Ceramic Gas Turbine Component Development and Characterization

Progress in Ceramic Gas Turbine Development, Volume 2

Mark van Roode
Mattison K. Ferber
David W. Richerson

NEW YORK ASME PRESS 2003

Ceramic gas turbine component development and characterization : progress in ceramic gas turbine development volume 2 / [edited by] Mark van Roode, Mattison K. Ferber, David W. Richerson.
p. cm.
Includes bibliographical references and index.
ISBN 0-7918-0197-7
1. Gas-turbines—Ceramic materials. I. van Roode, Mark. II Ferber, Mattison K. III. Richerson, David W., 1944-

TJ778.C385 2003
621.43'3—dc21 2003051835

The front cover depicts various stages in the evolution of processing and characterization of silicon nitride turbine gas turbine components. A vane is shown in the green state and in the final as-fired condition. The vane was manufactured by Honeywell Ceramic Components and is discussed in detail in Chapter 5. Characterization is represented by the scanning electron micrograph of the polished cross-section of an advanced silicon nitride. Also shown is a high-speed X-ray computed tomography image of a gelcast rotor fabricated of the same advanced silicon nitride. The rotor image was generated at Argonne National Laboratory (see Chapter 23 for details). The front cover design was developed by Matt Ferber of Oak Ridge National Laboratory.

Table of Contents

Foreword vii

Preface ix

Acknowledgements xi

About the Editors xiii

Chapter 1
An Overview of Ceramic Materials Development and Other Supporting Technologies 1
David W. Richerson, Takayoshi Iseki, Anatoly V. Soudarev, and Mark van Roode

Chapter 2
Ceramic Technology Development at Ford Motor Company from 1967 to 1986 31
John A. Mangels

Chapter 3
Ceramic Component Development at Saint-Gobain/Norton Advanced Ceramics 55
David W. Richerson, Eric Bright, and Robert Licht

Chapter 4
Silicon Carbide Component Development at Saint-Gobain/The Carborundum Company 77
Roger S. Storm, Martin R. Kasprzyk, and James W. MacBeth

Chapter 5
The Development of Monolithic Silicon Nitride Structural Ceramics at Honeywell 97
John P. Pollinger

Chapter 6
Technology Development at Kyocera Corporation 125
W. David Carruthers, Edwin H. Kraft, Makoto Yoshida, Koichi Tanaka, Toshifumi Kubo, Hirofumi Terazono, and Sazo Tsuruzono

Chapter 7
Research & Development of Ceramic Materials and Components for Gas Turbines 147
Tsutomu Yamamoto and Masaaki Masuda

Chapter 8
Sintering and Mechanical Properties of Silicon Carbide with Alumina Addition 165
Nobuhiro Shinohara

Chapter 9
The German National Program for Ceramic Materials and Component Development 175
Mark van Roode

Chapter 10
Hot Isostatic Pressing of Ceramic Turbine Components to Shape—Experience at Asea High Pressure Laboratory, Asea Cerama AB, ABB Cerama AB and AC Cerama AB 199
Hans T. Larker

Chapter 11
Ceramic Heat Exchanger Development by BOOSTEC SA 211
Marc Ferrato and Ginès Martinez

Chapter 12
Scientific Principles of Design and Production of Ceramic Parts for Engine Applications 229
Alexandr G. Romashin, Vladimir V. Vikulin, and Anatoly S. Shatalin

Chapter 13
Structural Alumino-Boron-Nitride Ceramic Material for Gas Turbine Engine Applications 245
Anatoly V. Soudarev, and Vladimir V. Grishaev

Chapter 14
Ceramic Matrix Composites Development at GE Power Systems Composites, LLC 259
Dennis J. Landini, Ali S. Fareed, Hongyu Wang, Phillip A. Craig, and Stan Hemstad

Chapter 15
Ceramic Matrix Composites for Gas Turbine Applications 277
Andrew Szweda, Timothy E. Easler, Robert A. Jurf, and Steven C. Butner

Chapter 16
Silicon Melt Infiltrated Ceramic Composites—Processes and Properties 291
Gregory S. Corman, Krishan L. Luthra, and Milivoj K. Brun

Chapter 17
Development of Hot Section Components Composed of Ceramic Matrix Composite (CMC) at the Research Institute of Advanced Material Gas-Generator (AMG) 313
Mikio Hiromatsu, Shigeto Nishide, Kozi Nishio, Ken-ichiroh Igashira, Go Matsubara, and Takeshi Suemitsu

Chapter 18
Development of Continuous Fiber-Reinforced Reaction Sintered Silicon Carbide Matrix Composite for Gas Turbine Hot Parts Application 331
Tsuneji Kameda, Shoko Suyama, Yoshiyasu Itoh, Katsutoshi Nishida, Isao Ikeda, Tsuneo Hijikata, and Takanari Okamura

Chapter 19
Mechanical Characterization of Monolithic Ceramics for Gas Turbine Applications 353
Mattison K. Ferber, Hua-Tay Lin, Michael G. Jenkins, and Tatsuki Ohji

Chapter 20
Development of High-Temperature Ceramics for Application in Gas Turbine Engine Components 397
Victor I. Trefilov, Oleg N. Grigoriev, Gennady V. Trunov, Georgy A. Gogotsi, and Pavel S. Kislyi

CHAPTER 21
Equipment and Methods for Ceramic Materials Evaluation 425
Youry A. Nozhnitsky, Youlia A. Fedina, and Alexander V. Fishgoit

CHAPTER 22
Properties of Continuous Fiber-Reinforced Ceramic Matrix Composites for Gas Turbine Applications 441
Edgar Lara-Curzio

CHAPTER 23
Nondestructive Evaluation/Characterization 493
William A. Ellingson, Yasushi Ikeda, and Jürgen Goebbels

CHAPTER 24
Test Standards and Design Codes for Structural Ceramics and Ceramic Matrix Composites 521
Michael G. Jenkins, Yutaka Kagawa, and Marc Steen

CHAPTER 25
Life Prediction of Structural Components 553
Stephen F. Duffy, Lesley A. Janosik, Andrew A. Wereszczak, Bjoern Schenk, Akihiko Suzuki, Jacques Lamon, and David J. Thomas

CHAPTER 26
Corrosion Issues for Ceramics in Gas Turbines 607
Nathan S. Jacobson, Dennis S. Fox, James L. Smialek, Elizabeth J. Opila, Peter F. Tortorelli, Karren L. More, Klaus G. Nickel, Takehiko Hirata, Makoto Yoshida, and Isao Yuri

CHAPTER 27
Coatings for Engineering Ceramics 641
Kang N. Lee, Holger Fritze, and Yuzuru Ogura

CHAPTER 28
Design of Impact-Resistant Ceramic Structural Components 665
Alonso D. Peralta, and Hiro Yoshida

CHAPTER 29
The Ceramic Gas Turbine—Retrospective, Current Status and Prognosis 693
David W. Richerson, Mattison K. Ferber, and Mark van Roode

CHAPTER 30
Future Needs and Developments in Structural Ceramic Materials for Gas Turbines 743
W. Dave Carruthers, John Pollinger, Paul F. Becher, Mattison K. Ferber, and Mark van Roode

Index 767

FOREWORD

The nation has been experiencing one of the longest periods of sustained economic growth in U.S. history. During this period, significant progress has been made in addressing some of the country's most pressing energy and environmental problems. For example, the U.S. Environmental Protection Agency reports that emissions of certain air pollutants have measurably declined since the 1980s. As a result, the once strongly held belief that it is not possible to improve the environment without slowing the economy has been proven false. One of the factors that has made these economic and environmental gains possible is greater energy efficiency, which has been brought about in part by the development and deployment of cleaner and more fuel efficient appliances, industrial equipment, transportation vehicles, and power generation devices.

As we have entered the new millennium, energy problems have once again become major public concerns. Energy price spikes, electric power supply constraints, power reliability problems, and air pollution are once again national news. In fact, the need for cleaner, reliable, affordable, and more fuel efficient energy devices has never been greater.

The U.S. Department of Energy (DOE) has a long-standing commitment to supporting technological innovation aimed at increasing energy efficiency, reducing dependence on energy imports, and lowering emission of air and water pollutants from energy generation, delivery, storage, and end-use consumption. A critical energy research and development (R&D) area has been that of advanced materials for improving the efficiency of combustion processes. In fact, for more than thirty years the DOE and other government agencies have been funding the design and development of new materials for energy generation systems to improve performance and lower emissions.

The oil crises of the 1970s stimulated considerable interest in the development of more energy efficient engines for passenger vehicles. More recently, increasingly stringent environmental emissions requirements have spurred interest in low-emission electric power plants. Recent growth in worldwide power demand has led to a surge in orders for electricity generation systems, particularly combustion turbines. In fact, the worldwide demand for combustion turbines is now so great that the major manufacturers are experiencing difficulties expanding production to keep pace with customer demand.

One of the most productive R&D pathways has been the development of advanced ceramic materials in collaboration with gas turbine manufacturers and producers of ceramic materials and components as part of a concerted strategy to combat the wasting of energy resources and to promote environmentally friendly engineering practices. The Federal government initiated major programs in the 1980s, including the Advanced Gas Turbine (AGT) program and its successor, the Advanced Turbine Technology Applications Project (ATTAP). These successful programs resulted in major technology breakthroughs in ceramic component design and materials processing and properties. The industrial participants in these programs – in collaboration with government laboratories and universities – developed the basic design, fabrication, and testing tools which led to the introduction of small ceramic gas turbines for automotive and related vehicular applications.

The great strides made in this endeavor are reflected in the advancements in silicon nitride ceramics, the strength of which have almost doubled since this material was first introduced in the 1970s. In fact, as a direct result of the AGT and ATTAP programs, silicon nitride nozzles have recently been certified for auxiliary power units for commercial aircraft. Lufthansa and Alaska Airlines planes are now flying with durable and reliable ceramic parts.

In the early 1990s, the DOE launched several new initiatives to advance the development of more energy efficient gas turbines through the application of advanced ceramic materials. These included the Ceramic Stationary Gas Turbine (CSGT) program and the Continuous Fiber-reinforced Ceramic Composites (CFCC) initiative. Both programs were aimed at applying the AGT and ATTAP improvements in the design and fabrication of advanced ceramic components to improve the design and performance of medium size industrial gas turbines.

The CSGT and CFCC programs have been extremely successful. Recent tests of silicon carbide-CFCC combustor liners have now been demonstrated for over 50,000 operating hours in industrial gas turbines at commercial sites. Tests have shown NO_x emissions at these sites dropped by more than 50 percent and CO emissions have been reduced 5 to 10 fold.

In addition, microturbines—small-scale gas turbines often less than 100 kilowatts in size – are being sold commercially and have the capability to revolutionize the global power generation industry over the next ten years. To be competitive with alternative energy generation devices, the fuel efficiency and air emissions profiles of microturbines must be improved. Ceramic materials for various microturbine subsystems and components will play a key role in reaching the performance levels required for commercial success. The DOE has recently launched a five year microturbine R&D initiative in collaboration with industry to meet those technological challenges and reach the performance targets.

This book provides important perspectives on the history of ceramic gas turbine development worldwide. There are lessons that have been learned that apply to a wide variety of industry-government R&D partnerships. The book is intended to highlight the excitement of successful R&D ventures in general and detail how researchers have responded to the challenges of designing ceramic turbines, how materials suppliers have improved their processes, how scientists have developed techniques to measure properties reliably and to establish life prediction and diagnostic technologies, and the key lessons learned in each of these endeavors.

Finally, innovations require the dedication of talented, visionary individuals. The three editors of this book are each a major contributor to the science, engineering and impact of advanced materials on the world. Each one is clearly among the best and brightest who have worked many additional hours, uncompensated, with the desire to leave more than they take in this world. Mark van Roode ran one of the best Public-Private partnerships that the Department of Energy was engaged in during the 1990s; work that contributed to his employer winning the Malcolm Baldrige Quality Award for excellence. This book is one of many major significant contributions to Ceramics and Education that Dave Richerson has given the world.

It is my hope that readers will find many useful insights that can be applied for further improvements in ceramic materials for gas turbines and in the development of other technologies for achieving cleaner, reliable, affordable, and more efficient energy generation technologies for the 21st century.

William P. Parks, Jr.
Director, Office of Distributed Energy and Electric Reliability
Office of Energy Efficiency and Renewable Energy
US Department of Energy

PREFACE

As ceramic components become viable design alternatives for commercial gas turbines the need will arise amongst the manufacturers of turbomachinery equipment, their component suppliers, and end users to become knowledgeable about the applicability of ceramics and their properties. For more than fifty years reports have appeared in the technical literature about design and materials development for specific gas turbine related projects. Taken together the published literature represents an impressive body of data that can be utilized to guide the materials process engineer, and the engine component and system designer.

Published conference proceedings discuss specific aspects of ceramics for gas turbine applications and represent the state-of-the-art at a certain point in time. However, at this time there exists no comprehensive text that addresses all the relevant issues involved in the application of ceramic gas turbine components and summarizes the evolution of the ceramic gas turbine development work. Ceramic gas turbines development programs have been conducted for a wide range of commercial and military applications for which the output power ranges from several tens of kilowatts to tens of megawatts in size. Large programs have been undertaken for automotive, industrial and utility size gas turbine development, each of which has its own peculiarities that need to be appreciated to obtain a complete picture of the development efforts. For example, data on component performance under typical operating conditions for automotive gas turbines is typically limited to durations from several hours to a few thousand hours at best, because the expected service life of an automotive gas turbine can be simulated by such short term testing. On the other hand, test data representing tens of thousands of hours are needed to effectively establish the life of components in industrial gas turbines for which the end users expect service lives on the order of 30,000 hours between overhauls. And such long term data is still very scarce and has only become available over the past decade.

These and other critical issues are very evident to those who have worked on ceramic gas turbine design, materials development and engine testing. Many valuable lessons have been and are still being learned, and the emerging knowledge base starts to provide a clearer picture of the potential and limitations of designing gas turbine hot section components with ceramics.

It is against this background that a publication that consolidates the data generated and summarizes the lessons learned over the more than fifty years of ceramic gas turbine development is timely and will be helpful for those involved with designing, testing, repairing, and marketing of turbomachinery equipment with ceramic parts, the suppliers of these parts, organizations involved in materials and component testing, and the end users of equipment.

The publication in front of you starts from the premise that the contributors to the various aspects of knowledge on ceramic gas turbines themselves are the best source of information. Ceramic gas turbine design is still a developing field, and there are different design approaches. On the other hand, similar probabilistic life prediction methodologies are widely employed for a variety of applications, although there are differences in the software routines used by the various workers in the field.

The editors of the publication discussed here, when contemplating the state of the technology, feel that the best approach to become familiar with the issues of ceramic gas turbine development is to consult directly the experts that have been intimately involved with ceramic gas turbine related work at their own organizations, whether these are the gas turbine manufacturers, suppliers of ceramic materials and components, or test laboratories.

This approach also has the advantage that it brings together the view points of the contributors that have learned valuable lessons from first hand experience. As a result the text will illustrate the various approaches in which the gas turbine community has embarked upon using ceramic materials. To guide the reader there are introductory chapters that will filter the cumulative experience and distill the key lessons learned, while highlighting still unresolved issues.

The large amount of design and materials data will be valuable for those with various degrees of exposure to the field. There will be information useful to the novice as well as to the experienced gas turbine designer and materials expert. The editors hope that the data presented will shorten the iterative development cycle and provide an up-to-date perspective on recent as well as historically more distant development work in ceramic materials and components. To provide a balanced view point the applications of ceramic matrix composites are being included alongside with those of the monolithic ceramics.

This publication on ceramic gas turbines draws on the accumulated experience of those closely involved with their development in the four regions of the world where active research has been conducted, the United States, Japan, Western Europe, and Russia and other countries that previously made up the Soviet Union. Some of the data has been accessible to those that cared to search for it. However, this information has been consolidated here in one English language text for the first time. Moreover, the perspective of a worldwide body of experts with its focus on relevance adds a novel dimension to the data.

The goal of this publication is to summarize the collective experience of the international gas turbine community with ceramic gas turbine development over the past couple of decades. Representatives from various organizations (equipment manufacturers, suppliers of ceramic parts, and scientists and engineers involved with materials evaluation) from four areas in the world (U.S., Japan, Western Europe, and Russia) have contributed chapters that summarize the projects in their organizations. The emphasis is on evolution of development projects, lessons learned, successes, problems, and potential for commercialization.

The text is published in two volumes. The first volume highlights the development work carried out by the gas turbine equipment manufacturers and emphasizes design and testing of ceramic components and engines. This volume is organized by each of the four regions and there is an introductory chapter for the work in the U.S., Japan, Western Europe, and Russia.

The second volume summarizes the work of the component suppliers and materials laboratories. The laboratory work is predominantly organized by interest area, i.e. monolithic materials and components development, ceramic matrix composite development, materials characterization, and special challenges. This volume is concluded with two summary chapters that highlight the overall perspective.

The editors hope that the two volumes of this text will spark interest of the gas turbine design and ceramic component fabricating communities, and within other organizations that relate to these industries. Universities that educate future engineers that will be working in the gas turbine related industries will find much in this text that is valuable and intellectually challenging.

The authors retain full responsibility for the material presented in their chapters and they deserve the credit for it. The editors are greatly indebted to their dedication in presenting their material; they have merely had the pleasure of structuring the overall text and providing editorial assistance where appropriate.

Mark van Roode, Mattison K. Ferber, David W. Richerson

ACKNOWLEDGEMENTS

The initial objective of this two-volume treatise was to document the worldwide efforts and progress involved in applying ceramic materials to gas turbine engines. Considering the diverse nature of the technologies involved and a timeline that goes back at least to the late 1960s, meeting the objective turned out to be quite a challenge. The editors are indebted to all of the authors who agreed to be part of this challenge, to the many individuals who helped us identify and recruit the authors, to the individuals and organizations who funded and managed the programs (many on a national scale), and to the hundreds of individuals who participated in the programs. In addition the editors thank ASME Press for their willingness to publish the treatise, the ASME editorial team, and all of the individuals who were willing to proofread draft chapters and provide feedback.

The editors especially acknowledge with great appreciation the support of the following individuals:

Mr. Masafumi Chikaraishi, Materials Manager of Kyocera Corporation, who was fulfilling an assignment with the Japanese Fine Ceramics Association, early on in the project invited representatives of the Japanese organizations that could contribute to this book for a meeting in Tokyo where one of the editors (Mark van Roode) was given the opportunity to explain the book project. Dr. Takane Itoh of Tokay University attended that meeting and was extremely helfpul in enlisting contributors for the various chapters, as well as being the prime author on the chapter in Volume I that introduces the Japanese ceramic programs. He further contributed as the prime author on the chapter reviewing the Japanese automotive ceramic programs. Without the help of Mr. Chikaraishi and Dr. Itoh, the Japanese contributions would not have been as numerous and complete.

Prof. Dr. Anatoly Soudarev, Assistant Managing Director at the Research Technological Institute for Power Engineering in St. Petersburg, Russia, very effectively coordinated the chapters from Russia and the Ukraine. He also organized the translation of the Russian chapter text into English, a job the editors were ill equipped to handle. The editors also recognize the helpful suggestions of Dr. Vladimir Shirmanov of Energomash, St.Petersburg, Russia, on the technical content of some of the chapters.

For the early work in Germany, the editors appreciate the comments of Dr. Wolfgang Bunk and Dr. Manfred Böhmer (retired from DFVLR), Dr. Jörgen Heinrich of the Technical University of Clausthal, Dr. Fritz Thümmler, Professor Emeritus of the University of Karlsruhe, Prof. Dr. Fritz Aldinger of the University of Stuttgart, Dr. Ernst Gugel, former Head of R&D of CFI, Dr. Gerhard Wötting of CFI, and Dr. Bjoern Schenk of Honeywell Engines, Systems & Services. The editors also thank Mr. Manfred Langer of Volskwagen AG, Dr. Klaus Mörgenthaler of Daimler-Chrysler AG, Mr. Gerhard Andrees of Motoren Turbinen Union, and Dr. Ernst Gugel for their comments on the chapter on German automotive development in Volume I. Dr. Eberhard Seitz of the Research Center Jülich, Ltd., reviewed the introductory chapter on ceramic gas turbine development in Western Europe. Dr. Peter Walzer supplied supplemental literature for the automotive gas turbine development work in Germany. Prof. Dr.-Ing. Helmut Pucher of the Technical University of Berlin provided several names of researchers that were contacted to review the German ceramic gas turbine development activities.

Special thanks also to Sandra L. Maldonado, formerly of Oak Ridge National Laboratory (ORNL), and Austin K. Ferber for organizing and scanning the hundreds of drawings and photographs associated with the various figures. The support of Dr. Michael Karnitz of ORNL throughout the book preparation project is also gratefully acknowledged.

Finally, the editors are much indebted to the support of Mary Grace Stefanchik, Manager, Books at ASME Press and Tara Smith, Production Coordinator at ASME Press who guided the project patiently through the publication preparation process.

Mark van Roode, Mattison K. Ferber, David W. Richerson

About the Editors

Mark van Roode has been active in scientific and technical research and development since he received his Ph.D. in Chemistry from the University of Calgary in 1975. His varied assignments in chemistry include an assistant professorship at Memorial University of Newfoundland, Sir Wilfred Grenfell College, Corner Brook, Newfoundland, Canada, and a post-doctoral fellowship in pre-biotic chemistry at the Salk Institute in San Diego, California. He became involved in industrial research in ceramics in 1981 when he joined the Ontario Research Foundation, Mississauga, Ontario, Canada. In 1984 Dr. van Roode joined Solar Turbines Incorporated as a development engineer. He worked for a number of years on the development of ceramic and metallic coatings for oxidation and corrosion protection of gas turbine hot section components. When Solar was awarded the contract for the Ceramic Stationary Gas Turbine (CSGT) program by the U.S. Department of Energy in 1992, Dr. van Roode became the program manager with responsibility for the programmatic and technical leadership of the program. Under this program Solar and its subcontractors designed ceramic blades, nozzles, and combustor liners for the Solar Centaur 50 engine. These components were extensively tested in in-house engine testing at Solar and in field testing at several industrial sites. He is currently the Manager of Development Engineering at Solar, with responsibility for contracts in advanced materials development and technology strategy for gas turbines. Dr. van Roode has been active in the scientific community and has published extensively in the area of ceramic gas turbine development. He served on the executive of the ceramics division of the International Gas Turbine Institute of ASME from 1994 to 1999.

Mattison K. Ferber is currently employed as a Senior Research Staff Member in the Metals and Ceramic Division at the Oak Ridge National Laboratory in Oak Ridge, Tennessee. Dr. Ferber received his B.S., M.S., and Ph.D. degrees in Ceramic Engineering at the University of Illinois. From 1981 until 1984 Dr. Ferber was employed at the Oak Ridge National Laboratory as a Research Staff Member in the Structural Ceramics Group. In 1987 he joined the Department of Ceramic Engineering at the University of Illinois in Urbana-Champaign as an Associate Professor. From 1987 until 1996 he was the Group Leader of the Mechanical Characterization and Analysis Group in the High Temperature Materials Laboratory. Dr. Ferber, who is currently a member of the Structural Ceramics Group, is author or coauthor on more than 160 publications including 130 technical papers. Dr. Ferber's honorary include Tau Beta Pi, Phi Kappa Phi, Keramos, and Bronze Tablet. He is currently affiliated with the Engineering and Basic Science Divisions of the American Ceramic Society. Dr. Ferber is a Fellow of the American Ceramic Society.

David W. Richerson has been one of the long-term pioneers in efforts to develop ceramics and supporting technologies for turbines. Following his undergraduate studies in Ceramic Engineering at the University of Utah and his graduate studies in Ceramic Science at Pennsylvania State University, Mr. Richerson joined Norton Company in Worcester, Massachusetts, in 1969. There he developed the first commercial silicon nitride in the U.S. (NC-132 hot pressed silicon nitride) and fabricated many prototype components such as stator vanes and bearings. He then moved to AiResearch Manufacturing Company of Arizona (presently Honeywell Engines, Systems & Services) in Phoenix, AZ and spent the next 11.5 years guiding efforts to introduce ceramics into turbine engines. This included materials and fabrication development, characterization, design and life prediction, NDE, rig and engine testing, and failure analysis. In addition he taught courses at Arizona State University, wrote the textbook *Modern Ceramic Engineering: Properties, Fabrication, and Use in Design*, participated in National Materials Advisory Board studies, and wrote a white paper for the ASME Gas Turbine ceramics committee that became the model for the DOE/ORNL Ceramic Technology for Advanced Heat Engines Project. Mr. Richerson temporarily left gas turbine development efforts in 1985 to become Director of R&D (and later Vice President) at Ceramatec, Inc. in Salt Lake City, Utah. However, in 1991 he established his own consulting business and again focused much of his efforts on ceramics for gas turbines. Some of these efforts included being a major participant in the Solar Turbines CSGT program, supporting the DOE ATS project, helping with ceramic matrix composites development under the CFCC program, and currently supporting the DOE Microturbines programs. In addition Mr. Richerson has been active in education. Efforts have included teaching at the University of Utah, teaching numerous short courses internationally, writing a home study course on ceramics for ASM, Int., writing a second edition of *Modern Ceramic Engineering* and chapters on ceramics and composites for numerous books, establishing an education committee for the American Ceramic Society and being its chair since 1991, and writing the education outreach book *The Magic of Ceramics*.

CHAPTER

1

An Overview of Ceramic Materials Development and Other Supporting Technologies

David W. Richerson
University of Utah, Salt Lake City, Utah, USA

Takayoshi Iseki
Department of Inorganic Materials, Tokyo Institute of Technology, Tokyo, Japan

Anatoly V. Soudarev
Boyko Research-Engineering "Ceramic Heat Engines" Center (NIZ-KTD), St. Petersburg, Russia

Mark van Roode
Solar Turbines Incorporated, San Diego, California, USA

THE CHALLENGE OF USE OF CERAMICS IN TURBINES

The hot section of a gas turbine engine is a challenging environment for materials to survive. The materials must withstand high mechanical and/or thermal stress at high temperature for thousands of hours, resist oxidation and corrosion by high velocity gases (often with entrained particles or condensed phases), avoid chemical reaction with adjacent components, be stable in high-cycle and low-cycle vibration conditions, and have acceptable creep and stress rupture life. Metals, coatings, and cooling schemes that can survive this combination of severe conditions have evolved steadily since the initial invention of the gas turbine engine to allow higher and higher temperature operation and increased efficiency. This evolution, however, has been slow, on the order of a few degrees temperature increase per year, and has become increasingly difficult as metal alloys have approached their melting temperatures.

Because of their high temperature capability and potential for enabling large increases in engine efficiency, ceramics were initially proposed to be used as gas turbine engine hot section components (rotor blades, turbine nozzles, and combustor liners) in the 1940s [1]. Unfortunately, the ceramic materials such as alumina and cermets that were available at that time did not survive the engine operating conditions [2,3]. By the late 1960s, though, silicon nitride and silicon carbide ceramics had been developed that appeared to have a suitable combination of mechanical, thermal, and chemical properties to be considered for turbine components [4–10].

Starting in the late 1960s, a series of turbine engine development programs were conducted to implement ceramics. These programs are reviewed in detail in Volume 1. At the time of the earliest engine programs, technology for ceramics for turbines was at an early level of development:

- Design methodologies for brittle materials had not been developed.
- Only reaction-bonded silicon nitride, hot pressed silicon nitride, and reaction-sintered silicon carbide were available, but as early-stage materials with room for considerable improvement and optimization for turbine use; generally, there wasn't even a high quality commercial source of starting powders.
- Very little material data were available and virtually no standardized tests, thus limiting the ability to make comparisons of different materials and limiting the ability to do analytical design.
- Ceramic manufacturers had no experience in fabricating the complex shapes required in turbines and little or no experience in fabricating materials that could be guaranteed to not have flaws above a critical size that would cause component failure in a gas turbine engine.
- Non-destructive evaluation (NDE) methods and equipment were not available to assure that a ceramic turbine component did not have a critical flaw, and proof test procedures had not been established.
- Life prediction methodology and analytical tools had not yet been developed.

To achieve commercial introduction of ceramics in turbines requires maturity in each of the above categories. Early engine programs had to work with the technologies that were available at the time or could be developed during the duration of the program. The engine programs became strong drivers of technology by identifying deficiencies and providing program planning and funding to help resolve the deficiencies. Thus, the chapters in Volume 1 have a wealth of information about the evolution of each technology and the plateau that each reached during the span of that program. However, those chapters focused on the details of the engine development and testing and not on the details of these other required supporting technology developments. The objective of Volume 2 is to focus on the details of the evolution of the supporting technologies: ceramic materials fabrication and properties improvement, test methods and database generation, standards and design codes, life prediction methodology, NDE, and machining.

The remainder of this introduction provides a brief summary of the evolution of the various supporting technologies in different regions of the world and of some of the companies that participated. There are two reasons for presenting this summary: (1) to provide the reader a quick overview and (2) to provide a more complete history of the evolution of ceramics for turbines by also acknowledging some of the companies that were involved at various times, but are either no longer involved or that did not participate in preparation of subsequent chapters. Of course, the companies with current active involvement in the development of the supporting ceramic technologies have been included in this summary.

SUMMARY OF EVOLUTION OF CERAMICS FOR TURBINES IN THE UNITED STATES

Materials Development

The ceramic materials available at the beginning of the 1970s in the United States were reaction-bonded silicon nitride (RBSN), hot pressed silicon nitride (HPSN), and reaction-sintered silicon carbide (RSSC), primarily based on technology developed in Great Britain during the 1960s [4–10]. The reaction-bonded silicon nitride had high porosity and relatively low strength (typically under 200 MPa as measured in 3-point bending), but retained the strength to over 1400°C. Because of the high open porosity, the early RBSN was very susceptible to oxidation and degraded with time at turbine operation temperatures. Ford Motor Company, Norton Company, and later AiResearch Casting Company developed improved grades of RBSN with decreased porosity and increased strength (>300 MPa as measured in 4-point bending) [11–14]. These improved RBSN materials were used for the turbine nozzles and other static structure components in the Ford and Garrett Turbine Engine Company programs throughout the 1970s. The Ford material was also used for rotor blades in their duo-density rotor in conjunction with a hot pressed hub.

The best RSSC material in the world by 1970 was fabricated by British Nuclear Fuels based on technology developed at the United Kingdom Atomic Energy laboratories at Harwell, Great Britain. Ford, Norton, and Carborundum each developed their own version of RSSC during the 1970s [15–17]. These materials were evaluated for combustor liners, stator vanes, and other static structure components. Because the RSSC ceramics had no open porosity, they had excellent oxidation resistance. They also had high thermal conductivity and retained strength of about 300 MPa up to the melting temperature of silicon (slightly above 1400°C).

The RBSN and RSSC ceramics were not strong enough to survive the stresses in a gas turbine engine rotor. The best candidate ceramic for a rotor in 1970 was hot pressed silicon nitride from Lucas in Great Britain [18]. This material had strength at room temperature in 3-point bending over 800 MPa, but a strength of less than 200 MPa at 1400°C and relatively short stress rupture life, even at 1200°C. It was suitable to start turbine rotor development, but did not offer long component life. The hot pressed silicon nitride from Lucas was made using MgO additions to silicon nitride powder synthesized by nitriding of silicon.

Initial hot pressed silicon nitride development in the U.S. was conducted at Ford Motor Company and at Norton Company using MgO sintering aid addition to silicon nitride powder [19–21]. Norton Company licensed the Lucas hot pressed silicon nitride technology and in 1972 made substantial improvements in the high temperature strength and stress rupture life by careful control of the chemistry of the grain boundary phase. Soon thereafter, Gazza at the Army Materials and Mechanics Research Center (AMMRC) demonstrated that further improvements in high temperature stability could be achieved using yttrium oxide sintering aid rather than MgO [22].

Hot pressed silicon nitride, especially the Norton NC-132 material, became the standard for fabrication of turbine rotor blades and large industrial turbine stator vanes throughout the 1970s. However, complex shapes could not be easily fabricated by hot pressing, so considerable diamond grinding was required. Terwilliger of Westinghouse demonstrated by 1974 that silicon nitride could be fabricated by pressureless sintering at 1650°C with 5% MgO addition [23]. Although the material he produced had greater than 10% porosity and low strength, it stimulated considerable interest and hope that complex shapes could be fabricated by conventional sintering and thus eliminate much of the costly final diamond grinding. During the remainder of the 1970s, extensive efforts worldwide focused on achieving near-theoretical density and high strength by pressureless sintering [24,25].

The initial attempt to increase the density and strength of sintered silicon nitride was to sinter at higher temperature. However, this resulted in decomposition of silicon nitride, which inhibited sintering. Several approaches were developed to minimize decomposition: packing the parts in a silicon nitride powder bed, using a closed crucible, and sintering in an overpressure of nitrogen [26–29]. Ultimately, procedures were developed that allowed sintering in the 1900–2000°C range and resulted in near-theoretically-dense material with excellent properties. Sintering at higher temperatures had an additional benefit. It allowed enough grain growth to form a highly intertwined microstructure of β-Si_3N_4 grains that resulted in substantially higher fracture toughness than for traditional ceramics [30–32].

Parallel to efforts to inhibit decomposition, there also were many studies to assess various sintering aids besides MgO. Examples included Y_2O_3, $Y_2O_3 + Al_2O_3$, spinel, Ce-O-Si-N compounds, yttria-magnesia-zirconia, alumina-ceria, and even ceria-magnesia-strontia-zirconia [33–35]. Much of the early work in the U.S. focused on yttria-alumina, including work at Stanford Research Institute, AiResearch Casting Company, and GTE. However, by the mid-1980s most of the sintered silicon nitride turbine components evaluated in U.S. engine programs were fabricated in Japan by Kyocera and NGK.

Another area of development that played an important role in achieving quality sintered silicon nitride was powder synthesis. Initial silicon nitride powders were prepared by nitriding of silicon and crushing and grinding the resulting silicon nitride "cake" into a fine powder. Initial silicon nitride powders were relatively coarse, severely contaminated, and not very reactive. Herman Starck Company in Germany achieved a series of refinements in direct nitriding of silicon nitride during the late 1970s and during the 1980s. GTE Sylvania in the U.S. developed a fine-grained, reactive, chemically derived silicon nitride using a gas-phase reaction of $SiCl_4$ and NH_3. Several excellent chemically derived silicon nitride powders were developed in Japan, as described later in this chapter. Other silicon nitride powders have been chemically synthesized at Ford Motor Company and at Dow Chemical, but are not commercially available.

Two other directions of development were pursued in the late 1970s to achieve decreased porosity and higher strength silicon nitride: sintered reaction-bonded silicon nitride (SRBSN) and hot isostatically pressed silicon nitride (HIPSN). SRBSN was fabricated by either nitriding silicon with a sintering aid present or adding a sintering aid by infiltration after nitriding [36,37]. The RBSN with the sintering aid was subsequently sintered in an overpressure of nitrogen in the 1800–2000°C range. SRBSN developments in the U.S. were conducted at Ford Motor Company. Initially, the development was directed towards turbocharger rotors. Mangels and co-workers at Ford focused on additions of yttria to achieve sintering of their RBSN.

Early efforts on hot isostatic pressing (HIP) of silicon nitride were conducted in the U.S. at Battelle and in Sweden at Asea [38,39]. HIP consists of preparing a powder preform of the desired shape, encapsulating the powder preform in a material that will deform but remain impermeable under high temperature and pressure. Battelle initially densified silicon nitride plus yttria encapsulated in a tantalum capsule and later demonstrated that they could use a high-silica glass composition instead of the tantalum. Asea developed a commercially viable procedure of applying powdered glass over the shaped silicon nitride powder preform (with a boron nitride coating between the silicon nitride and glass). The glass softened and flowed to form a gas-impermeable layer during the HIP cycle, then spalled off during cooling. Norton Company and AlliedSignal Ceramic Components both licensed the Asea technology during the 1980s and developed HIPSN materials that were evaluated for turbine components in the late 1980s and 1990s [40,41].

All of the materials discussed so far behave in a brittle fashion; they fracture with very little strain-to-failure. During the early 1980s, composite ceramics reinforced with high-strength ceramic fibers started to become available. One of the earliest was a SiC/SiC composite from the Société Européenne de Propulsion (SEP) in France. The early SiC-based ceramic matrix composites degraded at high temperature in an oxidizing environment, so were initially considered only for short life applications. However, by about 1992 the ceramic composite technology had progressed far enough that the "Continuous Fiber Ceramic Composite" program was started by the U.S. Department of Energy (DOE) to accelerate ceramic composite development. Several of the projects in this program focused on gas turbine components. Another DOE-funded program, the "Ceramic Stationary Gas Turbine" (CSGT) development program at Solar Turbines Inc., evaluated some of the resulting composite materials and other ceramic composite materials as combustor liners in combustor test rigs and later in actual engines. After several development iterations, SiC/SiC ceramic composite combustor liners have now survived more than 50,000 hours under normal engine operating conditions.

Characterization and Property Improvement

Very little property data were available for silicon nitride and silicon carbide candidate turbine materials in the U.S. in 1970, and there was little understanding of chemistry/processing/microstructure/property relationships. The Defense Advanced Research Projects Agency/Army Materials and Mechanics Research Center (DARPA/AMMRC) Ford/Westinghouse "Brittle Material Design, High Temperature Turbine" program stimulated a rapid learning curve in the early 1970s. Physical, thermal, and mechanical properties were screened for a number of materials over a broad temperature range. Scanning electron microscopy, energy dispersive x-ray analysis, transmission electron microscopy, and Auger spectroscopy were just emerging as analysis tools available to industry. By comparing the measured properties with observations of the microstructure and analyses of chemistry of grains and grain boundaries made possible by these new analysis tools, deficiencies such as decrease in strength with temperature began to be understood [42]. As failure modes were identified and the causes understood, the chemistry and microstructure of the materials were modified in efforts to achieve properties commensurate with the turbine engine requirements. It should be noted that 1975-vintage ceramics were able to survive in a turbine engine, but only for short times. The property improvement efforts since then have focused on increasing the life, reliability, and temperature capability.

Property characterization in the late 1960s and early 1970s was directed largely towards material development, such as comparing different process modifications, compositions, and fabrication lots. The primary modes of characterization were density and flexure strength. Flexure strength was measured in 3-point bending. As the needs of turbine engines and turbine designers became better known, characterization expanded dramatically to provide data suitable for design and life prediction. This included thermal diffusivity, thermal conductivity, strength, and elastic modulus as a function of temperature. By about 1973 most strength measurements were being switched over to 4-point bending with self-aligning fixtures, and the resulting data were presented in a Weibull format. Special SiC fixtures were designed for strength testing up to about 1400°C, and a high temperature database was beginning to accumulate. Need for time-dependent data was also recognized, resulting in initiation of creep, stress rupture, and oxidation testing. The importance of fracture toughness was recognized, so test methods were developed and in common use by the early 1980s.

As data identified property deficiencies, material developers would conduct research to improve their materials. Each new material or modification was given a new name or designation. For example, Norton Company dense silicon nitride evolved from HS-110 to NC-132 to NCX-34 to NT-154 to NT-164. Each subsequent material had improved high temperature properties, but required generation of a database suitable for turbine engine companies to conduct component design and life prediction.

Property measurements were initially conducted at the materials company or contracted out to a university or test lab. There were no standards for sample preparation, size of samples, test fixtures, method of measurement, or method of data analysis. Data for different materials or test sites could not be reliably compared. The Ford/Westinghouse program and later the DARPA/Navy/Garrett Turbine Engine Company "Ceramic Gas Turbine Engine Demonstration Program" gave strong encouragement and support to establishment of test standards and material standards [43–45]. AMMRC and the National Bureau of Standards also provided leadership and support [46–48].

The Air Force started a key program in 1979: "Property Screening and Evaluation of Ceramic Turbine Materials" at the Illinois Institute of Technology Research Institute (IITRI) to systematically characterize candidate turbine engine ceramics [49]. A vital part of this program was to make sure that all samples were surface ground in the same way and tested on standardized fixtures with specified test procedures. Also, the fractured or otherwise tested surfaces of the samples were systematically examined to learn as much as possible about the behavior. This included information such as the flaw causing fracture or whether slow crack growth had occurred in a high temperature test.

The mechanical property tests under the Air Force sponsored program were conducted in bending. By the early 1980s, designers were requesting uniaxial tensile data. DOE initiated programs through Oak Ridge National Laboratory (ORNL) that were conducted primarily at ORNL and at the University of Dayton Research Institute (UDRI) [50,51]. New tensile fixtures were designed and standardized and extensive tensile strength, creep, and stress rupture data generated. Tests are still continuing today [52]. Parallel testing has been conducted at the National Institute of Standards and Technology (NIST) and at some of the ceramics companies.

Early U.S. engine programs identified additional needs for materials characterization. One involved material response to high temperature oxidation in a combustion environment. NASA pioneered early efforts using static furnace exposures and burner rigs [53]. Later they funded a long term program to Garrett Turbine Engine Company to expose flexure test bars for up to 3500 hours in a burner test rig and measure the effect of the exposures on strength, dimensions, and near-surface microstructure and chemistry. This testing continued well into the 1990s [54,55]. During the late 1990s, engine field tests conducted by Solar Turbines Inc.

under the CSGT program and by Rolls-Royce Allison under its ATS (Advanced Turbine Systems) program identified that high water vapor pressure in an operating turbine engine could cause rapid surface recession of silicon-based ceramic matrix composites and monolithic ceramics [56,57]. Test facilities have been established at Oak Ridge National Laboratory and at General Electric Company–Corporate Research & Development (GE CRD, currently GE GRC-Global Research Center) that better simulate the moist conditions inside industrial turbines and have led to better understanding of the potential mechanisms of degradation [57]. The NASA Glenn Research Center, ORNL, and GE CRD combined information from engine tests and laboratory tests determined that an "environmental barrier coating (EBC)" would be necessary to protect the surface and provide acceptable component life.

An additional characterization need identified in early engine programs was contact stress. Extensive efforts were conducted at Garrett Turbine Engine Company (later AlliedSignal Engines, and presently Honeywell Engines, Systems & Services) to understand the mechanisms of contact stress fractures and to find design and materials solutions for both rotating and stationary turbine components [58,59].

Turbine Engine Component Fabrication

Turbine engine components are complex in shape and are exposed to high stresses. In 1970 ceramic manufacturers in the U.S. did not know how to economically make the complex shapes with a material that had enough strength to survive the turbine operating conditions, so they had to address two major challenges: (1) developing processes to achieve the shape and tolerances and (2) eliminating fabrication defects such as pores and inclusions to achieve acceptable strength.

Reaction-bonded silicon nitride and reaction-sintered SiC were selected for early component fabrication, partially because they exhibited very little dimensional change during densification. Ford Motor Company conducted extensive fabrication development of injection molding and slip casting during the 1970s [11,17]. They produced rotor blade rings and one-piece turbine stator rings. Additional shape-forming efforts were conducted with RBSN at Norton Company and at AiResearch Casting Company. The Norton effort focused on isostatic pressing and green machining, while the AiResearch effort focused on injection molding and slip casting [60].

The hot pressed materials were a different challenge. Hot pressing in the early 1970s could only be conducted uniaxially, so the only shape that could be fabricated initially was a flat plate. Complex shaped components such as rotor blades were then achieved by cutting and grinding the hot pressed silicon nitride with diamond coated tooling, which was a very time-consuming and expensive process. To add to the challenge, existing diamond grinding wheels were not effective at cutting silicon nitride. Furthermore, surface grinding caused tiny near-surface cracks that decreased strength. Through many iterations, the technical problems were overcome, but not the cost problems [61].

HIP with glass frit encapsulation was an important innovation that allowed hot pressed quality silicon nitride to be fabricated to complex shapes with a substantial reduction in diamond grinding [38]. As mentioned before, both Norton and AiResearch Casting Company (whose ceramic fabrication operation became AlliedSignal Ceramic Components and is presently Honeywell Ceramic Components) licensed the HIP technology from Asea and developed high quality HIPSN materials in the late 1980s. The Norton HIP material was designated NT-154 and later improved to NT-164 [62]. The AlliedSignal material was designated GN-10 [63].

Although HIP decreased final diamond grinding, efforts were still required to fabricate a powder preform of complex shape that was uniform enough to yield excellent properties and that was strong enough for coating with the glass frit and HIPing. Injection molding, slip casting, and combinations of cold isostatic pressing and green machining or slip casting and green machining were developed during the late 1980s and 1990s. By the late 1990s, gel-casting technology, initially developed at ORNL, was licensed to AlliedSignal Ceramic Components.

Conventional pressureless sintering and nitrogen overpressure sintering were developed parallel to HIP. Carborundum Company invented and commercialized a sintered α-SiC ceramic designated Hexoloy SA

[64]. GE CRD developed a sintered β-SiC. GTE developed a sintered silicon nitride material designated as AY-6 [65,66]. AlliedSignal achieved a high-toughness sintered silicon nitride designated AS-800 [32]. Extensive R&D was conducted with all of these materials to fabricate complex turbine shapes with a high degree of microstructure uniformity and resulting high strength. As part of the efforts, the ceramic fabrication companies worked to establish "fixed processes" where every step in the fabrication process was controlled by a specification and certification procedure to assure quality. For example, Norton Company used Taguchi statistical experiments under a DOE/ORNL program to help refine their process and implement quality control. All of the companies demonstrated a large increase in the uniformity and reliability of their materials. The Weibull modulus was increased from below 10 to routinely over 20.

The turbine component fabrication efforts at Ford, Norton, AiResearch Casting/AlliedSignal/Honeywell, and Carborundum are described in detail in subsequent chapters and in the literature cited in those chapters. Further details on the rig and engine evaluation of these components is detailed in Volume I. The GE CRD work on β-SiC was conducted to support an internal General Motors automotive development program and was never published. The effort lasted for nearly ten years and resulted in some high quality injection molded radial and axial rotors. Key investigators were G. Renlund, C. Johnson, and R. Giddings.

Material Inspection and Component Qualification

Initial silicon nitride and silicon carbide materials were fabricated on an R&D basis with no standards, specifications, or quality control procedures. In-process and post-process inspections evolved as the materials evolved. Initial quality control consisted of density measurement and usually fracture of test bars fabricated with the component. Ford Motor Company added dye-penetrant, conventional x-ray radiography and ultrasonic c-scan nondestructive inspections plus many in-process measurements such as particle size distribution, viscosity, and tooling temperature [67]. Garrett Turbine Engine Company explored microfocus x-ray radiography and introduced analysis of the film with computerized image-enhancement algorithms [68,69]. They also explored high frequency ultrasonics and later computed tomography (CT Scan or CAT Scan) [70,71].

Engine companies and materials companies applied non-destructive evaluation methods and were able to reject components with large defects. However, they did not feel confident that they could detect and interpret defects in the 50–200 μm range that could cause failure in a highly-stressed turbine component. As a result of this uncertainty, each engine company developed proof tests to stress actual turbine components to the engine design stress or an overstress to eliminate defective or marginal ceramic components. Many of these proof tests are described in Volume 1.

Renewed efforts on NDE were begun in the late 1980s and have continued to the present. Much of this effort has been conducted by Ellingson and co-workers at Argonne National Laboratory (ANL) including development of a laser scattering method for detecting surface and near-surface defects, thermal imaging equipment and techniques, and CT (for both defect detection and dimensional measurement) [72,73]. The team at ANL also has pioneered techniques for NDE of large shapes, inspection of composites, and methods to reduce inspection time and cost.

Brittle Material Design and Life Prediction Methodology

Although NASA had conducted some design work with cermets in the 1950s, no brittle materials design methodology was in place in 1970 in the U.S. Ford Motor Company and Westinghouse in their DARPA/AMMRC program during the early 1970s pioneered development and verification of brittle material design codes using finite element analysis [74]. During the mid-1970s, the Allison Division of General Motors (later Allison Turbine Engines and now part of Rolls-Royce) and AiResearch Manufacturing Company of Arizona (which later became Garrett Turbine Company) used the Ford methodology as a baseline and developed their own design and life prediction codes [75].

The 1970s design and life prediction codes were based primarily on fast fracture data and modes of failure and were adequate for the short-term demonstrations. Long life applications, though, would require a life prediction methodology that also accounted for time-dependent changes in the materials such as slow crack growth, creep, and oxidation. NASA began a long-term in-house program to accumulate property data and establish and verify versatile codes that were available to all engine companies [76]. The engine companies continued to evolve their own codes, often with substantial support from the DOE through the ORNL "Ceramic Technology for Heat Engines Project" [77]. This included analysis and testing of various shapes and volumes of ceramic material and assessment of various statistical methods of handling the data. These programs were closely linked to other programs that generated property data and helped to drive the acquisition of tensile data and time-dependent high temperature data.

Most of the engine programs in the U.S. during the 1980s were directed towards automotive turbines with a total life requirement of 3500 hours, but with peak stress conditions only being a small portion of that time. Thus, data generation was typically for 3500 hours or less. During the 1990s, with the start of the Solar Turbines program on industrial turbines, emphasis changed to 20,000–30,000 hour life. This required much more attention to time-dependent properties and led to long-term property measurements and extension of the design and life prediction codes.

SUMMARY OF EVOLUTION OF CERAMICS FOR TURBINES IN JAPAN

Introduction

The history of the development of ceramic gas turbine technology originated in the Advanced Gas Turbine Project which started in 1978. The project was operated under the energy saving program of the Japanese Ministry of International Trade and Industry (MITI—became the Ministry of Economy, Trade and Industry, METI, in 2001), the Moonlight Project. Other research and development projects followed, ending with the extraordinary results of the 300 kW CGT (Ceramic Gas Turbine) Project of the New Energy and Industrial Technology Development Organization (NEDO) by the year 1999. T. Itoh and A. Sugawara describe the histories of the ceramic gas turbine engine programs in chapter 12 of Volume I of this text.

One way to overview the advances in ceramics for gas turbines is to read the reviews on recent developments in ceramic science and engineering that were published about every 10 years by the Ceramic Society of Japan. The 1971 Yogyo Taikan [78] has no section concerning ceramic gas turbine materials. There is discussion of sintering of oxide ceramics, but no mention of silicon nitride or silicon carbide. The 1981 Yogyo Taikan [79] has significant sections on silicon nitride and silicon carbide and also includes the term "ceramic gas turbine." In 1991, in celebration of its centennial, the Ceramic Society of Japan published "21-Seiki E Habataku Seramikkusu—Ceramics 100th" [80]. This review included short administrative explanations of the main national projects including the Fine Ceramics Project, the 300 kW NEDO CGT Project, and the 100 kW PEC (Petroleum Energy Center) CGT Project. In addition non-oxide and turbine engine component sections also appeared.

Tremendous progress has been accomplished in Japan during the 30 years of effort on ceramics for gas turbine engines. Key milestones are shown in Table 1.1. The Fine Ceramics Project was particularly important in advancing ceramic materials technology.

The Fine Ceramics Project

The Fine Ceramics Project was launched in 1981 under the National Research and Development Project on Basic Technologies for Future Industries supported by MITI, so called "Jisedai Fine Ceramics" [82,83]. The progress is documented in two thick progress reports [84,85].

The research and development were carried out for 12 years from FY1981 to FY1992. The goal of the project was to establish key technologies for developing high performance ceramics as structural materials. This includes essential performance and reliability achievements, as well as development of design techniques for

TABLE 1.1 MILESTONES OF 30 YEARS OF DEVELOPMENT IN CGT CERAMIC MATERIALS IN JAPAN

1969	Base composition of Si_3N_4-rare earth oxide
1970	Sialon
1975	SiC continuous fiber
1978	Imide-decomposed Si_3N_4 powder
1981–1992	"Fine Ceramics" Project
1981	Ceramic plug for diesel engine
1983	First Int. Symp. on Ceramic Components for Heat Engines (at Hakone) [81]
1985	Ceramic turbocharger
1988–1998	300kW CGT Project (NEDO)
1990–1996	100kW CGT Project (PEC)

using brittle materials for industrial purposes. The research plan was divided into four development stages. Through the development of material and processing technology, the target performance was attained for monolithic materials in the form of test pieces in the 1st stage (FY1981–1983) and for monolithic materials as simple-shaped models in the 2nd stage (FY1984–1987). In the 3rd stage (FY1988–1990), new performance levels were achieved for the aforementioned materials including high toughness, surface strengthening, and complex-shaped models for practical use. In the final stage (FY1991–1992) overall results were to be verified by evaluation tests under practical circumstances for mechanical parts for ceramic turbines for coal gasification.

The 1st stage was for the development of the process fundamentals, the 2nd stage for the development of technology bases, and the 3rd stage for confirmation and verification. Table 1.2 shows the target values of the performance properties of fine ceramics in the initial stage [82,86]. The goals of the 3rd stage became more challenging. For example, the requirements increased to materials usable above 1400°C, with a minimum guaranteed strength of 400 MPa (40.8 kg/mm^2), a rate of rejection <20%, and a Weibull modulus $m > 20$.

A standard testing method for fine ceramics or advanced technical ceramics/engineering ceramics was first established in 1981 as Japanese Industrial Standard (JIS) JIS R-1601 "Testing Method for Flexural Strength of High Performance Ceramics" (revised in 1995). This was followed by other standards: R1602 (1986, revised in 1995) "Testing Methods for Elastic Modulus," R1603 (1987, revised in 1994) "Methods for Chemical Analysis of Silicon Nitride Powders," and R1604 (1987, revised in 1995) "Testing Method for Flexural Strength at Elevated Temperature." By 2000 more than 40 JISs for fine ceramics had been established. Additional international standards have been established for fine ceramics: ISO 14703 (2000) "Fine

TABLE 1.2 PERFORMANCE PROPERTY OBJECTIVES OF FINE CERAMICS (Adapted from ref. [86], with permission of the American Ceramic Society, www.ceramics.org, © 1982, all rights reserved. Also, ref. [86])

Classification	Objective	Values
High strength materials	≥1200°C, in air, after 1000hrs holding, average tensile strength	$\bar{\sigma} \geq 30kg/mm^2$
	Weibull modulus	$m \geq 20$
	1200°C, in air, after 1000hrs continuous loading, creep rupture strength	$\bar{\sigma} \geq 10kg/mm^2$
Corrosion-resistant materials	≥1300°C, in air, after 1000hrs holding, average tensile strength	$\bar{\sigma} \geq 20kg/mm^2$
	Weibull modulus	$m \geq 20$
	Corrosion resistance (weight gain)	$\leq 1mg/cm^2$
Wear-resistant materials	Room temperature	
	Wear resistance	$\leq 10^{-8}mm^3/kg\text{-}mm$
	Surface flatness	$R_{max} \leq 2\mu m$
	800°C, in air after 1000hrs holding, average tensile strength	$\bar{\sigma} \geq 50kg/mm^2$
	Weibull modulus	$m \geq 22$

$1kg/mm^2 = 9.8MPa$

ceramics (advanced ceramics, advanced technical ceramics) sample preparation for the determination of particle size distribution of ceramic powders," ISO 14705 (2000) "Test method hardness of monolithic ceramics at high temperature," ISO 14704 (2000) "Test method for flexural strength of monolithic ceramics at room temperature," and ISO 15490 (2000) "Test method for tensile strength of monolithic ceramics at room temperature." Several other ISO standards for fine ceramics will appear soon, and more than 10 are now at the preparatory or proposal stage. Japan has been the secretary country of technical committee ISO/TC206 and has been contributing to the implementation of these ISO standards since 1992.

The 300 kW NEDO CGT Project

The 300 kW NEDO CGT Project started in FY1988 as one of MITI's Moonlight Projects. There was a close relationship and much synergy between this CGT project and the Fine Ceramics Project. In the Fine Ceramics Project, a large number of ceramic manufacturers were engaged in the development of comprehensive ceramic fields such as materials, manufacturing technology, evaluation technology, etc. On the other hand, in the CGT project, only 3 or 4 ceramic suppliers with much experience in engineering ceramics participated by fabricating ceramic components for the engines. The objectives were to manufacture prototype components close to the practical application, and doing the performance testing in an actual engine.

The 300 kW Project was conducted at a time that it benefited from the precious experience of the European and USA gas turbine programs. It also benefited from the 100 kW automotive CGT project conducted concurrently from FY1990–1996 under management from the Petroleum Energy Center (PEC). The details of the two Japanese projects are described in chapters 13–16 in Volume I of this text, while research and development of ceramic materials is described in subsequent chapters in this volume. The fabrication development for the 300kW NEDO CGT Project, and the project objectives and target performance values for the ceramic materials are included in several of the chapters. A special issue for the 300 kW NEDO project was published in 1999 focusing on the ceramic materials development for the CGT [87].

Early design questions in the 300 kW Project included "Which components could be made of ceramics?" and "What should be the target values for the properties of the ceramics?" To help answer these questions in the preparatory stage of the Project, the common tests for high temperature strength and fracture toughness were carried out by several ceramic manufacturers to establish the present state of properties of each company's product in those days. The property values of Si_3N_4 and SiC materials were compared from the above tests and the improvements that would be needed in the subsequent five years were estimated.

The high-temperature bending test was actually performed by using test pieces about twice the size of the JIS specimens and close to the representative dimensions of gas turbine components. Representative data are shown in Table 1.3 [88]. For a pressureless sintered Si_3N_4, the high temperature bending strength at 1200°C was 960 MPa and the Weibull modulus was 42. It was a surprise to see the results in those days. The product was made in a very clean factory compared to a conventional factory, which resulted in the higher

TABLE 1.3 BENDING STRENGTH OF CANDIDATE MATERIALS [88]

Material	K_{IC}* ($MPa.m^{1/2}$)	Room temp. strength** (MPa)	High temp. strength** Temperature (°C)	Strength (MPa)	m
SiC	6.0	400	1350	260	9
	2.8	760		570	14
	3.0	340		600	6
Si_3N_4	7.0	790	1200	650	19
	5.9	900		960	42
Sialon	7.5	1030	1200	740	22

* Indentation method
**4-point bending, size 80 mm × 8 mm × 6 mm (JIS specimen size × 2)

strength and uniformity. However, the design targets for the ceramic components could not be satisfied until the final stage of the 300 kW CGT project, regardless of the fact that the material and technology of the above mentioned Si_3N_4 had practically already been developed at the starting point.

Another effort that was beneficial to the CGT program was the development and production of ceramic turbocharger rotors for automobiles between about 1985 and 1995. The success of the ceramic turbocharger was highly affected by the automobile designer's idea that the strength level of ceramics had been sufficient to use for this structural component. The ceramic turbocharger program clearly demonstrated that ceramics and ceramic designs could be highly reliable. The success of the CGT project further showed that the designer could use ceramics, and the designer recognized the increase in reliability of the ceramics and became very knowledgeable in using ceramics.

A very close collaboration between the designer and the ceramic manufacturer was required for both the turbocharger and gas turbine projects to be successful. By actually setting up the engine and carrying out the performance test, there was frank and severe discussion between both sides until the performance targets were achieved. This was certainly the primary motivating force for the success of this project. Actually, it took about three years before the designer and the ceramic fabricator began to understand and appreciate each other's technical language and perspective. In the mean time, it can also be said that three years are necessary to develop the same taste, even if they live under the same roof. In this 12-year project, if there was no laboratory visit from the other party for a while, a letter came saying, "It is a pretty lonely time without you." The research and development was carried out in this atmosphere.

It is needless to say that the success of the project was the fruit of the desperate effort of each member and close cooperation between the material side and the design side. Furthermore, a change of the plan was proposed at the conclusion of the evaluation at the intermediate stage, which involved a two-year time extension. This deserves special mention, for without this extension success would have been highly unlikely. If the economic depression in Japan had come earlier, the extension would not have been granted.

The two candidate ceramics for the CGT, Si_3N_4 and SiC, had almost the same capability for the application in the 1970s, each having merit and demerit. However at the beginning stage of the CGT project, it was recognized that the thermal shock was very severe at the start up and shut down of the CGT. In addition, the fairly rapid progress in the development of Si_3N_4 compared with that of SiC caused overwhelming use of Si_3N_4 since then. Figure 1.1 shows a history of the development of Si_3N_4 ceramics [89] and Figure 1.2 the improvement in 4-point bending strength of Si_3N_4 [90]. The details are documented in the Japanese literature and patents as well as in some publications in English [24,25,29–31,33,34,91–109].

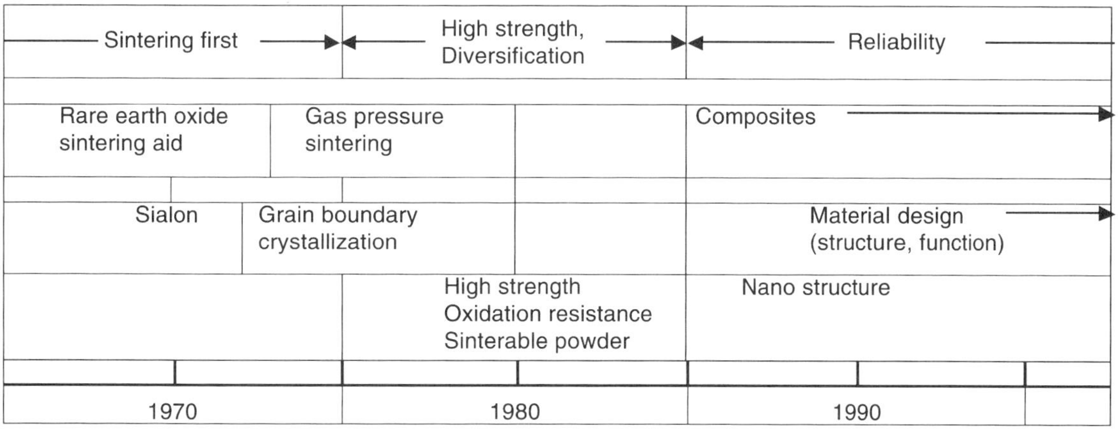

Figure 1.1 Development of Si_3N_4 Ceramics (Reprinted with permission from Prof. K. Komeya, ref. [89])

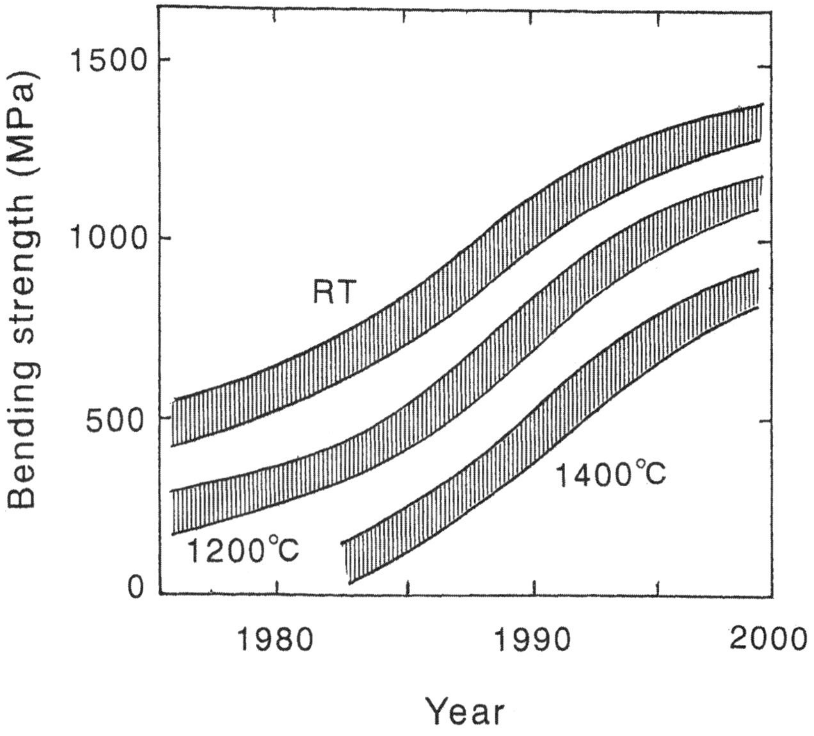

Figure 1.2 Improvement in 4-Point Bending Strength of Si_3N_4 Ceramics (Reprinted with permission from Dr. T. Sasa, ref.[90])

Ceramic Matrix Composite Materials

It is the invention of continuous or long SiC fiber that has contributed remarkably to the development of ceramic matrix composite (CMC) materials in Japan. The SiC long fiber made from a polycarbosilane precursor-polymer was invented by Prof. S. Yajima [110] of Tohoku University in 1975 and then Nicalon fiber was developed for the first time by Nihon Carbon Co., Ltd. This fiber is composed of Si, C and O. By heating above 1200°C SiO(g) and CO(g) are generated and the tensile strength decreases rapidly. In order to reduce the degradation of the fiber at high temperature, it is necessary to reduce the oxygen content introduced in the thermal oxidation curing process. The crosslinked polycarbosilane fiber was developed by the radiation curing process [111]. The ceramic fiber (Hi-Nicalon) obtained by this radiation curing process possesses less oxygen. Ube Industries Ltd. developed another competitive SiC-based ceramic fiber in 1978. The Si-Ti-C-O fiber (Tyranno) was continuously prepared from polytitano-carbosilane. Typical properties of the two fibers are listed in Table 1.4 [111]. Tonen Corporation also developed continuous Si_3N_4 fiber made from perhydropolysilazane in 1988 [112], but the manufacturing was regretfully ceased in 1996.

There were several CGT research projects in which CMCs were applied. In Volume I, the 100 kW Automotive CGT project operated by the Petroleum Energy Center (PEC) (see chapter 13), and in Volume II components for power generating gas turbines of Toshiba Corporation (chapter 18), and the project at the Research Institute of Advanced Material Gas Generator (AMG) (chapter 17) are included.

The PEC 100 kW Automotive CGT Development Project supported by MITI (FY1990–1996), one of the Japanese big two CGT Projects, had as one of its objectives the utilization of high toughness CMCs for the CGT. The materials development in this project was carried out in 3 groups. Each group consisted of petroleum companies as raw material manufacturers and ceramic companies as processed material and

TABLE 1.4 TYPICAL PROPERTIES OF POLYMER-DERIVED SIC-BASED FIBERS [111]

SiC fibers	Nicalon			Tyranno			
	NL-200	HI-Nicalon	HI-Nicalon-S	Lox M	Lox E	ZE	SA
Atomic composition	$SiC_{1.34}O_{0.36}$	$SiC_{1.39}O_{0.01}$	$SiC_{1.05}$	$SiTi_{0.02}C_{1.37}O_{0.32}$	$SiTi_{0.02}C_{1.59}O_{0.16}$	$SiZr_{0.01}C_{1.52}O_{0.05}$	SiC O, Al < 0.008
Tensile strength (GPa)	3.0	2.8	2.6	3.3	3.4	3.5	>2.5
Tensile modulus (GPa)	220	270	420	187	206	233	>300
Elongation (%)	1.4	1.0	0.6	1.8	1.7	1.5	0.8
Density (g/cm^3)	2.55	2.74	3.10	2.48	2.55	2.55	—
Diameter (mm)	14	14	12	11	11	11	10
Specific resistivity (Ω-cm)	10^3–10^4	1.4	0.1	30	0.8	0.3	—
Thermal expansion coeff. (10^{-6}/K)	—	3.5 (298–773K)	—	—	—	—	—

component manufacturers. By the end of the project, stationary components, such as a back plate and an orifice liner, were successfully made from short fiber or particle dispersed composite and the actual engine test was carried out for the performance test. These composites were made from Si_3N_4 or SiC matrix composite. By the time of the testing, it was realized that the dimensional accuracy and the resulting surface roughness were equivalent to those of monolithic ceramics and that the forming technology was almost established. Furthermore, the required component performance from a materials perspective was satisfied. As for the components made from long fiber CMC, testing of components in assemblies or engines was not performed, but in the thermal shock fracture tests and FOD (foreign object damage) tests, it was verified that the CMCs were superior to monolithic ceramics. In addition, 14 standards for test methods for CMCs were prepared in this project and have been widely proposed. T. Kagawa describes them later in this volume (see chapter 24).

SUMMARY OF EVOLUTION OF STRUCTURAL CERAMIC MATERIALS IN RUSSIA

Technologies, Properties and Facilities for Their Characterization

The initial interest in ceramics in Russia (USSR until 1990) manifested itself in the mid-1960s. The development of sputniks and other equipment for space exploration, research for rocket and aircraft engineering and, also, the refractory materials production on the basis of oxides (Al_2O_3, ZrO_2, MgO, etc), nitrides (AlN, BN, Si_3N_4), carbides (SiC, B_4C), and other materials served as a springboard for the application of ceramics as structural materials. The studies were routinely carried out at the institutes for internal combustion diesel machine applications.

Powders produced in Russia and the Ukraine were used as the starting materials. The quality of the powders was rather poor, they had a relatively high percentage of admixtures, including carbon, free oxygen, etc. So, the carbide- and nitride-based ceramics, mainly reaction-bonded and reaction-sintered, had flexure strength on the order of 200 MPa, porosity amounting to 30%, and with the oxidation processes being initiated from 800–900°C.

The oxide ceramics in the meantime featured good heat resistance (over 1300°C), were chemically sustainable, had a high toughness, and not a bad flexure strength (200–300 MPa), although with a rather high density.

In the late 1960s, the first ceramic parts were made of the oxide- and silicon nitride-based ceramics, though the tests of these parts did not yield any positive and optimistic results. The main causes of the failures were low flexure strength, low fracture toughness, and low thermal stability for the oxide materials and a fairly low heat resistance up to 1200°C and a high oxidizability at temperatures over 1000°C for the nitride- and carbide-based ceramic materials. The other no less important cause was that all the ceramic structures copied in detail the metal components and did not account for the specific properties inherent in ceramics. These first failures, on the one hand, promoted a dramatic increase in the interest in the ceramics for engine manufacture applications and, on the other hand, led to the emergence of special laboratories where the development and production of the novel structural ceramic materials (SCMs) for the manufacture of engines, including gas turbine engines (GTEs) were launched.

The first systematic research and the development projects under the industrial and state programs for the gas turbine applications were started in 1972. First and foremost was the Ministry of Aircraft Industry (MAP*) program, associated not only with the development of the novel SCMs fabrication techniques but also with the development of facilities to test SCM specimens under high-temperature conditions aimed at determining the mechanical and physical properties, the development of data processing methodologies, specimen geometries, etc. Actually, the activities were launched simultaneously at three Research Centers of the Soviet Union: in Moscow, Leningrad and Kiev. A number of the affiliates of MAP and the institutes of the Academy of Sciences joined them. Through the efforts by the scientists of Moscow (CIAM, VIAM, IVTAN, VNIIMET-MASH, etc), Leningrad (St.Petersburg) (VIO, GOI, VNIIMASH, Institute of Technologies), Kiev (IPM, ISM, IPP), Obninsk ("Technologiya"), Riga (CKTB of non-organic materials), etc, novel SCMs, technologies and facilities to produce SCMs and to test them were developed.

A strong incentive for the development of SCMs for gas-turbine manufacture applications was generated in 1975 by the Interbranch State Program "Composite Materials with Higher Impact Viscosity for Production of Parts and Elements of Gas Turbine Engines Operating Within 1300–1600°C Temperatures." It was by the late 1970s that such materials as OTM-904 (RBSN, "Technologiya"), OTM-907 (SRBSN, "Technologiya"), NKM-2 (RBSN, IPM), Tisinit (IPM), OTM-912 (SSC, "Technologiya"), etc. were developed. Over the same period, entire research entities to study SCM properties and to test SCMs thoroughly at high temperatures were established. The CIAM's complex involved facilities to carry out the mechanical tests of specimens in tension, compression, flexure and shear with temperatures as high as 1400–3000°C. Equipment for cold and hot spin testing was also established. At the IPP "Technologiya," the mechanical tests were carried out to identify the thermal-physical properties of ceramics, their structure, corrosion and erosion resistance.

At the same time, NDE methodologies for SCMs, including fluorescent dye penetrant inspection (FPI), X-ray and ultrasound control techniques as well as the codes for computing the stress-strain state, reliability, endurance and proof test regimens were progressing, at CIAM especially. By the middle of the 1970s, the level of the strength properties at room temperature for the oxide ceramics had increased to 700 MPa, and up to 800 MPa for the carbide- and nitride-based ceramics. Nevertheless, at temperatures above 1200°C, the flexure strength still did not exceed 300 MPa.

---

*MAP—Ministry of Aircraft Industry,

CIAM—Central Institute of Aviation Motor Manufacture;

VIAM—All-Russian Research-Technological Institute for Aircraft Materials,

IVTAN—High Temperature Institute of Academy of Sciences,

VNIIMETMASH—All-Russian Research Institute of Metal Engineering,

VIO—All-Russian Optic Institute,

GOI—State Optic Institute,

VNIIMASH—All-Russian Research Institute for Engineering;

IPM—Institute for Material Science Problems of the Ukraine,

ISM—Institute of Superhard Materials of the Ukraine,

IPP—Institute of Strength Problems of the Ukraine,

CKTB—Special Design Technological Bureau of Non-organic Materials.

In the early 1980s, many scientists believed that if they succeeded in producing SCMs with a flexure strength above 1000 MPa at 20°C, above 300 MPa at 1400°C, and with a density not exceeding 3.2 g/cm^3, and with a coefficient of linear expansion of about $3 \cdot 10^{-6}$, the task of GTE production was practically solved. By the mid 1980s, the positive results of tests of individual components of SCMs began to appear. The nozzle vanes for the GTE of Si_3N_4 had up to 20–30 operating hrs in test rigs and over 100 cycles under startup—shutdown conditions at temperatures from 20–1500°C. Attempts were made to manufacture all-ceramic rotors, and ceramic rotor blades in combination with a metal disc.

Projects for the development of the novel SCMs in the late 1980s and early 1990s in Russia (Soviet Union) gained in scope. Many new technologies for the manufacture of SCMs and products of SCMs were developed. As for the starting powders for SCM fabrication, the level of dispersivity was 10–100 nm for the chemical powders, 100 nm for the plasmachemical powders, and up to 500 nm for the mechanical powders while the purity of the starting powders was as high as 99.9999%.

The strength levels for individual SCMs exceeded 1000 MPa at 20°C, and were as high as 400–700 MPa at 1400–1500°C. Oxidizability of the nitride- and carbide-based SCMs at temperatures between 1200°C and 1300°C was actually zero.

An impressive contribution to the development of the novel nitride- and carbide-based SCMs was made by "Tekhnologhia". The company produced a complete series of various materials on the basis of a variety of technologies, such as packed reaction-bonded, hot-pressed brands of silicon nitride upgraded with other materials; sintered and upgraded silicon carbide and a number of oxide ceramics. Concurrent with the activities aimed at developing the novel SCMs, activities were undertaken to standardize and to introduce quality control procedures as well as to provide a prediction of the service life of the ceramics. The research labs of IPM and "Tekhnologhia" succeeded in developing methodologies for mapping out of and quality control for ceramic materials. The NDE techniques based on acoustic emissions were also developed.

In the mid-1980s, the operations were launched at an affiliate (Zaporozhye city) of the NIID Institute*) to develop a technology for the manufacture of a ceramic material on the basis of aluminum-boron-nitride compositions, but it was only by the middle of the 1990s that a new class of SCMs was developed in Russia (St.Petersburg, Research-Technological Institute for Power Engineering-NITI EM—"NIZ KTD") based on these compositions. Primarily, they are the zero-shrinkage structural ceramics "SUGRAV" and "ABN" which can be easily manufactured and have good weldability. They need no diamond machining after sintering, and lend themselves easily to the electro-milling treatment, etc. Their cost is relatively low and they are heat resistant up to 1600°C. Their fabrication techniques are based on the vacuum-compression treatment of the powder compositions, including impregnation by a melt metal.

Manufacture of Parts and Components of Gas Turbine Engines (GTE) of SCMs

The initial attempts to manufacture individual GTE parts of SCMs were undertaken in the 1970s. In particular, the aeroderivative GTE nozzle vanes ("Technologiya"), ceramic regenerative heat exchanger (CIAM), GTE flow passage parts (IPM) were manufactured of silicon nitride. In the early 1970s and in the mid-1970s, many organizations participated in collaborative activities to put the GTE ceramic projects into practice. Detailed studies into the nature of the cracking initiation process in SCMs and techniques for fracture toughness measurement (CIAM), the comparative analysis of the effect of various factors (temperature, heat treatment, etc) on the fracture toughness, etc., contributed greatly to the implementation of these projects. Somewhat later, a similar approach was used to study the cracking in composite ceramic materials, based on a mathematical method which considered the specific features of the composite structure and interactions between the reinforcement and the matrix during fluctuation in temperature and pressure.

In the mid-1980s, MAP started funding of a project to develop a ceramic aeroderivative engine with an output power (at take-off) up to 1300 hp. CIAM, "Technologiya" and the Rybinsk Motors Works joined the

*NIID—Research-Engineering Institute for Engines.

project. Under this project, two-shaft ceramic gas generators for the GTE, including the combustor, nozzle vanes, turbine wheel, bearings, and tip shroud seals were fabricated and tested. The gas generator design allowed the study of ceramic component operation both in test rigs and in an engine. Several versions of each ceramic element were studied. To manufacture them, various monolithic and composite SCMs and engineering approaches were employed. The impregnation materials, bearing cooling cycles, etc., varied. The turbine wheels survived the cold spin tests at 130% of the design speed and the hot spin tests. The nozzle vanes segments survived the proof tests at 1500°C. Regretfully, this project was not brought to an end since funding was ceased in the 1990s after the break-up of the USSR.

Nevertheless, despite the dramatic recession that all the above-stated entities in Russia and the Ukraine experienced, activities in the field of ceramic gas turbine manufacture continued; novel technologies to fabricate SCMs, including composites (IPM), and advanced techniques for their evaluation (IPP, CIAM) and testing ("Technologiya", CIAM) were developed. In the early 1990s, the government made a decision to establish the Research-Engineering Center "Ceramic Heat Engines" at the NITI EM (St.Petersburg). The Center was given the task to develop a ceramic stationary GTE for the gas industry in Russia. By the end of the 1990s, the above task was actually in the final stage of its implementation. All the main parts, components, facilities and high-temperature path systems (nozzle vanes, turbine wheel, combustor, airheater, scroll, hot gas ducts, tip shroud seals, etc) of this CGTE had been manufactured and tested individually. The engine demonstrator was scheduled to be tested as an assembly in 2001.

SUMMARY OF EVOLUTION OF CERAMICS FOR TURBINES IN WESTERN EUROPE

There is a common thread between the countries in Western Europe with major ceramic gas turbine programs. They had well established gas turbine engine manufacturers and they had an indigenous mature ceramics industry. The countries that have led the ceramic gas turbine development efforts in Western Europe were Great Britain, Germany, and Sweden, and, to a lesser extent, France and Italy.

A more recent development is the result of the increasing integration of Western Europe in the European Union (EU), a supra-national body which controls a substantial budget and which is actively promoting international collaboration between its member states. Over the past ten years ceramic gas turbine related development programs have been directly funded by the EU, and have stimulated activities in countries such as Italy, that were previously not very involved in this area.

This section will primarily focus on the ceramic development activities of the suppliers of monolithic ceramic components while the ceramic composite development will be discussed in a more summary manner.

Monolithic Ceramic Component Development

Great Britain. Much of the early developments and understanding of silicon nitride, sialon, and silicon carbide ceramics occurred in England by researchers at the Admiralty Materials Laboratory, the British Ceramics Research Association, the University of Newcastle upon Tyne, the University of Warwick, the Atomic Energy Laboratory at Harwell, the University of Leeds, and Joseph Lucas, Ltd. [113–115]. Subsequent fabrication efforts at Lucas, Advanced Materials Engineering (AME), and British Nuclear Fuels, Ltd. resulted in some of the first commercially-available products.

Collins and Gerby [4] reported in 1955 on fabrication of reaction-bonded silicon nitride by pressing, slip casting, and extrusion. They achieved 72% of theoretical density and conducted measurements of physical, thermal, electrical, and chemical properties. Parr and coworkers [5,116–119] and Popper and coworkers [7] provided further development and studies of reaction-bonded silicon nitride during the 1950s and 1960s. Davidge and coworkers [120,121] conducted property evaluations during the late 1960s and into the 1970s. During this same time, Godfrey and coworkers [122–124] explored engineering and design of reaction-bonded silicon nitride.

Reaction-bonded silicon, by nature of its fabrication process, contained substantial porosity that limited strength and oxidation resistance. Deely and coworkers [6] reported in 1961 that silicon nitride could be fully densified by adding MgO or Mg_3N_2 and hot pressing at 1850°C. Glenny and Taylor [125] compared some of the properties of the hot pressed material with those of reaction-bonded material. Material and hot pressing improvements were conducted at Lucas by Lumby and coworkers [18] and further studies were conducted at the University of Newcastle and at Leeds [126–128].

Parallel to the work on hot pressing of silicon nitride, Jack and coworkers [129–131] identified interesting compositions in the Si-N-O-Al system that became known as "Sialons." Some of these could be densified by pressureless sintering and led to important applications such as cutting tool inserts. Other efforts on sialons were conducted at Lucas [132,133] and at the University of Warwick [134].

Early efforts on dense SiC also were conducted in England. Popper [8] reported in 1960 about fabrication of dense SiC by reaction-sintering. Various improvements were made by Popper [135] and Forrest [136]. By the early 1970s, a high quality material referred to as "Refel" was being manufactured into combustor liners and other turbine shapes by British Nuclear Fuels, Ltd.

Germany. The ceramic materials, process and component development programs in the Federal Republic of Germany (FRG) mimicked similar efforts ongoing in the USA and Great Britain, albeit with a few years time lag. The first substantial program that accelerated the development work in the FRG was the "Keramische Komponenten für Fahrzeug-Gasturbinen" (Ceramic Components for Gas Turbines). This nine year program (1974–1983) was sponsored by the German Federal Ministry for Research and Technology with significant cost sharing by German industry. The program participants included two car manufacturers (Daimler-Benz AG, Volkswagen AG) and an engine manufacturer (MTU), six ceramic component suppliers (Annawerk-Ceranox, Feldmühle, Rosenthal Technik, ESK, SIGRI, Degussa), and six research institutes (Max Planck Institut—Stuttgart, Technical University—Berlin, Technical University—Clausthal-Zellerfeld, IzfP Saarbrücken, Technical University—Karlsruhe, and the German Aerospace Research Establishment (DFVLR—Cologne), providing supporting technologies.

The progress on the program was reviewed at three workshops in Bad Neuenahr, Germany, and the readers that are conversant with the German language and interested in the detailed results are referred to the proceedings [137–139]. English language summaries of the program are also available [140,141]. The work at the ceramic suppliers is described in chapter 9 of this volume, and the work at the gas turbine manufacturers in chapter 21 of Volume I.

Few reasonably well established ceramic fabrication technologies were available at the start of the program and much effort was expended in developing these technologies with eventual commercialization in mind. For low cost manufacturing, high-purity and very fine sinter-active powders are required, and a lot of effort went into improving the quality of powders for the two key materials, SiC and Si_3N_4. H.C. Starck, the leading German powder manufacturer, focused on new ways to process powders and establish correlations between powder characteristics, powder processing, and final product properties. Powder manufacturing was tailored to the various processes for the fabrication of the two major materials, silicon nitride and silicon carbide. The increase in specific surface area of high-grade α-Si_3N_4 powders is a good indicator for the improvement in powder quality. Until 1978 relatively coarse Si_3N_4 powders were produced (grade H1-specific surface area: 9 m^2/g), primarily for hot pressing or sintering with high amounts of sintering additives. By 1980 the finer LC10 powder became available (15 m^2/g), and by 1981 the very fine LC12 powder (24 m^2/g) [140].

The main ceramics available at the beginning of the German national program were reaction bonded silicon nitride (RBSN), hot pressed silicon nitride (HPSN), and siliconized silicon carbide (SiSiC). The improvement in material properties over the course of the "Ceramic Components for Gas Turbines" program was quite dramatic as witnessed by the marked increase in room temperature 4-point bending strength of RBSN and HPSN which almost doubled. Typical values for RBSN were slightly over 150 MPa in 1974, at the beginning of the program, and increased to about 300 MPa when the program ended in 1983. Comparable strength values for HPSN at the start and completion were about 500 MPa and 900 MPa, respectively.

Process modifications and new processes were also developed which improved the properties or offered enhanced shaping capabilities for component fabrication. The incorporation of a final sintering step in the RBSN process at Annawerk resulted in sintered RBSN which doubled the room temperature flexure strength of this material from about 300 MPa to 650–750 MPa. Replacing the sintering additive magnesium oxide with yttria in HPSN improved oxidation resistance and strength. Processes were developed for newer materials such as sintered silicon carbide (SSiC), hot pressed silicon carbide (HPSC), and hot isostatic pressed silicon carbide (HIPSiC). The latter two materials had good room temperature flexure strength, in the 550–600 MPa range.

The improvements in strength were accompanied by an increasing ability of the German ceramic manufacturers to fabricate components with shapes of varying complexity. Injection molding for small complex shaped parts and slip casting and isostatic pressing for larger thin walled parts were the main fabrication routes. Table 1.5 lists various components fabricated using the common forming techniques and processing routes.

The German National Program contributed to remarkable progress in materials and process capability at the German ceramic component manufacturers which continued under the "Structural Ceramics" program from 1983 to 1986, and which broadened the application range to include engines other than gas turbines. The shift in focus was in part the result of the realization that the gas turbine goals were extremely challenging and would not likely result in near term commercialization. Promising applications that were more near term were emerging to which the ceramic fabricators directed their attention.

Areas of interest were high temperature components for burners, furnaces and heat exchangers, insulating parts, wear parts, seals, bearings, cutting tool inserts, automotive components, catalyst carriers, implants, sensors, and various parts for the electronics industry, fiber-optic applications, and critical components for the nuclear industry [142]. The progress achieved under the German National Program did in fact contribute to the more rapid advancements in related areas since the high temperature strength, durability and reliability of SiC and Si_3N_4 had dramatically improved as a result of that program.

One rather successful application of monolithic ceramics for gas turbines is the use of combustor tiles in Siemens VX4.2 gas turbines. The tiles consist of high-alumina refractories based on a corundum-sillimanite mixture with a relatively high silica content. The combustor tiles have been used for many years and over 8

TABLE 1.5 VARIOUS COMPONENTS FABRICATED BY THE GERMAN CERAMIC MANUFACTURERS

Material	Process			
	Injection molding	Slip casting	Isostatic Pressing	Tape casting
RBSN	Turbine blades	Nose cones		Recuperator foils
	Blade rings	Radial turbine rotors		
	Nozzles			
	Multi-vane segments			
	Integral nozzle rings			
	Disk blanks			
SRBSN	Turbine blades			
HPSN	Turbine blades			
	Rotor castings			
SiSiC	Nozzles	Turbine blades	Combustors	
	Multi-vane segments	Radial turbine rotors		
	Integral nozzle rings	Combustors		
		Nozzle rings		
		Nose cones		
		Scrolls		
SSiC	Turbine blades	Combustors	Rings	
	Nozzles	Rotors	Balls	

million operating hours have been accumulated. More recently Siemens has been investigating alternative tile materials to meet the requirements for hotter combustor liner conditions. Chapter 28 in Volume I describes the Siemens experience in detail.

Sweden. The ceramic development programs in Sweden up to 1983 have been summarized by Kronogård and Malmrup [143]. Sweden had a significant and vibrant ceramics industry manufacturing electronic, sanitary, enamel, and household ceramics, and bricks and refractories, totaling several hundred million dollars in the early 1980s. Sweden also had a significant gas turbine industry. The large Volvo company had several divisions with engine manufacturing activities for cars, trucks, buses, and boats, for civil and military jet aircraft. One Volvo subsidiary, United Turbine, had developed an automotive gas turbine. Another automotive manufacturer was Saab-Scania. Stal Laval developed and manufactured gas and steam turbines for power generation, industrial use and ship propulsion.

Based on Sweden's background in engines, particularly in gas turbines and Sterling engines and with an indigenous ceramics industry, the Swedish Board for Technical Development (STU) in 1976 decided to financially support the work in ceramic materials for heat engines. Two programs encompassed two parallel activities: one aimed at material development, the other at engine development. The long-term goal of the material development program was the domestic production of ceramic heat engine parts. The materials and process development work was started at R&D institutes with the ceramic suppliers becoming progressively more involved. The engine development was performed by the engine companies. The development strategy was to procure parts from overseas to give the Swedish industry time to develop its own experimental parts.

The ceramic engine turbine development work in Sweden started later than in Germany and the focus was on more recent process technologies. The initial development work on high temperature ceramics was carried out at the Swedish Institute for Silicate Research in the 1979–1981 time frame, and the priority materials were pressureless sintered silicon carbide and silicon nitride. The studies progressed from the fabrication and testing of basic materials properties using test bars, to manufacturing studies using injection molding and slip casting as the basic forming processes, to sintering of simple shapes for which the properties were determined.

The premier ceramic fabricator on the Swedish National Program was Asea, primarily known as a manufacturer of electrical equipment and power stations. But Asea also produced equipment for hot isostatic pressing (HIP) and in 1965 the company set up a high pressure laboratory at Robersfors. A significant amount of development work had been carried out at the Robersfors laboratory prior to the start of the Swedish National Program. It was at this laboratory that a new HIP technique to process Si_3N_4 was developed [38]. The advantage of the HIP process was that it produced fully isotropic materials of complete density with minimal sintering aids. The process steps included initial powder preparation, the forming of the green body by any process, the encapsulation of the shaped body with a glass powder, followed by the evacuation of the capsule and hot isostatic pressing and sintering at temperatures in the 1700–1800°C range and >100 MPa pressure. The encapsulant was removed after the HIP sintering step, followed by any finish machining that was required to achieve the final shape and tolerances.

The HIP sintered Si_3N_4 had attractive properties: a flexure strength of 700 MPa at room temperature and 550 MPa at 1370°C with a Weibull modulus m = 20. The creep rate was 1.6×10^{-6}/hr at 70 MPa/1300°C. Shrinkage was totally uniform and predictable. Asea transitioned the technology from the laboratory experiment level to a process capable of handling volume part production [143]. The Asea HIP sinter development work has been described in detail in chapter 10 of this volume. With the process, complex parts such as integral automotive turbine rotors were fabricated. The fabricated parts were tested in rigs and gas turbines. A milestone was achieved when in March 1982 professor Kronogård drove a modified Volvo 760 powered by a MK1 turbine with a HIP sintered silicon nitride rotor on public roads. Asea licensed its HIP encapsulation technology to several other companies including Norton Company and AlliedSignal Ceramic Components during the 1980s. They subsequently incorporated the process for their HIPSN materials.

Asea's HIP sintered materials technology was also used in the European AGATA (Advanced Gas Turbines for Automobiles) program which was conducted from 1992–1998. This program which was a collaborative effort of a number of Swedish and French companies had the goal to develop three components for a hybrid electric vehicle gas turbine with a 60 kW output. These components were a ceramic radial turbine wheel, a catalytic combustor, and a ceramic heat exchanger. AC Cerama (the new entity that had emerged from the former Asea) was selected to fabricate the turbine wheel from its CSN 101 HIPed Si_3N_4. AC Cerama combined two green forming methods in a two-piece process. The wheel hub was formed by cold isostatic pressing (CIP) and green machining and the blade ring was injection-molded. The two pieces were then green-joined, glass-encapsulated and HIP sintered. Later AC Cerama developed a process to form a single piece component by injection molding. CSN 101 had improved properties over previous HIP sintered Si_3N_4 vintages with a flexure strength of 955 MPa at room temperature (m = 28) and 630 MPa at 1350°C [144].

Another CSN 101 component fabricated by AC Cerama for the AGATA project was the catalytic combustor afterburner cone.

France. France has a strong ceramic manufacturing sector. Over the past decade one of its major ceramic companies, Saint Gobain, acquired the Norton and Carborundum companies in the U.S. and AC Cerama in Sweden, and strongly redirected the efforts of these companies into profit making manufacturing, refocusing on their core competencies in materials processing and market niches, and away from government funded R&D in ceramic gas turbine development and related areas. Prior to this refocusing French companies participated in a number of programs, often collaborative efforts with partners elsewhere, notably in Sweden and Russia. France has made unique contributions to the development of fiber-reinforced composites and these will be mentioned in a separate section below.

One area where France has played a leading role with monolithic ceramic expertise is in the development of materials and fabrication technologies for ceramic recuperators. The efforts gained momentum when Céramiques et Composites (C&C) was formed in 1985 and continued at its spin-off, Boostec Industries. Specific recuperators can be configured with gas turbines, primarily for automotive applications, which makes them of interest to this book. Starting in 1983, C&C, in collaboration with the French National Aerospace Institute (ONERA) embarked on a program to develop a ceramic recuperator for a turbine cycle for aircraft engines. The heat exchanger was of the tubular type and initially fabricated from SSiC, but later Sialons were used because of their superior thermomechanical resistance.

C&C became a major ceramic supplier to Volvo Aero Corporation under the AGATA program for the development of a compact plate type heat exchanger designed to operate at temperatures up to 1027°C and pressures up to 4 bar. The recuperator elements consisted of stacked flat and corrugated plates, 0.2–0.3 mm thick, and were fabricated by extrusion and calendering. An extensive study was done to select the recuperator material, cordierite and SSiC being the two primary candidates. Cordierite was eventually selected because of thermomechanical stress considerations. C&C also supplied the mullite-zirconia (SM3) honeycomb substrates for the catalytic combustor. The development activities of C&C have been described in detail in chapter 11 of this volume.

Another French company that contributed to the AGATA program was Aerospatiale which fabricated the whisker-reinforced silicon nitride housing for the catalytic elements as well as the silicon nitride combustor housing.

Ceramic Matrix Composite Development

Western European companies have played a major role in developing the ceramic matrix composite (CMC) materials that have recently received increasing interest from the gas turbine design community. The developments of CMCs were stimulated by prior experience with polymer and intermetallic composites and much of the early theoretical framework for the development was derived from prior experience in those areas.

The French company SEP (Société Européenne de Propulsion—currently a division of Snecma Moteurs) was largely responsible for developing the chemical vapor infiltration process for carbon and silicon carbide fiber-reinforced ceramic matrix composites (C/SiC and SiC/SiC CMCs) which started in the early 1980s. Over time SEP's applications were mainly in the area of jet engine static components such as after-burner components (exhaust cones and flame-holders) and exhaust nozzles. An established SEP CVI (chemical vapor infiltration) SiC/SiC CMC (2.5 D) has a yield tensile strength of 360 MPa and 0.8% elongation at failure. The linear elastic limit is 80 MPa (all data at room temperature) [145].

Since 1984, SEP has participated in several R&D testing programs of SiC/SiC and C/SiC composites for rotating and stationary components of rocket and turbojet engines. These engines have relatively short mission lives, up to a few hundred hours and gas temperatures are on the order of 1100°C to 1400°C. Testing of C/SiC and SiC/SiC CMC rotors at peripheral (blade tip) speeds between 400 to 600 m/s gave component lives on the order of one hour. SEP indicated that the achievable speed level did not provide sufficient margin over the usual peripheral speeds of turbomachinery. Development efforts were therefore refocused on stationary components [145]. A SiC/SiC CMC (CERASEP®A410) with a self-sealing matrix was used to fabricate nozzle seals which were installed in a F100-PW-229 engine and tested for over 250 hours, reportedly without experiencing damage [146]. Materials property data for CERASEP®A410 have been reported [147].

A major developer of CMCs since the mid-1980s in Germany is MAN Technology AG. MAN modified the CVI process to include temperature and pressure gradients and developed the Liquid Polymer Impregnation (LPI) and Pyrolysis processes, essentially to reduce processing time compared to the conventional CVI process which takes place at atmospheric pressure. A gradient CVI SiC/SiC with 42–47 vol% fiber (SiC fiber not specified) had a room temperature flexure strength of 450–550 MPa and 0.5–0.7% strain to failure. A CVI SiC/SiC tube, simulating a combustor chamber was tested in a small gas turbine for 50 hrs under cyclic loading with peak temperatures up to 1500°C, apparently without damage [148].

Holmquist et al. [149] reported on potential materials combinations for all-oxide ceramic matrix composites as part of a broad four year study under a Brite/Euram project. Participants in the project were Rolls-Royce (UK), Snecma Moteurs and ONERA (France), Volvo Aero Corporation and AC Cerama (Sweden). A promising model composite incorporated sapphire fibers (Saphikon Inc., USA), a ZrO_2 interphase coating deposited by a sol-slurry aproach, and an alumina matrix, infiltrated either by gas pressure infiltration or tape casting. The prepregs were cut, stacked and laminated. Consolidation was done by either hot pressing or HIPing. Flexural strength exceeded 600 MPa for the best material. Preliminary test results indicated thermal stability of the composite at 1400°C for about 1000 hrs [149,150].

At Alstom in the UK, experimental combustor development is being carried out with a SiC based CMC and an oxide based thermal protection system (TPS). The TPS acts as a thermal barrier protecting the underlying CMC from thermal degradation. An overview over the approach can be found in chapter 27 of Volume I.

A relatively recent activity has been the development of a new borosilicon-nitride fiber by Bayer AG which reportedly has oxidation resistance up to 1600°C. The fiber is fabricated from a polymer precursor by a pyrolysis process. At high temperatures a dense SiO_2 layer combines with an additional Si-O-N-B layer to give the material its good oxidation resistance [151].

REFERENCES

[1] Probst, H. B., "Structural Ceramics in Heat Engines—The NASA Viewpoint," in *Proceedings of the First International Symposium on Ceramic Components for Engines*, Hakone, Japan, Oct. 17–19, 1983, S. Sōmiya, E. Kanai and K. Ando, eds., KTK Sci. Pub. Tokyo, Japan, 1984, and Elsevier Applied Science Publishers Ltd., London, UK, 1986. pp. 45–58.

[2] Conway, H. M., "The Possible Use of Ceramic Materials in Aircraft Propulsion Systems," NACA CB No. 4D10, 1944.

[3] Morgan, W. C., and Deutsch, G. C., "Experimental Investigation of Cermet Turbine Blades in an Axial-Flow Turbojet Engine," NACA TN-4030, Oct. 1957.

[4] Collins, J. G., and Gerby, R. W., "New Refractory Uses for Silicon Nitride Reported," J. Metals, 7[1], 612–615, 1955.

[5] Parr, N. L., "Silicon Nitride, a New Ceramic for High Temperature Engineering and Other Applications," Res. (Lon.) 13, 261–269, 1960.

[6] Deely, G. G., Herbert, J. M., and Moore, N. C., "Dense Silicon Nitride," Powder Met., 8, 145–151, 1961.

[7] Popper, P., and Ruddlesden, S. N., "The Preparation, Properties and Structure of Silicon Nitride," Trans. Brit. Ceram. Soc., 60, 603–626, 1961.

[8] Popper, P., "The Preparation of Dense Self-Bonded Silicon Carbide," in *Special Ceramics*, P. Popper, ed., Academic Press, New York, USA, 1960. pp. 209–219.

[9] Godfrey, D. J., and Taylor, P.G., "Designing with Brittle Materials," Eng. Mater. Design, 12 [9], 1339–1342, 1969.

[10] Godfrey, D. J., "The Use of Ceramics in Engines," Proc. Brit. Ceram. Soc., 26, 1–15, 1978.

[11] Mangels, J. A., "Development of Injection Molded Reaction Bonded Si_3N_4," in *Ceramics for High Performance Applications II, Proceedings of the Fifth Army Materials Conference,* Newport, RI, USA, March 21–25, 1977, J. J. Burke, E. N. Lenoe, and R. N. Katz, eds., Brook Hill Publishing Company, Chestnut Hill, MA, USA, 1978. pp. 113–130.

[12] Washburn, M. E., and Baumgartner, H. R., "High-Temperature Properties of Reaction-Bonded Silicon Nitride," in *Ceramics for High Performance Applications, Proceedings of the Second Army Materials Conference,* Hyannis, MA, USA, November 13–16, 1973, J. J. Burke, A. E. Gorum, and R. N. Katz, eds., Brook Hill Publishing Company, Chestnut Hill, MA, USA, 1974. pp. 479–491.

[13] Gersch, H., Mann, D., Rorabaugh, M., and Schuldies, J. J., "Slip-Cast and Injection-Molding Process Development of Reaction-Bonded Silicon Nitride at AiResearch Casting Company," in *Proc. DARPA/NAVSEA Ceramic Gas Turbine Demonstration Engine Program Review*, J. W. Fairbanks and R. W. Rice, eds., Battelle Columbus Laboratories, Columbus, OH, USA, Report No. MCIC-78-36, MA, USA, March 1978. pp. 313–340.

[14] Richerson, D. W., Smyth, J. R., and Styhr, K., "Material Improvement Through Iterative Process Development," Cer. Eng. and Sci. Proc., 4 [9–10], 841–852, 1983.

[15] Alliegro, R. A., and Coes, S. H., "Reaction Bonded Silicon Carbide and Silicon Nitride for Gas Turbine Applications," ASME paper 72–GT-20, presented at the Gas Turbine Conference, San Francisco, CA, USA, March 26–30, 1972.

[16] Lucek, J. W., Torti, M. L., Weaver, G. Q., and Olson, B. A., "Cast Densified Silicon Carbides," SAE Paper 790253, 1979.

[17] Whalen, T. J., Noakes, J. E., and Terner, L. L., "Progress on Injection-Molded, Reaction-Bonded SiC," in *Ceramics for High Performance Applications II, Proceedings of the Fifth Army Materials Conference,* Newport, RI, USA, March 21–25, 1977, J. J. Burke, E. N. Lenoe, and R. N. Katz, eds. Brook Hill Publishing Company, Chestnut Hill, MA, USA, 1978. pp. 179–189.

[18] Lumby, R. J., and Coe, R. F., "The Influence of Some Process Variables on the Mechanical Properties of Hot-Pressed Silicon Nitride," Proc. Brit. Ceram. Soc., 15, 91–101, 1970.

[19] Richerson, D. W., "Effects of Impurities on the High Temperature Properties of Hot-Pressed Silicon Nitride," Bull. Am. Ceram. Soc., 52 [7], 560–562, 1973.

[20] Baumgartner, H. R., and Richerson, D. W., "Inclusion Effects on the Strength of Hot-Pressed Si_3N_4," in *Fracture Mechanics of Ceramics*, Vol. 1, R. C. Bradt, D. P. H. Hasselman, and F. F. Lange, eds., Plenum Press, New York, NY, USA, 1974. pp. 367–386.

[21] Baker, R.R., Ezis, A., Hartsock, D.L., and Goodyear, M.U., "Developments in Press Bonding of Duo-Density Rotors," in *Ceramics for High Performance Applications II, Proceedings of the Fifth Army Materials Conference*, Newport, RI, USA, March 21–25, 1977, J. J. Burke, E. N. Lenoe, and R. N. Katz, eds., Brook Hill Publ., Chestnut Hill, MA, USA, 1978. pp. 207–230.

[22] Gazza, G. E., "Hot-Pressed Silicon Nitride," J. Am. Ceram. Soc., 56[12], 662, 1973.

[23] Terwilliger, G. R., "Properties of Sintered Silicon Nitride," J. Am. Ceram. Soc., 57[1], 48–49, 1974.

[24] Masaki, H., and Kamigaito, O., "Pressureless Sintering of Silicon Nitride with Additions of MgO, Alumina, and/or Spinel," Yogyo Kyokai Zasshi (J. Ceram. Soc. Japan), 84[10], 508–512, 1976 (in Japanese).

[25] Oda, I., Kaneno, M., and Yamamoto, N., "Pressureless Sintered Silicon Nitride," in *Nitrogen Ceramics, Proc. NATO-ASI Nitrogen Ceramics Conf.*, Canterbury, England, UK, Aug. 16–27, 1976. Noordhoff International Publishing, Leyden, The Netherlands, 1977, pp. 359–365.

[26] Greskovich, C., and Rosolowski, J. H., "Sintering of Covalent Solids," J. Am. Ceram. Soc., 59[7–8], 336–343, 1976.

[27] Priest, H. F., Priest, G. L., and Gazza, G. E., "Sintering of Silicon Nitride Under High Nitrogen Pressure," J. Am. Ceram. Soc., 60 [1–2], 81, 1977.

[28] Rowcliffe, D. J., and Jorgensen, P. J., "Sintering of Silicon Nitride," in *Proc. Workshop on Ceramics for Heat Engines*, Energy Res. & Dev. Adm., CONRT, Orlando, FL, USA, Jan. 24–26, 1977. pp. 191–196.

[29] Mitomo, M., "Pressure Sintering of Silicon Nitride," J. Mater. Sci., 11[6], 1103–1107, 1976.

[30] Tani, E., Umebayashi, S., Kishi, K., Kobayashi, K., and Nishijima, M., "Gas-pressure Sintering of Silicon Nitride with Concurrent Addition of Alumina and 5 wt.% of Rare Earth Oxide: High Fracture Toughness Silicon Nitride with Fiber-like Structure," Am. Ceram. Soc. Bull., 65[9], 1311–1315, 1986.

[31] Matsuhiro, K., and Takahashi, T., "The Effect of Grain Size on the Toughness of Sintered Silicon Nitride," Ceram. Eng. and Sci. Proc., 10[7–8], 807–816, 1989.

[32] Li, C. W., and Yamanis, J., "Super-tough Silicon Nitride with R-Curve Behavior," Ceram. Eng. and Sci. Proc., 10[7–8], 632–645, 1989.

[33] Komeya, K., Tsuge, A., and Nishida, K., "Silicon Nitride-based Refractory Composite," Japanese Patent No. 74,110,708, Oct. 22, 1974.

[34] Tsuge, A., Nishida, K., and Komatsu, M., "Effect of Crystallizing the Grain-Boundary Glass Phase on the High-Temperature Strength of Hot-Pressed Si_3N_4 Containing Y_2O_3," J. Am. Ceram. Soc., 58[7–8], 323–326, 1975.

[35] Smith, J. T., and Quackenbush, C. L., "A Study of Sintered Silicon Nitride Compositions with Yttria and Alumina Densification Additives," in *Proc. Int. Symp. on Factors in Densification and Sintering of Oxide and Nonoxide Ceramics*, Hakone, Japan, Oct. 3–6, 1978, S. Sōmiya and S. Sayto, eds., Association of Science Documents Information, Tokyo Institute of Technology, 1979. pp. 426–443.

[36] Mangels J. A., "Sintered Reaction-Bonded Silicon Nitride," Ceram. Eng. and Sci. Proc., 2[7–8], 589–593, 1981.

[37] Giachello, A., and Popper, P., Ceramurgia Int., 5, 110, 1979.

[38] Larker, H., Adlerborn, J., and Bohman, H., "Fabricating of Dense Silicon Nitride Parts by Hot Isostatic Pressing," SAE Publication 770335, 1977.

[39] Wills, R. R., Brockway, M. C., McCoy, L.G., and Niesz, D.E," Preliminary Observations on the Hot Isostatic Pressing of Silicon Nitride," *Ceram. Eng. Sci. Proc.* 1, 534–539, 1980.

[40] Pujari, V. K., Tracey, D. M., Corbin, N. D., Foley, M. R., Garg, A. K., Paille, N. I., Pelletier, P. J., Sales, L. C., Willkens, C. A., and Yeckley, R. L., "Reliability Improvements of High Strength Silicon Nitride Through Process Optimization and Control," in *Proc. Annual Automotive Technology Development Contractors' Meeting*, Dearborn, MI, USA, Oct. 28–31, 1991, SAE P-256, Society of Automotive Engineers, Warrendale, PA, USA, 1992. pp. 137–144.

[41] Pollinger, J. P., "Improved Silicon Nitride Materials and Component Fabrication Processes for Aerospace and Industrial Gas Turbine Applications," ASME paper 95-GT-159, presented a the International Gas Turbine and Aeroengine Congress and Exposition, Houston, TX, USA, June 5–8, 1995.

[42] Kossowsky, R., "Creep and Fatigue of Silicon Nitride as Related to Microstructures," in *Ceramics for High Performance Applications, Proceedings of the Second Army Materials Technology Conference,* Hyannis, Mass, USA, November 13–16, 1973, J. J. Burke, A. E. Gorum, and R. N. Katz, eds., Brook Hill Publ., Chestnut Hill, MA, USA, pp. 347–371, 1974.

[43] McLean, A. F., and Fisher, E. A., *Brittle Materials Design, High Temperature Gas Turbine,* AMMRC TR 81–14, Final Report, March 1981.

[44] Richerson, D. W., and Johansen, K. M., *Ceramic Gas Turbine Engine Demonstration Program,* Final Report DARPA/Navy Contract N00024-76-C-5352, May 1982.

[45] Fairbanks, J. W., and Rice, R. W., eds., *Proc. 1977 DARPA/NAVSEA Ceramic Gas Turbine Demonstration Engine Program Review*, MCIC Report MCIC-78-36, March 1978.

[46] Evans, A. G., and Wiederhorn, S. M., "Crack Propagation and Failure Prediction in Silicon Nitride at Elevated Temperatures," J. Mater. Sci., 9, 270–278, 1974.

[47] Evans, A. G., Russell, L. R., and Richerson, D. W., "Slow Crack Growth in Ceramic Materials at Elevated Temperatures," Met. Trans. A., 6A, 707–716, 1975.

[48] Quinn, G. D., *Characterization of Turbine Ceramics after Long-Term Environmental Exposure,* AMMRC-TR-80-15, Final Report, April 1980.

[49] Larsen, D. C., and Adams, J. W., *Property Screening and Evaluation of Ceramic Turbine Materials,* Final Technical Report AFWAL-TR-83-4141, April 1984.

[50] Chuck, L., Goodrich, S. M., Hecht, N. L., and McCullum, D. E., "High-Temperature Tensile Strength and Tensile Stress Rupture Behavior of Norton/TRW NT-154 Silicon Nitride," Ceram. Eng. and Sci. Proc., 11 [7–8],1007–1027, 1990.

[51] Ferber, M. K., Jenkins, M. G., and Tennery, V. J., "Comparison of Tension, Compression, and Flexure Creep for Alumina and Silicon Nitride Ceramics," Ceram. Eng. and Sci. Proc., 11[7–8],1028–1045, 1990.

[52] Cranmer, D. C., and Richerson, D. W., eds., *Mechanical Testing Methodology for Ceramic Design and Reliability*, Marcel Dekker, New York, 1998.

[53] Sanders, W. A., and Probst, H. B., "High Gas Velocity Burner Tests on Silicon Carbide and Silicon Nitride at 1200°C," in *Ceramics for High Performance Applications, Proceedings of the Second Army Materials Technology Conference,* Hyannis, MA, USA, November 13–16, 1973,

J. J. Burke, A. E. Gorum, and R. N. Katz, eds., Brook Hill Publishing Company, Chestnut Hill, MA, USA, 1974. pp. 493–531.

[54] Carruthers, W. D., Richerson, D. W., and Benn, K. W., "Combustion Rig Durability Testing of Turbine Ceramics," in *Ceramics for High Performance Applications III Reliability, Proceedings of the Sixth Army Materials Technology Conference,* Orcas Island, WA, USA, July 10–13, 1979, E. M. Lenoe, R. N. Katz, and J. J. Burke, eds., Plenum Press, New York, NY,USA, 1983. pp. 571–595.

[55] Schienle, J. L., *Durability Testing of Commercial Ceramic Materials,* Final Report DOE/NASA/0027-1, NASA CR-198497, Jan. 1996.

[56] More, K. L., Tortorelli, P. F., Ferber, M. K., Walker, L. R., Keiser, J. R., Brentnall, W. D., Miriyala, N., and Price, J. R., "Exposure of Ceramics and Ceramic Matrix Composites in Simulated and Actual Combustor Environments," ASME paper 99-GT-292, presented at the International Gas Turbine and Aeroengine Congress and Exposition, Indianapolis, IN, USA, June 7–10, 1999.

[57] Ferber, M. K., Lin, H. T., Parthasarathy, V., and Wenglarz, R. A., "Degradation of Silicon Nitrides in High Pressure, Moisture-Rich Environments," ASME paper 2000-GT-0661, presented at the International Gas Turbine and Aeroengine Congress and Exposition, Munich, Germany, May 8–11, 2000.

[58] Richerson, D. W., Finger, D. G., and Wimmer, J. M., "Analytical and Experimental Evaluation of Biaxial Contact Stress," in *Fracture Mechanics of Ceramics, Vol. 5, Surface Flaws, Statistics and Microcracking*, R. C. Bradt, A.G. Evans, D.P.H. Hasselman, and F. F. Lange, eds., Plenum Press, New York, NY, USA, 1983. pp. 163–184.

[59] Richerson, D. W., Carruthers, W. D., and Lindberg, L. J., "Contact Stress and Coefficient of Friction Effects on Ceramic Interfaces," in *Surfaces and Interfaces in Ceramic and Ceramic-Metal Systems*, J. Pask and A. G. Evans, eds., Plenum Press, New York, NY, USA, 1981. pp. 661–667.

[60] Rorabaugh, M. E., and Styhr, K. H., "Reliability of Slip Cast Silicon Nitride Components," in *Ceramics for High Performance Applications III Reliability, Proceedings of the Sixth Army Materials Technology Conference,* Orcas Island, WA, USA, July 10–13, 1979, E. M. Lenoe, R. N. Katz, and J. J. Burke, eds., Plenum Press, New York, NY, USA, 1983. pp. 309–322.

[61] Torti, M. L., "Processing Hot Pressed Silicon Nitride for Improved Reliability: HS-110 to NC-132," in *Ceramics for High Performance Applications III Reliability, Proceedings of the Sixth Army Materials Technology Conference,* Orcas Island, WA, USA, July 10–13, 1979, E. M. Lenoe, R. N. Katz, and J. J. Burke, eds., Plenum Press, New York, NY, USA, 1983. pp. 261–273.

[62] McEntire, B. J., Taglialavore, A. P., Heichel, D. N., Bright, E., Yeckley, R. L., Holowczak, J. E., and Quackenbush, C. L., "Silicon Nitride Component Development for Advanced Gas Turbine Engines," in *Proc. 27th Automotive Tech. Devt. Contractors' Coord. Meeting*, Dearborn, MI, USA, Oct. 23–26, 1989, SAE P-230, Society of Automotive Engineers, Warrendale, PA, USA, 1990. pp. 341–356.

[63] Pollinger, J. P., and Busovne, B. J., "Development of Pressure Slip Cast Silicon Nitride Rotors for ATTAP," ibid., pp. 357–364.

[64] Funk, J. E., Lawler, H. A., Ohnsorg, R. W., and Storm, R. S., "ATTAP SiC Ceramic Component Development Year 2," ibid., pp. 319–326.

[65] Neil, J., Bandyopadhyay, G., Sordelet, D., and Mahoney, M., "Fabrication of Silicon Nitride ATTAP Components at GTE Laboratories," ibid., pp. 303–310.

[66] Pasto, A. E. and Natansohn, S., "Advanced Processing of High-Performance Silicon Nitride Ceramics," ibid., pp. 197–206.

[67] Cassidy, D. J., "NDE Techniques Used for Ceramic Turbine Rotors," in *Ceramics for High Performance Applications II, Proceedings of the Fifth Army Materials Conference,* Newport, RI, USA, March 21–25, 1977, J.J. Burke, E. N. Lenoe, and R. N. Katz, eds., Brook Hill Publishing Company, Chestnut Hill, MA, USA, 1978. pp. 231–242.

[68] Schuldies, J. J., and Richerson, D. W., "NDE Approach, Philosophy, and Standards for the DARPA/NAVSEA Ceramic Turbine Program," in *Proc. 1977 DARPA/NAVSEA Ceramic Gas Turbine Demonstration Engine Program Review*, J. W. Fairbanks and R. W. Rice, eds., MCIC Rep. MCIC-78-36, Battelle Laboratories, Columbus, OH, USA, 1978. pp. 381–402.

[69] Schuldies, J. J., and Spaulding, W. H., "Radiography and Image Enhancement of Ceramics," ibid., pp. 403–428.

[70] Schuldies, J. J., and Derkacs, T., "Ultrasonic NDE of Ceramic Components," ibid., pp. 429–448.

[71] Kino, G. S., Khuri-Yakub, B. T., Murakami, Y., and Yu, K. H., "Defect Characterization in Ceramics Using High Frequency Ultrasonics," in *Proceedings of the ARPA/AFML Review of Progress in Quantitative Nondestructive Evaluation,* July 17–21, 1978, Scripps Institute of Oceanography, La Jolla, CA, USA, 1979. pp. 242–245.

[72] Ellingson, W. A., and Vannier, M. W., *X-Ray Computed Tomography for Nondestructive Evaluation of Advanced Structural Ceramics,* Argonne National Laboratory Report ANL-87-52, Argonne National laboratory, Argonne, IL, USA, 1987.

[73] Sun, J. G., Shirber, M., and Ellingson, W. A., "Laser-Based Optical Scattering Detection of Surface and Subsurface Defects in Machined Silicon Nitride Components," Ceram. Eng. and Sci. Proc.,18[4], 273–280, 1997.

[74] McClean, A. F., and Hartsock, D. L., "Design with Structural Ceramics," in *Structural Ceramics*, J. B. Wachtman, Jr., ed., *Treatise of Material Science and Technology Volume 29,* Academic Press, Inc., Boston, MA, USA, 1989. pp. 27–97.

[75] Comfort, A. M., and Cuccio, J. S., "Life Prediction Methodology for Ceramic components of Advanced Heat Engines," in *Proc. 27th Automotive Tech. Devt. Contractors' Coord. Meeting*, Dearborn, MI, USA, Oct. 23–26, 1989, SAE P-230, Society of Automotive Engineers, Warrenville, PA, USA, 1990. pp. 275–284.

[76] Gyekenyesi, J. P., "SCARE: A Postprocessor Program to MSC/NASTRAN for the Reliability Analysis of Structural Ceramic Components," Transactions of the ASME, Journal for Gas Turbines and Power, 108[3], 540–546, 1986.

[77] *Lifetime Prediction Methodologies and Data for Ceramic Materials*, ASTM STP 1201, Brinkman, C.R., and Duffy, S.F., eds, American Society for Testing and Materials, Philadelphia, PA, USA, 1993.

[78] *Yogyo Taikan—Hattensuru Seramikkusu*, (Comprehensive Review—Developing Ceramics), The Ceramic Society of Japan, Tokyo, Japan, 1971. pp.100–101 (in Japanese).

[79] *Yogyo Taikan—80 Nendai No Seramikkus*, (Comprehensive Review—Ceramics in the 1980s), The Ceramic Society of Japan, Tokyo, Japan, 1982. pp.174–176, 223–225 (in Japanese)

[80] *21-Seiki E Habataku Seramikkusu—Ceramics 100th*—The Ceramic Society of Japan, Tokyo, Japan, 1991. pp.30–31, pp.72–81, pp.334–338 (in Japanese).

[81] *Proc. 1st Int. Symp. on Ceramic Components for Engines,* Hakone, Japan, Oct. 17–19, 1983, S. Sōmiya, E. Kanai and K. Ando, eds., KTK Sci. Pub. Tokyo, Japan, 1984, and Elsevier Applied Science Publishers Ltd., London, UK, 1986.

[82] Nagahiro, A., *Present Status of the National Project of R and D of Fine Ceramics*, FC Annual Report, Japan Fine Ceramics Association, 1986. pp. 61–66.

[83] Kenney, G. B., and Bowen, H. K., "High Tech Ceramics in Japan: Current and Future Markets," Am. Ceram. Soc. Bull., 62[5], 590–596, 1983.

[84] *Fain Seramikkusu. Jisedai Kenkyu No Ayumi* (Fine ceramics: Steps of the Fine Ceramic Project), Japan Fine Ceramics Gijyutsu Kenkyu Kumiai (Engineering Research Association for High Performance Ceramics), Tokyo, Japan, 1988 (in Japanese).

[85] *Fain Ceramics. Jisedai Kenkyu Kaihatsu No Kiseki To Seika* (Fine Ceramics, Locus and Results of the Fine Ceramics Project), Fine Ceramics Gijyutsu Kenkyu Kumiai (Engineering Research Association for High Performance Ceramics), Tokyo, Japan,1993 (in Japanese).

[86] Mueller, J. I., "Handicapping the World's Derby for Advanced Ceramics," Am. Ceram. Soc. Bull., 61[5], 588–591,1982.

[87] *Seramikkus* (Ceramics Japan, Bull. Ceram. Soc. Japan), 34[12], 1999 (in Japanese).

[88] *Research Report for State of the Art Ceramic Gas Turbine*, Japan Fine Ceramics Association, 1988. p.66 (in Japanese).

[89] *21-Seiki E Habataku Seramikkusu—Ceramics 100th*—The Ceramic Society of Japan, Tokyo, Japan, 1991. p. 80. (in Japanese).

[90] Sasa, T., *Handbook of Ceramic Technology*, Revised Edition, The Ceramic Society of Japan (to be published—in Japanese).

[91] Hattori, Y., "High Density Silicon Nitride Products," Japanese Patent No. 78 46,307, 25 April, 1978.

[92] Kamigaito, O., "Ceramic Components for Engine in Japan," SAE paper 870019, 1987. Kamigaito, O. and Oyama, Y., "Method of Producing Ceramics of Silicon Nitride," US Patent No. 3,903,230, 2 September 1975.

[93] Kawamura, H., "Development of Engineering Ceramics for Engine," *International Workshop for Advanced Materials Technology, Ceramics III*, Nagoya, Japan, 1987.

[94] Komeya, K., "Ceramics for Gas Turbine Rotors," in *Proc. Int. Symp. Factors in Densification and Sintering of Oxide and Non-Oxide Ceramics*, S. Sōmiya and S. Saito, eds., Tokyo Gakujutsu Bunken Fukyu-kai, 1978. pp. 565–579.

[95] Kuroda, H., "Development of Ceramic Turbine Rotor for Automotive Turbochargers," *International Workshop for Advanced Materials Technology, Ceramics III*, Nagoya, Japan, 1987.

[96] Narita, K., and Mori, K., "Crystal Structures of Silicon Nitride," Bull. Chem. Soc. Japan, 32, 417–419, 1959.

[97] Nishida, K., and Nakamura, N., "Silicon Nitride Sintered Product," Japanese Patent No. 74 128,915, 10 December 1974.

[98] Niwa, S., Oki, K., Arakawa, Y., and Watanabe, T., "Refractories Containing Silicon Nitride," Japanese Patent No. 72 23,167, 28 June 1972.

[99] Ogawa, Y., Machida, M., Miyamura, N., Tashiro, K., and Sugano, M., "Ceramic Rocker Arm Insert for Internal Combustion Engines," presented at the 1986 SAE Annual Conference.

[100] Ohji, T., and Ito, S., "High Temperature Tensile Strength of Silicon Nitride," *International Workshop for Advanced Materials Technology, Ceramics II*, Nagoya, Japan, 1987.

[101] Oyama, Y., and Kamigaito, O., "Hot Pressing of Silicon Nitride—Alumina," Yogyo- Kyokai-Shi, 80 [8], 327–336, 1972 (in Japanese).

[102] Oyama, Y., "Solid Solution in the Ternary System, Si_3N_4-AlN-Al_2O_3," Japan J. Appl. Phys., 11, 750–751, 1972.

[103] Sato, T., Kanno, Y., Endo, T., and Shimada, M., "Corrosion of Silicon Nitride in Molten Alkali Sulfate and Carbonate," Adv. Ceram. Mat., 2, 3A, 228–231, 1987.

[104] Soma, T., Nasuda, M., Matsui, M., and Oda, I., "High Temperature Tensile and Torsional Strength of Sintered Silicon Nitride," *International Workshop for Advanced Materials Technology, Ceramics II*, Nagoya, Japan,1987.

[105] Suzuki, H., "The Synthesis and Properties of Silicon Nitride," Bull. Tokyo Inst. of Technology, 54, 163–177, 1963.

[106] Suzuki, K., "Gas Pressure Sintering Method," Nikkan Kogyo Shimbun, 19 March 1987. p. 32.

[107] Tani, E., Kishi, K., Umebayashi, K., and Nishijima, M., "Gas Pressure Sintering of Composite of Silicon Nitride—Silicon Carbide System," J. Ceram. Soc. Japan., 95, 917–920, 1987.

[108] Tani, S., Umebayashi, S., Kishi, K., Kobayashi, K., and Nishijima, M., "Effect of Size of Grains with Fibre-like Structure of Silicon Nitride on Fracture Toughness," J. Mat. Sci. Lett., 4[5], 1454–1456, 1985.

[109] Tsuge, A., Komeya, K., Inoue, H., Hashimoto, H., and Komatsu, M., "Microstructural Development and Densification During Sintering of Silicon Nitride," *Proc. Int. Symp. Factors in Densification and Sintering of Oxide and Non-Oxide Ceramics*, S. Sōmiya and S. Saito, eds., Tokyo Gakujutsu Bunken Fukyu-kai, 1978. pp. 503–516.

[110] Yajima, S., "Special Heat-Resisting Materials from Organometallic Polymers," Am. Ceram. Soc. Bull., 62[8], 893–898, 1983.

[111] Okamura, K., "Status Quo and Future Trend on R&D for High Temperature and High Performance Ceramic Fibers Derived from Polymers," Adv. Composite Mater., 8[1],107–115, 1999.

[112] Funayama, O., Arai, M., Tashiro, Y. Aoki, H., Suzuki, T., Tamura, K., Kaya, H., Nishii, H., and Isoda, T., "Tensile Strength of Silicon Nitride Fibers Produced from Perhydropolysilazane," J. Ceram. Soc. Japan (Seramikkusu Ronbunshi), 98[1],104–107, 1990.

[113] Popper, P., "Introductory Lecture," in *Nitrogen Ceramics*, F. L. Riley, ed., Noordhoff Int. Publishing, Leyden, The Netherlands, 1977. pp 3–19.

[114] Cutler, I. B., and Croft, W. J., "Powder Met. Review 7: Silicon Nitride (Part 1)," Powder Met. Int. 6, 92–96, 1974; "Part 2," Powder Met. Int. 6, 144–146,1974.

[115] *Nitrogen Ceramics,* F. L. Riley, ed., Noordhoff Int. Publishing, Leyden, The Netherlands, 1977.

[116] Parr, N. L., Sands, R., Pratt, P.L., May, E. R. W., Shakespeare, C. R., and Thompson, D. S., "Structural Aspects of Silicon Nitride," Powder Met., 8, 152–163, 1961.

[117] Parr, N. L., Martin, G. F., and May, E. R. W., "Preparation, Microstructure, and Mechanical Properties of Silicon Nitride," in *Special Ceramics*, P. Popper, ed., Academic Press, New York, NY, USA, 1960. pp. 102–135.

[118] Pratt, P. L., "The Microstructure and Mechanical Properties of Silicon Nitride," in *Mechanical Properties of Engineering Ceramics,* W. W. Kriegel and H. Palmour, eds. III, Interscience Publishers, New York, NY, USA, 1961. pp. 507–519.

[119] Noakes, P. B., and Pratt, P. L., "High-Temperature Mechanical Properties of Reaction- Sintered Silicon Nitride," in *Special Ceramics, Vol. 5, Proceedings of the 5th International Symposium on Special Ceramics,* held by the BCRA, Stoke-on-Trent, England, 14–16 July 1970, P. Popper, ed., British Ceramic Research Association, Manchester, England, UK, 1972. pp. 299–310.

[120] Davidge, R. W., "Mechanical Properties of Reaction Bonded Silicon Nitride," in *Nitrogen Ceramics*, F.L.Riley, ed., Noordhoff Int. Publishing, Leyden, The Netherlands, 1977. pp. 541–559.

[121] Davidge, R. W., Evans, A. G., Gilling, D., and Wilyman, P. R., "Osication of Reaction- Sintered Silicon Nitride and Effects on Strength," in *Special Ceramics, Vol. 5, Proc. 5th Int. Symp. on Special Ceramics*, BCRA, Stoke-on-Trent, England, 14–16 July 1970, P. Popper, ed., British Ceramic Research Association, Manchester, England, UK, 1972. pp. 329–344.

[122] Godfrey, D. J., and Taylor, P. G., "Designing with Brittle Materials," Eng. Mater. Design, 12[9], 1339–1342, 1969.

[123] Godfrey, D. J., and Lindley, M. W., "The Strength of Reaction-Bonded Silicon Nitride Ceramics," Proc. Brit. Ceram. Soc., 22, 229–252, 1973.

[124] Godfrey, D. J., and Mitchell, K. W., "Statistical and Practical Approaches to the Development of Design Criteria for Brittle Materials," Materials for Engineers Series, N. L. Parr, ed., The Engineer, 224, 737–738, 1967.

[125] Glenny, E., and Taylor, T. A., "Mechanical Strength and Thermal Fatigue Characteristics of Silicon Nitride," Powder Met., 8, 164–195, 1961.

[126] Wild, S., Grieveson, P., Jack, K. H., and Latimer, M. J., "The Role of Magnesia in Hot- Pressed Silicon Nitride," in *Special Ceramics, Vol. 5, Proc. 5th Int. Symp. on Special Ceramics,* BCRA, Stoke-on-Trent, England, 14–16 July 1970, edited by P. Popper, British Ceramic Research Association, Manchester, England, UK, 1972. pp. 377–384.

[127] Wild, S., Grieveson, P., and Jack, K. H., ibid., pp. 289–298.

[128] Bowen, L. J., Weston, R. J., Carruthers, T. G., and Brook, R. J., "Mechanisms of Densification During the Pressure Sintering of Alpha-Silicon Nitride," Ceramurgia Int., 2[4], 173–176, 1976.

[129] Jack, K. H., and Wilson, W. I., "Ceramics Based on the Si-Al-O-N and Related Systems," Nature Phys. Sci. (London), 238, 28–29, July 10, 1972.

[130] Jack, K. H., "Review: Sialons and Related Nitrogen Ceramics," J. Mater. Sci., 11, 1135–1158, 1976.

[131] Thompson, D. P., Korgul, P., and Hendry, A., "The Structural Characterisation of Sialon Polytypoids," in *Progress in Nitrogen Ceramics*, F. L. Riley, ed., Martinus Nijhoff Publishers, The Hague, The Netherlands, pp. 61–74, 1983.

[132] Lumby, R. J., North, B., and Taylor, A. J., "The Formation of Single Phase Sialon Ceramics," in *Nitrogen Ceramics*, F. L. Riley, ed., Noordhoff International Publishers, Leyden, The Netherlands, 1977. pp. 393–404.

[133] Lumby, R. J., North, B., and Taylor, A. J., "Properties of Sintered Sialons and Some Applications in Metal Handling and Cutting," *Ceramics for High Performance Applications II, Proceedings of the Fifth Army Materials Conference,* Newport, RI, USA, March 21–25, 1977, J. J. Burke, E. M. Lenoe, and R. N. Katz, eds., Brook Hill Publishing Company, Chestnut Hill, MA, USA, 1978. pp. 893–906.

[134] Drew, P., and Lewis, M. H., "The Microstructure of Silicon Nitride/Alumina Ceramics," J. Mater. Sci., 9, 1833–1838, 1974.

[135] Popper, P., US Patent 3,275,722, "Production of Dense Bodies of Silicon Carbide," Sept. 27, 1966.

[136] Forrest, C.W., "US Patent 3,495,939, "Manufacture of Dense Bodies of Silicon Carbide," Feb. 17, 1970.

[137] *Keramische Komponenten für Fahrzeug-Gasturbinen* (Ceramic Components for Vehicular Gas Turbines), W. Bunk and M. Böhmer, eds., Project Status Workshop, 18 and 19 April, 1978, Bad Neuenahr, Germany, Springer Verlag, Berlin, 1978 (in German).

[138] *Keramische Komponenten für Fahrzeug-Gasturbinen II* (Ceramic Components for Vehicular Gas Turbines II), W. Bunk and M. Böhmer, eds., Project Status Workshop, 24–26 November 1980, Bad Neuenahr, Germany, Springer Verlag, Berlin, 1981 (in German).

[139] *Keramische Komponenten für Fahrzeug-Gasturbinen III* (Ceramic Components for Vehicular Gas Turbines III), W. Bunk, M. Böhmer, and H. Kißler, eds., Project Status Workshop, 13–15 February 1984, Germany, Bad Neuenahr, Springer Verlag, Berlin, 1984 (in German).

[140] Bunk, W., "Overview on the German Program on Ceramic Components for Vehicular Gas Turbines," *Proceedings of the First International Symposium on Ceramic Components for Engines*, Hakone, Japan, Oct. 17–19, 1983, S. Sōmiya, E. Kanai and K. Ando, eds., KTK Sci. Pub. Tokyo, Japan, 1984 and Elsevier Applied Science Publishers Ltd., London, UK, 1986. pp. 9–20.

[141] Bunk, W., Gugel, E., and Walzer, P., "Overview of the German Ceramic Turbine Program," in *Ceramics for High-Performance Applications—III Reliability, Proceedings of the Sixth Army Materials Technology Conference,* Orcas Island, WA, USA, July 10–13, 1979, Edward M. Lenoe, R. Nathan Katz, and John J. Burke, eds., Plenum Press, New York, NY, USA, 1983. pp. 29–50.

[142] Thümmler, F., "State of the Art Engineering Ceramics," J. Eur. Ceram. Soc., 6, 139–151, 1990.

[143] Kronogård, S.O. and Malmrup, L., "Ceramic Engine Research and Development in Sweden," in *Ceramics for High-Performance Applications—III Reliability*, E. M. Lenoe, R. N. Katz, and J. J. Burke, eds., Plenum Press, New York, NY, USA, 1983. pp. 51–80.

[144] Lundberg, R., and Ferrato, M., "Ceramic Component Development for AGATA," ASME paper 99-GT-392, presented at the International Gas Turbine and Aeroengine Congress and Exposition, Indianapolis, IN, USA, June 7–10, 1999.

[145] Spriet, P., and Habarou, G., "Applications of Continuous Fiber Reinforced Ceramic Composites in Military Turbojet Engines," ASME paper 96-GT-284, presented at the International Gas Turbine and Aeroengine Congress and Exposition, Birmingham, UK, June 10–13,1996.

[146] Bouillon, E. P., Habarou, G., Spriet, P. C., Lecordix, J. L., Ojard, G. C., Linsey, G. D., Feindel, D. T., and Stetson, D. P., "Characterization and Nozzle Test Experience of a Self Sealing Ceramic Matrix Composite for Gas Turbine Applications," ASME paper GT-2002-30458, presented at the ASME TURBO EXPO Land, Sea & Air 2002, Amsterdam, The Netherlands, June 3–6, 2002.

[147] Bouillon, E., Lamouroux, F., Baroumes, L., Cavalier, J.-C., Spriet, P., and Habarou, G., "An Improved Long Life Duration CMC for Jet Aircraft Engine Applications," ASME paper GT-2002–30458, presented at the ASME TURBO EXPO Land, Sea & Air 2002, Amsterdam, The Netherlands, June 3–6, 2002.

[148] Wurtinger, H., and Mühlratzer, A., "Cost Effective Manufacturing Methods for Structural Ceramic Matrix Composite (CMC) Components," ASME paper 96-GT-296, presented at the International Gas Turbine and Aeroengine Congress and Exposition, Birmingham, UK, June 10–13, 1996.

[149] Holmquist, M., Lundberg, R., Razzell, A., Sudre, O., Molliex, L., and Adlerborn, J., "Development of Ultra High Temperature Ceramic Composites for Gas Turbine Combustors," ASME paper 97-GT-413, presented at the International Gas Turbine and Aeroengine Congress and Exposition, Orlando, FL, USA, June 2–5,1997.

[150] Razzell, A., Molliex, L., Holmquist, M., and Sudre, O., "Oxide/Oxide Ceramic Matrix Composites in Gas Turbine Combustors," ASME paper 98-GT-30, presented at the International Gas Turbine and Aeroengine Congress and Exhibition, Stockholm Sweden, June 2–5, 1998.

[151] Bornemisza, T.G., "Western European Status of Ceramics for Gas Turbines," Proceedings of the Advanced Turbine Systems Annual Program Review Meeting, U.S. Department of Energy, Abbie Layne and Patricia Hoffman, Washington, D.C., November 1996. DOE/OR-2048. pp. 659–663.

CHAPTER

2

Ceramic Technology Development at Ford Motor Company from 1967 to 1986

John A. Mangels

Director of Sales and Marketing, Ceradyne, Inc., Costa Mesa, California, USA

INTRODUCTION

A ceramic technology program at Ford Motor Company was initiated in 1967 in support of Ford's advanced gas turbine program. The goals of this program were to develop a 200-horsepower, regenerative, axial flow gas turbine engine with a turbine inlet temperature of 1371°C (2500°F). The goals of this program necessitated the use of ceramic materials. Ford's ceramic effort concentrated on the development of silicon nitride and silicon carbide materials, ceramic component fabrication technology and material characterization and nondestructive evaluation (NDE) techniques. The ceramic effort resulted in a number of application firsts including: the first uncooled gas turbine to run at 1371°C (2500°F) turbine inlet temperature, the first demonstration of a ceramic turbocharger and the first use of silicon nitride as a cutting tool material. However, most importantly, the success of the Ford gas turbine program lead to the "ceramic fever" which began in the mid 1970s and has spawned the development of non-oxide ceramics around the world. These materials are used as enabling technologies in many industrial, semiconductor processing and automotive applications.

The Beginning

Ford's ceramic program began in 1967 as an outgrowth of a high horsepower turbine engine program designed for heavy-duty truck and marine applications. A.F. McLean, lead design engineer on this program, realized that turbine efficiencies could be radically improved if the operating temperatures could be increased. He proposed that Ford develop a gas turbine for automotive application that would run at elevated temperatures. The initial goals of this program were to develop a 200-horsepower, regenerative, axial flow gas turbine that would operate at a turbine inlet temperature of 1371°C (2500°F).

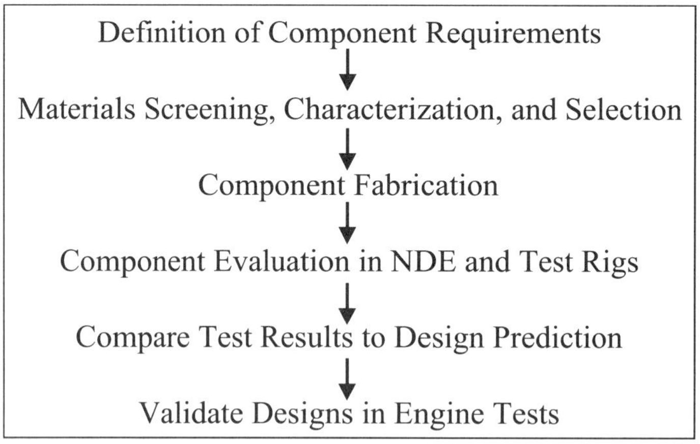

Figure 2.1 Structure of Ford's Ceramic Turbine Program

While the original program goals did not mention the use of ceramics, their use was certainly implied as McLean realized that they were the only materials that could withstand the temperatures, which would be necessary for program success. The proposal of such an ambitious program was visionary on McLean's part, especially considering the state of ceramic technology at the time.

In 1967, advanced technical ceramics as we now know them did not exist and the ceramic community did not have the infrastructure to develop the required materials. Likewise, the ceramic community did not have the fabrication technology to manufacture complex shaped turbine components. Material characterization, component NDE techniques, finite element modeling and life predictions methods commonly used today were not available to the Ford ceramic turbine team.

Program Structure

McLean and his team instituted a program structure (Figure 2.1), which is the model of many materials related programs today. The ceramic gas turbine team consisted of turbine aerodynamic and thermodynamic design engineers, engine test engineers together with ceramic engineers and material scientists. This diverse group of technologists worked together to create the world's first ceramic gas turbine engine.

The ceramic program, headed by E.A. Fisher, was structured (Figure 2.2), to initially select and later develop the materials and create the infrastructure required to produce the various turbine components using

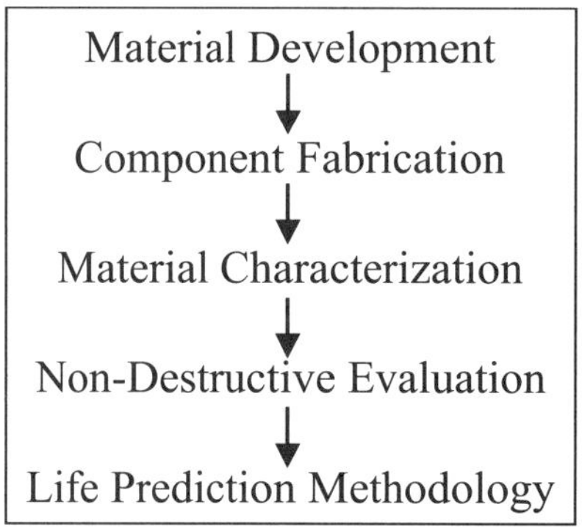

Figure 2.2 Structure of Ford's Ceramic Technology Program

TABLE 2.1 CHRONOLOGY OF THE FORD CERAMIC TURBINE PROGRAM

Years	Program	Contract
1967–1971	820 Gas Turbine Engine Program	IR&D Funding
1971–1976	Brittle Materials Design, High Temperature Gas Turbine	DOD–ARPA Funding–Contract No. DAAG 46-71-C-0162
1977	AGT 101 Design	DOE/NASA Funding–Contract No. 0037-79/1
1978–1984	AGT 101	DOE–Subcontractor to Garrett Turbine Engine Company
1978–1981	High Strength RBSN	DOE/NASA–Contract No. DEN 3-20
1983–1985	Ceramic Life Prediction Methodology	DOE–Contract No. MTL TR 86-16
1985–1987	Sintered SiC	NASA–Contract No. NAS3-24384
1977–1986	Various Technology and Application Development Programs	IR&D Funding

Notes: IR&D: Internal Research and Development; DOD: U.S. Dept. of Defense; DOE: U.S. Dept. of Energy; ARPA: U.S. Dept. of Defense Advanced Research Projects Agency; NASA: U.S. National Aeronautics and Space Administration

methods that could lead to high volume manufacturing. The original emphasis was on material development, fabrication and characterization. NDE and life prediction gained emphasis as the fundamental knowledge of materials behaviors increased.

Program Duration

The Ford ceramic turbine program began in 1967 and continued as a ceramic technology program until 1987. Table 2.1 provides a chronology of the Ford ceramic turbine effort.

The Original Materials

The initial screening tests consisted of strength, oxidation and thermal shock testing using simple bend bars. Successful materials were evaluated in a simple "burner" rig where the materials were subjected to a "gas turbine" environment with temperatures approaching 1370°C for incremental time periods. These tests showed that the advanced materials of the time, alumina, mullite, carbon and glassy carbon were all deficient in a variety of ways including thermal shock and oxidation resistance.

Based on the screening tests, original materials selected for the gas turbine program are summarized in Table 2.2.

It became apparent that silicon nitride was to be an important material for the axial flow parts. It further became apparent that the quality of this material was not adequate for structural components and there were no manufacturing sources available to produce components from silicon nitride.

A portion of the Ford turbine program was set up to characterize and monitor supplier material and component development for the regenerator, housing and combustor. However, the majority of the ceramic effort was to develop improved grades of silicon nitride and to develop the fabrication technology to produce axial flow turbine components from silicon nitride.

TABLE 2.2 ORIGINAL MATERIAL SELECTION FOR FORD'S CERAMIC GAS TURBINE

Component	Material	Comments
Regenerator, Housing	Lithium Aluminum Silicate	Manufacturing source in place, Corning Glass Works (now Corning, Inc.)
Combustor	Reaction Bonded Silicon Carbide (RBSC)	Manufacturing sources in place (Norton Company, Carborundum Company)
Stationary Flow Path Components, Turbine Rotor	Reaction Bonded Silicon Nitride (RBSN)	No manufacturing source for either the material or the components

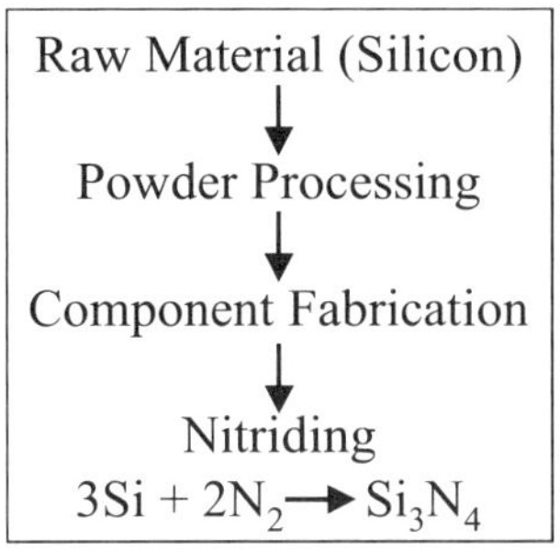

Figure 2.3 Process Flow for Reaction Bonded Silicon Nitride (RBSN)

MATERIAL DEVELOPMENT

Reaction Bonded Silicon Nitride

Reaction Bonded Silicon Nitride (RBSN) was first developed in the 1960s. The process flow for this material is summarized in Figure 2.3. There are a number of advantages of RBSN including a simple process, a readily available starting raw material and the fact that relatively high chemical purities can be attained. The high purity nature of RBSN results in a material that maintains its strength to 1400°C. However, the most important feature of this material is that there is no shrinkage during the firing (nitriding) process. This makes net shape fabrication of turbine components (including airfoil geometries) possible. The disadvantages of RBSN are its low density (70–80% of theoretical density), its low strength and wall thickness limitations.

The objective of the Ford RBSN material development program was to increase the strength of the material. Figure 2.4 shows the strength of RBSN versus time.

The strength increase was accomplished by increasing the bulk density of the material and by improving microstructure through process controls during nitriding. Figure 2.5 summarizes the strength—density relationship obtained for Ford's RBSN.

The high densities were obtained by increasing the green density of the silicon compacts. This was accomplished through the creation of an optimum particle size distribution. The optimal, broad particle size distributions (Figure 2.6(a)) were obtained by dry ball milling for various amounts of time [1]. The broad distributions also positively effected the fabrication processes, especially injection molding. The relationship between RBSN density and particle size distribution is shown in Figure 2.6(b). As the green densities

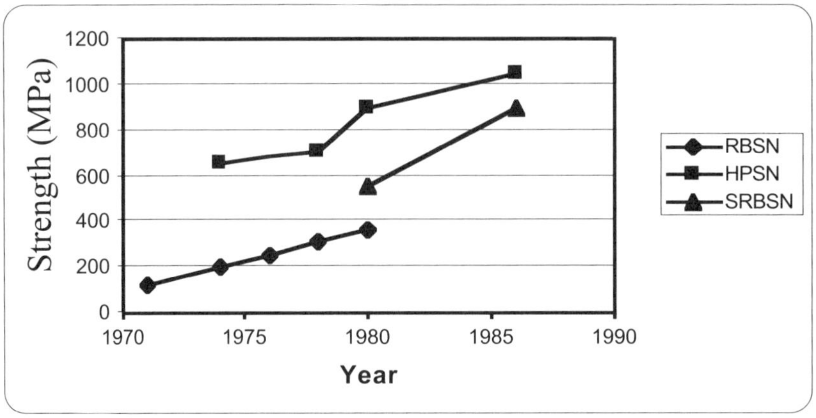

Figure 2.4 Strength of Ford's Silicon Nitride vs.Time

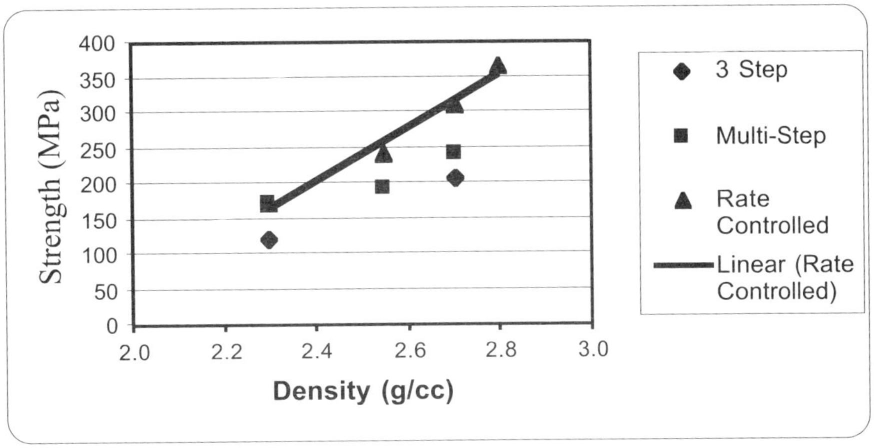

Figure 2.5 Strength-Density Relationship for Ford's RBSN

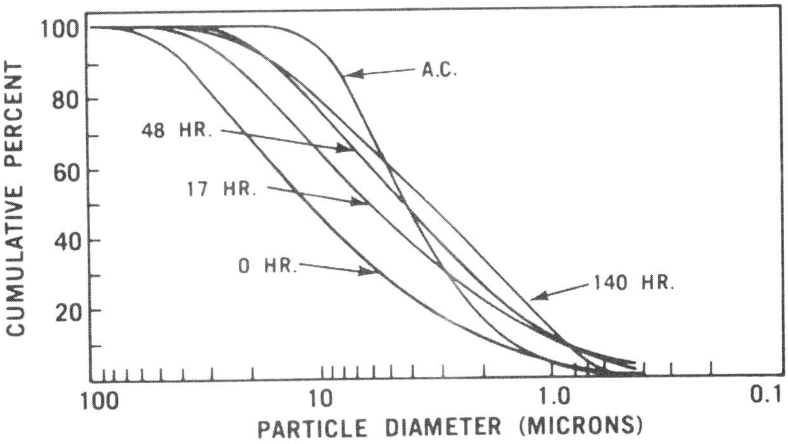

Figure 2.6(a) Silicon Particle Size Distributions for RBSN (reprinted from ref. [1])

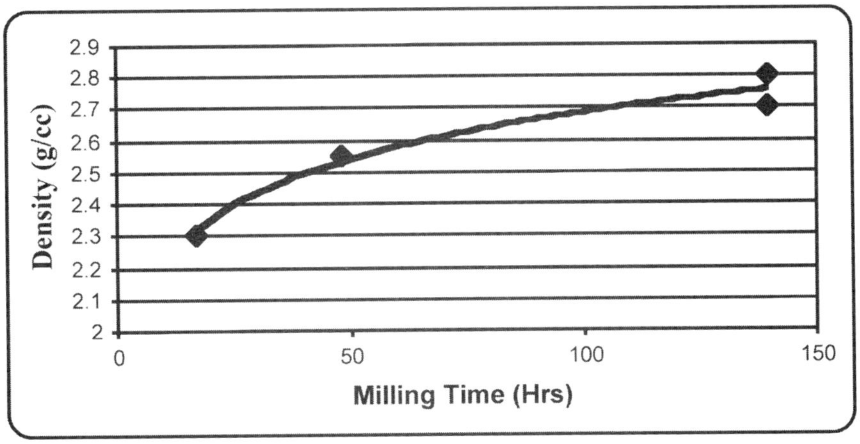

Figure 2.6(b) Relation between RBSN Density and Particle Size Distributions

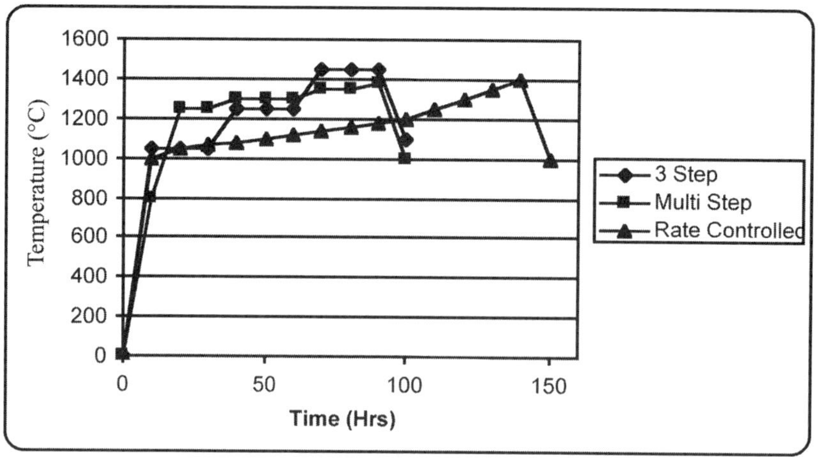

Figure 2.7 Nitriding Cycles for RBSN

increased, more care was required in the nitriding process [2–4]. Higher green densities make the transport of nitrogen into the green compact more difficult. Figure 2.7 summarizes typical nitriding cycles used during various stages of the Ford program.

The key factor in the design of a nitriding cycle is the control of the exothermic, nitriding reaction. Lack of adequate process controls, such as the "multi step cycle" can result in temperatures exceeding the melting point of silicon (1400°C) and result in a large grain sized microstructure (Figure 2.8(a)). The "rate controlled cycle" controls the exotherm and avoids the melting of silicon. This results in a more uniform, fine grained structure (Figure 2.8(b)), higher strength and a more uniform transport of nitrogen into the green compact.

Sintered Reaction Bonded Silicon Nitride

In the mid 1970s, Terwilliger [5,6] demonstrated that silicon nitride compacts could be sintered to full density with the addition of sintering aids such as magnesium oxide. This opened up the potential of achieving fully dense silicon nitride components with near net shape capability. However, the lack of commercially available grades of sinterable silicon nitride powder, along with the costs of sinterable powder and the technical difficulties associated with the fabrication of components from sinterable silicon nitride powder were major obstacles with this new technology.

Ford found that the addition of sintering aids (MgO, Al_2O_3, Y_2O_3) to silicon and proceeding with the typical RBSN process combined with the addition of a sintering step would result in a fully dense silicon nitride component [7–9]. Figure 2.9 describes the Sintered Reaction Bonded Silicon Nitride (SRBSN) process. Ford further demonstrated that gas pressure sintering (GPS) was a key to the development of a fully dense, high strength material (see Figure 2.4 for strength). A typical GPS sintering cycle is summarized in Figure 2.10.

Ceradyne, Inc. [10] commercialized the SRBSN* process with properties summarized in Table 2.3. This material has outstanding strength, reliability, fracture toughness and microstructure, which

*In 1986, Ceradyne, Inc. was selected to commercialize the silicon nitride technology developed during the Ford ceramic program. Ceradyne elected to commercialize the SRBSN process and began commercial production of this material in 1990. They first found application for this material in industrial wear products and later in automotive applications. See further details in the Applications section of this paper.

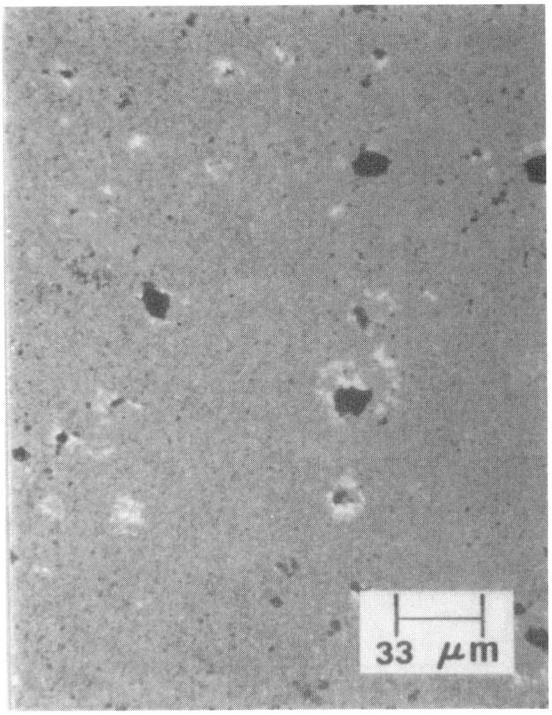

(a)

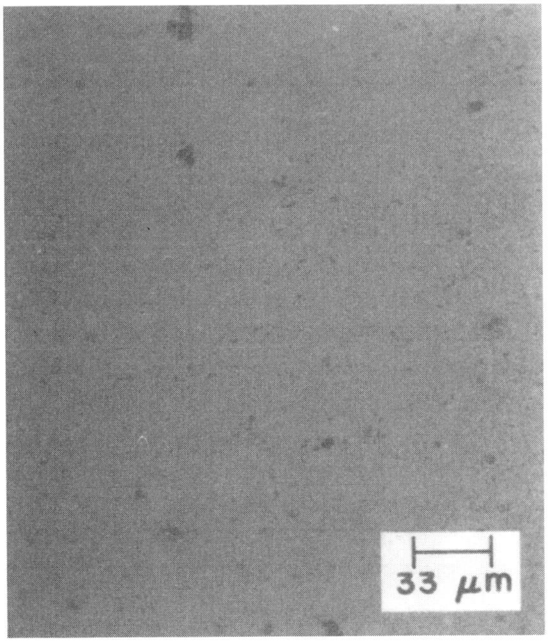

(b)

Figure 2.8 Microstructure of 2.7 g/cc RBSN a) Multi Step Cycle b) Rate Controlled Cycle (Reprinted from ref. [3], with permission of the American Ceramics Society, www.ceramics.org, © 1981, all rights reserved)

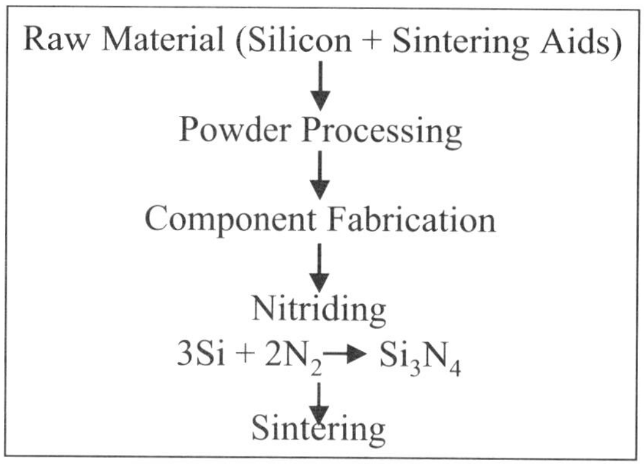

Figure 2.9 Process Flow for Sintered Reaction Bonded Silicon Nitride (SRBSN)

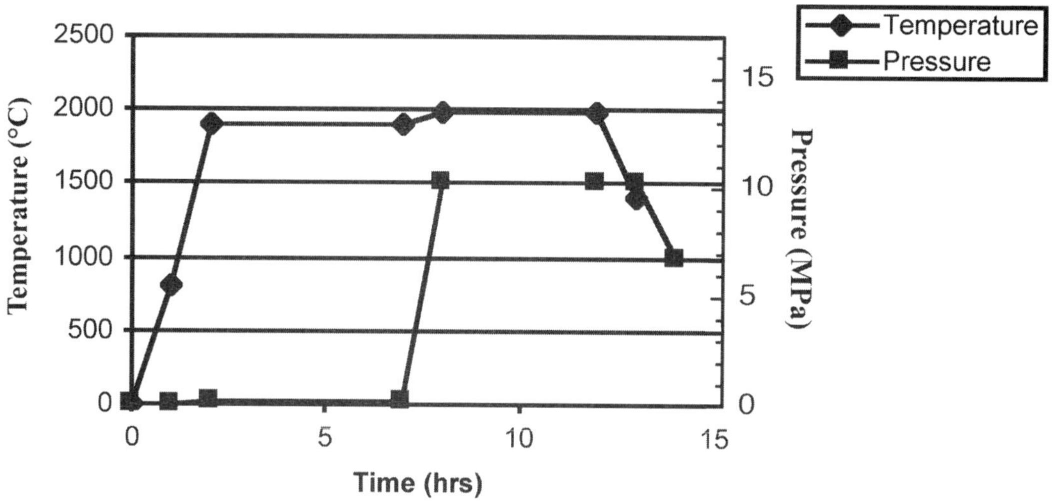

Figure 2.10 SRBSN—Gas Pressure Sintering Cycle

TABLE 2.3 PROPERTIES OF CERADYNE'S CERALLOY® 147-31N SRBSN

Grade	Ceralloy® 147-31N
Density	3.21 g/cm^3 (99.5% TD)
Strength	800 MPa
Weibull Modulus	>15
Fracture Toughness	5.5–6.0 MPa.m$^{1/2}$
Hardness	1550 kg/mm^2
Elastic Modulus	310 GPa
Poisson's Ratio	0.2
Thermal Conductivity	26 W/m.K

makes this material ideal for highly stressed mechanical applications such as cam rollers and other valve train components.

Hot Pressed Silicon Nitride

The goal of the hot pressed silicon nitride development program was to produce a material that could be used as the hub of a turbine rotor (as part of the duo-density turbine rotor described later). The specific material goal was to optimize the strength of the hot pressed silicon nitride at both room temperature and at 1100°C, the proposed operating temperature of the turbine hub.

The initial work focused on the use of MgO as a densification aid while investigating various processing parameters including powder type, MgO level, hot press temperature and time at temperature. This work showed that the principal variable affecting the room temperature strength was the hot pressing temperature (Figure 2.11). Figure 2.11 also shows that changing to a higher purity grade of powder resulted in further room temperature strength improvements. However, the high temperature strength of hot pressed silicon nitride produced using MgO resulted in significant strength reductions on the order of 25 percent at high temperatures (1100°C).

Alternate densification additives were evaluated to determine if the high temperature strength could be significantly improved. Results showed that Y_2O_3 additives resulted in room temperature strength and 1100°C strength of approximately 700 MPa with no significant reduction in strength at 1100°C. However, the hot pressing pressures necessary to achieve 99% density were 34.5 MPa (5000 psi), which were too high to successfully fabricate a duo-density turbine rotor.

In 1980, silicon nitride was found to be useful as a cutting tool material for the machining of gray cast iron [12]. (See later section on cutting tools). This invention prompted a large amount of work to optimize the properties of hot pressed silicon nitride for cutting tool applications. These properties include room temperature and high temperature strength, crystalline phase composition and hardness. This work resulted in an increase in the strength of Hot Pressed Silicon Nitride (HPSN) to >1050 MPa (see Figure 2.4) through the optimization of sintering additives and starting raw materials, especially with respect to purity, particle size and chemistry [13].

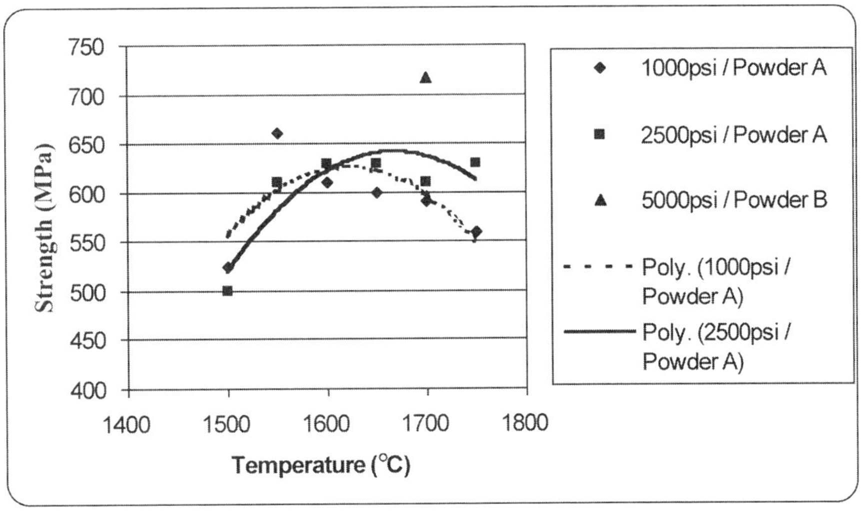

Figure 2.11 Room Temperature Strength of Hot Pressed Silicon Nitride-1psi = 6.9kPa (Reprinted from ref. [11])

COMPONENT FABRICATION DEVELOPMENT

The fabrication development programs were aimed at developing methods to produce turbine components from silicon nitride ceramics and, in fact, producing the components and supplying them for rig and engine testing. In manufacturing these components, the normal process engineering steps were taken to insure robust processes were developed which could produce components in high volume at low costs.

The processes investigated were ones that could produce turbine shapes, i.e. airfoil geometries to net shape. These include injection molding and later a novel slip casting process. Conventional slip casting was employed for simple flow path components. However, conventional ceramic fabrication processes such as isostatic pressing or dry pressing followed by green machining were not emphasized.

Injection Molding

Injection molding [14,15] was determined to be the natural process to produce multi-bladed airfoil components, such as monolithic turbine rotors and stators, as well as some of the other complicated hot flow path components shown in Figure 2.12.

The attraction of injection molding is its ability to produce complex shaped components to net shape at high production rates. The challenges faced with the development of ceramic injection molding technology

(a)

(b)

Figure 2.12 Injection Molded Silicon Nitride Turbine Components ((a) Reprinted from ref. [14], with permission of the American Ceramics Society, www.ceramics.org, © 1984, all rights reserved, (b) reprinted with permission from Ford Motor Company)

was the choice of a binder system which would allow for proper fluid flow at high solids loading, tooling design for the complex turbine components and the development of a rapid, reliable binder extraction method.

Many different binder types were utilized during the course of the Ford ceramic injection-molding program. The main characteristics of the binders were that they have a thermoplastic behavior and they have a low melt viscosity. The formulations selected were mixtures of binders that were chemically compatible and had different melting temperatures and vapor pressures to facilitate binder removal. Bi-polar, organo-metallic compounds were used as deflocculants to increase the solids loading to obtain high green densities. The final formulations consisted of mixtures of waxes having different melting points and zinc stearate as the deflocculant.

The injection molding formulations were evaluated with the use of spiral flow injection molding tests where the distance the molding compound traveled along the spiral die was a direct indication of the injection molding behavior of the material. Figure 2.13 shows typical spiral flow data for different particle size distributions of starting raw materials.

Figure 2.13(a) Spiral Flow Injection Molding Tool (Reprinted from ref. [11]).

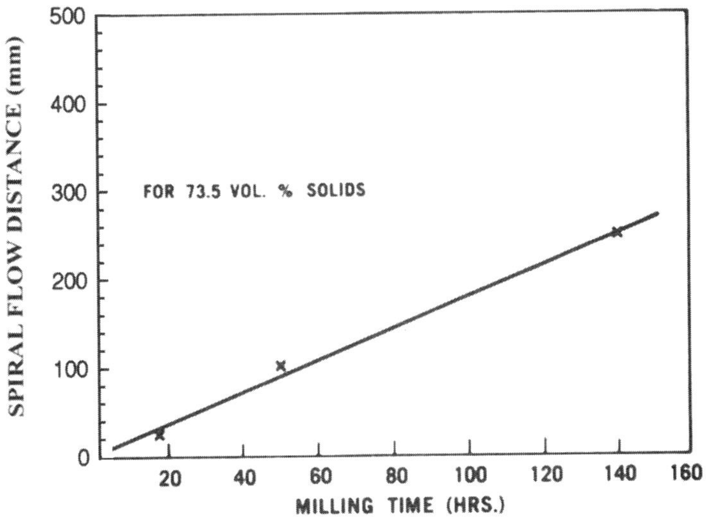

Figure 2.13(b) Typical Spiral Flow Injection Molding Results (Reprinted from ref.[11])

Figure 2.14 Typical Components Produced with Low Pressure Injection Molding. (Reprinted with permission from Ford Motor Company)

This data shows that the particle size distribution was critical in obtaining high solids loadings and consequently the high green densities required to obtain the mechanical properties of the fired ceramic component.

Removal of the injection molding binder can be accomplished in a number of ways: 1) thermal decomposition, 2) thermal extraction (wicking) or 3) chemical extraction. The Ford program utilized thermal extraction with the use of activated charcoal as the "wicking" medium. The parts were imbedded in the charcoal and then gradually heated above the melting point of the binders in either a vacuum or argon atmosphere. This process was greatly improved through the use of a high-pressure argon atmosphere, binder removal cycle. This process resulted in higher yields and a reduced cycle time.

Low-pressure injection molding [16] was found to be practical for the manufacture of thick cross section radial turbocharger rotors and turbine rotors that were typical of the later stages of the Ford program. Figure 2.14 shows examples of components produced with low-pressure injection molding.

Low-pressure injection molding has the advantages of processing parts at pressures of 690 kPa (100 psi) versus pressures of 6.90 to 69 MPa (1,000 to 10,000 psi) with conventional injection molding processes. This results in reduced tooling costs (aluminum vs. hardened steel tooling) and reduced residual stress in the molded component. The reduced stress level was found to be a contributing factor to improved yields after the binder removal process. Research indicated that components having solid loadings of up to 65 volume percent solids could be processed with low-pressure injection molding.

Slip Casting

Slip casting is a conventional ceramic fabrication technique used to make relatively simple shapes and was ideal for the fabrication of selected turbine components, as shown in Figure 2.15.

The main challenge faced with slip casting of RBSN components was the reactivity of silicon powder with water [17]. The hydrolysis reaction could be controlled by altering the pH of the system and by the addition of selected chemical deflocculants.

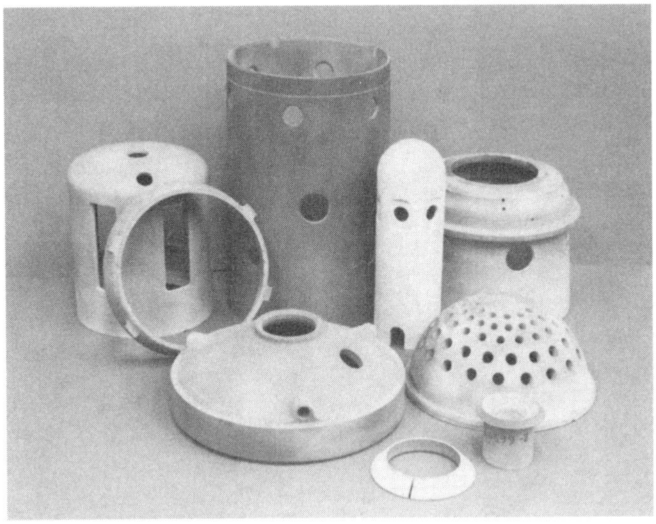

Figure 2.15 RBSN Components Produced by Conventional Slip Casting Techniques (Reprinted with permission from Ford Motor Company)

A novel slip casting technique was developed, fugitive wax slip casting [18–20] which was used to produce airfoil components. The flow chart for this technique is described in Figure 2.16.

The fugitive method was able to produce complex turbine and turbocharger rotors to net shape without the use of expensive injection molding tools (Figure 2.17). Aluminum injection molding tools, similar to those used to fabricate investment-casting patterns, were employed to make the original patterns. Investment casting waxes were used for the water-soluble and alcohol-soluble patterns.

Produce Master Pattern from Water Soluble Material
↓
Coat Pattern with Alcohol Soluble Material
↓
Dissolve Pattern Leaving a Shell Mold
↓
Mount Shell Mold into Plaster Mold Base
↓
Pour Slip into Shell Mold
↓
Monitor Casting Progress
↓
Remove Casting from Plaster Mold Base
↓
Dissolve Shell Mold in Alcohol
↓
Dry and Fire Casting

Figure 2.16 Ford Fugitive Wax Slip Casting Process

Figure 2.17 Components Produced by the Fugitive Wax Slip Casting Process (Reprinted with permission from Ford Motor Company)

Duo-Density Turbine Rotor Fabrication

The most difficult challenge in the Ford ceramic turbine program was the development of the ceramic turbine rotor. The Ford ceramic turbine rotor was an axial design with a center-mounting hole (Figure 2.18). This design resulted in very high stresses at the center bore, but relatively low stresses in the airfoil blades. The material challenge was to find an available material that had the strength to withstand the tensile stresses at the center of the rotor and had the high temperature strength to withstand the temperature and stresses in

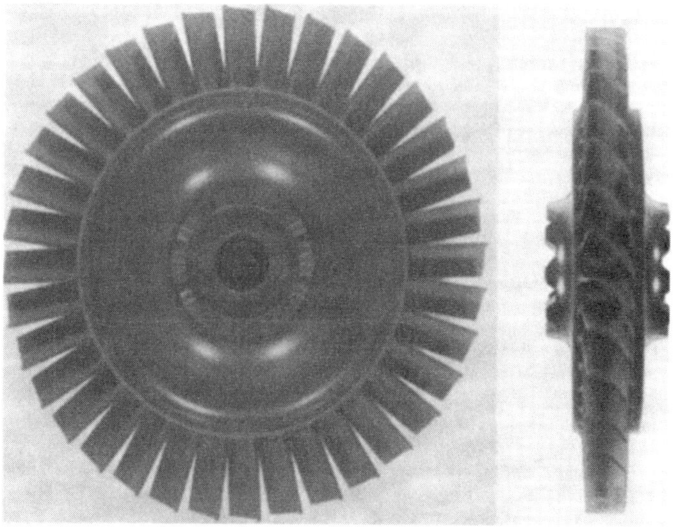

Figure 2.18 Ford Axial Duo-Density Ceramic Turbine Rotor (Reprinted from ref. [11])

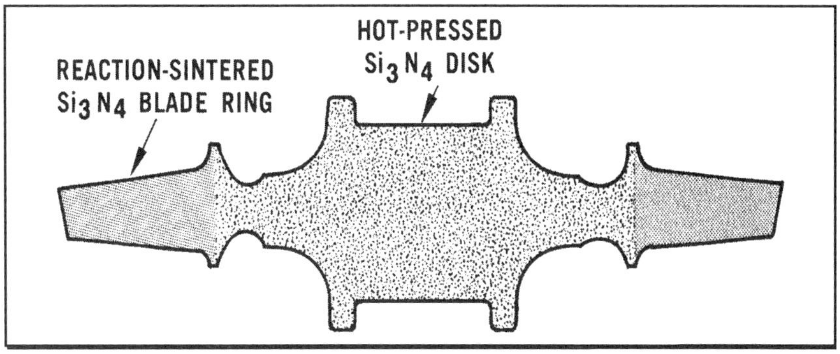

Figure 2.19 Graphic Depiction of the Ford "Duo-Density" Turbine Rotor (Reprinted from ref. [11])

the airfoil. (Remember this was before sintered silicon nitride was invented.) An equal challenge was to fabricate such a rotor from the appropriate materials.

The material choices were to use hot pressed silicon nitride for the rotor hub and to use reaction bonded silicon nitride for the airfoils. The two components were then diffusion bonded together to form a "Duo-Density" turbine rotor assembly [21] shown graphically in Figure 2.19. Ultimately the hot pressed hub was hot pressed to near net shape and diffusion bonded to the RBSN airfoils in one operation [22].

This rotor concept enabled Ford to meet its DOD program requirements. Rotors of this type were run in turbine engines at 1371°C (2500°F) turbine inlet temperature at 64,000 rpm over a 200-hour duty cycle. This accomplishment demonstrated to the world that ceramics could be successfully employed in structural applications.

NONDESTRUCTIVE EVALUATION (NDE)

At the beginning of the ceramic turbine program, the only NDE techniques available were visual dye penetrant examination using colored dyes combined with visual examination using various magnification techniques. These techniques examined only the surface of a component. However, due to the high stressed nature of various turbine components, NDE techniques were required to inspect for subsurface defects. A major effort was undertaken to apply emerging NDE techniques to ceramics to determine if they could be successful in identifying subsurface defects in complex shaped components.

X-ray Radiography

Conventional x-ray radiography was useful for the detection of gross flaws in injection-molded components having complex geometries. This technique was demonstrated to have the capability for detecting cracks or voids on the order of 1 mm and for detecting metallic inclusions on the order of 0.2 mm in diameter.

Micro focus x-ray radiography was a technique to magnify the defect size by altering the distance of the x-ray film from the component. This technique enhanced the ability to view small defects.

Ultrasonic Inspection

Ultrasonics techniques were used to determine the sonic properties of various ceramics. This information was used to calculate the elastic properties or to determine the density of a material.

Ultrasonic inspection were successful in measuring defects as small as 25 microns, but only in very simple shaped components, such as an MOR bar.

Acoustic Emission

An acoustic emission technique was useful in determining the presence of flaws in a material or the onset of failure of a stressed component. However, the technology was not useful in practical evaluation of flaws in a complex shaped component.

The results of the above investigations revealed that the techniques were not ready for implementation on a day-to-day basis for the inspection of ceramic components. The principal techniques used for NDE of ceramics were: 1) enhanced visual inspection, 2) dye penetrant inspection and 3) proof testing for known types of flaws which were experimentally determined to occur in the same area of a component. Figure 2.20a shows a crack at the blade root of a turbine blade by visual inspection while Figure 2.20b shows a test rig to measure the strength or proof test each blade on a turbine rotor.

SPIN-OFF PROGRAMS/COMPANIES/APPLICATIONS

"The major purpose of the DARPA program was the demonstration of brittle materials for demanding high temperature structural applications in order to encourage wider use of these materials in engineering systems" [23].

A number of spin-off programs and applications resulted from the initial Ford ceramic turbine program. Likewise, a number of companies were formed by persons who were associated with the Ford ceramic turbine program or with technology initially developed in the Ford ceramic turbine program. These are summarized below:

Programs

- DOE's AGT Programs with Garrett Turbine Engine Company and Allison Gas Turbine Company plus many related ceramic technology programs
- German Ceramic Gas Turbine Programs [24]
- Japanese Fine Ceramic Programs

Companies

- Ceramic Systems (Eldor Hermann, Founder)
- Industrial Ceramic Technology (John Schuldies, Founder)
- Honeywell Ceramic Components (formerly AiReseach Casting Company and then Garrett Ceramic Components and later AlliedSignal Ceramic Components)
- Cercom (Andre Ezis, Co-Founder)
- Ceradyne, Inc.—commercialization of Ford's silicon nitride technology

Applications

Silicon Nitride Cutting Tools. In 1980 Ford was awarded the first patent for silicon nitride cutting tools for the machining of gray cast iron [12]. Now, every cutting tool company in the world offers this product, which has become the industry standard for machining of gray cast irons. This material increased the productivity of the cast iron machining operations (both milling and turning) by enabling the machining processes to be conducted at higher speed, feeds and depths of cut while at the same time offering longer tool life.

Ceramic Turbocharger Rotors. In 1981 Ford first demonstrated that a ceramic turbocharger rotor could be used with a small internal combustion engine [25]. The ceramic turbocharger rotor offered improved "drivability" by reducing the "turbo" lag (Figure 2.21) because of the 53 percent reduction in the polar moment of inertia of the ceramic rotor versus the corresponding metal rotor. In the early to mid 1980s, Japanese companies offered this component for sale on cars sold in Japan.

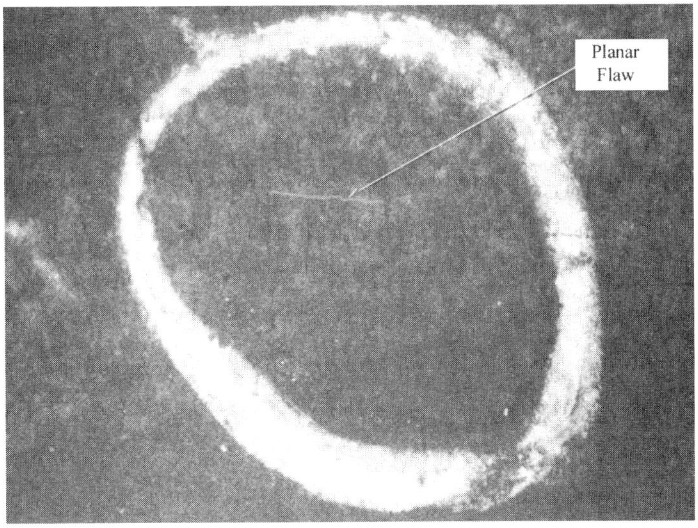

(a)

(b)

Figure 2.20 (a) Crack at a Turbine Rotor Blade Root Detected by Visual Examination and (b) Proof Test Rig to Measure the Strength of Each Turbine Blade (Reprinted from ref. [11])

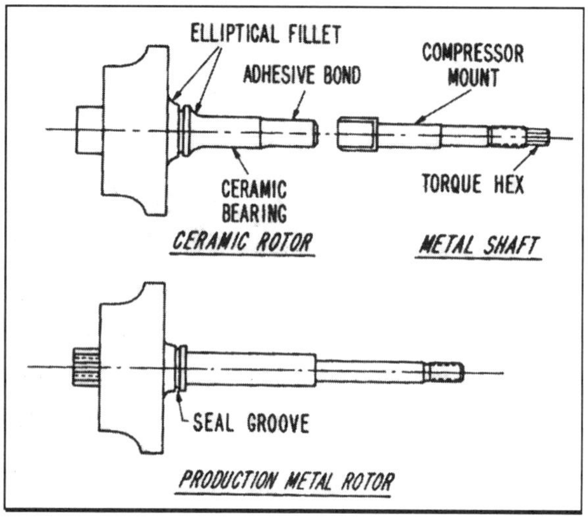

(a)

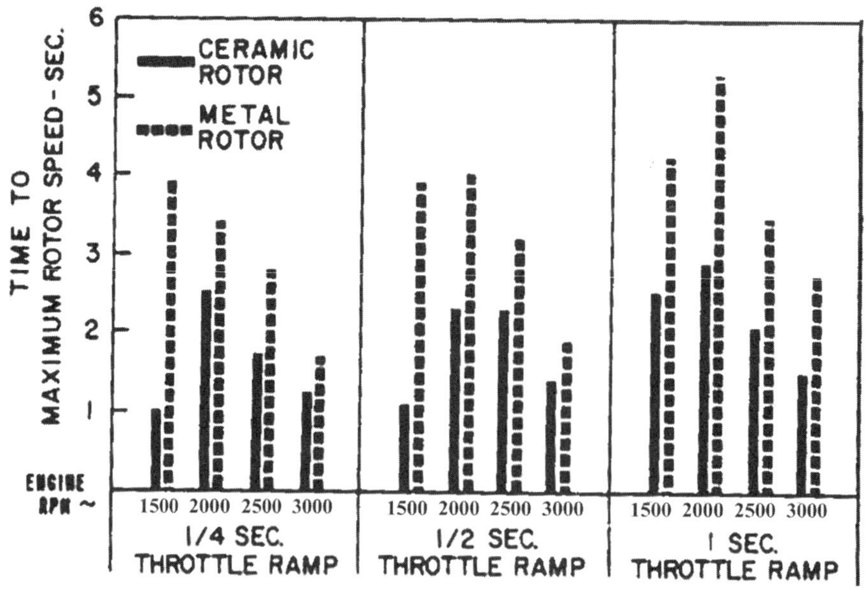

(b)

Figure 2.21 Ford Turbocharger Rotor and Its Performance Characteristics (a) Schematic of a Silicon Nitride Ceramic Turbocharger Rotor, (b) Comparative Response Times for Ceramic vs.Metal Turbocharger Rotors at Various Engine Conditions (Reprinted from ref. [25].)

Figure 2.22 Typical Wear Components Produced from SRBSN by Ceradyne, Inc. (Reprinted with permission from Ceradyne Inc.)

Wear Components. In 1986 Ford selected Ceradyne, Inc. [10] as a commercialization partner for its silicon nitride technology. Ceradyne concentrated its early efforts on the manufacturing scale up of the sintered reaction bonded silicon nitride (SRBSN) process. They developed a series of compositions under the generic designation of Ceralloy 147-3. The original markets for the wear products made of these materials which included cutting tools, were in the paper, mineral processing, metal processing, petro-chemical and oilfield industries (Figure 2.22). By 1996, they were selling over 600,000 pieces (over 27,000 kg) of SRBSN wear components annually. The advantage of silicon nitride over other ceramics was its combination of properties including hardness, toughness, thermal shock resistance and chemical durability in a variety of operating conditions.

Engine Components. Ford expanded its ceramic effort in the late 1970s to investigate the potential of ceramics for other engine applications, especially in the area of small adiabatic diesel engines [26]. Ford investigated the feasibility of a number of components from ceramics including tappet inserts, valve guides and engine valves [27]. The results of these programs showed that traditional engine components could be fabricated and successfully operated in piston engine applications.

During the 1980s and 1990s, a number of engine components had been commercialized using silicon nitride. These are summarized in Table 2.4.

In 1999, Ceradyne began producing components for engine applications on a production scale using the Ford SRBSN technology. Components include cam rollers for heavy-duty diesel engines, rollers for high-pressure fuel pumps and components for a variety of valve train components for racing applications

TABLE 2.4 SILICON NITRIDE ENGINE APPLICATIONS [28,29]

Component	Year Introduced	Benefits
Glo-Plugs	1981–85	Faster start-up
Precombustion chamber	1983–86	Lower emissions, lower noise, faster start-up
Rocker-arm pads	1984–87	Lower wear
Turbocharger	1985–89	Lower inertia lag
Fuel injector link	1989	Reduced wear, lower emissions
Cam roller follower	1991	Increased injection pressure, lower emissions
Fuel Injector check ball	1992	Reduced wear, lower emissions
Compression brake master	1993	Reduced wear on cylinder face
Exhaust gas control valve	1993	Increased acceleration, reduced gas leakage

(Figure 2.23). The common feature of these components is that they all operate under high Hertzian contact stresses, up to 2.4 GPa (350,000 psi), at low to intermediate temperatures. The silicon nitride provides increased life and durability versus the corresponding metal components as well as reduced operating friction. The decision to use ceramic instead of metal parts is often driven by cost considerations. Ceramic components are several times more expensive than the metal parts they replace. Customers will pay a premium for ceramics but only if necessary to solve a real problem or if they can provide a quantifiable improvement in performance.

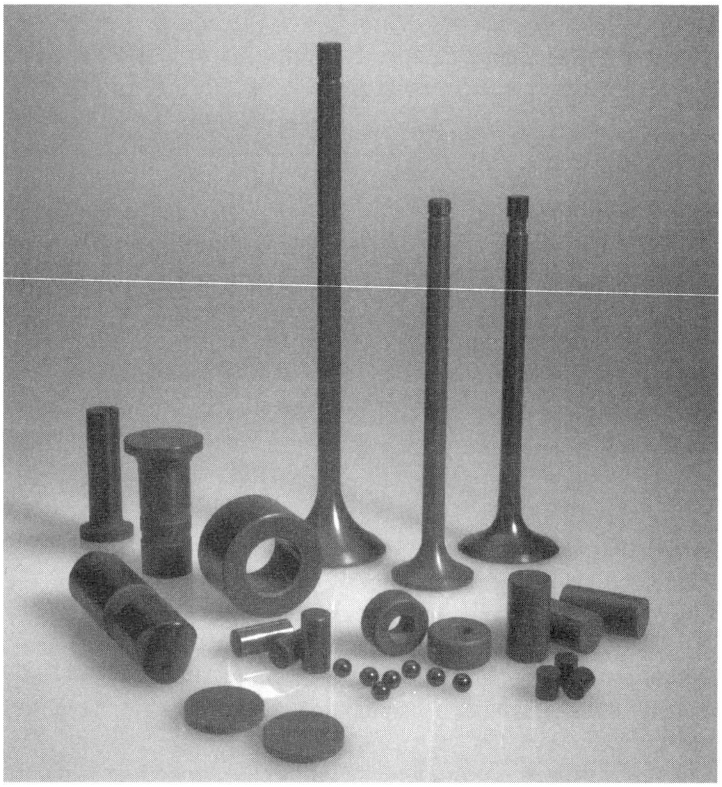

Figure 2.23 Typical Engine Components Produced from SRBSN by Ceradyne, Inc. (Reprinted with permission from Ceradyne Inc.)

SUMMARY

Ford's ceramic effort concentrated on the development of silicon nitride and silicon carbide materials, ceramic component fabrication technology and material characterization and non-destructive evaluation techniques. The ceramic effort resulted in a number of application firsts including: the first uncooled gas turbine to run at 1371°C (2500°F) turbine inlet temperature, the first demonstration of a ceramic turbocharger, the first use of silicon nitride as a cutting tool material. However, most importantly, the success of the Ford gas turbine program lead to the "ceramic fever" which began in the mid 1970s and has spawned the development of non-oxide ceramics around the world. These materials are now used as enabling technologies in many industrial, semiconductor processing and automotive applications. The Ford ceramic turbine program demonstrated beyond any doubt that ceramics could be used successfully in structural applications.

ACKNOWLEDGEMENTS

The author would like to thank A.F. McLean and E.A. Fisher for their leadership of the Ford ceramic turbine effort and for the help and guidance they gave me throughout his career at Ford. Also, thanks to the many persons involved in the Ford ceramic programs from 1967 to 1986 who made this program a technological success. Thanks also go to R.N. Katz, program manager for the ARPA Ceramic Gas Turbine Program, who had the patience to help manage, direct and fund this challenging program. Thanks also go to Richard Allor and Robert Baker who reviewed this paper, to Susan Marra for help in the preparation of the manuscript and to the Ford Motor Company and Ceradyne, Inc. for permission to publish this paper. I also especially thank two special individuals, G.J. Tennenhouse and F. Runkle, who had a profound effect on the author's career—may they rest in peace.

Authors Note

Mr. Mangels was a member of the Ford Motor Company's ceramics team from 1969 to 1986. In 1986, Mr. Mangels joined Ceradyne, Inc. to transfer the Ford ceramics technology to Ceradyne as part of a joint venture between the two companies. Ceradyne continued the development and commercialization of sintered reaction bonded silicon nitride (SRBSN) that was initially developed at Ford. Unless otherwise indicated the figures and tables in this chapter were originals prepared by the author.

BIBLIOGRAPHY

McLean, A. F., Fisher, E. A., *Brittle Materials Design, High Temperature Gas Turbine*, AMMRC TR 81–14, Contract No. DAAG 46-71-C-0162, March 1981.

Ceramics for High Performance Applications, Proceedings of the Second Army Materials Conference, Hyannis, MA, USA, November 13–16, 1973, J. J. Burke, A. E. Gorum, and R. N. Katz, eds. Brook Hill Publishing Company, Chestnut Hill, MA, USA, 1974.

Ceramics for High Performance Applications II, Proceedings of the Fifth Army Materials Conference, Newport, RI, USA, March 21–25, 1977, J. J. Burke, E. N. Lenoe, and R. N. Katz, eds. Brook Hill Publishing Company, Chestnut Hill, MA, USA, 1978.

Hartsock, D. L., "Ford's Development of the 820 High Temperature Ceramic Gas Turbine Engine, Chapter 3 in *Ceramic Gas Turbine Design and Test Experience—Progress in Ceramic Gas Turbine Development—Volume I,* M. van Roode, M. K. Ferber, and D. W. Richerson, eds., ASME Press, New York, USA, 2002.

REFERENCES

[1] Mangels, J. A., "Development of Injection Molded Reaction Bonded Silicon Nitride," in *Ceramics for High Performance Applications II, Proceedings of the Fifth Army Materials Conference,* Newport, RI, USA, March 21–25, 1977, J. J. Burke, E. N. Lenoe, and R. N. Katz, eds., Brook Hill Publishing Company, Chestnut Hill, MA, USA, 1978. pp. 113–130.

[2] Mangels, J. A., "Strength-Density–Nitriding Cycle Relationships for Reaction Sintered Si_3N_4," in *Nitrogen Ceramics*, F. L. Riley, ed., Noordhoff Int. Pub., Leyden, The Netherlands, 1977. pp. 569–576.

[3] Mangels, J. A., "Effect of Rate-Controlled Nitriding and Nitriding Atmospheres on the Formation of Reaction-Bonded Si_3N_4," Bull. Am. Ceram. Soc., 60[6], 613–617, 1981.

[4] Mangels, J. A. *Development of Moldable, High Density Reaction Bonded Silicon Nitride—Final Report,* NASA CR-165537, Contract No. DEN-3-20, September 1981.

[5] Terwilliger, G. F., "Properties of Sintered Silicon Nitride," J. Am. Ceram. Soc., 57 [1], 48–49,1974.

[6] Terwilliger, G. F., and Lange, F. F., "Pressureless Sintering of Silicon Nitride," J. Mater. Sci., 10, 1169–1174, 1975.

[7] Mangels, J. A. and Tennenhouse, G.J., "Densification of Reaction Bonded Silicon Nitride," Bull. Am. Ceram. Soc., 59[12], 1216–1218, 1980.

[8] Mangels, J. A., and Tennenhouse, G. J., "A Method of Densifying a Reaction Bonded Si3N4 Article," U.S. Patent No. 4,285,895, August 25, 1981.

[9] Mangels, J. A., "Sintered Reaction-Bonded Silicon Nitride," Ceram. Eng. and Sci. Proc., 2[7–8], 589, 603, 1981.

[10] Ceradyne Inc., 3169 Redhill Avenue, Costa Mesa, Ca., www.ceradyne.com

[11] McLean, A. F. and Fisher, E. A., *Brittle Materials Design, High Temperature Gas Turbine,* AMMRC TR 81–14, Contract No. DAAG 46-71-C-0162, March 1981.

[12] Samanta, S. K., Subramanian, K., and Ezis, A., "Method of Using Si3N4-Y2O3-SiO2 Ceramic System for Machining Cast Iron," U.S. Patent No. 4,227,842, October 14, 1980.

[13] Ezis, A., and Beckwith, E., "Method of Making High Strength Hot Pressed Silicon Nitride," U.S. Patent Application No. 629,921, July 1984.

[14] Mangels, J. A., and Trela, W., "Ceramic Components by Injection Molding," Forming of Ceramics," Advances in Ceramics, Vol. 9, J. A. Mangels, G. Messing, eds., American Ceramic Society, 1984. pp. 220–233.

[15] Mangels, J. A., and Williams, R. M., "Injection Molding Ceramics to High Green Densities," Bull. Am. Ceram. Soc., 62 [5], 601–606, 1983.

[16] Mangels, J. A., "Low-Pressure Injection Molding," Bull. Am. Ceram. Soc., 73[5], 37–41, 1994.

[17] Ezis, A., "The Fabrication and Properties of Slip-Cast Silicon Nitride," in *Ceramics for High Performance Applications, Proceedings of the Second Army Materials Conference,* Hyannis, MA, USA, November 13–16, 1973, J. J. Burke, A. E. Gorum, and R. N. Katz, eds., Brook Hill Publishing Company, Chestnut Hill, MA, USA, 1974. pp. 207–222.

[18] Ezis, A., and Nicholson, J. M., "Method of Manufacturing a Slip Cast Article," U.S. Patent 4,067,943, January 10, 1978.

[19] Ezis, A., and Nicholson, J. M., "Slip Cast Article Manufacturing Method," U.S. Patent No. 4,217,320, August 12, 1980.

[20] Ezis, A., and Neil, J.T., "Slip Casting of Silicon–Based Ceramics," Bull. Am. Ceram. Soc., 58[9], 883, 1979.

[21] Fisher, E. A., Goodyear, M. U., McLean, A. F., and Styhr, K. H., "Method of Making a Duo Density Silicon Nitride Article," U.S. Patent No. 4,087,500, May 2, 1978.

[22] Baker, R. R., "Apparatus for Forming Ceramics Bodies," U.S. Patent, 4,140,469, February 20, 1979.

[23] McLean, A. F., and Fisher, E.A., "Brittle Materials Design, High Temperature Gas Turbine," AMMRC TR 81-14, Contract No. DAAG 46-71-C-0162, March 1981. p.vii.

[24] *Ceramics for High Performance Applications II, Proceedings of the Fifth Army Materials Conference,* Newport, RI, USA, March 21–25, 1977, J. J. Burke, E. N. Lenoe, and R. N. Katz, eds., Brook Hill Publishing Company, Chestnut Hill, MA, USA, 1978.

[25] DeBell, G. C., and Secord, J. R., "Development and Testing of a Ceramic Turbocharger Rotor," ASME paper 81-GT-195, presented at the Gas Turbine Conference & Products Show, Houston, TX, USA, March 9–12, 1981.

[26] Wade, W. R., Havstad, P. H., Ounsted, E. J., Trinker, F. H., and Garwin, I. J., "Fuel Economy Opportunities with an Uncooled DI Diesel Engine," SAE 841286, 1984.

[27] Kabat, D. M., Garwin, I. J., and Hartsock, D. L., "Ceramics Valve Analysis, Reliability and Test Results," SAE 880670, 1988.

[28] Katz, R. N., "Applications of Silicon Nitride-Based Ceramics," Ind. Ceram. (Firenza, Italy), 17[3], 158–164, 1997.

[29] Riley, F. L., "Silicon Nitride Related Materials," J. Am. Ceram. Soc., 83[2], 245–265, 2000.

CHAPTER

3

Ceramic Component Development at Saint-Gobain/Norton Advanced Ceramics

David W. Richerson
University of Utah, Salt Lake City, Utah, USA

Eric Bright
Saint-Gobain Abrasives, Worcester, Massachusetts, USA

Robert Licht
Saint-Gobain Ceramics & Plastics, Inc., Northboro, Massachusetts, USA

INTRODUCTION

Norton Company (presently Saint-Gobain Ceramics & Plastics, Inc.) began development of silicon nitride and silicon carbide ceramics for turbine engine components in the early 1970s. Focus during the 1970s was on reaction bonded silicon nitride (RBSN), hot pressed silicon nitride (HPSN), reaction sintered (siliconized) silicon carbide (RSSC), and recrystallized silicon carbide. Focus shifted in the 1980s to hot isostatically pressed silicon nitride (HIP-SN). This chapter first briefly reviews the roots of the Norton silicon nitride and silicon carbide technologies and then describes the evolution, especially of silicon nitride materials, to the present time. Emphasis is on the efforts to improve the room temperature and high temperature strength, creep resistance, and stress rupture life; to increase the reliability by improving the fabrication process; and to learn to fabricate the complex shapes required for gas turbine engine components.

BASELINE TECHNOLOGIES

Table 3.1 identifies some of the baseline technologies that were present at Norton Company by the late 1960s when engine companies and government agencies began to have a strong interest in ceramics

TABLE 3.1 BASELINE TECHNOLOGIES AVAILABLE AT NORTON BY THE LATE 1960S

Key Technologies	Source of These Technologies
Reaction-bonded ceramics	• Nitride-bonded silicon carbide developed for refractory applications • Silicon oxynitride developed as a refractory for the aluminum melting industry
Recrystallized SiC	Porous high purity sintered SiC developed for heating elements, kiln furniture, and other industrial products synthesis of non-oxide powders, synthesis of silicon carbide in large quantities by the Acheson process for abrasives, heating elements, and other products
Hot pressing	Fabrication of boron carbide ceramic armor in sizes exceeding 50 cm × 50 cm
Complex shape fabrication processes	Experience with isostatic pressing and green machining and with slip casting
Diamond tooling design and fabrication; and grinding of hard ceramics	Division within Norton Company that conducted design and fabrication of diamond-tipped saw blades and grinding wheels

for turbine engines. These technologies provided an excellent starting point for adaptation and modification to develop prototype turbine engine components.

DEVELOPMENTS DURING THE 1970S

Developments during the 1970s focused on reaction-bonded silicon nitride (RBSN), reaction-sintered silicon carbide (RSSC), hot pressed silicon nitride (HPSN), and hot pressed silicon carbide (HPSC).

Reaction-Bonded Silicon Nitride Developments

RBSN is formed by first producing a compact of silicon powder to the desired shape and then reacting the silicon powder with nitrogen in a sealed high temperature furnace to yield silicon nitride. A big advantage of the process is that very little (about 0.1%) dimensional change occurs during the conversion from silicon to silicon nitride, which simplifies fabrication of complex shaped parts such as required by gas turbine engines.

RBSN materials had been successfully developed in England prior to 1970, but the residual porosity was quite high and the strength typically under 150 MPa. Washburn and coworkers [1,2] used experience gained in development of nitride-bonded SiC refractories and silicon oxynitride plus an improved nitriding furnace design [3] to develop RBSN with decreased porosity, improved uniformity of microstructure, and increased strength [4,5]. The resulting RBSN (designated NC-350) had average room temperature strength (of test bars with as-fabricated surfaces, tested in 4-point bending) of 214 MPa. and 1375°C strength of 391 MPa. A creep test with 4-point loading of 138 MPa at 1260°C exhibited no deformation after 260 hours of testing.

The NC-350 RBSN was fabricated into stator vanes and other turbine components using green machining of cold isostatic pressed silicon blocks. Considerable engine testing was successfully conducted, especially under the DARPA/Navy/Garrett Turbine Engine Company Ceramic Engine Demonstration program [6]. Later, NC-350 was used as the matrix for composites with SiC fiber reinforcement [7].

Silicon Carbide Developments

Norton pursued several SiC material options for development of gas turbine engine components in the 1970s. These included hot pressed SiC, recrystallized SiC, silicon infiltrated recrystallized SiC, and reaction-sintered SiC.

The hot pressed SiC technology was an extension of work done by Alliegro and coworkers [8] in the 1950s at Alfred University and by Alliegro, Weaver and others at Norton during the 1960s and early 1970s associated with commercialization of hot pressing of boron carbide armor. The resulting hot pressed SiC [9] (designated NC-203) was evaluated as stator vanes for a 30-MW size gas turbine engine by Westinghouse under the ARPA/AMMRC Brittle Material Design, High Temperature Turbine program [10]. This material had 3-point flexural strength of about 800 MPa at room temperature and nearly 500 MPa at 1400°C. It exhibited good oxidation resistance, but hot pressed silicon nitride exhibited better thermal shock resistance, so further work concentrated on silicon nitride.

The baseline Norton non-hot-pressed SiC evaluated for early turbine components, especially combustor liners, was Noralide NC-400 [11]. It was a relatively pure recrystallized SiC with a bimodal grain structure. It had an average flexural strength of about 165 MPa from room temperature to over 1400°C. Higher strength reaction-sintered (siliconized) SiC materials (designated NC-430 and NC-433) were developed during the 1970s [11–13]. The NC-430 had 3-point flexural strength of about 300 MPa at room temperature and 215 MPa at 1400°C. The NC-433 was a little stronger, i.e. 310 MPa at room temperature and about 295 MPa at 1400°C.

Considerable shape forming development was conducted to produce prototype combustor liners, turbine vanes and integral stators, rotor blades, turbocharger rotors, and heat exchanger components [13]. One combustor liner for an Army program conducted by AiResearch (now Honeywell Engines, Systems & Services) was 0.5 m in diameter and 0.75 m high [14].

Hot Pressed Silicon Nitride Development

Developments in England in the 1960s demonstrated that high strength silicon nitride could be achieved by adding sintering aids to silicon nitride powder and hot pressing [15,16]. For example, the hot pressed silicon nitride available from Joseph Lucas Limited in 1970 had average 3-point flexural strength of about 840 MPa at room temperature and about 415 MPa at 1200°C.

Norton Company began limited hot pressing trials on silicon nitride in 1970, but did not achieve properties comparable to those reported by Lucas. Norton decided to obtain a license to the Lucas hot pressed silicon nitride technology and subsequently proceeded to improve the properties and scale up size capability. Two directions were pursued to improve properties:

(1) Modify the chemistry of the starting silicon nitride powder and additives, especially to improve high temperature properties [17,18]; this later became known as "grain boundary engineering."
(2) Examine fracture surfaces optically and by scanning electron microscopy (SEM) to identify the defect or microstructure feature responsible for failure and conduct iterative fabrication developments to eliminate the source of the defects [19,20].

Figure 3.1 illustrates the improvements in strength achieved by October 1971. HS-110 designates the baseline Lucas hot pressed silicon nitride, and HS-130 designates the early improvements achieved at Norton. Both materials used MgO as the sintering aid. The improvement in high temperature strength was especially important for turbine applications. By reducing the impurities in the silicon nitride powder, especially calcium, the 3-point flexure strength at 1200°C was increased from 415 MPa to about 700 MPa [18]. At 1350°C, the strength was increased from about 140 MPa to over 310 MPa. The stress rupture life (in 4-point flexure) of the baseline HS-110 material was only 1 hour for an applied stress of 138 MPa at 1200°C. The stress rupture life of the HS-130 was 250 hour for a stress of 207 MPa at 1200°C. This work was an important step in the evolution of silicon nitride and set the stage for extensive grain boundary engineering during the next 20 years to further improve strength, stress rupture life, and creep resistance.

In parallel to the grain boundary chemistry studies, fractography efforts were conducted to identify the microstructure features that were controlling fracture. As the strength-limiting defects were identified, the fabrication process was modified in an attempt to either eliminate or decrease the size of these defects [19]. The results are illustrated in Figure 3.2. Note that the fracture stress values in this case were measured in 4-point flexure and reported in Weibull format. For comparison, the SN1 material (which was HS-130) had a mean strength of 966 MPa in 3-point flexure and 718 MPa in 4-point flexure.

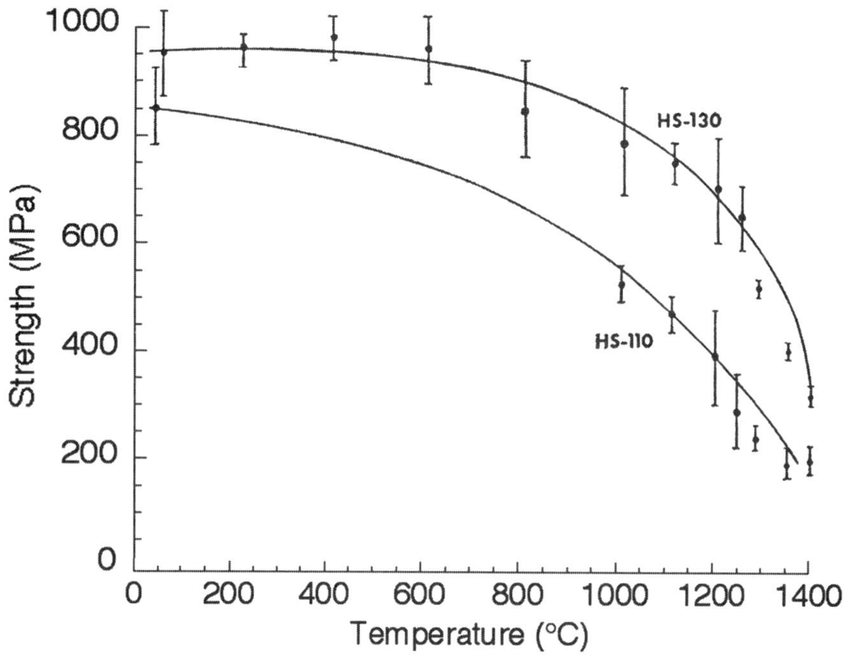

Figure 3.1 Early Improvements in Hot Pressed Silicon Nitride by Control of the Composition of the Grain Boundary Phase (Adapted from ref. [18] with permission from the American Ceramic Society, www.ceramics.org, © 1973, all rights reserved)

As can be seen from the curves in Figure 3.2, both the mean strength and the Weibull modulus of hot pressed silicon nitride were substantially increased. The SN2 material became the well-known NC-132 material that was the first truly commercial dense silicon nitride. NC-132 was fabricated in a pilot-production facility that included a positive pressure moderate clean room environment, detailed specifications for each step in the fabrication process (including raw materials specifications), magnetic separation and slurry filtration to eliminate inclusions and large particles, and in-process and post-process certification procedures to assure high quality.

NC-132 became a standard for dense silicon nitride throughout the 1970s and even into the 1980s and was the baseline for comparison of new silicon nitride materials as they were developed. It was characterized extensively [21–26] and was evaluated with considerable success for turbine blades by Pratt and Whitney [27], Westinghouse [10,28], and Garrett Turbine Engine Co. [6,29]; for industrial turbine stator vanes by Pratt and Whitney [30,31] and Westinghouse [10,32]; and for high-speed bearings [33,34].

Gazza [35] at AMMRC explored the use of yttrium oxide as the sintering aid for hot pressing of silicon nitride. He reported flexure strength of 814 MPa at room temperature and 400–475 MPa at 1315°C. Weaver and co-workers subsequently developed a hot pressed silicon nitride (designated NCX-34) at Norton using 8% yttrium oxide additions [36]. They reported 3-point flexure strength of 1069 MPa at room temperature, 897 MPa at 1200°C, and 517 MPa at 1400°C. All of these values were higher than for NC-132. In an independent study at Illinois Institute of Technology Research Institute (IITRI), Larsen [37] reported 4-point flexure strength of NC-132 of 703 MPa at room temperature and 324 MPa at 1400°C compared to 897 MPa at room temperature and 352 MPa at 1400°C for NCX-34. Other tests [24] suggested that NCX-34 had superior stress rupture life and oxidation resistance compared to NC-132 above 1000°C.

The NCX-34 looked like a promising improved silicon nitride material for blades and vanes. However, burner rig tests [38] revealed that severe degradation consisting of swelling and cracking occurred when the

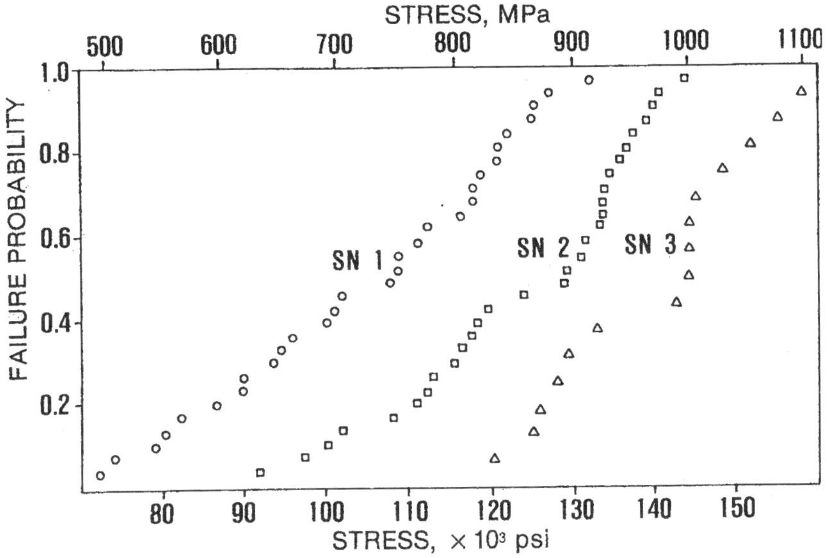

Figure 3.2 Improvements in the Strength and Weibull Modulus of Hot Pressed Silicon Nitride by Modifying the Fabrication Process to Decrease the Size and Frequency of Strength-Limiting Defects (Reprinted with permission from ref. [19])

material was exposed to oxidation at intermediate temperatures. Further study in a gradient furnace [38] determined that the degradation occurred at temperatures from 890°C down to at least 550°C. Lange et al. [39] reported similar degradation in silicon nitride containing 13% yttria that they attributed to oxidation and volume change of quaternary Si-Y-O-N compounds (J and H phases). NCX-34 was not further pursued for gas turbine engine components.

DEVELOPMENTS DURING THE 1980S

Norton Company continued development of silicon nitride and silicon carbide into the 1980s. Material development efforts focused on (1) achieving hot pressed silicon nitride compositions with yttrium oxide sintering aid that were stable at intermediate temperatures, (2) hot isostatic pressing (HIP) of silicon nitride, (3) whisker-reinforced and fiber-reinforced silicon nitride, and (4) improved siliconized silicon carbide. A joint venture was established with TRW (Norton-TRW Ceramics, i.e. NTC) to develop ceramic engine components. In parallel, efforts were conducted to establish cost-effective production of bearings based on the NC-132 silicon nitride.

Continuation of Monolithic Silicon Nitride Development

Phase equilibrium considerations and experimental hot pressing studies indicated that the Si-Y-O-N phases that caused the oxidation degradation in NCX-34 at intermediate temperatures could be avoided by decreasing the percentage of yttria sintering aid. However, this also increased the difficulty of achieving full densification. HIP was successfully explored as a means of achieving full densification with decreased sintering aid. NTC ultimately licensed the HIP and glass encapsulation technology from Asea [40]. Following substantial development and optimization, NTC settled on a HIP silicon nitride material containing 4%

TABLE 3.2 BASELINE PROPERTIES OF NT154 IN 1988 (ADAPTED FROM REF. [41])

Density (g/cm3)	3.23 ± 0.01
Elastic Modulus (GPa)	320
Shear Modulus (GPa)	126
Poisson's Ratio	0.273
Thermal Expansion Coefficient	3.93×10^{-6}/°C
Flexure Strength at room temperature (MPa)	920 ± 153
Weibull Modulus at room temperature	9.6 ± 0.3
Flexure Strength at 1370°C (MPa)	531 ± 46
Weibull Modulus at 1370°C	13.2 ± 1.3
Fracture Toughness (MPa.m$^{1/2}$)	3.6 ± 0.3
Thermal Conductivity (W/m.K)	
25°C	37.6
500°C	28.3
1000°C	19.5
1400°C	9.1

yttria. This material was designated NT154. The baseline properties for NT154 reported in 1988 are listed in Table 3.2 [41].

The baseline NT154 was selected in 1987 for evaluation and continued development under the DOE Advanced Turbine Technology Applications Project (ATTAP) under subcontract to the Garrett Auxiliary Power Division (GAPD) of AlliedSignal Aerospace Company. The objectives were (1) to further optimize properties through refinement of the raw materials and fabrication process steps and (2) to learn to cost effectively fabricate turbine components with properties comparable to the baseline established for test bars.

The efforts during the initial three years focused on evaluation of properties of samples fabricated using cold isostatic pressing, injection molding, and slip casting (especially pressure slip casting) as the shape-forming processes prior to HIP. Taguchi experimental design methods were effectively used to explore variables in powder beneficiation, forming, pre-HIP treatments, HIP, and post-HIP treatments [41].

An example of a Taguchi experiment design for initial screening of powder processing, HIP, and forming method is shown in Figure 3.3 [42]. Variables evaluated included powder blend (PB), milling time (MT), solids concentration in the slip (SC), HIP temperature (TP), HIP time (TM), cooling rate (CR) and cold isostatic pressing (CIP) versus slip casting. The experiment required 72 separate trials and thus was one-sixth of a full factorial.

Figures 3.4 and 3.5 illustrate the type of information that resulted from the Taguchi L9XL4 experiment [42,43]. The information gained from the initial Taguchi experiment was used to set up focused Taguchi experiments to better understand and optimize each critical process step. The following are some observations and conclusions reported [43]:

- Powder blend was a dominant factor for every response variable.
- Fine powder gave the highest strength, but a broader particle size distribution was necessary for acceptable forming behavior.
- Short mix/mill times and high solids content were favorable for slip casting.
- HIP time, temperature, and cooling rate all were determined to be critical.
- Modest improvements in fracture toughness resulted in substantial improvements in strength.
- Fine powders with high surface area (and relatively high oxygen content) resulted in high strength and toughness, but not optimum fatigue life at high temperature. Compromises were chosen for both particle size/distribution and HIP parameters to achieve superior high temperature properties at less than peak fast fracture strength.

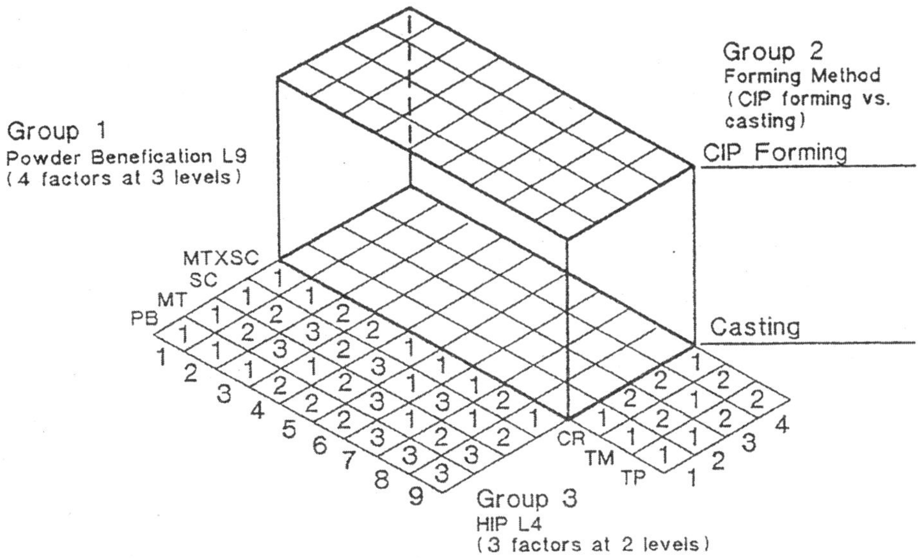

Figure 3.3 Schematic of a L9 × L4 Taguchi Experiment Design to Evaluate Powder Beneficiation, Forming Method, HIP parameters, and Interactions for Norton HIP Silicon Nitride (Reprinted from refs. [42,43]).

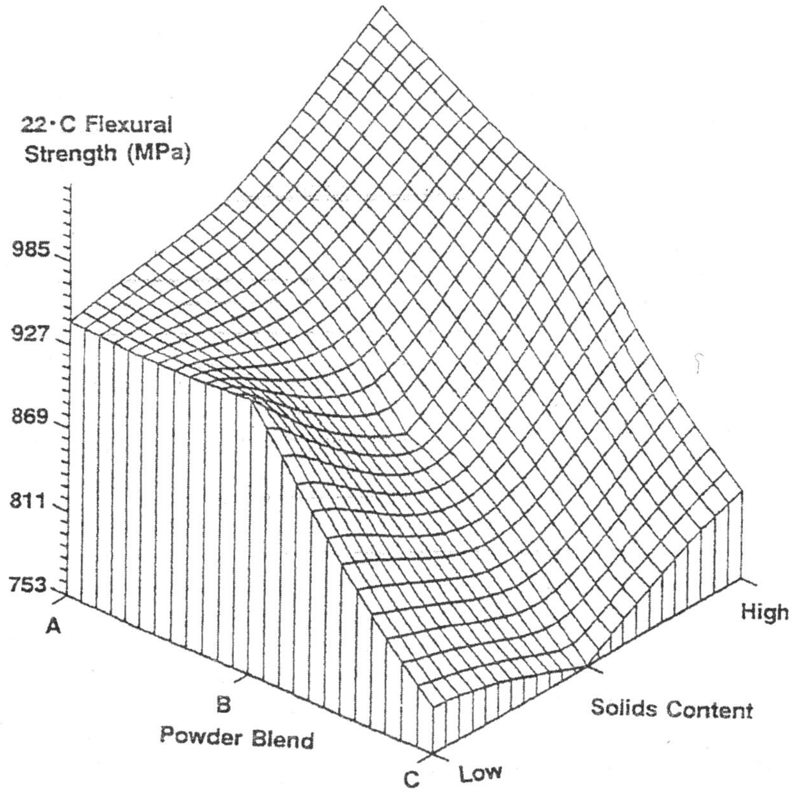

Figure 3.4 Example of the Output from a Taguchi Experiment Showing the Influence of Powder Blend and Solids Content Variations on the Room Temperature Strength (Reprinted from refs. [42,43])

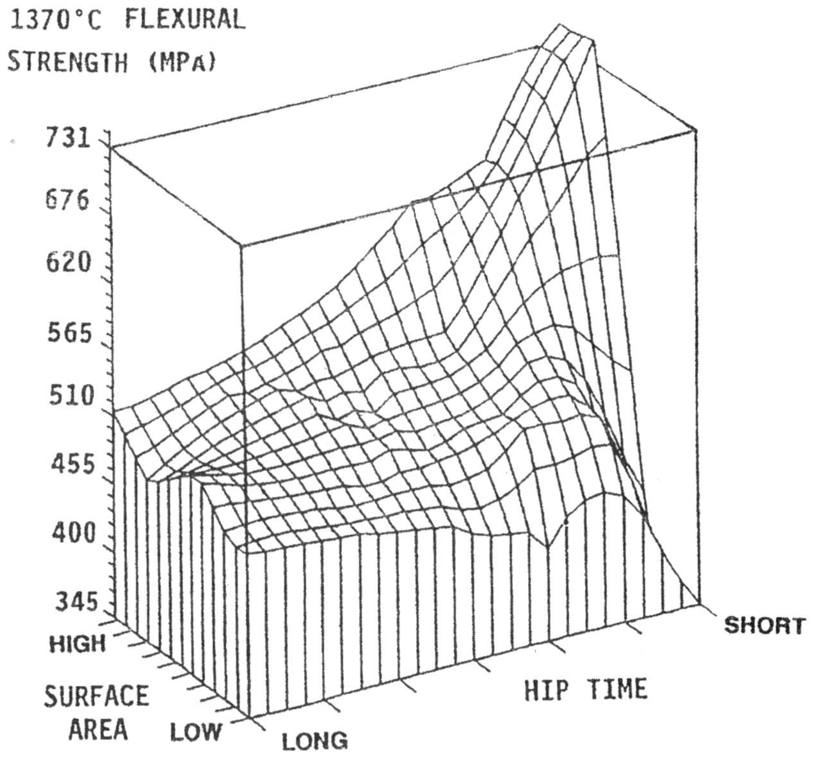

Figure 3.5 Example of the Output from a Taguchi Experiment Showing the Influence of Powder Surface Area and HIP Time on the High Temperature Strength of Silicon Nitride (Reprinted from refs. [42,43])

- A high temperature inert gas heat treatment (defined as degassing) prior to encapsulation for HIPing was identified to be beneficial. Compromise was again required since the optimum conditions for peak fast fracture strength were different than for time-dependent properties.
- Two heat treatments were established post-HIP. One was a high temperature exposure in an inert atmosphere to crystallize any residual glass. The second was an oxidation treatment to increase strength by decreasing the severity of surface defects.
- Samples prepared by pressure slip casting exhibited the highest strength and Weibull Modulus, followed by cold isostatic pressing and injection molding.

Fiber-Reinforced Silicon Nitride

During the early 1980s, Corbin, et. al. [7] conducted fabrication and property studies at Norton on SiC monofilament reinforcement of RBSN. The baseline NC-350 had a flexure strength of about 330 MPa and strain-to-failure less than 0.2%. A composite of NC-350 RBSN containing about 35 volume % unidirectional SCS-6 SiC monofilaments (~140 μm diameter) from AVCO Specialty Materials Division (later became Textron) had flexure strength of about 650 MPa and greater than 3% strain-to-failure.

Norton received a contract from ORNL in October 1986 under the Ceramic Technology for Advanced Heat Engines program to develop fully dense silicon nitride containing a dispersion of SiC whiskers [44,45].

Two fabrication routes were studied, both based on addition of 4% Y_2O_3 and on final densification by HIP with the Asea glass encapsulation technology. One fabrication route started with silicon powder to produce a RBSN part that was subsequently HIPed. The other route started with silicon nitride powder.

The one-year Phase 1 effort under the program explored a variety of SiC whiskers (Tateho, Huber, American Matrix, Kobe Steel, Nikkei, Kanebo, Arco, and Mitsui-Toatsu) plus coatings on selected whiskers. Coatings were deposited by CVD and consisted of carbon, BN, dual coatings of SiC or Si_3N_4 over BN or carbon. Composite samples containing typically 20% whiskers were fabricated and compared based on flexure strength from room temperature to 1370°C, fracture toughness from room temperature to 1200°C, and stress rupture life up to 200 MPa and 1370°C.

Samples fabricated with uncoated fibers had approximately the same fracture toughness as monolithic control samples (~3 MPa.m$^{1/2}$). Coating the whiskers with BN or a BN/SiC dual coating increased the toughness to nearly 5 MPa.m$^{1/2}$. Fracture toughness decreased at elevated temperature. Room temperature flexure strength was typically about 700 MPa and decreased to 450–480 MPa at 1370°C. Stress rupture life at 1370°C was greater than 200 hours at 200 MPa stress [44].

Whisker-reinforced HIP silicon nitride efforts continued under a 2-year Phase 2 program [45]. Much of the effort involved detailed study of different whiskers and attempts to understand chemical effects versus properties. The best toughness (8.2 MPa.m$^{1/2}$) was obtained for addition of 30% chopped carbon fibers (Union Carbide P55S) to silicon nitride with 5 weight % MgO sintering aid. The next best toughness (6.8–7.0 MPa.m$^{1/2}$) was for a composite that contained uniaxially aligned carbon-coated whiskers.

NT230 Siliconized SiC Development

Parallel to the HIP silicon nitride efforts under ATTAP, NTC conducted an internal R&D process development program to improve their existing NT235 Si-SiC material. The baseline NT235 had flexure strength of 245 MPa at room temperature, 274 MPa at 1260°C, and 251 MPa at 1370°C. The improved NT230 had flexure strength of 410 MPa at room temperature, 540 MPa at 1260°C and 500 MPa at 1370°C [43]. As with NT235, the NT230 exhibited no shrinkage during densification and was thus selected for fabrication of challenging shapes such as the Allison ATTAP scroll.

Turbine Component Development During the 1980s

An integral part of both silicon nitride and silicon carbide developments during the late 1980s was learning to fabricate complex turbine components [46]. NTC received substantial development funding from DOE under the ATTAP program through a subcontract from GAPD. Pressure slip casting was selected as the primary candidate for rotors and scrolls. Injection molding was selected initially for stators (nozzle guide vanes).

NT154 was selected for radial rotor development. Efforts during 1988 and 1989 focused on slip preparation, additions, and pressure casting of bladeless rotor hubs in porous plastic molds at greater than 3.45 MPa (500 psi). Uniform density, crack-free hubs were achieved and sectioned into test bars. Flexure strength at room temperature approached 1000 MPa and decreased to 740 MPa at 1200°C and 630 MPa at 1370°C. This demonstrated that engine size radial rotor cross sections could be produced with NT154 level of strength.

NT154 also was selected for development of injection molded stator vanes. Early efforts focused on binder selection and optimization, binder removal procedures, and fracture analysis to understand the source of strength-limiting flaws. More than 1500 stator vanes and many test bars were molded. Average flexural room temperature strength was under 750 MPa. The strength-limiting defects were high density inclusions traced to equipment and other defects traced to impurities in the binder raw materials. Filtration of the binder raw material was implemented and some of the equipment was modified, but much more extensive modifications of both the equipment and process were projected to be necessary. NTC decided to abandon injection molding and switch to slip casting.

DEVELOPMENTS DURING THE 1990S

Efforts expanded during the early 1990s on fabrication of HIPed silicon nitride components under ATTAP subcontracts to both GAPD (renamed AlliedSignal Engines; currently Honeywell Engines, Systems & Services) and Allison Gas Turbine Division of General Motors (now part of Rolls-Royce). Additional efforts supplied NT154 nozzles to AlliedSignal Engines under their internally funded and DOD funded programs, and diffusers, nozzles and monorotors to Sundstrand Power Systems (currently Hamilton Sundstrand division of United Technologies Corporation). Later during the 1990s Norton supplied components to the DOE Ceramic Turbine Engine Development Project (CTEDP) and to the DOE/Solar Turbines Ceramic Stationary Gas Turbine (CSGT) program.

Material property improvement continued in support of the ATTAP programs and under the DOE/ORNL Ceramic Technology Program. A new HIPed silicon nitride material (NT164) with improved high temperature stability was introduced and an experimental composition (NCX-5102) was demonstrated. In addition to turbine hot section materials, material development continued for bearings based on the lower temperature NC-132 material. After so many years of development, silicon nitride bearings finally were commercialized and a production plant established.

Other significant changes occurred during the 1990s. One key event was the acquisition of Norton by Saint-Gobain and the change in name to Saint-Gobain-Norton Industrial Ceramics Corporation (currently Saint-Gobain Ceramics and Plastics).

In the early 1990s, Saint-Gobain/Norton acquired the Torrington interest in the Cerbec ceramic bearing joint venture and continued with further silicon nitride development and commercialization. Saint-Gobain Advanced Ceramics in East Granby, CT, currently operates a successful Cerbec® commercial silicon nitride ball facility.

In 1993, Saint-Gobain acquired the TRW interest in Norton/TRW Ceramics (NTC) and formed the Norton Advanced Ceramics Division to continue development and commercialization of silicon nitride engine components and bearing components.

Continuation of Silicon Nitride Development

An effective partnership was established between Norton/TRW and the ORNL High temperature Materials Laboratory (HTML) during the late 1980s and into the 1990s, especially to study the high temperature behavior of NT154. Tensile creep studies identified that a life-limiting creep mechanism in NT154 at 1260 and 1370°C was cavitation along the grain boundaries. Studies were conducted to modify the grain boundary phase, particularly to decrease the amount of amorphous phase present. These studies were successful and resulted in an improved HIPed silicon nitride designated NT164. The comparison of fatigue life of NT164 with earlier silicon nitride materials at 1370°C is illustrated in Figure 3.6 [47,48]. Other properties of NT164 are compared with 1995-vintage NT154 in Table 3.3 [49].

Parallel studies, starting in 1989 under the DOE/ORNL Office of Transportation Technologies Ceramic Technology Project, were conducted to develop and improve advanced ceramics for advanced heat engines [50]. Norton was awarded one of two major contracts to develop and demonstrate methodologies for achieving reliability of high temperature silicon nitride [51]. The other major contractor, GTE Laboratories, exited the silicon nitride business before the end of the program. The Norton Improved Processing Program was considered the "flagship" program of the DOE/ORNL Ceramic Technology Project. Improved Processing evaluated the properties and reliability of near-net shape buttonhead tensile specimens made by both injection molding and pressure slip casting. In-process NDE methodologies, including Microfocus X-radiography, Computed Tomography and liquid dye penetrant inspection, were incorporated to detect and control critical flaws. The pressure slip casting process was chosen after the initial process down selection phase.

The Norton Improved Processing program developed NCX5102, an optimized silicon nitride with 4% yttria sintering aid using colloidal processing techniques [51–53]. Previous processing was done by milling the powder in alcohol, drying, screening, and then re-dispersing the powder in water. This resulted in powder

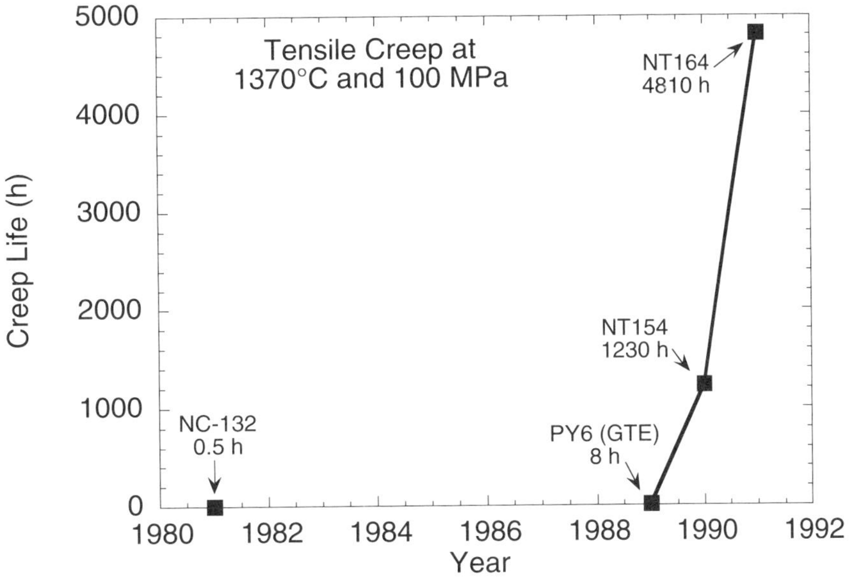

Figure 3.6 High Temperature Static Fatigue Comparison of NT164 with Earlier Silicon Nitride Materials

TABLE 3.3 COMPARISON OF THE PROPERTIES OF NT154 AND NT164 IN 1995 (ADAPTED FROM REF. [49])

Properties	Typical Values NT154	Typical Values NT164
Density (g/cm^3)	3.232 ± 0.004	3.190 ± 0.004
Elastic Modulus (GPa)	310–320	306 (1)
Shear Modulus (GPa)	126	N/A
Poisson's Ratio	0.273	N/A
Hardness (kg/mm^2)	1620	1891 (1)
Coefficient of Thermal Expansion	3.93×10^{-6}/°C	4.30×10^{-6}/°C
Thermal Conductivity (W/m.K)		
@ 25°C	37.6	N/A
@ 1400°C	15.8	N/A
22°C Bulk Surface Ground Mech. Props.		
Average Flexural Strength (MPa)	896 ± 55	887 ± 100
Characteristic Strength (MPa)	917	917
Weibull Modulus	8–12	9–12
Fracture Toughness ($MPa.m^{1/2}$)	6.0 (2)	5.6
Tensile Strength (MPa)	762 ± 112 (3)	546–700 (4)
22°C As Processed Surface Mech. Props.		
Average Flexural Strength (MPa)	490 ± 90	637 ± 62
Characteristic Strength (MPa)	524	669
Weibull Modulus	6	8
1370°C Mechanical Properties		
Average Flexural Strength (MPa)	641 ± 40	630 ± 49
Characteristic Strength (MPa)	669	664
Weibull Modulus	13.0	13.0
Fracture Toughness (1200–1400°C—$MPa.m^{1/2}$)	4.1 (5)	N/A
Tensile Strength (MPa)	240–520 (4)	388–567 (4)
1260°C Tensile Creep Rate (300 MPa)/s	1.9×10^{-8}	2.2×10^{-9}
1370°C Tensile Creep Rate (200 MPa)/s	8.7×10^{-7}	1.0×10^{-8}

Notes: (1) UDRI (University of Dayton Research Institute) data, (2) Controlled Flaw Method: 10 kg indent, (3) Includes CIPed and cast samples—UDRI and ORNL (Oak Ridge National Laboratory) data, (4) Loading rate dependent—UDRI and ORNL data, (5) Indentation Method

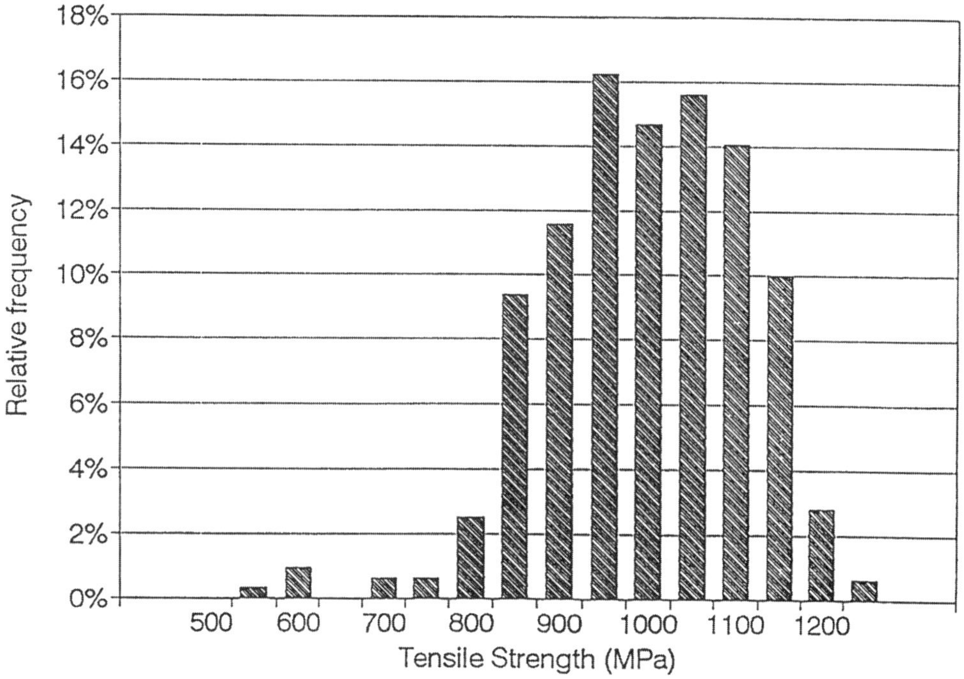

Figure 3.7 Histogram Showing the Tensile Strength Distribution for Over 320 NCX-5102 Silicon Nitride Samples (Reprinted from ref. [53])

agglomerates that were difficult to break down and that became defects in the final material. A "closed loop aqueous" process was developed that allowed passing the slurry through a 5μm filter to assure that agglomerates were minimized prior to pressure casting. Ube silicon nitride powder was used for these studies.

The HIPed silicon nitride samples fabricated under the program were evaluated both by flexure and tensile strength tests. Early samples in the program had tensile strength less than 500 MPa at room temperature. However, as the program progressed the material was steadily improved until average tensile strength of nearly 1000 MPa was achieved. The material was designated NCX-5102. Figure 3.7 shows a histogram based on 320 tensile tests. The average tensile strength was 997 MPa with a Weibull modulus of 10.4.

Another evaluation of the NCX-5102 involved fabrication of a 50 mm diameter cylindrical sample. The cylinder was machined into flexure strength test bars. Bars tested after a post-machining oxidation heat treatment had room temperature characteristic strength of 1018 MPa with Weibull modulus of 12. Bars without the heat treatment had room temperature characteristic strength of 873 MPa with Weibull modulus of 24. Strength at 1370°C was 600 MPa with Weibull modulus of 46. Fracture toughness (4-point indentation strength method) was 6.8 ± 0.36 MPa.m$^{1/2}$.

Demonstration of Cost-Effectiveness

Cost and reliability have long been considered two major barriers to advanced ceramics commercialization. The DOE/ORNL Ceramic Technology Project, after successfully demonstrating silicon nitride reliability in the Norton Improved Processing program and through other contracts, shifted emphasis to cost-effectiveness. Norton was again involved in a flagship program of this DOE initiative. Norton Advanced Ceramics led a contract team on the Advanced Ceramic Manufacturing Technology Program (ACMT) [54]. While this program developed cost-effective manufacturing processes for silicon nitride diesel engine valves, many of the lessons learned could be applied to high temperature silicon nitride for gas turbine applications. Norton Company Superabrasives Division also led a program with Norton Advanced Ceramics called Innovative Grinding Wheel for Cost-Effective Machining of Advanced Ceramics which led to the development of the Scepter® wheel for grinding silicon nitride [55].

Continuation of Turbine Component Development

Norton conducted during 1990 and 1991 pressure slip casting development of NT154 fully-bladed AGT101 radial rotors for GAPD under the ATTAP program. Three rotors were successfully spin tested at GAPD to above the proof speed of 105 krpm [56]. Examples of two finished AGT101 rotors are shown in Figure 3.8. An additional five engine quality rotors were delivered to GAPD in 1992 and passed the 105 krpm proof speed. Two were spin tested to failure at 127 krpm and 119 krpm [57]. A total of 189 engine quality AGT101 stator vanes of NT154 also were delivered to Garrett Auxiliary Power Division of AlliedSignal (GAPD) in 1992. Examples are illustrated in Figure 3.9.

Parallel component development efforts were conducted during the early 1990s in support of the Allison ATTAP program. Seven NT154 AGT-5 axial gasifier rotors were fabricated in 1991; four passed spin testing and were delivered to Allison. One of the rotors was rig tested for 200 hours of cyclic durability. This included more than 1850 start/stop cycles. During 1992 Norton refined the pressure casting process, including reducing casting time from >20 hours to <5 hours. Twenty-one AGT-5 rotors of NT154 and NT164 were fabricated. Seventeen of the 21 rotors survived the Allison 80 krpm proof speed. Two failed at 30 krpm at blade flaws and two failed at 75 and 78 krpm at hub flaws. Two others were spin tested to burst at 88 krpm and 92 krpm [57].

Silicon nitride components were also fabricated for other programs during the early 1990s. These included diffusers, nozzles and monorotors for the Sundstrand Power Systems TJ50 turbojet engine under Department of Defense (DoD) funding, a variety of tensile test specimens for the GAPD Life Prediction Methodology for Ceramic Components program, and nozzles for the GAPD GTCP85 auxiliary power unit [57]. Silicon nitride hardware for these and other programs are illustrated in Figure 3.10. The GTCP85 program resulted in some early engine testing of NT154 nozzle guide vanes. By 1992 more than 1000

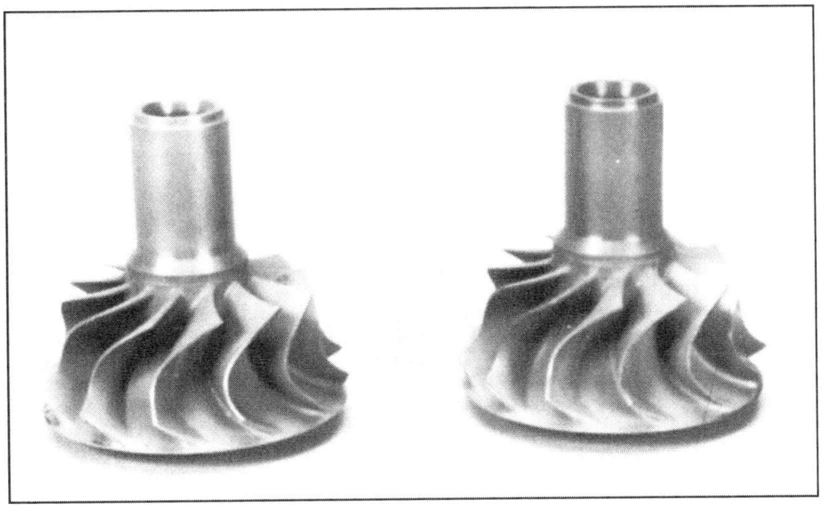

Figure 3.8 NT154 AGT101 Rotors Fabricated by Pressure Slip Casting (Reprinted from ref. [56])

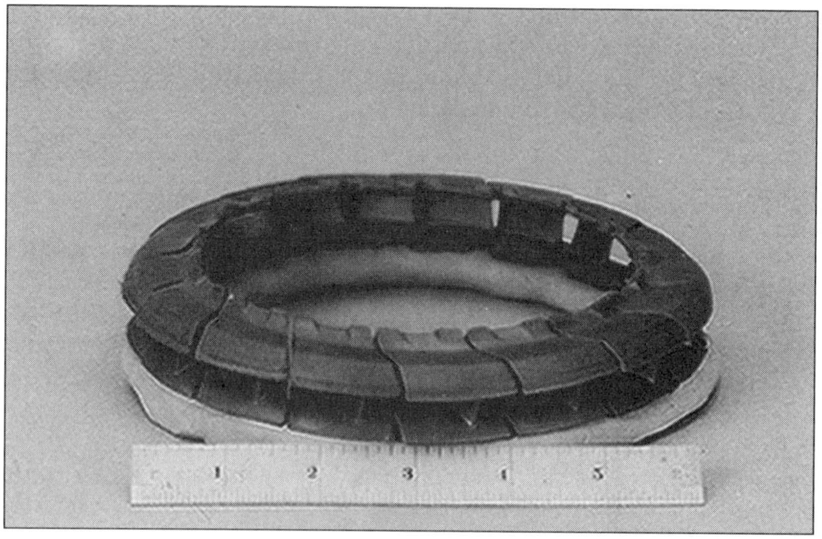

Figure 3.9 NT154 AGT101 Stator Vanes Fabricated by Pressure Slip Casting (Reprinted from ref. [57])

hours had been accumulated, including ~10 hours of sand ingestion. The NT154 nozzles exhibited no distress, while metal nozzles exhibited substantial wear, corrosion, and distortion.

Engine component development of Norton NT230 reaction sintered silicon carbide (Si-SiC) continued during 1990–1992 under the ATTAP and other programs parallel to the silicon nitride efforts. Components successfully fabricated included AGT101 transition ducts, AGT-5 scrolls and combustor liners, and miscellaneous

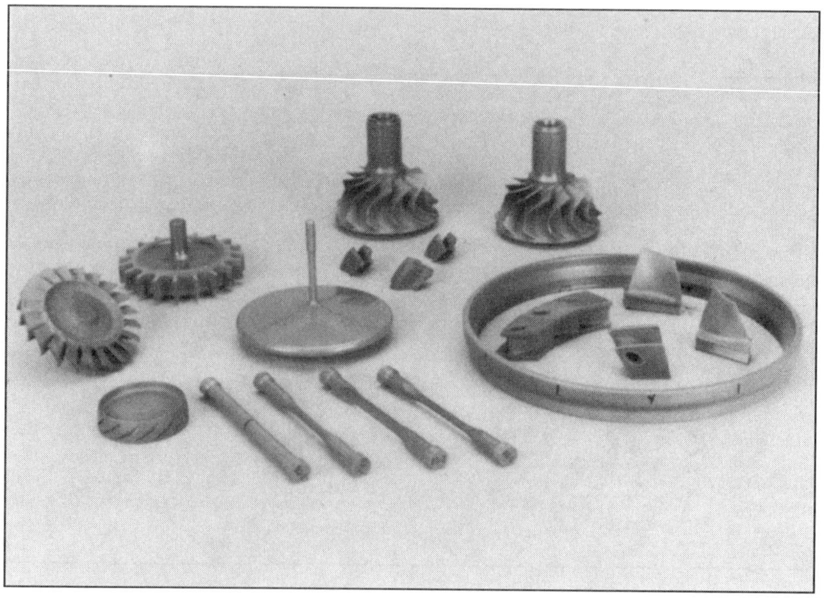

Figure 3.10 HIP Silicon Nitride Hardware Delivered During the Early 1990s to Various Turbine Engine Companies for Evaluation (Reprinted from ref. [57])

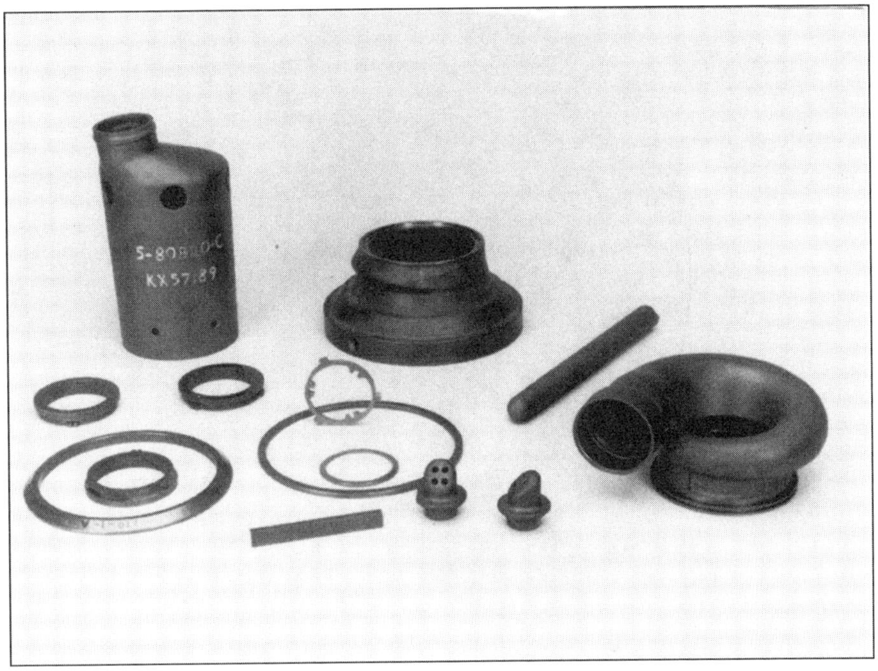

Figure 3.11 NT230 Si-SiC Gas Turbine Components (Reprinted from ref. [57])

seal rings. These and other NT230 parts are shown in Figure 3.11. The Allison AGT-5 scroll for the ATTAP program, shown in Figure 3.12, was especially challenging. A clam-shell mold of porous plastic was designed that allowed casting of a scroll in less than 80 seconds at a pressure greater than 2070 kPa (300 psi).

By the mid-1990s Norton Advanced Ceramics (NAC) was fabricating components for the DOE/Solar Turbines Incorporated Ceramic Stationary Gas Turbine (CSGT) program. In 1994–1995 time frame NAC delivered simulated blades and full size blades that were tested in spin rigs and in the Solar Centaur 50S engine. In August 1995 a full set of 62 NT164 first stage rotor blades was successfully tested for one hour in the Centaur 50S engine at the baseline turbine rotor inlet temperature (TRIT) of 1010°C. This was the first time silicon nitride components were tested in an industrial gas turbine under typical operating conditions. Post test evaluation indicated that the parts were in good condition. The test validated the blade dovetail design (see Figure 3.13). However, dynamic fatigue data and life prediction indicated that the NT164 blades would not have adequate life for the typical 30,000+ hrs expected time between overhaul (TBO) for an industrial gas turbine because the surface strength was compromised by the Asea HIP sintering process used for component consolidation. Subsequent testing was therefore conducted with blades fabricated using gas pressure sintering which don't suffer the deleterious surface strength degradation.

The promise of a commercialized automotive gas turbine did not materialize. The DOE Office of Transportation Technologies began to shift emphasis from ATTAP to ceramics for heavy vehicle diesel engines. Ceramic gas turbines remained a possibility in the development of a hybrid electric vehicle engine but this approach was soon abandoned. Most of the ceramic gas turbine development work was shifted to stationary power gas turbines under DOE Office of Industrial Technologies and later under the DOE Office of Power Technologies. As mentioned above Norton Advanced Ceramics supplied blades to Solar Turbines Incorporated for their DOE Ceramic Stationary Gas Turbine Program.

In 1995 Saint-Gobain/Norton made the decision to shelve the high temperature silicon nitride development work and suspend ceramic gas turbine activity. Saint-Gobain made this decision in order to

Figure 3.12 NT230 Scroll for the Allison AGT-5 Engine (Reprinted from ref. [57])

Figure 3.13 NT164 Blade Fabricated for the DOE/Solar Turbines Inc. CSGT Program (Reprinted with permission from Solar Turbines Incorporated)

concentrate on their Cerbec® commercial silicon nitride bearing component business and to further develop diesel engine components.

SUMMARY AND LESSONS LEARNED

Norton and the field of advanced ceramics for turbines have evolved together since the early 1970s. Efforts were initiated with Norton as a private company, progressed under a joint venture with TRW, and continued after Norton was acquired by Saint-Gobain. The ceramics efforts began with RBSN and hot pressed silicon nitride and progressed to HIP silicon nitride and a substantially improved version of Si-SiC. In 1996 Saint-Gobain decided to focus their resources on launching their silicon nitride bearing business and to cease their turbine component developments. The decision was made in 2001 to reactivate turbine component developments to support microturbine programs.

Many lessons have been learned over the years that have contributed to advances in the state of the art for advanced ceramics for turbines and other applications.

- Through the work on NC-132 silicon nitride, Norton identified the importance of grain boundary engineering in achieving improvements in high temperature properties.
- Also through the work on NC-132 silicon nitride, Norton was an early pioneer in the use of SEM and fracture analysis to identify strength-limiting microstructural defects and to apply the information gained to achieve process improvement.
- Later, through the work on NT154, Norton demonstrated the value of statistical experiments in optimization of raw materials and fabrication process steps to establish a reliable material with fixed-process controls.
- Norton demonstrated that the use of screening and filtration procedures during wet colloidal processing could limit flaw populations in the finished ceramic and assure uniform high strength and high reproducibility/reliability.
- Norton demonstrated that complex shaped turbine components such as scrolls and integral rotors could be fabricated by pressure slip casting using long-life porous plastic molds.

Although ceramic turbine components have not yet become a commercial success, the technologies developed in silicon nitride have resulted in major commercial successes in cutting tools, bearings, and wear-resistant components.

Even with the enormous improvements that have been accomplished in properties, complex shape fabrication and reliability, additional challenges must be overcome before advanced ceramics will be commercially successful in gas turbine applications. One big challenge is cost. Another, which currently contributes to high cost, is achieving a level of quality control that will assure acceptable performance and life of the ceramic component in the turbine application.

Some materials issues also must be addressed. For example, the properties of as-HIPed NT154 and NT164 silicon nitride are inferior to those of machined samples. This is due to physical and chemical effects associated with glass encapsulation. Improvement of the as-HIPed properties may be necessary for some turbine applications. Other areas where material improvements will be beneficial (and perhaps required for some applications) include increase in fracture toughness and improvement in oxidation resistance (or development of environmental barrier coatings).

ACKNOWLEDGEMENTS

Many individuals were responsible for the material advances accomplished at Norton, Norton/TRW, and Saint-Gobain Ceramics and Plastics. Most of the names are included in the references. In many cases the efforts of these individuals were made possible by collaboration with engineers in the gas turbine industry and by funding from various organizations within the DoD and the DOE.

REFERENCES

[1] Washburn, M. E., Norton Co., Worcester, MA, USA, "Production of Silicon Oxynitride," US Patent No. 3,356,513, Dec. 5, 1967.

[2] Washburn, M. E.,"Silicon Oxynitride Refractories," Bull. Amer. Ceram. Soc. 46[7], 667–671, 1967.

[3] Washburn, M. E., Norton Co., Worcester, MA, USA, "Control System in an Apparatus for Reacting Silicon with Nitrogen," US Patent No. 3,854,882, Dec. 17, 1974.

[4] Washburn, M. E., Fredriksson, J. I., and Alliegro, R. A., "Properties and Applications of Reaction Bonded Silicon Nitride," paper presented at the 74th Annual Meeting of the American Ceramic Society, Washington, D.C., May 6–11, 1972. [Abstract: Bull. Amer. Ceram. Soc., 51[4], 429, 1972].

[5] Washburn, M. E., and Baumgartner, H. R., "High-Temperature Properties of Reaction-Bonded Silicon Nitride," in *Ceramics for High Performance Applications*, J.J. Burke, A. E. Gorum, and R. N. Katz, eds., Brook Hill Publishing Company, Chestnut Hill, MA, USA, 1974. pp. 479–492.

[6] Richerson, D. W., and Johansen, K. M., *Ceramic Gas Turbine Engine Demonstration Program,* Final Report, January 1976–May 1981, Garrett Report 21-4410, prepared under DARPA/NAV-AIR contract N00024-76-C-5352, May 1982.

[7] Corbin, N. D., Rossetti, G. A., and Hartline, S. D., "Microstructure/Property Relationships for SiC Filament-Reinforced RBSN," Ceram. Eng. and Sci. Proc., 7[7–8], 958–968, 1986.

[8] Alliegro, R. A., Coffin, L. B., and Tinklepaugh, J. R., "Pressure-Sintered Silicon Carbide," J. Am. Ceram. Soc., 39[11], 386–389, 1956.

[9] Torti, M. L., Alliegro, R. A., Richerson, D. W., Washburn, M. E.., and Weaver, G. Q., "Silicon Nitride and Silicon Carbide for High-Temperature Engineering Applications," Proc. Brit. Ceram. Soc., 22, 129–146, 1973.

[10] Bratton, R. J., and Miller, D. G., "Brittle Material Design, High Temperature Gas Turbine—Stationary Turbine Project Summary," in *Ceramics for High Performance Applications-II, Proc. 5th Army Materials Tech. Conf.,* Newport, RI, USA, Mar. 21–25, 1977, J. J. Burke, E. N. Lenoe, and R. N. Katz, eds., Brook Hill Publishing Company, Chestnut Hill, MA, USA, 1978. pp. 689–730.

[11] Alliegro, R. A., and Coes, S. H., "Reaction Bonded Silicon Carbide and Silicon Nitride for Gas Turbine Applications," ASME paper 72-GT-20, presented at the Gas Turbine Conference, San Franciso, CA, USA, March 26–30, 1972.

[12] Alliegro, R. A., "Processing and Fabrication of Non-Hot-Pressed Silicon Carbide," in *Ceramics for High-Performance Applications, Proc. 2nd Army Materials Tech. Conf.,* Hyannis, MA, USA, Nov. 13–16, 1973, J. J. Burke, A. E. Gorum, and R. N. Katz, eds., Brook Hill Publishing Company, Chestnut Hill, MA, USA, 1974. pp. 253–263.

[13] Lucek, J. W., Torti, M. L., Weaver, G. Q., and Olson, B. A., "Cast Densified Silicon Carbides," SAE Paper 790253, 1979.

[14] Matuschak, P. E., "Ceramic Reheat Combustor," in *Ceramics for High Performance Applications-II, Proc. of the 5th Army Materials Tech. Conf.,* Newport, RI, USA, Mar. 21–25, 1977, J. J. Burke, E. N. Lenoe, and R. N. Katz, eds., Brook Hill Publishing Company, Chestnut Hill, MA, USA, 1978. pp. 923–943.

[15] Deely, G. G., Herbert, J. M., and Moore, N. C., "Dense Silicon Nitride," Powder Met., 8, 145–151, 1961.

[16] Lumby, R. J., and Coe, R. F., "The Influence of Some Process Variables on the Mechanical Properties of Hot-Pressed Silicon Nitride," Proc. Br. Ceram. Soc., 15, 91–101, 1970.

[17] Richerson, D. W., and Washburn, M. E., Norton Co., Worcester, MA, USA, "Hot Pressed Silicon Nitride," U.S. Patent 3,836,374, Sept. 17, 1974.

[18] Richerson, D. W., "Effect of Impurities on the High Temperature Properties of Hot-Pressed Silicon Nitride," Bull. Am. Ceram. Soc., 52 [7], 560–562, 569, 1973.

[19] Baumgartner, H. R., and Richerson, D. W., "Inclusions Effects on the Strength of Hot Pressed Silicon Nitride," in *Fracture Mechanics of Ceramics*, Vol. 1, R. C. Bradt, D. P. H. Hasselman, and F. F. Lange, eds., Plenum Press, New York, NY, USA, 1974. pp. 367–386.

[20] Richerson, D. W., Smyth, J. R., and Styhr, K., "Material Improvement Through Iterative Process Development," Cer. Eng. and Sci. Proc., 4 [9–10], 841–52, 1983.

[21] Evans, A. G., and Wiederhorn, S. M., "Crack Propagation and Failure Prediction in Silicon Nitride at Elevated Temperatures," J. Mater. Sci., 9, 270–278, 1974.

[22] Tighe, N. J., "Microstructure Aspects of Deformation and Oxidation of Magnesia-Doped Silicon Nitride," in *Nitrogen Ceramics*, F. L. Riley, ed., Noordhoff Inter. Pub. Co, Leyden, The Netherlands, 1977. pp. 441–448.

[23] Evans, A. G., Russell, L. R., and Richerson, D. W., "Slow Crack Growth in Ceramic Materials at Elevated Temperatures," Met. Trans. A. 6A, 707–716, 1975.

[24] Richerson, D. W., Yonushonis, T. M., and Weaver, G. Q., "Properties of Silicon Nitride Rotor Blade Materials," *Proc. 1977 DARPA/NAVSEA Ceramic Gas Turbine Demonstration Engine Program Review*, J. W. Fairbanks and R. W. Rice, eds., Battelle Columbus Laboratories, Columbus, OH, USA, Report No. MCIC-78-36, 1978. pp.193–217.

[25] Quinn, G. D., "Characterization of Turbine Ceramics after Long-Term Environmental Exposure," AMMRC-TR-80-15, Final Report, April 1980.

[26] Larsen, D. C., and Adams, J. W., "Property Screening and Evaluation of Ceramic Turbine Materials," AFWAL TR-83-4141, Final Technical Report AFWAL Contract F33615-79-C-5100, April 1984.

[27] Calvert, G. S., and Carruthers, W. D., "Ceramic Blade Attachments," *Ceramics for High Performance Applications-II*, Proc. of the 5th Army Materials Tech. Conf., Newport, RI, USA, Mar. 21–25, 1977, J. J. Burke, E. N. Lenoe, and R. N. Katz, eds., Brook Hill Publishing Company, Chestnut Hill, MA, USA, 1978. pp. 839–860.

[28] Andersson, C. A., Bratton, R. J., Sanday, S. C., and Cohn, A., "Progress on EPRI Ceramic Rotor Blade Program," ibid., pp. 783–804

[29] Richerson, D. W., Schuldies, J. J., Yonushonis, T. M., and Johansen, K. M., "ARPA/NAVY Ceramic Engine Materials and Process Development Summary," ibid., pp. 625–650.

[30] Nessler, C. G., "Prototype Ceramic Vanes," in *Ceramics for High Performance Applications, Proc. 2nd Army Materials Tech. Conf.*, Hyannis, MA, USA,13–16, 1973, J. J. Burke, A. E. Gorum, and R. N. Katz, eds., Brook Hill Publishing Company, Chestnut Hill, MA, USA, 1974. pp. 609–630.

[31] Hayes, C. W., Nessler, C. G., and Zabierek, D., "Turbine Vane Ceramic Endwall," ibid., pp. 749–761.

[32] Booher, C. R., Jr., "Thermal Transient Stress Analysis of an Industrial Combustion Turbine Ceramic Stator Vane and Correlation with Test Results," in *Ceramics for High Performance Applications-II, Proc. of the 5th Army Materials Tech. Conf.*, Newport, RI, USA, Mar. 21–25, 1977, J. J. Burke, E. N. Lenoe, and R. N. Katz, eds., Brook Hill Publishing Company, Chestnut Hill, MA, USA, 1978. pp. 731–781.

[33] Baumgartner, H. H., "Ceramic Bearings for Turbine Applications," ibid., pp. 423–443.

[34] Sundberg, D. V., "Ceramic Roller Bearings for High-Speed and High-Temperature Applications," SAE Publication 740241, 1974.

[35] Gazza, G.E., "Hot-Pressed Silicon Nitride," J. Am. Ceram. Soc., 56[12], 662, 1973.

[36] Weaver, G. Q., and Lucek, J. W., "Optimization of Hot-Pressed Si_3N_4-Y_2O_3 Materials," Bull. Am. Ceram. Soc., 57[12], 1131–1134, 1978.

[37] Larsen, D. C., *Property Screening and Evaluation of Ceramic Turbine Engine Materials,* Final Report AFML-TR-79-4188, Oct. 1979.

[38] Carruthers, W. D., Richerson, D. W., and Benn, K. W., "Combustion Rig Durability Testing of Turbine Ceramics," in *Ceramics for High Performance Applications III*, E. M. Lenow, R. N. Katz, and J. J. Burke, eds., Plenum Press, New York, NY, USA, 1983. pp. 571–595.

[39] Lange, F. F., Singhal, S. C., and Kusnicki, R. C., "Phase Relations and Stability Studies in the Si_3N_4-Y_2O_3 Pseudoternary System," J. Am. Ceram. Soc., 60[5–6], 249–252, 1977.

[40] Larker, H., Adlerborn, J., and Bohman, H., "Fabrication of Dense Silicon Nitride Parts by Hot Isostatic Pressing," SAE Paper 770335, 1977.

[41] McEntire, B. J., Taglialavore, A. P., Heichel, D. N., Johnson, J. W., Bright, E., and Yeckley, R. L., "ATTAP Ceramic Component Development Using Taguchi Methods," *Proc. 26th Auto. Tech. Dev't Contractors Coord. Meeting*, Dearborn, MI, USA, Oct. 24–27, 1988, SAE Publication P-219, 1989. pp. 289–300.

[42] McEntire, B. J., Taglialavore, A. P., Heichel, D. N., Bright, E., Yeckley, R. L., Holowczak, J. E., and Quackenbush, C. L., "Silicon Nitride Component Development for Advanced Gas Turbine Engines," *Proc. 27th Auto. Tech. Dev't Contractors Coord. Meeting*, Dearborn, MI, USA, Oct. 23–26, 1989, SAE Publication, April 1990. pp. 341–356.

[43] McEntire, B. J., Hengst, R. R., Collins, W. T., Taglialavore, A. P., Yeckley, R. L., Bright, E., and Bingham, M. G., "Ceramic Component Processing Development for Advanced Gas Turbine Engines," ASME paper 91-GT-120, presented at the Int. Gas Turbine and Aeroengine Congress and Exposition, Orlando, FL, USA, June 3–6, 1991. Transactions of the ASME, J. Engineering for Gas Turbines and Power, 115[1], 1–8, 1993.

[44] Corbin, N.D., and Wilkins, C.A., *Development of Toughened Silicon Nitride Composites by Glass Encapsulated Hot Isostatic Pressing,* Final Report ORNL/Sub/86-95906/1, Aug. 1988.

[45] Corbin, N.D., Sundberg, G.J., Siebein, K.N., Wilkens, C.A., Pujari, V.K., Rossi, G.A., Hansen, J.S., Chang, C.L., and Hammarstrom, J.L., *Material Development in the Silicon Nitride System Using Glass Encapsulated HIPing,* Final Report ORNL Sub. 86-95906/2, April, 1992.

[46] Hengst, R.R., Heichel, D.N., Holowczak, J.E., Taglialavore, A.P., and McEntire, B.J., "A Comparison of Forming Technologies for Ceramic Gas-Turbine Engine Components," ASME paper 90-GT-184, presented at the Gas Turbine and Aeroengine Congress and Exposition, Brussels, Belgium, June 11–14, 1990.

[47] Ferber, M.K., Oak Ridge National Laboratory, unpublished data.

[48] Ferber, M.K., Jenkins, M.G., Nolan, T.A., and Yeckley, R., "Comparison of the Creep and Creep Rupture Performance of Two HIPed Silicon Nitride Ceramics," J. Amer. Ceram. Soc., 77[3], 657–665, 1994.

[49] Bright, E., Burleson, R., Dynan, S. A., and Collins, W. T., "NT 164 Silicon Nitride Gas-Turbine Engine Turbine Blade Manufacturing Development," ASME Paper 95-GT-74, presented at the Int. Gas Turbine and Aeroengine Congress and Exposition, Houston, TX, USA, June 5–8, 1995.

[50] Johnson, D. R. and Schulz, R. B., "Ceramic Technology for Advanced Heat Engines", presentation at Material Conference 1993, North Carolina Agricultural and Technical State University, Ceramic Technology Newsletter, No. 41, October-December 1993, ORNL/DOE.

[51] Pujari, V. K., Amin, K. E., and Tewari, P. H., "Development of Improved Processing and Evaluation of Silicon Nitride," ASME paper 91-GT-317, presented at the Int. Gas Turbine and Aeroengines Congress and Exposition, Orlando, FL, USA, June 3–6, 1991.

[52] Pujari, V. K., and Tracey, D. M., "Processing Methods for High Reliability Silicon Nitride Heat Engine Components," ASME paper 93-GT-319, presented at the Int. Gas Turbine and Aeroengine Congress and Exposition, Cincinnati, OH, USA, May 24–27, 1993.

[53] Pujari, V. K., Tracey, D. M., Foley, M. R., Paille, N. I., Pelletier, P. J., Sales, L. C., Wilkens, C. A., and Yeckley, R. L., *Development of Improved Processing and Evaluation Methods for High Reliability Structural Ceramics for Advanced Heat Engine Applications,* Phase I, Final Report ORNL/Sub/89-SB182/1, August 1993.

[54] Pujari, V. K., Bright, E., Broderick, L. A., Burk, C.B., Collins, W. T., Garrett, J. M., Leo, T. P., Licht, R. H., Lynch, L. D., and McEntire, B. J., *Development of Advanced Ceramic Manufacturing Technology,* Final Report, Subcontract No. 86X-SP234C, published by Oak Ridge National Laboratory Ceramic Technology Project, Report Number ORNL/SUB/94-SP234/1, April 2001.

[55] Licht, R. H., Kuo, P., Liu, S., Murphy, D., Picone, J. W., and Ramanath, S., *Innovative Grinding Wheel Design for Cost-Effective Machining of Advanced Ceramics,* Phase II Final Report, Subcontract No. 86X-SU697V, published by Oak Ridge National Laboratory Ceramic Technology Project, Report Number ORNL/Sub/93-SM037/2, December 1998.

[56] Taglialavore, A. P., Hengst, R.R., Bright, E., Lynch, L. D., and McEntire, B. J., "Advanced Ceramics for Gas Turbines Applications," *Proc. Annual Auto. Tech. Development Contractors' Coord. Meeting 1991*, Oct. 28–31, 1991, Dearborn, MI, USA, SAE Publication P-256, April 1992. pp. 71–79.

[57] Taglialavore, A. P., Bright, E., McEntire, B. J., Collins, W. T., Dynan, S. A., Lynch, L. D., and Girvin, R. F., "Progress in Ceramic Engine Component Development," *Proc. Annual Auto. Tech. Development Contractors' Coord. Meeting 1992*, Nov. 2–5, 1992, Dearborn, MI, USA, SAE Publication P-265, May 1993. pp. 277–287.

CHAPTER

4

Silicon Carbide Component Development at Saint-Gobain/The Carborundum Company

Roger S. Storm
MER Corporation, Tucson, Arizona, USA

Martin R. Kasprzyk
SyCore Inc., Ransomville, New York, USA

James W. MacBeth
Saint-Gobain Advanced Ceramics, Niagara Falls, New York, USA

INTRODUCTION

Reflecting upon its one hundred year existence, a 1991 commemorative history describes how The Carborundum Company got its name: ". . . *Carborundum.* Edward Goodrich Acheson liked the rich, hard sound of the word. It was the spring of 1891 and he and a friend, William McCallister, were on a train traveling from Monongahela City, Pennsylvania to New York City. In Acheson's vest pocket was a vial containing a strange, crystalline powder. A short time before, Acheson had produced the mysterious crystals during the course of an experiment and found they were hard enough to scratch the surface of his diamond ring. He and McCallister were on their way to see if there was a market for the new product among the diamond cutters of New York City. No name had yet been given to the substance, since its composition was still unknown. The men agreed that it had to be named if they were to sell it in New York, so throughout the train ride numerous suggestions were considered and rejected. Finally, on the assumption that the contents of the vial were a chemical combination of carbon and corundum (natural aluminum oxide), the name *carborundum* was chosen. It was a name that would change the course of their lives and the ways of the industrialized world for the next 100 years. (It wasn't until a year later that a chemical analysis of the material revealed that the tiny "crystal stones" were, in fact, not carbon and corundum, but silicon carbide.) Acheson's trip to New York City resulted in the sale of his Carborundum crystals to a John Street diamond dealer for 40 cents a carat, or about $880 a pound. Acheson took the money from this first sale, about $60, and bought a microscope to study the structure of his material. So started the first cycle of research, discovery, product sale, and more research so successfully repeated throughout the company's history. . ." [1].

Carborundum retained that tradition of research and innovation for over 100 years. Silicon carbide was the basic building block for many of Carborundum's businesses including abrasives (bonded and coated), refractories (Carbofrax®, Refrax®), and electric products (Globar® heating elements). Carborundum also diversified into other ceramic products including oxide fibers (Fiberfrax®) and fused cast refractories.

For most of the 1960s and 1970s William Wendell was president of the Carborundum Co., which was still an independent company at that time. One of Wendell's goals was to grow the company through a significant effort in research. In the mid 1960s, the R&D Department had a worldwide staff of 600, with a budget equal to about 3% of sales. Many of these efforts were in non-traditional areas for Carborundum. A fire resistant fiber (Kynol®) was developed which was based on a phenolic precursor fiber. A family of high temperature polymers based on aromatic co-polyesters (Ekkcel®) was commercialized. Substantial effort was directed toward activated carbon and water treatment. These technologies turned out to be a significant departure from the predominantly ceramics based businesses Carborundum was most expert at. Eventually these businesses were sold off.

In the early 1970s however, a new technology was discovered which did provide a very strategic fit with Carborundum's strengths: sintered alpha silicon carbide (SASC). Carborundum had large manufacturing plants world-wide which made substantial quantities of SiC grain by the Acheson process. These SiC powders were ground and classified to various particle sizes for use in Carborundum products such as abrasives and refractories. A considerable quantity of powder came out of the backend of the process which was too fine to use in any existing products. The powder was sub-micron alpha SiC with a surface area above 10 m^2/g. Carborundum researchers in Niagara Falls found that the powder could be transformed into a dense body by pressureless sintering. Small amounts of boron carbide and carbon were added to the sub-micron SiC powder, and the mixture was compacted to a density of 50–60%, e.g. by isostatic pressing or dry pressing. The resultant green body would undergo a volumetric shrinkage of >50% during sintering at 2100–2200°C in argon or nitrogen. Densities of >3.05g/cm^3 were achieved (>95% of the theoretical density of 3.21 g/cm^3).

Up to that point, the densification of alpha SiC required the simultaneous application of heat and pressure, e.g. by hot pressing. While the strength and high temperature properties of hot pressed SiC were of great interest for turbine engine hot components, they could only be produced in simple shapes such as cylinders. In order to produce a turbine vane or blade, the cylinder had to be diamond ground, a very slow and very expensive process. The pressureless sintering process was recognized to offer a low cost route to the manufacture of complex ceramic components.

At about the same time as the discovery of the sinterability of alpha SiC, Carborundum had been working with the Detroit Diesel Allison (DDA) Division of General Motors (Allison) on the Ceramic Applications for Turbine Engines program, or CATE (discussed later in the chapter). This program was sponsored by the US Department of Energy. The goal of CATE was to raise the efficiency of a truck turbine engine by increasing the operating temperature. The key to increasing the temperature was the use of ceramic components. The stationary components such as vanes and rings were being developed in reaction bonded SiC (RBSiC). The process for RBSiC involved the formation of a green body containing SiC and carbon which was then infiltrated with molten Si metal. The Si reacted with the carbon to form SiC. The residual Si also filled the remaining porosity resulting in a fully dense body. As the fired body cooled to room temperature, the difference in CTE between the SiC and the Si caused the Si to exude from the dense body and collect on the surface forming a hard crust. It had to be subsequently removed, e.g. by grit blasting at room temperature.

The RBSiC had a maximum useful operating temperature of ~1100°C for long term exposure, hotter than uncooled metals could run. The Allison program manager for CATE was Gene Helms. He was a leading proponent for the value ceramics could bring to turbine engines, both in fuel efficiency and emissions. He was always looking for ways to increase turbine operating temperature. Helms became aware of the development of the SASC technology and started an effort under CATE to develop a net shape SASC axial turbine blade by injection molding to push the operating temperature of the ceramic turbine even hotter.

Meanwhile, in Washington D.C., the Department of Energy saw the progress being made under CATE and other programs to develop ceramic turbine engine components. Under the guidance of Al Chesnes, another strong proponent of the potential for ceramics in hot turbine components, the effort was expanded to a much larger effort to develop a gas turbine for a passenger vehicle. This initiative was called the Automotive Gas Turbine program, or AGT, and gathered attention from around the world. All the big three automotive manufacturers, plus the Garrett Turbine Engine Co. participated. The timely development of the SASC technology enabled Carborundum to play a major role in AGT.

The target engine efficiencies for AGT raised the temperature requirements to a new level, with components required to perform at 1260°C and above for 3500 hours with minimal loss in strength. The Allison temperature requirement was 1371°C. The two materials that were selected for the AGT program were SASC and sintered silicon nitride (SSN). SASC had a low fracture toughness, ~4.5 $MPa.m^{1/2}$, as determined by the single-edged notched bend (SENB) method, which made it very flaw sensitive and gave it limited strength. (As described later, the strength was improved to >600MPa by reduction of process flaw size.) The low fracture toughness resulted from the apparent absence of a grain boundary phase. This microstructural feature however gave SASC its strongest attribute for turbine applications: excellent high temperature strength with minimal creep and slow crack growth at 1371°C and above in air. Silicon nitride ceramics on the other hand were sintered with a considerable grain boundary phase. Fracture toughness was excellent (values of 7–10 $MPa.m^{1/2}$ by SENB were typical) contributing to high room temperature strength and resistance to catastrophic failure from foreign object damage or FOD. However the grain boundary phase led to creep, slow crack growth and considerable reduction in strength at elevated temperatures.

Allison under Gene Helms selected SASC as the predominant ceramic material for their AGT engine. In the mean time, ceramic fever was building in the U.S., and also worldwide, and there was talk of replacing cast iron engine blocks with ceramic parts. Bill Wendell, the President of Carborundum, was still looking for an investment that could have a major pay-back for Carborundum. He knew of the SASC technology and was following the turbine engine programs with interest. A meeting between Wendell and Helms helped convince him this was a real opportunity for Carborundum. In 1978, Wendell approved the construction of a new state of the art ceramic research and prototyping center for Carborundum at a cost of over $5 Million. It was named the William H. Wendell Technical Center in his honor.

Separate areas were provided for material preparation and spray drying, green body forming (isostatic pressing, dry pressing, slip casting, injection molding, aqueous and plastic extrusion), and furnacing. The forming labs all had Class 10000 walk-in hoods to minimize defect formation from airborne particles. The furnace room had ducting to collect excess heat which was then stored in an underground pool and was used to partially heat the building. A class 100 hood was provided in the laboratory area. Extensive destructive and nondestructive testing was provided including microfocus X-ray, fluorescent dye penetrant, scanning ultrasonics, and SEM (scanning electron microscopy).

Carborundum under Wendell provided the R&D funding to ensure his vision would be accomplished. Under the DOE AGT and subsequent ATTAP programs, Carborundum made major advances in the fabrication of very complex, demanding components. While the automotive ceramic turbine engine has not become a reality, the technology that was developed led to the use of ceramics in applications where other materials such as metals and plastics could not perform. Some examples of commercialization are included at the end of the chapter.

SiSiC COMPONENT DEVELOPMENT UNDER THE CERAMIC APPLICATIONS FOR TURBINE ENGINES PROGRAM

The 1973 oil crises created an interest in the U.S. to reduce the need for imported oil by improving the efficiency of automobile and truck engines. One efficiency improvement concept was to use turbine engines instead of diesel engines in large trucks. It was well understood that increases in the turbine inlet temperature would lead to improvements in fuel efficiency, and it was also known that a ceramic turbine could allow the highest temperature capability.

The U.S. Department of Energy (DOE), Office of Vehicle and Engine Research and Development, created the Ceramic Applications for Turbine Engines program to develop a ceramic turbine tractor trailer truck engine with the Detroit Diesel Allison Division of General Motors (Allison) as the prime contractor. Allison was already producing axial type light-weight turbine helicopter engines and the goal was to utilize this type of engine with some of its metallic alloy parts being replaced by ceramics having higher temperature capability. This was not a simple material substitution with existing designs. Allison established its own computer linear elastic finite element design capability to modify part designs to make them more reliable with ceramic materials.

In the mid 1970s the most suitable ceramic materials were either monolithic silicon carbide or silicon nitride types due to their high temperature strength, and resistance to creep, oxidation, corrosion, and thermal shock. They also had low density (important to reduce stresses in rotating parts and to reduce engine weight).

The Carborundum Company was chosen in 1973 to be one subcontractor to Allison due to its experience in developing silicon carbide (SiC) types of ceramic materials. One purpose of the CATE program was to develop an engine that was to be low enough in cost to be a commercially viable alternative to conventional diesel engines. Thus Carborundum was asked to supply silicon carbide parts that were made with processes that would be suitable for large volume production. Since silicon carbide can only be machined with expensive diamond grinding it meant that net shape and size forming capability was very important for ceramic parts.

The axial Allison turbine stationary stator vane assembly was made with individual airfoil shaped vanes held in place with inner and outer support rings, about 0.5 cm thick. The inside ring (about 25 cm OD) had blind airfoil shaped holes on its OD and the outside ring had airfoil shaped through holes. The individual airfoil vanes were inserted through the outer ring until they bottomed out in the blind holes in the inner ring. The vanes were held in place with a 0.5 cm thick vane retaining ring slid over the outer support ring.

These three rings had to be made with tolerances of only about 0.05 mm and, in order to avoid a large amount of final grinding, it was decided to make them out of a fine grain silicon infiltrated material that had no shrinkage in its final processing. The starting green material was a mixture of dry thermosetting phenolic resin powder and 1000 mesh (4.5 micron) SiC powder that had acetone added to it to dissolve the resin. After allowing the acetone to evaporate, the caked dry material was hammer milled back into a powder. The powder was then warm compression molded into oversized rings at several thousand kg/cm^2 in steel dies heated to 177°C in order to first melt the resin and then set it. The resin-bonded SiC rings were then baked in a vacuum induction furnace to completely decompose the resin leaving about half of it behind as a carbon residue which still held the SiC particles together. The rings typically had a shrinkage of about 1% during this bake out step. The carbon bonded rings were then easily lathe turned with a single point diamond tool to dimensions with only a few thousandths of an inch (1 inch = 2.54 cm) extra material. The vane holes in the rings were ultrasonically machined. The level of resin was selected to give the baked rings enough carbon residual bond strength to prevent breakage of the parts during the green machining operations.

The rings were then infiltrated at 1700°C with molten silicon in a vacuum induction furnace. The carbon reacted with the silicon to form a small amount of secondary silicon carbide, bonding the original SiC particles together. No shrinkage occurs in this step. The ring surfaces were then diamond ground to remove any excess silicon material and achieve a smooth surface finish and the exact size required.

The turbine vanes were also made from fine grained silicon infiltrated material that used phenolic resin binder but in this case they were transfer-molded into shape. The molding mix also had synthetic wax added to it to achieve sufficient flowability. In this process a small cylindrical "puck" of the molding mix powder was first cold compression molded in a steel die. This puck was then quickly microwave heated and dropped into a small piston which pushed (transferred) the material into a heated split steel mold having a vane shaped cavity. After about a half minute to set the resin the hard green vane was removed from the mold.

The green molded vanes were then silicon infiltrated in a vacuum induction furnace without any prior bake-out steps. The total shrinkage in this process was about 2% and the steel mold was made oversize so the final parts were the correct size. The silicon-infiltrated vanes were then lightly sand blasted to remove excess silicon from their surfaces and they were used without any diamond grinding. A photo of the rings and vanes is shown in Figure 4.1.

These fine grained warm compression molded and transfer molded materials, which were developed for the CATE program had some properties which were quite different from previously used commercial reaction-bonded SiC (RBSiC) materials. Most other applications of RBSiC materials were based upon silicon carbide's wear resistance due to its high hardness. All RBSiC materials have a residual second phase of elemental free silicon metal from the infiltration process which is quite soft so RBSiC producers traditionally tried to minimize this amount of free silicon in order to maximize wear resistance. A typical commercial RBSiC has a density of 3.15 g/cm^3 and only 7% by volume of silicon. Pure silicon carbide has a density of 3.21 g/cm^3; and pure silicon has a density of 2.33 g/cm^3, so the relationship between the RBSiC final density (D_f) and volume fraction of silicon (V_{si}) is: $D_f = 3.21 - 0.88V_{si}$.

Free silicon is minimized by using coarse-fine combinations of SiC particle sizes to achieve high green packing densities (D_i) and using a large green weight fraction (f) of very fine graphite powder which reacts to form new silicon carbide between the original SiC particles. It can be shown that the relationship between volume fraction of silicon and the initial baked green density (D_i) and the green part's carbon fraction (f) from resin carbon residue and graphite powder additions is: $V_{si} = I - (0.728f + 0.311)D_i$.

The Carborundum CATE silicon infiltrated materials had a final density of about 2.88 g/cm^3 and a free silicon volume fraction of 37.5% because the carbon fraction from the resin was only 8% with a low green density of 1.69 g/cm^3 from using only fine SiC particles which packed poorly. The microstructure had a continuous phase of free silicon as shown in Figure 4.2.

The fine SiC particle size was chosen to achieve high strength. Room temperature 4-point bend strengths were typically 350–400 MPa. The decision to use a large amount of free silicon had some

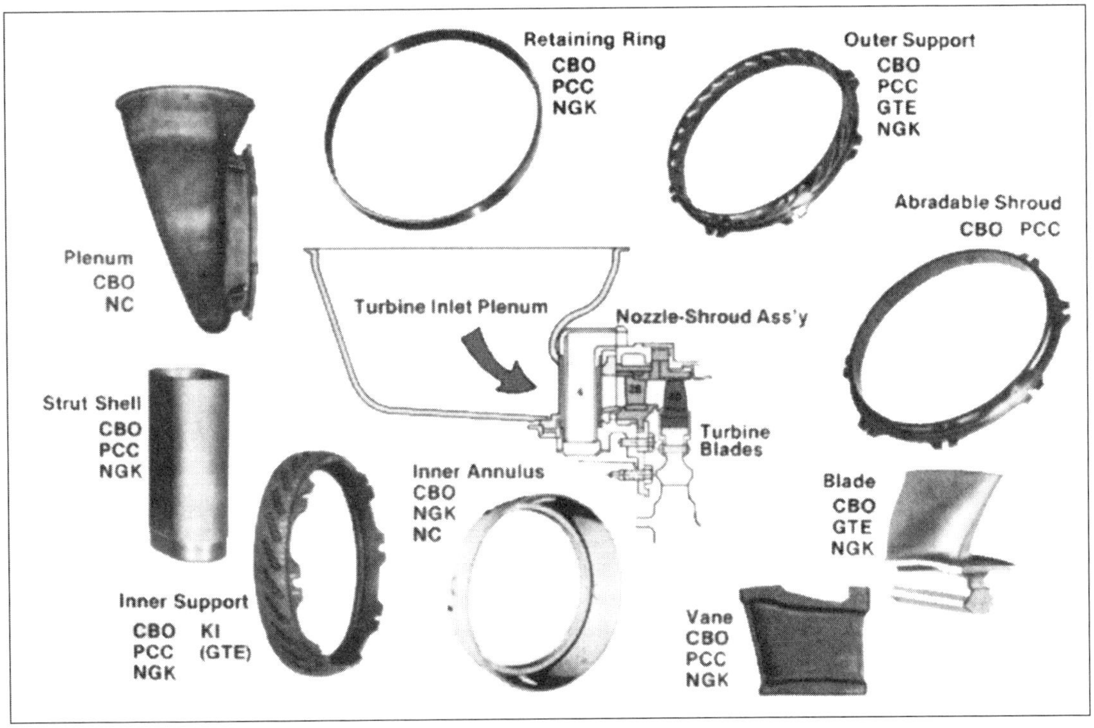

Figure 4.1 CATE Silicon Infiltrated SiC Turbine Components Made by Carborundum (Reprinted with permission of Saint-Gobain Advanced Ceramics).

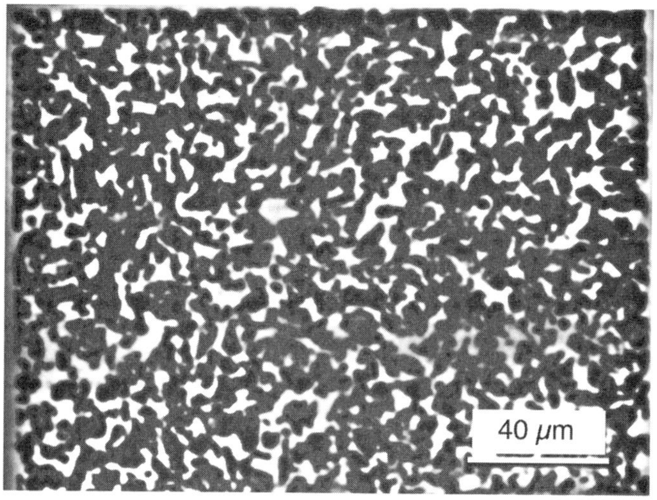

Figure 4.2 Microstructure of KX01 Silicon Infiltrated SiC (Reprinted with permission of Saint-Gobain Advanced Ceramics)

important influences on properties. Silicon is brittle at low temperature but it shows considerable plasticity above 800°C and the high temperature bend strength tests indicated nonlinear plastic deformation above 1000°C. The bend tests indicated strength increases at high temperatures as shown in Figure 4.3, but the strength increase was thought to be caused by a redistribution of stresses in the test bar from plastic deformation rather than an actual increase in the tensile strength. The Weibull modulus also increased at high

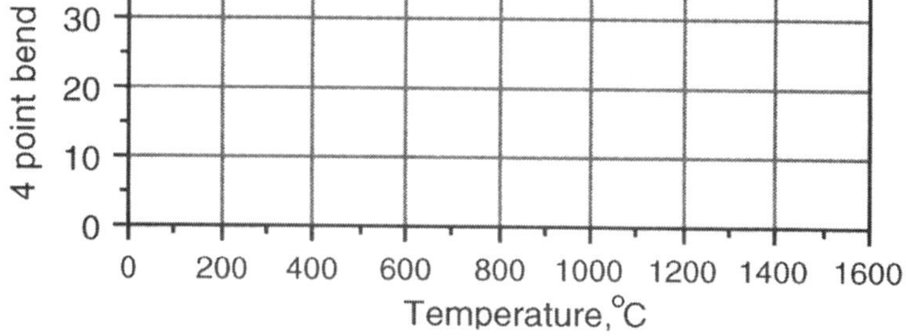

Figure 4.3 Four Point Bend Strength of KX01 Silicon Infiltrated SiC Material vs. Temperature (Reprinted with permission of Saint-Gobain Advanced Ceramics) Note that 1 psi = 6.90 kPa.

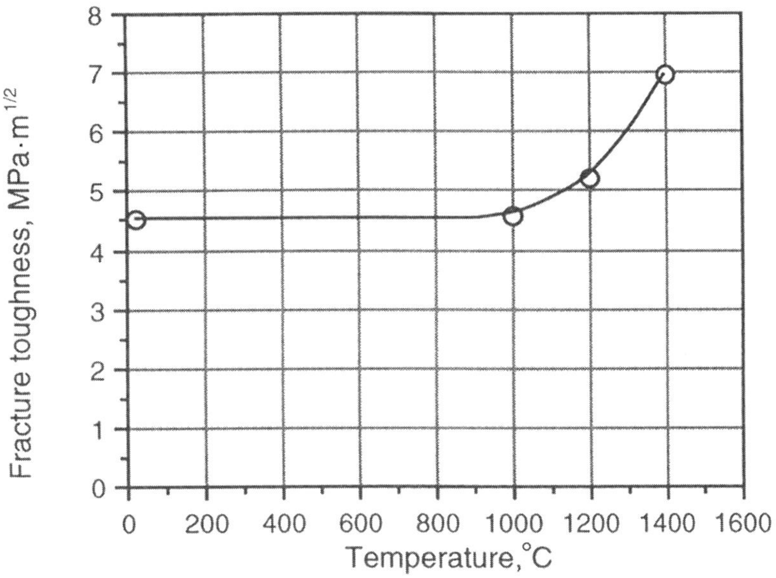

Figure 4.4 Fracture Toughness of KX01 Silicon Infiltrated SiC Material vs. Temperature (Reprinted with permission of Saint-Gobain Advanced Ceramics)

temperature: 10 at 25°C, 16 at 1000°C, and 19 at 1400°C, which seemed to indicate that the high temperature plastic deformation capability was reducing the influence of defects that cause strength scatter.

Silicon's plasticity may have also been responsible for an increase in the material's high temperature fracture toughness. The critical stress intensity factor K_{IC} was measured using a single edge notched beam crack in a four-point bend test and the results are shown in Figure 4.4.

The elastic modulus of pure silicon carbide is 406 GPa while pure silicon's is only 193 GPa so it is not surprising that the elastic modulus of the material was an intermediate value of 331 GPa. A coarser grain size material was made with 23 μm SiC powder and was similar to the 7 μm material but with strength of only 200 MPa. It was not used in the CATE program.

The low elastic modulus, high strength, and high fracture toughness all help the material's thermal shock resistance. The DDA turbine laboratory tests included a burner flame out test in which the fuel was turned off to the combustor so that cold air was suddenly blown through the hot engine. This caused thermal "down shock" where the ceramic part's outer surfaces were colder than their interiors thus putting the outer surfaces in tension which is usually undesirable for brittle ceramic parts. Because the stationary vane assembly is the first engine stage after the combustor, it experienced the most sudden cooling. This was a particularly severe test for the turbine vanes because their thin trailing edges (about 0.076 cm. thick) cooled much faster than the thicker portion of the airfoil shape.

Some Carborundum vanes did develop cracks perpendicular to their trailing edges but unlike other materials the cracks, surprisingly, did not propagate completely through the vanes and they stayed together in their support rings. It was believed that the material's plasticity from its silicon phase was arresting the crack propagation.

Allison did successfully demonstrate a CATE truck engine using these RBSiC vanes and support rings and many thousands of miles of road tests were completed successfully without any failures of these parts. Allison did not commercially produce the CATE truck engine but their ceramic design and testing experiences led to Allison's participation in the DOE-NASA Automobile Gas Turbine (AGT) program in the early 1980s. Fuel efficiency was still a high priority but there was also a growing need to reduce automobile NOx pollution emissions.

SINTERED ALPHA SILICON CARBIDE DEVELOPMENT UNDER THE AGT AND ATTAP PROGRAMS

As the CATE program was progressing it its effort to improve fuel efficiency in a truck turbine engine through the use of ceramics, DOE started a parallel program aimed at automotive turbines. The Automotive Gas Turbine (AGT) program was directed at a ceramic turbine engine of approximately 100 horsepower, much smaller than the CATE engine. The engine concepts for AGT from DDA and the Garrett Turbine Engine Company required much higher operating temperatures than the CATE engine, approaching 1371°C. None of the silicon nitride compositions available, or the RBSiC used for the CATE program had sufficient long-term stability to meet the target of 3500 hour life.

Carborundum had been developing a sintered SiC, which offered the potential to meet this goal of stability at 1371°C. The process for sintered alpha SiC (SASC) was first demonstrated at Carborundum in the 1970s. Submicron alpha SiC grain was readily available in tonnage quantities at low cost as a by-product from Carborundum's abrasive powder processing plants in Europe. The densification process required the addition of small amounts of carbon (for reduction of surface oxygen) and boron. Green bodies were formed by either dry pressing at moderate pressures (e.g. 700 kg/cm^2) or slip casting. Heating the green bodies to approximately 2100–2200°C in a N_2 or Ar atmosphere resulted in a volume shrinkage of ~50% and a dense body with a density of 3.05–3.15 g/cm^3, compared to a theoretical density of 3.21 g/cm^3. Grain size was typically 3–5 μm. High resolution TEM (transmission electron microscopy) did not show any evidence of a grain boundary phase, in contrast to the sintered Si_3N_4s which gained their strength and toughness from a substantial grain boundary phase. The microstructure for SASC resulted in a low fracture toughness (K_{IC} = 2.5 – 3 MPa.m$^{1/2}$ by indentation or 4.5 MPa.m$^{1/2}$ by SENB) with a transgranular fracture mode. The 4-point fracture strengths typically averaged ~350 MPa for the early materials. Strength values of 640 MPa were later achieved, as reported below, by systematic elimination of surface and volume flaws.

The apparent absence of a grain boundary phase gave rise to excellent retention of strength at high temperature. No decrease in fast fracture flexure strength was observed in air at temperatures of 1500°C and above. Slow crack growth was the lowest of any of the structural ceramics being considered for turbine engines. Cyclic durability testing of various structural ceramics by AFML showed excellent results for SASC [2]. Modulus of rupture (MOR) bars were cycled between room temperature and an elevated temperature for up to 3500 hours to assess lifetime in an automotive turbine engine. SASC showed no loss of strength when tested at 1371°C for 3500 hours.

This high temperature stability resulted in SASC being the exclusive material selected for the DDA AGT radial rotor which had a design operating temperature of 1371°C. However, a radial rotor was a very challenging component to produce in a monolithic ceramic at that time. Complex blade curvature would require a net shape process. Also any net shape ceramic process had not successfully demonstrated the considerable thickness of the hub section at that time. It was therefore decided to do the initial process development using a simpler shape. The fabrication process selected was injection molding, which had already been under development for several years at Carborundum. Individual axial turbine blades for the CATE axial truck turbine engine were selected as a complex component without the complication of a thick hub for the initial process development work (Figure 4.5). The design was such that individual ceramic blades were inserted into a metal hub.

The process developed at Carborundum involved preparing a blend of thermoplastic resins and plasticizers with alpha SiC powder [3]. This blend could then be formed into complex shapes using conventional plastic forming equipment. After molding, the parts were baked using an accurately controlled heat-up rate to remove the organics. During sintering, the parts experienced a linear shrinkage of approximately 18% producing a strong, single phase silicon carbide body with a density in excess of 95 percent of the theoretical value (3.21 g/cm^3). One of the goals of the AGT effort was to establish the degree to which this shrinkage could be controlled, especially with regard to minimizing distortion. An injection molding tool was built and parts were successfully fabricated in that blades were obtained which were free of major flaws or defects, and distortion was kept to a minimum.

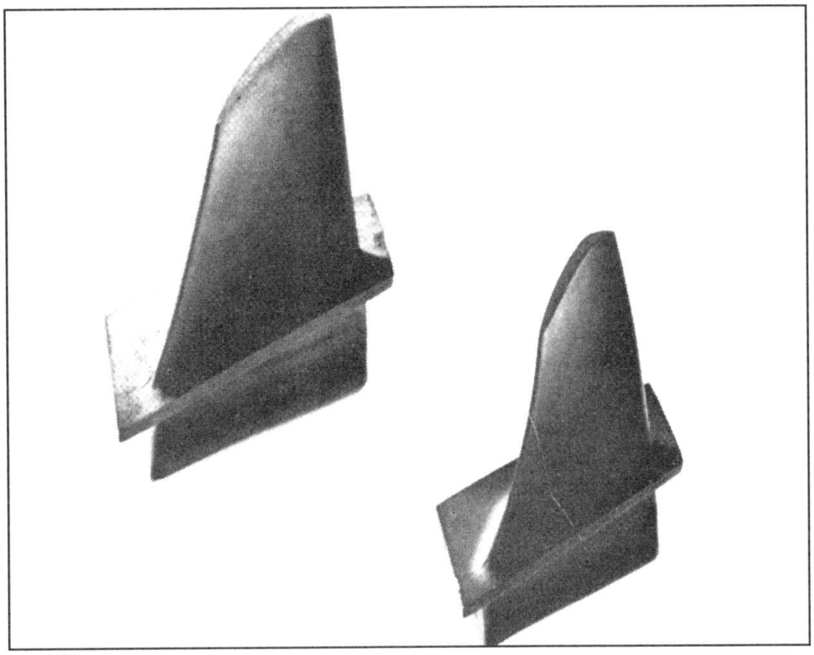

Figure 4.5 Injection Molded SASC Turbine Blade (Reprinted from ref. [3])

Based on the accomplishments in this initial phase, a new injection molding tool was constructed using a revised, optimized design which was fully analyzed by Allison to calculate thermal and mechanical stress distributions. This tool was a two-cavity mold: the turbine blade and an accompanying test bar. The two cavities were fed by a U-shaped runner and since the cavities were of differing dimensions, the system was not balanced; i.e., one cavity would fill while the other was only partially filled during injection.

An initial set of turbine blades was fabricated using a trial and error technique to product parts which had an optimum appearance based on the elimination of cracks and flow lines. Two types of injection molding gate were employed: a conventional circular end gate and a fan gate which spread the material flow out 90 degrees before the plastic was injected into the blade cavity. Parts were supplied to Allison after baking and sintering. Their dimensional analysis indicated that the standard gate provided blades of greater uniformity. The parameters used for processing these blades were then used as a starting point for establishing an extensive process variation matrix carried out under the CATE program.

The purpose of the blade process matrix study was to identify process parameters, which yielded the maximum strength of the sintered alpha silicon carbide blades, used in axial turbine wheels for the AGT 404 engine. The matrix consisted of 32 experiments with 13 process variables thought to have a direct impact on strengths and which were investigated at two levels. These included: binder composition, compounding method, sintering aid level, processing additive level, molding temperature and pressure, sintering temperature and heatup rate, furnace type (continuous vs. periodic), furnacing atmosphere and fixtures, grinding and post grinding treatment. One hundred fifty blade/bar pairs were molded and processed through sintering for each experiment. The sintered parts were then evaluated using NDE (nondestructive evaluation) techniques (microscopic visual examination at 8–40x, fluorescent dye penetrant, and micro-focus X-ray), as well as dimensionally to establish the conditions that resulted in the maximum yield of acceptable parts. These parts were delivered to Allison for further characterization and statistical analysis of the data produced.

Variables which were found to be significant in analyzing the results of this experiment were the use of a new binder, the sintering aid level and the processing additive level. Other variables were less significant within the ranges investigated [4]. This designed experiment, which involved a very large number of molded sintered parts allowed Carborundum to identify the optimum process conditions for yield and quality, and subsequent delivery of parts for rig and engine testing.

The goals of the program were achieved insofar as a process route was specified by which high strength blades were produced at a yield of 88%. Spin test results showed burst speed averaging 55,700 RPM which was 151% of the design speed of 36,700 RPM. As-fired test bars accompanying the blades had mean strengths of 456 MPa.

Based on the success in injection molding of individual turbine blades, the program moved on to the more difficult radial turbine rotor. Allison designed a prototype radial turbine rotor for the automotive AGT program to assess feasibility of producing this much more complex component by injection molding (Figure 4.6). Blade overlap at the exducer section was eliminated in the prototype design to facilitate fabrication of the injection molding die. Process conditions to produce crack free SASC rotors were quickly defined by Carborundum, and rotors delivered to Allison for spin testing. The design speed of 86,500 RPM was routinely demonstrated and provided the basis for moving on to the engine configuration rotor (ECR) which would be used for engine testing with other ceramic components. While the changes from the prototype to the ECR were not drastic, they were significant and necessary for functionality and aerodynamic efficiency. The hub configuration was essentially unchanged. The blade curvature was modified and lengthened such that the new mass was approximately 4% greater than that of the prototype. Blade overlap at the exducer section was accommodated by a more complex injection molding tool.

The first task addressed in the development of the ECR was to reproduce the performance achieved with the prototypes. From there it was intended to increase reliability and improve performance.

The initial group of ECRs were produced using essentially the same process route as that for the prototypes except that molding pressures were slightly lower. When the same molding pressures were used, molding compound was forced into the sliding mechanism of the more complex tool and the molded rotor could not be extricated. The net effect of using lower pressures appeared to be negligible since the spin data was approximately the same when a correction was made for the increased mass. The burst speed of 15 prototype rotors averaged 97,000 RPM whereas the burst speed of 7 ECRs of equivalent quality as determined by NDE was 95,000 RPM when corrected for additional blade mass [4].

Figure 4.6 Injection Molded, Sintered Alpha SiC Turbine Rotor (Reprinted from ref. [3])

Figure 4.7 Injection Molded Original and Thickened SiC Turbine Rotors (Reprinted from Ref. [5])

It should be noted that both the prototype and ECRs contained linear indications primarily on the hub surface between the blades. These flow lines developed during molding and resulted from the rheology of the compound, machine parameters, and the nature of the filling pattern. An intensive effort to improve surface quality was conducted in which a series of sprue/nozzle configurations were evaluated over a range of machine settings. These efforts were successful in reducing flow lines and improving surface quality; however, the lines were not completely eliminated. Over 1000 rotors were molded during this evaluation and about 200 rotors were selected for further processing through sintering, and 160 rotors were delivered.

A clean-up procedure was then instituted in which the flow lines were rubbed out after bake-out. Eleven rotors from this group with equivalent quality (as determined by NDE) to the previously described groups of prototype and ECR rotors were spun. The average burst speed was 99,330 RPM (110% of design speed) indicating an improvement over any group of rotors spun to that point. As a result of engine testing of SiC gasifier turbine rotors in which the inducer blade tips were damaged because of foreign object damage (FOD), a redesign was initiated in 1985 to thicken the inducer blade tips from 0.76 mm to 1.91 mm to improve the airfoil impact resistance. Both the thin-bladed and thick-bladed rotors are shown in Figure 4.7. Sintering cycle modifications increased average density to from 3.10 to 3.14 g/cm^3 and improved the microsctructure uniformity. A total of 23 thick-bladed SiC rotors achieved an average burst speed of 90,700 RPM with a Weibull modulus of 16.0. Hot isostatic pressing (HIP) processing trials on 30 thick-bladed rotors at Asea in Sweden resulted in an increase in average burst speed to 98,200 RPM with a Weibull modulus of 19.3 [5]. The thick-bladed rotors had improved impact resistance while the burst speed was in the range of the previous thin-bladed rotor design.

The DOE AGT Program had come to the end of its planned duration. It had made very significant advances in the design, fabrication, and testing of ceramic automotive gas turbines. Carborundum had progressed down the path of injection molding test bars and moved on to individual turbine blades and finally to complex radial rotors with a thick cross section. In order to move further down the path toward the ceramic engine however, the quality and reproducibility of the ceramic components needed to be improved. To address that need DOE initiated the Advanced Turbine Technology Applications Project (ATTAP) as a follow-on program. The goals of ATTAP were directed toward improving ceramic component performance and reliability.

At about the same time as ATTAP was being started up, GM announced that they had been developing an axial automotive turbine engine at the GM Tech Center, under the direction of Al Bell. GM had started the initial planning to move the automotive gas turbine down the commercialization path. Subsequently GM and Allison chose to merge the two programs, and Allison utilized the axial GM configuration for the ATTAP program.

The objective of Carborundum's ATTAP efforts was to fabricate defect-free sintered alpha SiC turbine components, which met ATTAP performance requirements by improving upon materials understanding, and

Figure 4.8 Allison AGT-5 Axial Rotor (Reprinted from ref. [6])

process development [6]. The activities applied to injection molding of rotating components included flow analysis, equipment instrumentation, compound evaluation, HIPing, and NDE, specifically computed tomography (CT).

The extensive process improvement programs carried out by Carborundum under ATTAP focused on the Allison axial AGT-5 axial rotor design (Figure 4.8). Flow modeling and analysis were conducted using Moldflow's "Moldflow" software to optimize specific molding variables which keep shear rates and shear stresses as low and uniform as possible. The approach was to first simulate filling using start-up molding factor levels to determine fill time, pressure, shear rate, shear stress, and temperature profiles and to identify potential molding problems. The control variables (melt temperature, mold temperature, velocity profile, sprue and gate configuration) could then be optimized to minimize response variable gradients. Flow modeling recommendations included a centroid gate location opposite the shaft, a higher melt temperature, a higher mold temperature, and, most importantly, a velocity profile to accommodate the thinnest regions or blades based on the uniform flow front velocity concept.

Injection molding equipment instrumentation was incorporated to monitor injection molding variables on a real-time basis to verify parameter levels and establish reproducibility. Both the injection process controller and mold were instrumented to monitor position of the screw, pressure and temperature.

Figure 4.9 shows the first two seconds of the injection molding cycle during which time the mold is filled. There is an initial pressure surge required to produce movement in the screw resulting in a high initial velocity.

A 2^{5-1} Taguchi L16 experiment (2 = number of levels, 5 = number of factors, 1 = fractionalizing element, 16 = number of treatment combinations) was then conducted on the axial rotor. Mold temperature was shown to be the most important factor from previous experiments and a velocity profile was run as suggested by flow analysis. The factors from the rotor hub experiment were set at their optimum levels. The experimental results are shown in Table 4.1. The two strongest factors, sprue size and mold temperature, explained 61% of the variability for flow lines. A, B, C, AxD, and BxC explained 84% of the variability.

When the temperature of the mold was heated to 16–27°C above the best previous temperature (with respect to flow lines), the flow lines virtually disappeared. Now sprue size could be increased and a velocity profile maintained to minimize shear rates and stresses. The resultant rotors had a more homogeneous surface appearance and no flow lines could be found in the green or sintered state.

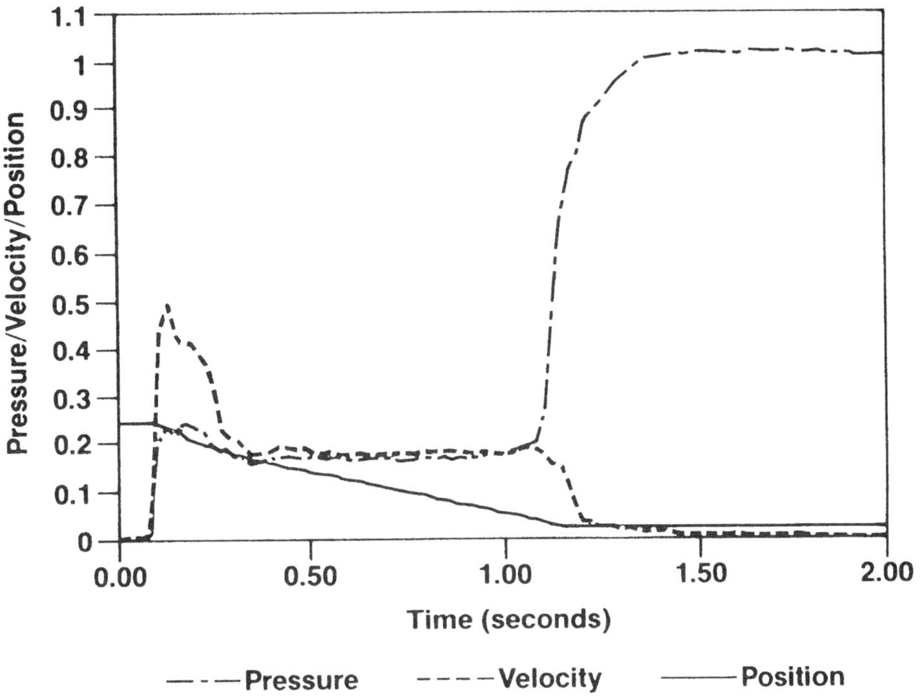

Figure 4.9 Injection Molding Cycle First Two Seconds (Reprinted from ref. [6])

Techniques to remove inclusions during compounding were then investigated. A linear filter and a screen pack were used to trap 0.25 mm and larger inclusions. The x-ray detectable inclusions were greatly reduced when a screen pack was used in compounding and virtually eliminated using a screen pack during molding. The impact of this process modification on strength is shown in Table 4.2. A 10-fold reduction in the number of inclusions resulted in a 25% improvement in MOR (Modulus of Rupture). After dismantling the screen packs and subsequent microscopic examination, the inclusions were found to be largely SiC particulates, originating from ball milling with SiC media.

Next, the effect of post hot isostatic pressing (HIPing) on SASC strength was investigated as a means to further reduce volume flaws. The axial AGT-5 rotors used for the study were fabricated with compound that had been processed using the screen pack filtration during compounding and molding described above. As shown in Figure 4.10, the as-fired rotor had a low-density core. This porosity was reduced to a very low level after the post HIPing operation. As a result, the burst speed for failure during spin testing (Table 4.3) increased from 107,500 RPM for the as sintered parts to 128,500 RPM after post HIPing. The corresponding 4-point MOR, calculated assuming a failure probability of 50%, increased from 398 MPa to 467 MPa.

TABLE 4.1 2^{5-1} TAGUCHI L16 EXPERIMENTAL MOLDED FLOW LINES [6]

Factor	Contribution	Best Level
Sprue Size (A)	Strong	Low
Mold Temperature (B)	Strong	High
Sprue Temperature (C)	Moderate	Low
Velocity Sliders 1–9 (D)	Very Weak	High
Velocity Slider 10 (E)	Very Weak	High

TABLE 4.2 CORRELATION OF BILLET MOR WITH AVERAGE NUMBER OF INCLUSIONS/BILLET IN GREEN BILLETS [6]

Mix	Inclusions/Billet	MOR (MPa)
A-2	2.00	428
B-1	2.75	396
A-3	10.50	358
B-3	14.25	330
B-2	22.00	341

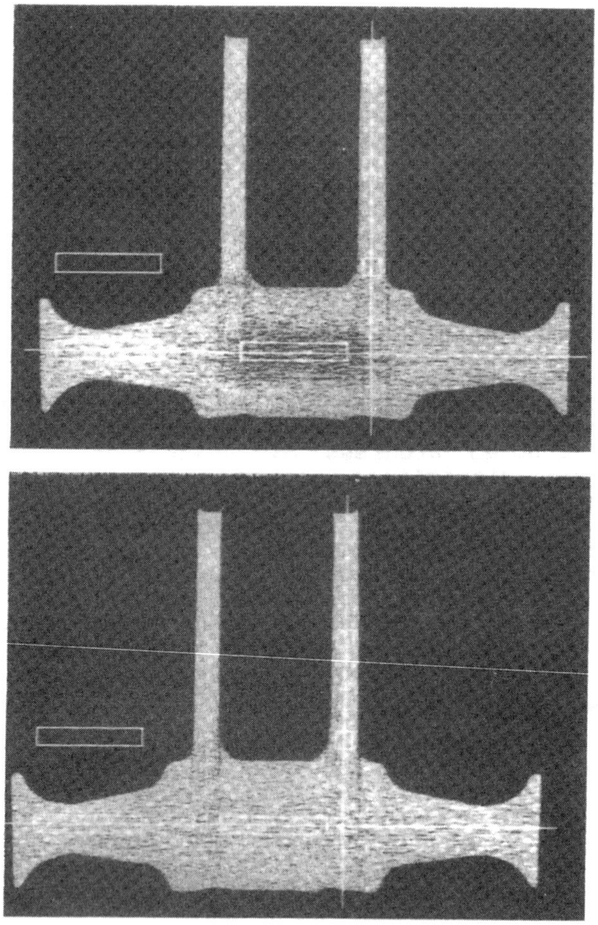

Figure 4.10 Computed Tomography Scans of As-Fired (top) and Post-HIPed (bottom) Rotor Hub (Reprinted from ref. [6])

TABLE 4.3 AS-FIRED AND POST-HIPED ROTOR HUB DATA [6]

	As-Fired (No. of Hubs)	Post-HIPed (No. of Hubs)
Density (g/cm^3)	3.150 (21)	3.163 (21)
Burst Speed (kRPM)	107.5 (2)	128.5 (7)
RT MOR (MPa)	398 (2)	467 (7)
CT Density Difference (%)	8.0 (21)	1.6 (21)

Finally, with the significant reduction in flaw population from optimized injection molding conditions and elimination of flow lines, combined with a reduction in inclusions (use of screen packs) and porosity (post HIPing), the grain size of the SiC became a strength limiting defect. Under the ATTAP program, both sintering and post HIPing of AGT-5 axial rotors were carried out at optimized lower temperatures. A 4-point MOR of 593 MPa was obtained for bars machined from the rotor hubs [7]. This was repeated with as fired injection molded test bars, and an MOR of 642 MPa was measured.

Overall, a strength improvement of ~100% was demonstrated under the AGT and ATTAP programs through a systematic flaw reduction effort. This was accomplished without changing the chemistry of the SASC. Thus the excellent resistance to high temperature creep and slow crack growth at turbine operating temperatures of 1371°C were maintained.

Both the Allison and Garrett automotive gas turbine programs required the development of stationary SASC components for the gas flow path. The radial turbine design of the Allison program generally required components with a uniformly thin wall. Carborundum focused on slip casting for these components which included a combustor body, scroll assembly and backplates. The axial Garrett design led to variable part thickness, with a need for thicker cross sections. Isostatic pressing and green machining were utilized for these components, in addition to slip casting and injection molding.

A large component from the Garrett Turbine Engine Company was also undertaken by injection molding [3]. Figure 4.11 shows the turbine shroud which was formed in a one piece injection molding. Maximum wall thickness was 3 cm and diameter was 35 cm. Part weight was approximately 5 kg. The turbine shroud for the AGT 101 Engine is the largest component Carborundum formed by injection molding. Because of its size, it was molded on a 1000T reciprocating screw injection press having a shot size of 4687 cm^3.

In the fabrication of this component, fold lines were frequently evident on the surface and were removed prior to sintering. In some cases, however, the folds were below the surface and undetectable by X-ray. By increasing the gate size, the incidence of these folds diminished and several good parts were molded. Another problem which was addressed early in the development of this component was flange distortion. By adjusting the sintering cycle, the problem was satisfactorily resolved. The close tolerances needed to accommodate mating components and to produce the correct stacking height was achieved by precision grinding after sintering.

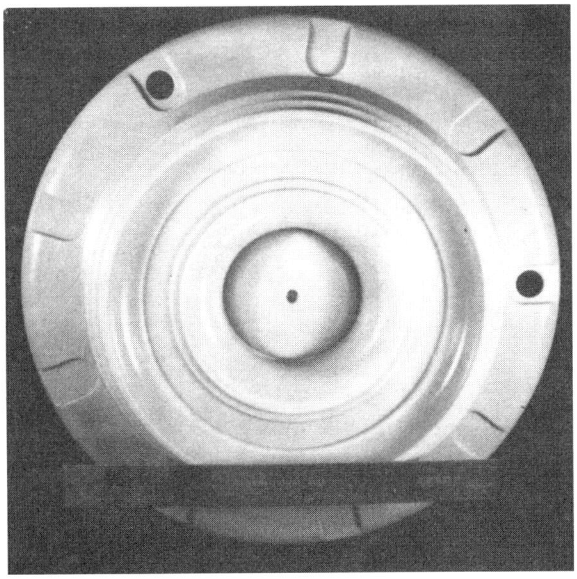

Figure 4.11 Injection Molded Wax Replica, Garrett Turbine Shroud (Reprinted from ref. [3])

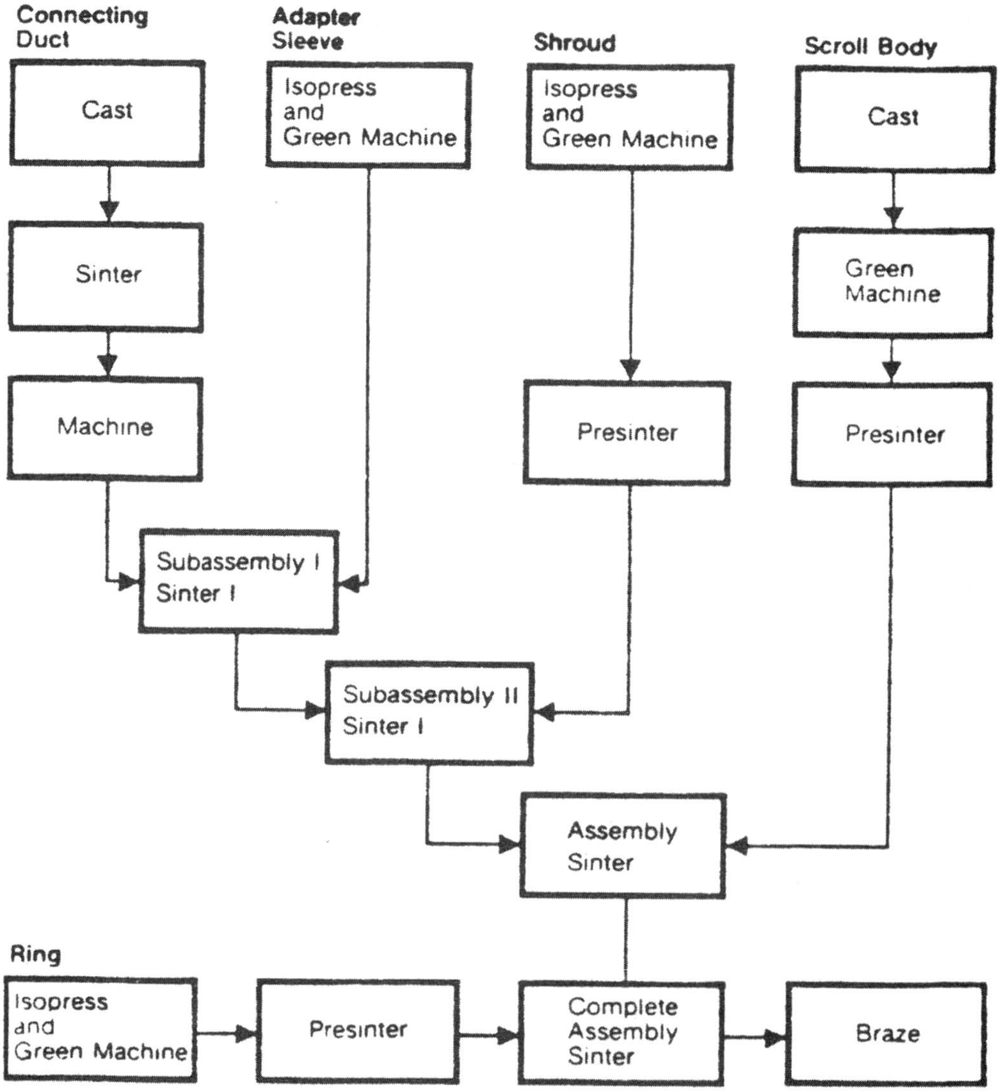

Figure 4.12 Schematic of Process for the Allison Gas Turbine Operation (AGTO) Scroll Assembly (Reprinted from refs. [4,5])

For the Allison AGT program, the gasifier turbine scroll assembly was a very complex geometry. The gasifier turbine scroll assembly consisted of five individual parts bonded together by sintering and/or brazing to form the finished component. The process is shown schematically in Figure 4.12 and a completed assembly with the major components is shown in Figure 4.13 [4,5].

For the initial prototype design, parts were slip cast using a very complex mold with multiple inserts, The scroll body was cast in a twelve-piece mold consisting of an outer shell, inner removable core sections and several inserts. Several mold designs were investigated before arriving at this configuration. The complex geometry made the demolding process difficult and very dependent on timing and slip properties. In general, the problems encountered in demolding resulted from drying shrinkage in the mold. If the part shrank too much, it tightened on the inner core; too little shrinkage and the part would not readily release

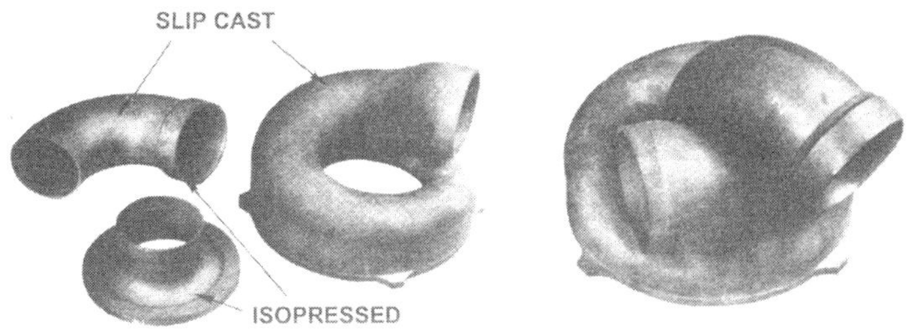

Figure 4.13 Completed AGT Scroll Assembly and Constituent Parts (Reprinted from ref. [5])

from the outer wall. Precise control of the drying shrinkage was found to depend on particle size distribution, degree of deflocculation, water content, and pH. Each lot of powder was screened for castability. Initially yields were very low on the scroll body; however, with experience and a better understanding of the critical parameters, demolding yields increased above 50% which was still well below the 90% which is typical for slip cast parts.

The joining of five components into a completed assembly required six firings. Extensive in-process NDE after each step was performed to eliminate any potentially defective parts before firing the final assembly.

Achieving the correct alignment and dimensional tolerances required accurate fixturing to accommodate shrinkage during the sintering process. The bond between the shroud and the scroll body was critical to maintain mechanical integrity as well as gas tightness. After the final assembly firing, the joint was inspected both visually and by X-ray. A final firing was conducted in which a high temperature silicon based braze was applied to the bond area to ensure proper filling and strength. While this process produced parts for testing, it was very intricate. In addition, the tight dimensional requirements were very challenging. Despite the complexity of this process, assemblies were produced (albeit in low yield) with acceptable dimensions.

Although, the AGT and ATTAP programs did not lead to the commercialization of ceramic automobile engines they did help provide, through successful demonstrations, direction to improve the ceramic manufacturing processes used and the insight to make them more reliable and robust. Additionally, much was learned about nondestructive testing, proof testing, finite element design, and the economics of producing reliable high strength components with complex shape.

COMMERCIALIZATION AT INEX INC.

One of the authors (M. Kasprzyk) founded Inex Inc. in 1984, which produced simple large diameter thin-walled SiC tubes (see Figure 4.14) which were used as gas fired radiant tubes in industrial heat treating furnaces. The radiant tubes had generous tolerances and the only strength requirement was that the part had to hold up its own weight at 1200°C without excessive creep and not fail due to thermal shock.

Inex Inc. developed a silicon infiltrated silicon carbide radiant tube material with 50% silicon (by volume) but with very low C-ring and O-ring strengths of about 598 kg/cm^2 due to its very large SiC particle grain size of 210 microns. C-ring tests at 1250°C and 1350°C had increases in strength to 809 kg/cm^2 and 774 kg/cm^2, respectively. The tensile test stress-strain curves were nonlinear with plastic deformations of 0.1% at 1250°C and 0.6% at 1350°C (Figure 4.15). This proved that indications of increases in bend strength at high temperatures for materials with large proportions of silicon and little SiC particle to particle bonding were due to stress redistribution in the bend test bars or C-rings rather than an increase in the material's actual tensile strength. A material made with a large amount of brittle silicon carbide bond phase shows less of this effect. Thus, the use the term "reaction bonded SiC" or "RBSiC" for this type of mate-

Figure 4.14 Schunk.Inex Corp. Si-SiC Tubes with Outside Diameters from 1 to 4.5 Inches (Reprinted with permission from Schunk-Inex Corporation)

rial and "SiSiC" for those with a large amount of silicon is preferred even though both types are made by a silicon infiltration method.

Inex Inc. chose to use large grain size for several reasons. Coarse SiC grain was less expensive than fine powder which is important for radiant tubes that weighed between 2.5 and 25 kg. Coarse SiC grain is also necessary for Inex's novel net size manufacturing process which relies upon dry powder gravity flow to form thin walled tubes [8]. Fine powders do not flow well and cannot be used.

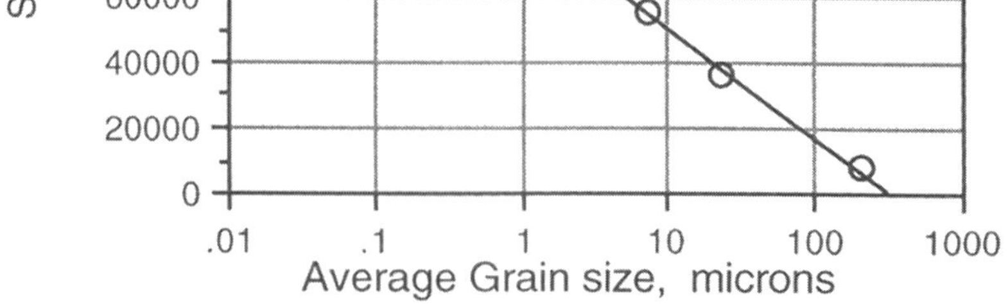

Figure 4.15 Room Temperature Strength vs. Grain Size for Si-SiC Materials (Reprinted with permission from Schunk-Inex Corporation) Note That 1 psi = 6.90 kPa.

The microstructure was also coarse enough that individual SiC grains could be seen with a naked eye. A material's strength limiting critical flaw size is often proportional to its grain size so the very coarse grain size produced defects which could be easily visually detected in order to eliminate the need for expensive NDE equipmentAbout 12,000 Inex Inc. (Schunk-Inex Corp. as of 1998) Si-SiC radiant tubes are now in use. The first ones were installed in 1989 and are still functioning perfectly despite having an intentionally low strength so it appears that they may have a very long life potential.

SINTERED SiC COMMERCIALIZATION AT CARBORUNDUM/SAINT-GOBAIN

Over the past twenty years Carborundum has developed sintered silicon carbide (SiC) materials having a high level of performance and reliability through fabrication methods with the potential to make complex shapes suitable for a broad range of applications. The AGT projects in which Carborundum was engaged in the 1970s and 1980s have aided more recent developments and the commercial growth of silicon carbide through greater understanding of fabrication requirements, including both net-shape and near net-shape processes, resulting in improved product reliability.

Emphasis was placed not only on material systems development but on manufacturing development as well. Critical objectives of these development efforts included using fabrication methods adaptable to mass production, increased understanding between component design and the method of fabrication, and most important, methods based on an economically sound approach.

Commercial success in newer product applications has been in part a consequence of maturity and improved reproducibility of key manufacturing processes, playing as important a role as the required material characteristics. Effective manufacturing manifests itself in consistent material properties which in turn translates into improved reliability and reduced cost.

As described above, SiC as a synthetic material was invented by Dr. Edward Acheson in 1891, and was the first man-made abrasive originally used in such applications as grinding wheels and sandpaper. By the mid 1970s, this versatile material opened up new opportunities when Carborundum invented a pressureless sintered alpha form of SiC, giving it the tradename Hexoloy®. Silicon carbide can best be defined as an advanced non-oxide ceramic offering a substantially higher level of performance, and a market value based on unique performance which justifies a higher added value than is characteristic of traditional ceramics.

Progress in both traditional ceramic and high technology industries including aerospace transportation (new vehicular propulsion concepts), electronics and semiconductor, medical and biotechnology industries is being advanced through the use of silicon carbide and other advanced ceramics. The advantages of silicon carbide ceramics include light weight, high hardness, resistance to wear and corrosion, strength retention at high temperature, excellent tribological behavior, and high thermal conductivity. Benefits of these outstanding properties and characteristics contribute to conservation of energy and natural resources, and improved products throughout broad market sectors.

Typically, silicon carbide serves as a substitution material for older technology materials, such as alumina, tungsten carbide, or exotic metal alloys. The unique physical, mechanical, and thermal properties of silicon carbide make it ideal for applications such as mechanical seals for chemical pumps and automotive water pumps, bearings, valves, blast and atomization nozzles, heat exchanger tubes, and ballistic armor protection.

Within the past several years Carborundum has commercialized newer versions of sintered silicon carbide, including a grade with the addition of spherical pores, and a grade containing some free graphitic carbon having high electrical conductivity. The controlled porosity version of sintered SiC, identified as Hexoloy® SP, has spherical pores that are discrete, non-interconnecting, and dispersed in a controlled manner throughout the body of the material. The spherical pores act as fluid or lubricant reservoirs helping to promote the retention of a fluid film at the interface of sliding component surfaces. This grade was designed specifically for optimum performance in sliding contact applications such as mechanical seal faces and product lubrication bearings. Seal designs incorporating self-mated hardfaces can particularly benefit from this grade of silicon carbide and the material has found extensive application replacing aluminum oxide faces in automotive water pump seals. The material has become a standard seal component material used by all automotive companies worldwide.

The electrically conductive version of sintered SiC, identified as Hexoloy® SG-90, has less than 5 ohm-cm bulk electrical resistivity. It was developed primarily for specialized applications as a dc magnetron sputtering target material to deposit silicon carbide thin films used in optical data storage disks. Further potential application areas include hard protective coatings on flat glass and eyeglass lens, and conductive or vapor/corrosion barrier coatings.

More recently, new product applications for sintered silicon carbide have been commercialized in the semiconductor industry, including plasma chamber components for semiconductor wafer processing, vacuum and electrostatic chucks, susceptors, RF windows, and chemical mechanical polishing blocks. These applications take advantage of silicon carbide's thermal expansion match to silicon, high elastic modulus, chemical inertness, allowing economic benefits of maintenance and reuse, and high thermal conductivity for even, rapid heating of the silicon wafer. Silicon carbide is suitable as a high elastic modulus material for silicon wafer support, being rigid and dimensionally stable, and able to be polished to a very high degree of flatness.

The commercial strength of silicon carbide as a high performance ceramic lies in its unique range of material properties. Recognition of the versatility of silicon carbide as a high performance material, and the capability to be economically and reliably manufactured in complex and useful shapes will be strongly valued in the marketplace. As technological advancements evolve, new markets open up, and expanded use in more traditional applications continues, the future commercial prospects for silicon carbide remain very strong.

REFERENCES

[1] "*The Carborundum Company, The First 100 Years 1891–1991*," Flower City Printing, Inc., Rochester, NY, USA, 1991. pp. 8–12. Reprinted with permission from Saint–Gobain Advanced Ceramics.

[2] Larson, D. C., *Property Screening and Evaluation of Ceramic Turbine Engine Materials,* Technical Report AFML-TR-79-4188, 1979.

[3] Storm, R. S., Ohnsorg, R. W., and Frechette, F. D., "Fabrication of Injection Molded Sintered Alpha SiC Turbine Components," ASME paper 81-G7-161, presented at the Gas Turbine Conference & Products Show, Houston, TX, USA, March 9–12, 1981.

[4] Sweeting, T. B., Frechette, F. J., and MacBeth, J. W., "Progress in Net Shape Fabrication of Alpha SiC Turbine Components," ASME paper 84-GT-273, presented at the International Gas Turbine Conference and Exhibition, Amsterdam, The Netherlands, June 4–7, 1984.

[5] *Advanced Gas Turbine AGT Technology Project Final Report,* Allison Gas Turbine Division, General Motors Corporation, Indianapolis, IN, USA. Contract DEN 3-168. DOE/NASA 0168-11, NASA CR-182127, EDR 13295, August 1988.

[6] Ohnsorg, R. W., Funk, J. R., and Lawler, H.A., "Fabrication of ATTAP Ceramic Turbine Components," ASME paper 90-GT-185, presented at the Gas Turbine and Aeroengine Congress and Exposition, Brussels, Belgium, June 11–14, 1990.

[7] Ohnsorg, R. W., Lau, S.K., TenEyck, M.O., and White, D.A., "Injection Molded and Isopressed SiC Components for Stationary Gas Turbine Engine Components," ASME paper 95-GT-75, presented a the International Gas Turbine and Aeroengine Congress and Exposition, Houston, TX, USA, June 5–8, 1995.

[8] Frechette, F. J., Kasprzyk, Martin R., McDowell, D. J., and Zanghi, J. S., "Composite Material of Discontinuous Silicon Carbide Particles and Continuous Silicon Matrix and Method of Producing Same," US Patent Number 4,795,673, January 3, 1989.

CHAPTER

5

The Development of Monolithic Silicon Nitride Structural Ceramics at Honeywell

John P. Pollinger

Honeywell International Inc., Ceramic Components, Torrance, California, USA

INTRODUCTION

Advances in the development of high temperature structural ceramic materials during the late 1960s and early 1970s in Great Britain and the United States led the Garrett Corporation to initiate efforts to fabricate components for a variety of their turbine applications, including aircraft auxiliary power units and propulsion engines, and automotive turbochargers. This effort was started in 1976 at the AiResearch Casting Company (ACC) division in Torrance, California. Over the last 25 years great progress has been made in both materials and component fabrication technology. This chapter will discuss the historical development of these materials, component fabrication technologies, applications, the lessons learned, and future technology needs and plans.

Although these efforts have occurred at the same organization since 1976, significant corporation name changes and mergers have occurred. To minimize confusion, a brief corporate history will be described to provide a simplified business nomenclature for the subsequent discussion. In 1976, the Garrett Corporation was part of Signal Oil Corporation. The divisions of Garrett included the Garrett Turbine Engine Company (GTEC), located in Phoenix, Arizona—the manufacturer of aircraft turboprop and turbofan propulsion engines and auxiliary power units (APUs), and AiResearch Casting Company (ACC), located in Torrance, California—a manufacturer of aluminum, steel, and superalloy components for turbine engines and turbochargers, including cast wheels, blades, nozzles, and other static structures. The ceramic component fabrication efforts were initiated at ACC in 1976. In 1985, Signal Oil merged with Allied Chemical Corporation with the new name of AlliedSignal Corporation. In 1986, the structural ceramics development effort within ACC was formed into a separate business center and named Garrett Ceramic Components (GCC). In 1995,

GCC was renamed AlliedSignal Ceramic Components (ASCC) and the Phoenix engine division was renamed AlliedSignal Engines (AE). In December 1999, AlliedSignal purchased Honeywell, and ASCC was renamed Honeywell Ceramic Components (HCC). To minimize confusion, HCC will be used as the acronym to denote the ceramic component manufacturing operations located in Torrance and the engine manufacturing operations in Phoenix will be denoted as Honeywell Engines, Systems & Services (HES&S).

Honeywell Engines, Systems & Services was and is a leader in the production of small gas turbines for aircraft propulsion and auxiliary power units. Recognizing the need for enhanced hot section materials performance compared to the limitations of current metals technologies in the early 1970s, HES&S initiated efforts to investigate the application of high temperature structural ceramics, with additional support from the U.S. Department of Defense (DOD), Department of Energy (DOE), and the National Aeronautics and Space Administration (NASA). Because of the general lack of high temperature structural ceramic component materials and fabrication capability at that time inside the United States and elsewhere in the world, Dave Richerson of HES&S proposed the setup of fabrication capabilities within Garrett. Facilities were set up in 1976 at the AiResearch Casting Company facilities in Torrance to develop both materials and component fabrication with guidance and support from HES&S. This was the start of the operations which eventually became HCC. The material efforts focused on silicon nitride, because of its promise of high strength and high temperature capability, while exhibiting excellent thermal shock resistance and light weight. After the merger with Allied Chemical in 1985, the Honeywell Laboratories (formerly the AlliedSignal Corporate Research Laboratories in Morristown, New Jersey) built a strong ceramic research group that developed advanced silicon nitride materials and forming processes for HCC and HES&S. A historical discussion of the development and maturation of silicon nitride materials and fabrication of components at HCC is presented for a variety of applications in gas turbines, highlighting progress made, lessons learned, and hurdles still to be overcome.

SILICON NITRIDE MATERIALS

HCC initiated its work in structural ceramics work in 1976 by developing reaction bonded silicon nitrides (RBSN), based on the work at Ford Motor Company in the early 1970s and earlier work in Britain on reaction sintering of silicon nitride and silicon carbide. Karsten Styhr in Torrance and John Schuldies in Phoenix were hired from Ford to utilize their experience with Ford's RBSN material and component fabrication development efforts. RBSN materials are formed by the nitriding of silicon powder. Silicon powder is first formed into a powder compact, with a packing density of 50–60 vol%. The nitridation reaction is carefully controlled at temperatures between 1000°C and 1400°C to fully convert the silicon to silicon nitride, without blocking off the interconnecting porosity in the powder compact, which would prevent full conversion. The result is a silicon nitride material with ~80–92% of theoretical density with residual porosity in the 8–20 vol% range. The nitridation reaction results in less than 0.5% shrinkage of the powder compact, thus nitridation can be a near-net shape forming process. HCC's first RBSN material, demonstrated in 1978, was RBN-101, formed by slipcasting –325 mesh (<45 μm particle size) silicon powder. Typical mechanical properties are presented in Table 5.1 and stress rupture conditions to achieve 1000 hours life are presented in Table 5.2. RBN-101 was a relatively low strength material, with low toughness and large scatter in strength (low Weibull modulus). Failure origins occurred at pores and at metallic (chrome, iron) inclusions. Due to internal oxidation because of the open porosity, RBN-101 also had relatively poor oxidation resistance.

HES&S and HCC improved the RBSN material properties by utilizing a much finer powder (<5 μm dia.) to reduce pore and grain size, magnetic filtering to reduce metallic contamination, improved powder processing to eliminate large voids, and flash oxidation to generate a protective silica oxidation layer on the material. The material improvement efforts were guided by extensive fractography and chemical analysis of the baseline RBN-101 material, supported by the Phase II Ceramic Components for Turbine Engines Program funded by the Air Force [1,2]. The resulting improved RBSN material, RBN-104, was introduced in 1980. Compared to RBN-101, the strength was improved from 257 MPa up to 345 MPa, and the highest

TABLE 5.1 SUMMARY OF PROPERTIES FOR HCC SILICON NITRIDES

	Flexure Strength (MPa)			Fracture	Weibull
	25°C	25°C	1204°C	Toughness	Modulus
Material	Machined	As-Sintered	Machined	(MPa$\sqrt{m}$)	
RBN-101	257	—	320	2.5	6–10
RBN-104	345	—	405	2.5	10–16
CODE 2 (Sintered)	547	—	325	5.3	6–10
CODE 2 (Sinter-HIPed)	670	—	405	6.1	10–15
GN-10	890	450	700	6.0	12–18
GN-10 w/SiC Whiskers	695	—	550	6.9	—
GS-44	930	705	—	8.0	20–35
AS800	760	660	590	8.5	20–35

Weibull modulus value from 10 up to 16 (Table 5.1). Although these were significant improvements, the desired applications for ceramics (turbine hot section nozzles, blades, wheels) still required higher strengths, reliability (Weibull modulus), toughness, durability and oxidation resistance.

Hot pressing of silicon nitride, using small amounts of oxide sintering additives to liquid phase sinter to >99% density, had been developed in the early 1970s and had shown promise as an approach to achieve high strength material with improved oxidation resistance compared to RBSN, but hot pressing obviously had limited shape forming capability. As a result, HCC focused on developing a low pressure sinterable version of hot-pressed silicon nitride in the early 1980s. Liquid phase sintering of silicon nitride required very fine silicon nitride powders (≤1 μm dia.) for reactivity due to the refractory nature of the second phase compositions and resulting low liquid phase activity. These powders began to become available in the early 1980s, coincident with the acceleration of silicon nitride liquid phase sintering development efforts. Through discussions with David Rowcliffe and Paul Jorgenson of Stanford Research Institute (who had performed some of the first work on sintering of silicon nitride with Y_2O_3 and Al_2O_3 additions) and through funding of preliminary research on silicon nitride sintering and microstructure development at the University of California Los Angeles with Ilhan Aksay, HES&S (Richerson [3]) helped define the sintering technology required. Code 2 silicon nitride was developed and utilized 6% Y_2O_3 and 2% Al_2O_3 as the sintering additives, and the new fine size SN-502 silicon nitride powder developed by GTE. Code 2 was sinterable at 0.7 MPa (100 psi) of nitrogen overpressure and 1800°C–2000°C. It was introduced in 1983 and exhibited a significant improvement in strength (547 MPa), compared to RBSN (345 MPa) materials, and an improvement

TABLE 5.2 COMPARISON OF TEMPERATURE AND STRESS CONDITIONS TO ACHIEVE 1000 HOUR LIFE FOR HCC SILICON NITRIDE MATERIALS

Material	Temperature (°C)	STRESS (MPa)
RBN-101	1100	170
CODE 2	1100	240
GN-10	1204	240
	1316	100
AS800	1204	380
	1316	200
	1370	140
AS950EXP	1370	170
	1400	140
	1460	40

in toughness (from 2.5 to 5.3 MPa$\sqrt{m}$). However, the liquid phase sintering process results in extensive shrinkage of the powder compact (40–50 vol%), and made the achievement of near-net shape fabrication much more difficult. In addition, processing of these much finer size powders was much more difficult (including the facts that these first generation silicon nitride powders had undesirable particle morphologies, including whiskers, and contained significant impurities), resulting in strength limiting defects such as voids, agglomerates, and inclusions. The Code 2 composition, selected for sinterability, also resulted in a second phase with relatively poor high temperature strength and stress rupture properties (Table 5.2).

Concurrently, researchers had been developing high pressure sintering techniques to gain the enhanced densification benefits of hot pressing without the shape forming limitations. Hot Isostatic Pressing (HIP) had been extensively used for metal components to reduce internal material flaws—such as pores, by applying very high pressures (up to 207 MPa) as a post-casting operation. For ceramic powder compacts, the technique was refined so that materials were first sintered to closed porosity (~92% density) in a furnace at low pressure (<0.7 MPa), then high pressure could be applied to the part to assist in final densification and void elimination. HCC developed sinter-HIPing of Code 2 silicon nitride and achieved a significant increase in strength and Weibull modulus by maximizing density and reducing pores as failure origins (Table 5.1). However, sinter-HIPed Code 2 still had the high temperature limitations of its second phase chemistry and starting silicon nitride powder impurity levels. More refractory compositions were unable to be sinter-HIPed to full density, so further improvements in densification technology were needed. Richerson and Wimmer from HES&S investigated a potential improvement in HIP processing by using Ta metal or fused silica glass tubes to encapsulate slipcast powder compacts [4]. The HIP work was performed at Battelle Columbus Laboratories. HES&S and HCC were able to demonstrate that RBSN compositions with small amounts of Y_2O_3, unable to be sintered or hot-pressed to full density, could be encapsulation HIPed to full density and achieve high strengths.

Under a NASA sponsored program to improve silicon nitride high temperature properties that ran from 1984 to 1987 [5], HCC evaluated a number of more refractory second phase compositions, using state-of-the-art improved morphology and higher purity silicon nitride powders (from Denka, H.C. Starck, and Ube) to improve silicon nitride properties. As expected, the new compositions proved very difficult to sinter to closed porosity and then to sinter-HIP to full density, so encapsulation-HIPing of powder in metal cans was explored. Encapsulation-HIP was very effective in achieving full density of the experimental compositions. HCC was able to demonstrate a material with significantly improved strength, toughness, Weibull modulus (Table 5.1), and high temperature stress-rupture life (Table 5.2). This material was designated GN-10.

Due to the fact that metal can or glass tube encapsulation HIP is limited to very simple shapes, Richerson [3] had investigated the Asea glass encapsulation-HIP technology developed during the 1970s. In this process, a viscous glass conformable envelope is generated around the ceramic powder compact before densification temperatures are reached. Thus, closed porosity is not required to apply full HIP pressure and there are no powder compact shape limitations. HCC evaluated the glass encapsulation-HIP technology and achieved full density and excellent mechanical properties on complex shaped parts (e.g. turbine wheels), eventually licensing the process from Asea (which later became AC Cerama, which, in turn, was purchased in 1998 by the Saint-Gobain Corporation) and implementing the process for production of GN-10 silicon nitride starting in 1987. Because the pressure can be applied early in the densification step, glass encapsulation HIPing results in significantly lower sintering temperatures (200°C–300°C) and densification times. This also results in a relatively fine microstructure, e.g. GN-10 silicon nitride has an average grain size of ~1 μm. The glass encapsulation process does have the limitation that the molten glass reacts with the silicon nitride powder compact surface, resulting in roughening of the surface and chemical interdiffusion of glass and silicon nitride sintering additives. This can result in a significant decrease in mechanical properties, and in the case of GN-10, decreases the as-HIPed surface strength by 45–50% (Table 5.1) compared to machined surface strengths.

Although the toughness of GN-10 was significantly improved over RBSN and early sintered silicon nitride materials, it was still considered too low, especially in engine applications where impact by foreign object damage (FOD) was a concern, and when compared against the toughness of existing metal materials. As a possible intermediate solution to continuous fiber-reinforced ceramic composites (CFCCs), which in the 1980s had the limitations of expensive and poor quality SiC fibers, undesirable fiber/matrix interactions, difficult formability, and very low strengths, HCC and and Honeywell Laboratories, with support from DOE, explored the use of SiC whiskers to improve the toughness of GN-10 [6]. SiC whiskers were also very expensive, a health hazard due to their small size and high aspect ratios, difficult to process, and also reacted with the silicon nitride second phase during densification. Improvement in toughness was achieved, from 6.0 to 6.9 MPa$\sqrt{m}$, but the small improvement was not worth the costs, health issues, and processing difficulties involved.

In the early 1970s Lange [7] had investigated liquid phase sintering of silicon nitride with oxide additives and had observed the unique grain nucleation and growth mechanisms of the α to β silicon nitride dissolution–reprecipitation mechanism and the fact that the β silicon nitride grains grow in a very columnar and acicular morphology. Honeywell Laboratories scientists and a number of other silicon nitride manufacturers began efforts in the mid-1980s to control and exploit this acicular grain growth to generate isotropically oriented microstructures of β silicon nitride grains with high aspect ratios to increase toughening by generation of tortuous intergranular crack paths, with wake zone phenomena, such as crack bridging, and resistance to grain pull-out imposing further restrictions on crack growth [8]. Honeywell Laboratories was very successful and developed a new class of silicon nitride materials with microstructures that AlliedSignal refers to as in-situ reinforced (ISR). A typical ISR silicon nitride microstructure is shown in Figure 5.1. A unique feature of these materials is that the controlled grain growth of the β grains results in significantly larger grain sizes than previous generation sintered and HIPed silicon nitrides. The much larger average grain size has significant impact on many properties and will be subsequently discussed. To achieve the acicular β silicon nitride grain growth, specific second phase compositions were developed to balance effective liquid phase activity to achieve strong grain growth during sintering, and resulting high temperature strength and stress-rupture life. As a result, gas pressure sintering (sintering below ~14 MPa) was used as the densification process. Glass encapsulation-HIP could not be used, as the higher temperatures required to achieve significant grain growth caused excessive reaction with the glass envelope. Gas pressure sintering (GPS) turned out to have the added benefits of achieving high as-sintered surface strengths and requiring significantly less expensive furnaces than HIPing.

Honeywell Laboratories developed two ISR silicon nitride materials. The first ISR silicon nitride, introduced in 1993 by HCC, was GS-44, designed for lower temperature applications up to 1000°C. GS-44 has a lower temperature second phase composition which is less amorphous. Typical GS-44 properties are shown in Table 5.1. AS800 ISR silicon nitride was developed with a more refractory second phase compo-

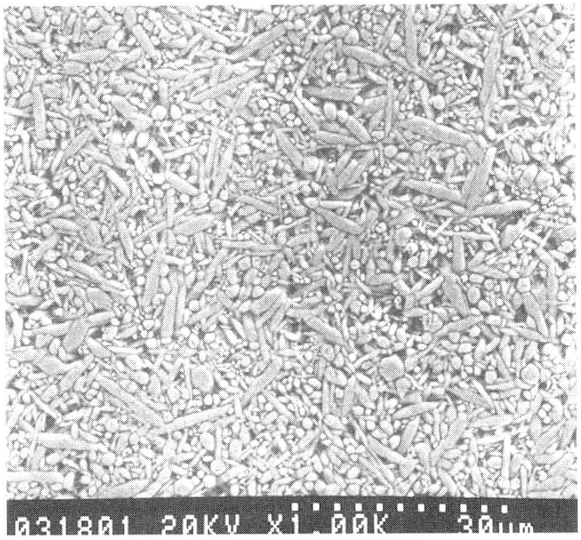

Figure 5.1 In-Situ Reinforced (ISR) Silicon Nitride Microstructure (AS800 Silicon Nitride) (Reprinted from ref. [9])

sition for high temperature applications up to 1370°C and was introduced in 1995. In addition, the AS800 second phase is carefully crystallized to maximize high temperature creep and oxidation resistance. Typical AS800 mechanical properties are shown in Table 5.1 and stress-rupture capability shown in Table 5.2. The ISR interlocked microstructure also improves slow crack growth and creep resistance in addition to improved second phase composition [9], resulting in the significant performance improvement over previous generation sintered and HIPed materials such as Code 2 and GN-10.

The ISR microstructures of GS-44 and AS800 result in a fracture toughness of 8–9 MPa$\sqrt{m}$ and typical Weibull modulus of 20–35. The unique characteristic of these materials is that, unlike earlier generation silicon nitrides with finer (~1 μm) and relatively equiaxed grain size, the relatively large grain size and interlocked acicular grains result in the grains themselves being the failure origins. With careful processing, the grains are approximately all the same size, so that very narrow strength distributions, and thus high Weibull values, are obtained. This result also makes powder processing much more robust in that the typical critical flaw size (large grains) is 75–150 μm, well above typical powder processing flaw sizes of 5–25 μm observed when using submicron silicon nitride powder. In a recent example, buttonhead tensile strength of twenty separate lots (separate powder processing, forming, and sintering runs) of AS800 were measured [10]. The total of 98 specimens exhibited a Weibull modulus of 33 (Figure 5.2).

The large grain size of AS800 silicon nitride, in combination with the rare-earth based composition of its second phase, also results in significant increase in thermal conductivity. AS800 has a room temperature thermal conductivity of 80 W/m.K, whereas conventional silicon nitrides have conductivities ranging from 15 to 30 W/m.K. The combination of very high thermal conductivity and flaw tolerance gives AS800 excel-

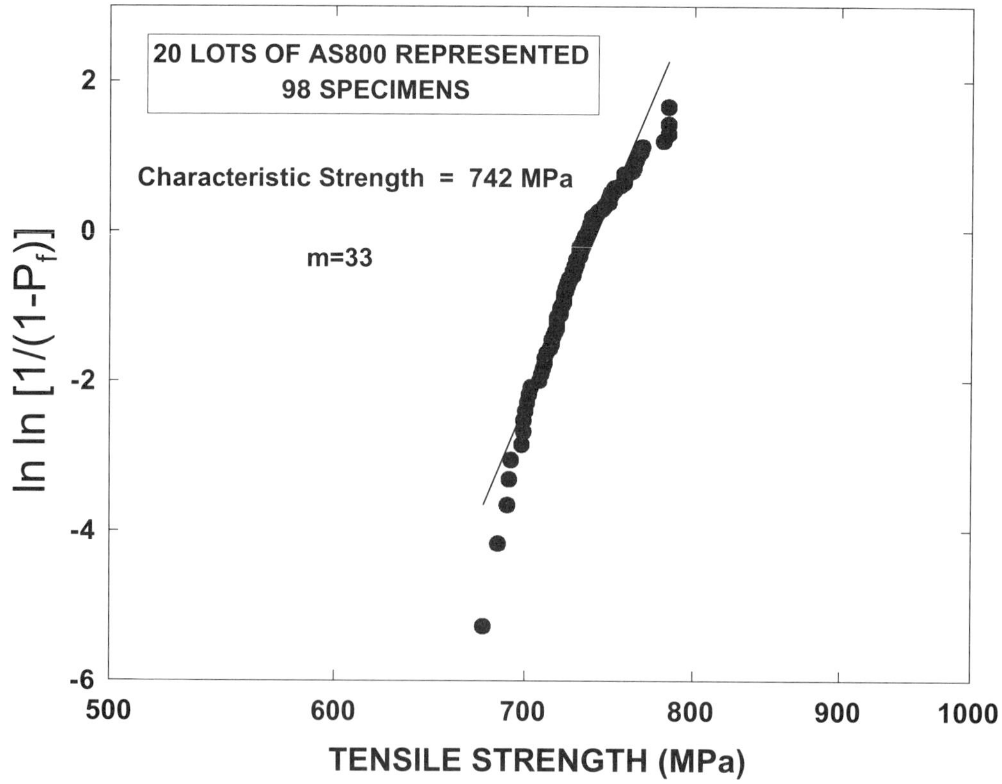

Figure 5.2 Reproducibility of AS800 Silicon Nitride Tensile Strength

lent thermal shock resistance. Tests at NASA Glenn Research Center with rocket engine exhaust impingement have shown that AS800 can withstand repeated 1700°C/s upshocks [11] and conventional thermal shock quench testing has shown ability to survive greater than 1200°C downshocks without strength loss.

The ISR microstructures can be controllably manipulated through careful control of the solution-reprecipitation and grain growth processes during sintering. In fact Honeywell Laboratories and HCC have been able to demonstrate ISR materials with toughnesses up to 12 MPa$\sqrt{m}$ and thermal conductivities up to 90 W/m.K, by significantly increasing grain size. Honeywell Laboratories and HCC have observed the trade-off that, with microstructures that exhibit toughnesses above ~9.0 MPa$\sqrt{m}$, strength starts to decrease due to the significantly larger grains that subsequently become failure origins. ASCC balances strength and toughness of its ISR materials to maximize both, thus GS-44 and AS800 silicon nitride both have been designed with microstructures that exhibit toughnesses in the 8–9 MPa$\sqrt{m}$ range.

Recently, a number of advanced gas turbine applications, such as the Defense Advanced Research Project Agency's (DARPA) Integrated High Performance Turbine Engine Technology (IHPTET) program, have identified the need for a higher temperature capable material for application in hot section components. Honeywell has further refined the AS800 second phase composition to extend its temperature capability and oxidation resistance to 1400°C. AS950EXP is an experimental material now being refined and characterized. The development of AS950EXP focused on improving creep resistance at 1400°C to achieve creep rates no greater than 10^{-10} s^{-1} at continuous stress loads of 70–100 MPa (typical of hot section 1st and 2nd stage turbine blades operating in advanced propulsion and small power generation gas turbines). The creep rate for AS950EXP at 1400°C has achieved the desired creep/stress target range and is approximately one order of magnitude improved over AS800 at 1400°C (Figure 5.3). Corresponding tensile stress rupture data (Figure

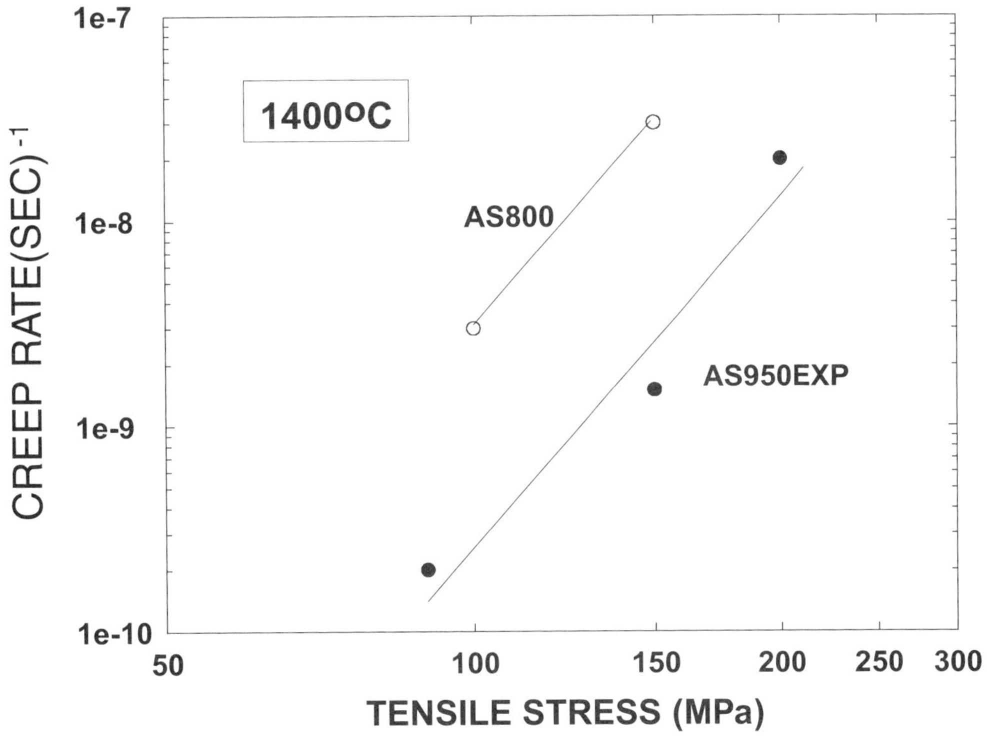

Figure 5.3 Comparison of AS800 and AS950EXP Tensile Creep Rate at 1400°C (Reprinted from ref. [10])

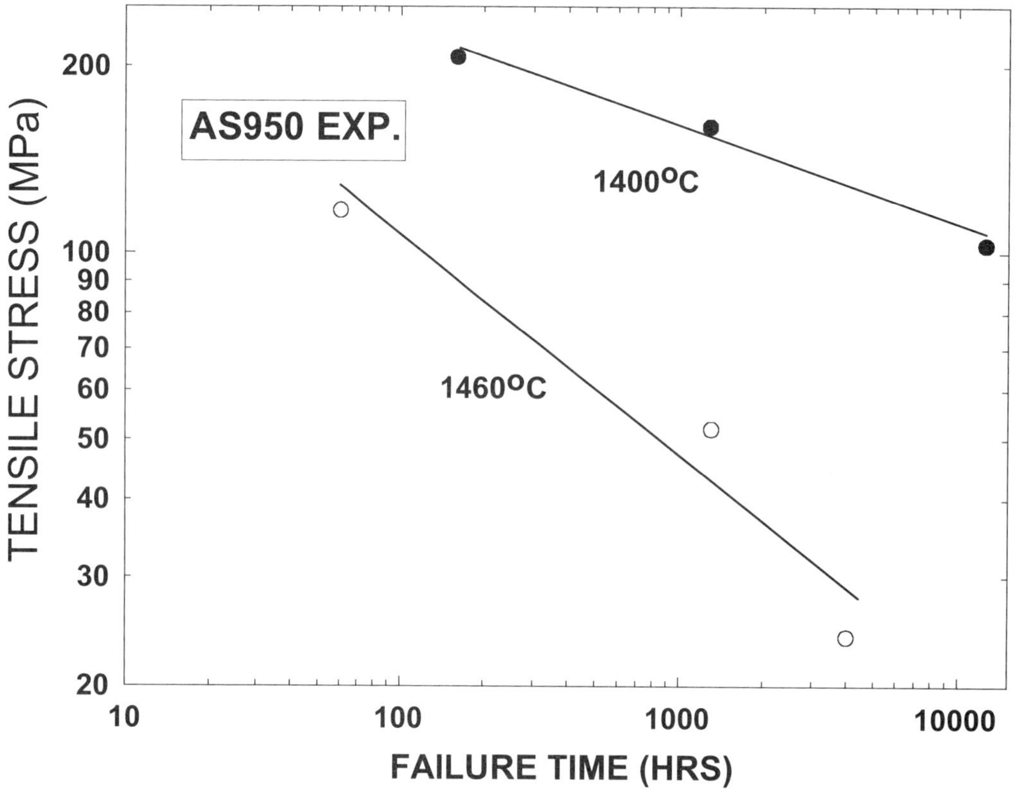

Figure 5.4 Tensile Stress Rupture Properties of AS950EXP Silicon Nitride (Reprinted from ref. [10])

5.4) indicates that suitable lifetimes can be achieved for stress levels of <150 MPa. Honeywell Laboratories and HCC believe that composition refinements to improve silicon nitride use temperatures are reaching their limits and thus do not expect to see further significant improvements in silicon nitride compositions to push uncooled use temperatures above 1400°C.

Over the past several years, two significant challenges have become apparent for silicon nitride applications in gas turbine hot sections. One is the volatilization of the protective silica oxidation layer on silicon nitride at high pressure, high water vapor content, high velocity, high temperature conditions, and the other is the issue of impact sensitivity due to foreign or domestic object damage (FOD/DOD). As progress has been made in the development and maturation of monolithic and composite non-oxide ceramic materials and component fabrication processes, a series of investigations at NASA [12] and Oak Ridge National Laboratory (ORNL) [13] have explored the durability of these materials in test environments simulating the combustion environment of steady-state medium and larger gas turbines. Previously, most environmental testing of these materials has been in either static air or simulated combustion environments at atmospheric pressure and at relatively low velocities. Under these conditions, silicon nitride materials generate a self-protecting passivating layer of silica, with higher temperatures accelerating the oxidation process and oxidized layer thickness. For a number of lower temperature (<1150°C) applications in relatively low pressure and low velocity environments and cyclic applications such as aircraft APUs and small automotive propulsion and turbogenerator engines, these tests adequately represented the material performance. Medium (>1 MW) and larger gas turbines continuously operate at much higher pressures and gas flow velocities, with significant amounts of water vapor, and at significantly higher temperatures (1200°C and higher in many cases).

Evaluations of silicon based non-oxide materials in test rigs and in actual engine tests under high pressure, high velocity, high water vapor, and high temperature conditions have demonstrated that the protective silica layer is volatilized, resulting in recession of material from the surface. This mechanism can result in aggressive surface recession and is a significant issue limiting application of silicon based non-oxide structural ceramics in gas turbines operating under these conditions. As an example, Rolls-Royce Allison completed an 800 hour field test of their 501 industrial gas turbine engine in 1999, retrofit with AS800 silicon nitride 1st stage nozzles. The 501 engine nozzles (Figure 5.5) performed well with no performance issues during the test, but inspections during and at the end of the 800 hours of testing showed that the nozzles were experiencing significant surface recession, up to 30–40% of trailing edge thickness loss after 800 hours [14]. Analyses of exposed nozzles at ORNL show that the recession is very much a surface effect, with mechanical properties and microstructure of the remaining bulk material in the airfoil sections equivalent to starting properties [15]. The amount of surface recession is much greater than allowable for desired component lifetimes for industrial power generation turbine applications (>10,000 hours) and thus a solution to the volatilization process is required.

A number of organizations are investigating possible solutions to silica volatilization of monolithic silicon nitride and silicon carbide based continuous fiber composites, including Honeywell. The primary path being pursued is to place a protective environmental barrier coating (EBC) on the ceramic to prevent oxygen and water vapor reaction with the silicon in the ceramic. To be effective and durable, the coatings must: a) be an oxygen/water vapor barrier, b) closely match the thermal expansion of the ceramic, c) have relatively low modulus to prevent spalling during cyclic loading, d) be oxidation resistant, e) non-silicon containing or self-healing to minimize volatilization, and f) be non-reactive with the ceramic base material at desired operating temperatures. The coatings must be able to be applied effectively to bond to the ceramic and are obviously desired to be low cost. Mullite [16] and a proprietary coating developed under the NASA Enabling Propulsion Materials (EPM) program [17] are currently being evaluated on both composite and silicon nitride monolithic materials and show some promise. Honeywell is developing an oxide coating system that does not contain silicon, and which is currently under evaluation at ORNL in their high-pressure high-water-vapor test rig on

Figure 5.5 AS800 Silicon Nitride 1st Stage Nozzles For Rolls-Royce Allison 501 Gas Turbine Engine (Reprinted from ref. [10])

AS800 silicon nitride test specimens. In addition, HCC has fabricated more AS800 501 engine nozzles, which were coated with the Honeywell oxide EBC, and were scheduled to be evaluated by Rolls Royce Allison in their field test engine in 2001 to help determine potential effectiveness of the EBC coating.

A number of applications of silicon nitride components have seen minimal or no FOD/DOD issues due to specific component robustness, design for impact resistance, cleanliness of engine, small size, and low rotational speeds. Larger gas turbines have less flexibility and catastrophic damage to ceramic parts has occurred, an example being the Ceramic Stationary Gas Turbine (CSGT) program Solar Centaur 50 engine with ceramic blades, where FOD events have resulted in catastrophic failure of the entire set of silicon nitride 1st stage blades [18]. Continuous fiber composite ceramics (e.g. SiC fiber reinforced SiC) have shown significant damage tolerance capability in preliminary impact testing [19], but monolithic silicon nitride is required for the blades due to the high stresses and temperatures of this application. Although minimization of FOD/DOD is primarily an engine and component design issue, efforts are also ongoing to improve the damage tolerance of monolithic silicon nitride. HCC is exploring a number of refinements to ISR silicon nitride to improve impact resistance. Impact resistance improvement from the material standpoint can be addressed by strength improvements and fracture toughness improvements. HCC has determined that there is no obvious way to significantly improve material strength without compromising ISR properties in silicon nitride materials, but is exploring a number of routes to improve toughness, including hierarchical reinforcement structures such as fibrous monoliths using ISR silicon nitride materials, continuous fiber composite/monolithic hybrids, and compressive coatings.

COMPONENT FABRICATION PROCESSES

As described above, HCC began its development of silicon nitride in 1976 by fabricating RBSN. The starting raw material for RBSN is silicon powder, which is then formed into a powder compact of desired shape. A key benefit of the reaction bonding process is that the nitridation of silicon takes place with less than 0.5% shrinkage, thus the starting powder compact shape and size does not change during nitridation, so that forming of RBSN can be a near-net-shape process. The silicon powder used for forming of RBN-101 was –325 mesh (<45 μm particle size). This is relatively coarse and low surface area powder and thus easy to form into powder compacts by traditional slipcasting techniques using aqueous slurries of powders and casting in porous plaster molds. One difficulty with silicon powder is that it reacts with water to form hydrogen gas, thus silicon powder slurries had to be carefully prepared to allow for outgassing and stabilization of powder surface chemistry to achieve stable slurries. HCC also used a variant of the lost core wax mold process, first demonstrated by Ford, to cast complex shape parts for applications where thin shapes and small features were required (such as turbine nozzles) but very difficult to cast and remove using plaster molds while achieving tight dimensional reproducibility. In the lost core technique, water-soluble wax is formed into the shape of the desired final component, usually by injection molding. The wax part (or core) is then dip coated with a water-insoluble wax (but soluble in an organic solvent). The water-soluble wax core is dissolved out in water, leaving a mold in the desired near-net shape. The mold has two sets of openings, one opening that is contacted to plaster to draw out water, and a second opening for filling with slurry. After casting, the wax mold is removed by immersion in organic solvent. This process was effectively developed and demonstrated for fabrication of the turbine nozzles, shown in Figure 5.6, under the DARPA-Navy funded Ceramic Gas Turbine Engine Demonstration (CGTED) program at HES&S for the TSE331C-1 engine in 1979 [20,21].

In addition to slipcasting, injection molding of silicon powder using wax binders was developed for RBSN applications starting in 1978. In this process, approximately 45 vol% wax was melt mixed with silicon powder and then pelletized into feedstock for use in screw injection molders with hardened metal screws and barrels. In the case of RBN-101, the coarse silicon powder used was able to be effectively molded and the binder subsequently removed by pyrolysis due to the large particle sizes and resulting powder compact large pore sizes. HCC was also able to demonstrate and deliver engine quality RBSN nozzles for the TSE331C-1 engine made by wax injection molding in 1979 [20,21].

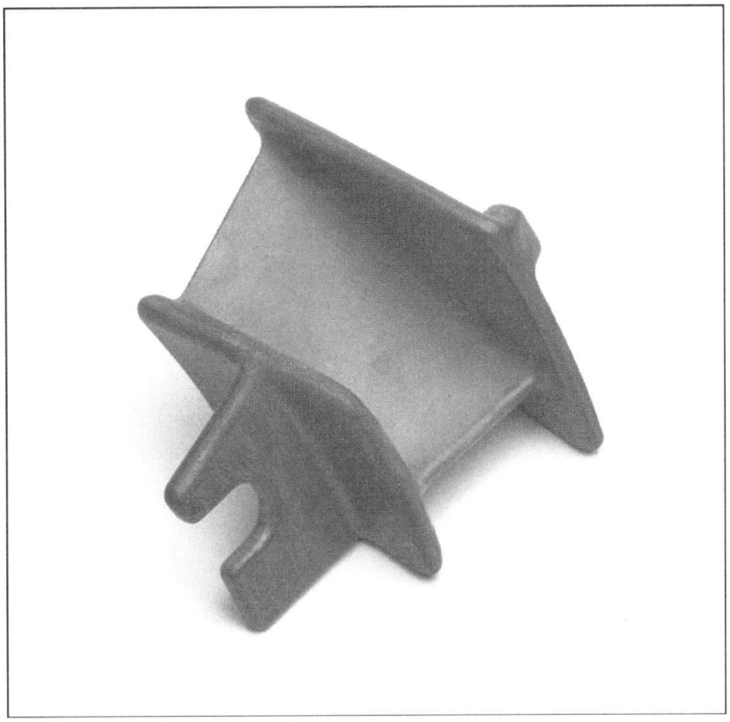

Figure 5.6 RBSN Nozzles for Honeywell TSE331C-1 Engine

When HCC developed RBN-104 with a much finer particle size (<5 μm) silicon powder to improve mechanical properties, powder processing and forming became more difficult. For slipcasting, the higher surface area silicon powder was more reactive in water, requiring up to a week to stabilize slurry surface chemistry before use. The finer powders were also more difficult to pack uniformly due to higher particle surface area and resulting increased colloidal forces. For injection molding, RBN-104 had a less desirable molding rheology and it was more difficult to pyrolyze the wax binder without cracking or bloating due to the smaller pore sizes in the powder compact.

With the advent of commercially available silicon nitride powders and liquid phase sintering concepts in the early 1980s, HCC transitioned to the development and forming of materials such as Code 2 silicon nitride. Code 2 used silicon nitride powders such as GTE's SN-502, which was approximately 1 μm in avg. particle size, and the additions of similar sized oxide sintering additives. The use of much finer powders and the fact that the liquid phase sintering process results in extensive shrinkage were new processing challenges. The primary forming processes developed for Code 2 were slipcasting and isopressing. Although much more stable in water than silicon powder, the silicon nitride and oxide powders still hydroxylate, requiring aging times of up to 24 hours to achieve slurries with stable desired rheologies. The slipcasting process allowed for achievement of relatively high powder packing densities, ~55% to 60 vol%, but the fine powders resulted in long casting times due to the small powder compact pore channels for dewatering, and in undesirable powder packing density gradients, which result in part distortion and potential cracking upon subsequent densification. The powder compacts shrink 40–45 vol% during liquid phase sintering, which requires careful determination of shrinkage factors to fabricate correct size tooling, and robust reproducible powder packing during casting to achieve consistent shrinkage. By 1984, HCC was able to fabricate components such as the AGT101 radial turbine wheel (Figure 5.7) by slipcasting of Code 2, but casting times were 60 hours and yields were low due to powder processing flaws.

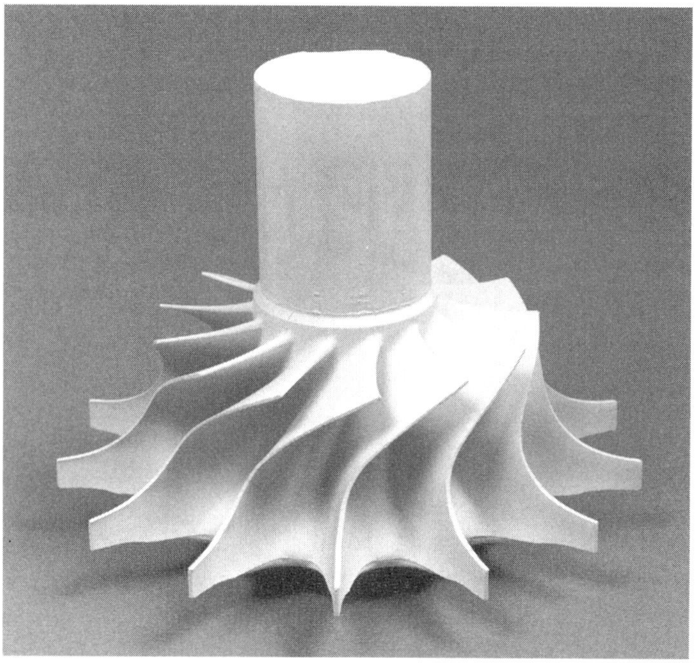

Figure 5.7 Code 2 Silicon Nitride Radial Design Turbine Wheel

Isopressing, another traditional ceramic powder compact forming process, was also developed and demonstrated for Code 2 silicon nitride simple shapes and for stock where extensive final grinding was planned to achieve final part geometries. In the isopressing process, the silicon nitride/sintering additive slurry is combined with a small amount of wax binder and then spray dried into large free flowing granules. The spray dried powder is then placed into flexible bag molds which are isostatically pressed to compact the granules together. The small amount of wax binder (2–4 wt%) holds the pressed part together. The small amount of wax is readily removed by subsequent pyrolysis through the powder compact open pore network. The isopressing process was used to make a number of simple Code 2 shapes such as tubes and rings.

Extensive efforts were made at HCC to develop wax injection molding of Code 2 silicon nitride in the early 1980s, including a significant iterative process development effort led by HES&S that was funded by the Army to develop silicon nitride turbocharger wheels by injection molding [22]. The small size of the starting powders made the molding and pyrolysis process much more difficult. Approximately 50 vol% wax binder was required to achieve moldable material rheology. The high binder content and small pore sizes made it extremely difficult to pyrolyze the binder without cracking or bloating of parts. In addition, the low powder packing densities resulted in difficulty in sintering injection molded Code 2 and the extensive shrinkage resulted in part distortion. The wax injection molding technology was further investigated throughout the 1980s by HCC, but the problems of low packing densities and pyrolysis, due to the need to use micron and submicron sized powders for Code 2 and GN-10 silicon nitride, were not overcome.

As GN-10 silicon nitride was being developed for applications such as turbine wheels and nozzles in the mid-to-late 1980s, it became obvious that improvements in slipcasting were needed to address powder packing nonuniformity and long casting times, especially with the fact the GN-10 used an even finer silicon nitride powder (~0.7 μm avg. dia.) than Code 2. HCC developed and demonstrated pressure slipcasting technology, where the slipcasting mold is enclosed and pressurized with 276–828 kPa (40–120 psi) of air such that the pressure is exerted on the slurry to help force the water component through the plaster dewatering surfaces. The pressure casting process resulted in a significant reduction in powder packing density gradients—from 3.5–4.0% down to 0.5–1.5% in large components such as the AGT101 turbine wheel, and also significantly reduced casting times (e.g. from 60 hrs to 8 hrs for the AGT101 wheel). In many cases, the lost wax mold technology described earlier was combined with pressure slipcasting to form complex near-net-shape parts such as the AGT101 impact resistant turbine (IRT) wheels, made by pressure slipcasting of GN-10 (Figure 5.8) and delivered to HES&S in 1991.

In the late 1980s, the improvement in silicon nitride materials, such as GN-10 and GS-44, led to the desire by a number of engine companies to evaluate prototype quantities of components. The immaturity, potential tooling costs, and long tooling lead times of the net-shape technologies such as pressure slipcasting using wax core technology and wax injection molding led HCC to develop a more rapid prototype component fabrication process—bisque machining, that utilized simple shape starting stock and did not require complex shape mold tooling. Bisque (or presinter) machining of simple shaped stock is the process where a ceramic powder compact of simple geometry is partially heat treated to provide enough strength for machining by typical turning, milling, and rapid grinding techniques, but not densified enough to require diamond grinding, as is needed for fully densified material. The bisque firing process results in an increase of only ~0.5–1.0% density, allowing carbide and other relatively low cost cutting tools to be used. HCC developed and demonstrated bisque machining of simple shaped slipcast stock to fabricate development quantities of complex shaped components. The bisque machining process proved very effective for fabrication of engine quality blades and nozzles, where the airfoils are bisque machined to net-shape and only tight tolerance attachment features such as dovetails, platform edges, and features such as blade tips are final machined. Examples include the Solar Turbines Centaur 50 1st stage turbine blade of AS800 silicon nitride (Figure 5.9) fabricated for Solar under the DOE Ceramics for Stationary Gas Turbines (CSGT) program. HCC also used this process to fabricate low volumes (500–1000/month) of simple shaped silicon nitride production turbine engine seal components by isopressing simple shaped tubes and bisque machining seal blanks from them. The bisque machining process

Figure 5.8 GN-10 Silicon Nitride Impact Resistant Design AGT101 Wheels for Honeywell Engines, Systems & Services

Figure 5.9 1st Stage Turbine Blades (Slipcast AS800 Silicon Nitride) for Solar Centaur 50 CSGT Industrial Turbine Engine (Reprinted from ref. [10])

is most appropriate for these parts since the relatively low volume (<1000 month) and multiple parts with slightly different sizes did not justify investment in near-net-shape forming tooling, and because all part surfaces must be final machined to meet tight tolerances (many at +/– 0.005 mm).

Although the bisque machined complex shaped components have performed well as engine quality hardware, this fabrication process has been demonstrated to be only capable of relatively low volumes and is high cost for complex shapes due to extensive bisque and final machining time using CNC machining. In general, turbine engine companies are currently not willing to pay a higher price for ceramics over metal superalloy components, even though in a number of cases the ceramics have demonstrated potential lifetimes 2× to 10× that of the equivalent metal components. That is, life cycle cost is currently not a consideration in most cases for the purchase of ceramic components for gas turbine applications. Currently, bisque machined complex shaped ceramic components (e.g. nozzles, blades) in volumes of 100s to the low 1000s per year cost 2× to 5× that of superalloy components. This has very much limited the implementation of ceramic components even where performance, durability, and life cycle cost advantages have been demonstrated. Thus, there still exists a strong need to have viable high-yield near-net-shape forming processes to minimize machining costs and meet production volume and cost target requirements.

In 1989, HCC began investigating the gelcasting powder compact forming technology developed by Oak Ridge National Laboratory [23] through a Cooperative Research and Development Agreement (CRADA) with ORNL. The gelcasting process exhibited excellent potential as a viable near-net-shape fabrication process for production of turbomachinery silicon nitride nozzles, blades, wheels, and thin-walled static components such as combustors and scrolls. The gelcasting process utilizes the concept of adding a small amount of organic monomer and crosslinker to the silicon nitride aqueous slurry. Upon heating, the monomer polymerizes and the resulting gel (which is ~90% water) rigidizes the silicon nitride slurry into the shape of the mold. Gelation times are as short as 15 minutes for turbine wheels. The gelled powder compact is then easily dried to remove the water, leaving a powder compact with an open pore network and 2–4 wt% dried gel. The gel is then easily removed by a short pyrolysis cycle. The gelcasting process demonstrated numerous advantages over pressure slipcasting as a net-shape forming process, including: being 50%–80% faster than slipcasting, generating more uniform powder packing, having a much higher part strength after forming and after drying, and requiring simpler molds (no need for dewatering surfaces). Most importantly, the gelcasting process is much more amenable to automation than pressure slipcasting of complex shapes. But the ORNL CRADA demonstrated that the gelcasting technology needed significant improvements to become viable for silicon nitride applications, including development of a low toxicity monomer, improvements in gel pyrolysis technology, and significant improvements in submicron powder dispersion.

HCC worked with ORNL from 1992 to 1995 to improve and refine the gelcasting process for silicon nitride under an Advanced Technology Program (ATP) sponsored by the National Institute of Standards and Technology (NIST). The program helped develop improved gel systems including a low toxicity monomer that pyrolyzed much more readily [24], and improved colloidal processing such that the submicron silicon nitride powders could be effectively used [25]. The improved gelcasting technology demonstrated great flexibility in silicon nitride shape forming capability, ranging from small thin-cross-section parts such as a commercial aircraft APU nozzle (Figure 5.10), thicker cross-section parts such as turbine wheels (Figure 5.11), and large thin walled shapes such as combustors (Figure 5.12). The gelcasting process has shown the limitation of low packing densities when used with fine (<1 μm powders). The gelcasting slurries must have low enough viscosity to be able to effectively fill molds, which in the example of silicon nitrides such as GS-44 and AS800, results in slurry solids contents of 45–50 vol%. Since the gelation process "freezes" in the slurry solids content, the powder compact packing densities range from 48– 53 vol% (2–3% shrinkage occurs during drying and pyrolysis). These relatively low packing densities can inhibit densification and the extensive shrinkage can cause part distortion.

DARPA and DOE have supported HCC in automating the gelcasting process to fabricate complex shaped components such as turbine wheels to meet expected cost targets and required production volumes for potential applications such as missile turbojet engines and microturbines for power generation. The result is the development of a first generation automated gelcasting system (shown in Figure 5.13) designed to fabricate 10,000 silicon nitride near-net-shape turbine wheels per year [26]. The Teledyne M304 turbojet wheel (shown in Figure 5.14) was the demonstration component. In 1999, the automatic gelcasting system demonstrated the capability of fabricating 4 parts per hour, meeting the 10,000/yr production rate.

Figure 5.10 Honeywell Engines, Systems & Services Series 85 APU Turbine Nozzle (Gelcast AS800 Silicon Nitride)

Figure 5.11 Honeywell Air Turbine Starter Wheels (Gelcast GS-44 Silicon Nitride) (Reprinted from ref. [10])

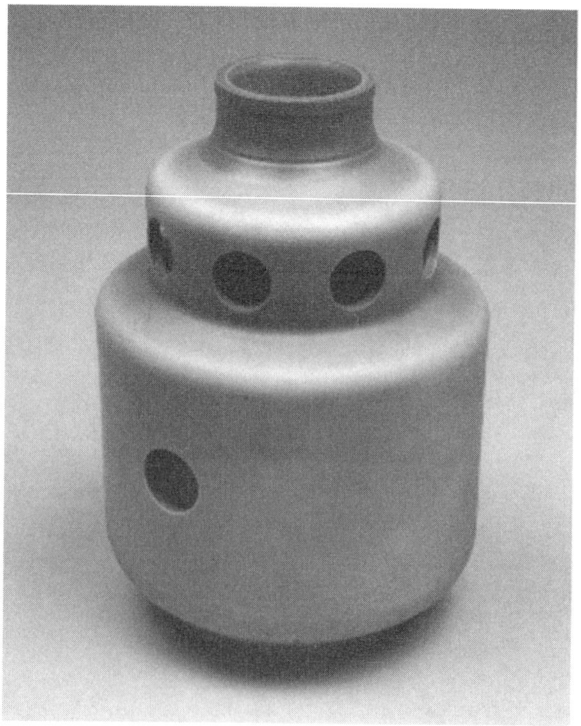

Figure 5.12 Gelcast AS800 Silicon Nitride Combustor for Rolls-Royce Allison AGT-5 Hybrid Vehicle Gas Turbine Engine

Figure 5.13 Automated Gelcasting System for Silicon Nitride Turbine Wheels (Reprinted from ref. [10])

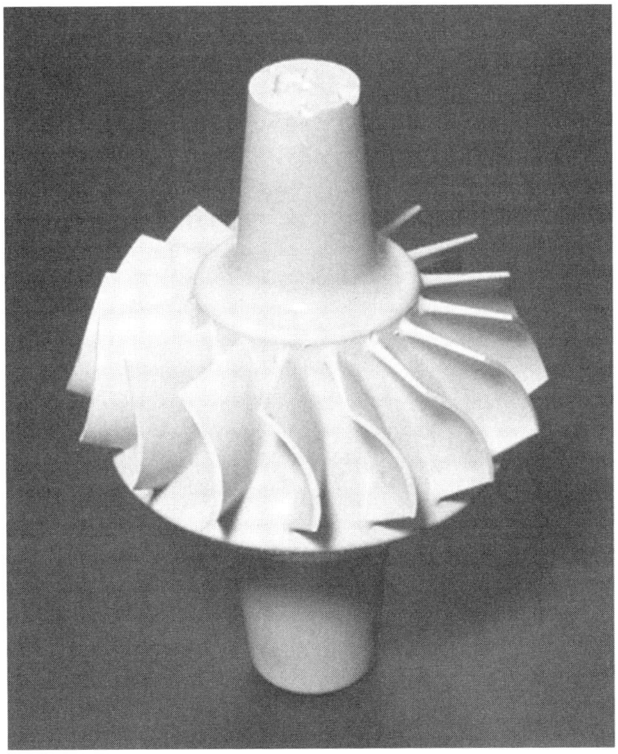

Figure 5.14 Teledyne M304 Turbine Wheel (Gelcast AS800 Silicon Nitride) (Reprinted from ref. [10])

The gelcasting process has also demonstrated the ability to effectively form large cross-section and size silicon nitride powder compacts, leading to recent evaluations of the ability to fabricate larger components, including the 190 mm dia. HES&S 331-200 APU turbine wheel previously formed from slipcast blanks, with support from DOE. Gelcast AS800 331-200 spin disks (Figure 5.15) were spun-to-burst in 1999, demonstrating equivalent performance to slipcast spin disks and wheels. Recently, further investigations have demonstrated the ability to gelcast 400 mm dia. AS800 silicon nitride blisk blanks weighing 20 kg, for potential integral blisk applications for advanced unmanned aircraft propulsion engines and turbopump applications [10].

In the late 1980s, Honeywell Laboratories developed a rapid and automatable near-net-shape forming technology based on aqueous injection molding (AIM) of ceramic material feedstock [27]. The feedstock is made moldable by preparing an aqueous mixture of the ceramic powder and agar—a polysaccharide derivative. The agar dissolves in the water at slightly elevated temperatures and results in a viscous but fluid material that is readily injectable into molds using standard thermoplastic injection molding equipment. The agar regells in the mold upon cooling, utilizing water as part of its gel structure. Molding cycle times of 5 minutes have been demonstrated for components such as small turbine blades and nozzles. AIM is a faster gelation process than gelcasting, but is more limited in thicknesses and sizes of parts that can be formed because it is a viscous filling process. Again, because water is ~90% of the gel structure, the agar/water binder is readily removed by first drying the part and then pyrolyzing the 2–4 wt% of remaining agar. HCC attempted to develop the AIM technology for silicon nitride under an ATP program from NIST that ran from 1994 to 1997. As in gelcasting, the submicron silicon nitride powder significantly limited the rheology of the AIM feedstock, requiring relatively low solids loadings (45–50 vol%) to achieve moldable feedstock. Because of the relatively weak gel strength of agar, silicon nitride parts would crack and distort during drying and binder pyrolysis due to extensive shrinkage because of low packing densities. Honeywell was very successful in applying AIM to coarser (>1 μm) ceramic and metal powder applications, such as alumina, zirconia, and stainless steel, and

Figure 5.15 Comparison of Net-Shape Gelcast 331-200 Spin Disk and Bisque Machined Slipcast Blisk of AS800 Silicon Nitride (Reprinted from ref. [10])

started a new business named at this time as AlliedSignal PowderFlo™ for fabrication of these materials by AIM in 1998. However, the requirement for submicron silicon nitride powder for GS-44 and AS800 silicon nitride has limited the application of the AIM technology to gas turbine components, until improvements in colloidal processing to increase effective powder loading and packing can be developed.

HCC is also evaluating solid freeform fabrication (SFF) techniques to address the needs to: 1) rapidly make functional near-net-shape prototype parts without expensive tooling or machining, 2) fabricate even more complex structures such as complex cooling passages in small turbine blades, and 3) build multiple materials and hybrid microstructures in a single component. HCC is working with Stanford Research Institute (SRI) to evaluate their Direct Photo Shaping (DPS) SFF technology for the Rolls Royce Allison 501 engine 1st stage nozzle, funded by DARPA. The DPS process combines photopolymerizable preceramic polymer with ceramic powder into a slurry, which is tapecast into layers and polymerized by projecting light in desired patterns onto each layer in sequence. The AS800 made by DPS has been demonstrated to have equivalent mechanical properties to benchmark slipcast AS800. The DPS nozzles exhibit very good dimensional control, and process refinements are now addressing the different scaling factors unique to DPS to generate parts meeting nozzle dimensional requirements. HCC is also working with Honeywell Laboratories and Rutgers University under DARPA funding to develop Fused Deposition of Ceramics (FDC) to fabricate silicon nitride turbine blades for HES&S advanced propulsion engines that contain complex cooling passages needed for ultra high operating temperatures (>1400°C). Additionally, the FDC process is being developed to generate hybrid microstructures/materials in single components. Examples include placing different modulus materials (e.g. silicon nitride and silicon carbide) strategically within the component to reduce stresses due to thermal gradients or mechanical loads. Fibrous monolith structures are also being developed for fabrication by FDC to potentially generate hybrid properties between monolithic and continuous fiber composite ceramics such as improved damage tolerance with high strength retention.

APPLICATIONS

Silicon nitride materials and components have been evaluated and produced by HCC for a variety of turbomachine applications including aircraft propulsion engines, APUs, starters, automotive direct and hybrid (turbogenerator) propulsion, small/medium size industrial turbines for power generation, and microturbines for hybrid propulsion and distributed power. The first component application at HCC was RBN-101 RBSN nozzles (Figure 5.6) for the DARPA-Navy CGTED program at HES&S, running from 1976 to 1981, [21,28] to retrofit an existing turboprop engine—the Garrett T76, redesignated the TSE331C-1 with ceramic components. The engine evaluation incorporated RBSN nozzles and shrouds, and hot pressed silicon nitride blades, and increased the combustion temperature from 1000°C to 1200°C. HCC successfully fabricated near-net-shape nozzles by both slipcasting and wax injection molding. The RBSN nozzles performed well, although significant limitations of the RBSN material were revealed—the low strength and toughness resulted in contact loading damage and high oxidation rates and poor stress-rupture properties placed limits on the ultimate life of the components.

In 1979, DOE initiated a program to develop direct propulsion automotive gas turbines. The Advanced Gas Turbine (AGT) program, managed for DOE by NASA, supported development of engines at Detroit Diesel Allison and at HES&S. The HES&S engine—the AGT101, was a derivative of the earlier Ford AGT820 gas turbine and was a unique single shaft radial turbine design. As a subcontractor to HES&S, HCC developed and demonstrated the fabrication of a number of components over the period from 1979 to 1984. A number of static structure components such as shrouds and ducts were made by slipcasting of RBN-104 RBSN. In addition, the individual RBN-104 turbine nozzles were successfully fabricated by wax injection molding. Fabrication of the radial turbine wheel marked the first application of sintered silicon nitride for HCC. Slipcasting of Code 2 silicon nitride was developed and demonstrated for the radial turbine wheel (shown in Figure 5.7) with the casting process taking 60 hours. Wheels were successfully proof-spun to 115,000 RPM but the one wheel evaluated failed in engine testing, indicating that both better material properties (strength) and improved forming (elimination of processing flaws) were required.

In 1984, the AGT program was refined to focus on key engine design improvements and on improving ceramic materials and fabrication processes to achieve components meeting engine requirements. The new program, running from 1984 to 1993, was designated the Advanced Turbine Technology Applications Project (ATTAP). HCC again supported HES&S on the AGT101 engine and also supported Rolls-Royce Allison on the AGT-5 automotive gas turbine engine. HCC focused on improving the fabrication of the AGT101 wheel for HES&S, developing pressure slipcasting, and using GN-10 glass-HIPed silicon nitride. HES&S also redesigned the AGT101 wheel for improved impact resistance. HCC successfully fabricated a number of the redesigned AGT101 wheels (Figure 5.8) which were successfully proof-spun and received limited rig and engine testing. HCC also developed a fabrication process for the AGT-5 hot section axial wheel for Allison, once again using pressure slipcasting of GN-10 silicon nitride. The axial wheel proved a more difficult fabrication challenge for the pressure slipcasting process due to the thinner sections and more complex geometry. HCC delivered prototype GN-10 wheels that did not meet engine quality by the end of that effort in 1993.

In the early 1990s, interest shifted from direct gas turbine propulsion of vehicles to hybrid propulsion where small gas turbogenerators would be used in conjunction with electric propulsion and battery energy storage. HES&S and DOE revised the program in 1993, retitled the Ceramic Turbine Engine Demonstration Project (CTEDP) to utilize the 331-200 commercial aircraft APU as a test bed to develop and demonstrate technologies for ceramic applications in small turbogenerators for hybrid vehicle applications. HES&S focused on replacing the hot section 1st stage nozzles and wheel with silicon nitride components, and HCC was tasked with fabricating these components. HCC developed and fabricated 331-200 blades and nozzles (Figure 5.16) for HES&S from 1994 to 1996 by bisque machining simple shaped AS800 silicon nitride slipcast stock. In subsequent engine testing, the nozzles performed very effectively in the 331-200 test engines with over 2500 hours on a single set and over 5100 hours on multiple sets. In early 2000, two Airbus commercial aircraft 331-200 APUs were equipped with AS800 nozzles by HES&S for field testing and are currently in passenger service. The AS800 blades were successfully engine tested to over 1080 hours with the development of effective dovetail and compliant layer technology by HES&S. Because of the cost

Figure 5.16 HES&S 331-200 APU Turbine Nozzle and Blade (Slipcast AS800 Silicon Nitride)

(ceramic and metallic hardware) and complexity of mounting 23 individual ceramic 331-200 blades (Figure 5.17), HES&S designed a monolithic ceramic turbine wheel for evaluation. HCC developed and fabricated a number of these wheels by bisque machining of simple shaped slipcast disks of AS800 in 1996 and 1997 (Figure 5.18). The blisks were tested in a 331-200 test engine and performed effectively again—over 750 hours as the only ceramic component, and over 300 hours with the AS800 nozzles in combination. To further evaluate the performance of the AS800 331-200 wheel, HES&S has embarked on a 10,000+ hour field test with the industrial version of the 331-200 APU, in an application where volatile organic compounds are burned off [29]. By May of 2003 in excess of 10,000 test hours had accumulated on a ceramic blisk in this field test.

Rolls-Royce Allison also revised its ATTAP program with DOE in 1993 to address hybrid vehicle propulsion. Thc program was retitled the Hybrid Vehicle Turbine Engine Technology Support Program (HVTE-TS). HCC supported Rolls-Royce Allison by developing fabrication of gelcast AS800 AGT-5 turbine wheels for evaluation in the AGT-5 turbine engine technology test bed, and developed and demonstrated the gelcasting fabrication of near-net-shape AS800 Si_3N_4 AGT-5 testbed combustors. In 1996, HCC fabricated and delivered prototype gelcast near-net-shape AGT-5 wheels and utilized a combination of near-net shape metal mold tooling and wax core technology to fabricate and deliver the AS800 gelcast combustor shown in Figure 5.12.

HCC also developed and demonstrated gelcasting of near-net-shape GS-44 silicon nitride impulse turbine wheels for commercial and military aircraft air turbine starters for HES&S—Tempe under NASA and DARPA funding support (1993–1999). The silicon nitride starter wheels, shown in Figure 5.11, allow reduced weight containment structures, and improved erosion and corrosion durability for cartridge starter applications. The gelcast wheels are currently in testing in starter units and have successfully completed over 1700 of 3000 planned start/stop cycles to-date. Cartridge starts were also successfully performed at –54°C, 22°C, and 49°C to simulate emergency start conditions in the field with no damage to the silicon nitride wheels or braze joint attachment.

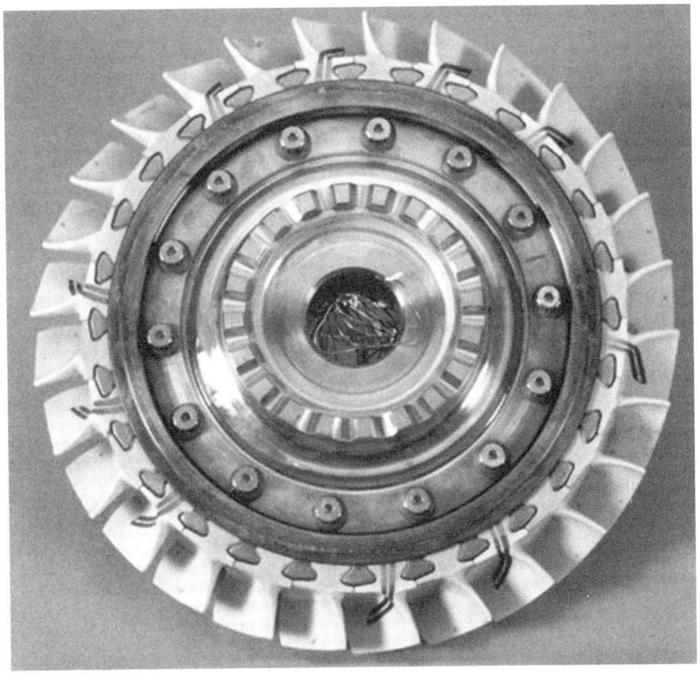

Figure 5.17 HES&S 331-200 APU 1st Stage Turbine Wheel with Inserted AS800 Silicon Nitride Blades (Reprinted from ref. [10])

Figure 5.18 HES&S 331-200 APU 1st Stage Turbine Blisk (Slipcast AS800 Silicon Nitride) (Reprinted from ref. [10])

HCC began producing seal runners of GS-44 silicon nitride, shown in Figure 5.19, using bisque machining of isopressed stock in 1996 for the Honeywell TFE 731 family of aircraft turbofan propulsion engines. The silicon nitride seals provide a much longer life oil seal for the turbine engine main shaft by providing a better thermal expansion match to the carbon component of the seal assembly and higher hardness than the metal seal rings they replace. The silicon nitride seal runners are being produced at a rate of over 4000 per

Figure 5.19 Seal Runner for Honeywell 731 Turbofan Propulsion Engine (GS-44 Silicon Nitride)

year and have accumulated over three million flight hours through 1999. In addition, HCC has been producing silicon nitride ring rotors for commercial aircraft APUs since 1994. The rotors are part of the APU oil pump and provide critical enhanced emergency starting of the APU during flight. The rotors, shown in Figure 5.20, are also produced by bisque machining of isopressed stock. Over 4000 a year are produced. The applications include APUs in the Boeing 777, the new generation Boeing 737 (–500,–600,–700,–800), and many of the Airbus range of commercial aircraft.

HCC developed and demonstrated the fabrication of GN-10 and AS800 industrial gas turbine blades, formed by presintering machining of slipcast stock, as a subcontractor to Solar Turbines Inc. under its DOE sponsored CSGT program, which started in 1992. HCC delivered 150 AS800 Centaur 50 1st stage turbine blades (Figure 5.9) to Solar in 1996. After 104 hours of in-house engine testing, Solar successfully ran full sets of 62 blades for 948 hours in a field test engine in 1998 before FOD damage caused catastrophic blade failure. Solar has halted further monolithic ceramic blade insertion efforts under the CSGT program until impact resistance of monolithic silicon nitride can be effectively improved by either engine design, component design, or material changes [18].

HCC also developed and fabricated AS800 nozzles in support of Rolls-Royce Allison for their 501 industrial gas turbine engine under the DOE Advanced Turbine Systems (ATS) program. The 1st stage nozzles, again fabricated by bisque machining of simple shaped slipcast stock were delivered in 1999 (Figure 5.5). The nozzles performed successfully for 800 hours in field test in 1999, but significant surface recession due to the high pressure, high velocity, high water vapor content combustion environment was observed, as described previously [14], causing Rolls-Royce Allison to halt the test to address the recession issue. HCC has since fabricated more AS800 nozzles with its developmental environmental barrier coating (EBC) for evaluation in the 501 field test engine to determine if the EBC coating can provide improved protection and component life.

HCC has also developed the gelcasting of near-net-shape prototype cooled vanes of AS800 for United Technology Research Center's (UTRC) FT-8 industrial gas turbine engine under DOE funding support [30]. The initial development parts, demonstrated in 1998 and shown in Figure 5.21, do not include the final design small trailing edge cooling passages. HCC is currently developing the fabrication process for these

Figure 5.20 Ring Rotors for Honeywell Commercial Aircraft APU's (GS-44 Silicon Nitride)

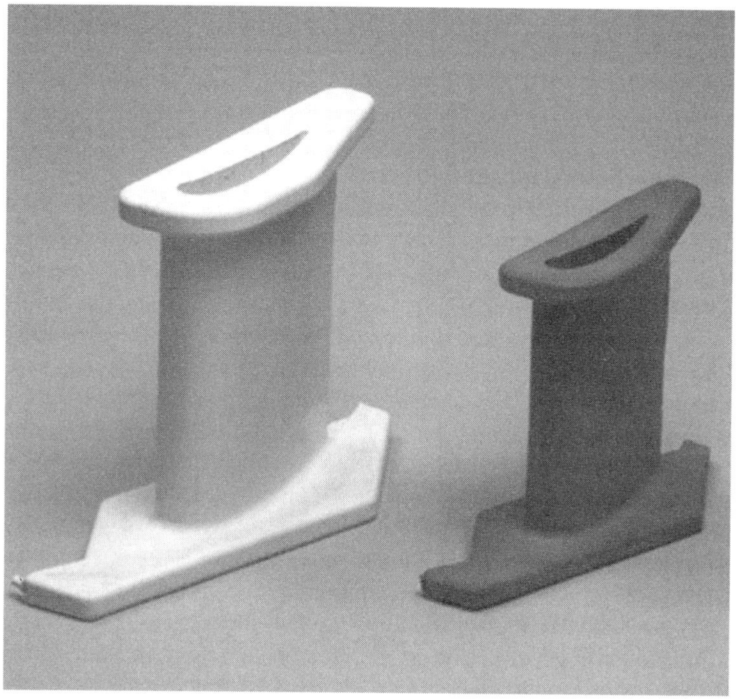

Figure 5.21 UTRC FT-8 Industrial Turbine Engine Nozzle Prototypes (Gelcast AS800 Silicon Nitride) As-Cast and Fully Dense (Reprinted from ref. [10])

cooling passages and demonstration of engine quality FT-8 silicon nitride vanes is scheduled under Phase II of UTRC's DOE-sponsored program.

Finally, as mentioned previously, HCC is developing automation of the gelcasting process to fabricate complex shaped components such as turbine wheels to meet expected cost targets and required production volumes for applications such as missile turbojet engines and microturbines for power generation. The ability to automatically gelcast turbine wheels at a rate of 10,000 per year was demonstrated in 1999 using the Teledyne M304 turbojet wheel (Figure 5.14). Development efforts sponsored by DOE have now transitioned to developing and demonstrating, in 2000, the ability to gelcast the significantly larger radial turbine wheel for the Honeywell 75 kW distributed power generation microturbine, out of AS800 silicon nitride.

SUMMARY

Extensive progress has been made at HCC over the last 25 years in the development of silicon nitride structural ceramic materials and fabrication of components for gas turbine applications. Silicon nitride materials have transitioned from low strength, low toughness RBSN to gas pressure sintered in-situ reinforced silicon nitrides such as AS800 that exhibit a combination of high toughness, high Weibull modulus, and high temperature creep and stress rupture resistance. Gas turbine applications are continuing to push the need for further advances in environmental durability (e.g. environmental barrier coatings) and impact resistance (e.g. hybrid toughening concepts such as fibrous monolith structures).

Complex shape forming technologies were initially effective due to the large powder particle sizes used. As the materials technology transitioned to very fine powders, significant improvements to powder compact forming technologies were needed to achieve engine quality, near-net-shape, and cost effective fabrication.

It has taken almost 20 years to develop forming approaches for submicron silicon nitride powders that are as capable as slipcasting and wax injection molding were for first generation RBSN materials. There is now a further drive to mature the state-of-the-art technologies such as gelcasting for production viability and cost effectiveness, and to develop new forming technologies to address complex features such as cooling passages and fibrous monolith microstructures.

Finally, the initial applications expected for silicon nitride structural ceramics in gas turbines have transitioned from aircraft propulsion engine hot sections, to automotive direct propulsion turbines, to hybrid vehicle turbogenerators and small power generation industrial turbines, and now to distributed power generation microturbines. Although the only engine parts currently in production are seal components for aircraft turbofans and APU oil pump components, the improvements in material properties, forming capabilities, and experience gained in engine and field testing are expected to lead to the production of nozzles and turbine wheels for aircraft APU's and turbine wheels for microturbines in the next 1–3 years. The 25 year promise of silicon nitride structural components for gas turbines may soon be realized.

ACKNOWLEDGMENTS

The great challenges overcome and progress made in advancing the technology and application of silicon nitride for gas turbines described in this review are the result of the highly skilled teams of people assembled at Honeywell Ceramic Components, Honeywell Engines, Systems & Services, and Honeywell Laboratories from 1976 to present. Much of the work presented in this review was performed under a number of programs supported by the Department of Energy, Department of Commerce, Defense Advanced Research Projects Agency, NASA, and under subcontract to Honeywell Engines, Systems & Services, Rolls-Royce Allison, Solar Turbines Inc., UTRC, and SRI. Unless otherwise indicated the figures and tables were supplied courtesy of Honeywell Ceramic Components.

REFERENCES

[1] Johansen, K. M., Richerson, D. W., and Schuldies, J. J., *Ceramic Components for Turbine Engines,* Phase II Technical Report for AFML/APL Contract No. F33615-77-C-5171, February 1980.

[2] Richerson, D. W., Smyth, J. R., and Styhr, K. H., "Material Improvement Through Iterative Process Development," Ceram. Eng. and Sci. Proc., 4 [9–10], 841–852, 1983.

[3] Richerson, D. W., Personal Communication.

[4] Richerson, D. W., and Wimmer, J. M., "Properties of Isostatically Hot Pressed Silicon Nitride," J. Am. Ceram. Soc., 66 [9], C173–176, 1983.

[5] Yeh, H. C. and Fang, H. T., *Improved Silicon Nitride for Advanced Heat Engines,* Final Report, NASA Report CR-182193, Contract NAS3-24384, April 1991.

[6] Yeh, H. C., et. al., *Development of Ceramic Matrix Composites for Applications in the Ceramic Technology for Heat Engines Program,* Final Report, ORNL Subcontract 85-22008-1, 1988.

[7] Lange, F. F, "Fabrication and Properties of Dense Polyphase Silicon Nitride," Am. Ceram. Soc. Bull., 62 [12], 1369–1374, 1983.

[8] Li, C-W, and Yamanis, J., "Super-Tough Silicon Nitride with R-Curve Behavior," Ceram. Eng. and Sci. Proc., 10 [7–8], 632–645, 1989.

[9] Pollinger, J. P., "Status of Silicon Nitride Component Fabrication Processes, Material Properties, and Applications," ASME paper 97-GT-321, presented at the International Gas Turbine and Aeroengine Congress and Exposition, Orlando, FL, USA, June 2–5, 1997.

[10] Newson, D. D., Pollinger, J. P., and Twait, D. J., "Status of Silicon Nitride Material Properties, Component Fabrication and Applications for Small Gas Turbines," ASME paper 2000-GT-0533, presented at the ASME Turbo Expo 2000, Land, Sea, & Air, Munich, Germany, May 8–11, 2000.

[11] Pollinger, J. P., "Improved Silicon Nitride Materials and Component Fabrication Processes for Aerospace and Industrial Gas Turbine Applications," ASME paper 95-GT-159, presented at the International Gas Turbine and Aeroengine Congress and Exposition, Houston, TX, USA, June 5–8, 1995.

[12] Opila, E. J., and Hann Jr., H. E., "Paralinear Oxidation of SiC in Water Vapor," J. Am. Ceram. Soc., 80[1], 197–205, 1997.

[13] Ferber, M. K., Lin, H. T., Parthasarathy, V., and Brentnall, W., "Degradation of Silicon Nitrides in High Pressure, Moisture Rich Environments," ASME paper 99-GT-265, presented at the International Gas Turbine and Aeroengine Congress and Exposition, Indianapolis, IN, USA, June 7–10, 1999.

[14] Wenglarz, R. A., and Kouns, K., "Ceramic Vanes for a Model 501-K Industrial Turbine Demonstration," ASME paper 2000-GT-0073, presented at the ASME TURBO EXPO 2000, Land, Sea, & Air, Munich, Germany, May 8–11, 2000.

[15] Ferber, M. K., Lin, H. T., Parthasarathy, V., and Wenglarz, R. A., "Degradation of Silicon Nitrides in High Pressure, Moisture Rich Environments," ASME paper 2000-GT-0661, presented at the ASME Turbo Expo 2000, Land, Sea, & Air, Munich, Germany, May 8–11, 2000.

[16] Lee, K. N., "Key Durability Issues with Mullite-Based Environmental Barrier Coatings for Si-Based Ceramics," ASME paper 99-GT-443, presented at the International Gas Turbine and Aeroengine Congress and Exposition, Indianapolis, IN, USA, June 7–10, 1999.

[17] Eaton, H. E., Linsey, G. D., More, K. L., Kimmel, J. B., Price, J. R., and Miriyala, N., "EBC Protection of SiC/SiC Composites in the Gas Turbine Combustion Environment," ASME paper 2000-GT-0631, presented at the ASME Turbo Expo 2000, Land, Sea, & Air, Munich, Germany, May 8–11, 2000.

[18] Price, J. R., Jimenez, O., Parthasarathy, V., and Miriyala, N., "Ceramic Stationary Gas Turbine Development Program—Sixth Annual Summary," ASME paper 99-GT-351, presented at the International Gas Turbine and Aeroengine Congress and Exposition, Indianapolis, IN, USA, June 7–10, 1999.

[19] van Roode, M., Jimenez, O., McClain, J., Price, J., Parthasarathy, V., Poormon, K. L., Ferber, M. K., and Lin, H.-T., "Ceramic Gas Turbine Materials Impact Evaluation," ASME paper GT-2002-30505, presented at the ASME TURBO EXPO Land, Sea & Air, Amsterdam, The Netherlands, June 3–6, 2002.

[20] Gersch, H., Mann, D., Rorabaugh, M., and Sculdies, J. J., "Slipcast and Injection Molding Process Development of Reaction Bonded Silicon Nitride at AiResearch Casting Company, "Ceramic Gas Turbine Demonstration Engine Program Review, J. W. Fairbanks and R. W. Rice, eds., MCIC Report No. MCIC-78-36, 313–40, 1978.

[21] Richerson, D. W., and Johansen, K., *"Ceramic Gas Turbine Engine Demonstration Program Final Report,* January 1976–May 1981," DARPA/Navy Contract N00024-76-C-5352, May 1982.

[22] Richerson, D. W., Smyth, J. R., and Styhr, K., *Low-Cost Net-Shape Ceramic Radial Turbine Program,* Final Report AMMRC TR 85-2, Prepared for AMMRC Under Contract DAAG46-81-C-0006, May 1985.

[23] Omatete, O. O., Janney, M. A., and Strehlow, R. A., "Gelcasting—A New Ceramic Forming Process," Am. Ceram. Soc. Bull., 70[10], 1641–1649, 1991.

[24] Janney, M. A., Omatete, O. O., Walls, C. A., Nunn, S. D., Ogle, R. B., and Westmoreland, C. G., "Development of Low-Toxicity Gelcasting Systems," J. Am. Ceram. Soc., 81[3], 581–591, 1998.

[25] Omatete, O. O., and Nick, J. J., "Improved Gelcasting Systems," Ceram. Eng. and Sci. Proc., 20 [3], 241–248, 1999.

[26] Nick, J. J., Newson, D. D., Masseth, M., and Monette, S., "Gelcasting Automation for High Volume Production of Silicon Nitride Turbine Wheels," Ceram. Eng. and Sci. Proc., 20[3], 225–231, 1999.

[27] Fanelli, A. J., Silvers, R. D., Frei, W. S., and Marsh, G. B., "New Aqueous Injection Molding Process for Ceramic Powders," J. Am. Ceram. Soc., 72[10], 1833–1836, 1989.

[28] Richerson, D. W., "Evolution in the U.S. of Ceramic Technology for Turbine Engines," Am. Ceram. Soc. Bull., 64[2], 282–286, 1985.

[29] Schenk, B., "Ceramic Gas Turbine Technology Development," ASME paper 99-GT-315, presented at the International Gas Turbine and Aeroengine Congress and Exposition, Indianapolis, IN, USA, June 7–10, 1999.

[30] Holowczak, J. E., Bornemisza, T. G., and Day, W. H., "Sector Testing of Cooled Silicon Nitride Ceramic HPT Vanes for the FT8 Aeroderivative Engine," ASME paper 2000-GT-0662, presented at the ASME Turbo Expo 2000, Land, Sea, & Air, Munich, Germany, May 8–11, 2000.

CHAPTER

6

TECHNOLOGY DEVELOPMENT AT KYOCERA CORPORATION

W. David Carruthers
Dave Carruthers & Associates, Battle Ground, Washington, USA

Edwin H. Kraft
EHK Technologies, Vancouver, Washington, USA

Makoto Yoshida, Koichi Tanaka, Toshifumi Kubo, Hirofumi Terazono and Sazo Tsuruzono
R&D Center, Kagoshima, Kyocera Corporation, Kokubu, Kagoshima, Japan

INTRODUCTION

Kyocera Corporation, a leading producer of advanced ceramics for automotive, industrial, and electronic applications has its global headquarters in Kyoto, Japan. For over 30 years, Kyocera has been on the forefront of ceramic technology, aimed at efficient, durable and low polluting heat engines. Since road testing of its own all-ceramic automotive diesel prototype engine in the mid-1980s, Kyocera has supported numerous development projects for ceramic valves, rollers, turbine rotors, combustors, and other components for numerous diesel and gas turbine programs.

Kyocera's production operations provide silicon nitride components for diesel and gasoline reciprocating engines, including cam follower rollers, pre-combustion chambers, rocker arm tips, and "instant start" glow plugs, ceramic turbocharger rotors, and exhaust gas bypass valves. The high performance turbocharger rotors are produced on an automated production line using unique fabrication techniques developed by Kyocera. Rotors are provided with a metal shaft attached, in a fully balanced, proof tested condition. The cam roller followers are produced in Kyocera's Vancouver, WA facility (KICC). Upon its inception the cam roller production represented the first high volume production of a Si_3N_4 structural ceramic part in the U.S. Its success was followed by the production of a silicon nitride based seal system in a popular turbofan propulsion engine used in many business aircraft.

Kyocera's support of ceramics for gas turbine applications has included participation in Japanese projects including the NEDO 300 kW CGT (ceramic gas turbine for cogeneration) and 100 kW automotive engine proj-

Figure 6.1 Sintered Silicon Nitride Parts Provided to Support the Early Japanese and U.S Ceramic Gas Turbine Programs: Scroll and Turbine Rotors for the 300kW CGT, and the Allison (Now Rolls-Royce Indianapolis) and, AlliedSignal (Now Honeywell) AGT Turbine Scrolls and Rotors.

ects [1–11], and the Toshiba 15 MW and the Tokyo Electric Power Company (TEPCO) 20 MW projects. Support of U.S projects includes the Dept. of Energy funded programs including the Honeywell Engines, Systems & Services' (formerly Garrett Auxiliary Power Division of AlliedSignal Aerospace Corporation) Advanced Gas Turbine (AGT) project [12] and Rolls-Royce Indianapolis (formerly Allison Engine Company, Inc., and prior to that Allison Gas Turbine Division of General Motors Corporation) AGT, Advanced Turbine Technology Applications (ATTAP), and Hybrid Vehicle Turbine Engine Technology Support (HVE-TS) projects [13–15], Solar Turbines Incorporated's Ceramic Stationary Gas Turbine Project (CSGT) [16], Honeywell Engines, Systems & Services Ceramic Technology Engine Demonstrator Project [17], and United Technologies FT-8 vane program [18,19]. Components supplied included turbine blades, turbine nozzles, shrouds, scrolls and combustor components. These were fabricated from Kyocera's SN252, and more recently SN281 and SN282 materials.

Parts typical of those provided in the 1970s, 1980s, and early 1990s are illustrated in Figure 6.1. These components were of moderate temperature capability, by the standards in 2000, and were fabricated using low yield, low production quantity approaches, at very high cost. These components were typically used in programs aimed at establishing the feasibility of successful use of ceramics in the gas turbine environment, which was accomplished. More recent efforts have extended both the materials characteristics and fabrication options, and have provided components aimed at demonstrating high engine performance, durability, and cost.

MATERIAL PROPERTIES

Silicon Nitride Material Development

Kyocera has been developing gas turbine grade ceramics for many years. Although silicon carbide and silicon nitride were both of interest and SiC's oxidation resistance was initially better than that of Si_3N_4, the

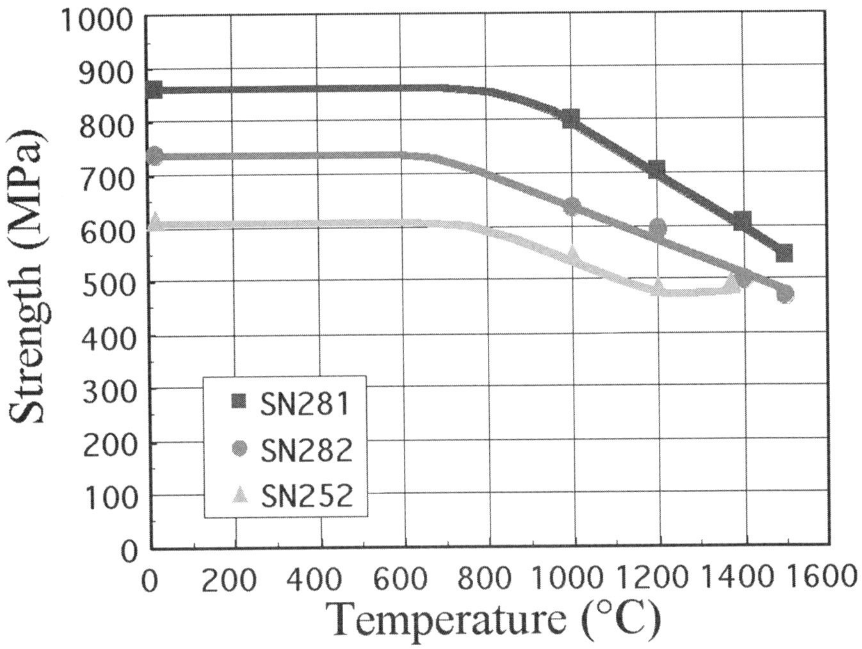

Figure 6.2 Four Point Bending Flexural Strengths of SN281, SN282, and SN252 (Reprinted from ref. [6])

higher fracture toughness of silicon nitride made it the preferred engine candidate. Initial materials, such as SN220 were limited to use at temperatures well below 1000°C. Kyocera has continued its efforts to provide materials for use at temperatures up to 1400°C. Toward that effort, Kyocera previously developed SN252 and SN253, and more recently, SN281 and SN282, using a rare earth oxide with small ionic radius as a sintering additive [6]. These materials have a fine acicular microstructure with high melting point oxide grain boundary phases. These materials are suitable for use above 1000°C as summarized below. All flexural test data were developed using 3x4 mm cross section bars with inner and outer spans of 10 and 30 mm, respectively, per JIS R-1601 and R-1604.

Flexural Strength

Figure 6.2 shows the flexural strength of SN281, SN282 and SN252 up to 1500°C. The higher strength of SN281 is attributed to: 1) the fine grain structure of the matrix realized by HIP (hot isostatic press) processing. The grain growth has been restrained as a result of controlling the alpha to beta phase transformation of silicon nitride, and the internal defects are reduced by high pressure HIP; 2) A grain boundary phase with higher melting point obtained from optimized sintering additive compositions. SN282 has medium strength among these three materials. This material has the same chemical composition as SN281, but is not HIPed.

Stress Rupture and Creep Rate

The stress rupture results at 1200°C and 1400°C are shown in Figure 6.3 and Figure 6.4, respectively. The characteristics were measured per JIS R-1612. For SN281, the allowable stress for 100hrs without

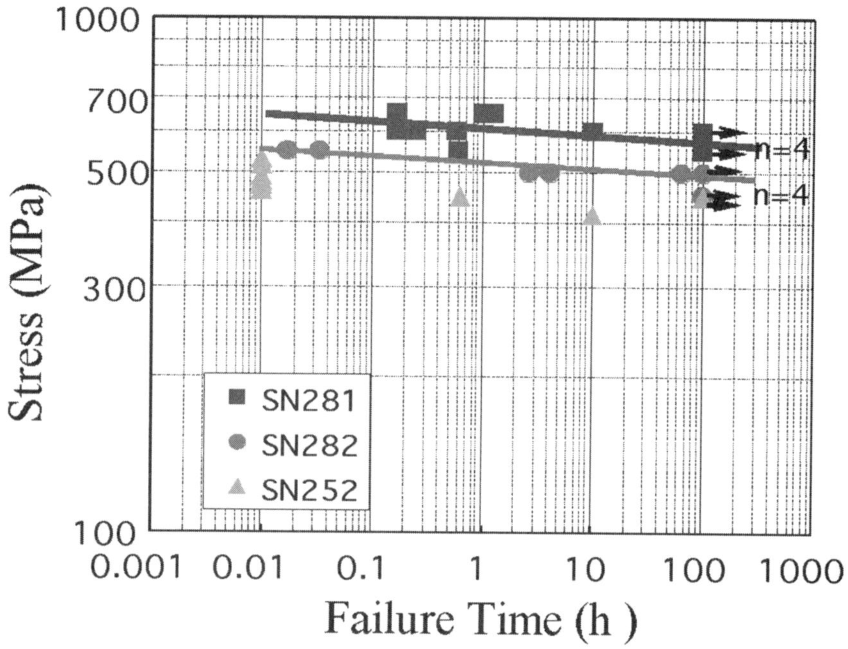

Figure 6.3 Four-Point Bending Stress Rupture Results at 1200°C (Reprinted from ref. [6])

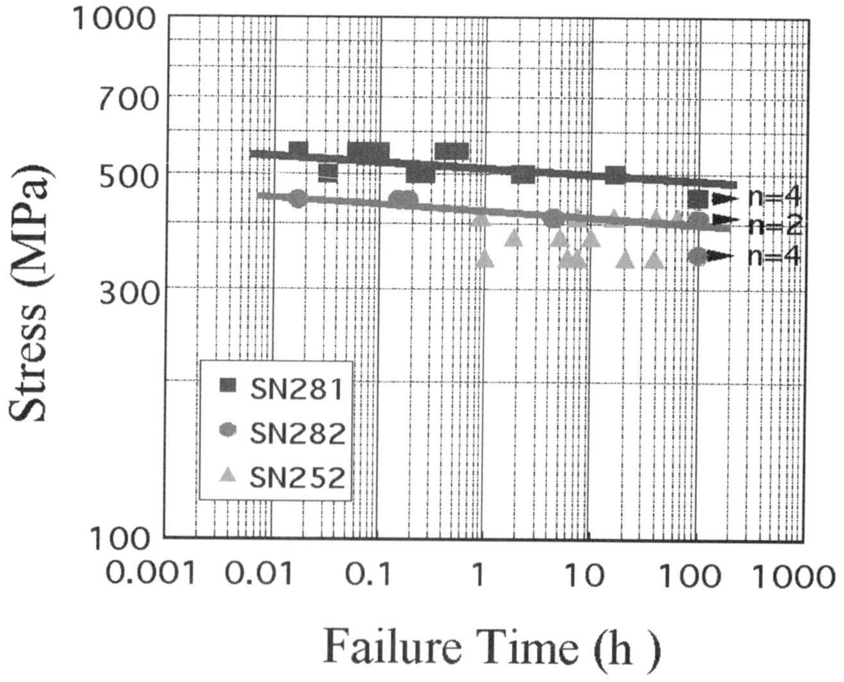

Figure 6.4 Four-Point Bending Stress Rupture Results at 1400°C (Reprinted from ref. [6])

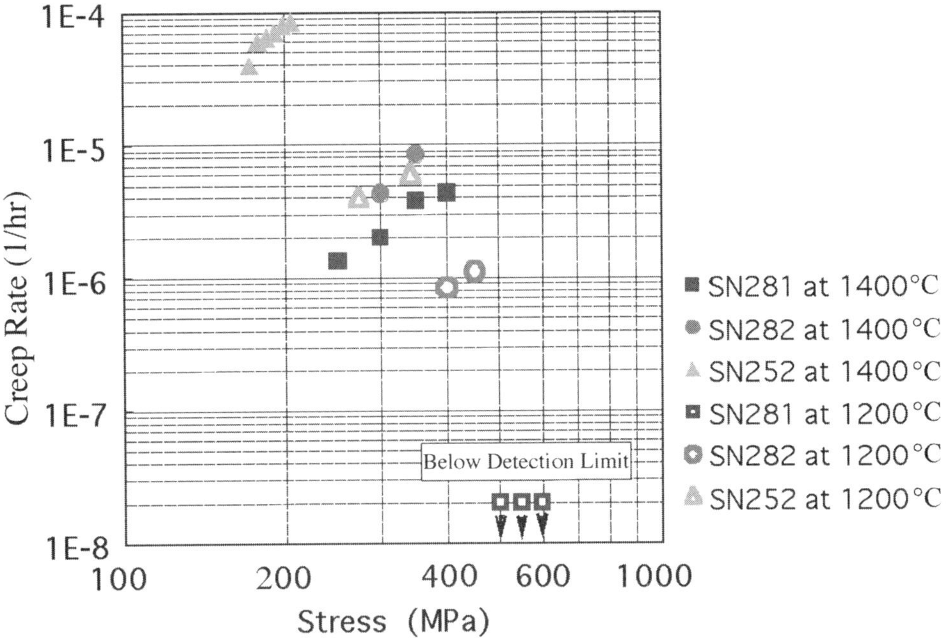

Figure 6.5 Four-Point Bending Creep Rate of SN281, SN282, and SN252 (Reprinted from ref. [6])

failure is 550MPa at 1200°C, and 450MPa at 1400°C. These stress rupture strengths are about 150MPa higher than those of SN252.

Figure 6.5 shows the creep rates of SN281, SN282 and SN252. The creep rate of SN281 at 1200°C is lower than the detection limit. The creep rates of SN281 and SN282 at 1400°C are approximately the same as those of SN252 at 1200°C. Testing performed at the U.S. National Institute of Standards and Technology found SN282 to have a creep resistance several orders of magnitude greater than that of previously examined silicon nitrides [20].

Fracture Toughness

Fracture toughness as measured by the SEPB (single-edge precracked beam) method per JIS R-1607 and R-1617 is shown in Figure 6.6. High temperature testing was performed in a N_2 atmosphere. The fracture toughnesses of SN281 and SN282 at room temperature are 6.2 MPa.m$^{1/2}$ and 6.4 MPa.m$^{1/2}$, respectively. These values are almost the same as those of SN252. The toughness decreases with increasing temperature.

Oxidation Resistance

The weight gain during oxidation at 1500°C in an air atmosphere is presented in Figure 6.7. SN281 and SN282 show similar oxidation resistance because of their similar grain boundary composition. The oxidation rates of these two materials are similar to that of silicon carbide. The high oxidation resistance of SN281 and SN282 is due to the grain boundary phase being an oxide that does not contain any nitrogen.

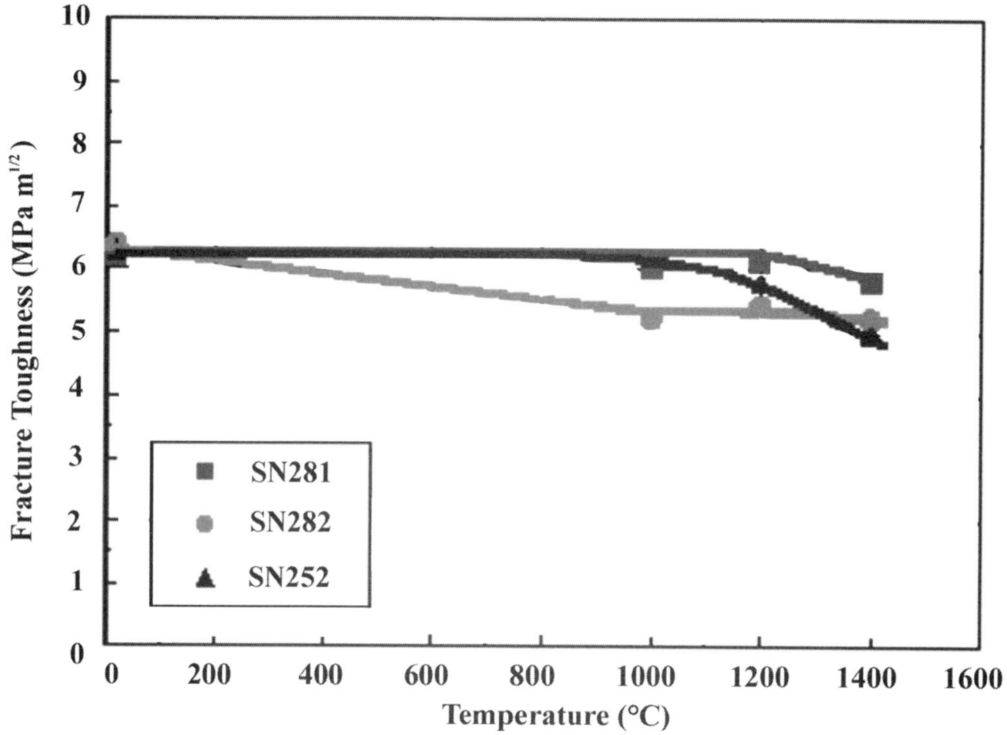

Figure 6.6 Fracture Toughness (SEPB Method) of SN281, SN282 and SN252 (Reprinted from ref. [6])

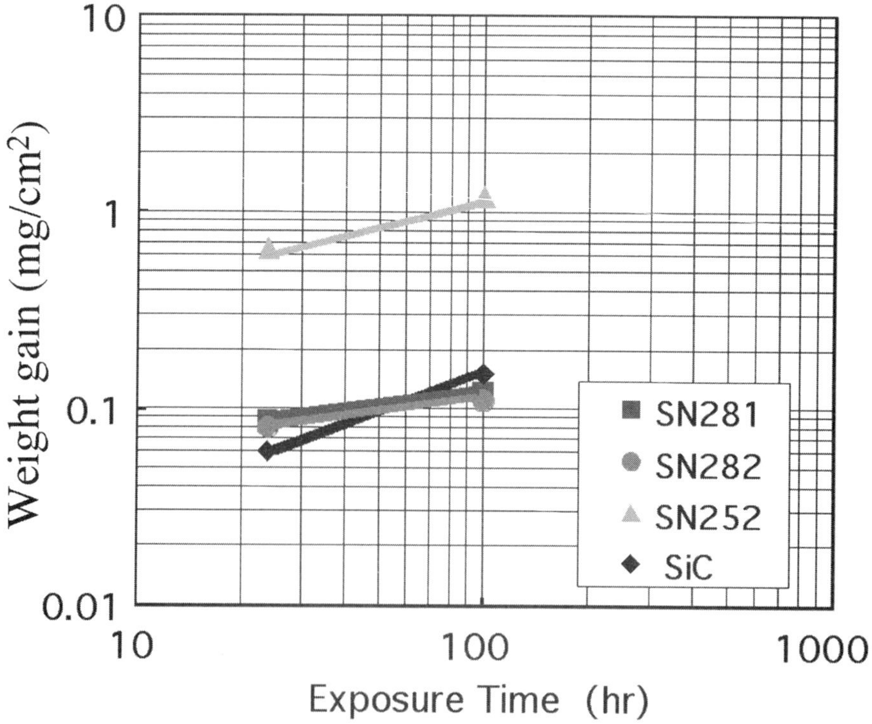

Figure 6.7 Weight Gain After Oxidation at 1500°C (Reprinted from ref. [6])

FABRICATION PROCESSES

For production of gas turbine components, Kyocera has developed a variety of forming processes, including cold isostatic pressing (CIP), hybrid molding and drain casting. Additionally, bisque machining can be used in conjunction with all of these processes to enhance the ability to form net- or near net-shape preforms. Selection of the fabrication process should consider the operating conditions, the part complexity and the production quantity.

All parts are inspected at appropriate in-process steps and in the final condition using real-time and film radiography, fluorescent dye penetrant, and coordinate measuring machine dimensional inspections. Rotors and other selected parts are also qualified by ultrasonic inspection and spin or other proof testing.

Hybrid Molding

The Hybrid Molding process utilizes a low viscosity suspension, or "slip" which is injected into a metal mold. The use of metal molds provides the desired reproducible dimensional control. The slip is a combination of silicon nitride powder, sintering aids, liquid suspension agents and binders. Once the slip is introduced into the mold, solidification occurs under controlled thermal conditions. Subsequently, the binding agents are removed, and the part is sintered to full density. Shrinkage and, possibly, distortion accompany densification. To the extent that this shrinkage and distortion are controlled, the goal of hybrid molded parts is to leave the typically complex gas path surfaces in the as-sintered condition. A mold iteration is typically required to achieve the demanding tolerances of gas turbine components. Other surfaces, such as attachment surfaces typically require final grinding to meet the tighter tolerances.

Kyocera's Hybrid Molding process is well suited for high volume production of small, complex parts. It has been utilized in the production of automotive turbocharger rotors since 1986, as shown in Figure 6.8. In support of gas turbine efforts, hybrid molding has been used to fabricate turbine blades, nozzle segments, and integral rotors as illustrated in Figures 6.9–6.13. For most of these parts, only the attachment and other select features that require close tolerances and surface finishes, need to be ground after sintering.

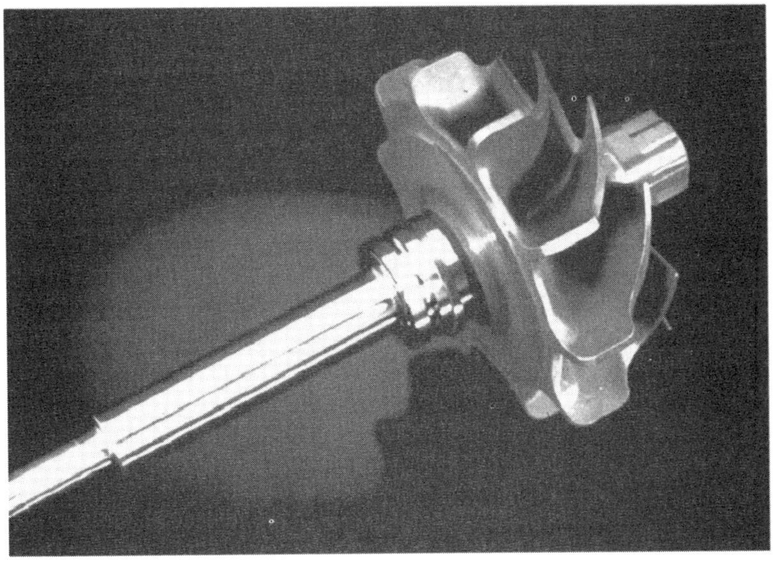

Figure 6.8 Ceramic Automotive Turbocharger Rotor

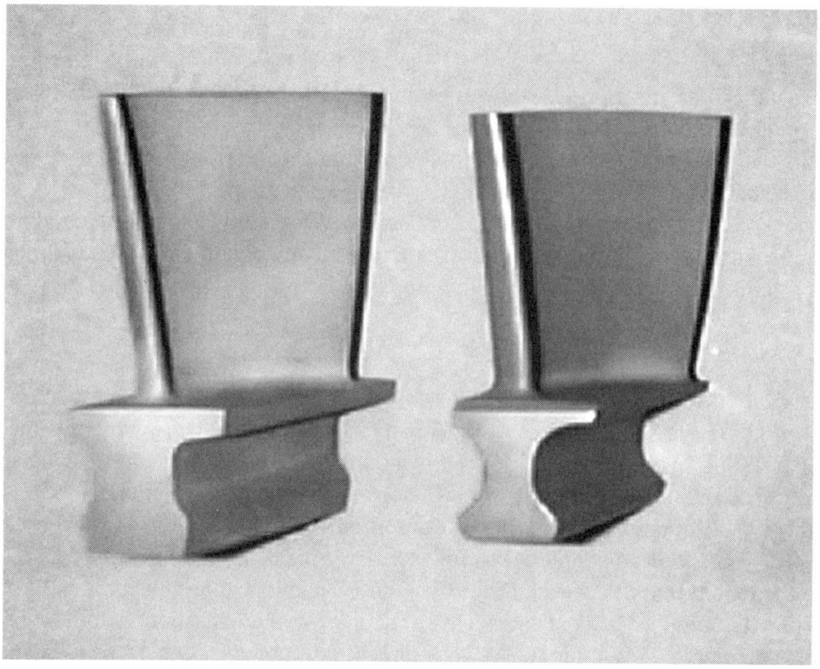

Figure 6.9 SN281 Dovetail Turbine Blades for Use in a Solar Turbines Incorporated Industrial Turbine Engine

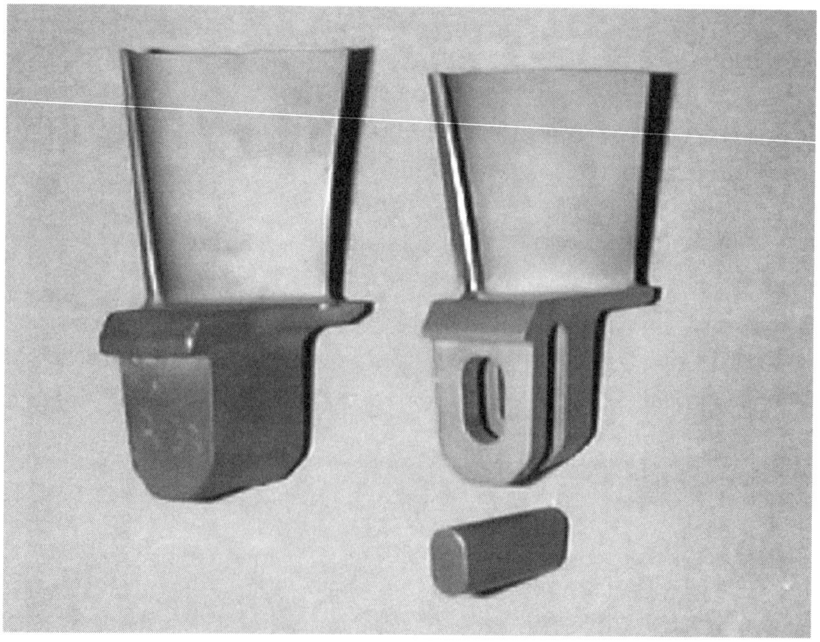

Figure 6.10 SN253 Pinned Root Turbine Blades for Solar Turbines Incorporated Industrial Turbine Engine

Figure 6.11 SN253 Turbine Rotor for Use in Rolls-Royce's AGT-5 Automotive Engine (Reprinted from refs. [14,15])

Gasifier Rotor Power Turbine Rotor

Figure 6.12 Hybrid Molded Integral Monolithic Rotors for the CGT 302 Engine, with Metallic Shafts Attached (Reprinted from refs [6,7])

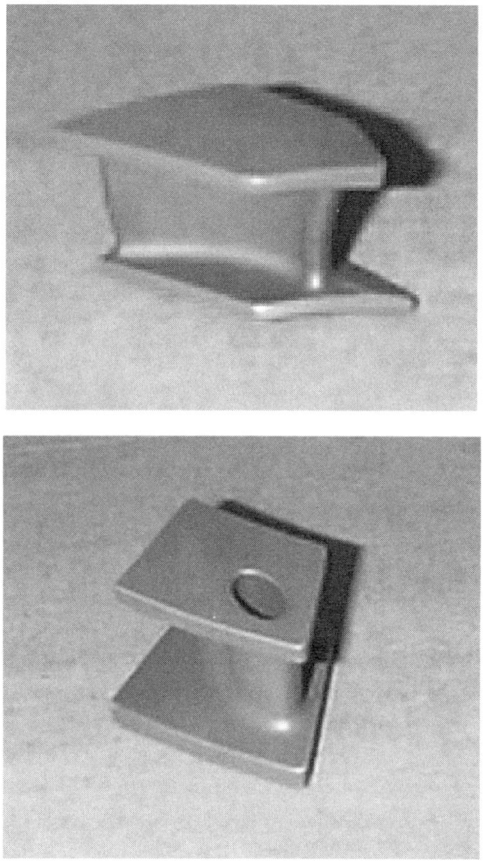

Figure 6.13 SN252 (Top) and SN282 (Bottom) Turbine Nozzles Fabricated by Hybrid Molding for Honeywell's 331 APU Engine

The pinned root blade, Figure 6.10 was a particularly difficult fabrication challenge. A significant portion of the root required grinding, and the success of this design depended upon strict control of the dimensions in the oblong holes to +/– 10 microns. Those dimensional requirements, as well the overall complexity of the part required approximately 40 hours of grinding time for the prototype parts. Such complex designs should be avoided to keep the cost of ceramic components as low as possible.

Green and Bisque Machining

Processing of low quantity, prototype components for development programs can be a lengthy and costly effort if high volume production methods are used. For example, the hybrid molding process is ideal for high volume production, but requires the iterative fabrication of metal molding dies. The process can exceed one year for process optimization. Thus, Kyocera sought to develop bisque firing and machining as a means of providing prototypes or low volume production in significantly less time. The process utilizes a bisque firing process to partially sinter a ceramic preform. In the partially sintered condition, the preform can be machined by conventional computer numerical control (CNC) techniques with carbide cutting tools. This process has been applied to integral nozzles and rotors, and individual nozzle segments as illustrated in Figures 6.14–6.16. Most of these parts utilized CIPed preforms, but the process has also been used in conjunction with hybrid molding and slip casting.

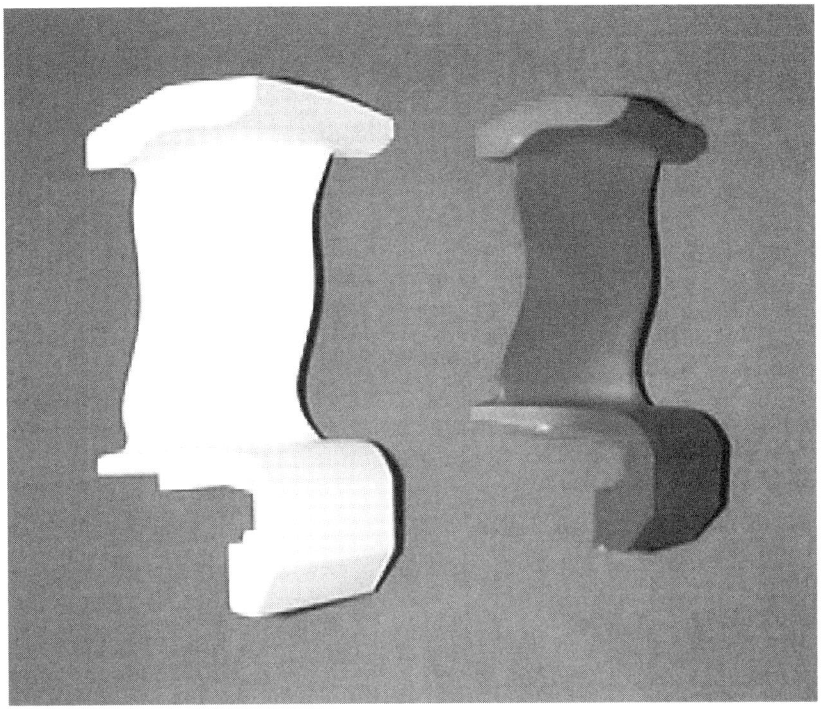

Figure 6.14 Turbine Nozzles Fabricated by Green and/or Bisque Machining. As-Bisque Machined and as-Sintered SN282 Turbine Nozzle Segments for Use in Solar Turbines Centaur 50S Engine

Figure 6.15 A 17.5-cm Bisque Processed SN 281 Integral Rotor Designed by Honeywell for the 331-200 APU. One Blisk was Fully Processed and Spin Tested to Burst at 49,512 rpm.

Figure 6.16 Hamilton Sundstrand Nozzle for TJ-50 Engine Fabricated by Bisque Machining

Prototype deliveries for components fabricated by this process have been completed in as short as 3 months. The process also allows the fabrication of parts with negative or undercut features that would not be fabricable from "hard" molds or tools. This allows the design of airfoils without the restriction on twist or lean.

Solid Slip and Drain Casting

Slip casting involves mixing silicon nitride and sintering aids in a liquid, which is then poured into a porous mold, such as plaster. The liquid is drawn into the mold surface, depositing the powder on the walls. Pouring the center liquid out before it solidifies creates hollow parts (drain casting). Allowing complete solidification to occur produces solid parts (solid slip casting). Among the many forming processes, the drain casting process has the highest yield of powder. The recovered drained slurry may be reused. This process is particularly effective for large or hollow components with uniform wall thickness. Kyocera's drain casting process for silicon nitride material results in high yield for turbine scrolls and combustor liners, bladed rotors, and is also suitable for hollow, air cooled turbine nozzles as shown in Figures 6.17–6.19.

For the scrolls and combustor liners only the gas inlet, outlet, and mounting surfaces of the parts were ground. A successfully cast and machined part is illustrated in Figure 6.18. Polycarbonate mold patterns fabricated by solid free form processing can also be utilized to reduce the lead-time for mold procurement. The wall thickness of these parts is typically 3 to 4 mm. Turbine scrolls and combustor liners are typically some of the largest ceramic components used, up to 600mm with a wall thickness of 4 to 6mm. Figure 6.1 (top left) shows a large scroll for the 300 kW CGT302 engine.

Air-cooled FT8 Vane Fabrication

The hollow, air-cooled Si_3N_4 vane segments (Figure 6.17) for use in the Pratt & Whitney (P&W) FT8 industrial turbine engine were fabricated from SN282 using combined drain casting and bisque machining techniques. The vanes incorporated an internal cavity in the airfoil and trailing edge cooling passages. The hollow interior of the airfoil was achieved by casting the airfoil against the mold walls and draining the mold prior to complete solidification. The thicker, solid portions of the platform were also cast in the same operation. A particular challenge was to obtain a 3 mm-thin wall in the hollow airfoil location, while also achieving up to approximately 10 mm of thickness in the platform.

After casting, the vane segments were bisque fired. United Technologies Research Center (UTRC) pursued laser machining trailing edge (TE) vent holes of an oval, tapered configuration; however, this approach resulted in cracks around the holes. As an alternate approach, KICC machined round TE holes using conventional carbide drills. This approach proved successful, but the round holes adversely affect the cooling of

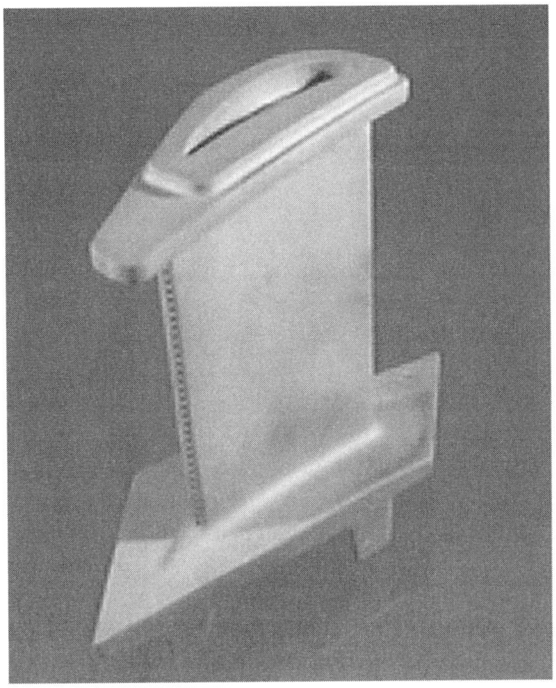

Figure 6.17 As-Sintered, Drain Cast Hollow SN282 Turbine Nozzle Segment with Cooling Holes in a Truncated Trailing Edge. Trailing Edge Holes were Bisque Machined.

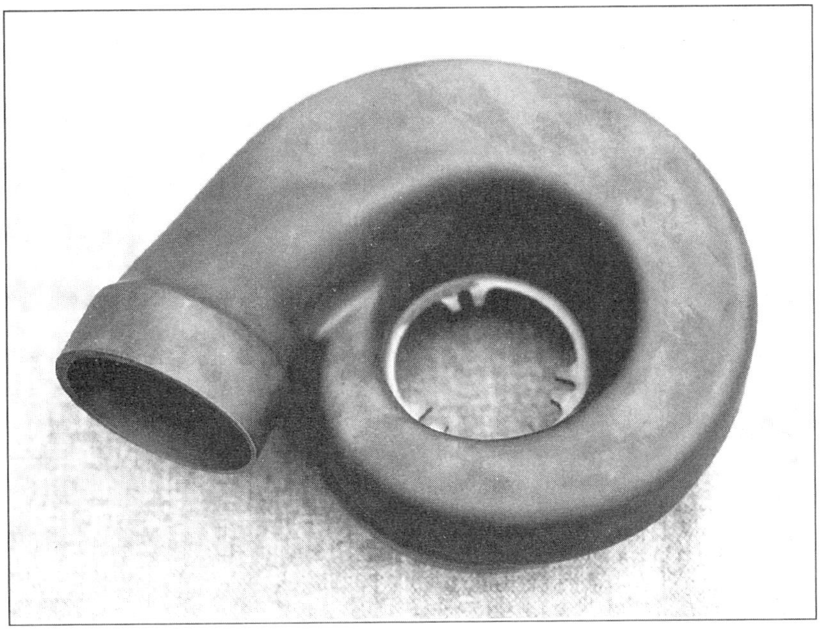

Figure 6.18 Drain Cast Turbine Scroll for the Rolls-Royce AGT-5 Engine (Reprinted from refs. [14,15])

Figure 6.19 Drain Cast SN282 Combustor Liner for the Rolls-Royce AGT-5 Automotive Engine (Reprinted from ref. [15])

the TE. To compensate for the cooling loss a portion of the trailing edge was cut back. Bisque machining was followed by sintering and final machining. Completed vanes were successfully exposed to combustion exhaust conditions by UTRC [18,19].

JAPANESE 300KW INDUSTRIAL CERAMIC GAS TURBINE

Successful Engine Application of Ceramic Components

In 1988 the Japanese New Energy and Industrial Technology Development Organization (NEDO) started the "300kW Industrial Ceramic Gas Turbine Research and Development Project" under contract to the Agency of Industrial Science and Technology of the Ministry of International Trade and Industry (MITI) [1]. The objective was to achieve 42% thermal efficiency at 1350°C TIT (turbine inlet temperature).

In support of this program, Kyocera developed ceramic components for the CGT302 engine in cooperation with Kawasaki Heavy Industries (KHI) and Sumitomo Precision Products [2,3]. The CGT302 is a regenerated two-shaft axial flow gas turbine. Ceramic materials were used for the combustor liner, scrolls, turbine nozzles and turbine rotors. The combustor is a can of the premixed lean burn type to achieve low NOx emissions. Engine specifications are summarized in Table 6.1. In 1995, engine tests at 1200°C TIT (the "Basic" CGT target) were successfully carried out by using ceramic components made of SN252 [4]. In 1996, the group started the evaluation of the engine with the final configuration (the "Pilot" CGT) to achieve the final target of 1350°C TIT [5]. To achieve the 1350°C operation, SN281 and SN282 were utilized. All ceramic components for the CGT302 were successfully fabricated using these new materials. For the pilot CGT, the rotational speed was reduced to 64,000 rpm from 76,000 rpm to prevent the rotor blades from chipping caused by foreign object damage (FOD).

Table 6.2 shows the selected material and the forming process for the pilot CGT302 components, and Figure 6.20 shows the principal ceramic components of the CGT302 engine. The nozzle assembly is unique in that the assembly was wrapped with SiC fibers bonded by CVD SiC.

TABLE 6.1 CGT 302 ENGINE SPECIFICATION TARGETS (ADAPTED FROM REF. [10])

	Unit	Basic CGT	Pilot CGT
Max Power	kW	140	300
Thermal Efficiency	%	30	42
Turbine Inlet Temp	°C	1200	1350
Air Flow Rate	kg/s	0.68	0.94
Pressure Ratio		5.9	8.0
Gas Generator Speed	rpm	68,400	64,000
Power Turbine Speed	rpm	51,300	47,800
Compressor Efficiency	%	78	82
Turbine Efficiency	%	82.2	85.5
Heat Exchanger Efficiency	%	80	80

TABLE 6.2 MATERIAL AND FORMING PROCESSES FOR THE PILOT CGT COMPONENTS (ADAPTED FROM REF. [6])

Component	Material	Fabrication Process
Combustor Liner	SN282	Drain cast
Support Ring, Combustor	SN282	CIP
Scroll Parts	SN282	Drain cast
Gas Gen Nozzle Segment	SN282	Hybrid mold
GGT Nozzle Support Structures	SN282	CIP
PT nozzle Segments	SN282	Hybrid mold
PT Nozzle Ducts	SN282	CIP
GGT Rotor	SN281	Hybrid mold
PT rotor	SN282	Hybrid mold
PT shroud	SN282	CIP
Seal Ring	SN282	Drain cast

Figure 6.20 Ceramic Components of the Pilot CGT302 Engine (Reprinted from ref. [6])

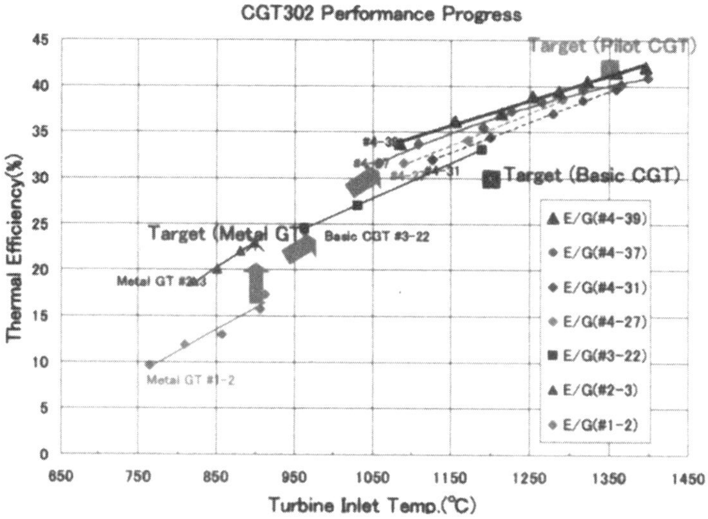

Figure 6.21 Historical Trend of the Thermal Efficiency of the Pilot CGT302 Engine (Reprinted from ref. [6])

Spin proof testing of GGT rotors was performed to evaluate the capability of the fabrication process. The maximum principal stress is 266MPa and high stress occurs throughout the disk.

Photographs were taken at the moment of fracture and the failure speed of the turbine rotor was measured. Fifteen rotors attained the proof speed of 91,200 rpm and the average burst speed was nearly 100,000 rpm.

Power Turbine Rotor. The power turbine (PT) rotor for the basic CGT was a hybrid component, fabricated by fitting 42 ceramic blades in a metallic disc. In order to reduce tip clearances, however, a monolithic ceramic rotor design was selected for the PT rotor for the pilot CGT. The rotor was 192mm in diameter. SN282 material and the hybrid molding process were used to fabricate the PT rotor.

Spin proof testing of PT rotors was also performed to evaluate the fabrication process. Five rotors attained the proof speed of 57,400rpm, and the average burst speed was 30% higher than the rated rotating speed of the PT rotor.

Figure 6.12 shows the two integral ceramic rotors used in the CGT302 engine.

Engine Test Results

The durability of the CGT302 was demonstrated in 2100 hours of cumulative operation between 1200 and 1396°C TIT from 1995 through 1999. A thermal efficiency of 42.1% was obtained at 1396°C [8,10,21]. To achieve the peak performance, a reduced tip clearance and abradable shroud were utilized in the power turbine. The accomplishments are summarized below:

- All ceramic components were successfully fabricated from SN282 and SN281.
- 20 hours of operation were achieved at 1335°C, and 1000 hrs were achieved at 1200°C.
- Over 2100 hours of cumulative operation.
- A thermal efficiency of 42.1.0% and a power output of 322 kW were achieved at 1396°C TIT.
- No defects in any parts were detected in post-test FPI (fluorescent dye penetrant inspection).
- Combustor testing at 1350°C TIT produced less than 30 ppm (16% O_2) NOx emissions.
- Geometric precision of the pilot GGT rotor made of SN281 achieved the design targets.
- The 192 mm dia SN282 PT rotor was successfully fabricated by hybrid molding.

The relationship of these accomplishments is illustrated in Figure 6.21.

PROCESS DEVELOPMENT FOR AUTOMOTIVE HYBRID VEHICLE ENGINE

Kyocera supported the automotive hybrid vehicle engine developments by providing components and applying low cost, high volume manufacturing processes to the parts. The engine used as a test bed for these technologies was Rolls-Royce Allison's AGT-5 [15].

Parts fabricated included:

- SN253 Rotors by slip casting (Figure 6.11)
- SN252 Scrolls by drain casting (Figure 6.18)
- SN282 Combustor liners by drain casting (Figure 6.19)

Although the Kyocera Corporation in Japan had previously provided SN252 AGT-5 gasifier rotors under the previous AGT program, a second iteration effort was conducted to provide rotors from Kyocera's Vancouver, WA facility using Kyocera's SN253 material. Rotors were cast to near net shape to provide as-processed blades and minimal machining of the hub, shaft, and blade tips. Figure 6.11 illustrates the rotor in the as-HIPed condition.

MOR bars cut from rotor hubs exhibited strengths above both the in-house room temperature and elevated temperature specifications.

COST MODEL DEVELOPMENT

Kyocera developed a detailed and comprehensive model of the manufacturing costs for its ceramic components. As input variables the model utilizes material and tooling, labor, machine use, and yield for each step, and production quantity. The model also incorporates overhead burdens, and profit. The output of the model is both a graphic and tabulated cost breakdown for the cost of each step and the total price. The model also summarizes the cost of each step by elements of labor, material, and machine use. The model is capable of utilizing amortization of tooling or other capital expenses. This model was subsequently used by Kyocera in several cost reduction, and manufacturing scale up efforts as discussed below.

Production Scale-up and Cost Reduction for a Ceramic Turbine Nozzle

Kyocera used its detailed cost model to demonstrate cost reducing process improvements for a hybrid molded ceramic turbine nozzle segment. The nozzle was designed by Honeywell for the first stage of the 331-200 APU engine. The objective of this effort was to reduce the cost to below $200/part for a production level of 500 parts/month. The effort was performed during 1995–1998 [17,22,23]. The turbine nozzle segment is illustrated in Figure 6.13.

Vanes were fabricated from SN 252 and SN282 materials. Each segment incorporated 9 edges, and 2 cylindrical surfaces to be ground. Multiple grinding fixtures and inspection steps are required to support, position and confirm dimensional results during grinding. Inspections included real time x-ray radiography in the green state, film x-ray radiography, florescent penetrant inspection (FPI), 10x microscopy, visual, and dimensional.

The approach for cost reduction was to 1) establish a baseline process, 2) identify primary cost drivers via Kyocera's cost model, and 3) focus on improvements of cost drivers to achieve the cost goals.

A baseline process capable of prototype quantities of parts was first developed. Vanes were processed to establish both the baseline process and costs. Inspections were performed to characterize and quantify the causes and rates of rejection.

The cost model highlighted grinding and inspection as primary cost drivers at 44% and 32% of the parts cost, respectively. The baseline process suffered the greatest yield loss from shroud distortion during sintering which also contributed to high costs during dimensional inspection and grinding steps. The cost model directed process improvements in the following areas:

- Characterizing of, and reducing the sintering distortion.
- Developing procedures for quick dimensional inspection screening after sintering.
- Developing ganged grinding fixtures capable of multiple grinding operations and in-process dimensional inspection without part removal.

Baseline parts were characterized for shroud distortion, and, alternate approaches to setting the parts were evaluated. The alternate approaches provided a 70% reduction in the distortion compared to the baseline method.

Additionally, the grinding and inspection processes were further analyzed for yield and cost to better refine the cost driving operations. Improved fixtures that minimize setups and allow on-machine, in-process dimensional inspections were fabricated to demonstrate the improved concepts. Use of the improved grinding and inspection tooling resulted in a timesaving of 65% for the most expensive grinding operation, and similar benefits to all other grinding steps.

When the distortion, inspection, and grinding improvements were incorporated into the cost model, it showed that the combined grinding and dimensional inspection costs were reduced 45%, and an overall part cost reduction of 52% was achieved due to improvements in yield. The model also projected that an overall cost reduction of 76% was feasible with "next generation" higher efficiency tooling. The cost target of $200/part would nearly be met with the projected improvements. Figure 6.22 illustrates the cost breakdown for baseline, demonstrated improvements, and projected results [23].

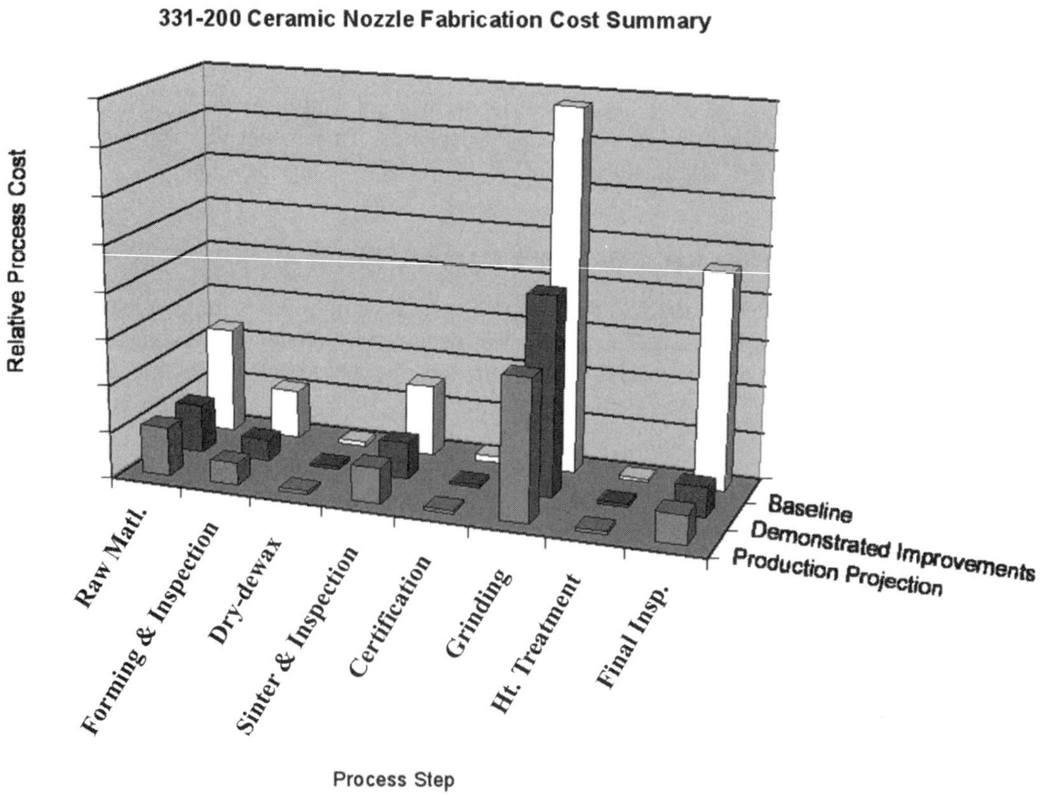

Figure 6.22 Cost Model Summary for Baseline and Improved Process (Adapted from ref. [22])

LESSONS LEARNED/HURDLES TO COMMERCIALIZATION

- Careful cost analysis/modeling at an early stage of development is necessary in order to direct activity and make decisions on project viability.
- Major cost drivers for advanced engine ceramic components are raw materials, grinding and inspection.
- Early stage prototypes should be made by prototype methods rather than production methods; forming methods that offer seamless transition from prototype to production should be used where possible.
- Early stage design for manufacturability should be used to shorten development time and cost and improve probability of commercial success.
- Focus by users on first-cost, rather than life cycle costing or cost of ownership, limits adoption of this and other advanced technologies.
- Cost burden for technology development for such speculative markets should not be placed on the component supplier if there is little corollary financial benefit.
- User and Government supported central materials database development would reduce cost and improve quality and availability of data.
- Continued materials development is required, but cannot be justified by expected return on investment by the materials suppliers.

ACKNOWLEDGEMENTS

1) Thanks to the U.S DOE and Honeywell Engines, Systems & Services for program support under contract DEA0296EE50454, formerly DEN3-335.
2) Thanks to the U.S. DOE and Solar Turbines Incorporated for program support under contract DE-AC02-92CE40960.
3) Thanks to the U.S. Dept of Defense and United Technologies Research Center for program support under contract N00014-98-C-0355.
4) Thanks to the Agency of Industrial Science and Technology of MITI/METI and NEDO for support of the CGT programs in Japan
5) Thanks to Kawasaki Heavy Industries, Ltd., Sumitomo Precision Products Co., Ltd., Honeywell Engines, Systems &Services, Rolls-Royce Indianapolis and United Technologies Corporation for their support in their projects.
6) Unless otherwise indicated figures and tables in this chapter were provided courtesy of Kyocera Corporation.

REFERENCES

[1] Arakawa, H., Suzuki, T., Saito, K., Tamura, S., and Kishi S., "Research and Development of 300kW Class Ceramic Gas Turbine Project in Japan," ASME paper 97-GT-87, presented at the International Gas Turbine and Aeroengine Congress and Exposition, Orlando, FL, USA, June 2–5, 1997.

[2] Yoshida, M., Tsuruzono, S., Ono, T., and Gejima, H., "Development of Silicon Nitride Rotors for Gas Turbine Applications," ASME paper 94-GT-332, presented at the International Gas Turbine and Aeroengine Congress and Exposition, The Hague, The Netherlands, June 13–15, 1994.

[3] Tanaka, K., Tsuruzono, S., Kubo, T., and Yoshida, M., "Development and Fabrication of Ceramic Gas Turbine Components," ASME paper 96-GT-446, presented at the International Gas Turbine and Aeroengine Congress and Exposition, Birmingham, U.K., June 10–13, 1996.

[4] Tatsumi, T., Takehara, I., and Ichikawa, Y., "Current Status of CGT302 Ceramic Gas Turbine," *Proceedings of the 1995 Yokohama International Gas Turbine Congress*, GTSJ, Yokohama, October 23–27, 1995. 95-YOKOHAMA-IGTC-125, Vol. I, I-153/I-156, 1995.

[5] Ichikawa, Y., Tatsumi, T., Nakashima, T., Takehara, I., and Kobayashi, H., 1997, "Current Status of CGT302 (Progressing to Final Phase)," ASME paper 97-GT-461, presented at the International Gas Turbine and Aeroengine Congress and Exposition, Orlando, FL, USA, June 2–5, 1997.

[6] Yoshida, M., Tanaka, K., Kubo, T., Terazono, H., and Tsuruzono, S., "Development of Ceramic Components for Ceramic Gas Turbine Engine (CGT302)," ASME paper 98-GT-398, presented at the International Gas Turbine and Aeroengine Congress and Exhibition, Stockholm, Sweden, June 2–5, 1998.

[7] Tatsumi, T., Takehara, I., and Ichikawa, Y., "Development Summary of the 300 kW Ceramic Gas Turbine CGT302, ASME paper 99-GT-105, presented at the International Gas Turbine and Aeroengine Congress and Exhibition, Indianapolis, IN, USA, June 7–10, 1999.

[8] Takehara, I., Tatsumi, T., and Ichikawa, Y., "Development Summary of CGT302 Ceramic Gas Turbine," *Proceedings of the International Gas Turbine Congress 1999 Kobe,* T. Tamaru, ed.. GTSJ, Kobe, Japan, November 14–19, 1999. IGTC'99Kobe-101, Vol. I, pp. 57–64, 1999.

[9] Terazono, H., Tanaka, K., Kubo, T., Tsuruzono, S., and Yoshida, M., "Development and Fabrication Process for Ceramic Gas Turbine Components," *Proceedings of the International Gas Turbine Congress 1999 Kobe,* T. Tamaru, ed.. GTSJ, Kobe, Japan, November 14–19, 1999. IGTC'99Kobe-104, Vol. I, pp. 81–86, 1999.

[10] Takehara, I., Tatsumi, T., and Ichikawa, Y., "Summary of CGT302 Ceramic Research and Development Program," ASME paper 2000-GT-0644, presented at the ASME TURBO EXPO 2000, Land, Sea & Air, Munich Germany, May 8–11, 2000.

[11] Tanaka, K., Yoshida, M., Kubo, T., Terazono, H., and Tsuruzono, S., "Development and Evaluation of Ceramic Components for Small Gas Turbine Engine, ASME paper 2000-GT-0531, presented at the ASME TURBO EXPO 2000, Land, Sea & Air, Munich Germany, May 8–11, 2000.

[12] *Advanced Gas Turbine (AGT) Technology Development Project*, Final Report, Garrett Auxiliary Power Division, Allied-Signal Aerospace Co., Contract DEN 3-167, DOE/NASA/0167-12, NASA CR-180891, December 1987.

[13] *Advanced Gas Turbine AGT Technology Project,* Final Report, Allison Gas Turbine Division, General Motors Corporation, Contract DEN 3-168, DOE/NASA/0168-11, NASA CR-182127, EDR 13295, August 1988.

[14] *Hybrid Vehicle Turbine Engine Technology Support Project*, 1995–1996 Annual Report, Allison Engine Company, Inc., Contract DE-AC02-96EE50453, DOE 0336-8, EDR 18370.

[15] *Advanced Turbine Technology Applications Project (ATTAP)/Hybrid Vehicle Turbine Engine Technology Support Project (HVTE-TS)*, Final Summary Report, Rolls-Royce Allison, Contract DE-AC02-96EE50453, formerly Contract DEN 3-336, DOE 0336-9F, EDR 18635, December 1998.

[16] Price, J.R., Jimenez, O., Parthasarath, V., and Miriyala, N., "Ceramic Stationary Gas Turbine Development Program—Sixth Annual Summary," ASME paper 99-GT-351, presented at the International Gas Turbine & Aeroengine Congress & Exhibition, Indianapolis, IN, USA, June 7–10, 1999.

[17] Kinney, T.R., and Easley, M.L., "Ceramic Gas Turbine Technology Development," ASME paper 97-GT-465, presented at the International Gas Turbine & Aeroengine Congress & Exhibition, Orlando, FL, USA, June 2–5, 1997.

[18] Day, W.H., Kouris, K., Holowczak, J. E., Blair, M. F., Hale, S. E., and Bornemisza, T., "Development of Cooled, Ceramic First Stage Vanes for the FT8 Aeroderivative Gas Turbine," ASME paper 98-GT-133, presented at the International Gas Turbine and Aeroengine Congress and Exhibition, Stockholm Sweden, June 2–5, 1998.

[19] Holowczak, J.E., Bornemisza, T.G., and Day, W.H., "Sector Testing of Cooled Silicon Nitride Ceramic HPT Vanes for the FT8 Aeroderivative Engine," ASME paper 2000-GT-662, presented at the ASME TURBO EXPO 2000, Land, Sea, & Air, Munich, Germany, May 8–11, 2000.

[20] Lofaj, F., Wiederhorn, S.M., Long, G.G., Jemian, P.R., Ferber, M.K., "Cavitation Creep in the Next Generation Silicon Nitride," *Proceedings of the 7th International Symposium, Ceramic Materials and Components for Engines*, J.G. Heinrich and F. Aldinger, eds., Goslar, Germany, June 19–21, 2000, Wiley-VCH Verlag GmbH, Weinheim, Germany, 2001. pp. 487–493.

[21] Tatsumi, T., "Development of the 300 kW Ceramic Gas Turbine CGT302," Chapter 15 in *Ceramic Gas Turbine Design and Test Experience, Volume I of Progress in Ceramic Gas Turbine Development,* M. van Roode, M.K. Ferber, and D.W. Richerson, eds., ASME Press, July 2002, New York, NY, USA. pp. 331–360.

[22] Easley, M.L., Schenk, B., and Cai, H., "Ceramic Gas Turbine Technology Development,"ASME paper 98-GT-554, presented at the International Gas Turbine and Aeroengine Congress and Exhibition, Stockholm Sweden, June 2–5, 1998.

[23] Schenk, B., "Ceramic Gas Turbine Technology Development," ASME paper 99-GT-315, presented at the International Gas Turbine and Aeroengine Congress and Exhibition, Indianapolis, IN, USA, June 7–10, 1999.

CHAPTER

7

RESEARCH & DEVELOPMENT OF CERAMIC MATERIALS AND COMPONENTS FOR GAS TURBINES

Tsutomu Yamamoto and Masaaki Masuda

Corporate R&D Group, NGK Insulators, Ltd., Nagoya, Aichi, Japan

INTRODUCTION

NGK Insulators, Ltd. (NGK) was established in 1919. Major business activities are manufacture and sale of electrical insulators, industrial ceramic products, beryllium-copper products, and plant engineering. In the 1970s, NGK began developing high strength silicon nitride and silicon carbide for use as structural ceramics applicable to reciprocating engine components or gas turbine components. At the same time, NGK participated in several ceramic gas turbine (CGT) programs and concentrated efforts on the research and development of the component fabrication process technology.

First, during the 1980s, NGK supplied many components made of silicon nitride for CGT projects in the USA. As a result, the base of the fabrication process technology for CGT components using silicon nitride was established. Subsequently, in the late 1980s, research and development of a power generating CGT by Tokyo Electric Power Company (TEPCO), and a Japanese government CGT program for co-generation and automotive applications was initiated. As a result of the progress on the Japanese CGT programs, NGK developed a new silicon nitride material having a higher heat resistance that is equivalent to that of silicon carbide. For component design methodology development, mechanical and fatigue property data for the silicon nitride materials were accumulated. Also, a new forming process capable of fabricating complex-shaped components was studied and established.

This chapter describes the status of research and development of silicon nitride materials and component fabrication process technologies.

MATERIAL RESEARCH AND DEVELOPMENT

Silicon nitride ceramics for heat engine components and gas turbine components have been studied with several kinds of sintered additives, such as the CeO_2-MgO-SrO-ZrO_2, the Y_2O_3-MgO-ZrO_2, and the Y_2O_3-Yb_2O_3 dopant systems [1,2]. The densification and the α to β transformation process have been studied to develop a high strength material. The grain boundary phase has been controlled to create a high temperature material. Figure 7.1 shows the improvement of strength of NGK's silicon nitride over the last 20 years. The properties of the silicon nitride materials after the above research are shown in Table 7.1.

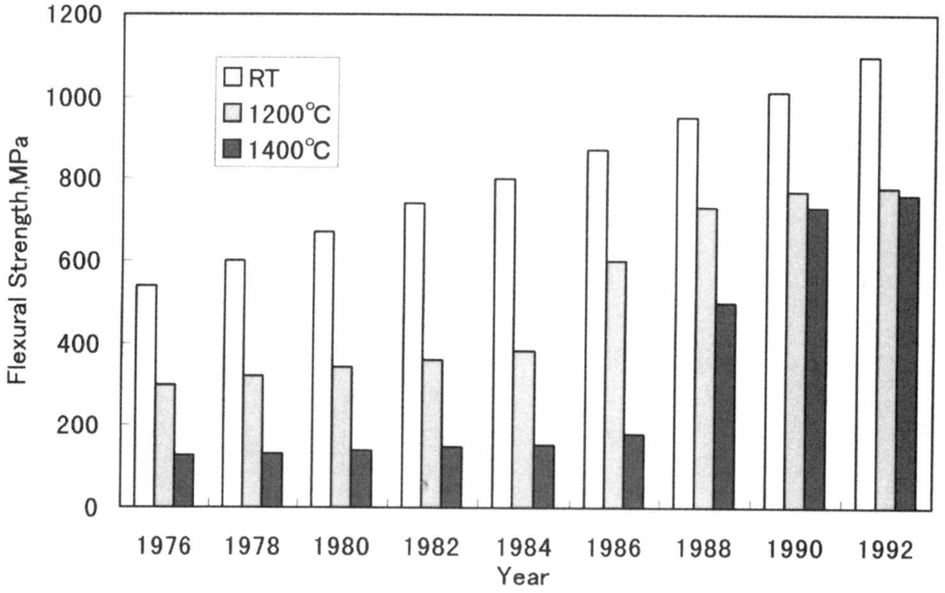

Figure 7.1 Improvement of Strength of Silicon Nitride at NGK

TABLE 7.1 PROPERTIES OF SILICON NITRIDE MATERIALS

Material code	SN-55	SN-73	SN-84	SN-88
Density (g/cm^3)	3.2	3.2	3.2	3.5
Bending strength (MPa, 4-point, JIS R1601)				
RT	850	1150	860	860
800°C	750	950	800	—
1000°C	380	600	760	770
1200°C	—	410	780	770
1400°C	—	—	—	760
Elastic modulus (GPa)	260	290	280	260
Poisson ratio	0.27	0.27	0.27	0.27
Fracture toughness (MPa.m$^{1/2}$, SEPB, JIS R1607)	7	7.5	6	7
Hardness (GPa, Knoop, 300 kg)	15	14	15	15
Coefficient of thermal expansion (40–800°C, K^{-1})	3.4	3.4	3.7	3.5
Thermal conductivity (RT, W/m.K)	35	45	40	75
Heat capacity (kJ/kg.K)	0.7	0.7	0.7	0.7
Thermal shock resistance (°C)	800	1000	1000	1200
Oxidation resistance				
1000°C, 1000 hrs, (mg/cm^2)	0.2	0.2	<0.1	<0.1

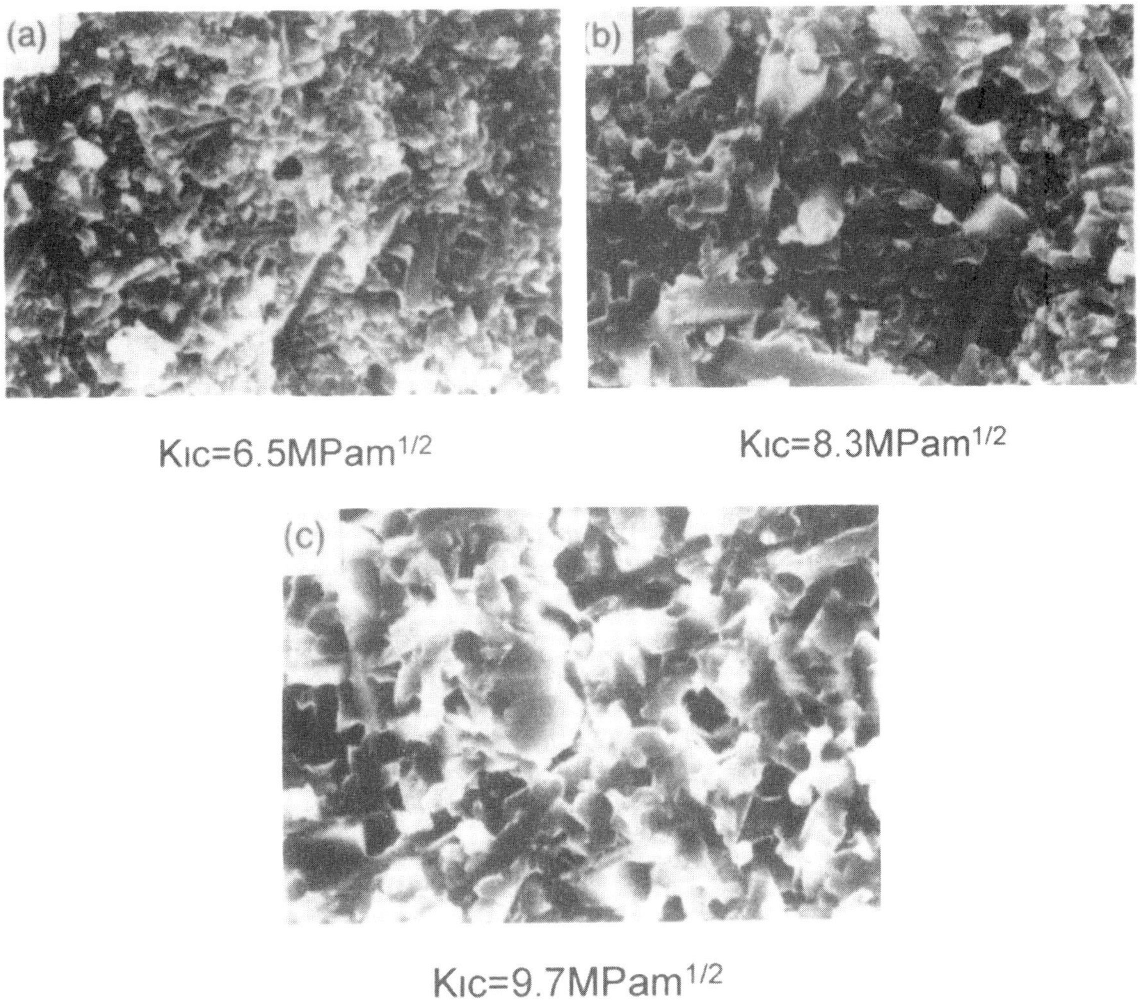

Figure 7.2 Microstructure of Various Silicon Nitrides with Different Fracture Toughness Values [3]

The microstructures of the materials with high fracture toughness shown in Figure 7.2 have been discussed [3]. The pictures show high strength Si_3N_4 sintered materials produced by different chemical processes, starting powders and/or sintering conditions. Their K_{IC} values are 6.5, 8.3, and 9.7 $MPa.m^{1/2}$,respectively. The bodies with a coarser microstructure exhibit greater toughness. Grain size analysis results show that β–Si_3N_4 grains having a diameter larger than 1 μm are stronger. The relationship between the grain population of each classified diameter range and K_{IC} is illustrated in Figure 7.3 [3]. Populations of grains having a diameter between 1.0 and 1.5 μm, and especially larger than 1.5 μm, contribute the most to toughness. Grains with a diameter larger than 1 μm primarily increase toughness, while grains with a diameter smaller than 1.0 μm have a negative effect on toughness.

The mechanical properties and stability of SiO_2 doped silicon nitride were investigated up to 1600°C [4]. The material showed almost constant strength around 600MPa between room temperature to 1600°C, as shown in Figure 7.4 [4]. Weight gain during soaking time in air at 1600°C for 100 hours was 0.2 mg/cm^2, and no change in the crystal phase inside the specimen was detected. The oxidation resistance at 1600°C is equivalent to that of the rare earth oxide doped silicon nitride ceramic at 1400°C. The fracture toughness of

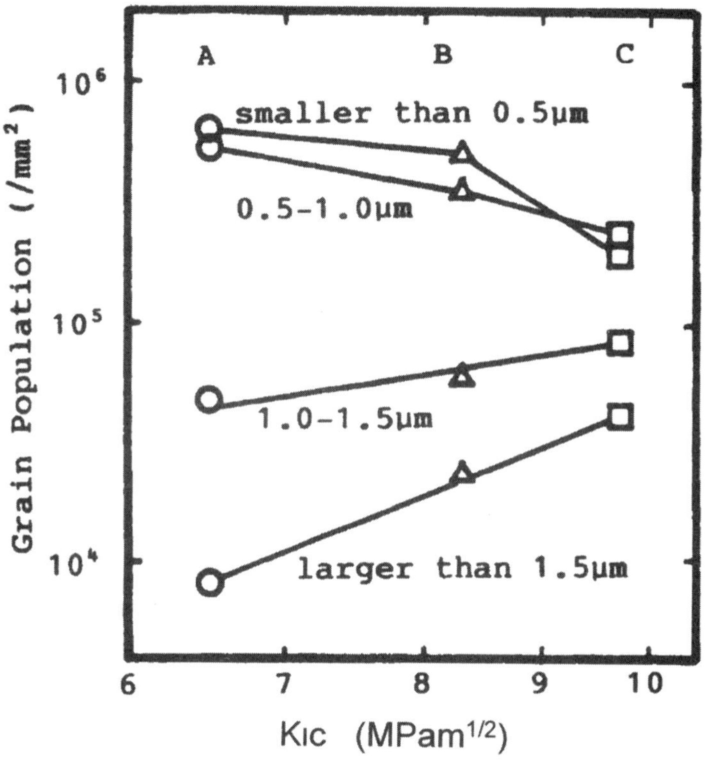

Figure 7.3 Classified Grain Diameter Population vs. K_{1C} Relation [3]

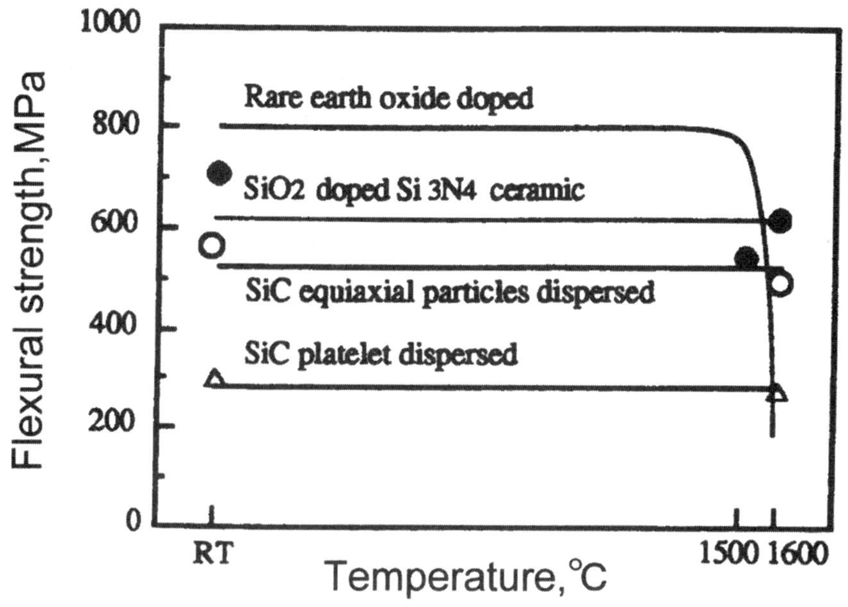

Figure 7.4 Temperature Dependence of Bending Strength of Various Ceramics [4]

SiO_2 doped silicon nitride is 2 MPa.m$^{1/2}$ because of the rigid grain boundaries. Figure 7.4 shows the strength of a material with 50wt%, 5 μm diameter SiC equiaxial particles, and of a material with 40wt%, 30 μm diameter dispersed platelets. The fracture toughness of the former material was 3 MPa.m$^{1/2}$, while that of the latter material was 5 MPa.m$^{1/2}$. The bending strength decreases with increasing fracture toughness.

STUDY OF LIFETIME PREDICTION METHOD FOR SILICON NITRIDE

To clarify the effect of grain size based on the Weibull statistical theory, the fracture strengths of various types of silicon nitride specimens were measured. Figure 7.5 [5] shows the experimental results. Each effective volume was calculated using the uniaxial Weibull distribution function and only the maximum principal stress was taken into account. Weibull moduli which were obtained using 4-point bending test results are almost the same as with values calculated from the slope in Figure 7.5. This result shows that the fracture strength of components within 5 orders of magnitude can be predicted based on the Weibull statistical theory [6].

Tension compression biaxial stress tests have been carried out by measuring the 6 mm diameter gauge portion of a button-head type specimen. Figure 7.6 [5,7] shows the fracture specimens. The surface fracture in each case was perpendicular to the direction of the maximum principal stress, so the maximum principal stress controlled the multiaxial stresses within the fracture. Figure 7.7 [5] shows the experimental results from tests of multiaxial stresses, and the predicted results of fracture theories from multiaxial stresses. The experimental results and the predictions are normalized so that tensile strength is unity. The experimental data agreed with the two curves predicted by the maximum principal stress and Weibull's multiaxial theories.

The cyclic fatigue results using tapered beam specimens at frequencies from 0.03Hz to 3kHz are shown in Figure 7.8 [8]. The cyclic fatigue behavior of silicon nitride at room temperature appears to depend prin-

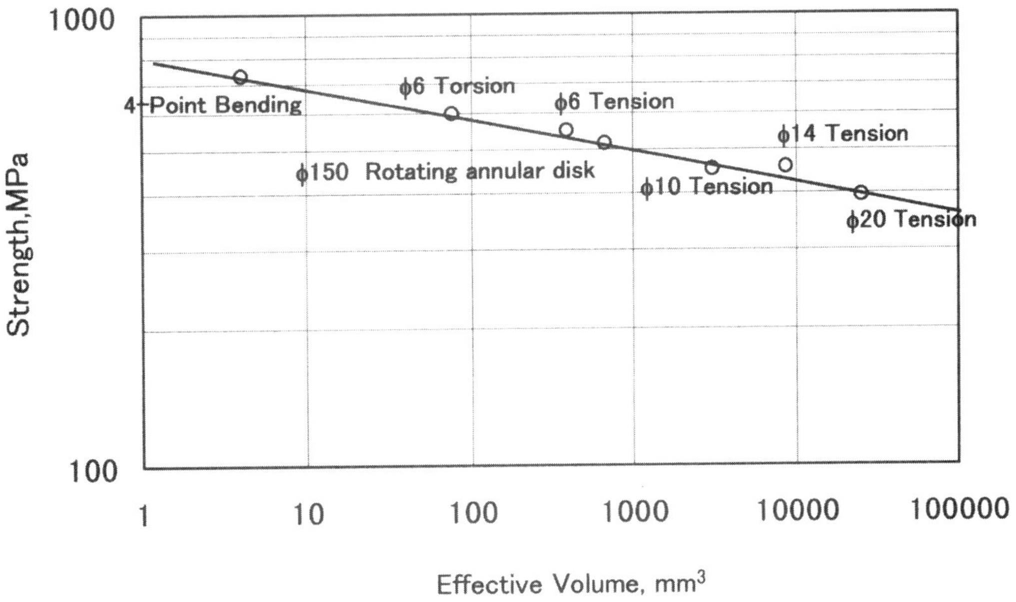

Figure 7.5 The Relationships Between Mean Strength and Effective Volume for Silicon Nitride [5]

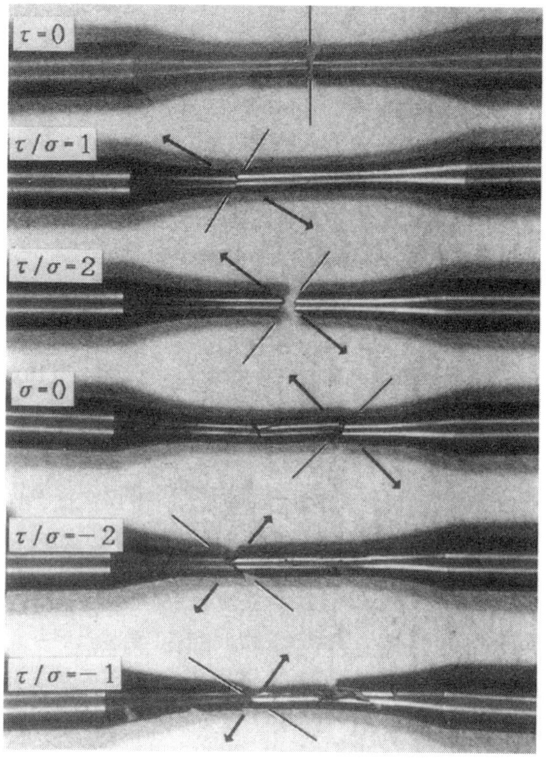

Figure 7.6 Typical Fractured Silicon Nitride Specimens in Multiaxial Load [5]

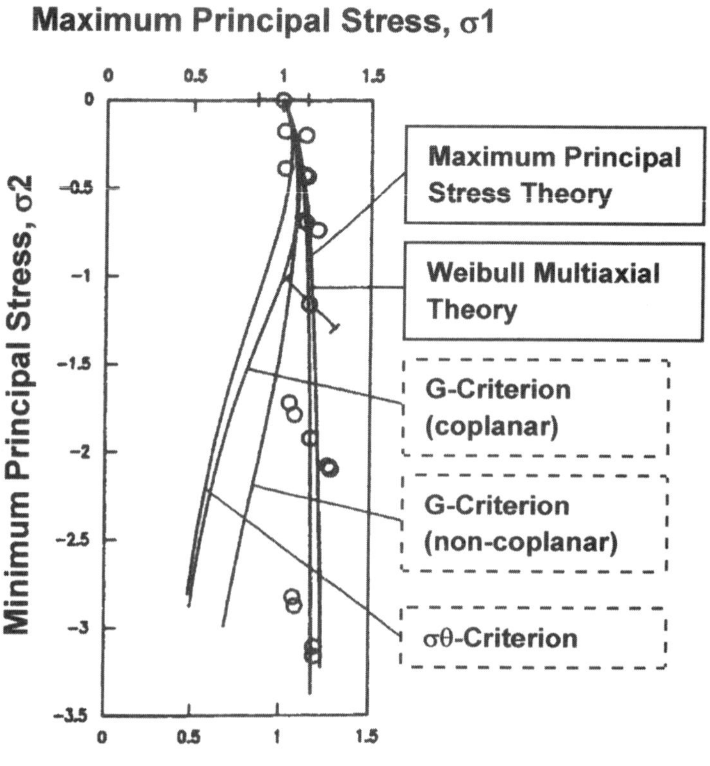

Figure 7.7 Experimental and Predicted results under multiaxial stresses [5].

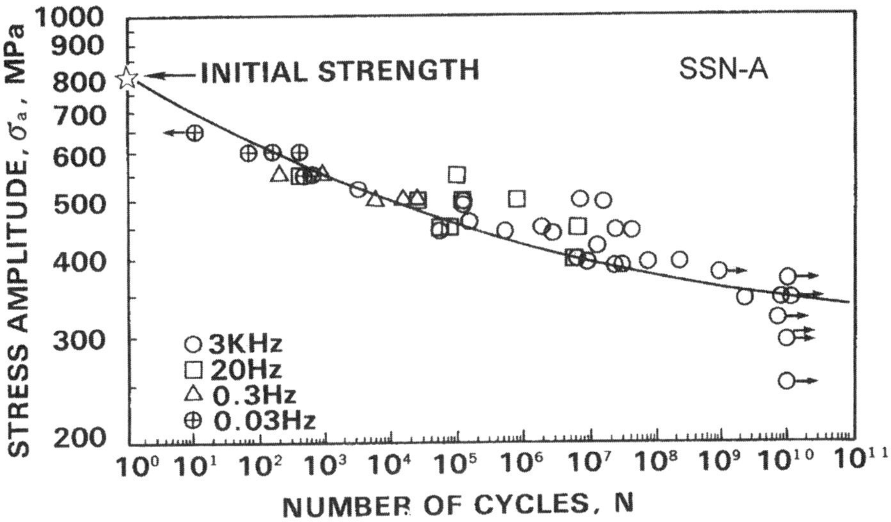

Figure 7.8 S-N Curve of Cantilever Beam Specimen Under Alternating Load [8]

cipally on the number of cycles rather than on time, because the fatigue strength at various frequencies did not show any discontinuities compared to the number of cycles. The solid curve suggests a decrease in the fatigue limit to about 40% of its initial strength.

Figure 7.9 [9] shows the mean stress and stress amplitude diagram of a material at room temperature. On the horizontal axis σ_t represents the mean tensile strength of 6-mm diameter button-head specimens.

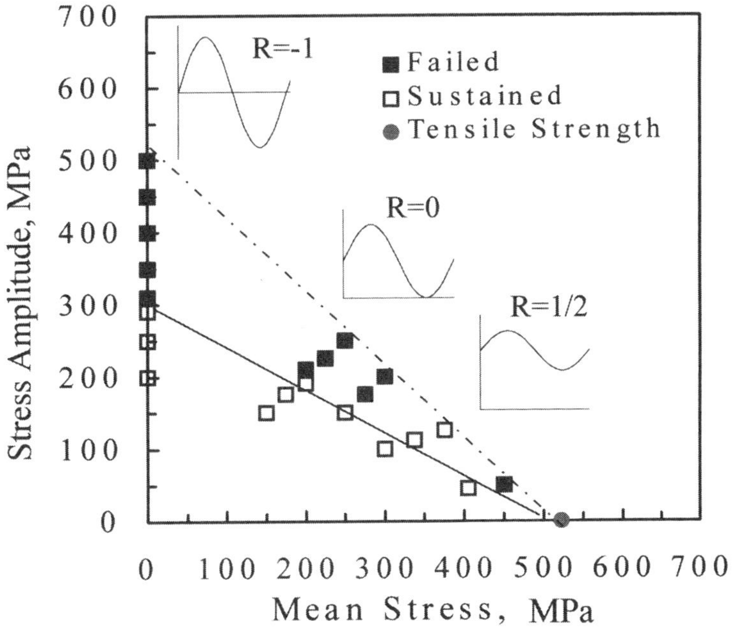

Figure 7.9 High Cycle Fatigue Test Result [9]

Open squares represent specimens surviving at 10^7 cycles, and closed squares represent the failed specimens. The dotted line represents tensile strength. Alternating load decreased strength to a larger extent than pulsating load. This diagram shows more clearly the safe stress regions. The solid line from σ_e to σ_t represents a modified Goodman line [10], which is expressed by Eq. (1) and is used as a design criterion for metal components:

$$\sigma_a = \sigma_e (1 - (\sigma_m/\sigma_t)) \tag{1}$$

where σ_a is stress amplitude, σ_m mean stress, σ_t tensile strength and σ_e alternative fatigue strength at a stress ratio of R = –1. The modified Goodman line agreed well with the results of the present fatigue test. The design methodology proposed for metal fatigue is applicable to the silicon nitride examined in our study.

The fatigue life of ceramic materials has been predicted by the power laws or exponential laws of crack propagation. The following power law crack propagation equation is widely used to predict the lifetime of ceramic materials:

$$da/dt = AK_I^n \tag{2}$$

where A is a constant, n is the slow crack growth exponent, and a is crack length. Based on Eq. (2), the static, dynamic and cyclic fatigue life of three kinds of silicon nitride have been discussed. The sintering additives were SrO, MgO, CeO_2, and ZrO_2 for SSN-A; Y_2O_3, MgO, and ZrO_2 for SSN-B; and Y_2O_3 and Yb_2O_3 for SSN-C. The grain boundary phase is glassy in SSN-A, and crystallized in SSN-B and SSN-C. These are the original materials for SN-73, SN-84 and SN-88, respectively. A comparison of lifetimes of SSN-A under dynamic, static and cyclic load is shown in Figure 7.10 (a)[11]. Equivalent times were calculated using n values of 40 at 800°C and 11 at 1000°C, based on the static fatigue test results. The equivalent lifetimes of dynamic fatigue and cyclic fatigue agree with the static fatigue data at 800°C. At 1000°C, on the other hand, the equivalent lifetimes of dynamic fatigue also agree with the static fatigue data. However, the equivalent lifetimes of cyclic fatigue at 1000°C tended to shift to the higher-stress, longer-life side than the static fatigue data. Eq.(2) is not applicable to cyclic fatigue under high frequency (e.g., 20Hz) cyclic stress. Another mechanism must be considered for cyclic fatigue at elevated temperatures [12].

The relationship between stress and equivalent time to failure for SSN-B and SSN-C are shown in Figure 7.10(b). The equivalent times of tensile strength data were calculated using the n value of the static fatigue data. The results for SSN-B and SSN-C are represented by two straight lines divided into two regions of stress. It is possible to apply the power law crack growth formulation to the high stress region. However, Eq.(2) is based on slow crack growth from a preexisting flaw, and tends to overestimate lifetime in low stress regions. It is therefore necessary to use another procedure to predict behavior in the creep rupture region.

The following Larson-Miller parameter P is widely used to predict lifetime of creep rupture in metallic materials:

$$P = f(s) = T (C + \log t_r) \tag{3}$$

where T is absolute temperature, t_r is time to creep rupture, and C is a constant. The C value of 20 in Eq.(3) is widely used in the case of metallic materials. The relationship between applied stress and the Larson-Miller parameter depends on the temperature [13]. However, a C value of 20 is unsuitable for the silicon nitride used. Figures 7.11(a) and (b)[11] show the master Larson-Miller curve, where C = 44 for SSN-B and C = 30 for SSN-C are used. The results for both SSN-B and SSN-C are represented by a straight line. The effect of temperature on the creep rupture strength of SSN-B was significantly more severe than for SSN-C. The C value in Eq.(3) depends on the material. The higher the C value, the more sensitive is the influence of temperature on creep rupture time.

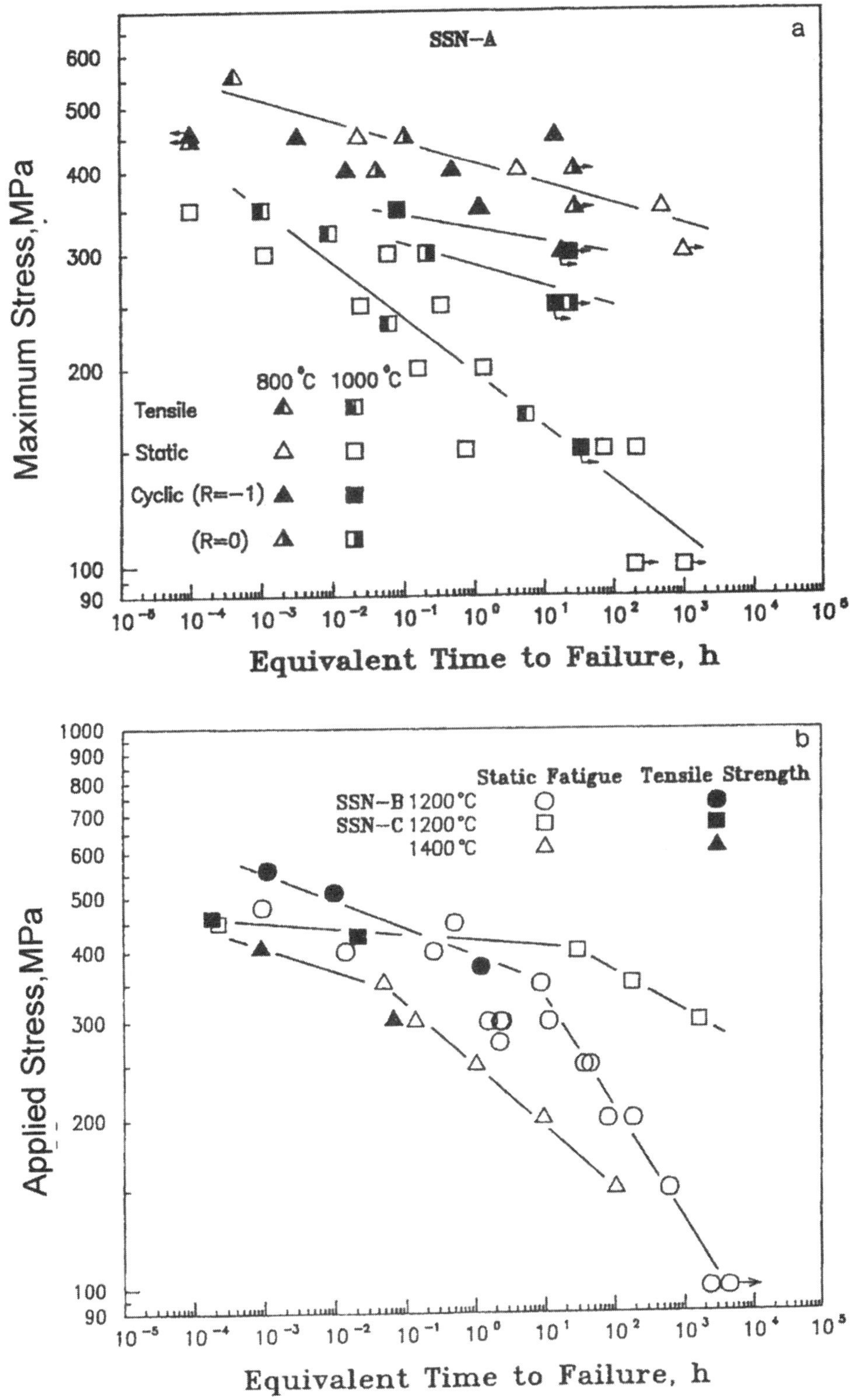

Figure 7.10 Comparison of Lifetime Under Dynamic, Static and Cyclic Load [11]

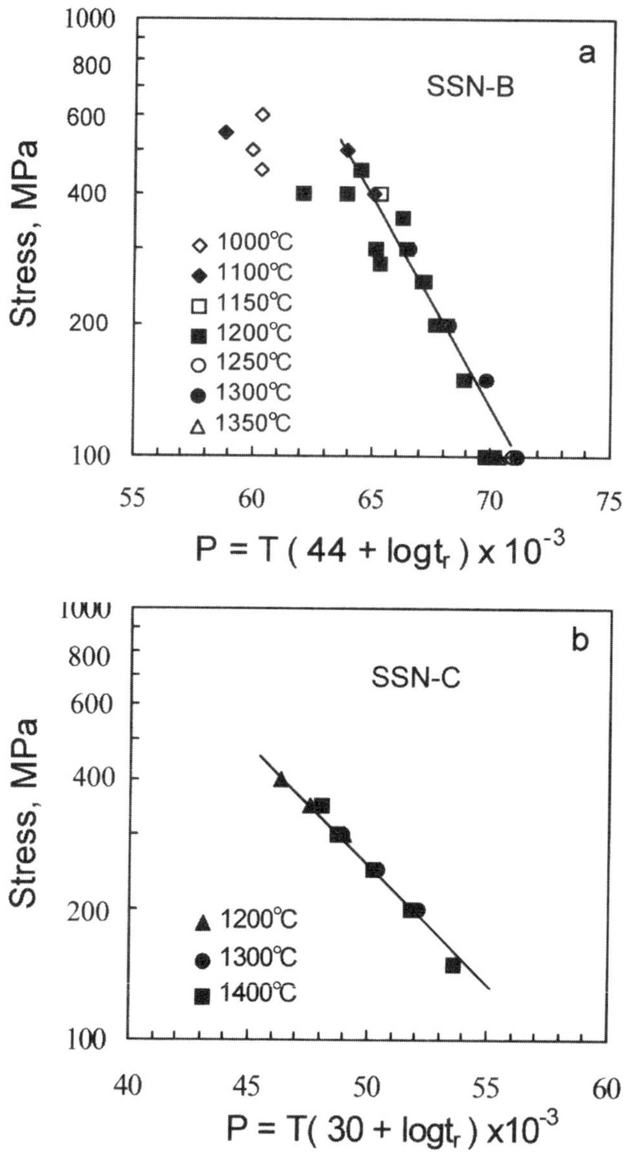

Figure 7.11 Master Larson-Miller Curve for Silicon Nitride [11]

Figure 7.12 [5] shows the design concept for high heat resistant ceramic components. First, the initial design is determined, including such factors as geometric designs, material selection, and operating conditions. From this information, FEM techniques analyze stress and temperature distributions in the component. On the other hand, mechanical properties such as strength, Weibull modulus, slow crack growth (SCG) behavior, creep failure and oxidation fatigue, are evaluated to obtain the allowable stress-temperature critical conditions for each failure mode. A graph of stress vs. temperature, the Fracture Map, shows the stress-temperature condition for each mode. Comparing these variables in the Fracture Map provides either the required material properties for the component or new information for design. If necessary, the same analysis is performed on new materials and designs.

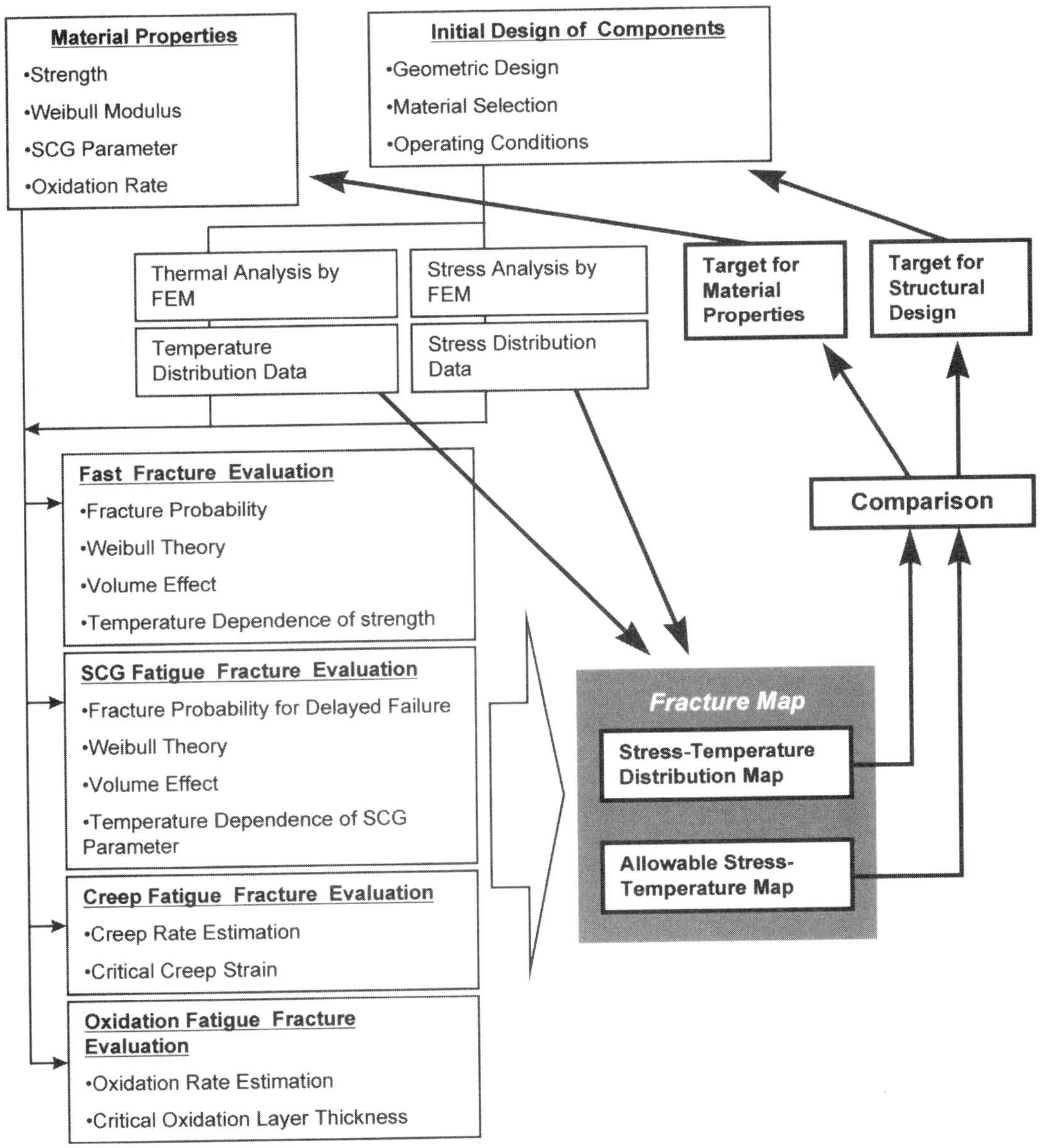

Figure 7.12 Design Concept for Ceramic Components [5]

COMPONENT FABRICATION PROCESS DEVELOPMENT

In order to fabricate various ceramic component shapes, several fabrication processes have been developed through participation in government and in-house CGT programs. The critical tasks were to establish the near-net and complex-shape forming technologies.

Injection Molding Process [14]

The injection molding process is one of the best techniques for producing a complex shape correctly for high production volumes. This technique has been successfully used for ceramic turbocharger rotors for

Figure 7.13 AGT101 Turbine Rotor Made of SN-84

automotive use. However, a large radial turbine rotor such as the AGT101 rotor (Figure 7.13) for Garrett Auxiliary Power Division and the automotive 100kW CGT rotor, having an outside diameter of about 130mm (Figure 7.14), for the Japan Automobile Research Institute, Inc.(JARI) are difficult to fabricate by the injection molding process due to the limitations of de-waxing. A section component thicker than about 50mm cannot be de-waxed properly based on our experience. Consequently, the complex shaped blade section was fabricated by the injection molding process and the shaft section attached by cold isostatic pressing (CIP). Both pieces were combined into one by CIP, and then sintered to obtain a body comprised of an

Figure 7.14 Automotive 100kW CGT Turbine Rotor Made of SN-88

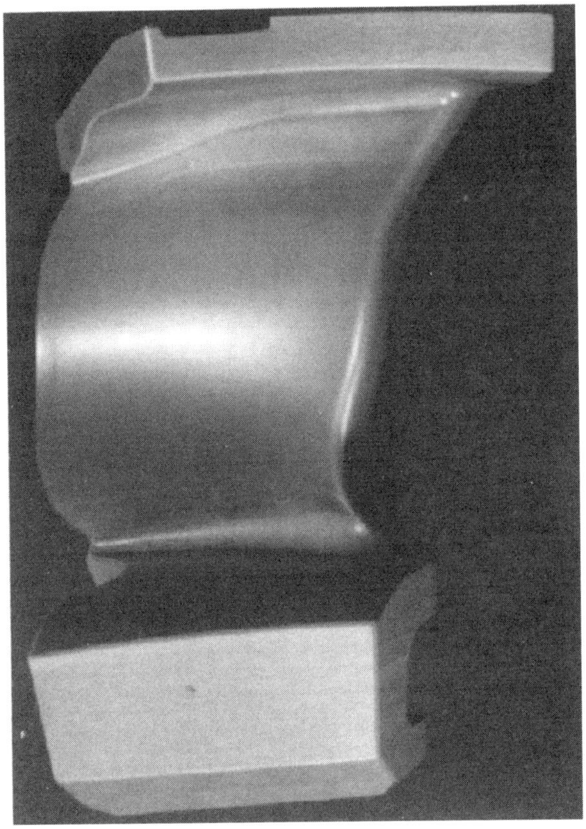

Figure 7.15 Turbine Nozzle Made of SN-88

integral radial turbine rotor. The JARI 100kW CGT rotor made of SN-88 silicon nitride survived a high temperature engine test (TIT 1350°C, 100hours, a tip speed of 665m/s) intact, which verified the reliable strength of the ceramics.

Calcined Block Machining

In this process a calcined block formed by CIP is machined with a five-axes computer numerically controlled grinding machine [15]. This technique is very effective for low volume manufacturing applications and for manufacturing complex-shaped components which cannot be fabricated by the injection molding process. The first stage nozzles for the Centaur 50S of Solar Turbines Incorporated (Figure 7.15) and the CGT301 turbine nozzles fabricated for Ishikawajima-Harima Heavy Industries, Ltd, were manufactured by calcined block machining. Figure 7.16 shows the air-cooled nozzles for a power generating CGT of the TEPCO 20MW CGT program [16]. An air-cooled nozzle vane was designed so that the maximum surface temperature can be maintained below 1300°C by a small amount of cooling air. A one-piece nozzle body made of SN-88 was fabricated by calcined block machining, and the cooling slits having a width of 0.5mm were formed by ultrasonic machining. Cascade tests were carried out for the air-cooled nozzles to evaluate the validity of the design. All tests were successfully conducted at a pressure of 6 atm and temperature up to 1500°C.

Figure 7.16 Air-Cooled Nozzles Made of SN-88

CIP Joining Process

Combuster liners and gas-path components are comparatively large complex shapes. To fabricate such components, the CIP joining process was studied [17]. For example, the nose cone of the CGT301 engine was fabricated using the CIP joining process. As shown in Figure 7.17, three components were formed separately, then combined into one body by CIP. Next, the integrated green body was sintered and finally machined. Thanks to the CIP joining process, manufacturing of large ceramic components has become much easier.

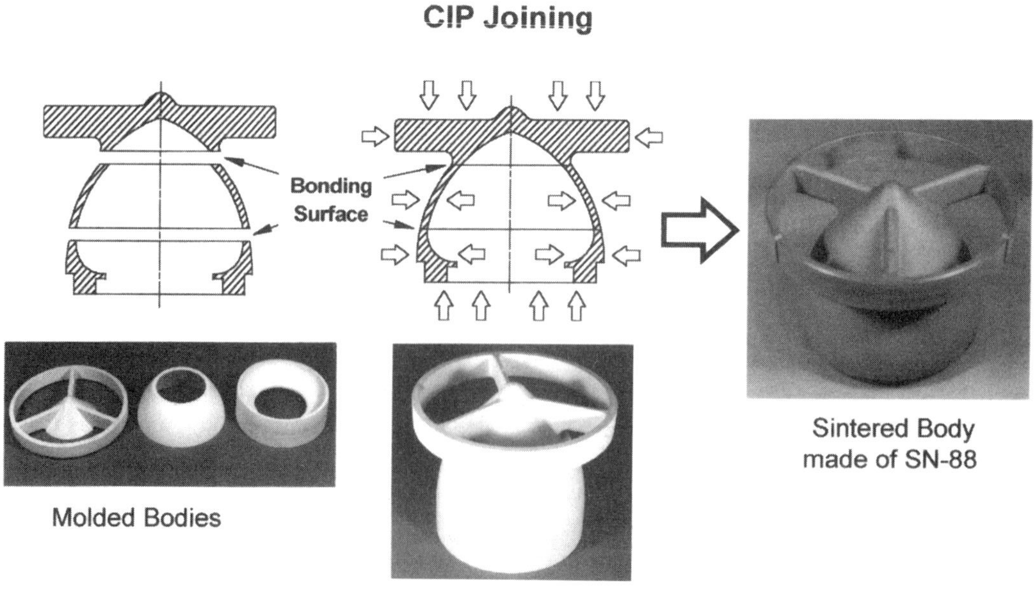

Figure 7.17 An Example of the CIP Joining Process

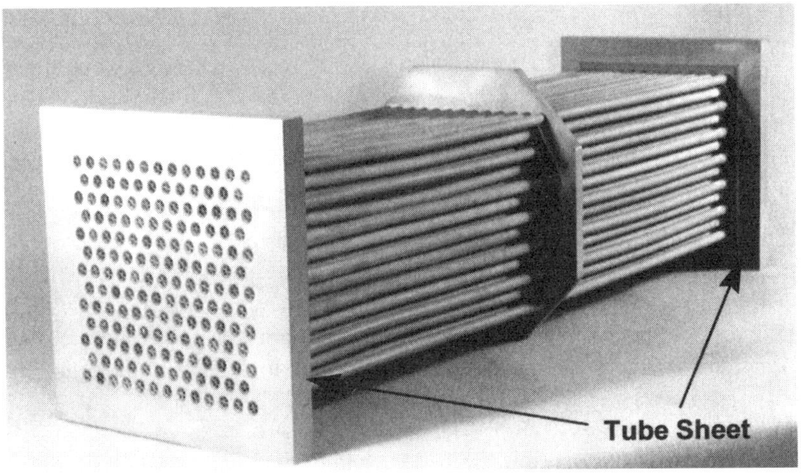

Figure 7.18 Ceramic Heat Exchanger (195mm x 185mm x 630mm)

Sintering Linking Process

The ceramic heat exchanger for the CGT301 engine (Figure 7.18) was very difficult to manufacture. To develop the ceramic tube-bundle, a fabrication process for the ceramic heat transfer tubes and tube sheet as well as a bonding technique between tubes and tube sheet had to be established [17]. The ceramic-tube bundle required both airtightness and strong bonds between the tubes and tube sheet. As shown in Figure 7.19, a new technique was developed which used the shrinkage of the tube sheet during re-sintering after the sintered tubes had been mounted in the holes of green tube sheet. The optimum allowable shrinkage was selected for both airtightness and bonding strength between the tube and tube sheet.

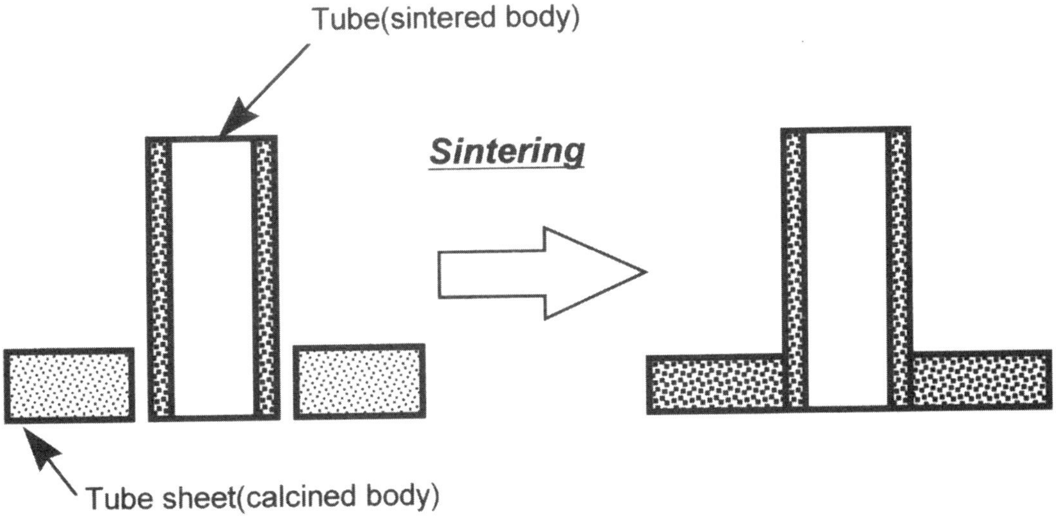

Figure 7.19 Sintering Linking Process

FURTHER TASKS

The material properties and the component fabrication process technology for silicon nitride ceramics have been dramatically improved in the last 20 years. Ceramic component reliability was verified by the 1200°C TIT CGT engine under the Japanese NEDO (New Energy and Industrial Technology Development Organization) sponsored program. This test showed that the ceramic parts had a durability of about 1,000 hours. However, there remain obstacles that need to be overcome before the CGT engine can be commercialized.

1. *High corrosion-resistant material*: The CGT component requires a guaranteed lifetime of at least 10,000 hours. However, testing of engines incorporating monolithic ceramic components have not been performed for such long periods of time, and it is not entirely clear what problems with the ceramics might arise. The gas turbine is operated in a high-pressure/moisture-rich environment. These are very severe conditions for ceramics. Therefore, research must continue to develop a material having not only high temperature strength but also excellent long-term durability.
2. *Ceramic component design methodology:* The material property data base and the standardization of the testing methods for ceramics have progressed considerably, but the design methodology for ceramic components is still insufficient compared with that of metallic materials. Further cooperation between ceramics manufacturers and turbine manufacturers will be essential to establish better design methodologies.
3. *Cost effective component fabrication process:* It will be necessary to establish a revolutionary cost effective near-net-shape forming process for the mass production of ceramic components. Also, continuing development of fabrication processes for large-sized, complex-shaped components is very important.

ACKNOWLEDGEMENT

All figures and tables were provided courtesy of NGK Insulators, Ltd.. Some of these have appeared in previous publications which have been referenced.

REFERENCES

[1] Matsuhiro, K., and Takahashi, T., "Physical Properties of Sintered Silicon Nitride Controlled by Grain Boundary Chemistry and Microstructure Morphology," MRS Int'1. Mtg. On Adv. Mats. Vol.5, 11–15,1989.

[2] Takahashi, T., "The Sintering Behavior of Y_2O_3-MgO-ZrO_2 Doped and Y_2O_3-Yb_2O_3 Doped Silicon Nitride," Ceramic Transactions, 7, 674–686,1990.

[3] Matsuhiro, K., and Takahashi, T., "The Effect of Grain Size on the Toughness of Sintered Si_3N_4," Ceram. Eng. and Sci. Proc.,10 [7–8], 807–816,1989.

[4] Takahashi, T., Isomura, M., Endoh, Y., and Furuse, Y., "SiO_2-Doped Si_3N_4 Ceramics," Key Engineering Materials, 89–91, 225–228,1994.

[5] Matsui, M.,"Utilization Techniques for Structural Ceramics," in *Proceedings of the 6th International Symposium on Fracture Mechanics of Ceramics*, D. Munz, ed., Karlruhe, Germany, 18–20 July, 1995, University of Karlsruhe Publications, Germany, 1995. Vol.11, 1996.

[6] Weibull, W.A., "A Statistical Theory of the Strength of Materials," Ingeniors Vetenskaps Akademien Handlinger," No.151,1939.

[7] Oda, I., Matsui, M., Soma, T., Masuda, M., and Yamada, N., "Fracture Behavior of Sintered Silicon Nitride Under Multiaxial Stress States," J. Ceram. Soc. Japan, 96[5], 539–545,1988. (in Japanese). *J. Ceram. Soc. Japan. Inter. Ed.,* 96[5], 523–530, 1988.

[8] Masuda, M., Yamada, N., Soma, T., Matsui, M., and Oda, I., "Fatigue of Ceramics (Part 2)—Cyclic Fatigue Properties of Sintered Si_3N_4 at Room Temperature-," J. Ceram. Soc. Japan, 97[5], 520–524,1989 (in Japanese). *J. Ceram. Soc. Japan. Inter. Ed.,* 97[5], 509–514, 1989.

[9] Masuda, M., Soma, T., Matsui, M., and Oda, I., "Fatigue of Ceramics (Part 1)—Fatigue Behavior of Sintered Si_3N_4 under Tension-Compression Cyclic Stress-," J. Ceram. Soc. Japan, 96[3], 278–283,1988 (in Japanese). *J. Ceram. Soc. Japan. Inter. Ed.,* 96[3], 275–280, 1988.

[10] Goodman, J., *Mechanics Allied to Engineering*, London, Longmans, Green & Co.,1899.

[11] Masuda, M., Makino, T., Nakasuji, Y., and Matsui, M., "Fatigue Behavior of Non-Oxide Ceramics at Elevated Temperature," Fracture Mechanics of Ceramics, Vol. 9, 481–491, 1992.

[12] Masuda, M., Soma, T., Matsui, M., and Oda, I., "Fatigue of Ceramics (Part 3)—Cyclic Fatigue Behavior of Sintered Si_3N_4 at High Temperature," J. Ceram. Soc. Japan, 97[6], 612–618, 1989 (in Japanese). *J. Ceram. Soc. Japan. Inter. Ed.,* 97[6], 601–607, 1989.

[13] Masuda, M., Soma, T., Matsui, M., and Oda, I., "Fatigue in Ceramics (Part 4)—Static Fatigue Behavior of Sintered Silicon Nitride under Tensile Stress," J. Ceram. Soc. Japan, 98[1], 83–92, 1990 (in Japanese). *J. Ceram. Soc. Japan. Inter. Ed.,* 98[1], 86–95, 1990.

[14] Watanabe, K., Masuda, M., Ozawa, T., Matsui, M., and Matsuhiro, K., "Research and Development of Ceramic Turbine Wheels," ASME paper 92-GT-295, presented at the 37th International Gas Turbine and Aeroengine Congress and Exposition, Cologne, Germany, June 1–4, 1992. Transactions of the ASME, Journal of Engineering for Gas Turbines and Power, 115[1], 36–31, Jan. 1993.

[15] Yamada, S.,Yamamoto, T., Watanabe, K., and Masuda, M., "Development of Ceramic Turbine Rotors for the 100KW Automotive Ceramic Gas Turbine Program," *Proceedings of the 6th International Symposium on Ceramic Materials and Components for Engines*, K. Niihara, S. Kanzaki, K. Komeya, S. Hirano, and K. Morinaga, eds., October 19–23, 1997, Arita, Japan. Keirin, Tokyo, Japan, 1998. pp.283–287.

[16] Tsuchiya, T., Furuse, Y., Yoshino, S., Chikami, R., Tsukagoshi, K., and Mori, M., "Development of Air-cooled Ceramic Nozzles for a Power Generating Gas Turbine," ASME paper 95-GT-105, presented a the International Gas Turbine and Aeroengine Congress and Exposition, Houston, TX, USA, June 5–8, 1995.

[17] Hattori, M., Yamamoto, T., and Watanabe, K., "Development of Ceramic Turbine Rotors for the CGT301 Engine" *Proceedings of the 6th International Symposium on Ceramic Materials and Components for Engines,"* K. Niihara, S. Kanzaki, K. Komeya, S. Hirano, and K. Morinaga, eds., October 19–23, 1997, Arita, Japan. Keirin, Tokyo, Japan, 1998. pp. 222–227.

CHAPTER

8

SINTERING AND MECHANICAL PROPERTIES OF SILICON CARBIDE WITH ALUMINA ADDITION

Nobuhiro Shinohara

Asahi Glass Co., Ltd., Yokohama, Kanagawa, Japan

INTRODUCTION

Silicon Carbide (SiC) ceramics have been known as one class of candidate materials for high temperature structural applications because of their excellent performance in high temperature environments. Asahi Glass Co., Ltd. (AGC) started with the development of the SiC material in 1978. At that time, it was known that the addition of boron (B) and carbon (C) was effective for the pressureless sintering of SiC materials [1]. Through the investigation of various sintering agents for densifying SiC, we found that the addition of alumina (Al_2O_3) enabled the SiC materials to densify with pressureless sintering [2–4]. Because the dense, sintered SiC-Al_2O_3 materials had high strength, high fracture toughness, and high fatigue resistance, it was an attractive goal to apply the SiC materials to structural components in gas turbine engines.

Typical properties of the SiC-Al_2O_3 materials developed by AGC are listed in Table 8.1. Materials fabricated from commercially available and high pure β-SiC powders have been included for comparison. It is notable that the fracture toughness is 5.5–5.8 MPa.m$^{1/2}$, which is almost twice as high as that of other SiC materials such as the SiC materials fabricated with B and C additions. The high fracture toughness is ascribed to the crack deflection or bridging effects caused by the unique microstructure of the sintered body, a microstructure composed of fine, interlocking plate-like grains as shown in Figure 8.1.

The application of SiC-Al_2O_3 materials for the fabrication of gas turbine components has been performed in several programs, including the Japanese national "Research and Development of Fine Ceramics" program (1981–1993), the 100 kW Automotive CGT (ceramic gas turbine) program (1990–1997), and the TEPCO (Tokyo Electric Power Company) 20MW CGT program (1984–1995). The mechanical and thermal properties of SiC-Al_2O_3 ceramics have been improved through these programs, and they have been used for

TABLE 8.1 MECHANICAL AND THERMAL PROPERTIES OF PRESSURELESS SINTERED AND ISOSTATICALLY HOT PRESSED SiC-Al_2O_3 MATERIALS MADE FROM COMMERCIAL AND HIGH PURITY SiC POWDERS

Starting SiC Powder			Commercial	High Purity
Young's Modulus	GPa	R.T.	406	426
		1000°C	385	405
		1400°C	376	395
Poisson's Ratio		R.T.	0.14	0.15
Coefficient of Thermal Expansion	K^{-1}	R.T.	2.7×10^{-6}	2.6×10^{-6}
		1000°C	4.5×10^{-6}	4.5×10^{-6}
		1400°C	4.7×10^{-6}	4.7×10^{-6}
Thermal Conductivity	W/m.K	R.T.	73	70
		1000°C	41	40
		1400°C	31	32
Specific Heat	J/g.K	R.T.	0.64	0.64
		1000°C	1.30	1.30
		1400°C	1.33	1.34
Fracture Toughness (SEPB Method)	$MPa.m^{1/2}$	R.T.	5.8	5.5
3-point bending strength sample size: 3 × 4 × 40 mm span: 30 mm	MPa	R.T.	605	740
		1000°C	580	650
		1200°C	560	620
		1400°C	510	605
Vicker's Hardness	GPa	R.T.	24	24
Oxidation Resistance (Weigh gain by oxidation at 1300°C)	mg/cm^2	10 hrs	0.06	0.03
		100 hrs	0.18	0.10
		1000 hrs		
3-point bending strength after oxidation at 1300°C for 100 hrs	MPa	before	618	750
		after	606	735
Thermal Shock Resistance (quenching in water) sample size: 3x4x40 mm	°C		500	600

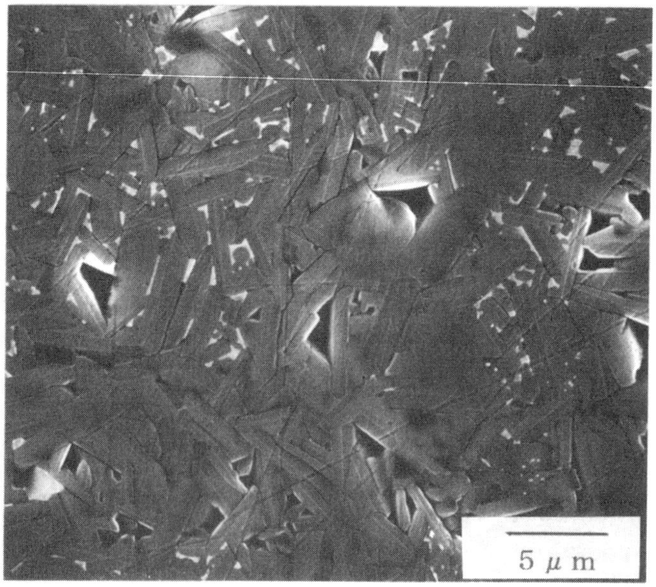

Figure 8.1 Microstructure of the Sintered SiC Material with the Addition of Al_2O_3 (Reprinted with permission from Asahi Glass Co., Ltd.)

the fabrication of static parts used at high temperature (>1200°C), such as stator vane models and combustor liners.

In this chapter, first the manufacturing processes and the properties of SiC-Al_2O_3 ceramics are being introduced. Subsequently, the results of applying the material to CGT components in the "Research and Development of Fine Ceramics" and the "100 kw Automotive CGT" are being discussed.

FABRICATION PROCESS OF SiC-Al_2O_3 CERAMICS

Forming of components

SiC ceramics with the addition of Al_2O_3 can be produced with conventional powder processing techniques. Using fine powders with an average grain size of 0.2–0.6 μm and thoroughly mixing of both SiC and Al_2O_3 powders are important for obtaining high performance sintered bodies. In our system, the pH of the mixed slurry of SiC and Al_2O_3 (5 mass %) powders is controlled to around 9–10 in order to mix both powders uniformly and ensure a sufficiently co-dispersed state during ball-milling.

Various forming techniques generally used for the fabrication of ceramics, such as dry pressing, cold isostatic pressing (CIP), slip casting, injection molding, etc., are applicable to the SiC-Al_2O_3 system. For example, CIP was selected as a favorable forming technique for the fabrication of combustor liners. Green compacts for the combustor liner were made by cold isostatic pressing the granules, that were prepared by spray drying the mixed slurry of SiC and Al_2O_3 powders, in a rubber mold at a pressure of 150–200 MPa, subsequently followed by machining to near-net component shape. Stator vane models were formed by using an injection molding technique. For injection molding, 52.5 vol% of a dried powder mixture of SiC and Al_2O_3 were compounded with binder (polyethylene and polystyrene), plasticizer (octyl palmitate) and lubricant (stearic acid and stearyl alcohol) in a kneader, and then injection molded in the metal mold. These green bodies were dewaxed by heat-treating in air before sintering.

Sintering of SiC with the addition of Al_2O_3

Densification of SiC with the addition of Al_2O_3 is enhanced by mass transport through the liquid phase formed by the reaction between added Al_2O_3 and nominal oxide impurities contained in the starting SiC powders, such as SiO_2. Because the microstructure development and densification kinetics in the SiC-Al_2O_3 system is strongly dependent on the characteristics of the starting SiC and Al_2O_3 powders, purity, particle size, phase (α- or β-SiC), etc., and the amount of added Al_2O_3, it is important to examine the densification behavior of the material during sintering in order to control the microstructure and resultant mechanical properties.

Figure 8.2 shows the typical densification behavior of the SiC-Al_2O_3 system made with high purity SiC powder (β-SiC, average particle size: 0.6 μm) which was developed in the "Fine Ceramics" program. The amount of added Al_2O_3 was controlled to be 5.5 mass % of the total amount of SiC and Al_2O_3 powders. Sintering was performed by pressureless sintering in an argon atmosphere (0.1MPa) at various temperatures for 5 hrs. With the characterization of the sintered bodies fabricated at each temperature, it was concluded that the densification of SiC-Al_2O_3 materials proceeded through the following three stages:

a) Initial stage (1700–1800°C)—Rearrangement of SiC grains governs the slow densification in this stage. Formation of bridges, like necks, between SiC grains is considered responsible for the rearrangement.
b) Intermediate stage (1800–1950°C)—The formation of a viscous glassy phase promotes the rapid densification of the material. The solution-precipitation process of SiC grains contributes to both the densification and the growth of SiC grains accompanied by the formation of plate-like α-SiC grains. The starting β-SiC phase in powder particles transforms to various α-SiC polytypes, mostly 6H, during this sintering stage.
c) Final stage (>1950°C)—Densification is dominated by the growth of plate-like silicon carbide grains. Residual pores are entrapped among the interlocking plate-like grains, and further densification ceases.

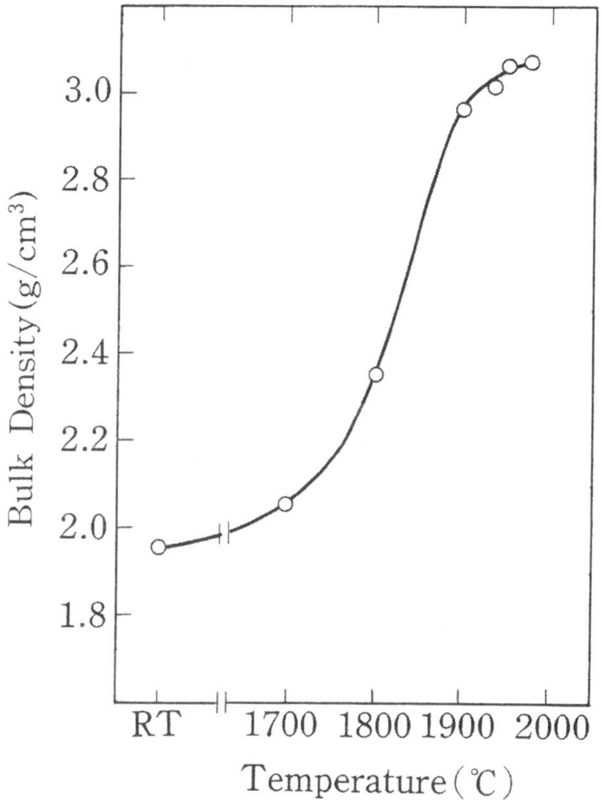

Figure 8.2 Dependence of the Bulk Density on Sintering Temperature for SiC with 5.5 Mass% Al_2O_3 Addition (Reprinted with permission from Asahi Glass Co., Ltd.)

The important feature in the microstructure development of SiC-Al_2O_3 materials during sintering is the growth of uniformly distributed, fine, plate-like interlocking SiC grains with a high aspect ratio. These plate-like grains are composed of mainly well ordered 6H lamellae and of other polytypes, 15R, 4H and 3C [5]. These grains have a clean basal plane along both of their long sides with an extremely thin second phase layer on the plane [5]. It is believed that the unique feature of SiC grains in the SiC-Al_2O_3 system contributes to the deflection of the fracture front along the basal plane, as demonstrated in Figure 8.3. The high fracture toughness of the present SiC-Al_2O_3 materials can, therefore, be explained by crack deflection and bridging mechanisms due to the unique microstructure of sintered bodies. It is also possible to form similar microstructures in sintered bodies by using commercial fine β- and α-SiC starting powders.

Hot isostatic pressing (HIP) or gas pressure sintering (GPS) are well known techniques for the increase in bulk densities and for the improvement of mechanical properties of ceramics. Figure 8.4 shows the change in densities of pressureless sintered high pure SiC materials with Al_2O_3 addition, subsequently followed by hot isostatic pressing at 1900°C for 1 hr under the pressure of 190 MPa argon. Almost full density could be achieved by applying the HIP technique. Post-sintering under the high gas pressure was an indispensable technique in the manufacturing processes of SiC-Al_2O_3 materials for CGT components in order to satisfy the high performance requirements.

In SiC-Al_2O_3 ceramics, the added Al_2O_3 forms extremely thin oxide films at the grain boundaries although the thickness is not affected by the amount of added Al_2O_3. The majority of the excess amount of Al_2O_3 is located between SiC grains, e.g. at multiple junction points, or it evaporates as gaseous species (AlO, Al_2O, SiO, CO, etc.), caused by the reaction with SiC at high temperatures. Therefore, the increase in

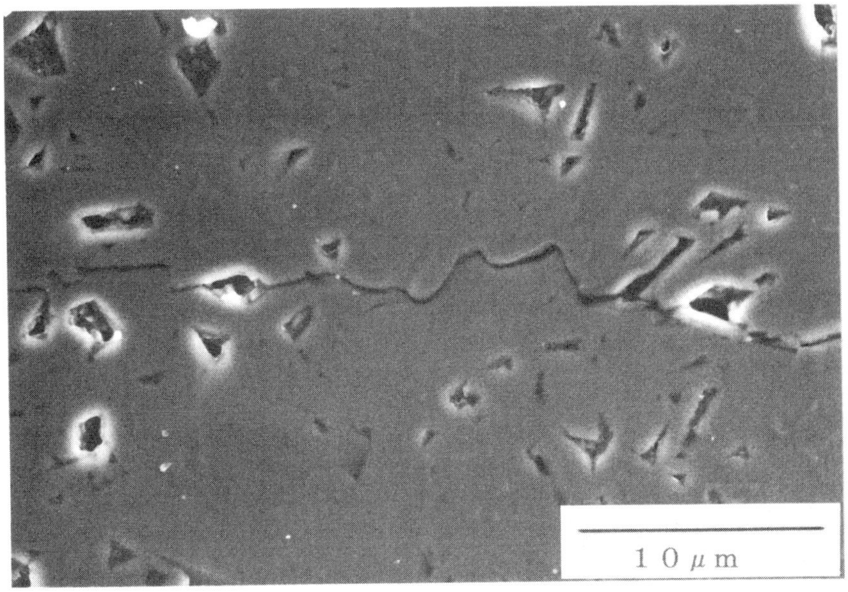

Figure 8.3 A SEM Micrograph of Crack Propagation in the SiC-Al_2O_3 Material (Reprinted with permission from Asahi Glass Co., Ltd.)

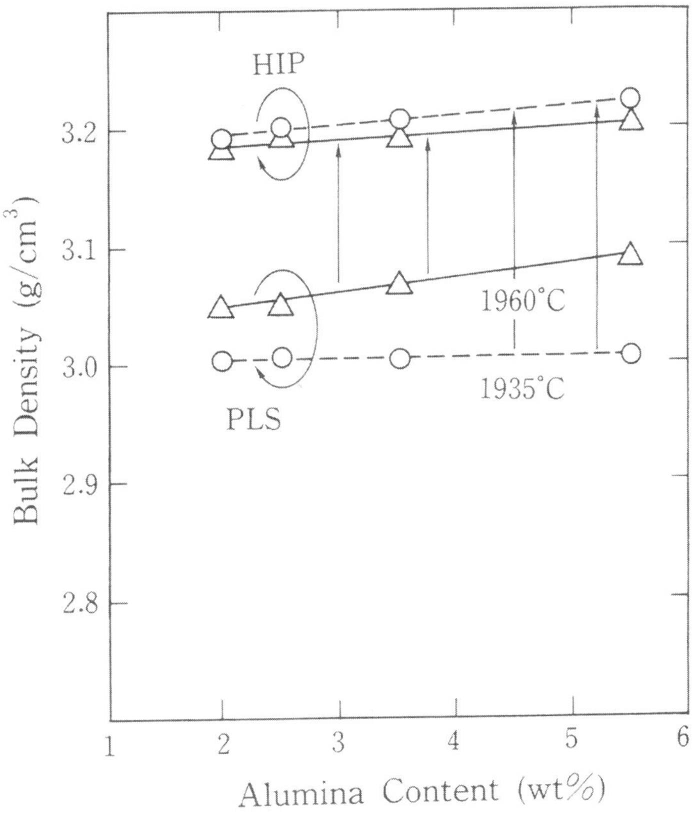

Figure 8.4 The Change in Densities of Pressureless Sintered SiC-Al_2O_3 Materials After Hot Isostatic Pressing (1900°C-1hr in 190 MPa Ar) (Reprinted from ref. [3], with permission from Asahi Glass Co., Ltd.).

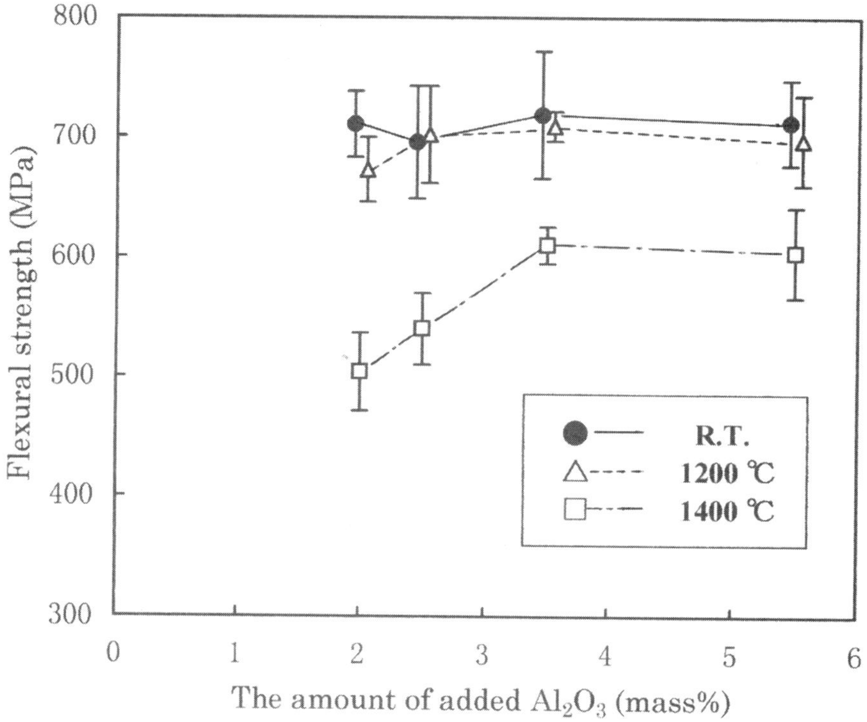

Figure 8.5 Dependence of the Flexural Strength (3-Point Bending Strength) on Temperature for SiC Materials with the Addition of Various Amounts of Al_2O_3 (Reprinted with permission from Asahi Glass Co., Ltd.)

the amount of added Al_2O_3 is not so detrimental to high temperature flexural strength in the present SiC-Al_2O_3 system. High temperature strength is rather improved with the increase of the added Al_2O_3 amount, as shown in Fig.8.5.

DEVELOPMENT OF HIGH PERFORMANCE SiC-Al_2O_3 CERAMICS

Fine Ceramics Program

The improvement of the mechanical properties of SiC-Al_2O_3 ceramics in order to enable their application for the fabrication of CGT components was performed in the national project, "Research and Development of Fine Ceramics" (1981–1993) [6,7]. In this project, the following target values were established for the application of the materials to stator models:

- Flexural strength of 400 MPa at 1400°C
- Failure probability: 20%
- Creep rupture strength: 250 MPa at 1400°C for 10,000 hrs
- Fracture toughness: 3 MPa.m1/2

As presented in Table 8.1, several high temperature mechanical properties of SiC-Al_2O_3 ceramics are improved depending on the purity of the starting SiC powders. Therefore, the use of highly pure starting SiC powder was required to achieve high performance at high temperature.

AGC successfully developed the SiC-Al_2O_3 materials satisfying the target values by using high purity β-SiC powders that were also developed in this program by Showa Denko Co. Ltd.. Figure 8.6 shows the integrated stator-nozzle model fabricated with SiC-Al_2O_3 materials [7]. The injection molding technique was

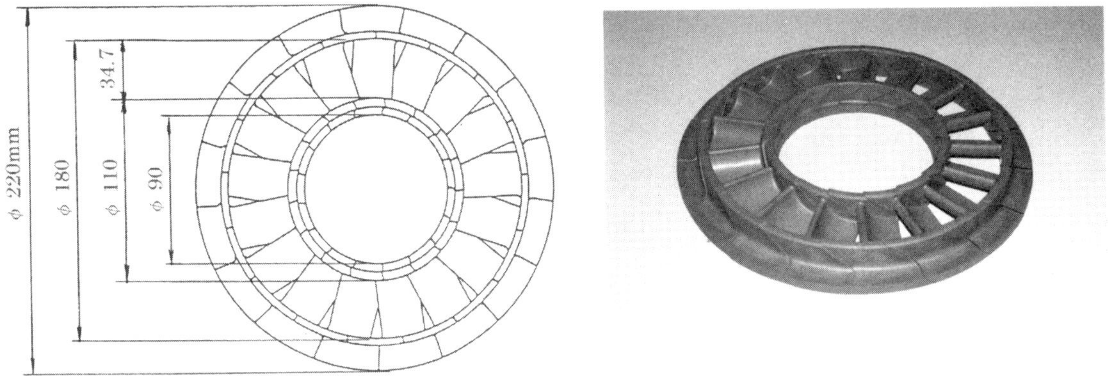

Figure 8.6 Integrated Stator-Nozzle Model Fabricated with SiC-Al_2O_3 Materials (Reprinted from ref. [7], with permission of the Fine Ceramics Research Association)

used to form the stator vane models with complex shape. Hot isostatic pressing (HIP) was applied following pressureless sintering in order to achieve full densification and to improve mechanical properties of the materials. The results that were obtained by testing the developed materials were as follows.

The high temperature mechanical properties of SiC-Al_2O_3 materials were improved by using the high purity SiC starting powder. Although the results are not shown here, the SiC-Al_2O_3 materials fabricated from high purity SiC powders exhibit higher fatigue resistance than those made from commercial grade SiC powders. Figure 8.7 shows the creep rupture diagram plotted with the estimated stresses for the effective volume

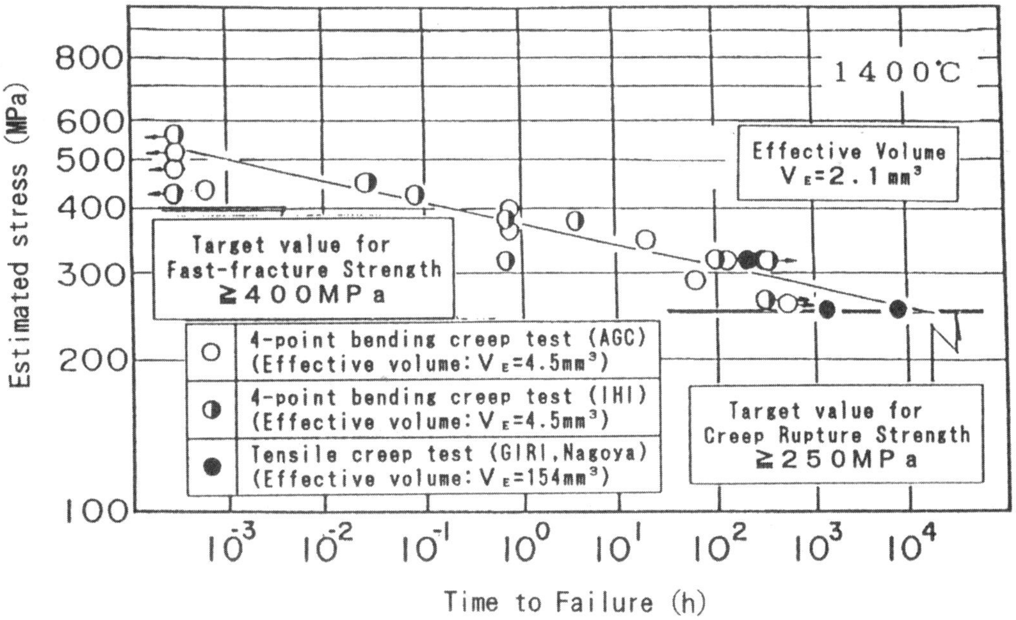

Figure 8.7 Creep Rupture Diagram Plotted with the Estimated Stresses for the Effective Volume of Stator Vane Models from the Results of 4-Point Bending and Tensile Creep Tests Measured at 1400°C (Reprinted from ref. [7], with permission of the Fine Ceramics Research Association)

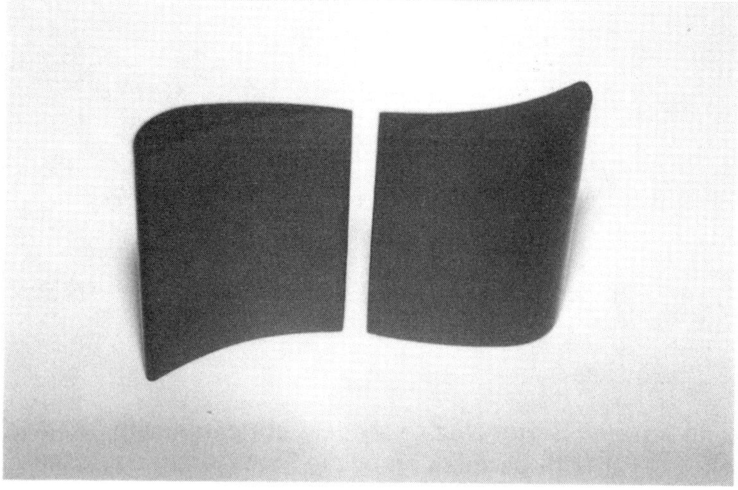

Figure 8.8 Simple Stator Vane Models for Thermal Shock Tests (Reprinted with permission from Asahi Glass Co., Ltd.)

of stator vane models (effective volume $V_E = 2.1mm^3$) from the results of 4-point bending ($V_E = 4.5mm^3$) and tensile ($V_E = 1549mm^3$) creep tests conducted at 1400°C. It can bee seen that the target value for creep rupture strength, 250 MPa at 1400°C for 10,000 hrs, was satisfied for these materials. For the SiC-Al_2O_3 materials, both made from high purity and commercial grade SiC powders, creep deformation was not observed below 1400°C, although slow crack growth was observed at the fracture surfaces of tensile tested specimens.

The results of thermal shock tests for the simple stator vane models (Figure 8.8) made from high purity SiC powders are shown in Table 8.2. The tests were conducted by blowing the cooling air with various flow rates, 107–164 m/s, to the models that were pre-heated at 1400°C. Thermal stresses generated were calculated from the temperature distribution on the surface of the models which was measured with a scanning radiation thermometer. As presented in Table 8.2, it was estimated that thermal stresses of 490–568 MPa

TABLE 8.2 THERMAL SHOCK TEST RESULTS FOR STATOR VANE MODELS OF PRESSURELESS SINTERED AND ISOSTATICALLY HOT PRESSED SiC-Al_2O_3 MATERIALS (ADAPTED FROM REF. [6], WITH PERMISSION FROM THE FINE CERAMICS RESEARCH ASSOCIATION).

No.	Cooling Air V(m/s)	Temperature at the beginning of cooling (°C)*			Temperature within 3 s after cooling (°C)*			Maximum stress generated (MPa)
		MT	TE	LE	MT	TE	LE	
1	107	1380	1250	—	1320	1160	—	490
2	107	1380	1250	—	1320	1160	—	490
3	109	1370	1200		1300	1130		490
4	109	1370	1200		1300	1130		490
5	112	1370	1210		1340	1150		510 (fractured)
6	136	1340	1210		1280	1170		529
7	144	1360	1180	800	1310	1025	720	549
8	164	1370	1200	770	1320	1100	750	568
9	164	1370	1150	780	1330	1080	760	568

*MT: Region of maximum thickness; TE: Trailing edge; LE: Leading edge

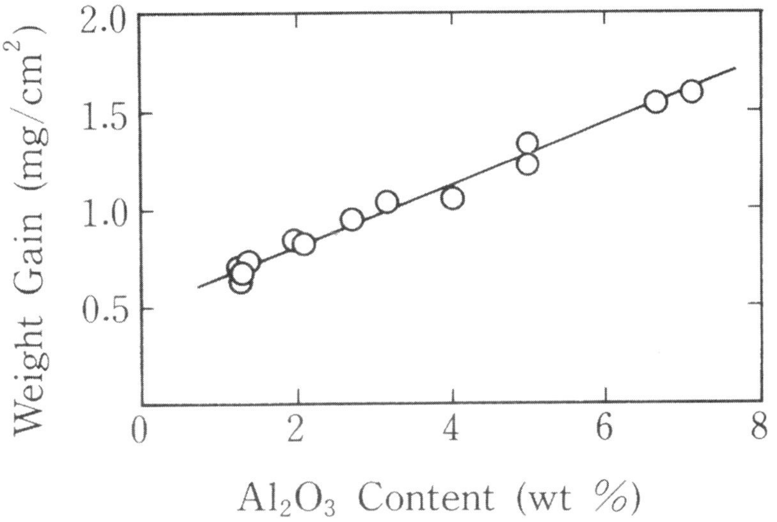

Figure 8.9 The Increase in Weight of SiC-Al_2O_3 Materials with <1 % Porosity After Oxidation at 1300°C for 1000 hrs in Air (Reprinted with permission from Asahi Glass Co., Ltd.)

were generated in the models. Although one model fractured under a stress of 510 MPa, other models survived, suggesting that the results of fast-fracture tests (3-point bending strength at 1400°C, 595MPa (n=30), Weibull modulus: 20.1) with test specimens were applicable for component design.

The oxidation resistance, which is also an important property of SiC ceramics for high temperature applications, was influenced by the amount of added Al_2O_3. Figure 8.9 shows the results of the increase in weight of the SiC-Al_2O_3 materials with <1 % porosity after oxidation at 1300°C for 1000 hrs in air. As presented, oxidation resistance deteriorates with an increase in the amount of added Al_2O_3.

The final evaluation test was carried out for the integrated stator-nozzle model (Figure 8.6) using 17 stator vane models that passed the guarantee test for each component. The integrated stator model was assembled in a rig test apparatus and the inlet gas temperature was elevated from 700°C to 1400°C in 30 seconds by using propane gas as a fuel. After holding for 120 seconds at 1400°C, the gas temperature was decreased to 400°C in 120 seconds. It was found that the maximum thermal stress, 156MPa, was generated in the components in 2 seconds after the decrease of the gas temperature from 1400°C. However, no damage was observed in the models, showing the potential of using SiC-Al_2O_3 materials for CGT components.

100 kW Automotive CGT Program

In the Japanese 100 kW Automotive CGT development program (1990–1997), AGC supplied SiC-Al_2O_3 combustor liners. All parts were inspected by fluorescent dye penetrant inspection and X-ray transmission methods before applying them for testing. Commercially available β-SiC powders were used for part fabrication. The combustion liners made of the SiC-Al_2O_3 materials demonstrated good performance in the actual component testing conducted in the program.

Problems with the application of the SiC-Al_2O_3 ceramics to actual CGT components

The SiC-Al_2O_3 materials have the potential to be used for the fabrication of actual CGT components. One of the important factors that enables the material to be used for the CGT components is the high fracture toughness in $SiC\text{-}Al_2O_3$ ceramics which results from the unique microstructure. However, it may still be difficult to apply the material for dynamic components such as a turbine rotor due to the insufficient strength and fracture toughness compared to silicon nitride ceramics or continuous fiber-reinforced ceramic composites. Relatively low thermal shock resistance may also be the crucial problem for using SiC ceramics under circumstances in which a rapid temperature change occurs in the actual operation of the CGT. Therefore, it would be required to design the component such that only a low thermal stress is generated under the expected operating conditions. In order to achieve high performance in $SiC\text{-}Al_2O_3$ ceramics, the overall process needs to be enhanced through the use of highly pure starting powders, pressure-assisted densification, machining with high accuracy, application of various NDE techniques for all products, etc.. These requirements may contribute to an increase in manufacturing cost, which is a serious concern.

Although AGC has focused on the research and development of SiC ceramics with the addition of Al_2O_3 for the structural application, SiC-B-C materials may have superior mechanical properties in the temperature range >1400°C. In that temperature regime SiC-B-C materials exhibit higher strength, higher creep deformation and oxidation resistance than $SiC\text{-}Al_2O_3$ materials.

In recent years, AGC has manufactured several types of high temperature valves, parts for wear resistance, etc. with the $SiC\text{-}Al_2O_3$ materials made from commercially available SiC powders. Application of SiC materials to the electronics industrial field has been our concern. Extremely pure SiC-Si composite materials developed by AGC have been widely used in semiconductor related industrial fields. Various types of tubes, boats, trays, etc. are manufactured with the SiC-Si materials for the diffusion and oxidation treatments in processing of Si wafers.

Production of the $SiC\text{-}Al_2O_3$ materials for CGT components has been discontinued since the end of the 100kw CGT program. Further development of the materials and the improvement of production processes for the fabrication of the components would depend on the expectation of future demands for the CGT.

REFERENCES

[1] Prochazka, S., "The Role of Boron and Carbon in the Sintering of Silicon Carbide," in *Special Ceramics 6,* P. Popper, ed., British Ceramic Research Association, Stoke-on-Trent, U.K., 1975. pp. 171–181.

[2] Suzuki, K., and Sasaki, M., "Pressureless Sintering of Silicon Carbide," presented at the Japan-US Seminar on Fundamental Structural Ceramics, Seattle, WA, USA, 1984; *Fundamentals of Structural Ceramics*, S. Sōmiya and R. C. Bradt, eds., 1987. pp. 75–87.

[3] Shinohara, N., Suzuki, K., and Kanno, T., Reports of the Research Laboratory, Asahi Glass Co., Ltd., 41[1], 25–40, 1991 (in Japanese).

[4] Shinohara, N., Suzuki, K., and Kanno, T., Reports of the Research Laboratory, Asahi Glass Co., Ltd., 43[1], 21–34, 1993 (in Japanese).

[5] Shinozaki, S.S., Hangas, J., Carduner, K.R., Rokosz, M.J., Suzuki, K., and Shinohara, N., Correlation between Microstructure and Mechanical Properties in Silicon Carbide with Alumina Addition,"J. Mater. Res., 8[7], 1635–1643, 1993.

[6] Kanno, T., Shinohara, N., and Suzuki, K., "Development of Monolithic Silicon Carbides with the Addition of Alumina," in *Fine Ceramics,* edited by the Fine Ceramics Research Association, 1993. pp. 53–68 (in Japanese).

[7] Kanno, T., and Shinohara, N., "Development of Integrated Stator-nozzle Models by Injection Molding of Silicon Carbide with the Addition of Alumina," in *Fine Ceramics,* edited by the Fine Ceramics Research Association, 1993. pp. 586–601 (in Japanese).

CHAPTER

9

The German National Program for Ceramic Materials and Component Development

Mark van Roode

Solar Turbines Incorporated, San Diego, California, USA

INTRODUCTION

Large scale ceramic materials and component development in Germany was conducted between 1974 and 1983 under a program entitled "Ceramic Components for Gas Turbines" ("Keramische Komponenten für Fahrzeug-Gasturbinen"). This nine year program was sponsored by the German Federal Ministery for Research and Technology (Bundesministerium für Forschung und Technologie—BMFT), and was significantly cost shared by German industry. The BMFT program was managed by the German Aerospace Research Institute (Deutsche Forschungs- und Versuchsanstalt für Luft- und Raumfahrt e.V.—DFVLR). The participants included car manufacturers and an engine manufacturer, various ceramic materials and component suppliers and research institutes. Table 9.1 lists the automotive and gas turbine equipment manufacturers and suppliers on the program and the components they were responsible for, and Table 9.2 lists the research institutes providing supporting technologies. The program followed on the heels of the ceramic gas turbine projects at Ford and Westinghouse in the USA initiated with government support in 1971. The program was conducted in three phases; Phase I was conducted between 1974–1978, Phase II from 1978 to 1980, and Phase III between 1980 and 1983. Details on the program's progress can be found in the proceedings from three program review workshops in Bad Neuenahr, Germany [1–3], and in summaries by Bunk et al. [4,5]. The work at the companies was 50% funded by the West-German government while the research institutes were fully funded. It has been estimated that approximately 97 million DM was spent by government and industry under the program [6].

There were several drivers behind the BMFT program [7]. First, there was the desire to substitute strategic metallic elements for which the raw materials have to be imported with easily accessible ceramic

TABLE 9.1 ACTIVITIES OF THE PARTICIPANTS IN THE BMFT PROGRAM [7]

	Combustor	Stator	Rotor	Recuperator
Materials Suppliers	Annawerk Rosenthal SIGRI	Annawerk Rosenthal SIGRI	Annawerk Degussa ESK Feldmühle	Rosenthal
Gas Turbine Manufacturers	MTU Volkswagen	MTU Volkswagen	Daimler-Benz MTU Volkswagen	Daimler-Benz

TABLE 9.2 ACTIVITIES OF RESEARCH INSTITUTES [7]

Institute	Activities
Max Planck Institute—Stuttgart (Powder Metallurgy Laboratory)	Basic research
Technical University—Berlin	Powder Technology
Technical University—Clausthal-Zellerfeld	Joining
Fraunhofer-IzfP Saarbrücken	Nondestructive testing
Technical University—Karlsruhe	Creep and oxidation properties
DFVLR Cologne	HIP Material characterization
	Project management

materials that can be produced in almost limitless quantities. Secondly, the superalloys used for aircraft turbines which incorporate these strategic metals, were limited to a use temperature of about 1000°C, and could therefore not be considered for vehicular engines.

A third major factor was the 1974 oil crisis which led to a steep increase in the price of gasoline almost overnight, and forced the Western countries to devise strategies to lessen the dependence on foreign oil. The new political imperative of energy conservation translated into support for programs to improve the fuel efficiency of automobiles and industrial processes. The desired improvement in performance of power generating equipment necessitated the higher temperature capability of hot section parts made of the more durable ceramics compared to those made of superalloys.

A fourth factor was the growing public awareness of the pollution from automobiles and industrial processes. As a result there was an interest in finding new ways for the generation of automotive power that would ensure both improved energy efficiency and lower emissions of NOx, CO and unburned hydrocarbons (UHC).

The gas turbine became the focus for the advanced material and power generating equipment development because of its efficient operation and the relatively low levels of exhaust emissions compared to the internal combustion engine. The goal became to improve its performance further by operating it at high firing temperatures through the incorporation of high temperature ceramics as structural materials for hot section components such as combustor liners, rotating blades, stationary vanes, and recuperators.

At this early stage of the development there were many challenges. First, there was the materials aspect. Materials processing and fabrication technologies had to be developed for components for engines with 1350°C operating capability using multiple fuels. Novel conceptual designs for demanding dynamic and temperature conditions had to be developed using the latest analytical design methodologies. Furthermore these designs had to reconcile the properties of the interfacing metallic and ceramic components in the gas turbine.

This chapter summarizes the materials and component development activities of the suppliers of ceramic parts. The development work by the gas turbine manufacturers has been described in Volume I, chapter 21.

TABLE 9.3 PROCESSES USED FOR THE MANUFACTURE OF Si_3N_4 AND SiC GAS TURBINE COMPONENTS

Material	Process	Acronym	Companies
Si_3N_4	Reaction Bonded Si_3N_4	RBSN	Annawerk, Feldmühle, Degussa, Rosenthal
	Sintered RBSN	SRBSN	Annawerk
	Hot Pressed Si_3N_4	HPSN	Annawerk
	(Sintered Si_3N_4	SSN	under development)
SiC	Hot Pressed SiC	HPSiC	ESK
	Hot Isostatically Pressed SiC	HIPSiC	ESK
	Siliconized SiC	SiSiC	Annawerk, Rosenthal, SIGRI
	Sintered SiC	SSiC	Annawerk, ESK

CERAMIC PROCESSING

The early ceramic materials and processing development under the BMFT program has been summarized by Thümmler [8]. While Si_3N_4 and SiC are brittle materials at higher temperatures, they have some real advantages because of an unusual combination of properties: low density, low coefficient of thermal expansion, high strength (including creep strength), and good oxidation resistance in the 1100–1400°C range. Also, based on their elastic modulus and thermal conductivity, Si_3N_4 and SiC have acceptable thermal shock resistance. They are particularly attractive candidate materials for use in the 1300–1350°C range, well above the 1000°C use temperature of 1970's superalloys. The basic materials properties of Si_3N_4 and SiC were very promising but many challenges had to be addressed in the areas of powder processing, materials process development and property optimization, and the various aspects of component development. At the end of the BMFT program many of these challenges had been successfully met by several promising materials, processing, and fabrication technologies, but much development was still needed. In particular, the properties of the components fabricated were often still marginal in meeting the design requirements of the gas turbine equipment fabricators.

Some of the key challenges were to develop materials that:

- are dense and homogeneous in microstructure since these properties correlate with strength.
- retain their strength up to high temperatures (1400°C), and that have good resistance against slow crack growth, fatigue, creep, and oxidation.
- have reproducible properties with a narrow spread, and, consequently, high Weibull modulus.
- are easily formed into complex shapes
- can be incorporated into processes that are economical and scalable to large volumes of parts and to larger component sizes.

It is understandable that these many criteria are not easily met by a single material or process, so a number of different materials systems and processes were developed to meet a favorable combination of these requirements. Table 9.3 lists the main processes used by the manufacturers of the Si_3N_4 and SiC gas turbine components under the BMFT program. The principal features of the processes have been described in chapter 1 of this text.

CERAMIC MATERIALS PROPERTIES

Table 9.4 gives a snapshot of the properties of some of the key Si_3N_4 and SiC materials as of 1978, the end of Phase I of the BMFT program. Table 9.5 shows the variation in room temperature flexure strength for Si_3N_4 and SiC materials prepared by different processing routes. The processes for reaction bonded Si_3N_4

TABLE 9.4 TYPICAL PROPERTIES OF Si_3N_4 AND SiC (ADAPTED FROM REF. [8])

	Si_3N_4		SiC	
Structure	Hexagonal 2 Mod. ($\alpha \rightarrow \beta$)		Hexagonal (α) Cubic (β)	
Density (g/cm^3)	3.19		3.2	
Melting Point (°C)	1900 (dissociation)		2830 (peritectic point)	
	RBSN	HPSN	SSiC	HPSiC
Elastic Modulus (GPa)	160	300	400	400
Open Porosity (%)	15–20	<0.5		<0.5
Thermal Conductivity (20–1100°C) (W/m.K)	10	15	84 (500°C) 39 (1200°C)	30
Coefficient of Thermal Expansion (20–1100°C) ($K^{-1}.10^{-6}$)	2.7	3.3	4.3	4.5
Flexure Strength (MPa) (RT)	200–300	650–900	400–500	550–800
Flexure Strength (MPa) (1300°C)	200–300	300–700	400–500	400–650
Maximum Use Temperature (°C)	1400	1400	<1400	1600

TABLE 9.5 ROOM TEMPERATURE FLEXURE STRENGTH OF Si3N4 AND SiC (ADAPTED FROM [8])

Material	Process	Density (g/cm^3)	σ_f (average) (MPa)*	m	K_{IC} (MPa.m$^{1/2}$)
RBSN	Injection molding	2.5–2.75	250–300	8–11	
RBSN	Slipcasting	2.5–2.75	250	—	2.1–4
RBSN	Isostatic pressing	2.50–2.70	255–281	11–20	
SRBSN	Hot densified	–	570	~20	8
HPSN	Hot pressed	3.18–3.20	~600–840	16–22	3.5–7 (20°C) 10–12 (1400°C)
HPSiC	Al-doped	3.20	500–700	5–11	4.4–5.7
HPSiC	Al_2O_3-doped	3.22	480	16	
SiSiC	8–15 vol.% Si	3.0–3.10	220–400	6–11	2–5
SSiC	Presssureless Sintered	2.9–2.95	200–400	—	
SiC	Recrystallized	—	235	14	

(RBSN), hot pressed Si_3N_4 (HPSN) and siliconized silicon carbide (SiSiC) were available at the outset of the program. Other processes were at a lesser stage of development at the start of the program or were developed during the program.

CERAMIC POWDER DEVELOPMENT

One of the cornerstones of the BMFT program was to develop advanced and economical powder manufacturing techniques. The first requirement became to develop clean powders with a high surface area. The premier powder manufacturer was H.C. Starck in Berlin. The company had focused on new ways to process powders and had found good correlations between powder characteristics, powder processing and final product properties. A detailed description of the work at H.C. Stark was presented by Schwier [9,10]. The work has also been summarized by Bunk [4].

The various processes for silicon nitride and silicon carbide fabrication have their own requirements for powder characteristics. The low-cost pressureless sintered components require high-purity powders with a fine particle size distribution, that are homogeneous and have constant physical and chemical characteristics.

TABLE 9.6 GRADES OF SILICON NITRIDE AND SILICON CARBIDE POWDERS (ADAPTED FROM REF. [4])

Grades of α-Si_3N_4 Powders

	H2	H1	LC1	LC10	LC12
Application	HPSN	SSN HPSN	SSN HPSN	SSN	SSN
Typical Powder Data					
FINENESS					
BET (m^2/g)	5	9	10	15	24
FSSS (μm)	1.0	0.6	0.6	0.5	0.4
Tap Density (g/cm^3)	0.8	0.6	0.6	0.6	0.8
PURITY					
Ca (%)	0.01	0.008	0.008	0.007	0.006
Fe (%)	0.09	0.02	0.03	0.02	0.02
Al (%)	0.08	0.08	0.05	0.05	0.05
C (%)	0.30	0.40	0.18	0.15	0.16
O (%)	1.3	1.2	1.2	1.3	1.2
N (%)	38.3	38.4	38.4	38.4	38.4
PHASES					
β-Si_3N_4(%)	4	4	2	2	2
Free Si (%)	<0.7	<0.1	<0.1	<0.1	<0.1

Grades of SiC Powders

	A1	A10	B10
Application	SSiC	SSiC	SSiC
Typical Powder Data			
FINENESS			
BET (m^2/g)	8	16	17
FSSS (μm)	1.2	0.7	0.7
Tap Density (g/cm^3)	1.0	0.7	0.8
PURITY			
Ca (%)	0.010	0.004	0.003
Fe (%)	0.01	0.01	0.01
Al (%)	0.02	0.03	0.03
O (%)	0.2	0.4	0.6
N (%)	0.02	0.03	0.04
C (%)	30.1	30.1	30.5
PHASES			
SiC Type	α-	α-	β-
Free Si (%)	<0.1	<0.1	<0.1

The manufacturing process for high-grade α-Si_3N_4 powder is based on the nitriding of suitable silicon powders. The produced α-Si_3N_4 is milled to the desired powder fineness and subsequently purified by a wet chemical process. Until 1978 relatively coarse silicon nitride powders were produced (grades H1 and H2 with surface area of 5–9 m^2/g, mainly for hot pressing or for sintering with high amounts of sintering additives). Since 1980 the finer sinter-powder LC10 became available with a surface area of 15 m^2/g, and since 1981 the fine and very uniform powder LC12 with a surface of 24 m^2/g. Data for Si_3N_4 and SiC powders are shown in Table 9.6. Morphology and fineness of some of the SiC powders produced by H. C. Starck can be seen in Figure 9.1.

Support work at the universities was primarily directed at powder fabrication on a laboratory scale. The Technical University of Berlin developed sinter-active SiC and Si_3N_4 powders. An example was a pure β-Si_3N_4 with a high specific surface area of 21 m^2/g made by the thermal decomposition of methylsilane at 1600°C. Laboratory scale powder processing and characterization was also conducted at the Max Planck Institute.

COMPONENT FABRICATION BY THE CERAMIC SUPPLIERS

In this section the efforts in materials, processing, and fabrication development of the suppliers involved in the BMFT program have been summarized with an emphasis on the evolution of the know-how and capabilities over the duration of the program.

Annawerk-Ceranox

Annawerk-Ceranox in Rödenthal produced a large number of materials and components both from Si_3N_4 and SiC, using a variety of raw materials, processes, and forming methods. Its scope of work was broader than that of any of the other suppliers [11–15].

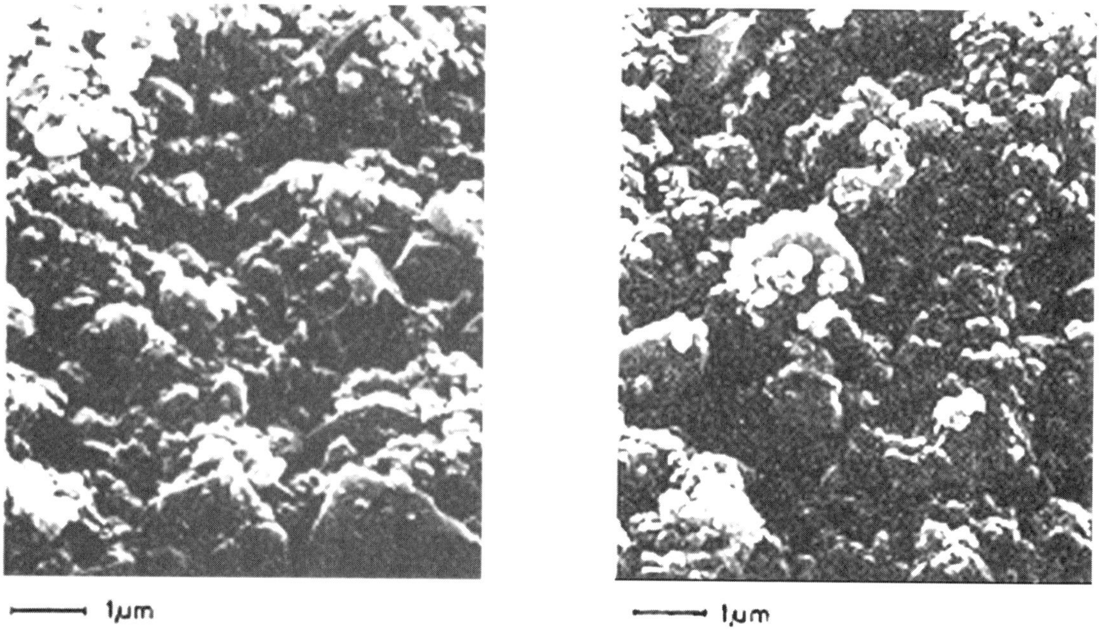

β-SiC Quality: B10 α-SiC Quality: A10

Figure 9.1 Morphology and Fineness of Some SiC Powders (Reprinted with permission from ref.[9])

RBSN. Controlling the density of the silicon preform was found to be critical for the strength properties of RBSN. A density of RBSN in the 2.60–2.65 g/cm^3 range was found to be too high for effective nitriding of the preform. Lowering the density to 2.55–2.60 g/cm^3 resulted in improved room temperature flexure strength (around 300 MPa). However, lowering the preform density still further, resulted in poorer oxidation resistance. The detrimental impact of impurities (mainly calcium) on the time dependent properties demonstrated the need for high purity silicon powder. Using fine-grained silicon powder improved the mechanical properties further.

Initially, Annawerk collaborated with Degussa on the fabrication of RBSN components, injection molding being the primary fabrication route. When Degussa abandoned the RBSN component fabrication activities in 1978 the injection molding process was transferred to Annawerk. Isostatic pressing and slip casting were also used for component forming. For high quality, injection molded RBSN parts need to be small (e.g. blades) or thin walled. A post-sintering step after nitriding greatly improved the strength of the material. During this step the component underwent 5–6% linear shrinkage which resulted in a significant densification and a substantial improvement in room temperature flexure strength. Values in the 650–750 MPa range were achieved for SRBSN compared to about 300–320 MPa for RBSN. Thus, while RBSN turbine blades could not meet the spin test target rpm, SRBSN blades could be tested in the 87,000–90,000 rpm range, easily surpassing the minimum requirements. Slip casting became the preferred processing technique for relatively large thin walled parts (e.g. nose cones) and small compact components (e.g. radial turbine rotors), in part because of economic factors (i.e. cost of mold fabrication). Figure 9.2 shows various slip cast RBSN radial turbine rotors, fabricated towards the end of the program. These components were not tested because of time constraints.

Figure 9.2 Slip Cast RBSN Radial Turbine Rotors (Reprinted with permission from ref. [13])

HPSN. The use of MgO as an additive gave a very homogeneous product with average ambient flexure strength in the 700–750 MPa range and Weibull moduli of 20–30. Average flexure strength dropped to 400–500 MPa at 1200°C. Castings were fabricated of rotor blades and of integrally bladed rotors. The latter parts required an enormous effort in machining which made the method very expensive compared to SRBSN and SSN. HPSN with an yttria additive was found to give superior oxidation resistance and strength and was used to fabricate monolithic rotors for Daimler-Benz. Average room temperature flexure strength was in the 750–850 MPa range with Weibull moduli between 15 and 20. Flexure strength dropped to 550–600 MPa at 1400°C. The materials had good slow crack growth resistance and a low creep rate.

A great challenge for the program was the fabrication of larger bladed Si_3N_4 rotors. Processing of HPSN is not practical, because of the substantial machining effort required, and RBSN does not have sufficient strength. The solution pursued was to fabricate the hub of HPSN and a blade ring of RBSN. Joining of the two parts was investigated using various approaches including bonding with a high temperature cement or glue, nitriding of a RBSN blade ring preform which was pressed against the HPSN hub, and hot pressing of the RBSN blade ring against the hub. The latter method proved to be the preferred route as bonded integrally bladed rotors were successfully spin tested.

SSN. Sintered silicon nitride was a relatively new material. Initial materials studies were carried out in collaboration with the University of Karlsruhe [14]. This material was attractive because of its good mechanical properties and potential cost advantage over HPSN. Room temperature flexure strengths in the range of 600–800 MPa were achieved. Various forming techniques were available such as isopressing and slip casting. A number of component shapes were fabricated including nozzle vanes, small integral rotors, disk shapes and balls.

SiSiC. This material had already undergone considerable development prior to the start of the BMFT program, and the focus of the work became fabrication development. One area was the reduction of the amount of residual carbon in the finished component which was a source of defects and greater than desirable spread in the strength values. Another effort was the reduction in grain size of the SiC microstructure which also resulted in improved strength. As a result an average flexure strength of 431 MPa was achieved. Figure 9.3 compares the strength of Annawerk's SiSiC with that of various other SiSiC materials over a wide range of temperatures. The strength of the material falls somewhere in between that of the two Norton SiSiC materials, NC-430 and NC-435. Note that there is a substantial spread in the data points for a specific temperature measurement. The forming process mostly used was slip casting, although isostatic pressing was

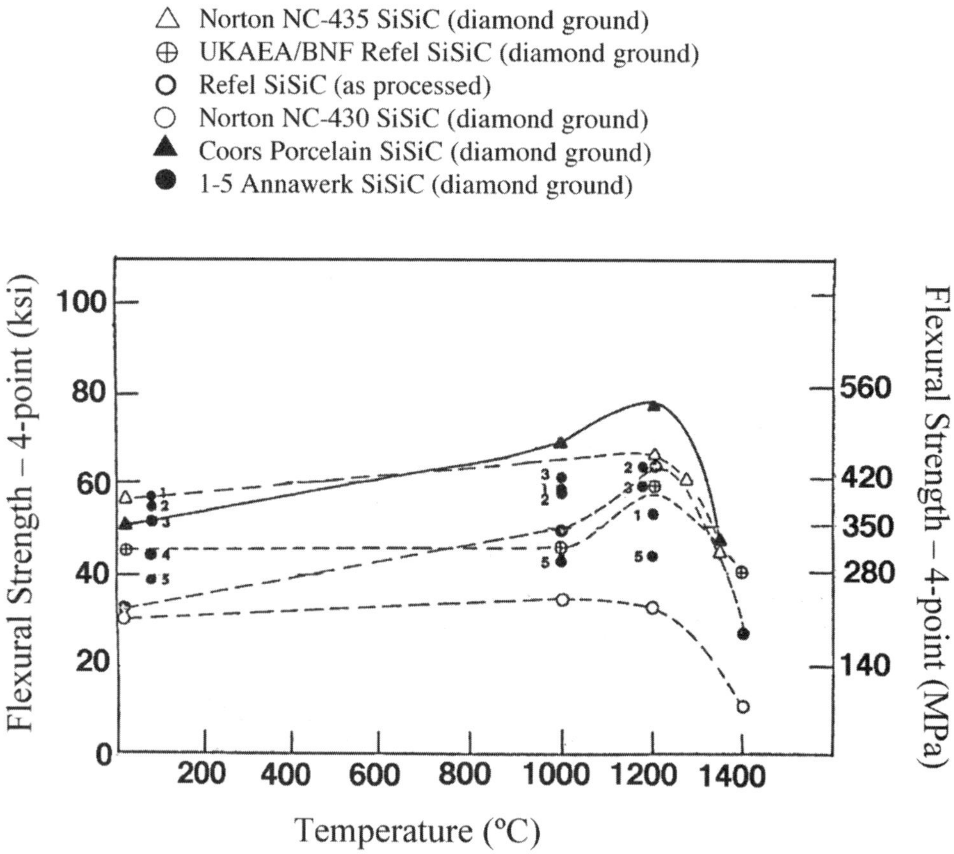

Figure 9.3 Flexure Strength of SiSiC Materials from Various Suppliers (Adapted with permission from ref. [15])

also employed. Components fabricated included radial turbine rotors, combustors, nose cones and nozzle rings. An integral SiSiC nozzle ring fabricated for MTU is shown in Figure 9.4. The technique was also used for other applications, such as parts for wear protection. A slip ring with 360 mm diameter was fabricated using the technology developed under the program.

SSiC. This was a new material for the BMFT program and a substantial effort was devoted to all aspects of the materials and process development. Powder development was carried out at H.C. Starck with support from the Technical University of Berlin. A density of 3.17 g/cm^3 was achieved with this material and an average room temperature room temperature flexure strength of 339–381 MPa and Weibull modulus of about 13. The process was used to fabricate injection molded turbine rotor blades, isopressed rings and balls, slip cast combustor chambers and small integral rotors. Figure 9.5 shows various stages of the development of SSiC combustor liners fabricated for Detroit Diesel Allison.

Feldmühle AG

Both Feldmühle Aktiengesellschaft [16–18] in Plochingen and Degussa in Hanau (to be discussed in the next section) worked on the development of the injection molding process for the fabrication of RBSN

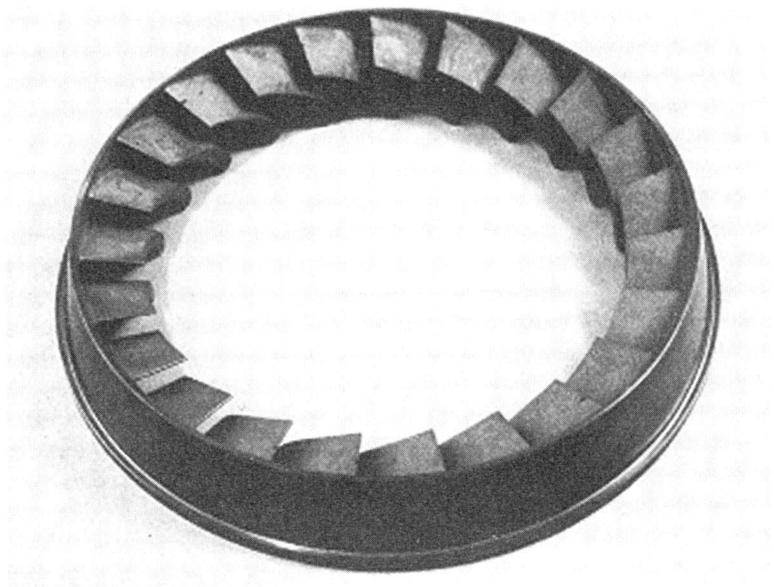

Figure 9.4 Integral SiSiC Nozzle Ring for MTU Gas Turbine (Reprinted with permission from ref.[15])

Figure 9.5 Component Development for SSiC Combustor Chamber for Detroit Diesel Allison's AGT 100 Program (Reprinted with permission from ref. [15])

TABLE 9.7 COMPARISON OF VARIOUS FABRICATION PROCESSES FOR TURBINE BLADES AND INTEGRAL TURBINE BLADE RINGS (ADAPTED FROM REF. [16])

		Injection molding		Slip casting		Dry pressing	
		Blade	Ring	Blade	Ring	Blade	Ring
Material	Strength	+	+	+	0	0	
	Oxidation resistance	+	+	+	+	0	
	Dimensional stability	+	+	0	0	+	
	Quality for large volumes	+	+	0	–	0	Part is
Fabrication	Suitability for mixing	+	+	–	–	0	technically
Process	Cost small volumes	Medium	High	Low	Medium	Medium	not feasible
	Cost large volumes	Low	High	Low			

and SRBSN silicon nitrides (SSN was not available at the time), alhough the work at these companies focused on different size components. Feldmühle directed its attention to materials optimization and initially to the fabrication of small components while Degussa concentrated on larger components such as nozzle rings. Feldmühle also started to fabricate larger parts towards the end of the program. The decision to use injection molding was based on a comparison of this process with other candidate fabrication techniques such as slip casting and dry pressing for small parts (turbine blades) and larger components (turbine blade rings). Table 9.7 summarizes the results of the assessment.

In 1974 when the BMFT project started injection molded RBSN had relatively modest mechanical properties. Densities were less than 2.2 g/cm^3 and the room temperature flexure strength was about 120 MPa. Internal oxidation and creep were not yet recognized as important degradation mechanisms. Feldmühle AG focused on the development of powder processing, optimization of the mechanical properties, including a narrowing of the spread in the strength values, optimization of the oxidation behavior and long term stability of the material. An essential task became increasing the density of the material, and ensuring that it was homogeneous throughout the material. Density was increased to 2.5–2.75 g/cm^3 during the development.

Careful studies were performed to correlate the overall density with the particle size range and distribution, and with the temperature of the nitriding process. It was found that a very fine silicon powder (maximum particle size <10 μm) was needed to ensure that the silicon reacted optimally in the nitriding process with a minimum of residual porosity. The result was a material that had a room temperature maximum flexure strength around 300–350 MPa, minimal creep up to 1250°C, and a retention of strength under conditions of thermal cycling in air for up to 3000 hrs. The creep behavior of RBSN was shown to be very dependent on internal oxidation as a comparison of creep deformation in air and in vacuum demonstrates (see Figure 9.6) A correlation with porosity and pore radius was also observed. Thümmler and Grathwohl have discussed the the creep and oxidation behavior of RBSN extensively [19].

Feldmühle successfully fabricated relatively small components such as turbine blades for hybrid metal disk/ceramic blade rotors (see Figure 9.7) as well as much larger blanks for disks (see Figure 9.8). While the smaller components were largely free of defects imperfections were observed in the larger components, primarily density gradients near the surface of the part. Because of this no integral rotors were fabricated that could survive planned spin testing experiments.

Degussa

As mentioned above Degussa in Hanau also developed process and fabrication technology for injection molded RBSN with the emphasis on larger shaped components [16,20]. Degussa worked closely with Annawerk-Ceranox during Phase I of the BMFT program (1974–1978). The two companies prepared a joint program report for this period. Degussa did not participate in the final phase of the BMFT program and its

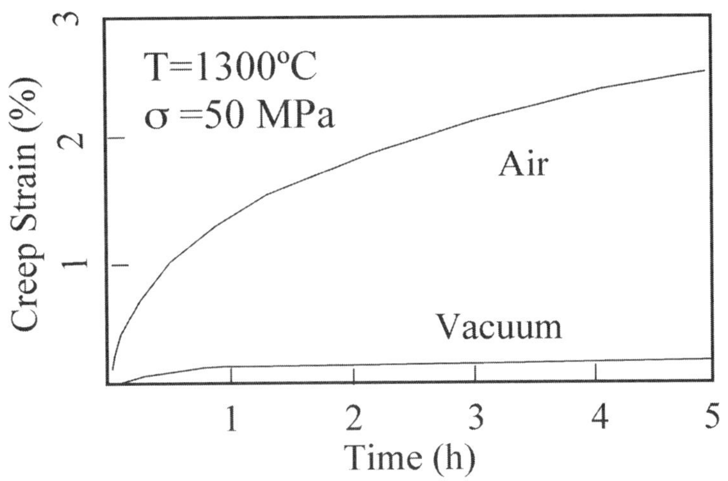

Figure 9.6 Creep Deformation of RBSN in Air and in Vacuum (Adapted with permission from ref. [18])

Figure 9.7 Experimental Injection Molded RBSN Blades in Metallic Disk (Reprinted with permission from Ref. [18])

Figure 9.8 Injection Molded Silicon Preforms for Integral RBSN Turbine Rotors (Reprinted with permission from ref. [18])

reports only extend to 1980, the year of the second review workshop in Bad Neuenahr. Annawerk took over Degussa's injection molding work in Phase III (1980–1983).

The emphasis of the work between 1978–1980 was not so much on improving the density of the material, but more on the narrowing down of the spread in the strength values in a series of experiments, i.e. on improving the Weibull modulus. For materials with room temperature bending strength values of 250 MPa, Weibull moduli of about 20 were achieved. However there was a real challenge for strength around 300 MPa were the Weibull modulus was only about 8 and the goal became to increase it to 15. This implied that, assuming a probability of failure (POF) of 0.1%, the maximum load on a component could be raised from 130 MPa to values in the 180–200 MPa range. Table 9.8 summarizes the status of the development at Degussa

The strategy at Degussa for component fabrication became the establishment of minimal values for key properties such as flexure strength and density with a definite failure rate. A reasonable criterion at the time was to reject a component if the failure rate was somewhere between 0.01 and 0.1%. Low rejection rates are only achievable when the components have a narrow spread in their properties, i.e. a high Weibull modulus. Therefore, apart from materials development the development of appropriate manufacturing processes for volume scale components play a decisive role.

Figure 9.9 shows the relationship between the POF and flexure strength for components for different Weibull moduli. It is assumed that the component strength is in excess of 200 MPa and that a guaranteed

TABLE 9.8 FLEXURE STRENGTH AND WEIBULL MODULI FOR INJECTION MOLDED RBSN PARTS AT DEGUSSA (ADAPTED FROM REF. [20])

	1978		1980	
Density (g/cm^3)	2.5–2.7		2.5–2.7	
Flexure strength (MPa)	210	280	250	310
Weibull modulus	15	8	20	15
Max. component stress For POF = 10^{-3} (10^{-4})	137 (118)	127 (95)	182 (162)	203 (168)

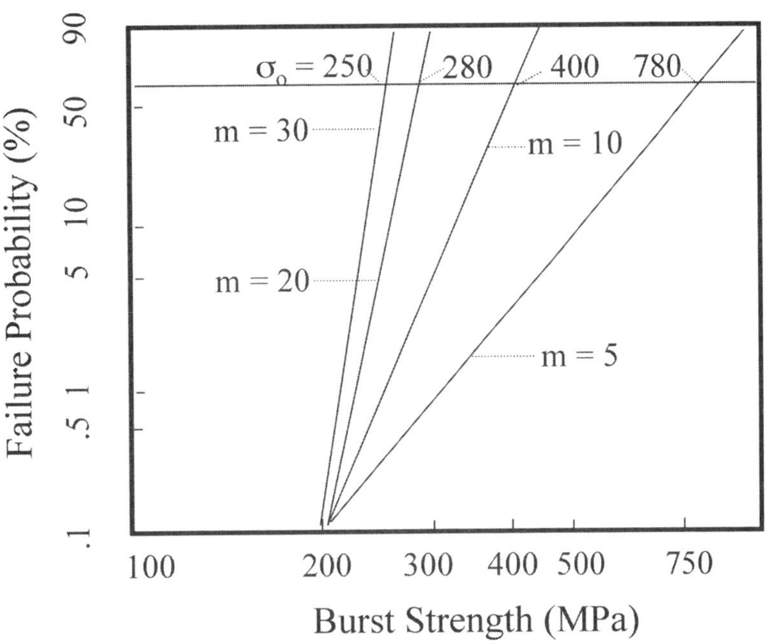

Figure 9.9 Relationship Between Flexure Strength, Weibull Modulus, and Probability of Failure (Adapted with permission from ref. [20])

POF of 0.1% or less has to be achieved. Thus for m = 20 a flexure strength of 280 MPa will be required while for the much lower m = 5 the flexure strength needs to be at least 780 MPa. In 1980 components with maximum allowable service stress values of 200 MPa were fabricated at Degussa (see Table 9.8).

Small component (rotor blades, nozzles) development was also carried out with injection molding (see Figure 9.10). To demonstrate the quality control procedures large numbers of parts (in some cases >1000 parts) were fabricated. The injection molding process was largely automated. These turbine blades were evaluated in cold spin testing at various rotor speeds. Table 9.9 compares the stress calculated based on the rotor speed at blade failure with the flexure strength of co-processed flexure bars.

Figure 9.10 Injection Molded RBSN Turbine Rotor Blades and Nozzles for a Car and Truck Design (Reprinted with permission from ref. [20])

TABLE 9.9 COMPARISON OF FLEXURE STRENGTH OF FLEXURE AND THE STRESS IN INJECTION MOLDED RBSN ROTOR BLADES DURING COLD SPIN TESTING (ADAPTED FROM REF. [20])

	4-pt Flexure strength (MPa)	Rotor speed (rpm) at rupture	Blade stress (MPa) at rupture
Truck design speed	—	40,000	258
D-017	252 (±17)	27,840	132
		39,980	258
		41,055	272
		45,450	333
Car design speed		64,000	200
D-018	264 (±18)	65,200	207
		55,175	149

Failure occurred over the range of 132–333 MPa in the truck design where the material had an average flexure strength of 252 MPa. The failure range for the car design was 149–207 MPa for an average flexure strength of 264 MPa. Apparently, some components actually failed at stresses well below the average flexure strength of co-processed flexure bars.

Proof tests were also conducted for single vane nozzles (hollow and un-cored) and multi-vane segments. The hollow single vane nozzles were tested for 50 cycles at 1200°C and 1400°C.

The blade ring of a dual-density rotor was also fabricated by injection molding. It was estimated that the burst strength of the ring would be 175 MPa at a rotor speed of 36,000 rpm. The average of 6 components failed at 168 MPa and 33,200 rpm. This burst speed would have corresponded to a burst failure stress of 151 MPa. These data are all very close indicating there is little excess margin over the design values. Rotor disk preforms were also fabricated with wall thickness on the order of 20 mm. It was however realized that for parts of this methods adapted to high thicknesses needed to be developed.

Rosenthal Technik AG

The emphasis of the R&D at Rosenthal Technik Aktiengesellschaft in Selb was on the establishment of production techniques for ceramic components for gas turbine applications using RBSN and SiSiC [21–25]. Using these methods combustors, nose cones, nozzles and heat exchanger components were developed. RBSN was also used for the fabrication of integrally bladed rotors. Forming techniques included isostatic pressing, injection molding, and slip casting. For recuperative heat exchangers a tape technology was evaluated and RBSN tapes were fabricated with a wall thickness as low as 0.2 mm.

For RBSN component fabrication critical process parameters were established to be silicon particle size, particle size distribution and grain shape, chemical purity, and reproducible fabrication. Conditioning of commercially available silicon became an integral aspect of the development work.

Reports from Rosenthal at the first two review meetings of the BMFT program were exclusively devoted to the heat exchanger development efforts conducted in collaboration with Daimler-Benz. The important parameters for a recuperator ceramic are strength, imperviousness to gases, thermal shock resistance and oxidation resistance. Besides RBSN, SiC and cordierite were considered for recuperator materials.

The initial requirements were summarized as follows:

- Materials: RBSN, SiC, cordierite
- Concept: Daimler-Benz recuperator
- Foil thickness: 0.4 mm (initially)—0.2 mm (goal)
- Pressure ratio: 5:1
- Leak rate: 0.5% at 10^{-8} kg/s.cm^2
- Application temperature: 1000°C (initially)—1200°C (goal)
- Test duration: 10 hrs

A variety of materials were evaluated in collaboration with Daimler-Benz, the primary end user of the recuperator materials. Initial work started with RBSN. It was an attractive material, and its major drawback, residual porosity, could be addressed by infiltrating it with mineral compounds. Various fabrication technologies were investigated starting with a stacked structure of corrugated foils, next an extruded structure, and finally a structure based on slip cast laminated foils. Figure 9.11 shows the assembly line for foil production at Rosenthal. The foils were shaped by a die-and-punch process to a thickness of 1.2 ± 0.2 mm.

Figure 9.12 shows two cross sections of RBSN heat exchange modules. The densities and specific surface areas were 2.05 g/cm^3 and 2.7 m^2/g for the module on the left, and 2.25 g/cm^3 and 1.2 m^2/g for the mod-

Figure 9.11 RBSN Foil Production Assembly Line (Reprinted with permission from ref. [22])

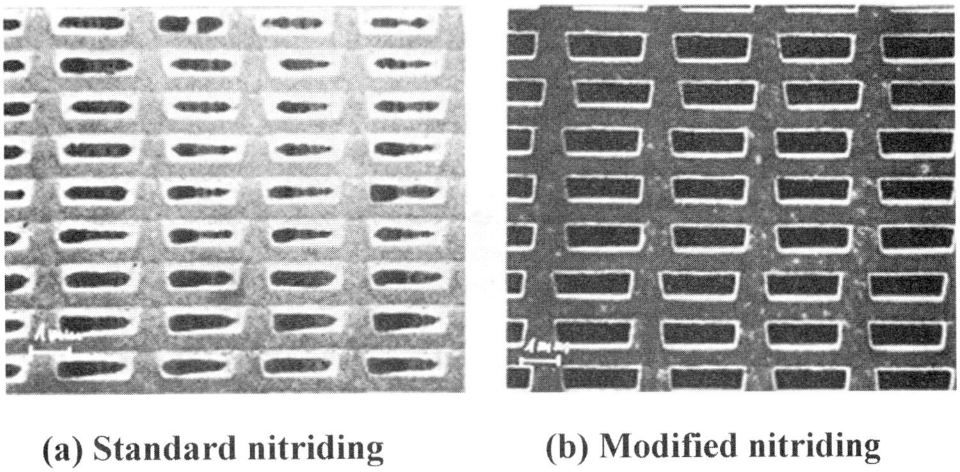

Figure 9.12 Cross Section of RBSN Heat Exchange Modules—Effect of Firing Regime (Reprinted with permission from ref. [22])

TABLE 9.10 TYPICAL MATERIALS PROPERTIES ACHIEVED FOR SiSiC AT ROSENTHAL TECHNIK AG (ADAPTED FROM REF. [24])

	Isostatic pressing	Tape casting	Injection molding	Slip casting
Density (g/cm^3)	3.07 ± 0.02	2.96 ± 0.02	3.05 ± 0.02	2.95 ± 0.02
Flexure strength (MPa)	380	260	275	260
Weibull modulus	12	10	12	12
Flexure strength (MPa) after 200 hrs/1260°C	420	260	300	260

ule on the right, respectively. The differences were achieved by modifying the nitriding process. Evidently, the low density high surface area structure was very prone to oxidation during processing.

Efforts were expended to close the residual porosity of the RBSN for the recuperator application. Extensive testing of modules fabricated by Rosenthal was carried out at Daimler-Benz. Small sections were subjected to burst testing, and larger cubular modules (100 mm × 100 mm × 100 mm) were subjected to thermal shock testing following a stepwise temperature profile up to 1100°C and a 3:1 pressure ratio. The test modules met the customer's specifications.

Component development similar to that for RBSN was explored for SiSiC. Critical process parameters for the SiSiC formings process are a consistent quality of the raw materials including well defined grain morphology and grain size distribution, well defined carbon content of the SiC and low content of impurities. A homogeneous distribution of the silicon carbide, carbon carrier, and of additives in the forming slurry or paste are critical for good component quality. Table 9.10 summarizes the materials properties of SiSiC achieved with various forming processes. Figure 9.13 shows SiSiC combustor liners fabricated by isostatic pressing for MTU. Injection molding and slip casting were used for complex components such as nozzle segments.

Figure 9.13 SiSiC Combustors for MTU (Reprinted with permission from Ref. [24])

ESK

Elektroschmelzwerk Kempten (ESK) of Munich concentrated on pressure sintering of SiC using the HPSiC and HIPSiC approaches [26–29]. Initial process development work started on HPSiC in collaboration with Volkswagen AG of Wolfsburg and MTU of Munich.

HPSiC with good mechanical properties (e.g. a flexure strength of 550 MPa), a density of 3.21 g/cm^3 (close to theoretical density), and good oxidation resistance was already available at the start of the program. An early application was the fabrication of rotor disks (without blades). The transfer of the technology for the fabrication of test specimens and thin sheets to the manufacture of a heavy thick part such as a rotor disk was not simple. One of the problems was that the density of the disk decreased from the outer surface to the interior. Cracks developed in the inner region of the hub as a result of thermal stresses generated during cooling after sintering. A further problem was the difficulty in maintaining dimensional control during hot pressing as the densification under the influence of a temperature gradient started at the bottom and subsequently at the top leading to an asymmetric component. Improvements were achieved over time, especially with respect to thermal shock resistance, but further progress in high temperature strength and fracture behavior was still needed.

Dense HIPSiC parts were fabricated without sintering additives under typical HIP conditions of 2000°C, 1900 bar, and 2 hrs holding time. Room temperature flexure strength was on the order of 600 MPa. The strength dropped by about 12.5–17% at 1370°C. Adding 0.2 wt. % Al improved the strength somewhat. Regardless of actual chemical composition, HIPed test specimens all had the same homogeneous fine-grained microstructure with maximum grain size around 1.5 μm and densities in the 3.21–3.22 g/cm^3 range, i.e. around 100% of theoretical density. Figure 9.14 compares the flexure strength as a function of tempera-

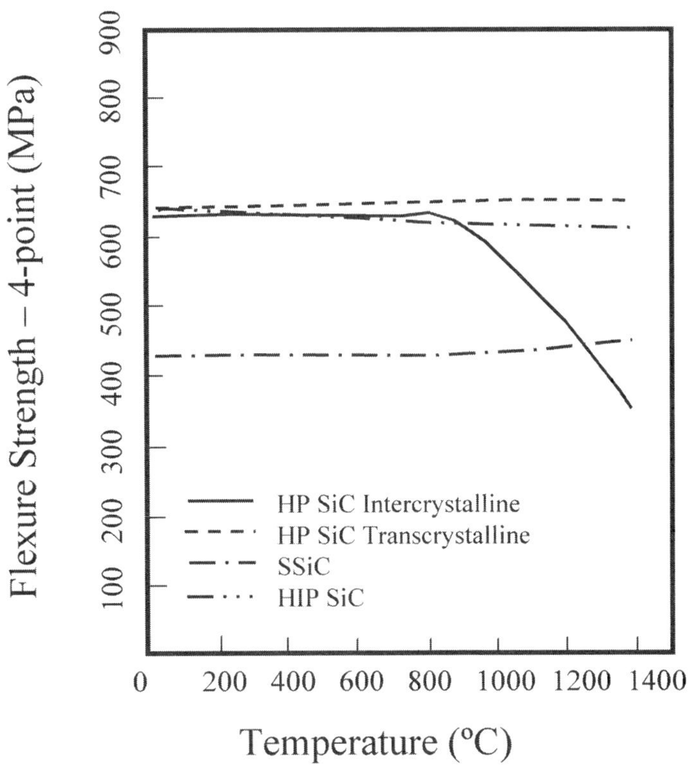

Figure 9.14 Flexure Strength as a Function of Temperature for Various Types of SiC (Adapted with permission from ref. [28])

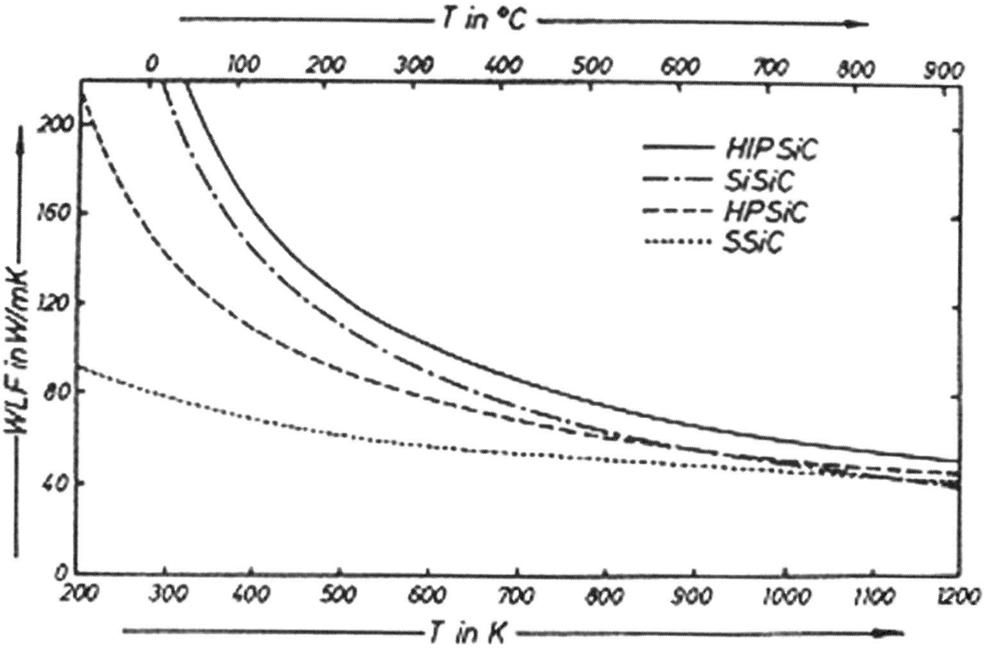

Figure 9.15 Temperature Dependence of the Thermal Conductivity of Various SiC Materials (Adapted with permission from ref. [28])

ture for various silicon carbides. It can be seen that HIPSiC is almost as good as HPSiC and definitely superior to SSiC. However, it is also much more expensive.

Increasing the thermal conductivity of SiC improves the thermal shock resistance. The room temperature thermal conductivity is in the range of 70–100 W/m.K for SSiC and about 200 W/m.K for SiSiC. Standard HPSiC has a thermal conductivity of about 100 W/m.K which towards the end of the program had increased to about 140 W/m.K. Figure 9.15 illustrates the thermal conductivity as a function of temperature for various SiC materials. It can be seen that HIPSiC has thermal conductivity properties that are comparable to those of the best SiSiC materials with a value of 210 W/m.K at room temperature and 55 W/m.K at 800°C.

It was also considered to apply the HIP method to prepare mixed compositions of SiC and Si_3N_4 over a wide compositional rage without any sintering additives. It was hoped that by selecting a specific composition and capitalizing on the superior thermal conductivity and oxidation resistance of SiC, and the better elastic properties and generally higher flexure strengths of Si_3N_4, a combination of desirable properties could be achieved. However, this concept was not further developed.

The HIP technology was very useful for joining SiC and Si_3N_4 parts. An example is a fabricated rotor for an axial gas turbine in which the hub is joined to the inserted blades by a diffusion bonding method. The joint needs to have the same high temperature strength as the matrix material.

SIGRI ELEKTROGRAPHIT GMBH

SIGRI, based in Meitingen, worked on the development, fabrication, and demonstration of stationary gas turbine components of SiSiC, capitalizing on its more than 50 years of expertise in the manufacture of silicon carbide heating elements [30–33]. The material was trademarked as Silit SK. The process development initially focused on slip casting, a technique which lends itself well to complex component fabrication, with injection molding being developed later when the interest shifted to volume component fabrication. Figure 9.16 shows a scroll fabricated by SIGRI from SiSiC.

SIGRI in collaboration with the Max Planck Institut für Metallforschung in Stuttgart and the Fraunhofer Institut für Werkstoffmechanik in Freiburg studied the behavior of the material under long term static and dynamic loading conditions at elevated temperatures. Engineering data were developed for temperatures up to 1200°C for life under static load, fatigue under dynamic load, and volume dependence of strength.

Flexure strength (4-point bending) increases to a maximum of about 400 MPa at 1200°C then shows a sharp drop at 1400°C as the silicon phase (8–12%) starts to weaken. The Weibull modulus improves from about 10 at ambient to about 30 at 1200°C, mainly because of the healing of surface cracks through oxidation, and possibly relaxation of frozen-in residual stresses. The fracture toughness is about 4 $MPa.m^{1/2}$ at room temperature and about 2.5–3.0 $MPa.m^{1/2}$ at 1200°C. The crack velocity coefficient, n, was determined at 200 (±20) at room temperature, and at 115 (±15) at 1200°C, indicating that the material was fairly resistant to subcritical crack growth. This assumption was borne out in proof testing where samples subjected to 1.09 times the service load over a 12 day period failed at stresses well above the proof test load. Fatigue at ambient and elevated temperatures was found to be negligible in cyclic testing at ambient and 1200°C.

One of the problems SIGRI admitted towards the end of the program was the lack of reproducibility of the materials properties which was attributed to a lack in the ability to control surface and volume defects such as protrusions, pores, or inclusions. These defects were reflected in relatively low Weibul moduli which only towards the end of the program improved to a more acceptable level. The initial efforts were primarily directed towards improving the bending strength of the material which was increased from an initial level of about 160 MPa in 1976 to about 300 MPa in 1981. The Weibull modulus was increased from 3 to 8 in the same period. In the final phase of the development the efforts were directed towards improving the reproducibility and by 1983 the Weibull modulus had increased to 15 while the bending strength had only been modestly improved to 320 MPa. These were all room temperature values. SIGRI

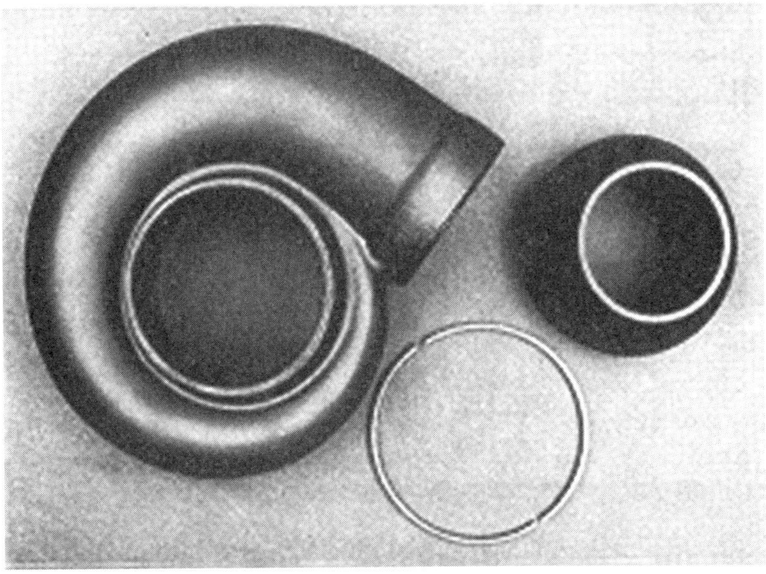

Figure 9.16 Scroll Fabricated from SiSiC by SIGRI (Reprinted with permission from ref. [32])

claimed particular success in improving the high temperature properties of test specimens and simple components with a bending strength of 350 MPa and Weibull modulus of 20 at 1200°C.

Retrospective and Future Perspective

A large number of components were fabricated over the nine years duration of the BMFT program. Testing of some of these components has been described in some detail in chapter 21 of Volume I. At the concluding workshop DLFVR program manager, Manfred Böhmer, summarized some of the main achievements of the BMFT program [6]:

Stationary parts (combustor, inlet scroll, nose cone, nozzle, turbine ring) were all fabricated from various materials and tested in a 200 hour simulated road test at a maximum combustor outlet temperature of 1352°C (1625 K) and a combustor inlet pressure of 5 bar.

A silicon nitride based recuperator was fabricated with a wall thickness of 0.2 mm, a 5:1 pressure ratio, an allowed leakage rate of 0.5% and a maximum inlet temperature of 1202°C (1475K). The recuperator was evaluated in a 10 hr test.

Hybrid metal-ceramic rotors were fabricated for passenger cars and trucks and tested for 200 hours in a road test. The turbine inlet temperature was 1252°C (1525K). Integral ceramic rotors were also fabricated and tested in a 50 hr road test at a turbine inlet temperature of 1352°C (1625K). These road tests demonstrated that ceramics could be produced in complicated shapes with properties that withstand the conditions in a high temperature gas turbine!

Böhmer also presented a retrospective and future perspective at the conclusion of the third and final workshop for the BMFT program [34]. The challenges at the start of the program were probably not well appreciated and the capabilities of the ceramic materials to meet the rigorous design and service requirements of the automotive gas turbine were overestimated. Especially, the time necessary for materials development and properties optimization was underestimated. Nevertheless, significant advancements had been achieved, e.g the almost 100% improvement in the flexure strength of two of the major silicon nitride materials, RBSN and HPSN (see Figure 9.17).

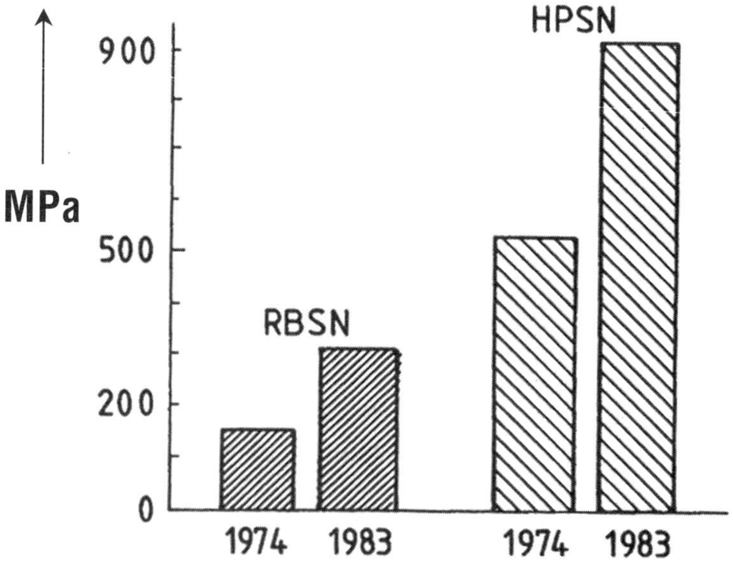

Figure 9.17 Improvement in the Flexure Strength (4-point Bend Test Data) of RBSN and HPSN During the German National Program (Adapted with permission from ref. [34])

Significant progress was also made in ceramic processing, e.g. the development of process technologies for pressureless sintered silicon carbide (SSiC) and silicon nitride (SSN), post-sintered reaction bonded silicon nitride (SRBSN), and hot isostatic pressing of silicon carbide. Also, because of the progress made on the program various "spin off" applications emerged, such as sealing rings, cutting tools, fuel nozzles, ball bearings, valves, pump components, etc.

Still major problems remained to be solved. These related to reliability, spread in test data, and reproducibility in fabrication, as well as economic factors pertaining to weight and cost, all of which can be grouped under the general heading of the ability for mass production of large part volumes for commercial application. It was felt at the time that only major support from the government could address these challenges adequately, and a new follow-on program emphasizing ceramic manufacturing was initiated. This three-year program, entitled: "Strukturkeramik" (Structural Ceramics") was carried out between 1983 and 1986 and was funded with about DM 36.9M. In addition to silicon nitride and silicon carbide this follow-on program also included development work for other ceramics such as zirconia and an expansion of the applications into engines other than gas turbines.

The German ceramic component suppliers underwent a number of changes in ownership and product focus since the BMFT program [35–37]. Annawerk sold its Ceranox division to Feldmühle in 1983. This sale ended Annawerk's activities in engine components. All Ceranox equipment was transferred to Feldmühle following the sale. Annawerk itself was acquired by the French company Saint-Gobain in 1999. Feldmühle continued the engine component fabrication for a while to satisfy contract requirements but eventually stopped this activity altogether. Around 1984/1985 Rosenthal Technik was bought by Hoechst and became its CeramTec AG. Interest in engine components continued for some time with the support of a strong R&D group in the Hoechst R&D laboratory in Frankfurt. CeramTec AG was later taken over by Metallgesellchaft AG which also acquired the advanced ceramics part of Feldmühle. Both entities were merged in the 1993–1995 time frame and operated under the CeramTec AG name.

SIGRI acquired Great Lakes Carbon and merged to SGL, an independent company which today produces (among others) ceramic brake discs for Porsche. ESK has always retained an interest in engine components. They belong to the Wacker group of companies and had not been acquired by another company as of 2000. Degussa sold its ceramics Degussit division to Friatec AG (a sister company of Annawerk in the Cremer Group—see below) around 1978. Degussa then shut down all of its ceramic activities which have never been resumed since.

Another company that has had ongoing engineering ceramics activities is CFI (formerly the Cremer Forschungsinstitut, currently Ceramics For Industry). CFI is the central R&D facility for the Cremer group of companies, one of which was Annawerk. CFI established the basis for the non-oxide activities of Annawerk's Ceranox division. CFI became a joint venture of Bayer and Cremer in 1986 with Bayer holding 75% of the shares. In 1999 Bayer acquired 100% of the CFI shares. CFI was a leader in the fabrication of automotive ceramic components in Europe. CFI evaluated silicon nitride valves in Mercedes Benz cars. In 1999, H.C. Starck which is also owned by Bayer took over all ceramic activities from Bayer. In 2000, H.C.Starck acquired the company TeCe (Technical Ceramics) from Shell, being active in SiC components. Also in 2000, the decision was made, to merge CFI and TeCe to a new company, H.C. Starck Ceramics, with new combined fabrication and R&D-facilities in Selb, Germany. Current products include engine and engineering components, and bearing balls made from silicon nitride and silicon carbide.

ACKNOWLEDGEMENTS

The author is indebted to Dr. Fritz Thümmler, Professor Emeritus of the University of Karlsruhe, Germany, Dr. Fritz Aldinger, Professor of Ceramics at the University of Stuttgart, Dr. Ernst Gugel, former Head of R&D at CFI, Dr. Gerhard Wötting, Head of Materials R&D at CFI, and to Dr. Bjoern Schenk, Manager Strategic Technology at Honeywell Engines, Systems & Services, Phoenix, Arizona, USA, for their review of the chapter. Dr. Gugel, Dr. Wötting, and Dr. Aldinger also provided information about the history of the ceramic component manufacturers.

REFERENCES

[1] *Keramische Komponenten für Fahrzeug-Gasturbinen* (Ceramic Components for Vehicular Gas Turbines), W. Bunk and M. Böhmer, eds., Project Status Workshop, 18 and 19 April, 1978, Bad Neuenahr, Germany, Springer Verlag, Berlin, 1978 (in German).

[2] *Keramische Komponenten für Fahrzeug-Gasturbinen II* (Ceramic Components for Vehicular Gas Turbines II), W. Bunk and M. Böhmer, eds., Project Status Workshop, 24–26 November 1980, Bad Neuenahr, Germany, Springer Verlag, Berlin, 1981 (in German).

[3] *Keramische Komponenten für Fahrzeug-Gasturbinen III* (Ceramic Components for Vehicular Gas Turbines III), W. Bunk, M. Böhmer, and H. Kißler, eds., Project Status Workshop, 13–15 February 1984, Germany, Bad Neuenahr, Springer Verlag, Berlin, 1984 (in German).

[4] Bunk, W., "Overview on the German Program on Ceramic Components for Vehicular Gas Turbines," in *Proceedings of the First International Symposium on Ceramic Components for Engines*, Shigeyuki Sōmiya, Eizo Kanai, and Ken-ichiro Ando, eds., Hakone, Japan, October 17–19, 1983. pp. 9–20.

[5] Bunk, W., Gugel, E., and Walzer, P., "Overview of the German Ceramic Turbine Program," in *Ceramics for High-Performance Applications—III Reliability*, *Proceedings of the Sixth Army Materials Conference*, Edward M. Lenoe, R. Nathan Katz, and John J. Burke, eds., Plenum Press, New York, NY, USA, 1983. pp. 29–50.

[6] Böhmer, M., "Bericht der Projektträgerschaft über das Förderungsprogramm "Keramische Bauteile für Fahrzeug-Gasturbinen" (Project Management Report for the Development Program "Ceramic Components for Vehicular Gas Turbines"), in ref. [3], pp. 27–40 (in German).

[7] Bunk, W., "Bericht der Projektbegleitung über das Programm" (Program Management Report), in ref. [1], pp. 3–10, (in German).

[8] Thümmler, F., "Die Keramischen Hochtemperaturwerkstoffe Si_3N_4 und SiC" (The Ceramic High Temperature Materials Si_3N_4 and SiC), in ref. [1], pp. 45–79 (in German).

[9] Schwier, G., "Herstellung von Siliciumnitrid- und Siliciumcarbid Pulvern zum Sintern und Heiβpressen" (Fabrication of Silicon Nitride and Silicon Carbide Powders for Hot Pressing), in ref. [2], pp. 255–263 (in German).

[10] Schwier, G., "Entwicklung und Herstellung verbesserter Alpha-Siliciumnitrid Pulver für das Heisspressen und Sintern" (Development and Fabrication of Improved Alpha-silicon Nitride Powder for Hot Pressing and Sintering), in ref. [3], pp. 55–69 (in German).

[11] Gugel, E., Kessel, H., Müller, N., and Lange, E., "Herstellung von Siliziumnitrid-Verbundbauteilen für die Gasturbine" (Fabrication of Joined Silicon Nitride Components for the Gas Turbine), in ref. [1], pp. 141–160 (in German).

[12] Steinmann, D., Gugel, E., and Kessel, H., "Heiβpressentwicklung von Siliziumnitrid" (Development of Hot Pressed Silicon Nitride), in ref. [2], pp. 321–355 (in German).

[13] Gugel, E., and Leimer, G., "Entwicklung von Gasturbinenbauteilen auf der Basis von Siliziumnitrid" (Development of Silicon Nitride-based Gas Turbine Components), in ref. [3], pp. 133–157 (in German).

[14] Hausner, H., Landfermann, H., Wötting, G., Gugel, E., and Leimer, G., "Sintern von SiC und Si_3N_4" (Sintering of SiC and Si_3N_4), in ref. [2], pp. 265–296 (in German).

[15] Leimer, G., and Gugel, E., "Entwicklung von Gasturbinenbauteilen auf der Basis von Siliziumkarbid" (Development of Silicon Carbide-based Gas Turbine Components), in ref. [3], pp. 255–283 (in German).

[16] Lange, E., Müller, N., Olapinski, H., and Stannek, W., "Turbinenbauteile aus spritzgegossenem Si_3N_4" (Turbine Components from Injection Molded Si_3N_4), in ref. [1], pp. 183–200 (in German).

[17] Dworak, U., and Olapinski, H., "Gasturbinenbauteile aus spritzgegossenem RBSN—Materialenentwicklung" (Gas turbine Components from Injection Molded RBSN—Materials Development), in ref. [2], pp. 385–395 (in German).

[18] Dworak, U., Olapinski, H., and Finerle, D., "Herstellung von Bauteilen aus reaktionsgesintertem Siliciumnitrid im Spritzgießverfahren" (Fabrication of Components from Reaction Sintered Silicon Nitride by the Injection Molding Process), in ref. [3], pp. 113–132 (in German).

[19] Thümmler, F., and Grathwohl, G., "High Temperature Oxidation and Creep of Si_3N_4- and SiC-Based Ceramics and Their Mutual Interaction," MRS International Meeting on Advanced Materials, Vol.4, 237–253, 1989.

[20] Lange, E., and Müller, N., "Gasturbinenbauteile aus spritzgegossenem RBSN—Material und Bauteilenentwicklung" (Gas Turbine Components from Injection Molded RBSN—Materials and Component Development), in ref. [2], pp. 397–410 (in German).

[21] Pohlmann, H.-J., Maier, H.R., and Krauth, A., "Entwicklung und Herstellungstechnologien von keramischen Wärmetauschern" (Development and Fabrication Technologies for Ceramic Heat Exchangers), in ref. [1], pp. 201–217 (in German).

[22] Keller, H., Maier, H.R., and Krauth, A., "RBSN-Recuperatoren in Folien-Laminiertechnik" (RBSN-Recuperators Using Foil Laminating Techniques), in ref. [2], 411–429 (in German).

[23] Berroth, K., and Kraut, A., "Entwicklung von Siliciumnitrid-Bauteilen für den Gasturbinenbau" (Development of Silicon Nitride Components for Gas Turbines Manufacture), in ref. [3], pp. 93–112 (in German).

[24] Krauth, A., and Berroth, K., "Entwicklung von hochwarmfesten, korrosionbeständigen Siliciumcarbid-Formkörpern für die Verwendung im Gasturbinenbau" (Development of High Temperature Corrosion Resistant Silicon Carbide Castings for Application in Gas Turbine Manufacture), in ref. [3], pp. 227–253 (in German).

[25] Krauth A., "Recuperative Ceramic Heat Exchanger for Gas Turbine and Other Applications," in *Proceedings of the First International Symposium on Ceramic Components for Engines*," Shigeyuki Sōmiya, Eizo Kanai, and Ken-ichiro Ando, eds., Hakone, Japan, October 17–19, 1983. pp. 110–119.

[26] Kriegesmann, J., Reinmuth, K., and Lipp, A., "Die Herstellung und Eigenschaften von heissgepresstem und heissisostatisch gepresstem SiC" (The Fabrication and Properties of Hot Pressed and Hot Isostatically Pressed SiC), in ref. [1], pp.219–233 (in German).

[27] Hunold, K., and Grellner, W., "Möglichkeiten und Grenzen beim Heisspressen und Heissisostatpressen von Siliciumcarbid" (Potential and Limitations in Hot Pressing and Hot Isostatic Pressing of Silicon Carbide), in ref. [2], pp. 499–516 (in German).

[28] Hunold, K., and Grellner, W., "Heißisostatischen Verdichten von Siliciumcarbid" (Hot Isostatic Densification of SiC), in ref. [3], pp. 305–328 (in German).

[29] Schwetz, K.A., "Injection Molded Sintered Turbine Components of Aluminum Doped Alpha Silicon Carbide," in *Proceedings of the First International Symposium on Ceramic Components for Engines*, Shigeyuki Sōmiya, Eizo Kanai, and Ken-ichiro Ando, eds., Hakone, Japan, October 17–19, 1983. pp. 583–594.

[30] Böder, H., and Heider, W., "Eigenschaften von reaktionsgesintertem Siliciumcarbid und Anwendungsmöglichkeiten in der Fahrzeug-Gasturbine" (Properties of Reaction Sintered silicon Carbide and Potential Applications for Vehicular Gas Turbines), in ref. [1], pp. 235–250 (in German).

[31] Heider, W., and Böder, H, "Fortschritte bei der Entwicklung statischer Turbinenbauteile aus Si-infiltriertem Siliciumcarbid" (Progress with the Development of Stationary Turbine Components from Si-infiltrated Silicon Carbide), in ref. [2], pp. 457–471 (in German).

[32] Heider, W., and Böder, H., "Zur Eignung von SiSiC für GT-Bauteile" (Development of SiSiC for Gas Turbine Components), in ref. [3], pp. 285–303 (in German).

[33] Böder, H., and Heider, W., "Engineering Data for Silit SK®", in *Proceedings of the First International Symposium on Ceramic Components for Engines*, Shigeyuki Sōmiya, Eizo Kanai, and Ken-ichiro Ando, eds., Hakone, Japan, October 17–19, 1983. pp. 573–582.

[34] Böhmer, M., "Förderungsprogramm "Keramische Bauteile für Fahrzeug-Gasturbinen"—Zusammenfassung—Rückblick—Ausblick" (Development program "Ceramic Components for Vehicular Gas Turbines"—Summary—Retrospective—Future Perspective), in ref. [3], pp. 677–686.

[35] Gugel, E., private communication, 2000.

[36] Aldinger, F., private communication, 2000.

[37] Wötting, G., private communication, 2000.

CHAPTER

10

Hot Isostatic Pressing of Ceramic Turbine Components to Shape—Experience at Asea High Pressure Laboratory, Asea Cerama AB, ABB Cerama AB and AC Cerama AB

Hans T. Larker

Before retirement affiliated with AC Cerama AB, Robersfors, Sweden
(Since July 1998 in the Saint-Gobain Corporation)

HISTORICAL INTRODUCTION

In 1965 Asea AB, the leading Swedish heavy electrical company, (in 1988 merged with the Swiss company Brown Boveri to ABB Ltd,) inaugurated the Asea High Pressure Laboratory at Robertsfors in the north of Sweden. The production of synthetic diamonds, developed by Asea in the early 1950s, independent of the better known US efforts, had started at Robertsfors a couple of years earlier. The goal of the Asea High Pressure Laboratory was to develop products and processes utilizing higher pressures than common in the industry at the time (but lower than the extreme pressures used in synthetic diamond production). Processing and furnace equipment for hot isostatic pressing (HIP) became one of the main development areas. Projects in cooperation with other Swedish industries led, for example, around 1970 to the production by HIP of large cemented carbide rolls by Sandvik and of high speed tool steels from powder by STORA (ASP-steels).

By the end of the 1960s silicon nitride was (and still is) considered a leading candidate material for highly stressed ceramic turbine components. Development, predominantly in Great Britain, was focused on reaction bonded silicon nitride (RBSN), with the potential of making complex shapes, however with low strength, and on hot pressed silicon nitride (HPSN), produced under uniaxial pressure to high density and strength but limited to very simple shapes, requiring costly diamond machining. The author, head of the Asea High Pressure Laboratory at that time, put forward the idea to develop the HIP technology for silicon nitride with a *vision of achieving both shape and strength.* Gas turbine components were seen as a major potential application area. Much of the work from 1977 on was carried out in a very fruitful cooperation with the gas turbine component developer/user United Turbine/Volvo. Their ceramic gas turbine experience is presented in chapter 22 in Volume I of this publication [1].

It should be noted that Asea, contrary to most other potential gas turbine component makers, came into the area with a heavy manufacturing/processing technology focus rather than an orientation towards the development of materials. The R&D efforts were also entirely company/industry funded, i.e. there was no governmental or similar financing until in the early 1980s, when a major organizational change was made.

Asea Cerama AB was formed in 1984 (with Volvo as one of the four minority shareholders). Licensing then commenced of the glass encapsulation and HIP technology (and in some cases the injection molding know-how) to companies both in Europe, Japan and the USA. Several of the licensees used the technology for materials development and for supplying products in important gas turbine programs. Some of these materials are AlliedSignal Ceramic Components' (currently Honeywell Ceramic Components) GN-10, GTE Labs' PY6 and Norton's NT154, NT164 and NCX-5102. Some financial support for technology development was from that time on obtained from STU (the Swedish National Board for Technical Development). From 1988 the company was one of the about 1100 in the ABB group (ABB Cerama AB). Early 1993 there was a management buy-out and name-change to AC Cerama AB. Saint-Gobain S.A., the worlds largest industrial ceramics group, acquired the company on July 1, 1998.

PROCESSING PRINCIPLES AND EARLY EXPERIENCE

Around 1970 the common shaping method in isostatic pressing, cold or hot, was to use a container with suitable interior shape, carefully fill it with powder, seal the container and apply isostatic pressure on the container at a selected temperature [2]. The fill density of the fine silicon nitride powders to be used would, however, be far too low for a reasonable probability to avoid heavy shape distortion during HIP. Furthermore, a high processing temperature would be needed, around 1700°C, which would require refractory container materials like molybdenum or tantalum. The processing route of 'Direct HIP in a Pre-Shaped Container' was thus very early judged not feasible, for these and other reasons, including high cost.

The processing route of 'Direct HIP with Shape Determined by Green Body' [3] was instead considered the way to go. Like in the manufacturing of RBSN products, a powder body should determine the shape of the product. An at the densification temperature resilient but gas impermeable encapsulation, upon which the gas pressure in the HIP process could act, would then enable three-dimensional hot pressing of the powder body to high density, *ideally without changing its shape, just reducing its size.* As the HIP processing would require temperatures around and above 1700°C, high-melting glass was identified as a possible candidate for the encapsulation. The author got support for these ideas from Asea's top management and could, early in 1971, employ a glass specialist, Dr. Jan Adlerborn.

By the end of that year the first samples were HIPed, using cold isostatic pressing (CIP) of silicon nitride powder and a conformable high silica glass container (Figure 10.1). Remarkably good shape tolerances were obtained of cylinders prepared by CIP plus green machining, and then HIPed to high density using such an oversize-glass-ampoule technique. It was found that minor constituents of the glass, like boron, had a marked effect on glass penetration and on reaction with the sample. A recent Ph.D. thesis [4] throws light on this early observation. In TEM (transmission electron microscopy) micrographs crystalline boron nitride (BN) was found in an approximately 0.5 to 0.8 micrometer deep layer, near the original silicon nitride green body surface and more or less covering it. The textured BN in Figure 10.2

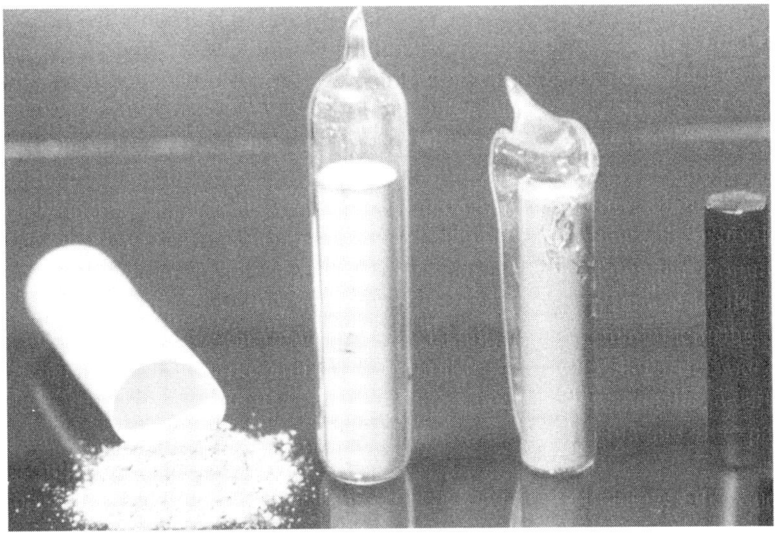

Figure 10.1 The Encapsulated HIP Process Using a High Silica Glass Container (Ampoule). Silicon Nitride Powder is CIPed and Machined to a Cylindrical Body and then Slipped into a Glass Ampoule, Evacuated and Sealed, Left and Center Left. After Heating the Glass to Softening (At About 1300°C) and Allowing the Glass to Conform to the Cylinder, Center Right. Fully Dense Cylinder after HIP and Glass Removal by Sand Blasting, Right. (Reprinted with permission from AC Cerama AB)

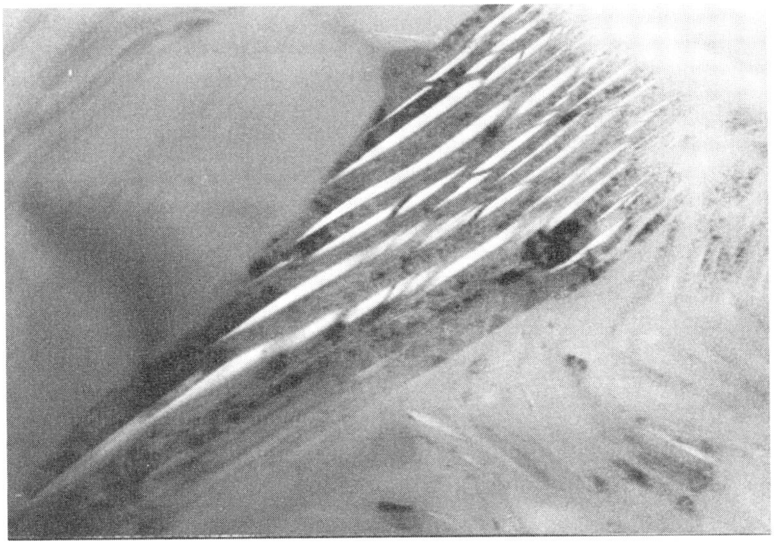

Figure 10.2 TEM Micrograph of Textured, Crystalline BN Formed Near an Original Silicon Nitride Green Body Surface Facing Encapsulation Glass Containing Boron Oxide. Adjacent Grains Now Consist of Silicon Oxynitride. Total Width of the Figure is Approximately 0.6 Micron. (Reprinted with permission from A.-K. Westman)

was facing encapsulation glass containing boron oxide during HIP and as in this micrograph, the apex of a wedge-shaped BN area usually points inwards. Adjacent grains now consist of silicon oxynitride. BN was not present before HIP, but is assumed to result from a reaction between a boron containing encapsulation glass and silicon nitride particles. If such boron nitride forms in former pore channels, when the glass starts to penetrate the small pores of the green body, the flow will be hindered. Even a low viscosity glass, which would create only insignificant shear stresses on the surface of the ceramic body, could thus be used without risk for gross glass infiltration.

The experimental conditions can be particularly well defined with Vycor® glass, due to its softening behavior, which is more suitable than that of quartz glass. The method is still well suited for material samples. It is, however, unsuited for products with weak, protruding parts, as the viscous glass container may break these off the green body when conforming to its shape. Single-blade dummies of crude shape were tried but with very low yield. Glass encapsulation development gradually shifted towards methods better suited for shaped parts.

DEVELOPMENT OF MATERIALS AND PROCESS

Commercial silicon nitride powders were used as experimental materials. They were alloyed, prepared, glass encapsulated and HIPed at different sets of parameters. It was soon found that a lower additive content in the silicon nitride alloy was needed than possible, even with uniaxial hot pressing (HP). A commercial British silicon nitride powder grade, AME CP85, was—as a matter of fact—HIPed to full density without the use of any additives at all in 1974.

A first goal was to obtain similar mechanical properties as obtained in materials made by hot pressing (HP). Contacts with Lucas Ltd. were initiated in 1974 and in 1976 tests were conducted in parallel. A Lucas proprietary powder blend lot (grade HS110, with magnesia-alumina additives) was split into two parts, one densified by Lucas using their HP process, and the other HIPed by Asea. Mechanical testing by Lucas confirmed the same strength level of the samples densified by HIP as the HP samples, when tested in their "strong" direction (perpendicular to the hot pressing direction). However, the HIPed material had uniform strength, independent of direction, while the HP material showed the typical strength reduction of about 20% parallel to the pressing direction.

The R&D efforts were gradually increased. One direction was towards silicon nitride alloy development with relatively low additive content, and based on yttria rather than on magnesia—for improved short term and long-term high temperature properties, vital for gas turbine applications. Another direction was glass encapsulation development, focusing on creating the encapsulation starting from glass particles, rather than the cumbersome and expensive glass ampoule technique. The Asea team pioneered the use of glass for HIP of silicon nitride and a comprehensive patent portfolio on glass encapsulation and ceramics was built up from 1971 and forward (see e.g. [5–8]).

The ongoing R&D efforts were for the first time made public in March 1977 at the annual SAE conference in Detroit [9], where some early results were presented. Achieved density was for example shown as function of pressure, after glass encapsulated HIP at 1750°C for 1hr, of a silicon nitride powder with 0.5 to 10 wt% yttria additive. The silicon nitride powder used had a metal impurity level of 0.3 wt% (0.2% Fe, <0.1% Al). For example, with only 0.5 wt% yttria, 250 MPa was required to achieve full density, while about 100 MPa was sufficient with 10 wt% additive. Also, a 3-point bend strength result at 1370°C of 554 MPa with an m-value of 26 (m = Weibull constant) was reported. Steady-state creep rates at 70 MPa of 1.6×10^{-6}/hr at 1300°C and 1×10^{-5}/hr at 1400°C were shown in a diagram together with literature data.

BILATERAL COOPERATION

The first cooperation and license agreement, directed towards gas turbine components for the KTT-concept, was signed with United Turbine AB (later Volvo) in 1977 (KTT is short for Kronogård Turbine Transmission [10]).

A breakthrough was made in 1978 in the development of encapsulation using glass particles but the method to make the green body, which determined the shape, was still CIP plus green machining. Wheels with straight and flat blades were made. A shape deviation of ≤0.1 mm on a 100 mm diameter hub was typical after HIP and glass removal. A common material composition for turbine parts was H.C. Starck silicon nitride grade H1 with 1–5 wt% yttria as the only additive.

NC-machining from a CIPed block was used to shape the first axial turbine wheels with 24 blades (straight airfoils), attached to a hub. After encapsulation, using glass particles, HIPing, gentle glass removal by sand blasting and shaft attachment, such an axial turbine wheel of silicon nitride was driving the compressor and tested as the first stage of a self-running KTT/MK1 turbine in 1981. The operating conditions were modest however, about 1000°C and 47,000 rpm. Aerodynamically improved 33-blade wheels, which were similarly made but green machined, using five-axis NC grinding, were similarly processed. Such a wheel was cold spin tested to 80,000 rpm and used in the MK1 turbine/transmission powering a modified Volvo 264, driven on public roads by Prof. Kronogård in March 1982 [11].

By the end of the 1970s injection molding was identified as a potential method to accurately shape complex green ceramic bodies, like integrated turbine wheels. The team at Robertsfors was trusted with the development of a binder/binder removal system. Airfoil shapes with remarkable form consistency after glass encapsulation and HIP were made soon thereafter. United Turbine and Asea made an advanced injection-molding tool for a 37-blade axial turbine wheel with high aerodynamic efficiency. The above-mentioned three generations of axial turbine wheels—a wheel with 24 blades (straight airfoils) and an aerodynamically improved 33-blade wheel, both formed by green machining of CIPed blocks, using four- and five-axis NC grinding, respectively, and finally a 37-blade axial turbine wheel with high aerodynamic efficiency, green formed by injection molding—are shown in Figure 10.3. The injection-molded wheel exhibited a linear size change of about 14%, in all directions, during the shrinkage from green body to fully densified turbine wheel.

Shape consistency clearly exceeded the author's highest expectations at the outset, a decade earlier. No shape adjustment of the airfoils was needed. For example, tolerances were well within specifications for a series of single blades with dump-bell hub attachment, made for and measured by Rolls-Royce. Airfoil shape requirements were also met for the above-mentioned 37-blade integrated wheel, which had as injection-molded plus HIPed blade trailing edges of just 0.3 mm thickness!

United Turbine developed a balancing/stub-shaft machining/shrink fitting method to the metal shaft, avoiding the common balance machining of the hub. The gravity center was determined by weighing and the

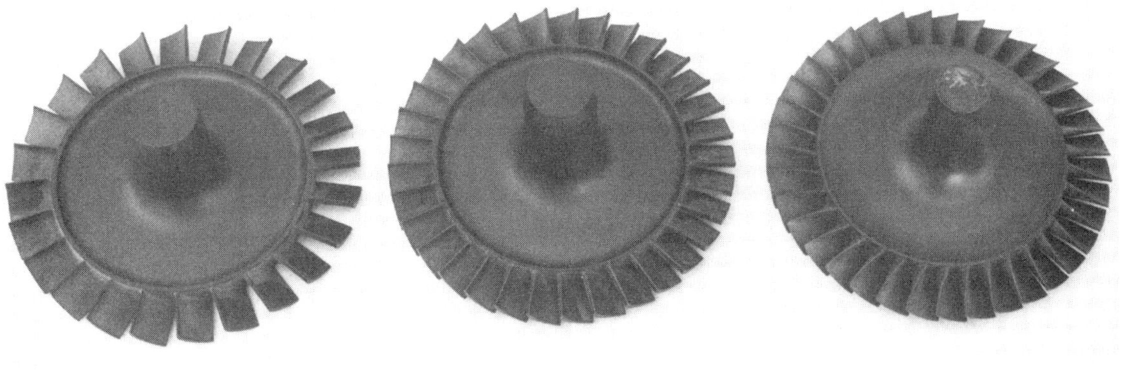

Figure 10.3 Three Generations of Silicon Nitride Axial Turbine Wheels, HIPed to Full Density Using Glass Powder Encapsulation of Shaped Green Bodies. Left and Center 24- and 33-Blade Wheels with Green Bodies Prepared by CIP and Green Machining. At Right a 37-blade Wheel with the Green Body made by Injection Molding (Reprinted with permission from Volvo Aero Corporation)

stub-shaft outer diameter was diamond ground. Finally, the blade tips were adjusted by cylindrical grinding, for proper clearance to the shroud. Different generations of 37-blade wheels were used in cold and hot spin testing and in the full ceramic hot section of the GT110, a further development of the MK1 gas turbine.

Unfortunately, injection molded wheels often failed from defects, which the available quality assurance methods had failed to detect. Spin testing to failure is very expensive and the attempts to find the nature of the cause of such premature failures in the shattered debris of a fractured wheel is very time-consuming, and is more often than not, unsuccessful. A method for low-cost tensile testing—the ASCERA® method and equipment—was therefore developed, allowing large volumes of injection-molded and HIPed ceramic material to be fractured in tension [12] at low costs, and investigated for strength-critical defects. The ability to identify the origin of such defects, usually of metal pick-ups from different equipment used in the preparation and molding of the products, was very important for the work to eliminate the most critical defects. As much as practical, HIPed silicon nitride material was used to replace metals, e.g. in the lining and the Z-shaped rotor of the mixing equipment and in the liner of the screw-type injection molding machine. Mechanical properties of the turbine wheels were gradually improved.

RADIAL TURBINE WHEELS AND AGATA

Radial turbine wheel development was for many years directed towards turbochargers (e.g. for Volkswagen). An early experience was however (in connection to a response to a tender request) to make a hub of the size needed in the radial turbine wheel for the Garrett/Ford AGT 101 single shaft automotive gas turbine. It was injection molded and HIPed (the binder removal time exceeded one month, however). This 0.8 kg heavy "door-knob" with 90 mm diameter looked sound but was never further tested. The experience certainly influenced a later invention [13], allowing the use of an injection molded hollow blade ring and a CIPed hub, green machined to fit into the blade ring. After assembling and some sealing preparation at the joint line, the two components grow into one piece in the subsequent glass encapsulated HIP process.

This radial wheel experience was certainly of use in the European AGATA (Advanced Gas Turbine for Automobiles) project from 1992 through 1998. The interested reader is referred to chapter 22 in Volume I with its references for further detail of the project from a gas turbine manufacturers point of view [1]. AC Cerama's participation in AGATA was in the development and manufacture of the catalytic combustor after-burner exit cone and of the turbine wheel. The work was done in close cooperation with Volvo.

Material

An improved silicon nitride grade, CSN 101, was selected for both components. The silicon nitride grade CSN 101 was developed by AC Cerama to meet the demands of high strength and fracture toughness coupled with good creep resistance at 1350°C and good thermal shock resistance. It is a 3 wt% Y_2O_3 composition based on high purity Si_3N_4 powder (with ~ 0.01 wt% metal impurities) and can be green formed either by injection molding or by CIP. The densification is then performed using the proprietary glass encapsulation and HIP technique discussed above and in more detail in reference [3].

Typical properties of this material are given in Table 10.1.

Tensile creep rupture tests in air at temperatures up to 1350°C and 50 to 200 MPa tensile stress for up to 1700 hrs are presented in detail in ref. [1]. So are also long term oxidation tests on bend test specimens both in static air and in a high velocity burner gas up to 1350°C. The average weight change after exposure was limited to fractions of mg/cm^2. A drop in bend strength occurred, mainly in the first 50 hrs but then remained approximately unchanged up to 1000 hrs. Particularly encouraging was the observation that burner rig exposure caused lower strength change than air exposure (for details see [1]).

4-point flexural Stepped-Temperature-Stress-Rupture (STSR) tests going from 1000°C up to 1400°C were also performed. Both test bars with machined surface, with as-HIPed surface (standard glass encap-

TABLE 10.1 PROPERTIES OF AC CERAMA CSN 101 SILICON NITRIDE

Property	Value
Density	3.22 g/cm^3
Hardness	19 GPa
Flexural strength (RT, 4-point bending)	955 MPa
Weibull modulus (RT)	28
Flexural strength (1350°C, 4-point bending)	630 MPa
Fracture toughness (Chevron notch)	6 MPa.m$^{1/2}$
Coefficient of thermal expansion (K^{-1})	3.7 × 10^{-6} (20–800°C)
Thermal conductivity (RT)	40 W/m.K

sulation) and with *as-HIPed surface using a novel interlayer* between glass encapsulation and silicon nitride, were tested. As can be seen in Table 10.2, the novel interlayer technique gave material with high temperature properties as good as the bulk material, and at the level of the best silicon nitride materials reported in the literature [14].

The novel interlayer makes it thus possible to produce an as-HIP-processed-surface with stress rupture properties that are fully comparable to a carefully diamond machined part (in spite of a significantly smoother surface of the latter). A drawback is, however, the higher sensitivity at room temperature (similar to the sensitivity of a machined surface) for sand blasting, the method usually used for finish removal of the encapsulation glass. A reason for this is the silicon oxynitride, formed just at the surface in contact with glass [4]. Since its thermal expansion coefficient is slightly lower than silicon nitride's a favorable stress state for withstanding sand blasting develops at the surface.

In interpreting the results in Table 10.2, it may be noted that the *time to stress rupture* of the as-HIPed, just glass encapsulated surface at 250 MPa stress is at the same level as the time to stress rupture of the other two surfaces at 300 MPa in this STSR-test.

TABLE 10.2 STEPPED-TEMPERATURE-STRESS-RUPTURE (STSR) RESULTS FOR CSN 101 WITH DIFFERENT SURFACE CONDITIONS

STSR Bend stress (MPa)	Time at temperature				
Machined surface	1000°C	1100°C	1200°C	1300°C	1400°C
200	24hrs	24hrs	24hrs	24hrs	>24hrs
250	24hrs	24hrs	24hrs	24hrs	>24hrs
300	24hrs	24hrs	24hrs	24hrs	2–24hrs
350	24hrs	24hrs	24hrs	24hrs	0.3–4hrs
As-HIPed surface (glass encapsulated)	1000°C	1100°C	1200°C	1300°C	1400°C
200	24hrs	24hrs	24hrs	24hrs	>24hrs
250	24hrs	24hrs	24hrs	24hrs	2–15 hrs
300					
350					
As-HIPed surface (novel interlayer+glass)	1000°C	1100°C	1200°C	1300°C	1400°C
200	24hrs	24hrs	24hrs	24hrs	>24hrs
250	24hrs	24hrs	24hrs	24hrs	>24hrs
300	24hrs	24hrs	24hrs	24hrs	2hrs
350	24hrs	24hrs	24hrs	24hrs	0.1hr

Figure 10.4 Ceramic Combustor Exit Cones of HIPed Silicon Nitride (CSN 101). End Faces (only) will be Ground and Polished (Reprinted with permission from Volvo Aero Corporation)

CATALYTIC COMBUSTOR AFTERBURNER

AC Cerama used the CSN 101 grade silicon nitride to manufacture the exit cone of the catalytic combustor for the AGATA project. This component was at the prototype scale manufactured by cold isostatic pressing, followed by green machining, glass encapsulation and HIP. In a future larger production volume such products could be formed by injection molding. The outer and inner surfaces of the combustor component are used with as-HIPed surface finish. The novel interlayer technique was used.

Two afterburner cones, prior to delivery for catalytic section assembly (for details see [1]), are shown in Figure 10.4. The full scale testing of the combustor was successfully completed. The afterburner cones showed no evidence of mechanical damage or oxidation and withstood transient tests and steady state tests at 1350°C.

TURBINE WHEEL

CSN 101 Si_3N_4 was also the material chosen for the AGATA turbine wheel. AC Cerama's technique of combining two green-forming methods in a two-piece approach, joined as a single piece in the glass encapsulated HIP process [13], was further developed. The hub section of the wheel was CIPed and green machined, while the hollow blade-ring was injection molded (see Figure 10.5). Following binder burnout, the hub and blade ring are green-joined, glass encapsulated and HIPed, the advantage being that binder removal is achieved much more quickly and easily with two smaller components (and can initially be kept very low in the CIPed part), compared to when the component is injection molded as a single piece.

AC Cerama has also evaluated injection molding of the wheels as a single piece. After some initial problems with large internal defects this route was judged to be also a feasible way for producing good quality wheels (see Figure 10.6). The novel interlayer technique was only in limited use due to the risk of surface damage in finish removal of the encapsulation glass. The results of non-destructive and destructive testing of the wheels are discussed in [1]. Unfortunately results from spin testing came in too late in the program to

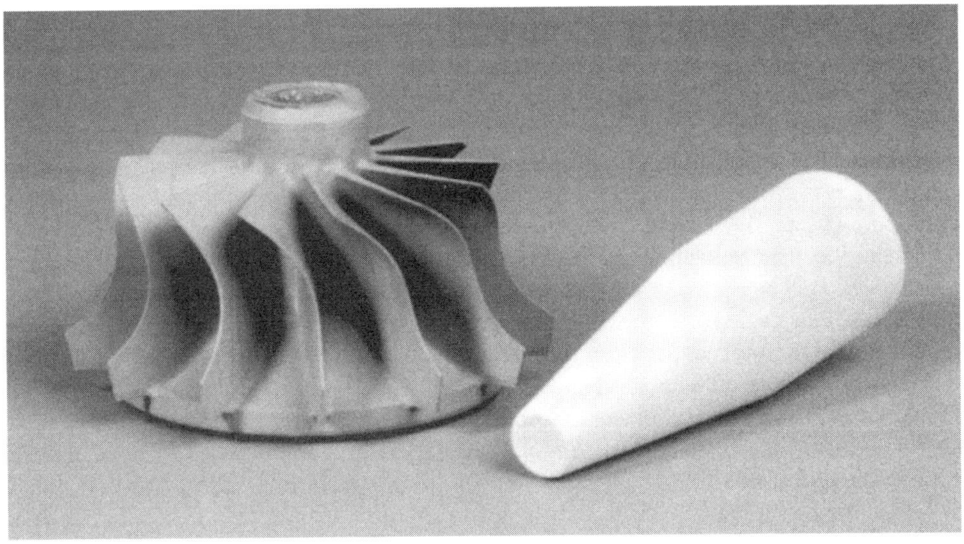

Figure 10.5 Injection Molded Blade Ring and Cold Pressed Hub Green bodies, prior to glass encapsulation and HIP (Reprinted with permission from Volvo Aero Corporation)

allow necessary iterations in order to fulfill program targets. One cause to of premature failure was that green body surface cracks (from injection molding and binder removal) did not heal properly but became glass-filled during HIP. A contributing factor might be that the wheels were "brown fired" to facilitate handling, before glass encapsulation and HIP. The general experience has been that fine green body surface cracks heal more easily if pre-sintering before HIP is avoided (if not, even a slight pre-sintering before HIP significantly reduces surface crack healing).

Figure 10.6 HIPed Turbine Wheels for AGATA of CSN 101 Grade Silicon Nitride. The Back Face and Stub Shaft Only are Finish Machined (Reprinted with permission from Volvo Aero Corporation)

OTHER CERAMIC MATERIALS

The initial reason for focusing on silicon nitride was the high inherent potential for high temperature properties and its relative toughness, coupled with an expected high effectiveness of isostatic pressure in the processing. Other potential ceramic materials for high temperature structural applications, like silicon carbide and oxide ceramics, also appeared to us as having a better developed processing technology. Furthermore, these materials were not found to benefit from the favorable glass penetration blocking mechanism [4] observed with borosilicate glasses and silicon nitride. That mechanism appears to work the very best with undoped silicon nitride, but also well with moderate rare earth oxide additions, like yttria. Alumina additions, for example, must be kept at a low level, or additional interlayers on the green body will be required.

Silicon carbide can readily be processed, using interlayers. AC Cerama produces fine grain material with high strength, mainly for wear parts. *Boron carbide*, with superior wear resistance, but also more expensive, has surpassed silicon carbide for some applications.

In the late 1970s, there was a strong bias towards silicon carbide for gas turbines, and away from silicon nitride. Some environmental stability results at high temperature for silicon nitride, which were very discouraging, had been published. Silicon nitride as such has, however, intrinsic creep and oxidation resistance properties, well at the level of silicon carbide even at temperatures as high as 1550°C! To the author's satisfaction, that was already shown in 1975 in tests on CVD (chemical vapor deposited) silicon nitride [15]. Also undoped, high purity polycrystalline silicon nitride, made by glass encapsulated HIP, exhibits excellent high temperature stability, superior to state-of-the-art silicon carbide at temperatures up to 1440°C [16]. It is the additives, enriched at grain boundaries, which give rise to the problems. Silicon nitride with 3–4 wt% yttria, depending on minor impurities in the basic powder, at present appears to offer satisfactory environmental stability. AC Cerama's CSN 101 is one example. The Norton experimental grade NCX-5102 illustrates the potential at hand with careful powder and green body processing [17].

Whisker, long fiber, platelet or particle reinforcement in a matrix of silicon nitride or oxide ceramic has successfully been fully densified with glass encapsulated HIP. Silicon carbide platelets in undoped silicon nitride were shown to improve fracture toughness while maintaining low creep rates up to 1400°C [18]. Combustor tiles were made from *single crystal oxide fiber reinforced oxide*, using glass encapsulation and HIP at AC Cerama in a Brite/EuRam supported program with French, Swedish and British companies [19]. Cost projections for production of long fiber combustor tiles by HIP were favorable in comparison with other available processes. The feasibility to make curved panels is particularly interesting.

SPIN-OFF APPLICATIONS

Some spin-off applications have already been mentioned but the single most important production application is silicon nitride balls for bearing applications, meeting the highest bearing producer standards. To some extent they are used in turbine/turbocharger bearings. For upper grade sports equipment and quality machine tools, for example lathes, they are quickly becoming state-of-the art.

KEY LESSONS LEARNED AND FUTURE POSSIBILITIES

In looking back the author realizes that a more open attitude towards universities and researchers would have brought the glass encapsulated HIP to a more favorable position relative to competing processes. Just look at what one PhD student could accomplish [4]! A "supercritical" mass of brains for further development and better understanding of the process was lacking. The secrecy around the technology, needed with the chosen commercialization strategy, and the desire at AC Cerama and its predecessors to be self-financing, have both contributed to the generally prevailing opinion of prohibitive processing costs for the glass encapsulation and HIP technology. However, as an example, the common powder beds in sintering of silicon nitride are more expensive than the encapsulation glass! And a simple textile industry wear part of a silicon carbide /silicon nitride powder composite has been in profitable production for over a decade in direct competition with sintered alumina. The key factor in that case is the superior shape precision achieved with injection molding and glass encapsulated HIP. For some reason most of our licensees of the technology appear not to have fully focused on the shaping potential of the process. Interlayer technology may have to be used for some products. High pressure is a great help for achieving final densification but a favorable microstructure may be better developed at more moderate pressures. A concerted effort may very well make ceramic turbine rotors (of silicon nitride, of course, but very possibly modified into a composite) to a viable (and cost effective) proposition!

ACKNOWLEDGEMENTS

Many of the views put forward in this chapter are only those of the author, but I would like to thank many friends and associates particularly at AC Cerama, Volvo/United Turbine, ABB/Asea and Luleå University of Technology. Thanks also to STU/NUTEK (the Swedish National Board for Technical Development) for sponsoring of different projects.

REFERENCES

[1] Lundberg, R., and Gabrielsson, R.,"Ceramic Components for Automotive Gas Turbines—Experience at Volvo," in *Ceramic Gas Turbine Design and Test Experience, Progress in Ceramic Gas Turbine Development, Vol. I,* M. van Roode, M.K. Ferber, and D.W. Richerson, eds., ASME Press, New York, USA, 2002. Chapter 22, pp. 499–512.

[2] Larker, H.T., and Larker, R. "Direct HIP in a Pre-Shaped Container," in *Processing of Ceramics, Part 2*, R.J. Brook, ed., in *Materials Science and Technology, a Comprehensive Treatment,* VCH Publishers, 1996. Vol. 17B, Sec. 16.3.1.1., pp. 152–153

[3] Larker, H.T., and Larker, R., "Direct HIP with Shape Determined by Green Body," in *Processing of Ceramics, Part 2*, R.J. Brook, ed., *in Materials Science and Technology, a Comprehensive Treatment*, VCH Publishers, 1996. Vol. 17B, Sec. 16.3.1.2, pp.153–157.

[4] Westman, A-K., *Hot Isostatic Pressing of Ceramics and Influence of Glass Encapsulation,* PhD thesis, Luleå University of Technology, Luleå, Sweden, LTU-DT-98/11, 1998.

[5] Isaksson, S-E., and Larker, H.T., "Method of Manufacturing a Sintered Powder Body," US Patent 4,339,271,1971.

[6] Adlerborn, J., and Larker, H.T., "Method of Manufacturing Bodies of Silicon Nitride," US Patent 4,455,275, 1974.

[7] Adlerborn, J., and Larker, H.T., "Method of Manufacturing an Object of Silicon Nitride," US Patent 4,112,143, 1976.

[8] Adlerborn, J., Larker, H.T., Mattsson, B., Nilsson, J., German Patent 2,950,158, 1978, and "Method of Manufacturing an Object of Metallic or Ceramic Material," US Patent 4,478,789 1984.

[9] Larker, H.T., Adlerborn, J., and Bohman, H., "Fabricating of Dense Silicon Nitride Parts by Hot Isostatic Pressing," SAE Technical Paper No. 770335, presented at the International Automotive Engineering Congress and Exposition, Detroit, MI, USA, 28 February–4 March, 1977.

[10] Kronogård, S.O., Automotive Turbine—Advantages of Three Shaft Configurations," Gas Turbine International, Nov.-Dec. 1975.

[11] Press release, March 27, 1982.

[12] Nilsson, J., and Mattson, B., "A New Tensile Test Method for Ceramic Materials," in *Ceramic Materials and Components for Engines, Proceedings of the Second International Symposium*, W. Bunk and H. Hausner, eds., Lübeck-Travemunde, Federal Republic of Germany, April 14–17, 1986, Verlag Deutsche Keramische Gesellschaft, Bad Honnef, 1986. pp. 651–656.

[13] Adlerborn, J., Larker, H.T., Mattsson, B., and Nilsson, J., (1981). SE Patent 8,101,509-1, and JP Patent 1,636,429.

[14] Pollinger, J.P., "Progress in Fabrication of Silicon Nitride Structural Components for Turbomachinery Applications," ASME paper 96-GT-347, presented at the International Gas Turbine and Aeroengine Congress and Exposition, Birmingham, UK, June 10–13,1996.

[15] Greskovich, C., Rosolowski, J.H., and Prochazka, S., *Ceramic Sintering*, Final Report SRD-75-084, General Electric Corp. R&D, 1975.

[16] Tanaka, I., Pezzotti, G. "Evaluation of Slow Crack Resistance in Ceramics for High Temperature Applications," J. Am. Ceram. Soc., 75[4], 772–777, 1992.

[17] Pujari, V.K., and Tracey, D.M., "Processing Methods for High Reliability Silicon Nitride Heat Engine Components," ASME paper 9-GT-319, presented at the International Gas Turbine and Aeroengine Congress and Exposition, Cincinnati, Ohio, USA, May 24–27, 1993.

[18] Pezzotti, G., "Si_3N_4/SiC-Platelet Composite without Sintering Aids: A Candidate for Gas Turbine Engines", J. Am. Ceram. Soc., 76[5], 1313–1320, 1993.

[19] Razzell, A.G., Holmquist, M., Molliex, L., and Sudre, O., "Oxide/Oxide Ceramic Matrix Composites in Gas Turbine Combustors," ASME paper 98-GT-30, presented at the International Gas Turbine and Aeroengine Congress and Exhibition, Stockholm Sweden, June 2–5, 1998

CHAPTER

11

CERAMIC HEAT EXCHANGER DEVELOPMENT BY BOOSTEC SA

Marc Ferrato and Ginès Martinez

BOOSTEC SA, Bazet, Hautes-Pyrénées, France

INTRODUCTION

CGEC (Compagnie Générale Electro Céramiques) which was founded in 1922 as a subsidiary of CGE (Compagnie Générale d'Electricité) started its production with the manufacturing of electric insulators made of bone china and alumina powders. In 1971, the French ceramic company Sediver joined CGEC and became Céraver in the CGE goup. This company became involved with nuclear applications by manufacturing of porous alumina tubes for the enrichment of uranium by gaseous diffusion, employing a staff of two thousand. In 1985 Céramiques et Composites (C&C), a subsidiary of Rhône-Poulenc in association with Aérospatiale was established to develop advanced ceramic parts based on non-oxide materials. The R&D team of Céraver joined C&C. BOOSTEC SA was founded in 1999 as a spinoff of C&C. The mechanical seals activity for automotive and chemical applications of C&C joined the Wacker Chemie Group and became, in the beginning of 2002, Wacker Ceramics France company, while BOOSTEC SA is an independent company focused on new demanding markets for ceramics, like space optics, electronics, automotive and chemical engineering.

For more than twenty years the companies Céraver, Céramiques et Composites (C&C) and, more recently, BOOSTEC SA, in cooperation with a number of partners, have been developing ceramic heat exchangers, mostly incorporating silicon carbide but also, in special cases, ceramic oxides. This chapter will focus on the ceramic heat exchanger manufacturing for which, now, BOOSTEC SA is the rightful heir of these two decades of experience at Bazet.

Energy recuperation with ceramic heat exchangers is predominantly of interest to chemists or petrochemists who are interested in using the technology to improve the performance of their processes, and more specifically, to protect materials against oxidation and corrosion. The introduction of our heat exchangers in gas turbines started in the mid-1980s, essentially because of two European programs which were partially funded through public funds.

Between 1986 and 1990, the French National Aerospace Research Institute (ONERA—Office National d'Etudes et Recherches en Aérospatiales) and Céramiques et Composites with the support of the French Ministry of Defense conducted a research and development program focused on a ceramic heat exchanger for a gas turbine cycle for aircraft engines.

The second major project, conducted under the European Union Eureka umbrella, was the AGATA (Advanced GAs Turbine for Automotive) program which was conducted between 1987 and 1998. The goal of the AGATA program was to develop a gas turbine that could power a conventional passenger car. The AGATA project was redefined in 1992 to take advantage of a gas turbine in an hybrid electric vehicle, specifically from an environmental point of view. The revised AGATA project began in early 1993 and spanned a five year, three month period until March 1998.

CERAMIC RECUPERATION

For many years, suitable clay ceramics have been used for thermal applications. The transportation channels for hot water to ancient roman houses are some of the oldest examples. During the industrial period, the first large-scale utilization of ceramics for heat exchange was primarily recuperators for the glass industry. Since the beginning of the twentieth century, this application has been improved steadily, particularly when considering the thermal and corrosion aspects, and today it is a very efficient technology.

Other applications have followed the emergence of new ceramic materials. For example, one of the first applications of silicon carbide combined with more traditional clay ceramics was the manufacturing of refractory components for high temperature furnaces with a controlled atmosphere. These products for thermal management were manufactured by the U.S. based Carborundum Company.

It is only during the 1970s that some projects were initiated in which the primary focus of the new refractory ceramics became heat exchange devices. There are two main reasons for this evolution. The first one is the 1973 oil crisis that resulted in strong public support for development in the area of energy management [1,2]. At that time industry was the major energy consumer, and so it was natural that all efforts concentrated on industrial applications of recuperation. Thermal recovery of exhaust gas from high temperature processes was a potentially important route for energy recovery (see Figure 11.1).

The second reason originated in the successes obtained during this period with military and civil nuclear and aerospace activities, and also in the fields of electrical engineering and data-processing. Some examples were a heat shield for ballistic missiles and space shuttles, porous tubes for the enrichment of uranium that subsequently led to the manufacturing of large volumes of advanced ceramic parts, and thermal processing furnaces with controlled atmospheres, necessary for the manufacturing of silicon substrates for the microprocessor industry.

The Western countries, particularly the U.S., Great Britain, France and Germany strongly contributed to these programs which strengthened their industrial ceramic companies. The Japanese industry rapidly joined the progress with integrated and vertical development in companies such as Kyocera Corporation and NGK Insulators, Ltd.

Two basic concepts have been proposed for ceramic heat exchangers: a cyclic recuperator (C1) and a stationary recuperator with direct heat transfer (C2). The first concept uses the ceramic component for its thermal capacity and is alternatively heated and cooled through counterflow contact between the hot and cold fluids. There are two basic options for this concept.

C1V1: Cyclic recuperator consisting of two heating compartments. The first one is heated by hot gas while the second chamber yields its enthalpy during the preheating of cold air. The next cycle is an inversion of the previous mechanism through the action on a set of sluices. This is the primary option for large furnaces requiring several tens of megawatts of power. In the early 1980s the British company Hotwork, on the basis of this principle, commercialized compact thermal recuperators directly integrated with burners for metal processing applications. The required thermal capacities were attained with two chambers filled with alumina marbles.

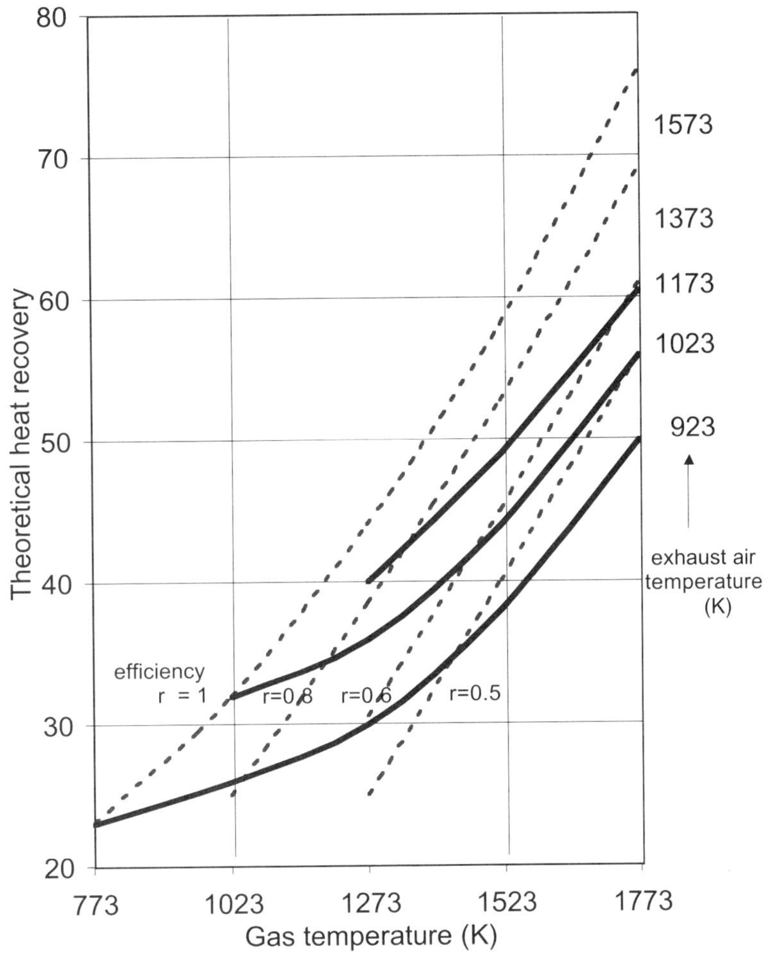

Figure 11.1 Graph Showing Heat Recovery as a Function of the Temperature of the Industrial Exhaust Gas and the Efficiency of the Recuperator

C1V2: Rotating recuperator consisting of one disk with axial permeability. The flat faces of the rotating disk are in contact with two collector/distributor parts which control the fluid distribution. The cold fluid is heated as it passes along the hot ceramic walls. The rotation of the disk establishes continuous thermal cycling of the ceramic disk. The best known fabrication method is based on the extrusion process. In the 1980s, companies such as Corning in the USA and Céraver in France proposed honeycomb structures with cell densities in the range of 93–124 cells/cm^2 (500–800 cell.in^{-2}) (Figure 11.2)

This concept was selected for the first European automotive ceramic turbine programs. It has the advantage that its geometry offers good thermomechanical behavior, compatible with process manufacturing requirements. But a major problem is to maintain airtightness between the rotating ceramic core and the inlet/outlet collectors. Notwithstanding time consuming development efforts, this problem has not been satisfactorily resolved.

Concept C2 involves a stationary recuperator with direct heat transfer. Considering all current materials and applications, this is the most widely applied concept. There are three main options for ceramic recuperators:

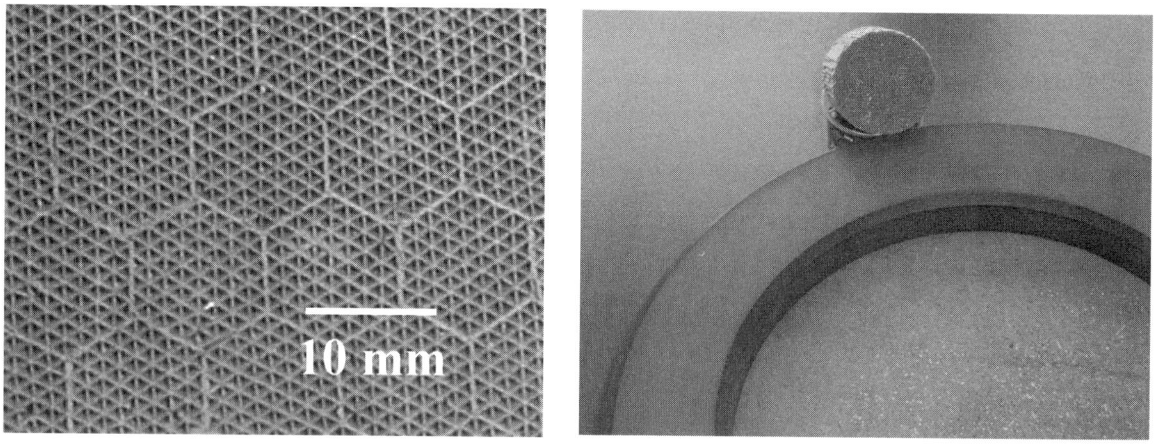

Figure 11.2 Ceramic Rotating Recuperator (>300mm in Diameter) with Cell Unit

C2V1—Tubular heat exchanger with convection and radiation heat transfer used for high power modules such as power stations for fuel or heavy coal burning. The exhaust gas is very corrosive and fouling of the equipment a major challenge [3,4]. Large diameter (2.5 to 5.0 cm) tubes, 2–4 meters in length are connected to the inlet/outlet air zone. The exhaust gas circulates inside a chamber outside the tubes, with parallel and/or counter flow arrangement. One limitation of this kind of equipment is the collector interface because its large size and complex shape offers poor thermomechanical resistance.

C2V2—Tubular recuperator with convection heat transfer. The tubes, 1.25 cm in diameter, are smaller than for the previous type (C2V1). The fluid is generally not corrosive or fouling to the equipment. These units offer higher specific thermal power and a better thermomechanical performance than the C1V1 option. The constituting tubes are also easier to manufacture.

Céramiques et Composites (C&C) manufactured tubular recuperators with sintered silicon carbide tubes, PTFE plates and an enameled stainless calender [5]. The first industrial application was for a hydrogen fluoride acid (HF) production process based on the chemical reaction between sulphuric acid (H_2SO_4) and fluorspar (CaF_2). Due to the very high chemical resistance of silicon carbide, the recuperator also allows excellent thermal exchange between sulphuric acid on the one hand and a mixture of hydrogen fluoride and sulfuric acid (HF and H_2SO_4) on the other hand. This equipment appreciably increases the process efficiency and moreover offers substantial energy savings. This successful development improved the lifetime up to 8 years without the need to replace any critical components. On the other hand, a metal recuperator, e.g. based on Hastelloy X alloy, was not reliable because the components did not last for more than 5 to 6 weeks.

A high temperature (800 to 1200°C) gas/gas version was proposed by C&C with sintered silicon carbide (SSiC) components (plates, tubes and deflectors) as shown in Figure 11.3.

Airtightness between ceramic plates, tubes and deflectors can be controlled as follows:

(1) During sintering through self bonding (SBS) as shown in Figure 11.4

Mineral additives are introduced in the starting raw materials in order to create, at high temperature, a molten glass which can migrate and create a bond between the different elements. This process is used for cordierite based recuperators. For SSiC, no additive is necessary. The coefficient of thermal expansion coefficient is controlled during sintering shrinkage to bond the constituent components [6]. The bonding is homogeneous (see Figure 11.5)

(2) After sintering, through the use of cements or glass ceramics, or (in the case of silicon carbide) through a brazing operation with silicon (Figure 11.6).

The homogeneous method is preferred for recuperators for operation in a corrosive environment.

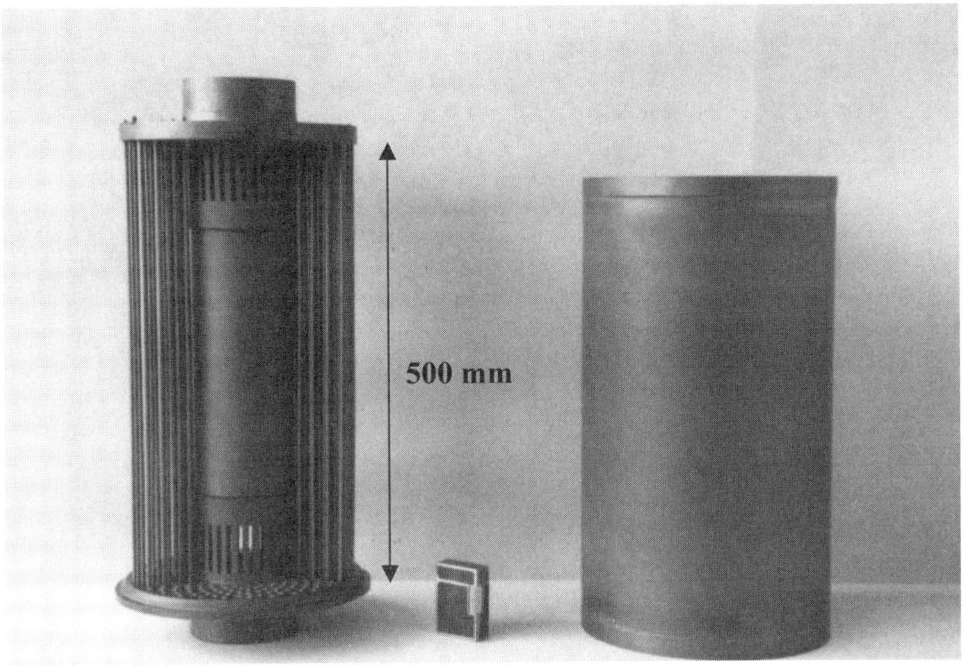

Figure 11.3 Sintered SiC Tubular Recuperator

Figure 11.4 Self Bonding During Sintering (AGATA Program) (Reprinted from ref. [9])

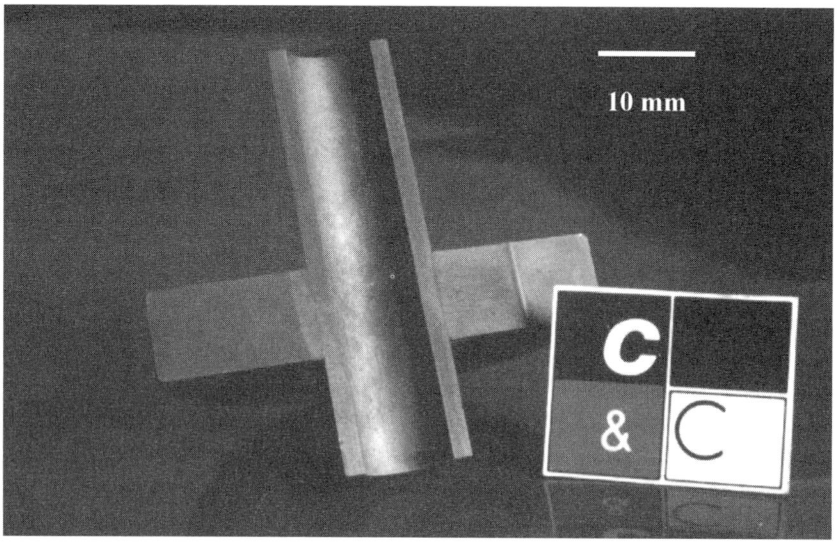

Figure 11.5 Homogenous Bonding During Sintering

C2V3—Compact recuperator. Greater thermal effectiveness can be achieved by increasing recuperator compactness [7]. Two main design paths have focussed on:

(1) a fluid arrangement in order to increase the thermal exchange coefficients ($W.K^{-1}.m^{-2}$)
(2) a matrix design with large specific surface exchange values ($m^2.m^{-3}$)

The first approach has been widely used for fluids operating with impingement jets [8]. Céraver in collaboration with the French company Bertin has designed and manufactured this kind of recuperator with a silicon carbide plate structure. This application concerned the thermal recovery for a high temperature process (see Figure 11.7). The recuperation is mainly achieved by means of impact jets which impart very high overall exchange coefficients (150 $W.K^{-1}.m^{-2}$) [8]. The concept is modular. An assembly of basic modules, each with a capacity of about 30 kW, results in industrial output power of several hundred kW.

Successful laboratory trials, conducted with a module at operating conditions as high as 1200°C, validate this technology.

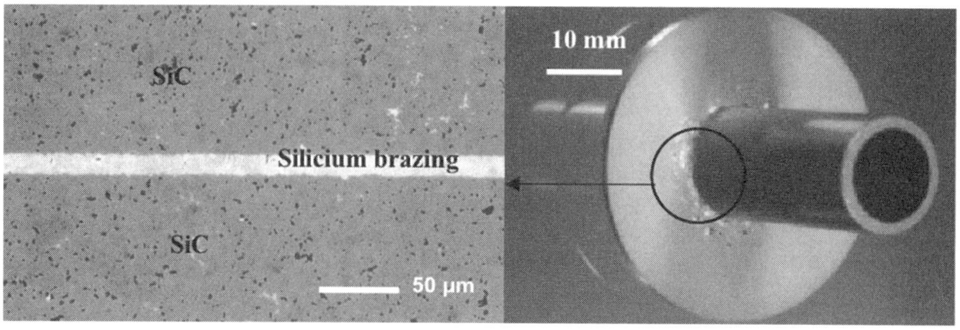

Figure 11.6 Bonding through Silicon Brazing

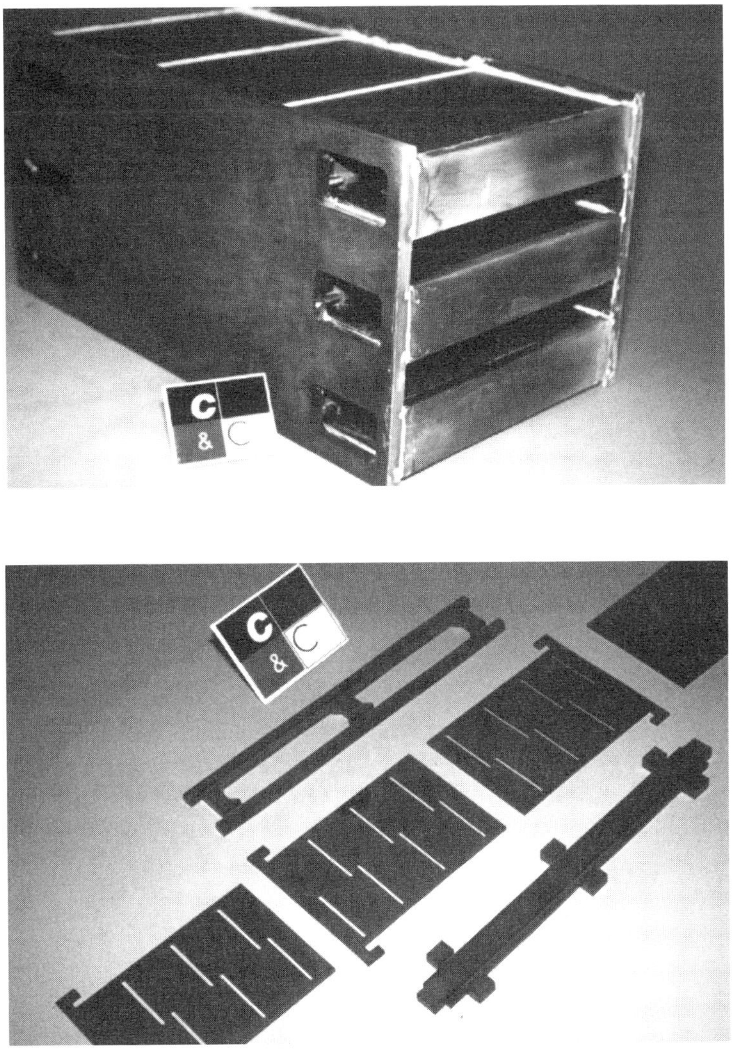

Figure 11.7 Bertin/Céraver Compact Recuperator (200mm × 200mm × 400mm)

The flat geometry of the exchange surface associated with the mechanical bonding between ceramic parts limits this application to gases at atmospheric pressure without any overpressure in order to prevent leakage. This design is unsuitable for the air pressure levels of a turbine cycle.

From the mid-1980s ONERA and Céramiques et Composites with the support of the French Ministry of the Defense led a research program for a ceramic recuperator for a turbine cycle for aircraft engines [5]. The basic element of the heat exchanger consisted of a multi-channel tube, or to be more precise, three concentric cylindrical tubes with fin-type connections (see Figure 11.8). This results in a hydraulic diameter of typically a millimeter with a structure that is resistant to pressures of 8–10 bars and with fins that are very compact. Each of the bayonet type tube assemblies consists of one half of a heat exchanger, in which two tube sheets connect the various heat exchanger tubes. Homogeneous bonding technology was used to ensure air tightness. With this approach it is mandatory that the joint keeps the assembly rigid. This is difficult

Figure 11.8 Heat Exchanger Component for Aircraft Engine Turbine

because the time-temperature behavior of the elements is very different, and the thermomechanical performance was drastically decreased which directly impacted service life.

Finally, l'Institut Français du Pétrole (IFP) and Céramiques et Composites have co-developed a recuperator/reactor for a very high temperature petrochemical process with the financial support of the Direction à la Recherche de l'Union Européenne (DG XII). Based on a SSiC honeycomb monolith, half the layers are plugged with ceramic paste. The resultant structure consists of alternating reactant gas layers separated by the hot gas layers in a counter flow arrangement (Figure 11.9). This concept was applied for the gas filter of Diesel engines using porous ceramic material as a filtering medium.

Figure 11.9 Sintered Silicon Carbide Honeycomb Heat Exchanger.

CERAMIC RECUPERATOR FOR THE AGATA PROGRAM

The highest degree of compactness for a C&C recuperator was achieved for the AGATA program (Advanced GAs Turbine for Automobiles). The objectives of the AGATA program were to develop three critical ceramic components—a catalytic combustor, a radial turbine wheel, and a static heat exchanger—for a 60 kW gas turbine for a hybrid electric vehicle. AGATA is a Eureka joint project conducted by the commercial companies and research institutes in France and Sweden listed in Table 11.1.

As conceived in 1987, the original aim was to develop a gas turbine to drive a conventional car power train. To take account of the advantages of a gas turbine in an hybrid electric vehicle, specifically from the environmental point of view, the AGATA project was redefined in 1992. The revised AGATA project began in early 1993 and spanned a five year and three months period until March 1998.

C&C focussed on the development of the compact ceramic heat exchanger designed to operate at high temperatures (up to 1027°C/1300 K) and moderate pressures (4 bars). This chapter will focus primarily on the heat exchanger fabrication development aspect of the AGATA program. The testing of the heat exchanger is described in more detail in the chapter by ONERA in Volume I (chapter 23) of this text. C&C also became involved with the fabrication of monolithic ceramics with honeycomb structures for the catalytic combustor. The catalytic combustor development is described in the chapter by Volvo Aero Corporation (Volume I, chapter 22).

The extensive discussions between the design partners and C&C resulted in a high efficiency heat exchanger compatible with the ceramic molding process [9]. Tubular and plate compact recuperator concepts were compared, on the basis of calculations and manufacturing tests, in order to value their characteristics with regards to the following parameters.

—thermal effectiveness of the heat exchanger
—weight
—volume
—manufacturing
—thermomechanical strength

Due to the volume and mass criteria, a plate heat exchanger was selected with ceramic parts that had to be as thin as possible. The compact plate heat exchanger was made by stacking corrugated plates separated by thin flat sheets. The layers are alternatively hot gas and cold air paths. Because of the high ther-

TABLE 11.1 COMPANIES AND INSTITUTES IN THE (EUREKA) AGATA PROGRAM (1993–1998)

Technology Area	Company
Ceramics	AC Cerama, Sweden
Ceramics	Aerospatiale, France
Catalytic materials	AlliedSignal Cat. Env. (ACE), France
Turbines	AlliedSignal Turbo, France
Materials	CEA/CEREM, France
Ceramics	Céramiques et Composites (C&C), France
Vehicles	PSA, France
Vehicles	Renault, France
Turbines, ceramics	Volvo Aero Corporation, Sweden
Technology Area	Research Institute
Heat exchangers	GRETh, France
Combustion	Institut Français du Petrole, France
Turbines, combustion, heat exchangers	ONERA, France

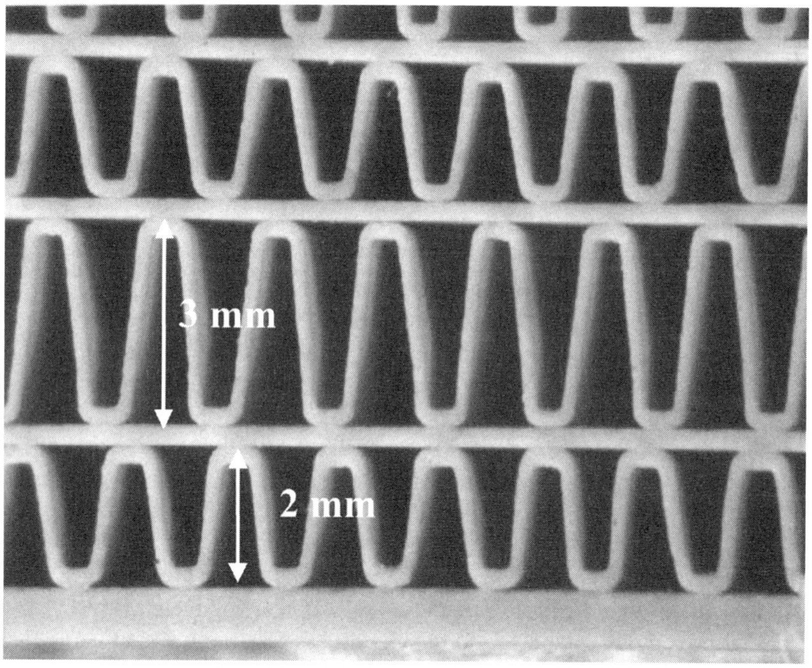

Figure 11.10 Cross Section of the Active Ceramic Matrix (Adapted from ref. [9])

mal effectiveness target a counter flow arrangement of the hot and cold flows was required with a perfect tightness between them. A radial hot gas inlet/outlet was selected based on the cold air flow.

The following design characteristics were specified:

• Flat plate thickness	0.3 mm
• Corrugated plate thickness	0.2 mm
• Layer height (cold air flow)	2 mm
• Layer height (hot gas flow)	3 mm
• Sinusoidal folds period (cold air flow)	1.6 mm
• Sinusoidal folds period (hot gas flow)	1.6 mm

Figure 11.10 is a cross section of a assembly of flat and corrugated sheets showing the sizes of the different constitutive components.

Processing Challenges

The manufacturing technology. The compact recuperator poses the greatest manufacturing challenges, particularly the manufacturing of large flat sheets (0.3 mm in thickness) and thin corrugated folds (0.2 mm in thickness). For the AGATA program, different industrial forming processes were investigated: casting, pressing, extrusion, and calendering. Casting results in non-homogeneous thickness, and is not well-suited for sinusoidal folds. Pressing was not selected because of the specified low thickness and large surface of the thin sheets, and because it is not a convenient technology for corrugated folds. On the other hand, extrusion and calendering techniques give good results, both in terms of the quality of the green molded parts, and also from a cost perspective.

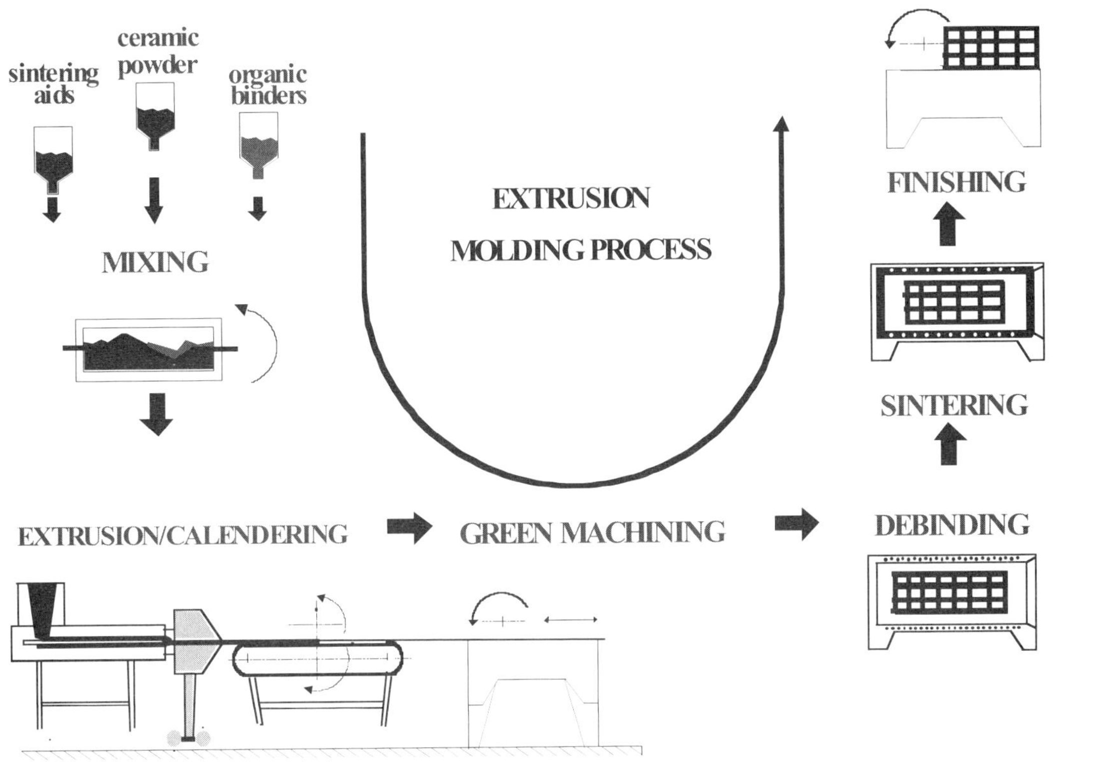

Figure 11.11 Manufacturing of the Ceramic Parts by the Extrusion/Calendering Process

The plastic forming technique: extrusion and calendering. The equipment consists of a screw extruding machine with a special die. In the case of the plate recuperator, the final geometry of the green ceramic parts is achieved by calendering. For a tubular recuperator or a honeycomb unit, the same extrusion process is used without the final calendering step.

This molding process, schematically represented in Figure 11.11, involved the processing of a thermoplastic paste, which is a mixture of the ceramic powder and a binder system. The latter consists of a primary binder and processing aids such as plasticizers, lubricants and surfactants to control the rheological behavior of the compound paste.

The thermoplastic paste, after having been properly mixed up to 100°C, is forced through the die which confers to the part its exact shape. These green components are then cut with the use of a high speed water cutting system which gives, with high accuracy, the final geometry of the parts. Figure 11.12 shows the constituent components for a compact plate recuperator, just before assembly.

Compact heat exchangers are made of a great number of different elements, which have to be joined together in order to ensure tightness between the hot gas and cold air flows, with a good contact to ensure the best heat exchange effectiveness. The joining can be performed following the two processes described as previously for the C2V2 tubular recuperator, during sintering or after sintering.

Manufacturing tests demonstrated that the SBS process gives the best results for cordierite (Figure 11.4). It ensures homogeneous bonding for all surfaces which have to be in contact to optimize thermal effectiveness. These observations are related to the strength of the bonding. Single-layer samples are glued onto two plates, and installed on a mechanical testing machine, with which the rupture strength is measured. The higher strengths are those which exhibit rupture inside of the sinusoidal sheet, instead of taking place at the interface between flat and sinusoidal sheets.

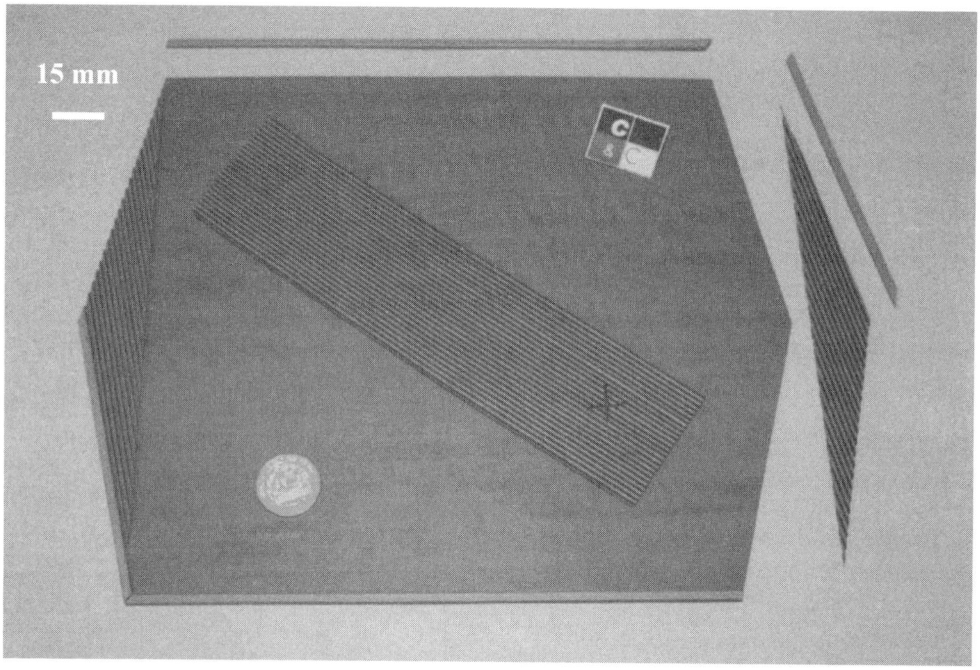

Figure 11.12 Constituent Elements of One Gas Layer from Quarter Scale Compact Exchanger (AGATA Program) (Reprinted from ref. [9])

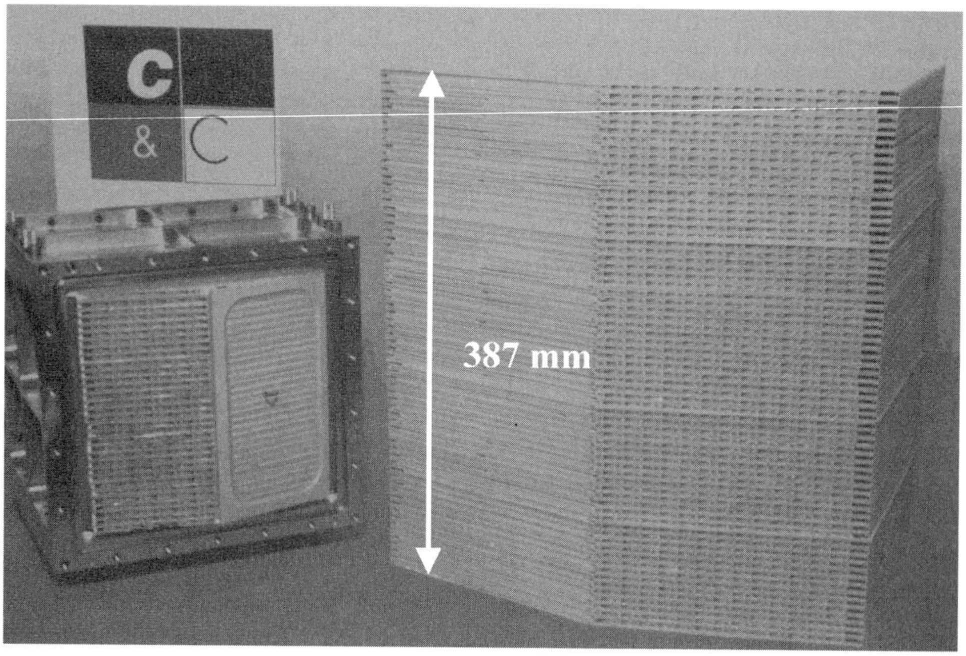

Figure 11.13 One Quarter Scale and Full Scale Compact Heat Exchangers After Sintering (Adapted from ref. [9])

TABLE 11.2 CHARACTERISTICS OF THE AGATA COMPACT RECUPERATOR

Characteristics	Cordierite Matrix As-Manufactured
Number of active gas layers	65
Number of active air layers	64
Upper layer circuit	Gas
Lower layer circuit	Gas
Length	190 mm
Width	376 mm
Module height	387.2 mm
Module weight	30.3 kg

The SBS process is more cost-effective than brazing because it is a very simple in-line process which fewer processing steps. Prior to sintering the binder system is removed during a thermal treatment in air at a temperature up to 500°C [10–12]. Final sintered ceramic recuperators are shown in Figure 11.13.

The final characteristics of the fabricated heat exchanger are given in Tables 11.2 and 11.3. The main geometrical objectives (thickness, wavelength) met the AGATA requirements. A micrograph of the active ceramic matrix is shown in Figure 11.10.

Materials Challenges

Ceramic materials are of interest for thermal applications, especially heat exchangers, because of an interesting combination of properties compared to those of metallic materials:

- Superior thermomechanical properties
- Better corrosion resistance in humid and dry gas
- Lower specific mass which is of interest for mobile equipment

From a geopolitical point of view, the abundance of ceramic raw materials was very much appreciated by the French government during the sensitive period before the end of the Cold War.

TABLE 11.3 CONTROL OF FINAL CHARACTERISTICS OF FULL-SCALE AGATA CERAMIC CORE

Air Layer Characteristics	As-Defined by the Designers	Cordierite Matrix As-Manufactured	Mean Variation
Flat plate thickness	0.3 mm	0.30 –0/+0.04	+5%
Corrugated folds thickness	0.2 mm	0.20 –0/+0.02	+5%
Corrugated folds period	1.6 mm	1.74	+9%
Layer height	2 mm	2.06	+3%
Hydraulic diameter Dh	0.923	0.96	+4%
Gas Layer Characteristics	**As-Defined by the Designers**	**Cordierite Matrix As-Manufactured**	**Mean Variation**
Flat plate thickness	0.3 mm	0.30 –0/+0.03	+5%
Corrugated folds thickness	0.2 mm	0.20 –0/+0.02	+5%
Corrugated folds period	1.6 mm	1.65	+3%
Layer height	3 mm	3.06	+2%
Hydraulic diameter Dh	1.0	1.077	+8%

Nevertheless, these materials have well-known limitations that must be reconciled during the design phase. An important limitation for monolithic ceramics is their brittleness which is reflected in the fracture toughness ($1 < K_{IC} < 10$ MPa.m$^{1/2}$), and the mechanical property distribution characterized by the Weibull modulus ($3 < m < 20$). The ceramist and the designer will operate with a specific methodology and computational tools in order to make the design compatible with the brittle material. In ceramic component design some basic rules must be followed such as:

- equilibrium of ceramic cross sections
- ceramic components must be as small as possible (In ceramics "small is beautifull")
- stress distribution at the interfaces
- smooth surfaces, rounded fins, etc.

Also, the molding and bonding processes for components are less versatile than for metallic alloys.

Silicon carbide and cordierite-based materials are well known for high temperature applications because of their refractoriness and their behavior under severe thermal shock conditions. Cordierite is an ancient material that has contributed to automotive development as a catalyst substrate for exhaust gases. Because of this application, the extrusion process has been developed for this material on an industrial scale. Moreover, cordierite-based materials exhibit good thermomechanical properties, especially their ageing at high temperature (see Table 11.4). However for temperatures approaching 1000°C and above, no commercial powders exhibit the required properties and material development is necessary. This was one of the objectives of the AGATA program as described below.

The quality of cordierite-based material is strongly dependent on the raw materials which have been selected for a minimal content of impurities. Careful chemical analysis is required to establish the precise chemical composition of the starting materials. This information is needed to control the final stoichiometry of the cordierite-based material. Starting powder characteristics, such as particle size distribution and specific surface area must be known in order to control all the processing steps.

During the AGATA program, C&C has developed a high performance cordierite on a pre-industrial scale. The properties of the mature material are given in Table 11.4 together with the properties of other materials used for heat exchange application.

Silicon carbide materials offer good thermomechanical properties, especially sintered silicon carbide (SSiC), which can be used at very high temperature with excellent oxidation and creep resistance. Typical characteristics of these materials are summarized in Table 11.5. SSiC exhibits high temperature strength 2–3 time greater than nitride bonded, siliconized or recrystalized SiC materials.

As Table 11.4 shows, cordierite has low thermal conductivity, typically ~ 2 W.m^{-1}.K^{-1} Cordierite recuperators are designed with the objective to decrease the heat conduction transfer impact on the thermal effectiveness factor. Designing more compact components is a good solution. The focus of the designer is to obtain a fluid/wall transfer coefficient which is lower than the wall conductivity. This ratio is the Biot coefficient. Thin walls (0.1 to 1 mm in thickness) with short fins (0.5 to 2 mm in length) lead to a Biot number <1 in the case of a gas/gas recuperator. For other types of heat exchange, such as liquid/liquid or gas/liquid, cordierite is not suitable.

On the other hand, silicon carbide exhibits a high thermal conductivity, close to values for metallic alloys, typically exceeding 100 W.m^{-1}.K^{-1}. This material does not offer significant heat transfer resistance even for recuperators with large exchange values [7]. However, due to the high value of the elasticity modulus and the absence of any intergranular vitreous phase, SiC is more sensitive than cordierite-based material from the point of view of flaw propagation. Particular attention must be given to the geometry of the parts and the connecting areas by designing shapes as simple as possible. Knowledge of the thermal shock resistance and thermal cycling is critical.

Based on the recuperator design, the required thermal effectiveness, and thermomechanical considerations, these materials must be compared not only on the basis of their intrinsic properties but also from the point of view of:

TABLE 11.4 SUMMARY OF PROPERTIES OF DIFFERENT CANDIDATE MATERIALS FOR RECUPERATORS

Material		Units	BOOSTEC Sintered SiC (SSiC)	AGATA Cordierite	BOOSTEC Aluminum Nitride (AlN)	Sintered Silicon Nitride (Si_3N_4)	Aluminum Alloy (12% Si)	Nimonic Refractory Alloy
Apparent bulk Density		10^3 kg.m^{-3}	>3.10	2.50	3.00	3.20	2.8	8.3
Apparent porosity		%	0	0	0	0	0	0
Mechanical strength	20°C	MPa	450	200	360	700	300	1100
(3-point bending)	1200°C	MPa	450	200 (1000°C)	—	700	70 (300°C)	850 (1000°C)
Modulus of elasticity	20°C	GPa	420	120	330	300	75	210
	1200°C	GPa	370	100	—	250	—	—
Thermal conductivity at 20°C		W.m^{-1}.K^{-1}	180	2.5	200	30	160	20
Coefficient of thermal Expansion	(20–1000°C)	10^{-6} °C^{-1}	4.6	1.7	4.3 (20–400°C)	3	22 (20–300°C)	15
Maximum working temperature in air		°C	1750	1200	1350	1600	300	900

TABLE 11.5 SUMMARY OF PROPERTIES OF DIFFERENT SILICON CARBIDES

Material		Units	BOOSTEC Sintered SiC (SSiC)	Recrystalized SiC (ReSiC)	Siliconized SiC (SiSiC)	Reaction Bonded SiC (RBSiC)
Apparent bulk Density		10^3kg.m^{-3}	>3.10	2.70	3.00	2.80
Apparent porosity		%	0	16	0	10
Mechanical strength	20°C	MPa	450	150	350	300
(3-point bending)	200°C	MPa	450	150	330	280
	1600°C	MPa	460	150	—	—
Modulus of elasticity	20°C	GPa	420	280	300	200
	1200°C	GPa	370	190	200	200
Thermal conductivity	20°C	W.m^{-1}.K^{-1}	180	100	120	70
	1200°C	W.m^{-1}.K^{-1}	35	20	25	15
Coefficient of thermal expansion	20–1000°C	10^{-6}°C^{-1}	4.6	5.5	4.8	4.9
Maximum working temperature		°C	1750	1600	1350	1450

- feasibility and the quality of the bonding between the plates
- manufacturing process for the ceramic parts
- the heat exchanger design
- thermomechanical stress calculations.

For example, the AGATA requirements, especially thermomechanical stress considerations, led to the selection of cordierite-based material rather than silicon carbide. It has been clearly demonstrated that the safety factors, expressed as the ratio of the mechanical strength over the induced stress, are higher in the case of the cordierite material. This result is surprising because of the known higher mechanical strength of silicon carbide. However, due to a lower elastic modulus and coefficient of thermal expansion, the thermal stresses induced in the cordierite material are lower than those induced in the silicon carbide.

KEY LESSONS LEARNED AND CONCLUSIONS

Initially, the recuperation of energy from industrial exhaust gases was the objective of these developments. Subsequently, the enormous industrial gas turbine market estimated at billions of dollars, justified the development projects. Nevertheless, today we note that the optimistic forecasts of the 1980s have not yet been realized, because the developments have been much slower than were envisioned.

This slowdown is primarily related to production challenges which, initially, were underestimated, but also to the communication between ceramists and end users which was not always constructive. The end user requirements, in terms of performance and cost, often led to designs that were too complex and did not take into account ceramic technologies. One of the lessons learned is the need for the ceramist to acquire an adequate understanding of the complete system of which he is a part, to be an effective participant in the sometimes deeply involved redesign that implies the application of ceramics. The end user must also accept this requirement for redesign even if, initially, system performance may need to be sacrificed. The iterative trial approach which inevitably accompanies the production of complex ceramic systems is like the acrobatics of high stunt-flying. Initially, one modifies the material, then the component design, then one returns to the material, and so on. Therefore, these developments are involved projects which require many long term tests which are expensive.

However, these difficulties should not overshadow the progress made. The ceramic materials, in particular silicon carbide and cordierite, have now matured, and are being extensively used in the industrial environment (automotive and chemical applications, etc.). The production processes are controlled, and the equipment, designed specifically for heat exchanger components, is operational. The assembly processes have also evolved, and new techniques have been tested successfully. All this progress has resulted in a body of know-how and powerful equipment, which enables us to consider the start of pre-commercialization programs.

ACKNOWLEDGEMENT

Unless otherwise indicated, all figures are being reprinted with permission from BOOSTEC SA.

REFERENCES

[1] McDonald, C.F., "The Role of the Ceramic Heat Exchanger in Energy and Resource Conservation," ASME paper 79-GT-106, presented at the Gas Turbine Conference and Exhibit and Solar Energy Conference, San Diego, CA, USA, 12–15 March, 1979. Transactions of the ASME, Journal of Engineering for Power, 102[2], 303–315, April 1980.

[2] Drouet, C., Martinez, G., and Minjolle, L., "Echangeurs céramiques" (Ceramic Heat Exchangers), Journées Techniques la Maîtrise de l'Energie dans l'Industrie, MEI 87, Paris, France, 31 March-3 April, 1987. pp. 289–307. (in French).

[3] Mori, Y., " Future Development of High Temperature Heat Exchangers," *Proceedings of the XVIIth International Symposium on High Temperature Heat Exchangers*, Y. Mori, A.E. Sheindlein, and N. Afgan, eds., Dubrovnik, Yougoslavia, August, 1985. Hemisphere Publishing Corporation, 1986. pp. 3–14.

[4] Laws, W.R. and Reed, G.R., "Ceramic Heat Exchangers for Coal Fired and Dirty Gas Applications," ibid., pp. 550–566.

[5] Minjolle, L., "Réalisation d'échangeurs de chaleur en céramique: problèmes mécaniques et thermiques," (Development of Ceramic Heat Exchangers: Mechanical and Thermal Problems), Matériaux Mécanique Electricité, 422, 23–27, Sept.–Oct., 1987 (in French).

[6] "Procédé d'assemblage d'objets en carbure de silicium et assemblages ainsi obtenus" (The Process for Arranging Silicon Carbide Objects into Assemblies), French Patent FR 90 12240 du 01/10/90 (Publication No.: FR 2667591), European Patent EP 0479657 (in French).

[7] Ferrato M. and Thonon B. "A Compact Ceramic Plate-fin Heat exchanger for Gas Turbine Heat Recovery," *Compact Heat Exchangers for the Process Industries*, R.K. Shah, ed., Begell House, Inc., New York, NY, USA 1997. pp.195–199.

[8] "Echangeur de chaleur à impact de jets" (Heat Exchanger with Jet Impact), French Patent FR 88 08639, 28 June, 1988 (Publication No.: FR 2633379) (in French).

[9] Lundberg, R. and Ferrato, M., "Ceramic Component Development for AGATA," ASME paper 99-GT-392, presented at the International Gas Turbine and Aeroengine Congress and Exposition, Indianapolis, IN, USA, June 7–10, 1999.

[10] Chartier, T., Ferrato, M., and Baumard, J.F., "Influence of the Debinding Method on the Mechanical Properties of Plastic formed Ceramics," J. Eur. Ceram. Soc., 15[9], 899–903, 1995.

[11] Chartier, T., Ferrato, M., and Baumard, J.F., "Supercritical Debinding of Injection Molded Ceramics," J. Am. Ceram. Soc., 78[7], 1787–1792, 1995.

[12] Ferrato, M., Chartier, T., Baumard, J.F., and Coudamy, G., "Der Bindemittelabgang in keramischen Scherben" (The Issue of Bonding of Broken Ceramics), CFI Ceram. Forum Int., 71[1–2], 8–12, 1994 (in German).

CHAPTER

12

Scientific Principles of Design and Production of Ceramic Parts for Engine Applications

Alexandr G. Romashin, Dr/Sc, Acad. of MIA
Vladimir V. Vikulin, Dr/Sc, Acad. of MIA
Anatoly S. Shatalin, Dr/Sc
OR&PE Technologiya, Kaluga Region, Russia

INTRODUCTION

In 1975, extensive scientific research and applied development of structural ceramic materials and components of gas turbine engines started in Russia, including work at OR&PE "Technologiya"*. The OR&PE "Technologiya" was established in 1959 to develop and produce the materials and products of glass, ceramics, polymers and polymer-composite materials for the aircraft industry. By 1975 it gained expertise in the field of development and production of ceramic insulators and pressure-seal feed-throughs for the thermoemissive conversion of nuclear power into electric power, crucibles for casting GTE (gas turbine engine) blades of chrome-nickel alloys, ducts for burning the ceramic cores forming the cooling channels in GTE's metal blades, radiotransparent fairings, fairings for C-300 type rockets and other ceramic products. Materials and technology development, design and testing were studied within the framework of the interdepartmental program GKNT "High toughness composite materials for the production of parts and assemblies for gas turbine engines operating at firing temperatures in the 1300–1600°C (1573K–1873K) temperature range" and a number of departmental programs. There were two extensive programs for the production of engines

*OR&PE "Technologiya"—The State Research Center of Russia Obninsk Research and Production Enterprise "Technologiya".

with ceramic parts and components operating at high temperatures which had been approved by the Ministry of Aviation Industry. It was envisioned that in these programs ceramic materials, technology, and GTE hot section component development would follow the sequence: small-size, short-life, (~20 hours), medium-life, (hundreds of hours), and finally a demonstration of an engine operating at full power.

Silicon nitride- and carbide-based materials were developed and studied thoroughly during the performance period of the programs by OR&PE "Technologiya". Selected properties are shown in Table 12.1. The techniques of manufacture (flow sheets and basic parameters) of the reaction-bonded silicon nitride (RBSN) OTM-904, extra sintered RBSB (SRBSN) OTM-907, sintered silicon carbide (SSiC) OTM-912 and silicon nitride (SSN) OTM-913, hot-pressed silicon nitride (HPSN) and the products of the above materials are well known. The properties of the materials are mainly determined by the facilities available, original materials and some proprietary know-how of the developers. The materials indicated in Table 12.1 were initially developed for application in gas turbine engines, i.e. for the fabrication of components operating at high temperatures, mechanical stresses and in the fuel combustion environment. The reaction-bonded material OTM-904 is intended for the fabrication of complex shaped products such as nozzle vanes which are characterized by a low (around 0.5%) and steady shrinkage at sintering. The reaction-bonded material OTM-907 is consolidated at higher temperatures, and the sintered materials OTM-912 and OTM-913, are used for the fabrication of nozzle vanes and rotor blades, combustor liners, fuel nozzles and flame holders. The hot-pressed material OTM-914 is intended for the fabrication of rotor blades, inner and outer support rings of nozzle vanes, and hybrid and all-ceramic bearings. All the materials were synthesized from high-purity parent powders. They contain the minimum required amount of sintering agents and have superior oxidation resistance in air and in the gas turbine combustion environment. Their structure consists essentially of elongated grains as can be seen in Figure 12.1 (except for the materials designated OTM-904 and OTM-912) which favors reasonably high values of strength within the 1200°C–1500°C (1473K–1773K) temperature range and good fracture toughness, K_{IC}.

What distinguishes the ceramics as a structural material is that all ceramic parts must be fabricated instantaneously and in one piece. There is at present no practical way to manufacture them by welding or molding of a preform. You cannot assemble them using bolts, screws or rivet them as you do with the fabrication of metal structures. Because of these limitations there is a rigid interdependence between product design, fabrication technology, and material properties, all of which must be directed towards the function

TABLE 12.1 PROPERTIES OF Si_3N_4 AND SiC-BASED MATERIALS

Properties (Units)		OTM-904, RBSN, Si_3N_4	OTM-907, SRBSN, Si_3N_4	OTM-912, SSiC, SiC-AlN-Y_2O_3	OTM-913, SSN, Si_3N_4-Y_2O_3-Al_2O_3	OTM-914, HPSN, Si_3N_4-Y_2O_3
Density (g/cm^3)		2.6	3.2	3.2	3.3	3.4
Open porosity (%)		13	0.5	0.25	0.15	0
4-point bending	20°C (293K)	250	500	400	670	830
strength (MPa)	1200°C (1473K)	240	400	380	430	760
	1400°C (1673K)	220	500	400	—	700
Dynamic elasticity modulus (GPa)		164	248	320	294	294
Microhardness (kg/mm^2)		2200	—	—	1550	1840
Fracture toughness-K_{IC} (MPa·m$^{0.5}$)		2.5	5.5	5.4	7.5(6.4)	7.9
Thermal conductivity (W/m·K)		4–7	16–11	—	—	16–11
Heat capacity (kJ/kg·K)		0.7–1.4	0.7–1.0	—	0.9	0.8–1.3
Coefficient of thermal expansion (10^{-6}/K-R.T. – 927°C)		2.7	3.2	4.6	3.2	3.7
Maximum operating temperature (°C)		1500	1300	1500	1300	1400

Figure 12.1 Typical Material Structure with Elongated Grains 2400 × Magnification

and duty cycle of the product. For example, under actual operating conditions, the structural integrity of a design can be significantly compromised due to residual stresses, and design deficiencies by not considering specific ceramic properties, even if a high strength material was used.

Most of the current scientific publications are dedicated to the problems of producing materials, and research into their properties, but much less attention is paid to the test results for the ceramic parts. There have been only sporadic attempts at summarizing scientific approaches applied to the design of thermally stressed products of brittle ceramic materials [1].

We have over 30 years of experience with the production of thermally stressed high strength ceramic and glass structures. This experience enabled us to develop the following complex approach: the iterative process of designing and testing of a component or assembly, and its material properties and fabrication technology must be performed with the understanding that they are continually and mutually interdependent, i.e. a design must be selected while taking into account material properties and fabrication technology and, vice versa, material properties and process technology must be selected keeping in mind the effects on design and operating conditions of the component.

Based on studies of the microstructures and the specific properties of ceramic materials, experimental studies of the stress-strain state of various designs, development experience, and service performance criteria for thermally stressed ceramic products for various applications, the scientific principles for design and production of ceramic parts and components of engines described below evolved.

SCIENTIFIC PRINCIPLES OF CERAMIC COMPONENT DESIGN AND FABRICATION

In contrast to metallic materials where plasticity resulting from sliding dislocations reduces the level of the stresses in the areas where they are concentrated to levels below the ultimate yield strength, ceramics, in fact, do not relax stresses. The lesser (by 2–4 orders of magnitude) basic density of the dislocations, the higher stiffness of interatomic links, the greater complexity of stress fields in the grid, the preponderance of covalent bonding, the lower incidence of dislocation sliding systems, all these restrict the mobility of dislocations, and the rate of dislocations solidity increases with stress and rise in temperature [2]. Therefore, first and foremost, localized stresses must be reduced to improve ceramic component designs subject to stress-strain conditions.

Principle of Minimization of Stress Concentrations, Uniform Distribution of Loads

Essentially, this principle can be summarized as follows: a ceramic part must have no stress concentrators, especially within the highly loaded zone. To show how important the principle of minimizing stress concentrations is can be demonstrated for a ceramic cowl when considering how it should be joined to a metal support structure. This component after thermal cycling is shown in Figure 12.2. All the cracks here cross the holes in multiple ways, or to be precise, the holes themselves are the stress concentrators. All sorts of leaks, sharp shoulders, thickness variations, sharp angles are highly undesirable since they are prone to becoming stress concentrators.

Packings. A material to be used at the interface of a ceramic part should be selected with the lowest possible modulus of elasticity. Thus, assuring that non-parallelism of cone ceramic washers is less than 0.05 mm, and applying packings of a niobium alloy facilitated fabrication of parts for fixturing components. They can be operated between –130°C and 1200°C (143K and 1473K) in air under vacuum and survive with loads as high as 1,200 kg and with vibration and impact overloads as well (Figure 12.3). These packings were applied to the "Buran" aerospace craft.

A skewed "dovetail" type lock was selected at the joint between a ceramic rotor blade inserted in a metal disk. Similar joints, more often with more complex profiles of the mating surfaces, have been widly used to connect the metal blades with the disk. We have selected the simplest version of the joint design as a reverse wedge on the blade, mounted at the side into the recess of the same profile in the metal disc. The

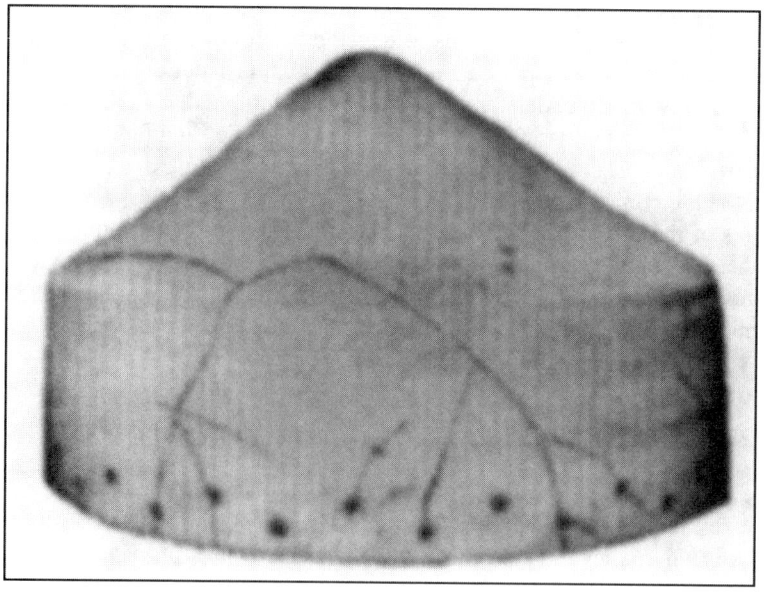

Figure 12.2 Ceramic Cowl After Thermocyclic Tests

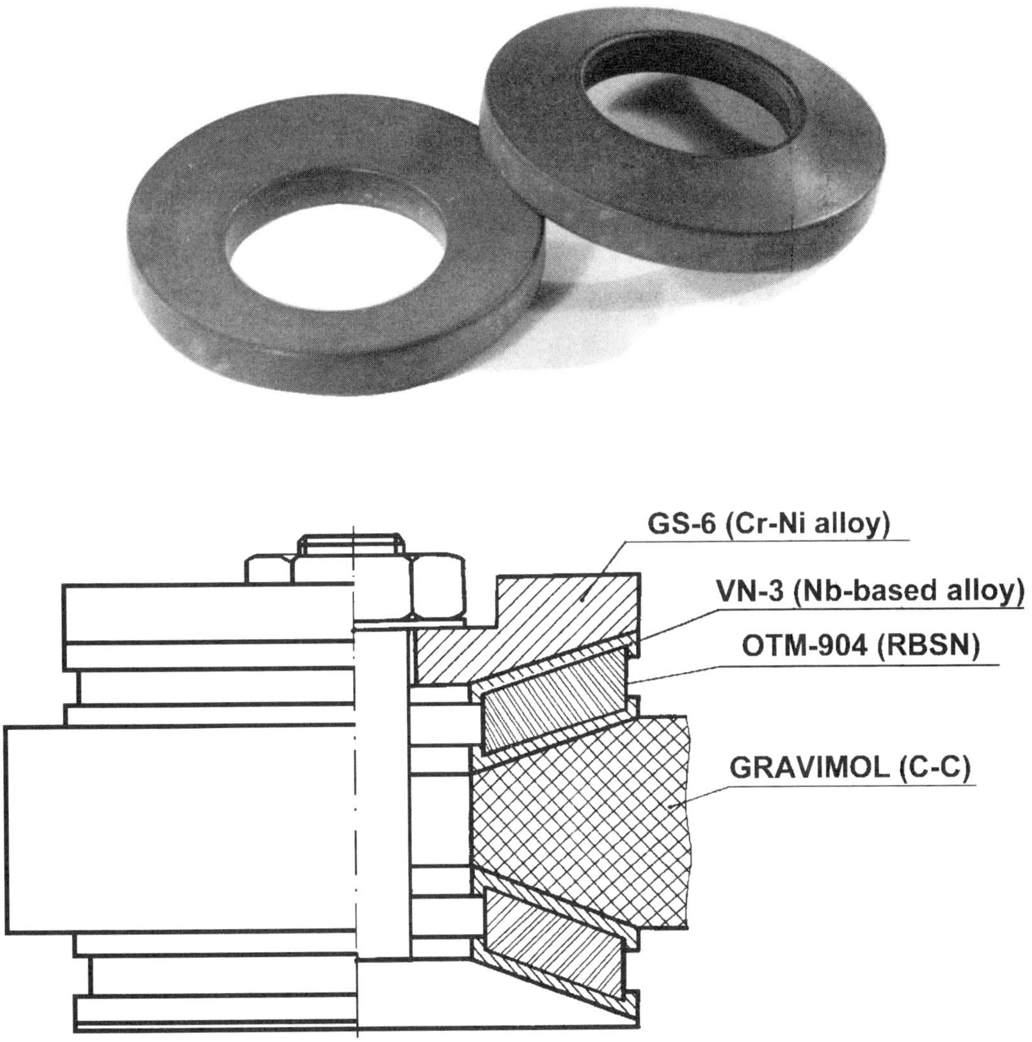

Figure 12.3 Ceramic Conic Washers and Fixturing Component with Niobium Packings

joint design was iterated computationally and experimentally by the angle α (Figure12.4, top left), transition radius R, contact length, and its remoteness from the blade neck. In addition, the effect of the thin compliant layers of nickel, copper and other metals mounted between the contacting blade and the disk surfaces was studied. The stress distribution within the blade dovetail (Figure 12.4, top right) is shown for the version without the compliant layer. The blades were manufacturerd of sintered RBSN OTM-907 and HPSN OTM-914. The tests were performed on the spin test rig at temperatures up to 170°C, on the gas generator BT-60 at temperature up to 1050°C, and on a gas-dynamic test bed at gas temperatures up to 1250°C. The exposure time at the maximum speed and temperature varied from 0 to 25 hours. The main benefit of such a lock applied to low plasticity materials is the availability of two planes of contact which permits a reasonably uniform distribution of contact loads. A calculation of the stress-strain state of the blade dovetail on the basis of the opening angle and contact length due to the action of centrifugal forces demonstrated that a reduction in the maximum principal tensile stress could be achieved by opening up the dovetail angle, decreasing the dovetail contact line length, and its maximum possible displacement from the neck, and by varying the radius of rounding of the neck width. Calculations of the stress-strain state for a ceramic rotor blade design, to be more precise for the dovetail (Figure 12.4) demonstrated that full contact between the ceramic dovetail and the metal disk was available. The application of a packing reduced the friction between the ceramic dovetail and the metal disk, and as a result the stress was lowered by 20–44%. Illustrations of distribution of the main tensile stresses in the blade dovetail tail are shown in Figure 12.4. By applying the above-mentioned conditions service-ready blades were produced. They were tested at a speed up to 40,000 rpm within the 1050°C–1250°C (1323K–1523K) gas temperature range [3].

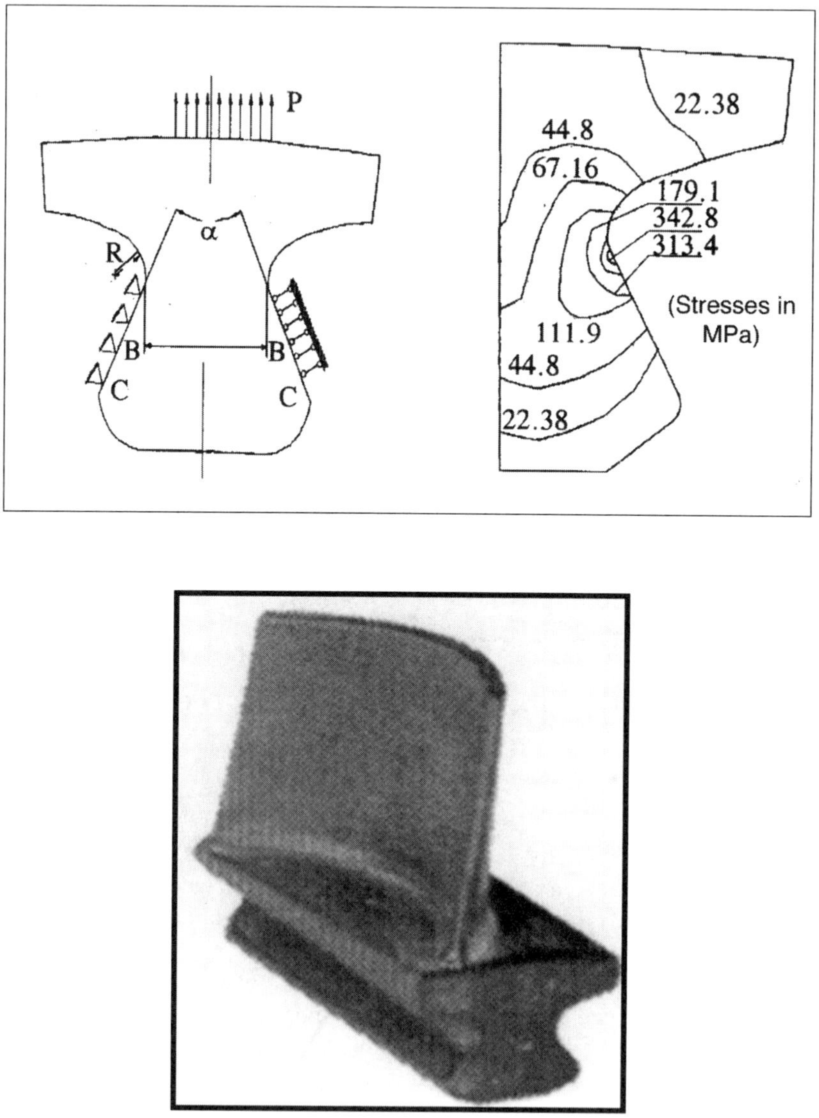

Figure 12.4 Schematic of Loading and Distribution of Main Tensile Stresses in Ceramic Rotor Blade Dovetail

Principle of Free Thermal Expansion of a Ceramic Element

The thermal expansion of all the elements joined to the ceramics must closely match the deformation of the ceramic component in service. A mismatch between the thermal deformation of the support structure and the ceramic shell caused the component failure shown in Figure 12.2. There are essentially two ways for applying the free temperature deformation principle in assemblies, either make an appropriate selection of the coefficients of linear thermal expansion for the ceramics and the metal or use elastic compensating elements.

Development of ceramic nozzle vanes including a set of separate vanes, and three internal and external support ceramic rings, and use of spring-loaded elements ensured free deformation of the vanes under heating and cooling conditions. Short-life GTE nozzle vanes designed using the above principle survived the tests in rigs and in engines with the gas stream temperature level ranging from 1000°C to 1400°C (1273K to 1673K). The duration of the test was several hundred hours (Figures 12.5 and 12.6). The nozzle vanes were fabricated of the reaction-bonded silicon nitride OTM-904, the inside and outside support rings of the nozzle boxes with a maximum diameter of around 280 mm were of the sintered reaction-bonded silicon nitride

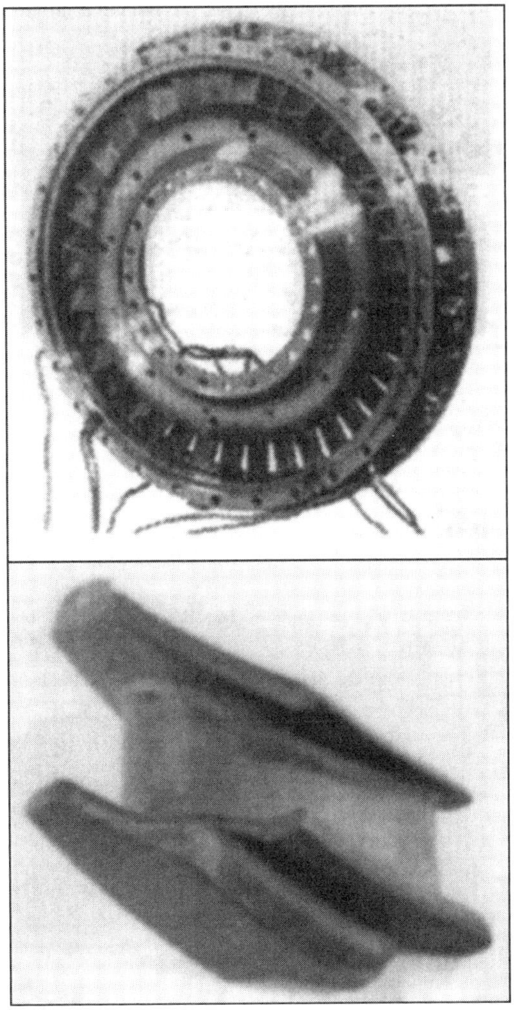

Figure 12.5 Ceramic Nozzles and Support Rings

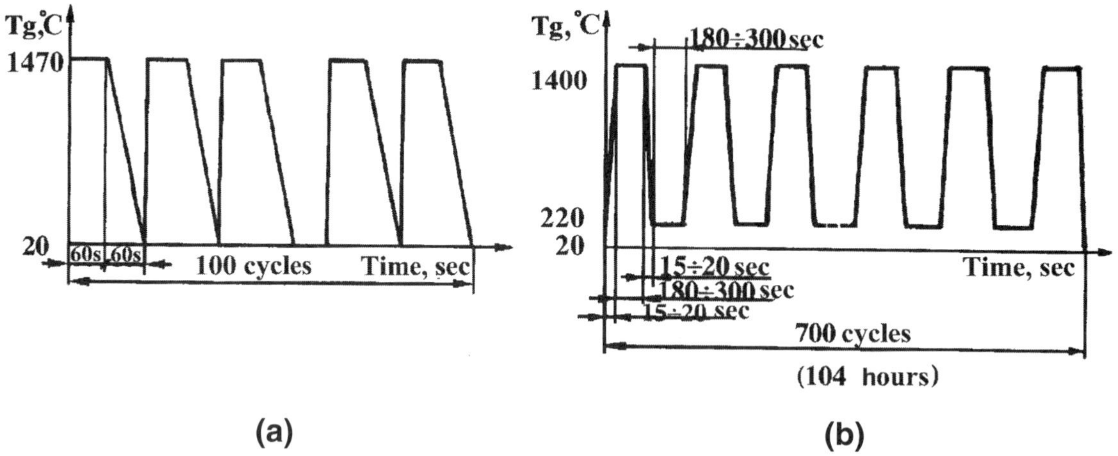

Figure 12.6 Diagrams of Thermocyclic Tests for Nozzle Vanes (a) and Nozzle Vanes with Ceramic Support Rings (b) Before the Components Were Installed in a Test Rig or Engine

OTM-907 and hot-pressed OTM-914. The design maximum stresses did not exceed 80 MPa for the vanes and 180 MPa for the outside support ring.

The Principle of Segmentation of a Complex Component Along the Highest Stressed Sections

This principle underlies the fabrication of segmented nozzle vanes. A segmented design generates much lower thermal stresses during heating and cooling than an integral design. Stresses are further mitigated because of non-uniformity of the pattern factors compared to those for integral nozzle vanes.

Using the above principle, a catalytic igniter component for a booster engine was designed and fabricated of reaction-bonded silicon nitride, OTM-904. The gas temperature at the catalytic igniter was 600°C (873K) and 1700°C (1973K) at the exhaust. A solid structure did not survive the thermal stresses and failed during initial thermal cycling. At the same time, a component segmented in three parts (Figure 12.7) which was part of the catalytic igniter underwent successfully 1,000 cycles in the test rig and 784 cycles in the engine with the exhaust gas temperature being 1700°C (1973K). Application of this concept was necessary for operation of the experimental combustor liner which is a tube of about 180 mm diameter, around 360 mm high, with a wall thickness of about 6 mm. Holes to supply fuel and air were made in the liner wall. The monolithic liner failed during the initial testing. The segmented liner which splits in the middle of the holes, survived over 20 heating cycles with 20 hours of total exposure at 1600°C without failure.

The same concept was used to develop the experimental ceramic tip seal. Segmentation of the monolithic ring of around 300 mm diameter with a wall thickness of around 4 mm fabricated of sintered silicon nitride segments provided stability during temperature cycling from 20°C to 1200°C while the application of hot-pressed laminated material of Si_3N_4 and BN with increasing BN content towards the surface contacting the rotor blades ensured the required fitting of the seal [4].

Principle of Thermal Uniformity of Ceramic Elements

This principle is based on the condition that the temperature field of the ceramic component is uniformly distributed along the perimeter and the surface. Violation of this principle may result in additional thermal stresses limiting the ability of the component to perform its intended function.

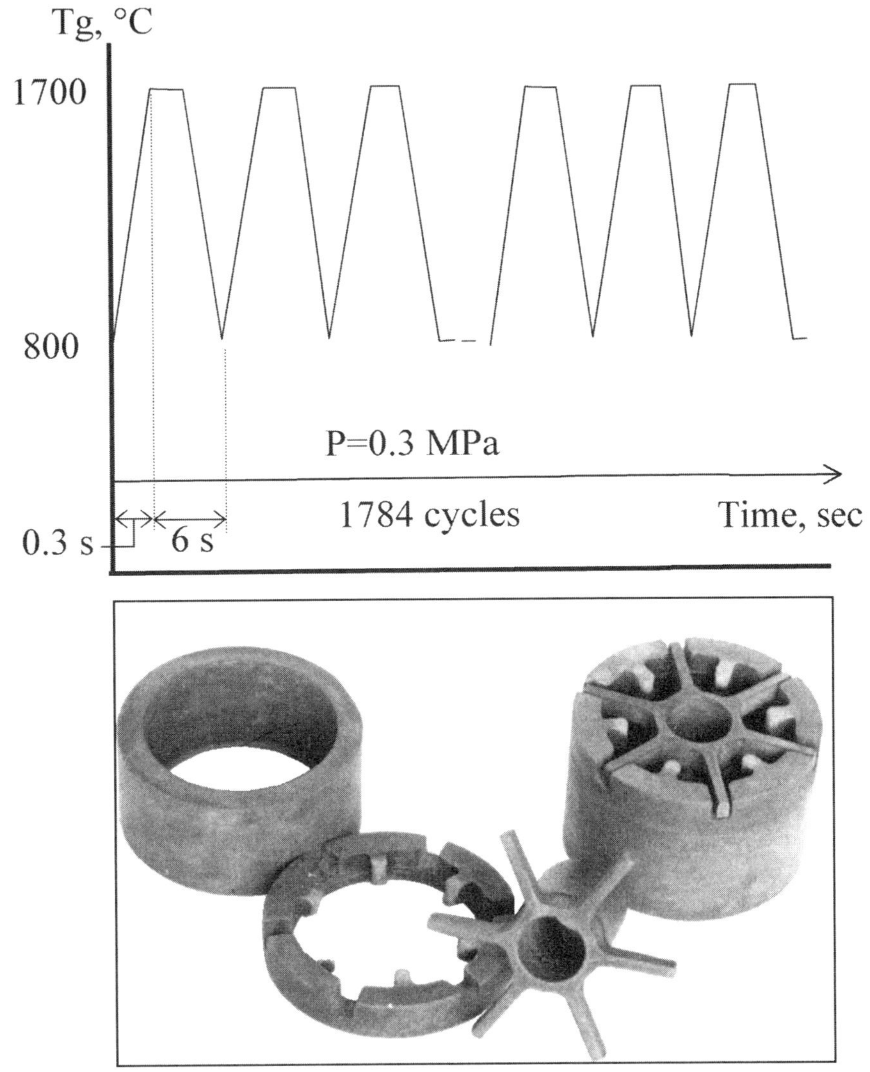

Figure 12.7 Catalytic Igniter of Booster Chamber and Diagram of Its Thermocyclic Tests (1000 Cycles on Test Rig, 784 Cycles in Engine, Gas Temperature at Inlet –600°C, at Outlet –1700°C)

Figure 12.8 shows some designs which support (a) and violate (b) the thermal uniformity principle. In Figure 12.8 (a) the ceramic cylinder mated with a metal along its inner surface is externally heated, and the heating proceeds uniformly lengthwise. For the case presented in Figure 12.8 (b), the metal sleeve provides thermal insulation for a portion of the cylinder which generates a difference in the average temperature at the top and bottom portions of the ceramic element, which results in thermal stresses. A similar situation arises for heating a plate or for heating a rotor blade (Figure 12.8(c)) where thicknesses vary substantially. Application of a composite structure of two cones (Figure 12.8(d)) ensured uniform pattern factors and durability of the part while being heated and under thermal loads in service. At the same time, the design version in Figure 12.8 (e), which shows a cone with boring within the lip zone, would fail during heating at the boring area, due to breaching of the thermal uniformity principle. Furthermore, the boring is the stress concentrator.

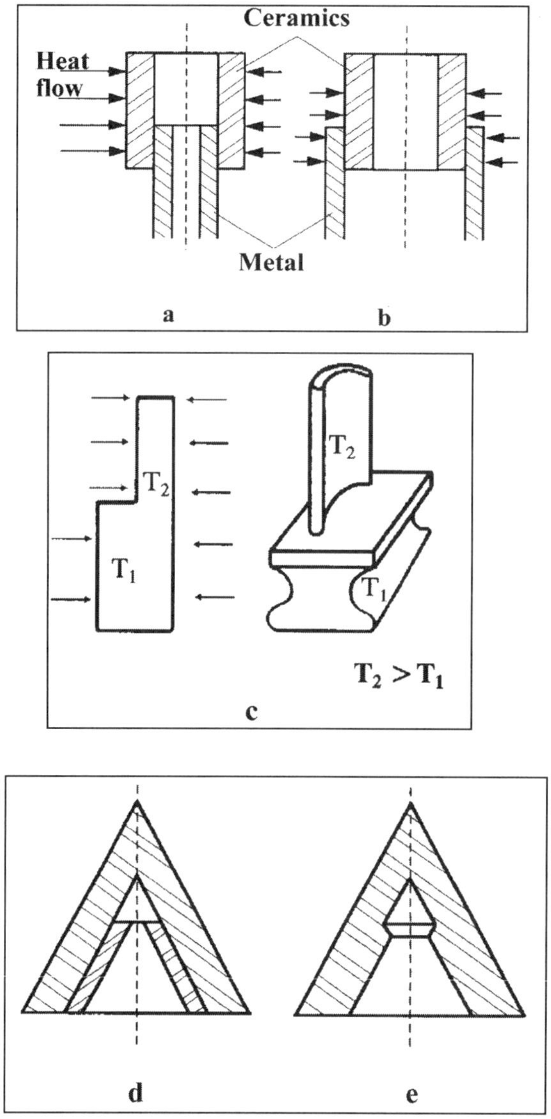

Figure 12.8 Examples of Structures with (a, d) and without (b, c, e) Thermal Uniformity

There are additional important principles: deliberately generated compressive stresses in ceramic elements, and geometric dispersal of stresses generated by fixturing of ceramic parts and localized stress concentrators, in particular, at the end faces of ceramic elements.

The basic design principles described above do not guarantee that a certain design solution for a ceramic engine component is absolutely risk free, although analysis of specific designs may indicate to what extent they match the above principles and will help avoiding solutions which are doomed to be a failure from the very beginning.

Strength is often the fundamental criterion applied for the development of structural ceramic materials for various designs of engine components. However, calculations of the stress-strain state of a component and the results of experimental trials have shown that the performance of a component in service will depend

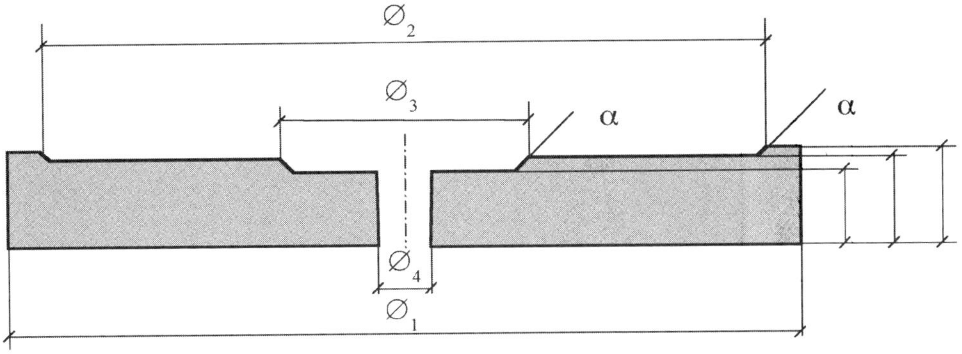

Figure 12.9 Schematic of Ceramic Sleeve on Metal Disc

on the specific application and conditions in the engine which are not only determined by strength but, to a greater extent, by a combination of mechanical and physical properties of the materials, such as thermal conductivity, specific heat, coefficient of thermal expansion, thermal stability, ceramic-to-metal contact stresses, erosion and corrosion resistance at elevated temperatures in air and in the combustion environment, chemical resistance to corrosive media and to other materials. The materials considered for the development must retain their mechanical and physical properties reproducibly and reliably in components, and these properties must be maintained during long-term thermocyclic loading [3,5].

The application of the statistical approach [6] provides a quantitative way to distinguish various component design versions on the basis of the criterion of minimum probability of failure (POF). This approach enables the designer to determine the requirements for ceramic materials under specific loading conditions or, conversely, helps the designer to establish ultimate allowable loads based on the known material properties, the component's geometry, and the acceptable probability of failure.

The advantage of materials which have a desirable combination of properties compared to materials whose only distinctive feature is their strength was demonstrated by the calculation of the stress-strain state of a ceramic sleeve on a metal disk using a central bolt (Figure 12.9). The thermal and mechanical stresses are critical for this design. At the same time, the physical and mechanical properties of the materials affect the stresses as well. Therefore, the selection of the materials for the fabrication of the sleeve cannot be reasonably supported by a high value of only one parameter such as average strength. The calculation code "Gradient" [7,8] includes properties of 4 types of ceramic materials that have superior strength, solidity, thermal conductivity, and coefficient of thermal expansion. The values of the main tensile stresses in the strap section obtained from a solution to the non-stationary problem of thermal conduction, the elastic theory of an axisymmetrial component are presented in Table 12.2.

TABLE 12.2 VALUES OF THE MAIN TENSILE STRESSES IN A CERAMIC SLEEVE ON A METAL DISK

Material	Density (g/cm^3)	Bending strength σ_{bnd} (MPa)	Maximum thermal stress σ_m (MPa)	σ_{bnd}/σ_m
OTM-904 (Si_3N_4)	2.6	250	31	8.1
EC-131 (Si_3N_4)	3.35	1000	127	7.9
INA-99 (Al_2O_3)	3.94	600	292	2.0
ITZ-20 (ZrO_2)	4.90	1000	425	2.4

TABLE 12.3 RESULTS OF CALCULATION OF STRESS-STRAIN STATE OF A CERAMIC SLEEVE ON A METAL DISK

Material of the Sleeve	Principal maximum stress (MPa)	Target strength (MPa) POF = 10^{-1}	POF = 10^{-2}	POF = 10^{-3}
EC-111, (Si_3N_4)	46	95	121	150
EC-131, (Si_3N_4)	151	312	395	497
ITZ-10, (ZrO_2)	370	764	969	1217

The lowest thermal stresses are generated in OTM-904, a reaction-bonded silicon nitride [5] and, despite a considerably lower strength compared to other materials, it has a bigger safety margin expressed as the ratio of strength to maximum thermal stress. Table 12.3 shows the results of calculations for the sleeve for several different materials. The calculations demonstrate the need for high strength (~1200 MPa) to ensure a POF of 10^{-3} for the zirconia-based ceramic material. On the other hand, the reaction-bonded silicon nitride which has good strength (see Table 12.2) and also an optimum combination of physical and mechanical properties is a good candidate material given the relatively low principal maximum stress of 151 MPa.

Figures 12.10 and 12.11 show the results of calculations for the ceramic swirler of a combustor liner. Properties of three types of silicon nitride-based materials, namely OTM-904 (reaction-bonded Si_3N_4), OTM-907 (densified reaction-bonded Si_3N_4) and OTM-908 (reaction-bonded silicon nitride-based on the system Si_3N_4-BN) are incorporated in the calculation [3,5,6]. The thermal process in this part under transient stress conditions is analyzed in terms of the probability of failure. The maximum values of the main tensile stresses σ_1^{max}, load factor K_L and POF are presented in Table 12.4.

A comparative analysis of the predicted service life of ceramic components under specified loading conditions, material properties and assuming a specific design leads to the following conclusions. Regardless of the level of the principal maximum stresses for the material, it being higher for OTM-907 compared to OTM-904 and OTM-908, the design load factor is notably lower during the entire startup period. Therefore, the

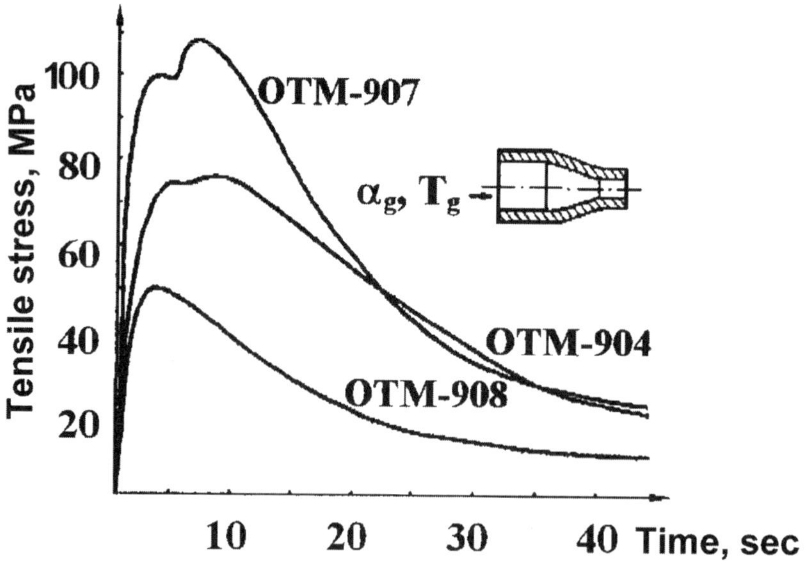

Figure 12.10 Variation of Principal Maximum Stress in Flame Swirler Under Engine Startup Conditions

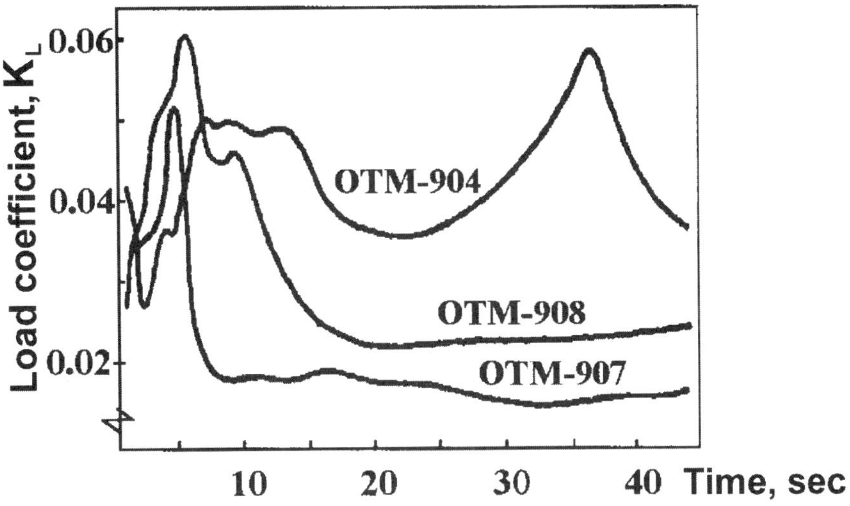

Figure 12.12 Variations of Load Factor with Ceramic Material Properties for Flame Swirler Under Engine Startup Conditions

TABLE 12.4 RESULTS OF COMPUTATION OF THE STRESS-STRAIN STATE AND POF FOR A CERAMIC SWIRLER

	Type of material		
Properties	**OTM-904**	**OTM-907**	**OTM-908**
Load factor for design, K_L	0.06	0.0513	0.0614
Time of emergence of maximum stresses (s)	9.1	3.45	4.45
Main maximum stress σ_1^{max} (MPa)	121.2	124.9	54.7
Probability of failure (POF)	$0.647 \cdot 10^{-1}$	$0.66 \cdot 10^{-4}$	$0.345 \cdot 10^{-3}$

POF of this design must be lower. Thus, the predicted service life of a ceramic component is determined by the combination of its properties and all aspects of its operation in service.

TEST RESULTS

The lower conditions and test results for nozzle vanes and rotor blades for GTEs designed for short-term and intermediate term operation, catalytic igniters of boosters and, also, successful tests of swirlers and flame holders, ceramic inserts of blade tip seals (see Table 12.5) have verified the correctness of the complex approach to the development of thermally stressed engine components, and the efficacy of the application of the principles of ceramic engine component design outlined above.

DISCUSSION

There is no doubt that ceramic materials could be regarded as very promising for application to gas turbine engines. This is mostly because these materials enable an increase in the turbine operating temperature,

TABLE 12.5 TEST RESULTS FOR CERAMIC PARTS AND COMPONENTS FOR GTES

Type of Engine	Material	Component	Test Temperature, °C (K)	Gas Pressure, (MPa)	Operating Hours	Number of Thermal Cycles, Startups
	OTM-904	Nozzle vane	1600 (~1870), rig			100
Short-life	OTM-904	Nozzle vane	1100 (~1370), rig	0.6	16	25
	OTM-904	Flame holder	1400 (~1670), rig	0.26	19	294
	OTM-904	Nozzle vane	1600 (~1870), rig			100
	OTM-904	Support rings of	1400 (~1670), rig		40	250
Intermediate	OTM-914	nozzle vanes				
Life	OTM-904	Nozzle vane	1400 (~1670), rig	2	200	700
	OTM-914		1150(~1420),engine		16	20
	OTM-907	Rotor blade	1400 (~1670), rig		104	100
			1130 (~1300), engine		27	150
	OTM-904	Catalytic igniter of	1700 (~1973), rig	0.3		1000
		booster chamber	1700 (~1973), rig	0.3		784
Demonstration	OTM-908	Insert of blade tip seal	1210–1250 (~1580–1620), engine		8	7
	OTM-910	Flame swirler	1400 (~1670), rig	0.3	20	200
			1100 (~1370), rig	0.3	16	400
	OTM-912	Flame holder	1530 (~1800), rig		20	20

an increase in fuel efficiency, and a reduction in exhaust emissions. The state-of-the art of materials properties and the application of a complex design-engineering approach enables the fabrication of ceramic gas turbine components subjected to thermal stresses during engine operation. Presently, ceramic materials with K_{IC} up to 8 MPa.m$^{0.5}$ are being produced. However, to ensure the actual wide-spread application of ceramics as gas turbine hot section parts, many challenges still need to be addressed. Here, we will single out only two of them.

First, there is the problem of predicting failure of a specific ceramic component. This problem has two aspects. On the one hand, there has been a lot of effort in predicting target life. On the other hand, many researchers, designers and test engineers have regularly encountered incidences of an unpredictably long operation of a ceramic component and even an increase in the material strength after having completed a test. And they have observed an easing of slow crack growth or change of direction of its evolution at elevated temperatures when a mass transfer, crystallization of an intergranular phase, or phase and structure changes take place, including those changes dependent on the stress state of a component. In this case, these changes tend to be observed more noticeably in locations of stress concentrations, or in the vicinity of a crack initiation site. For example, the observation that crack growth may come to a halt, and the initial strength level may be maintained and even be exceeded after long-term tests (up to 300 hours) under a load amounting to 0.7 of the ultimate strength at temperatures of 1200°C, 1300°C, and 1400°C have been reported [2,9]. The only explanation for this effect of "strengthening" observed at lower temperatures is a likely decrease in the level or direction of internal stresses during tests. In any case, we are faced with the real challenge of finding methods for imposing a physical effect on a ceramic part aimed at increasing its life under specified conditions.

The second challenge concerns increasing the fracture toughness (K_{IC}) values to ensure reliable operation of ceramic products. The observed gradual improvement of K_{IC} values for silicon nitride- and silicon carbide-based materials is due to the improvement of their structure, composition, and processing technology. It seems that it is feasible to increase K_{IC} for ceramic materials to the level of 12–15 MPa.m$^{0.5}$ in the near future. This might be achieved by producing fine-grained isotropic self-reinforced structures with elongated grains and a purposeful direction of the composition and crystallinity properties of the intergranular phase. Even higher values of K_{IC}, in excess of 15 MPa.m$^{0.5}$ have been obtained in composite ceramic materials, reinforced by SiC and C fibers. The practical application of these reinforced composite materials is still a remote prospect since the high-temperature property stability problems still need to be solved and the problems of thermodynamic and thermomechanical stability of the systems, especially at the fiber-matrix interface tend to become more complicated compared to issues with unreinforced, i.e. monolithic ceramic materials. In Russia studies focused on increasing K_{IC} values for the monolithic sintered Si_3N_4-based ceramics by the formation of a structure with elongated grains of β-Si_3N_4 and hot-pressed composites with a Si_3N_4–based matrix reinforced by single-crystal whiskers and SiC fibers [2,10,11] are currently in progress. K_{IC} up to 11 MP.m$^{0.5}$ and flexure strengths up to 1,000 MPa have been achieved. These results provide hope that after additional tests and studies these composites will eventually find wide application in gas-turbine engines.

ACKNOWLEDGEMENT

All figures and tables were provided courtesy of OR&PE Technologiya.

REFERENCES

[1] Sasa, T., "Design of Ceramic Parts," Ishikawajima-Harima Gicho, 20[5], 362–367, 1980.

[2] Shatalin, A.S., "Some Feature of the Fracture Behavior of Structural Ceramics," *Proceedings of the Conference New Materials and their Applications*, The University of Warwick, The Institute of Physics, Coventry, England, UK, 10–12 April, 1990. pp. 509–519.

[3] Vikulin, V.V., Romashin, A.G., and Leshchuk, T., "Si_3N_4-Si Composite Ceramics and Chemically Toughened SiC for Engine Components," *Proceedings of the Institute of Energy Conference, Second International Conference on Ceramics in Energy Applications,* London, UK, 20–21 April, 1994. pp. 21–32.

[4] Kelina, Yu., Ershova, N.I., and Chasovskoi, E.N., "Special Features of the Thermal Shock Behavior of Composite Materials Based of Silicon and Boron Nitrides," Refractories and Industrial Ceramics, (Ogneupori i Technicsheskia Keramika), 39[11–12], 444–449, 1998 (in Russian).

[5] Vikulin, V.V., "Manufacturing of Tailor-made Structural Ceramic Materials Based on Si_3N_4 and SiC," *Proceedings of the II European Ceramic Society Conference*, G. Ziegler and H. Hausner, eds., Augsburg, Germany, September 11–14, 1991. Deutsche Keramische Gesellschaft, Cologne, Germany, 1991. pp. 753–758.

[6] Vikulin, V.V., Kurskaja, I., and Leshchuk, T., "Si_3N_4- and SiC-based Structural Ceramic Materials for Operating Temperatures Above 1400°C," *Proceedings of IRC'92- Materials for High Performance Processing, Properties and Applications of Metallic and Ceramic Materials,* M.H. Loretto and C.J. Beevers, eds., Birmingham, England, UK, Sept. 7–10, 1992. The Processing, Properties and Applications of Metallic and Ceramic Materials Advisory Ltd. (formerly MCE Publications Ltd.), ISBN 0947817557, 1992, pp. 327–332..

[7] Postnikov, A.A., Selynsky, A.A., Chasovsky, E.N., "Analysis of Strength and Reliability of Ceramic Parts of Machine," Mashinovedenie, 3, 36–41, 1986 (in Russian).

[8] Vikulin, V.V., Postnikov, A.A., and Chasovsky, E.N., "Assessment of Serviceability of Heat-Stressed Engine Elements," Oborudovanie, Materialy, Prozessy, M., VIMI, 3, 47–53, 1989 (in Russian).

[9] Shatalin, A.S., and Romashin, A.G., "Ceramics are the Materials of the XXIst Century," Science to Production, (Nauka Proizvodstvu), 9, 4–7, 1999 (in Russian).

[10] Kelina, Yu., Drobinskya, V.A., and Plyasunkova, L.A., "Microstructure and Properties of Silicon Nitrid-matrix Ceramic Composites," Plenum Pablishing Corporation, Refractories and Industrial Ceramics, 39[1–2], 22–25, 1998 (in Russian).

[11] Kelina, I. Yu., Ershova, N.I., Pljasunkova, L.A., and Yakovenko, E.I., "Advanced Hot-Pressed Ceramic Matrix Composites (CMC) in SiC_w, SiC_f, C_f/Si_3N_4 Systems," in *Ceramic Materials and Components for Engines*, J.G. Heinrich and F. Aldinger, eds., 7th International Symposium, Goslar, Germany, 19–21 June, 2000. pp. 641–646.

CHAPTER

13

Structural Alumino-Boron-Nitride Ceramic Material for Gas Turbine Engine Applications

Anatoly V. Soudarev and Vladimir V. Grishaev

JSC "Boyko Research-Engineering Ceramic Heat Engines Center" (NIZ KTD), St. Petersburg, Russia

INTRODUCTION

The practical application of the ceramic gas-turbine engine (CGTE) is, first and foremost, related to the development of the structural ceramic materials (SCMs). The existing metal oxide-, silicon nitride- and silicon carbide-based SCMs developed in many countries worldwide give reason to believe that sooner or later the manufacture of the CGTE will become a reality. To date, the first production CGTEs have already been manufactured in Japan [1–3], in the USA [4], and in Russia [5,6] and at present, engine testing is in progress.

The joint-stock company "Boyko Research-Engineering Ceramic Heat Engines Center" (NIZ KTD) was established in 1988 as a part of the Research-Technological Institute for Power Engineering (NITI EM), St. Petersburg, Russia. Its major objective was the development of techniques for the fabrication of SCMs and ceramic components for the CGTEs. The leading specialists of this company have been engaged in studying the feasibility of applying various SCMs and their technologies to the CGTEs since 1990. The experience and expertise of various companies in Japan, the USA, Germany, etc. have been analyzed. Many companies in Russia specializing in the development of SCMs were involved with the development of CGTE components of nitride-, carbide- and oxide-based ceramics.

NIZ KTD's own fabrication development work for SCMs has been conducted in the area of the application of special and intensive technologies, including:

- production of fine and nanophase silicon carbide-, silicon nitride- and oxide-based composite powders by mechanical, chemical and plasmochemical techniques
- molding of products by high hydrostatic pressures up to 20 kbar

- sintering of ceramic parts by hot pressing, hot isostatic pressing and other techniques and, also, thermal sintering by means of an electron beam with energies of 1 to 15 MeV
- study of the potential of the vacuum-compression impregnation technology

The above studies were conducted in close cooperation with the designers of gas turbines, combustors and heat exchangers under the program described in chapter 32 of Volume I of this text [7]. In September 2001, NIZ KTD abandoned NITI EM. At present, it is an independent legal body named after A.M.Boyko, a leading specialist of the joint-stock company OAO GAZPROM who was a special supervisor of all the activities in the field of the CGTU developments. Its activities are focused on upgrading and modification of alumino-boron-nitride based SCMs using the vacuum-compression treatment technology (VCT).

The ceramic materials "SUGRAV" [8] and "ANB" manufactured by the vacuum-compression treatment technique (VCT) using an alumino-boron-nitride composition are described in this chapter. Actually, all ceramic components of the CGTE hot section [5,6] and a number of devices operating under thermally loaded conditions were manufactured from these SCMs. It was done in collaboration with the French National Aerospace Research Institute (Office National d'Etudes et Recherches en Aérospatiales—ONERA).

The ceramics "SUGRAV" and "ANB" feature the following technological advantages: zero shrinkage at sintering (with a 0.5–0.8% volume increase), the feasibility of machining in the "green" state without the use of a diamond or hard-alloy tool, the possibility of diffusive welding of various parts made of these SCMs into all-ceramic components, high heat resistance (1600°C), fairly satisfactory thermal stability (1200°C), a low rate of oxidation, etc.

Also, the ceramic "ANB" has the following properties:

- a defined chemical composition with controlled phases
- ceramic materials with a different fine-grain structure
- 1.5–2 times the strength of "SUGRAV"
- it can be used to fabricate composite ceramic materials with fiber or tissue reinforcing elements.

This chapter provides information about the fabrication processes for the "SUGRAV" and "ANB" ceramics, and also information about grain structure and microstructure. The major physical and mechanical properties are also described. Snapshots of the ceramic components for the CGTE with a description of their fabrication, testing and operating conditions provide an excellent insight into the potential of the "SUGRAV" ceramics.

MAIN REQUIREMENTS FOR THE DEVELOPMENT OF SCMS

Increasing the efficiency of gas-turbine engines (GTE) is primarily dependent upon an increase in the maximum temperature of the working media. For metal GTEs, this is accomplished both through the development of advanced high-temperature metal alloys and by sophisticated complex cooling schemes for metal hot section components. The current metal GTEs can operate at temperatures up to 1300°C. This is the practical limit since the efforts required for component cooling above this operating temperature, which is close to the limit for metals, cannot be offset by the efficiency gains of the GTEs.

The application of SCMs to the entire hot section of a 2.5 MW CGTE, with a turbine inlet temperature (TIT) of 1350°C, enables an increase in the thermal efficiency from 28 to 45%, while with a TIT of 1500°C, the efficiency can be increased to 48%. Based on the above, the initial objective was to develop the CGTE design and, subsequently, produce SCMs for the fabrication of components for reliable operation in this type of engine.

It would be beneficial to use several types of SCMs since the operating conditions within specific zones of the hot section (combustor, guide vanes, turbine wheel, inlet and exhaust piping, heat exchanger) are different. Based on an analysis [9] a minimum level of physical and mechanical properties for application to various CGTE components is required (see Table 13.1).

It follows from Table 13.1 that one SCM could be applied for all components of the CGTE. The differences in thermal conductivity requirements can be solved through different design approaches. Comparative

TABLE 3.1 MINIMUM LEVEL OF MAIN PROPERTIES OF SCMS FOR DESIGNING MAJOR COMPONENTS AND DEVICES FOR STATIONARY GTE AT A TURBINE INLET TEMPERATURE (TIT) OF 1350°C

	Elements of CGTE*			
Properties	**TW**	**GV**	**CC**	**HE**
1 Heat-resistance, °C	≥1500	≥1500	≥1500	≥1100
2 Density, g/cm^3	2–4	2–4	2–4	2–4
3 Flexure strength, MPa	400	200	100	100
4 Compressive strength, MPa	800	400	200	200
5 Fracture toughness, K_{IC}, MPa.m$^{1/2}$	15	15	15	15
6 Weibull modulus	25	25	25	25
7 Thermal conductivity, W/m.K	<10	5–10	<6	>15
8 Coefficient of thermal expansion, 10^{-6}/°C	8–10	8–10	8–10	8–10
9 Thermal stability, T_{max}, °C	1350	1350	1350	1100
10 Oxidation weight change, 10^{-3} (mg/cm^{2}h)	<1.0	<1.0	<1.0	<1.0

*TW—turbine wheel; GV—guide vanes; CC—combustion chamber, HE—heat exchanger—air heater.

properties of metal oxide-, silicon nitride- and silicon carbide-based SCMs developed and currently applied in various countries have been described in the literature [9]. Analysis of these properties indicates that there is no SCM available at present that could meet all the requirements presented in Table 13.1. The most important requirements are high-temperature strength, fracture toughness and stability against oxidation.

The development of SCM structures operating in compression, but not in tension or bending, has resulted in a relaxation of the requirement for fracture toughness. The above approach of GTE designers and materials specialists has led to the commercialization of the VCT technology for the fabrication of ceramic materials on the basis of the alumino-boron-nitride compositions. These compositions have a number of important properties.

PECULIAR FEATURES AND TECHNIQUES OF MANUFACTURE OF THE "SUGRAV" CERAMICS

The feasibility of the fabrication of components with a complex configuration using the unsintered material by various machining techniques without the use of hard-alloy or diamond tools and, also, the zero shrinkage during subsequent sintering was the basis for a definite interest in the vacuum-compression impregnation technique. However, this approach was not used further since the ceramic material produced by this method had low flexure strength and used to be prone to very strong oxidation even at temperatures of 1100–1150°C. At temperatures over 1400°C, oxidation would result in destruction of the material.

Focused research and explorations (1990–1993) into the VCT and ceramics synthesis processes aimed at the study of the mechanisms of formation of the alumino-boron-nitride composition-based ceramic material, and at increasing the strength characteristics in terms of higher heat-resistance with reduced susceptibility to oxidation resulted in the development of a ceramic material [10], subsequently identified as "SUGRAV" [8] that had heat-resistance up to 1600°C without oxidation, while its strength actually increased by about 10%. The studies of the mechanisms of the phase composition and microstructure of the "SUGRAV" have been described in detail in ref. [11].

This ceramic material had a number of advantages, namely

- zero shrinkage at sintering
- good fabricability prior to sintering
- no need for machining
- an established technology for diffusive joining of parts of this ceramic into all-ceramic components

- high heat-resistance (over 1600°C)
- good thermal stability (on the order of 1200°C)
- good oxidation resistance

The efforts of the designers focused on the development of low-load ceramic structures, i.e. special ceramic components operating in compression but not in tension or bending. The developers of SCMs and GTEs took a special interest in the behavior of the ceramic for operation under high temperature conditions (1350–1500°C and higher). The French companies "ONERA" and "EDF" (Electricité de France) as the patentees of the ceramic "SUGRAV" [8], in collaboration with INSA (Institut National des Sciences Appliquées) of Lyon, France, conducted studies of the high-temperature properties of the ceramic "SUGRAV" using samples provided by the Joint-Stock Company "C.T.S.". Parallel explorations were conducted for the SN-235P Si_3N_4 ceramic using samples provided by Kyocera Corporation and an AlN ceramic using samples from the company "BOOSTEC". Table 13.2 shows some high-temperature properties of the ceramics "SUGRAV", Si_3N_4 and AlN, using the data provided by "INSA" de LYON (C.Ouguay, C. Olagnon, and G. Fantozzi).

As is evident from Table 13.2, there is a noticeable reduction in the properties with increasing temperature for Si_3N_4 and AlN, in the range of 1400–1500°C, which could become critical in some cases. This type of change of mechanical properties is typical for all SCMs of this class. A degradation in the above properties from about 500–700°C could also be observed for "SUGRAV", although there is a gradual improvement again, including at temperatures of 1400 and 1500°C. The actual values of the properties of "SUGRAV" are considerably less than the corresponding values for Si_3N_4 and AlN, although at high temperatures, the difference is quite insignificant. Since "SUGRAV"'s density is 10–15% lower, the corresponding stresses are reduced as well.

The more recently developed "ANB" ceramic has the same technological advantages as "SUGRAV", but the flexure strength of "ANB" approaches 250–300 MPa, which enables new applications for this class of ceramic materials for gas turbine manufacture and for other industries. Furthermore, the fabrication techniques for this ceramic enables the development of several new SCMs of the "SUGRAV" family through:

- development of ceramics having a fine-grain structure
- use of controlled additives in the initial powder composition
- development of composite ceramic materials with the 'SUGRAV" matrix using whiskers, fibers and tissue of SiC, Al_2O_3, etc. as the reinforcing elements.

The structure of "ANB" enables the fabrication of parts of the unsintered material by the electro-milling treatment technique (this method is impracticable for the ceramic "SUGRAV") which ensures a high qual-

TABLE 13.2 COMPARATIVE DATA ON SOME PROPERTIES* OF CERAMIC "SUGRAV", Si_3N_4 AND AIN (DATA REPRINTED WITH PERMISSION FROM "INSA" DE LYON)

		"SUGRAV" "NIZ KTD"	AlN "BOOSTEC"	SN-253P Si_3N_4 Kyocera
ρ(g/cm^3)		2.6–2.8	3.1	3.25
E (GPa)	20°C	132	294	295
	1200°C	91	274	237
	1400°C	~100	~270	—
σ_n bending (MPa)	20°C	122	349	632
	1350°C	55	131	169
	1500°C	~63	—	~115
m	20°C	14.8	—	12.5
K_{IC}, (MPa·m$^{1/2}$)	20°C	4.1	—	8.1
	1200°C	—	—	6.8
$\frac{\Delta m/S\ (1300°C)}{\Delta m/S\ (1400°C)}$ (g/cm^{2}**)		1.16	—	2.46

*The high-temperature properties of "ANB" are being studied at present.
**Δm/S—variation of sample mass referred to its surface, unit (g/cm^2).

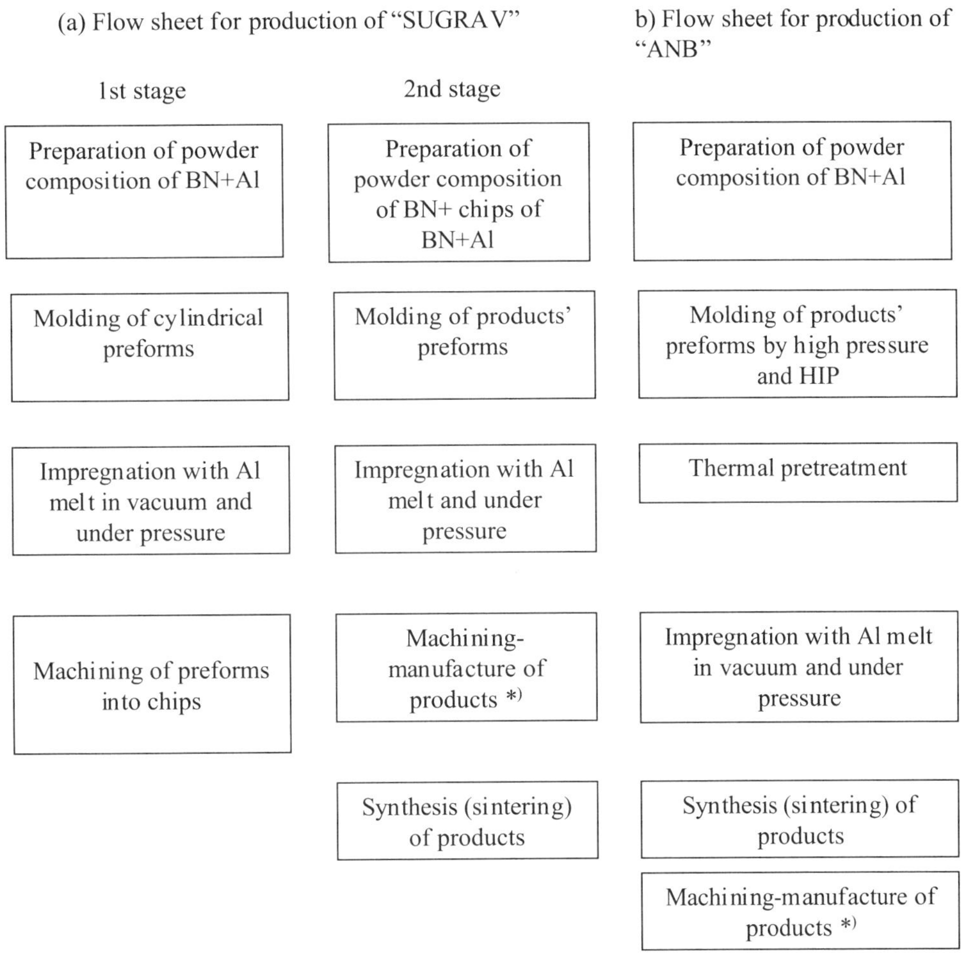

*) If required, prior to synthesis complex sectioned products are assembled of separate elements to enable welding during synthesis.

Figure 13.1 Flow Sheets for Production of the Ceramics "SUGRAV" (a) and "ANB" (b)

ity of ceramic parts. The technology of manufacture of the ceramic "SUGRAV" is presented in a number of publications [8,10–12].

The vacuum-compression impregnation of powder preforms with liquid aluminium is the major and determining process; the boron nitride decomposition into nitrogen and boron with the subsequent formation of an aluminium-nitrogen-boron compound which results in formation of the AlN phase and appropriate phases of AlB_n (n = 2,10,12) is the major chemical process. Of special importance for the entire process is the optimization of the initial ratio of the boron nitride powder and the liquid aluminium introduced, which eliminates oxidation during the processing of the ceramic and during its operation under high temperature conditions.

Figure 13.1 shows the processing flow sheets for the "SUGRAV" and "ANB". As can be seen the flow sheet for "ANB" is single-staged. However, the application of a more active technology is required to mold the products' preforms, which means higher power consumption with VCT. The cost of "ANB" fabrication

Figure 13.2 Microstructure of "SUGRAV" Ceramic

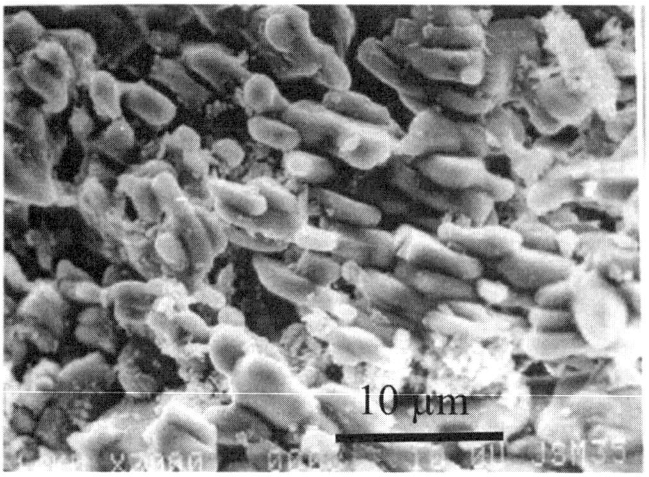

Figure 13.3 Microstructure of "ANB" Ceramic. Mag. 2000×

at the same time exceeds that for "SUGRAV" by no more than 20%. Figures 13.2 and 13.3 show microstructures of "SUGRAV" and "ANB".

SPECIFIC FEATURES OF MANUFACTURE OF CERAMIC COMPONENTS FOR GTE APPLICATION

As was noted above, the fabrication of ceramic parts of "SUGRAV" and "ANB" is carried out using VCT preforms. The strength of the unsintered material is around 50 MPa and 120 MPa for "SUGRAV" and "ANB", respectively; so, the manufacture of a variety of support fixtures for the machining of parts is required. These holders vary from part to part depending on the design. In fact, all the developed and manufactured ceramic parts have a minimum required thickness (ranging from 1 to 10 mm) since the basic goal

Figure 13.4 Ceramic Element and Complete Shell of "SUGRAV" (CGTE-2.5) Rotor Blade

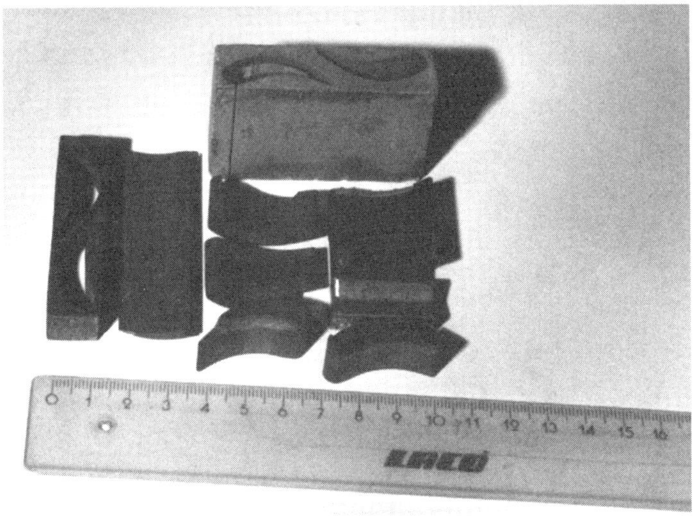

Figure 13.5 Ceramic Shells and Preforms of "ANB" (CGTE-2.5) Rotor Blade

of the ceramics is to take advantage of the high temperature effect (1350–1500°C) thereby ensuring the desired strength of the design and also its reliability during operation.

The limitations of various machining techniques in combination with requirements posed by the design of a concrete part constrain fabrication. Figure 13.4 shows separate elements of the rotor blade ceramic shell of the ceramic "SUGRAV" and the solid shell itself fabricated from these elements by a diffusive joining technique. The minimum radius of the inside profile of the shell is 0.5 mm which restricts the height of the fabricated shells to 6–7 mm using a milling cutter of 1.0 mm diameter. To fabricate a solid shell, 30 mm high, joining needs to be performed after the individual elements are cut.

Figure 13.5 shows the rotor blade ceramic shells of "ANB",15–17 mm high, and preforms, 60 mm high, from which these shells are fabricated by the electro-milling technique using a wire that is 0.1–0.3 mm in diameter. In this particular case, there is no need to apply the joining technology since the quality of the

Figure 13.6 Turbine Wheel of Metallic-Ceramic Turbine for CGTE-2.5 Engine Application

Figure 13.7 All-Ceramic Scroll for 2.5 MW CGTE

surface treatment is higher compared to that in the former case. The fabrication of these shells of conventional monolithic ceramic materials (SiC, Si_3N_4, etc) is, in fact, impractical.

Figure 13.6 shows the metal turbine disk with the inserted ceramic shells. Figure 13.7 shows the all-ceramic scroll for the 2.5 MW GTE manufactured from 18 separate elements, using the same diffusive joining technique. Note that the scroll is made up of elements (segments) joined to another by diffusive welding.

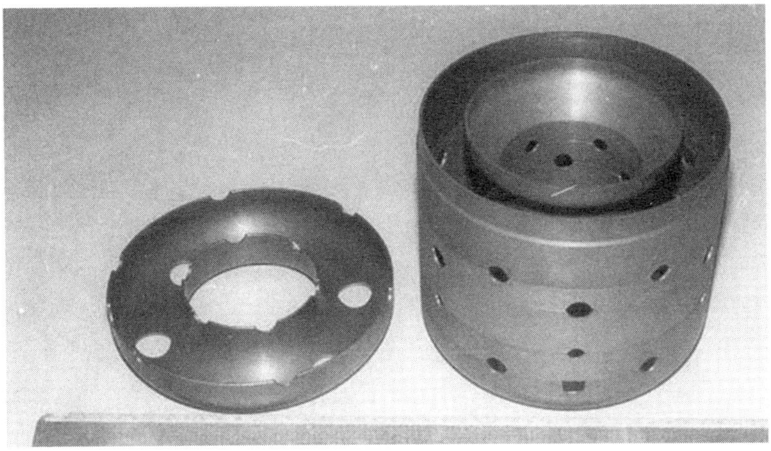

Figure 13.8 Annular Liquid Fuel Firing Combustor for Separate Ceramic Engine Application, TIT=1350°C

The objective of fabricating two types of scrolls is to assess which of the two designs is more reliable in terms of operating in the GTE during transient conditions (startup, shutdown, etc). The effect of a concrete design of ceramic parts under the impact of thermal loads will be discussed below.

Figure 13.8 shows the annular combustor as an assembly and its separate rings of "SUGRAV" with the wall thickness being 2.0 mm. Figure 13.9 shows the design of the inlet tubing of the high pressure turbine CGTE-2.5 which is made up of separate plates of ceramic lining secured on the elastic metal carrier membranes. The same concept was applied to the fabrication of the elements of the guide vanes (GV) (Figure 13.10) and exhaust diffuser (Figure 13.11) for the 2.5 MW GTE application.

TESTS OF CERAMIC COMPONENTS

Tests are an essential requirement for metallic GTEs as a final validation of the overall development efforts, and for the commercialization of both the components and the GTE as a whole. Tests are major challenge for ceramic GTEs when appropriate SCMs are still unavailable or when there are no mature components and structures with ceramic parts as yet.

The test program for SCMs, the ceramic components of the CGTE, and the CGTE as a whole, developed by engineers and designers include the following stages:

1) Evaluation of the major properties of the SCMs (from 20–1500°C) (density, porosity, phase composition, heat capacity, thermal conductivity, coefficient of thermal expansion, flexure and compressive strength, fracture toughness, heat resistance, oxidation resistance). Some data are presented in Table 13.2 and in the literature [10,11].
2) Tests of SCM samples for heat-resistance for temperatures up to 1600°C, thermal stability up to 1400°C, thermal cycling resistance up to 50–100 cycles, and oxidation resistance at temperatures of 1100–1600°C for 1000 hours. The main findings have been published elsewhere [10,13].
3) Tests of separate ceramic elements and parts under relevant operating loads, static and dynamic strength, thermal stability and thermal cycling resistance.
4) Tests of metallic-ceramic parts and component assemblies with relevant thermal and mechanical loads, 1.5 times those predicted under the rated conditions.

Figure 13.9 Inlet Metallic-Ceramic Tubing from Combustor to Guide Vanes (CGTE-2.5)

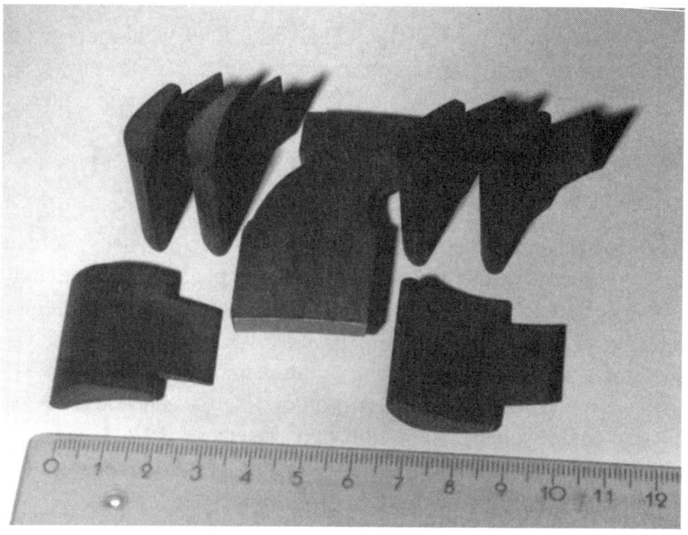

Figure 13.10 Elements of Ceramic Guide Vanes for CGTE-2.5 Engine Turbine

Figure 13.11 Inlet Metallic-Ceramic Diffuser for CGTE-2.5 Turbine Stage

5) Tests of separate units of the CGTE (heat exchange modules, combustors, inlet and exhaust tubing, guide vanes, turbine wheels, etc).
6) Test of the CGTE as a whole.

The first two tests are, in fact, focused on the study of materials. There will be no advancements in the development of ceramic parts and components if the properties of the ceramic materials do not meet specified requirements. The direct involvement of GTE designers is required at these two stages to make corrections and to assist with the comparisons between the existing and anticipated properties of the SCMs.

The third stage of testing is decisive for the design of zero-shrinkage components for a CGTE since the effect of a design varies from part to part and its thermally stressed state is different from that of ceramic samples. Thus, the original design of the ceramic stator element of the "dovetail" type with dimensions of 40 mm × 30 mm × 4 mm used to fail in the first or second thermal cycle (20°C–1400°C–20°C) while the samples with dimensions of 80 mm × 50 mm × 10 mm would last for more than 50 cycles. Only after the "dovetail" design was made "open", was component breakdown eliminated. Or to give another example: the ceramic combustor rings of 220 mm diameter, 7 mm thick, and 70 mm high used to fail in thermal cycling tests (20°C–1200°C–20°C) while ceramic rings of 285 mm diameter, 7 mm thick and 50 mm high would not fail. Analytical computations and subsequent check-ups indicated that the ratio for diameter, thickness, and height of a product were factors of primary importance for the design of closed annular ceramic structures.

The fourth stage tests enable the evaluation of interactions between the metallic and ceramic parts, and the identification of benefits and deficiences in the design of a concrete metallic-ceramic component. Consecutive tests of the ceramic shells (Figures 13.4 and 13.5) were carried out which showed that they would survive all the thermal and static tests as specified for them, although when spin tests of the shells along with the metal core were conducted, failure of the ceramics would occur. The preliminary and subsequent analytical computations did not reveal any excessive mechanical loads. The initial characterization of the failures indicated the emergence of off-design flexure stresses in the dynamic tests. When "ANB" ceramic shells were substituted for the "SUGRAV" shells ("ANB" has higher strength), the flexure stresses disappeared, though other stresses, related to flexure (during rotation) of the metallic core, emerged. Only after the ceramic shell design was changed (to two sections that were not joined—see Figure 13.5), and after the inside gap of the shell's bottom area was enlarged, were positive test results obtained.

In the fourth stage of the tests, the metallic-ceramic designs of the inlet tubing and exhaust diffusers of module devices were also verified. These verifications confirmed that the computations were correct and established options for the designs of these components.

To date, some metallic-ceramic parts, e.g. the combustor assembly, for the 2.5 MW GTE [14], and the prototype turbine, and the metallic-ceramic heat exchanger module [14–18] have undergone the fifth stage tests, namely rig tests under conditions close to the engine operating conditions. Some components have been in service for around 400 hours at 200 cycles simulating the GTE startup and shutdown duty cycles. Testing of the guide vanes and the turbine wheel of the 2.5 MW GTE are close to being completed in the 2000 timeframe. Final tests of the prototype turbine stage as an assembly will be subsequently performed. Testing of a demonstration unit of the 2.5 MW GTE is scheduled to be concluded in 2003 and 2004.

CONCLUSIONS

The results of research to-date on the developed "SUGRAV" and "ANB" SCMs as well as tests results for the ceramic and metallic-ceramic designs using these SCMs give confidence that the development of the CGTE will be completed. The authors are aware that this is only the beginning of a long path where many errors and failures are still to be expected. Nevertheless, the production of a reliable, low-NO_x and efficient ceramic gas-turbine engine is becoming a real possibility in the near future.

ACKNOWLEDGEMENT

All artwork in this chapter has been provided courtesy of the authors' organization. The important contributions of our colleagues Souryanov, A.A., Molchanov, A.S., Podgorets, V.Ya., and Soudarev, B.V. for the work reported are acknowledged.

REFERENCES

[1] Niwa, K., "100 kW Automotive CGT and Its Application," *6th International Symposium on Ceramic Materials and Components for Engines*, K. Niihara, S. Kanzaki, K. Komeya, S. Hirano, K. Morinaga, eds., Arita, Japan, October 19–23, 1997. pp. 212–217.

[2] Hattori, M., Yamamoto, T., and Watanabe, K., "Development of Ceramic Gas Turbine Components for the CGT 301 Engine," ibid., pp. 222–227.

[3] Nakashima, T., Tatsumi, T., Takehara, I., Ichikawa, Y., and Kobayashi, H., "Research & Development of CGT302," ibid., pp. 233–236.

[4] van Roode, M., Brentnall, W.D., Smith, K.O., Edwards, B.D., Mc.Clain, J., and Price, J.R., "Ceramic Stationary Gas Turbine Development Program - Fourth Annual Summary," ASME paper 97-GT-317, presented at the International Gas Turbine and Aeroengine Congress and Exposition, Orlando, FL, USA, June 2–5, 1997.

[5] Soudarev, A.V., Soudarev, B.V., Grishaev, V.V., Molchanov, A.S., Podgorets, V.Y., Surjaninov, A.A., and Avran P, "History of Development of Metal-ceramic Axial Turbine Stage for Highly Effective Gas Turbine Engine," VDI-Gesellschaft Energietechnik (VDI-GET) Universität, GU, Siegen, Germany, 10–11 September, 1996. p. 527.

[6] Soudarev, A.V., Soudarev, B.V., Molchanov, A.S., Grishaev, V.V., Podgorets, V.J., and Souryaninov, A.A., "Program of Development of Metal-Ceramic Stationary Gas-Turbine 2.5 MW Unit and the Initial Results of its Implementation," ASME paper 97-GT-356, presented at the International Gas Turbine and Aeroengine Congress and Exhibition, Orlando, FL, USA, June 2–5, 1997.

[7] Soudarev, A.V., and Tikhoplav, V.Yu., "Research-Engineering Center Ceramic Heat Engines (NIZ KTD) at the Research-Technological Institute for Power Engineering (NITI EM)," in *Ceramic Gas Turbine Design and Test Experience - Progress In Ceramic Gas Turbine Development, Vol. I*, M. van Roode, M.K. Ferber, and D.W. Richerson eds., ASME Press, New York, USA, 2002. Chapter 32, pp. 683–707.

[8] Avran, P., Soudarev, A.V., and Grishaev, V.V.. French patent N 9616135, 27.12.96 PCT N FR 97/02042.

[9] Grishaev, V.V., and Soudarev, A.V. "Materials Science Aspects of Development of Stationary Gas-Turbine Engine," Turbines and Compressors, No. 8,9 [1,2], 82–92, 1999 (in Russian).

[10] Grishaev, V.V., and Soudarev, A.V., "Expansion of High Temperature Range of Alumo-boron-nitride Ceramics Application," International Conference on SHAPING of Advanced Ceramics, Mol, Belgium, April 25–27, 1995. Extended Abstract. pp. 321–323.

[11] Grishaev, V.V., Soudarev, A.V., and Okatova, G.P. "Research into Mechanism of Formation of Alumo-boron-nitride Ceramics: A Structural Material for GTE Parts," ASME paper 97-GT-323, presented at the International Gas Turbine and Aeroengine Congress and Exhibition," Orlando, FL, USA, June, 2–5, 1997.

[12] Soudarev, A., Vinogradov, E., Zakharov, Yu., Avran, P., and Grishaev, V., "Test Bed Development of Design of Ceramic Combustor for Gas-Turbine Engine," *Proceedings of the International Gas Turbine Congress 1999 Kobe*, Vol.II", T.Tamaru, ed., Kobe, Japan, 14–19 November, 1999, The Organizing Committee, International Gas Turbine Congress 1999 Kobe, c/o Gas Turbine Society of Japan. pp. 679–683.

[13] Soudarev, A., and Grishaev, V., "Shrinkage-free, Easily Adaptable to Manufacture High-temperature Ceramic Material ANB,"—*Trends in Materials Science and Technology, Proceedings of the Third International Workshop on Materials Science (IWOMS'99)*, Hanoi, Vietnam, November 2–4, 1999, Part I, pp.1a–1b.

[14] Soudarev, A.V., and Soudarev, B.V., "Heat Exchange and Hydraulic Resistance Inside the Inner Path of the Ceramic Airheater for GTU," *6th UK National Conference on Heat Transfer*, Edinburgh, U.K., September 15–16, 1999, Book Abstracts. pp. 503–508.

[15] Avran, P., and Boudigues, S., "Ceramic Development at ONERA," in *Ceramic Gas Turbine Design and Test Experience - Progress In Ceramic Gas Turbine Development, Vol. I*, M. van Roode, M.K. Ferber, and D.W. Richerson, eds., ASME Press, New York, USA, 2002. Chapter 23, pp. 513–537.

[16] Soudarev, A.V., Soudarev, B.V., and Grishaev, V.V., "Development and Study of Compact High-Temperature Ceramic Turbular Airheaters," *Proceedings of the Fourth ISHMT—ASME Heat and Mass Transfer Conference and Fifteenth National Heat and Mass Transfer Conference*, M.S. Loknath, S.P. Venkateshan, B.V.S.S.S. Prasad, B. Basu, Vishwanath Prasad, eds., Puna, India, Jan. 12–14, 2000, Tata McGraw-Hill Publishing Company Ltd., New Delhi, India, 2000. pp. 1033–1038.

[17] Soudarev, A.V., Soudarev, B.V., and Kondratiev, V.V., "Convective Heat Transfer and Hydraulic Resistance at Air Flow Through Ceramic Tubes with Heat Exchange Intensifiers," *Proceedings of the Fourth International Symposium on Experimental and Computational Aerothermodynamics of Internal Flows*, Dresden, Germany, August 31–September 2, 1999, Vol.II. pp. 208–216.

[18] Soudarev, A.V., Soudarev B.V., Soudarev V.B., Avran, P., and Leclair, A., "Shell-and-tube Ceramic Heat Exchangers for High-temperature GTEs applications," *Proceedings of the 11th International Heat Transfer Conference, Kyongju*, Korea, August 23–28, 1998, Vol.6. pp. 255–260.

CHAPTER

14

Ceramic Matrix Composites Development at GE Power Systems Composites, LLC

Dennis J. Landini, Ali S. Fareed, and Hongyu Wang
GE Power Systems, Newark, Delaware, USA

Phillip A. Craig
Consultant, Rochester Hills, Michigan, USA

Stan Hemstad
Powdermet, Landenburg, Pennsylvania, USA

INTRODUCTION

Ceramic matrix composites (CMCs) have been pursued for more than 20 years as the next-generation materials that are of higher temperature capability than the current state-of-the art metallic superalloys and tougher than the traditional monolithic ceramics. In this over 20-year-long journey, a 15-year-old company now called GE Power Systems Composites, LLC (PSC) stands high amongst its peers in the research, development and prototype-production of ceramic composite materials. This chapter intends to present to you the history of GE PSC and its journey in developing many of the currently used material systems as a testimony of the great efforts the CMC community has made in introducing this revolutionary new material.

Fiber reinforced CMCs have a unique failure mechanism versus monolithic ceramic materials; they exhibit a graceful failure mode.As the matrix cracks under thermal or mechanical stresses, load is transferred from the matrix to the fibers. The fibers can slide in the matrix thanks to a debond layer called interface coating allowing load transfer and energy absorption. This permits multiple micro-cracks to exist prior to fracture, resulting in increased toughness. The ability of the material or component to carry load while cracked is known as graceful failure. This characteristic provides confidence to designers for using CMCs as structural parts that are generally not highly flaw sensitive, and are therefore resistant to brittle failure.

The superior high temperature capability of CMCs makes them desirable in applications such as gas turbine hot-gas-path components of combustors, transition liners, shrouds, vanes and vane edges, afterburners, jet pipes, exhaust flaps and seals. CMC have also proven utility in industrial applications other than turbine engines, such as petrochemical furnaces, heat exchangers, and in combustion of coal and municipal solid waste fuels, where in addition to the mechanical characteristics, the matrix phase provides good corrosion resistance.

The acquisition of Honeywell Advanced Composites, Inc. (predecessor to PSC) by GE in December 2001 symbolized GE's determination in using this next-generation high temperature material in its world-leading power-generating gas turbine and aircraft engine products. The investment by GE in PSC and its effort in bringing the technology and material systems from R&D to commercial production will undoubtedly accelerate the overall CMC development and realization of the dreams of thousands in the materials processing and engineering community.

GE Power Systems Composites, LLC and its predecessor companies have been on the forefront of CMC developments with state-of-the-art materials processing and component fabrication technologies. PSC currently can fabricate SiC/SiC and C/SiC composites by chemical vapor infiltration (CVI SiC/SiC and CVI C/SiC), and SiC/SiC composites by melt infiltration (MI SiC/SiC). Following an introduction to the history of PSC, sections will be devoted to a more detailed description of these material systems and their potential applications.

HISTORY OF GE POWER SYSTEMS COMPOSITES, LLC

GE Power Systems Composites, LLC started in 1987 from two entities devoted to ceramic matrix composites development—DuPont Lanxide Composites and DuPont Advanced Composites. The former focused on the development of the direct metal oxidation technology, known as DIMOX, and the latter on chemical vapor infiltration technology. Table 14.1 summarizes the timing of key events, milestones, and accomplishments.

Early Efforts at DuPont

In the mid 1980s DuPont began a broad effort in developing ceramic matrix composites. This effort was concentrated at DuPont's corporate research and development center—the DuPont Experimental Station, Wilmington, DE but also involved other DuPont research groups. A New Business Development group in the Fibers Department utilized DuPont scientists and outside consultants to review research efforts and help identify future business opportunities. One such effort suggested that DuPont had many of the core technologies needed for metal matrix composite and ceramic matrix composite materials.

This was supported by DuPont management, because it was in alignment with the history of the DuPont organic composite efforts, which grew from research work in Fibers and Engineering Polymers Departments. The organic composites division grew from a similar New Business Development effort in the early 1980s, into a free-standing business, although it was sold by DuPont in 1996.

Contributions from three independent research activities were brought together. Ceramic fibers had been in research as a higher temperature extension of the Fiber business beyond the aramids (Kevlar™ and Nomex™). Ceramic solutions and precursors were researched for oxide and non-oxide ceramic matrices for monolithic filters and small structural pieces such as yarn guides by the Fibers and Engineering departments. Separately, CVD coatings and ceramic solutions were studied by the Chemicals and Pigments groups for corrosion-resistant coatings for processing equipment. Part of the ceramic solutions work came from the Pigments department, active in titanium dioxide and other ceramic materials for improved products such as paint. Oxide ceramic matrix development was based primarily on colloidal alumina and mostly reinforced with DuPont's FP™ alumina fiber and later reinforced with S-Glass fiber, which became PRD-66™. The effort also included laboratory scale chemical vapor deposition of interface coatings including pyrolytic carbon, boron nitride, and a matrix of silicon carbide. A development effort in FP™ fiber reinforced aluminum

TABLE 14.1 SUMMARY OF THE EVOLUTION OF CMC MATERIALS AT PSC

Time	Key Events, Milestones, Accomplishments
Pre-1980 [1]	• Work by Professor R. Naslain funded by Société Européenne de Propulsion (SEP) and DRET (French military research agency) on chemical vapor infiltration (CVI) of a carbon fiber perform with a SiC-C mixture. • DuPont work to make oxide matrix composites using colloidal alumina and FP™ alumina fibers and glass fibers.
1980–1985 [1–5]	• Development by SEP of higher strain-to-failure (0.5–1.0%) C/SiC-C composites and substitution of Nicalon™ SiC-based fibers for the C fibers. • Development at Lanxide of the directed metal oxidation (DIMOX™) process and early trials of reinforcement with Nicalon™ fibers.
1987	• Acquisition by DuPont of an exclusive license to the SEP CVI CMC technology.
1989	• Installation by DuPont of two CVD reactors in Newark, DE along with all of the support facilities to produce CVI SiC/SiC composites.
Early 1990s	• Formation of three joint ventures between DuPont and Lanxide and combination of one of these with the SEP technology to establish DuPont Lanxide Composites, Inc. (DLC) in Newark, DE. • Award of two major contracts under the DOE Continuous Fiber Ceramic Composites program to optimize CVI SiC/SiC and DIMOX alumina/SiC composites and to fabricate and evaluate prototype components for industrial applications.
Mid-1990s [6]	• Identified key limitations of their composites (DIMOX™) and conducted extensive developments to overcome these limitations, especially achieving increased life at high temperature in an oxidizing atmosphere.
1994–1997	• DLC conducted development and evaluation of melt infiltrated (MI) SiC/SiC under the NASA Enabling Propulsion Materials (EPM) program. • Conducted composite development using improved SiC fibers (Hi-Nicalon™ and Sylramic™) and interface coatings.
1997	• Enhanced CVI SiC/SiC composite combustor liner survived 1048 hours in a Solar Turbines Centaur 50 engine field test, but exhibited substantial oxidation degradation.
1998	• Acquisition of DLC by AlliedSignal; renamed Allied Signal Composites, Inc. (ASCI)
1999	• Merger between AlliedSignal and Honeywell; ASCI became Honeywell Advanced Composites, Inc. (HACI)
1999/2000	• CVI SiC/SiC and MI SiC/SiC combustor liners with environmental barrier coatings (EBCs) survived 13,937 hours in a Solar Turbines Centaur 50 engine in field tests at a ChevronTexaco site.
December 2001	• Acquisition of HACI by GE; renamed GE Power Systems Composites, LLC (PSC).
2000/2002	• CVI SiC/SiC combustor liner with environmental barrier coating survived 15,144 hrs in a Solar Turbines Centaur 50 engine at the Malden Mills site in Lawrence, MA.
2002	• Improved robustness and capability of CMC processes, transitioning a new CMC process from GE Global Research Center, and invested millions of dollars in capital equipment for establishing CMC production readiness.

and/or magnesium metal matrix composites (MMCs) yielded components such as automotive engine connecting rods, main bearing caps, pistons, and industrial components such as die casting press parts. A principle hurdle existed in the MMC activity as it was seen to be one of the many materials that functioned well in the temperature range of the aluminum matrix. The MMC activity was ended and transitioned into the ceramic composite joint venture with Lanxide, called DuPont Lanxide Composites Inc. The ceramic composite materials, however, seemed to have a clear position in the niche of extremely high temperature operation.

SEP License

Simultaneously, and with the learnings of the FP™ fiber /aluminum MMC work, DuPont made the decision to pursue the CMC market in earnest. The missing piece of core technology seemed to be the ceramic matrix, as the matrices available via DuPont technology were primarily based on PRD-66 and other oxides, which were porous and weak. To overcome this shortcoming, DuPont intentionally took two approaches, one more near-term commercial (more expensive), and another more long-term R&D (and less expensive a venture).

For the commercial approach, DuPont obtained an exclusive license of technology from the French Société Européenne de Propulsion (SEP) for use in the United States and formed a group—DuPont Ceramics, staffed with about 50 volunteer scientists, engineers, technicians and operators. The licensed technology included a pioneering "process" patent for making a ceramic matrix composite using a carbon or boron nitride interface coating, chemical vapor deposited (CVD) on the fibers, and a CVD deposited matrix, principally of silicon carbide. Terms of the license allowed for technology transfer between the two companies as well as future patents awarded to SEP. Also included in the license was the design of full-scale CVD reactors for carbon and silicon carbide, with on-site installation and operations training at the French and the new U.S. site. Two initial reactors, one carbon, one silicon carbide, were co-located with DuPont's organic matrix composite business at the Pencader Plant, Newark, Delaware in 1989. A second SiC reactor was added soon after plant start-up. The carbon reactor is 1.4 meters inside diameter, and the SiC reactors are 1.0 and 1.4 meter inside diameter. The facility included supporting infrastructure for the size and expected volume of parts flowing through these reactors, including preform areas, tooling and CMC machine shops, NDE and engineering offices.

The initial license included three basic methods for fabricating fiber reinforced ceramic composites, 2-D fabric laminate SiC/SiC, 2-D fabric laminate C/SiC, and 3-D C/SiC composites utilizing SEP's Noveltex™ non-woven quasi three dimensional (3D) preform. The SiC/SiC materials were considered ready for military applications as flaps and seals for USAF fighter aircraft, C/SiC materials for short lifetime engine parts, such as cruise missiles, and the Noveltex composites for rocket thrust chambers. All the materials were proven and demonstrated in application testing, and have been improved with refinements for oxidation control and high temperature durability.

Formation of DuPont Lanxide Composites, Inc.

The second thrust into ceramic composites was considered more speculative. Also in Newark, Delaware, Lanxide, a materials technology company was growing rapidly with a business philosophy of joint venturing and partnering to expand the presence of their new MMC and CMC technologies. Three joint ventures were formed between DuPont and Lanxide, two were Lanxide controlled and the third, DuPont Lanxide Composites, Inc (DLC) was with a DuPont majority. The two minority ventures sought applications in market areas of electronics and armor. DLC sought applications in rockets, missiles, turbine engine and high temperature heat exchanger market segments. In contrast to the significant capital investment made in DuPont Ceramics with the SEP license, DLC was started with a simple purchase order.

Lanxide had developed and patented Directed Metal Oxidation (DIMOX™) methods of making ceramic matrix composites. The most promising of those was silicon carbide fiber reinforced alumina (SiC_f/Al_2O_3), but included other fibers and matrices such as silicon nitride and silicon carbide matrix composites. The process patented and practiced was to deposit a boron nitride interface coating on fibers in a tooled preform followed by a silicon carbide CVD deposited layer over the boron nitride to protect the BN from the subsequent matrix "growth" steps. The porous preform with its dual layer of BN and SiC was then melt infiltrated with metal or metal alloys and the metal allowed to oxidize. In the case of aluminum it was allowed to oxidize to alumina or aluminum nitride. In the case of silicon, it was allowed to react with free carbon to silicon carbide.

In 1993 DuPont merged its former effort based on the SEP CVD technology and its joint venture with Lanxide to form DuPont Lanxide Composites (DLC). DuPont contributed the SEP technology and Pencader facilities and PRD-66, and Lanxide donated the DIMOX™ process patents, Ceraset™, a pre-ceramic polymer patented originally by Hercules and licensed to Lanxide, and its facilities for making DIMOX™ at its Marrow's road facility in Newark, DE to the new Joint Venture.

Acquisition by AlliedSignal/Honeywell

In 1996, DuPont elected to divest many of their advanced materials efforts including organic matrix composites (OMCs) and ceramic matrix composites. In 1997, DuPont sold the OMC business at the Pencader site to Fiberite, which was subsequently acquired by Cytek who moved the business to Havre De

Grace, Maryland, in 1999. In 1998, DuPont sold their portion (so did Lanxide) of the DLC joint venture business to AlliedSignal Inc. and it was renamed AlliedSignal Composites Inc. (ACI), a Delaware Corporation, reporting to AlliedSignal Aerospace's Aircraft Landing Systems business unit. The acquisition was a good technical and business fit as AlliedSignal Engines was a principal customer of DLC and several of the other AlliedSignal business units including Aircraft Landing Systems and Power Management and Generating Systems would have benefited from the technology available at DLC. Finally, AlliedSignal merged with Honeywell at the end of 1999, adopted its well-recognized name for all of the new combined business, and hence Honeywell Advanced Composites Inc.

Formation of GE Power Systems Composites, LLC

In late 2001 as the deal for GE to acquire overall Honeywell fell through, a number of Honeywell business units were made available to GE, including Honeywell Advanced Composites Inc.. Because CMC had been considered as a strategic "game-changer" technology by a synergy team of GE Power Systems, GE Aircraft Engines and GE Global Research Center, and power-generation represented a near-term product introduction opportunity, GE acquired HACI through its Power Systems business and named the new addition as Power Systems Composites, LLC. The new GE business has established a mission of becoming the "Supplier of Choice" for GE and non-GE customers in the development and manufacturing of advanced ceramic matrix composite materials and components. Its primary markets lie in turbine engines, rocket/missile hot-gas-path components, and thermal protecting hot structures for aerospace applications.

GE PSC has a unique advantage of leveraging the synergy in CMC resources within GE. Its joint efforts and technical exchange with GE Global Research Center have brought the technical capability of PSC to a brand new level. In the first year after the GE acquisition, with a significant investment by GE Power Systems in capital equipment and infusion of a rigorous Six-Sigma methodology and discipline, PSC made remarkable advances in developing process robustness and establishing production readiness, which will undoubtedly accelerate the introduction of CMCs in commercial products and will ultimately benefit PSC's many GE and non-GE customers.

CMC MATERIALS DEVELOPMENT

In the history of PSC, a number of CMC materials and processes have been studied, developed, licensed and used to make a variety of CMC components for different applications. Three of the material systems, schematically shown in Fig. 14.1 will be described in details in the following sub-sections.

SiC or C Fiber-Reinforced CVI SiC Matrix Composites (CVI SiC or C/SiC)

Figure 14.2 illustrates the general fabrication processes used by PSC to fabricate CMC components by the chemical vapor infiltration (CVI) process. A preform of SiC or C fiber is first prepared by braiding or laying up woven fabric on a graphite mandrel. The mandrel provides support for the preform, establishes dimensions, and controls CVI gas flow to component surfaces. The common procedure is to wrap layers of a 2-D woven fabric around the mandrel until the desired thickness is achieved. The second step is to deposit a thin layer of pyrolytic carbon on the SiC or C fiber. Alternatively, the preform may be coated with a thin layer of boron nitride. These interlayers regulate the bond between the fiber and matrix. The SiC matrix is then deposited by a sequence of CVI cycles. Machining is conducted as needed between cycles and following the final cycle to achieve the desired dimensions.

The baseline CVI SiC/SiC composite licensed from SEP was fabricated using a 0/90 balanced layup of CG Nicalon™ fabric with a pyrolytic carbon coating on the fibers. Typical properties reported in 1988 for this baseline material are listed in Table 14.2.

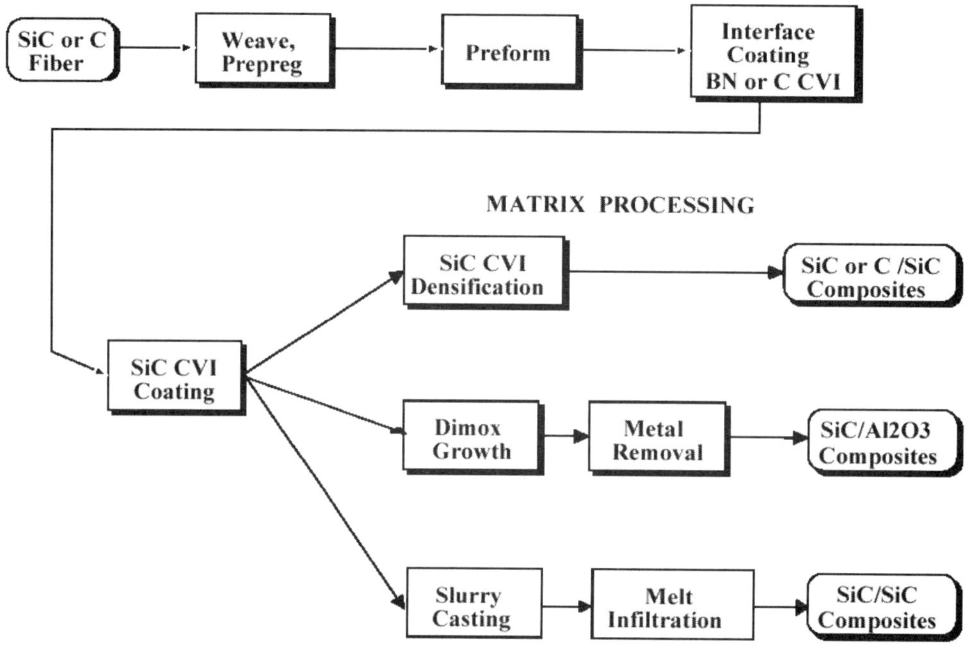

Figure 14.1 Schematic of Three CMC Processes at PSC

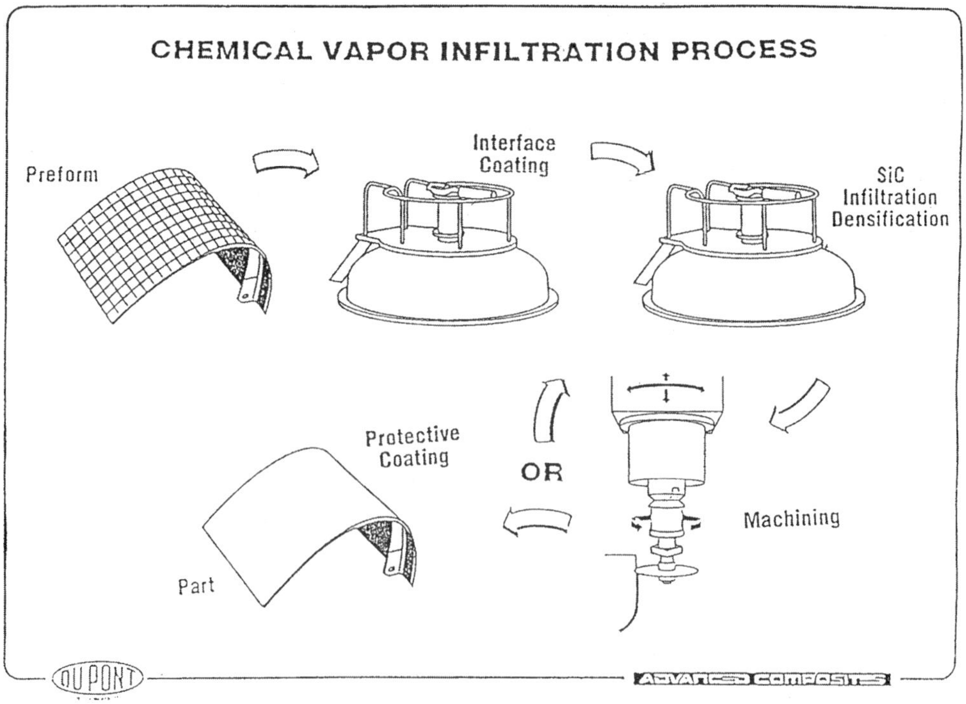

Figure 14.2 Schematic of PSC's CVI Process

TABLE 14.2 1988 BASELINE PROPERTIES OF THE CVI SiC/SiC CMC WITH NICALON™ FIBER (SOURCE: DUPONT PRELIMINARY ENGINEERING DATA INFORMATION SHEET H-10627-4)

		Temperature		
Property	**Units**	**23°C**	**1000°C**	**1400°C**
Fiber Content	%	40	40	40
Specific Gravity	g/cm^3	2.5	2.5	2.5
Porosity	%	10	10	10
Tensile Strength	MPa	200	200	150
Flexural Strength	MPa	300	400	280
Tensile Elongation	%	0.3	0.4	0.5
Initial Young's Modulus	GPa	230	200	170
Poisson's Ratio				
V12		0.05		
V13		0.18		
Compressive Strength	MPa			
In Plane		580	480	300
Thru the Thickness		420	380	250
Shear Strength (interlaminar)	MPa	40	35	25
Coefficient of Thermal Expansion	10^{-6}/K			
In Plane			3	
Thru the Thickness		1.7	3.4	
Thermal Conductivity	W/m.K			
In Plane			15.2	
Thru the Thickness		9.5	5.7	

The baseline CVI SiC/SiC material exhibited reasonable strength retention to at least 1000°C (based on short term tests in both air and an inert atmosphere), but the strength decreased as the temperature increased. Also, when the composite was tested in air under an applied stress (stressed oxidation), the material degraded rapidly. The pyrolytic carbon fiber coating oxidized and was replaced by silica [7]. Slip could no longer occur along the fiber-matrix interface, so the material fractured in a brittle manner.

Extensive efforts were conducted at PSC during the 1990s to address the stressed oxidation problem and achieve longer life at higher temperature and higher stress [6]. The following are some of the approaches that were explored:

- BN fiber coating
- Layered BN/SiC coating
- Si-BN coating
- Higher temperature fibers: ceramic grade Nicalon™, Hi-Nicalon™, and Sylramic™ fibers
- Modification of the matrix chemistry by adding proprietary "oxygen getters" to produce "Enhanced" CVI SiC/SiC
- Overcoating with CVD SiC
- Overcoating with oxide "Environmental Barrier Coatings" (EBCs)

By the end of 1993 substantial progress had been achieved. Table 14.3 shows data reported in a DOE program review in September 1993 that compares CVI SiC/SiC composites fabricated from Nicalon™ and Hi-Nicalon™ fibers with and without matrix modification (enhancement). The number of data points were limited, but it indicated that the newer materials had comparable fast fracture properties to the baseline material. To explore high temperature stability in an oxidizing atmosphere, stepped stress rupture tests (using 4-point flexure samples) were conducted for each material at 100 MPa stress. The baseline Nicalon™/SiC

TABLE 14.3 MECHANICAL PROPERTY COMPARISON FOR VARIOUS CVI SiC/SiC CMCS

Property	Nicalon™/SiC	Hi-Nic/SiC	Nicalon™/E-SiC	Hi-Nic/E-SiC
Density (g/cm^3)	2.5	2.7	2.35	2.6
Ambient Temperature				
Tensile Strength (MPa)	193	214	227	
Elongation (%)	0.22	0.23	0.4	
Elastic Modulus (GPa)	232	266	138	221
Elastic Limit Stress (MPa)	76	90	55	104
Elastic Limit Strain (%)	0.033	0.034	0.039	0.047
Flexural Strength (MPa) S:D=30:1*	276	310	345	428
1000°C				
Tensile Strength (MPa)	239		255	
Elongation (%)	0.35		0.6	
Elastic Modulus (GPa)	232		145	
Elastic Limit Stress (MPa)	124		90	
Elastic Limit Strain (%)	0.053		0.80	

*S:D is the Span to Depth (sample thickness) ratio.

failed in less than one hour at 1000°C. The Hi-Nicalon™/SiC failed at 1250°C in about 3.3 hours. The Hi-Nicalon™/Enhanced-SiC survived at 1400°C for 7.0 hours. During other stress rupture tests at 1300°C and 170 MPa, Nicalon™/Enhanced-SiC survived 100 hours, but exhibited extreme creep. The Hi-Nicalon™/Enhanced-SiC survived 288 hours, and exhibited much less creep. These tests verified that PSC was working in the right direction to achieve improved CMC life.

Based on the promising results, development and testing of Enhanced CVI SiC/SiC variations continued. By the end of 1995 tensile stress rupture tests demonstrated sample survival of >4500 hours at 1200°C and 80 MPa and >1200 hours at 1200°C and 90 MPa. However, at 1200°C and 100 MPa samples fractured between 17 and 63 hours, showing a significant stress limitation. By 1998 a sample with Hi-Nicalon™ fibers in Enhanced CVI SiC/SiC survived 540 hours at 1204°C and 110 MPa.

Solar Turbines Inc. identified a new problem in 1997 during engine testing of Enhanced CVI SiC/SiC combustor liners. The high water vapor pressure in the engine contributed to volatilization of SiC and resulted in a high rate of surface recession of the composite [8–12]. Increased thickness of the CVD SiC overcoat and deposition of an oxide environmental barrier coating (EBC) have addressed this problem. These mitigation strategies are discussed later in this chapter.

In parallel to improving the properties of CVI SiC/SiC, PSC has worked on improving reproducibility and on cost reduction. Parts have been successfully fabricated with a wide range in size and shape: (1) simple cylinders from pencil size to a 18 cm diameter by 107 cm long tube, (2) integrally-stiffened panels of 61 × 61 cm, (3) turbine blades and integral rotors with complex airfoil shapes and (4) rocket thrusters. Present CVI furnaces are capable of producing parts up to 137 cm in diameter by 213 cm high.

DIMOX™ Composites

The DIMOX™ (directed metal oxidation) process was invented at Lanxide Corporation. DuPont Lanxide Composites used the DIMOX™ technology to fabricate fiber-reinforced ceramic matrix composites for aerospace and turbine engine applications.

In this process, the first step in fabrication is preform making, which typically, though not always, utilizes 2-D woven fabric. The preforms are then coated via a chemical vapor infiltration (CVI) process. The coating consists of a duplex layer of boron nitride (with or without silicon doping) followed by silicon carbide. Boron nitride provides a weak fiber-matrix interface, which is essential to achieving composite behavior. The outer layer of silicon carbide protects the underlying boron nitride coating and fiber during subsequent matrix infiltration. An aluminum oxide matrix is grown through the coated fiber preform. by placing the preform in contact with a molten aluminum alloy in an air atmosphere at temperatures in the 900–1100°C range. A three-dimensionally interconnected matrix of alpha alumina grows into the fiber preform, starting from the alloy/preform interface. An interconnected network of microscopic metal channels of aluminum that are at most a few microns in diameter is present in the matrix phase. The molten alloy wicks from the reservoir through these microchannels to the growth front, where it reacts to continue the oxide matrix growth process. The reaction is terminated when it comes in contact with a gas-permeable barrier layer, which is normally applied to one or more external surfaces of the preform, enabling fabrication of components to the desired shape.

The matrix of the processed composite is ceramic rich with some residual metal, most of which is in the microscopic, interconnected channels. For applications requiring elevated temperature exposure and use, the presence of residual aluminum, which would become molten, is undesirable. Hence, an aluminum metal removal process is used to reduce the residual aluminum content in the matrix from ~7 to 10 wt.% to ~1 wt.%. Fabricated composites typically contain about 35 vol.% of continuous fibers in a 2-D weave [2–5,13].

Attractive features of the directed metal oxidation process include the fabrication of components to net or near-net shape using a low cost matrix densification process. The process is versatile in that various types of fiber reinforcements have been incorporated in different matrices. These included reinforcements of silicon carbide, aluminum oxide and mullite fibers and matrices of aluminum oxide, aluminum nitride and silicon nitride. To date, the majority of the DIMOX™ directed metal oxidation composites fabricated have used a silicon carbide fiber reinforcement in an aluminum oxide matrix.

Mechanical properties for the DIMOX™ SiC_f/alumina composite, using CG Nicalon™ fibers as a reinforcement, are reported in Table 14.4 [5]. Further tests also indicated good resistance to thermal shock. Water quench from 1000°C resulted in a small (7%) decrease in strength. Five cycles of thermal shock lowered the composite strength by 17%, which was equivalent to the decrease from one thermal shock cycle from a higher temperature of 1200°C. Thermal shock testing was conducted in air.

Although SiC_f/alumina composites exhibited some strength decrease due to high temperature exposure and to thermal shock, this loss was small enough to justify the fabrication of prototype components for engine testing at Williams International. SiC_f/alumina composite was compared with CVI SiC_f/SiC and SiC_f/SiC processed by a polymer precursor approach as part of the engine test.

Component design and testing was conducted by Williams International under an ARPA and U.S. Air Force Wright Laboratory/Materials Directorate funded program. The project goal was to evaluate the performance of CMC composite hardware at operating temperatures from 1093°C to 1371°C (2000°F to 2500°F) for potential use in expendable gas turbine engines. Each combustor set was to be tested to maximum temperatures appropriate for the material. The goal was to reach a 10-hour life in a Williams International WR-24-7 engine test at the Williams International Walled Lake facility in Michigan.

The WR24-7 combustor consisted of two components, a combustor cover and a primary plate. Combustion occurred in the gap between the cover and the primary plate. Holes were machined in the

TABLE 14.4 1995 PROPERTIES OF DIMOX™ NICALON™/ALUMINA CCOMPOSITES

Property	Room Temperature	1000°C	1200°C
4-pt. Flexure Strength (MPa)	477	398	351
Fracture Toughness by Chevron Notch ($MPa.m^{1/2}$)	21	23	18
Interlaminar Shear Strength (MPa)	62		
4-pt. Flexure Strength after 24 h in air at 1200°C (MPa)	310		

components to provide inlets for air into the combustion chamber. Composite toughness was a key criterion for ensuring component survivability during the test. Williams International performed all combustion and combustor design and provided hole patterns for the composite fabrication vendors.

Each of the three materials selected for the engine test demonstrated a satisfactory level of performance based on material properties from flat plates. Physical properties obtained from flat plates included flexure and tensile properties. Flexural specimens were then sectioned from one of the complex combustor components, to obtain some level of comparison of component properties. The complex shaped flexure bars were not straight, but an order of magnitude ranking was obtained. The data for the SiC_f/Al_2O_3 material was 372 MPa from the flat plate, and 352 MPa from the components at room temperature tests. Hence, the flexural strengths demonstrated on flat panels were replicated in the fabricated component.

Of the three materials tested, the DIMOX™ material showed the highest overall durability. Even though minor cracking was observed between air inlet holes, the SiC_f/Al_2O_3 combustors were the only components to successfully meet the engine test goal. The tests were run with engine stops about every $1^1/_2$ hours, during which borescope exams were made. Engines were run until severe cracking appeared on the combustor liners. All combustors that were tested suffered minor cracking between air inlet holes. Running times prior to severe cracking for two SiC_f/SiC composites were a maximum of 2 hours 3 minutes. Temperatures ranged from 1093°C to 1260°C (2000°F to 2300°F) for these parts. Both combustor sets of the SiC_f/Al_2O_3 DIMOX™ material ran for more than 10 hours each at temperatures up to 1482°C (2700°F).

This set of tests provided valuable experience in materials development and component fabrication. The combustor cover proved sufficiently complex with the 90 degree bend at the outer edge. To address fabric drapability, cut seams, and effective consolidation to a specific fiber volume fraction, vacuum bag techniques were developed. The outer edges of the fabricated parts were diamond ground, and the air inlet holes were laser machined. Several iterations of laser machining were conducted to establish the pulsed-CO_2 laser technique that was used for many future parts. From a materials viewpoint, most of the mechanical testing up to this point in time was tensile stress rupture testing at constant stress, and limited low-cycle fatigue testing. Analyses of the fractured specimens from these mechanical tests had shown the presence of an internal crack sealing mechanism for oxidation protection. The performance of the combustor sets in the presence of cracks served as an added indicator of an oxidation protection mechanism. Further detailed testing determined that this oxidation protection mechanism was time-limited. However, the expendable engine combustor mission profile and a relatively short time requirement for very high performance seemed to be a reasonable match to the properties of the DIMOX™ SiC_f/alumina CMC.

The main drawbacks of the SiC_f/alumina composite for longer term turbine engine applications were a microcracked matrix in the as-fabricated state due to a relatively large fiber-matrix CTE mismatch, and an inherently low thermal conductivity matrix. The development of melt infiltrated (MI) SiC_f/SiC composites resulted in a system, which utilized the duplex BN/SiC coating approach invented for DIMOX™, but with a dense and higher thermal conductivity matrix. Furthermore, MI SiC_f/SiC composite fabrication did not require the additional processing step used in DIMOX™ to remove the lower melting residual aluminum metal. Reduction of a processing step has significant implications from a manufacturing perspective with regard to cost and yield. Hence, development efforts for DIMOX™ SiC_f/alumina were substantially reduced for turbine applications, and a major shift was made to MI SiC_f/SiC. However, considering that the DIMOX™ aluminum oxide matrix is an ideal candidate for oxide-oxide CMCs, the ability to resurrect this system still exists at PSC in the event that a suitable, high temperature, commercially viable oxide fiber does become available.

SiC Fiber-Reinforced Melt-Infiltrated SiC Matrix Composites (MI SiC/SiC)

Between 1994 and 1999, PSC conducted earnest development of silicon melt infiltrated (MI) SiC/SiC composites, partially funded by the Enabling Propulsion Materials (EPM) program and with technical guidance from the EPM team of NASA, General Electric and Pratt and Whitney. The MI process is described in detail in Chapter 16 of this volume and the process steps used at PSC are shown in Figure 14.1. As with other techniques, the MI process begins with a preform of SiC fibers prepared by typical preform fabrication methods

TABLE 14.5 PROPERTIES OF PSC MI SiC/SiC COMPOSITES

Property	Sylramic™	Hi-Nicalon™
Density (g/cm^3)	2.86	2.7
23°C		
UTS (MPa)	357	359
Elongation (%)	0.37	0.74
Elastic Modulus (GPa)	223	196
Elastic Limit Stress (MPa)	117	120
1204°C		
UTS (MPa)	307	271
Elongation (%)	0.46	0.52
Elastic Modulus (GPa)	182	145
Elastic Limit Stress (MPa)	166	130

such as lay-up of 2D woven cloth or filament winding. As in CVI SiC and DIMOX™ processing, the fiber preform is first coated with an interface or de-bonding layer such as BN or pyrolytic carbon. The preform with interface coating is next coated with a layer of CVI SiC or Si_3N_4 to protect the fiber and interface during the melt infiltration process. The final steps of the process are designed to completely densify the matrix. This is typically accomplished by infiltrating any remaining porosity with either carbon or silicon carbide particulate followed by infiltration with molten silicon. When carbon is present in the preform, the silicon infiltrant reacts with the carbon to form additional SiC. Any remaining pores are filled with silicon

The MI SiC/SiC composite has special characteristics. Unlike the CVI SiC/SiC and the DIMOX™ SiC/alumina, the MI SiC/SiC contains little porosity. As a result it has higher yield strength (stress at which microcracking first occurs), greater resistance to oxidation, and higher thermal conductivity. The high thermal conductivity significantly reduces thermally induced stresses in turbine components such as combustor liners and shrouds. The mechanical properties of MI SiC/SiC composites fabricated at PSC using Sylramic™ and Hi-Nicalon™ SiC fibers are summarized in Table 14.5.

Current PSC Material Systems and Potential Applications

To meet the demands in its primary markets of turbine engines, rocket/missile hot-gas-path components, and thermal protecting hot structures for aerospace applications, PSC is developing a number of material systems. Table 14.6 summarizes examples of current PSC material offerings and potential applications. These include materials with reinforcement of SiC fiber, C fiber (T-300) and non-woven quasi 3-D preform (Novoltex™), and matrix by chemical vapor infiltration and melt infiltration. Characteristics of applications range from short time, high temperature to long time, but relatively low temperatures.

TABLE 14.6 CURRENT PSC CMC MATERIALS OFFERING AND EXAMPLE APPLICATIONS

Material Systems	MI SiC/SiC	CVI C (Novoltex™)/SiC	CVI C (T-300)/SiC
Primary Application	Turbine Engine Hot Section	Rocket Hot Gas Path	RLV/SOV Air Frame
Temperature	1204–1316°C (2200–2400°F)	1927–2482°C (3500–4500°F)	1538–1649°C (2800–3000°F)
Life	10,000+ hrs	10–1000 s	100+ hrs
Requirements Density	2.80 g/cm^3	2.16 g/cm^3	2.16 g/cm^3
Other Attributes	Thermally Conductive	Erosion Resistant	High Specific Strength and Stiffness

COMPONENT FABRICATION AND ENGINE TESTING

PSC has actively pursued opportunities for component fabrication and testing throughout its history. Figure 14.3 shows some of the components PSC has made in support of various customer programs. The fabrication of combustor liners for the DOE/Solar Ceramic Stationary Gas Turbine (CSGT) program is highlighted in the following sub-sections.

Fabrication and Testing of Subscale Solar Turbines Inc. Combustor Liners

Fabrication of subscale liners for the DOE/Solar CSGT program was performed in 1993–1994. These subscale cylindrical liners were the initial CMC test pieces for the Solar ceramic combustor liner concept to be evaluated in a subscale combustor rig which simulated key features of the full scale combustor in a simplified but representative configuration. The nominal dimensions of the subscale liner were 20 cm long × 20 cm diameter × 0.25 cm thick. Based on these dimensions PSC acquired graphite tooling. Four cans and co-processed burst-rings of Enhanced SiC/SiC with a proprietary coating were fabricated and delivered. The fiber used was ceramic grade Nicalon, and the CVI process was used for matrix densification. Figure 14.4 shows some of the subscale components and rings delivered to Solar.

Prior to testing the subscale liners were evaluated nondestructively at PSC and Argonne National Laboratory (ANL) [14]. The subscale liners provided an opportunity for ANL to develop the basic nondestructive evaluation (NDE) methodology that was subsequently used for the full-scale CMC liners. ANL's NDE data, collected with infrared thermal diffusivity and X-ray Computed Tomography, correlated well with water-coupled C-scan ultrasonic transmission data gathered at PSC. One of the liners showed minor delamination, a condition that did not worsen during subscale combustor testing, but PSC made some process changes which eliminated this problem in subsequent fabrication runs. Further details on ANL's NDE methodology can be found in Chapter 23 of this volume.

Solar's standard subscale combustor test program in its single injector can rig involved successive tests of 1, 10, and 100 hrs. The test rig conditions resulted typically in a maximum liner temperature of about 1200°C, and the rig pressure was around 7 bar. Stresses on the liner were low, the maximum being around

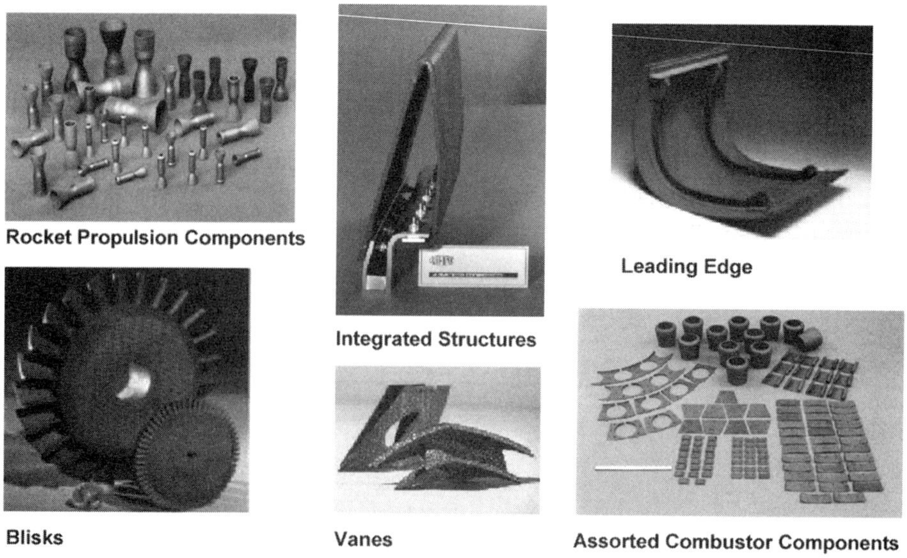

Figure 14.3 Examples of CMC Components Fabricated by PSC

Figure 14.4 Various Subscale Components and Test Specimens Fabricated for Solar Turbines Inc. Rig Tests

30 MPa. The liners survived all three tests with no marked changes to the CMC microstructure based on NDE observations. Burst ring testing showed that the tested liner had retained on average 89% of the as-received strength. The strain to failure increased about 19% compared to the as-received. The favorable subscale liner test data contributed to Solar's selection of the PSC CVI SiC/SiC CMC for full-scale liner fabrication and subsequent testing in the Centaur 50 S engine.

Fabrication and Engine Testing of Full-Scale Solar Turbines Combustor Liners

Several full-scale liner sets were shipped to Solar over the course of the CSGT program. An example is shown in Figure 14.5. Each set consisted of an outer liner, about 77 cm in diameter and an inner liner approximately 33.8 cm in diameter. Both liners were about 20.5 cm long and the target thickness was about 2.7 mm.

Figure 14.5 Full-Scale CMC Inner and Outer Liners Fabricated for Solar Turbines Engine Tests

Actual as-fabricated thickness was considerably greater and the target thickness was achieved through grinding. In the initial trials fiber volume was less than the desired 35%, but over time the CMC fabrication process was improved and the target fiber volumes were obtained. The first two sets of full scale liners fabricated by PSC for Solar used ceramic grade Nicalon fiber. Subsequent sets were fabricated using Hi-Nicalon fiber. Up to 1998, the standard process for matrix infiltration was CVI. In later fabrication runs the MI process was also used to fabricate the inner liners. Details of the fabrication of the liners have been reported [15].

Several challenges were faced during the scale up from subscale to full-scale liner fabrication. To avoid problems with wrinkling while compacting the plies, each ply of the outer liner was applied as three 120° arc segments of fabric. The inner liner was preformed with one full-circumference piece of fabric per ply. Seams were staggered so that there was at least 7.1 cm between seams. Co-processed coupons were fabricated with a seam to simulate the seams in the liners. The seam was placed in the middle of the gauge sections. All full-scale liners in Phase II of the program were supplied with a SiC seal coat applied by chemical vapor deposition (CVD). NDE at PSC and ANL was used to evaluate the liner integrity and to ensure that the liners were free from delaminations.

Full-scale PSC SiC/SiC CMC combustor liners have been tested at Solar Turbines Inc. and at industrial gas turbine test sites at ChevronTexaco (formerly Texaco, prior to that ARCO) in Bakersfield, California, and Malden Mills in Lawrence, Massachusetts, under the CSGT program. The selection of the CMC materials and the progression of testing is described in detail in Chapter 7 of Volume I of this text. The highlights of the testing will be summarized here briefly.

PSC full scale enhanced SiC/SiC CVI full-scale liners were first tested in atmospheric and high pressure rigs at Solar to establish their durability in short term engine exposure. These tests were typically short term (up to 50 hrs) and included cyclic operation. Maximum liner temperature was about 1200°C. Stress levels were similar to those expected in actual engine operation, the maximum predicted level being between 78 and 91 MPa. The liners were then tested in the Centaur 50S engine at Solar. The standard qualification test prior to long term engine durability testing at field sites was 100 hrs. All PSC SiC/SiC liners tested met the 100 hr qualification goal. A slight amount of glass formation was usually detected at the liner surfaces upon completion of the tests, but this was not believed to be harmful for subsequent long term testing.

Table 14.7 summarizes the engine field testing at the ChevronTexaco Bakersfield and Malden Mill sites in which PSC CMC liners were used. Borescope inspection revealed that these liners in the absence of a protective environmental barrier coating showed significant degradation after about 500 hrs of exposure in the gas turbine combustion environment. An outer liner that had been exposed for 5,028 hrs had recession to a depth of 80% of the thickness of the composites.

Recent engine field testing has shown a dramatic improvement in CMC durability when an EBC was applied [16]. At the conclusion of the test at the end of November 2000 13,937 cumulative hours had been tallied for a set of CMC liners in the Solar Centaur 50S engine at the ChevronTexaco (Texaco and ARCO at the time the testing was conducted) site. While the EBC is showing evidence of degradation, surface recession of the CMC appears to be limited to the areas where the EBC has spalled.

FUTURE CHALLENGES TO CMC COMMERCIALIZATION

Application of the CMCs is still in its early stage. Many challenges still need to be overcome before CMCs can gain widespread commercial use in gas turbine engines and other advanced applications. The first and possibly most constraining, is the issue of cost. CMC components are currently several times more expensive than components fabricated from conventional high temperature superalloys with thermal barrier coatings (TBC). An accurate assessment of life-cycle benefit of CMC is critical in justifying the initial component cost [15].

Manufacturing Resources, Inc. (MRI) was contracted through Oak Ridge National Laboratory (ORNL) to work with PSC to study the potential manufacturing costs at PSC for CMC liners and recommend initiatives to reduce costs to accelerate the commercialization of CMC technology [17]. The assessment began with the

TABLE 14.7 SUMMARY OF FIELD TESTING AT THE TEXACO (CURRENTLY CHEVRON/TEXACO) AND MALDEN MILLS SITES

Test Site/Date	Test Duration	PSC Liners	Seal Coat or EBC	Observations/ Comments
Texaco(ARCO)—1997	1048 hrs*	Outer and inner liners CG-Nicalon/Py-C/E-SiC CVI	Thin SiC CVD seal coat	Glass formation—significant on inner liner
Texaco (ARCO)—1998/1999	2266 hrs (inner) 5028 hrs (outer)	Outer and inner liners Hi-Nicalon/Py-C/E-SiC CVI E-SiC CVI	Thick CVD seal coat	Glass formation—significant on both liners
Texaco (ARCO)—1999/2000	13,937 hrs**	Outer liner Hi-Nicalon/Py-C/E-SiC CVI Inner liner Hi-Nicalon/BN/SiC MI	Thick CVD seal coat + EBC	Surface degradation where EBC spalled
Malden Mills I—1999/2000	7,238 hrs**	Outer liner Hi-Nicalon/Py-C/E-SiC CVI	Thick CVD seal coat + EBC	Surface degradation where EBC spalled
ChevronTexaco (continued from Malden Mills I)	5,135 hrs			EBC was refurbished after Malden Mills test
Malden Mills II—2000/2002	15,144 hrs**	Outer liner Hi-Nicalon/Py-C/E-SiC CVI	Thick CVD seal coat + EBC	Surface degradation where EBC spalled, some cracking in liners
Malden Mills II	> 4,600 hrs***	Inner and outer liner Tyranno ZMI/BN/SiC MI	No seal coat + EBC	

* Includes 100 hrs of qualification testing at Solar
** At completion of test
*** As of end of January, 2003

development of a set of functional diagrams that describe the present CVI and MI processes for combustion liners. From these functional descriptions and on-site analyses a series of bottom-up costs models was developed for both processes for fabricating 120 liners/year. The results are shown in Figure 14.6. Raw materials was determined to be the greatest factor and was dominated by the cost of SiC fibers. The other factors were distributed roughly equally. Going from production of 120 liner sets per year to 360 sets per year was estimated to reduce the total cost by 50% and reduce the raw materials portion from 38 % to 32 %. Much greater increase in production would be required to significantly reduce the cost of the fibers.

Other factors that were evaluated in the study were life-cycle cost, time to market, and potential for refurbishment. Value-in-use factors suggested that ceramic combustor initial acquisition prices may be as high as twice of the cost of the metal liners, when cost is determined at the point of parts in stores ready for engine assembly.

Other challenges include (1) limited data on life expectancy under field service operating conditions and the lack of tools for life prediction, (2) limited property data available to designers and lack of experience in designing with these new materials, and (3) lack of mature production process and facility in producing commercial quantity and quality of CMC components that bring satisfactory values to CMC end-users. Like the introduction of any new materials, these challenges are being overcome as time progresses. PSC is playing a major role in all aspects of this challenging and yet exciting endeavor, in particular, we are making a significant contribution in establishing a mature CMC production base.

CLOSING REMARKS

GE Power Systems Composites, LLC and its predecessor companies have been involved in development of ceramic matrix composites since around 1980 and in the application of CMC to gas turbine engines

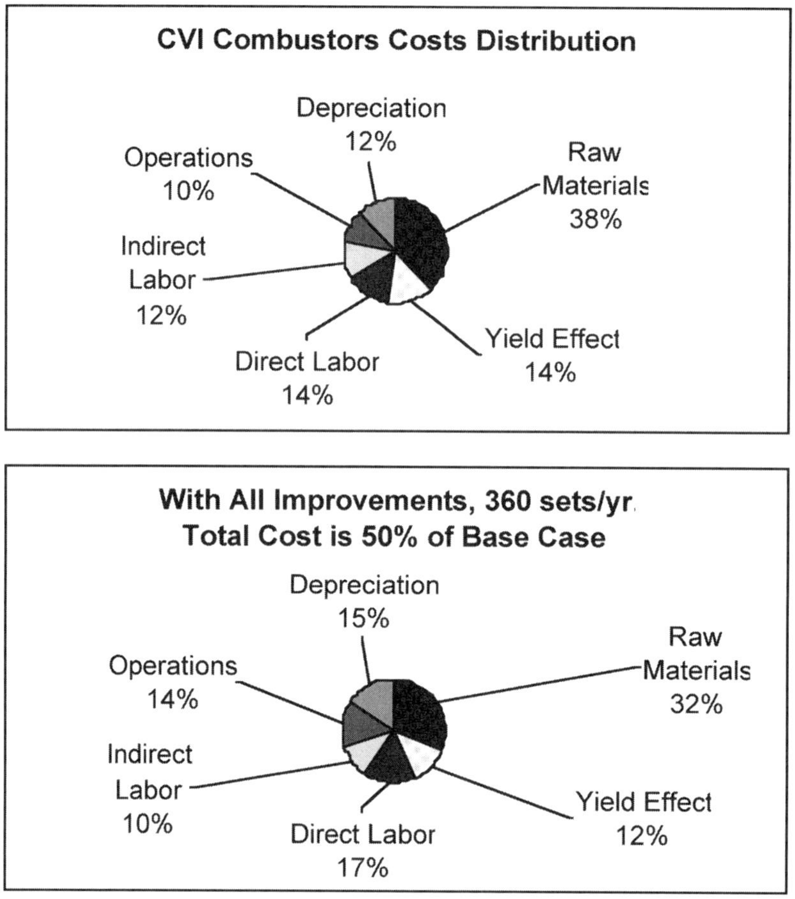

Figure 14.6 Results of a Manufacturing Cost Analysis (Reprinted from ref. [17])

since the early 1990s. The PSC CMC of choice for turbines are based on a silicon carbide fiber-reinforced silicon carbide matrix system (SiC/SiC). Two advanced processing routes are used to produce SiC/SiC composites, chemical vapor infiltration and melt infiltration. Using these techniques, PSC has contributed to many successful testing programs of this revolutionary material.

The technical success of these efforts has been the result of ongoing diligent materials and processing R&D efforts and partnership with customers. PSC has brought together various key CMC technologies and integrated them to meet demanding technical challenges. This has resulted in an ability to fabricate increasingly larger and more complex shapes with a high degree of reliability. Vital elements in the CMC process evolution have been the use of superior fiber materials, the development of effective matrix infiltration technologies, and the application of sophisticated fiber/matrix interface coatings. Long lifetimes in service were obtained through further improvements including the incorporation of matrix enhancements to suppress oxidation, and the application of oxidation resistant surface and environmental barrier coatings to resist high-pressure steam environments.

Looking forward to the future, PSC has been engaging in a great effort to improve its process capability so that it can produce quality products consistently, and reduce the cost of CMCs through process yield improvement, cycle-time reduction, and lean manufacturing. PSC is working with various fiber suppliers in improving the fiber quality and reducing its cost, as it is the biggest contributor to the overall CMC cost. Six-Sigma methodology and rigor are being practiced by PSC technology and production staff in their daily activities to achieve process robustness and cost reduction. Existing processes are being scaled up and new and improved processes are being developed in synergy with GE Global Research Center and government laboratories. Large-scale production equipment is being planned, designed, fabricated and installed. With the support from GE, government funding agencies and its valued customers, PSC is overcoming challenges and bringing the CMC technologies across the goal-line from R&D to commercialization.

ACKNOWLEDGEMENTS

The authors thank all of the individuals who contributed to the CMC efforts at PSC and its predecessors and the Government agencies that provided funding. Unless otherwise indicated figures and tables were provided courtesy of GE PSC.

REFERENCES

[1] Lamicq, P.J. and Jamet, J.F., "Thermostructural CMCs: An Overview of the French Experience," in *High-Temperature Ceramic-Matrix Composites I: Design, Durability, and Performance*, A.G. Evans and R. Naslain, eds., American Ceramic Society, Westerville, OH, USA, 1995. pp. 1–11.

[2] Newkirk, M.S., Urquhart, A.W., Zwicker, H.R., and Breval, E., "Formation of Lanxide™ Ceramic Composite Materials," J. Mater. Res., 1 [1], 81–89, 1986.

[3] Newkirk, M.S., Lesher, H.D., White, D.R., Kennedy, C.R., Urquhart, A.W., and Claar, T.D., "Preparation of Lanxide™ Ceramic Matrix Composites: Matrix Formation by the Directed Oxidation of Molten Metals," Ceram. Eng. and Sci. Proc., 8[7–8], 879–885, 1987.

[4] Andersson, C.A., Barron-Antolin, P., Fareed, A.S., and Schiroky, G.H., "Properties of Fiber-Reinforced LANXIDE™ Alumina Matrix Composites," in *Proc. Int. Conf. On Whisker- and Fiber-Toughened Ceramics*, Bradley, R.A., Clark, D.E., Larsen, D.C., and Stiegler, J.O., eds., ASM, Metals Park, OH, USA, 1988. pp. 209–215.

[5] Fareed, A.S., Sonuparlak, B., Lee, C.T., Fortini, A.J., and Schiroky, G.H., "Mechanical Properties of 2-D Nicalon™ Fiber-Reinforced LANXIDE™ Aluminum Oxide and Aluminum Nitride Matrix Composites," Ceram. Eng. and Sci. Proc., 11[7–8], 782–794, 1990.

[6] Garnier, J., Headinger, M., Ueda, K., and Hatton, K., "Advanced Fiber Reinforced CMCs Fabricated by CVI and DIMOX™ for Turbine Engine Applications," in *Proceedings of the 1995 Yokohama International Gas Turbine Congress*, GTSJ, Yokohama, Japan, October 23–27, 1995. 95-YOKOHAMA-IGTC-113, Vol. I, I-73/I-80, 1995.

[7] Richerson, D.W., *Modern Ceramic Engineering*, Marcel Dekker, Inc., New York, 1992. pp. 785–787.

[8] Robinson, R.C., *SiC Recession Due to Scale Volatility Under Combustor Conditions*, NASA Contractor Report 202331, March 1997.

[9] Smialek, J.L., Robinson, R.C., Opila, E.J., Fox, D.S., and Jacobson, N.S., "SiC and Silicon Nitride Recession Due to Volatility under Combustor Conditions," Adv. Composite Mater., 8 [1], 33–45, 1999.

[10] Ferber, M.K., Lin, H.T., Parthasarathy, V., and Wenglarz, R.A., "Degradation of Silicon Nitrides in High Pressure, Moisture-Rich Environments," ASME Paper 2000-GT-0661, presented at the ASME TURBO EXPO 2000, Land, Sea & Air, Munich, Germany, May 8–11, 2000.

[11] Miriyala, N. and Price, J.R., "The Evaluation of CFCC Liners after Field-Engine Testing in a Gas Turbine—II," ASME paper 2000-GT-0648, presented at the ASME TURBO EXPO 2000, Land, Sea & Air, Munich, Germany, May 8–11, 2000.

[12] More, K.L., Tortorelli, P.F., Ferber, M.K., Walker, L.R., Keiser, J.R., Brentnall, W.D., Mirayala, N., and Price, J.R., "Exposure of Ceramics and Ceramic Matrix Composites in Simulated and Actual Combustor Environments," ASME paper 99-GT-292, presented at the International Gas Turbine and Aeroengine Congress, Exposition and Users Symposium, IN, Indiana, USA, June 7–10, 1999.

[13] Fareed, A.S., Landini, D.J., Johnson, T.A., Patel, A.N., and Craig, P.A., "High-Temperature Ceramic Matrix Composites by the Directed Metal Oxidation Process," SME paper EM92-216, presented at the High Temperature Composites Clinic, Los Angeles, CA, USA, August 12–13, 1992.

[14] Ellingson, W., *Nondestructive Evaluation (Characterization) Development for Stationary Gas Turbine Ceramic Components*, Argonne National Laboratory CSGT Final Report prepared for Solar Turbines Inc., under DOE Contract Number DE-AC02-92CE40960, February 1999.

[15] Hatton, K., Landini, D., Hemstad, S., and Robinson, R.C., "SiC Fiber Reinforced Ceramic Matrix Composites for Turbine Engine Applications," ASME paper 2000-GT-72, presented at the International ASME TURBO EXPO 2000, Land, Sea & Air, Munich, Germany, May 8–11, 2000.

[16] Eaton, H.E., Linsey, G.D., More, K.L., Kimmel, J.B., Price, J.R., and Miriyala, N., "EBC Protection of SiC/SiC Composites in Gas Turbine Combustion Environment," ASME paper 2000-GT-0631, presented at the ASME TURBO EXPO 2000, Land, Sea & Air, Munich, Germany, May 8–11, 2000.

[17] Craig, P.A. and Godfrey, D., "Reductions in Acquisition Costs for State-of-the-Art Fabrication of CFCC Turbine Engine Combustor Liners," ASME paper 99-GT-352, presented at the International Gas Turbine and Aeroengine Congress, Exposition, and Users Symposium, Indianapolis, IN, USA, June 7–10, 1999.

CHAPTER

15

CERAMIC MATRIX COMPOSITES FOR GAS TURBINE APPLICATIONS

Andrew Szweda, Timothy E. Easler, Robert A. Jurf, and Steven C. Butner

COI Ceramics, Incorporated, San Diego, California, USA

INTRODUCTION

Gas turbine development has progressed to the point where high temperature, lightweight Ceramic Matrix Composite (CMC) materials are needed to replace currently used metal alloys that are operating close to the limits of their thermal and mechanical properties. The use of CMCs will enable gas turbines to operate at higher temperatures, improving performance and efficiency. Over the years there has been significant interest in these materials from both the military gas turbine and the power generation gas turbine communities [1].

During the past ten years, the Dow Corning Corporation and Composite Optics, Incorporated have independently been developing non-oxide and oxide based CMCs, respectively, as high temperature structural materials for use in both land-based and aircraft gas turbines [2,3]. In 1999 these two companies formed a subsidiary known as COI Ceramics, Incorporated (COIC) to continue the development and commercialization of these CMC materials, and a brief review of their progress is given.

MATERIALS AND PROCESSING

The two main CMC systems that have been evaluated for gas turbines are based on a non-oxide (Si-N-C) CMC and an oxide (both alumina and alumino-silicate) CMC. The non-oxide Si-N-C CMC is derived from a pre-ceramic polymer infiltration and pyrolysis (PIP) process, which involves multiple infiltration and pyrolysis cycles to achieve a specified densification level, typically greater than 95%. The oxide CMC requires only a single infiltration step since it relies on matrix porosity of about 22% to provide a fiber pullout mechanism. While a variety of different fibers and matrices can be utilized, the most applicable versions are chosen within each system to meet specific application requirements.

Non-Oxide CMC Material and Process

At the beginning of the non-oxide CMC process, ceramic fibers are coated with a thin interface layer to control the degree of bonding of the fibers to the matrix. This is required in order to obtain the desired crack deflection behavior, which translates to a graceful failure mode. A ceramic filler material is typically added to the matrix phase to modify properties and processing behavior. A small amount of residual porosity and matrix cracks are also present, the latter being formed during conversion of the matrix precursor to a ceramic phase.

The CMCs described are fabricated using a Polymer Impregnation and Pyrolysis (PIP) process. This process is widely recognized as a versatile method to fabricate large, complex shaped structures. It consists of the impregnation of ceramic fabric with the matrix precursor, pyrolysis to convert the polymeric precursor to a ceramic phase, and then repeated reimpregnation and pyrolysis of the component until the desired level of density is attained. In comparison to other ceramic composite fabrication processes, the PIP process offers significant flexibility. By utilizing low-temperature forming and molding steps typically used in polymer matrix composites, the PIP approach allows the use of existing equipment and processing technology developed for those materials. This includes the ability to machine details into components before they are converted to the ceramic phase, greatly reducing the cost and time for machining. As shown in Figure 15.1,

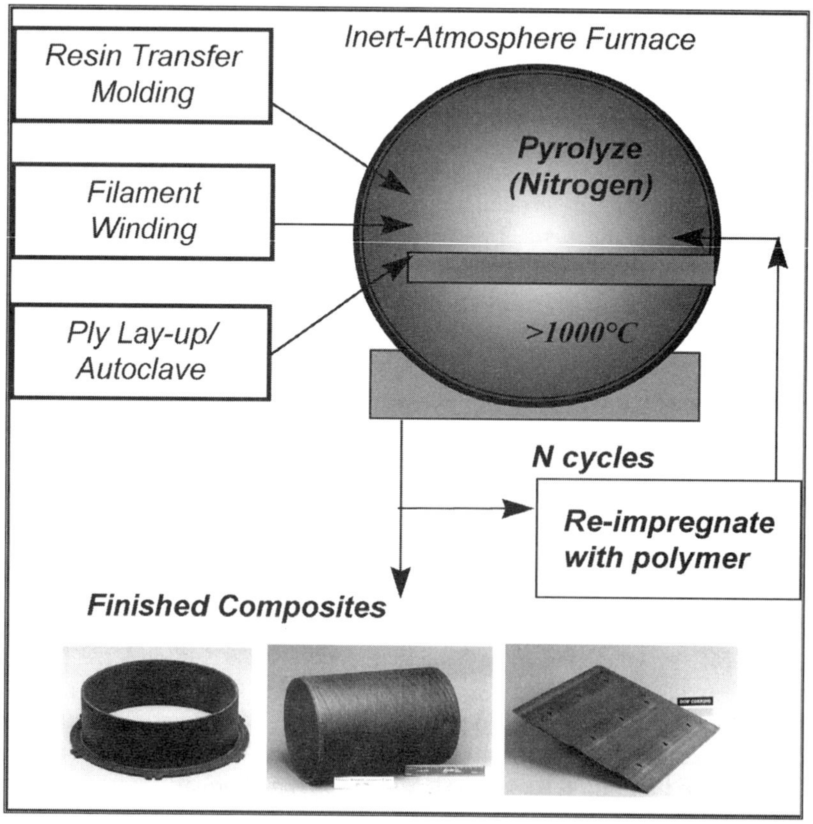

Figure 15.1 Schematic of the Non-Oxide CMC Process

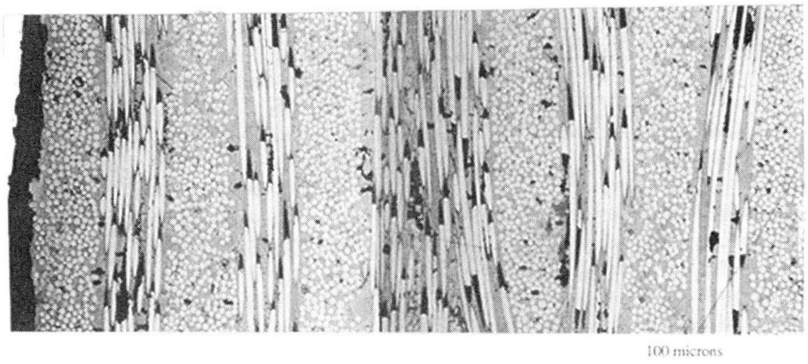

Figure 15.2 Microstructure of a Non-Oxide CMC

an inert-atmosphere furnace is used to pyrolyze the matrix precursor at temperatures of >1000°C, which converts the polymer to a ceramic phase. During this process, there is a reduction in the volume of matrix material as it is converted from polymer to ceramic and increases in density. The resultant void space is filled by reimpregnating the composites with additional polymer, followed by further pyrolysis. This process loop is repeated until the desired levels of density and porosity are achieved. A bulk density of 2.1–2.2 g/cm^3 and an open porosity of <5% are typical.

The non-oxide CMC has a versatile chemistry and can consist of various fibers, interface coatings, and matrix chemistries. The systems developed are divided into 3 classes of materials denoted as follows:

S200: various interface coatings (typically boron-containing) on CG NICALON™ fiber in an amorphous SiNC matrix. These materials can be used at temperatures up to 1200–1250°C in an oxidizing or inert environment, and can tolerate temperatures up to 1400°C for limited lifetime applications.

S300: various interface coatings on SYLRAMIC™ SiC fiber in an amorphous SiNC, or crystalline Si_3N_4/SiC or SiC matrix for higher temperature applications up to 1400°C for long duration applications.

S400: various interface coatings on carbon fiber in a crystalline Si_3N_4/SiC or SiC matrix for higher temperature applications up to 1600°C in inert environments.

A typical microstructure of a non-oxide CMC is shown in Figure 15.2. The individual fibers are approximately 15 microns in diameter. The fibers are surrounded by the matrix phase that is continuous except for a small number of voids and cracks that form during processing. Filler particles are less than one micrometer in size and therefore not visible at this magnification.

Oxide CMC Material and Process

The baseline oxide CMC fabrication approach is a sol-gel derived alumino-silicate or alumina matrix that can be combined with a variety of commercially available fiber reinforcements such as Nextel 550, Nextel 610, and Nextel 720. The baseline oxide-oxide system relies on controlled matrix porosity for toughness, thereby eliminating the need for fiber coatings. Recent findings support the conclusion that high strength, damage tolerant oxide-oxide CMCs can be made without the use of fiber coatings [4,5].

A range of oxide based CMCs can be produced using a variety of compatible reinforcing fibers available from 3M Corporation [6]. Some of these materials are shown in Table 15.1 along with the range of physical and mechanical properties that can be obtained.

The process for manufacturing oxide-oxide CMCs is simple and low cost. A conventional approach to fabricating a complex shaped component is shown in Figure 15.3. The matrix is a viscous slurry that is prepregged into the fabric and staged to a tacky consistency. Components are laid-up using organic

TABLE 15.1 FEATURES OF DIFFERENT OXIDE CMC FABRIC REFINFORCEMENTS

	Oxide CMC Properties with Various Oxide Fabrics				
	Astroquartz	**Nextel 312**	**Nextel 550**	**Nextel 720**	**Nextel 610**
Composite Density (g/cm^3)	1.50	2.30	2.41	2.60	2.83
Nominal Fiber Volume (%)	50	48	38	48	51
RT Tensile Modulus (GPa)	22 (est)	31.0	40.0	76.5	124.1
RT Tensile Strength (MPa)	46 (est)	124.8	147.8	179.0	366.1
CTE (ppm/°C)	0.40	4.85	5.40	6.30	7.92

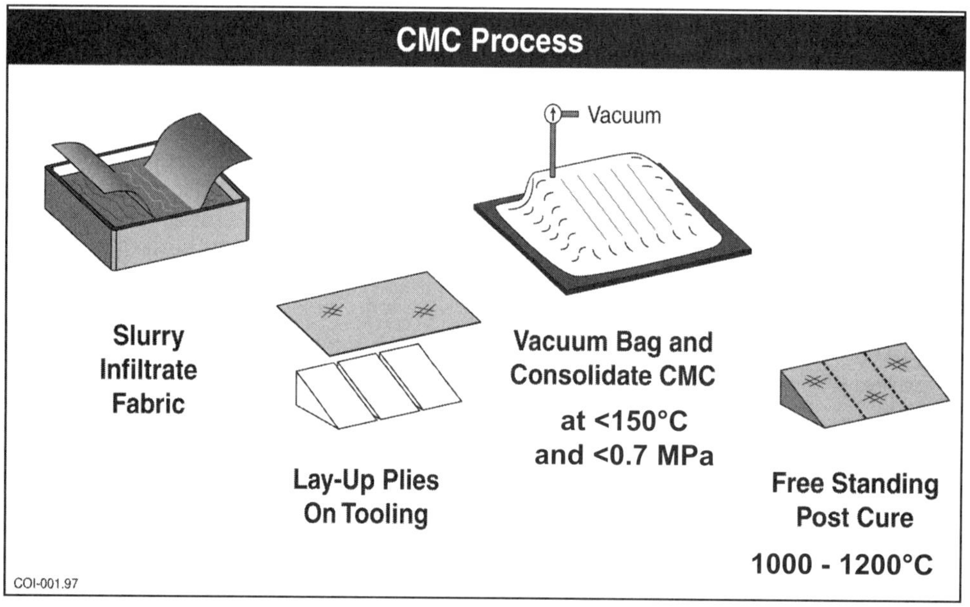

Figure 15.3 Oxide CMC Fabrication Process

composite techniques and cured (dried) using low pressures (<0.7 MPa) and temperatures (<150°C). Composite parts have sufficient strength at this point to be demolded. The final step required is a free standing pressureless sinter of about 1150°C to consolidate the matrix to final density.

CMC Component Fabrication

Both of the CMC processes employed have some similarities and can utilize the same equipment. Also, the ability to fabricate preforms having the desired shape and architecture is of primary importance for ceramic composites. Gas turbine applications require versatile approaches to produce the wide range of structural shapes, wall thickness and tolerances. Techniques such as filament winding, resin transfer molding, and ply lay-up followed by autoclaving can be used to fabricate a variety of shapes and sizes of components. When the CMC process is completed, net shape parts can result that require a minimum of final machining. When final machining is required, conventional diamond tooling as well as waterjet, laser and ultrasonic machining can be used.

MECHANICAL PROPERTIES

Properties of Non-Oxide CMCs

The typical mechanical properties of a S200 CMC at 20°C and 1000°C are shown in Table 15.2. There is about a 15% increase in tensile strength at the elevated temperature, which is attributed to the redistribution of residual stresses. Concurrently, there is a decrease in shear strength with temperature. The latter behavior is a matrix-dominated property. In this case, the higher temperature condition results in a decrease in stiffness but an increase in tensile strength. This stress-strain behavior is shown graphically in Figure 15.4 for tensile data at three temperatures.

The thermal conductivity of S200 as a function of temperature is shown in Figure 15.5. This data shows that thermal conductivity is a relatively isotropic property and is only slightly affected by temperature over the range from room temperature to 1200°C. The constant nature of this property in the x, y, and z planes in the CMC occurs because both the fiber and matrix are amorphous and have similar conductivities.

TABLE 15.2 MECHANICAL PROPERTIES OF S200 CMCS

	Tensile Strength MPa (ksi)	Shear Strength MPa (ksi)	Flexure Strength MPa (ksi)	Compressive Strength MPa (ksi)
20°C	262 (38)	38 (5.5)	379 (55)	434 (63)
1000°C	300 (43)	24 (3.5)	414 (60)	NA
	Tensile Modulus MPa (ksi)	**Proportional Limit MPa (ksi)**	**Proportional Strain %**	**Strain-to-Failure %**
20°C	96 (14)	96 (14)	0.09	0.5
1000°C	90 (13)	90 (13)	0.08	0.7

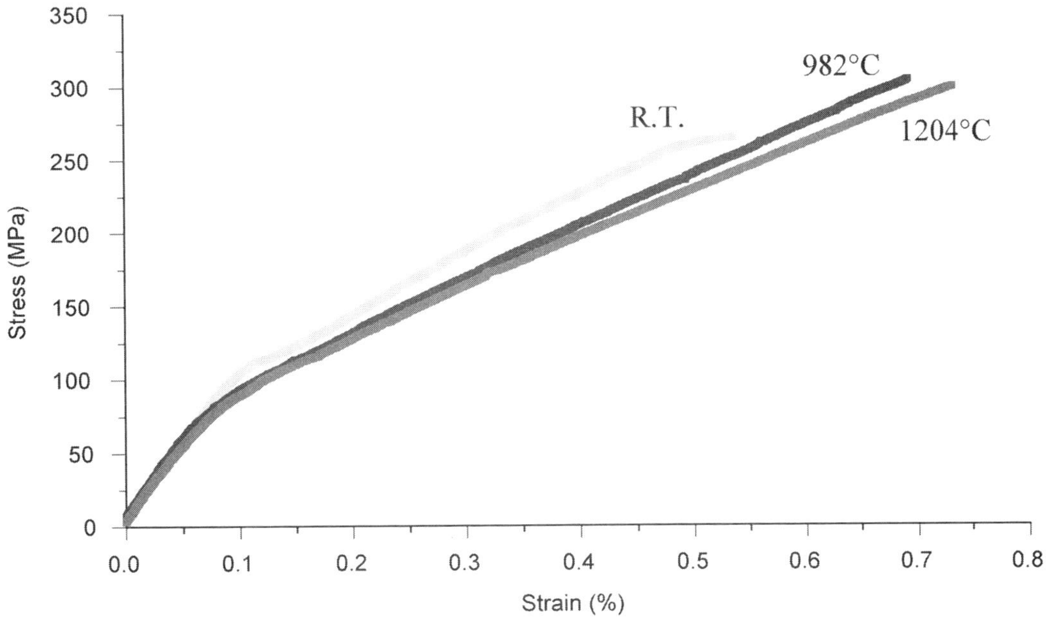

Figure 15.4 Tensile Stress-Strain Behavior vs. Temperature for an S200 CMC

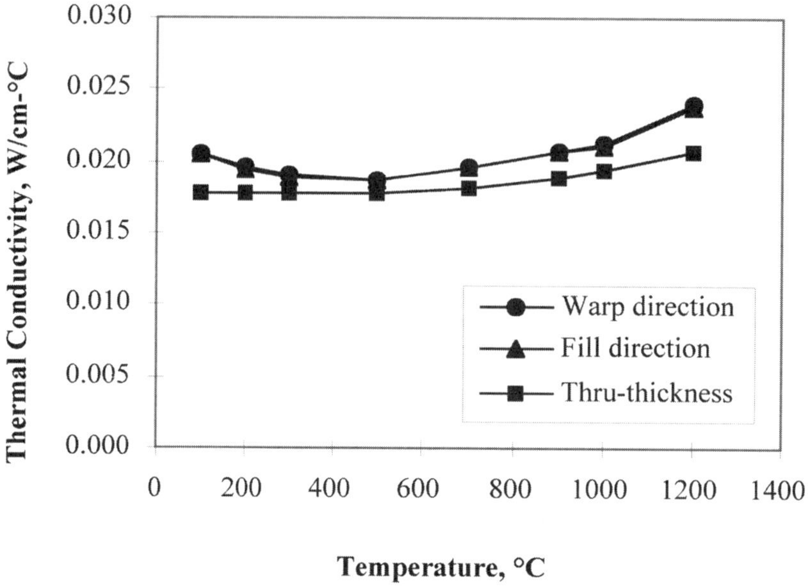

Figure 15.5 Thermal Conductivity of S200 as a Function of Temperature (Reprinted from ref. [3])

Properties of Oxide CMCs

Mechanical testing has been an ongoing part of the maturation process for this CMC system. Table 15.3 summarizes basic mechanical strength and modulus values compiled to date for the Nextel 720 aluminosilicate system.

Creep and Fatigue Properties. Resistance to creep rupture and fatigue is an important material characteristic for both military and commercial gas turbine engines. The Air Force Research Laboratory (AFRL) Materials and Manufacturing Directorate, Wright-Patterson AFB, Ohio, USA, has characterized the Nextel 720 alumino-silicate CMC at 1000°C and 1100°C [8]. The results are summarized in Table 15.4. The residual strengths after fatigue are equal to, or greater than, those of the as-fabricated material, suggesting that repeated loading may have a work-hardening/toughening effect. The creep rupture tests yield residual strengths of 209 MPa after 100 hours at a maximum stress of 150 MPa.

TABLE 15.3 TYPICAL PROPERTIES OF NEXTEL 720 REINFORCED ALUMINOSILICATE CMC

	Temperature	Strength (MPa)	Modulus (GPa)	Strain (%)
Tension	RT	179	76.5	0.30
Tension (after 100 hours at 1000°C)	RT	187	82.1	0.29
Tension [7]	1000°C	170	73.1	0.27
Tension (after 100 hours at 1000°C)	1000°C	176	78.6	0.28
Flexure	RT	216	97.9	0.22
Compression	RT	186	80.0	0.22
In-Plane Shear	RT	31	14	0.49
Interlaminar Shear Strength	RT	11.5	—	—

TABLE 15.4 RESIDUAL TENSILE STRENGTHS AFTER FATIGUE AND CREEP LOADINGS IN AIR

	Test Temperature (°C)	Stress Level (MPa)	Loading Conditions	Residual Strength (MPa)
Fatigue	1000	125	100,000 cycles	193
	1000	150	100,000 cycles	211
	1000	160	100,000 cycles	195
	1100	150	100,000 cycles	168
Creep Rupture	1000	150	100 Hours	209

Environmental Resistance. NAVAIR Materials Division at Patuxent River, Maryland, USA, completed preliminary hot corrosion studies of Nextel 550 and 720 reinforced CMCs [9]. The first test run on the material was a 500 cycle oxidation resistance test to 900°C in air only. Each cycle consisted of 55 minutes at 900°C and 5 minutes at room temperature. The CMC was extremely inert to the oxidation exposure; there was no weight change or visible change in the CMC samples.

The second test run on the material was a hot corrosion thermal exposure. When salt water is ingested into aircraft turbine engines and mixed with sulfur residue, corrosive salts are deposited on nozzle parts, greatly reducing their useful operating life. To model this behavior, Na_2SO_4 is deposited on the CMC coupon surfaces at 5 g/cm^2. The 500 cycle thermal exposure to 900°C is repeated. After completing 500 cycles of cycling, no visible glass formation and no significant weight changes were reported. Initial analysis gave no indication of any sodium silicate layer formation. The surface topography and morphology looked unchanged. The results showed that these materials were well behaved at 900°C and are very promising.

Effects of Long-Term Aging. Siemens Westinghouse Power Corporation has carried out long-term aging studies on the Nextel 720 alumino-silicate CMC system. Samples were exposed for up to 1000 hours at temperatures ranging from 1000°C to 1200°C (Figure 15.6) in air. These data indicate that the properties of the Nextel 720 CMC are stable for very long-term exposures (1000 hours) up to 1000°C, then degrade rapidly beginning at 1100°C. Along with this degradation in strength with temperature exposure above 1100°C, there is a transition from a fibrous to brittle failure (Figure 15.7). While the mechanism for the strength/microstructural degradation has not been fully characterized, it is believed that the primary loss in strength is a result of additional matrix densification rather than fiber degradation. As the matrix is further densified (through sintering) it shrinks, becomes stronger, and bonds more extensively to the reinforcement fibers thereby degrading the toughness mechanism.

APPLICATION EXPERIENCE

A number of prototype components were fabricated for evaluation in both aerospace and industrial applications and some were subsequently tested in simulated application environments.

Figure 15.8 illustrates a number of CMC exhaust flaps for a General Electric F-110 aircraft engine. These flaps (approx. 15 cm × 60 cm) are being evaluated as replacements for metal flaps that have limited life capability under the severe thermal and oxidative environment of the engine exhaust [10]. Figure 15.8 shows S200 flaps, as well as similar flaps fabricated by General Electric, HITCO, and DuPont Lanxide. These flaps were tested in a GE F110-129 engine for up to 120 hours. The HITCO and S200 flaps survived well, whereas the General Electric and DuPont Lanxide flaps exhibited some cracking and wear.

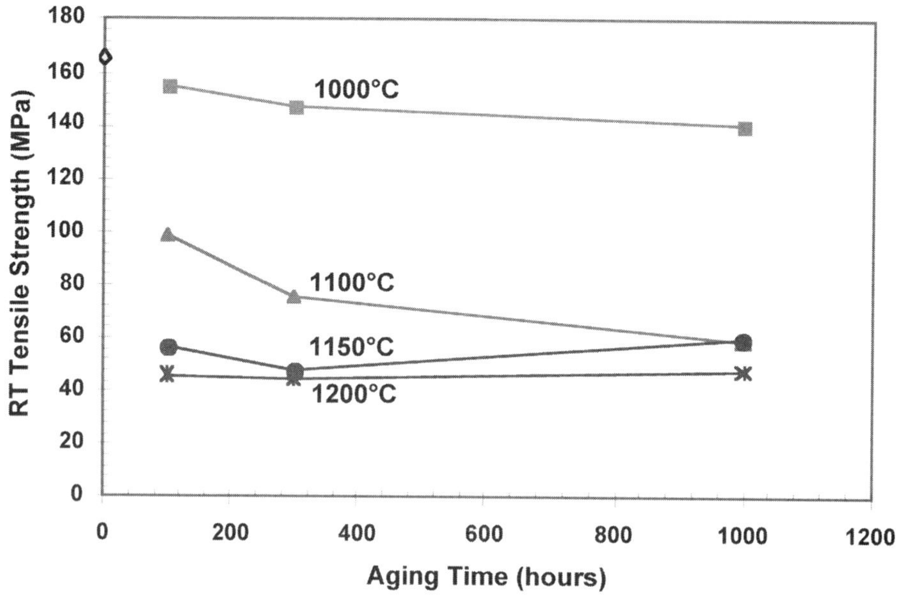

Figure 15.6 Tensile Strength Retention after 1000 Hours of Exposure up to 1200°C (Reprinted from ref. [2])

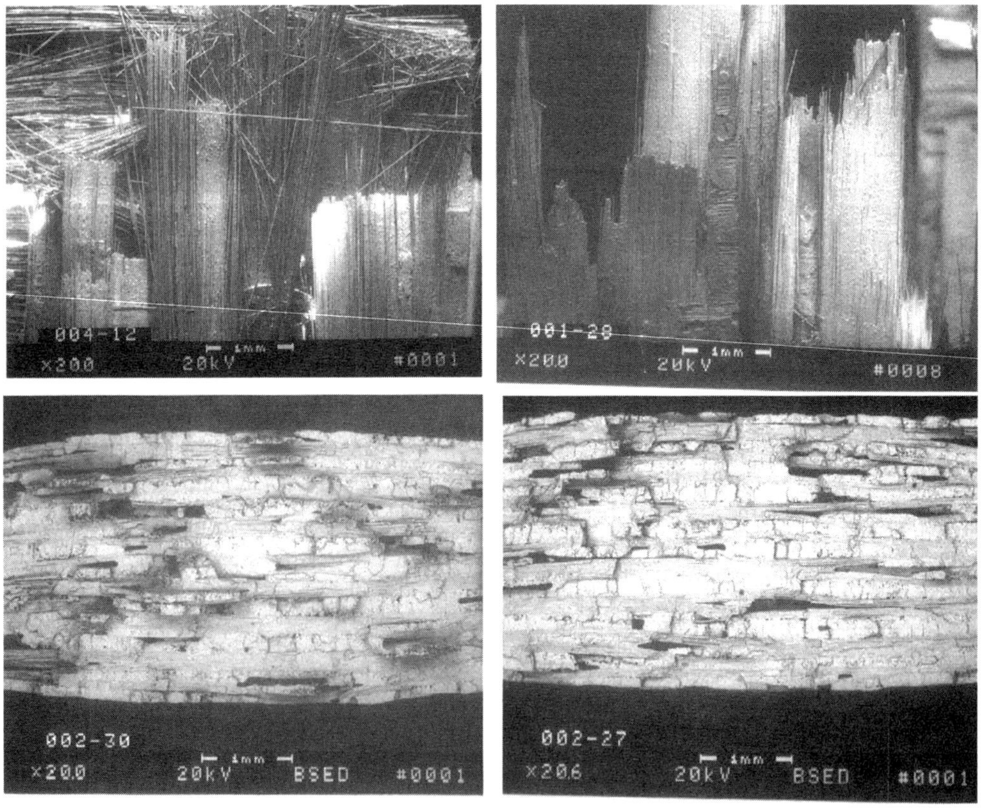

Figure 15.7 Fracture Surfaces as Fabricated (top-left), and After 1000 Hour Exposure to 1000°C (Top Right), 1100°C (Bottom Left) and 1200°C (Bottom Right) (Reprinted from ref. [2])

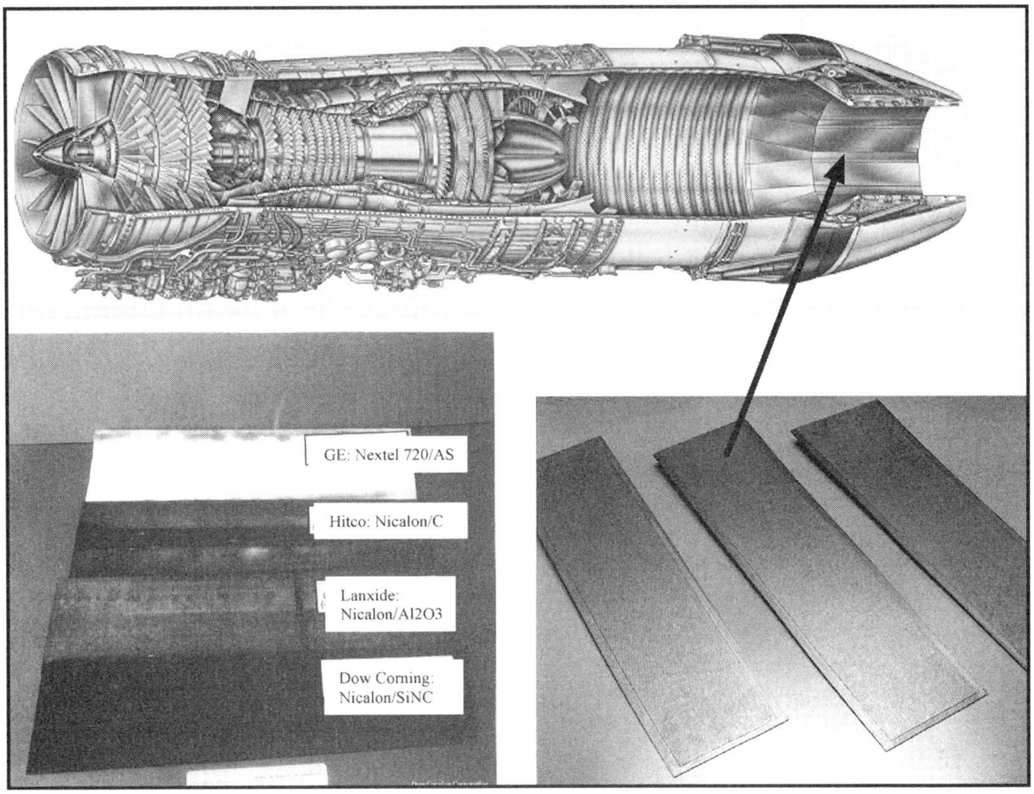

Figure 15.8 Exhaust Flaps for General Electric F-110 Aircraft Engine (Reprinted from ref. [3])

Figure 15.9 shows an exhaust panel for the proposed NASA High-Speed Civil Transport (HSCT) aircraft [11]. These S200 CMC tiles were to line the exhaust nozzle to provide high temperature structural stability to the acoustic damping system. The tile shown in Figure 15.9 has six fastener locations as well as a large number of perforations that permit the system to attenuate the acoustic energy. A four panel array was successfully tested in a hot acoustic rig at the Air Force Research Laboratory [11].

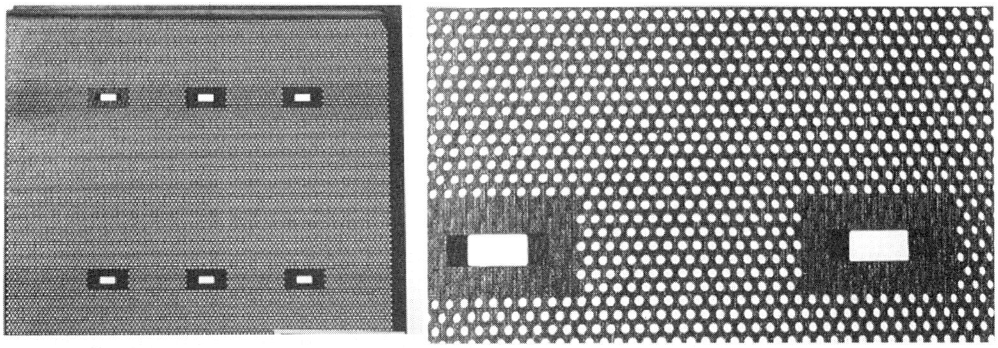

Figure 15.9 NASA-HSCT (High-Speed Civil Transport) Exhaust Panels (Reprinted from ref. [3])

Prototype sub-scale components have been fabricated for land-based gas turbines also. Figure 15.10 shows a cross section of a typical land-based gas turbine and the location of components that have been evaluated. These components were selected because they represent the need in the turbine for higher temperature capability and more durable materials. Use of these materials will enable significantly improved efficiency and reduced emissions for these turbines [12].

Both S200 and S300 CMC sub-scale combustor liners (~20 cm dia. × 20 cm long × 0.25 cm thick) have been fabricated and tested for up to 100 hours in the Solar Turbines subscale combustor test rig. The S200 liner was found by non-destructive evaluation to have sustained some damage [13]. However, Figure 15.11 shows the S300 sub-scale outer combustor liner that was evaluated in a 100-hour combustor rig by Solar Turbines Incorporated. Non-destructive evaluation of this liner indicated that the liner was defect free after the

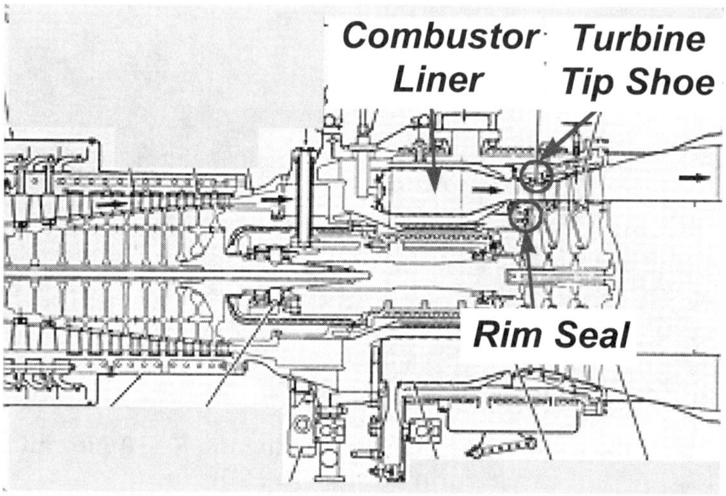

Figure 15.10 Location of CMC Components for Land-Based Gas Turbines (Adapted with permission from Solar Turbines Incorporated)

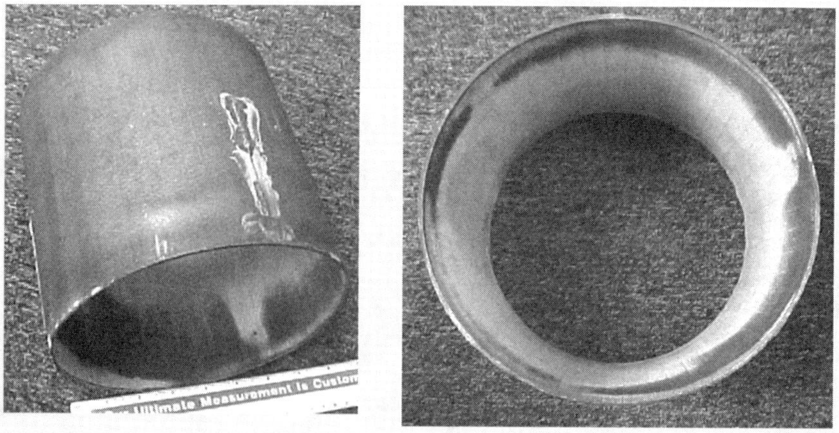

Figure 15.11 Post 100-Hour Rig-Tested S300 Sub-Scale Combustor Liner (Reprinted from ref. [13])

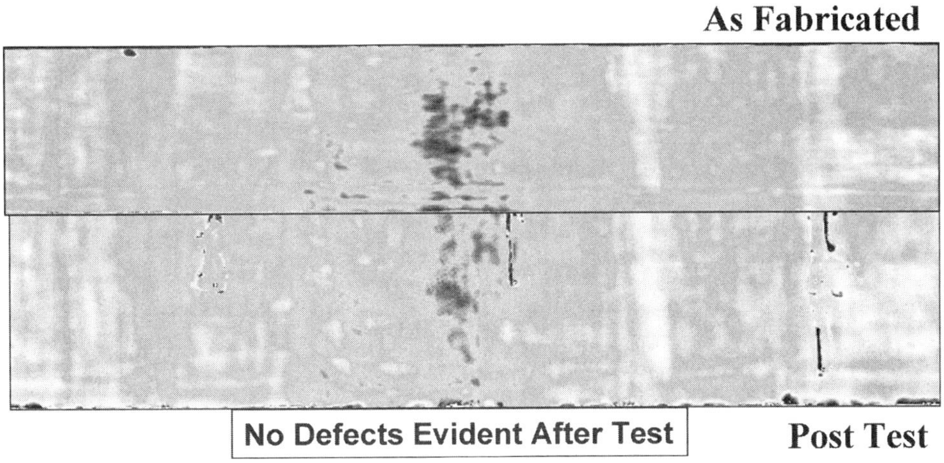

Figure 15.12 Thermal Diffusivity Maps for As-Fabricated and 100-Hour Rig Tested S300 CMC (Reprinted with permission from Argonne National Laboratory)

100-hour rig test (Figure 15.12). Note that the variations in gray scale in these NDE thermal diffusivity image data sets are not defects but rather represent reasonable variations in the materials. Note that the lower image also shows the thermocouples that were added after the upper image data were taken. These are not to be confused with defects. The sub-scale combustor liner tests also indicated that these materials were susceptible to rapid oxidation and surface recession in high temperature, high pressure steam environments. Subsequent evaluation of environmental barrier coatings (EBC) for these materials, such as the United Technology Research Center's multilayer coating based upon BSAS and mullite, and our own NZP-based coating, in the Oak Ridge National Laboratory's "Keiser Rig" has shown that significant protection of both S200 and S300 is obtained for up to 1500 hours of exposure. NZP is a low thermal expansion material based upon sodium zirconium phosphate chemistry where ionic substitution can allow tailoring of the thermo-physical properties.

Figure 15.13 shows the process steps involved in the fabrication of a sub-scale combustor liner using the oxide CMC process. Figure 15.14 shows a prototype sub-scale rim seal and turbine tip shoe that were fabricated (both approx. 20 cm diameter). These S200 prototypes were made to demonstrate the ability to fabricate thick-section (~1.0 cm) rings using the PIP process. Both of these components have a need for surface abradability to allow rotating knife-edges to cut into the surface and create a seal.

CONCLUSIONS

Both non-oxide and oxide CMCs are materials that are viable for gas turbine applications. The fabrication processes for both of these types of ceramic composite are versatile methods for producing both simple and complex net-shaped components. These CMCs have excellent mechanical properties at both room and elevated temperatures. A number of prototype components have been fabricated and validated as both full scale and sub-scale components in gas turbines. These materials will continue to be evaluated in both aerospace and industrial applications.

ACKNOWLEDGEMENTS

Support for some of the work detailed above came from the Air Force Research Laboratory under SBIR Contract F33615-98-C-5210, Dow Corning and the Department of Energy, Office of Industrial Technologies

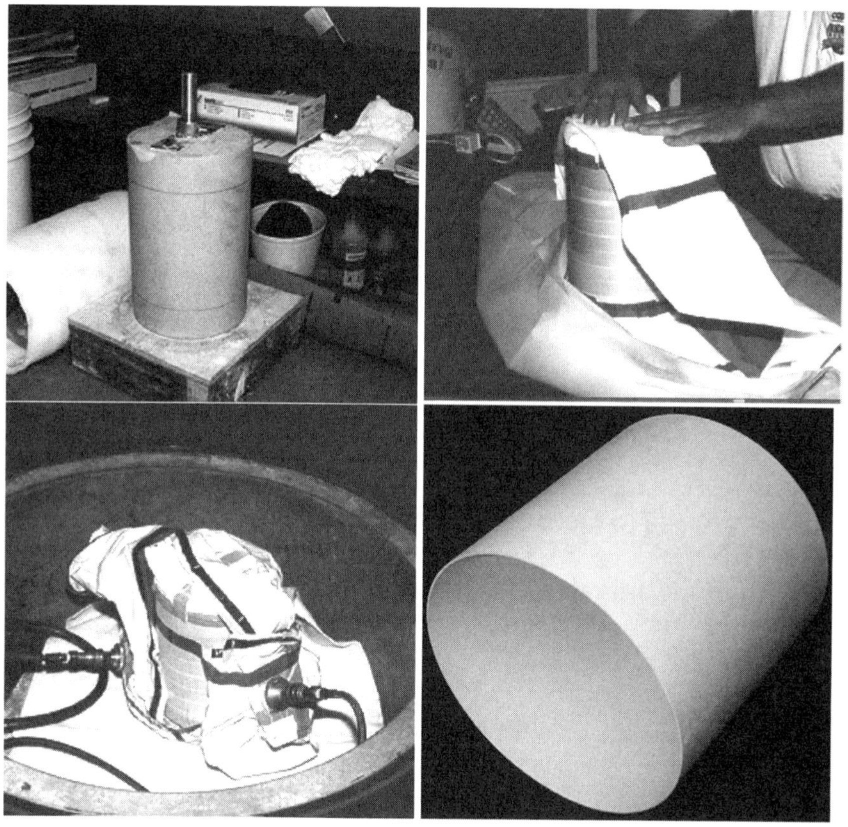

Figure 15.13 Tooling (Top-Left), Ply Lay-Up (Top-Right), Bagging Process (Bottom Left) And Finished Oxide CMC Part (Bottom Right) (Reprinted from ref. [2])

(a) (b)

Figure 15.14 Fabricated Sub-Scale S200 Prototype Components; (a) Rim Seal, (b) Tip Shoe

under Cooperative Agreement No. DE-FC02-92CE4093, and NASA. Unless otherwise indicated figures and tables were supplied courtesy of COI Ceramics.

REFERENCES

[1] Simpson, J. F., Parthasarathy V., and Fahme, A., "Testing of Monolithic Ceramics and Fiber Reinforced Ceramic Composites for Gas Turbine Combustors," Ceram. Eng. and Sci. Proc., 18[4], 229–238, 1997.

[2] Jurf, R. A and Butner, S.C., "Advances in Oxide-Oxide CMCs," ASME paper 99-GT-190, presented at the International Gas Turbine and Aeroengine Congress, Exposition, and Users Symposium, Indianapolis, IN, USA, June 7–10, 1999.

[3] Szweda, A., Easler, T.E., Petrak, D.R., and Black, V.A., "Continuous Fiber Ceramic Matrix Composites for Gas Turbine Applications," ASME paper 99-GT-189, presented at the International Gas Turbine and Aeroengine Congress, Exposition, and Users Symposium, Indianapolis, IN, USA, June 7–10, 1999.

[4] Tu, W.C., Lange F. F., and Evans A., "Concept for a Damage-Tolerant Ceramic Composite with "Strong" Interfaces," J. Am. Ceram.Soc., 79[2], 417–424, 1996.

[5] Cazzato A., Colby M., Daws D., Davis J., Morgan P, Porter J., Butner S. and Jurf R., 1997, "Monazite Interface Coatings in Polymer and Sol-Gel Derived Ceramic Matrix Composites," Ceram. Eng. and Sci. Proc., 18[3], 269–278, 1997.

[6] 3M Ceramic Fiber Technical Data Brochure, 1998.

[7] Personal Communication with J. French, Boeing, St. Louis, MO, USA.

[8] Personal Communication with L. Zawada, AFRL, Dayton, OH, USA.

[9] Personal Communication with R. Kowalik, NAVAIR, Patuxent River, MD, USA.

[10] Staehler, J.M., and Zawada, L.P., "Performance of Four Ceramic-Matrix Composite Divergent Flap Inserts Following Ground Testing on an F110 Turbofan Engine," J. Am. Ceram.Soc., 83 [7], 1727–1738, 2000.

[11] Pacia, A.B., Crumbacher, R., Wentz, K.R., Camden, M.P., "Combined Thermal/Acoustic Testing of Ceramic Matrix Composite Engine Nozzle Exhaust Components," *Electronic Proceedings of the 69th Shock & Vibration Symposium,* Oct 12–16, 1998, The Shock & Vibration Information Analysis Center, Minneapolis/St. Paul, MN, USA, 1999.

[12] Price, J.R., "Ceramic Stationary Gas Turbine," *Proceedings of the 1997 Advanced Turbine Systems Annual Program Review Meeting,* Abbie Layne, Patricia Hoffman, Conference Technical Coords, DOE/FETC-98/1057, Morgantown, WV, USA, October 28–29, 1997. Section 5.4.

[13] Szweda, A., and Parthasarathy, V., "Liner Rig Test," U.S. Dept. of Energy CFCC News, 9, 8–11, Aug. 1997.

CHAPTER

16

SILICON MELT INFILTRATED CERAMIC COMPOSITES—PROCESSES AND PROPERTIES

Gregory S. Corman, Krishan L. Luthra and Milivoj K. Brun

GE Global Research Center
Niskayuna, New York, USA

INTRODUCTION

GE Global Research Center (GEGRC) (formerly Corporate Research & Development—CRD) is one of the world's largest and most diversified industrial laboratories. GEGRC performs research and development efforts in support of all GE businesses, including GE Power Systems (GEPS) and GE Aircraft Engines (GEAE). Development of advanced ceramic materials for potential gas turbine engine applications has been pursued at GEGRC for over 30 years. Efforts in the early- to mid-1970s focused on "pressureless" fabrication of engineering ceramics. Monolithic SiC, fabricated by either reaction bonding, i.e. the "Silcomp" process [1], or sintering [2], and monolithic sintered Si_3N_4 [3] materials were successfully developed. Although these materials showed excellent engineering properties for their day, they did not find use in gas turbines because of concerns over brittle fracture and reliability. In the mid-1980s the focus of material development switched to fiber reinforced ceramic matrix composites (CMCs) in order to deal with these concerns. Building on prior experience with monolithic Silcomp, the first melt infiltrated ceramic matrix composite (MI-CMC) was invented at GEGRC. Extensive development of this material through the 1990s has lead to successful rig and engine testing of MI-CMC shroud and combustor liner components. This chapter reviews the GE efforts on melt infiltrated ceramics, particularly the development and evolution of MI-CMCs at GE.

"MONOLITHIC" SiC-Si CERAMICS

The fabrication of SiC-Si ceramic bodies via silicon infiltration into porous, carbon-containing preforms is a technology that is now over 40 years old. Such infiltrations can be done using silicon liquid [4] or vapor [5]. Such materials are commonly collectively referred to as "reaction bonded SiC", and the processing, microstructures and properties of these materials have been described [6–11].

During the early to mid 1970s, researchers at GEGRC developed a type of reaction bonded SiC material referred to as Silcomp™. This material is made by infiltrating molten silicon into a porous preform containing carbon in a fibrous form. The infiltrating silicon reacts with the carbon in the preform to form a fully dense body of silicon carbide and silicon. Details of the fabrication, properties and potential applications for monolithic Silcomp have been discussed previously [1,12–15].

The fabrication of monolithic Silcomp involves two major steps: fabrication of a porous, carbon-containing preform, and melt infiltration with silicon or a silicon alloy. Preforms utilizing continuous carbon fibers can be prepared using techniques analogous to those for fabrication of polymeric composites. Preforms made from chopped carbon fiber and other particulates can be processed using conventional powder processing techniques. This flexibility in the fabrication of the preform enables one to control the content and architecture of carbon in the preform, which in turn controls the microstructure and properties of the final infiltrated material. For instance, the amount of SiC in the infiltrated body can be controlled over a range of 25% to 90% by tailoring the microstructure of the preform. During infiltration, the porous preform is heated in vacuum to slightly above the melting point of silicon (1410°C). Molten silicon is then introduced, which readily wets the carbonaceous preform and is pulled into the porous body by capillary pressure. Infiltration of typical preforms, several centimeters in size, is complete within a few minutes. The molten silicon also reacts with the carbon, forming continuous, interpenetrating networks of SiC and Si. Overall, the infiltration process is net shape (it causes no changes in preform shape or size), and results in a fully dense body.

The silicon-carbon reaction during infiltration consumes any unprotected carbon fiber. Silcomp is thus not a reinforced composite in the traditional sense but rather behaves as a typical brittle ceramic. Nevertheless, it has many highly desirable engineering properties, such as high thermal conductivity, thermal shock resistance and oxidation resistance. For instance, the bend strength of Silcomp is relatively independent of temperature up to about 1200°C, and then drops off gradually up to the melting point of silicon (1410°C). However, even above that temperature Silcomp still retains roughly 30% of it room temperature strength, as shown in Figure 16.1.

The flexibility of the fabrication process has allowed the manufacture of various pieces of hardware for turbine engine evaluations, some of which are shown in Figure 16.2. Despite outstanding performance in

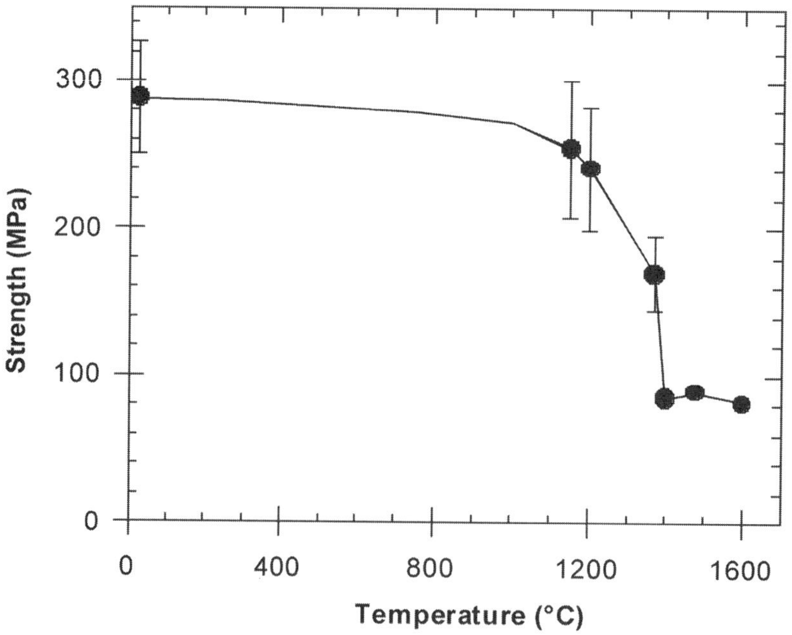

Figure 16.1 Bend Strength as a Function of Testing Temperature for Monolithic Silcomp Showing Roughly 30% Strength Retention Above the Melting Point of the Silicon Phase

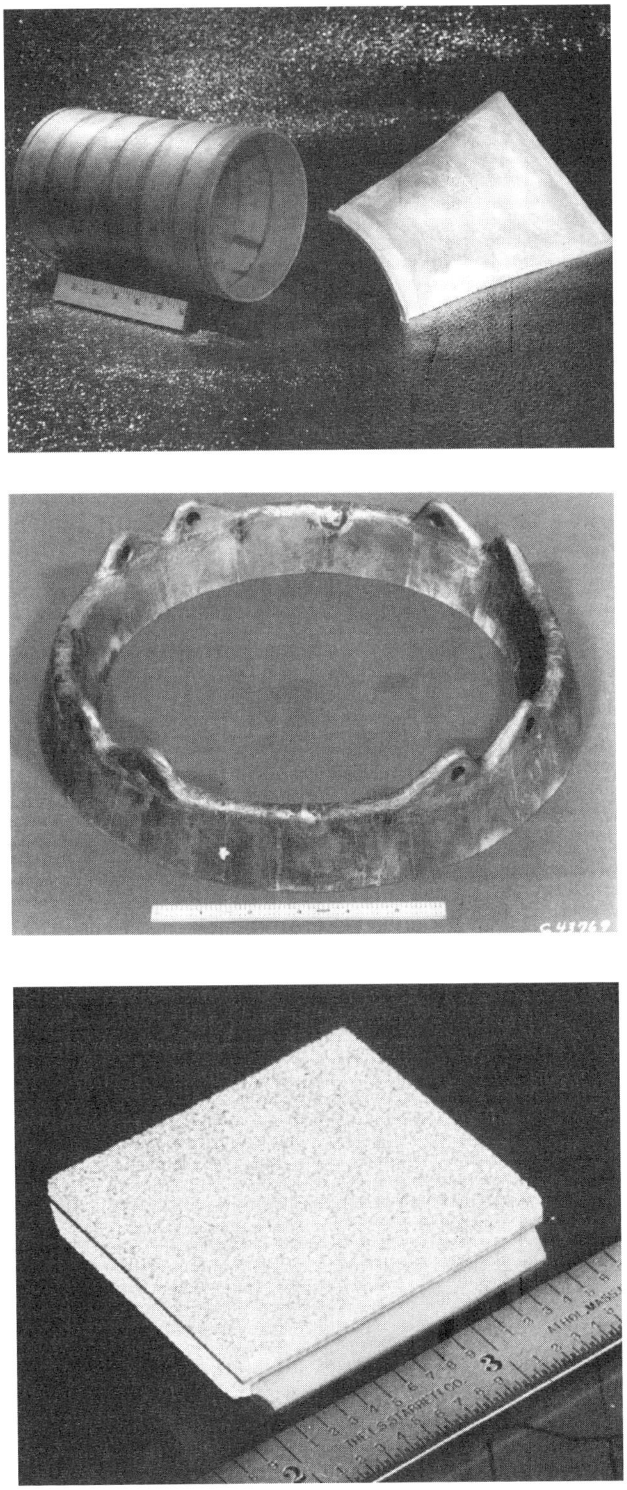

A

B

C

Figure 16.2 Examples of Gas Turbine Hardware Made from Monolithic Silcomp™. A: Industrial Turbine Segmented Combustor Liner and Section of Transition Piece. B: Jet Engine Afterburner Flame Holder. C: Industrial Turbine Shroud Segment these evaluations, Silcomp has not been utilized commercially as a high temperature structural material because of the reliability questions inherent with any brittle material.

Melt Infiltrated Ceramic Matrix Composites (MI-CMCs)

Many processes have been developed for making ceramic-matrix composites. Conventional ceramic-processing techniques, such as sintering, introduce a dimensional change during consolidation. Densification of the matrix around already dense fibers introduces matrix cracks. Although this problem can be alleviated with pressure-assisted sintering (hot pressing or hot isostatic pressing), the shape and size limitations associated with pressure-assisted processes are significant.

Most methods of consolidating ceramic composites use externally supplied matrix materials to fill a pre-shaped, porous structure containing the reinforcing fibers, or preform. These methods include gas transport [16] (i.e., chemical vapor infiltration, or CVI), simultaneous oxidation and capillary transport of molten metal leading to the formation of an extended oxide matrix [17] (Lanxide Dimox™ process), pressure transfer molding using molten glass [18], reactions of solids and gases (e.g., reaction-bonded silicon nitride) [19,20], infiltration with molten matrix phase [21] or infiltration with pre-ceramic resins or polymers followed by pyrolysis [22]. The silicon melt infiltration (MI) process uses molten silicon to infiltrate a porous preform containing C and/or SiC. Any carbon present reacts with the silicon melt to form additional SiC, and residual silicon fills any remaining porosity. The MI process is net-shape, i.e. there are no dimensional changes during infiltration, and, unlike many of the other composite fabrication processes, results in a fully dense, crack-free matrix. The low residual porosity of the matrix gives MI composites superior mechanical and thermal properties, making them particularly suited to gas turbine engine applications.

For this article, the term "melt infiltrated ceramic matrix composite" (MI-CMC) will refer only to continuous fiber composites whose matrices are formed by molten silicon (or alloy) infiltration into a porous SiC- and/or C-containing preform. Originally such materials were referred to as "Toughened Silcomp" [23], but GE now prefers the more generic MI-CMC designation. GE holds numerous patents on the composition and fabrication of these materials, only a few of which are listed in reference [24]. Such composites can be made from a variety of constituents and processes, as described below.

MI-CMC Constituents

The three major constituents of any continuous fiber ceramic matrix composite are the reinforcing fibers, matrix and fiber-matrix interphase, usually included as a fiber coating. In some cases, an external oxidation protection or thermal barrier coating may also be used.

Fiber. The choice of reinforcement fibers is strongly dependent upon the processing and use temperatures and environments. For gas turbine applications, the payoff in terms of energy savings is significant only if the composites are used for periods of several thousand hours in oxidizing environments at temperatures of over 1100–1200°C. The fiber must have useful strength properties under these conditions. In addition, it must survive the melt infiltration conditions of ~1425°C for 15 minutes to 2 hours.Other desirable characteristics for the fiber include a thermal expansivity close to that of the matrix to prevent harmful residual stresses, high stiffness for efficient load transfer to the fiber from the matrix, low density, high thermal conductivity, and creep resistance. One material that meets these requirements is SiC. It has a thermal expansivity (5×10^{-6}/°C from 0 to 1400°C) very close to that of the SiC-Si MI matrix, a low density (3.2 g/cm^3), high modulus (440 GPa), high thermal conductivity (270 W/m.K at room temperature), and excellent creep resistance in the absence of oxide impurities.

The original development of MI-CMC material (or "Toughened Silcomp" as it was referred to) was done using primarily monofilament SiC fibers, such as Textron's SCS-6 [23,25]. However, problems with environmental stability, shape fabrication capability, and cost have driven the adoption of fine diameter (<15 μm), tow-based SiC fibers. Such fibers with oxygen contents above 5% generally lack sufficient thermal stability to survive composite fabrication or to be useable at 1100°–1200°C for extended periods. Most development effort has therefore focused on low-oxygen, but C-rich, SiC fibers, such as Nippon Carbon's Hi-Nicalon™, UBE's Tyranno™ ZE, or on stoichiometric SiC fibers, such as Hi-Nicalon-S, Tyranno SA or Dow Corning's Sylramic™.

Fiber-Matrix Interface. As with most other ceramic composite systems, a fiber coating is necessary to prevent chemical attack of the fibers during processing and to provide for a weak mechanical interface between the fiber and matrix for enhanced toughness and graceful failure. The coating most widely used is boron nitride (BN) applied by CVD. Unfortunately, BN coatings can be degraded by contact with molten silicon during the melt infiltration process, and consequently an over-coating of SiC or Si_3N_4 is commonly used to protect the BN layer.

Unfortunately, the BN interphase is susceptible to oxidative degradation in the presence of matrix cracks, as may occur from foreign object damage or transient over-stressing. This is particularly true for turbine applications where the oxidation product, boron oxide, is highly volatile in the high water vapor pressure atmospheres present in the combustion gasses. The removal of the B_2O_3 oxidation product by volatilization can lead to continuous interface degradation and embrittlement of the composite large distances from the original damage site. For this reason the currently preferred fiber coating is silicon-doped BN, which forms a combined B_2O_3–SiO_2 oxidation product. The presence of the SiO_2 greatly slows oxide volatilization and thus greatly reduces overall oxidation rates of the fiber coating. The superior oxidation resistance of silicon-doped BN compared to pure BN has been demonstrated for both dry and wet atmospheres, as shown in Figure 16.3.

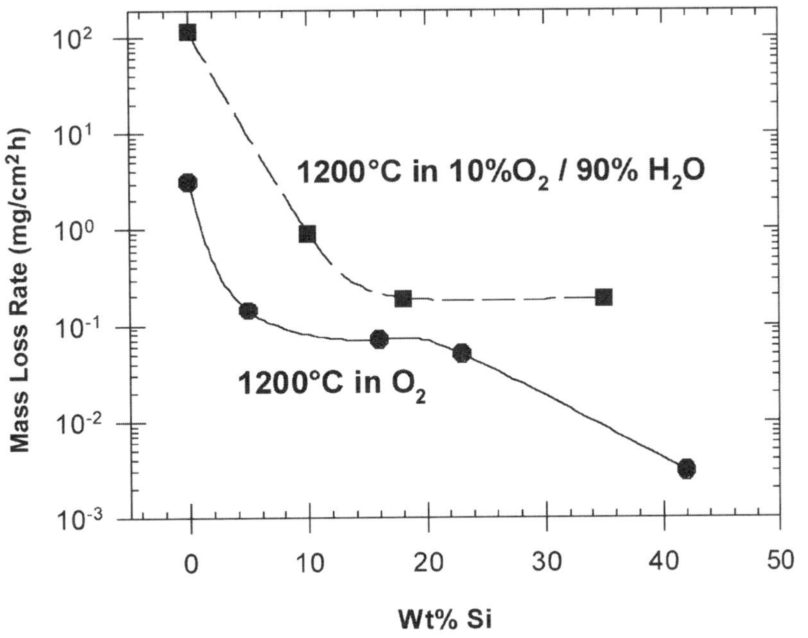

Figure 16.3 Oxidation/Volatilization Rates of Pure and Silicon-Doped BN at 1200°C in Dry (Pure O_2) and wet (10% O_2/90% H_2O) Atmospheres. Additions of 10-15 wt% Si Reduce the Mass Loss Rate by About 2 Orders of Magnitude Compared to Pure BN

Matrix. By definition, the matrices of MI composites all consist of SiC plus Si with occasionally small amounts of residual carbon. The coefficients of thermal expansion (CTEs) of Si and SiC differ only slightly such that relatively wide ranges of matrix composition are possible while still providing an overall match to the CTE of the SiC-based fibers. Typically MI composites have residual silicon levels of 10 to 30 vol%. The details of the microstructure of the matrices depend upon the specific fabrication process used, as discussed below.

Processing of MI-CMCs

The overall preparation of MI composites involves two main process steps: preparation of the porous composite preform and silicon infiltration. Preform fabrication itself typically involves many process steps, many of which are similar to those used for fabrication of polymeric, carbon-carbon or other types of ceramic composites. The most commonly used fabrication processes can be broken down into a combination of steps as shown in Figure 16.4.

The original MI-CMC fabrication process invented by GEGRC utilized individual lengths of coated monofilament fibers that were laid up into a mold and the matrix was slurry cast into the fiber bed (path 1 in Figure 16.4) [23]. Subsequent process scale-up utilized fiber alignment and tape casting [23,25] to form unidirectional prepreg tapes that were then laminated to give controlled fiber architectures (path 2 in Figure 16.4). Later development of continuous fiber coatings for both monofilament [26] and tow-based fibers [27] allowed the use of a wet drum winding process for formation of the unidirectional prepreg tapes (path 3 in Figure 16.4). Because the fibers are coated and impregnated with matrix slurry before preform shaping the matrix slurry has the chance to penetrate all inter-fiber voids. Consequently this "prepreg MI" process (path 3 in Figure 16.4) yields composites with very low (<3%) residual porosity. Typical microstructures of monofilament and fine fiber reinforced composites generated by the drum winding process are shown in Figure 16.5.

GE also played a lead role in the development of the "slurry casting" approach under the NASA-sponsored High Speed Civil Transport (HSCT) program [28], in cooperation with several material vendors. This

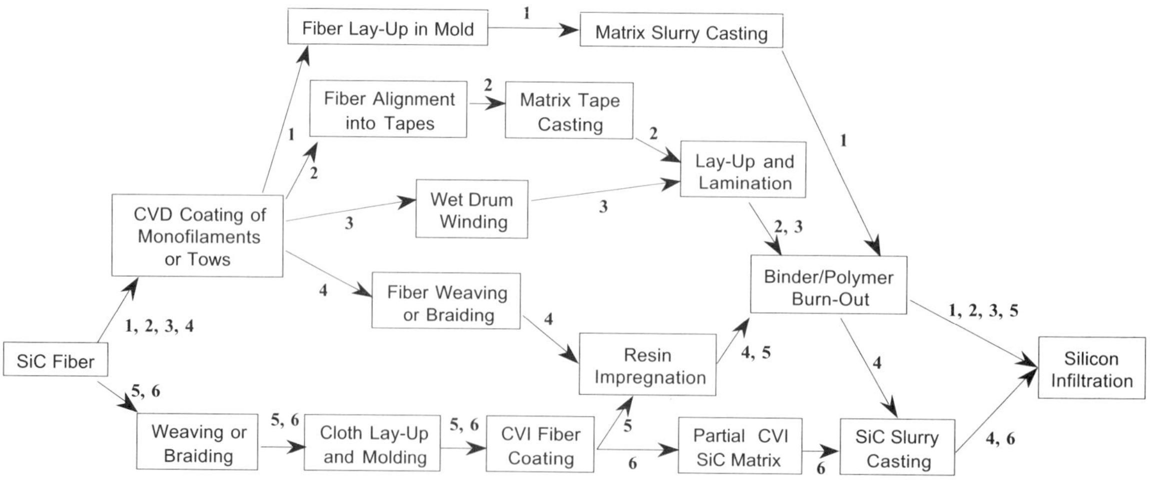

Figure 16.4 Process Diagram Showing the Individual Process Steps Used in the Various MI CMC Fabrication Processes. All Processes Start with the Incoming Continuous SiC Fiber and End with the Silicon Melt Infiltration Step. The Numbers Beside the Arrows Indicate the Paths for the Processes Described in the Text

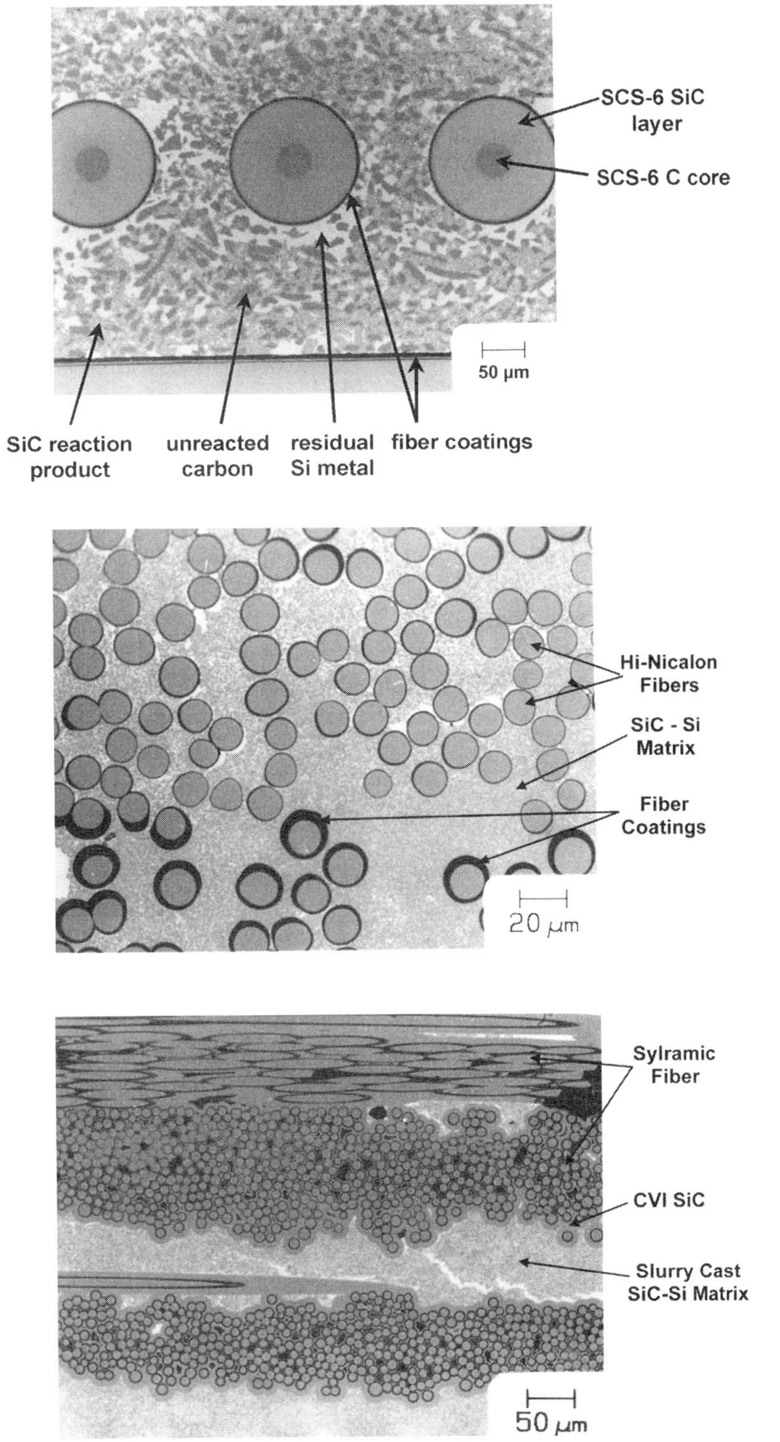

Figure 16.5 Typical Microstructures of Different MI-CMC Materials: Top—Prepreg (Path 3 in Figure 16.4) with SCS-6 Monofilament Fibers and Coarse Matrix. Middle—Prepreg with Hi-Nicalon™ Fiber and Fine Matrix. Bottom—Slurry Cast (Path 6 in Figure 16.4) with Sylramic™ Fiber

approach builds upon the process capability developed for chemical vapor infiltrated (CVI) SiC-SiC composites. The first four steps of the process (path 6 in Figure 16.4) are identical with the starting process steps for the preparation of CVI composites. However, the SiC matrix deposition step is cut short to give a composite preform with roughly 30vol% porosity. Typical ceramic slip casting technology is then used to partially fill the matrix porosity with SiC particulate. The body is then infiltrated with silicon to fill the remaining open porosity in the matrix. While inter-tow porosity is generally filled during the slurry casting and melt infiltration steps, intra-tow porosity that is trapped during the fiber coating and SiC CVI steps still remains, giving the overall composite roughly 6 vol% residual porosity. Although the slurry casting process uses more steps than the CVI SiC-SiC composite process, the simplicity and speed of the slurry casting and melt infiltration steps keep the overall cost of the process comparable to that for CVI composites. A typical microstructure produced by the slurry casting process is also shown in Figure 16.5.

Additional preform processes have been developed for making MI composites. Again as part of the HSCT program, NASA drove the development of process referred to as CRM, which is a combination of CVI, Resin injection molding, and Melt infiltration (path 5 in Figure 16.4) This process begins with fiber weaving, lay-up and coating identical to that for the CVI or slurry cast MI composite processes. Rather than slurry cast particulate into the preform porosity following fiber coating, the CRM process impregnates the preform with a mixture of organic resins [29]. During resin cure and burnout a carbon phase is formed within the preform porosity that has continuous, micron-sized pores. Silicon infiltration into this microporous carbon phase completes the matrix densification.

Another approach, pursued by Toshiba Corp., has also been shown to yield promising composite properties [30]. In this approach (path 4 in Figure 16.4) the reinforcing fiber is coated as individual monofilaments or tows and then woven to form cloths for lay-up. Resin impregnation and pyrolysis steps are used to add carbon char primarily into intra-tow porosity. The inter-tow porosity is then filled with a SiC and C particulate slurry. Final densification and conversion of the C to SiC is accomplished by silicon melt infiltration. As with the prepreg MI approach, this Toshiba approach has the potential to produce composites with very low residual porosity levels; however, weaving or braiding of the coated fiber can lead to physical damage of the fiber coatings.

Currently, the two main MI-CMC fabrication processes being used by GE are the prepreg (path 3 in Figure 16.4) and slurry casting (path 6 in Figure 16.4) approaches, both utilizing fine diameter SiC fibers. Schematic process flow diagrams for these two processes are shown in Figure 16.6. Although

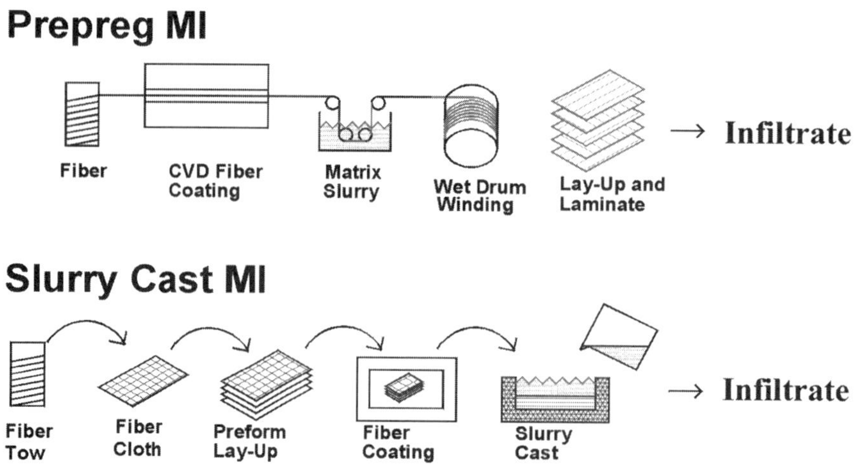

Figure 16.6 Schematic Representations of the Prepreg MI-CMC and Slurry Cast MI-CMC Fabrication Processes Currently Being Optimized by GE

most composite development to date has concentrated on laminated 2D structures, both processes are being modified to allow the use of 3D fiber architectures.

Although the various techniques described above can be used for preform fabrication, the silicon infiltration step is practically identical for all of the processes. In this step the composite preform is simply heated in vacuum above 1410°C while in contact with a source of silicon or silicon alloy. Because silicon readily wets the SiC and C phases present in the preforms, the molten silicon is drawn into the preform porosity by capillary action. Any carbon in the preform reacts with the silicon to form SiC and any remaining porosity is filled with residual silicon metal. Depending on the size of the carbon phase, the composition of the silicon melt, and the reaction times and temperatures used some residual carbon may remain after infiltration.

COMPOSITE PROPERTIES

A large amount of process development and material characterization work has been done on slurry cast MI-CMC under the HSCT program. However, discussions of material properties in this paper will be restricted to measurements made on GE's Continuous Fiber Ceramic Composite (CFCC) program, sponsored by the U.S. Department of Energy, Office of Industrial Technologies and Office of Power Technologies, which is an ongoing program started in 1992.

SiC Monofilament Reinforced Composites

Through the development history of MI-CMCs several combinations of monofilament fiber, fiber coating and fabrication process have been investigated. Some of these combinations are listed in Table 16.1 along with representative tensile strength data at room temperature. Additional details regarding the processing and properties of these composites is available elsewhere [31].

All of the monofilament fibers evaluated were obtained from Textron Specialty Materials (TSM). These fibers are formed by CVD deposition of SiC onto a 33μm diameter, continuous carbon filament. The SCS-6 fiber is a standard product used for reinforcement of metal-matrix and ceramic-matrix composites. It is nominally 142μm in diameter, has a roughly 4 GPa room temperature tensile strength, and consists of several distinct SiC layers as well as a multi-layered, 3μm thick carbon surface coating [32]. The SCS-9 fiber is smaller (79μm diameter) than the SCS-6 fiber, but is similar in structure, including the carbon-rich surface coatings and has similar strength and thermal stability properties. SCS-Ultra is a newer fiber developed by Textron that has higher strength than SCS-6 (roughly 6.9 GPa). The particular fiber used by GE for MI-CMC evaluations (Ultra-First Pass) was pulled from the manufacturing process before the entire outer carbon layers were fully deposited, and thus had only a 0.5 μm thick carbon layer compared to the 3 μm layers on SCS-6 and SCS-9. Attempts were also made to utilize SCS-0 fiber, which has no surface carbon coating at all. Although small quantities of composites were produced for evaluation purposes, the relatively low and highly variable strength of this fiber made it impractical to use for large samples or components.

Most of the fiber coatings used with monofilaments were based on BN; however, because of the carbon surface layers intrinsic to the SCS-6 fiber the deposition of an extra de-bonding layer is not necessary. A protective coating over the intrinsic carbon layers, such as Si_3N_4, is sufficient to give fiber-matrix debonding and tough composite behavior. The source of the fiber coatings listed in Table 16.1 also varied depending on the fabrication process. For composites made by the tape casting prepreg approach short fiber lengths can be used, so that coatings for this approach were done in a batch CVD reactor at GEGRC. The drum winding prepreg approach requires continuous lengths of coated fiber, so that fiber coatings for this process were deposited in a continuous CVD fiber coater at TSM.

Elevated temperature tensile strength behavior was measured for all of the SCS-6 and SCS-9 reinforced composites listed in Table 16.1. The matrix cracking stresses were relatively constant with temperature, as

TABLE 16.1 SEVERAL VARIATIONS OF MONOFILAMENT FIBER-REINFORCED MI-CMCS AND TYPICAL ROOM TEMPERATURE TENSILE STRENGTH BEHAVIOR*

Fiber	Fiber Coating	Process†	Initial Modulus (GPa)	Matrix Cracking Stress (MPa)	Ultimate Strength (MPa)	Strain at Ultimate Strength (%)
SCS-6	BN	Tape	221	97	207	0.80
SCS-6	BN	Drum	200	124	296	0.80
SCS-6	Si_3N_4	Tape	207	145	352	0.75
SCS-6	Si_3N_4	Drum	228	159	393	0.90
SCS-9	BN	Drum		100	317	0.95
SCS-UFP‡	BN	Tape	241	131	179	0.50
SCS-0	BN	Tape		96	152	0.25

*All composites are 0–90° crossply lay-ups with nominally 20–23 vol% fiber unless otherwise noted

†"Tape" refers to a tape casting prepreg approach (path 2 in Figure 16.4) and "Drum" refers to a wet drum winding prepreg approach (path 3 in Figure 16.4).

‡Composites made with SCS—"Ultra first pass" fiber contained nominally 13 vol% fiber.

was the composite initial modulus up to about 1100°C. Above 1100°C the modulus tends to drop off, which is believed to be caused by increasing plasticity in the residual silicon phase.

The ultimate tensile strengths decreased with testing temperature above 800°C, closely following the strength behavior of the reinforcing fiber. Normalized ultimate strength values (ultimate strength at test temperature divided by average ultimate strength at room temperature) are plotted as a function of test temperature in Figure 16.7. The composite strength decrease with temperature follows the fiber strength behavior very well up to 1200°C, as would be expected for a fiber-dominated property like ultimate strength. Above 1200°C the composite strengths are actually better than the measured fiber strengths, suggesting that surface oxidation damage during the fiber tests was contributing to the observed loss in fiber strength. Since the fibers in the composites are protected from such oxidation damage they would not be subject to such oxidative degradation. Strain at ultimate strength values showed a similar decrease with test temperature as did the ultimate strength, with the strain values at 1200°C being roughly 40% of the room temperature values.

One of the main advantages of MI-CMC materials over other ceramic composites is their relatively high thermal conductivity. Thermal conductivity values for several variations of MI-CMC are compared to those of several typical gas turbine alloys in Figure 16.8. The thermal conductivity of MI-CMCs at their projected use temperature (1100–1200°C) are comparable to those of the metallic alloys at their typical use temperatures (800°–1050°C). Another advantage is relatively high interlaminar properties. For instance, the interlaminar shear strength of monofilament-reinforced MI-CMC is ~65 MPa at room temperature.

There are primarily two main technical limitations with the use of monofilament fibers. The first limitation is that the large fiber diameter and high stiffness of the fiber greatly limit allowable fiber bend radius, thus limiting the component geometry that can be fabricated. To address this problem GE has developed techniques for joining composites in either the preform or finished stage. Subsequent infiltration (or re-infiltration) yields a single piece joined composite structure. Moreover, if the joint geometry is well engineered the resulting joint will display tough fracture behavior. Figure 16.9 shows tensile stress-strain behavior for several joined composite samples demonstrating that tough joints with 70% of the un-joined composite strength can be fabricated.

The second major limitation with current SiC monofilament fibers is the poor oxidation resistance imparted by the carbon-rich surface layers. As long as the composite matrix remains intact and the fiber surfaces are isolated from the external environment the thermal stability of monofilament composites is very good. However, in real-world gas turbine applications there are always unintentional events, such as foreign object damage or unusual engine transients, that will temporarily strain a CMC component above the matrix cracking stress. In such cases the fiber-matrix interface will be exposed to the environment through the

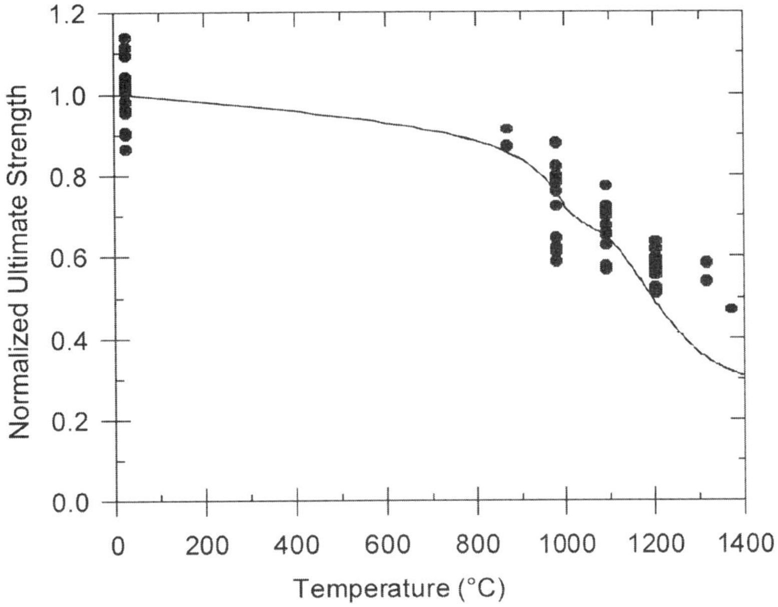

Figure 16.7 Normalized Ultimate Tensile Strength as a Function of Test Temperature for a Variety of SiC Monofilament Reinforced MI-CMC Materials. The line, representing the strength behavior of SCS-6 fiber tested in air (unpublished data from GE Aircraft Engines), is shown for comparison.

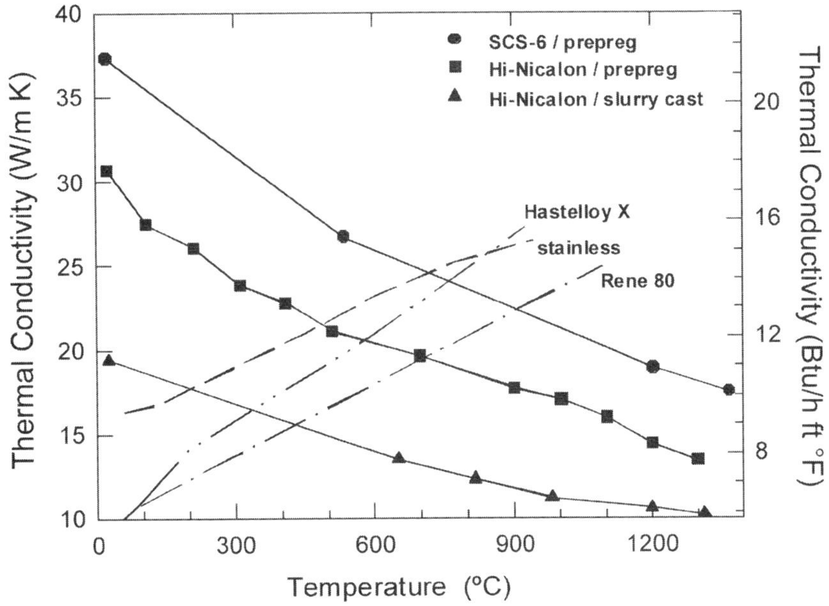

Figure 16.8 Through-Thickness Thermal Conductivity of Various Types of MI-CMC Composites Compared to Typical Gas Turbine Alloys. In-Plane Conductivity of the MI-CMC Materials is Generally 30% Higher Than the Through-Thickness Value

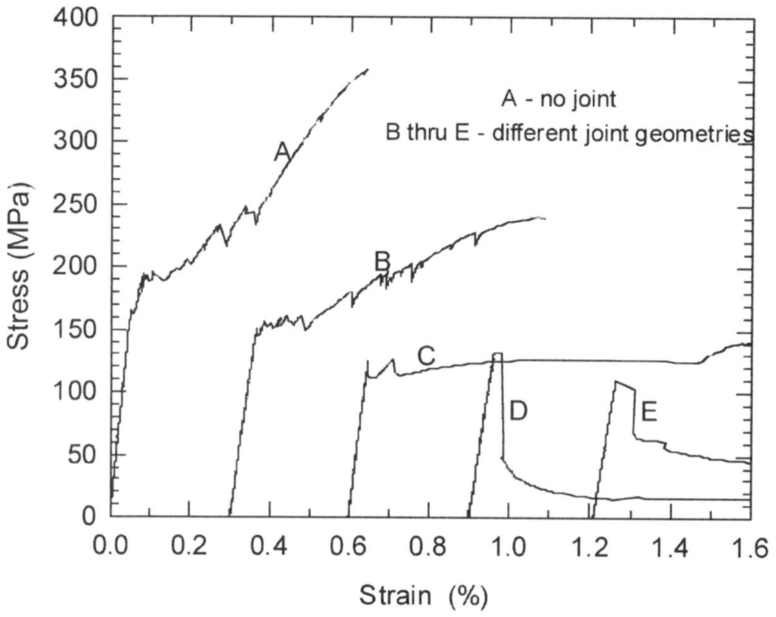

Figure 16.9 Tensile Stress-Strain Behavior of Joined, Monofilament-Reinforced MI-CMC Samples Having Different Joint Geometries. Joint Configuration B Gives a Joined Composite with 70% of the Strength and All of the Strain Capability as an Un-Joined Composite (Curves are offset for clarity.)

matrix cracks, and thus inherent stability of the interface is needed until the matrix cracks can be sealed by oxidation. Unfortunately, the rapid oxidation of the surface carbon layers on current monofilament fibers leads to rapid weakening and composite embrittlement following matrix cracking.

This environmental degradation is demonstrated by the oxidation test data in Table 16.2. Composites reinforced with SCS-6 fiber, with both BN and Si_3N_4 fiber coatings, were exposed in air at 1200°C for 100 hours. Composites exposed in the as-fabricated state (i.e. un-cracked) showed excellent property retention; however, samples that were exposed after cracking the matrix to 0.2% strain were very weak and nearly brittle. This severe property degradation following only 100 hours of interface exposure would be unacceptable for gas turbine applications. Similar degradation was seen for the SCS-UFP fiber samples. Composites with SCS-0 fibers, which do not have a surface carbon layer, should not be susceptible to such degradation; however, SCS-0 reinforced composites could not be readily fabricated because of low, highly variable fiber strengths. Because of these results GE switched emphasis of composite development away from monofilament fibers to using small diameter, multi-filament tow fibers which do not have a carbon surface coating.

TABLE 16.2 OXIDATIVE DEGRADATION OF SCS-6 FIBER-REINFORCED PREPREG MI-CMC AFTER 100 HOURS EXPOSURE AT 1200°C IN AIR

Sample Condition	Matrix Cracking Stress (MPa)	Ultimate Strength (MPa)	Strain at Ultimate Strength (%)
As Fabricated	141	363	0.85
Oxidized Un-cracked	110	460	0.87
Oxidized Pre-cracked	52	70	0.06

Hi-Nicalon™ Reinforced Composites

Representative tensile properties of Hi-Nicalon reinforced prepreg and slurry cast MI-CMC are listed in Table 16.3. The sources of these materials were GEGRC, GE Power Systems Composites (PSC—formerly Honeywell Advanced Composites, Inc.) and Goodrich Corporation (GC—formerly BF Goodrich Aerospace). These data represent average values for 5 or more test samples that were fabricated as witness coupons for a turbine shroud component, and thus represent what one might expect from actual component material. Nevertheless, the properties are from a limited sample and substantial batch-to-batch variation in properties is common. One should note that the prepreg MI-CMC has similar tensile properties to the two slurry cast varieties even with 40% less fiber.

The temperature dependence of the matrix cracking and ultimate strengths and strains are shown graphically in Figure 16.10 for the GEGRC prepreg MI-CMC material. The ultimate strength behavior of slurry cast MI-CMC with Hi-Nicalon fiber is similar, with roughly a 25% drop in ultimate strength at 1200°C compared to room temperature. However, the matrix cracking stress of slurry cast material decreases slightly with increasing temperature in contrast to the prepreg MI-CMC, which shows constant or slightly increasing cracking stress with increasing temperature. Interlaminar shear strengths of both prepreg and slurry cast

TABLE 16.3 REPRESENTATIVE TENSILE STRENGTH PROPERTIES OF HI-NICALON—REINFORCED MI-CMC

Source*	Process	Initial Modulus (GPa)	Matrix Cracking Stress (MPa)	Ultimate Strength (MPa)	Strain at Ultimate Strength (%)
GEGRC	prepreg	235 ± 24	163 ± 22	301 ± 23	1.02 ± 0.14
PSC	slurry cast	163 ± 6	125 ± 10	387 ± 34	0.75 ± 0.09
GC	slurry cast	255 ± 19	157 ± 8	292 ± 21	0.68 ± 0.06

*Sources of material were GE Global Research Center (GEGRC), GE Power Systems Composites (PSC—formerly Honeywell Advanced Composites, Inc.—HACI) and Goodrich Corporation (GC—formerly BF Goodrich Aerospace—BFG). The GEGRC prepreg composite had nominally 20–25 vol% fiber whereas the PSC and GC slurry cast materials had 35–40 vol% fiber.

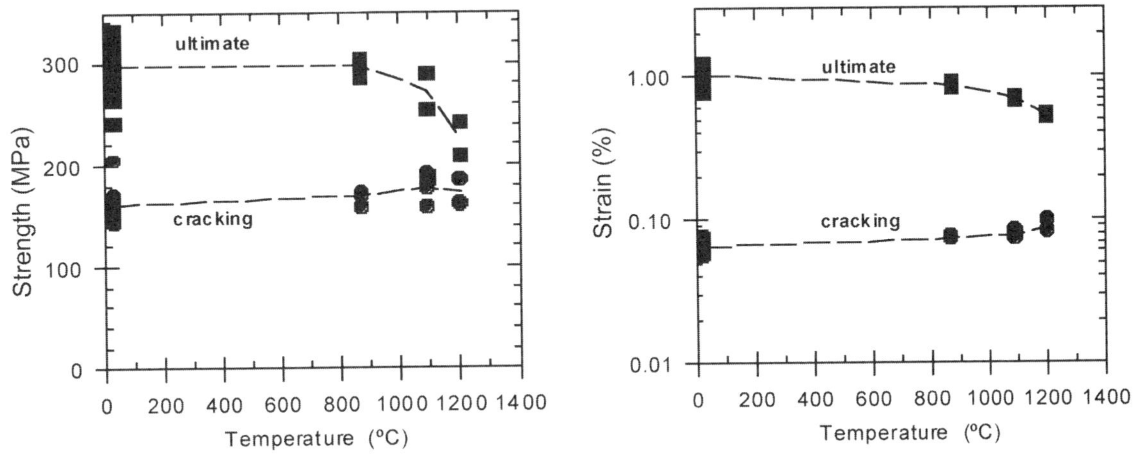

Figure 16.10 Effect of Temperature on the Tensile Fast Fracture Properties of Prepreg MI-CMC Reinforced with Hi-Nicalon Fiber. "Cracking" Refers to the Matrix Cracking Stress and Strain and "Ultimate" Refers to the Failure Strength and Strain

materials with Hi-Nicalon fiber are about 48 MPa at room temperature, with prepreg showing an increase to 66 MPa at 1100°C and slurry cast decreasing to 41 MPa.

Thermal conductivity data for MI-CMC with Hi-Nicalon fiber is included in Figure 16.8. The thermal conductivity of this system is lower than for the SCS-6 reinforced system primarily because of the lower fiber conductivity and higher volume fraction of matrix coatings in the Hi-Nicalon reinforced system. Thermal conductivity for slurry cast MI-CMC with Hi-Nicalon fiber is another 30% lower than for prepreg MI-CMC with Hi-Nicalon. The lower conductivity of the slurry cast material is caused by its higher volume fraction of both fiber and fiber coatings (both Hi-Nicalon and the BN interface coatings have lower thermal conductivity than the SiC-Si matrix) and by its somewhat higher porosity level (roughly 6vol% for slurry cast compared to <3vol% for prepreg).

It should be noted that all three MI-CMCs discussed here are relatively immature materials and that, in GE's experience, each displays substantial batch-to-batch variability in mechanical properties. The data presented here typically represents limited measurements on one to two "batches" of materials fabricated in the late 1997 to early 1998 time frame. Additional property measurements are needed over numerous material "batches" in order to better characterize the true property averages and variability. Also, as each fabricator (GEGRC, PSC and GC) continues to modify their material and/or process one should expect changes in the average properties with time.

The thermal stability of uncracked MI-CMC with Hi-Nicalon fiber and BN fiber coatings is quite good, although the details of the behavior are still being determined. Preliminary data on samples exposed in air at 1200°C for up to 2000 hours show only very minor changes in tensile strength and strain to failure, as shown in Figure 16.11. More recent data suggests a roughly 20% reduction in strength and strain to failure values following 500 hours of exposure at 1200°C in either air or steam, as indicated by the data in Table 16.4. Similar reductions are seen in slurry cast MI-CMC from PSC, and somewhat poorer behavior in slurry cast MI-CMC from GC. No further degradation was observed in prepreg composites up to 2000 hours exposure

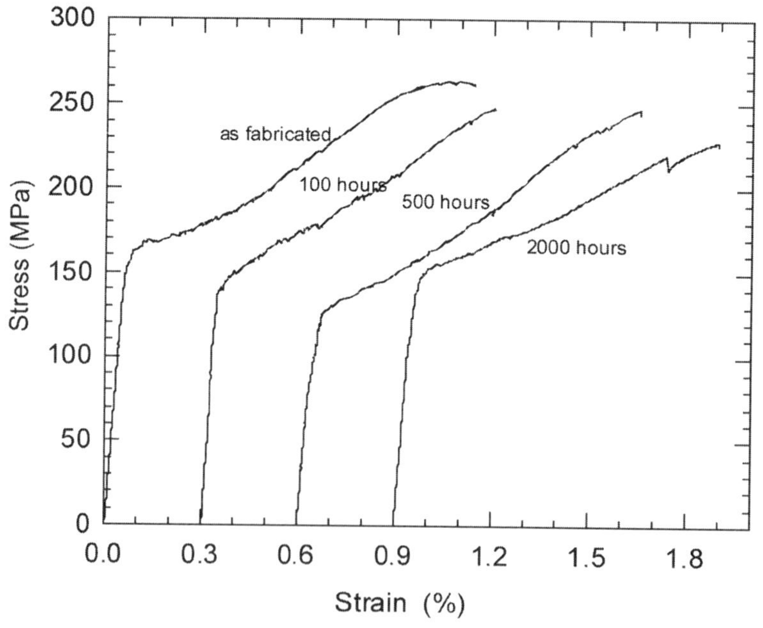

Figure 16.11 Room Temperature Tensile Stress-Strain Behavior of Hi-Nicalon/Prepreg MI-CMC Samples Following Exposure in Air at 1200°C for the Indicated Times (Curves are offset for clarity.)

TABLE 16.4 STRENGTH RETENTION BEHAVIOR OF UNCRACKED MI-CMC FOLLOWING LONG-TERM THERMAL/ENVIRONMENTAL EXPOSURE AT 1200°C

	Environmental Exposure Conditions					
	500 hrs, Air		500 hrs, Steam		2000 hrs, Steam	
Material	%σ	%ϵ	%σ	%ϵ	%σ	%ϵ
GEGRC Prepreg	78	100	79	75	76	74
PSC Slurry Cast	74	83	78	82	54	53
GC Slurry Cast	80	35	67	24	57	25

%σ and %ε values are the percentages of the as-fabricated ultimate strength and strain to failure, respectively, that is retained following the given environmental exposure. Steam environment is 90% H_2O–10% O_2.

whereas the slurry cast materials did show continuing degradation. The increased degradation of the slurry cast material is believed to be caused by the interconnected residual porosity in this system.

Thermal/environmental stability of uncracked composites is important, but it is likely that limited matrix cracking of components will occur in service. The potential for material degradation following matrix cracking is much greater, particularly for high water vapor pressure environments such as gas turbine combustion gasses, where oxidation/volatilization of the fiber coatings can occur. This is shown by the data in Figure 16.12, where 500 hour steam oxidation exposure on pre-cracked samples caused a 20% to 40% loss of ultimate strength in samples with pure BN fiber coatings. The degradation caused by such exposures can

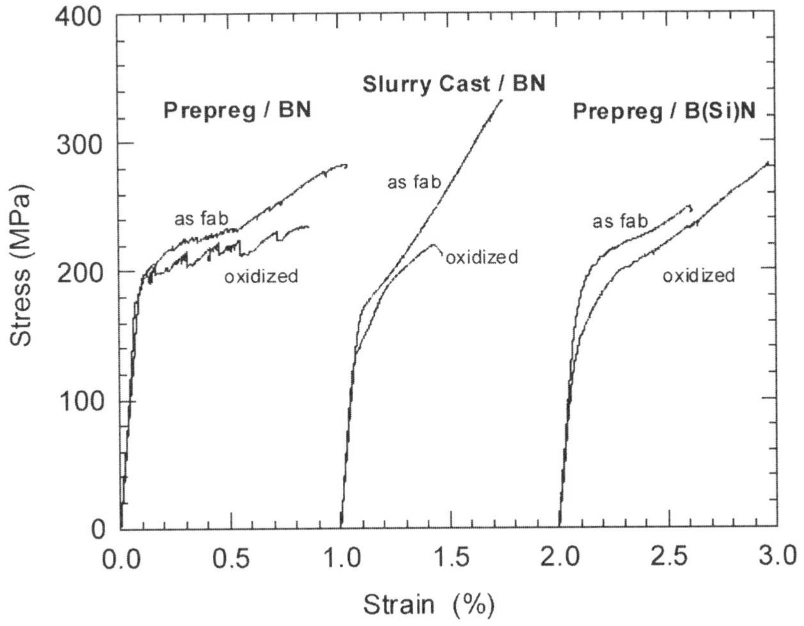

Figure 16.12 Effect of Steam Oxidation (90% H_2O + 10% O_2, 1200°C for 500 Hours) on the Tensile Strength Behavior of Pre-Cracked Hi-Nicalon Reinforced MI-CMC with Pure BN and Silicon-Doped BN Fiber Coatings (Curves are offset for clarity.)

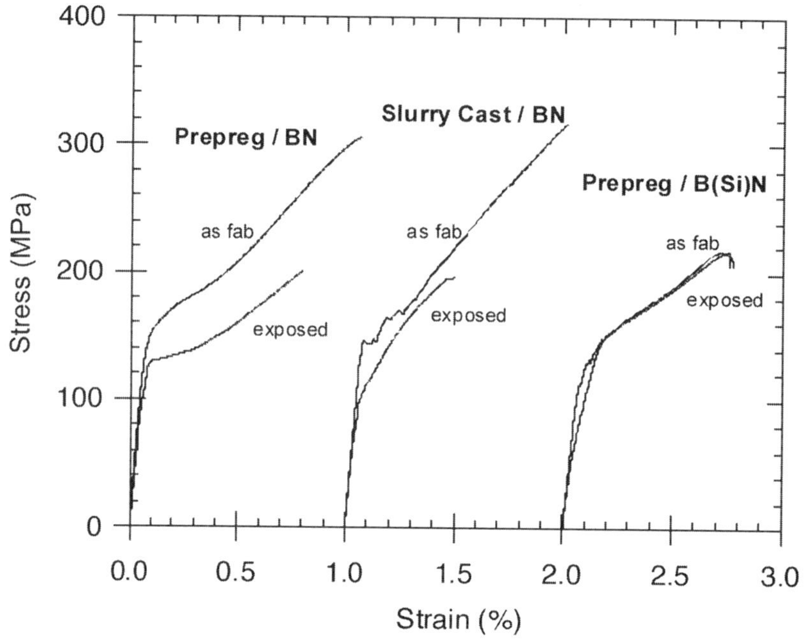

Figure 16.13 Effects of High Pressure Combustion Rig Exposure (100 Hours at Roughly 1300°C and 7 Bars in Natural Gas Combustion Atmosphere) on the Tensile Strength Behavior of Precracked MI-CMC Test Bars (Curves are offset for clarity.)

be largely overcome by the use of more oxidation resistant fiber coatings, such as silicon-doped boron nitride. Although the samples made with silicon-doped BN fiber coatings had somewhat lower as-fabricated strengths than the samples with pure BN coatings, these composites did not show any degradation in ultimate strength or strain following pre-cracked steam oxidation exposures. This same trend was seen with samples exposed in a high pressure combustion rig for 100 hours, as shown by the data in Figure 16.13. GE is continuing to work on the optimization of silicon-doped BN fiber coatings for both prepreg and slurry cast composites. Similarly, long-term, high temperature mechanical behavior, such as low cycle and high cycle fatigue, creep and stress rupture, of these materials is currently being determined.

Stoichiometric SiC Fiber Reinforced Composites

Stoichiometric SiC fibers, such as Sylramic™, Hi-Nicalon S™ and Tyranno SA™ offer several advantages and disadvantages compared to low-oxygen, carbon-rich fibers like Hi-Nicalon™. One of the main advantages is a higher thermal conductivity for the composite [28]. Data from the HSCT program also indicates that MI-CMC with stoichiometric fiber has a higher fatigue limit stress and shows less mechanical degradation during fatigue testing than similar materials with Hi-Nicalon fiber [28].

Disadvantages of the stoichiometric fiber include reduced strain to failure of the composites and increased cost. The reduced strain to failure is partly caused by the higher modulus, yet similar strength, of the stoichiometric fibers, giving a lower fiber strain to failure. However, composite failure strains are significantly lower than the fiber failure strain, typically ranging from 0.3 to 0.4% at room temperature and even lower at high temperature. The full impact of this low strain to failure on the damage tolerance of the composites still needs to be evaluated.

RIG AND ENGINE TESTING OF MI-CMC COMPONENTS

GE has conducted high-pressure combustion rig and engine testing of several MI-CMC turbine shroud and combustor liner components. Rig and engine testing of turbine shroud components are discussed elsewhere [33–35]. Similar rig testing has been performed on sub-scale and full-scale combustor liners made of MI-CMC composites.

A high-pressure combustion test rig was designed and constructed for rig testing of turbine shroud components under realistic engine exposure conditions. The rig was originally assembled using a cooled stainless steel liner. However, during initial shroud cyclic exposure testing at a rig firing temperature of 1200°C the stainless liner was severely damaged and had to be pulled from service after less than 50 hours total exposure at the 1200°C firing temperature. Subsequently two 10 cm diameter by 15 cm long prepreg MI-CMC liners were fabricated and used in the rig. The MI-CMC liners were found to withstand much higher rig firing temperatures (up to 1650°C), longer exposure times (up to 150 hours) and more thermal cycles (over 400) with no observable damage. One of the MI-CMC liners after 150 hours of rig testing is shown in Figure 16.14, and the failed stainless steel combustor liner is also shown for comparison.

Larger combustor liner components, scaled to fit an MS5002B (25MW) engine, have also been tested at GEGRC. The MI-CMC liner sections were full engine size in diameter (28 cm), but were only 25 cm long compared to the full 91 cm metallic liners. This size was selected so that two liner sections, one of prepreg MI-CMC and one of slurry cast MI-CMC, could be rig tested simultaneously. Both liners have undergone 150 hours of exposure testing, both under air cooled and uncooled conditions, with no observable damage. The rig combustion gas temperature for these tests was roughly 1480°C, well in excess of the normal 1065°C

Figure 16.14 Post-Exposure Comparison of Stainless Steel (Left) and MI-CMC (Right) Combustor Liners from the Shroud Test Rig. The intact MI-CMC liner had roughly 3× the exposure time as the failed stainless liner.

A B

Figure 16.15 Photographs of 28 cm Diameter Combustor Liners Following 150 Hours of High Pressure Combustion Rig Testing. A. Prepreg MI-CMC Made by GEGRC. B. Slurry Cast MI-CMC Made by GE PSC (Formerly HACI).

firing temperature of an MS5002B engine. A photograph of the liners following 150 hours of rig testing is shown in Figure 16.15.

Slurry cast MI-CMC materials with Hi-Nicalon fiber is also being used for the inner combustor liner on a Solar Turbines Centaur 50 engine under the Ceramics for Stationary Gas Turbines program [36]. Such liners are currently in engine testing with very promising results. Similar material with Sylramic fiber, being developed primarily for aircraft engine applications, has also undergone combustion rig testing under the HSCT program [28].

REMAINING BARRIERS

While GE has had considerable success to date with rig and engine testing of MI-CMC turbine components, considerable challenges remain before these components can be commercialized. For industrial gas turbine applications some of the most difficult challenges relate to component life. Typical combustor liner and shroud inspection intervals are 8,000 hours and 24,000 hours, respectively, and there is considerable commercial incentive to push these to even longer times. Ensuring component reliability over such long times in high-pressure, high-temperature, high vibration combustion environments is therefore necessary for successful adoption of these materials.

There are two main mechanisms of environmental (oxidative) degradation that threaten to limit the life of the MI-CMC materials. The first is volatilization of the normally protective oxidation product layer. All of the silicon-based non-oxide ceramics (SiC, Si_3N_4, Sialon, $MoSi_2$, etc.) rely on the formation of a stable, low-permeability SiO_2 film to limit material oxidation. In high pressure combustion gasses, which contain a substantial water vapor pressure, the SiO_2 film is found to volatilize from the formation of volatile Si-O-H species. This phenomenon has been described in detail elsewhere in this book [33].

The solution to this silica volatility problem is the use of an Environmental Barrier Coating (EBC). Development of EBCs for MI-CMC was initiated under the HSCT program where several candidate oxide coatings were developed, the most promising of these being an Air Plasma Sprayed (APS) Barium-Strontium Alumino-Silicate (BSAS) top coat with a silicon/mullite bond layer. GE is also investigating alternatives to coating the MI-CMC as part of the DOE-CFCC program.

Another life issue for MI CMCs relates to the oxidation of the fiber coatings following matrix cracking by foreign object damage (FOD) or temporary overstressing. Cracking of the matrix allows for direct contact of the environment with the fiber coatings, and coating oxidation leads to a loss of composite strength and toughness. A potential solution to this problem is to slow the fiber coating oxidation/volatilization process so that sealing of the matrix cracks via oxidation can occur before significant interfacial degradation can occur [37]. This approach is most likely to succeed if the fiber coating is itself resistant to oxidation. The most oxidation resistant fiber coating known, which still gives good fiber-matrix debonding and tough composite behavior, is silicon-doped BN. GE has patented the use of such coatings with MI-CMC and is optimizing such coatings for both prepreg and slurry cast MI-CMC.

Reproducibility of material properties, or lack thereof, is also a major concern of those wishing to utilize CMC materials. Ceramic composites, including MI-CMC, are relatively new materials that are made by relatively immature and yet complex fabrication processes. This situation is compounded by the fact that most CMC producers are still continually modifying the manufacturing processes in order to improve material performance. Consequently the reproducibility of material properties from batch to batch has in general been rather poor. The development of robust component designs, especially considering the long component life requirements of industrial gas turbines, requires that the material behavior be well characterized and reproducible. Some improvement in reproducibility is expected as the processes mature and vendors gain additional fabrication experience.

Another barrier to implementation of MI-CMC is cost. The cost of MI-CMC is significantly higher than that of the materials it is intended to replace, partially because current production volumes are very low. The current cost of MI-CMCs is of the order of \$20,000 to \$30,000 per pound. However, GE cost projections show that the finished cost of MI-CMC components may reach the levels of TBC-coated, single crystal superalloy components at moderate commercial production levels. In addition, when a cooled component is replaced by an uncooled component, elimination of coolant delivery hardware, seals and other ancillary components can reduce the overall sub-module cost even after the higher cost of the MI-CMC components is taken into account. Therefore the cost of MI-CMC components must be weighed against the benefit provided by their utilization.

SUMMARY AND CONCLUSIONS

The development of melt infiltrated SiC-Si ceramics for gas turbine applications has been a 30+ year effort. Over the past 12 years this development has concentrated on continuous fiber reinforced composites using monofilament, non-stoichiometric and stoichiometric SiC fibers. Current materials have demonstrated good performance in a variety of combustion rig and engine tests of gas turbine hot section components.

ACKNOWLEDGEMENTS

The authors would like to thank P. Meschter, H. Wang, J. Steibel, H. McGuigan, G. Buczkowski and A. Dean for contributing to the technical work described in this paper. The described work was partially funded by the U.S. Department of Energy, Office of Industrial Technologies under contract DE-FC02-92CE4100. The financial support of DOE and the technical input of the contract monitor, J. Mavec, are greatly appreciated. All tables and figures in this chapter were supplied courtesy of GE Global Research Center.

REFERENCES

[1] Hillig, W.B., Mehan, R.L., Morelock, C.R., DeCarlo, V.J. and Laskow, W., "Silicon/Silicon Carbide Composites," Am. Ceram. Soc. Bull., 54[1], 1054–1056, 1975.

[2] Prochazka, S. "The Role of B and C in the Sintering of SiC," in *Special Ceramics 6,* P. Popper, ed., British Ceramic Research Association, Stoke-on-Trent, U.K., 1975. pp. 171–182.

[3] Greskovich, C., "Preparation of High-Density Si_3N_4 by a Gas-Pressure Sintering Process," J. Am. Ceram. Soc., 64[12], 725–730, 1981.

[4] Popper, P., "The Preparation of Dense Self-Bonded Silicon Carbide," in *Special Ceramics,* P. Popper, ed., Heywood and Company, London, U.K., 1960. pp.209–219.

[5] U.S. Patent No. 2,938,807, "Method of Making Refractory Bodies," Anderson, J.C., to the Carborundum Company, August 13, 1957.

[6] Forrest, C.W., Kennedy, P. and Shennan, J.V., "The Fabrication and Properties of Self-Bonded Silicon Carbide Bodies," in *Special Ceramics 5,* P. Popper, ed., British Ceramic Research Association, Stoke-on-Trent, U.K., 1972. pp. 99–123.

[7] Alliegro, R.A., "Processing and Fabrication of Non-Hot-Pressed Silicon Carbide," *Ceramics for High Performance Applications, Proceedings of the Second Army Materials Conference*, Hyannis, MA, USA, November 13–16, 1973, J.J. Burke, A.E. Gorum and R.N. Katz, eds., Brook Hill Publishing Company, Chestnut Hill, MA, USA, 1974. pp. 253–263.

[8] Sawyer, G.R. and Page, T.F., "Microstructural Characterization of "REFEL" (Reaction-Bonded) Silicon Carbides," J. Mater. Sci., 13, 885–904, 1978.

[9] Larsen, D.C. and Adams, W.J., *Property Screening and Evaluation of Ceramic Turbine Materials, Final Technical Report,* AFWAL-TR-83-4141, Air Force Wright Aeronautical Labs, April 1984.

[10] Whalen, T.J., "Processing and Properties of Structural Silicon Carbide," Ceram. Eng. and Sci. Proc., 7[9–10], 1135–1143, 1986.

[11] Chiang, YM., Messner, R.P., Terwillliger, C.D. and Behrendt, D.R., "Reaction-Formed Silicon Carbide," Mat. Sci. Eng., A144, 63–74, 1991.

[12] Mehan, R.L., "Anisotropic Behavior of Si/SiC Ceramic Composites," Am. Ceram. Soc. Bull., 56[2], 211–212, 1977.

[13] Hillig, W.B., "Tailoring of Si/SiC Composites for Turbine Applications," in *Ceramics for High Performance Applications II, Proceedings of the Fifth Army Materials Conference,* Newport, RI, USA, March 21–25, 1977, J. J. Burke, E. N. Lenoe, and R. N. Katz, eds., Brook Hill Publishing Company, Chestnut Hill, MA, USA, 1978. pp. 989–1000.

[14] Mehan, R.L., "Effect of SiC Contend and Orientation on the Properties of Si/SiC Ceramic Composites," J. Mater. Sci., 13, 358–366, 1978.

[15] Mehan, R.L., Hillig, W.B., and Morelock, C.R., "Si/SiC Ceramic Composites: Properties and Applications," Ceram. Eng. and Sci. Proc., 1[7–8], 405–418, 1980.

[16] Stinton, D.P., Caputo, A.J., and Lowden, R.A., "Synthesis of Fiber-Reinforced SiC Composites by Chemical Vapor Infiltration," Am. Ceram. Soc. Bull., 65[2], 347–350, 1986.

[17] Newkirk, M.S., Urquhart, A.W., Zwicker, H.R., and Breval, E., "Formation of Lanxide™ Ceramic Composite Materials," J. Mater. Res., 1[1], 81–89, 1986.

[18] Prewo, K.M., Brennan, J.J., and Layden, G.K., "Fiber Reinforced Glasses and Glass—Ceramics for High Performance Applications," Am. Ceram. Soc. Bull., 65[2], 305–313, 1986.

[19] Bhatt, R.T., "Mechanical Properties of SiC Fiber-Reinforced Reaction-Bonded Si_3N_4 Composites," in *Tailoring Multiphase and Composite Ceramics, Proceedings of the 21st University Conference on Ceramic Science.* R. E. Tressler, G. L. Messing, C. G. Pantano, and R. E. Newnham., eds., Materials Science Research Vol. 20, Plenum Press, New York, NY, USA, 1986. pp. 675–686.

[20] Washburn, M.E., and Coblenz, W.S., "Reaction-Formed Ceramics," Am. Ceram. Soc. Bull., 67[2], 356–363, 1988.

[21] Hillig, W.B., "Melt Infiltration Approach to Ceramic Matrix Composites," J. Am. Ceram. Soc., 71[2], C96–99, 1988.

[22] Easler, T.E., Petrak, C.R., and Szweda, A., "Sylramic Ceramic-Matrix Composites Processing and Properties," Ceramic Transactions, 96, 113–122, 1999.

[23] Luthra, K.L., Singh, R.N., and Brun, M.K., "Toughened Silcomp Composites—Process and Preliminary Properties," Am. Ceram. Soc. Bull., 72[7], 79–85, 1993.

[24] U.S. Patent No. 4,889,686 (December 26, 1989), "Composite Containing Coated Fibrous Material," R.N. Singh and A.R. Gaddipati; U.S. Patent No. 5,015,540 (May 14, 1991), "Fiber Containing Composite," M.P. Borom, W.B. Hillig, R.N. Singh, W.A. Morrison, and L.V. Interrante; U.S. Patent No. 5,294,489 (March 15, 1994), "Protective Coating with Reactive Interlayer on Reinforcement in Silicon Carbide Composite," K.L. Luthra, M.K. Brun, and G.S. Corman; U.S. Patent No. 4,442,352 (September 3, 1996), "Silicon Carbide Composite with Coated Fiber Reinforcement," M.K. Brun, K.L. Luthra, and R.N. Singh; U.S. Patent No. 5,952,100 (September 14, 1999), "Silicon-Doped Boron Nitride Coated Fibers in Silicon Melt Infiltrated Composites," G.S. Corman and K.L. Luthra.

[25] Luthra, K., Singh, R., and Brun, M., "Toughened Silcomp Composites," in *High Temperature Ceramic Composites*, R.Naslain, J. Lamon, and D. Doumeing, eds., *Proceedings of the 6th European Conf. On Composite Materials,* 20–24 September 1993, Bordeaux, France, Woodhead Publishing Ltd., Cambridge, U.K., 1993. pp. 429–436.

[26] Corman, G.S., Heinen, J.T. and Goetz, R.H., "Ceramic Composites for Industrial Gas Turbine Engine Applications: DOE CFCC Phase 1 Evaluations," ASME paper 95-GT-387, presented at the 40th ASME International Gas Turbine and Aeroengine Congress and Exposition, Houston, TX, USA, June 5–8, 1995.

[27] Corman, G.S., Brun, M.K. and Luthra, K.L., "SiC Fiber Reinforced SiC-Si matrix Composites Prepared by Melt Infiltration (MI) for Gas Turbine Engine Applications," ASME paper 99-GT-234, presented at the 44th ASME Gas Turbine and Aeroengine Technical Congress, Exposition and Users Symposium, Indianapolis, IN, USA, June 7–10, 1999.

[28] LaChapelle, D.G., Noe, M.E., Edmondson, W. G., Stegemiller, H.J., Steibel, J.D. and Chang, D.R., "CMC Materials Applications to Gas Turbine Hot Section Components," AIAA Paper 98-3266 (A98-35122), *34th AIAA/ASME/SAE/ASEE Joint Propulsion Conference & Exhibit*, Cleveland, OH, USA, July 13–15, 1998.

[29] Singh, M., Dickerson, R.M., Olmstead, F.A. and Eldridge, J.I., "SiC (SCS-6) Fiber Reinforced–Reaction Formed SiC Matrix Composites: Microstructure and Interfacial Properties," J. Mater. Res., 12[3], 706–713, 1997.

[30] Suyama, S., Kameda, T., Amiji, N., Umezawa, V. and Ichikawa, H., "Fabrication and Properties of Dense Silicon Carbide Matrix Composites," Ceram. Eng. and Sci. Proc., 17[4], 118–124, 1996. See also Volume II, chapter 18 of this book.

[31] Corman, G.S., Luthra, K.L., Brun, M.K. and Meschter, P.J. *Toughened Silcomp Composites for Gas Turbine Engine Applications,* DOE CFCC Phase 1 Contract Final Report, DOE/CE/41000-2, July, 1994.

[32] Ning, X.J. and Pirouz, P., "The Microstructure of SCS-6 Fiber," J. Mater. Res., 6[10], 2234–2248, 1991.

[33] DiMascio, P.S., Orenstein, R.M., Schroder, M.S., Corman, G.S. and Dean, A.J., "Ceramic Gas Turbine Programs at GE Power Systems," Chapter 9 in *Ceramic Gas Turbine Design and Test Experience, Progress in Ceramic Gas Turbine Development, Volume I,* van Roode, M., Ferber, M.K., and Richerson, D.W., eds., ASME Press, New York, USA, 2002. pp. 225–236. See also, N. Jacobson, E. Opila, J. Smialek, P. Tortorelli, K.G. Nickel, M. Yoshida and T. Hirata, "Environmental Response of Ceramics," Chapter 26 in *Ceramic Gas Turbine Component Development and Characterization, Progress in Ceramic Gas Turbine Development Volume II,* van Roode, M., Ferber, M.K., and Richerson, D.W., eds., ASME Press, New York, USA, 2003; and K. Lee, H. Fritze and Y. Ogura, "Protective Coatings for Ceramics," ibid., Chapter 27.

[34] Dean, A.J., Corman, G.S., Bagepalli, B., Luthra, K.L., DiMascio, P.S. and Orenstein, R.M., "Design and Testing of CFCC Shroud and Combustor Components," ASME paper 99-GT-235, presented at the 44th ASME Gas Turbine and Aeroengine Technical Congress, Exposition and Users Symposium, Indianapolis, IN, USA, June 7–10, 1999.

[35] Corman, G.S., Dean, A.J., Brabetz, S., Brun, M.K., Luthra, K.L., Tognarelli, L. and Pecchioli, M., "Rig and Engine Testing of Melt Infiltrated Ceramic Composites for Combustor and Shroud Applications," ASME paper 2000-GT-0638, presented at the ASME TURBO EXPO 2000, Land, Sea & Air, May 8–11, 2000, Munich, Germany.

[36] Miriyala, N., Simpson, J.F., Parthasarathy, V.M. and Brentnall, W.D., "The Evaluation of CFCC Liners After Field-Engine Testing in a Gas Turbine," ASME paper 99-GT-395, presented at the 44th ASME Gas Turbine and Aeroengine Technical Congress, Exposition and Users Symposium, Indianapolis, IN, USA, June 7–10, 1999.

[37] Luthra, K.L., "Oxidation-Resistant Fiber Coatings for Non-oxide Ceramic Composites," J. Am. Ceram. Soc., 80[12], 3253–3257, 1997.

CHAPTER

17

Development of Hot Section Components Composed of Ceramic Matrix Composite (CMC) at the Research Institute of Advanced Material Gas-Generator (AMG)

Mikio Hiromatsu, Shigeto Nishide

Research Institute of Advanced Material Gas-Generator Co., Limited, Tokyo, Japan

Kozi Nishio, Ken-ichiroh Igashira, Go Matsubara

Research Institute of Advanced Material Gas-Generator Co., Limited, Tokyo, Japan

Technical Institute of Kawasaki Heavy Industries Limited (KHI), Akashi, Hyogo, Japan

Takeshi Suemitsu

Technical Institute of Kawasaki Heavy Industries Limited (KHI), Akashi, Hyogo, Japan

INTRODUCTION

The Research Institute of Advanced Materials Gas-Generator Co., Ltd. (AMG) was jointly established by the Japan Key Technology Center and 14 firms in Japan in March 1993. There were three gas turbine manufacturers, five materials, four mechanical component and two electronic control system companies amongst the 14 investors. The research and development activity is planned to finish in early 2002.

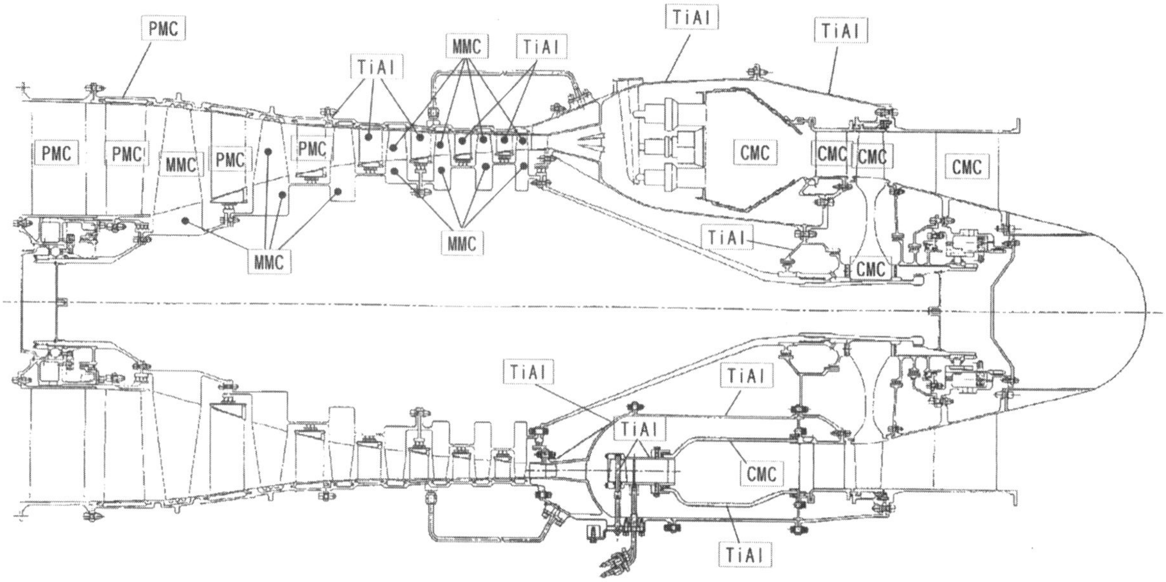

Figure 17.1 Concept of an Innovative Gas-Generator and the Advanced Materials (Reprinted with permission of AMG, ref. [1])

AMG has been conducting technological studies on an innovative gas- generator that will use 20% less fuel, weigh 50% less, and emit 70% less NOx than the conventional gas-generator through the use of advanced materials. The baseline of the performance is the most advanced land-use gas turbine engine or aircraft turbofan engine as of 1993 when the company started the activity.

Figure 17.1 shows AMG's concept of an innovative gas-generator with candidate advanced materials [1]. Selected materials are high temperature polymer matrix composites (PMC), metal matrix composites (MMC), TiAl intermetallic compounds (IMC) and ceramic matrix composites (CMC). CMCs were selected as the key materials for turbine hot section components because of their thermal stability, lower weight and higher fracture toughness or superior impact resistance compared to the corresponding properties of monolithic ceramics. AMG set a target of a 1600°C (1873K) turbine inlet temperature (TIT) to achieve higher thermal efficiency and lower weight with no cooling or less cooling. The company is also investigating the feasibility of applying the advanced gas-generator to gas turbines used in land, marine and air applications.

Because of funding and time scale limitations, the program is concentrating on the basic key technology areas. Therefore, demonstrations of the AMG developed components in gas turbine engines will be conducted in a separate future program.

DEVELOPMENT OF A COMBUSTOR LINER

Introduction

Continuous SiC fiber-reinforced SiC composite (SiC^F/SiC) was selected as the most suitable CMC for the combustor liner because of its thermal stability and formability. An evaluation of the applicability of the SiC^F/SiC to the combustor liner on the basis of material properties and stress analysis of the combustor liner was carried out. Trial SiC^F/SiC combustor liners, the largest of which was 500 mm in diameter, were fabricated by braiding and a combination of the CVI (Chemical Vapor Infiltration) and PIP (Polymer Impregnation and Pyrolysis) methods using the technology for producing high performance SiC^F/SiC developed by Kawasaki Heavy Industries Limited (KHI). Using SiC^F/SiC liners built to the actual dimensions, non-cooling combustion tests were carried out up to 1600°C (1873K) gas temperature at the outlet of the liner.

Materials Production Technology and Evaluation Technology (Developed at KHI)

The process developed for producing the SiC^F/SiC is shown in Figure 17.2. This process consists mainly of two stages. The first stage is the fiber coating for the formation of the interface between the fibers and the matrix with a SiC layer (approximately 1 μm in thickness) and a C layer (thinner than 50 nm) using the CVI method, which we shall refer to as the CVIed-C/SiC Process. The second stage is the synthesis of the matrix using the PIP method with a slurry containing polycarbosilane and fine SiC particles. By repeating the impregnation and pyrolization treatment for six cycles, the SiC^F/SiC with a residual pore volume of 10 % and bulk density of about 2.5 g/cm^3 was obtained.

The characteristics of the developed material were evaluated in a flexural test at room temperature and another after an oxidation test for 500 hours at various temperatures ranging from 900°C (1173K) to 1200°C (1473K). The stress strain diagram for the room temperature flexural test is shown in Figure 17.3. The test piece exhibited extremely high strength and large fracture strain confirming the efficacy of the CVIed-C/SiC interface for improving strength characteristics.

Through the evaluation of strength after heating, we confirmed that the effect of temperature of heating on residual strength is linear while that of time is logarithmic. Therefore, we attempted to estimate the durability of the test materials by using the Larson Miller parameter [LMP = T × (C + logt); T: absolute temperature (K), t: exposure time (hr), C: constant], which is used for evaluating a metallic material's creep characteristics. In Figure 17.4, the residual strength after heating in air and in vacuum is correlated with LMP. The value of the constant C in LMP will vary by material. For the test material, linearity was obtained by setting C at 10. Based on this result, we decided that an accelerated test would be possible by raising the test temperature.

The materials production and evaluation technology (developed at KHI) has been described in the literature [2–4].

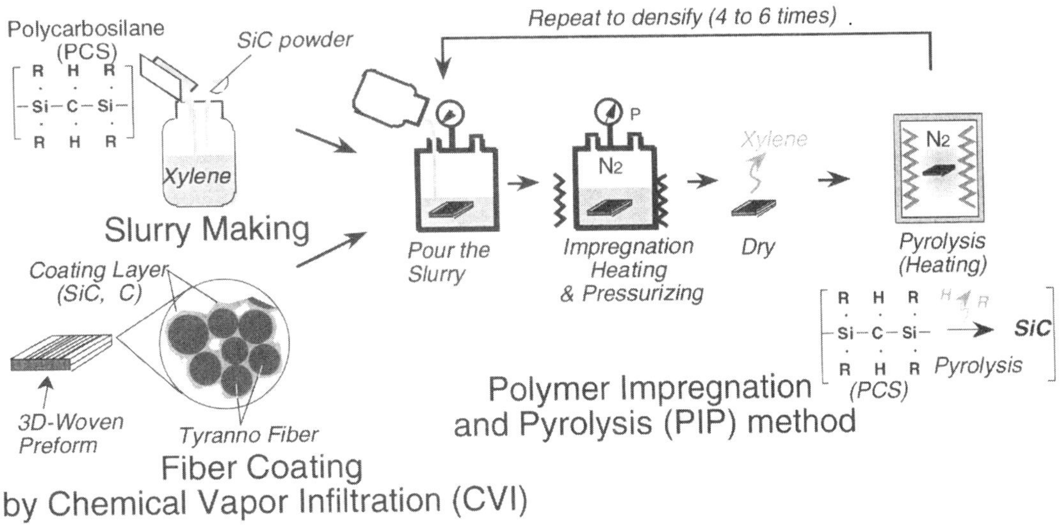

Figure 17.2 Schematic Illustration of the Fabricating Procedure of SiC^F/SiC (Reprinted with permission of KHI)

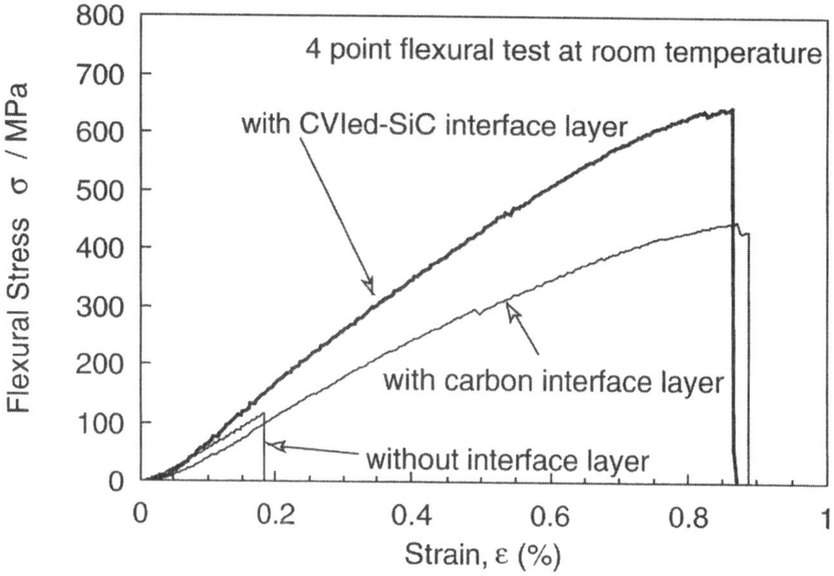

Figure 17.3 Stress Strain Curves of SiC^F/SiC Composites with Different Fiber/Matrix Interface Layers (Reprinted with permission of KHI)

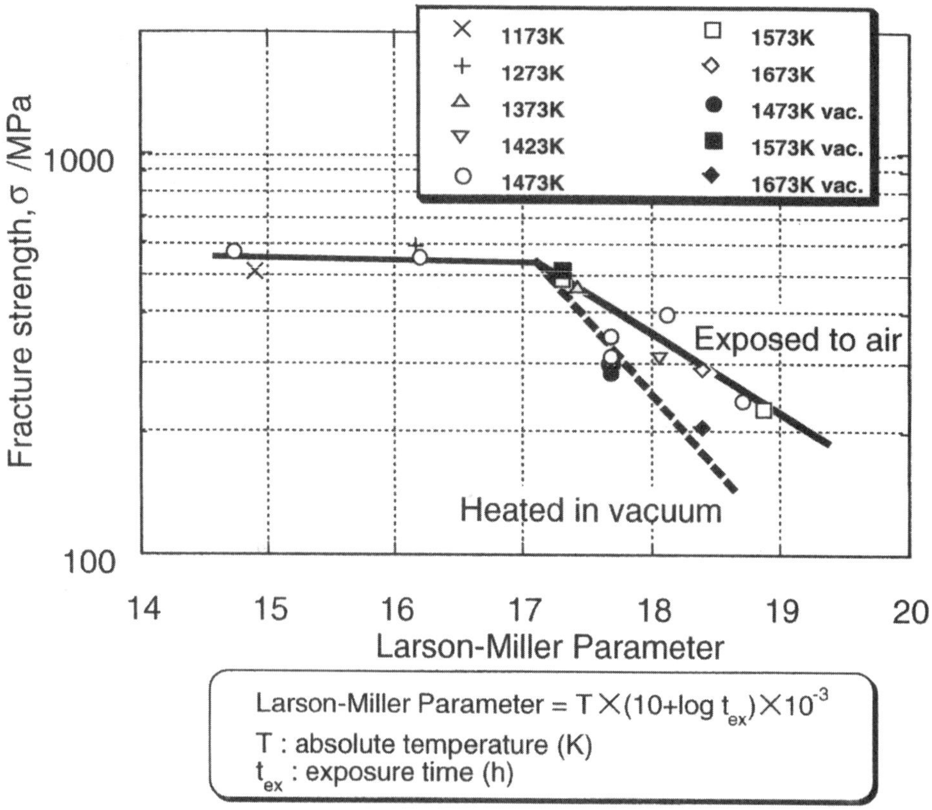

Figure 17.4 The Relationship Between Remaining Flexural Strength of SiC^F/SiC Composite and Larson-Miller Parameter (Reprinted with permission of KHI, ref. [4])

Application of CMC to Parts

In the AMG project, the plan is to apply SiC^F/SiC to the combustor liner and TiAl to the swirler. The conceptual configuration of the combustor is shown in Figure 17.5. As the first step in the R&D of the CMC combustor liner, we chose the filament winding and lay up of unidirectional fiber prepregs as the CMC combustor liner molding method, and evaluated the combustor liners from trial manufacturing. As the second step in the program, to reduce the thickness of the liner, we chose the braiding molding method, which makes it fairly easy to adjust the orientation angle of the reinforcing fiber and permits near-net molding over a broad range of sizes by simply replacing the molding forms. CMC application to components has been detailed in the literature [5–8].

Material Properties of the CMC. For evaluating the material properties of the CMC combustor liner, we conducted evaluation tests of the CMC fabricated using the same method and the same raw material as that for the liner. We conducted flexural tests and tensile tests at room temperature to evaluate basic characteristics of the proposed combustor liner material. As the evaluation test of durability in actual use, we conducted fatigue tests at room temperature and at the liner's expected operating temperature of 1300°C (1573 K) assuming changes in load when the gas generator was started and stopped. The fatigue test was conducted using the tension/tension method at a load stress ratio of 0.1 and frequency of 1 Hz.

As examples of the evaluation tests of the material properties, maximum stress-cycles to failure (S-N) diagrams are shown in Figure 17.6. The low cycle fatigue strength of the combustor liner, at about 100 thousand cycles, is believed to be around 200 MPa at room temperature and around 100 MPa at 1300°C (1573 K). Our separate flexural strength measurements of the experimental material after oxidation tests for 500 hours at an exposure temperature of 1100°C (1373 K) showed virtually no deterioration in strength.

Based on these results, we surmised preliminary criteria for the application of the SiC^F/SiC to the combustor liner. The maximum stress of the liner should be less than 100 MPa at 1300°C (1573 K), and less than

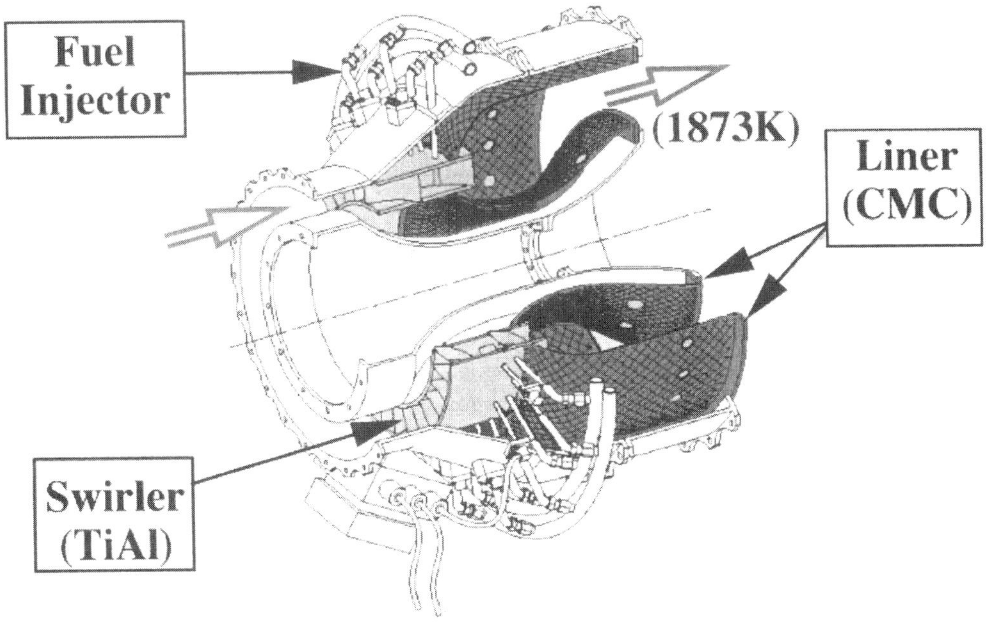

Figure 17.5 The Conceptual Configuration of the AMG Combustor (Reprinted with permission of AMG, refs. [1,7])

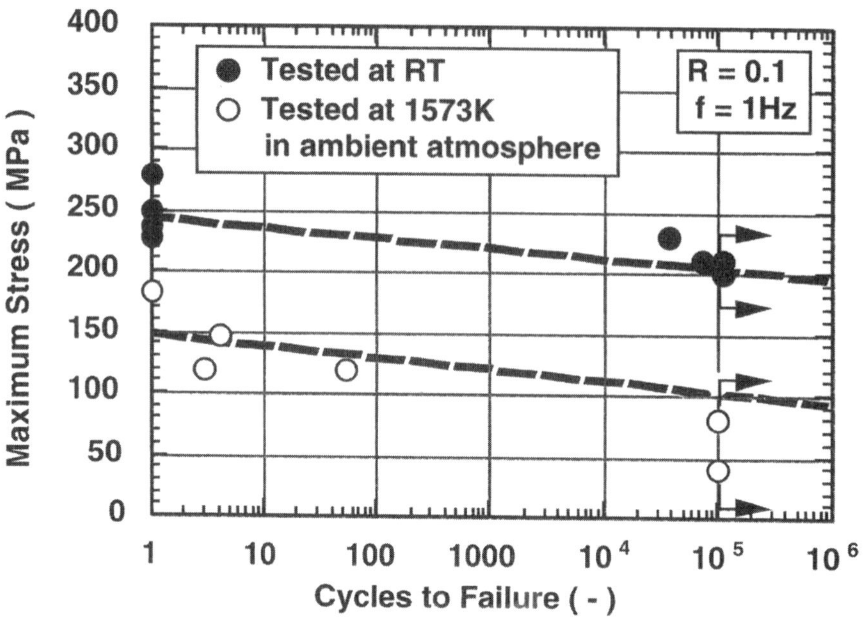

Figure 17.6 Relationship Between Maximum Cycle Stress and Cycles to Failure in the Low-Cycle Fatigue Test for Trial Products (Reprinted with permission of AMG, ref. [8])

200MPa at a temperature lower than 1100°C (1373 K), and not more than 100 MPa to 200 MPa in the range between these two temperatures. In addition, in the actual design of the full scale combustor, the acceptable stress levels must be below the proportional limits of the materials.

Thermal Stress Analyses for the CMC Combustor Liner. The thermal stress of the CMC combustor liners in operation was estimated by the analysis of heat transfer and thermal stress using ABAQUS (trade mark of Hibbit, Karlsson & Sorensen, Inc.) analysis software, based on the material property data of the trial products. The combustor liner model for the analysis was created using axi-symmetric solid elements. For the tie-in part between the liner and the swirler, a spring-type fitting structure was used, and a model was produced using linear spring elements. In the analysis, we used a general spring constant of 330 N/mm. The model used for analysis is shown in Figure 17.7.

The analysis of heat transfer and thermal stress in the CMC combustor liners at steady state conditions indicated that the maximum thermal stress occurred near the tie-in part between the liner and the swirler, and that it was hoop stress. The maximum thermal stresses in the inner and outer liners were around 150 MPa and 170 MPa, respectively. The temperature at the location where the maximum stress occurred was about 1000°C (1273 K).

We also analyzed the thermal stress under transient conditions when the combustion state shifted from idle to design point combustion, and from design point to idle. The heat transfer analysis under transient conditions confirmed that the liner temperature reached the highest level at the location where the combustion gas was assumed to reach the maximum temperature.

The thermal stress analysis under transient conditions revealed that the peak thermal stress appears several seconds after the combustion state shifts for both the inner and outer liners. Examples of the thermal stress analysis under transient conditions, specifically changes in thermal stress with time at the maximum

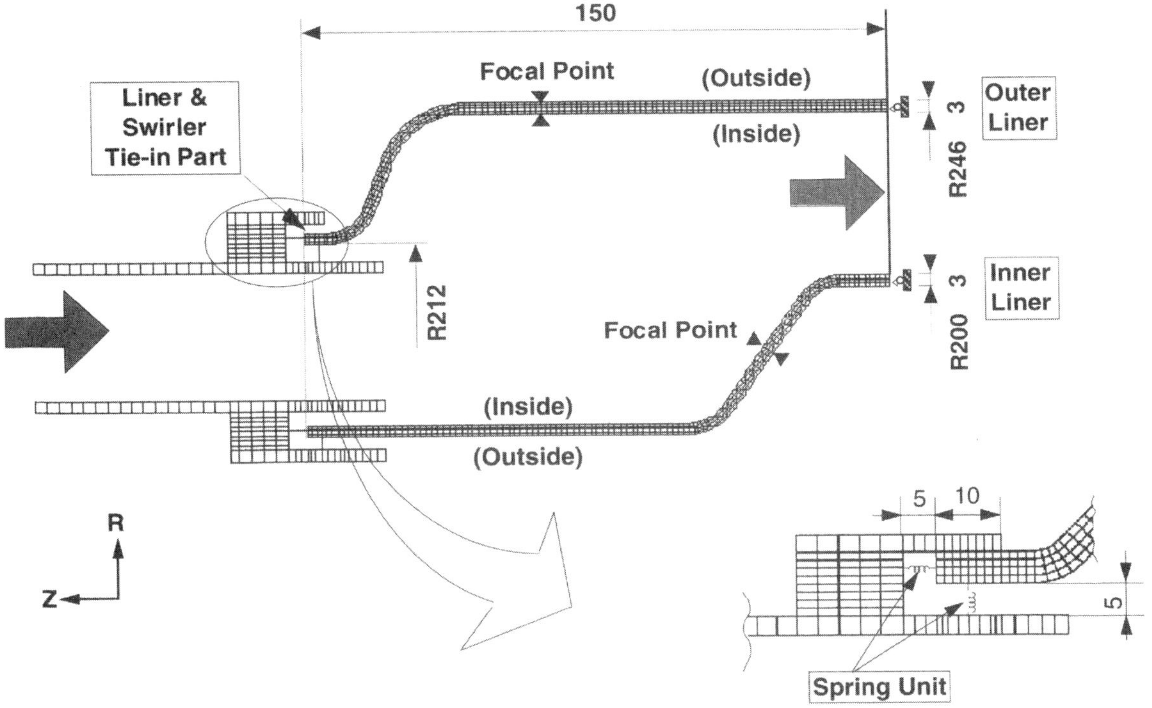

Figure 17.7 Thermal Stress Analysis Model for the CMC Combustor Liner (Reprinted with permission of AMG, refs. [7,8])

temperature location of the inner liner, are shown in Figure 17.8. The peak thermal stress value did not exceed 100 MPa, and the material temperature of the part where the peak thermal stress occurred was 1220°C (1493 K).

The stress values obtained by the steady state and the transient analyses were within the range of the preliminary criterion of application of the proposed SiC^F/SiC material for the combustor liner. These results suggest the SiC^F/SiC to be suitable as combustor liner material.

Evaluation of the CMC Combustor Liner in Combustion Testing. The inner and the outer liners were produced by braiding molding, the fiber coating by using the CVIed-C/SiC Process, and the matrix by the PIP method. The reinforcement fiber used was a Si-Ti-C-O-based low-oxygen fiber (product name: Tyranno ZMI from Ube Industries, Ltd.). For the matrix precursor polymer, we used polycarbosilane (PCS). Braided fibers run around the liner at 80° inclined or declined to the axial direction.

To evaluate the prototype CMC combustor liners, we assembled them inside the combustor housing and a combustion test was conducted. Kerosene was used as the fuel. The combustion conditions were: an inlet gas temperature of 525°C (798 K), air flow rate of 1.0kg/s, air pressure of 0.2MPa, air/fuel ratio of 30 and outlet gas temperature of 1600°C (1873 K), which was the AMG target temperature. The combustion schedule consisted of two four-hour spans of continuous combustion at the design point, and 110 cycles between 30 seconds at the design point and 30 seconds at idling conditions. The retention time at high temperature reached about 10 hours as the cumulative total. The temperature of the liner during the combustion test was measured with a thermocouple fitted near the dilution hole on the inner surface (combustion gas side) of the outer liner and with thermal paint coated on the outer surface of the liners. Upon completion of the test, the

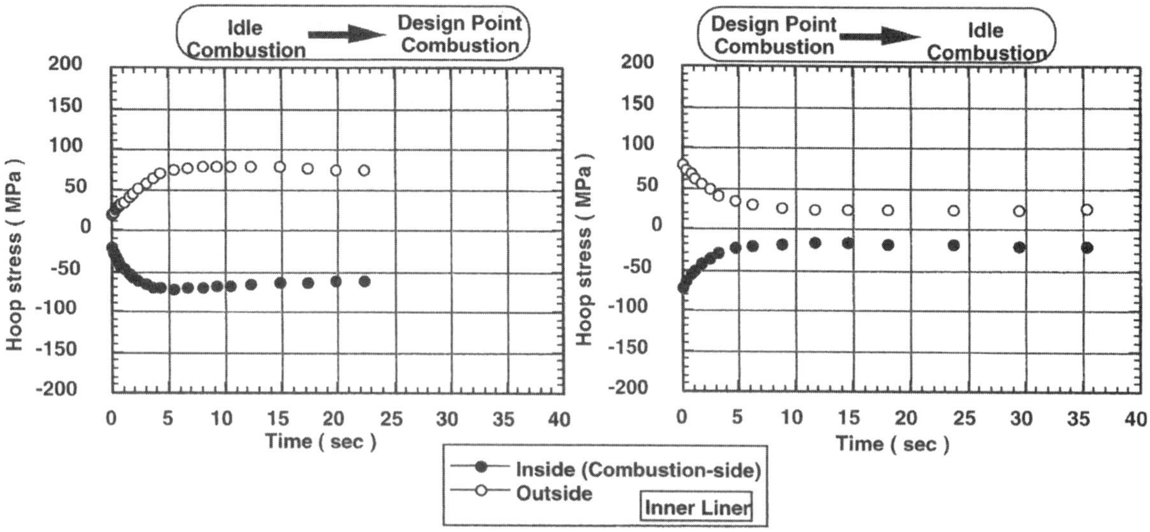

Figure 17.8 Results of Transient Thermal Stress Analysis of the CMC Liner for AMG (Reprinted with permission of AMG, refs. [7,8])

external appearance was observed and the liner's dimensions and weight were measured in order to analyze changes from the pre-test condition. The changes in external appearance are shown in Figure 17.9. The inspection revealed that no abnormalities such as cracking or delamination had occurred, but the surface on the combustion gas side had turned blue due to the formation of an oxide layer.

Regarding dimensions, the measuring precision is believed to be ±0.5 mm due to undulations caused by the reinforcement fiber flux appearing on the surface. When the results of the dimensional measurements are compared, taking into account the measuring precision, the amount of change during the test falls within the

(Before Test) **(After Combustion Test)**

Figure 17.9 Results of Combustion Test (External Appearance) (Reprinted with permission of AMG, ref. [8])

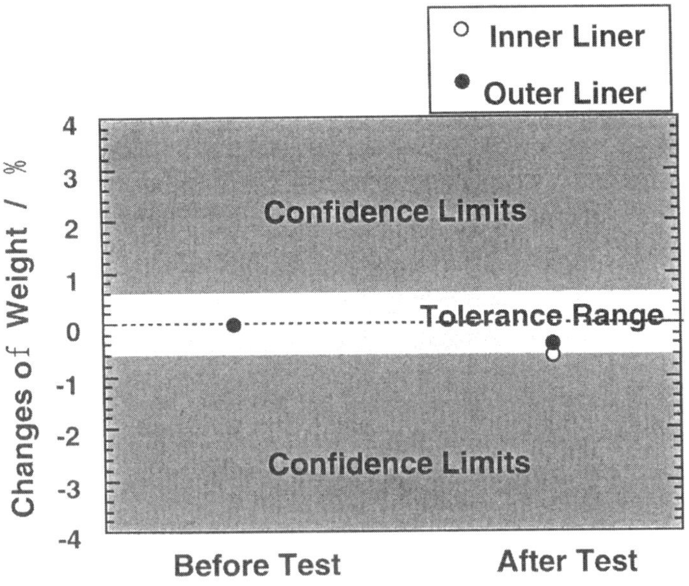

Figure 17.10 Weight Changes in the Combustion Test (Reprinted with permission of AMG, ref. [8])

measurement tolerance range. Therefore, it can be said that there was little or no deformation during the combustion test. The changes in weight are shown in Figure 17.10. Although there was a slight (10 g) change in weight, we believe the weight changed very little by the combustion test. There may have been some influence on the measurement from the thermal paint coating and the sample being weighed in a dry state.

Although the test period was brief, the SiC^F/SiC liner suffered no changes in characteristics during combustion using kerosene. Therefore, we believe that the applicability of SiC^F/SiC for the combustor liner was confirmed by this first-phase, short-term combustion test using liquid fuel, conducted to gauge the performance of the prototype low NOx combustor. Preparations are being made for the second-phase long-term combustion tests.

DEVELOPMENT OF A TURBINE BLISK

Introduction

The design, manufacturing and application technologies for continuous SiC fiber-reinforced SiC ceramic matrix composites (CMC) for the turbine blisk (bladed disk) are being developed in the research program of AMG [9]. The specifications for the high performance CMC turbine blisk are [10]:

—Material temperature: 1400°C (1673K)
—Blade tip speed: 570m/s

Columnar symmetrical three-dimensional (3-D) woven fabrics with a high volume fraction of SiC fiber were developed and densified by CVI followed by PIP.

Key issues to be addressed are the weaving technology and its specifications, CMC component manufacturing technologies that are capable of fabricating a blisk of the desired thickness, material data acquisition, and stress analysis for the blisk, including an understanding of failure modes, and joining between the metallic shaft and the CMC blisk.

TABLE 17.1 SIC FIBERS DERIVED FROM POLYMER PRECURSORS (REPRINTED WITH PERMISSION FROM REF. [11])

Chemical Composition (wt%)	LoxM	LoxE	ZMI	AM	SA	Nicalon	Hi-Ni	Hi-NiS
Si	55	59	57	53	68	57	62	69
C	32	37	35	34	31	32	37	31
O	10	3	8	12	0.3	11.7	0.5	0.2
Ti	2	2	—	—	—	—	—	—
Zr	—	—	1	—	—	—	—	—
Al	—	—	—	<1	<1	—	—	—
C/Si(at%)	1.37	1.47	1.44	1.49	1.08	1.32	1.40	1.05
Properties								
Fiber diameter (μm)	11	11	11	11	10	14	14	12
SiC beta size (nm)	1.4	—	2.0	1.8	38.0		5.0	10.9
Density (g/cm^3)	2.5	2.6	2.5	2.4	3.0	2.6	2.7	3.1
Thermal conductivity (W/m.K)	1.4	2.4	2.5		65			
Tensile strength (GPa)	3.3	3.3	3.4	2.8	2.8	3.0	2.8	2.6
Tensile modulus (GPa)	187	220	200	160	420	220	270	420
Elongation (%)	1.8	1.5	1.7	1.6	0.7	1.4	1.0	0.6

The structural integrity of the CMC blisk was validated by cold spin testing at a maximum peripheral speed of 644 m/s, and it will be ultimately evaluated by hot gas stream spin testing.

Materials

Candidate reinforcement SiC fibers for the CMC blisk are listed in Table 17.1 [11]. Tyranno Lox-M (Ube Industries, Ltd.) was selected for the initial study on the manufacturing and design feasibility of the CMC blisk because of its relatively high strength and flexibility for the complex weavings. 3-D weavings of various specifications were designed and manufactured for CMC disks and blisks, where fiber yarns were arranged following the cylindrical coordinate system (r-θ-z), suitable to withstand the centrifugal stresses during the high speed spin testing of the blisk.

The manufacturing process for the CMC is shown in Figure 17.11. A thin carbon interface layer (approx. 100 nm thick) is deposited on the fiber surface of the 3-D weavings. Subsequently the SiC matrix is formed by a hybrid process which is a combination of CVI and PIP. After machining, a thin SiC CVD coating is applied to the surface of the CMC to protect the carbon interface from oxidation. The CVI/PIP dual phase matrix CMC has advantages for the turbine blisk application, because it has higher strength than an all-CVI matrix CMC and higher thermal conductivity than an all-PIP matrix CMC. A typical stress-strain curve for the CVI/PIP dual phase matrix CMC, and comparative curves for an all-CVI and an all-PIP matrix CMC are shown in Figure 17.12.

Development of a CMC Blisk with a 3-D Woven Preform

To obtain a CMC thick enough to machine the blisk, a CMC block, 25 mm thick, was manufactured by the hybrid process, and the uniformity of the mechanical properties through the thickness was evaluated. The variation in the strength and modulus of the 25mm thick CMC was in the acceptable range, though the modulus and density were slightly lower for specimens sectioned from the center than those for specimens cut near the surface (Figure 17.13 and Figure 17.14). The development of the CMC blisk has been detailed in the literature [12,13].

The stress analysis data for the CMC blisk were confirmed by burst spin testing of thin CMC disks (Figure 17.15). The stress distribution was analyzed by finite element modeling (FEM) using material

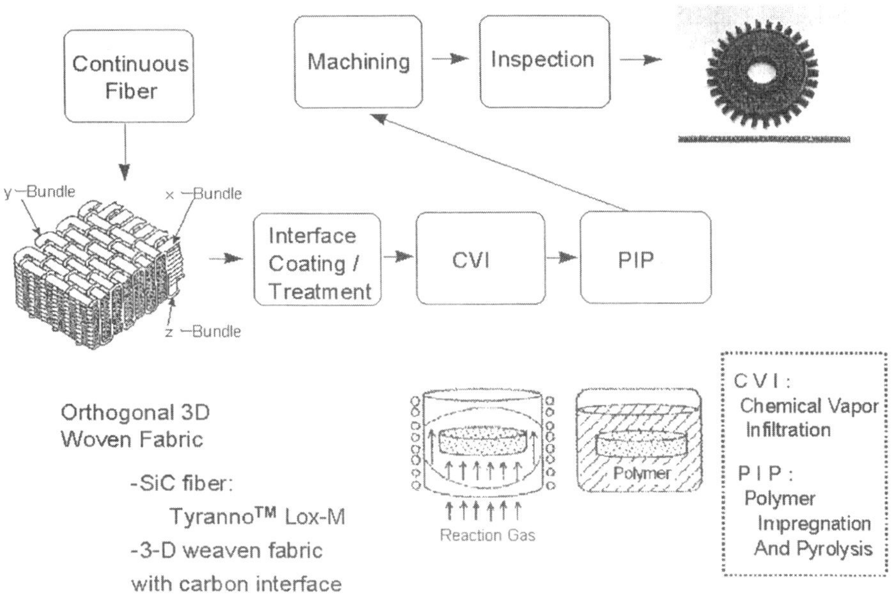

Figure 17.11 Manufacturing Process for the CMC Blisk (Reprinted with permission of AMG, refs. [10,12])

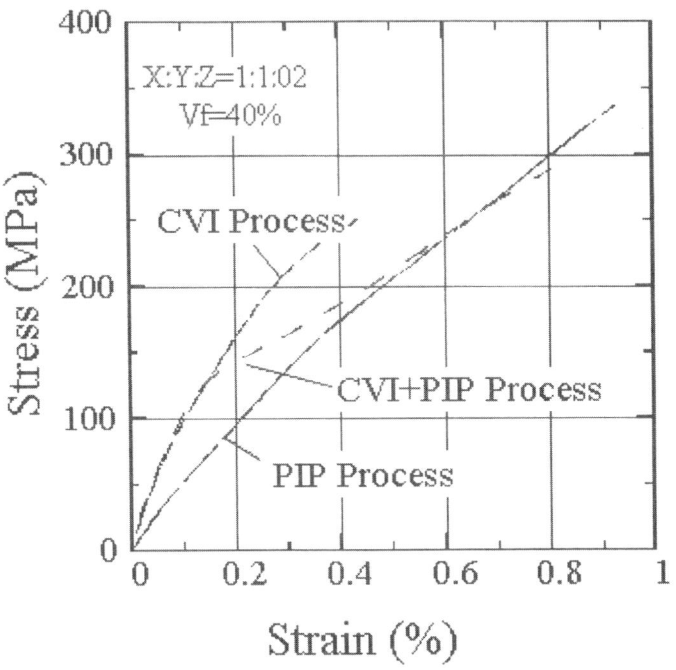

Figure 17.12 Typical Stress-Strain Curve for the 3-D CMC (Reprinted with permission of AMG, ref. [13])

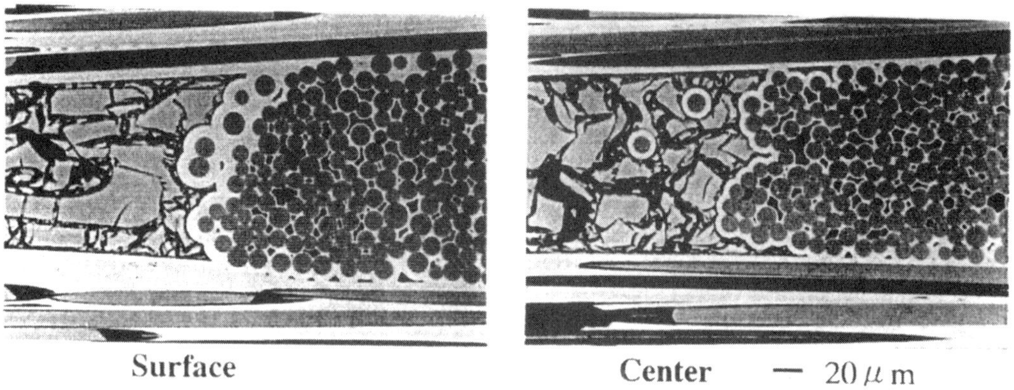

Figure 17.13 Microstructure of 25 mm Thick CMC (Reprinted with permission of AMG, ref. [13])

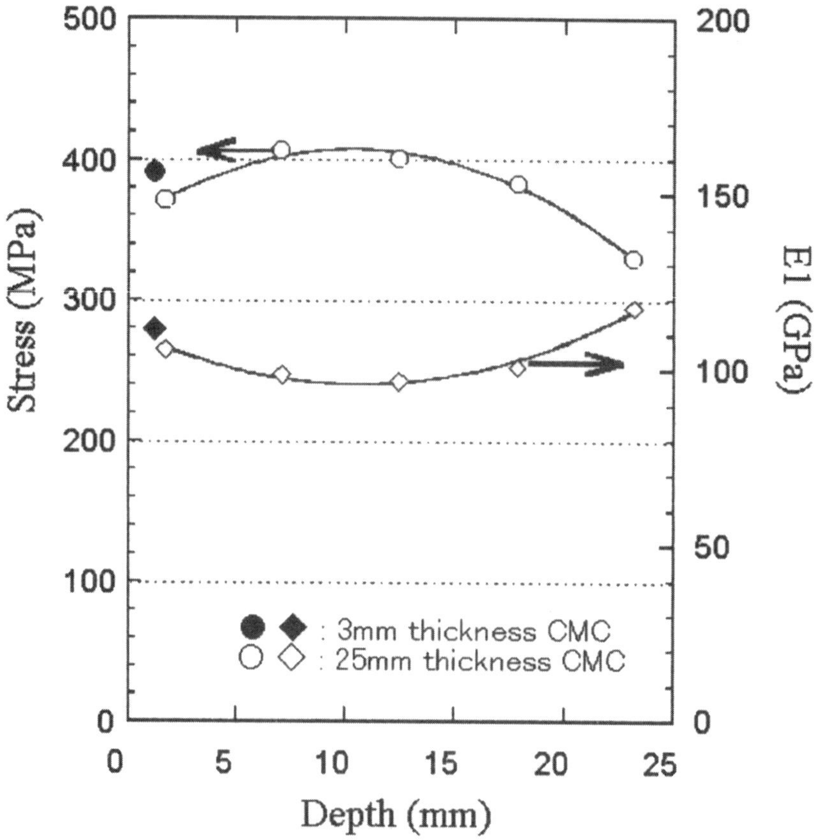

Figure 17.14 Variation of Mechanical Properties Distribution with Thickness (Reprinted with permission of AGM, ref. [13])

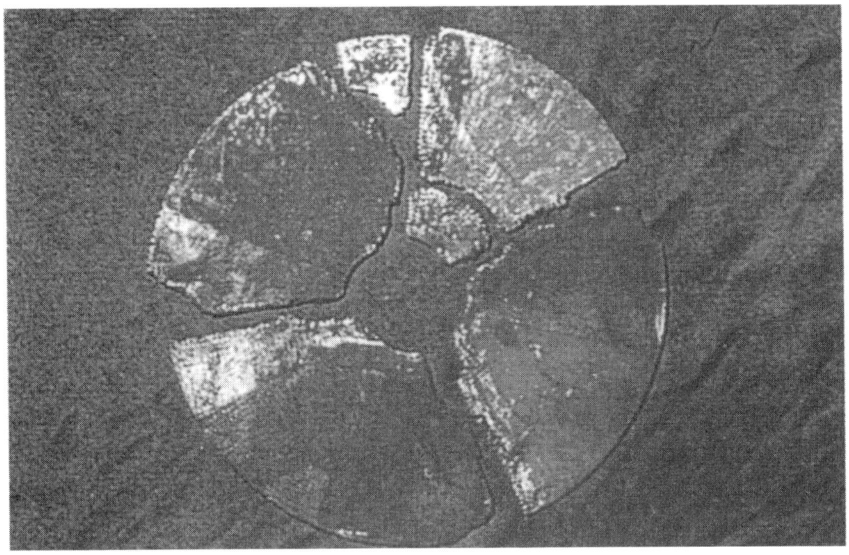

Dimensions: ϕ 250- ϕ 50- t 7mm
Burst speed: 36200 RPM

Figure 17.15 Typical CMC Disk After Burst Test (Reprinted with permission of AMG, ref. [12])

TABLE 17.2 THE SCATTER OF THE BURST SPEEDS OF CMC DISKS

Test No.	1	2	3	4	5	6
Actual burst speed: N_t (rpm)	36200	37900	40200	36100	37800	37800
Predicted burst speed: N_a (rpm)	35600	37500	38900	38400	39500	38200
$N_a/N_t \times 100$ (%)	98	99	97	106	104	101

constants, which were calculated by the unit cell model corresponding to the structure of 3-D weavings. The burst strength was predicted based on the maximum hoop stress criterion. The test CMC disks had an outer diameter of 250 mm, an inner diameter of 50 mm, and 7 mm thickness. Each disk was attached to a shaft driven by an air turbine and spun to burst. The scatter of the burst speeds of the six CMC disks was small as shown in Table 17.2, and agreed well with the analysis.

The results of the analysis and testing of the blade segment of the model tests were as follows. The stress concentration at the fillet R of the blade root section has a negligible effect on the blade strength. The average radial stress criterion can be applied to the blade root of the CMC blisk. Measurement of the natural frequency of each blade cut from the CMC disk confirmed that there was no destructive blade vibration through the hot gas stream test.

Curvic couplings and a spring mechanism were designed and applied to resolve technical issues related to the joint between the CMC blisk and the metallic shaft, one of them being a difference of thermal expansion coefficients.

Current Status of the Evaluation of the CMC Blisk

CMC blisks with 246mm blade tip diameter, 23mm blade height, approximately 23mm chord, and 28mm maximum thickness in the disk portion (Figure 17.16) have been manufactured and evaluated by cold

Figure 17.16 CMC Blisk and Metal Adapter (Reprinted with permission of AMG, ref. [10,12])

spin testing and preliminary hot gas stream spin testing. For the cold spin test, the 3-D CMC blisk was accelerated to 50,000 rpm at room temperature and the blade tip speed reached 644m/s. There was no evidence of damage on the disk surface and blade root portion after the test. Preliminary hot gas stream spin tests were conducted to evaluate thermal effects on the joint mechanism, and to check the performance of the test apparatus. The preliminary hot gas stream spin test was carried out at 700°C (973K).

In the hot gas tests, pressurized air heated by the combustor was supplied through the turbine nozzle made of SiC whisker-reinforced Si_3N_4 ceramic and then to the blade section of the CMC blisk to rotate it. The blisk was joined to the metallic shaft, of which the other end was connected to a hydraulic dynamometer to absorb the power output. The CMC blisk was rotated without any vibration and without any damage up to 30,000 rpm at 700°C (973K), which was the maximum test temperature condition planned (Figure 17.17). A uniform and proper contact mark was observed on the curvic coupling showing that the newly designed CMC-metal joint mechanism worked properly.

PROSPECTS FOR THE FUTURE

Through our studies, we have confirmed the applicability of the selected CMC for the combustor liners of the gas turbine engine. However, it is necessary to expand the experience for the long service life expected for gas turbines in the high temperature combustion gas environment.

The feasibility of the design methods and manufacturing process for the turbine blisk were confirmed using a CMC with the Lox-M fiber at this stage. But it cannot survive under high temperature creep conditions. Hot gas stream spin testing up to 1400°C (1673K) will be conducted using creep resistant fibers such as SA (Ube Industries, Ltd.) (Figure 17.18). It is necessary to conduct further intensive studies to accumulate experience and data to realize CMC rotational components for the gas turbine.

Technologies developed for the CMC combustor liner and blisk can be applied to other hot section components and gas path parts in the turbine, such as turbine blades, vanes and shrouds. Remaining issues for the CMC components are a detailed understanding of the mechanical behavior of this material, especially in

Figure 17.17 CMC Blisk After Hot Gas Stream Spin Test (Reprinted with permission of AMG, ref. [12])

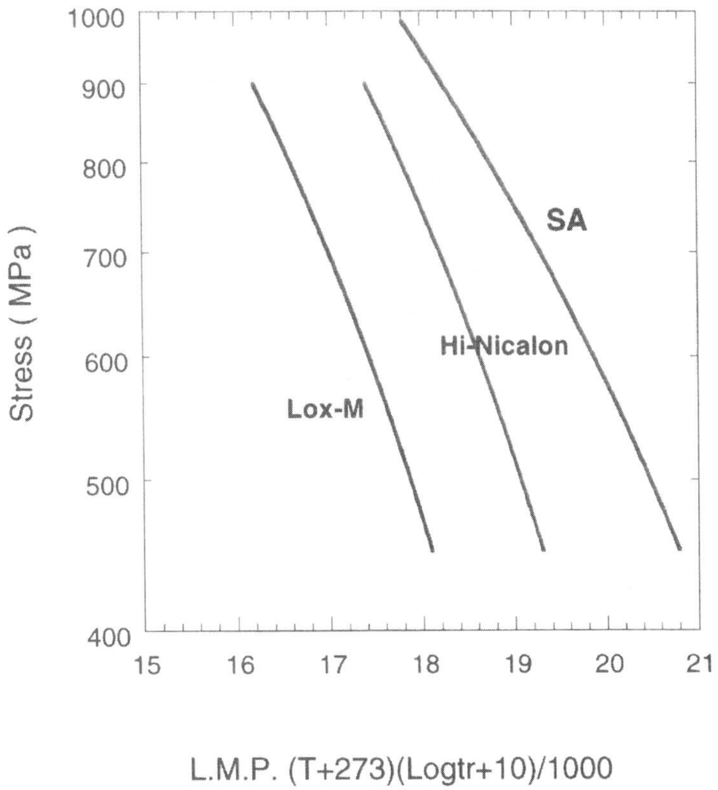

Figure 17.18 Creep Properties of SiC Fibers (Reprinted with permission of AMG)

long term operation, quality assurance methods, and cost. Those shall be addressed through ongoing research and development efforts.

The project has been bridging a gap between the materials industry and engine manufacturers effectively. It is hoped that an engine program will be organized to demonstrate a high performance, lightweight and low-NOx emission gas turbine using AMG developed CMC components in the near future.

SUMMARY

The Advanced Material Gas-Generator(AMG) program has been ongoing in Japan since 1993 to develop the key technologies for applying the various advanced materials to the gas-generator. Design methodology, including stress analysis, for the CMC combustor liner and turbine blisk have been studied and developed.

Different processes such as a filament winding molding or a lay-up of unidirectional fiber prepregs were evaluated for trial manufacture of the combustor liners. It was found that the liners manufactured by the braiding molding method using Tyranno ZMI fiber coated with carbon interface and SiC by means of CVI, and the matrix infiltrated by PIP had desirable features such as a relatively thin wall thickness and manufacturability. The liners made by this process were tested in the AMG low NOx combustor for a short period (approximately 10 hours). The results encouraged AMG to enter into extended duration combustion tests.

Three dimensionally (3-D) woven Tyranno Lox-M fiber reinforced disks were made with various combinations of fiber distribution, and spun to burst to evaluate and improve strength for a turbine blisk application. The CVI and PIP dual phase matrix was found to be the most suitable for the blisk. A CMC blisk of 244 mm blade tip diameter with curvic couplings was tested at room temperature up to 50,000 rpm. The result showed that the strength was sufficient for proceeding to the hot gas stream spin test. The preliminary hot gas spin test was conducted. As the Lox-M fiber has limited creep resistance, further development of the blisk should be conducted using some other fibers.

ACKNOWLEDGEMENT

Unless otherwise indicated figures and tables in this chapter were originally supplied and are owned by the Research Institute of Advanced Materials Gas-Generator. Their appearance in previous publications has been referenced.

REFERENCES

[1] Hiromatsu, M., and Seki, S., "Status of Advanced Material Gas-Generator Research and Development Project," *Proceedings of the 1995 Yokohama International Gas Turbine Congress*, GTSJ, Yokohama, October 23–27, 1995. 95-YOKOHAMA-IGTC-125, Vol. I, I-203/I-210, 1995.

[2] Igashira, K., Nishio, K., Hino, H., and Okazaki, S., "High Temperature Endurance and Degradation Mechanism of Si-Ti-C-O Fiber/SiC Composite," Journal of the Japan Institute of Metals, Vol. 60[12], 1229–1235, 1996 (in Japanese).

[3] Igashira, K., Nishio, K., Hino, H., and Okazaki, S., "Fabrication and Mechanical Properties of Si-Ti-C-O Fiber/SiC Composites with CVD-SiC Layer between Fiber and Matrix," Journal of the Japan Institute of Metals, 62[8], 766–773, 1998 (in Japanese).

[4] Igashira, K., Nishio, K., and Suemitsu, T., "Development of the combustor liner composed of ceramic matrix composites," *Proceedings of the 8th International Conference on Composite Materials (ECCM-8)*, I. Crivelli Visconti, ed., 3–6 June 1998, Naples, Italy, Woodhead Publishing Ltd, Cambridge, U.K., Vol.2, 1998. pp.39–46.

[5] Nishio, K., Igashira, K., Take, K., and Suemitsu, T., "Development of Combustor Liner Composed of Ceramic Matrix Composite (CMC)," Journal of the Japan Gas Turbine Society, 25[98], 94–98, 1997 (in Japanese).

[6] Nishio, K., Igashira, K., Take, K., and Suemitsu, T., "Development of Combustor Liner Composed of Ceramic Matrix Composite (CMC)," Journal of the Japan Gas Turbine Society, 26[101], 102–106, 1998 (in Japanese).

[7] Nishio, K., Igashira, K.-I., Take, K., and Suemitsu, T., "Development of a Combustor Liner Composed of Ceramic Matrix Composite (CMC)," ASME paper 98-GT-104, presented at the International Gas Turbine and Aeroengine Congress and Exhibition, Stockholm, Sweden, June 2–5, 1998. Transactions of the ASME, Journal of Engineering for Gas Turbines and Power, 121[1], 12–17, 1999.

[8] Nishio, K., Igashira, K., Matsubara, T., Suemitsu, T., "Development of Combustor Liner Composed of Ceramic Matrix Composite (CMC)," *Proceedings of the International Gas Turbine Congress 1999 Kobe*, T. Tamaru, ed., 14–19 November 1999, Kobe, Japan, © Gas Turbine Society of Japan, Vol. II, 1049–1056, 1999. IGTC'99 Kobe TS-110.

[9] Hiromatsu, M., Seki, S., "Research and Development Status of Advanced Material Gas-generator(AMG) Project," ASME paper 95-GT-287, presented at the International Gas Turbine and Aeroengine Congress and Exposition, Houston, TX, USA, June 5–8, 1995.

[10] Nojima, Y., Araki, T., Saida, Y., Suzumura, N., Funatogawa, O., Nishide, S., and Yahsuhira, K., "Development of CMC Bladed Disk for Advanced Gas-Generator," *Proceedings of the International Gas Turbine Congress 1999 Kobe*, T.Tamaru, ed. 14–19, November 1999, Kobe, Japan, ©Gas Turbine Society of Japan, Vol. II, 1031–1036, 1999. IGTC'99 Kobe TS-107.

[11] Masaki, S., Yamamura, T., "High Performance Tyranno (Si-C-M-O) Fibers for Composites," High Temperature Ceramic Matrix Composites-3, CSJ Series, 3, 67–72, 1998.

[12] Natsumura, T., Nojima, Y., Suzumura, N., Araki, T., Moriya, K. and Yasuhira, K., "Component design of CMC and MMC rotor for turbine engine applications," *Proceedings of 31'st International SAMPE Technical Conference (Advanced Materials & Processes: Preparing for the New Millennium)*, J.E. Green and D.D. Howell, eds., Chicago, IL, USA, 27–31 October 1999, SAMPE Publishing, Vol.31, 130–141, 1999.

[13] Araki, T., Szumura, N., Masaki, S., Natsumura, T., Onozuka, M., Ohnabe, H. and Yasuhira, K., "Manufacturing of Ceramic Matrix Composite Rotor for Advanced Gas-Generator," Ceram. Eng. and Sci. Proc., 19[4], 241–248, 1998.

CHAPTER

18

Development of Continuous Fiber-Reinforced Reaction Sintered Silicon Carbide Matrix Composite for Gas Turbine Hot Parts Application

Tsuneji Kameda, Shoko Suyama, Yoshiyasu Itoh, Katsutoshi Nishida, Isao Ikeda, Tsuneo Hijikata and Takanari Okamura

Power & Industrial Systems Research & Development Center, Toshiba Corporation, Yokohama, Kanagawa, Japan

INTRODUCTION

Toshiba Corporation has been developing ceramics targeted for application in the gas turbine hot section. The development was aimed at improving the gas turbine efficiency due to the reduction of cooling air, and the potential reduction of NOx emissions by keeping the combustion temperature low. These objectives are to be achieved by incorporating ceramic parts in the gas turbine hot section which have high strength and oxidation resistance at high temperature. Figure 18.1 shows the timeline for the development of the ceramic parts at Toshiba Corporation. In the early stage of the development, some monolithic ceramics were investigated as materials for combustor, turbine stator, and rotor blades. Those tests for monolithic parts were conducted successfully, but on the other hand, there remained a reliability problem, concerning the application for large-scale

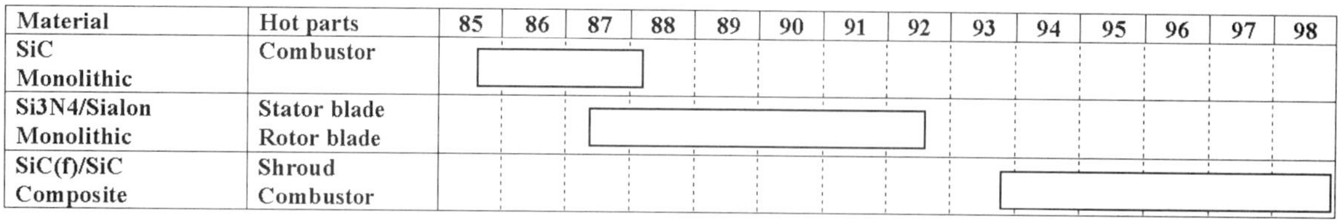

Figure 18.1 Timeline of Ceramic Parts Development for Gas Turbine Applications at Toshiba Corporation (Reprinted with permission of Toshiba Corporation)

land-based gas turbines. So, in the next stage of the development, the focus was on fiber-reinforced ceramic matrix composites, because materials from this class of ceramics have improved fracture toughness compared to monolithic ceramics, and therefore, have improved slow crack growth and impact resistance.

Fiber-reinforced ceramic matrix composites are some of the most promising candidate materials for high temperature structural applications. SiC matrix composites are expected to be used as materials for gas turbine hot section parts, because of their high temperature stability [1–6]. With regards to protection of fibers and interfaces from the oxidizing environment of the gas turbine, a dense matrix is preferred. The recent development of a low oxygen content SiC fiber (Hi-Nicalon), which shows both excellent high temperature stability and weaving ability, makes it possible to apply various matrix consolidation processes at high temperature. The reaction sintering process is one of the processes suitable for forming a dense matrix without firing shrinkage [7–12].

In this chapter, a study of the application of SiC fiber-reinforced reaction sintered SiC matrix composites for gas turbine hot parts is being reported. The reaction sintered SiC matrix composites reinforced by Hi-Nicalon braided fabric and uni-directional fiber were developed and their mechanical and fracture properties were evaluated. Combustor and shroud model parts were fabricated, and their reliability was evaluated under atmospheric pressure combustion test conditions.

MATERIAL DEVELOPMENT

Fabrication Method for Composite Test Samples

The fabrication method for composite test samples is shown in Figure 18.2 [13]. Filaments of SiC fiber (Hi-Nicalon/Nippon Carbon Co., Ltd.) were coated with BN or BN/SiC by chemical vapor deposition (CVD), to provide the appropriate interfacial weak bonding and to prevent the interfacial reaction. The BN and SiC coating thicknesses were about 0.4 μm and 0.3 μm, respectively. The fiber yarn consisted of 500 filaments, and the filament diameter was about 14 μm. After 4 yarns were gathered into one fiber bundle unit, flat braided preforms made of 13 bundle units were fabricated. Alternatively, the fiber yarns were gathered uni-directionally. For the uni-directional samples, a 1.0 μm thick BN coating was deposited. The fibrous preforms were set in a porous mold and the green composites were obtained by pressure impregnating and casting of the matrix slurry. The slurry was a mixture of SiC and carbon powders, and water with some dispersant. The green composites were dried and then reaction sintered at 1450°C (1723 K) in vacuum for 18 ks for braided samples, and at 1420°C (1693 K) in vacuum for 0.3 ks for uni-directional samples. The infiltrating metals were 99.9 mass% pure silicon metal and 5 mass% boron-doped silicon metal. The sintered bodies were machined into 1 mm × 10 mm × 40 mm test bars.

Effect of Coating Structure and Composition of Infiltrating Material

In previous studies on the reaction between SiC and molten Si [14,15], it was reported that a small amount of SiC dissolved into molten silicon, and then precipitated. Figure 18.3 shows a scanning electron

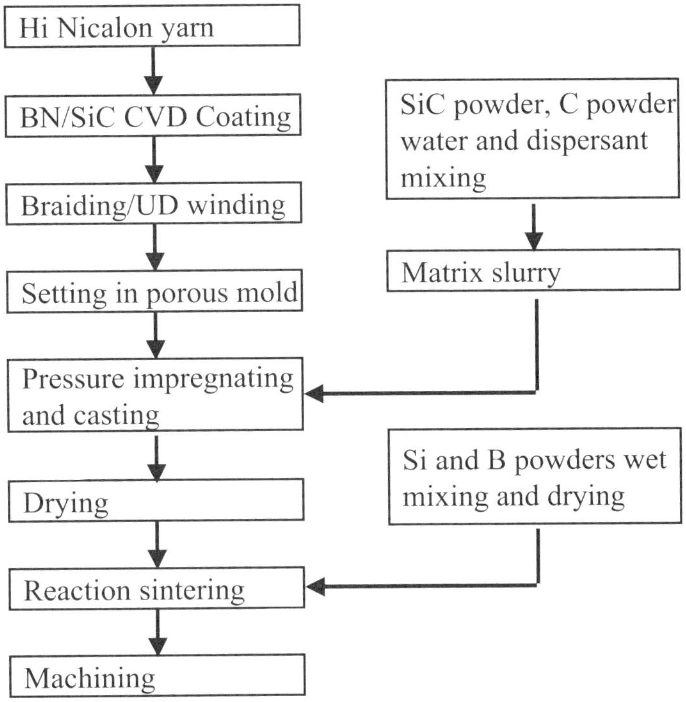

Figure 18.2 **Flow Chart of the Fabrication Process for the Composite Test Samples (Reprinted with permission from refs. [10,13])**

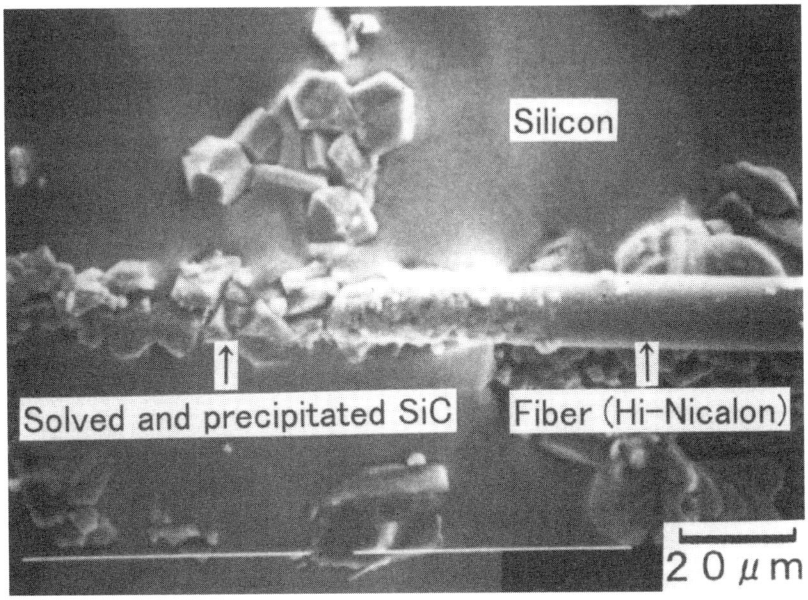

Figure 18.3 **SEM Microphotograph of the Contact Region of Hi-Nicalon and Silicon After Heat Treatment at 1420°C (1693K) in Vacuum (Reprinted with permission from ref. [10])**

microscope (SEM) microphotograph of the contact region of Hi-Nicalon and silicon after heat treatment at 1420°C (1693K) in vacuum. Material in the contact region of the fiber had dissolved and then reprecipitated. This remarkable morphology change was believed to have caused the drastic degradation of the fiber strength. Therefore controlling the contact and the reaction between fiber and molten silicon in the fabrication process is believed to be important for obtaining desirable mechanical properties. So, the effect of the process conditions on two kinds of interface structures, BN and BN/SiC, and two kinds of infiltrating metals, silicon and boron doped silicon, were investigated.

Figure 18.4. shows the typical load-displacement curves for braided samples in a 3-point bending test at room temperature, for BN coated fiber and Si infiltrated sample (a), for BN/SiC coated fiber and Si infiltrated sample (b), for BN coated fiber and Si-B infiltrated sample (c) and for BN/SiC coated fiber and Si-B infiltrated sample (d) [13]. The crosshead speed was 0.5 mm/min. SEM microphotographs of corresponding fracture surfaces are shown in Figure 18.5. For BN coated fiber and Si infiltrated sample (a), the fracture behavior was relatively brittle. The fracture surface showed little fiber pull-out. This was thought to be due to the degradation of the mechanical properties of the fibers, which were in contact with molten silicon during sintering. To prevent the reaction between the fibers and the molten silicon, a SiC overlay coating (b) and boron doping of the silicon were applied, respectively. High strength and high fracture energy properties were obtained by selecting the BN/SiC coated fiber and the Si/B infiltrated condition (d). The fracture surface showed obvious fiber pull-out. It was believed that the SiC overlay coating on the BN layer was effective in preventing contact of molten silicon with the fibers, and that boron doping might reduce silicon reactivity.

Figure 18.6 shows the EPMA (Electron Probe Micro Analysis) composition analysis for the polished section of the BN/SiC coated fiber and Si-B infiltrated sample. Bright contrast regions were observed for some

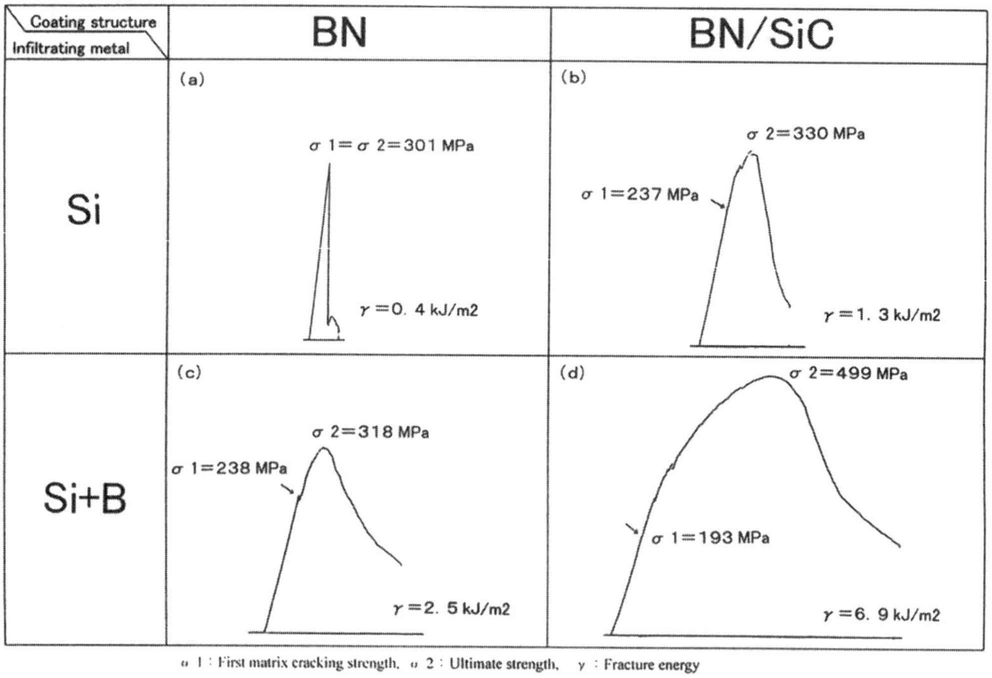

Figure 18.4 Typical Load-Displacement Curves for Braided Samples in a 3-Point Bending Test at Room Temperature, for BN Coated Fiber and Si Infiltrated Sample (a), for BN/SiC Coated Fiber and Si Infiltrated Sample (b), for BN Coated Fiber and Si-B Infiltrated Sample (c) and for BN/SiC Coated Fiber and Si-B Infiltrated Sample (d). (Reprinted with permission from refs. [10,13])

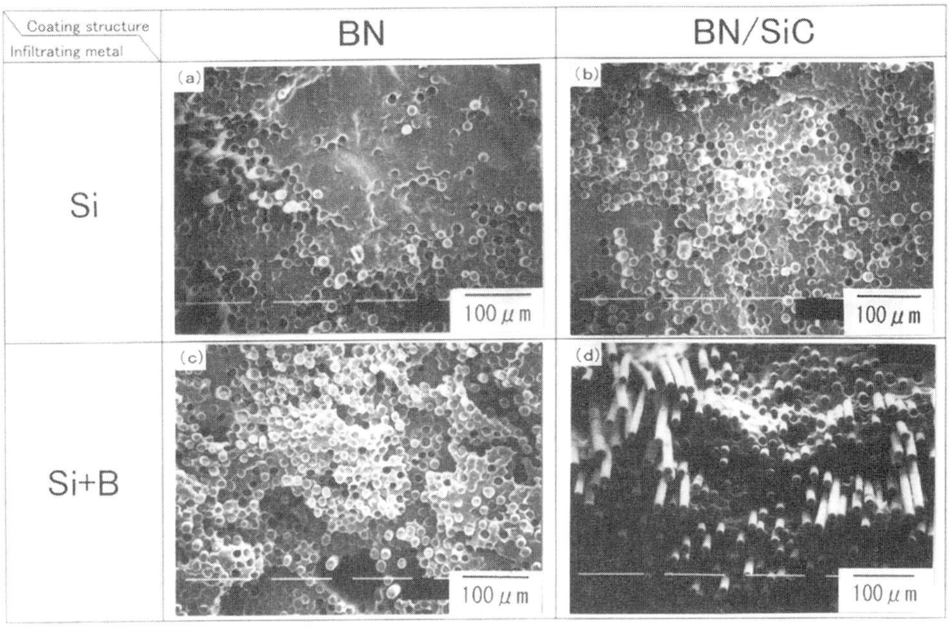

Figure 18.5 SEM Microphotographs of Fracture Surfaces After 3-Point Bending Test at Room Temperature, for BN Coated Fiber and Si Infiltrated Sample (a), for BN/SiC Coated Fiber and Si Infiltrated Sample (b), for BN Coated Fiber and Si-B Infiltrated Sample (c) and for BN/SiC Coated Fiber and Si-B Infiltrated Sample (d) (Reprinted with permission from refs. [10,13])

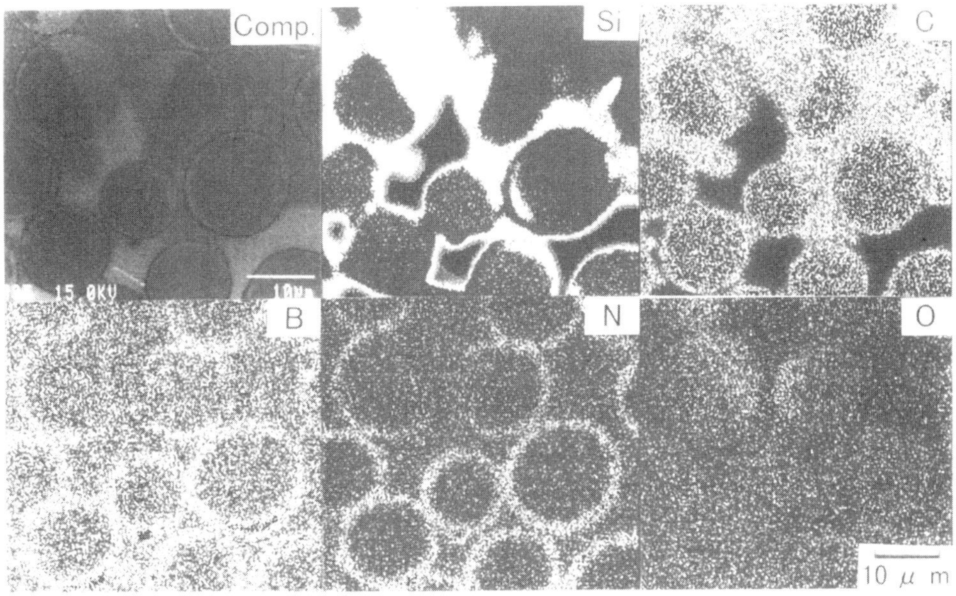

Figure 18.6 EPMA Composition Analysis for the Changed Contrast Region of the Composite Fibers (Reprinted with permission from ref. [10])

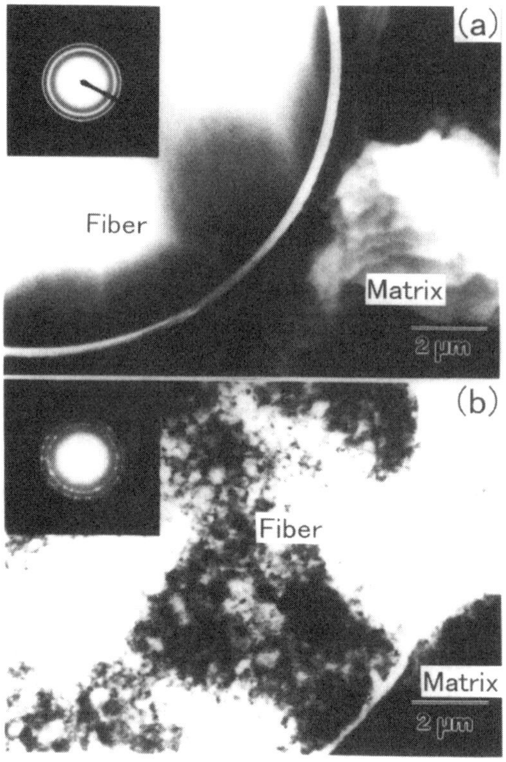

Figure 18.7 Typical Microstructure at the Fiber and Matrix Interface Observed by TEM Bright Field Image and Electron Diffraction Patterns, for Normal Fiber Region (a) and Reacted Fiber Region (b). (Reprinted with permission from ref. [10])

fibers in the composition image. They corresponded to the silicon-rich and carbon-poor regions. The original Hi-Nicalon fiber has a carbon-rich composition ($Si/C < 1$) [16]. It is possible that the bright contrast regions were the fiber regions that had reacted with molten silicon during sintering. Figure 18.7 shows a typical microstructure at the fiber and matrix interface observed by transmission electron microscope (TEM) bright field image and electron diffraction patterns, for a normal fiber region (a) and a reacted fiber region (b). Some fibers were recognized to be damaged even for the BN/SiC coated fiber and Si-B infiltrated sample.

Effect of Sintering Condition and Coating Thickness

The sintering condition was optimized in order to reduce the fiber damage. It was observed that the lowest sintering temperature for obtaining a uniform melt of infiltrating metal was 1420°C (1693K), and the minimum sintering time for forming a dense matrix was found to be 0.3 ks. This sintering condition of 1420°C (1693K) for 0.3 ks was employed for the test samples for which the data are given below.

Figure 18.8 shows the ultimate strength in room temperature tensile testing for uni-directional samples with BN(0.4μm), BN(0.4μm)/SiC(0.3μm), BN(1.0μm), and BN(1.0μm)/SiC(0.3μm) coated fibers [13]. The crosshead speed for tensile testing was 0.1 mm/min. It was clear that the ultimate tensile strength depended on the structure of the interfacial layer. The higher strengths were obtained for specimens with the thicker BN and BN/SiC coatings. SEM microphotographs of corresponding fracture surfaces are shown in Figure 18.9. For the sample with the BN(0.4μm) coated fiber, many fibers fractured at the same surface as the matrix were observed. The length of the pull-out fibers was small. For the samples with the

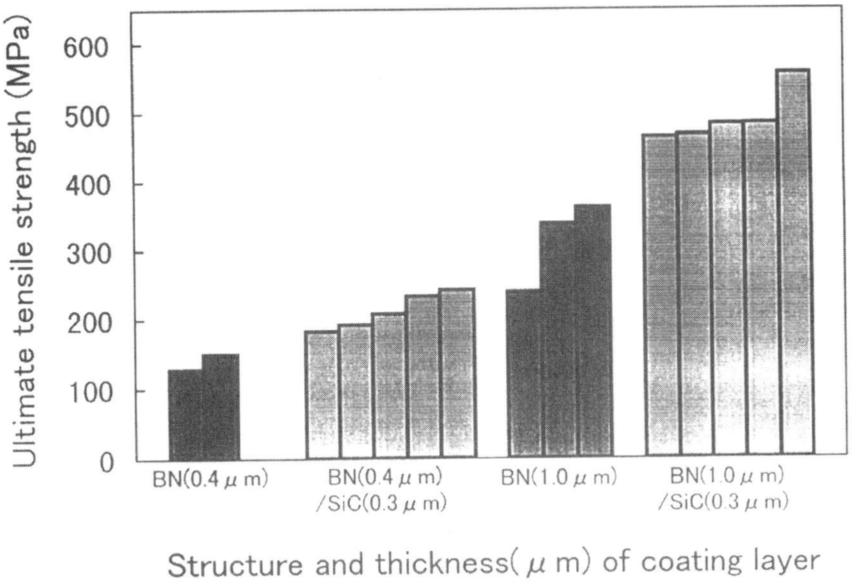

Figure 18.8 Ultimate Strengths in Tensile Testing for Uni-Directional Samples with BN(0.4μm), BN(0.4μm)/SiC(0.3μm), BN(1.0μm), and BN(1.0μm)/SiC(0.3μm) Coated Fibers. The sintering condition was Si-B infiltrating at 1420°C (1693K) for 0.3 ks (Reprinted from ref. [13])

Figure 18.9 SEM Microphotographs of Fracture Surfaces after Tensile Testing for Uni-Directional Samples with (a) BN(4μm), (b) BN(0.4μm)/SiC(0.3μm), (c) BN(1.0μm), and (d) BN(1.0μm)/SiC(0.3μm) Coatings. The sintering condition was Si-B infiltrating at 1420°C (1693K) for 0.3 ks. (Reprinted with permission from Toshiba Corporation)

BN(0.4μm)/SiC(0.3μm) and BN(1.0μm) coated fibers, the pull-out fiber length showed a large distribution. For the samples with the BN(1.0μm)/SiC(0.3μm) coated fibers, the length of the pull-out fibers was relatively small and of uniform distribution. Taking into consideration also the tensile strength properties, the reaction between fiber and molten silicon was believed to be mostly prevented in the sample with the BN(1.0μm)/SiC(0.3μm) coated fiber.

EVAUATION OF MICROSTRUCTURE AND MECHANICAL PROPERTIES

Density and fiber volume fraction of the fabricated composite test samples were evaluated by the Archimedes' method and observation of polished cross sections, respectively. They were about 3.0 g/cm^3 and 30 vol%, both for the braided and uni-directional samples.

3-Point bending strength properties at room temperature and at 1300°C (1573K) were evaluated for the braided samples fabricated with the BN/SiC coated fiber and Si-B infiltrated condition. The bending strength at room temperature after oxidation at 1300°C (1573K) for 360 ks in air was also evaluated for some samples. Figure 18.10 shows the typical load-displacement curves for the BN/SiC coated fiber and Si-B infiltrated samples in 3-point bending at room temperature (a), at 1300°C (1573K) in air (b), at 1300°C (1573K) in argon (c), and at room temperature for specimens oxidized at 1300°C (1573K) for 360 ks (d). The ultimate strengths in bending testing did not change significantly for any of the test conditions.

Figure 18.11 shows the stress-strain behavior in tensile testing for the sample with BN(1.0μm)/SiC(0.3μm) coated fiber [13,17]. The elastic modulus was 242 GPa, and the stress increased almost linearly with the increase of the strain, up to about 400 MPa. The matrix micro-cracking started at a stress of about 170 MPa. For further strain increase, the stress showed a plateau. The stress increased again at around 0.4 % strain and the ultimate tensile strength of 556 MPa was reached at about 0.82 % strain. The failure elongation was about 0.93 %. Figure 18.12 shows a SEM microphotograph of micro-cracks in the matrix after the tensile test. Multiple micro-cracks perpendicular to the tensile direction can be clearly observed in this figure [13,17].

To evaluate the constitutive properties of a uni-directional sample with BN(1.0μm)/SiC(0.3μm) coated fiber, the fracture surfaces of the pull-out fibers were evaluated. In situ fiber strength was related to fracture mirror size [18]. For about 150 pieces of pull-out fibers, the fracture surfaces were observed by SEM. The

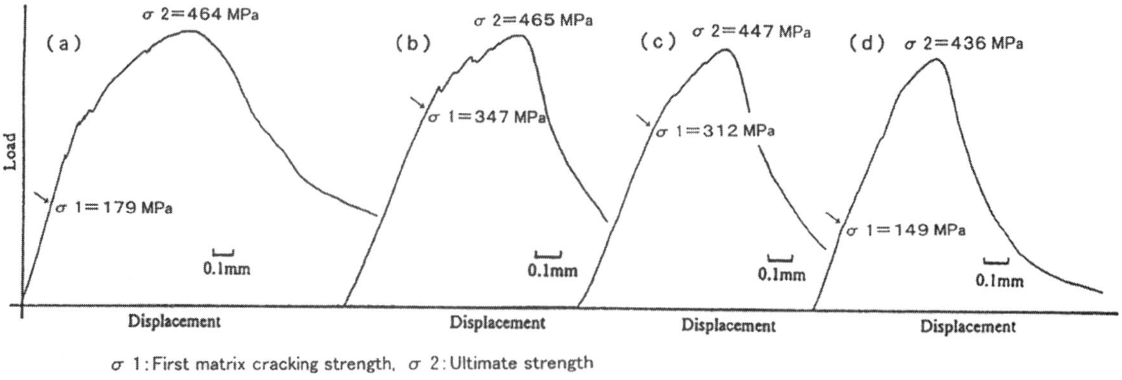

Figure 18.10 Typical Load-Displacement Curves in 3-Point Bending Testing at Room Temperature (a), at 1300°C (1573K) in Air (b), at 1300°C (1573K) in Argon (c), and at Room Temperature for Specimens Oxidized at 1300°C (1573K) for 360 ks (d) (Reprinted with permission from refs. [7,10])

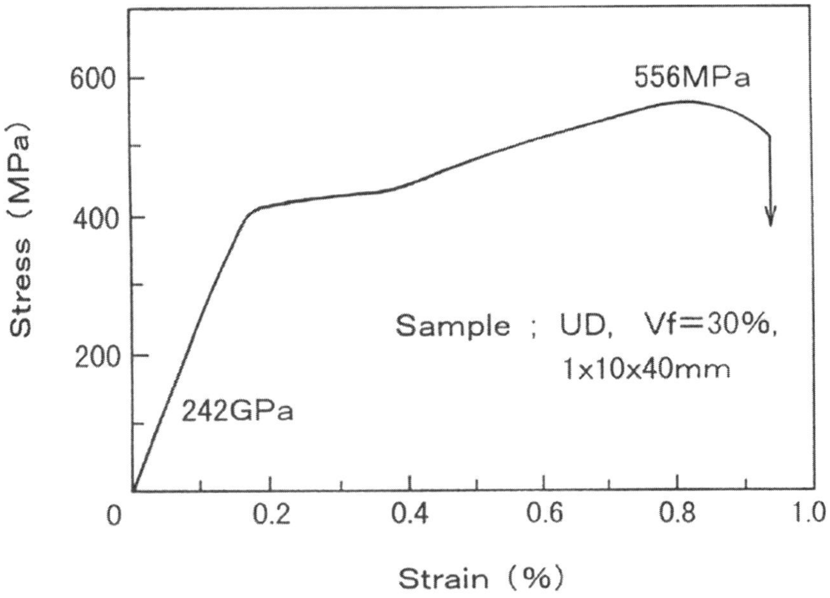

Figure 18.11 Typical Stress-Strain Behavior Obtained in Tensile Testing for the Uni-Directional Sample with BN(1.0μm)/SiC(0.3μm) Coated Fiber. The sintering condition was Si-B infiltrating at 1420°C (1693K) for 0.3ks (Reprinted with permission from refs. [13,17])

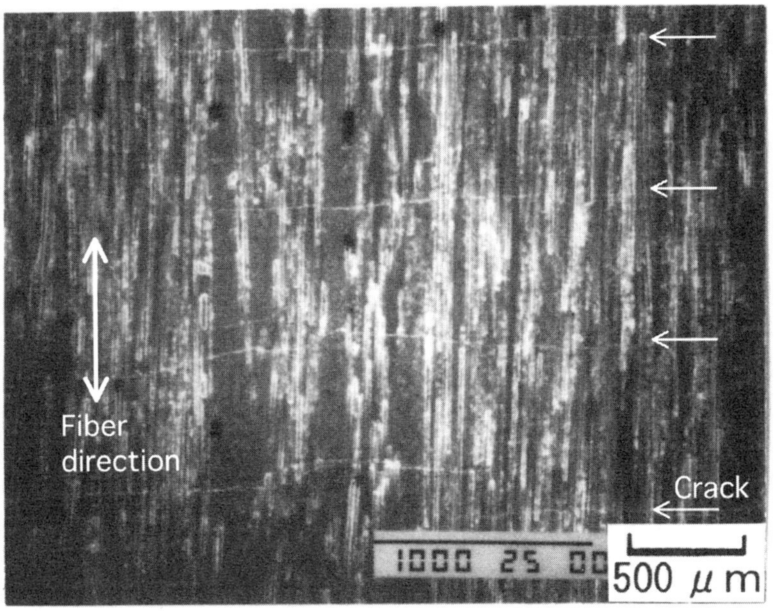

Figure 18.12 SEM Microphotograph of Micro-Cracking in the Matrix after Tensile Testing for the Uni-Directional Sample with BN(1.0μm)/SiC(0.3μm) Coated Fiber (Reprinted with permission from refs. [13,17])

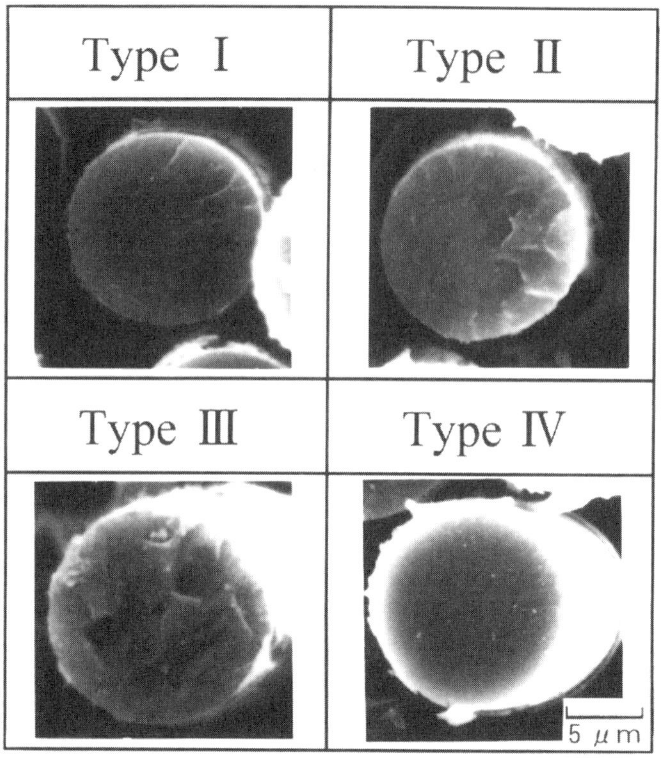

Figure 18.13 SEM Microphotographs Showing Four Types of Fracture Surfaces of Pull-Out Fibers after Tensile Testing (see Table 18.1) for the Uni-Directional Sample with BN(1.0μm)/SiC(0.3μm) Coated Fiber.
The sintering condition was Si-B infiltrating at 1420°C (1693K) for 0.3ks.
(Reprinted with permission from refs. [13,17])

morphology of the fracture surface was divided into 4 types. Figure 18.13 shows typical fracture surfaces illustrating the 4 types. They were Type I—the fracture origin was on the fiber surface and the mirror and hackle zones were distinguished, Type II—the fracture origin was inside of the fiber and the mirror and hackle zones were distinguished, Type III—the fracture origin was unclear and the overall surface was rough, and Type IV—the overall surface was flat (only mirror zone). The evaluated percentages of fracture surface types for pull-out fibers are summarized in Table 18.1 [13,17]. From the measurement of mirror radius for

TABLE 18.1 OBSERVED PERCENTAGE OF FRACTURE SURFACE TYPES FOR PULL-OUT FIBERS (REPRINTED FROM REF. [13])

Fracture Surface Type	I	II	III	IV
Observed (%)	56.6	9.6	14.5	19.3

I: Fractured from surface flaw. Mirror plane observed.
II: Fractured from inner flaw. Mirror plane observed.
III: Complex fracture surface.
IV. Almost flat fracture surface.

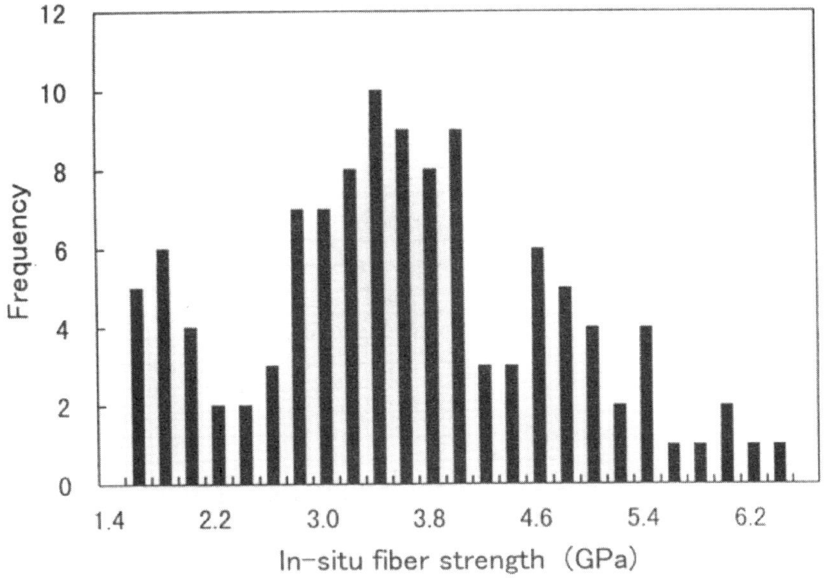

Figure 18.14 In Situ Fiber Strength Distribution from Fracture Mirror Size in Tensile Testing at Room Temperature (Reprinted with permission from refs. [13,17])

Type I and Type II, in situ fiber strengths were calculated. In this analysis, 1.48 MPa.m$^{1/2}$ was used for the K_{IC} value based on the experimental results for the relation between the tensile strength and the mirror radius of BN coated Hi-Nicalon monofilaments [19]. The in situ fiber strength distribution is shown in Figure 18.14 [13,17]. From the Weibull plot of these strength data shown in Figure 18.15, the values of s_c = 3.87 GPa and m = 3.65 were obtained. It is not believed that the fibers which failed in Type IV mode related to the composite ultimate strength, because they might have failed at a lower stress level. So the effective fiber volume fraction was corrected to 30 × (1 – 0.193) = 24.2 %. The constitutive properties of the sample are summarized in Table 18.2 [13].

Next, impact stress resistance was evaluated by foreign object collision and thermal shock tests. Figure 18.16 shows the fracture morphologies after foreign object damage testing for four kinds of braided samples, BN coated fiber and Si infiltrated samples (a), BN/SiC coated fiber and Si infiltrated samples (b), BN coated fiber and Si-B infiltrated samples (c) and BN/SiC coated fiber and Si-B infiltrated samples (d). The sample size was 10 mm × 20 mm × 1 mm. The collision object was a ϕ 2.38 mm steel ball, and the collision speed was 113m/s. For sample (a), some cracks through the sample were observed. On the other hand, for the samples (b), (c) and (d), the damages were limited to small regions. Figure 18.17 shows the fracture morphologies after foreign object damage testing for a Sialon monolithic ceramic fabricated by Toshiba Corp., which has about 800 MPa of 3-point bending strength and about 4 MPa.m$^{1/2}$ of fracture toughness, and the composite. The composite sample was fabricated with the BN/SiC coated fiber braided preform and Si-B infiltrated condition. The fiber volume fraction was 17.8%. And the sample size was 20 mm × 20 mm × 2 mm for both samples. The Sialon monolithic ceramic sample showed catastrophic failure after the ϕ 3mm steel ball's 31 m/s collision. On the other hand, the composite sample showed through hole failure, not catastrophic, after the same steel ball's 141m/s collision. Figure 18.18 shows the fracture morphologies after thermal shock testing in a 900°C (1173K) to room temperature water bath. The samples were filament wound and braided composites fabricated with the BN/SiC coated fiber and Si-B infiltrated condition. A reaction sintered SiC monolithic ceramic fabricated by Toshiba Corp., which has about 600 MPa of 3-point bending

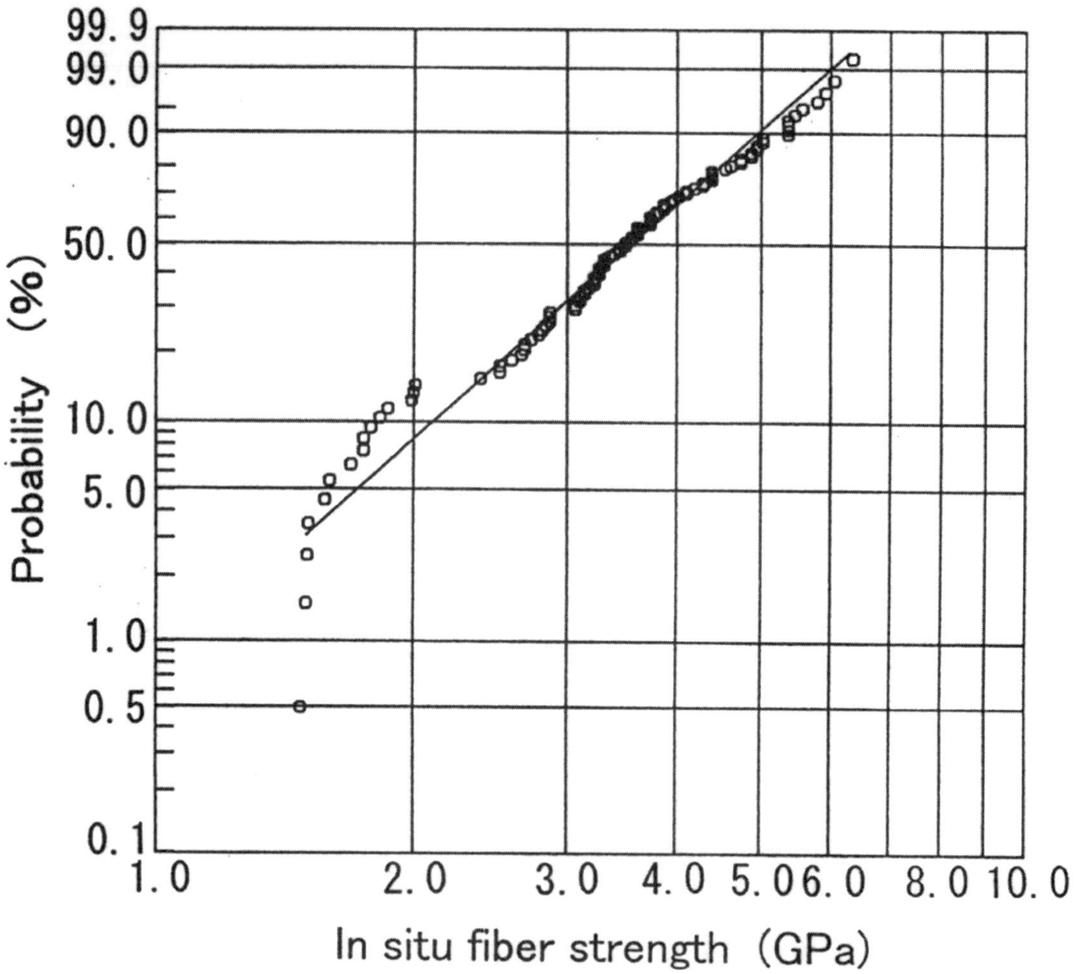

Figure 18.15 Weibull Plot for in Situ Fiber Strengths from Fracture Mirrors in Composites Tested in Tensile Mode at Room Temperature (Reprinted from ref. [13])

TABLE 18.2 CONSTITUTIVE PROPERTIES OF UNI-DIRECTIONAL SAMPLE (REPRINTED FROM REF. [13]).

V_f (%)	f_l (%)	σ_c (GPa)	m	σ_{uts} (MPa)
30	24.2	3.87	3.65	556

V_f: Fiber volume fraction for specimen
f_l: Effective fiber volume fraction for ultimate tensile strength
σ_c: In situ fiber mean strength
m: Weibull modulus for in situ fiber strength
σ_{uts}: Ultimate tensile strength

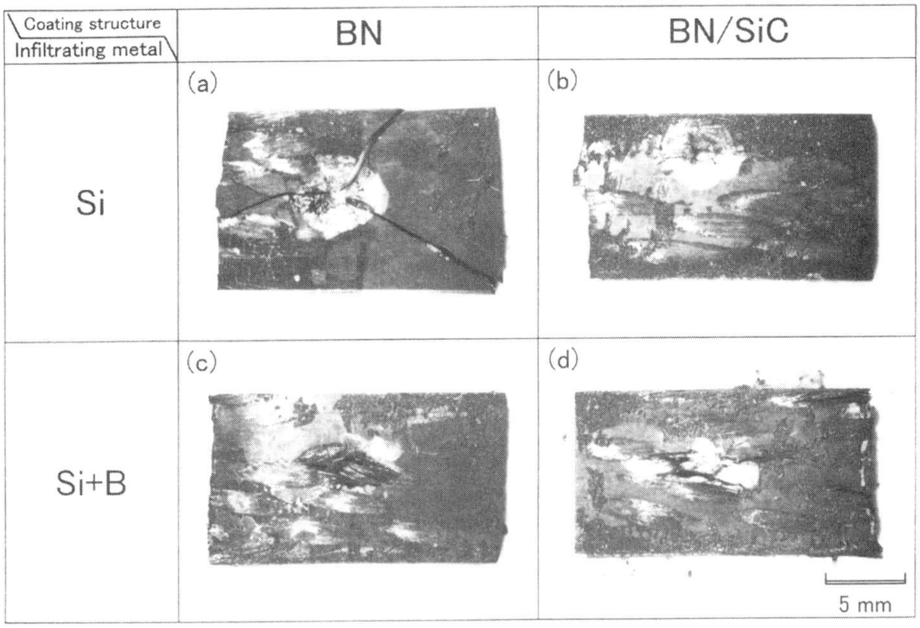

Figure 18.16 Fracture Morphologies After Foreign Object Damage Test at Collision Speed of 113m/s, for BN Coated Fiber and Si Infiltrated Sample (a), BN/SiC Coated Fiber and Si Infiltrated Sample (b), BN Coated Fiber and Si-B Infiltrated Sample (c) and BN/SiC Coated Fiber and Si-B Infiltrated Sample (d) (Reprinted with permission from Toshiba Corporation)

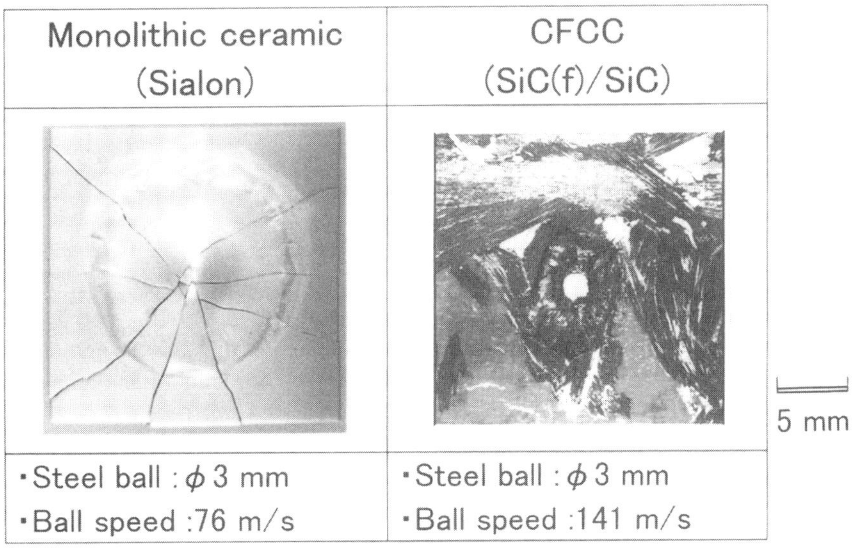

Figure 18.17 Fracture Morphologies After Foreign Object Damage Test for Sialon Monolithic Ceramic and Composite (Reprinted with permission from Toshiba Corporation)

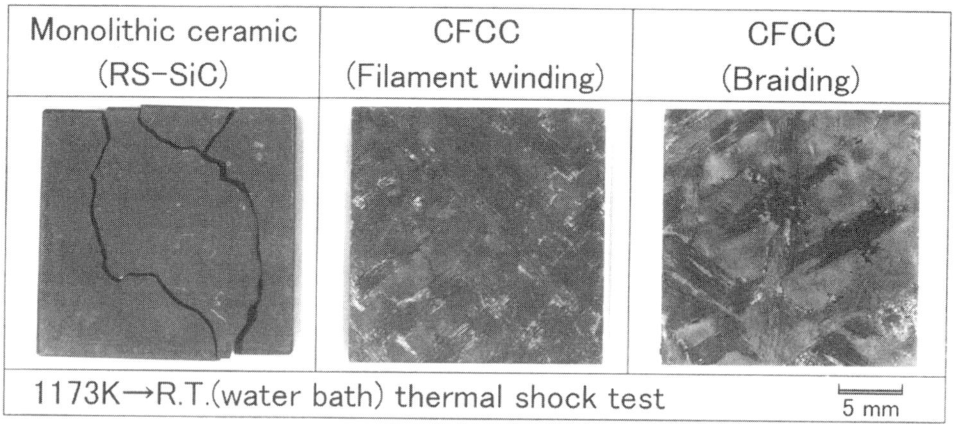

Figure 18.18 Fracture Morphologies after Thermal Shock Test from 900°C (1173K) to Room Temperature in a Water Bath, for Reaction Sintered SiC Monolithic Ceramic, Filament Wound and Braided Samples (Reprinted with permission from Toshiba Corporation)

strength and about 3 $MPa.m^{1/2}$ of fracture toughness, was used for reference. The fiber volume fraction values of the composites were 30.4% and 17.8%, respectively. The sample size was 20 mm × 20 mm × 2 mm. The reaction sintered SiC monolithic ceramic fragmented into small pieces, but no large crack was observed for both composite samples. This composite demonstrated improved impact resistance compared to the conventional monolithic ceramics.

In Table 18.3, some mechanical and thermal properties of the developed reaction sintered SiC matrix composite and other typical SiC matrix composites have been summarized. The reaction sintered SiC matrix composite had low porosity, and it was believed to have good oxidation resistance. The high Young's modulus and high thermal conductivity based on the dense matrix were also thought to be favorable for gas turbine hot section applications.

GAS TURBINE HOT SECTION COMPONENT DEVELOPMENT

Fabrication of Combustor Liner and Shroud Segment

The combustor liner and the shroud segment were fabricated applying the process conditions described for (d), i.e. a BN/SiC fiber coating and Si-B infiltrated matrix (see Figures 18.4 and 18.5). Figure 18.19 shows their dimensions. The combustor was a subscale component and the shroud segment was a full-scale part for a 15 MW gas turbine. The fiber preforms were fabricated by a filament winding technique. The fabricated parts are shown in Figure 18.20. As illustrated in Figure 18.19, some notches were machined to evaluate crack propagation properties in the combustion test. Some properties of the applied material have been summarized in Table 18.4 [13].

PARTS EVAUATION IN COMBUSTOR TEST

Crack propagation behavior for the combustor liner and the shroud segment were evaluated in the atmospheric combustion thermal cycle test. Figure 18.21 shows a schematic illustration of the combustion test apparatus. The combustion gas was pre-heated to 400°C (673 K) in the pre-combustor, then the pre-heated gas

TABLE 18.3 SOME MECHANICAL AND THERMAL PROPERTIES OF DEVELOPED REACTION SINTERED SiC MATRIX COMPOSITE AND OTHER TYPICAL SiC MATRIX COMPOSITES

Process	**RS**		**PIP**	**CVI**
Structure				
Fiber	Hi-Nicalon	Hi-Nicalon	Nicalon	Nicalon
Fiber volume fraction	30% (Uni-directional)	30% (Braiding)	30% (Cloth laminated)	40% (Cloth laminated)
Density	3.0 Mg/m^3	3.0 Mg/m^3	1.8 Mg/m^3	2.3 Mg/m^3
Porosity	2%	2%	23%	8–12%
Properties				
Heat resistant temperature	1623K	1623K	1273K	$\geqq$1478K
Young's modulus	240 GPa (Y-direction)	260 GPa (Y-direction)	50 GPa (Y-direction)	140 GPa (Y-direction)
Tensile strength	550 MPa (Y-direction)	—	110 MPa (Y-direction)	200 MPa (Y-direction)
Poisson's ratio	—	0.2	—	—
Thermal conductivity		50 W/mK(RT) 30 W/mK (1273K) (Z-direction)	0.56 W/mk (773K) 0.73 W/mK (973K) (Z-direction)	6.9 W/mK (RT) 6.2 W/mK (1473K) (Z-direction)
Thermal expansion coefficient	—	4.9 × 10^{-6}/K (Y-direction) 4.7 × 10^{-6}/K (X-direction)	3.6 × 10^{-6}/K (Y-direction)	3.0 × 10^{-6}/K (Y-direction) 2.5 × 10^{-6}/K (Z-direction) 
Characteristics				
Productivity	Good	Good	Not bad	Not good
Complex shape	Possible	Possible	Possible	Possible
Cost	Low	Low	Not high	High

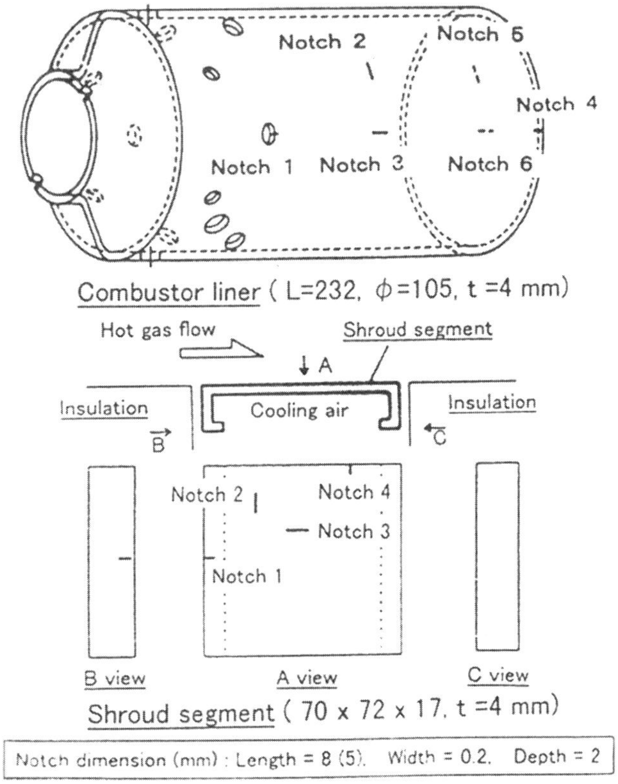

Figure 18.19 Combustor Liner and Shroud Segment for Combustion Test (Reprinted from ref. [13])

Combustor liner　　　　Shroud segment

Figure 18.20 Photographs of Fabricated Combustor Liner and Shroud Segment (Reprinted from ref. [13])

TABLE 18.4 MATERIAL PROPERTIES OF COMBUSTOR LINER AND SHROUD (REPRINTED FROM REF. [13])

Material		
Fiber/matrix	SiC/SiC (Hi-Nicalon / RS-SiC)	
Fiber volume fraction V_f	30%	
Pre-forming technique	Filament winding	
Matrix forming method	Reaction sintering	
Properties	**45°/–45° in plane**	**Through thickness**
3-point bending strength (at R.T.)	First matrix cracking strength	130 MPa
	Ultimate strength	209 MPa
Young's modulus (at 1000°C)	225 GPa	9.7 GPa
Poisson ratio (at 1000°C)	0.26	0.008
Thermal conductivity (at 1000°C)	30.4 W/m.K	16.7 W/m.K
Thermal expansion coefficient (at 1000°C)	5.1×10^{-6}/K	5.4×10^{-6}/K

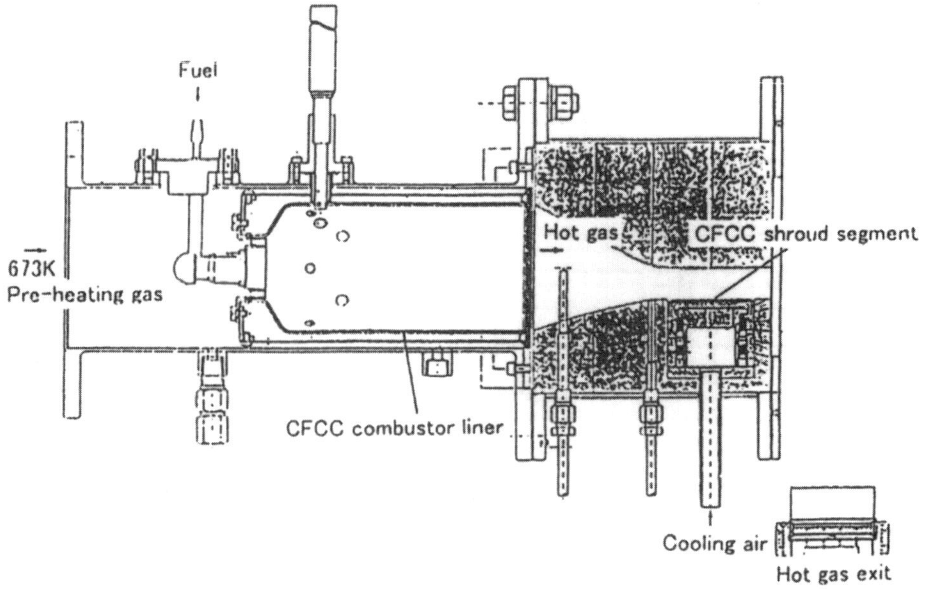

Figure 18.21 Schematic Illustration of Atmospheric Combustion Test Apparatus (Reprinted from ref. [13])

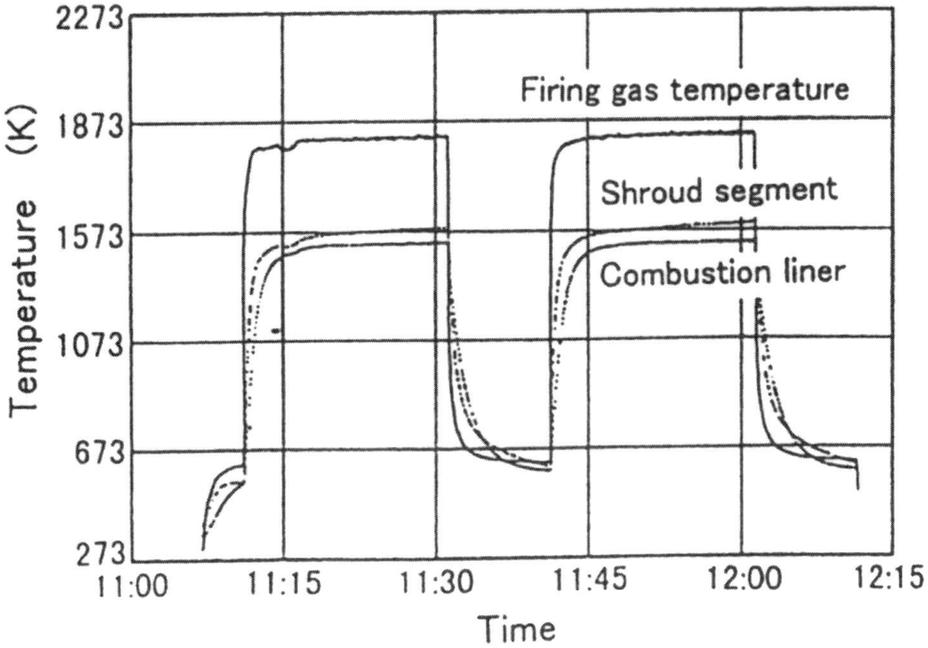

Figure 18.22 Temperature Measurement Results for 1500°C (1773K) Combustion Test (Reprinted from ref. [13])

flowed to the main combustor with the test liner. At the exit of the main combustor, the hot gas path section with the test shroud segment was located. Shutting off the fuel supply to the main combustor provided the thermal cycle. The fuel and air supply to the pre-combustor were constant. The temperature of the combustion gas was measured at the exit of the main combustor. The temperatures of the combustor liner and the shroud segment test parts were measured by attached thermocouples.

The parts were exposed to the thermal cycle test for 51 cycles of 1300°C/400°C (1573 K/673 K) for 15 min/10 min, and subsequently 166 cycles of 1500°C/400°C (1773 K/673 K) for 20 min/10 min. The measured temperature profiles are shown in Figure 18.22. The temperatures were 1500°C (1773 K) for combustion gas and about 1200°C (1473 K) for the test parts. After the thermal cycle test, no crack propagation was observed at the machined notches of the combustion liner. For the shroud segment, some crack propagation was observed from the notches No.1 and 4. The crack propagation from notch No.1 is shown in Figure 18.23. Obvious crack propagation was observed at the early stages of each test condition, but the crack propagation was saturated at higher cycle times. The microphotographs of the crack growth after 69 cycles and 132 cycles are shown in Figure 18.24. Photographs of the shroud segment after the thermal cycle test are shown in Figure 18.25. Catastrophic failure was not observed at all. The tested parts showed good durability.

Finite Element Stress Analysis

Figure 18.26 shows the finite element stress analysis for the shroud segment under the 1500°C (1773 K) steady state combustion condition. The largest tensile stress was at notches nos.1 and 4. In the transient analysis, about 1.2 times higher stress was observed at the same notches within 10 s after a fuel trip. This stress distribution analysis corresponds with the crack propagation locations observed in the combustion test.

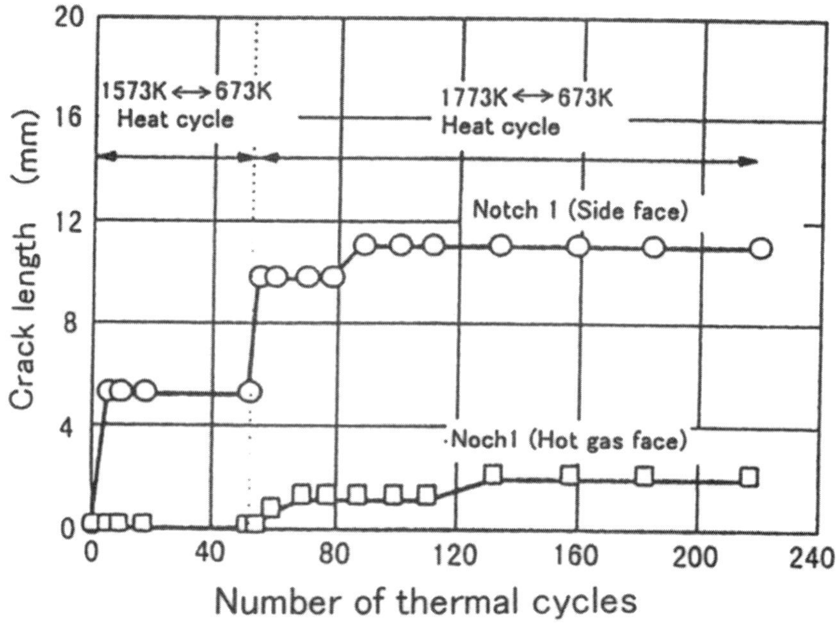

Figure 18.23 Crack Propagation versus Thermal Cycles for Notch 1 of the Shroud Segment (Reprinted from ref. [13])

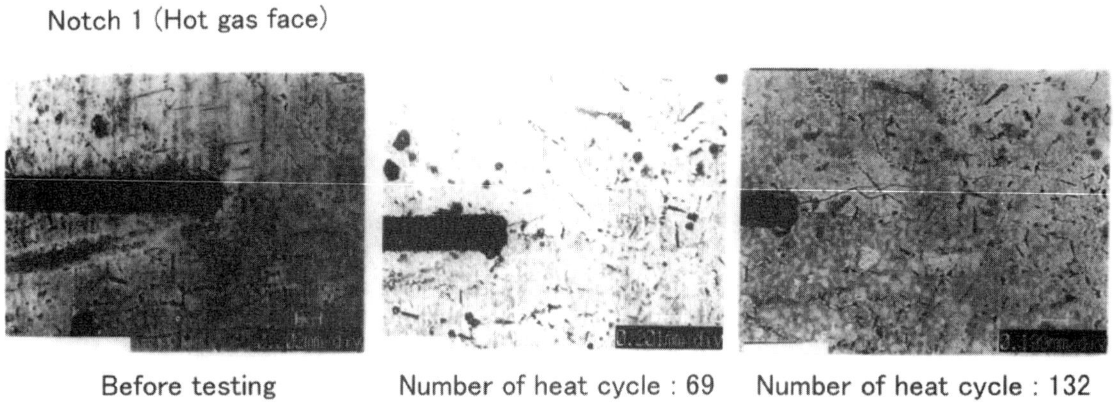

Figure 18.24 Optical Microphotographs of Crack Growth from Notch 1 of Shroud Segment after 69 and 132 Thermal Cycles (Reprinted from ref. [13])

CONCLUSIONS

(1) A Dense SiC matrix composite reinforced by Hi-Nicalon fibers was developed by slurry impregnation and subsequent reaction sintering process.

(2) By selecting appropriate processing conditions for the BN/SiC dual coating and Si-B metal infiltration for this composite, high strength and large fracture energy properties were achieved.

(3) The in situ fiber strength of a high tensile strength sample was evaluated from the fracture mirror size.

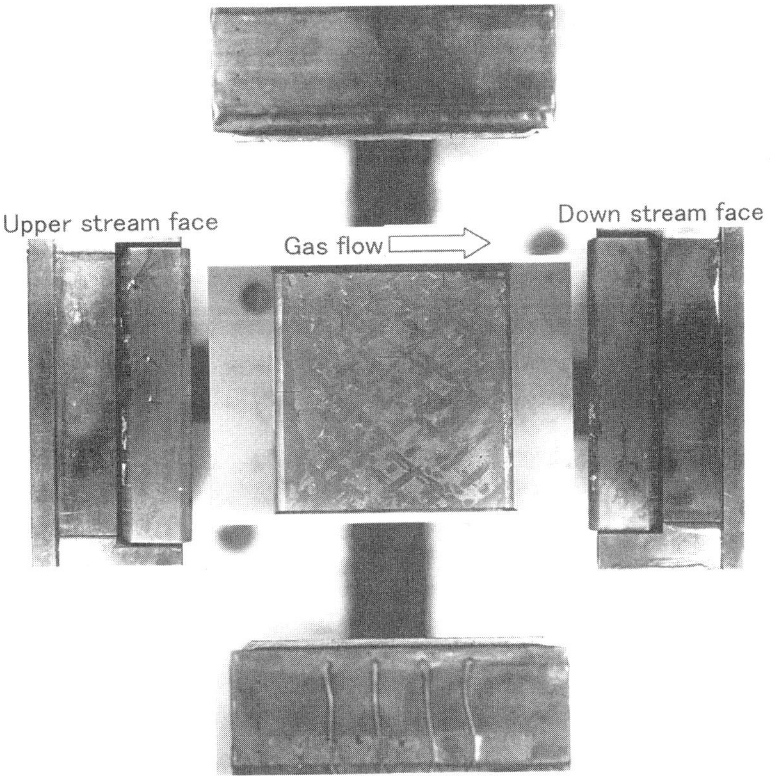

Figure 18.25 Photographs of the Shroud Segment after Combustion Test (Reprinted from ref. [13])

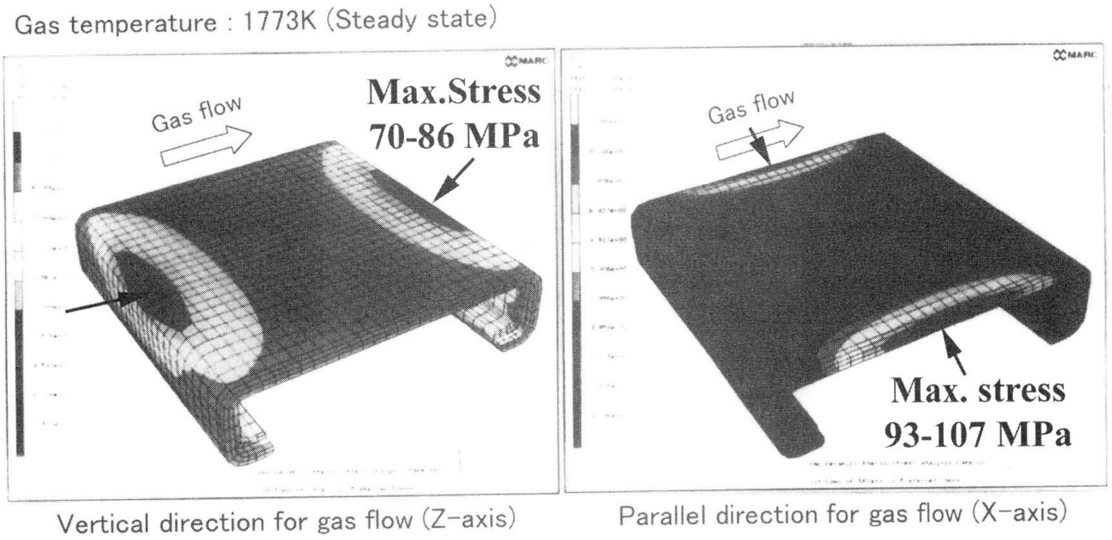

Figure 18.26 FEM Stress Analysis for the Shroud Segment Under the 1500°C (1773 K) Steady State Combustion Condition (Adapted from ref. [13])

(4) Improved impact stress resistance was evaluated by foreign object collision and thermal shock tests.
(5) High Young's modulus and high thermal conductivity properties were obtained with a dense matrix structure.
(6) The subscale combustor liner and the full scale scale shroud segment for a 15 MW gas turbine were successfully manufactured.
(7) The crack propagation behavior for the combustor liner and the shroud segment were evaluated in the combustion thermal cycle test. Crack propagation was observed at the early stage of each shroud test condition, but crack propagation was saturated at higher cycle times. The tested parts show good durability.
(8) Finite element stress distribution analysis corresponds with the crack propagation locations observed in the combustion thermal cycle test.
(9) In the future development, it is thought to be necessary to evaluate and also to improve the long-term durability of the CMC (Ceramic Matrix Composite) parts in realistic gas turbine environments.

NOMENCLATURE

σ 1	First matrix cracking strength in bending test
σ 2	Ultimate strength in bending test
γ	Fracture energy in bending test
CFCC	Continuous Fiber-reinforced Ceramic Composite
V_f	Fiber volume fraction
f_l	Effective fiber volume fraction for ultimate tensile strength
σ_c	In situ fiber mean strength
m	Weibull modulus for in situ fiber strength

REFERENCES

[1] Corman, G.S., Heinen, J.T., and Goetz, R.H., "Ceramic Composites for Industrial Gas Turbine Engine Applications: DOE CFCC Phase 1 Evaluations," ASME paper 95-GT-387, presented at the 40th ASME International Gas Turbine and Aeroengine Congress and Exposition, Houston, TX, USA, June 5–8, 1995.

[2] Van de Voorde, M.H. and Nedele, M.R., "CMCs: Research in Europe and the Future Potential of CMCs in Industry," Ceram Eng. and Sci. Proc., 17[4], 3–21, 1996.

[3] Luthra, K.L., Singh, R.N., and Brun, M.K., "Toughened Silcomp Composites—Process and Preliminary Properties," Am. Ceram. Soc. Bull., 72 [7], 79–85, 1993.

[4] Szweda, A., Easlar, T.E., Petrak, D.R., and Black, V.A., "Continuous Fiber Ceramic Matrix Composites for Gas Turbine Applications," ASME Paper 99-GT-189, presented at the 44th International Gas Turbine and Aeroengine Congress, Exposition and Users Symposium, Indianapolis, IN, USA, June 7–10, 1999.

[5] Corman, G.S., Brun, M.K. and Luthra, K.L., "SiC Fiber Reinforced SiC-Si matrix Composites Prepared by Melt Infiltration (MI) for Gas Turbine Engine Applications," ASME paper 99-GT-234, presented at the 44th ASME Gas Turbine and Aeroengine Technical Congress, Exposition and Users Symposium, Indianapolis, IN, USA, June 7–10, 1999.

[6] Dean, A.J., Corman, G.S., Bagepalli, B., Luthra, K.L., DiMascio, P.S. and Orenstein, R.M., "Design and Testing of CFCC Shroud and Combustor Components," ASME paper 99-GT-235, presented at the 44th ASME Gas Turbine and Aeroengine Technical Congress, Exposition and Users Symposium, Indianapolis, IN, USA, June 7–10, 1999.

[7] Kameda, T., Sayano, A., Amiji, N., Ichikawa, H., Hamada, H., Fujita, A., and Uozumi, T., "Fabrication and Mechanical Properties of Reaction Sintered Silicon Carbide Matrix Composite," Ceram. Eng. and Sci. Proc., 18[3], 419–426, 1997. Figure 3 in this reference was reprinted with permission of the American Ceramic Society, www.ceramics.org. © 1999. All rights reserved.

[8] Itoh, Y., Kameda, T., Nishida,K., Umezawa, M., Imai, K, and Ichikawa, H., "Effect of CVD-BN Coating on Mechanical Properties of Continuous Silicon Carbide-Fiber," J. Ceram. Soc., Japan. Inter. Ed., 106[8], 830–834, 1998.

[9] Itoh, Y., Kameda, T., Suyama, S., and Nagata, K., "Material Design for High Strength and Toughness of Continuous Fiber-Reinforced Ceramic Matrix Composites," J. Ceram. Soc. Japan. Inter. Ed., 106[12], 1190–1195, 1998.

[10] Kameda, T., Suyama, S., Itoh, Y., and Goto, Y., "Development of Continuous SiC Fiber-Reinforced Reaction Sintered SiC Matrix Composite," J. Ceram. Soc. Japan. Inter. Ed., 107[4], 327–334, 1999.

[11] Itoh, Y., Kameda, T., and Suyama, S., "Material Design for High Strength of Continuous Fiber-Reinforced and Reaction—Sintered SiC Matrix Composites by Paying Attention to Fiber-Pullout Behavior," J. Ceram. Soc. Japan. Inter. Ed., 107[5], 425–432, 1999.

[12] Kameda, T., Suyama, S., Itoh, Y., and Nishida, K. J., "Development of Reaction Sintered Silicon Carbide Matrix Composite and Tensile Strength Properties," J. Ceram. Soc. Japan Inter. Ed., 107[7], 622–626, 1999.

[13] Kameda, T., Itoh, Y., Hijikata, T., and Okamura, T., "Development of Continuous Fiber Reinforced Reaction Sintered Silicon Carbide Matrix Composite for Gas Turbine Hot Parts Application," ASME paper 2000-GT-67, presented at the ASME TURBO EXPO 2000, Land, Sea, & Air, Munich, Germany, May 8–11, 2000.

[14] Li, J.-G. and Hausner, H., "Reactive Wetting in the Liquid-Silicon/Solid-Carbon System," J. Am. Ceram. Soc., 79[4], 873–880, 1996.

[15] Fitzer, E. and Gadow, R., "Investigations of the Reactivity of Different Carbons with Liquid Silicon," in *Ceramic Components for Engines: Proceedings of the First International Symposium, Hakone, Japan, October, 1983",* Shigeyuki Sōmiya, Eizo Kanai, and Ken-ichiro Ando, eds., October 17–21, 1983, Hakone, Japan. Elsevier Applied Science Publishers, Ltd., Barking, Essex, England, UK, 1986. pp. 561–572.

[16] Ichikawa, H., "Development of High Heat Resistant Silicon Carbide Fiber Using E-Beam Curing," Hosyasenkagaku, 58, 21–29, 1994 (in Japanese).

[17] Kameda, T. and Itoh, Y., "Fabrication and Mechanical Properties of Reaction Sintered Silicon Carbide Matrix Composite," Key Engineering Materials, Vols. 164–165, 95–98, 1999.

[18] Rice, R.W., "Microstructure Dependence of Mechanical Behavior," in *Treatise on Materials Science and Technology,* Academic Press, New York, USA, 11, 199–381, 1978.

[19] Umezawa, M., Nippon Carbon Co. Ltd., private communication, 1998.

CHAPTER

19

Mechanical Characterization of Monolithic Ceramics for Gas Turbine Applications

Mattison K. Ferber and Hua-Tay Lin
Oak Ridge National Laboratory, Oak Ridge, Tennessee, USA

Michael G. Jenkins
University of Washington, Seattle, Washington, USA

Tatsuki Ohji
National Institute of Advanced Science and Technology (AIST)
Synergy Materials Research Center, Nagoya, Aichi, Japan

INTRODUCTION

Over the last 30 years, a number of programs in Russia, Europe, Japan, and the United States have sought to introduce monolithic ceramic components into gas turbines with the goals of increasing efficiency and lowering emissions. Overviews of these efforts are summarized in Volume I (Chapters 1, 2, 12, 20, and 30) of this book series. High performance silicon nitride and silicon carbide ceramics typically have been leading candidates for use in these applications [1–10]. These materials offer the following advantages over their metallic counterparts: (1) low density, (2) refractoriness, (3) adequate short-term strength, (4) good oxidation resistance at ambient pressures, and (5) their shape fabrication technology is fairly well established.

The gas turbine components, where ceramics are typically inserted, include combustors, transition sections, hot gas ducts, turbine vanes and shrouds, turbine blades, and integrally bladed rotors (blisks). All but turbine blades and blisks are classified as stationary components. The materials for the stationary component require moderate hot strength, good thermal shock resistance, and shape stability. The rotating turbine blades

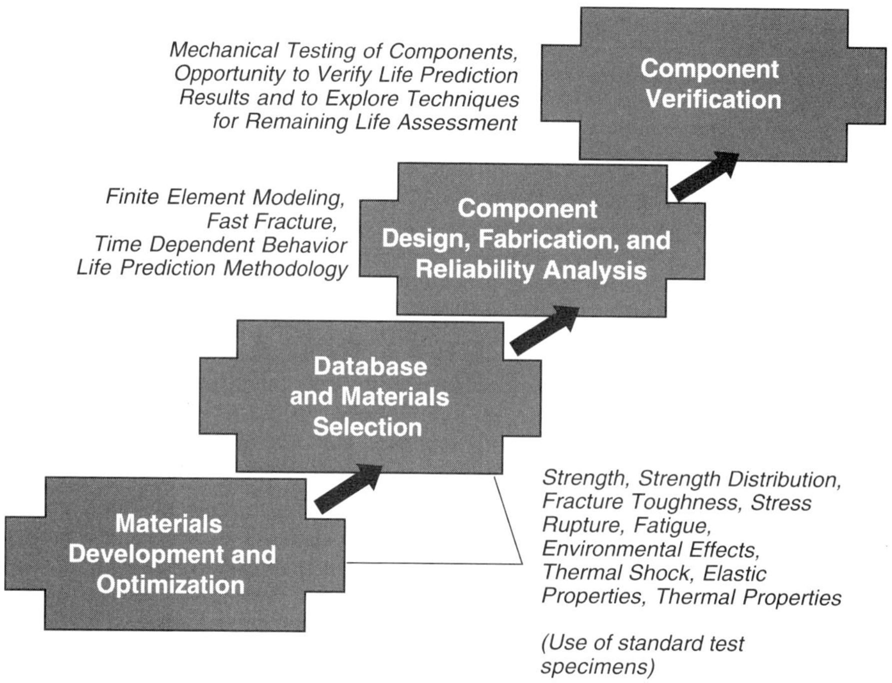

Figure 19.1 Key Factors Associated with the Application of Monolithic Ceramics to Gas Turbine Applications

call for higher strength materials and tolerance to impact. These components must also operate reliably for extended periods (>30,000 h). Ceramic components operating with surface temperatures in excess of 1000°C must also be resistant to the attack by combustion gases in the gas turbine hot section.

This chapter, which addresses mechanical characterization of monolithic ceramics for gas turbine applications, is divided into two sections. The first section discusses the various roles of mechanical characterization in the development and application of the monolithic structural ceramics in gas turbines in the context of the key elements shown in Figure 19.1. Particular focus is given to the important lessons learned in areas of material development and optimization, database generation, long-term mechanical reliability, and evaluation of the mechanical integrity of components.

The second section provides a summary of mechanical property data and illustrates the improvements in mechanical reliability achieved over the last 20 years. The primary focus is on commercially available materials since properties of experimental materials are continually evolving. Table 19.1 summarizes the materials and suppliers referenced in this chapter.

ROLE FOR MECHANICAL CHARACTERIZATION

The role of mechanical characterization in the application of monolithic ceramics to gas turbine applications can be described in terms of the key evolutionary elements required for successful use of these materials (Figure 19.1). *Materials Development and Optimization* is the first step in the process. Here the primary goal is to develop materials with improved toughness, higher strength, and better reliability (as reflected by a larger-valued Weibull modulus). Once these materials mature and become candidates for specific gas turbine components, a database of short-term material properties is required for purposes of *Database Generation and Materials Selection*. Such data allows designers to make preliminary estimates of the service stress and tem-

TABLE 19.1 SUMMARY OF SILICON NITRIDES AND SILICON CARBIDES CONSIDERED IN THE EVALUATION OF MECHANICAL PROPERTIES

Densification Mechanism	Material	Supplier	Comments
Hot-Pressed	NC-132 HP-Si_3N_4 (1% MgO)	Saint-Gobain/Norton Industrial Ceramics, Northboro, MA	
Hot-Pressed	NCX-34 HP-Si_3N_4 (8% Y_2O_3)	Saint-Gobain/Norton Industrial Ceramics, Northboro, MA	
Hot-Pressed	HP-Si_3N_4 (10% CeO_2)	Harbison-Walker, Pittsburgh, PA	
Hot-Pressed	SN-3 HP-Si_3N_4 (4% MgO, 5% Al_2O_3)	Kyocera, Kagoshima, Japan	
Hot-Pressed	HP-Si_3N_4 (4–16 wt% ZrO_2)	Naval Research Laboratory, Washington DC	
Hot-Pressed	Developmental HP-Si_3N_4 (CeO_2 and BN Additives)	AFML	
Hot-Pressed	Ceralloy 147A, HP-Si_3N_4 (1% MgO)	Ceradyne, Costa Mesa, CA	
Hot-Pressed	Ceralloy 147Y, HP-Si_3N_4 (15% Y_2O_3)	Ceradyne, Costa Mesa, CA	
Hot-Pressed	Ceralloy 147Y-1, HP-Si_3N_4 (8% Y_2O_3)	Ceradyne, Costa Mesa, CA	
Hot-Pressed	HP-Si_3N_4 (4% MgO)	Fiber Materials	
Hot-Pressed	HP-Si_3N_4 (4% Y_2O_3 3% Al_2O_3)	Toshiba, Tokyo, Japan	
Hot-Pressed	HP-Si_3N_4 (3% Y_2O_3 4% Al_2O_3 SiO_2)	Toshiba, Tokyo, Japan	
Hot-Pressed	HP-Si_3N_4 (4% Y_2O_3 SiO_2)	Westinghouse, Pittsburgh, PA	
Hot-Pressed	HP-Si_3N_4 (10% ZrO_2)	NASA/AVCO/Norton,	
Reaction Sintered	NC-350 Si_3N_4	Saint-Gobain/Norton Industrial Ceramics, Northboro, MA	
Reaction Sintered	Ford Injection Molded Si_3N_4	Ford, Dearborn, MI	
Reaction Sintered	RBN-101 Slip Cast Si_3N_4	AiResearch, Torrance, California	
Reaction Sintered	RBN-122 Injection Molded Si_3N_4	AiResearch, Torrance, California	
Reaction Sintered	RBN-104 Si_3N_4	AiResearch, Torrance, California	
Reaction Sintered	Raytheon Isopressed Si_3N_4	Raytheon, Lexington, MA	
Reaction Sintered	HPSN, Slip Cast or Injection Molded RBSN	Annawerk, Germany	
Reaction Sintered	Injection Molded RBSN	Degussa, Germany	
Reaction Sintered	Injection Molded RBSN	Feldmühle, Germany	
Sintered	SN-205 Si_3N_4 (5% MgO, 9% Al_2O_3)	Kyocera, Kagoshima, Japan	
Sintered	SN-201 Si_3N_4 (4% MgO, 7% Al_2O_3)	Kyocera, Kagoshima, Japan	
Sintered	AiResearch Si_3N_4 (8% Y_2O_3 4% Al_2O_3)	AiResearch, Torrance, California	
Sintered	Rocketdyne SN-50 Si_3N_4 (6% Y_2O_3 4% Al_2O_3)	Rocketdyne, Canoga Park, CA	
Sintered	Rocketdyne SN-104 Si_3N_4 (14% Y_2O_3 7% SiO_2)	Rocketdyne, Canoga Park, CA	
Sintered	Rocketdyne SN-46 Si_3N_4 (14% Y_2O_3 7% SiO_2)	Rocketdyne, Canoga Park, CA	
Sintered	SN88 Si_3N_4	NGK Insulators, LTD, Nagoya, Japan	Self reinforced
Sintered	AS800 Si_3N_4	Honeywell Ceramic Components, Torrance, CA	Self reinforced
Sintered	SN253 Si_3N_4	Kyocera, Kagoshima, Japan	Self reinforced
Sintered	SN281/282 Si_3N_4	Kyocera, Kagoshima, Japan	Self reinforced SN281 is HIPed
Hot Isostatic Pressed	HIP-Si_3N_4 (5% Y_2O_3)	Battelle	
Hot Isostatic Pressed	CSN 101 (3 wt% Y_2O_3)	AC Cerama, Sweden	
Hot Isostatic Pressed	GN10 Si_3N_4	Honeywell Ceramic Components, Torrance, CA	Formerly AlliedSignal
Hot Isostatic Pressed	PY6 Si_3N_4 (6% Y_2O_3)	GTE Laboratories, Inc., Waltham, MA	
Hot Isostatic Pressed	NT154 Si_3N_4 (4% Y_2O_3)	Saint-Gobain/Norton Industrial Ceramics, Northboro, MA	
Hot Isostatic Pressed	NT164 Si_3N_4 (4% Y_2O_3)	Saint-Gobain/Norton Industrial Ceramics, Northboro, MA	
Hot Pressed	NC-203 SiC (~2% Al_2O_3)	Saint-Gobain/Norton Industrial Ceramics, Northboro, MA	
Hot Pressed	Ceralloy 146A SiC (2% Al_2O_3)	Ceradyne, Costa Mesa, CA	
Hot Pressed	Ceralloy 146I SiC (2% B_4C)	Ceradyne, Costa Mesa, CA	
Hot Pressed	HPSiC	ESK, Germany	
Hot Pressed	HPSiC	Sigri, Germany	
Sintered	Sintered β-SiC	General Electric Schenectady, NY	
Sintered	Sintered α-SiC	Carborundum Niagara Falls, NY	1977 vintage
Sintered	SC-201 Sintered α-SiC	Kyocera, Kagoshima, Japan	1980 vintage
Sintered	SASC (Hexoloy SX-05)	Carborundum, Niagara Falls, NY	1981 vintage
Sintered	ESK Sintered α-SiC	ESK, Germany	
Sintered	Slip Cast Sintered SiC	Annawerk, Germany	
Sintered	NT-230	Saint-Gobain/Norton Industrial Ceramics, Northboro, MA	
Siliconized	NC-435 Si/SiC	Saint-Gobain/Norton Industrial Ceramics, Northboro, MA	
Siliconized	Refel Si/SiC (diamond-ground and as-processed)	ITKAEA/BNF	
Siliconized	NC-430 Si/SiC	Saint-Gobain/Norton Industrial Ceramics, Northboro, MA	
Siliconized	SC-1 Si/SiC	Coors Ceramics, Golden, CO	1979 vintage
Siliconized	SC-2 Si/SiC	Coors Ceramics, Golden, CO	1981/1982 vintage
Siliconized	Silcomp Grade CC Si/SiC	General Electric, Schenectady, NY	

perature distributions. Once a particular material composition is chosen, *Component Design, Fabrication, and Reliability Analysis (Life Prediction)* become critical factors. Design iterations are generally required to insure that the service stresses are sufficiently low to avoid time-dependent failure within the desired service life of the component. Such predictions of component lifetime also rely on the availability of long-term stress rupture and fatigue data. The manufacturing process must also be capable of producing a material having properties equivalent to those generated from the evaluation of standard test specimens. For example, differences in the mechanical response of the component (compared with standard test specimen) can arise from machining variations. The final factor is that of *Component Verification*. Techniques such as nondestructive evaluation and proof-testing are often used to insure that the component performance is consistent with lifetime predictions. The role of mechanical characterization in each of these elements is described in the sections that follow.

Materials Development and Optimization

In *Materials Optimization*, the major emphasis is placed upon developing materials with improved properties. Generally these properties are dictated by the specific application. For example, Table 19.2 lists target properties prescribed in the Japanese Fine Ceramic Project which was launched in 1981 under the National Research and Development Project on Basic Technologies for Future Industries [7,8].

From this list it is apparent that mechanical properties including the measurement of the room and high temperature strength, strength distribution, stress rupture, and fracture toughness are key indicators of the success of the materials optimization. Furthermore the ability to reproduce the materials process requires control of microstructural features such as grain size, grain boundary phase, and impurity level. Mechanical properties are generally quite sensitive to variations in these features. Several examples are provided below.

One early effort to use mechanical strength and its variability as a means to quantify the material processing is given in ref. [11]. In this study, the strength and its associated Weibull distribution were used to monitor the performance of hot-pressed, slip-cast/reaction bonded, and injection-molded sintered silicon nitride as function of the processing iterations. In the case of the hot-pressed silicon nitride, these iterations

TABLE 19.2 OBJECTIVE OF PERFORMANCE PROPERTIES OF FINE CERAMICS (Adapted from ref. [8] with permission of the American Ceramic Society, www.ceramics.org, © 1982, all rights reserved. Also ref. [7])

Classification	Objective	Values
High strength materials	≥1200°C, in air, after 1000h holding.	
	Weibull modulus	$m \geq 20$
	average tensile strength	≥30kg/mm^2
	creep rupture strength after 1000h continuous loading at 1200°C in air	≥10kg/mm^2
Corrosion-resistant materials	≥1300°C, in air, after 1000h holding	
	Weibull modulus corrosion	$m \geq 20$
	resistance (weight gain)	≤1mg/cm^2
	average tensile strength	≥20kg/mm^2
Wear-resistant materials	Room temperature.	
	wear resistance	≤10^{-8}mm^3/kg·mm
	surface flatness	$R_{max} \leq 2$ μm
	800°C, in air after 1000h holding.	
	Weibull modulus	$m \geq 22$
	average tensile strength	≥ 50kg/mm^2

1kg/mm^2 = 9.8MPa

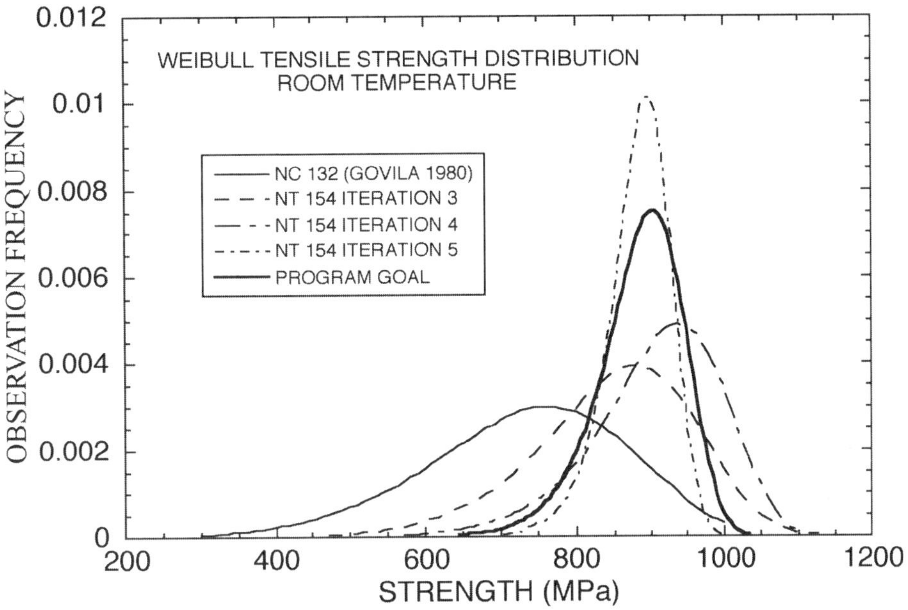

Figure 19.2 Use of Tensile Fracture Strength Distribution to Describe the Processing Evolution of a HIPed Silicon Nitride (Designated as HIP-SN). Data for Norton's NC 132 are provided for comparison.

focussed on controlling the level of impurities, which were assumed to control the refractoriness of the grain boundary phase and thus, the high-temperature strength.[1] For the slip-cast/reaction-bonded silicon nitride three iterations emphasizing the starting powder, powder processing method, slip preparation, and presintering treatment were evaluated. Iterations for the injection-molded material included injection molding parameters, sintering procedure, and chemical composition. This iterative process development approach resulted in significant improvements in mechanical performance of all three types of silicon nitrides.

A second example of process optimization is provided in ref. [12]. In this case the goal was to improve the strength distribution of a hot isostatically pressed (HIPed) silicon nitride. The improvements in strength characteristics (measured using button-head tensile specimens) highlighted in Figure 19.2 were due to a number of factors. The first involved the elimination of contamination, primarily iron, which was either present in the starting powder or introduced during green-state processing. The role of iron in creating large strength limiting flaws during densification is well established. Another critical factor in achieving better mechanical reliability was improved slip control and better casting procedures. Better dispersants and understanding the role of slip pH on agglomerate strength yielded a more uniform green body. The uniform green microstructure translated into a uniformly dense material with reduced porosity, improved glass distribution and decreased interactions with the HIP environment. These process improvements not only increased the strength reliability but were an important foundation for the ability to manufacture gas turbine components with properties comparable to the mechanical test samples.

Other studies in the area of *Materials Optimization* have focused on the improvement of the long-term mechanical reliability by attempting to better understand the effect of sintering additives (grain-boundary phases) and grain size distribution on the creep behavior of silicon nitride ceramics. Early work in this area

[1] Strength measurements were supplemented by extensive fractography. Other process monitoring techniques including measurement of particle size, viscosity, and chemistry were also applied.

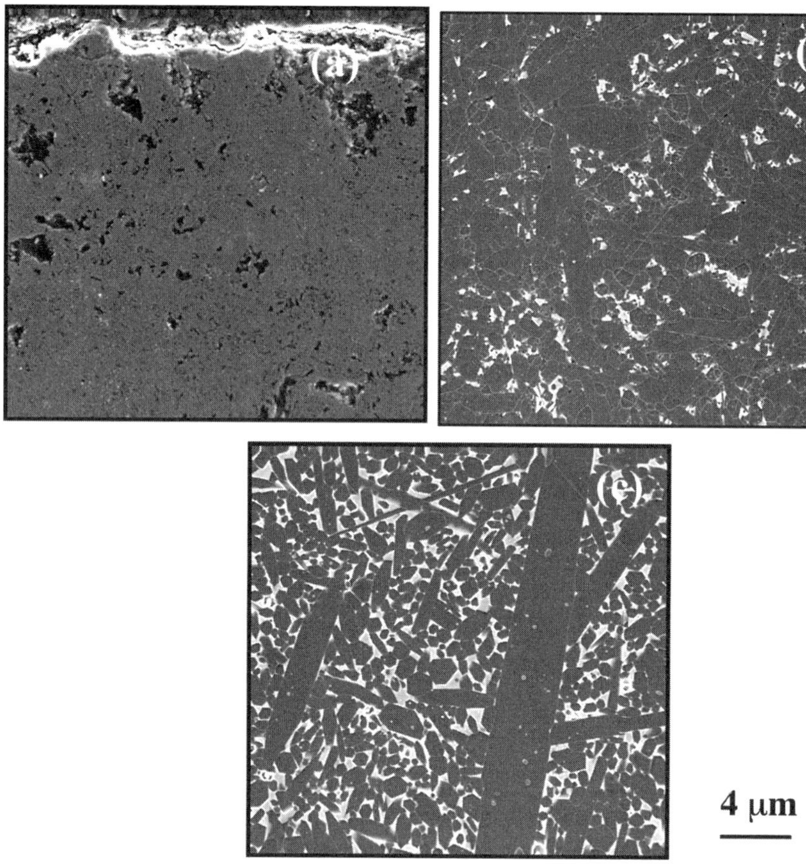

Figure 19.3 Evolution of Microstructures of Silicon Nitride Ceramics: (a) Reaction Sintered (Bonded), (b) Hot-Isostatically Pressed, and (c) Self-Reinforced

sought to minimize the detrimental effects of the secondary intergranular phases by hot isostatic pressing with a variety of refractory sintering aids [13–15]. Examples include silicon nitride materials such as PY6, GN10, and NT154/NT164 (Table 19.1).

More recently, improvements in the long-term mechanical reliability have been realized through the development of self-reinforced silicon nitride ceramics. The microstructures of these materials consist of elongated grains which are grown using special heat treatments (see Figure 19.3 for reference). In addition to improving the fracture toughness, this microstructure also increases creep resistance. The extent to which these properties are enhanced depends strongly upon the grain size distribution and composition of the intergranular phase [16–30]. For example, Figure 19.4 shows the effect of high Y:Al ratio upon the fracture resistance of a silicon nitride processed with large elongated seeds [23,24]. The addition of these beta silicon nitride seeds provided control over the grain size distribution. Significant improvement in the R-curve behavior was observed in the silicon nitride processed with a high Y:Al ratio. This effect was attributed to more extensive debonding at grain boundary interfaces.

Although the utility of mechanical characterization in *Materials Optimization* has been demonstrated, a number of cautionary notes are worthy of mention. First one must be sure that the properties which are monitored are sufficiently sensitive to unintended variations in the materials process. For example, flexure strength may not be sensitive to volume defects. Second, mechanical property measurements alone are not sufficient. Fractography, phase analysis, and microstructural characterization must also be used to determine the underlying reasons for measured changes in performance. In case of fractography [26] it is necessary to

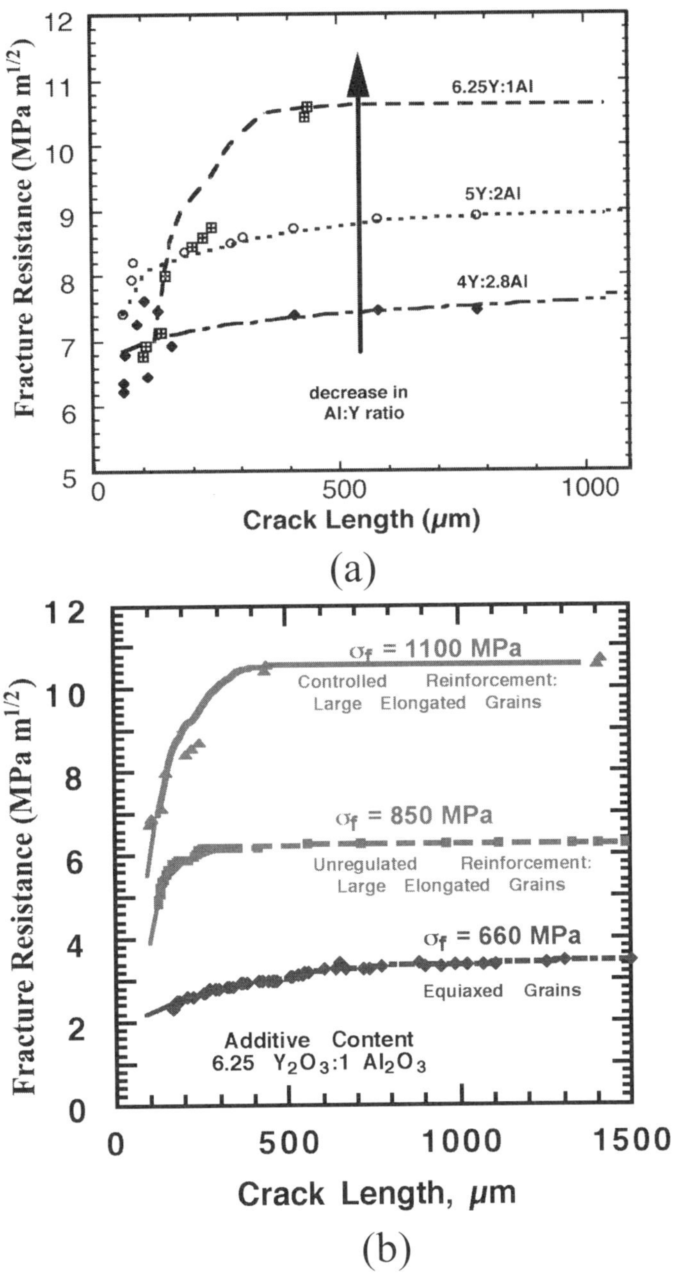

Figure 19.4 Effect of Intergranular Phase Composition (a) and Grain Morphology (b) Upon Fracture Toughness (Adapted from ref. [23] with permission from the American Ceramic Society, www.ceramics.org, © 1998, all rights reserved)

determine the nature (inclusion, large grain, porosity, etc.), the size (shape), and the location of the strength-limiting defect. Such information provides insights into what processing stage is responsible for limiting the mechanical performance. For example the presence of Fe inclusions in a HIPed silicon nitride was traced to powder contamination while blister formation in a sintered silicon nitride was attributed to reactions with the furnace insulation during high-temperature densification. Phase analysis is required to insure that grain boundary phases have the correct crystalline structure. Phases that are not stable at temperature can transform, leading to disruptive volume changes [31,32].

Finally, in using mechanical properties such as strength to monitor the consistency of a fixed materials process, the materials supplier must carefully examine trends in the data. In particular, the data may indicate that the strength is decreasing with time although the absolute value is still within specified tolerance bounds. This result may signal that the process is out of control.

Database Generation and Materials Selection

In order to evaluate the potential of a given material composition as a gas turbine component, the designer needs key thermal and mechanical property data. Specific properties of interest include thermal conductivity, thermal expansion, density, elastic modulus, Poisson's ratio, strength, and Weibull modulus. The temperature sensitivities of these parameters must generally be known as well. This information allows one to perform preliminary finite element analysis to establish the stress and temperature distributions within the component.

Typically, the design of a given component involves an iterative process in which finite element modeling is combined with probabilistic analysis to estimate and minimize the probability of faillure (P_f). If this value is too high the component design must be modified to lower the maximum stress. This can be achieved in various ways such as altering the component geometry, or lowering overall component temperature and/or reducing temperature gradients in the component at high stress locations. The ultimate goal of this process is to insure that the service stresses are sufficiently low to avoid time-dependent failure within the desired service life of the component. Such predictions of component lifetime obviously rely on the availability of long-term tensile stress rupture and fatigue data.

Over the years there has been a number of barriers that have limited the generation of the data required for materials selection and component life prediction. These barriers include the (1) lack of established test methodology and hardware for measuring tensile properties, (2) lack of dedicated test facilities, (3) insufficient data concerning long-term behavior, and (4) inability to measure properties in realistic turbine environments. Each of these issues is discussed in the sections that follow.

Test Methodology and Hardware. In many of the early ceramic gas turbine programs, the use of uniaxial stress tests (tension or compression) was limited because of the need for elaborate specimen preparation, specialized testing equipment (including specimen grips), and the difficulty of achieving the necessary uniform stress state. Therefore, the flexure bar remained the popular testing arrangement for ceramics, because of the ease of fabrication of the specimen geometry, the efficient use of material, the simplicity of gripping and loading, and the seemingly straightforward analysis. However, use of the flexure bar for unequivocally characterizing a material behavior is limited, especially at elevated temperatures, because of the variation of the stress state over the cross section of the bar as well as the relatively small, uniformly stressed volume. Moreover, designers generally demand that key mechanical properties be measured from uniaxial tensile specimens.

Because it is recognized that the limiting failure stress of brittle materials is tensile, it is necessary to test the material in a uniform tensile field to properly characterize the true, engineering, mechanical behavior. Thus tensile tests of ceramics in uniform stress fields are necessary to (1) determine strength distributions due to inherent flaws in large, stressed volumes; (2) characterize the tensile stress-strain behavior for engineering design purposes; (3) unambiguously quantify the effects of cyclic fatigue loading, and (4) elucidate the tensile creep behavior at elevated temperatures.

In terms of hardware, tensile test fixtures can be considered to be composed of two major parts. The first part is the actual specimen grip which contains the interface between the ceramic tensile specimen and the coupling to the loading frame. The second part is the attachment between the actual specimen grip and the test machine. Examples of the specimen/grip interface include single and multiple pin loading, shoulder loading, straight-shanks gripped by hydraulically-actuated jaws, and adhesively-bonded hydraulic piston loading.

Once the specimen has been properly gripped, the question of maintaining uniaxial alignment so as to promote uniform tension in the gage section is paramount. In most cases, the grip/machine attachments must be such so as to remove sources of eccentricity from the load train. Generally, two attachment methods have been employed: flexible, self-aligning and fixed, alignable.

Primary considerations in the design of a tensile specimen beyond the geometry of the specimen/grip interface, are the elimination of stress risers in the gage section, minimization of thermal gradients within in the gage section, and proper amounts of effective volume and effective surface area appropriate for the statistics of the design engineering needs. Stress raisers occur because of geometrical transitions. Although ideal dimensional cases exist for eliminating stress raisers in various geometrical transitions [33] in tensile specimens, the final specimen dimensions are often dictated by fabrication considerations. Examples of tensile specimens which have been successfully applied to strength and creep testing are shown in Figure 19.5 [34].

Dedicated Facilities. In order to establish a comprehensive mechanical properties database for materials selection, a number of properties must be measured including (1) strength and strength distribution, (2) fracture toughness, (3) elastic properties, (4) slow crack growth, (5) creep, (6) fatigue, and (7) thermal shock. These properties are generally determined as a function of temperature. It is also desirable to measure thermal properties such as thermal diffusivity and thermal expansion. Finally to better understand the underlying cause of property variability one should perform analyses of the microstructure, phases, chemistry, and strength-controlling defect. Unfortunately, the resources and facilities for making these extensive measurements have been limited. As a result when evaluating a given material, designers have often had to rely on multiple sources (materials supplier, university, federal laboratory, etc) for property data. In a few ceramic gas turbine programs this issue was addressed by establishing dedicated programs to measure key mechanical properties. These programs are discussed below.

In the United States, one of the first programs to focus on complete mechanical characterization of structural ceramics for gas turbine applications was funded by the US Air Force. This program, which was entitled, "Property Screening and Evaluation of Ceramic Turbine Materials," evaluated impurity content, phase ID, X-ray radiographic analysis, strength, strength distribution, fracture toughness, thermal shock water quench-residual strength, dynamic elastic modulus, slow crack growth by dynamic fatigue, stress rupture, and creep behavior for a number of candidate structural ceramics. The results of these screening studies are detailed in [35].

A more recent materials characterization effort was initiated in the early 1980s at the University of Dayton Research Institute (UDRI). This project entitled, "Environmental Effects in Toughened Ceramics" was funded by the Ceramic Technology for Advanced Heat Engines Program [6] and has provided extensive material property data for a variety of structural ceramics [36–39]. The goal of this program was to investigate the effects of environment on the mechanical behavior of commercially-available ceramics for heat engine applications. Mechanical studies conducted under this effort included: (1) flexure and tensile strength, (2) fatigue (dynamic, static, and cyclic), and (3) fracture toughness. Structural ceramics were also characterized in terms of microstructure, fractography, chemistry, and physical properties. When the Ceramic Technology for Advanced Heat Engines Program ended in the early 1990s, the program at UDRI continued under the DOE Advanced Turbines Systems (ATS) Program.

An important finding resulting from the UDRI work concerns the validity of flexure specimens for evaluating time-dependent properties at temperatures where creep processes are dominant. For example, flexural dynamic fatigue testing, which consists of measuring the strength as a function of stressing rate, is often used to quickly assess a material's susceptibility to slow crack growth. In general as the stressing rate is lowered, time-dependent degradation processes have more time to operate leading to a decrease in the measured fracture strength. The relationship between strength and stressing rate can be used to estimate the stress rupture behavior (see discussion below). However, the UDRI data show that when creep

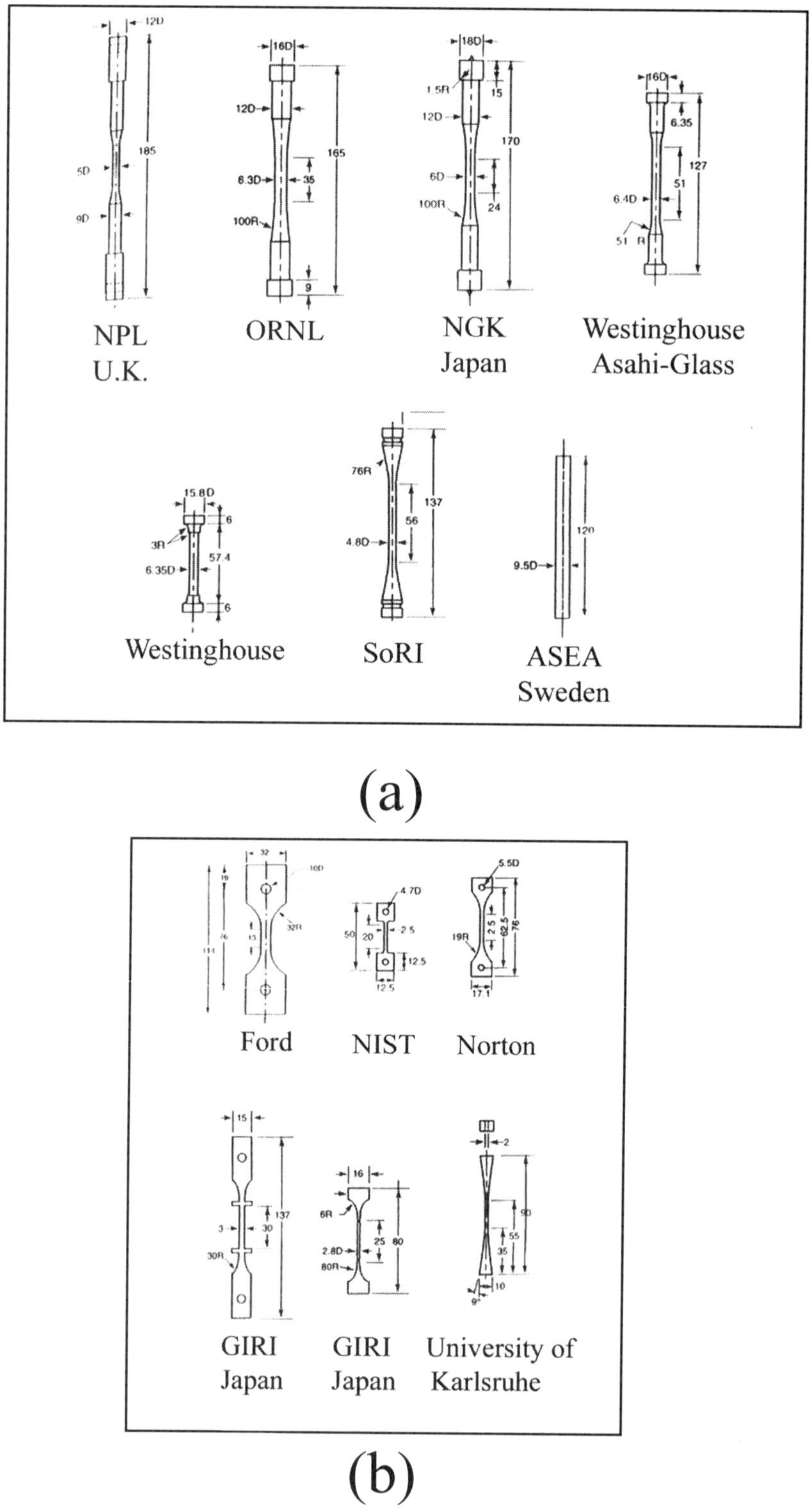

Figure 19.5 Circular (a) and Flat (b) Geometries Used for Tensile Testing of Monolithic Ceramics

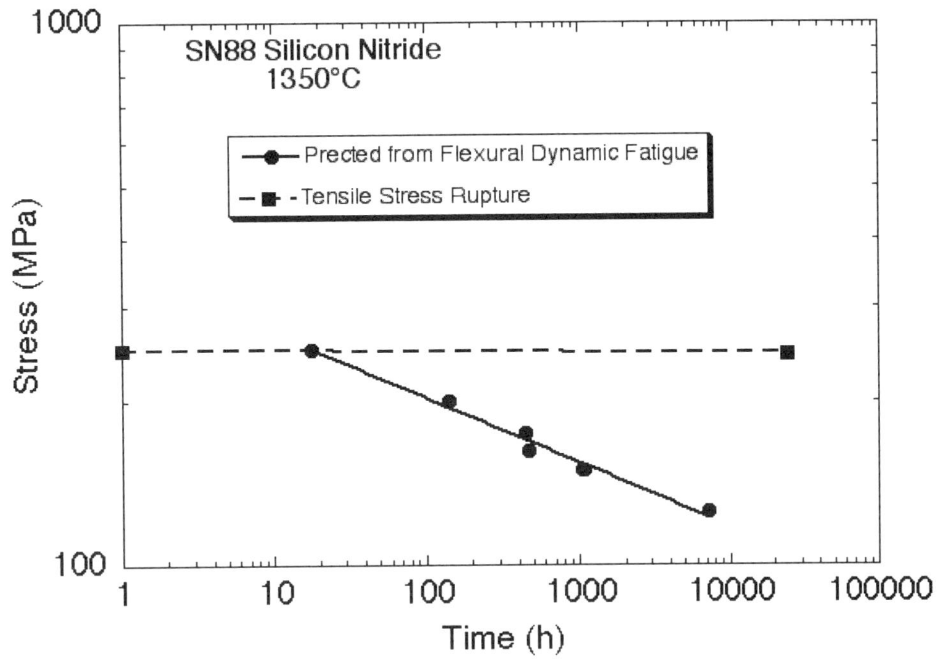

Figure 19.6 Comparison of Stress Rupture Curves of a Silicon Nitride Ceramic Predicted Using Flexural Dynamic Fatigue and Measured Directly Using Tensile Specimens

deformation occurs, the formula used to calculate the flexure strength will be in considerable error due to differences in the tension and compression creep response occurring in flexure specimen. Since the creep rate in tension generally exceeds that in compression for equivalent magnitudes of stress, the initial elastic tensile stresses in the flexure specimen are relaxed below the value given by the strength of material expression. This will tend to reduce the measured stressing rate dependence of strength and thus lead to an underestimation of the stress rupture susceptibility. This effect is illustrated in Figure 19.6 which compares stress (creep) rupture data predicted from flexural dynamic fatigue with that measured directly using tensile specimens. Therefore, flexural dynamic fatigue data should not be used to evaluate the long-term mechanical reliability of materials operating under creep conditions. For similar reasons, the application of static flexure tests to measure creep behavior tends to overestimate the creep rupture life.

Long-Term Testing. Efforts to predict the lifetime of ceramic components have been limited by lack of data describing time-dependent behavior. A number of mechanisms can lead to a time-dependent reduction in strength. The first is slow crack growth, which involves the subcritical extension of pre-existing defects under the combined effects of environment and stress. The effect of this growth on strength can be evaluated by first assuming that for a given flaw size, *a*, the associated fracture strength, S, is given as,

$$Y\ S\ (a)^{1/2} = K_{1C}, \tag{1}$$

where Y is a constant and K_{1C} is the fracture toughness. More generally, when the applied stress, σ, is less than the fracture strength, Eq. (1) becomes

$$Y\ \sigma\ (a)^{1/2} = K_1, \tag{2}$$

where K_I is the stress intensity factor. When slow crack occurs the crack length increases in accordance with the power law equation,

$$da/dt = V = A\ K_I^N, \tag{3}$$

where A and N are constants, V is the crack velocity, t is time, and K_I is the stress intensity factor. The K_I term in Eqs. (2) and (3) can be eliminating yielding the expression,

$$da/dt = V = A\ (Y\ \sigma\ (a)^{1/2})^N, \tag{4a}$$

or

$$da/(a)^{N/2} = A\ (Y\ \sigma)^N\ dt. \tag{4b}$$

After integrating each side one has that

$$[(a_f)^{-N/2+1} - (a_i)^{-N/2+1}]/(-N/2+1) = A\ Y^N \int \sigma^N\ dt \tag{5a}$$

or after rearrangement

$$[(a_i)^{-N/2+1} - (a_f)^{-N/2+1}] = (N - 2)/2\ A\ Y^N \int \sigma^N\ dt, \tag{5b}$$

where a_i and a_f are the initial and final crack lengths, respectively. These crack lengths can be related to the fracture strengths as follows

$$Y\ S_i\ (a_i)^{1/2} = K_{IC}, \tag{6a}$$

and

$$Y\ S_f\ (a_f)^{1/2} = K_{IC}. \tag{6b}$$

Consequently, Eq. (5) can be rewritten as

$$[(Y\ S_i / K_{IC})^{N-2} - (Y\ S_f / K_{IC})^{N-2}] = (N - 2)/2\ A\ Y^N \int \sigma^N\ dt \tag{7a}$$

or

$$(S_i)^{N-2}\ [1 - (S_f/S_i)^{N-2}] = (N - 2)/2\ A\ (K_{IC})^{N-2}\ Y^2 \int \sigma^N\ dt. \tag{7b}$$

As the strength, S_f, decreases due to the application of the time-dependent stress, σ, the quantity, $(S_f/S_i)^{N-2} << 1$ such that

$$(S_i)^{N-2} = (N - 2)/2\ A\ (K_{IC})^{N-2}\ Y^2 \int \sigma^N\ dt. \tag{7c}$$

Because the value of N dictates a materials susceptibility to time-dependent failure, a number of standard tests are generally used to estimate N. The most straight-forward method is to measure time-to-failure under a static stress. In this case the integral on the right hand side of Eq. (7b) is σ^N t, yielding the following expression,

$$(S_i)^{N-2} = (N - 2)/2\ A\ (K_{IC})^{N-2}\ Y^2\ \sigma^N\ t. \tag{8}$$

Equation (8) can be expressed in the log-linear relationship, ln (t) = –N ln (σ) + constant, which in the usual expression for time-to-failure under a static stress. The second approach for estimating N is to measure the failure stress as a function of stressing rate (dynamic fatigue). In this case, σ = (dσ/dt) t and Eq. (7b) becomes:

$$(S_i)^{N-2} = [(N\text{-}2)/(2(N+1))]AY^2(K_{IC})^{N-2}(d\sigma/dt)^{-1}\sigma^{N+1}. \tag{9}$$

The stress at failure is related to the stressing rate through the relationship, ln (σ) = 1/(N + 1) ln (dσ/dt) + constant, which is the usual relationship used to represent dynamic fatigue data.

The second mechanism that can be responsible for time-dependent failure is cyclic fatigue [40–57]. In the case of monolithic ceramics it is often assumed that the loss of strength arising from cyclic loading is simply a result of same slow crack growth process described above. If the cyclic stress is represented as $\sigma = \sigma_{max}$ f(t) where f(t) is a periodic function, then the expression for the cyclic fatigue becomes

$$(S_i)^{N-2} = (N-2)/2\ A\ (K_{IC})^{N-2}\ Y^2\ (\sigma_{max})^N\ N \int f(t)\ dt. \tag{10}$$

where N is the number of cycles to failure and the integral is evaluated over one cycle. Attempts to verify that cyclic fatigue for silicon nitride and silicon carbide ceramics is due to slow crack growth have met with limited success. For temperatures less than 1000°C, the fact the N values are relatively high means that there will be a large uncertainty in the lifetimes measured under static or cyclic loading. As a result it may not be possible to say with statistical certainty that a slow crack growth process is responsible for both mechanisms. For temperatures above 1000°C, other processes such as creep and oxidation may lead to differences in the behavior under static and cyclic loading.

The slow crack growth mechanism described above assumes that the subcritical extension of pre-existing flaws control failure. It does not account for generation of new flaws or time-dependent changes in toughness. Under conditions of creep, existing flaws can blunt while new flaws are generated via the formation and coalescence of cavities. Likewise, the toughness may change with time due to transformation of the intergranular phases. Cyclic loading can also reduce the toughness. For example, self-reinforced silicon nitrides rely on interlocking grains which inhibit crack growth. Cyclic loading lowers the sliding resistance between grains leading to a time-dependent loss in toughness.

The third and final mechanism responsible for time-dependent failure is creep. In the case of the silicon nitride ceramics, the stress rupture behavior (for temperatures above 1100°C) can be attributed to creep cavitation. For example, Figure 19.7 illustrates cavities observed for NT154 using transmission microscopy. The multi-grain junction cavities are generally much larger and account for as much as 95% of the measured creep strain. With time these cavities coalesce into a damage zone which weakens the material and causes failure. An example of one such damage zone (light colored region) is shown in the top micrograph in Figure 19.8. The surface cracks observed along the gage section (micrograph on the bottom) indicate that damage zone formation occurred at several points along the gage surface. The fact the damage zone originates from the specimen surface suggests that oxidation plays a key role in its formation. This has been confirmed by a number of studies [58,59].

One of the major findings from the creep rupture studies concerns the modeling of the stress and temperature sensitivities of the lifetime. At least four approaches have been used to describe the stress and temperature sensitivities of the measured lifetime in the creep regime. The first is the power-law model in which the lifetime, t_f, is described by the equation [60],

$$t_f = B_o\ (\sigma_a/\sigma'_o)^{-N} \exp(Q_f/RT), \tag{11}$$

where N and Q_f determine the stress and temperature dependencies of the dominant creep mechanism, B_o is a pre-exponential factor, σ_a is the applied stress, R is the universal gas constant, T is the absolute temperature, and σ'_o is a normalizing parameter (i.e. σ'_o= 1 MPa). For case where failure is controlled by diffusional creep for ideal viscous flow, N will be equal to 1. However, most tensile creep rupture data yield

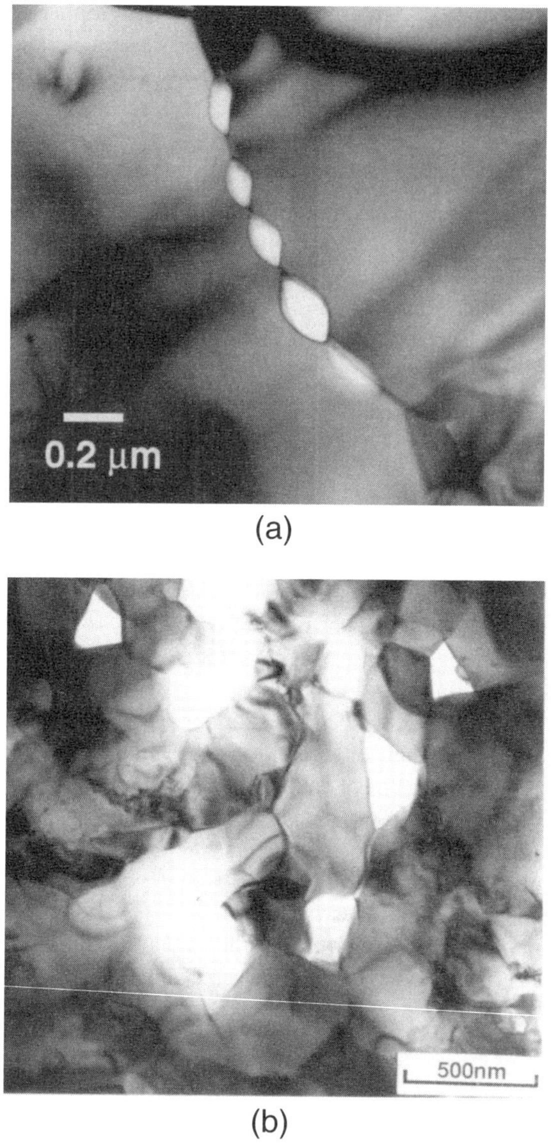

Figure 19.7 Cavities Observed in NT154 Silicon Nitride After Creep Testing at 1370°C: (a) Two-Grain Junction Cavities and (b) Multi-Grain Junction Cavities

much higher estimates of N (>4). Cavitation within the intergranular phase is generally cited as the key reason for this behavior.

The second approach is based upon the assumption that the steady-state or minimum creep strain rate, $d\varepsilon_s/dt$, reflects the rate of damage accumulation. Because lifetime should be inversely proportional to this rate, one can write,

$$t_f = C_o\,(d\varepsilon_s/dt)^{-m} \tag{12}$$

where C_o is a constant. This expression, which holds for a number of materials, is termed the Monkman-Grant relationship [61].

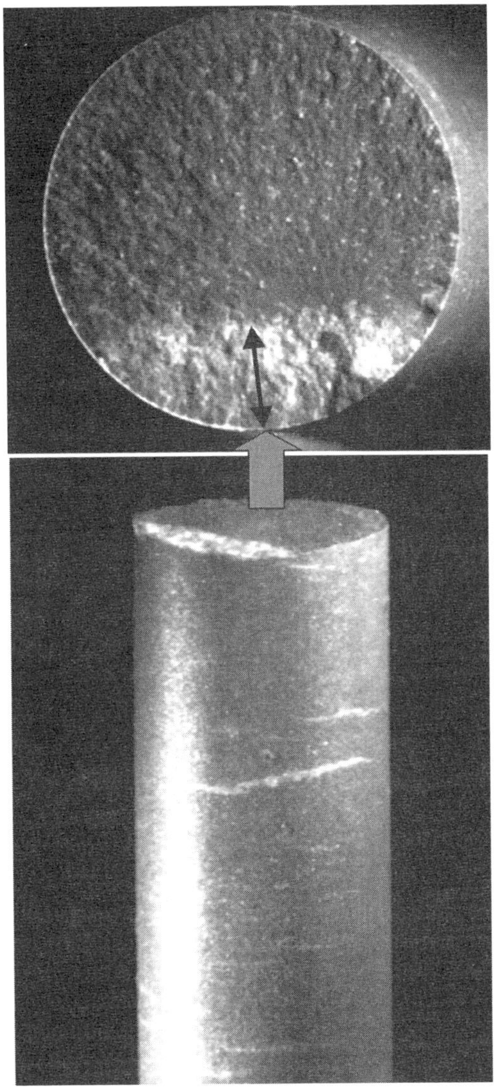

Figure 19.8 Fracture Characteristics of SN88 Button-Head Tensile Specimen Tested at 1350°C and 125 MPa. Arrows Highlight Damage Zone.

The third approach involves the use of parametric equations. Two commonly used parameters, the Larson-Miller and Sherby-Dorn, are given by the respective equations [62],

$$(20 + \log(t_f))\, T/1000 = f(\sigma) \tag{13}$$

and

$$t_f \exp(-Q_f/RT) = g(\sigma), \tag{14}$$

where $f(\sigma)$ and $g(\sigma)$ are empirical functions of stress only. In principle the application of either these time-modified expressions should collapse the experimental data to a single line or curve which is a function of stress only.

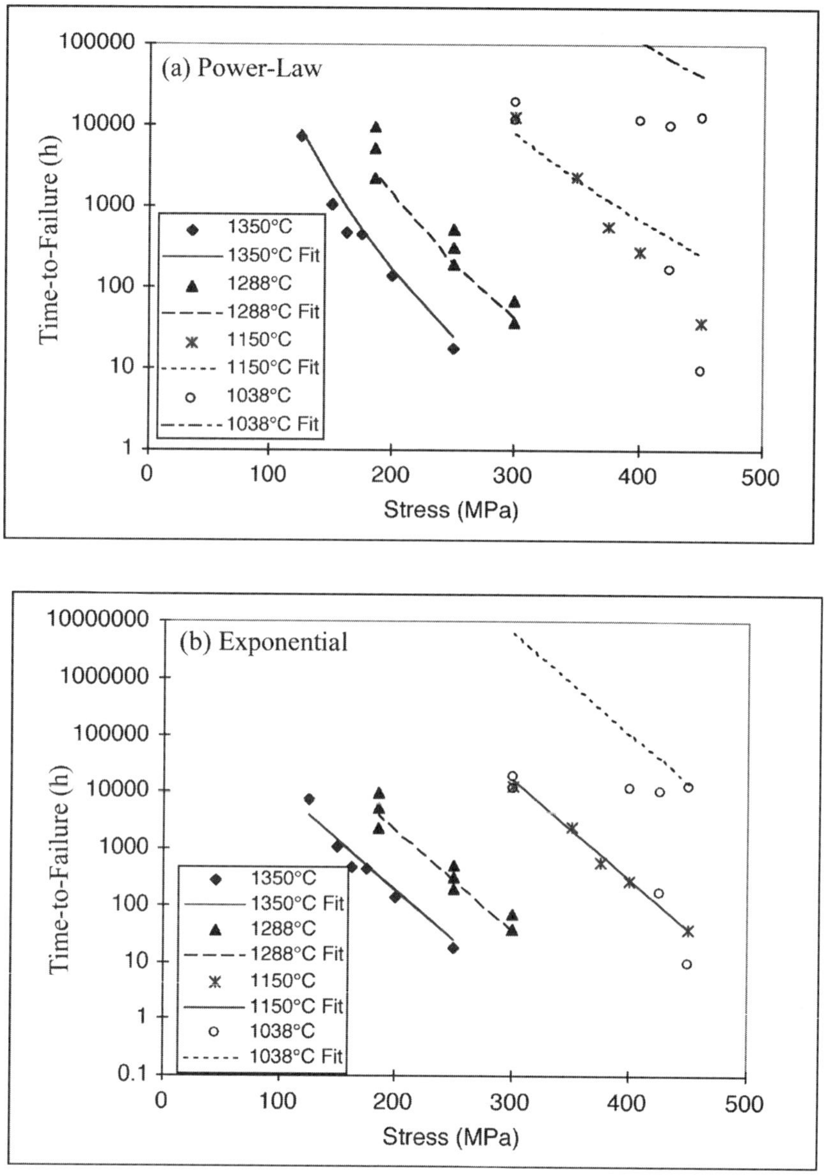

Figure 19.9 Comparison of the (a) Power-Law and (b) Exponential Fits to Creep Life Data for SN88 Silicon Nitride

A major problem with these three approaches is that they fail to adequately describe the data over a broad temperature range. A much better description of the experimental data is obtained using the exponential expression,

$$t_f = B \exp(b\sigma) \exp(Q_f/RT), \qquad (15)$$

where experimental stress dependency is characterized by the parameter, b. Figures 19.9 and 19.10 compare the power-law and exponential fits obtained for two silicon nitrides. In both cases, the exponential model provides a superior description of the stress and temperature dependencies.

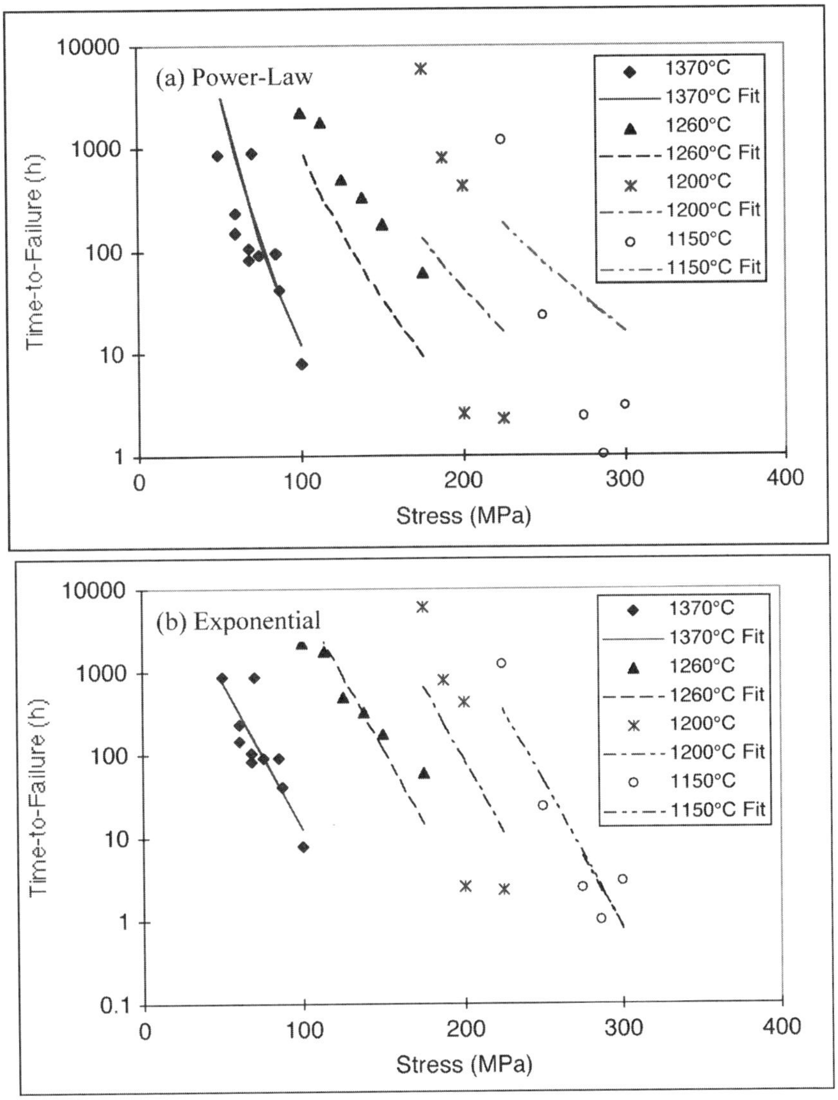

Figure 19.10 Comparison of the (a) Power-Law and (b) Exponential Fits to Creep Life Data for PY6 Silicon Nitride

An exponential creep-rate model has also been proposed by a group at the National Institute of Standards and Technology (NIST) [63–65]. This model is given by the expression,

$$d\varepsilon/dt = A_N\, \sigma \exp(B_N\sigma) \exp(-Q/RT), \qquad (16)$$

where A_N and B_N are constants. The exponential dependence of creep rate on stress stems from the kinetics of the secondary phase redistribution from growing cavities. Increasing stress increases the number of sites that can cavitate, which in turn increases the creep rate. The exponential stress term, $\exp(B_N\, \sigma)$, is found to approximate the stress dependency of the number of cavitating sites.

In terms of materials selection, engine designers have emphasized the need for lifetime data covering at least 10,000 hours. Efforts to characterize the long-term mechanical reliability of monolithic ceramics for gas turbine applications have been underway in several laboratories in the United States and Japan. The

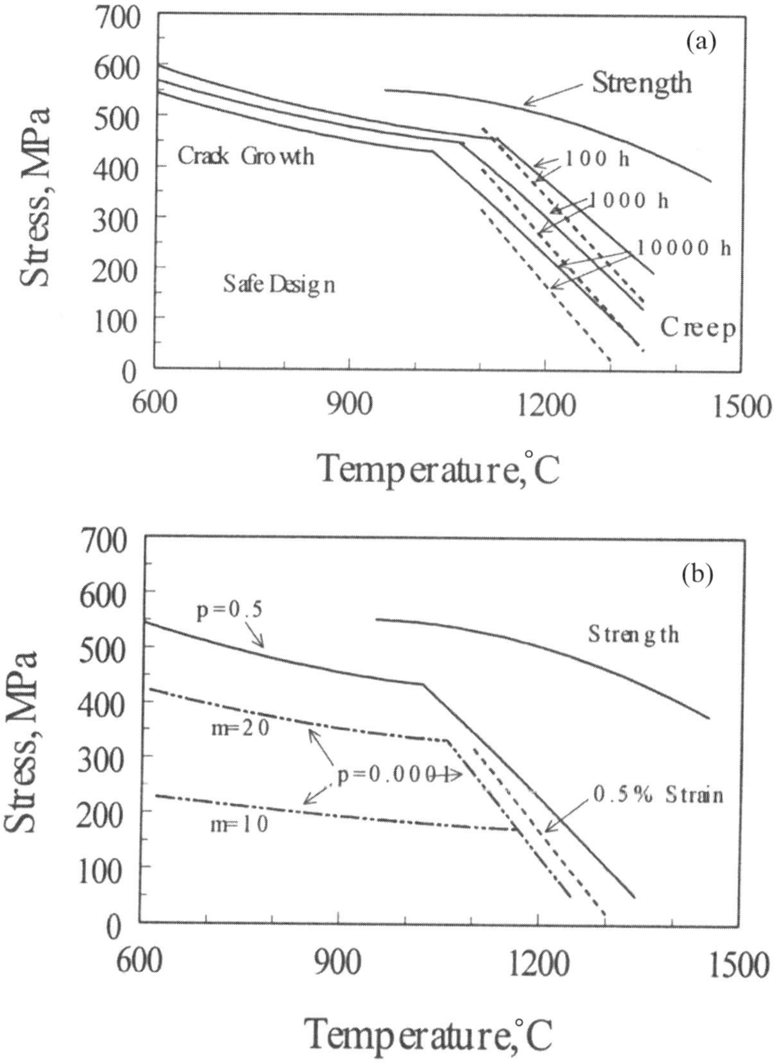

Figure 19.11 Fracture Mechanism Map for SN88 Silicon Nitride Showing (a) Curves for Average Lifetimes of 100, 1000, and 10,000 h and (b) Effect of Failure Probability and Weibull Modulus (Reprinted from ref. [74] with permission from Wiley-VCH Verlag GmbH)

results of these characterization efforts are described in the literature [35–59,66–74]. Specific examples of the evolution of the long-term reliability of structural ceramics will be provided in the section, Mechanical Property DataA for Selected Monolithic Ceramics.

An important finding arising from these efforts is that the dominant failure mechanism changes with temperature. At low temperatures subcritical crack growth from preexisting defects is generally assumed to be the main cause of failure. At high temperatures and high stresses slow crack growth may again prevail; however as the stress is lowered, creep can lead to loss of function due to excessive dimensional changes or failure arising from the evolution of creep damage.

Researchers at NIST have developed fracture mechanism maps [74] which account for both slow crack growth and creep controlled failure. Figure 19.11a illustrates the stress-temperature curves for three values of component lifetimes, 100, 1000, and 10,000 h. These lifetimes represent mean values (probability, p, = 0.5) as established by the experimental data. In order to establish a margin of safety, the lifetimes for more acceptable failure probabilities (p = 0.0001) should be used. Figure 19.11b illustrates the curves for a 10,000 h lifetime at two probabilities of failure (p = 0.5 and p = 0.0001). The effect of Weibull modulus upon the p = 0.0001 curve is also shown. At a temperature of about 900°C, a Weibull modulus of 20 reduces the allowable stress from the median value of about 450 MPa to about 350 MPa. For a Weibull Modulus of 10, the allowable stress is reduced to less than 200 MPa. These results demonstrate how important m is to component performance, since an allowable stress of less than 200 MPa lies very close to the design stresses in some ceramic components. The higher value of m yields a much greater flexibility for safe design.

Environmental Effects. In spite of the fairly extensive database describing the time-dependent behavior of monolithic ceramics, most of the data have been generated in a laboratory environment. Consequently the effects of the gas turbine environment upon the long-term mechanical reliability is poorly understood. Recent evaluations [32, 75] of various ceramic components exposed in field tests have shown that three processes can impact performance of the monolithic ceramics: (1) localized corrosion due to the presence of reactive species in the environment, (2) environmentally-induced de-stabilization of the intergranular phases, and (3) rapid recession of the silicon nitride (or silicon carbide) due to loss of the silica scale by direct reaction with water vapor. Each of the processes is discussed below. A more detailed discussion of corrosion issues can be found in Chapter 26 of this volume.

A decrease in strength can occur when metal impurities picked up by the combustion stream deposit on and react with the ceramic component. Figure 19.12 illustrates one such example in which the deposition of Fe resulted in the formation of a surface pit on a silicon nitride vane exposed for 68 h at 1288°C in an engine test stand [32]. It is possible that the Fe contamination lowers the viscosity of the silica scale thereby promoting active oxidation.

Environmental attack can also result in subsurface phase transformations that give rise to stresses large enough to nucleate cracks. This kind of failure was observed recently in a silicon nitride nozzle being tested in a gas turbine [32]. In this case, one of the phases present at the grain boundaries, the J-phase ($Yb_4Si_2O_7N_2$) decreased in volume by approximately 64% upon oxidation to $Yb_2Si_2O_7$. Cracks were observed to form in the environment-affected zone and to propagate almost completely across the nozzles.

Oxidation of both silicon nitride and silicon carbide is increased by (1) the replacement of oxygen by water vapor and (2) an increase in the pressure of the oxidant. In addition, the high velocities and presence of water in the environment can lead to the volatilization of the normally protective silica layer. Researchers at NASA Glenn Research Center [76–78] have shown that the presence of water vapor leads to the formation of a gaseous $Si(OH)_4$ species via a reaction with the silica layer. The rate of formation, k, of this species and thus the rate at which the SiC (or Si_3N_4) is consumed by its continued oxidation is determined by the expression,

$$k \propto v^{1/2}\, P(H_2O)^2/(P_{total})^{1/2} \tag{16}$$

where v is the gas velocity, $P(H_2O)$ is the pressure of the water vapor, and P_{total} is the total pressure. For lean burn conditions, k for a number of SiC based materials is found to be well described by the expression,

$$k\ (mg/cm^2\ h) = 2.04 \exp(-108\ kJ/mol/RT)\ v^{1/2}\ P_{total}{}^{3/2} \tag{17}$$

where T is the absolute temperature (K), P_{total} is the total pressure (atm), and v is the velocity (m/s). In addition to the volatility issues, which may lead to loss of function due to excessive changes in component dimensions, the effect of the environment upon the mechanical stability must be understood as well.

The validity of this model was recently confirmed in a study of the reliability of silicon nitride vanes exposed in an industrial gas turbine [79]. The vanes examined in this effort were fabricated from the AS800 silicon nitride. The first stage ceramic vanes were designed for retrofit into a Model 501-K turbine (Rolls-Royce, Indianapolis, IN). Following a 22 h shakedown run in a test turbine, the first stage vane assembly was mounted in a Model 501-K turbine at a commercial site (Exxon-Mobile, AL). During the field tests at Exxon, the average temperature and pressure of gas entering the vanes were approximately 1066°C

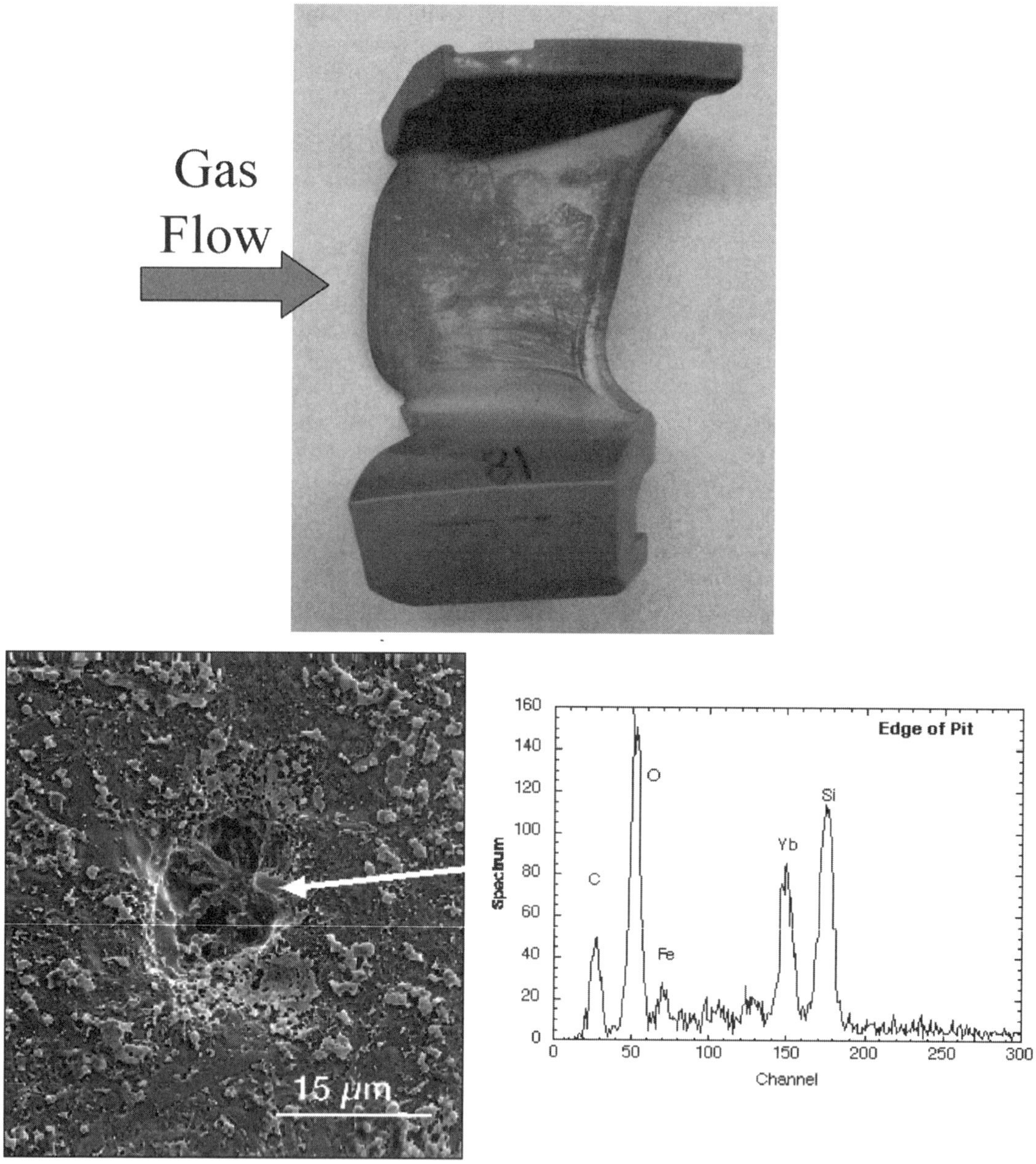

Figure 19.12 Example of Oxidation Pit in a SN88 Silicon Nitride Ceramic Vane Created by Deposition of Fe

(1950°F) and 883 kPa (128 psia), respectively. However, due to the combustor temperature pattern, the mid-span gas temperature could have been as high as 1260°C (2300°F) at the "hot spot". The inlet gas velocity at vane mid-span was about 162 m/s and the gas accelerated to about 573 m/s at the vane exit. The mole fraction of water vapor for the gas entering the vanes was calculated to be 0.101.

The engine sustained power and maintained performance throughout operation. It was subjected to the rigors of commercial operation including emergency shutdown (highest stress condition) and full power water wash. The engine was periodically shutdown for pre-planned inspections and removal/replacement of selected vanes for analyses. Although no vane failures occurred during the field test, dimensional measurements

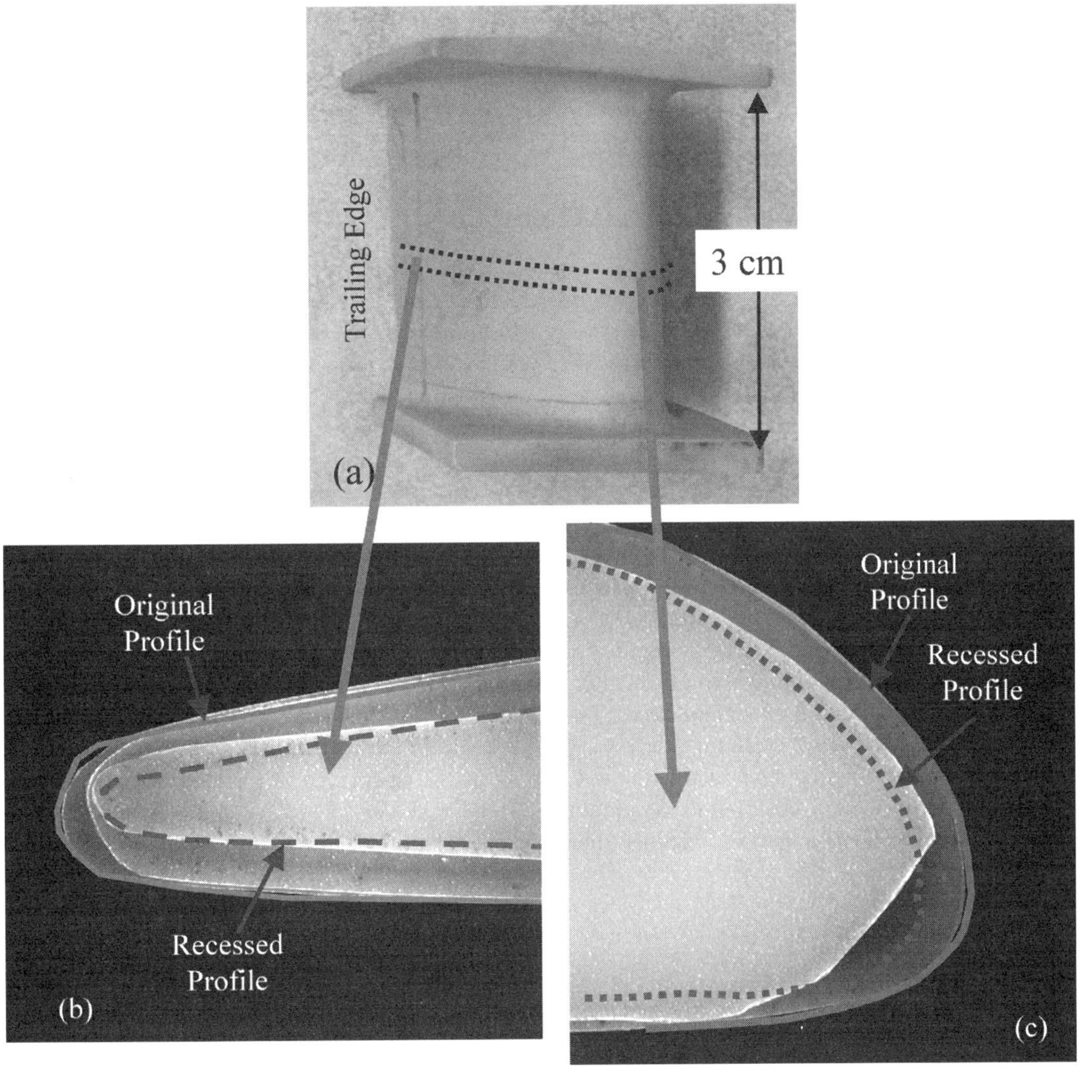

Figure 19.13 Photograph of (a) Uncoated AS800 Prior to Field Testing. The Dotted Lines in (a) Show the Sections That Were Machined for Subsequent Polishing and Evaluation by Scanning Electron Microscopy. Typical Sections for an As-Received Vane and a Vane Exposed for 815 h are Compared in (b) Trail Edge and (c) Leading edge. The Dotted Lines are Predicted Recession Profiles.

indicated that recession of the silicon nitride was a problem. Consequently, the engine test was terminated after 793 h (815 h total time including shakedown test). The analyses of the dimensional measurements are presented in the paragraphs that follow. For these measurements, the thickness of the trailing edge was measured at the tip, mid-span, and vane hub using a micrometer. Following completion of the field test all remaining vanes were measured in this manner.

The applicability of the recession model to predict the change in profile was examined by using Equation (17) to calculate the recession rate as function of position. The values of the pressure and velocity along the airfoil surface (required for this calculation) were obtained from the turbine manufacturer. These data were used to estimate the recession rate at each point along the airfoil surface. The change in the surface profile was calculated by assuming that the recession was normal to the surface at the point in question and that the magnitude was equal to the product of the rate and time. As shown in Figure 19.13, the prediction profile for the 815 h vane was in good agreement with the experimentally determined profile.

Currently, efforts are underway to develop coatings that will provide an effective barrier to the transport of water vapor to the surface of the silicon nitride or silicon carbide. Chapter 27 in this volume summarizes these coating efforts.

Component Design, Fabrication, and Life Prediction

Design issues are generally the responsibility of the engine manufacturer (see Volume I) while the ceramic suppliers must deal with the fabrication process (see Volume II). The most successful ceramic gas turbine programs are those in which there are strong interactions between the designers and suppliers.

Figure 19.14 illustrates the major steps associated with life prediction methodology. The first step involves estimating the temperature and stress distributions for the component in question. If creep processes are active, the prediction of valid stress profiles requires the use of appropriate finite elements. Unfortunately, designers often use linear elastic elements to reduce computation time. The resulting stress profiles do not account for effects such as stress relaxation, which occurs when the component creeps under fixed displacement conditions, or the residual stresses, which arise when the crept component is cooled to room temperature. It is also questionable whether some other important stress components such as contact stress are taken into consideration. The second step in the life prediction analysis is to use existing strength and time-dependent data (typically generated in the laboratory of standard specimens) to estimate appropriate Weibull parameters. The life prediction code then uses the Weibull parameters in conjunction with the finite element output to estimate the probability of survival at the desired operating life for the component. If this value is too low, the component geometry must be modified to lower the maximum stress. The ultimate goal of this process is to insure that the service stresses are sufficiently low to avoid time-dependent failure within the desired service life of the component. Further discussion on life prediction methodology can be found in Chapter 25 of this volume.

Although the procedure for lifetime prediction is well-established, there is often a significant discrepancy between the predicted values of lifetime/reliability and the actual component lifetimes. Narrowing this

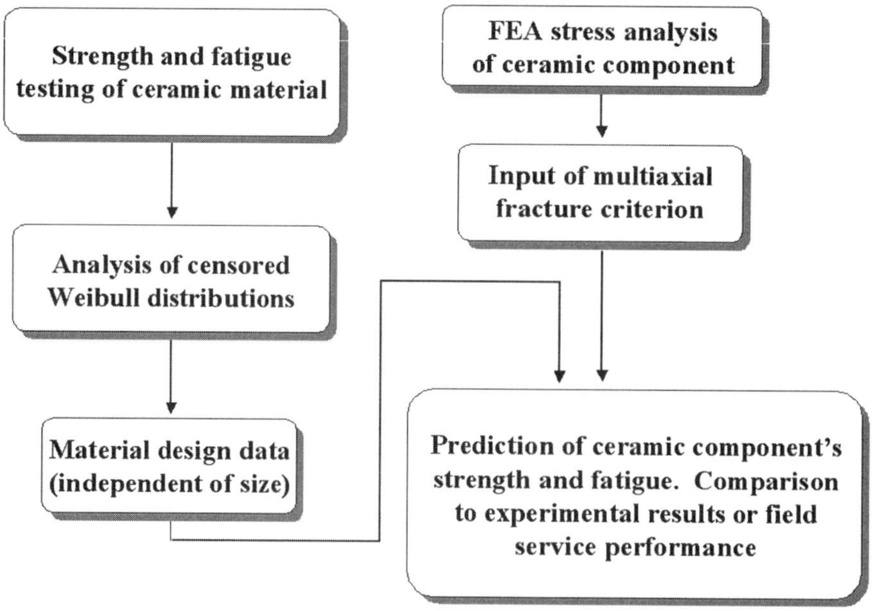

Figure 19.14 Key Steps Involved in Application of Life Prediction Methodology to a Ceramic Component

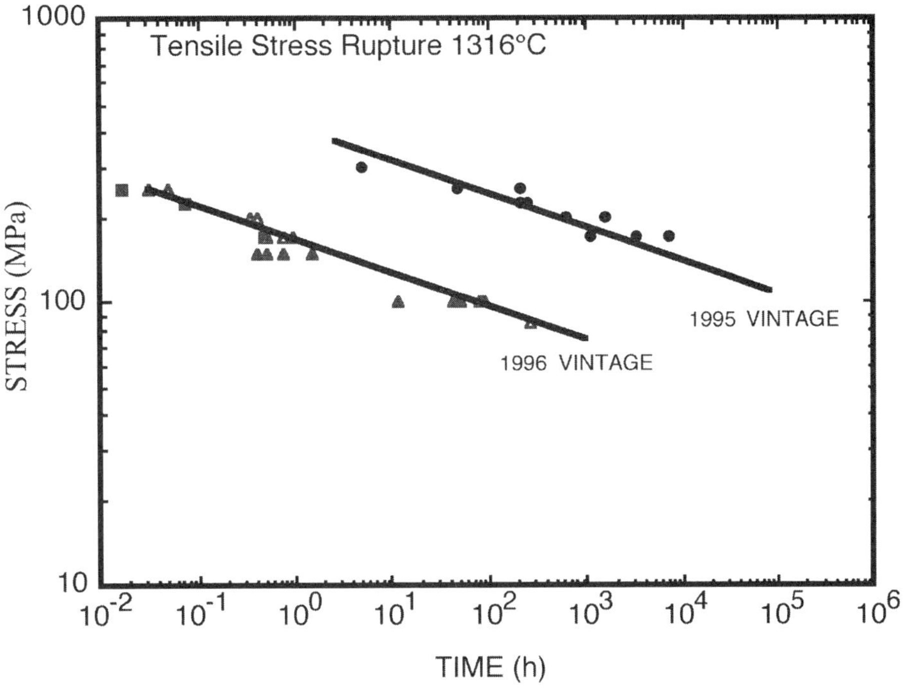

Figure 19.15 Comparison of Creep Deformation Behavior of Silicon Nitride Tensile Specimens Fabricated from Two Different Billets

gap to an acceptable level is necessary in order for ceramic components to be more extensively employed in gas turbines. Possible explanations for this discrepancy are:

- Inconsistent surface conditions. The surface state of the test specimens may be different from the component due to differences in the values of the machining parameters used. This calls for continued development of machining science and a better understanding of the effect of machining and evaluation of machined surface under engine conditions.
- Changing surface flaw population. In service, additional flaws may be introduced due to contact damage, particle impact, etc. This is exactly the case when creep conditions prevail.
- Difficulty of transferring meaningful data from test specimens. For example, surface fracture may be more frequent in the specimens with relatively high surface area to volume ratios, as compared with that of an engine component. Consequently strength measurements utilizing standard test specimens may not reflect the defects responsible for failure in the actual component.
- The stress state of the component may not be accounted for properly. In many instances finite element modeling of the component does not account for stress relaxation arising from creep deformation. While such relaxation lowers the maximum tensile stress at temperature, unacceptable residual tensile stresses may occur when the component cools to room temperature.
- Lack of information pertaining to environmental effects.
- Properties of the component differ from those in the standard test specimens used to establish the life prediction database. Such differences generally arise from unintentional changes in the processing. For example, Figure 19.15 shows stress rupture data obtained for two different vintages of a silicon nitride ceramic. Specimens fabricated from the 1996 vintage had much shorter lifetimes (and much higher creep rates than did specimens machined from billets fabricated in 1995. Preliminary x-ray data indicated that the billets having the poorer creep resistance contained an undesirable intergranular phase.

Component Verification

Once components are fabricated, component verification is traditionally accomplished through the use of thermal and mechanical proof testing, non-destructive evaluation, and limited engine testing. A somewhat under-utilized evaluation procedure involves the destructive characterization of a limited number of components. For example, the availability of the small dog-bone tensile specimen [32,79–81] has provided for the measurement of the tensile stress rupture properties of silicon nitride blades (Figure 19.16a). As shown in

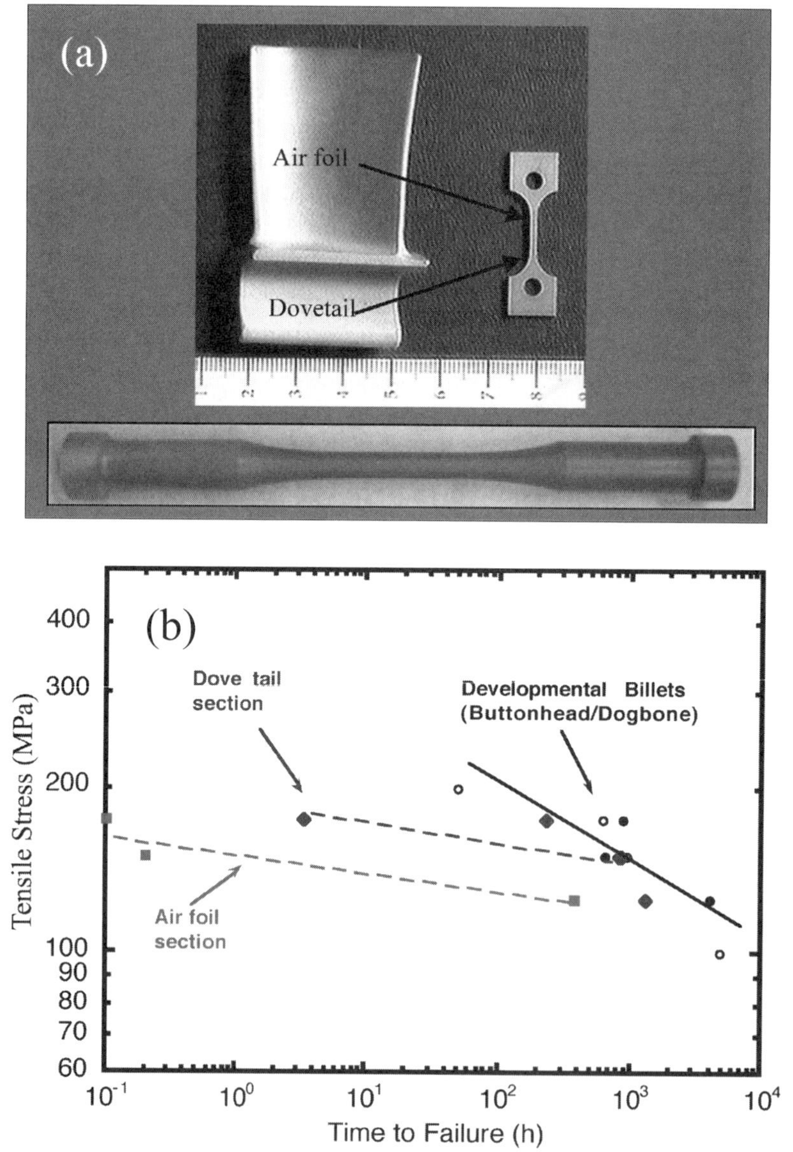

Figure 19.16 (a) Example of a Small Dog-Bone Tensile Specimens Fabricated from the Airfoil and Dovetail Regions of a Silicon Nitride Blade. A Standard Buttonhead Specimen Fabricated from Large Billets is Shown for Comparison. (b) Comparison of Stress Rupture Data Generated for Various Sections of a Ceramic Blade with that Obtained from Developmental Billets

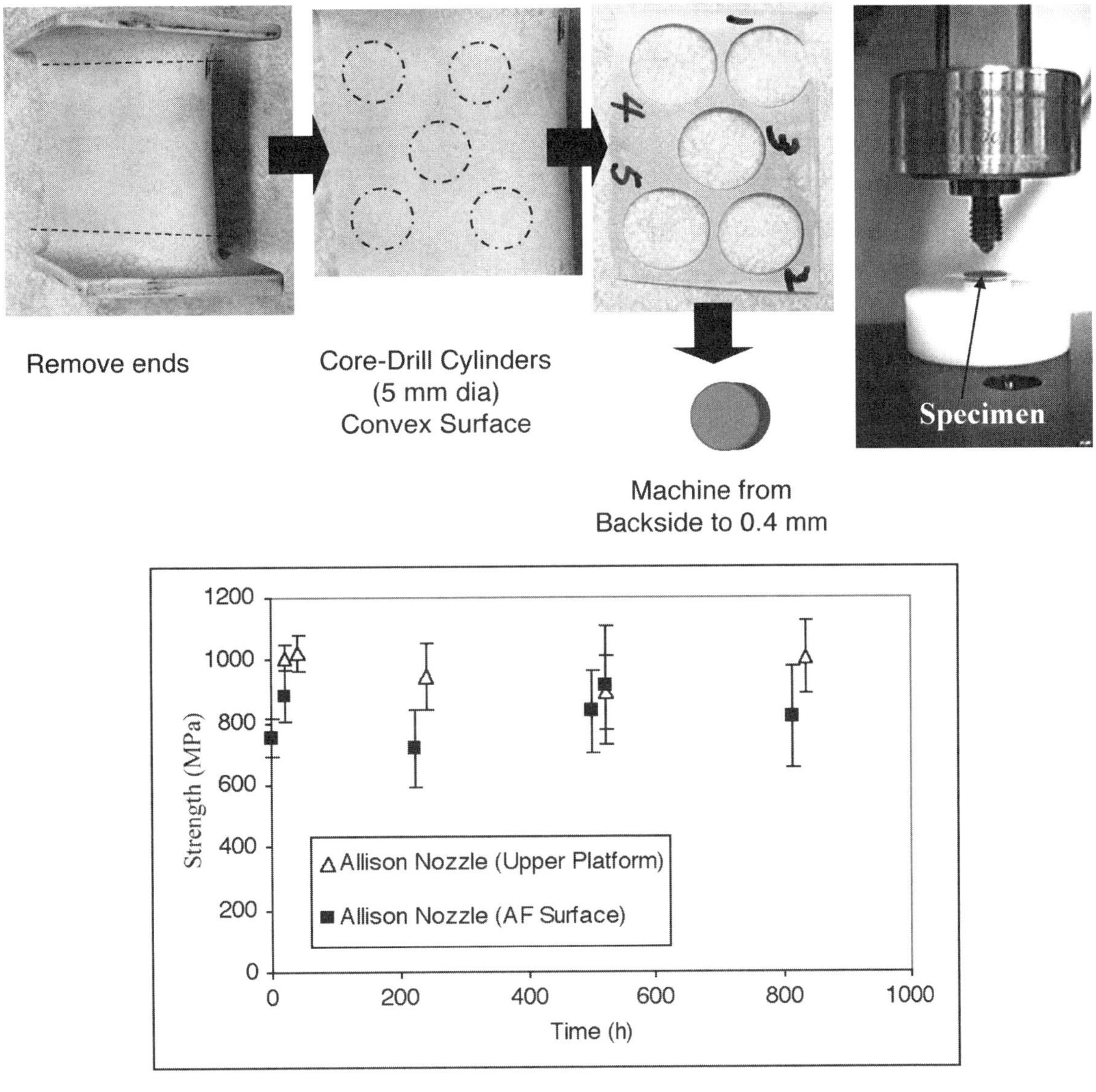

Figure 19.17 Measurement of Retained Strength for AS800 Vanes Exposed Is an 815 h Field Test Described in ref [79]. The Strength Values for Both the Airfoil and Platform Regions Are Provided in the Graph at the Bottom.

Figure 19.16b, there are significant differences in the stress rupture characteristics obtained for various parts of the ceramic blade. More importantly the data obtained from testing specimens prepared from the billets greatly over-estimate the lifetime.

It is also possible to use this approach to evaluate the properties of components after engine testing. For example, in ref. [79], a miniature biaxial flexure specimen (Figure 19.17) was used to measure surface strength as a function of exposure time. The strengths of the specimens removed from the airfoil platform were consistently higher than those removed from the airfoil surface. This difference could in part be attributed to the better surface quality of the airfoil ends (machined versus as-processed surfaces).

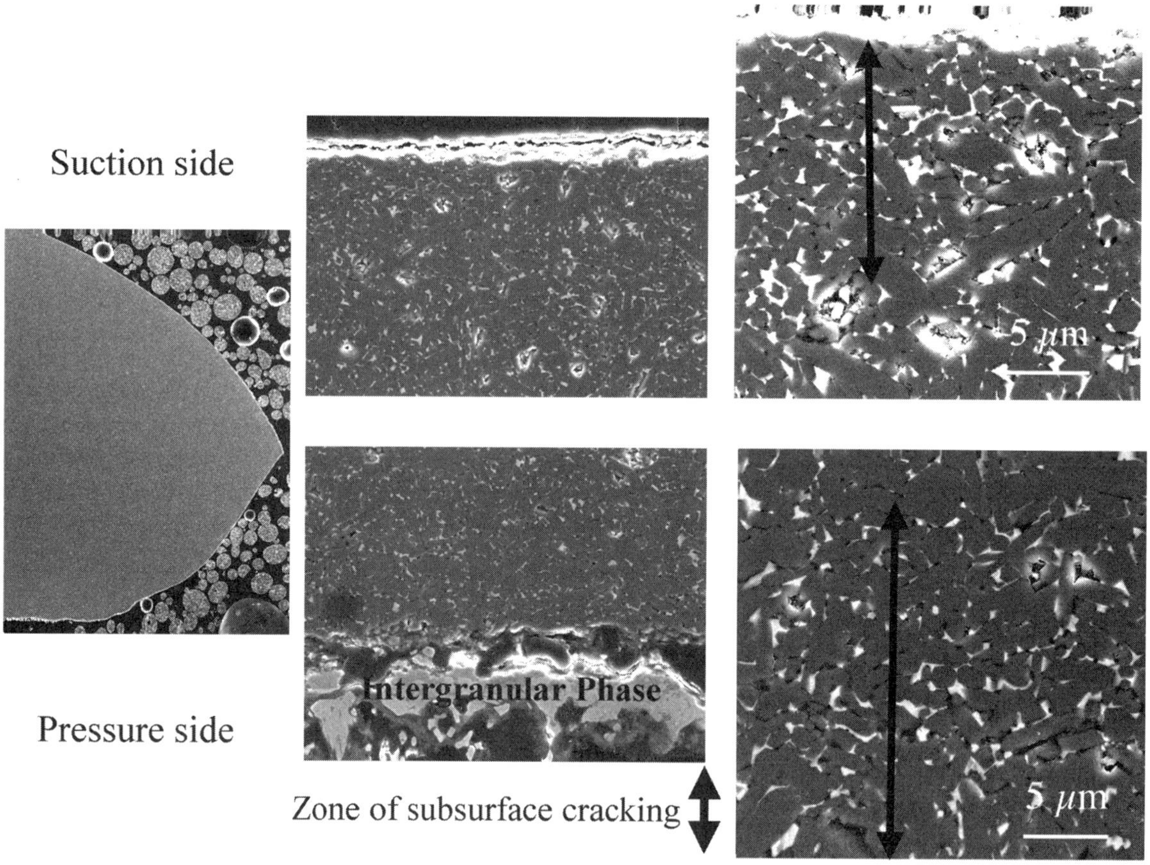

Figure 19.18 SEM of the Polished Cross-Section of Midspan Region of an Uncoated AS800 Vane After 815 h of Field Testing

There was very little change in average strength with time, although there was a small increase in the standard deviation in the strength of the specimens taken from the airfoil surface. These results indicate that the recession process effectively removed any environmentally-induced surface damage thereby preventing strength degradation. Note that although the strength was approximately constant, the probability of failure of the vanes most likely increased with time due to loss of material (and thus load-bearing capability) arising from the recession. The reduction in vane cross-section would be expected to increase the stresses generated during engine operation.

The strength measurements were supplemented with extensive dimensional, microstructural, and x-ray diffraction analyses. Scanning electron microscopy (SEM) of the airfoil surfaces showed that the light-colored scale noted above was actually the intergranular phase, which was composed primarily of La, Si, O, and trace amounts of Y. The appearance of this phase on the surface was most likely due to selective recession of the silicon nitride. SEM of the polished cross-sections of exposed vanes revealed that after 815 h (Figure 19.18) extensive recession resulted in significant changes in the shape of the airfoil. There was also a slight increase in the amount of subsurface cracking of the silicon nitride grains in the tested vane. The extent of this cracking was usually greater along the pressure surface. There was also a tendency for the pressure side to be slightly rougher than the suction side. Furthermore, the intergranular phase was often observed along the pressure surface.

MECHANICAL PROPERTY DATA FOR SELECTED MONOLITHIC CERAMICS

A comprehensive listing of relevant mechanical property data for SiC and Si_3N_4 based ceramics is not possible.[1] The purpose of this section is to illustrate the improvements in mechanical reliability of these materials over the last 30 years using specific examples. The primary focus is on commercially available materials (Table 19.1) since properties of experimental materials are continually evolving. The strength, toughness, and creep behaviors of these materials are described in the sections that follow. It should be noted that there are potential problems in comparing data from different sources. In the case of strength, for example, variations in the measured value can result from differences in the test procedures (geometry, machining conditions, etc.) which are not always adequately described. Second, even commercial materials exhibit properties which depend upon vintage. Finally the properties of a particular commercial material can depend strong on the green-state processing route (slip casting, injection molding, iso-pressing, etc.).

As indicated in Table 19.1 the materials are categorized in accordance with the primary densification mechanism. In the case of the silicon nitrides, these categories are hot-pressed, reaction sintered (bonded), sintered, and hot isostatically pressed. Hot-pressing, which is not generally considered as a viable method for component manufacturing, involves the simultaneous application of pressure (25–30 MPa uniaxial) and temperature to powder compacts held in high temperature dies. For silicon nitride, oxide additives such as MgO, Y_2O_3, CeO_2, ZrO_2 are required to achieve full density. The final product is predominantly β-Si_3N_4 that is close to 100% dense and exhibits a fine grain microstructure (grain size typically in the range of 0.5 to 2 μm) with a small fraction of grains being slightly elongated.

Reaction sintering involves nitridation of precompacted silicon powder perform, which can be.prepared by slip casting, dry pressing, flame spraying, injection molding, or some other forming process. The resulting green preform is then nitrided in an atmosphere of pure N_2 or N_2 and H_2 at a temperatures of up to 1400°C. The nitridation reaction must be carefully controlled to prevent blocking off the interconnecting porosity in the powder compact. Because this process involves little shrinkage, near-net shape component fabrication is possible. Compared with hot-pressed silicon nitride, reaction sintered materials require fewer additives, have lower densities (75–80% of theoretical are typically), and exhibit lower strengths (controlled by large pores or pore clusters).

Silicon nitride powder can also be densified via liquid phase sintering at modest nitrogen pressures. Early work in this area focussed on developing sinterable versions of the hot-pressed silicon nitride compositions. Fine reactive powders and a sufficient amount of oxide additives were required to achieve full density compacts with adequate properties. In general the high temperature properties of the sintered (and the hot pressed) silicon nitrides were determined by the refractoriness of the intergranular phase. In the early 1980s, materials suppliers began to exploit the α to β silicon nitride dissolution—reprecipitation mechanism to generate microstrutures consisting of acicular and interlocking grains. The toughness of these materials was significantly higher than that of their conventionally sintered counterparts due to the tortuous intergranular crack paths coupled with wake zone phenomena, such as crack bridging. These silicon nitrides are commonly referred to a self-reinforced or in-situ reinforced.

Hot isostatic pressing (HIPing) consists of applying high isostatic pressures (100s of MPas) during heating of powder compacts. Samples must either be pre-sintered to eliminate open porosity or be sealed in a gas

[1] Detailed data can be found in many of the references cited in this chapter as well as the following databases:

http://www.udri.udayton.edu/handbook/title.html

http://www.ceramics.nist.gov/srd/scd/scdquery.htm

http://www.ceramicindustry.com/FILES/HTML/ReferenceIndex/0,2796,,00.html

http://www.matweb.com/

http://www.jahm.com/

http://www.keramverband.de/eng/mechanische_eigen.html

impermeable envelope. Early HIP methodology utilized refractory metal cans, which limited the shape forming capability. More recently, glass encapsulation has been used to densify relatively complex parts. HIPing generally requires fewer additives to achieve full density. The resulting microstructures consist of a fine grain matrix (1 μm) with some elongated grains. Compared with conventional sintering, HIPing requires significantly lower sintering temperatures (200°C–300°C) and shorter densification times. Potential problems can arise due to reaction between the molten glass with the silicon nitride powder.

The silicon carbides are classified as hot-pressed, siliconized (reaction sintered), or sintered. Like the silicon nitrides, the hot-pressed silicon carbides require additives such as Fe_2O_3, Al_2O_3 or B_4C to promote the formation of dense materials. The major role of the Al_2O_3 is to suppress grain growth. Typical hot pressing conditions consist of pressures in the range of 25–30 MPa, temperature of up to 2150°C, and time of 1–2 hours.

In siliconized (or reaction sintered) SiC, SiC and C mixtures are first formed into a shaped compact using processes such as extrusion and dry pressing. This compact is then exposed at high temperatures to molten Si, which infiltrates into the pore structure and reacts with C to form SiC. The resulting material consists of the original SiC grains, newly formed SiC, and residual Si.

The SA SiC is a sintered silicon carbide which is fabricated in the green-state by injection molding. Small amounts of carbon and boron are added to assist the densification during sintering. Post-sinter HIPing is also used to improve density and reduce defect size. The microstructure of this material typically consists of silicon carbide grains ranging in size from 2 to 8 μm.

In both the United States and Europe, early ceramic gas turbine programs utilized reaction bonded and hot-pressed silicon nitrides. In the late 1970s and early 1980s, the HIPed materials having superior strength and toughness were primary candidates for components requiring high strength. By the late 1980s, most of research in Europe on structural ceramics for gas turbines ended (refer to Chapter 20 in Volume I). In the United States in the 1990s, the self-reinforced silicon nitrides became the material of choice for ceramic gas turbine components. Additional historical details of the evolution of these materials are provided in Chapter 1 of this volume.

In Japan, most ceramic gas turbine programs have utilized pressureless sintered silicon nitride containing rare earth oxides. Much of the early work focused on enhancing the properties of these materials by (1) optimizing the green-state processing, (2) improving the starting powder, and (3) optimizing the intergranular phase.

In Russia, hot-pressed, reaction-sintered, and sintered versions of both silicon carbide and silicon nitride have been developed for fabrication of gas turbine components (see Chapter 12 in this volume). In addition, emphasis has been placed on lowering cost by using less expensive ceramics such as the alumino-boron-nitride ceramic material "SUGRAV" (see Chapter 32 in Volume I and Chapter 13 in Volume II). Because of the relatively low strength of this material, design philosophy is to insure that the components are subjected primarily to compressive stresses.

In the sections that follow, representative data are provided for the strength (and distribution), fracture toughness, and creep of silicon nitride and silicon carbide ceramics.

Strength and Strength Distribution

Figure 19.19 compares the temperature dependency of the flexure strength for (a) reaction sintered (RSSN), (b) hot pressed (HPSN), (c) sintered (early vintages-SSN), (d) HIPed, and (e) self-reinforced (SSN-SR) silicon nitrides. The strength of the reaction sintered material is the lowest of all types of silicon nitride. Although much improved, the strength of the hot-pressed silicon nitride decreases rapidly as the temperature exceeds a critical value (1000–1200°C). The exact value depends upon the amount and type of secondary phase. The strengths of the HIPed and self-reinforced silicon nitrides, which represent the most recent materials of this class, exhibit further improvements in mechanical strength over the complete temperature regime. There is also a tendency for the scatter in strength data to be much lower for these newer materials suggesting much greater consistency.

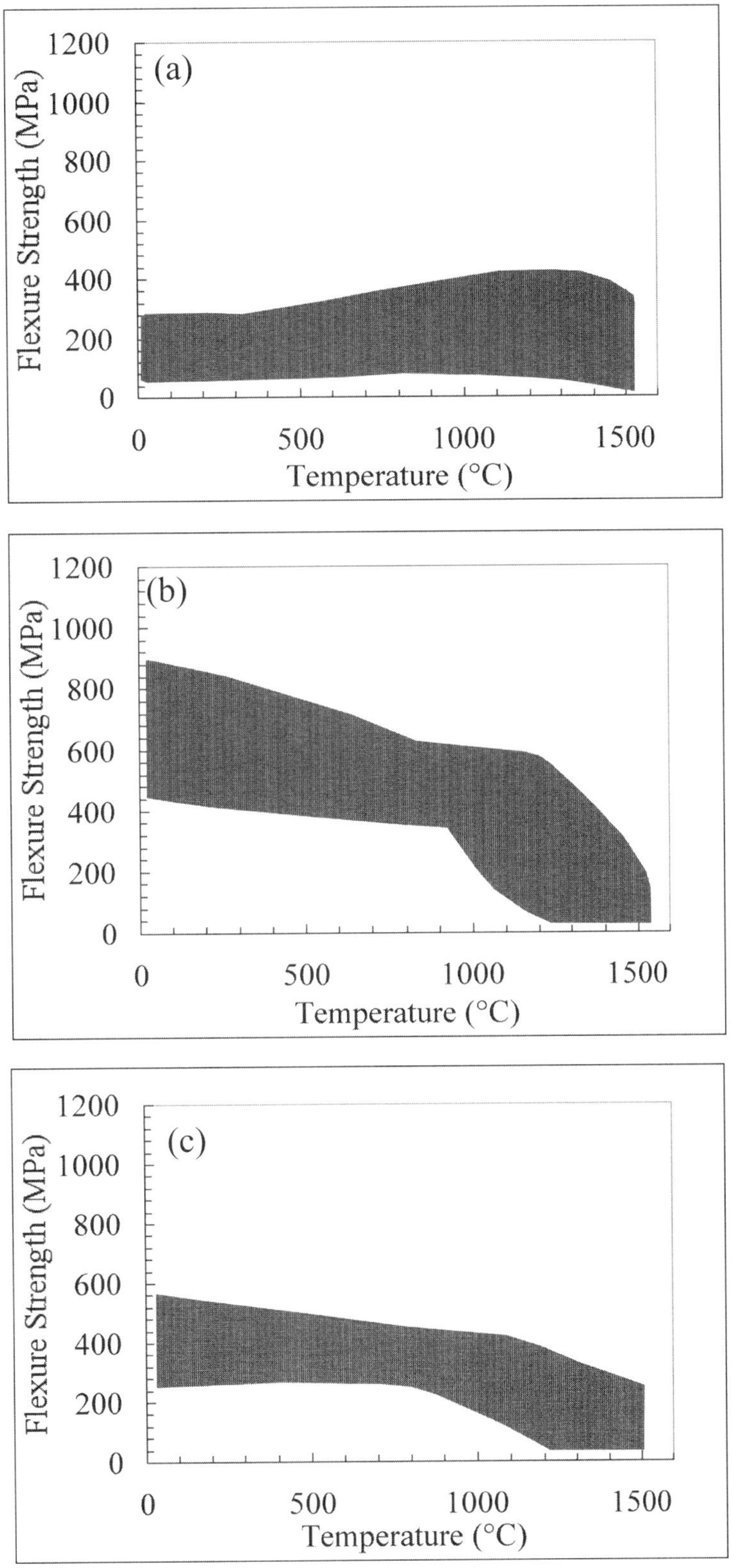

Figure 19.19 Comparison of Flexure Strength as a Function of Temperature for (a) Reaction Sintered, (b) Hot-Pressed, (c) Sintered (Early Vintages) ***(Figure continues)***

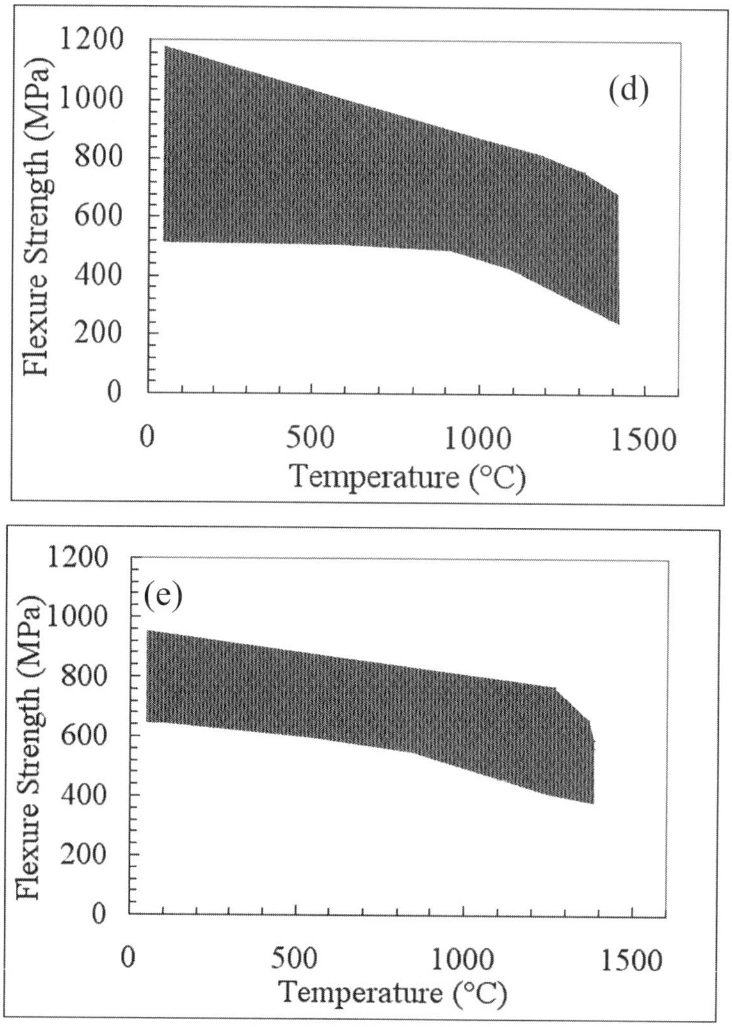

Figure 19.19 *(Continued)* (d) HIPed, and (e) Self-Reinforced Silicon Nitrides

Figure 19.20 illustrates similar strength-temperature data for (a) reaction sintered (RSSiC), (b) hot pressed (HPSiC), and (c) sintered silicon carbides (SSiC). The reaction sintered materials, which all contain free silicon, exhibit an abrupt drop in strength as the temperature approaches the melting point of free Si (1423°C). Like their silicon nitride counterparts, the strengths of HPSiCs are relatively high at low temperatures but decrease as the temperature is increased above a critical value. Finally the strengths of the SSiC are fairly independent of strength.

Comparisons of the strengths of the silicon nitride and silicon carbide materials at room and elevated temperatures are provided in Figures 19.21 (a) and (b), respectively. The strengths of the silicon carbides are comparable to those of the RSSN and SSN materials. It is also noteworthy that the spread in the high temperature strength values is much greater for the early materials (RSSN, HPSN, SSN, and all of the silicon carbides) than for the more recent HIPSN and SSN-SR silicon nitrides. This difference can be attributed to the fact that many of the early ceramic gas turbine programs (see Chapter 9 in Volume II, and

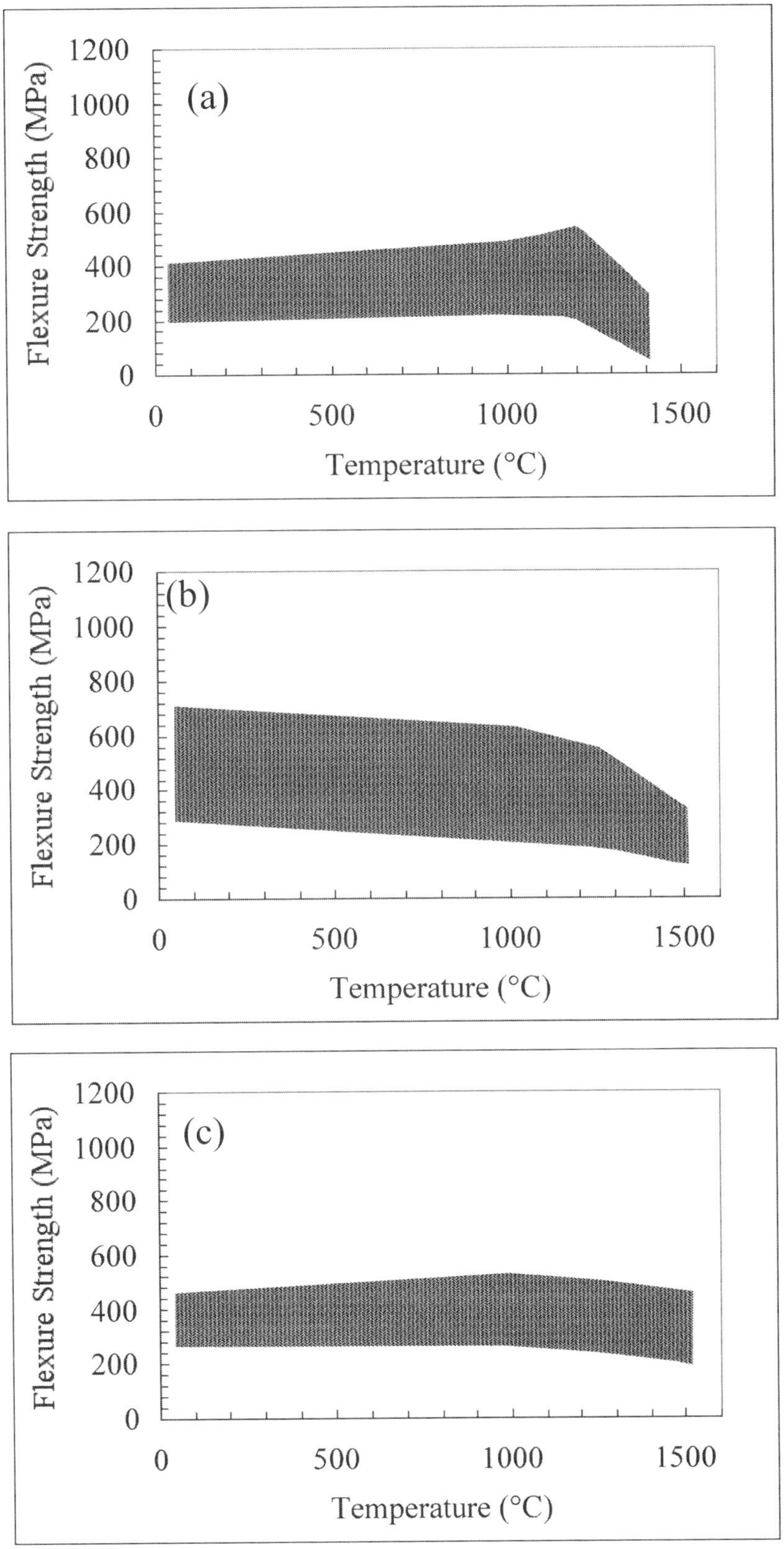

Figure 19.20 Comparison of Flexure Strength as a Function of Temperature for (a) Reaction Sintered, (b) Hot-Pressed, and (c) Sintered Silicon Carbides

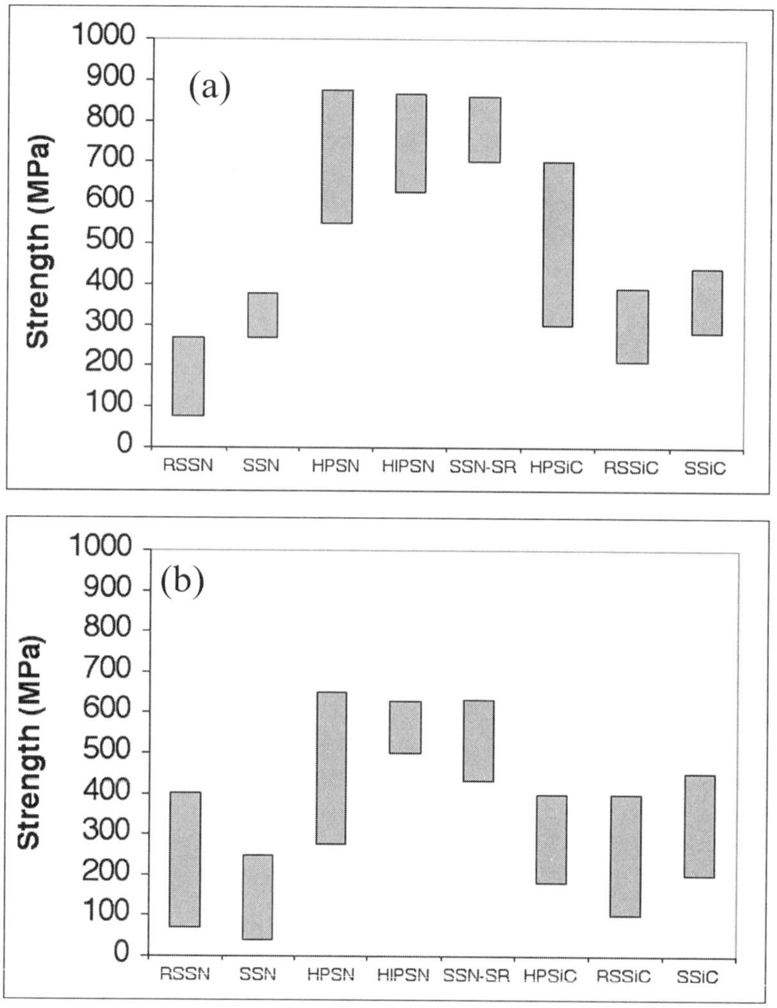

Figure 19.21 Comparison of Flexure Strength of Various Silicon Nitrides and Silicon Carbides at (a) Room and (b) Elevated Temperatures (1350–1400°C)

reference [82]) focused on the improvement of the mechanical behavior of the former materials. Figure 19.22 illustrates data from reference [82]. Another excellent example of the evolution of strength of silicon nitride is provided in Chapter 5 in this volume. The room-temperature strength data for the various silicon nitride compositions examined in this chapter are shown in Figure 19.23. The AS800 and GS44 are classified as self-reinforced materials, the GN10 is HIPed, the Code 2 materials are sintered, and the RBSN materials are reaction sintered. Representative data for the Japanese materials (primarily silicon nitride) are provided in Chapter 1 in this volume.

A similar comparison of the strength distributions as reflected by the Weibull parameters is complicated by a number of factors. First, the uncertainty in the estimates of these parameters cited in the literature is often largely due to the relatively small number of specimens tested. Secondly, when failure originates from the surface, the Weibull parameters exhibit a significant dependency on the surface preparation (machining, annealing treatment, etc.). Finally, many studies of the strength distributions do not include fractographic information, which is required to relate the Weibull parameters to a specific defect type(s). In spite of these problems, a review of selected data (obtained from machined flexure specimens) can be used to illustrate trends in Weibull

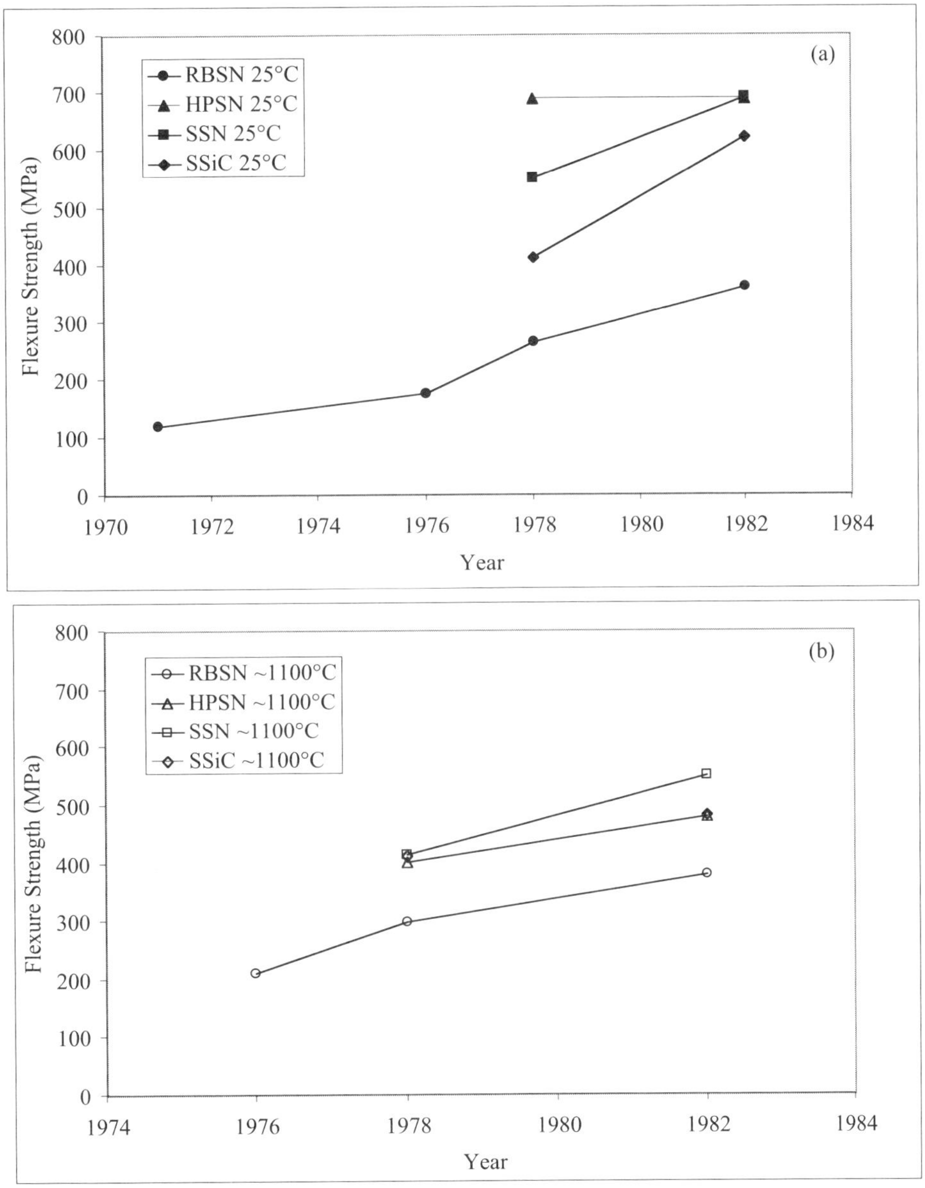

Figure 19.22 Evolution of the (a) Room-Temperature and (b) High-Temperature Strengths of Early Silicon Nitride and Silicon Carbide Materials [82]. The Creep and Oxidation Resistances of these Materials Improved as Well. [Designators—RBSN: Reaction Bonded Silicon Nitride, HPSN: Hot-Pressed Silicon Nitride, SSN: Sintered Silicon Nitride, and SSiC: Sintered Silicon Carbide]

modulus with material composition. As shown in Figure 19.24(a), m values for RSSN and all of the silicon carbide ceramics are lower than those for HPSN, HIPSN, SSN, and SSN-SR. More importantly, the self-reinforced silicon nitrides exhibit the greatest values of m indicating potential for excellent mechanical reliability. The data in Chapter 5 in this volume can also be used to illustrate the improvement in Weibull modulus with time (Figure 19.24(b)). The m values for the early materials were in the 5 to 15 range while the most recent self-reinforced silicon nitrides (AS800 and GS44) have exhibited m values as high as 35.

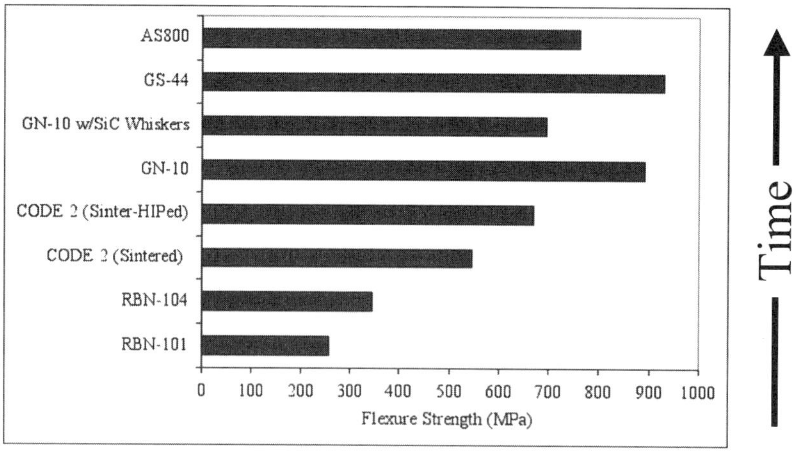

Figure 19.23 Evolution of the Strength of Silicon Nitride Developed by Honeywell Ceramic Composites [Chapter 5 in this volume]

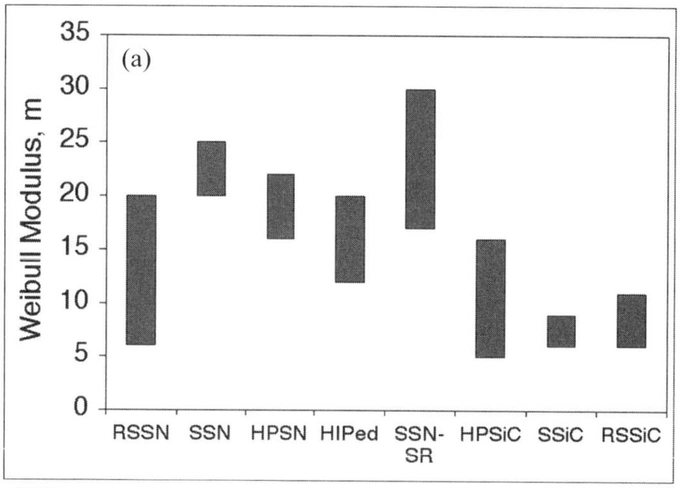

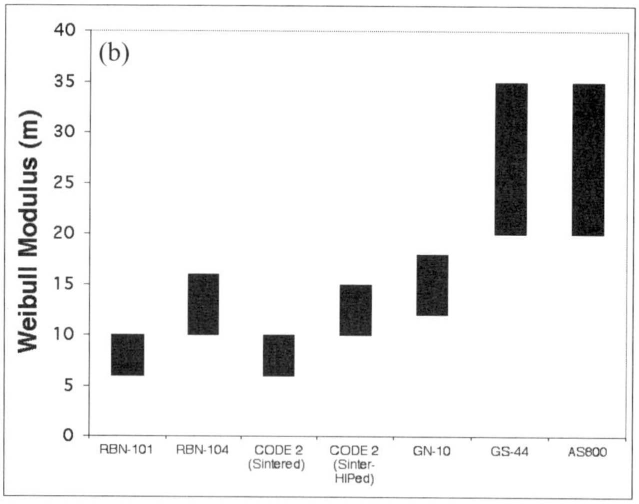

Figure 19.24 Weibull Modulus for Silicon Nitride and Silicon Carbide Ceramics Obtained from Flexure Strength Data for Machined Specimens (a). Data Ranges Obtained in Chapter 5 in This Volume Are Shown in (b).

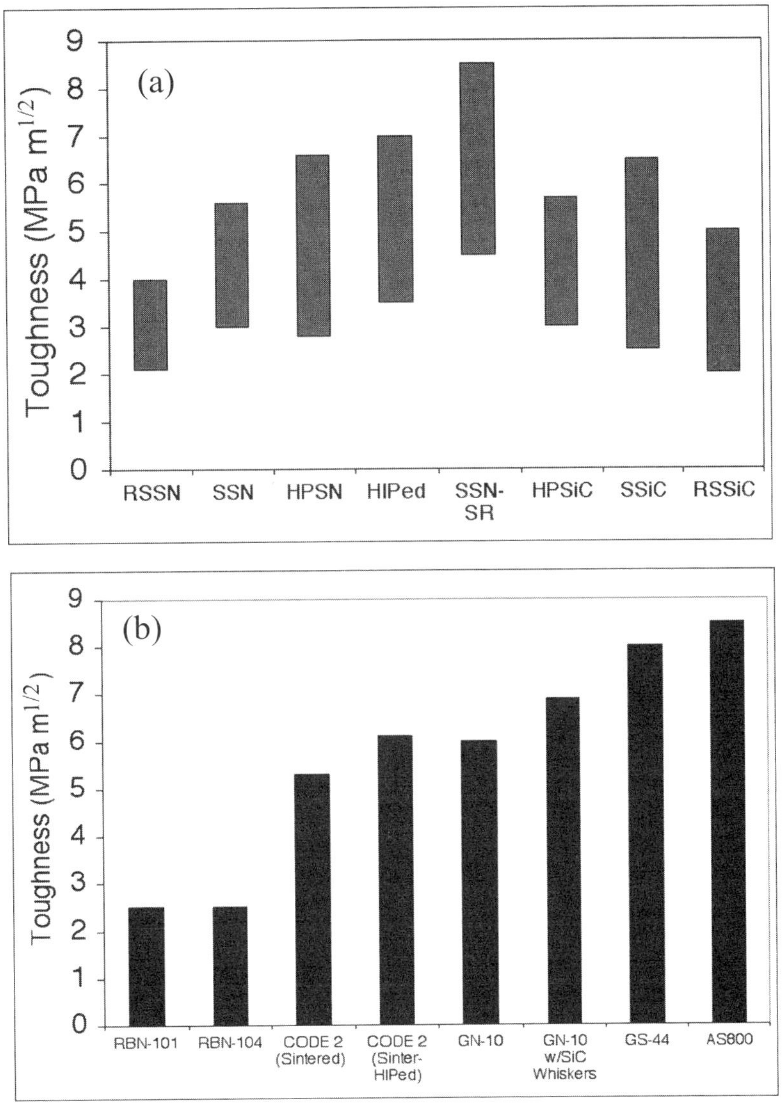

Figure 19.25 Fracture Toughness for Selected Silicon Nitride and Silicon Carbide Ceramics Listed in Table 19.1 (a). Data Obtained in Chapter 5 in this Volume Are Shown in (b).

Fracture Toughness

The fracture toughness is an important parameter for a number of reasons. First, it reflects the damage tolerance of a material in that for a desired level of strength, a higher fracture toughness means that larger flaws can be tolerated. The presence of larger flaws will also facilitate their detection by non-destructive methods. Toughness is also a critical parameter in determining impact resistance. For a detailed discussion the reader is referred to Chapter 28 of this volume. Finally as indicated in Equation (7), knowledge of the fracture toughness is required to fully define the slow crack growth behavior.

Typical values of fracture toughness for the silicon nitride and carbide material classes are given in Figure 19.25(a). As expected, the self-reinforced materials exhibit the highest values of toughness. Figure 19.25(b) illustrates the evolution of in the fracture toughness described in Chapter 5 in this volume. Although the maximum fracture toughness in these two figures approaches a respectable value of 9 MPa $m^{1/2}$, recent field experience (see Chapter 7 in Volume 1) has indicated that even higher values are required to eliminate foreign object damage in rotating components.

Time-Dependent Behavior

As discussed above, time-dependent strength degradation can occur from slow crack growth, cyclic fatigue, and creep damage. The slow crack growth susceptibility of monolithics is reflected by the values of A and N (Equation (3)), which are determined from stress rupture, dynamic fatigue, or direct V-K_I measurements. Of these two parameters, 1/N, which is the slope of a logarithmic plot of applied stress versus time-to-failure, is directly related to a materials susceptibility to degradation by slow crack growth. For example in the case where the applied stress is 80% of the fast-fracture strength, the failure times are 87 s, 2.1 h, 655 d, and 156 y for N values of 20, 40, 60, and 100 respectively.

Table 19.3 summarizes the N values determined from flexural dynamic fatigue measurements provided in ref. [83]. In general, the SiC based ceramics are less susceptible to slow crack growth than either the HIPed or SSN ceramics due in part to the relatively small amounts of intergranular phases present in the SiC materials. There is also a tendency for the N values to decrease with increasing temperature. However, it must be emphasized that the N values measured at elevated temperatures may not reflect the true slow crack growth behavior for several reasons. First, if creep deformation processes are active, then it is likely that stress redistribution within the flexure bar can invalidate the relationship used to calculate the maximum tensile stress. Secondly, these same creep processes can cause blunting of existing flaws which limits their ability to grow subcritically. In this respect, the slow crack growth mechanism can no longer dominate time-dependent behavior.

Cyclic fatigue is the second mechanism leading to time-dependent strength loss. Crack propagation behaviors under cyclic loading have been investigated at high temperatures, in tension and flexure, under various test conditions including frequency, stress ratio, stress wave form, atmosphere, and test temperatures [40–57]. A major question addressed by these studies is whether the strength degradation measured under cyclic loading can be attributed to the same slow crack growth mechanism active under static loading conditions. If so, the values of N and A in Eqs. (8) and (10) should be identical. The major findings from these cyclic fatigue studies are as follows:

(1) Materials having limited glassy phases at the grain boundaries exhibit cyclic fatigue lives that agree with those predicted from static and dynamic data.
(2) For those materials having a significant amount of intergranular phase, the cyclic fatigue lives measured at temperatures below the softening point of this phase are shorter than those predicted from static and dynamic fatigue lives. This behavior has been attributed to two cyclic loading mechanisms: (a) an increase in microcracking and (b) the destruction of grain bridging which reduces the effective fracture resistance.

TABLE 19.3 SUMMARY OF N VALUES CALCULATED FROM THE FLEXURAL DYNAMIC FATIGUE DATA GIVEN IN REF. [83]

	SiCs		HIPed SN		Early SSN	
T (°C)	Min N	Max N	Min N	Max N	Min N	Max N
20	47	167	23	104	96	>200
1000	>200	>200	70	>200	37	48
1200	>200	>200	30	46	14	>200
1250	77	>200	40	40	26	26
1350	>200	>200	24	24	22	22
1400	40	200	8	9	11	>200

(3) Creep rupture results for silicon nitride ceramics exposed to cyclic loading at temperatures where creep deformation is active indicate that at sufficiently high frequencies, the evolution of creep damage can be inhibited leading to much longer lifetimes.

The third mechanism responsible for time-dependent strength degradation is creep. In many silicon nitride ceramics creep deformation (strain) is controlled by the nucleation, growth, and coalescence of cavities located at two-grain and multi-grain junctions. Consequently, the damage accumulation rate is directly related to the measured creep rate. In the case of SiC and Si_3N_4 ceramics, creep behavior is largely affected by the refractoriness of the grain boundary phase, which depends on the sintering additives and processing conditions of the material.

The evolution in creep performance of silicon nitride ceramics over the last 20 years is nicely illustrated by the Monkman-Grant relationship between minimum or steady-state creep rate and creep life. As shown in Figure 19.26, the Monkman-Grant plot for a number of silicon nitride ceramics illustrate an inverse relationship between creep rate and time-to-failure in accordance with Eq. (12). The early hot-pressed silicon nitrides exhibited lifetimes of <1 h for an applied stress of 100 MPa and temperature of 1350°C. These lifetimes improved by 2 to 4 orders of magnitude for the HIPed materials (PY6, GN10, NT154, and NT164). The current self reinforced silicon nitrides (AS800 and SN282) exhibit creep lives that are up to 6 orders of magnitude higher than those of the hot-pressed materials.

Other Properties

Other properties of interest include elastic modulus, thermal expansion, thermal conductivity, oxidation resistance, thermal shock, and corrosion resistance. Typical values for the first three of these properties are provided in the databases cited in the footnote above and ref. [35, 36, and 83]. Although thermal shock resistance is also an essential property required for gas turbine applications, this property has not been studied so

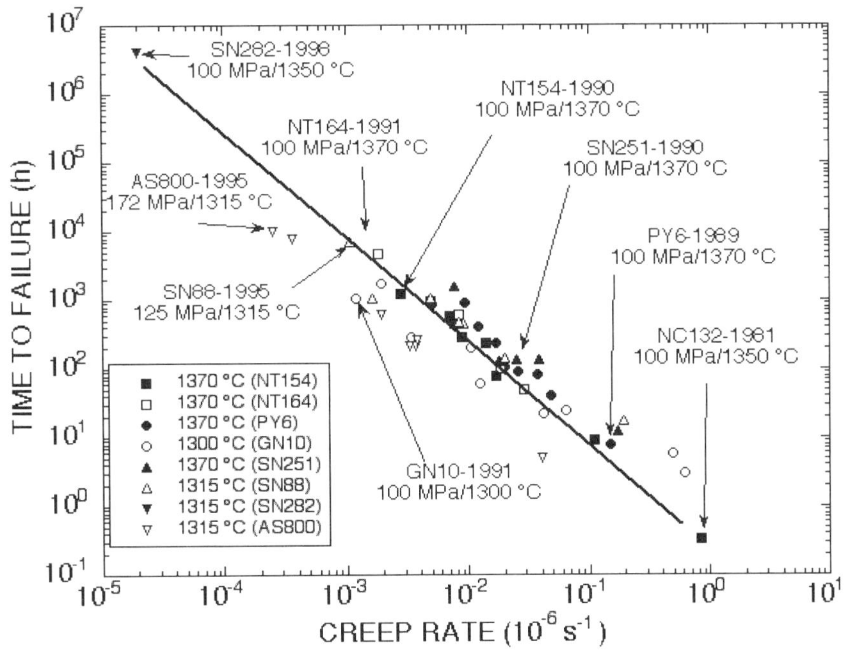

Figure 19.26 Comparison of Creep Rupture Behavior of Selected Silicon Nitrides

intensively due in part to the fact that it's strong dependence on specimen geometry greatly limits the availability of standard tests. Instead engine manufacturers often develop thermal shock tests for specific components (see Chapters 4 and 7 in Volume I). It is possible to predict this property from other properties like strength, elasticity, and thermal expansion coefficient.

The oxidation and recession characteristics of non-oxides in gas turbine environments are extremely important properties given the recent experience cited in refs [75–79]. The NASA model cited above and discussed in Chapter 26 of this volume provides a first order estimate of the expected behavior for many silicon carbide and silicon nitride ceramics. These predictions should be confirmed by burner rig studies conducted using conditions that closely mimic those of the gas turbine.

Key properties are also required for the successful application of environmental barrier coatings (EBCs) to inhibit surface recession of non-oxide ceramics. For example, thermal and elastic properties of the coating and substrate will dictate the level of residual stress. The mechanical strength of coated ceramics is also important since the application of a thick EBC may modify the critical flaws which are present. Finally, because most EBCs are oxide-based, they may enhance the materials susceptibility to time-dependent failure. A critical need in this area is the development of suitable standard mechanical tests to evaluate these properties.

SUMMARY

Although significant progress has been made in addressing the factors listed in Figure 19.1, there are still a number of needs which must be considered to insure the successful utilization of monolithic ceramics in gas turbines. First, the effect of cyclic fatigue must be better understood. As indicated above, the available data for many non-oxides show that the application of cyclic loading may either increase or decrease lifetime measured under static loading. The influences of microstructure and temperature upon these effects are poorly understood.

Current life prediction methodology does well in dealing with existing flaw populations which change only as a result of slow crack growth. However, processes such as creep, oxidation, recession and corrosion can introduce new flaw populations which are not easily accounted for in current design methodology. Mechanical characterization of components tested in the field coupled with extensive microstructural analysis may provide insights into how these flaw populations evolve.

The effect of environment upon both the short and long-term mechanical reliability of monolithic ceramics having EBCs must be more thoroughly evaluated. To be successful, the EBC must (1) inhibit oxidation, (2) not adversely affect the strength, slow crack growth, or creep behaviors of the underlying substrate, and (3) be stable in a gas turbine environment. Significant mechanical property and oxidation/recession testing are required to insure these goals are achieved. The development of NDE methods for assessing the integrity of EBCs on actual components is also needed. Life prediction methodology for EBC reliability is warranted especially in those cases where loss of the EBC could lead to rapid degradation of the underlying monolithic.

Finally, components should be routinely characterized to insure that the material properties are equivalent to those reflected in the database.

ACKNOWLEDGEMENT

Unless otherwise indicated the figures and tables in this chapter have been provided courtesy of the authors' organizations.

REFERENCES

[1] Anson, D., Ramesh, K. S., and DeCorso, M., *Application of Ceramics To Industrial Gas Turbines, Volume 1- Technology Background; Volume 2 Economics, Market Analysis, and R&D Plan*, DOE/CE/40878-1&2; Battelle Columbus, OH, USA, March 1991.

[2] Helms, H. E., Johnson, R. A., and Groseclose, L. E. "AGT 100-Advanced Gas Turbine Technology Development Project," in *Proceedings of the Twenty-Third Automotive Technology Development Contractors' Coordination Meeting* P-165, Society of Automotive Engineers,Warrendale, PA, USA, March, 1986. pp. 137–155.

[3] van Roode, M., "Ceramic Retrofit Program," in *Proceedings of the Joint Contractors Meeting: FE/EE Advanced Turbine Systems Conference—FE Fuel Cells and Coal-Fired Heat Engines Conference*, DOE/METC-93/6132, August 1993. pp. 77–93.

[4] Carruthers, D. and Lindberg, L., "Critical Issues for Ceramics for Gas Turbines," in *Third International Symposium on Ceramic Components and Materials for Engines*, V. J. Tennery, ed., American Ceramic Society, Westerville, OH, USA, 1988. pp 1258–1272.

[5] Haley, P. J, Holtman, R. L., Groseclose, L. E., Hilpisch, S. J., and Bell III, A. H., "Advanced Turbine Technology Applications Project (ATTAP)," in *Proceedings of the Twenty-Ninth Automotive Technology Development Contractors' Coordination Meeting*, Society of Automotive Engineers, Warrendale, PA, USA, June, 1992. pp. 19–29.

[6] *Ceramic Technology for Advanced Heat Engines Program Plan 1983–1993*, ORNL/M-1248, Oak Ridge National Laboratory, Oak Ridge, TN, USA, July, 1990.

[7] Nagahiro, A., *Present Status of the National Project of R and D of Fine Ceramics, FC Annual Report*, 1986. pp. 61–66.

[8] Mueller, J. I., "Handicapping the World's Derby for Advanced Ceramics," Am. Ceram. Soc. Bull., 61[5], 588–591, 1982.

[9] Kronogård, S. O., and Malmrup, L., "Ceramic Engine Research and Development in Sweden," in *Ceramics for High-Performance Applications III Reliability*, E. M. Lenoe, R. N. Katz, and J. J. Burke, eds., Plenum Press, New York, USA, 1983. pp. 51–80.

[10] Bunk, W., Gugel, E., and Walzer, P., "Overview of the German Ceramic Gas Turbine Program," in *Ceramics for High-Performance Applications III Reliability*, E. M. Lenoe, R. N. Katz, and J. J. Burke, eds., Plenum Press, New York, USA,1983. pp. 29–50.

[11] Richerson, D. W., Styhr, K. H., and Smythe, J. R., "Material Improvement Through Iterative Process Development," Ceram. Eng. and Sci. Proc, 4[9–10], 841–852, 1983.

[12] Pujari, V. K., Tracey, D. M., Corbin, N. D., Foley, M. R., Garg, A. K., Paille, N. I., Pelletier,P. J., Sales, L. C., Willkens, C. A., and Yeckley, R. L., "Reliability Improvements of High Strength Silicon Nitride Through Process Optimization and Control," in *Proceedings of the Annual Automotive Technology Development Contractors' Meeting*, P-256, Dearborn, Michigan, October 28–31, 1991, Society of Automotive Engineers, Warrendale, PA, USA, 1992. pp. 137–144.

[13] Ferber, M. K. and Jenkins, M. G., "Evaluation of the Strength and Creep-Fatigue Behavior of a HIPed Silicon Nitride," J. Amer. Ceram. Soc., 75[9], 2453–2462, 1992.

[14] Ferber, M. K., Jenkins, M. G., Nolan, T. A., and Yeckley, R., "Comparison of the Creep and Creep Rupture Performance of Two HIPed Silicon Nitride Ceramics," J. Amer. Ceram. Soc., 77[3], 657–665, 1994.

[15] Cranmer, D. C., Hockey, B. J., Wiederhorn, S. M., and Yeckley, R., "Creep and Creep-Rupture of HIP-ed Si3N4,"Ceram. Eng. and Sci. Proc., 12[9–12], 1862–1872, 1991.

[16] Li, C. W. and Yamanis, J., "Super-Tough Silicon Nitride with R-Curve Behavior," Ceram. Eng. and Sci. Proc., 10[7–8], 632–645,1989.

[17] Kawashima, T., Okamoto, H., Yamamoto, H., and Kitamura, A., "Grain Size Dependence of the Fracture Toughness of Silicon Nitride Ceramics," J. Cer. Soc. Japan, Int. Ed., 99[4], 310–313, 1991.

[18] Mitomo, M., “Toughening of Silicon Nitride Ceramics by Microstructural Control,” in *Proc. Science of Engineering Ceramics*, S. Kimura and K. Niihara, eds., Ceram. Soc. Japan, Tokyo, 1991. pp. 101–107.

[19] Becher, P. F., “Microstructural Design of Toughened Ceramics,” J. Am. Ceram. Soc., 74[2], 255–269, 1991.

[20] Krämer, M. and Hoffmann, M. J., “Grain Growth Studies of Silicon Nitride Dispersed in an Oxynitride Glass,” J. Am. Ceram. Soc., 76[11], 2778–2784,1993.

[21] Pollinger, J. P., “Improved Silicon Nitride Materials and Component Fabrication Processes for Aerospace and Industrial Gas Turbine Applications,” ASME paper 95-GT-159, presented a the International Gas Turbine and Aeroengine Congress and Exposition, Houston, TX, USA, June 5–8, 1995.

[22] Hwang, C. J. and Tien, T. Y. “Microstructural Development in Silicon Nitride Ceramics,” Mater. Sci. Forum, 47, 84–109, 1989.

[23] Becher, P. F., Sun, E. Y., Plucknett, K. P., Alexander, K. B., Hsueh, C., Lin, H.-T., Waters, S. B., Westmoreland, C. G., Kang, E. S., Hirao, K., and Brito, M. E.,”Microstructural Design of Silicon Nitride with Improved Fracture Toughness, Part I: Effects of Grain Shape and Size,” J. Am. Ceram. Soc., 81[11], 2821–2830, 1998.

[24] Sun, E. Y., Becher, P. F., Plucknett, K. P., Hsueh, C., Alexander, K. B., Waters, S. B., Hirao, K., and Brito, M. E., “Microstructural Design of Silicon Nitride with Improved Fracture Toughness, Part II: Effects of Additives,” J. Am. Ceram. Soc., 81[11], 2831–2840, 1998.

[25] Lin, H.-T., unpublished data, Oak Ridge National Laboratory, March 1998.

[26] C1322-96a, *Standard Practice for Fractography and Characterization of Fracture Origins in Advanced Ceramics*, Annual Book of ASTM Standards, Volume 15.01, ASTM International, West Conshohocken, PA, USA, 2001.

[27] Hirao, K., Nagaoka, T., Brito, M. B., and Kanzaki, S.. “Microstructure Control of Silicon Nitride by Seeding with Rodlike β-Silicon Nitride Particles,” J. Am. Ceram. Soc., 78 [7], 1857–1862, 1995.

[28] Hirao, K., Ohashi, M., Brito, M. B., and Kanzaki, S., “Processing Strategy for Producing Highly Anisotropic Silicon Nitride,” J. Am. Ceram. Soc., 78 [6], 1687–1690, 1995.

[29] Kondo, N., Suzuki, V., and Ohji, T., ‘Superplastic Sinter-Forging of Silicon Nitride with Anisotropic Microstructure Formation,” J. Am. Ceram. Soc., 82 [4], 1067–1069, 1999.

[30] Kondo, N., Brito, M. E., and Ohji, T., “Improved Creep Resistance in Anisotropic Silicon Nitride,” J. Mater. Res., 16 [8], 2182–2185, 2001

[31] Lang, F. F., Singhal, S. C., and Kuznicki, “Phase Relationships in the Si3N4-SiO2-Y2O3 Pseudoternary System,” J. Am. Ceram. Soc., 60[5–6], 2249–2252, 1977.

[32] Lin, H.T., Ferber, M.K., and van Roode, M., “Evaluation of Mechanical Reliability of Si_3N_4 Nozzles after Exposure in an Industrial Gas Turbine,” *7th Int. Symp. Ceramic Materials and Components for Engines*, J.G. Heinrich and Aldinger, F., eds., Goslar, Germany, June 19–21, 2000, Wiley-VCH Verlag GmbH, Weinheim, Germany, 2001. pp. 97–102.

[33] Rudnick, A., Marschall, C. W., Duckworth, W. H., and Emrich, B. R., *The Evaluation and Interpretation of Mechanical Properties of Brittle Materials*, AFML-TR-67-316, Air Force Materials Laboratory, Wright Patterson Air Force Base, OH, USA, 1968.

[34] Ferber, M. K, Jenkins, M. G., and Wereszczak, A. A., “Fracture Strength,” in *Mechanical Testing Methodology for Ceramic Design and Reliability*, D.C. Cranmer and D.W. Richerson, eds., Marcel Dekker, 1998. pp. 91–170.

[35] Larson, D. C. and Adams, J. W., *Property Screening and Evaluation of Ceramic Turbine Materials, Final Technical Report*, AFWAL-TR-83-4141, April 1983.

[36] Hecht, N. L., *Environmental Effects in Toughened Ceramics*, Final Report, UDR-TR-91-15, February 1991.

[37] Osborne, N. R., Graves, G. A., and Ferber, M. K., " Dynamic Fatigue Testing of Candidate Ceramic Materials for Turbine Engines to Determine Slow-Crack-Growth Parameters," Transactions of the ASME, Journal of Engineering for Gas Turbines and Power, 119[1], 173–178, 1997.

[38] Hecht, N. L., McCullum, D. E., and Graves, G. A., "Investigation of Selected Silicon Nitride and Silicon Carbide Ceramics," Ceram. Eng. and Sci. Proc., 9[9–10], 1313–1332, 1988.

[39] Hecht, N. L., McCullum, D. E., Graves, G. A., and Jang. S. D., "Environmental Effects in Toughened Ceramics," Ceram. Eng. and Sci. Proc., 8[7–8], 892–909, 1987.

[40] Lin, C.-K. J., Jenkins, M. G., and Ferber, M. K., "Cyclic Fatigue of Hot Isostatically Pressed Silicon Nitride at Elevated Temperatures," J. Mat. Sci., 29, 3517–3526, 1994.

[41] Hohjo, H., Yamada, A., and Saruki, K., "Fatigue Strength of Silicon Nitride Ceramics at High Temperatures," J. Soc. Mater. Sci. Japan, 44[496], 128–132, Jan. 1995.

[42] Masuda, M., Soma, T., Matsui, M., and Oda, I., "Cyclic Fatigue Behavior of Sintered Si_3N_4 at High Temperatures," J. Cer. Soc. Japan, 97, 601–607, 1989.

[43] Masuda, M., Soma, T., Matsui, M., and Oda, I., "Fatigue Behavior of Sintered Si_3N_4 Under Tension-Compression Cyclic Stress," J. Cer. Soc. Japan, 96, 275–280, 1988.

[44] Wereszczak, A. A., Ferber, M. K., Kirkland, T. P., and Lin, C. -K. J., "Effect of Cyclic Loading on the Creep Performance of Silicon Nitride," ASME Paper 95-GT-309, presented at the International Gas Turbine and Aeroengine Congress and Exposition, Houston, TX, USA,June 5–8, 1995.

[45] Masuda, M., Yamada, N., Soma, T., Matsui, M., and Oda, I., "Cyclic Fatigue Properties of Sintered Si_3N_4 at Room Temperature," J. Cer. Soc. Japan, 97, 509–514,1989.

[46] Makino, T., Nakasuji, Y., Masuda, M., and Matsui, M., "Fatigue Behavior of Non-Oxide Ceramics Under Static and Cyclic Stresses," in *Proc. 1st Int'l. Symp. Sci. Eng. Cer.*, S. Kimura and K. Niihara, eds., Ceramic Society of Japan, Tokyo, 1991. pp. 189–194.

[47] Bushby, A., Guiu, F., Knowles, J., and Reece, M. J., "Cyclic Fatigue of Silicon Nitride at 1000°C," Key Eng. Mat., 89–91, 673–677, 1994.

[48] Tajima, Y., Urashima, K., Watanabe, M., and Matsuo, Y., "Static, Cyclic and Dynamic Fatigue Behavior of Silicon Nitride," in *Third International Symposium on Ceramic Components and Materials for Engines*, V. J. Tennery, ed., American Ceramic Society, Westerville, OH, USA, 1988. pp. 719–728.

[49] Kawakubo, T. and Komeya, K., "Static and Cyclic Fatigue Behavior of a Sintered Silicon Nitride at Room Temperature," J. Am. Cer. Soc., 70[6], 400–405, 1987.

[50] Carlsson, L., *Cyclic Fatigue of Hot-Isostatically-Pressed Silicon Nitride*, SP Report 1989:43, Swedish [40] National Testing Institute, Borås, Sweden, 1989.

[51] Ritchie, R. O. and Dauskardt, R. H., "Cyclic Fatigue of Ceramics: A Fracture Mechanics Approach to Subcritical Crack Growth and Life Prediction," J. Cer. Soc. Japan, 99, 1047–1062, 1991.

[52] Ueno, A., Kishimoto, H., Kawamoto, H., and Asakura, M., "Crack Propagation Behavior of Sintered Silicon Nitride Under Cyclic Loads of High Stress Ratio and High Frequency," Eng. Fract. Mech., 40, 913–920, 1991.

[53] Gilbert, C. J., Dauskardt, R. H., Steinbrech, R. W., Petrany, R. N., and Ritchie, R. O., "Cyclic Fatigue in Monolithic Alumina: Mechanisms for Crack Advance Promoted by Frictional Wear of Grain Bridges," J. Mat. Sci., 30, 643–654,1995.

[54] Cornelissen, B. E., Dauskardt, R. H., Ritchie, R. O., and Thomas, G., "Cyclic Fatigue Behavior and Fracture Toughness of Silicon Nitride Ceramics Sintered with Rare Earth Oxides," Acta Metall. Mater., 42, 3055–3064,1994.

[55] Dauskardt, R. H., "Cyclic Fatigue-Crack Growth in Si_3N_4-Ceramics," in *Tailoring of Mechanical Properties of Si_3N_4 Ceramics*, M. J. Hoffman and G. Petzow, eds., NATO ASI Series. Series E, Applied Sciences, Vol 276, Kluwer Academic Publishers, 1994. pp. 365–378.

[56] Jacobs, D. S. and Chen, I-W., "Mechanical and Environmental Factors in the Cyclic and Static Fatigue of Silicon Nitride," J. Am. Cer. Soc., 77[5], 1153–1161,1994.

[57] Ewart, L. and Suresh, S., "Crack Propagation in Ceramics Under Cyclic Loads," J. Mat. Sci., 22, 1173–1192, 1987.

[58] Wereszczak, A. A., Ferber, M. K., Kirkland, T. P., Foley, M. R., and Yeckley, R. L., "Evolution of Stress Failure Resulting from Stress/Oxidation Damage in a Hot Isostatically Pressed Silicon Nitride at Elevated Temperatures," J. Amer. Ceram. Soc., 78[8], 2129–2140, 1995.

[59] Wereszczak, A. A., Ferber, M. K., Kirkland, T. P., and More, K. L., "Evolution of Oxidation and Creep Damage Mechanisms in HIPed Silicon Nitride Materials," in *Plastic Deformation of Ceramics*, R. C. Bradt, ed., Plenum Press, New York, USA, 1995. pp. 457–466.

[60] Norton, F. H., *The Creep of Steel at High Temperature*, McGraw Hill, New York, USA, 1929.

[61] Monkman, F. C. and Grant, N. J., "An Empirical Relationship between Rupture Life and Minimum Creep Rate in Creep-Rupture Tests," Proc. ASTM, 56, 593–620, 1956.

[62] Riedel, H., *Fracture at High Temperatures*, Springer-Verlag, Berlin, Heidelberg, Germany, 1987.

[63] Krause, R. F., Jr., Luecke, W. E., French, J. D., Hockey, B. J., and Wiederhorn, S. M., "Tensile Creep and Rupture of Silicon Nitride," J. Am. Cer. Soc., 82[5], 1233–1241, 1999.

[64] Luecke, W. E. and Wiederhorn, S. M., "A New Model for Tensile Creep of Silicon Nitride," J. Am. Cer. Soc., 82[10], 2769–2778,1999.

[65] Lofaj, F., Olada, T.A., and Kawamoto, H., "Cavitational Strain Contribution to Tensile Creep in Vitreous Bonded Ceramics," J. Am. Ceram. Soc., 80[6], 1619–1623, 1997.

[66] Ohji, T. and Yamauchi, Y., "Long-Term Tensile Creep Testing for Advanced Ceramics, "J. Am. Ceram. Soc., 75[8], 2304–2307, 1992.

[67] Murata, M., Takahara, W., Sato, J., and Deguchi, T., "High Temperature Fracture and Creep-Fatigue Characteristics for Structural Design of Silicon Nitride, "J. Soc. Mater. Sci. Japan, 46[8], 963–968, 1997.

[68] Tada, N., Kitamura, T., Ohtani, R., Kikura, M., and Tanaka, T., "Fracture of Sintered Silicon Nitride in Tension, Creep, Fatigue and Creep-Fatigue at High Temperatures, "J. Soc. Mater. Sci. Japan, 48[2], 173–180, 1999.

[69] Tanaka, T., Nakayama, H., Okabe, N., Yamamoto, S., and Fukui, S., "Creep Rupture Map of Engineering Fine Ceramics, " in *Fracture Mechanics of Ceramics, Vol.10*, Plenum, New York, NY, USA, 1992. pp. 473–491.

[70] Masuda, M. and Matsui, M., "Fatigue in Ceramics. 4. Static Fatigue Behavior of Sintered Silicon Nitride under Tensile Stress, "J. Ceram. Soc. Japan, 98[1], 83–92, 1990.

[71] Ohji, T. and Yamauchi, Y., "Tensile Creep and Creep Rupture Behavior of Monolithic and SiC-Whisker-Reinforced Silicon Nitride Ceramics," J. Am. Ceram. Soc., 76[12], 3105–3112, 1993.

[72] Ohji, T., "Long-Term Tensile Creep Behavior of Highly Heat-Resistant Silicon Nitride for Ceramic Gas Turbines," Ceram. Eng. and Sci. Proc. 22[3], 159–166, 2001.

[73] Ferber, M. K. and Graves, G. A., "Long-Term Testing," in *Proceedings of the Advanced Turbine Systems Annual Program Review*, November 9–11, 1994, DOE/OR-2025. pp. 121–138.

[74] Wiederhorn, S. M., "Lifetime Prediction for Silicon Nitride, *7th Int. Symp. Ceramic Materials and Components for Engines*, J.G. Heinrich and F. Aldinger, eds., Goslar, Germany, June 19–21, 2000, Wiley-VCH Verlag GmbH, Weinheim, Germany, 2001, pp. 109–114.

[75] Ferber, M. K., Lin, H. T., Parthasarathy, V., and Wenglarz, R. A., "Degradation of Silicon Nitrides In High Pressure, Moisture Rich Environments," ASME paper 2000-GT-661, presented at the ASME TURBO EXPO Land, Sea & Air, 2000, Munich, Germany, 8–11 May 2000.

[76] Opila, E.J. and Hann, Jr., R.E., "Paralinear Oxidation of SiC in Water Vapor," J. Am. Ceram. Soc., 80[1], 197–205, 1997.

[77] Opila, E.J., "Variation of the Oxidation Rate of Silicon Carbide with Vapor Pressure," J. Am. Ceram. Soc., 82[3], 625–636, 1999.

[78] Smialek, J.L., Robinson, R.C., Opila, E.J., Fox, D.S., and Jacobson N, "Recession Due to SiO_2 Scale Volatility under Combustor Conditions," Advanced Comp. Mater., 8[1], 33–45, 1999.

[79] Lin, H. T.,Ferber, M. K., Westphal, W., and Macri, F., "Evaluation of Mechanical Reliability of Silicon Nitride Vanes After Field Tests in an Industrial Gas Turbine," ASME 2002-GT-30629, presented at the TURBO EXPO Land, Sea, and Air 2002, June 3–6, Amsterdam, The Netherlands.

[80] French, J.D. and Wiederhorn, S. M., 1996, "Tensile Specimens from Ceramic Components," J. Amer. Ceram. Soc., 79[2], 550–552, 1996.

[81] Lin, H. T., Becher, P. F., Ferber, M. K., and Partasarathy, V., 1998 "Verification of Creep Performance of a Ceramic Turbine Blade," in (a) *Proceedings of 2nd International Symposium of the Science of Engineering Ceramics (EnCera'98),* K. Niihara, T. Sekino, E. Yasuda, and T. Sasa, eds., September 6–9, 1998, Osaka, Japan, Trans Tech Publications, Ltd., Switzerland, 1999. (b) Key Engineering Materials Vols. 161–163, 671–674, 1999.

[82] Lenoe, E. M., "Recent Accomplishments and Research Needs in Structural Ceramics," in *Ceramics for High-Performance Applications III Reliability*, E. M. Lenoe, R. N. Katz, and J. J. Burke, eds., Plenum Press, New York, USA,1983. pp. 3–18.

[83] Hecht, N. L., Graves, G. A., McCullum, D. E., Berens, A. P., Goodrich, S., Wolf, J. D., Hoenigman, J. R., Vaney, P., Grant, D., and Hilton, S., *The Evaluation of Environmental Effects in Toughened Ceramics for Advanced Heat Engines Investigation of Selected SiC and Si_3N_4 Ceramics*, Final Report," UDRI-TR-88-66, January 1987–August 1987.

CHAPTER

20

DEVELOPMENT OF HIGH-TEMPERATURE CERAMICS FOR APPLICATION IN GAS TURBINE ENGINE COMPONENTS

Victor I. Trefilov, Oleg N. Grigoriev, and Gennady V. Trunov
Institute for Problems of Materials Science NASc. of Ukraine, Kiev, Ukraine

Georgy A. Gogotsi
Institute for Problems of Strength NASc. of Ukraine, Kiev, Ukraine

Pavel S. Kislyi
Institute of Superhard Materials NASc. of Ukraine, Kiev, Ukraine

INTRODUCTION

The application of ceramics in engines is usually associated with the advantage of significant increases in the temperature of the operation process with a corresponding efficiency increase (as high as 25–30%). In general, the level of technological and structural development of materials science in this direction so far has prevented high-temperature ceramics from overcoming market barriers.

Secondly, ceramics can be applied to reduce friction losses and wear of bearings and friction pairs, and to improve the wear resistance of parts exposed to gas erosion. Appropriate materials and parts are finding practical engineering applications where they not only provide an efficiency increase (as high as 10–12%), but also improved durability and reliability.

And finally, a third application where ceramics and its parts have overcome market barriers, are environmentally friendly engine components (exhaust gas silencers, neutralizers, and soot filters).

The results of research for manufacturing processes and properties of high-temperature ceramics of gas turbine engine (GTE) elements are introduced in this chapter. The research cycle started in 1972 at the

Institute for Problems of Materials Science of the Ukraine Academy of Sciences. Silicon nitride and carbide based materials were studied in order to establish that advanced ceramics meet the requirements of high-temperature structural materials, i.e. that they have adequate oxidation and corrosion resistance, and high-temperature strength. The objective of the research was the development of ceramic manufacturing technology for nozzle guide vanes, monolithic ceramic and composite nozzles, combustors, rotor blades, heat exchangers, and exhaust gas after-burners of gas turbine engines with output power in the 300 kW to 10 MW range. This work was performed in accordance with state programs in collaboration with design and manufacturing machine engineering companies.

MANUFACTURING TECHNOLOGY FOR HIGH-TEMPERATURE MONOLITHIC CERAMICS AND CERAMIC MATRIX COMPOSITES

Traditionally, silicon nitride was considered the most advanced ceramic material for GTE components. Recently, sufficient interest in a wide range of ceramic matrix composites based on oxides (mullite, reinforced directionally crystallized eutectics), silicon carbide based materials, and other materials has emerged. We present the research results for silicon nitride ceramics and ceramic matrix composites bases on silicon nitrides and carbides (systems Si_3N_4-SiC and SiC-MeB_2, where Me = Ti, Zr). The generic term silicon nitride used here includes a whole class of materials that differ in composition, structure and properties. Reactive sintered (reaction bonded) silicon nitride, which is used for the manufacturing of stationary GTE parts (nozzles, combustors) is an example of this class of materials. Its manufacturing technology includes the following operations: forming of a preform with required dimensions and shape from fine dispersed silicon powder by pressing, slip casting or flame spraying, followed by heating of the preform in nitrogen or ammonia according to special process steps. During this process silicon is converted into silicon nitride with high heat resistance (strength does not decrease up to 1500°C) and low shrinkage, suitable for a product of the required design.

Raw Materials

A wide range of refractory powder compounds supplied by different manufacturers were used: 1) Donetsk Plant of Chemical Reagents (DPCR)—powders of Si_3N_4, rare metal oxides, Ti and Zr borides made by furnace synthesis; 2) Zaporozhye Abrasive Combinat (ZAC)—SiC, B_4C and BN powders made by furnace synthesis; 3) Riga Institute of Inorganic Chemistry (IIC)—aluminum and silicon nitride powders produced by plasma chemical synthesis; 4) Institute of Chemical Physics (ICPh, Tchernogolovka, Russia)—Si_3N_4 powders, self-propagating high-temperature synthesis (SHTS); Institute for Problems of Materials Science (IPMS)—powders of different carbides and nitrides, produced by the carbothermal reduction method. Also powders of other companies such as H.C. Stark (Germany) and Kema-nord (Sweden) were used during laboratory investigations. Depending on the manufacturing method powder characteristics varied widely. So, DPCR supplies Si_3N_4 with α-phase content up to 80–90% and specific surface area of 6–7 m^2/g, while Si_3N_4 manufactured at IIC is amorphous with specific surface area of 70–90 m^2/g.

The powders supplied by the above manufacturers, with a few exceptions, can be regarded as low quality products. They contain a high amount of free carbon, oxygen and other impurities, not classified by particle size (Table 20.1). It is well known that the technological properties of different fractions of a powder are essentially different. The use of polydispersed powders is accompanied by difficulties during forming and sintering, while the use of mono-dispersed powders during the compaction process leads to structures with regular geometry and also a high Weibull modulus. For this reason, we used different methods of purification (to remove impurities), milling, classification, granulation and low temperature annealing in active media. Milling conditions were varied according to types and regimes of the subsequent processing steps. So, for hot pressing, prepared charges were milled up to particle sizes $D_{50} = 0.6 - 0.7$ µm and $D_{max} = 4$ µm, in millers with silicon nitride linings and milling balls.

TABLE 20.1 CHEMICAL COMPOSITION OF POWDERS OF REFRACTORY COMPOUNDS

Compound	Basic Elements (%)		C (%)	Total Impurities (%)	Powder Source
Boron carbide	B—78.8	C—20.9	0.8	0.9	DPCR
	B—69.8	C—27.7	8.4	10.0	ZAC
Silicon carbide	Si—67.7	C—29.7	1.0	4.8	ZAC
Aluminum nitride	Al—67.8	N—31.5	0.25	3.4	DFCR
	Al—62.9	N—31.5	—	4.2	Riga, IIC
	Al—65.0	N—33.2	—	1.8	Tchernogolovka, ICPh
Silicon nitride	Si—60.8	N—37.3	0.1	1.8	DFPR
	Si—59.3	N—34.5	0.7	6.5	Riga, IIC

Preforming

Parts are fabricated by different methods. Preforms are formed using plasticizers by the following methods: pressing, hydrostatic compaction, cold and hot slip casting, injection molding, and machining. After forming by any of the above methods a basic shape of solid bonded particles is formed in the preform. The free space between particles is filled by air or plasticizer, which is removed during low-temperature heating before sintering. Thermoplastic slip casting was widely used in the work presented here. In essence, the method consists of the preparation of a homogeneous powder suspension in molten thermoplastic binder (paraffins with wax addition). The casting is made from the molten mass by casting under pressure in a metallic compression mold.

Typical casting conditions are:

- slip temperature: 80–90°C
- form temperature 20°C
- compressed air pressure 5–10 atm.
- endurance under pressure 20–60 s
- slip viscosity 50–120 poise

Binder removal involves a programmed slow preform heating in alumina charging from 20 up to 220°C with an average rate of 10°C/hour

Sintering

The following main manufacturing methods were used during sample investigations and component fabrication processes:

- reactive sintering under low nitrogen pressure (P ~ 1 atm.)
- activated sintering under low nitrogen pressure
- compression activated sintering (P ~ 100 atm.)
- hot pressing
- hot isostatic pressing (HIP)

In some cases a combination of methods was used (e.g. reactive + activated sintering, reactive sintering + hot pressing as a variant of quasi-hydrostatic loading, etc.).

Silicon Nitride Densification During Sintering

We have observed sintering peculiarities under low and high nitrogen pressures. Pressureless sintering of preforms from powders is one of the most important fabrication methods for dense products. This process is spontaneous, accompanied by a decrease in free surface, in the surface of boundaries, in the quantity of macro- and micro-defects, as well as with a decrease in the internal energy of the system. Different molecular processes also promote the sintering process: chemical interactions at the surface and interfaces, diffusion processes at surface, interfaces and in the bulk, and crystal lattice microstrain relaxation.

Silicon nitride sintering has its peculiarities, one of them being dissociation during heating with mass losses at high temperatures:

$$Si_3N_4 \rightarrow 3Si + 2N_2\uparrow \tag{1}$$

The partial pressure of nitrogen during dissocation is determined by the temperature:

$$\ln P_N = A - B/T, \tag{2}$$

where T is the temperature in Kelvin, A and B are constants, equal to 8.54 and 1.925×10^4, respectively [1].

It is known that sintering effectively takes place at temperatures approximately coinciding with some characteristic temperature t* [2], which is equal to $0.5.T_m$ for metals, $0.6 - 0.8.T_m$ for refractory compounds with a mixed character of interatomic bonds, and $0.85.T_m$ for covalent crystals. In theoretical evaluations silicon nitride is melted at temperatures of about 2700°C and, consequently, it shrinks enough only at very high temperatures (2100–2500°C) during pressureless sintering. Under these conditions dissociation is characterized by considerable pressure of the silicon vapor and there will be significant contribution to product formation from mass transfer processes through the gas phase.

Growth of the contact isthmus between particles in this case is determined by the difference in vapor pressure at the surface of the particles (positive radius of curvature) and the nearby contact surface (negative radius of curvature). An emerging concentration gradient and directed mass transfer leads to the growth of the contact isthmus without essentially altering the proximity of the centers of the particles. Thus, sintering is not accompanied by shrinkage and a decrease of porosity. Considerable mass losses start at temperatures in excess of 1200°C with low shrinkage during pressureless sintering, sintering under pressure, and/or activated sintering in the presence of special oxides introduced into the material.

The dependence on nitrogen pressure and temperature, at which mass losses don't exceed one percent is given by the expression:

$$P_N = A \exp(BT), \tag{3}$$

where A and B are constants, equal to 2.52×10^{-9} and 1.34×10^{-2}, respectively [3].

Values of nitrogen pressure at which silicon nitride is thermally stable, calculated by (3) are given in Table 20.2.

Under high nitrogen pressure regimes the equilibrium of the silicon nitride forming and dissociation processes is shifted to the side of higher temperatures and sintering is activated. A silicon nitride preform looses 1% of mass within 40 min. at 1680°C under a nitrogen pressure of 0.2 MPa, and under a pressure of 100 MPa at 2050°C. The thermal stability of silicon nitride is also observed when fillers are used, and decomposition occurs during heating with separation of silicon and nitrogen. The shrinkage of preforms synthesized in a furnace from coarse grain powders and from nano-grain powders synthesized by a plasma chemical process is essentially different. It was discovered, that preforms from silicon nitride powders

TABLE 20.2 NITROGEN PRESSURES AT WHICH SILICON NITRIDE IS THERMALLY STABLE

T (°C)	1600	1700	1800	1900	2000	2100
P_N (MPa)	0.5	2.0	7.5	29	110	419

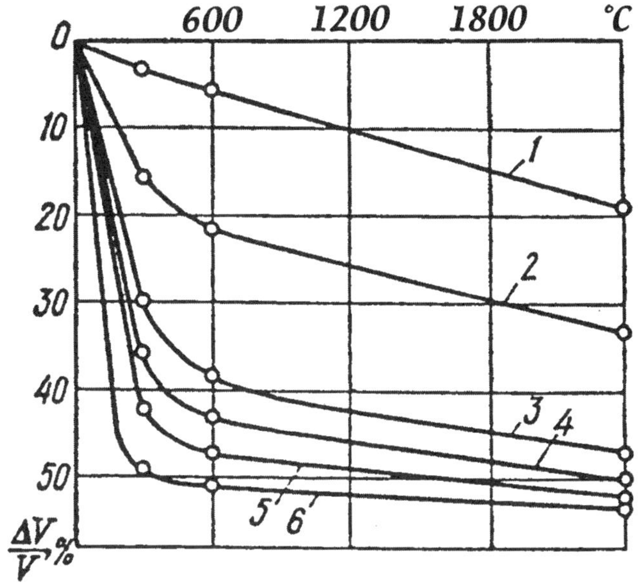

Figure 20.1 Dependence of Silicon Nitride Shrinkage versus Isothermal Exposure at Pressure of Nitrogen 100 MPa: 1—1700°C; 2—1800°C; 3—1900°C; 4—1950°C; 5—2000°C; 6—2050°C

manufactured at DPCR with a specific surface area of 7 m^2/g are not densified after sintering under nitrogen pressure and retain a residual porosity of about 50%. At the same time preforms from ultra-dispersed powders are densified under isothermal conditions in the same way as preforms from metal-like refractory compounds (Figure 20.1). This involves obviously a two-stage process with different densification speeds. At high temperatures all shrinkage is observed during the first stage. A structural investigation of the sintered materials showed, that during the first stage of sintering the shrinkage takes place by activated grain boundary slipping of ultra-dispersed powder particles. During the second stage the growth of elongated crystals by means of recrystallization from the gas phase, and from the liquid phase (because of the appearance of free silicon) is observed.

Introduction of oxide mixtures (MgO, Al_2O_3, Y_2O_3, etc.) and strengthening phases (SiC, TiN) into the charge is accompanied by transfer to liquid phase sintering with shrinkage actively taking place within the temperature range of 1500–1800°C. Self-reinforced structures from elongated silicon nitride needle-shaped grains are formed at the same time.

Reactive sintering of component preforms from silicon nitride and Si_3N_4-SiC composites was performed in a furnace with boron nitride and silicon nitride fillers in a nitrogen atmosphere. Sintering conditions were defined by the charge composition and for the composite Si_3N_4-SiC, where the charge composition was: α-Si_3N_4-40%, SiC-30%, Si-30%, the sintering temperature was equal to 1550°C, and the isothermal holding time was 3 hours.

Within the framework of the present investigations a high-speed hot pressing technology for products from silicon nitride and carbide based ceramics was developed. Hot pressing was performed in pilot installations with induction heating in graphite molds without a protective atmosphere. The isothermal holding temperature was in the 1750–2150°C range, depending on the sintered material under pressure (26–30 MPa), isothermal holding time (7–20 minutes), and heating and cooling speed (up to 100°C/min). After isothermal holding the axial pressure was released, the ceramics were cooled in the mold at the rate of 35°C/min., down to T = 1200 – 1500°C. Then, the ceramics together with the mold were placed in a graphite container for

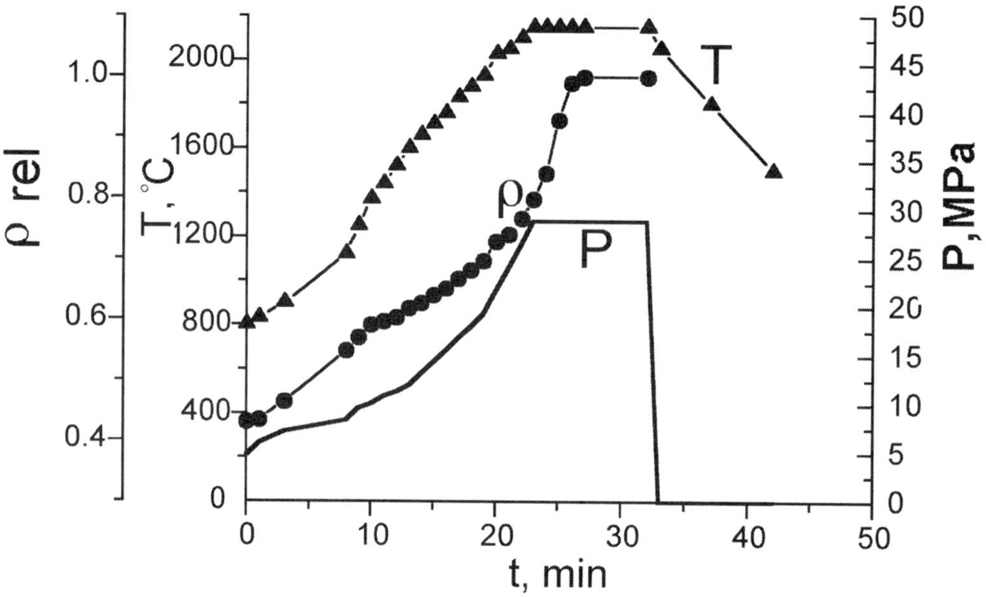

Figure 20.2 Change of Temperature (T), Pressure (P) and Density (ρ_{rel}) During Hot Pressing of Ceramics SiC-10%TiB_2

final cooling to room temperature. The shrinkage kinetics were fixed during the hot pressing process. An example of a typical hot pressing process is given in Figure 20.2.

It is necessary to note the peculiarities of the hot pressing process that was used. Compared to vacuum hot pressing of ceramics, sintering without a protective atmosphere takes place under conditions of a partially reducing atmosphere of CO-CO_2, derived from the graphite tooling. For example, during the hot pressing process for composites of the SiC-MeB_2 (Me = Ti, Zr)-B_4C system a number of chemical interactions take place in which the residual oxide phases of the charge (SiO_2, TiO_2, B_2O_3), boron carbide (reducing oxides), and also CO and CO_2 participate. Experimental data show that in the process of refining ceramics involving the formation of secondary TiB_2, the B_4C quantity decreases, and considerable gas evolution in the form of oxygen containing volatile compounds (SiO, B_2O_3, CO_2, etc.) takes place during the sintering process. Thus, solving the problems of high pressure optimization involving generation of gases is very important. The latter is achieved by isothermal holding stages at intermediate temperatures, the use of maximum pressure during the final process stages, etc.. As a result ceramic components have been fabricated with a volume of up to 300 cm^3 with 5–10% of structural and thermo-mechanical non-homogeneities.

PROPERTIES OF THE CERAMICS

Reaction Bonded Silicon Nitride Ceramics

The developed sintered materials have a wide range of physical and mechanical properties depending on their composition and the fabrication process parameters. Ceramic composite materials in the Si_3N_4-SiC system with additions of Mg compounds (ceramic materials SNCM) and boron (SNCMB) have been mostly approved and certified [4]. Representative materials properties are shown in Table 20.3.

TABLE 20.3 MAIN PROPERTIES OF THE SNCM(B) MATERIAL

Property (unit)	Value
Density (g/cm^3)	2.5–2.8
Porosity (%)	12–21
Ultimate bending strength (MPa)	
at 20°C	200–350
at 1400°C	150–180
Poisson's ratio	0.23
Ultimate tensile deformation in bending, 10^{-2} % (at 20°C)	10–16
Static modulus of elasticity (GPa)	
at 20°C	187
at 800°C	150
at 1000°C	146
at 1400°C	136
Dynamic modulus of elasticity (GPa)	198
Compressive strength (MPa)	865
Weibull ratio	10–40
Fracture toughness (MPa.m$^{1/2}$)	
at 20°C	4.07
at 1000°C	3.46
at 1200°C	3.42
at 1400°C	2.81
Thermal conductivity (W/m.K)	
at 20°C	11.5
at 200°C	12.6
at 600°C	12.2
at 900°C	10.8
Specific heat (kJ/kg.K)	1.97
Critical temperature difference (°C)	470
Coefficient of thermal expansion (10^{-6}/°C)	3.1
Tensile creep strength (1350°C, 500 hrs, MPa)	60.0

The composite is composed of a matrix with sufficiently developed silicon nitride prismatic grains and inclusions of silicon carbide grains, having an isometric or lamellar form (Figure 20.3). Statistical aspects of strength at room and high temperatures, non-elasticity and slow crack growth, thermal and corrosion resistance were studied [5–8]. This group of materials with additives of magnesium oxide and boron nitride showed the best strength and operating properties at temperatures up to 1500°C. These materials were deemed suitable for industrial production, and they have been used for the fabrication of GTE parts. The relatively low strength of these materials (bending strength ≤350 MPa) limits the range of applications to parts with low stresses under the design operating conditions.

Silicon Nitride Ceramics Sintered under Pressure (Compression Sintering and Hot Pressing)

Sintered and hot pressed composites of Si_3N_4-Al_2O_3-Y_2O_3, Si_3N_4-ZrO_2-Y_2O_3, α/β-sialons and other compositions, practically without porosity, were developed, based on Si_3N_4. Materials with bending strength of 400–600 MPa were made by slip casting of thermal plastic masses with further sintering under normal nitrogen pressure, as well as by injection molding and sintering in a vacuum-compression furnace under nitrogen pressure of up to 100 atm. When using hot pressing the bending strength of these materials at room temperatures was in the 700–1000 MPa range. In the case of self-reinforced ceramics based on

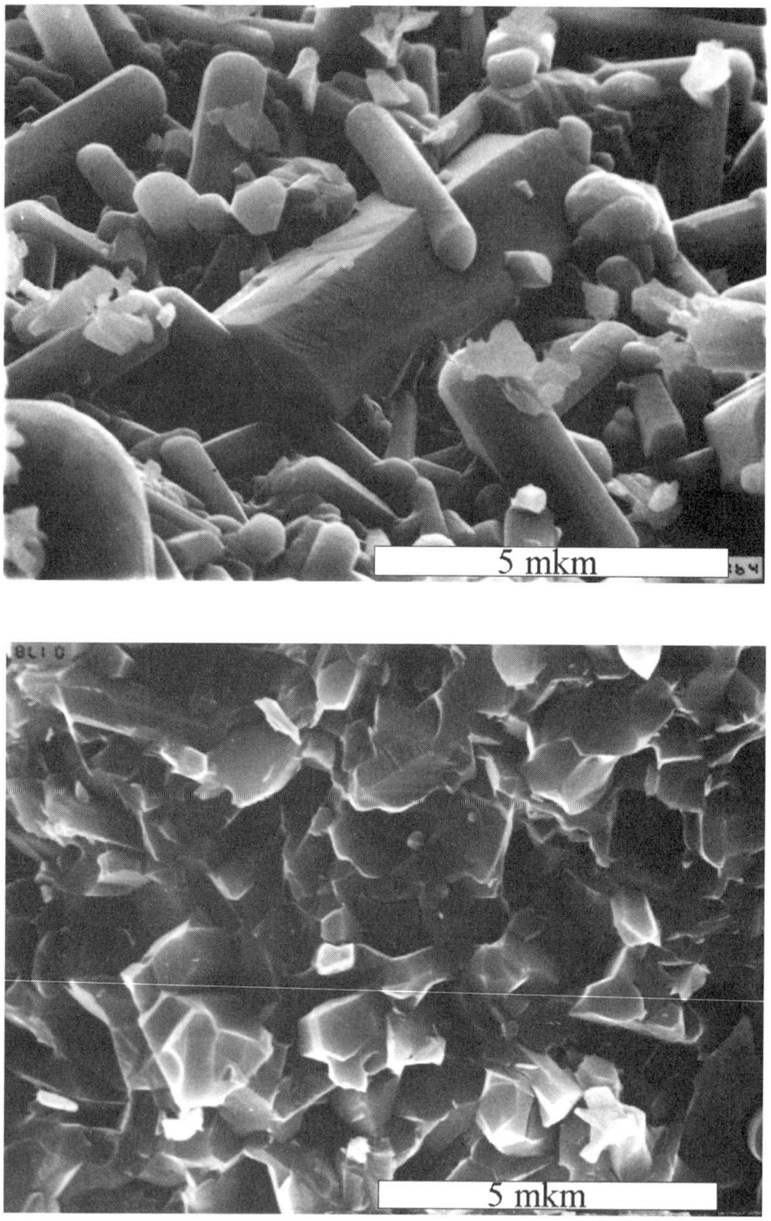

(a)

(b)

Figure 20.3 Grain Structure (a) and Fracture Surface (b) of NKM—2 Ceramics (mkm = μm)

β-Si_3N_4 in the system Si_3N_4-Y_2O_3-Al_2O_3 high strength values (up to 1000 MPa) and crack resistance (up to 9 MPa.$m^{1/2}$) were obtained, mainly through the mechanism of crack arresting by needle-like grains, formed during the $\alpha \rightarrow \beta$ transformation process in silicon nitride during sintering (Figure 20.4). But, ceramics made by activated sintering contain grain boundary amorphous phases, which can significantly reduce high-temperature strength. The study of the mechanical behavior of ceramics over a wide temperature range is therefore extremely important.

Figure 20.4 Structure of Si_3N_4-Y_2O_3-Al_2O_3 Ceramics (mkm = μm)

Strength at Low Temperatures and Optimization of Structure and Composition of Ceramics

In practically all actual development programs the ceramics are not single-phase but represent some form of a ceramic matrix composite. For example, to overcome the technological difficulties during sintering, it is necessary to add to the silicon nitride powders up to 10% of mixtures of oxides, as well as second phase particles (TiN, SiC) to strengthen the ceramics. Thus, there are problems of the physico-chemical and thermo-mechanical compatibility of components. The chemical interaction of the components results in the formation of new phases. Elastic interaction of phases at temperature and pressure changes during production and under external thermo-mechanical effects results in a composite stress-strain state of a material, which determines the features of its mechanical behavior. Therefore, the optimization of the structure of ceramics is reduced to the optimization of its internal stress fields.

An increase of composite fracture toughness and strength can be achieved by the selection of constituting components, by controlling their volume fractions and grain sizes, so that the matrix, subjected to the effect of residual thermal stresses, will be in compression, and isolated secondary phase inclusions in tension. The crack then will be forced to propagate in a compressed matrix, circumventing the secondary phase, due to the peculiarities of the non-homogeneous field of stresses in their vicinity.

A mathematical formulation of fracture toughness criteria for the composite with ceramic matrix and optimization methods for composite composition and structure have been presented by Galanov and Grigoriev [9].

In simple cases, composite fracture toughness is expressed by the ratio:

$$\tilde{K}_I = \tilde{K}_{IC} = \min_i\left(a_i^{-1}\left(K_{IC}^i - D\sigma^{ri}\right)\right) \quad (4)$$

and optimum composite composition is determined by the criteria:

$$\tilde{K}_{IC}^{opt} = \underset{Y}{MAX}\ \underset{i}{MIN}\left(a_i^{-1}\left(K_{Ic}^i - D\sigma^{ri}\right)\right); \quad (5)$$

$$\alpha_1 = \frac{G_i|Y_i|}{\sum_{i=1}^{N} G_i|Y_i|} \leq 1;\ D = \frac{2\sqrt{C}}{\pi} \quad (6)$$

Composite parameters, that were used in calculation of K_{IC}, are elastic phase characteristics and the parameter D.

Figure 20.5 shows the results of typical calculations. As can be seen from Figure 20.5, introduction of a high-α component into the composite is accompanied by an increase in fracture toughness. The maximum fracture toughness is shifted to lower second phase contents (~10–30%) with an increase of CTE (coefficient of thermal expansion) mismatch, the elastic characteristics, and grain sizes of the composite phases. At high concentrations of the second phase, the minimum K_{IC} with the value of $K_{IC} \rightarrow 0$ is found. It is caused by spontaneous failure under the effect of thermal stresses with flaw size approximately equal to the grain size (≈10 μm):

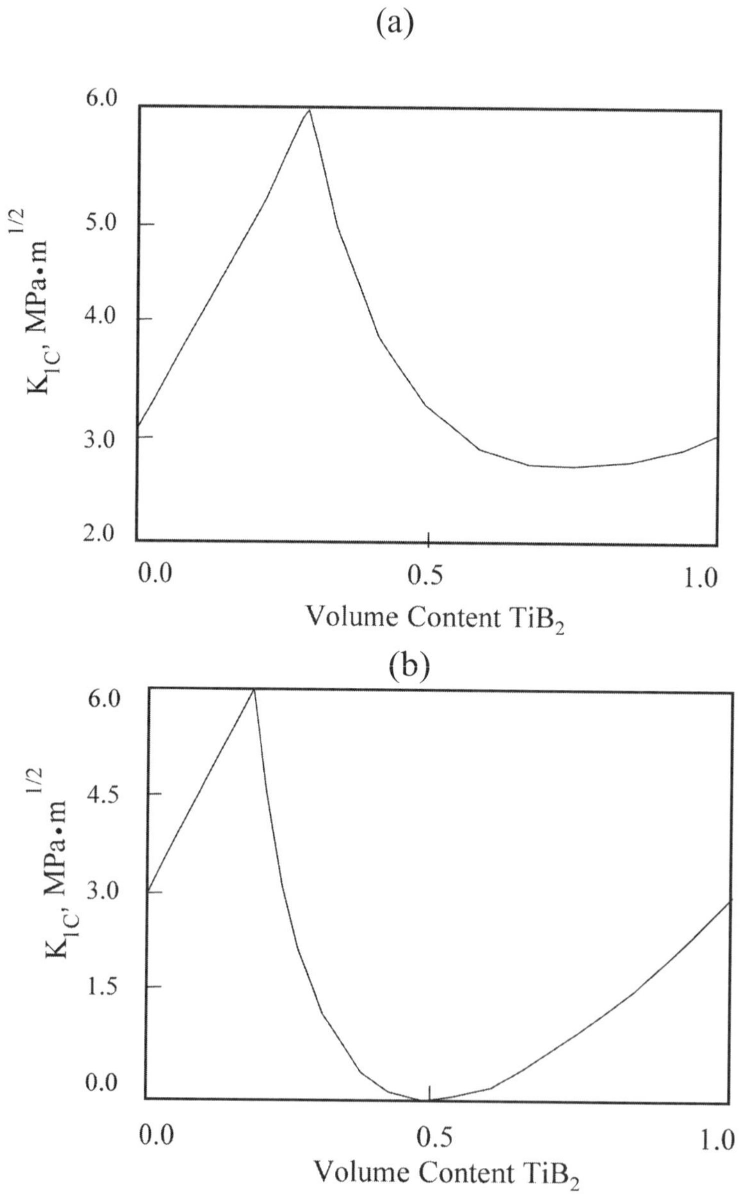

Figure 20.5 Calculated Dependencies of SiC-TiB_2 Composite Fracture Toughness on Composition; a) -grain size—2.5 μm; b) -grain size—10 μm

$$2C_{CR} = 2\,\underset{i}{MIN}\left[\left(\frac{\alpha_i^{-1} K_{IC}^i}{(2/\sqrt{\pi})\sigma^{ri}}\right)^2\right] \tag{7}$$

Generally, for the SiC-TiB_2 system, an improvement of composite properties can be expected in the range of 20–30 vol. % of TiB_2, and grain size of ~5–10 μm.

The present analysis considers thermal stresses, in part generated by the difference in coefficients of linear thermal expansion (α) of the constituting phases of the composite. Taking into account the material's macroscopic homogeneity and isotropic distribution of phase components, one can make an assumption for the hydrostatic nature of tensile and compressive stresses in each of the phases

$$(\sigma_{11})_i = (\sigma_{22})_i = (\sigma_{33})_i = \bar{\sigma}_i \tag{8}$$

governed by the conditions of stress equilibrium $\Sigma Y_i \bar{\sigma}_i = 0$.

Calculation of mean thermal stresses in composites was performed on the base of the statistical approach developed by L. P. Khoroshun [10] using the following ratios:

$$\langle \sigma_{ij} \rangle_1 = Y_2 \eta \delta_{ij};\ \langle \sigma_{iij} \rangle_2 = -Y_1 \eta \delta_{ij}; \tag{9}$$

$$\eta = \frac{\gamma}{K_1 K_2 + \langle K \rangle \gamma}\{(K_1 - K_2)p_s - 3K_1 K_2[(\alpha_1 - \alpha_2)\Delta T + (l_1 - l_2)]\}; \tag{10}$$

$$\gamma = (4/3)\langle G \rangle;\ \langle G \rangle = Y_1 G_1 + Y_2 G_2;\ \langle K \rangle = Y_1 K_1 + Y_2 K_2; \tag{11}$$

Intensity of internal stresses can be characterized by internal energy, which relates to the unit of micro volume,

$$\langle U \rangle = V_1 \langle U_1 \rangle + V_2 \langle U_2 \rangle; \tag{12}$$

The internal energies of components are defined as:

$$\langle U_1 \rangle = \frac{\eta^2 Y_2^2}{2K_1};\ \langle U_2 \rangle = \frac{\eta^2 Y_1^2}{2K_2} \tag{13}$$

For known values of the elastic and thermal characteristics of the composite phases, the value of the internal stresses is determined by ΔT—the difference between the temperature of the viscous-elastic transition T_{ve} and the final temperature, down to which the material is being cooled after hot pressing (typically, room temperature). Calculated stress and energy values for ΔT = 2000°C versus composite composition are given in Figure 20.6. As can be seen in Figure 20.6, the matrix of a SiC composite is in a compressive stress state, and TiB_2 particles are in hydrostatic tension with stresses up to 1.8 GPa. The stress calculations give only an estimation of stress levels in phases, without taking into account the viscoelastic relaxation. Therefore there is an important role for experimental methods for the determination of internal stresses. Usually, X-ray and neutron diffraction methods are used. In our investigations by X-ray methods the components of the deformation tensor in the i-phase $(\varepsilon_{33})_i$, related to the relevant thermal stresses (σ_i^t) were determined experimentally by the expression:

$$(\varepsilon_{33})_i = \frac{1 - 2\nu_i}{E_i}\sigma_i^t; \tag{14}$$

The procedure is identical to the one used for the Si_3N_4-TiN system [11].

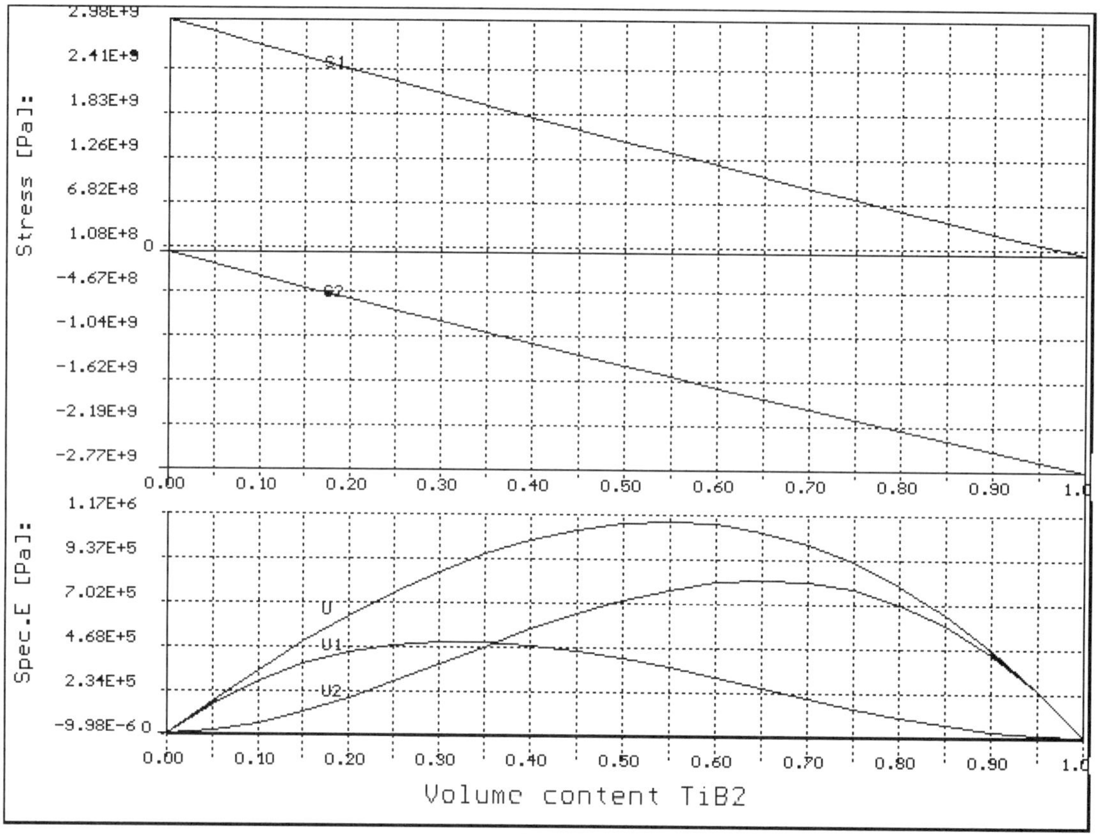

Figure 20.6 Changes of Calculated Thermal Stresses and Energy of Elastic Deformation of TiB_2 (S1) and SiC (S2) Phases Depending on TiB_2 Content, T = 2000°C

The results of the experimental measurement of stresses in the SiC-TiB_2 composite are presented in Figure 20.7. The experimental determination of thermal stresses not only gives quantitative information on internal stresses in phases, but also data on viscoelastic relaxation and the state of grain boundaries. A comparison of the experimental and theoretical values of internal stresses shows good agreement between theory and experiment only if we use a rather small ΔT value ($\Delta T \approx 1300°C$) in the calculation. It means that the temperature of the viscous-elastic transition is fairly low ($T_{ve} \approx 1300°C$) and the composite possesses significant relaxation ability at high temperatures ($T \geq 1300°C$). High relaxation of the material at $T \geq 1300°C$ could be explained by the eutectic nature of the system and the presence of impurities forming soft grain-boundary layers.

High Temperature Strength of Ceramics

The high-temperature behavior of ceramics (including also single crystals of refractory compounds) depends on the type of its structure. Deformation of ceramics made by solid phase sintering which do not have amorphous lamellas at grain boundaries, as well as single crystals, follow regularities of the dislocation mechanisms of crystal deformation with a high Peierl's barrier. For ceramics, made by liquid phase sintering with residual amorphous grain boundaries, lamellas have a peculiar dominating influence on visco-elastic

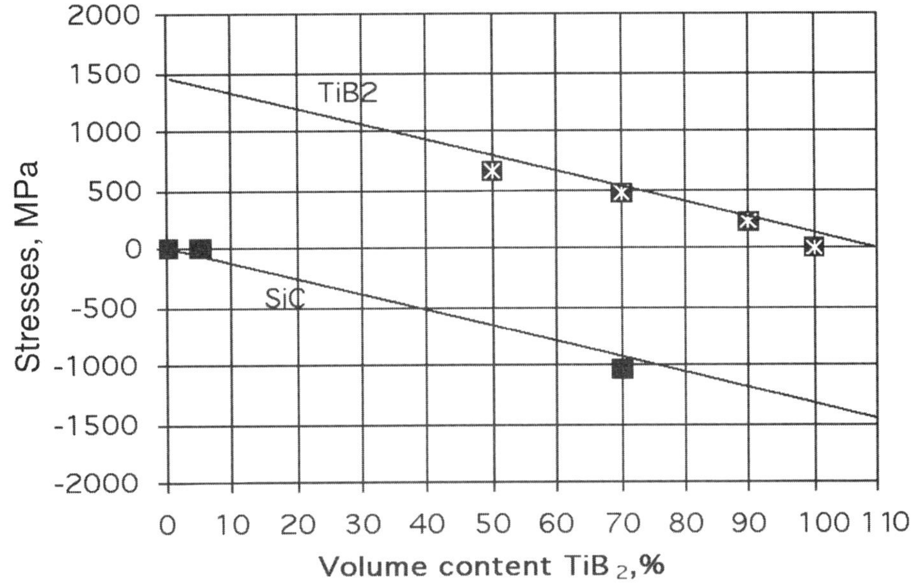

Figure 20.7 Dependence of Experimentally Measured Thermal Stresses in Composite Components versus TiB_2 Content

behavior. It is common that at high temperature brittle fracture of these materials takes place even after significant plastic pre-deformation and stages of slow crack growth. It is evidenced by the known high speed sensitivity of the yield point of refractory compounds with covalent bonds, described by the ratio:

$$(\varepsilon_{33})_{\mathrm{i}} = \frac{1-2\nu_i}{E_i}\sigma_i^t; \tag{15}$$

where the m parameter has low values (~ 2–5). For example, for BCC (body centered cubic)—metals the value of m is of the order of 30–40.

So, at the tip of the developing crack, where the deformation speed essentially exceeds nominal, at low m in accordance with the Yoffe scheme, temperature-speed conditions of brittle fracture are established [12]. A strong dependence of the yield point of materials with a low value of the parameter m, and temperature of the brittleness-plasticity transformation (T_b) from loading speed, essentially limits the temperature area of ceramics plastic deformation.

Studies of ceramics, sintered with additions of oxides [13,14] showed that, in spite of the fact that the quantity of the glass phase is usually very small, it has an important, even dominating influence on the mechanical properties at high temperature. In the presence of a glass phase, tension zones appear in ternary points during high-temperature grain boundary slipping, which leads to the formation of pores in the glass phase (cavitation) with further microcrack formation. The model developed by Tsai and Raj [14] forecasts, that crack growth during creep is possible when the damaged zone size exceeds some critical value. At low temperatures fracture is brittle and loading speed makes no difference. At high temperature a damaged zone of critical size can be formed very fast, because of the low toughness of the glass, and slow crack growth with strength degradation is possible. But at an intermediate temperature the toughness of some material leads to blunting of the microcracks, but the size of the damaged zone is not sufficiently large to provoke the slow microcrack growth. So, microcracking increases strength, if the damaged zone size is sub-critical, but it will weaken the material, if the damaged zone size exceeds the critical value. Because of this, investigations into the mechanical behavior of ceramics considering the amounts and types of introduced additives and impurities, and also processing parameters, have made significant progress over the past years.

As an example, let's take the high-temperature behavior of self-reinforced dense ceramics in the Si_3N_4-Y_2O_3-Al_2O_3 system. Results of strength and fracture toughness determinations with varying loading rates are presented in Figure 20.8. With increasing temperature the ultimate strength first decreases, up to 1000°C, next it increases, achieving a maximum at 1200–1300°C, and then decreases again up to the final temperature of 1600°C. At temperatures in excess of 1200°C there is noticeable non-elastic behavior of

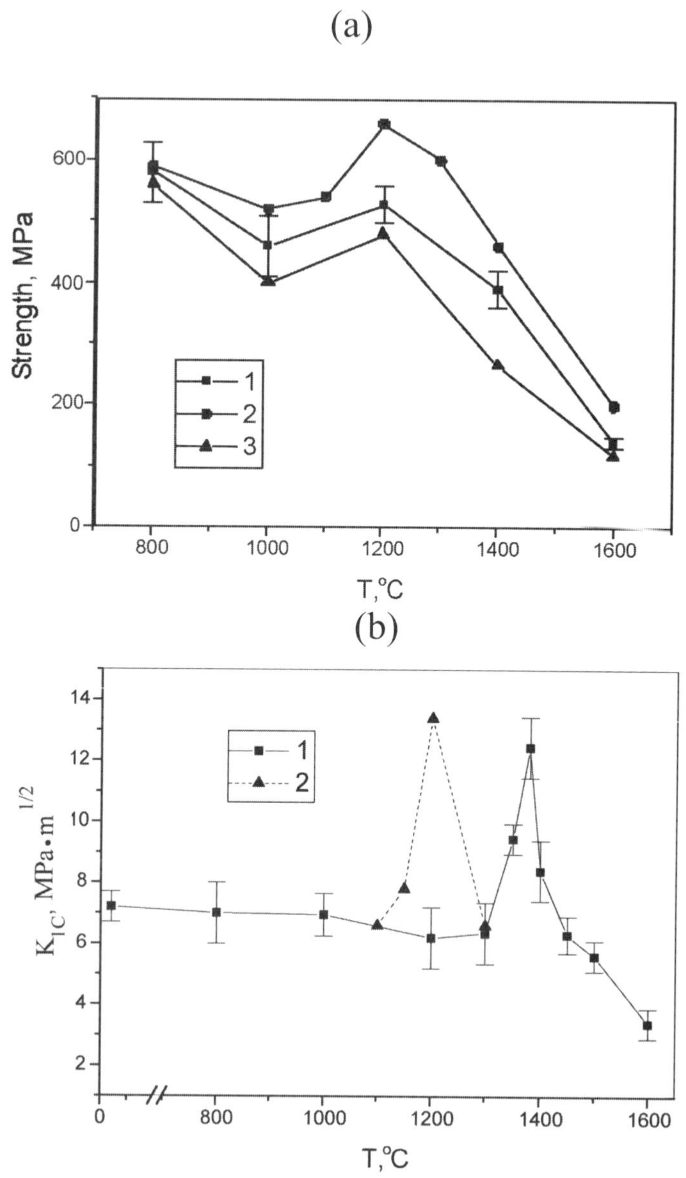

Figure 20.8 a) Temperature Dependence of Strength of Ceramics: Si_3N_4-7%Y_2O_3-2% Al_2O_3 (1); Si_3N_4-7% Y_2O_3-1%Al_2O_3 (2); Si_3N_4-7%Y_2O_3-3% Al_2O_3 (3) b) Temperature Dependence of Fracture Toughness of Si_3N_4-7%Y_2O_3-2% Al_2O_3 Ceramics; Travel Rate of Active Support 0.002 mm.min^{-1} (1) and 1 mm.min^{-1} (2).

specimens, characterized by a shift in the stress—strain curve from linear. The fracture toughness is slightly decreased with increasing temperature up to 1200°C. With a further increase in temperature it goes up, achieving a maximum value at 1370°C, and then it sharply decreases in the temperature range of 1400–1600°C. The decrease in the rate of deformation displaces the maximum K_{IC} to the side of higher temperatures.

Sample fracture surfaces, tested at 20°C, differ by developed wavy (or ridge) relief. The fracture character is mainly transgranular. With an increase in test temperature up to 1000°C the fracture surfaces, normal to the applied stress become flatter, and the proportion of intergranular fracture is increased. At high temperatures (above 1400°C) fracture is fully intergranular with a clearly visible granular material structure. At the fracture surfaces of the samples, tested at 1400–1600°C, areas of sub-critical crack growth are detected. Non-elastic deformation and areas of sub-critical growth become more pronounced with increasing temperature. The contour of the fracture surface, which corresponds to sub-critical crack growth, is characterized by a new structural peculiarity—many secondary grain cracks at the fracture surface, formed before the front of the slowly growing primary crack.

The results of precision measurements of lattice spacing and widths of X-ray diffraction lines, taken at the fracture surfaces of samples showed, that for high-temperature deformation the lattice spacing of silicon nitride grains is slightly enlarged, and that the line width becomes smaller. Consequently, evidence of grain deformation hardening is absent, and the observed changes of lattice spacing and widths of curves, apparently, are related to stress relaxation because of microcracking. Intensive microcracking is revealed during microscopic analysis of fracture surfaces. Examination of X-ray diffraction analysis and fractography, and also of literature data on electron-microscopic analysis, shows an absence of dislocating structures in Si_3N_4 grains after high-temperature deformation, which leads to the conclusion that all non-elastic deformations are localized in glassy phases.

The difference in the mechanical behavior of ceramics, which vary in their Al_2O_3 content is qualitatively described in terms of the model of a critical defect zone at the tip of the crack, proposed by Tsai. Under loading grain boundary shear leads to cavitation in ternary points and further growth of intergrain microcracks with a rate defined by the viscosity of the liquid in the intergranular channel.

The rate of the growth of a separate microcrack, and of the defect zone in general, and of sub-critical growth of the main crack, $\dot{a}$, depends on structural factors:

$$\dot{a} \propto \frac{\delta^3 E}{d^2 \mu} \tag{16}$$

It may be of interest to examine the temperature range where strength and K_{IC} increase. The decrease in the rate of deformation from 1 mm/min. down to 0.002 mm/min. displaces the curve $K_{IC} = K_{IC}(T)$ to the high temperature side. The value of the temperature shift determines the effective energy of the process which equals 720 kJ/mol, close to the activation energy of creep of Si_3N_4-Y_2O_3 ceramics, which is controlled by visco-elastic and diffusion mechanisms. This value is between 710 and 920 kJ/mol for slow crack growth in commercial silicon nitride ceramics [15,16]. It can be assumed that during the initial stage of the creep process the non-elastic deformation defect zone at the crack tip promotes an increase of ultimate strength and measured fracture toughness by means of crack tip blunting. If during this process the size of the defect zone does not achieve critical size, then strength and K_{IC} will be increased with an increase in temperature. As the temperature rises, the creep process accelerates, and as a result a new structural process starts which is characterized by intergranular coagulation of microcracks in elongated intergranular cavities together with a growth of the defect zone, and an increase in the crack tip radius. If the last process predominates, then the interaction of the crack with cavities will lead to a decrease of strength and fracture toughness with increasing temperature.

For Si_3N_4-Y_2O_3 ceramics it is assumed that the volume and the layer thickness δ of the grain boundary phase increases with increasing Al_2O_3 content. Consequently, the rate of sub-critical crack growth in Si_3N_4-7%Y_2O_3-3%Al_2O_3 ceramics will be greater than in Si_3N_4-7%Y_2O_3-1% Al_2O_3 ceramics, and the interval of significant lowering of strength will shift to a lower temperature range. This is indeed observed experimentally. Thus, additions of Al_2O_3, while promoting the stabilization the grain boundary phase in the amorphous state, lead to a significant drop in creep resistance of Si_3N_4-7%Y_2O_3 ceramics.

An increase in creep resistance can be achieved by suppressing one or several sequential processes: grain boundary slipping, cavitation nucleation and growth in ternary parts, and the growth and coagulation of grain boundary microcracks. Attempts to improve the heat resistance of nitride ceramics by introducing increasing amounts of dispersed-reinforcing additions of TiN was studied [17]. Test results with 2% Al_2O_3 and 5% Y_2O_3 additions are presented in Figure 20.9. Strength measurements were performed in air (curve 1) and in vacuum (curve 2). While the test results in vacuum show a high strength of the ceramics up to 850°C, the test in air gives a noticeable reduction is strength in the the intermediate temperature range (400–600°C). This reduction is related to oxidation of unstable yttrium silicates of the $Y_5(SiO_4)_3N$ type in intergranular phases, which starts at temperatures of around 400°C (see curve 3 in Figure 20.9a).

Introduction into the processing batch of additional amounts of TiN does not only suppress embrittlement at the intermediate temperature range, by elimination of the formation of unstable oxynitrides, but also, by means of precipitation strengthening, improves the short term strength of ceramics at temperatures above 1200°C (Figure 20.9b). It must be noted that during compressive creep testing (T = 1600°C, σ = 100 MPa) the creep rate of the materials containing TiN was rather high ($\dot{\varepsilon} = 3 - 6 \times 10^{-3}$ s^{-1}), but it was lower than in structures from isometric Si_3N_4 grains that did not contain glassy phases.

So, self-reinforced structures of silicon nitride materials with high-temperature grain boundary phases can be useful for high-temperature ceramics manufacturing. By comparing Figures 20.9a and 20.9b it is clear that the introduction of TiN suppressed embrittlement of ceramics in the intermediate temperature range, and gave a strength level of ~1000 MPa up to 1000°C. By examining the phase composition changes it was shown that this effect is related to the refining influence of TiN additions, to changes in the equilibrium of the reaction with carbon, and supression of the formation of unstable yttrium apatites.

Another way to increase the high-temperature strength of silicon nitride ceramics is the transition to sialons. This group of materials has improved corrosion resistance and thermodynamic stability compared to silicon nitrides. Powder synthesis can be performed using cheap raw alumino-silicates (kaolin, etc.), and the sintering process can be designed so that the liquid phase escapes. The resulting structure will not contain amorphous grain boundary layers.

A study was performed of the structure and properties of one of the advanced materials of this class, heterogeneous ceramics based on a mixture of α-β sialons in the Si_3N_4-Y_2O_3-AlN system [18]. Prismatic grains of the β-phase and more isometric grains of the α'-phase, which provide less reinforcement compared to the Si_3N_4-Y_2O_3-Al_2O_3 system, are present in the structure of this sialon. The temperature dependence of the strength of α-β sialons is shown in Figure 20.10. It is clear that weakening of the α-β sialons starts only at temperatures in excess of 1200°C, at the same time as in previously examined materials which weakened in the range of 900–1000°C. The high creep resistance of α-β sialon based ceramics, is directly related to the smaller amount of grain boundary phases and/or their greater fracture toughness.

Finally, it should be mentioned that the parameters affecting sub-critical crack growth are of paramount importance with regards to the application of engineered ceramics. Slow crack growth over a wide temperature range is described by the so-called $K_I - V$ diagrams. Analytical curves are presented by expressions of the type

$$V = A \cdot K_I^n \tag{17}$$

where A and n are empirical constants, characterizing sub-critical crack growth. For silicon nitride materials studied, the average value of n is about 100 at 20°C and this slope of the $K_I - V$ curves is preserved up to ~ 1000°C. With further temperature increase, n sharply decreases, and n = 3 at a temperature of 1400°C. The values of A and n can be used for evaluation of t, the "life" time of ceramics in a field of applied external stress σ_a:

$$t = \frac{2C^{\frac{2-n}{2}}}{Y^n \cdot \sigma_a^n \cdot A(n-2)}, \tag{18}$$

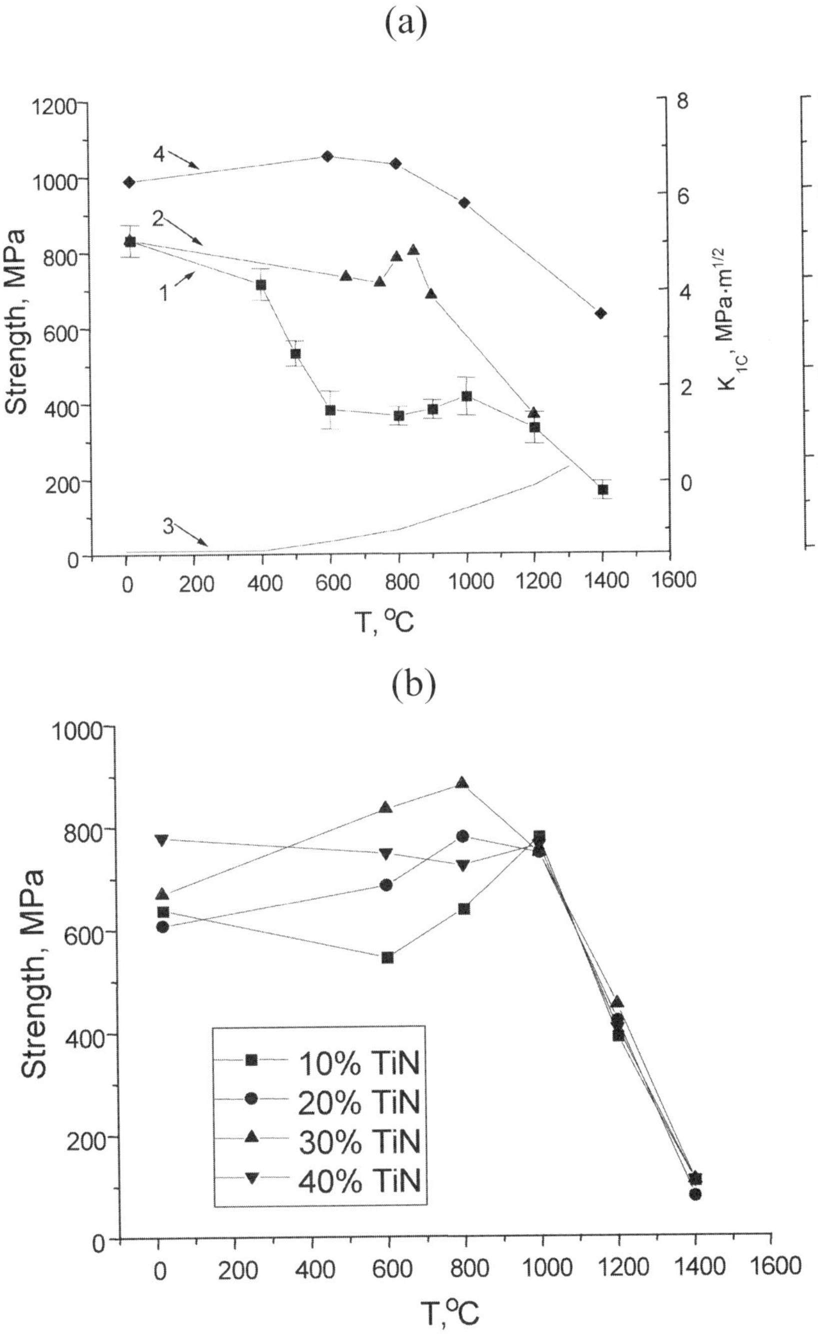

Figure 20.9 a) Dependence of Strength (in Air (1), Vacuum (2)) and Fracture Toughness (4) of Si_3N_4-5%Y_2O_3-2% Al_2O_3 Ceramics versus Temperature; (3) Thermogravimetric Curve b) Temperature Dependencies of Strength for Materials of the Si_3N_4-5%Y_2O_3-2% Al_2O_3-TiN System

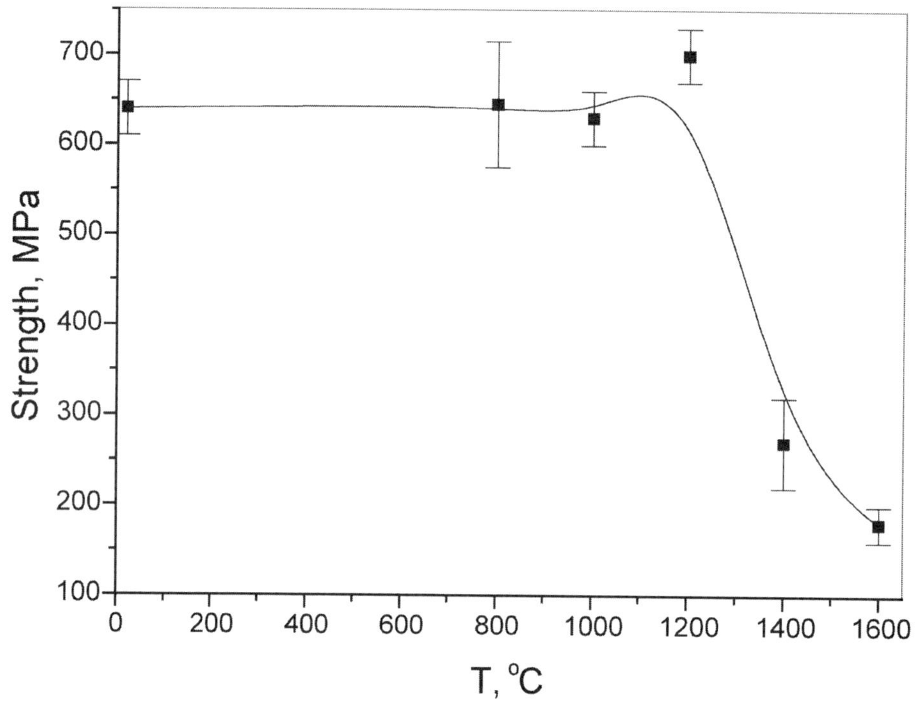

Figure 20.10 Temperature Dependence of Strength of α/β-Sialons (Si_3N_4-Y_2O_3.9AlN compositions)

Silicon Carbide Based Ceramic Matrix Composites

Ceramic matrix composites based on the SiC-TiB_2 eutectic system with eutectic temperature T_{eut} ~ 2200°C, containing strengthening particles of the second phase in amounts up to 40 vol.%, have been studied recently. Transition from monolithic ceramics to SiC-TiB_2 ceramic matrix composites is accompanied by an improvement in the mechanical properties of the materials at room temperature (fracture toughness, strength, hardness, wear resistance) [19–21].

Under ambient conditions the use of SiC-15% TiB_2 ceramics is limited to about 1200°C, where the oxidation rate is basically similar to that of SiC ceramics [19]. However, at 1400°C, the intense oxidation leads to the degradation of the materials' strength. Ceramics containing zirconium diboride have better oxidation resistance and are used for various high-temperature applications. The high oxidation resistance is determined by both the enhanced resistance of zirconium diboride itself, and by the fact that ZrO_2 is formed during oxidation. Accelerated solubility of SiO_2 in B_2O_3 forms a rather uniform protective film with high SiO_2 content. The influence of ZrB_2 additives, weak in the temperature range of 1000–1100°C, results in the reduction of weight loss at high temperatures (>1200°C).

Research was conducted on specimens of hot-pressed ceramics over a wide compositional range of MeB_2 (Me = Ti, Zr; 0–100%). To activate the sintering process, a charge was doped with not less than 5 vol.% of B_4C. The total content of boron carbide, accounting for milled material, and the contribution from grinding balls could be increased up to 15–17 wt.% (20–25 vol.%). Thus, this study considers not a two-phase system but rather the three-phase system SiC-MeB_2-B_4C.

Figure 20.11 shows typical microstructures of representative composites. In hot-pressing, the SiC particle size was 5–10 μm, the TiB_2 grain size remained in the 3–5 μm range. In some cases, the TiB_2 grains aggregate up to 10–15 μm in size, causing spontaneous microcracking as a result of thermal stresses. The

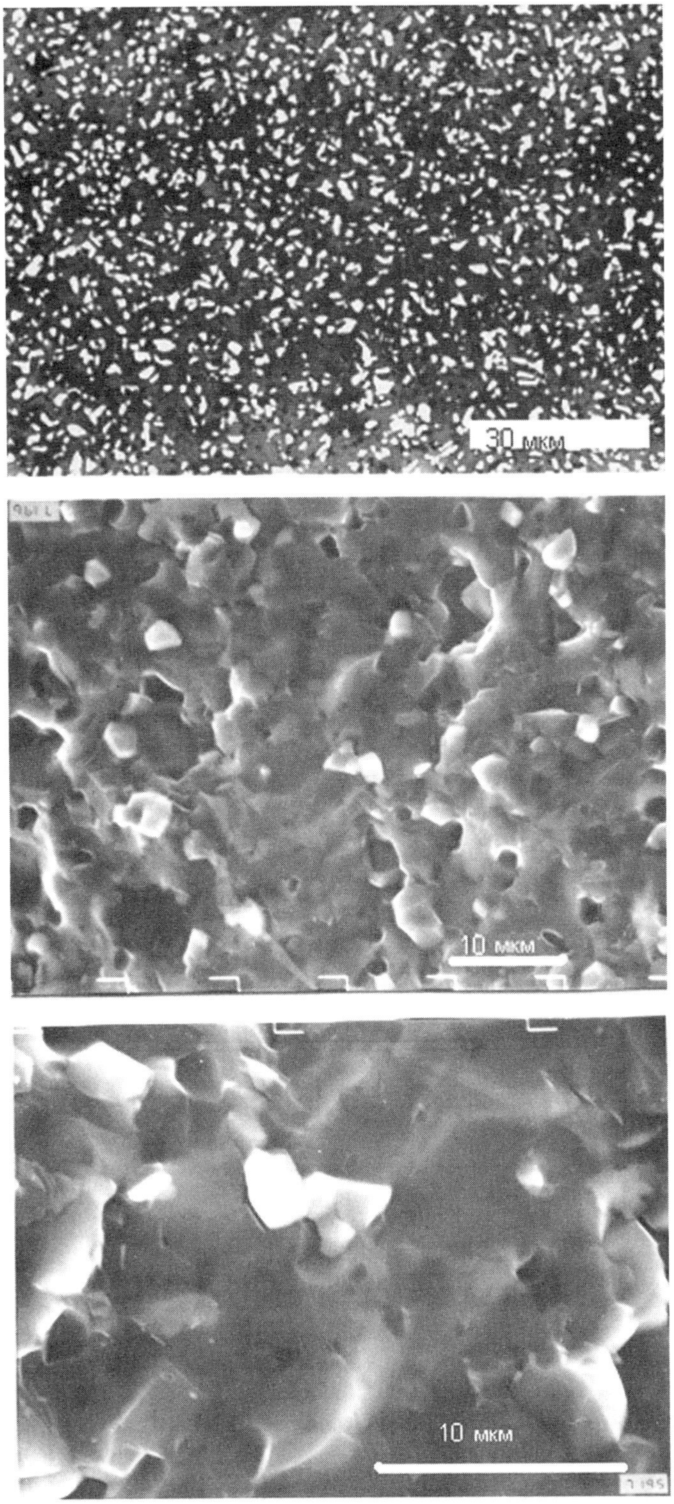

(a)

(b)

(c)

Figure 20.11 Microstructure SiC-TiB_2 Ceramics: a) –32 vol. % TiB_2; b), c) –15 vol. % TiB_2, Fracture Surface (mkm = μm)

size of the detected intergranular pores was in the 5–10 μm range. Fractographic analysis shows intercrystalline fracture.

Addition of 10–30 vol.% of titanium diboride to SiC was accompanied by a strong influence on all mechanical characteristics. Fracture toughness values reach a maximum (~7–8 MPa.m$^{1/2}$) at 15–20 vol.% of TiB_2 which subsequently decreases down to ~4 MPa.m$^{1/2}$) at 30% of TiB_2. This fracture toughness behavior is in agreement with the theoretical one calculated by the model, which accounts for redistribution of internal stresses existing between the phases. Results of the fracture surface analysis shows that the two mechanisms of toughening typically discussed for ceramic matrix composites are involved here as well: crack tilting around second phase particles and microcracking, initiated by stress fields at the crack tip, which contribute to the growth of fracture toughness.

The maximum strength, related to the analogous maximum fracture toughness, is achieved, in the range of 10–40 vol.% of TiB_2. With further increase in TiB_2 content, the values of strength and Young's modulus decrease, as evidenced by increasing porosity and an increase in internal stresses and structural non-homogeneity. The minimum values of strength and Young's modulus were observed for 50 vol.% of TiB_2 and were attributed to spontaneous microcracking in the neighborhood of coarse particles being subjected to high thermal stresses.

SiC–B_4C ceramics which do not contain titanium diboride, show a strength of 500 MPa at room temperature and at 1400°C (Figure 20.12). However, the strength decreases at temperatures slightly above 1000°C. Decrease of strength at 1000°C could be due to stress corrosion, which requires additional study. Strength degradation at 1000°C is suppressed in ceramics containing 20 vol.% TiB_2. However, at 1000–1400°C softening accompanied with strength degradation from 600 to 250 MPa occurs. Ceramics with 10 vol.% TiB_2 show the best high-temperature behavior: a drop in strength at 1000°C is less noticeable than for SiC-B_4C ceramics, and between room temperature and 1400°C the strength is more than 500 MPa. Thus, ceramics of the SiC-TiB_2 system, in addition to strengthening at room and moderate temperatures, can have high strength at elevated temperatures ($T > 1200$°C). To achieve a high level of strength at elevated temperatures, it is necessary to meet the condition of two-phase systems, the content of impurities, causing grain-boundary phases, needs to be minimized.

The oxidation resistance of ceramics can be qualitatively estimated from micrographs of the samples subjected to high-temperature bend tests. Surfaces of the ceramics with 20 vol.% TiB_2 contain oxidation products (TiO_2, B_2O_3) in the form of uniform yellow layers and bubbles, which appear at 1000–1200°C, and 1400°C, respectively. The former one was due to rutile formation upon oxidation of TiB_2. The latter one was due to evaporation of B_2O_3, formed upon oxidation of TiB_2 and B_4C, above 1200°C. Maximum oxidation resistance, as experimentally shown, belongs to the ceramics doped with zirconium diboride as the second phase.

The fabrication of of SiC-ZrB_2 composites is qualitatively similar to that of SiC-TiB_2 composites, but it is technologically very difficult. Additives of either ZrB_2 or B_4C introduced into the SiC powders in an amount of 20 vol.%, did not allow us to obtain ceramics with strength greater than 300 MPa. The simultaneous introduction of ZrB_2 and B_4C (up to 20 vol.% each) to a charge, results in ceramics with strength up to 400 MPa.

Compared to SiC-TiB_2 composites the microstructure of SiC-ZrB_2 composites is less uniform. Agglomerates of ZrB_2 can reach a size of 30 μm, and the average size of ZrB_2 grains exceeds the size of TiB_2 grains in SiC-TiB_2 composites by a factor of 2 – 3. The volume fraction of the second phase in micro- and in macro-volumes fluctuates much more than for the SiC-TiB_2 system. The appearance of large agglomerates in the studied system can be due to increased differences in density of SiC-ZrB_2-B_4C vs. SiC-TiB_2-B_4C, which may cause powders to delaminate during the drying stage of the batch due to the high density of ZrB_2. Moreover, these ceramics show greater apparent porosity in microscopic investigations which is the result of material pitting in abrasive processing. This suggests a decrease in strength of interphase boundaries and/or an increased level of interphase thermal stresses that directly follows from the larger CTE difference between SiC and ZrB_2 vs. SiC and TiB_2 phases. Therefore, to ensure higher strength in SiC-ZrB_2 ceramics, they should have better dispersion and structural homogeneity. Extending the milling time with rigorous quality control of

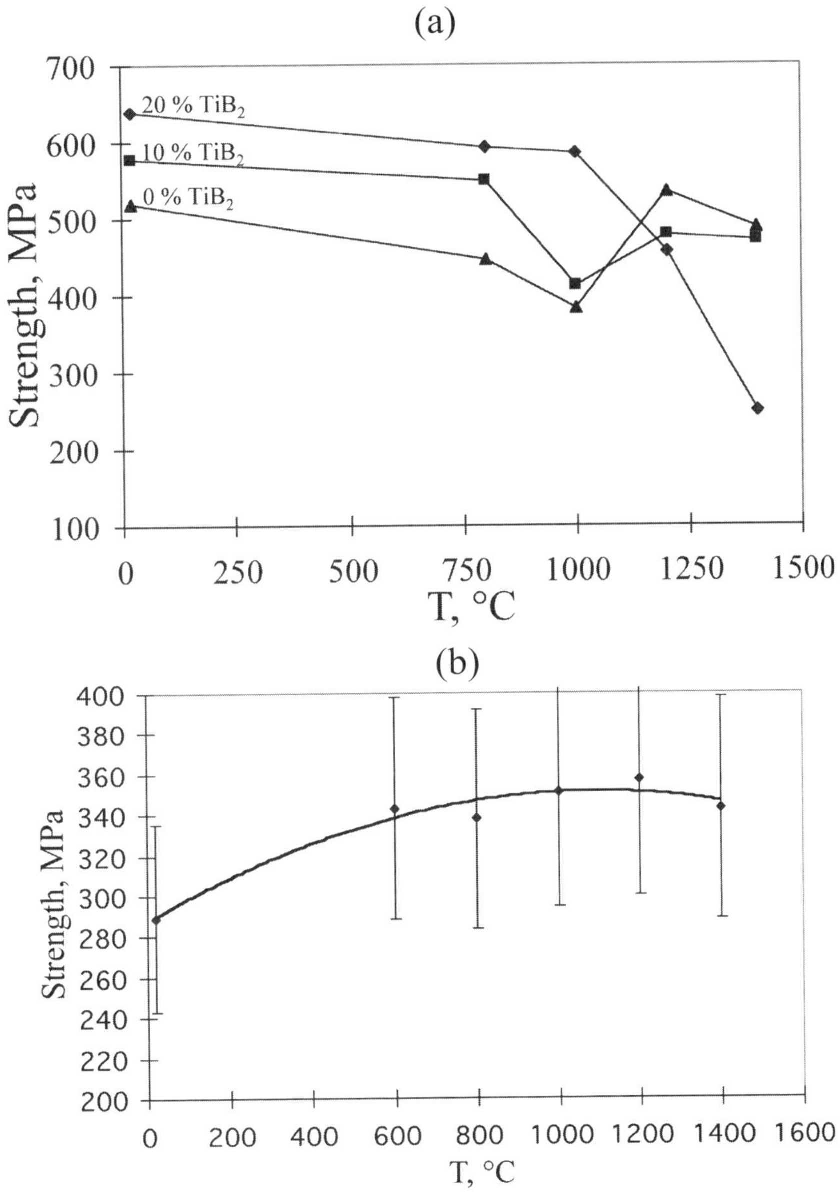

Figure 20.12 Temperature Dependencies of Strength for SiC with Different Volume Content of TiB_2 (a) and SiC-10 vol. % ZrB_2 (b) Composites

the milling process may lead to the necessary structural homogeneity and dispersion, similar to those for the SiC-TiB_2-B_4C system, and to 4-point bending strengths for samples with high ZrB_2 content, of 530–570 MPa.

In terms of high-temperature strength and oxidation resistance, materials containing moderate contents of the secondary phases are of the greatest interest. For investigations of mechanical properties over a wide temperature range, samples having a zirconium diboride content of 5–20 vol.%, and boron carbide of ~20 vol.% were selected. As can be seen in Figure 20.12b the SiC-ZrB_2-B_4C system, in contrast to SiC-TiB_2-B_4C, has a lower strength over a wide temperature range with fracture toughness values between 2.3–3.5 MPa.m$^{1/2}$.

However, SiC-ZrB_2-B_4C ceramics did not show strength degradation over the range of testing temperatures. Perhaps, it can be associated with the greater oxidation resistance of these ceramics and differences in the composition of grain boundary phases.

GAS TURBINE ENGINE COMPONENTS

Developed materials were evaluated during the fabrication of a wide range of GTE components for different applications—for surface transport, for energy supply in independent stations, and for marine and other applications. Some parts are shown in Figure 20.13.

Reaction sintered ceramics were used for the production of stationary parts (combustion chambers, nozzle assemblies, etc.). Working blades were made out of both dense sintered and hot pressed silicon nitride ceramics. SiC-MeB_2 ceramics were used for the manufacturing of working blades and to line a combustion chamber for gas turbine engines with a limited lifetime (~1 hour).

While using a hot pressed ceramic the parts of a given geometry were obtained by diamond machining. Upon the treatment they were annealed in pure nitrogen to obtain a homogeneous structure and to cure surface cracks as well.

Combustors and nozzles were made either as single pieces or as composite components. In the latter case different parts and pieces were separately formed, sintered and machined. Areas of components to be

Figure 20.13 Ceramic Parts for Engine Applications

joined were coated with an emulsion of dispersed silicon including activating additions prior to assembly. The composite structure was exposed to drying and heating to high temperature in nitrogen under pressure under conditions necessary for sintering.

Ceramics were joined to metal using brazing. Intermediate layers of a low modulus spring type, or composite ceramic intermediate layers (e.g. Si_3N_4-TiN, Si_3N_4-ZrO_2) with different contents of the secondary phase, dependent on the structure and function of the brazed parts, were used to reduce differences in coefficients of linear thermal expansion of the materials. Developmental compositions of adhesion-active brazes provided a shear strength of 15 KN for a brazing area of 15 cm^2 and at temperatures up to 900°C.

Let's observe more thoroughly the application of the developed ceramic materials, considering a GTE nozzle assembly. The first turbine was designed for land based applications and the second one for marine conditions at a maximum service temperature of 1400°C. The maximum calculated stresses were about 60 MPa based on the designers' estimates. A multi-faceted program was supposed to include:

1. Investigation of a deformation diagram of ceramics with identification of the brittleness measure χ (relation of elastic energy that the material accumulates to the moment of fracture versus the total specific energy spent on deformation; $\chi = 1$—for a brittle ceramic, $\chi < 1$—for a non-elastically deformed ceramic).
2. Definition of the changes of ultimate strength within an interval of working temperatures as well as temperature dependencies of fracture toughness.
3. Definition of the Weibull modulus, aimed at evaluating the level of allowable stresses and the reliability of the ceramic parts.
4. Definition of the characteristics of subcritical crack growth and durability of the ceramics under static and cyclic loading.
5. Determination of the resistance of the ceramic to thermal stresses and the evaluation of thermal-shock resistance.
6. Investigation of how the ceramic strength changes under the influence of oxidation, and corrosion by compounds such as Na_2SO_4, formed in a marine environment with sulfur containing fuel combustion.

Mechanical properties of an actual nozzle assembly and model constructions of the ring type under tensile stress loads, and the properties of specimens sectioned from entire components were determined. Both the composition of the ceramics, and the processing and fabrication technology of ceramic components that meet the performance specifications outlined previously in this chapter were optimized in a relatively short time span between 1978 and 1984. Reactive sintered ceramics (SNCM-78, 79, 80, 81, 83 and SNCMB-84) were used. Optimization of structure and ceramics properties was performed with the help of the methods described previously. Table 20.4 shows some properties of the materials at room temperatures.

The first problem to be addressed was governed by the statistical nature of strength, and by a low Weibull modulus which was evidence of insufficient homogeneity of the material. The evaluation of the early vintage materials (SNCM-78, 79) immediately revealed insufficient correlation between the materials'

TABLE 20.4 PROPERTIES OF THE REACTIVE SINTERED CERAMICS SNCM(B)

Ceramics	Density (g/cm^3)	Porosity (%)	Dynamic Elastic Modulus E (GPa)	Static Elastic Modulus E (GPa)	Brittleness Measure, χ	Ultimate Strain, $\varepsilon \times 10^4$	Flexure Strength, (MPa)	Weibull Modulus, m
SNCM-78	2.46	25	158	159	0.83	9.5	142	7–11
SNCM-79	2.49	24	178	163	1.0	9.9	165	7–8
SNCM-80	2.55	23	198	187	1.0	11.6	225	9–10
SNCM-81	2.53	23	—	167	0.86	11.5	190	10–15
SNCM-83	2.52	24	196	193	0.95	14.0	270	15–25
SNCMB-84	2.44	25	160	147	0.88	16.0	240	20–40

strength measured on samples (See Table 20.4), the strength of ring type model constructions (30–50 MPa), and the strength of actual nozzle assemblies (10–20 MPa). As a result of research and development efforts the materials SNCM-83 and SNCMB-84 with a relatively high Weibull modulus and a sufficient level of structural strength of the nozzle assembly (up to 50–60 MPa) were obtained.

Concurrently, efforts to optimize the composition and microstructure of the ceramic aimed at meeting specific life requirements under conditions of static and cyclic loading were undertaken. Figure 20.14 and Table 20.5 show some mechanical properties of the ceramics at 1400°C. Analysis of K_I – V and deformation diagrams as well as a creep rate revealed that at temperatures ~1400°C the lowest creep rate and rate of sub-critical crack growth was found for ceramics with a maximum value of χ. When considering all relevant high temperature properties (rate of subcritical crack growth, creep rate, ultimate strength, fracture toughness, and the thermal shock behavior) then we may conclude that SNCM-83 and SNCMB-84 ceramics fully meet the technical requirements.

All the ceramic materials have thermal shock resistance within a narrow range (500 ± 30°C), and if the coefficients of thermal expansion are about equal (~3.2 × 10^{-6}/°C), then an experimental assessment of the

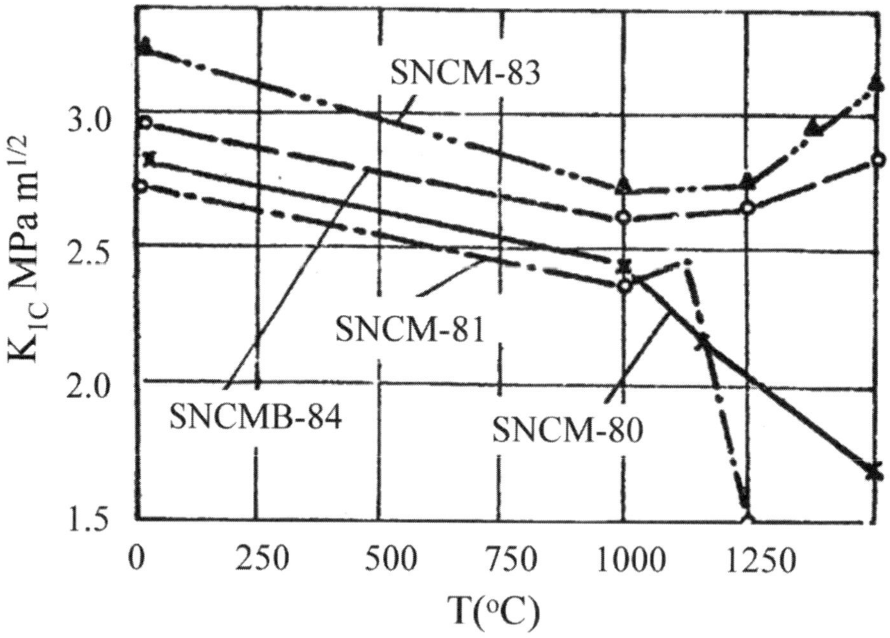

Figure 20.14 Temperature Dependence of Fracture Toughness of SNCM(B) Ceramics

TABLE 20.5 PROPERTIES OF CERAMICS AT T = 1400°C

Ceramics	Flexure Strength (MPa)	Brittleness measure (χ)	Fracture toughness, K_{IC} (MPa.m$^{1/2}$)
SNCM-78	20–30	—	—
SNCM-79	20–30	—	—
SNCM-80	150	0.12	1.7
SNCM-81	170	0.09	<1
SNCM-83	180	0.33	3.2
SNCMB-84	185	0.46	2.8

thermal shock resistance corresponds to the ultimate strain (See Table 20.4). The corrosion resistance rig tests were conducted at the design bureau "Mashproect" on an "RITM" gas-dynamic test stand in combustion gas flow with sea salt additions (15 ppm) at 1300°C for 300 hrs. An investigation of the characteristics and mechanisms of the oxidation of ceramics blades in various environments led to the development of a ceramic (SNCM-83) for which the surface layers practically do not degrade under the thermal cyclic conditions in aggressive gas flows. Corrosion and erosion resistance in a combustion environment and under the conditions of salt attack were found to be higher than for high-temperature structural metal alloys.

Working parts were bench tested at the design bureau "Turbine" of the Ministry of Defense of the USSR under the following test requirements: gas temperature before blades, number of launch—stop cycles, speed and density of gas stream, and vibration loads. There were different regimes of rig tests: –1) under continuous combustion gas flow at 700°C (0.5 hr.) → 1000°C (0.5 hr.) → 1330°C (0.5 hr.) → 1450 ± 20°C (1.5 hr.); –2) thermal cycling at special modes –a) start and increase of temperature up to 1250°C during 20 s, cooling down to 750°C during 15 s, holding at 750°C (120 s), blast refrigeration down to 80°C during 20 s and holding at 80°C (120 s); –b) start and increase of temperature up to 1420°C with ramp up time 12 s, holding at 1250°C (48 s) and blast refrigeration down to 100°C with ramp down time 120 s. In general materials and parts demonstrated durability with safe part life on the order of 200 hours and for up to 500 thermal cycles, representative of launch-stop conditions. They proved their efficiency, showed an absence of defects and got a certificate of approval for use.

Parts made of dense sintered and hot pressed silicon nitrides ceramic passed the same R&D program as reactive sintered materials. Nozzle assemblies made of dense silicon nitride ceramics for tank GTEs with a power of 1200 kW got a certificate to reliably work within 600 hours at a gas temperature of 1370°C, and 600 launch—stop cycles. The combustion chamber was awarded a use certificate for 600 hours at 1400°C, and 1200 launch—stop cycles. Yet, the working turbine wheel (HPSN hub with inserted blades) passed the test at a temperature of 1250°C and 48,000 rpm over only 140–147 hours, and failed to gain the certificate.

The prototype of a rotating ceramic heat exchanger with cellular honeycomb structure (silicon nitride ceramics, cell size –2 mm, wall thickness 0.2 mm, prototype's diameter 120mm) was tested at a gas stream temperature of 600°C and air temperature of 200°C. While in the experiment no visible defects were observed, the time and quantity were not sufficient to draw a positive conclusion.

For now new ceramic materials have been not introduced into the production of GTEs because the research efforts were terminated at the end of the 1980s.

ACKNOWLEDGEMENT

All figures in this chapter were supplied courtesy of the authors' organizations.

NOMENCLATURE

T	temperature
t^*	characteristic deformation temperature
$_DT$	temperature difference
T_{ve}	temperature of viscous—elastic transition
T_b	temperature of brittleness-plasticity transition
T_m	melting temperature
i	phase number
σ^{ri}	residual thermal stresses,
G_i	shear modulus of i- phase
E	Young's modulus
ν	Poisson's ratio
Y_i	volume fraction of i- phase

Y	geometric factor
C	typical size of initial defects (cracks)
C_{CR}	critical grain size for spontaneous failure under the effect of thermal stresses
α_i	coefficients of linear thermal expansion
K_i	modulus of bulk compression of i- phase
l_i	specific volume of i-component
p_s	external pressure at sintering
U_i	internal energy of elastic strain of i- phase
K_I	stress intensity factor for normal fracture (opening mode)
K_{1C}^i	fracture toughness of i-component of composite
σ_{ij}	components of tensor of residual (internal) stresses
$\dot{\varepsilon}$	rate of deformation
ε_{ij}	components of tensor elastic deformation
τ	yield stress
δ	thickness of grain boundary interlayers
μ	toughness of grain boundary interlayers
d	grain size
V	rate of macroscopic crack growth
σ_a	applied external stress

REFERENCES

[1] Deshman, C, *Scientific Fundamentals of Vacuum Technique*, Moscow, Mir Publ., pp. 286, 1964 (in Russian).

[2] Trefilov, V.I., Milman, Yu. V., and Grigoriev, O.N. "Deformation and Rupture of Crystals with Covalent Interatomic bonds," Prog. Crystal Growth and Charact., 16, 225–277, 1988.

[3] Kisly, P.S. and Kryl, Ya. A., "Massentransport in Siliziumnitrid unter hohem Druck," ("Mass Transport in Silicon Nitride Under High Pressure,") VII Int. Pulvermet. Tagung, Dresden, pp. 149–160, 1981 (in German).

[4] Sleptsov, V.M., Shcherbina, O.D., Pilyankevich, A.N., and Oleinik, G. S., "Strength and Microstructure of Silicon Nitride Based Materials," Powder metallurgy, 8, 83–87, 1976 (in Russian).

[5] Gogotsi, G.A. and Groushevsky, Ya.L., "Statistical Studies of the Strength of Inelastic Ceramics," Ceramics International, 8[1], 21–26, 1982.

[6] Gogotsi, G.A., Zavada V.P., and Shcherbina, O.D., "Strength and Crack Resistance of Ceramics (2)—Silicon Nitride Ceramic," Strength of Materials, 16[12], 1656–1659, 1984.

[7] Gogotsi, G.A., Groushevsky, Ya.L., Zudin, N.N., Trunov, G.V., and Shcherbina, O.D., "Composition Effect on Mechanical Properties of Silicon Nitride Based Materials," Powder Metallurgy, 2, 93–97, 1986 (in Russian).

[8] Gogotsi, G.A., Zavada, V.P., and Traskovsky, V.V., "Stress Corrosion of Silicon Nitride Based Ceramics," Ceramics International, 15[5], 305–310, 1989.

[9] Galanov, B.A. and Grigoriev, O.N., "Fracture Criterion of Composites with Ceramic Matrix," Problemy Prochnosti, 10, 30–40, 1993 (translated from Russian).

[10] Khoroshun, L.P., "Theoretical Methods of Random Functions in the Macroscopic Properties Problem of Microheterogeneous Solids," Applied Mechanics, Moscow, 14[2], 3–13, 1978 (in Russian).

[11] Grigoriev, O.N., Krivoshey, G.S., and Khoroshun, L.P., "Internal Stresses in Si_3N_4-Based Composites," Zavodskay Laboratoriya, Moscow, 10, 37–39, 1992 (in Russian).

[12] Aptekman, A.A., Bakun, O.V., Grigoriev, O.N., and Trefilov, V.I., "Deformation and Fracture of Diamond in Bending Tests in the Temperature Range 300–2100K," Reports of Academy of Sciences of USSR, 290[4], 845–848, 1986 (in Russian).

[13] Richerson, D.W., "Effect of Impurities on the High Temperature Properties of Hot-Pressed Silicon Nitride," Amer. Ceram. Soc. Bull., 52[7], 560–562, 1973.

[14] Tsai, R.S. and Raj, R., "Overview No. 18: Creep Fracture in Ceramics Containing Small Amounts of a Liquid Phase," Acta Metall., 30[6], 1043–1058, 1982.

[15] Evans, A.G. and Wiederhorn, S.M., "Crack Propagation and Failure Prediction in Si_3N_4 at Elevated Temperatures," J. Mat. Sci., 9[2], 270–278, 1974.

[16] Evans, A.G., Russell, L.R., and Richerson, D.W., "Slow Crack Growth in Ceramic Materials at Elevated Temperatures," Met. Trans., 6A[4], 707–716, 1975.

[17] Gogotsi, G.A, Grigoriev, O.N., and Jaroshenko, V.P., "Mechanische Eigenschaft von heissgepresstem Siliziumnitrid bei hohen Temperaturen," ("Mechanical Properties of Hot Pressed Silicon Nitride at Elevated Temperatures,") Silikattechnik, 41[3], 156–160, 1990 (in German).

[18] Grigoriev, O.N., Khomenko, G.E., and Perepelitsa, N.I., "Structure and Properties of α-Sialon-Based Ceramics," Powder Metallurgy, 9/10, 85–91, 1995, (in Russian).

[19] Janney, M.A., "Mechanical Properties and Oxidation Behavior of a Hot Pressed SiC-15 vol.%TiB_2 Composite," Amer. Ceram. Soc. Bull., 66[2], 322–324, 1987.

[20] Toshihiko, T. and Shigetaka, W., "Properties of Hot pressed SiC Matrix Composites with Internally-Synthesized TiB_2," J. Cer. Soc. Jap. Int. Ed., 97, 1188–1191, 1989.

[21] Kovalchuk, V.V., Yuga, A. I., and Grigoriev, O.N., "Physical, Mechanical and Tribological Properties of SiC-MeB_2 Materials," Powder Metallurgy, Kiev, 2, 95–99, 1992 (in Russian).

CHAPTER

21

Equipment and Methods for Ceramic Materials Evaluation

Youry A. Nozhnitsky, Youlia A. Fedina, Alexander V. Fishgoit

Central Institute of Aviation Motors (CIAM), Moscow, Russia

INTRODUCTION

Investigations on non-metallic small turbines for engines of different applications have been performed at the Central Institute of Aviation Motors (CIAM) under a program aimed at developing the enabling technologies for a new generation of engine components. A great number of silicon nitride and silicon carbide based monolithic ceramics, and fiber- and whisker-reinforced ceramic matrix composites as well as C/C and C-ceramic materials have been considered.

Advanced structural ceramics subjected to temperatures of 1400–1650°C can be used in designs for short life (the gas turbine engine part life for expendable vehicles or turbopump units of liquid-propellant rocket engines does not exceed several hours or even minutes). For the designs for long service life the ceramic part maximum operating temperature must be 200–300°C lower (the required part life of gas turbine drives for installations used for electric power generation or mechanical compressor drive is tens of thousands of hours). Therefore, the development of ultra-high temperature non-metallic materials is being vigorously pursued at present. The ultimate operating temperature of C/C materials is determined by the ability to protect the material against oxidation. The development of equipment and testing methodology must be carried out in parallel with the development of new materials, otherwise their full potential will not be realized in the near future.

While conducting high-temperature mechanical tests two challenges are faced [1–5]. The first challenge is related to the need to heat specimens up to the required temperature (which can exceed 1400°C), creating and maintaining the required temperature field, developing serviceable grips, providing load and alignment, and performing the high temperature measurements themselves. The second challenge stems from the mechanical, physical, and technological properties of the materials under investigation. Such material properties as low thermal conductivity, electrical non-conductivity, high temperature oxidation, interaction with

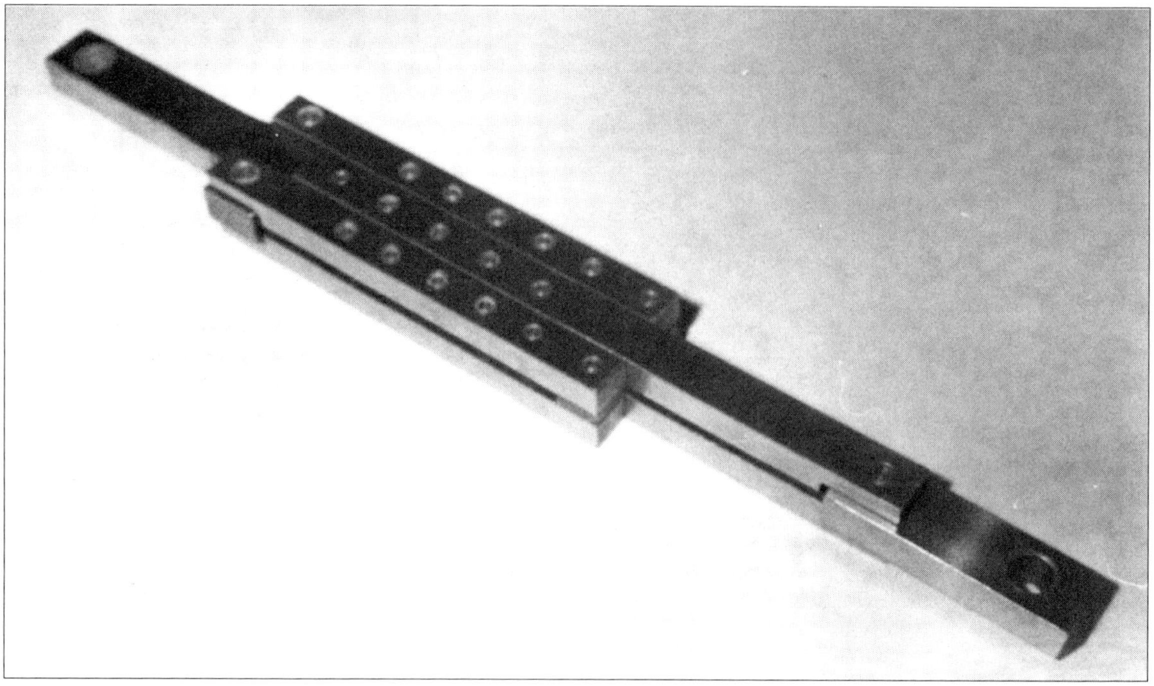

Figure 21.1 Tungsten Grip for Five Specimens for Bending Tests

grip material, and volatility must be considered while working out the procedures for high temperature testing (methods for heating, test environment, grip materials, etc.).

Anisotropy and brittleness of materials under investigation impact the selection of specimen size and geometry and methods for processing test results. It is also necessary to take into account the technological aspects of specimen fabrication, and the difficulties associated with extending the test results for specially fabricated specimens to full-scale components, because of differences in fabrication processes, reinforcement scheme (for composites), and scale factor.

Multipurpose equipment was used since a variety of materials were being tested in the laboratory. CIAM has extensive experience with conducting high-temperature tests using different test specimens or different methods for part heating: heating in furnaces and in rigs using gas flows, heating by conducting an electric current directly through a specimen under investigation, and high frequency current heating. Based on this experience and the analysis of the known methods of testing for ceramic and composite materials, facilities have been developed at CIAM to test ceramic monolithic and composite materials in tension, compression, bending and shear at temperatures in the range of 1400–3000°C.

The mechanical properties of the materials are mainly determined using resistance furnaces. For testing in vacuum, all tooling (heaters, shields, grips, extensometer elements subjected to high temperatures) must be manufactured using refractory alloys. The tungsten grips allow the tests to be conducted at maximum temperatures of 2300°C in vacuum and molybdenum grips allow testing up to 1800°C. For example, Figure 21.1 shows the tungsten grip intended for flexure testing of five specimens.

When conducting the tests in a protective medium it is advisable to use tooling made of carbon-graphite materials. Figure 21.2 shows the carbon-carbon grips for flexure and tensile tests. A dilatometer with carbon-graphite tooling is shown in Figure 21.3. The results of the measurements of coefficients of linear thermal

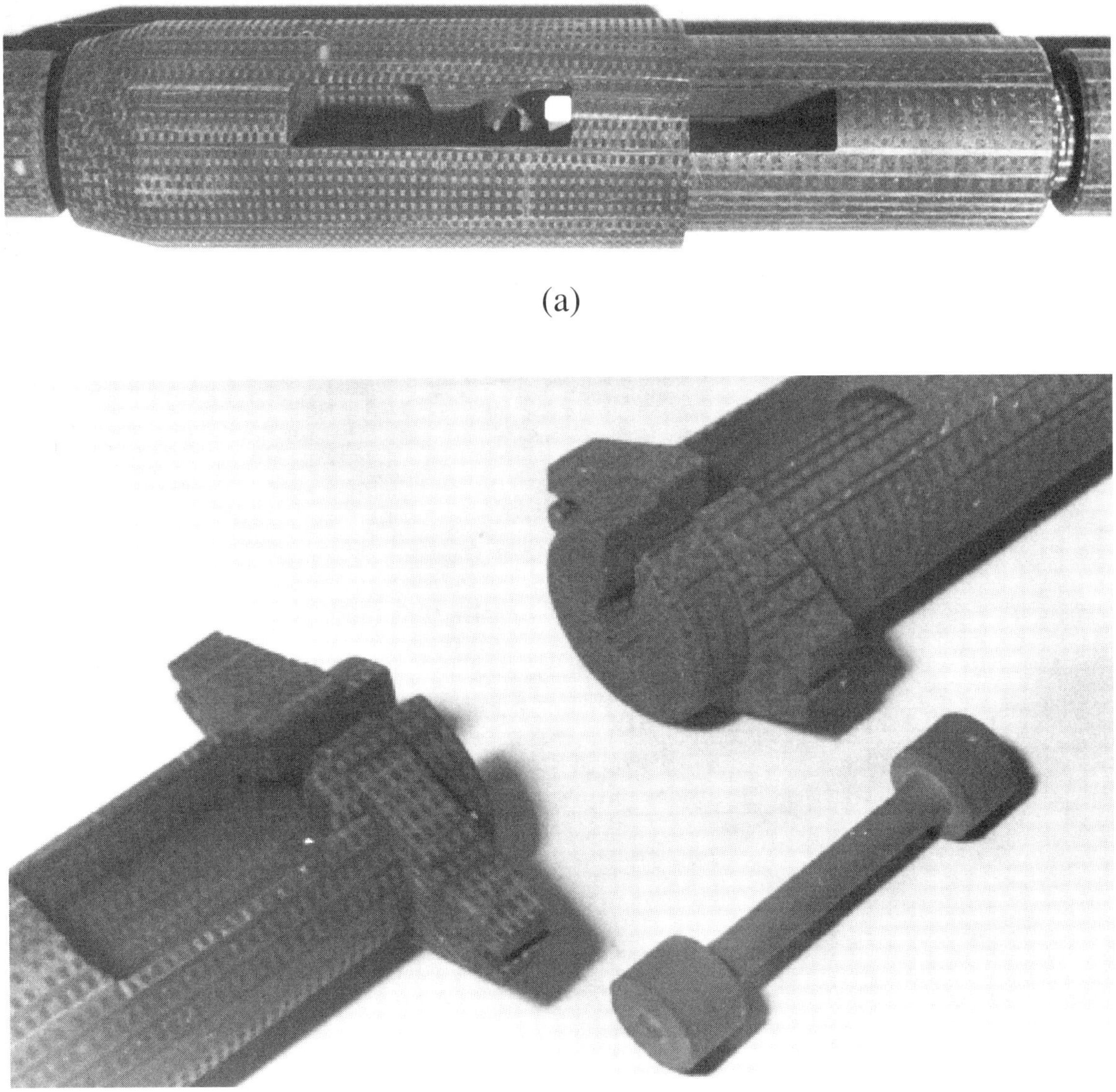

(a)

(b)

Figure 21.2 Carbon-Carbon Grips for Flexure (a) and Tensile (b) Tests

expansion are shown in Table 21.1. The carbon-graphite tooling enables testing in a protective environment at temperatures up to 3000°C.

Furnace testing of specimens in air is the most complicated because of the need to protect the grips. In this case molybdenum (for up to 1700°C), or tungsten alloys with coatings (up to 2200°C), chromium alloys (up to 1600°C), $MoSi_2$ (up to 1600–1700°C), SiC (up to 1600°C) and $LaCrO_3$ (up to 1600–1800°C) can be used to fabricate the heating elements. Higher heating temperatures can be achieved using heating elements made of oxides, primarily ZrO_2 or ThO_2. However, in the latter case the heaters have extremely high resistance at low temperatures and need to be started by means of removable additional heaters (e.g.

Figure 21.3 The Dilatometer with Carbon-Graphite Tooling and Specimens

TABLE 21.1 RESULTS OF THERMAL EXPANSION MEASUREMENTS

Temperature Interval (°C)	Coefficient of Thermal Expansion $\times 10^6 (K^{-1})$	
	$ZrO_2 - Al_2O_3$	Carbon/Carbon
200–300	8.3	0.11
300–500	9.1	0.2
500–800	10.4	0.3
800–1000	10.7	0.8
1000–1300	11.4	2.13
1300–1500	13.0	3.2
1500–1800	13.5	1.33
1800–2500		1.2

made of $LaCrO_3$) which can be located inside the high temperature heaters. At CIAM ceramic tooling made of non-oxide ceramics is used for bending tests at temperatures up to 1400°C. The use of ceramic grips for tension tests causes some difficulties as the strength of the grips that operate in tension appear to be insufficient and it is difficult to provide alignment by using reverse grips that operate in compression while the specimen is in tension. Therefore, tensile tests in air should be conducted in "short" furnaces and "long" specimens should be placed in such a way that the grips are located outside the furnace. For this purpose INSTRON tooling is used at CIAM. It allows tests to be conducted at temperatures up to 1600°C (Figure 21.4). For tests at higher temperatures (including thermocycling) the use of "short" reflective furnaces appears to be beneficial. For testing of "long" specimens made of material with low thermal conductivity, heating of the grips may be necessary in order to reduce thermal stresses in the specimen.

The choice of specimen design depends on the material. Specimens of fiber-reinforced composites, including laminated and 3-D reinforced materials, have a flat section. For monolithic ceramic tests there may be a preference for round specimens without sharp edges. Minimizing stress concentrators in the

(a)

(b)

Figure 21.4 The Tooling for the Tension Tests in an Air Medium: a) INSTRON Tooling with Resistance Furnace; b) "Short" Reflective Furnace and "Long" Specimen

specimen of a brittle material is of great importance. For composite material tests, because of the great influence of the edge effect (external layer overloading and internal layer underloading) it is necessary to increase the length of the specimen section where the load is applied and, therefore, to increase the length of the specimen heads. For tests with buttonhead specimens made of unidirectional composites the head length must take into consideration the low strength of the materials in interlaminar shear, and for tests with unidirectional C/C materials in 3-point bending, it is necessary to maintain the relation $l/h > 30$ (l—distance between outer supports; h—specimen thickness).

The methods and equipment developed at CIAM have been used for property determination of different carbon-carbon, carbon-ceramic and ceramic materials. Some results are presented below.

INVESTIGATION OF CRACK RESISTANCE OF CERAMICS

Gas turbine ceramic component life is often limited by crack growth due to initial defects. Therefore, the characteristics of material static and cyclic crack resistance are extremely important. In spite of the substantial amount of experimental work performed (e.g. the round robin VAMAS study) definite problems still remain. One of them is that the various methods of fracture toughness determination do not give the same or similar results. As the value of fracture toughness is used for strength calculations the question arises as to which method of its determination is the most applicable for the corresponding operating conditions. Gas turbine parts operate at high temperature and this restricts the range of the applicable methods for fracture toughness determination since there is a temperature limitation for some of these methods. The extent to which the conventional equations are applicable is also an issue.

The analysis of the equations for K_{IC} calculation shows that the elimination of residual stresses after indentation and reduction of a crack depth (healing) can significantly increase fracture toughness. This can be achieved by annealing in air or vacuum at temperatures that provide residual stress relaxation. Hence, it has been suggested to use annealing as a final step to eliminate the various surface defects that may have appeared during ceramic part manufacture. However, to validate this recommendation requires a thorough experimental test.

There are several methods for fracture toughness determination at room temperature [6].

- indentation with determination of indentation crack length (indentation fracture method—IF)
- bending of a specimen with an indentation crack (indentation strength method—IS)
- bending of a specimen with a macrocrack (single-edge precracked beam method—SEPB)
- bending of a specimen with a macronotch (single-edge notch beam method—SENB)

The SEPB method widely used for metallic materials is the most developed with respect to methodology. However, the existence of large cracks in real ceramic parts is not realistic. The IF method which generates a large amount of data from a single specimen is more suitable but it does not completely reflect actual conditions of loading and cannot be used at high temperatures. The IS method corresponds better to actual operating conditions. However, there is reason to doubt the validity of the results. Some of the questions are:

- It is not clear to what extent the form of a crack approaches the ideal semicircular shape used for the calculations.
- Is there the possibility of subcritical crack growth between indentation and bending as well as during bending, and what would be the rate of such subcritical crack growth?
- The equation for fracture toughness calculation by the IS-method assumes the presence of indentation residual stresses. Absolute or partial residual stress elimination takes place at high temperatures. For high temperature tests the temperature limitation of the applicability of the equations is not clear.
- Does healing of indentation cracks take place at high temperatures?

Because of these considerations fracture toughness needs to be determined by different methods and the results for various materials need to be compared for the selection of a method to be used for the calculation for real parts.

During service ceramic parts are subjected to constant and cyclic loads. To estimate the life of these parts information needs to be gathered about the crack growth rate at static and cyclic loads at room and elevated temperatures. It is desirable to get a relationship of the type:

$$dC/dt = F_1(K) \text{ for static loading}$$

and

$$dC/dN = F_2(\Delta K) \text{ for cyclic loading.}$$

With these relationships the number of loading cycles (or time) before a catastrophic failure can be established.

Materials and Types of Tests

The methodology developed in this study is applicable to a variety of ceramic materials. The properties of silicon nitride and silicon carbide ceramics with strengths varying from 330–1040 MPa and of silicon carbide whisker-reinforced silicon nitride composites have been investigated. The composition of materials, 3-point bending strength values at 20°C, and specimen dimensions are given in Table 21.2. Each material is designated by a letter to facilitate the discussion.

The following experiments were conducted:

- Determination of material hardness (H_v) and fracture toughness (K^i_{Ic}) by an indentation method (IF) at 20°C
- Determination of material strength by the three-point bending method in the 20–1500°C temperature range
- Determination of fracture toughness (K^b_{Ic}) by the three-point bending method for indented specimens (IS) in the 20–1400°C temperature range
- Determination of fracture toughness (K^n_{Ic}) by the three-point bending method for a specimen with a macronotch (SENB) in the 20–1400°C temperature range
- Investigation of indentation cracks growing during room temperature exposure and their healing at annealing
- Investigation of the kinetics of crack growth under cyclic and static loading

TABLE 21.2 COMPOSITIONS OF THE INVESTIGATED MATERIALS, THEIR STRENGTHS AND SPECIMEN DIMENSIONS

No	Designation	Composition	Dim. (mm)	Strength (MPa)	Comments
		Monolithics			
1	A	$SiC+30\%TiB_2$	50 × 5 × 3	350	Hot pressed materials.
2	B	$SiC+20\%TiB_2$	50 × 5 × 3	400	TiB_2 and AlN additions for better
3	C	SiC+20%AlN	50 × 5 × 3	330	densification
4	D	SiC+30%AlN	50 × 5 × 3	380	
5	E	Si_3N_4	50 × 5 × 5	430	Sintered materials.
6	F	$Si_3N_4+11\%Al_2O_3+TiN$	50 × 5 × 5	480	TiN technological addition
7	G	$Si_3N_4+5\%Y_2O_3+2\%Al_2O_3$	50 × 5 × 3.4	900	Hot pressed materials. Y_2O_3 and Al_2O_3 additions for better densification
8	H	$Si_3N_4+5\%Y_2O_3+2\%Al_2O_3$	50 × 5 × 3.4	1040	
		Composites			
9	K1	$Si_3N_4+Y_2O_3+20\%SiC_W$	50 × 5 × 3.5	820	Y_2O_3 additions for better densification, SiC_W reinforcement wiskers (diameter 0.3–0.5μm, length 5–10μm)

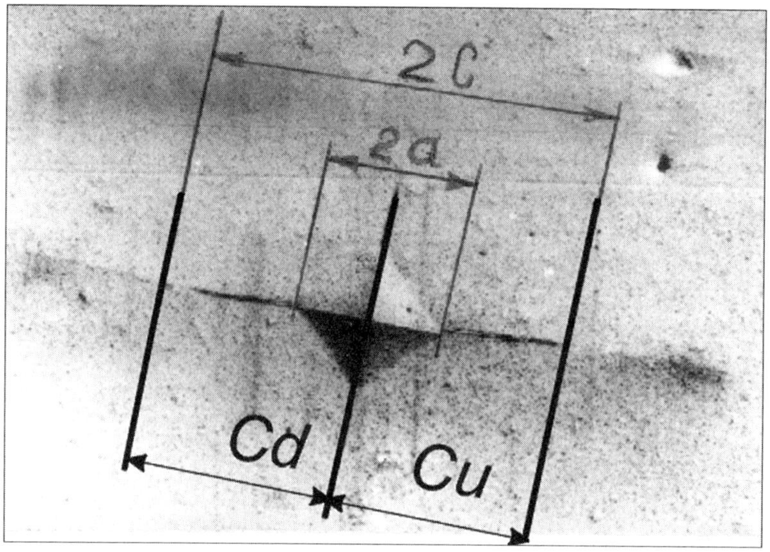

Figure 21.5 Imprint of the Indentor and Crack Originating at the Indent

Hardness, Strength and Crack Resistance against Static Loads at Room Temperature

For the determination of hardness, strength and crack resistance by the IF method the specimens were polished and indented by a Vickers diamond pyramid at 50–400 N loads. After indentation the dimensions of an indentor imprint diagonal $2a$ and length of cracks coming from impression angles $2C$ were measured by an optical microscope (Figure 21.5).

The hardness was calculated according to the equation

$$H_v = 0.4636\frac{P_i}{a^2}, \text{ [7]} \tag{1}$$

The fracture toughness was calculated according to the equation

$$K_{1c}^{i} = \eta_v^R \sqrt{\frac{E}{H_v}}\frac{P_i}{C^{3/2}}, \text{ [8]} \qquad \eta_v^R = 0.018 \text{ [7]} \tag{2}$$

For Si_3N_4 ceramics E = 290 GPa. For SiC ceramics E = 300 GPa.

The material strength at room temperature was determined by a three-point bending method and calculated by the equation:

$$\sigma_b = \frac{3P_c l}{2bh^2}, \tag{3}$$

To determine fracture toughness by the IF method the specimen was indented in the middle of the largest side at a load of 200N. Then the bending tests were performed. The fracture stress (σ_c) was determined according to equation (3). The fracture toughness was calculated without considering indentation residual stresses according to the equation

$$K_{1c}^{bw} = 1.26\sigma_c\sqrt{C}, \tag{4}$$

TABLE 21.3 HARDNESS AND FRACTURE TOUGHNESS VALUES FOR VARIOUS MATERIALS AT ROOM TEMPERATURE

Material Designation	H_V, MPa,	σ_C, MPa	K^{i}_{1c}, MPa$\sqrt{m}$	K^{bw}_{1c}, MPa$\sqrt{m}$	K^{b}_{1c}, MPa$\sqrt{m}$	K^{n}_{1c}, MPa$\sqrt{m}$
A	12600 ± 4700	230 ± 40	9.8 ± 2	3.8 ± 0.6	6.3 ± 1	6 ± 0.3
B	12300 ± 2000	280 ± 130	6.6 ± 1	5.2 ± 2.2	7.6 ± 2.5	—
C	14200 ± 1400	200 ± 30	5.7 ± 1.1	3.6 ± 0.6	5.7 ± 0.6	6.3 ± 0.6
D	12900 ± 2300	166 ± 23	6.3 ± 1.6	3 ± 0.4	4.9 ± 0.5	—
E	10400 ± 600	130 ± 6	7.3 ± 0.6	2.2 ± 0.4	3.5 ± 0.13	—
F	14000 ± 2000	140 ± 16	6 ± 1	2.2 ± 0.3	3.6 ± 0.3	—
G	17900 ± 2500	250 ± 50	6 ± 0.3	4.4 ± 0.3	5.7 ± 0.25	—
H	15700 ± 900	234	7.3 ± 0.6	3.8 ± 0.16	5.4	—
K1	13800 ± 1700	357 ± 40	8.5 ± 0.6	3.5 ± 0.5	8.3 ± 0.8	—

The factor 1.26 was calculated for a semicircular crack based on literature data [9]. The fracture toughness at bending was also calculated according to the equation where indentation residual stresses were taken into consideration [7].

$$K^{b}_{1c} = 0.59\left(\frac{H}{H_v}\right)^{1/8}\left(\sigma_c P_i^{1/3}\right)^{3/4}, \tag{5}$$

To determine fracture toughness by the SENB method a notch with depth equal to ~.5 *b* was inserted in the specimen, perpendicularly to the smallest side, by means of a diamond wheel. The notch width was ~0.5 mm. The specimens were tested by three-point bending at different temperatures. The fracture toughness was calculated with traditional equations for fracture mechanics [9]. The values for hardness and fracture toughness obtained by various methods at room temperature are given in Table 21.3.

It is worth noting that the results of fracture toughness determination by the various methods are close for the B, C, G, H and K materials. At the same time the values of K^{i}_{Ic} and K^{b}_{Ic}, for unknown reasons, differ significantly for the A, D, E and F ceramics. In our opinion, for strength calculations one needs to use the data from the IS method, since the loading conditions are closer to real operating conditions for a part with a small surface crack. It should be noted that the irregularity of the results for the SiC ceramics is essentially greater than for the Si_3N_4 ceramics. This may be explained by the significant uniformity of the SiC ceramics properties at the microscopic level.

The causes of coincidence or difference of fracture toughness determined by different methods are not clear to the authors. The experiments on various ceramics were performed by the same persons using the same equipment. At first the experiments on Si_3N_4 ceramics were conducted. The experiments demonstrated that the values of K^{i}_{1c} and K^{b}_{1c} of the K1 composite and ceramics of high strength G and H coincided but these values of low strength ceramics E and F were different. However the subsequent experiments on the SiC-based ceramics A, B, C and D have not confirmed this conclusion. The IS method is suggested for practical use as it simulates fractures during engine operation better than the other methods.

Influence of Room Temperature Exposure and Annealing on the Behavior of Indentation Cracks and Characteristics of Crack Resistance

To determine crack growth during exposure and crack healing following annealing, the crack lengths of some specimens (materials E, F and G) were measured immediately after, after 2 hours, and after 2 days following indentation. To evaluate the influence of annealing on crack healing, cracks (materials G and H) were measured immediately after indentation, and after annealing in air at 1070°C for 2 hours. To determine the influence of annealing on material crack resistance, specimens (materials G and H) were tested by the IS method in this regime in the following sequence:

(1) Without annealing
(2) Indentation, annealing, bending
(3) Annealing, indentation, bending

The results of the crack length measurements at different periods of time after indentation at 200 N loading did not show significant differences. The annealing after indentation of the G ceramic resulted in shortening of the average length of the cracks from 0.38 mm to 0.35 mm, but there was no noticeable effect on fracture toughness values (K^b_{1c} increased from 6.3 to 7.6 MPa$\sqrt{m}$). The effect of annealing after indentation on the H ceramic was much more dramatic with a very substantial increase in fracture toughness, K^b_{1c}, from 6 to 12 MPa$\sqrt{m}$. The reason could be more intensive crack healing for the H ceramic compared to the G material, and more complete elimination of residual stresses, or both. Annealing before indentation did not change the fracture toughness results. At the time of writing of this chapter a more detailed investigation of ceramic parts reliability improvement by annealing, possibly at the final stage of manufacture and component assembly, is being carried out.

Tests at Elevated Temperatures

The results of the measurement of material crack resistance in the range of 20–1400°C are given in Figures 21.6 through 21.8. At relatively low temperatures (up to 1200°C) the fracture toughness of the notched specimens (SENB method) coincides rather well with the fracture toughness by the IS method calculated according to (5) (which accounts for indentation residual stresses). At higher temperature it approaches the fracture toughness calculated according to (4) (which does not consider indentation residual stresses). Evidently, this can be explained by assuming that residual stresses are eliminated at elevated temperatures.

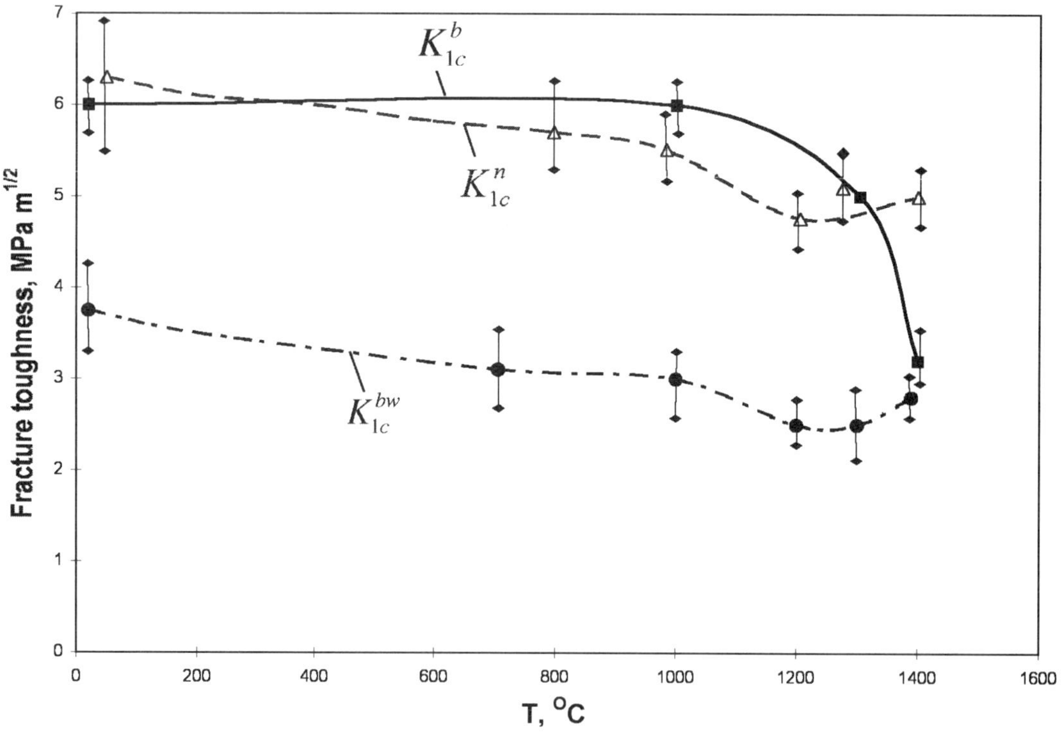

Figure 21.6 Fracture Toughness of Material A. Calculation by Equation (4) $-K^{bw}_{1c}$; Calculation by Equation (5) $-K^b_{1c}$; Fracture Toughness Determined on the Notched Sample $-K^n_{1c}$

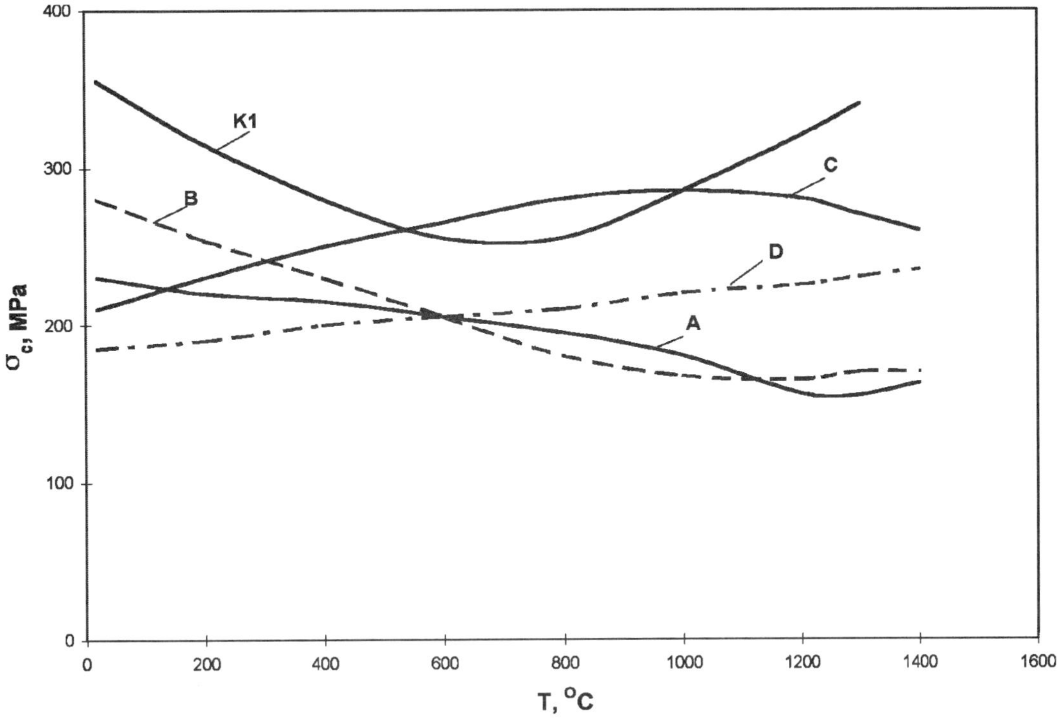

Figure 21.7 Comparison of the Fracture Stresses for Indented Samples Made from Different Materials

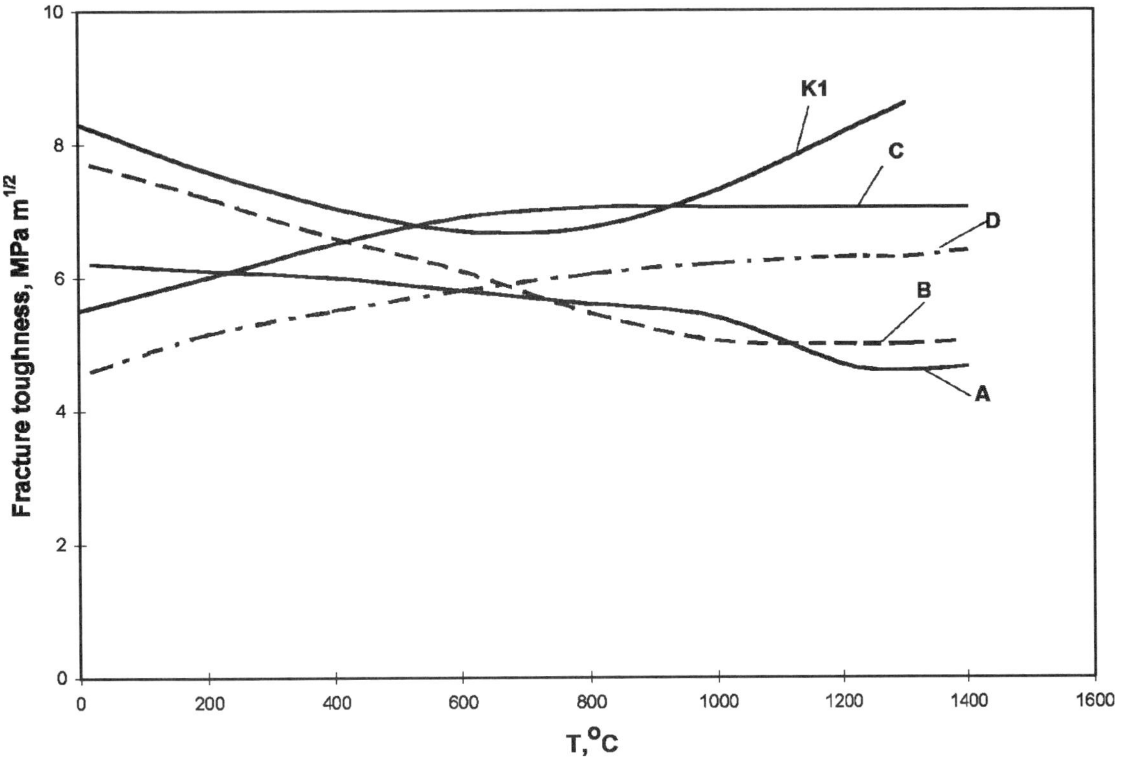

Figure 21.8 Comparison of the Fracture Toughness (K_{1c}^{b}) of Different Materials (see Table 1.2 for the materials tested)

The fracture stress (σ_c) and fracture toughness (K_{IC}) decrease rather monotonously with increasing temperature for materials A and B. On the other hand, the fracture stress (σ_c) and fracture toughness (K_{IC}) increase with temperature for materials C and D. The fracture stress and fracture toughness for the K1 material have minimum values around 700°C (Figures 21.7 and 21.8). Apparently, several processes influence the material behavior: strength reduction with increasing temperature, increasing plasticity and surface crack healing. The fracture stress σ_c for the indented specimen and fracture toughness values at room and operating temperatures (1000–1400°C) appear higher for the K1 ceramic composite (Figures 21.7 and 21.8) than for the A, B, C, and D silicon carbide ceramics.

Crack Growth under Conditions of Static and Cyclic Loading

The measurement of the crack growth rate under static loading was carried out in the "STARK" facility [10]. It is a multi-purpose rigid cell, a rectangular frame of high rigidity, the loading of which is carried out by a test machine. The specimen was loaded by the displacement of its central section according to a 3-point bending scheme. The displacement was defined by the deformation of the rigid frame under the external load of the test machine. The specimen deflection was measured by a variable-induction pickup and the loading on the specimen by a dynamometer. The measurement error for both values is less than 0.2%. The specimen compliance was calculated based on the load and deflection, and the crack depth was determined based on the compliance of the specimen. The pickup dynamometer signals are transmitted to a PC. The software allows the determination of the ultimate strength, modulus of elasticity, fracture toughness, I-integral, R-curve and the $dl/dt = F(K)$ diagram. In the latter case measurements of the average and instantaneous crack rates are possible. An example of the crack rate determination is shown in Figure 21.9.

It is worth noting that the data obtained from different specimens have a great deal of scatter. The interval of subcritical crack growth is not great (4–6) MPa.$\sqrt{m}$ and the rate is rather high (10^{-4}–10^{-1} mm/s). It is questionable whether determining slow crack growth at constant loading is useful for strength calculations. It seems that the value of the crack propagation threshold K_{th} is more easily used than the value of fracture toughness K_{IC}, and the difference between them is probably insignificant and within the expected range of error in the data.

The equipment for the observation of crack growth in ceramic specimens is a modified high-temperature metallographic microscope that has a device for cyclic loading. This microscope allows testing

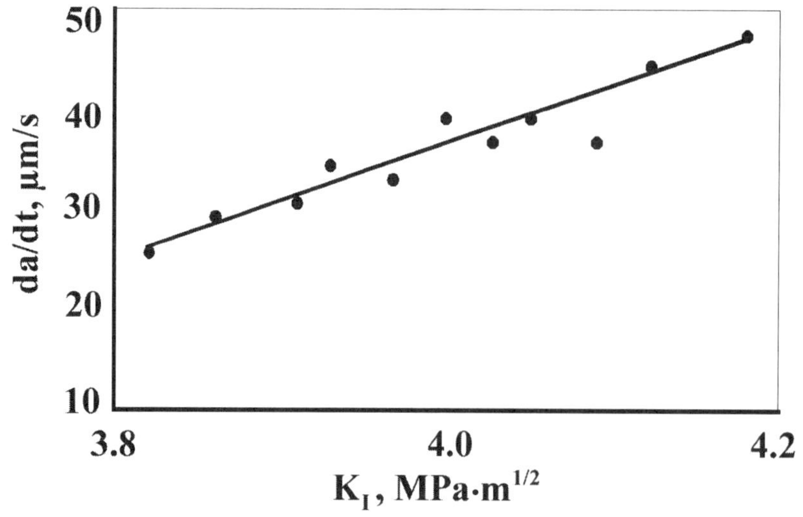

Figure 21.9 Crack Rate versus Stress Intensity Factor

of specimens in three-point bending under a load of 10–500 N in a chamber with air, under vacuum, or in a variable environment.

The cyclic load was applied to the specimen with a preliminary traced Vickers indentation in accordance with the three-point bending scheme. During loading, cracks originating from the imprint angles were observed by means of an optical microscope. The magnification of the optical system was up to 1000×. The frequency of loading was ~0.1 Hz. The specimens were periodically removed and investigated under a scanning microscope.

From the test results the relationship between crack length on the specimen surface (2*C*) and number of loading cycles was plotted [10,11]. Then these curves were replotted as the relationship between crack growth rate, *dC/dN*, and stress intensity factor range. The latter was calculated according to the equation $\Delta K = 0.89\sigma\sqrt{2C}$, where σ is the stress due to bending on the specimen surface.

The investigation of the crack growth kinetics was performed for the G and H monolithic ceramics, and for the K1 composite. The relation between crack length and number of loading cycles is shown in Figure 21.10. The uneven character of the crack growth attracts attention. The crack either does not progress, or quickly covers a 0.01–0.1 mm distance, after which it stops again. The growth of a crack can be arrested for up to several thousands of cycles. This type of crack propagation is evidently connected with the presence of local non-uniformities (0.01–0.1 mm) which are significantly greater than the coagulated particle dimensions (~1–5 μm).

In order to obtain the relationship between the crack growth rate, *dC/dN*, and the stress intensity factor, the curve representing the relation between the crack length and number of loading cycles was smoothed out. Subsequently, the differential relationship was obtained from the smoothed curve. It should be noted that due to the small number of increments for each curve there could be some degree of subjectivity in the data. The kinetic diagrams for the *dC/dN* and ΔK relationship are shown in Figure 21.11.

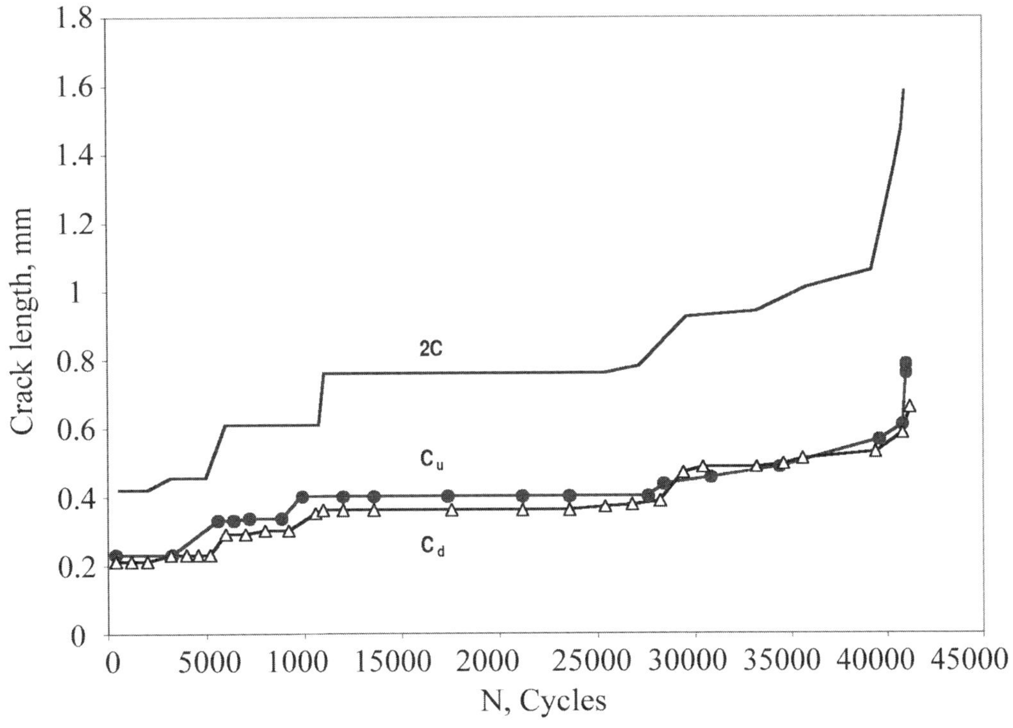

Figure 21.10 Crack Growth Rate in the H Material $-\sigma_{max}$ = 187 MPa

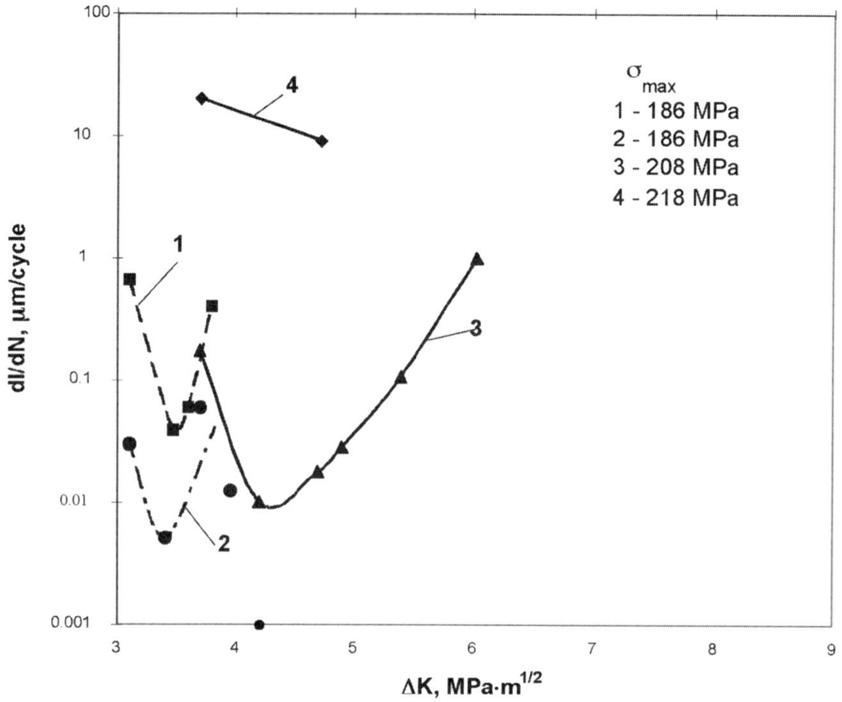

(a)

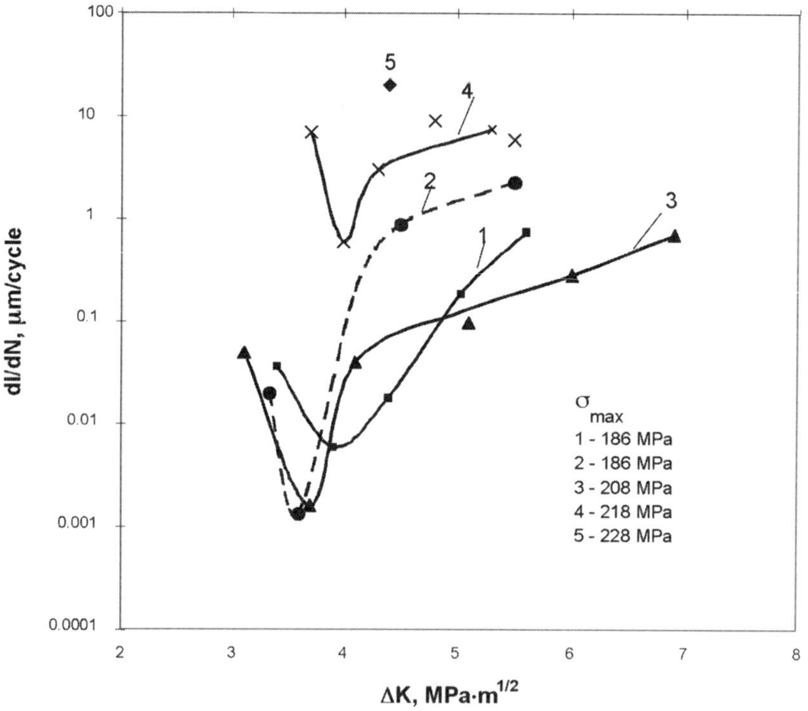

(b)

Figure 21.10 Kinetic Diagrams for the G (a) and H (b) Material

The same general shape of these relationships, which have a minimum at mean ΔK values, is noteworthy. The assumption has been made that the high crack growth rate at small values of ΔK is caused by large residual stresses due to indentation. The greater the effect of these stresses, the lesser the crack length. As the crack grows the influence of residual stresses from the indentation decreases and that results in a reduction in the crack growth rate. When the crack reaches a definite length, the growth rate increases, as the stress intensity factor increase connected with increasing crack length assumes a major role.

SUMMARY

1. Equipment and methodology for high-temperature mechanical tests for non-metallic materials have been developed.

2. An investigation of the crack resistance of some silicon carbide and silicon nitride ceramics and silicon carbide whisker-reinforced composites with a silicon nitride matrix has demonstrated the following:

- it has been shown experimentally that the crack resistance determination by IF and IS methods does not give similar or close results for all ceramics. For some ceramics the results obtained by the various methods differ significantly. It has been shown that annealing at 1070°C for 2 hours after indentation can result in a significant increase in fracture stress and fracture toughness based on three-point bending tests of a specimen with indentation cracks.
- The crack growth in the G and H ceramics and K1 composite has been investigated. The crack growth in all investigated materials at cyclic loading is of uneven character (with a jump length of 0.01–0.1 mm). The crack growth rate and stress intensity factor relationships are not monotonous: they have a minimum at the mean ΔK value. The relationship of the crack resistance and temperature can be of a different character.

ACKNOWLEDGEMENT

The figures and tables in this chapter were provided courtesy of CIAM.

NOMENCLATURE

a	half of indentor imprint diagonal
b	specimen width
C-C	carbon-carbon
C	half of indentation crack length on the sample surface, depth of semicircular crack
E	Young's modulus
F_1, F_2	functions designation
H_v	hardness
K	stress intensity factor
K_{Ic}^{b}	fracture toughness, determined by the three-point bending of an indented specimen method
K_{Ic}^{bw}	fracture toughness, determined by three-point bending of an indented specimen method, and calculated without taking into account residual stresses of the indentation
K_{Ic}^{i}	fracture toughness determined by the indentation method
K_{Ic}^{h}	fracture toughness determined by three-point bending of a specimen with macronotch
ΔK	range of stress intensity factor
l	distance between outer supports in the bending test
N	number of loading cycles
P_i	indentation load
P_c	fracture load
T	temperature

t	time
η_v^R	constant
σ	stress
σ_b	ultimate strength
σ_c	fracture stress

REFERENCES

[1] Negovskii, A. N., Drozdov, A. V., Kutnyak, V. V., Noshnitsky, Y. A., Menshikov, A. N., Qiao, S., and Bai, S., "Experimental Equipment for the Evaluation of the Strength Characteristics of Carbon-Carbon Composite Materials Within the Temperature Range 20–2200°C," Strength of Materials, 31[3], 319–325, 1999. Also in Problemy Prochnoste, 31[3], 319–325, 1999 (in Russian).

[2] Nozhnitsky, Y., Smirnov, L., Egorov, S., Markov, A., and Sakovich, V., "Experimental Investigation of Ceramic Materials and Turbine Rotor Components Strength, *Proceedings of the 4th International Symposium on Ceramic Materials and Components for Engines*, R. L. Carlson, T. Johansson, and L. Kahlman, eds., Göteborg, Sweden, 10–12 June, 1991. Kluwer Academic Publishers, Dordrecht, The Netherlands. pp. 1025–1034.

[3] Nozhnitsky, Y. A., Fishgoit, A. V., Smirnov, L. E., Rosanov, M. A., and Chercasova, S. A., "Investigation of High-temperature Monolithic Ceramic and Composite Materials Structural Strength," paper presented at the *4th European East-West Conference and Exibition on Materials and Process*. St-Petersburg, Russia, October 17–21, 1993. Prometey, Symposium B, No. B-III-3, p. 56.

[4] Nozhnitsky, Y. A. and Smirnov, L. E., "Ceramic, Carbon-Carbon and Other Composite Materials Tests at High Temperature," in *Ultra High Temperature Mechanical Testing*, R.D.Lohr and M.Steen, eds., Woodhead Publishing Limited, Cambridge, England, UK, 1995. pp. 184–192.

[5] Nozhnitsky, Y. A., Fedina, Y. A., Rekin, A. D., and Petrov, N. I., "Development and Investigation of Ceramic Parts for Gas-Turbine Engines," ASME paper 97-GT-157, presented at the International Gas Turbine and Aeroengine Congress and Exposition, Orlando, FL, USA, June 2–5, 1997.

[6] *ESIS TS 6 Round Robin on Fracture Toughness*, EMPA, Report No. 155, August, 1995.

[7] Lawn, B. R. and Marshall, D. B., "Hardness, Toughness, and Brittleness: An Indentation Analysis," J. Am. Ceram. Soc., 62 [7–8], 347–350, 1979.

[8] Lawn, B. R. and Faller, E. R., "Equilibrium Penny-Like Cracks in Indentation Fracture," J.Mat. Sci., 10[12], 2016–2024, 1975.

[9] Broek, D., *Elementary Engineering Fracture Mechanics*, Noordhoff International Publishing, Leyden, The Netherlands, 1975.

[10] Borovik, V. G., Kovalev, S. P., and Chushko, V. M., "Crack Resistance of Ceramic Materials with Controlled Crack Growth," Strength of Materials, 26 [5], 1994. Also in Problemy Prochnoste, 26[5], 69–77, 1994.

[11] Fishgoit, A. V., Nozhnitsky, Y. A., Fedina, Y. A., Rozanov, M. A., and Demidov, A. G., "The Strength and Crack Resistance of Ceramics and Composites with Ceramic Matrix at the Small Surface Defects Presence," paper E.346, presented at the *International Conference on Advanced Materials, Symposium A: "Engineering of Composites: Investigations, Technologies and Perspectives*," Kiev, Ukraine, October 4–7, 1999.

[12] Fishgoit, A. V., Nozhnitsky, Y. A., Rozanov, M. A., Demidov, A. G., and Teplova, S. V., 1993, "Small Cracks in Silicon Nitride Base Ceramics," *Proc. 7th Int. Conf. on Mechanical Behaviour of Materials*, May 28–June 2, 1995, The Hague, The Netherlands. Delft University Press, Delft, The Netherlands. pp. 613–614.

CHAPTER

22

Properties of Continuous Fiber-Reinforced Ceramic Matrix Composites for Gas Turbine Applications

Edgar Lara-Curzio

Metals & Ceramics Division, Oak Ridge National Laboratory, Oak Ridge, Tennessee, USA

INTRODUCTION

Improvements in the efficiency of gas turbine engines have been achieved by increases in turbine inlet gas temperature (TIT) through remarkable progress in component cooling and engine materials [1]. Because metallic materials currently used in the hot section of engines operate at 90% of their incipient melting temperatures, further improvements in efficiency will only be possible through the use of ceramic materials. It has been projected that fully optimized gas turbine engines incorporating ceramics would have significant increases in thermal efficiency and output power compared to all metal-engines with air-cooled components. Furthermore, such new gas turbine designs would offer increased performance, improved control of emissions and the potential for longer service lives.

Ceramics encompass all solids that are neither organic nor metallic. Compared with metals, ceramics have superior wear resistance, high temperature strength, and chemical stability. They also generally have lower thermal conductivity, thermal expansion, and lower toughness. The lack of toughness causes them to fail catastrophically, which is an undesirable trait if they are to be used in gas turbines engines. Several approaches have been followed to improve the toughness of ceramics. These include microstructural design and toughening by the addition of dispersed ceramic particulates, whiskers, or fibers to form ceramic matrix composites (CMCs). Among the various types of CMCs, Continuous Fiber-reinforced Ceramic Matrix Composites (CFCCs) are particularly attractive for gas turbine engine applications because of their ability to alleviate elastic stress concentrations at holes and notches by redistributing stresses through various inelastic deformation mechanisms [2].

As with all composite materials, the thermomechanical behavior and physical properties of CFCCs are determined by the properties of their constituents (i.e., fiber, fiber coating, matrix, matrix coating and their interfaces), by the architecture of the reinforcing fibers, and by a number of micromechanical mechanisms that include matrix cracking and fiber debonding and sliding to bridge the wake of advancing cracks in the matrix. The most widely used fiber architectures include woven fabrics (plain and satin weaves) but CFCCs also have been densified with simpler (e.g., stacking unidirectional layers) or more complex (e.g., 2-D polar, 2.5-D and 3-D) fiber architectures. Because most of the CFCCs that have been used for gas turbine engine applications are reinforced with 2-D continuous fiber architectures, the bulk of the data presented in this chapter corresponds to the properties of those materials.

Although most of the data available for CFCCs intended for gas turbine applications is for composites reinforced with SiC-based fibers and SiC matrices, data have been reported for CFCCs reinforced with carbon and oxide fibers (e.g. alumina, mullite) and CFCCs with carbon, glass-ceramic and oxide matrices. By far, carbon has been the most widely used fiber coating for CFCCs. The role of the fiber coating in CFCCs is to protect fibers during matrix synthesis and to facilitate fiber debonding and fiber sliding in the wake of advancing cracks in the matrix. However, the lack of oxidation resistance of carbon at temperatures above 300°C has prompted extensive work on the development of oxidation-resistant fiber coatings for CFCCs. Some of these fiber coatings include BN, Si-doped BN, multilayered coatings of C and SiC, pseudo-porous SiC and oxide fiber coatings [3].

Silicon carbide-based CFCCs are by far the most mature among CFCC systems with non-oxide matrices. This has been the result of more than 30 years of continuous R&D of SiC-based fibers and matrices. However, the lack of long-term environmental stability of SiC-matrix CFCCs in combustion-related environments (high-temperature, high-pressure, high-concentration of water vapor) has prompted the development of protective coatings (environmental barrier coatings or EBCs) and growing interest in all-oxide CFCCs because of the intrinsic stability of the latter in oxidizing environments. In contrast to the mechanisms responsible for providing damage tolerance in SiC-based CFCCs, damage tolerance in oxide/oxide CFCCs without fiber coatings is achieved by a network of fine pores in the matrix [4]. The mechanical characteristics of these materials are similar to those of carbon/carbon CFCCs but with superior oxidation resistance. Tensile fracture of all-oxide CFCCs occurs with substantial fiber pull-out although it doesn't conform to the classical model of individual fibers debonding and sliding out of their mating sockets in the matrix. Instead it involves fragmentation of the matrix and rearrangement of the surrounding matrix particles as the fibers slide [4]. However, the long-term thermal stability between the fibers and matrix in interphaseless all-oxide CFCCs remains a subject of concern, which has prompted significant efforts towards the development of fiber coatings for these CFCCs [5]. The use of all-oxide CFCCs is currently limited by the microstructural stability of the reinforcing fibers, which are less resistant to creep deformation than SiC fibers. Oxide fibers are prone to grain growth at high-temperatures, which results in loss of strength with time [6–10]. Although some single-crystal oxide fibers (e.g.—sapphire, YAG) possess good creep and stress-rupture resistance [6], their large diameter and high cost will limit their use to special applications with simple geometries.

Various routes have been established for the synthesis of CFCCs [11,12]. In the case of silicon carbide matrix composites, most systems are synthesized by chemical vapor infiltration (CVI), by the continuous impregnation and pyrolysis of polymer precursors (PIP) or by the infiltration and reaction of molten metals and alloys (MI). In the case of oxide-oxide composites, most systems are synthesized by sol-gel, powder processing and directed metal oxidation.

Most of the initial work to develop CFCCs for gas turbine engine applications took place in the UK during the 1950s and 1960s with the development of glass and glass ceramic matrix composites reinforced with carbon and silicon carbide fibers [13–18].

In subsequent years several programs were implemented in the U.S., Europe and Japan, to further research and develop CFCCs for gas turbine applications. In the U.S., these programs include, for example, U.S. Department of Defense Programs to develop fiber-reinforced glass and glass-ceramic matrix composites [19]; the U.S. Department of Energy's Continuous Fiber-Reinforced Ceramic Matrix Composites Program to address the long-term R&D priorities identified by industry as the key to the development of CFCCs for Industrial Applications, including stationary gas turbine engines [20]; the U. S. Department of Energy's Ceramic Stationary Gas Turbine Development Program to improve the performance of stationary gas turbines in cogeneration through the selective replacement (retrofit) of metallic hot section parts with uncooled ceramic components [21]; NASA's Enabling Propulsion Materials (EPM) Program, to develop a 1200°C combustor liner for high-speed civil transport [22]; the Ultra Efficient Energy Technologies (UEET) Program to develop SiC/SiC CFCCs for hot-section gas turbine engine components [23]; the Integrated High Performance Turbine Engine Technology (IHPTET) Program, which is a joint DoD/NASA/Industry effort to develop advanced engine technologies, including C/SiC turbine rotor and nozzles, and SiC/SiC inter-turbine transition ducts [24]; and the NASA Advanced High Temperature Engine Materials Technology (HITEMP) Program [25,26].

SEP (Société Européenne de Propulsion) and DRET (Direction de Recherches, Etudes et Techniques), the French military research agency, sponsored some of the first programs to research and develop CFCCs for use in gas turbines in France, in particular to fabricate afterburner components (exhaust cone and flameholder) and exhaust nozzles for the SNECMA M88 engine [27,28]. In The Netherlands a consortium sponsored by the Agency for Energy and the Environment worked towards the development of CFCCs for nozzles and a small turbine wheel for the Dutch OPRA and AMT turbines [29,30]. In Europe, the Commission of the European Communities sponsored the Human Capital and Mobility Program to develop and characterize carbon- and SiC-based CFCCs for advanced gas turbines [31].

In Japan, a consortium of government agencies and private organizations made extensive use of monolithic ceramics, CMCs and CFCCs to develop a 100 kW gas turbine with thermal efficiency over 40% at a turbine inlet temperature of 1350°C as part of the Automotive Ceramic Gas Turbine (CGT) Development Project [32]. The Research Institute for Advanced Materials Gas Generator, which is a joint effort by Japan's Key Technology Center and 14 Japanese corporations, implemented R&D programs to develop and use CFCCs for the manufacture of gas turbine engine stationary and rotating components [33,34] and more recently the CREST-ACE Program to research and develop SiC-based CFCCs with sound life cycles for advanced energy systems [35].

In the next section, examples of gas turbine engine components manufactured with CFCCs are discussed. This is followed by a compilation of CFCC properties that are relevant to gas turbine engine applications. Directions for future developments are also discussed.

COMPONENTS

Combustor Liners

The most common pollutants generated by gas turbines include NOx, CO and unburned hydrocarbons. Burning the fuel after premixing it with excess air is particularly effective at reducing NOx. Excess air in the reaction zone of the combustor acts as a heat sink that lowers the flame temperature to the point where no NOx is produced in the combustor. If the combustion liner is film cooled, as is necessary when using conventional superalloys, a good fraction of combustor air is not available for premixing with the fuel. Consequently it is impossible to operate the combustor lean enough to remain below the NOx formation threshold [1,36]. Combustion liners constitute the most widely explored application of CFCCs for gas turbine engines components because CFCCs have the potential of withstanding the temperature of the hot gas in the reaction zone without film air cooling. Furthermore, in this application mechanical loads are minimal and stresses arise mostly from thermal gradients. Using the available air for lean premixing, rather than for film air cooling, results in a reduction of NOx emissions. CO and unburned hydrocarbon emissions would also be reduced because the hot surface of a CFCC combustor liner could be operated at a temperature high enough to ensure that the fuel oxidation reactions were not quenched at the combustor walls.

In the U.S., the Department of Energy has sponsored various programs (e.g. the Ceramic Stationary Gas Turbine Development Program, the Continuous Fiber Ceramic Composites Program, and the Advanced Turbine Systems Program) to develop and demonstrate the use of CFCC combustor liners [20–21,37–38]. For example, as part of the Ceramic Stationary Gas Turbine (CSGT) Development Program, Solar Turbines Inc., has evaluated inner and outer SiC-based CFCC combustor liners in an industrial gas turbine engine with nominal power output of 4 MWe [21,39]. To date SiC-based CFCC combustor liners have accumulated more than 55,000 hours of engine operation. These engine runs were preceded by screening tests in which CG-Nicalon™/SiC, Al_2O_3/Al_2O_3 and CG-Nicalon™/Al_2O_3 subscale combustor liners were evaluated in a burner rig that simulates actual engine temperatures and pressures [40]. These screening tests showed that CG-Nicalon™/SiC exhibited the best performance of the liners evaluated. Figure 22.1 shows pictures of Hi-Nicalon™/CVI-SiC combustor liners for a Solar Centaur 50S gas turbine engine and a Hi-Nicalon™/MI-SiC combustor liner for an industrial-grade gas turbine engine. Several CFCC combustor liners have also been evaluated for military aeroengines [41].

In Japan, several programs have been focused on the development, fabrication and demonstration of SiC/SiC and C/SiC CFCC combustor liners for gas turbines used in the Super/Hypersonic Transport Propulsion System (HYPR), for the Advanced Material Gas-Generator (AMG), for a 100 kW automotive ceramic gas turbine and for helicopter turboshaft engines [32,34,42–44].

Programs in both Europe and the US have also been focused on using all-oxide CFCCs to manufacture combustor liners as an alternative to SiC-based CFCCs. Although all-oxide CFCCs have the potential of exhibiting improved stability in high-temperature oxidizing environments, they possess lower thermal conductivity than their SiC-based counterparts, which could result in higher thermal gradients and stresses. Also, the lower resistance of oxide fibers to creep deformation and the loss of fiber strength with time is a potential limitation of these materials. In Europe the four-year long program "Novel Oxide Ceramic Composites (NOCC)" was implemented to develop processing routes to fabricate an all-oxide CFCC combustor liner and to demonstrate it in a combustor rig at conditions representative of a gas turbine engine [45–48]. The all-oxide CFCC, which was synthesized by tape casting followed by hot pressing, consisted of single crystal alumina monofilaments, an alumina matrix and a zirconia fiber coating. Figure 22.2 shows a photograph of the all-oxide combustor tiles produced during this program. A tile was successfully tested in a rig at temperatures above 1260°C, but the major issues that remain to be addressed included high cost, delamination dur-

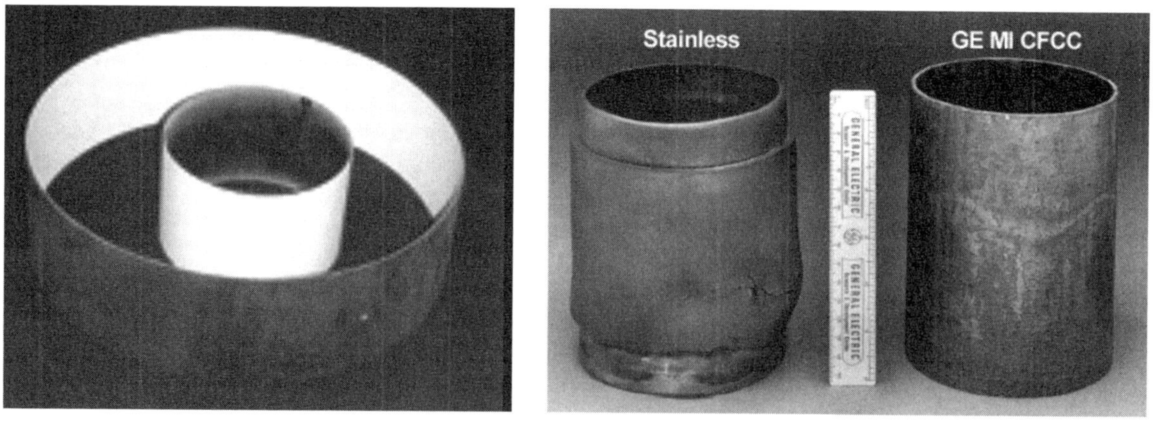

Figure 22.1 (a) Set of EBC Coated Liners Tested by Solar for 13,937 hrs—Inner Liner—Hi-Nicalon™/MI-SiC; Outer Liner—Hi-Nicalon™/SiC E-CVI (Reprinted from ref. [39]). (b) Hi-Nicalon™/MI-SiC CFCC Combustor Liner and Stainless Steel Counterpart (Reprinted from ref. [38]).

Figure 22.2 Combustor Can with All-Oxide CFCC Tiles (Reprinted with permission from ref. [46])

ing manufacture, and consistency in properties. Other programs in Europe have also investigated the potential use of C/C, SiC/SiC, and C/SiC CFCC combustor liners [49].

Flaps and Seals

Another popular application of CFCCs in gas turbine engines is for flaps and seals [50–54] Several CFCC systems (e.g.—CG-Nicalon™/Carbon, CG-Nicalon™/SiNCO, CG-Nicalon™/Al_2O_3 and Nextel™ 610/aluminosilicate) have been evaluated and their use has been demonstrated for these applications. While some of these CFCCs exhibited through-the-thickness cracking or excessive wear after screening tests, CG-Nicalon™/SiNCO and CG-Nicalon™/C were able to withstand successfully the conditions of this application [51].

Rotors and Blades

CFCCs have also been considered for the manufacture of rotating parts in gas turbine engines. In contrast to combustor liners or flaps, where stresses arise from thermal gradients and wear, in rotating parts the magnitude of stresses generated during service is significantly higher and proportional to the rotational speed. To optimize the resistance of CFCCs to stresses associated with high rotational speeds, special fiber architectures have been developed, which align the fibers along the direction of maximum stress (e.g.—fabrics with 2-D polar weaves for example). Cold spinning tests have shown that CFCCs are capable of withstanding the stresses associated with rotating gas turbine engine components and several prototypes of rotating disks have been fabricated using carbon and CG-Nicalon™ fibers [55–60]. For example, C/C,

C/PIP-SiC and Hi-Nicalon™/PIP-SiC rotors have been subjected to cold spinning tests and foreign object impact at speeds over 600 m/s, which correspond to the maximum speed of the blade tips [61]. C/SiC CFCC turbine rotors for the 100 kW automotive ceramic gas turbine project in Japan have withstood 90,000 rpm in cold spin tests [32]. Although the ultimate strength of many CFCCs is higher than the peak stress in these applications, the propensity of non-oxide CFCCs to environmental attack at stresses larger than the matrix cracking stress in combustion environments, and the expectancy of tens of thousands of hours of service life most likely will prevent their use for this application.

Shrouds

Components in the first stage of gas turbine engines (e.g.—shrouds) are exposed to some of the highest gas temperatures in the engine. Currently shrouds consist of coated nickel-based superalloys and the maximum capability of these materials limits the shroud temperature to below 1000°C, therefore requiring the use of cooling air. This results in substantial penalty in efficiency. CFCCs may satisfy the environmental and mechanical requirements, while at the same time substantially reducing the need for cooling air, resulting thus in improved engine efficiency. Programs in the U.S. and Japan have demonstrated the use of CFCCs for the fabrication of shrouds [32,62]. Figure 22.3 shows a SiC/MI-SiC shroud compared to its metallic counterpart.

TEST METHODS

In many instances the design and manufacture of CFCC components for gas turbine engines has followed the process illustrated in Figure 22.4. This process involves the use of standardized test methods to evaluate the physical and mechanical properties of CFCCs; the assembly of data bases with data generated through the

Figure 22.3 MI-CFCC Shroud and Metallic Counterpart (Adapted from ref. [62])

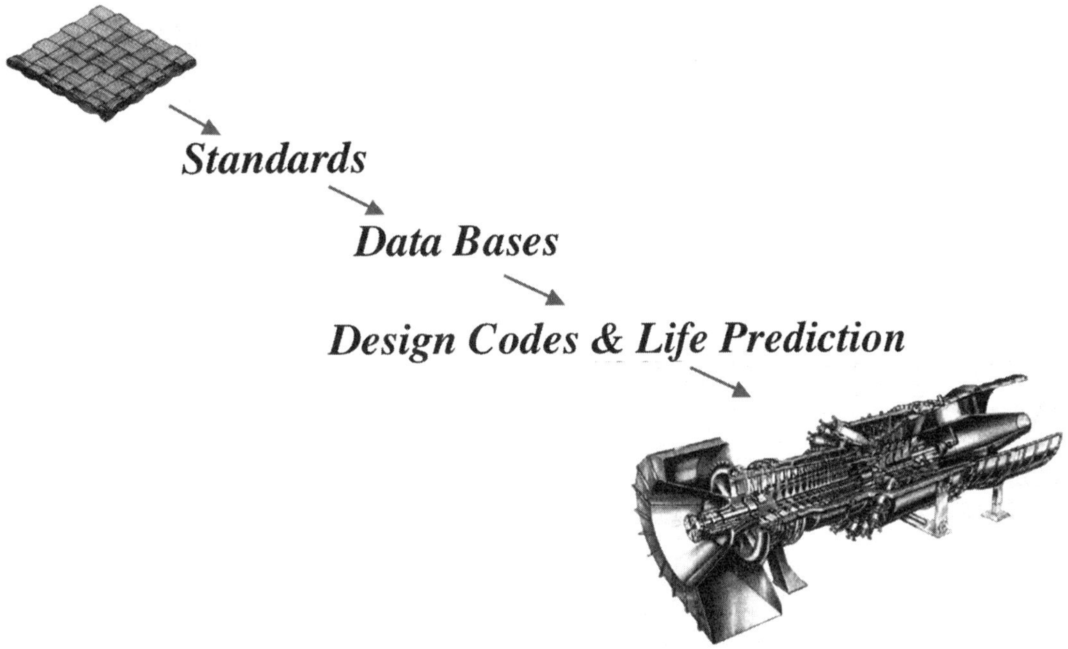

Figure 22.4 Process for the Design and Implementation of CFCCs in Gas Turbine Applications

use of full-consensus standardized test methods; and design codes. These design methodologies address the statistical nature of the strength of the CFCC constituents and are aimed at predicting their durability and reliability. Design methodologies usually start with the determination of stresses and strains through finite-element stress analyses. This requires knowledge of the properties of candidate materials for that design, which after being obtained through standardized test methods, would be available in data bases. Issues associated with fatigue, stress-rupture and environmental effects on mechanical and physical properties then can be addressed. In the U.S. an effort is currently under way to develop a design handbook for using CFCCs in gas turbines and industrial applications, as part of the DOD MILITARY HANDBOOK 17. The design methodologies contained therein depend strongly on the availability of data obtained according to full-consensus standardized test methods along with results from the evaluation of materials under special simulated or actual conditions of temperature and environment. Similar efforts are undergoing in Europe and Japan [63].

In the United States, the American Society for Testing and Materials (ASTM) has spearheaded the widespread introduction of standard test methods for advanced ceramics and ceramic composites [64,65]. In Europe, technical committee 184 of the Committee for European Normalization (CEN TC-184) is the organization responsible for the development of full-consensus test standards for CFCCs. In Japan, the Petroleum Energy Center has been one of the organizations responsible for sponsoring the preparation of several industry-accepted test standards for CFCCs [66]. Recently, international harmonization activities have resulted in the drafting of four international standards for CFCCs (tension, compression, in-plane shear and interlaminar shear) within the jurisdiction of technical committee 206 on Advanced Technical Ceramics of the International Standards Organization (ISO TC-206) [65]. Work is still needed to standardize the nomenclature related to CFCC and data reporting practices. For example, general references to "SiC/SiC composites" are meaningless considering the many forms and types of SiC-based fibers, fiber coatings, fiber architectures, and forms of SiC matrices that are currently available. Additional information on standardization of CFCCs can be found in chapter 24 of this volume.

In general, each gas turbine engine component has different material requirements. For example, combustor liners require materials with good thermal conductivity, good corrosion resistance and tolerance to foreign object impact. Materials for rotating components must have high tensile strength, high creep and fatigue resistance, whereas materials for nozzle flaps and petals must be resistant to erosion, corrosion and acoustic fatigue. In the following sections the properties of candidate CFCC materials for gas turbine applications are reviewed. Experimental results are presented for elastic constants, tensile and shear strength in various material planes, fracture resistance, performance-related behavior (e.g., resistance to mechanical fatigue, creep, stress-rupture, and thermal shock) thermal and physical properties (e.g., thermal conductivity and thermal expansion). Environmental effects (e.g., oxidation, corrosion) on mechanical performance and physical and mechanical properties of these materials are also reviewed.

MECHANICAL PROPERTIES

Figure 22.5 shows a typical tensile stress-strain curve for a 2-D CG-Nicalon™/PIP-SiNCO CFCC with a 0.1-μm thick fiber coating of BN and a fiber volume fraction of 40%. Also included in the plot is a record of the acoustic emissions generated by the material upon successive loading/unloading and reloading cycles to increasingly larger levels of stress. This stress-strain curve in Figure 22.5 exhibits well-defined regimes associated with: linear elastic behavior up to the proportional limit stress, σ_{pl}, and non-linear inelastic behavior for larger stresses. It can also be observed that the Felicity ratio, which may be used as a measure of damage and is defined as the ratio between the value of tensile stress at which acoustic emissions start upon reloading, divided by previous maximum tensile stress, decreases with increasing peak stress [67]. The hysteresis that occurs during unloading-reloading cycles relates to the magnitude of the interfacial shear stress. The three mechanisms responsible for the non-linear stress-strain behavior exhibited by many CFCCs are [68]: (1) matrix cracking; (2) changes in the residual stress distribution as a result of matrix

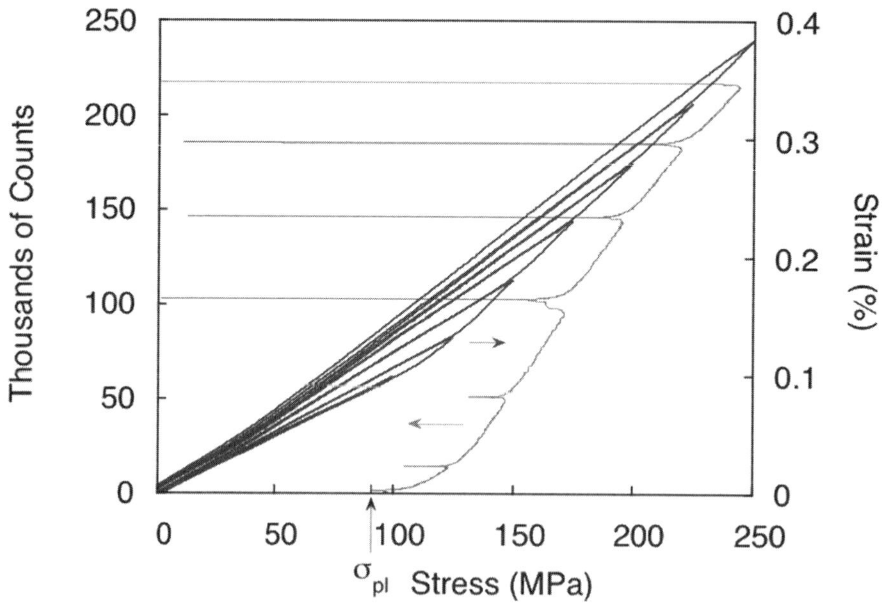

Figure 22.5 Tensile Stress-Strain Curve for CG-Nicalon™/BN/Si-N-C-O CFCC Obtained After Multiple Loading-Unloading Cycles. Also Enclosed are Acoustic Emissions Recorded During the Test. The Onset of Acoustic Emissions Indicates the Onset of Cracking in the Matrix.

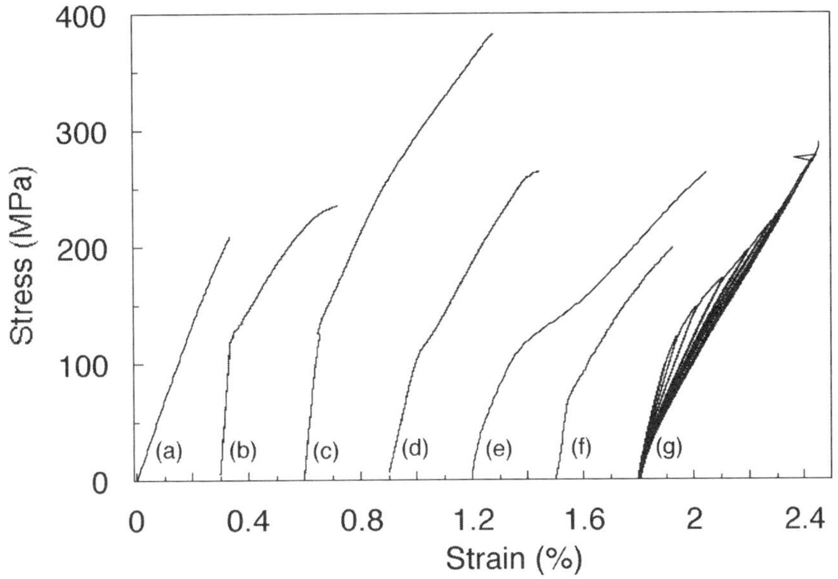

Figure 22.6 Typical Tensile Stress-Strain Curves of CFCCs. (a) 8-Harness Satin Weave Nextel 610/Sol-Gel $SiO_2Al_2O_3$ (b) 8-Harness Satin Weave Hi-Nicalon™/CVI SiC with Multilayered C/SiC Interfacial Coating (c) 5-Harness Satin Weave Hi-Nicalon™/CVI SiC with BN Interfacial Coating (d) 8-Harness Satin Weave CG-Nicalon™/ Polymer Impregnation and Pyrolysis Si-N-C-O with BN Interfacial Coating (e) 5-Harness Satin Weave CG-Nicalon™/ Directed Metal Oxidation Al_2O_3 with Dual BN/SiC Interfacial Coating (f) Plain Weave CG-Nicalon™/CVI SiC with Carbon Interfacial Coating (g) Plain Weave CG-Nicalon™/BMAS Glass-Ceramic with Dual BN-SiC Interfacial Coating.

cracking; (3) frictional dissipation that occurs at the fiber/matrix interfaces and that is dominated by the fiber coating, the fiber morphology and the topography of the fiber's surface. By varying the magnitude of the interfacial shear stress, the prevalent damage mechanism and the resultant non-linear stress-strain behavior can be dramatically modified. Figures 22.6 and 22.7 show a collection of tensile stress-strain curves for various types of CFCCs. Note that the shape of the curves is affected by the type of fiber, matrix and fiber architecture. The stress-strain curves obtained from the tensile evaluation of 0/90 oxide-oxide composites exhibit very slight nonlinearity that increases with increasing stress. In contrast, the stress-strain curves obtained from the evaluation of all-oxide composites with ±45° fiber architecture exhibit lower elastic modulus and ultimate strength, but greater capacity to inelastic straining (Figure 22.7).

Elastic Properties

Various techniques have been used to determine the elastic properties of CFCCs. These include the recording of stresses and strains within the elastic range during mechanical testing, and ultrasonic and small-amplitude vibration methods. Ultrasonic techniques have the advantage of being non-destructive and by aligning a particular direction of the material with the direction of propagation of the ultrasonic pulse, the elastic property of the material in that direction can be obtained. Ultrasonic techniques can be applied at elevated temperatures and therefore can be used to detect changes that may result from reactions between the material and the environment. For example, when CFCCs with SiC-based fibers and carbonaceous fiber coatings are exposed to air at elevated temperatures, the fiber coating will oxidize, resulting in loss of stiffness. This initial loss in stiffness is followed by an increase in stiffness as a result of the formation

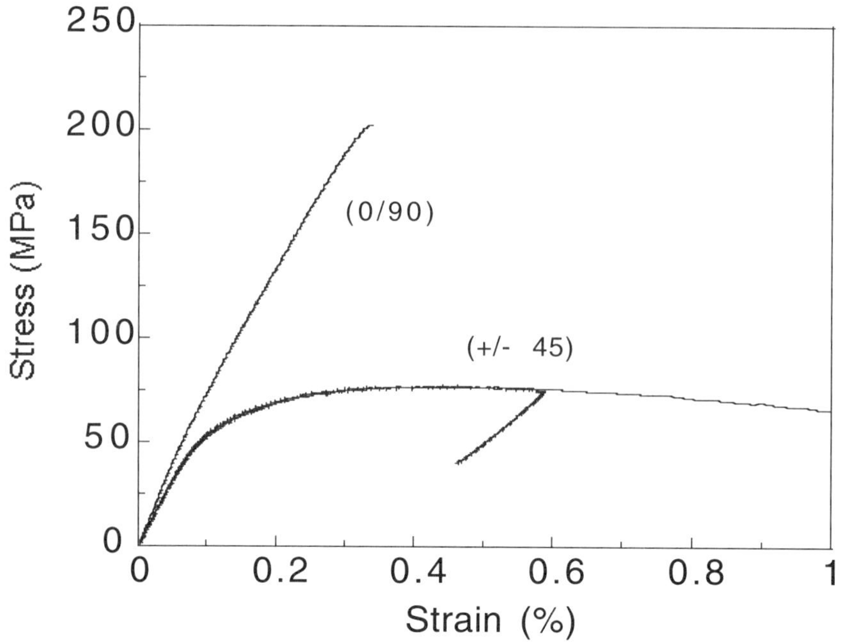

Figure 22.7 Tensile Stress-Strain Curves for Nextel™ 720/AS with (0/90) or (±45°) Fiber Orientation. Two Curves are Presented for the Latter to Illustrate the Reproducibility of Response

of silica on the surface of the fiber, and the filling of the gap previously occupied by the fiber coating with silica [69, 70]. Table 22.1 lists the elastic properties of several CFCCs along with the technique used for their determination.

In-plane Tensile Behavior

In 1999 a round-robin testing program was completed in the U.S. to determine the precision of ASTM test standard C1275 "Standard Test Method for Monotonic Tensile Strength Testing of Continuous Fiber-Reinforced Advanced Ceramics with Solid Rectangular Cross-Section Specimens at Ambient Temperatures" [82]. From this study coefficients of variation for repeatability and reproducibility were obtained for elastic modulus, proportional limit stress, ultimate tensile strength and strain at fracture of CG-Nicalon™/PIP SiNCO. However, no statement of bias was given because no acceptable reference standard material exists at this time. The elastic properties obtained for this CFCC as part of this study are listed in Table 22.1. Figure 22.8 shows a summary of the tensile strength results obtained by the nine laboratories that participated in the study, with each laboratory testing ten tensile specimens. The plot shows the great mean value and one standard deviation with respect to this value.

Limited studies have been carried out to assess the effect of specimen geometry, mode of loading (force versus displacement control), and rate of testing on the tensile strength and stress-strain behavior of CG-Nicalon™/CVI-SiC CFCCs [83]. It was found that for the range of values investigated, neither the test mode (displacement vs. force control) nor test rate (50 N/s vs. 500 N/s; 0.003 mm/s vs. 0.03 mm/s) had an effect on the proportional limit stress or ultimate tensile strength of the material. Although the geometry of the test specimens did not have an effect on the proportional limit stress, it was found that contoured tensile specimens with the smallest gauge length, for the range of values investigated, had the largest tensile strength. The in-plane tensile strength of CFCCs is size-dependent and this effect needs to be accounted for when designing components that are of different size from the standardized test specimens used for obtaining data [84,85].

TABLE 22.1 ELASTIC PROPERTIES OF CFCCs

Fiber	Matrix	V_f (%)	Fiber Architecture	Interface	Density (g/cm^3)	Porosity (%)	Technique	E (GPa)	G (GPa)	ν	Ref
PAN-carbon	CVI-SiC	40	2D PW	1 μm PyC	2.0	10–15	ultrasonic	118 (4)			[71]
								115 (12)			[72]
CG-Nicalon™	CVI-SiC					14	ultrasonic	248[1]			[73]
						30		130[1]			
C	CVI-SiC	40	2D PW		2.0	10–15	tensile testing	140 (20°C)			[74]
								145 (500°C)			
								135 (1000°C)			
								152 (1200°C)			
								144 (1400°C)			
								120 (1600°C)			
CG-Nicalon™	CVI-SiC + oxidation inhibitors	40	2D PW	PyC	2.58	9.7		234.9–235.4			[70]
CG-Nicalon™	CVI-SiC	33	8 plies 8HSW 0/90	PyC	2.54		tensile testing	240.6 (14.5)	75.8 (25.5)	0.19 (0.07)	[75]
			8 plies 8HSW 0/90±45					219.3 (20)	92.4 (21.4)	0.19 (0.04)	
			8 plies 8 HSW ±45					182.7 (24.1)	101.4 (15.2)	0.20 (0.02)	
CG-Nicalon™	CVI-SiC	41	5HSW with angle interlock between layers	0.5–1.0 PyC		10–15		237 (20°C)			[76]
								258 (1200°C)			
								60 (1550°C)			
T300	CVI-SiC		0/90				tensile testing	76			[77]
			0/±45/90					57			
			±45					56			
CG-Nicalon™	CVI-SiC		0/90	PyC			tensile testing	211			[77]
			0/±45/90					220			
			±45					210			
Tyranno™	CVI-SiC							190–210			[78]
Carbon	CVI-SiC							90–100			
Carbon	PIP-SiC							60–80			
Nextel™ 610	porous alumina-mullite	37–42	8HSW (0/90)	—		23.7–24.9	tensile testing	96		.05[2]	[79]
			8HSW (±45)					50		.55[2]	
CG-Nicalon™	PIP Si-N-C-O	42–48	8HSW (0/90)	0.5 μm BN			tensile testing	90.7 (3.7)	33 (2)	0.12 (0.01)	[80]
Nextel™ 720	AS	48	8HSW (0/90)	—	2.6			78			[81]
								72 (980°C)			
CG-Nicalon™	CVI-SiC						ultrasonic	235	78.8[3]		[70]
									68.3[4]		

Notes: 1–In-plane modulus; 2–ν_{12}; 3–Gzx; 4–Gxy, Gyz.

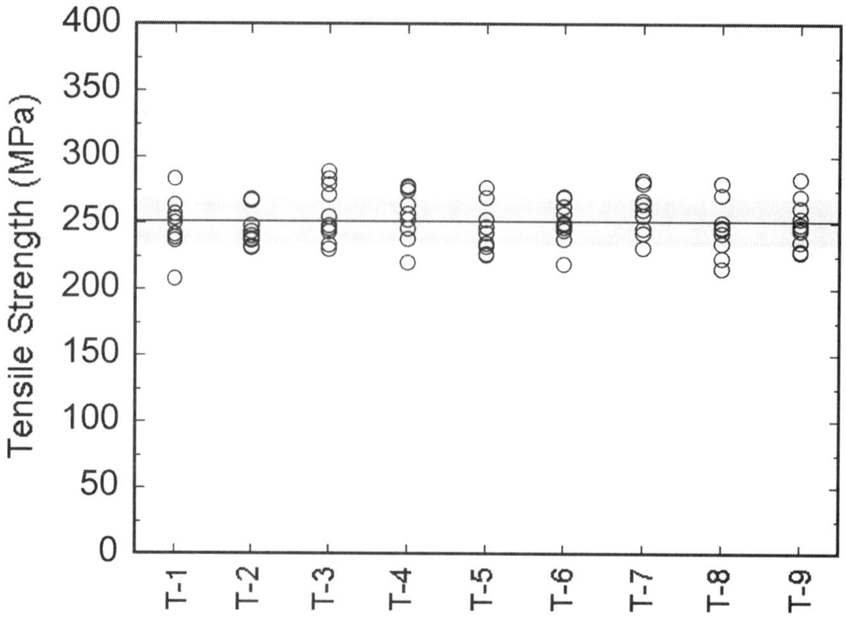

Figure 22.8 Summary of Tensile Strength Results for CG-Nicalon™/BN/SiNCO CFCC Obtained from Round-Robin Testing Program [80]

Figures 22.9 and 22.10 present tensile strength results for several continuous fiber-reinforced ceramic matrix composites as a function of temperature. Included in these graphs are data for composites reinforced with CG-Nicalon™ fibers, Hi-Nicalon™ fibers, T-300 graphite fibers, Nextel™ 610 and Nextel™ 720 fibers, Sumitomo® alumina-silica fibers, Bestfight™ carbon fibers, and Si-Ti-C-O Tyranno fibers. For composites with comparable fiber concentration and fiber architecture, CVI-SiC matrix composites reinforced with graphite fibers exhibit the highest tensile strength, and in inert environment, the best retention of strength at elevated temperatures. The effect of fiber stiffness (200 GPa for CG-Nicalon™ versus 270 GPa for Hi-Nicalon™) on the stiffness and on the magnitude of the proportional limit stress of CFCCs is evident in the tensile stress-strain curves presented in Figure 22.11 for SiNCO matrix CFCCs reinforced with CG-Nicalon™ and Hi-Nicalon™ fibers.

Even at room temperature, but particularly at elevated temperatures, the stress-strain behavior of CFCCs, including the magnitude of the matrix cracking stress and ultimate tensile strength, depend on the loading rate and loading mode [89–91]. At room temperature, the increase in the ultimate tensile strength of CFCCs can be attributed to an increase in interfacial shear stress with loading rate, whereas the load rate-dependence of the matrix cracking stress that has been observed for glass-ceramic matrix composites can be related to stress-corrosion cracking [90]. At high temperatures, as the loading rate increases—under force control—the ultimate tensile strength increases, but the opposite trend is observed when tensile tests are carried out under a constant displacement rate. These results suggest that the time-dependent deformation (e.g.—creep) and time-dependent redistribution of internal stresses in CFCCs [92] play a significant role. Figures 22.12–22.14 show results that illustrate the effect of loading rate on the ultimate tensile strength of CG-Nicalon™ fiber-reinforced CFCCs at high temperatures.

Transthickness Tensile Behavior

Because of the availability of ceramic fibers in tows and fabric form, most fiber preforms consist of stacked layers of woven fabric. As a consequence of the resulting laminated structure, 1-D and 2-D CFCCs exhibit large anisotropy between their in-plane and through-the-thickness physical and mechanical properties.

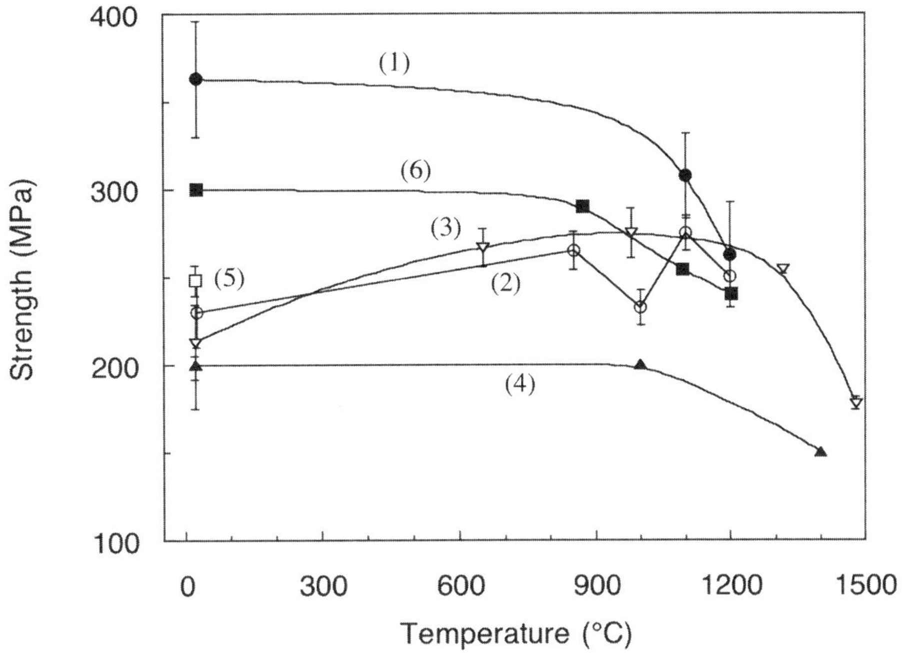

Figure 22.9 Tensile Strength Results for CFCCs: (1) Hi-Nicalon™/C/CVI-SiC [77]; (2) CG-Nicalon™/C/CVI-SiC [77]; (3) CG-Nicalon™/C/CVI-SiC [87]; (4) CG-Nicalon™/C/CVI-SiC [86]; (5) CG-Nicalon™/C/CVI-SiC [75]; (6) Hi-Nicalon™/BN/MI-SiC [38].

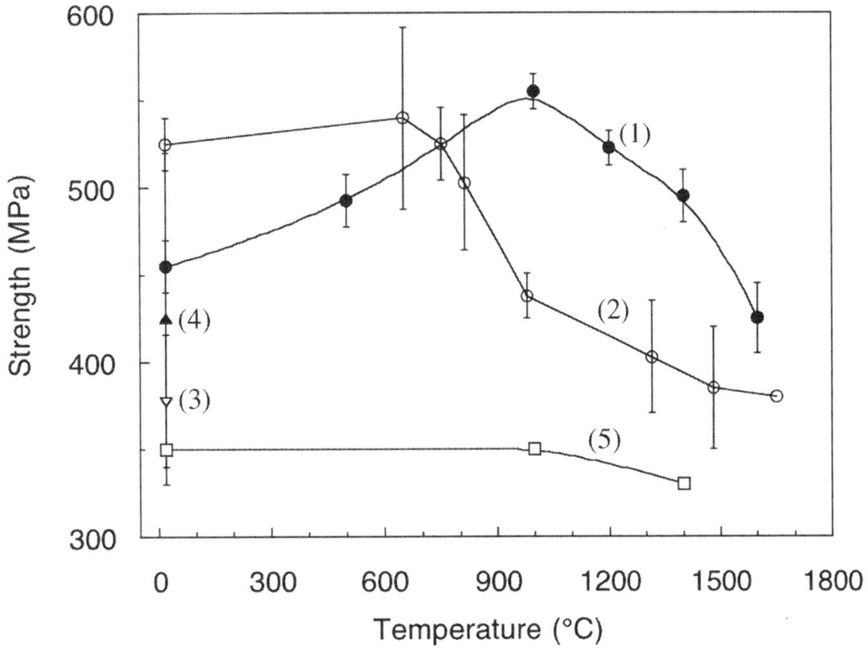

Figure 22.10 Tensile Strength Results for Carbon Fiber-Reinforced CVI-SiC. (1) [74]; (2) [88]; (3) [87]; (4) [77]; (5) [86].

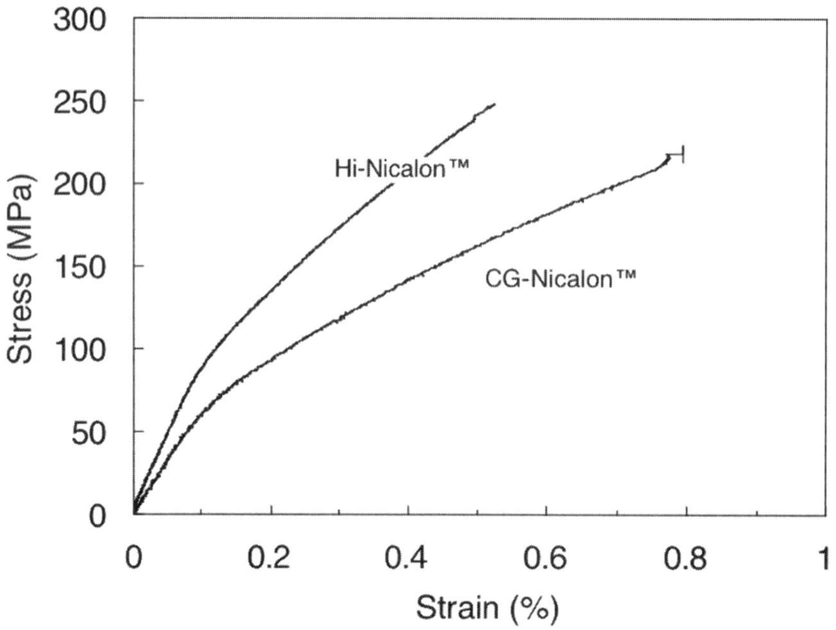

Figure 22.11 Effect of Fiber Type (CG-Nicalon™ versus Hi-Nicalon™) on the Monotonic Tensile Behavior of SiNCO Matrix CFCCs in Air at 1200°C

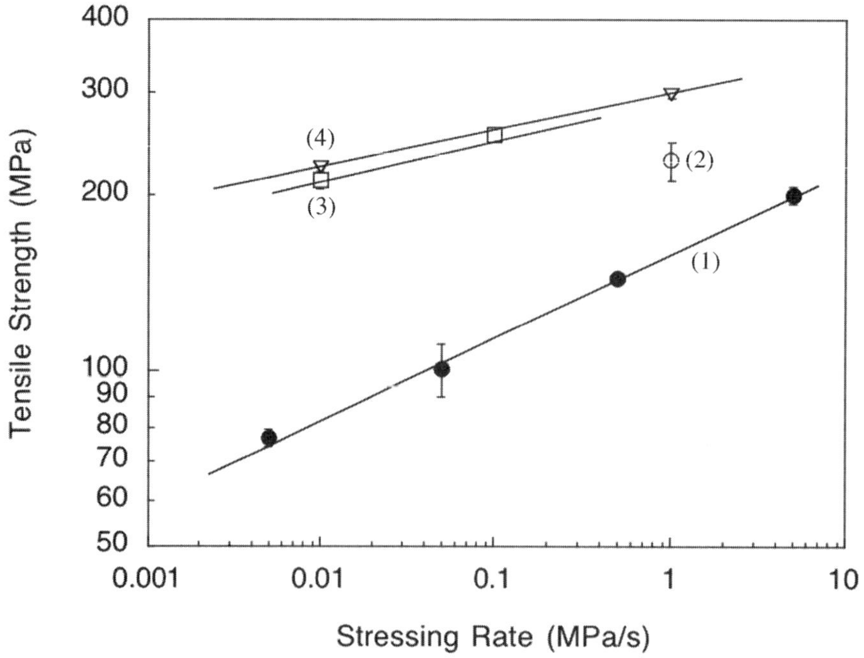

Figure 22.12 Effect of Loading Stress Rate and Temperature on the Tensile Strength of CFCCs: (1) CG-Nicalon™/C/CVI-SiC at 1200°C [89]. Open Symbols for CG-Nicalon™/BN/SiNCO [91]: (2) 23°C; (3) 1000°C; (4) 1200°C.

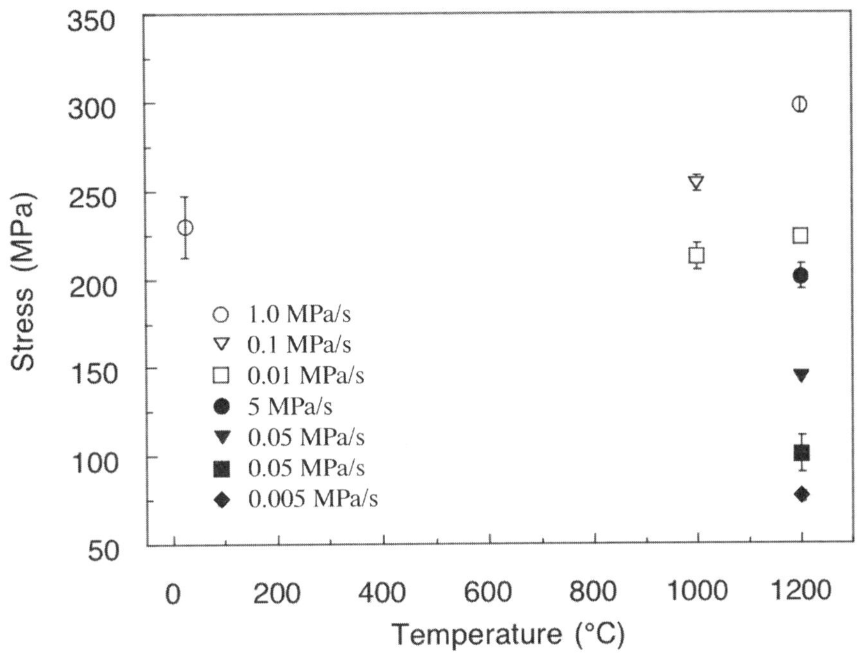

Figure 22.13 Effect of Loading Stress Rate and Temperature on the Tensile Strength of CFCCs: Open Symbols for CG-Nicalon™/BN/SiNCO [91] and Close Symbols for CG-Nicalon™/C/CVI-SiC [89].

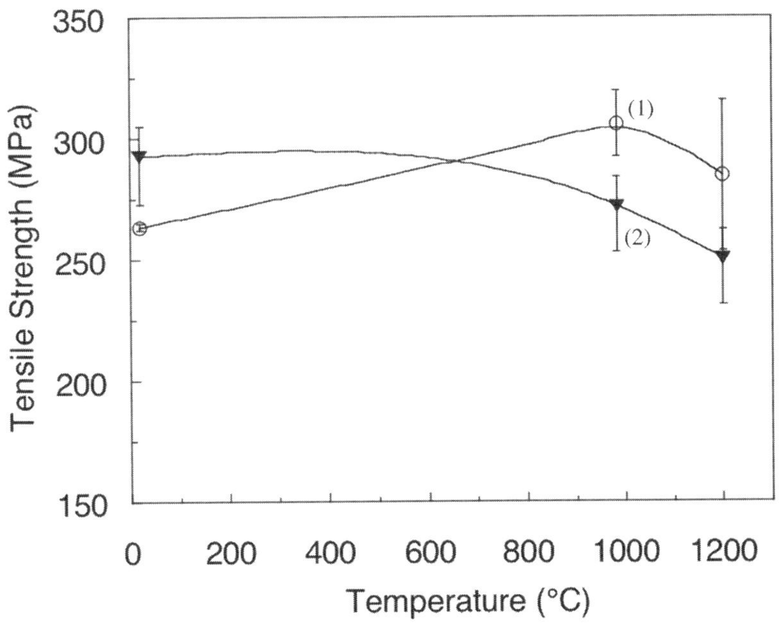

Figure 22.14 Effect of Temperature and Cross-Head Displacement Rate on the Tensile Strength of CG-Nicalon™/BN/SiNCO: (1) 25 μm/min and (2) 25 mm/min [91].

These differences have been accentuated due to efforts to optimize the in-plane tensile properties of CFCCs, which have occurred at the expense of the interlaminar and transthickness properties. In laminated CFCCs, the interlaminar shear and transthickness tensile modes of failure are dominated either by the properties of the matrix or by those of the fiber/matrix interface.

The optimization of the in-plane tensile properties of CFCCs has been achieved through the use of "weak" fiber-matrix interfaces, which promote crack deflection and fiber debonding and sliding. However, these "weak" interfaces become the weakest link for the propagation of interlaminar and transthickness cracks when CFCCs are subjected to tensile transthickness or shear loading. Figure 22.15 shows a series of stress-strain curves obtained from the transthickness tensile evaluation of unidirectional Hi-Nicalon™/CVI-SiC matrix composites with pyrocarbon fiber coating, in a direction normal to the fiber direction. The specimens were square in size (10 mm × 10 mm) and 2.3 mm thick. In this case, failure occurred along a plane that coincides with the fiber/matrix interface at a stress level lower than the matrix cracking stress associated with the onset of non-linear behavior during tensile loading. For this material, the ratio of the transthickness tensile strength to the tensile strength along the fiber direction was found to be 0.025, which illustrates the large degree of anisotropy in the strength of the material. Figure 22.16 summarizes transthickness tensile strength results for various CFCCs obtained from the evaluation of circular test specimens of diameters between 15 and 25 mm, and square test specimens of size between 10 and 17 mm [93,94].

A new test method is currently being considered for standardization within ASTM's sub-committee C28.07 on Ceramic Matrix Composites to determine the transthickness tensile strength of 1-D and 2-D CFCCs. This test method is based on the diametral compression of circular specimens as illustrated in Figure 22.17. The advantage of this test method is that it can be carried out at elevated temperatures, in contrast to the currently accepted ASTM Test Method C1468-00 "Standard Test Method for Transthickness Tensile Strength of Continuous Fiber-Reinforced Advanced Ceramics at Ambient Temperature," [95] which is limited by the properties of the adhesive used to transfer the load to the specimen.

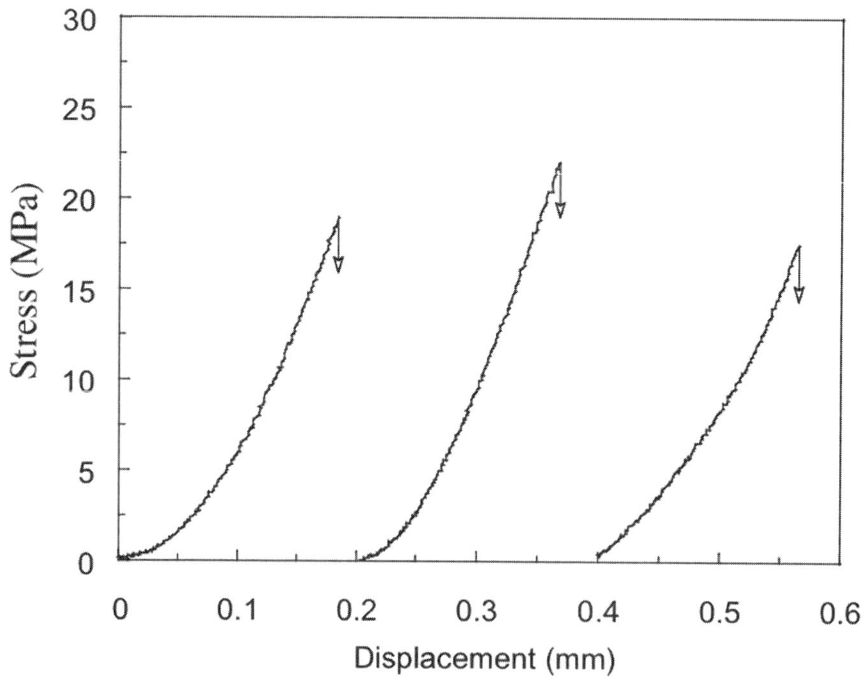

Figure 22.15 Typical Transthickness Stress versus Cross-Displacement Curve for 1-D Hi-Nicalon™/C/CVI-SiC [91]

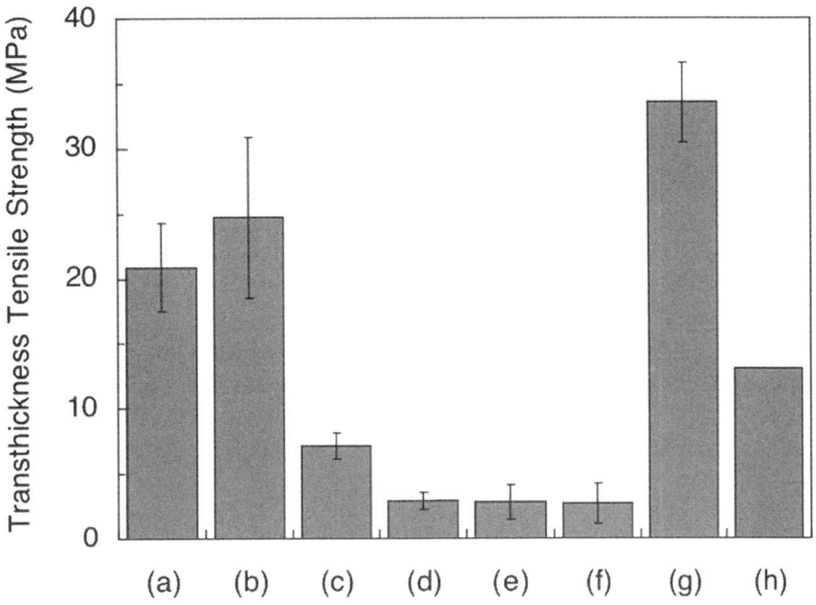

Figure 22.16 Transthickness Tensile Strength for 1-D and 2-D CFCCs. (a) CG-Nicalon™/MAS-5; (b) CG-Nicalon™/BN/SiNCO; (c) Nextel™610/AS (GE); (d) Nextel™610/AS (3M); (e) Nextel™720/AS (GE); (f) Nextel™720/AS (COI); (g) SCS-6/Si_3N_4 (Unidirectional) [93]; and (h) CG-Nicalon™/C/CVI-SiC [94]

Shear Properties

The matrix cracking that occurs in 2-D CFCCs when subjected to shear loading depends on the loading orientation and the properties of the matrix. For engineering applications, two dominant shear loading orientations are of interest: in-plane shear and interlaminar shear. For the case of interlaminar shear, matrix cracks evolve without significant interaction with the fibers. Conversely, for in-plane shear loading, matrix

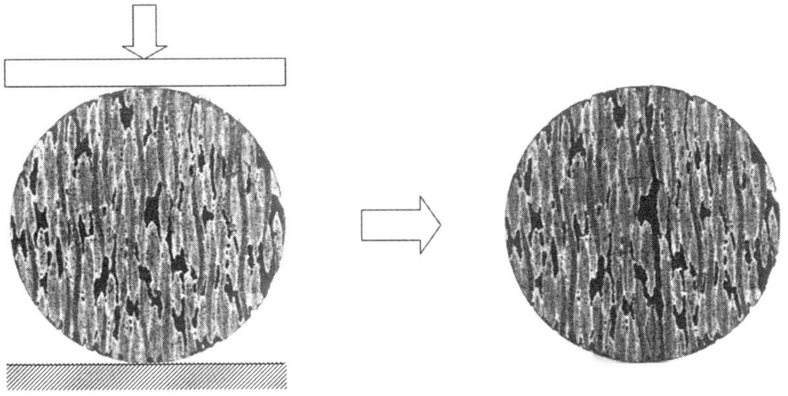

Figure 22.17 Schematic of Brazilian Test (Diametral Compression) to Determine the Transthickness Tensile Strength of CFCCs. CG-Nicalon™/CVI-SiC CFCC Specimen (a) Before and (b) After Testing

cracks must interact with the fibers which impede crack development. Consequently, the in-plane shear strength always exceeds the interlaminar shear strength [68].

Analysis of experimental shear strength data indicates that the matrix has a major influence on the shear strength and the shear ductility of continuous fiber-reinforced ceramic matrix composites [68]. Among the various matrices available, CVI-SiC matrix composites exhibit some of the largest in-plane shear strengths among CFCCs. The in-plane shear strength of CFCCs can be ranked using a parameter given by

$$W = \sqrt{\frac{\Gamma_m}{rG}} \tag{1}$$

where Γ_m is the fracture energy of the matrix, G is the shear modulus of the matrix, and r is the fiber radius.

ASTM standard test method C1292 "Standard Test Method for Shear Strength of Continuous Fiber-Reinforced Ceramics at Room Temperature" was first introduced in 1996 to determine the in-plane and interlaminar shear strength of CFCCs based on the Iosipescu test method, and interlaminar shear strength by the compression of double-notched specimens. Figure 22.18 shows a photograph of an Iosipescu test fixture with specimens for in-plane and interlaminar shear strength determination, along with schematics of the Iosipescu and compression of double-notched specimen test methods.

In 1999 a round-robin testing program was completed in the U.S. to determine the precision of ASTM test standard C1292 [80]. From this study coefficients of variation for repeatability and reproducibility were obtained for in-plane and interlaminar shear strength of CG-Nicalon/PIP SiNCO. Figures 22.19 and 22.20 summarize the results obtained by the various laboratories that participated in this testing program. These plots also

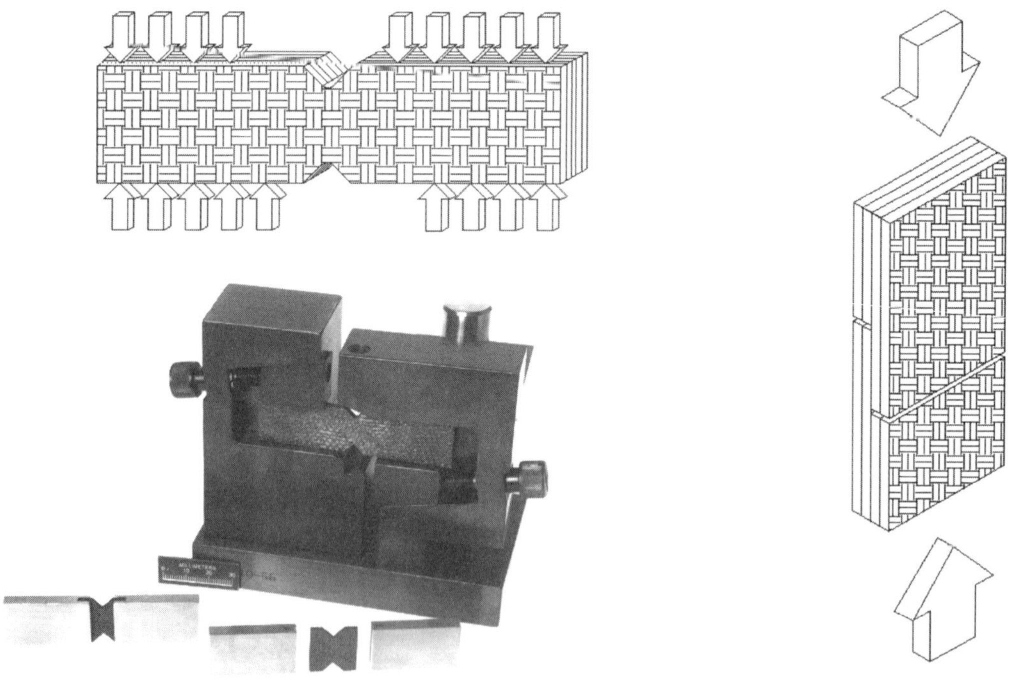

(a) (b)

Figure 22.18 (a) Schematic and Fixture for Iosipescu Test for Determination of In-Plane and Interlaminar Shear Strength. (b) Schematic of Compression of Double-Notched Specimen for Determination of Interlaminar Shear Strength.

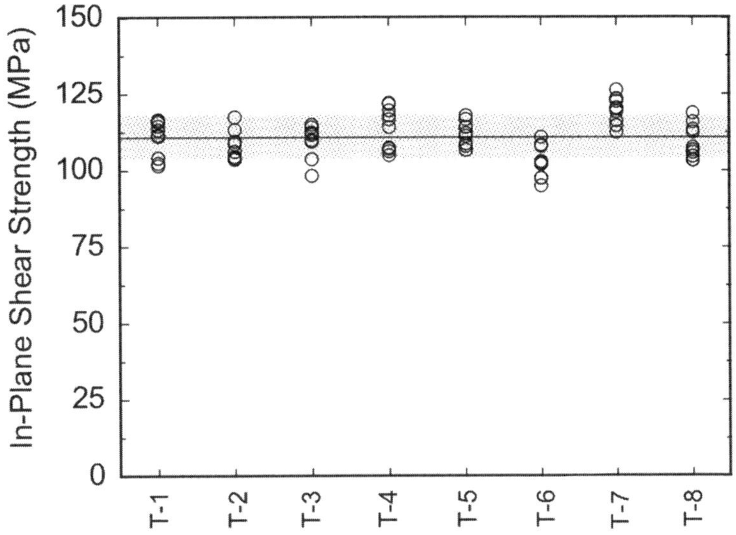

Figure 22.19 Summary of In-Plane Shear Strength Results for CG-Nicalon™/BN/SiNCO CFCC Obtained from Round-Robin Testing Program by Iosipescu Test Method. Notch Separation Was 11 mm [80].

show the value of the great mean and one standard deviation with respect to that value. Although the in-plane shear strength results show little variability, the interlaminar shear strength results exhibited significant variability. This large degree of variability is associated with the variation in the microstructure of the material and with the relative location of the roots of the notches with respect to the interlaminar region of the material. Figure 22.21 shows photographs of a double-notched specimen, used for the determination of interlaminar shear

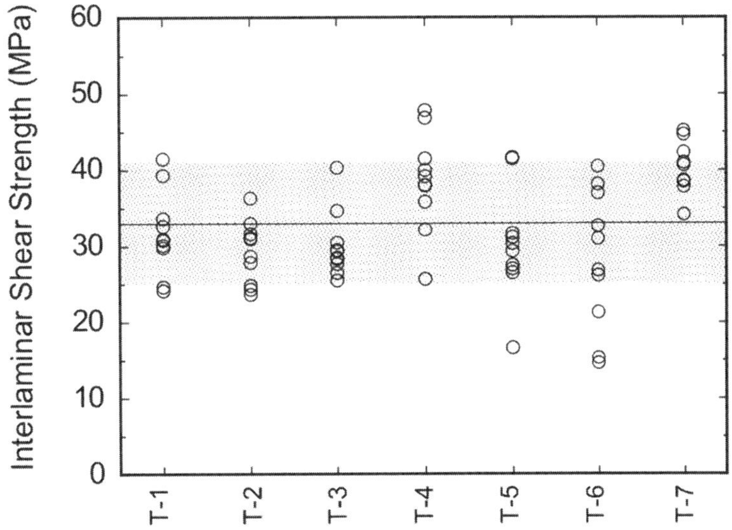

Figure 22.20 Summary of Apparent Interlaminar Shear Strength Results for CG-Nicalon™/BN/SiNCO CFCC Obtained from Round-Robin Testing Program by Compression of Double-Notched Specimens. Notch Separation Was 6 mm [80].

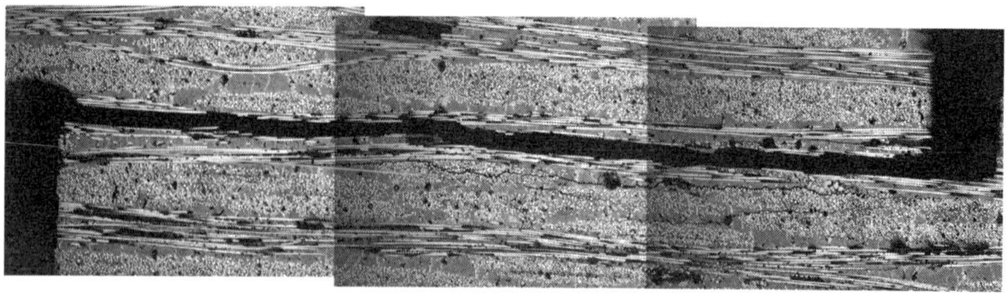

(a)

(b)

Figure 22.21 (a) Micrograph of CG-Nicalon™/BN/SiNCO CFCC Double-Notched Specimen After Test Illustrating the Propagation of Interlaminar Crack between Notches. (b) Micrograph of CG-Nicalon™/BN/SiNCO CFCC Iosipescu Specimen After Test [96].

strength, and of an Iosipescu specimen used for the determination of in-plane shear strength, after testing. Figures 22.22 and 22.23 contain interlaminar shear strength results for various CFCCs at ambient temperature as a function of notch separation and elevated temperature, respectively, while Tables 22.2 and 22.3 list additional results. In the case of Hi-Nicalon™/BN/MI-SiC CFCCs, the decrease of interlaminar shear strength with temperature shown in Figure 22.23 is associated with degradation of the fiber/matrix interfacial region [99].

Fracture Resistance

The main reason why CFCCs are preferred over monolithic ceramics for use in structural applications is the ability of the former to redistribute stresses around notches [101]. Failure mechanisms and processes in CFCCs during stable fracture are not merely confined to a localized region at the crack tip but include crack bridging, crack deflection, multiple matrix cracking and fiber pull-out over a larger fracture process zone [102, 103]. As a consequence, the nature of the fracture process in these materials precludes the application of linear elastic fracture mechanics (LEFM). In spite of this, many authors have made, and continue making use of existing LEFM formalisms for the analysis of fracture resistance of CFCCs. Others have made use of concepts such as equivalent elastic crack and R-curves, which characterize a material's resistance to fracture during slow, stable crack propagation and are dependent on the nature of the process zone extending from the crack tip in the material. Table 22.4 lists fracture resistance parameters obtained from the evaluation of various CFCCs.

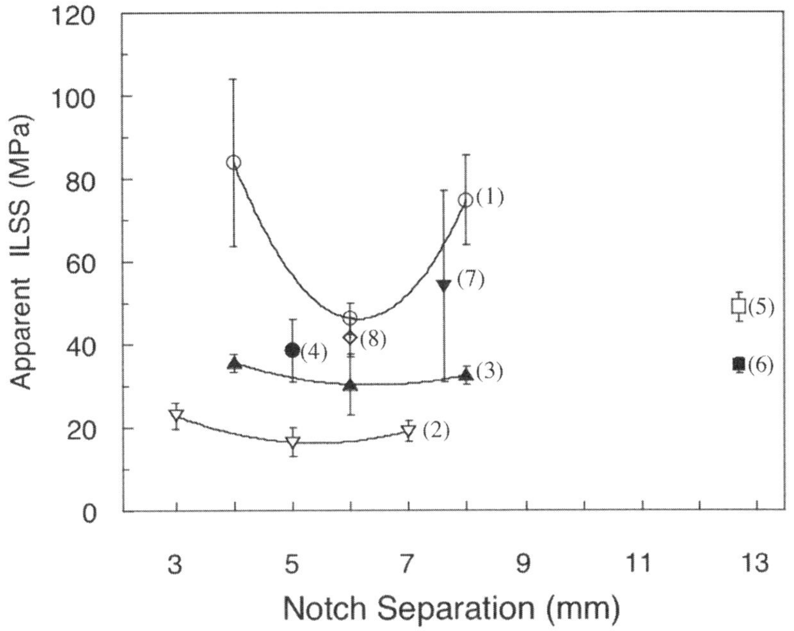

Figure 22.22 Effect of Notch Seperation on the Interlaminar Shear Strength of 2-D CVI-SiC Matrix Composites: (1) CG-Nicalon™, 0.3 μm Carbon Fiber Coating [97]. (2) CG-Nicalon™, 1.1 μm Carbon Fiber Coating [97]. (3) CG-Nicalon™, Boron Nitride Fiber Coating, SiNCO Matrix [96]. (4) CG-Nicalon™ [75]. (5) CG-Nicalon™, 0.1 μm Carbonaceous Fiber Coating [88]. (6) T300 Carbon Fiber 0.1 μm Carbonaceous fiber coating [88]. (7) CG-Nicalon™, 0.1 μm Carbonaceous Fiber Coating [98]. (8) Hi-Nicalon™, Dual Boron Nitride-CVD-SiC Fiber Coating, MI-SiC Matrix [99].

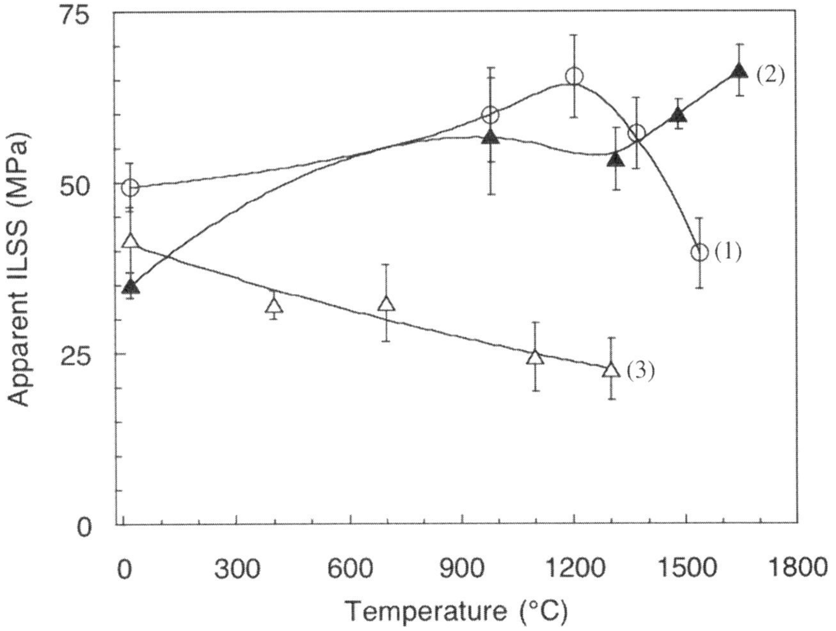

Figure 22.23 Effect of Temperature on the Interlaminar Shear Strength of: (1) CG-Nicalon™/C/CVI-SiC in Helium. (2) T300 carbon/C/CVI-SiC in Helium [88]. (3) Hi-Nicalon™/BN/MI-SiC in Air [99].

TABLE 22.2 IN-PLANE SHEAR STRENGTH OF CFCCs

Fiber	Matrix	Interface	V_f (%)	Fiber Architecture	Method	Porosity (%)	Density (g/cm^3)	Shear Strength (MPa)	Proportional Limit Stress (MPa)	Failure Shear Strain (%)	Reference
CG-Nicalon™	CVI-SiC	0.3 μm PyC	40	PW (0/90)	Iosipescu	15–20		101 (4.7)*			[100]
CG-Nicalon™	PIP SiNCO	0.5 μm BN	42–48	8HSW (0/90)	Iosipescu			110.8 (6.8)			[80]
CG-Nicalon™	CVI-SiC	PyC	33	8 plies 8HSW (0/90)	Iosipescu		2.54	95.8 (11.0)	39.3 (6.2)	0.62 (0.05)	[75]
				8 plies 8HSW (0/90/±45)			2.54	114.5 (11.7)	40.7 (9.)	0.55 (0.03)	
				8 plies 8HSW (±45)			2.54	124.1 (8.3)	44.1 (9.0)	0.50 (0.02)	
Nextel™ 720	AS		48				2.6	30.9		0.49	[81]

* Specimens 76 mm long, 19 mm wide and the separation between notches was 11 mm.

TABLE 22.3 INTERLAMINAR-PLANE SHEAR STRENGTH OF CFCCs

Fiber	Matrix	Interface	V_f (%)	Fiber Architecture	Method	Porosity (%)	Density (g/cm^3)	Shear Strength (MPa)	Reference
CG-Nicalon™	CVI-SiC	0.3 μm PyC	40	12 layers PW (0/90)	DNS	15-20			[91]
Nicalon™	PIP Si-N-C-O	0.5 μm BN	42–48	8HSW (0/90)	DNS			33.0 (8.0)**	[80]
CG-Nicalon™	CVI-SiC	PyC	33	8 plies 8HSW (0/90)	DNS		2.54	38.6 (7.6)*	[75]
				8 plies 8HSW (0/90/±45)				40.7 (5.5)	[75]
				8 plies 8HSW (±45)				41.4 (9.0)	
Nextel™ 720	AS		48				2.6	11.5	[81]
T300	CVI-SiC	PyC	45				2.1	19–35***	[98]
CG-Nicalon™			40				2.5	32–77***	
Hi-Nicalon™	MI-SiC	BN	40	8HSW 0/90	DNS				[99]

* compression of double-notched specimens (50 mm × 20 mm) with a notch separation of 5 mm using an anti-buckling fixture.
** Round robin testing program. 80 tests by 8 laboratories using double notched specimens
***The dimensions of the specimens were: 20.3 mm in length and 10.15 mm in width, while the notch separation was 7.62 mm.

TABLE 22.4 FRACTURE BEHAVIOR OF CFCCs

Fiber	Matrix	Interphase	V_f (%)	Fiber Architecture	Method	Porosity (%)	Density (g/cm^3)	Fracture Parameter	Reference
CG-Nicalon™	CVI-SiC	PyC		PW 0/90	Chevron-notched bend specimens			2.6–2.8 kJ/m^3	[103]
CG-Nicalon™	CVI-SiC	PyC			SENB			30–35 MPa.m$^{0.5}$	[104]
CG-Nicalon™	CVI-SiC	PyC	32	3-D braided	3-pt SENB ASTM E399	11	2.6	29.8 (2.5)[1] MPa.m$^{0.5}$	[105]
								18.1 (0.7)[2‡] MPa.m$^{0.5}$	
								13.3 (1.0)[1] MPa.m$^{0.5}$	
								10.0 (1.2)[2‡] MPa.m$^{0.5}$	
			35	PW 0/30/60		15	2.3	16.2 (1.7)[1] MPa.m$^{0.5}$	
								11.9 (0.7)[2‡] MPa.m$^{0.5}$	
								8.2 (0.5)[1] MPa.m$^{0.5}$	
								6.8 (1.0)[2‡] MPa.m$^{0.5}$	
CG-Nicalon™	CVI-SiC	PyC			SENT			22.1 (33.5) MPa.m$^{0.5}$	[106]
T300	CVI-SiC	PyC		3D	3-pt SENB	15–20		11.4–16.5*	[107]
T300	CVI-SiC		40	PW 0/90	ASTM E-813		2.6	30.9	[108]
CG-Nicalon™ **	CVI-SiC	ML C/SiC		PW 0/90	CT J-Integral			12–30 kJ/m^2	[109]
CG-Nicalon™ **	CVI-SiC				CT			10	[109]
CG-Nicalon™	CVI-SiC	PyC	33	8 plies 8HSW (0/090)	SENB		2.54	27.7 (1.0) MPa.m$^{0.5}$	[75]
				8 plies 8HSW (0/90/±45)				29.9 (1.1) MPa.m$^{0.5}$	
				8 plies 8HSW (±45)				27.6 (0.6) MPa.m$^{0.5}$	
CG-Nicalon™	CVI-SiC	PyC	40	11 plies PW (0/90)	SENT		2.58	14	[110]
				11 plies PW (±45)			2.58	14	
CG-Nicalon™	CVI-SiC	PyC		8 plies PW (0/90)	CT	10		14-27***	[111]
								12–18	[112]

1. In 1-3 plane
2. In 1-2 plane
‡ After heat treatment at 1200°C for 100 hours in air.
* Specimens 76 mm long, 19 mm wide and the separation between notches was 11 mm.
** Proprietary fiber surface treatment
*** 1200°C in air.

The resistance of CFCCs to crack propagation can also be evaluated by the notch sensitivity test. In this case, test specimens with notches or holes of different depth or diameter, are evaluated in tension and the results are compared with those of test specimens without notches or holes. The tensile strength values as a function of notch depth or hole diameter allow the determination of the range of notch sizes for which the material is sensitive to the presence of notches. Figure 22.24 contains a photograph of a Nextel™720/AS specimen with a hole, which was used as part of a study to assess the notch sensitivity of this material. Figure 22.25 contains a plot of normalized strength as a function of the ratio of hole diameter (or notch length) to specimen width for CG-Nicalon™/CVI-SiC and Nextel™610/AS. The line associated with notch insensitivity is also included in that plot. These results suggest that both of these CFCCs exhibit mild notch sensitivity and that in the case of the all-oxide CFCCs, it becomes more notch-sensitive at high-temperature. Figure 22.26 contains a plot of tensile strength results versus the ratio of hole diameter to width, for Nextel™720/AS CFCCs. These results suggest that this CFCC is effectively notch insensitive in the as-processed condition and that its strength exhibits a mild dependence on volume (specimen width). However, after an ageing treatment at 1100°C for 100 hrs the strength of this material decreases and the material becomes notch sensitive. Figure 22.27 presents results of the net section strength versus hole diameter for a Nextel™610/mullite-alumina CFCC, which suggest that this CFCC exhibits mild notch sensitivity and modest dependence of strength with hole size [79].

The evaluation of notched test specimens is also of interest because some gas turbine engine components (e.g.—combustor liners, flaps and seals) contain cooling holes and bolted attachment points [115].

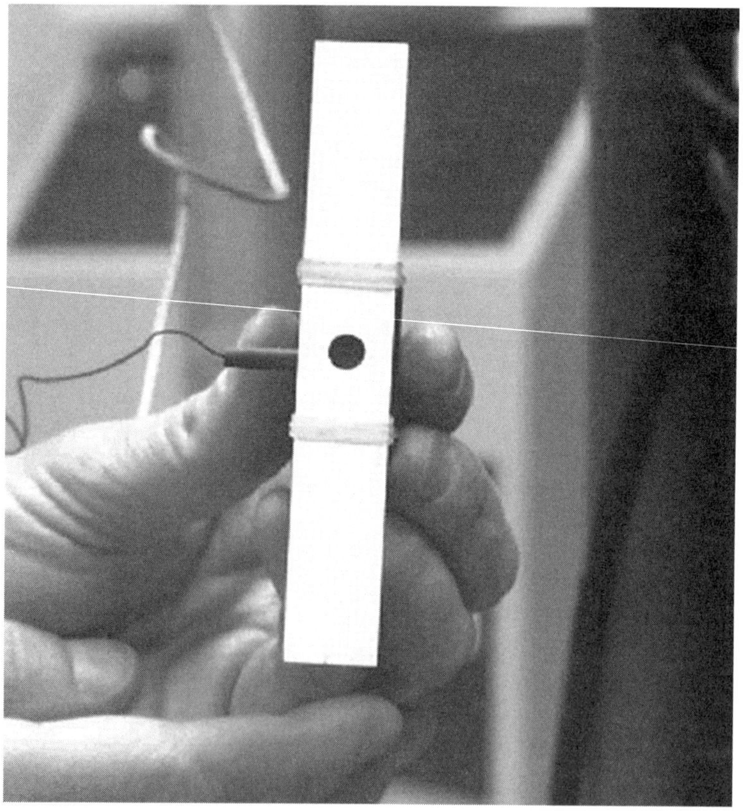

Figure 22.24 Photograph of All-Oxide Tensile Specimen with Hole Utilized for Determination of Notch Sensitivity of CFCCs

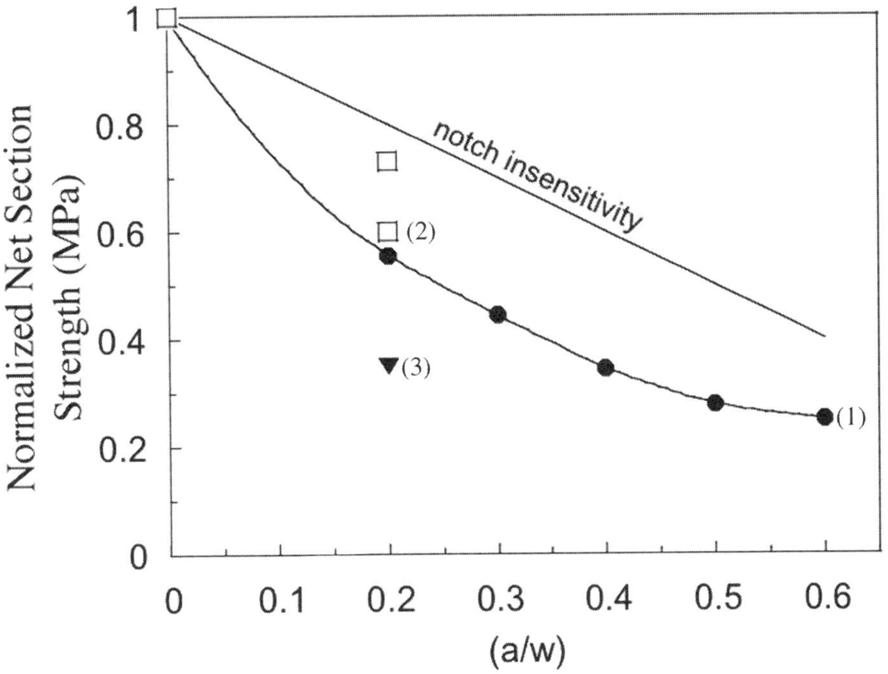

Figure 22.25 Notch Sensitivity of: (1) CG-Nicalon™/C/CVI-SiC [106], (2) Nextel™610/AS (23°C), and (3) Nextel™610/AS (950°C) [113].

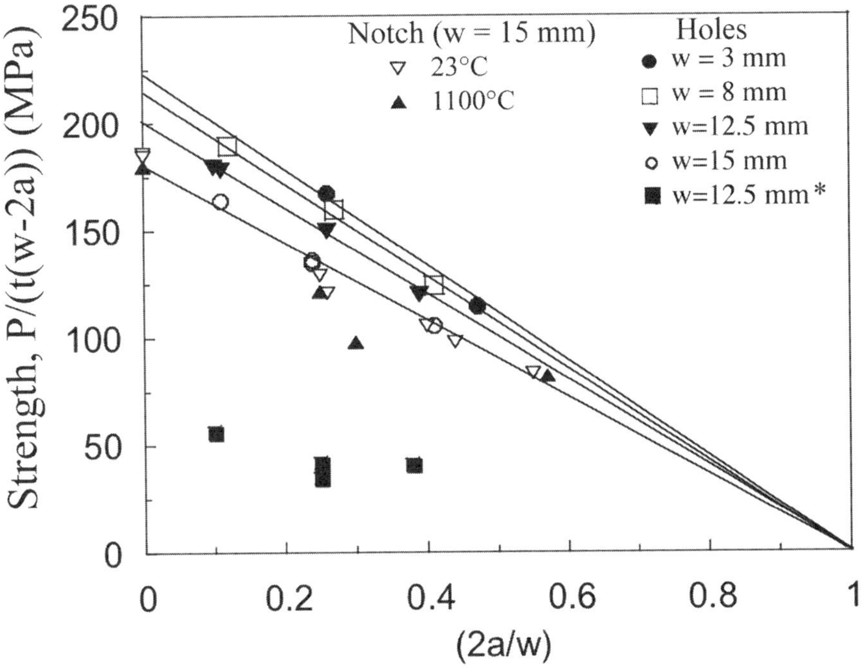

Figure 22.26 Notch Sensitivity Behavior of Nextel™720/AS: Specimens with Holes of Diameter 2a: * Indicates that Specimen was Tested after Aging at 1100°C for 100 hrs [114]. Double-Notched Specimens of w = 15 mm and Notches of Length a [115]. The Solid Lines Correspond to Notch Insensitive Behavior for Each Specimen Width.

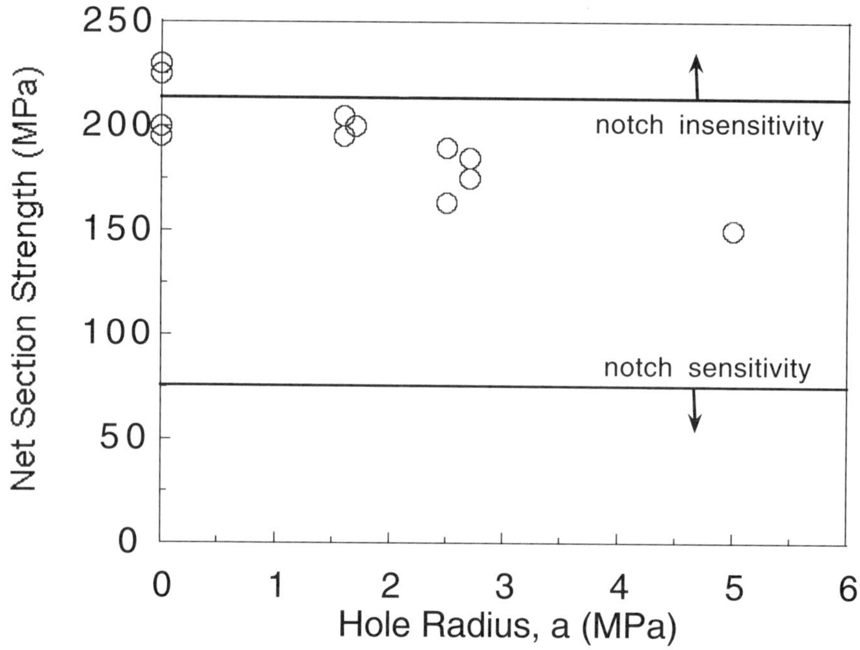

Figure 22.27 Effect of Hole Size on the Notch-Sensitivity of Center-Hole Tensile Specimens of Nextel™610/Mullite-Alumina CFCC [79]

PHYSICAL PROPERTIES

Transport Properties

When used in gas turbine engines CFCCs will be subjected to high temperatures, high heat fluxes and thermal gradients. Thermal gradients will induce thermal strains and consequently stresses. Among the various material properties that affect the magnitude of these thermal strains and stresses, thermal conductivity is the most important one [116]. In general, high values of thermal conductivity are desirable to minimize thermal stresses and strains, but the opposite would be true for applications requiring energy conservation. Although the thermal conductivity of CFCCs is independent of the environment, it becomes dependent when fiber-matrix debonding occurs and the environment can ingress in the interfacial region. It has also been found that the thermal diffusivity parallel to the fiber direction of composites exceeds the corresponding values for heat flow perpendicular to the fibers [116].

Figure 22.28 presents data for the temperature-dependence of thermal conductivity of CG-Nicalon™/CVI-SiC and values for Sylramic/MI-SiC composite with or without CVI-BN interface. Table 22.5 lists results for other CFCCs.

Thermal Expansion

Thermal expansion is one of the most important properties for the selection of materials for high-temperature applications. Volumetric changes with temperature become a prime consideration when materials with dissimilar thermal expansion values are incorporated in designs because thermally-induced stresses are proportional to the difference in the coefficients of thermal expansion. Table 22.6 lists values of thermal expansion for various CFCCs.

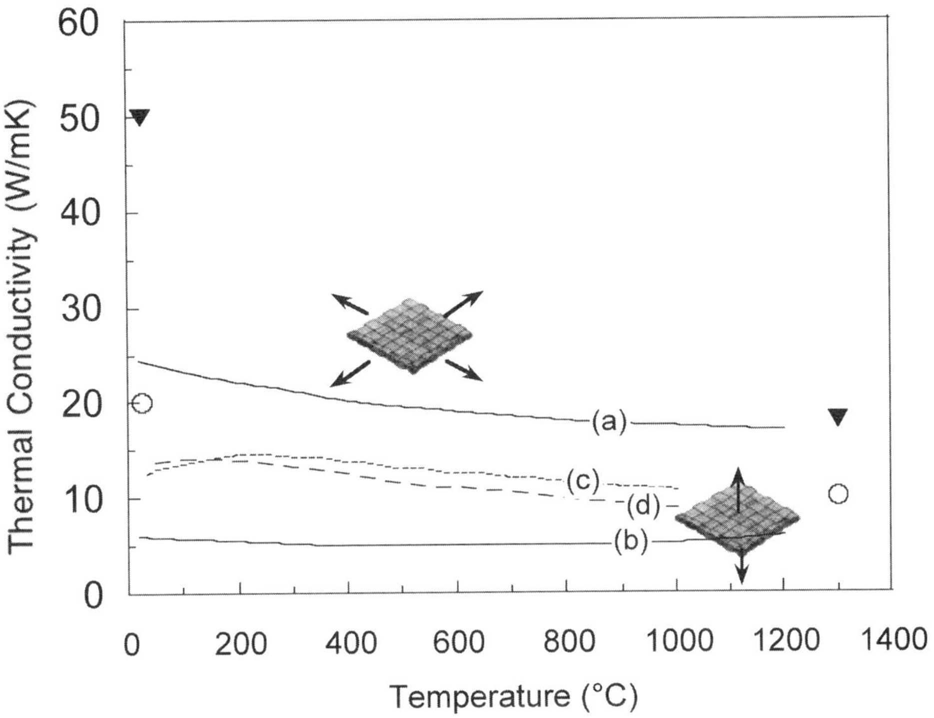

Figure 22.28 Thermal Conductivity of 2-D CG-Nicalon™ Fiber-Reinforced CVI-SiC Matrix Composites. (a) Parallel to Fiber Direction (b) Perpendicular to Fiber Direction [86]. (c) Parallel to Fiber Direction with a Carbonaceous Fiber Coating, 1 μm Thick; (d) Parallel to Fiber Direction with a Carbonaceous Fiber Coating, 0.1 μm thick [117]; Open Circle: Through-the-Thickness Thermal Conductivity of Sylramic/MI-SiC Composite with CVI-BN Interface; Closed Triangle: Through-the-Thickness Thermal Conductivity of Sylramic/MI-SiC Composite Without CVI-BN Interface [118].

Thermal Shock Resistance

Shutdowns for either scheduled or unscheduled maintenance will result in a decrease in the temperature of, and stress build-up in, hot-section components of gas turbine engines. The magnitude of these stresses will depend on the cooling rate and on the magnitude of thermal gradients. To be used in the hot section of gas turbine engines, CFCCs must possess adequate resistance to thermal shock in order to survive transients (e.g.—flame outs, which would subject hot components to thermal shock by incoming cold compressed air in the combustor) [124]. Traditionally, thermal shock conditions can be simulated in the laboratory using either water or air quenching after exposure to a furnace environment. However, more extreme thermal shock conditions have been obtained using stationary rocket engines [125].

In general CFCCs exhibit good thermal shock resistance. Damage resulting from thermal shock in CG-Nicalon™/SiC has been associated with microcracks that initiate from pores in the matrix, growth of cracks inside and between bundles in the transverse plane, and saturation of microcracks after repeated quenching [126, 127]. C/SiNCO and SiN-SiNCO CFCCs, which were used to fabricate scrolls in a 100 kW automotive gas turbine engine, have demonstrated excellent thermal shock resistance after being subjected repeatedly to high-temperature (1350°C) and high-speed (100 m/s) gas flow, followed by cold gas flow (20 m/s) [32].

TABLE 22.5 TRANSPORT PROPERTIES

Fiber	Matrix	Interface	V_f (%)	Fiber Architecture	Method	Porosity (%)	Density (g/cm³)	Thermal Diffusivity (10^{-6} m²s)	Thermal Conductivity (Wm⁻¹K⁻¹)	Reference
CG-Nicalon™	CVI-SiC	0.03–1.0 μm PyC	39.5–43.7	PW (0/30/60)	Xenon flash lamp	15–20	2.22–2.65	8.4–12.9*		[119]
Hi-Nicalon™	CVI-SiC	0.3 μm PyC	40	PW (0/30/60)	Xenon flash lamp	15–20	2.22–2.65		7.4 ≡ 25.3 ⊥	[120]
CG-Nicalon™				PW (0/30/60)					6.8 ≡ 24.7 ⊥	
Amoco K100, P55		—	44**	3D					214 ≡ 12 ⊥	
CG-Nicalon™	CVI-SiC	PyC		PW (0/30/60)					5.6 ± 0.6 ***	[121]

*The transverse thermal conductivity (which was calculated using the estimated composite specific heat obtained from the application of the rule of mixtures) increased with increasing carbon coating thickness from 8.4 Wm⁻¹ K⁻¹ for 0.03 μm to 12.9 Wm⁻¹ K⁻¹ for thickness of 1 μm while the thermal conductivity for a material without fiber coating was found to be 13.4 Wm⁻¹K⁻¹.

** P55 fibers were woven in the x and y directions with 2000 fibers/tow, while six tows (1000 fibers/tow) of K1100 fibers were woven through the orthogonal weave in the perpendicular direction (z). The overall fiber volume fraction was 44% of which 7% was in the x and y directions and 86% was in the z-direction.

*** Thermal diffusivity values were obtained after specimens were subjected to thermal shock (in water) for ΔT = 0, 200, 400, 600, 800 and 1000°C. Another specimen was water-quenched 4 times at DT = 800°C. The thermal diffusivity was found to remain constant (within scatter) regardless of quench history.

TABLE 22.6 THERMAL EXPANSION

Fiber	Matrix	Interphase	V_f (%)	Fiber Architecture	Method	Porosity	Density g/cm³	Coefficient of Thermal Expansion 10^{-6} K⁻¹	Reference
CG-Nicalon™	CVI-SiC		0.26	4HSW	Dual Dilatometer		2.47	5.30 ± 0.03	[122]
Nextel™ 312		0.34	8HSW				1.98	4.77 ± 0.04	
FP® alumina		0.31	8HSW				1.90	4.45 ± 0.07	
Nextel™ 720	AS	—	40	8HSW	Extensometer			5.3 (200°C–700°C) 6.6 (700°C–1000°C)	[10]
T300	MI-SiC	C	38	8HSW	Extensometer			2.5 (20°C–600°C) 2.8 (400°C–800°C) 3.2 (600°C–1000°C)	[123]

DURABILITY AND RELIABILITY

Matrix Cracking

Matrix cracking of CFCCs is very dependent on the microstructure of the composite, fiber properties, fiber architecture, porosity and residual stresses. Because matrix cracks in non-oxide matrix CFCCs can serve as avenue for the ingress of the environment into the interior of the composite, which may result in degradation and loss of strength, the matrix cracking stress is considered as the maximum allowable design stress for these CFCCs.

The various stages of matrix cracking in 2-D CVI-SiC CFCCs during tensile testing have been well identified [128, 129]. Microcracks tend to initiate first at singularities around macropores at 0.025% deformation, and then propagate across neighboring longitudinal yarns to the nearest macropores. Saturation of macropore-originated microcracks occurred at 0.12% deformation. At larger deformation levels, microcracks form in the transverse yarns and in the narrow strips of interply matrix located between macropores. The formation of these microcracks stops at 0.2% deformation, but microcracks appear randomly in the matrix surrounding the longitudinal bundles for deformations above 0.2% [130].

For many gas turbine engine components, resistance to foreign object impact damage is of critical concern, because catastrophic failure of components can lead to catastrophic engine failure. Limited studies have been focused on studying the resistance of CFCCs to foreign object impact. However, existing results indicate that the tensile strength of CG-Nicalon/CVI-SiC remains unchanged after impact with 1.0 mm diameter zirconia spherical particles at speeds of 600 m/s [32, 131]. Only a spalling crater and radial cracks were caused by the impact and the impact site was not the origin of the composite failure. However, at elevated temperatures in aggressive environments, formation of cracks resulting from foreign object impact could serve as avenues for the ingress of the environment into the material, which in turn would lead to degradation of properties and failure.

Mechanical Cyclic Fatigue

During service in gas turbine engine CFCCs will be subjected to variable thermal and mechanical loading. However, the microstructural features and damage mechanisms that are responsible for the tough behavior of CFCCs (i.e., low interfacial shear stress, long fiber debond lengths and matrix microcracking) often conflict with the microstructural requirements for optimal fatigue resistance [132].

In general, fatigue-induced failure of continuous fiber-reinforced ceramics is governed by interface wear, fiber damage and matrix crack growth [133]. Moreover, these damage processes are enhanced at elevated temperatures, particularly when the fibers and fiber coating are susceptible to corrosion/oxidation. It has been found that at ambient temperature the fatigue limit of CFCCs decreases with increasing frequency [134] and that the residual monotonic tensile strength of some CFCCs increases after reaching fatigue endurance. This phenomenon has been explained as the result of decreasing stress concentrations near the cross-over points between the 0° and 90° bundles.

Figure 22.29 contains a plot of peak stress versus number-of-cycles plot containing results for various CFCCs at various temperatures. In most cases, the life of CFCCs decreases with increasing peak fatigue stress. At peak fatigue stresses below the proportional stress limit, CFCCs exhibit endurance [138]. Limited work has been carried out to evaluate the thermomechanical fatigue of CFCCs [139, 140], but information obtained from this type of tests will be necessary to develop a thorough understanding of the behavior of CFCCs during transient operation of gas turbine engines.

Creep, Stress-Rupture and Stress Relaxation

When materials and structures are subjected to constant tensile loads at high homologous temperatures, they will experience dimensional changes with time. Although the resulting elongation of CFCCs is often referred to as "creep" deformation, it arises not only from the viscous deformation of the composite

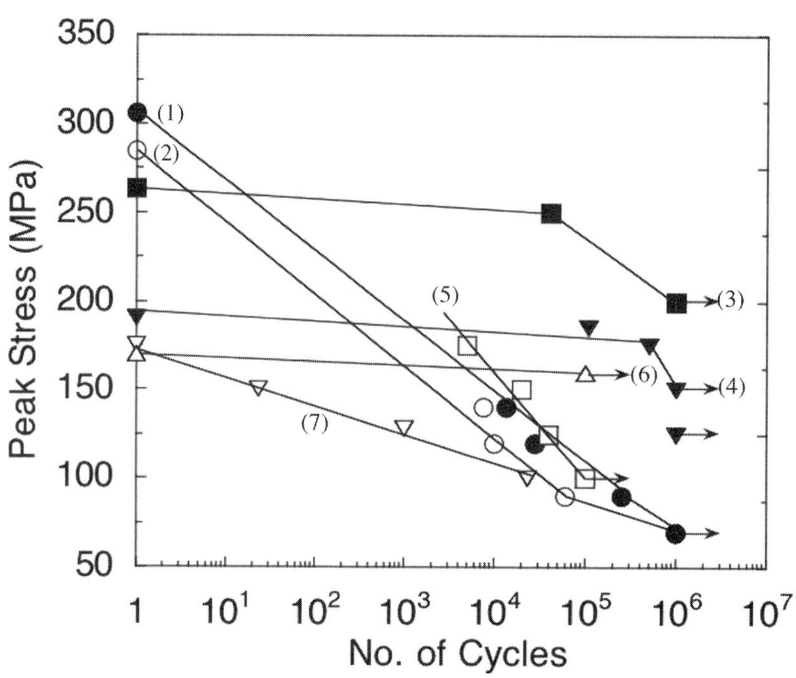

Figure 22.29 Fatigue Behavior of CFCCs in Air. (1) CG-Nicalon™/BN/(PIP-SiNCO + 25% SiC), 982°C [135]. (2) CG-Nicalon™/BN/(PIP-SiNCO + 25% SiC), 1204°C [135]. (3) CG-Nicalon™/BN/(PIP-SiNCO + 25% SiC), 23°C [135]. (4) CG-Nicalon™/BN/PIP-SiNCO, 23°C [136]. (5) CG-Nicalon™/BN/PIP-SiNCO, 1000°C [136]. (6) Nextel™720/AS, 1000°C [137]. (7) Nextel™720/AS, 1200°C [137].

constituents (e.g.—creep deformation), but also from matrix cracking, fiber debonding, fiber sliding and fiber failure. At elevated temperatures, time-dependent matrix cracking [141,142] (with the consequent compliance changes) can occur at composite stresses lower than the monotonic proportional limit stress at that particular test temperature as a result of dissimilar creep resistance of the constituents [92,143–145].

Figure 22.30 contains a series of tensile creep curves for CG-Nicalon™/PIP-SiNCO at 1200°C. In general, the amount of deformation and the rate of deformation of CFCCs increases with temperature and applied tensile stress. The curves in Figure 22.30 exhibit continuously decreasing rate of deformation with increasing time/deformation and in most cases the rate of deformation never reaches a true minimum value. Exceptions occur when due to cumulative fiber failure the effective load bearing capability of the CFCC in the plane of the critical crack is reduced. In this case CFCCs exhibit accelerated deformation analogous to "tertiary creep" in metals [146–148].

The magnitude of the rate of deformation of CFCCs is influenced by the rate of deformation of the constituents, the rate of load redistribution among them, which results from their differential creep resistance, and changes in compliance due to matrix cracking and fiber failure. Figure 22.31 illustrates the effect of fiber type on the creep deformation of PIP-SiNCO matrix composites. It can be observed that PIP-SiNCO matrix composites reinforced with Hi-Nicalon™ fibers are two orders of magnitude more resistant to creep deformation than PIP-SiNCO matrix composites reinforced with CG-Nicalon™ fibers. Hi-Nicalon™ fibers are processed through e-beam curing and therefore contain a lower amount of oxygen (below 0.5% wt) than CG-Nicalon™ fibers. As a result of this modification in processing, Hi-Nicalon™ fibers are stiffer, much more creep resistant than CG-Nicalon™ fibers and experience little of the strength degradation experienced by CG-Nicalon™ fibers at high-temperatures due to the decomposition of the oxycarbide phase.

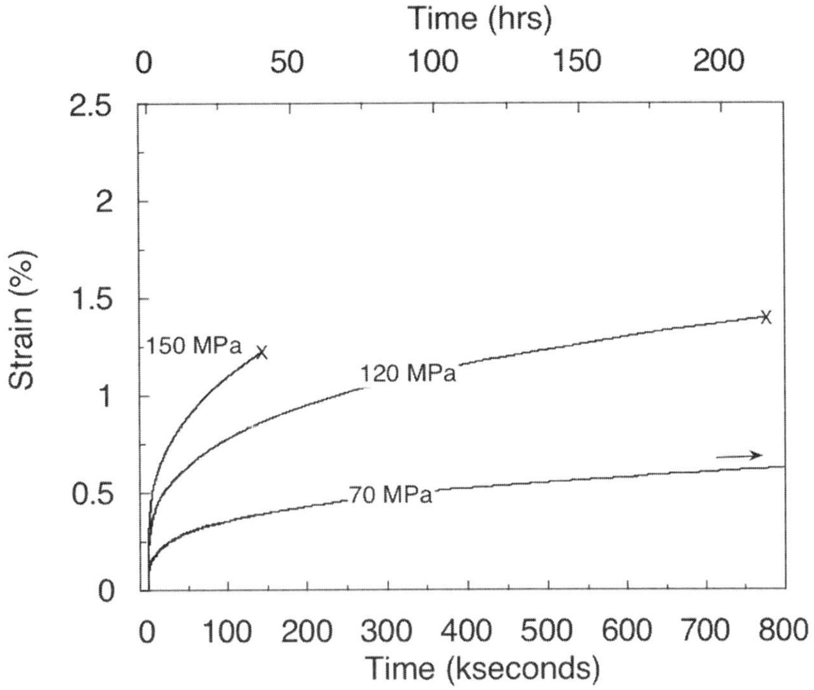

Figure 22.30 Tensile Creep Strain in Air at 1200°C for CG-Nicalon™/BN/SiNCO

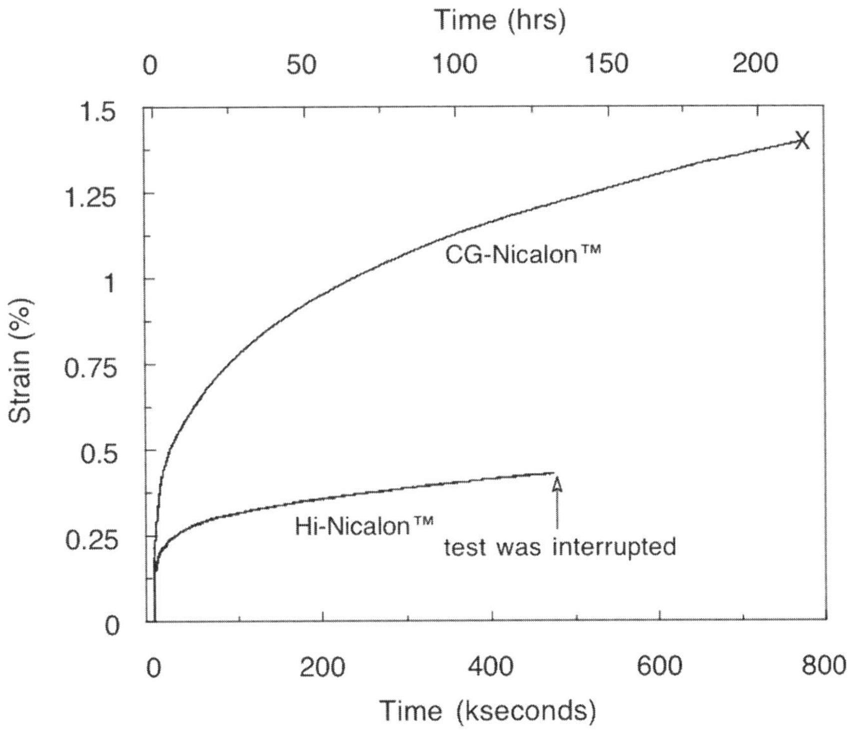

Figure 22.31 Effect of Fiber Type (CG-Nicalon™ versus Hi-Nicalon™) on the Creep Resistance of SiNCO Matrix CFCCs in Air at 1200°C. It is Effectively One Order of Magnitude Increase in Creep Resistance [97].

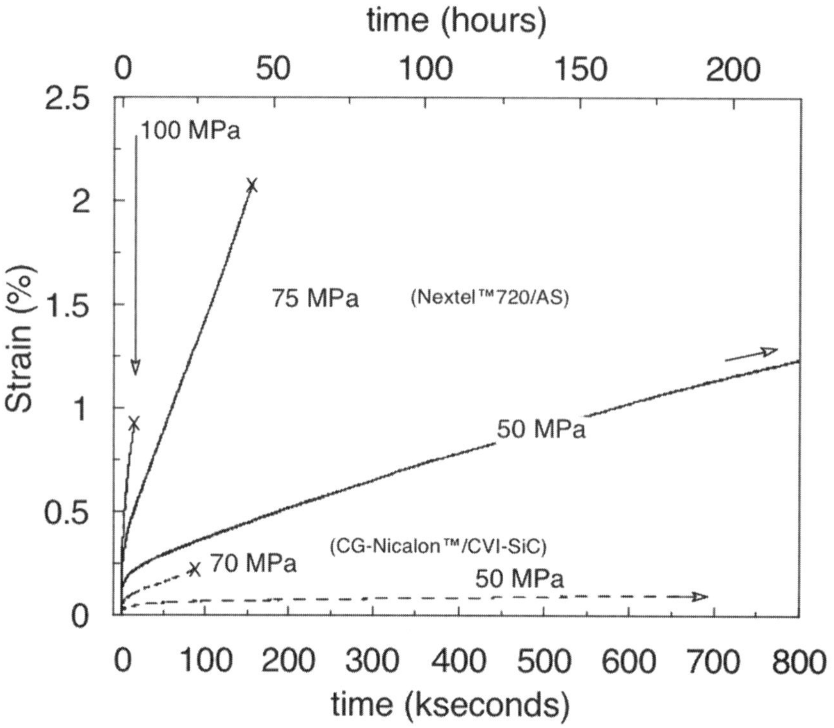

Figure 22.32 Tensile Creep Strain in Air at 1200°C for CG-Nicalon™/C/CVI-SiC and Nextel™720/AS [150]. Test at 50 MPa for CG-Nicalon™/C/CVI-SiC was Interrupted.

In general, SiC-based fibers are stronger than oxide fibers at room temperature and retain more of their strength at elevated temperatures. Oxide fibers are stiffer than many SiC-based fibers at room temperature, but they lose their stiffness advantage above 1000°C [149]. Although oxide fibers are more stable than SiC-based fibers in oxidizing environments at high temperatures, they experience grain growth and loss of strength due to softening of their grain boundaries[7–10, 149].

Figure 22.32 contains a plot with creep curves for CG-Nicalon™/CVI-SiC and Nextel™ 720/AS CFCCs at 1200°C in air. These results demonstrate that under the same conditions, the former is at least one order of magnitude more resistant than the latter. However, at stresses above the matrix cracking stress of CG-Nicalon™/CVI-SiC, its life is significantly shorter than that of Nextel™ 720/AS. This is the result of degradation in the interior of CG-Nicalon™/CVI-SiC from oxidation of the fiber coating and fibers, which leads to fiber failure and rupture of the CFCC.

An analysis of stress-rupture results for CG-Nicalon™/CVI-SiC CFCCs at temperatures between 600°C and 1000°C has shown that strength of this material decreases with time according to [151,152]

$$\sigma = C_3 t_f^{-0.25} \tag{2}$$

where σ denotes the tensile strength, C_3 is a constant and t_f is the time-to-failure. This time-dependence of CFCC strength is consistent with models based on the oxidation of the reinforcing fibers and the nucleation of notches or flaws in the oxide layer [146,148,150]. Similar models have been formulated to explain the stress-rupture of all-oxide CFCCs as a result of grain growth in the fibers and loss of fiber strength [10]. These models have also been able to predict the time-dependent changes in material compliance during

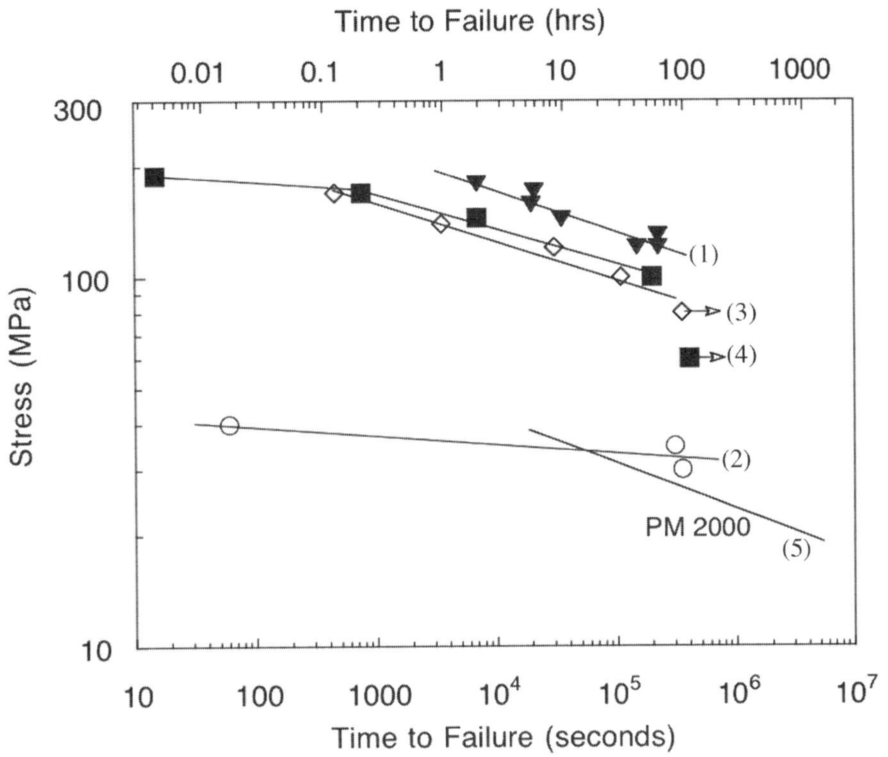

Figure 22.33 Plot of Stress (MPa) versus Time-to-Rupture for Tests Conducted in Air at 1300°C for: (1) 3-D Tyranno Lox-M/C/Si-Ti-C-O [153]. (2) 2-D, CG-Nicalon™/CVI-SiC [154]. (3) 2-D, Hi-Nicalon™/CVI-SiC [154]. (4) 2-D, CG-Nicalon™/"Enhanced" SiC [154]. (5) Data for PM2000.

stress-rupture conditions, which in addition to creep (e.g.—viscous deformation) result from the progressive oxidation of the fiber coating, the redistribution of stresses in the fibers bridging matrix cracks, the oxidation of the fibers and ultimately fiber failure [150].

In general, the stress-rupture behavior of CFCCs currently available is better than that of many metallic materials (other than single crystals) currently used in gas turbine engine applications. Figures 22.33–22.35 present stress-rupture data for several CFCCs in air at temperatures between 1100°C and 1300°C. In general, at these temperatures the strength of CFCCs decreases with time and temperature[1]. For comparative purposes, data are also included for several metallic alloys.

In many gas turbine engine applications, stresses are thermally-induced. To assess the potential of candidate CFCCs for these applications it is more appropriate to carry-out tests under strain-controlled conditions, rather than under load-controlled conditions (e.g.—creep, stress-rupture). Figure 22.36 illustrates the stress-relaxation behavior of Sylramic®/PIP-SiNCO when subjected to increasingly larger strains values at 1300°C in air. These results show that the material is able to relax as much as 80% of the stress associated with the

[1]At intermediate temperatures (e.g.—between 600°C and 950°C) many SiC-based CFCCs exhibit "anomalous" behavior, i.e.—they are less resistant to stress-rupture in air at those temperatures than at 1100°C. When plotting failure time versus temperature at constant stress, the resulting curve resembles an inverted "horse shoe".

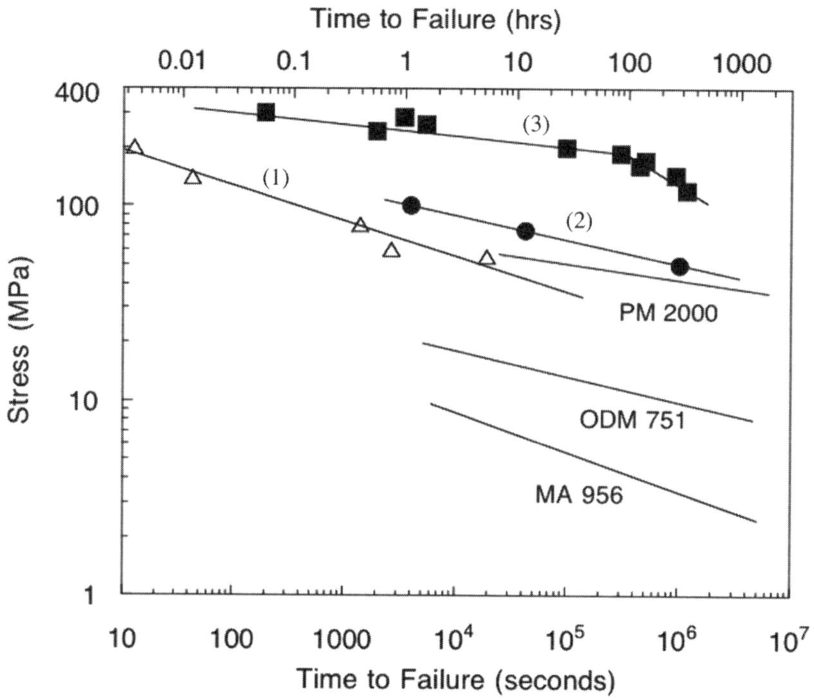

Figure 22.34 Plot of Stress (MPa) versus Time-to-Rupture for Tests Conducted in Air at 1200°C for: (1) 2-D, PW, CG-Nicalon™/CVI-SiC [89]. (2) 2-D, 8HSW, Nextel™720/AS [124]. (3) 3-D Tyranno Lox-M/C/Si-Ti-C-O [153]. For Comparative Purposes, Data are Provided for PM2000, MA 956 and ODM 751 at 1100°C in Air.

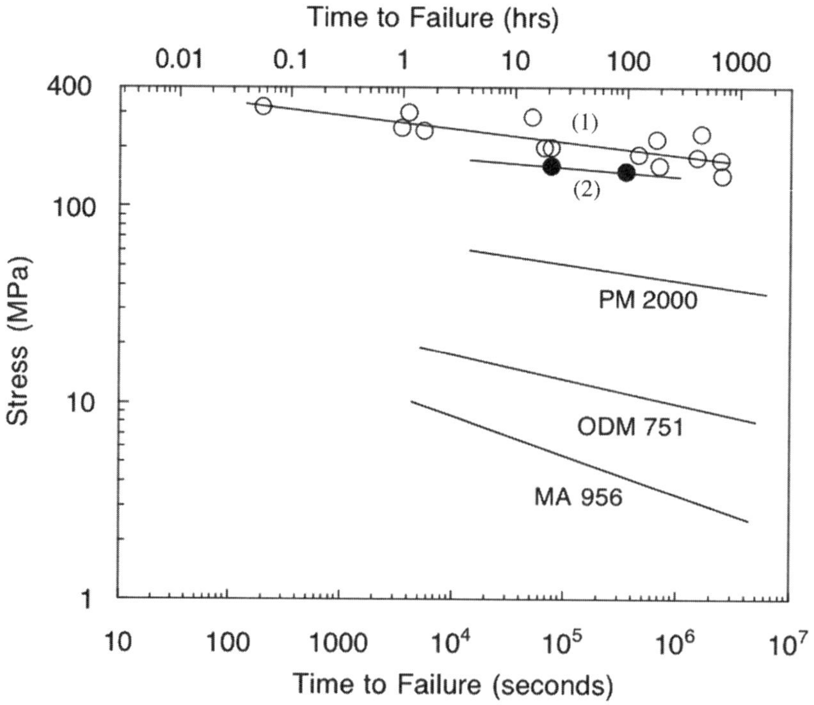

Figure 22.35 Plot of Stress versus Time-to-Rupture for Tests Conducted in Air at 1100°C for: (1) 3-D Tyranno Lox-M/C/Si-Ti-C-O [153]. (2) 2-D, 8HSW, Nextel™720/AS [115]. For Comparative Purposes, Data for PM2000, MA 956 and ODM 751 at 1100°C in Air.

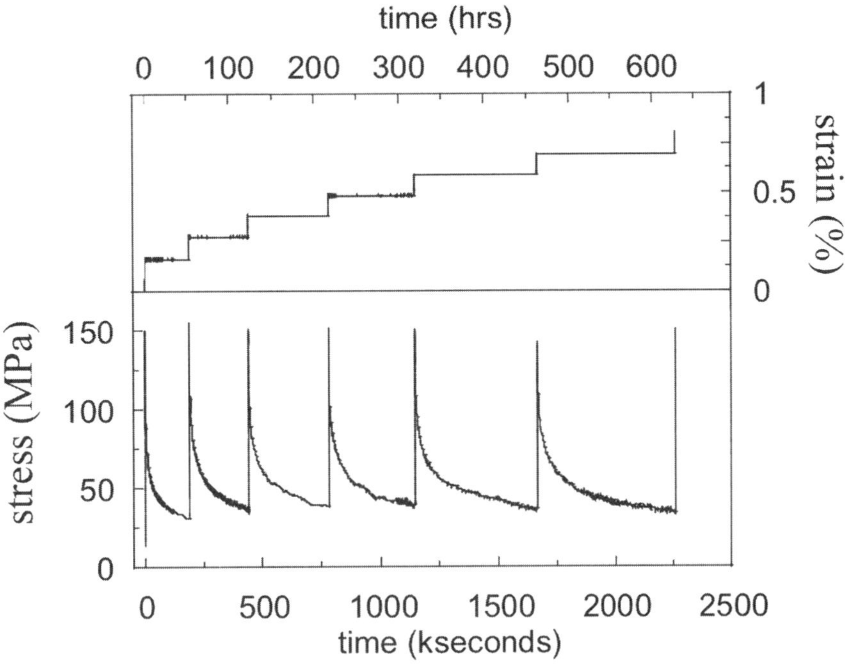

Figure 22.36 Stress-Relaxation Behavior of Sylramic®/PIP-SiNCO CFCC at 1300°C in Air after Successive Straining at Increasingly Larger Strain Levels

applied strain. The relaxation of stress in CFCCs at elevated temperatures is related to viscous deformation of the composite constituents as well as to damage resulting from the application of strain. In the case of C/MI-SiC CFCCs in air at 1000°C, when the environment can ingress to the interior of the material, stress relaxation results from the oxidation and failure of the reinforcing fibers (Figure 22.37) [123].

ENVIRONMENTAL EFFECTS ON PHYSICAL AND MECHANICAL PROPERTIES

Oxidation and Corrosion

One of the properties that make SiC attractive for applications in air at elevated temperatures is its excellent resistance to oxidation. The mechanism of passive oxidation of SiC is based on the formation of a protective layer of silica according to the following reactions:

$$SiC + (3/2)O_2 \rightarrow SiO_2 + CO$$
$$SiC + 2O_2 \rightarrow SiO_2 + CO_2$$

The oxidation behavior of SiC-matrix CFCCs in air has been analyzed extensively and values have been reported for the parabolic rate constant as a function of temperature [155–162]. However, despite the remarkable resistance of CVI-SiC to oxidation (in air) at elevated temperatures and ambient pressure, there are certain conditions of stress and temperature for which CVI-SiC matrix CFCCs exhibit severe degradation of mechanical properties as a result of oxidation.

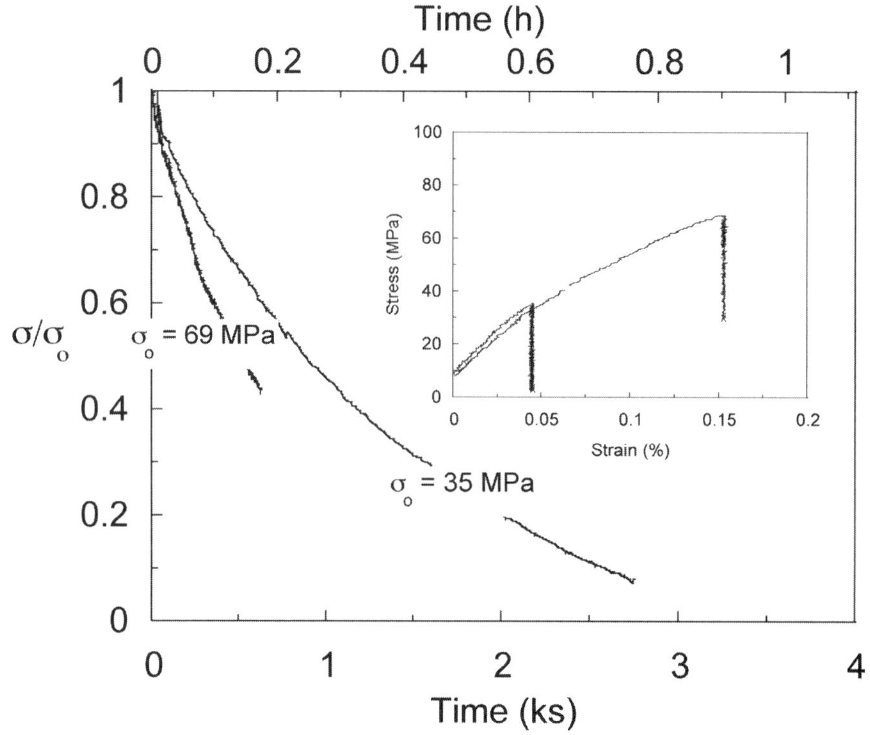

Figure 22.37 Stress-Relaxation Behavior of C/MI-SiC CFCC at 1000°C in Air at Two Different Initial Stresses. The Inset Shows the Corresponding Stress-Strain Curves [123].

The degradation of the physical and mechanical properties of continuous fiber-reinforced CVI-SiC composites as a result of oxidation, results from the ingress of the environment to the interior of the composite through matrix cracks, open machined surfaces, and fiber ends [88]. For carbonaceous fiber coatings, oxidation results in the removal of the fiber coating, leaving a gap between the fibers and the matrix that will result in the local redistribution of stresses, because load is transferred between the fibers and the matrix through the fiber coating. Once the fibers are exposed to the environment, the fibers will oxidize (in the case of non-oxide) and lead to loss of fiber strength.

Thermokinetic chemical analyses have been formulated to model the oxidation behavior of SiC-based and carbon fiber-reinforced CVI-SiC matrix CFCCs [158,159,160–162]. These analyses, which have been formulated taking into consideration the effect of partial pressures of oxygen and temperature, have provided the thickness of the fiber coating that is critical to promote the sealing of channels, created by the removal of the fiber coating, by matrix oxidation.

To assess the environmental durability of CFCCs in a particular application, it is necessary to evaluate candidate materials under conditions that are as realistic as possible. For example, data available from the evaluation of non-oxide CFCCs in air would not had predicted the degradation that these materials would experience in combustor rigs or actual engine operation [163–167]. In combustion environments water is present in amounts related to the composition of the fuel. In the case of Si-based CFCCs, water vapor will react with silica resulting in accelerated oxidation [165–167]. Chapter 26 presents a detailed discussion of environmental effects on the properties of monolithic ceramics and CFCCs. As a result of the interaction between SiC-based CFCCs and combustion environments, several environmental barrier coating concepts have been developed to protect CFCCs in combustion environments [168–172].

In marine environments, gas turbine components are subjected to even more severe conditions due to the presence of alkalis, which are known to attack SiC [173, 174].

COMPONENTS AND SPECIAL CONFIGURATIONS

To validate methodologies for designing components with CFCCs, it is necessary to proof-test or evaluate entire components, or sections of components, under loading conditions that are representative of those that are expected during service. Scaled combustor liners of CG-Nicalon™/PIP-SiNCO have been evaluated and their properties compared to those of standardized test specimens of the same material [175]. The scaled combustor liner, which was 145 mm in diameter, 2.87 mm in wall thickness and 12.75 mm tall, was subjected to internal pressurization using an elastomeric insert. The scaled combustor liner consisted of a convoluted structure in which segments of CFCC with 8HSW fiber architecture were overlapped at regular angles around the circumference. Figure 22.38 presents tensile stress-strain curves obtained from the evaluation of an ASTM C1275-size test coupon of the same material. Also included in that plot is the hoop stress versus hoop strain curve obtained from the internal pressurization of the scaled liner section. Although the proportional stress limit and the elastic modulus that were obtained from both tests are the same, the ultimate strength is not. This discrepancy results from the fact that the scaled combustor liner failed at a matrix-rich joint region, which would be equivalent to the weldment in a metallic cylindrical structure. Additional variations between results obtained from the evaluation of standardized test specimens and larger components would result from volumetric effects [84,85], from edge effects and constraints or from multiaxial states of stress not reproduced during the evaluation of standardized test specimens.

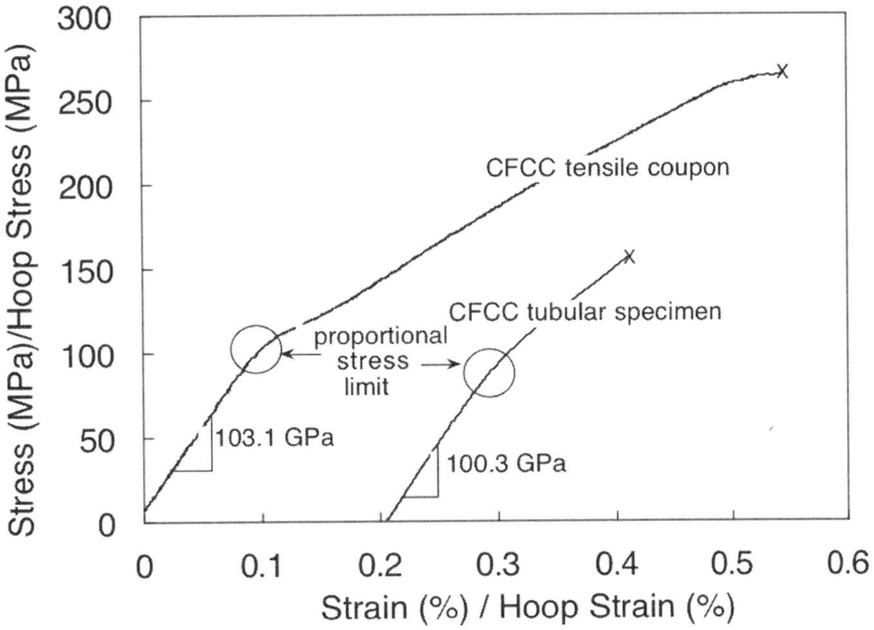

Figure 22.38 Tensile Stress-Strain Behavior of CG-Nicalon™/BN/PIP-SiNCO CFCC Test Coupon (200 mm Long with a Gauge Section 50 mm Long) and Tensile Hoop Stress-Hoop Strain Behavior of CG-Nicalon™/BN/PIP-SiNCO CFCC Tubular Specimen 145 mm Inner Diameter, 2.87 mm Thick and 12.75 mm Tall [175].

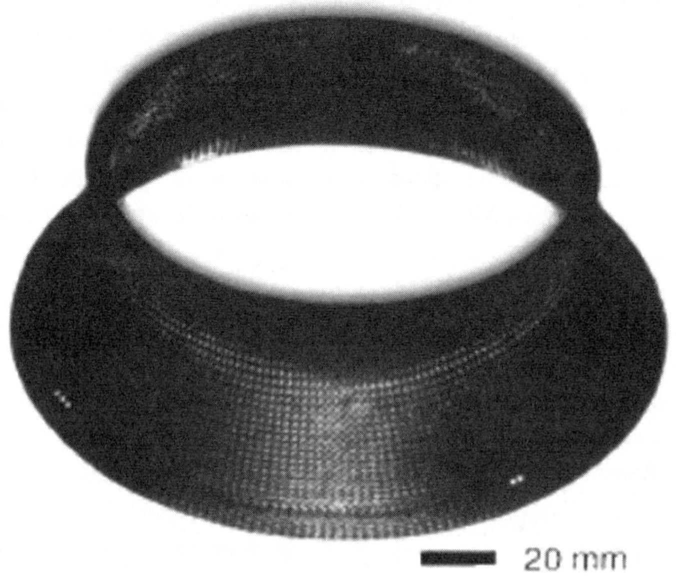

Figure 22.39 CG-Nicalon_/Si-N-C-O CFCC Inner Scroll Support of 100kW Automotive Ceramic Gas Turbine Development Project (Reprinted with permission from ref. [176])

A CG-Nicalon™/SiNCO CFCC inner scroll support for a 100 kW automotive ceramic gas turbine was evaluated by internal pressurization using oil, and instrumented with strain gauges to determine the distribution of strains during the test (Figure 22.39). The CFCC inner scroll support was subjected to increasingly larger values of internal pressure until failure, which occurred at a tangential stress of 180 MPa. The evaluation of tensile test specimens of the same material yielded lower tensile strengths [176] and these discrepancies suggest differences in the structure of the material between test specimen and inner scroll support and errors in the determination of stresses in the latter during pressurization.

In contrast to the applications illustrated above, where the state of stress is relatively simple, in other applications components will be subjected to multiaxial states of stress. This in turn will require the evaluation of special test specimen configurations including the evaluation of components under actual and simulated conditions to develop failure maps and multiaxial strength criteria [94].

One important aspect of the methodology for designing with CFCCs and validating life prediction models consists in determining the residual properties of components. Ideally it would be desirable to use nondestructive techniques that could provide a measure of residual properties as a function of the service history of the component. In the case of combustor liners, their residual mechanical properties (e.g. – tensile stress-strain behavior, interlaminar properties) have been determined after being removed from service operation. Figure 22.40 shows a plot containing stress-strain curves from the tensile evaluation of axial strips of Hi-Nicalon/BN/MI-SiC obtained from combustor liners that had accumulated 13,000+ hours at a peak service temperature of 1260°C [164]. The insets correspond to a photograph of one of those strips, illustrating the degree of corrosion experienced by the material, and a micrograph of a typical cross-sectional area. The stress values in the curves in Figure 22.40 were determined using the overall dimensions of the CFCC, including the thick oxide products. However, when the stresses are computed using the effective remaining cross-sectional area of the CFCC it is found that it only experienced a small loss of strength and most importantly, the CFCC still exhibited composite-like behavior and adequate failure strain comparable to that of the material in the as-processed condition.

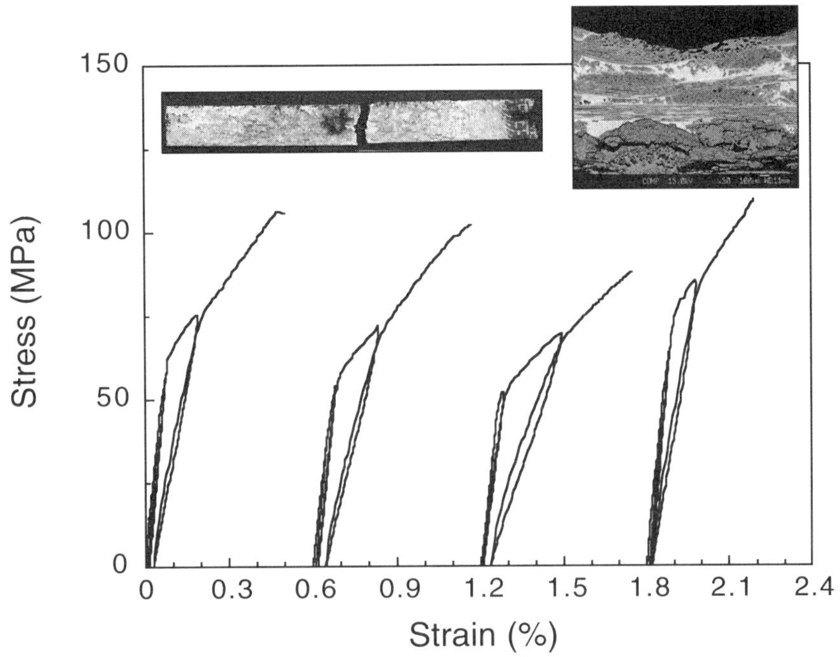

Figure 22.40 Tensile Stress-Strain Curves for Tyranno™-SA/BN/MI-SiC CFCC Strips Obtained from a Combustor Liner after 13,000+ Hours of Operation. The Insets Correspond to a Photograph of One of Those Strips Illustrating the Amount of Oxidation, and a Micrograph of the Cross-Sectional Area of the Material (Reprinted with permission of K.L. More, Oak Ridge National Laboratory).

SUMMARY

CFCCs exhibit remarkable physical and mechanical properties and the ability to alleviate elastic stress concentrations at holes and notches by redistributing stresses through various inelastic deformation mechanisms, which are attractive for the manufacture of components for gas turbines and many other industrial applications. CFCCs and their constituents have undergone substantial and significant developments over the last 25 years particularly in the area of reinforcing fibers and matrices. CFCCs currently available exhibit high tensile strength, damage tolerance, exceptional strength retention at elevated temperatures, and resistance to creep and stress-rupture at temperatures as high as 1300°C. Additional work is still needed to address the susceptibility of non-oxide CFCCs to environmental attack in combustion-related environments. Although developments in the area of EBCs are encouraging, dependence on their use does impact the cost and reliability of CFCCs.

All-oxide CFCCs have experienced rapid developments in recent years. There are currently various all-oxide CFCC systems available, which exhibit good mechanical properties and excellent damage tolerance and resistance to stress-rupture and mechanical fatigue. Although their use is currently limited by the microstructural stability of polycrystalline oxide fibers up to 1100°C, efforts are currently under way to develop small-diameter polycrystalline fibers with improved creep resistance and stability at high-temperatures [177].

The need to continue improving the efficiency of power generation systems and industrial processes and meeting increasingly more stringent environmental regulations on emissions remains the driving force for the further development and implementation of CFCCs in upcoming years.

ACKNOWLEDGEMENTS

The author gratefully acknowledges the support of the U.S. Department of Energy through various research programs at Oak Ridge National Laboratory. The support of program managers Debbie Haught, Merrill Smith and Mike Karnitz is sincerely appreciated. The author wishes to thank all those whose work has provided the material for this book chapter. Unless otherwise indicated all figures and tables were provided by the author.

NOMENCLATURE

Al_2O_3	aluminum oxide
AS	aluminosilicate
BN	boron nitride
C	carbon
CFCC	continuous fiber-reinforced ceramic matrix composites
CG	ceramic grade
CGT	ceramic gas turbine
CMC	ceramic matrix composite
CT	compact tension
CVD	chemical vapor deposition
CVI	chemical vapor infiltration
CO	carbon monoxide
DNS	double-notched shear specimen
HSW	harness satin weave
ILSS	interlaminar shear strength
kW	kilowatt
MI	melt infiltrated
ML	multilayer
MW	megawatt
N	nitrogen
NOx	nitrogen oxides
PIP	polymer infiltration and pyrolysis
PW	plain weave
SENB	single edge notch beam
SENT	single edge notch tensile
Si	silicon
SiC	silicon carbide
Si_3N_4	silicon nitride
SiNCO	silicon oxy-carbide-nitride
σ_{PL}	proportional limit stress
TIT	turbine inlet temperature
v_f	fiber volume fraction

REFERENCES

[1] *Ceramic Fibers and Coatings, Advanced Materials for the 21st Century*, The National Academy of Sciences Press, Washington, D.C., 1998.

[2] Miller, S., "Advanced Materials Mean Advanced Engines," Mater. World, 4[8], 446–449, 1996.

[3] Kerans, R. J., Hay, R. S., Parthasarathy, T. A., and Cinibulk, M. K., "Interface Design for Oxidation-Resistant Ceramic Composites," J. Am. Ceram. Soc., 85[11], 2599–2632, 2002.

[4] Levi, C. G., Zok, F. W., Yang, J-Y, Mattoni, M. and Lofvander, J. P. A., "Microstructural Design of Stable Porous Matrices for All-Oxide Ceramic Composites," Z. Metallkd., 90, 1037–1047, 1999.

[5] Lewis, M.H., Tye, A., Butler, E., and Al-Dawery, I., "Development of Interfaces in Oxide Matrix Composites," Key Engineering Materials, 164[16], 351–356, 1998.

[6] Tressler, R. E. and Pysher, D. J., "Mechanical Behaviour of High Strength Ceramic Fibres at High Temperatures," *Proc. Satellite Symp. Advanced Structural Inorganic Composites, 7th Int.Meeting on Modern Ceramics Technologies*, Montecatini Terme, Italy, June 27–30, 1990, Materials Science Monographs, 68, Elsevier Science Publishing Co. Inc., Amsterdam, The Netherlands, 1990. pp. 3–18.

[7] Wilson, D. M., Lieder, S. L., and Lueneburg, D. C., "Microstructure and High Temperature Properties of Nextel 720 Fibers," Ceram. Eng. and Sci. Proc., 16[5], 1005–1014, 1995.

[8] Milz, C., Goering, J., and Schneider, H, "Mechanical and Microstructural Properties of Nextel 720 Relating to its Suitability for High Temperature Application in CMCs," Ceram. Eng. and Sci. Proc., 20[3], 191–198, 1999.

[9] Goering, J., Kanka, B., Steinhauser, U., and Schneider, H., "Thermal Barrier Coated Nextel 720 Fiber/Mullite Matrix Composites: Their Potential for Long-term High Temperature Use in Gas Turbines," Ceram. Eng. and Sci. Proc., 21[3], 613–618, 2000.

[10] Lara-Curzio E. and More, K. L., "Stress-Rupture of Nextel™ 720 Fiber-Reinforced Aluminosilicate Composites: The Role of Microstructural Evolution," presented at the 24th Annual Cocoa Beach Conference and Exposition, Cocoa Beach, FL, USA, January 25, 2000.

[11] *Handbook on Continuous Fiber-Reinforced Ceramic Matrix Composites*, Lehman, R. L., El-Rahaiby, S. K., and Wachtman, J. B., eds., The American Ceramic Society, Westerville, OH, USA, 1995.

[12] Lackey. W.J. and Starr, T.L., "Composites by Chemical Vapor Infiltration: Processing, Structure and Properties," in *Fiber Reinforced Ceramic Composites: Materials Processing and Technology*, Mazdiyasni, K. S., ed., Materials Science and Process Technology Series, Noyes Publications, Park Ridge, NJ, USA, 1990. pp. 397–450.

[13] Clarke, D.A., "Fabrication Aspects of Glass Matrix Composites for Gas Turbine Applications," in *New Materials and their Applications*, University of Warwick, UK, 10–12 Apr. 1990, IOP Publishing, IOP Redcliffe Way, Bristol BS1 6NX, UK, 1990. pp. 173–183.

[14] Dominy, J., "Structural Composites in Civil Gas Turbine Aero Engines," Compos. Manuf. (UK), 5[2], 69–72, June 1994.

[15] Dominy, J., Evans, W. J., and Duff, S. L., "Ceramics in Aero Gas Turbines. An Engineer's View," *Materials Congress '98, Frontiers in Materials Science and Technology*, Cirencester, UK, 6–8 April 1998, Institute of Materials, Materials for High Power Generation and Process Plant Applications, UK, 2000. pp. 335–343.

[16] Ruffles, C., "Applications of Advanced Composites in Gas Turbine Aero Engines," *ICCM/9. Metal Matrix Composites. Vol. I*, Madrid, Spain, July 12–16, 1993. pp. 123–130.

[17] Bennett, A, "Overview of the Advanced Ceramics for Turbines (ACT) Programme," Advanced Materials Journal, 1[3], 21–22, 1990.

[18] Beesley, C. P., "Application of CMCs in High Integrity Gas Turbine Engines," Key Engineering Materials, 127–131, Part 1, 165–174, 1997.

[19] Nardone, V. C., and Prewo, K. M., "Tensile Performance of Carbon Fiber-Reinforced Glass," J. Mater. Sci., 23, 168–180, 1988.

[20] Karnitz, M. A., Craig, D. F., and Richlen, S., L., "Continuous Fiber Ceramic Composites Program," Am. Ceram. Soc. Bull., 70[3], 430–435, 1991.

[21] van Roode, M., Brentnall, W.D., Norton, P.F., and Boyd, G.L., "Ceramic Stationary Gas Turbine Development Program—First Annual Summary," ASME paper 94-GT-313, presented at the International Gas Turbine and Aeroengine Congress and Exposition, The Hague, The Netherlands, June 13–16, 1994.

[22] Brewer, D., "HSR/EPM Combustor Materials Development Program," Materials Science and Engineering A, 261[1–2], 284–291, 1999.

[23] DiCarlo, J.A., Yun, H. M., Morscher G. N., and Ogbuji, L. U. T., "Progress in SiC/SiC Composites for Engine Applications," in *High Temperature Ceramic Matrix Composites*, Krenkel, W., Naslain R., and Schneider, H., eds., 4th International Conference on High Temperature Ceramic Matrix Composites (HT-CMC4), Munich, Germany October 1–3, 2001, Wiley-VCH, 2001. pp. 777–782.

[24] Chung, V., Balis, C., Sohi, M., and van Erp, C., "Ceramic Matrix Composite Materials in Advanced Manned Gas Turbine Engine Combustors," in *Processing, Fabrication & Application of Advanced Composites*, Long Beach, California, USA, 9–11 Aug. 1993, ASM International (USA), 1993. pp. 91–97.

[25] Stephens, J. R., "Intermetallic and Ceramic Matrix Composites for 815–1370°C (1500–2500°F) Gas Turbine Engine Applications," NASA Tech. Memo., NASA TM-102326, *The Status of NASA's Advanced High Temperature Engine Materials Technology Program* (HITEMP), 1990. p.9.

[26] Levine, S.R., *Ceramic Composites—Enabling Aerospace Materials*, NASA TM 105599 (1992) for paper presented at the 37th International Gas Turbine and Aeroengine Congress and Exposition, Cologne, Germany, June 1–4, 1992.

[27] Lamicq, P. J., and Jamet, J. F., "Thermostructural CMCs: An Overview of the French Experience," Ceramic Transactions, 57, 1–12, 1995.

[28] Spriet, P. and Habarou, G., "Applications of Continuous Fibre Reinforced Ceramic Composites in Military Turbojet Engines," in *Proceedings of the 5th Conference and Exhibition of the European Ceramic Society (ECERS V)*, P. Abelard, J. Baxter, D. Bortzmeyer, et al., eds., Versailles, France, June 22–26, 1997, Key Engineering Materials, 132–136[3], 1930–1933, 1997.

[29] Terpstra, R. A., Gubbels, F., Roman, Y., van der Heijde, H., and Brinkman, E.," Ceramic Components in Gas Turbines," Klei/Glas/Keram., 17[6/7], 14–15, 1996.

[30] Gubbels, G. H. M., van der Heijde, J. C. T., Brinkman, H. W., Linden, J. L., and Terpstra, R. A., "Development of Ceramic Components for Gas Turbine Application," in *Proceedings of the 5th Conference and Exhibition of the European Ceramic Society (ECERS V)*, P. Abelard, J. Baxter, D. Bortzmeyer, et al., eds., Versailles, France, June 22–26, 1997, Key Engineering Materials, 132–136[3], 2096–2099, 1997.

[31] Iffländer, K., Gualco, G. C., Rosatelli, F., Hampshire, S., Van Brug, E. H., Ferrière, A., Rodriguez, G. P., Fordham, R. J., Dupel, P., Mason, N., and Nedele, M., "Selective Material Characterization of Ceramic Matrix Composites for Advanced Gas Turbines," Ceram. Eng. and Sci. Proc., 18[3], 625–632, 1997.

[32] Kaya, H., "The Application of Ceramic-Matrix Composites to the Automotive Ceramic Gas Turbine," Composites Science and Technology, 59, 861–872, 1999.

[33] Nishio, K., Igashira, K. I., Take, K., and Suemitsu, T., "Development of a Combustor Liner Composed of Ceramic Matrix Composite (CMC)," Transactions of the ASME, Journal of Engineering for Gas Turbines and Power, 121[1], 12–17, 1999.

[34] Nishio, K., Igashira, K., and Suemitsu, T., "Applications of SiC/SiC to the Components of a Gas Turbine," Ceram. Jap., 34[4], 286–288, 1999.

[35] Kohyama, A. and Katoh, Y, "Overview of CREST-ACE Program for SiC/SiC Ceramic Composites and their Energy System Applications," Ceramic Transactions, 144, 3–18, 2002.

[36] Smith, K.O. and Fahme, A., ""Experimental Assessment of the Emissions Benefits of a Ceramic Gas Turbine Combustor," ASME paper 96-GT-318, presented at the International Gas Turbine and Aeroengine Congress and Exposition, Birmingham, England, U.K., June 10–13, 1996.

[37] Corman, G., and Luthra, K., "Improved Properties of Gas Turbine Engine Components Utilizing CFCC and Melt Infiltration," Ind. Heat, 63[12], 39–40, 1996.

[38] Corman, G.S., Brun, M.K., and Luthra, K.L., "SiC Fiber Reinforced SiC-Si Matrix Composites Prepared by Melt Infiltration (MI) for Gas Turbine Engine Applications," ASME paper 99-GT-234, presented at the International Gas Turbine and Aeroengine Congress and Exposition, Indianapolis, IN, USA, June 7–10, 1999.

[39] Price, J., Jimenez, O., Parthasarathy, V., Miriyala, N., and Leroux, D., "Ceramic Stationary Gas Turbine Development Program—Seventh Annual Summary," ASME Paper 2000-GT-0075, presented at the ASME TURBO EXPO, Land, Sea & Air 2000, Munich, Germany, May 8–11, 2000.

[40] Simpson, J. F., Parthasarathy, V. M., Fahme, A," Testing of Monolithic Ceramics and Fiber Reinforced Ceramic Composites for Gas Turbine Combustors," Ceram. Eng. and Sci. Proc., 18[4], 229–236, 1997.

[41] Fohey, W. R., Battison, J. M., Nielsen, T. A., and Hastings, S, "Ceramic Composite Turbine Engine Component Evaluation," Ceram. Eng. and Sci. Proc., 16[4], 459–466, 1995.

[42] Kameda, T., Sayano, A., Amiji, N., Ichikawa, H., Hamada, H., Fujita, A., and Uozumi, T., "Fabrication and Mechanical Properties of Reaction Sintered Silicon Carbide Matrix Composite," Ceram. Eng. and Sci. Proc., 18[3], 419–426, 1997.

[43] Matsumoto, M., Fujioka, N., Hayakawa, T., Kawamura, N., and Sato, K., "Development of a Combustion Chamber Liner Utilizing a Long-Fiber Reinforced Composite Material Made Using the Polysilazane Impregnation Method and the Chemical Vapor Deposition Method," Key Engineering Materials, 164, 49–52, 1998.

[44] Itoh, Y., Kameda, T., Suyama, S., Nishida, K., and Okamura, T., "Development of Continuous Fiber Reinforced Reaction Sintered Silicon Carbide Matrix Composite and Fabrication of Gas Turbine Hot Parts," Materia Japan, 39[1], 63–65, 2000.

[45] Holmquist, M., Sudre, O., Lundberg, R., Molliex, L., Razzell, T., and Adlerborn, J., "Development of Ultra High Temperature Ceramic Composites for Gas Turbine Combustors," ASME paper 97-GT-413, presented at the 42nd International Gas Turbine & Aeroengine Congress & Exposition, Orlando, FL, USA, June 2–5, 1997.

[46] Holmquist M., Lundberg, R., Sudre, O., Razzell, A. G., Molliex, L., Benoit, J., and Adlerborn, J.," Alumina/Alumina Composite with a Porous Zirconia Interphase. Processing, Properties and Component Testing," J. Eur. Ceram. Soc., 20, 599–606, 2000. © 2000, Elsevier, All rights reserved.

[47] Sudre, O., Razzell, A. G., Molliex, L., and Holmquist, M.," Alumina Single-Crystal Fiber Reinforced Alumina Matrix For Combustor Tiles," Ceram. Eng. and Sci. Proc., 19[4], 273–280, 1998.

[48] Razzell, A. G., Molliex, L., Holmquist, M., and Sudre, O., "Oxide/Oxide Ceramic Matrix Composites in Gas Turbine Combustors," ASME paper 98-GT-30, presented at the ASME International Gas Turbine and Aeroengine Congress and Exhibition, Stockholm, Sweden, June 2–5, 1998.

[49] Filsinger, D., Münz, S., Schulz, A., Wittig, S., and Andrees, G.," Experimental Assessment of Fiber Reinforced Ceramics for Combustor Walls," ASME paper 97-GT-154, presented at the 42nd International Gas Turbine & Aeroengine Congress & Exposition, Orlando, FL, USA, June 2–5, 1997.

[50] Zawada, L. P. and Lee, S. S., "Evaluation of Four CMCs for Aerospace Turbine Engine Divergent Flaps and Seals," Ceram. Eng. and Sci. Proc., 16[4], 337–339, 1995.

[51] Staehler, J.M. and Zawada, L.P., "Performance of Four Ceramic-Matrix Composite Divergent Flap Inserts Following Ground Testing on an F110 Turbofan Engine," J. Am. Ceram. Soc., 83[7], 1727–1738, 2000.

[52] Das, S., Harlan, N., Lee, G., Beaman, J.J., Bourell, D.L., Barlow, J.W., Fuesting, T., Brown, L., and Sargent, K, "Direct SLS Processing for Production of Cermet Composite Turbine Sealing Components," Mater. Manuf. Processes, 13[2], 241–261, 1998.

[53] Morrison, J. A. and Kauth, K. M.," Design and Analysis of a CMC Turbine Blade Tip Seal for a Land-based Power Turbine," Ceram. Eng. and Sci. Proc., 19[4], 249–256, 1998.

[54] Khandelwal, P., Shepard, S. M., and Ahmed, T., "Thermal Wave Nondestructive Evaluation of Ceramic Matrix Composite Components," Ceram. Eng. and Sci. Proc., 18[4], 425–433, 1997.

[55] Jarmon, D. C. and Prewo, K. M., "Rotor for a Gas Turbine Engine," US Patent Number 4808076, 28 Feb. 1989.

[56] Huynh, T., Burkland, C. V., and Bustamante, B, "Densification of a Thick Disk Preform with Silicon Carbide Matrix by CVI," Ceram. Eng. and Sci. Proc., 12[9–10], 2005–2014, 1991.

[57] Klacka, R. W., Headinger, M. H., Moschelle, W. R., and Farr, J. D., *C/SiC Turbine Rotor Development*, NASA Conference Publication No. 3235, Pt. 2, 1994. pp. 855–862.

[58] Natsumura, T., Nojima, Y., Suzumura, N., Araki, T., Moriya, K., and Yasuhira, K., "Component Design of CMC and MMC Rotor for Turbine Engine Applications," in *Advanced Materials & Processes: Preparing for the New Millenium*, Society for the Advancement of Material and Process Engineering, USA, Oct. 1999. pp. 130–141.

[59] Okura, A., Tanatsugu, N., Naruo, Y., and Tachibana, M, "On the Test of C/C Composite Turbine Blade," *Advanced Composites '93: International Conference on Advanced Composite Materials*, Wollongong, Australia, Feb 15–19, 1993, The Minerals, Metals & Materials Society, USA, 1993. pp. 147–153.

[60] Araki, T., Suzumura, N., Masaki, S., Natsumura, T., Onozuka, M., Ohnabe, H., and Yasuhira, K, "Manufacturing of Ceramic Matrix Composite Rotor for Advanced Gas-Generator," Ceram. Eng. and Sci. Proc., 19[4], 241–248, 1998.

[61] Sanokawa, Y., Ido, Y., Sohda, Y., Nakazawa, N., and Kaya, H., "Application of Continuous Fiber Reinforced Silicon Carbide Matrix Composites to a Ceramic Gas Turbine Model for Automobiles," Ceram. Eng. and Sci. Proc., 18[4], 221–228, 1997.

[62] DiMascio, P. S., Orenstein, R. M., Schroder, M. S., Tognarelli, L., Corman, G. S., and Dean, A. J., Chapter 9 in *Ceramic Gas Turbine Design and Test Experience, Progress in Ceramic Gas Turbine Development, Volume I,* van Roode, M., Ferber, M. K., and Richerson, D. W., eds., ASME Press, New York, NY, USA, 2002. pp. 225–236.

[63] *The 9th International Symposium on Ultra-high Temperature Materials '99*, Japan Ultra-high Temperature Materials Research Center, Tajimi City, Japan, 1999.

[64] Lara-Curzio, E. and Jenkins, M. G., "Development of Test Standards for CFCCs in the US," Composites Part A, 30[4], 561–567, 1999.

[65] Lara-Curzio, E., "Status of Standardization Activities for Ceramic Matrix Composites in the United States," Ceramic Transactions, 144, 191–206, 2002.

[66] Mizuno, M., Cao J-W., Nagano, Y., and Kaya, H, "Development On Standard Test Methods For Ceramic Matrix Composites," Ceram. Eng. and Sci. Proc., 19[3], 541–548, 1998.

[67] Lara-Curzio, E., in "Continuous Fiber Ceramic Composite Program Task 2: Supporting Technologies," in *Bimonthly Progress Report for October-November 1994*, Oak Ridge National Laboratory, Oak Ridge, TN, USA, 1994. pp. 25–30.

[68] Evans, A. G., Zok, F. W., and Mackin, T. J., "The Structural Performance of Ceramic Matrix Composites," in *High Temperature Mechanical Behavior of Ceramic Composites*, Nair, S. V. and Jakus, K., eds., Butterworth-Heinemann, Newton, MA, USA, 1995. pp. 3–86.

[69] Huger, M., Fargeot, D., and Gault, C., "Ultrasonic Characterization of Oxidation Mechanisms in Nicalon/C /SiC Composites," J. Am. Ceram. Soc., 77[10], 2554–2560, 1994.

[70] Sakata, M. and Ohnabe, H., "Measurement of Orthotropic Elastic Constants of Ceramic Matrix Composite from Impact Sound," in *Thermal and Mechanical Test Methods and Behavior of Continuous-Fiber Ceramic Composites, ASTM STP 1309*, Jenkins, M. G., Gonczy, S. T., Lara-Curzio, E., Ashbaugh, N. E., and Zawada, L. P., eds., American Society for Testing and Materials, West Conshohocken, PA, USA, 1997. pp. 291–305.

[71] Camus, G., El Bouazzaoui, R., and Baste, S, "Development of Damage in a 2D Woven C/SiC Composite Under Mechanical Loading. II. Ultrasonic Characterization," Composites Science and Technology, 56[12], 1373–1382, 1996.

[72] Camus, G., Guillaumat, L., and Baste, S, "Development of Damage in a 2D Woven C/SiC Composite Under Mechanical Loading. I. Mechanical Characterization," Composites Science and Technology, 56[12],1363–1372, 1996.

[73] Liaw, P. K., Yu, N., Hsu, D. K., Miriyala, N., Saini, V., Snead, L. L., McHargue, C. J., and Lowden, R. A., "Moduli Determination of Continuous Fiber Ceramic Composites (CFCCs)," J. Nucl. Mater., 219[1–3], 93–100, March 1995.

[74] Camus, G. and Barbier, J. E., "Tensile Behavior of a 2D Woven C/SiC Composite at Ambient and Elevated Temperatures," Ceramic Transactions, 57, 407–412, 1995.

[75] Steffier, W. S., "The Orthotropic Mechanical Behavior of Nicalon® Fiber-Reinforced SiC-Matrix Composites," Ceram. Eng. and Sci. Proc., 14[9–10], 1045–1057, 1993.

[76] Chulya, A., Gyekenyesi, J. Z., and Gyekenyesi, J. P., "Failure Mechanisms of 3-D Woven SiC/SiC Composites Under Tensile and Flexural Loading at Room and Elevated Temperatures," Ceram. Eng. and Sci. Proc., 13[7–8], 420–432, 1992.

[77] *Data Sheet*, Allied Signal Composites, Inc., Newark, DE, USA.

[78] Wurtinger, H. and Mühlratzer, A, "Cost Effective Manufacturing Methods for Structural Ceramic Matrix Composite (CMC) Components," ASME Paper 96-GT-296, presented at the 1996 International Gas Turbine and Aeroengine Congress and Exhibition, Birmingham, UK, June 10–13, 1996.

[79] Heathcote, J. A., Gong, X-J., Yang, J. Y., Ramamurty, U., and Zok, F. W., "In-Plane Mechanical Properties of an All-Oxide Ceramic Composite," J. Am. Ceram. Soc., 82[10], 2721–2730, 1999.

[80] Jenkins, M. G., Lara-Curzio, E., Gonczy, S. T., and Zawada, L. P., "Multiple-Laboratory Round-Robin Study of the Flexural, Shear and Tensile Behavior of a Two-Dimensionally Woven Nicalon™/Sylramic™ Ceramic Matrix Composite," in *Mechanical, Thermal and Environmental Testing and Performance of Ceramic Composites and Components, ASTM STP 1392*, Jenkins, M. G., Lara-Curzio, E., and Gonczy, S. T., eds., American Society for Testing and Materials, West Conshohocken, PA, USA, 2000. pp. 15–30.

[81] *Data Sheet*, COI Ceramics Inc., San Diego, CA, USA, 2003.

[82] C1275-00, *Standard Test Method for Monotonic Tensile Behavior of Continuous Fiber-Reinforced Advanced Ceramics with Solid Rectangular Cross-Section Specimens at Ambient Temperatures*, American Society for Testing and Materials, West Conshohocken, PA, USA, 1990.

[83] Piccola, J. P., Jenkins, M. G., and Lara-Curzio, E., in *Thermal and Mechanical Test Methods and Behavior of Continuous-Fiber Ceramic Composites, ASTM STP 1309*, Jenkins, M. G., Gonczy, S. T., Lara-Curzio, E., Ashbaugh, N. E., and Zawada, L. P., eds., ASTM, West Conshohocken, PA, USA, 1997. pp. 3–15.

[84] Nozawa, T., Hinoki, T., Katoh, Y., Kohyama, A., and Lara-Curzio, E., "Influence of Specimen Geometry on Tensile Properties of 3-D SiC /SiC Composites," Ceramic Transactions, 124, 351–362, 2001.

[85] Nozawa, T., Katoh, Y., Kohyama, A., and Lara-Curzio, E., "Specimen Size Effect on the In-Plane Shear Properties of SiC/SiC Composites," Ceramic Transactions, 139, 127–138, 2002.

[86] Lacombe, A. and Bonnet, C., "Ceramic Matrix Composites: Key Materials for Future Space Plane Technologies," Paper AIAA-90-5208 presented at the *AIAA Second International Aerospace Planes Conference,* October 29–31, 1990, Orlando, FL, USA.

[87] Halbig, M. C., Brewer, D. N., Eckel, A. J., and Cawley, J. D., "Stressed Oxidation of C/SiC Composites," Ceram. Eng. and Sci. Proc., 18[3], 547–554, 1997.

[88] Effinger, M. R., Tucker, D. S., and Barnett, T. R., "Tensile and Interlaminar Shear Evaluation of DuPont Lanxide CMCs," Ceram. Eng. and Sci. Proc.,17[4], 316–323, 1996.

[89] Choi, S. R. and Gyekenyesi, J. P., "Effect of Load Rate on Tensile Strength of Various CFCCs at Elevated Temperatures—An Approach to Life-Prediction Testing," Ceram. Eng. and Sci. Proc., 22[3], 597–606, 2001.

[90] Sørensen, B. F. and Holmes, J. W., "Effect of Loading Rate on the Monotonic Tensile Behavior of a Continuous-Fiber-Reinforced Glass-Ceramic Matrix Composite," J. Am. Ceram. Soc., 79[2], 313–320, 1996.

[91] Lara-Curzio, E., Unpublished Results, 2003.

[92] Lara-Curzio, E. and C. M. Russ, "On the Matrix Cracking Stress and the Redistribution of Internal Stresses in Brittle-Matrix Composites," J. Mater. Sci. Eng., A250, 270–278, 1998.

[93] Zawada L. P. and Goecke, K. E., "Testing Methodology for Measuring Transthickness Tensile Strength for Ceramic Matrix Composites," in *Mechanical, Thermal and Environmental Testing of Ceramic Composites and Components, ASTM STP 1392*, Jenkins, M. G., Lara-Curzio, E., and Gonczy, S. T., eds., American Society for Testing and Materials, West Conshohocken, PA, USA, 2000. pp. 62–85

[94] Percival, M. J. L., Claxton, E., Gabelotaud, S., Dambrine, B., and Maire, J. F., "Material Testing for Multiaxial Macroscopic Modeling of Ceramic Matrix Composites," Key Engineering Materials, 127–131, 791–798, 1997.

[95] C1468-00, *Standard Test Method for Transthickness Tensile Strength of Continuous Fiber-Reinforced Advanced Ceramics at Ambient Temperature*, ASTM, American Society for Testing and Materials, West Conshohocken, PA, USA, 1990.

[96] Lara-Curzio, E., and M. G. Jenkins," Multi-Laboratory Round Robin Study of the Shear Behavior of a Continuous Fiber-Reinforced Ceramic Matrix Composite," presented at the 23rd Annual Cocoa Beach Conference and Exposition, Cocoa Beach, Florida January 28, 1999.

[97] Lara-Curzio, E., and Boisvert R. P., "Time-Dependent Deformation of Polymer-Derived CFCCs," CFCC News, 9, 13–15, August 1997,

[98] Fang, N. J.-J. and Chou, T.-W., "Characterization of Interlaminar Shear Strength of Ceramic Matrix Composites," J. Am. Ceram. Soc., 76[10], 2539–2548, 1993.

[99] Lara-Curzio, E. and Singh, M, "High-Temperature Interlaminar Shear Strength of Hi-Nicalon Fiber-Reinforced MI-SiC Matrix Composites with BN/SiC Fiber Coating," J. Mater. Sci. Letters, 19, 657, 2000.

[100] Lara-Curzio, E. and Ferber, M. K., "Shear Strength of Continuous Fiber Reinforced Ceramic Composites," in *Thermal and Mechanical Test Methods and Behavior of Continuous-Fiber Ceramic Composites, ASTM STP 1309*, Jenkins, M. G., Gonczy, S. T., Lara-Curzio, E., Ashbaugh, N. E., and Zawada, L. P., eds., American Society for Testing and Materials, West Conshohocken, PA, USA, 1997. pp. 31–48.

[101] Mackin, T. J., Purcell, T. E., He, M. Y., and Evans, A. G., "Notch Sensitivity and Stress Redistribution in Three Ceramic-Matrix Composites," J. Am. Ceram. Soc., 78[7], 1719–1728, 1995.

[102] Gomina, M., Themines, D., Chermant, J. L. and Osterstock, F., "An Energy Evaluation for C/SiC Composite Materials," International Journal of Fracture, 34, 219–228, 1987.

[103] Stull, K. R. and Parvizi-Majidi, A., "Fracture Toughness of Fiber-Reinforced Glass-Ceramic and Ceramic Matrix Composites," Ceram. Eng. and Sci. Proc., 12[7–8], 1452–1461, 1991.

[104] Heraud, L. and Spriet, P., "High-Toughness C/SiC and SiC/SiC in Heat Engines," in *Whisker- and Fiber-Toughened Ceramics*, Bradley, R. A., Clark, D. E., Larsen, D. C. and Stiegler, J. O., eds., ASM International, Oak Ridge, TN, USA, 1988. pp. 217–224.

[105] Yang, J. M., Lin, W., Shih, C. J., Kai, W., Jeng, S. M., and Burkland, C.U., "Mechanical Behaviour of Chemical Vapour Infiltration-Processed Two- and Three-Dimensional Nicalon/SiC Composites," J. Mater. Sci., 26, 2954–2960, 1991.

[106] Goto, K. and Kagawa, Y. "Fracture Behavior and Toughness of a Plane-Woven SiC Fiber-Reinforced SiC Matrix Composite," Materials Science and Engineering A, 211, 72–81, 1996.

[107] Xu, Y., Cheng, L., Zhang, L., and Yan, D., "Mechanical Properties and Microstructural Characteristics of Carbon Fiber Reinforced Silicon Carbide Matrix Composites by Chemical Vapor Infiltration," Key Engineering Materials, 164–165, 73–76, 1999.

[108] Shen, J. Y., Hirth, J. P., Zok, F. W., and Heathcote, J. A., "Effect of Notch Root Radius on the Initiation Toughness of a C-fiber/SiC-matrix Composite," Scripta Materialia, 38, 15–19, 1998.

[109] Droillard, C., Voisard, P., Heibst, C., and Lamon, J., "Determination of Fracture Toughness of 2-D Woven SiC Matrix Composites Made by Chemical Vapor Infiltration," J. Am. Ceram. Soc. 78[5], 1201–1211, 1995.

[110] Mizuno, M., Nagano, Y., Usami, H., and Kagawa, Y., "Effect of Specimen Shape on Fracture Toughness and Effective Fracture Energy in SiC-SiC Composite," Ceram. Eng. and Sci. Proc., 15[5], 859–866, 1994.

[111] Nair, S. V. and Wang, Y. L., "Failure Behavior of a 2-D Woven SiC Fiber/SiC Matrix Composite at Ambient and Elevated Temperatures," Ceram. Eng. and Sci. Proc., 13[7–8], 433–441, 1992.

[112] Nair, S. V. and Wang, Y. L., "Toughening Behavior of a Two-Dimensional SiC/SiC Woven Composite at Ambient Temperature: I, Damage Initiation and R-Curve Behavior," J. Am. Ceram. Soc., 81[5], 1149–1156, 1998.

[113] Kramb, V. A., John, R., and Zawada, L. P, "Notched Fracture Behavior of an Oxide/Oxide Ceramic-Matrix Composite," J. Am. Ceram. Soc., 82[11], 3087–3096, 1999.

[114] Antti M., and Lara-Curzio, E., "Effect of Notches, Specimen Size, and Fiber Orientation on the Monotonic Tensile Behavior of Continuous Fiber-Reinforced Oxide/Oxide Composites at Ambient and Elevated Temperatures," Ceram. Eng. and Sci. Proc., 22[3], 643–650, 2001.

[115] John, R., Buchanan J. R. and Zawada, L. P., "Notch-Sensitivity of a Woven Oxide/Oxide Ceramic Matrix Composite," in *Mechanical, Thermal and Environmental Testing of Ceramic Composites and Components, ASTM STP 1392*, Jenkins, M. G., Lara-Curzio, E., and Gonczy, S. T., eds., American Society for Testing and Materials, West Conshohocken, PA, USA, 2000. pp. 172–181

[116] Hasselman, D. P. H. and Donaldson, K. Y., "Thermal Conductivity of Continuous Fiber-Reinforced Ceramic Matrix Composites," in *Handbook on Continuous Fiber-Reinforced Ceramic Matrix Composites*, Lehman, R. L., El-Rahaiby, S. K. and Wachtman, J. B., Eds, The American Ceramic Society, Westerville, OH, USA, 1995. pp. 547–583.

[117] Youngblood, G. E., Senor, D. J., Jones, R. H., and Graham, S. "The Transverse Thermal Conductivity of 2-D SiC/SiC Composites," Composites Science and Technology, 62, 1127–1139, 2002.

[118] Mital, S. K., Bhatt, R. T., and Murthy, P. L. N., "Influence of Constituents on the Properties of Melt-Infiltrated SiC/SiC Composites," paper AIAA 2002–1686, 2002.

[119] Beecher, S. C., Dinwiddie, R. B., and Lowden, R. A., "The Thermal Conductivity of Carbon Coated Silicon Carbide Fibers Embedded in a Silicon Carbide Matrix," in *Thermal Conductivity*, Tong, T. T., ed., Technomic Publishing Co., Inc., Tempe, Arizona, USA, 22, 324–334, 1994.

[120] Snead, L. L. and Schwarz, O. J., "Advanced SiC Composites for Fusion Applications," Journal of Nuclear Materials, 219, 3–14, 1995.

[121] Ahuja, S., Ellingson, W. A., Steckenrider, J. S., and Koch, S. J., "Infrared-Based NDE Methods for Determining Thermal Properties and Defects in Ceramic Composites," in *Thermal and Mechanical Test Methods and Behavior of Continuous-Fiber Ceramic Composites, ASTM STP 1309*, Jenkins, M. G., Gonczy, S. T., Lara-Curzio, E., Ashbaugh, N. E., and Zawada, L. P., eds., American Society for Testing and Materials, West Conshohocken, PA, USA, 1997. pp. 209–218

[122] Eckel, A. J. and Bradt, R. C., "Thermal Expansion of Laminated, Woven, Continuous Ceramic Fiber /Chemical-Vapor-Infiltrated Silicon Carbide Matrix Composites," J. Am. Ceram. Soc. 73[5], 1334–1338, 1990.

[123] Lara-Curzio, E. and Singh, M., "The Time-Dependent Deformation of Carbon Fiber-Reinforced Melt-Infiltrated Silicon Carbide Matrix Composites: Stress-Rupture and Stress-Relaxation Behavior in Air at 1000°C," in *Mechanical, Thermal and Environmental Testing of Ceramic Composites and Components, ASTM STP 1392*, Jenkins, M. G., Lara-Curzio, E., and Gonczy, S. T., eds., American Society for Testing and Materials, West Conshohocken, PA, USA, 2000. pp. 216–228

[124] Lara-Curzio, E., "Stress-Rupture, Overstressing and a New Methodology to Assess the High-Temperature Durability and Reliability of CFCCs," in *Mechanical, Thermal and Environmental Testing of Ceramic Composites and Components, ASTM STP 1392*, Jenkins, M. G., Lara-Curzio, E., and Gonczy, S. T., eds., American Society for Testing and Materials, West Conshohocken, PA, USA, 2000. pp. 107–117

[125] Eckel, A. J., Herbell, T. P., Generazio, E. R., and Gyekenyesi, J.Z., "Thermal Shock Fiber-Reinforced Ceramic Matrix Composites," Ceram. Eng. and Sci. Proc., 12[7–8], 1500–1508, 1991.

[126] Kagawa, Y., "Thermal Shock Damage in a Two-Dimensional SiC/SiC Composite Reinforced with Woven SiC Fibers," Composites Science and Technology, 57, 607–611, 1997.

[127] Wang, H., Singh, R, N., and Lowden, R. A., "Thermal Shock Behavior of Unidirectional, 0/90 and 2-D Woven Fiber-Reinforced CVI SiC Matrix Composites," J. Mater. Sci., 32[12], 3305–3313, 1997.

[128] Guillaumat, L. and Lamon, J., "Multi Fissuration de Composites SiC/SiC," (Multi-cracking of SiC/SiC Composites) in *Microstructure, Comportements Thermomèchaniques et Modélisation des Composites Céramique-Céramique á Fibres*, Vol. 3., Chermant, J. L. and Fantozzi, G., eds., Hérmes, Paris, France, 1993. pp. 159–172 (in French).

[129] Gasser, A., Ladeveze, P., and Poss, A., "Damage Mechanisms of a Woven SiC/SiC Composite: Modelling and Identification," Composites Science and Technology, 56, 779–784, 1996.

[130] Guillaumat, L. and Lamon, J., "Model of the Nonlinear Stress-Strain Behavior of a 2D-SiC/SiC Ceramic Matrix Composite (CMC)," Ceramic Transactions, 57, 215–220, 1995.

[131] Akimune, Y., Ogasawara, T., Akiba, T., and Hirosaki, N., "Spherical Particle Impact Damage Behavior of a SiC Fiber-Reinforced Chemical Vapor Infiltrated SiC Composite," J. Mater. Sci. Letters, 18, 682-692, 1991.

[132] Holmes, J. W. and Sorensen, B. F., "Fatigue Behavior of Continuous Fiber-Reinforced Ceramic Matrix Composites," in *High Temperature Mechanical Behavior of Ceramic Composites*, Nair, S. V. and Jakus, K., eds., Butterworth-Heinemann, Newton, MA, USA, 1995. pp. 261–326.

[133] Reynaud, P., "Cyclic Fatigue of Ceramic-Matrix Composites at Ambient and Elevated Temperatures," Composites Science and Technology, 56[7], 809–814, 1996.

[134] Shuler, SF., Holmes, JW., Wu, X., and Roach, D, "Influence of Loading Frequency on the Room-Temperature Fatigue of a Carbon-Fiber/SiC-Matrix Composite," J. Am. Ceram. Soc., 76[9], 2327–2336, 1993.

[135] Lara-Curzio, E., Ferber, M. K., Boisvert R., and Szweda, A., "The High Temperature Tensile Fatigue Behavior of Polymer-derived Ceramic Composites," Ceram. Eng. and Sci. Proc., 16[4], 341–349, 1995.

[136] Lee, S. S., Zawada, L. P., Staehler, J. M., and Folsom, C. A., "Mechanical Behavior and High-Temperature Performance of a Woven Nicalon/Si-N-C Ceramic-Matrix Composite," J. Am. Ceram. Soc., 81[7], 1797–1811, 1998.

[137] Steel, S. G., Zawada, L. P., and Mall, S, "Fatigue Behavior of a Nextel 720/Alumina (N720/A) Composite at Room and Elevated Temperature," Ceram. Eng. and Sci. Proc., 22[3], 695–702, 2001.

[138] Kagawa, Y., Zhu, S., Mizuno, M., and Mutoh, Y, "Monotonic Tension, Fatigue and Creep Behavior of SiC-Fiber-Reinforced SiC-Matrix Composites: A Review," Composites Science and Technology, 59[6], 833–851, May 1999.

[139] Butkus, L. M., Holmes, J. W., Nicholas, T., "Thermomechanical Fatigue Behavior of a Silicon Carbide Fiber-Reinforced Calcium Aluminosilicate Composite," J. Am. Ceram. Soc., 76 [11], 2817–2825, 1993.

[140] Weinberg, D. J., Myers, F. K., and Holmes, J. W., "Thermal-Mechanical Fatigue of Fiber-Reinforced Ceramics," Ceram. Eng. and Sci. Proc. 18[3], 555–569, 1997.

[141] Evans, A. G., Zok, F. W., McMeeking, R. M., and Du, Z. Z., "Models of High-Temperature, Environmentally Assisted Embrittlement in Ceramic Matrix Composites," J. Am. Ceram. Soc., 79[9], 2345–2352, 1996.

[142] Jones, R. H., Henager, C. H., and Windisch, C. F., "Stress-Corrosion Cracking of Silicon Carbide Fiber/Silicon Carbide Composites," J. Am. Ceram. Soc., 83[8], 1999–2005, 2000.

[143] Lara-Curzio E. and Ferber, M. K., "Evolution of Internal Stresses in Composites During Creep," Ceram. Eng. and Sci., Proc., 16[5], 791–800, 1995.

[144] Holmes, J. W. and Wu, X., "Elevated Temperature Creep Behavior of Continuous Fiber-Reinforced Ceramic Matrix Composites," in *High Temperature Mechanical Behavior of Ceramic Composites*, Nair, S. V. and Jakus, K., eds., Butterworth-Heinemann, Newton, MA, USA, 1995. pp. 261–326.

[145] Widjaja, S., Jakus, K., Ritter, J. E., Lara-Curzio, E., Sun, E. J., Watkins, T. R., and Brennan, J. J., "Creep-Induced Residual Stress Strengthening in Nicalon™ Fiber-Reinforced BMAS Glass-Ceramic Matrix Composites," J. Am. Ceram. Soc., 82[3], 657–664, 1999.

[146] Lara-Curzio, E., "Stress-Rupture of Nicalon™/SiC Continuous Fiber Ceramic Composites in Air at 950°C," J. Am. Ceram. Soc., 80[12], 3268–3272, 1997.

[147] Lara-Curzio, E. and Ferber, M. K., "Stress Rupture of Nicalon™/SiC CFCCs at Intermediate Temperatures," J. Mater. Sci. Letters, 16, 23–26, 1997.

[148] Lara-Curzio, E., "Oxidation-Induced Stress-Rupture of Fiber Bundles," J. Eng. Mater. and Techn., 120, 105–109, April 1998.

[149] Pysher, D. J., Goretta, K. C., Hodder, R. S., Jr., and Tressler, R.E., "Strengths of Ceramic Fibers at Elevated Temperatures," J. Am. Ceram. Soc., 72[2], 284–288, 1989.

[150] Lara-Curzio, E., "Analysis of Oxidation-Assisted Stress-Rupture of Continuous Fiber-Reinforced Ceramic Matrix Composites at Intermediate Temperatures," Composites Part A, 30[4], 549–554, 1999.

[151] Verrilli, M. J., Calomino, A. M., and Brewer, D. N., in *Thermal and Mechanical Test Methods and Behavior of Continuous Fiber Ceramic Composites*, *ASTM STP 1309*, Jenkins, M. G., Gonczy, S. T., Lara-Curzio, E., Ashbaugh, N. E., and Zawada, L. P., eds. American Society for Testing and Materials, 1997. pp. 158–175

[152] Lipetzky, P., Stoloff, N. S., and Dvorak, G. J., "Atmospheric Effects on High-Temperature Lifetime of Ceramic Composites," Ceram. Eng. and Sci. Proc., 18[4], 355–362, 1997.

[153] Ogasawara, T., Ishikawa, T., and Suzuki, N., "Tensile Creep Behavior of 3-D Woven Si-Ti-C-O Fiber/SiC-based Matrix Composite with Glass Sealant," J. Mater. Sci., 35, 785–793, 2000.

[154] Mizuno, M., Zhu, S., Kagawa, Y., and Kaya, H, "Stress, Strain and Elastic Modulus Behavior of SiC/SiC Composites During Creep and Cyclic Fatigue," J. Eur. Ceram. Soc., 18[13], 1869–1878, 1998.

[155] Kleykamp, H., Schauer, V., and Skokan, A., "Oxidation Behavior of SiC Fiber-reinforced SiC," Journal of Nuclear Materials, 227, 130–137, 1995.

[156] Fox, D. and Nguyen, Q. N., "Oxidation Kinetics of Enhanced SiC/SiC," Ceram. Eng. and Sci. Proc., 16[5], 877–884, 1995.

[157] De Gaudenzi, G. P., Taut, C., Iffländer, K., and Corcoruto, S," Environmental Resistance of Ceramic Matrix Composite Materials for Gas Turbine Applications," Key Engineering Materials, 132-136[3], 1637–1640, 1997.

[158] Filipuzzi, L. and Naslain, R., "Oxidation Mechanism and Kinetics of 1D-SiC/C/SiC Composite Materials: II, Modelling," J. Am. Ceram. Soc. 77[2], 467–480, 1994.

[159] Filipuzzi, L., Camus, G., Naslain, R., and Thebault, J., "Oxidation Mechanism and Kinetics of 1D-SiC/C/SiC Composite Materials: I: An Experimental Approach," J. Am. Ceram. Soc., 77[2], 459–466, 1994.

[160] Lamouroux, F. and Camus, G., "Oxidation Effects on the Mechanical Properties of 2D Woven C/SiC Composites," J. Eur. Ceram. Soc., 14, 177–188, 1994.

[161] Lamouroux, F., Bertrand, S., Pailler, R., Naslain, R., and Cataldi, M., "Oxidation-Resistant Carbon-Fiber-Reinforced Ceramic-Matrix Composites," Composites Science and Technology, 59, 1073–1085, 1999.

[162] Lamouroux, F., Camus, G., and Thebault, J. "Kinetics and Mechanisms of Oxidation of 2D Woven C/SiC Composites: I, Experimental Approach," J. Am. Ceram. Soc., 77[8], 2049–2057, 1994.

[163] Ogbuji, L.U., "Degradation of SiC/BN/SiC Composite in the Burner Rig," Ceram. Eng. and Sci. Proc., 19[4], 257–264, 1998.

[164] Miriyala, N., Kimmel, J., Price, J., More, K. L., Tortorelli, P., Eaton, H., Linsey, G., and Sun, E., "The Evaluation of CFCC Liners After Field Testing in a Gas Turbine," ASME paper GT-2002-30585, presented at the ASME Turbo Expo 2002, Land, Sea & Air, Amsterdam, The Netherlands, June 3–6, 2002.

[165] Jacobson, N. S., Smialek, J. L., Fox, D. S., and Opila, E. J, "Durability of Silica-Protected Ceramics in Combustion Atmospheres," Ceram.Trans., 57, 157–170, 1995.

[166] Opila, E. J. and Hann, R. E., "Paralinear Oxidation of CVD SiC in Water Vapor," J. Am. Ceram. Soc., 80[1], 197–205, 1997.

[167] Opila, E. J., Fox, D. S., and Jacobson, N. S., "Mass Spectrometric Identification of Si-O-H(g) Species from the Reaction of Silica with Water Vapor at Atmospheric Pressure," J. Am. Ceram. Soc., 80[4], 1009–1012, 1997.

[168] Webster, J.D., Hayes, F.H., and Taylor, R," Thermodynamic Modeling and Experimental Studies in the Design of Integrated Oxidation Protection Systems for Ceramic Matrix Composites," Key Engineering Materials,127–131[2AB], 1225–1232, 1997.

[169] "Lee, K. N, "Current Status of Environmental Barrier Coatings for Si-Based Ceramics," Surface and Coatings Technology, 133–134, 1–7, Nov. 2000.

[170] Lee, K. N., Jacobson, N. S., and Miller, R. A., "Refractory Oxide Coatings on SiC Ceramics," MRS Bulletin, 35–38, October1994.

[171] Eaton, H. E. and Linsey, G. D, "Accelerated Oxidation of SiC CMCs by Water Vapor and Protection Via Environmental Barrier Coating Approach," J. Eur. Ceram. Soc., 22[14–15], 2741–2747, 2002.

[172] Kimmel, J., Miriyala, N., Price, J., More, K., Tortorelli, P., Eaton, H., Linsey, G., and Sun, E, "Evaluation of CFCC Liners with EBC After Field Testing in a Gas Turbine," J. Eur. Ceram. Soc., 22[14–15], 2769–2775, 2002.

[173] Richardson, G. Y., Kowalik, R. W.," Oxidation and Hot Corrosion of Dupont Lanxide Enhanced SiC/SiC and Hitco SiC/C Composites," Ceram. Eng.and Sci. Proc., 18[3], 599–606, 1997.

[174] Kumar, A, Oppici, M.A., Fox, A. G., and Wang S, "Effects of Hot Corrosion on the Microstructure of a Silicon Carbide Fiber-Reinforced Calcium Aluminosilicate," Ceram.Trans., 57, 455–458, 1995.

[175] Lara-Curzio, E., "Continuous Fiber Ceramic Composite Program Task 2: Supporting Technologies," in *Bimonthly Progress Report for December-January 1996*, Oak Ridge National Laboratory, Oak Ridge, TN, USA, 1996. pp. 25–29.

[176] Sato, K., Tezuka, A., Funayama, O., Isoda, T., Terada, Y., Kato, S., and Iwata, M., "Fabrication and Pressure Testing of a Gas-Turbine Component Manufactured by a Preceramic-Polymer Impregnation Method," Composites Science and Technology, 59 853–859, 1999. © 2000, Elsevier, All rights reserved.

[177] Wilson, D. M., and Visser, L. R., "Nextel 650 Ceramic Oxide Fiber: New Alumina-Based Fiber For High Temperature Composite Reinforcement," Ceram. Eng. and Sci. Proc., 21[4], 363–373, 2000.

CHAPTER

23

NONDESTRUCTIVE EVALUATION/ CHARACTERIZATION

William A. Ellingson
Argonne National Laboratory, Argonne, Illinois, USA

Yasushi Ikeda
Japan Fine Ceramics Center, Nagoya, Aichi, Japan

Jürgen Goebbels
Bundesanstalt für Materialforschung und Prüfung (BAM), Berlin, Germany

INTRODUCTION

Advanced structural ceramic materials, such as monolithic Si_3N_4, SiC, or ceramic matrix composites (CMCs), e.g., SiC/SiC, C/SiC, and oxide/oxide have been under development to meet the demanding environmental conditions of advanced gas turbines [1]. Indeed, the evolution of ceramic materials for turbine engine applications can be traced to the early 1970s when initial programs were developed in the U.S. [2] and Germany [3], and in the early 1980s, in Japan [4]. One historical issue with monolithic materials has been their relatively low fracture toughness and corresponding small critical flaw size, which, in the early materials, were considered to be of the order of 10–30 μm [5–9]. The small size of the flaws presented challenges to the available nondestructive-evaluation (NDE) technologies; hence, several new programs were initiated to advance the state of the art of the technology [10–14]. The challenge was so strong that many initial studies [15,16] of nondestructive evaluation/characterization technology (NDE/C) found little or no correlation between flaws detected and any significant mechanical property, e.g., 4-point bend strength or tensile strength. At that time, the available NDE methods, which lacked computer control and high-capability digital signal and image processing, were those that had been developed for the metals industry. They included conventional film-based through-transmission X-ray imaging with relatively large focal-spot X-ray sources, pulse-echo and through-transmission water-coupled C-scan ultrasonics, conventional dye penetrants, and

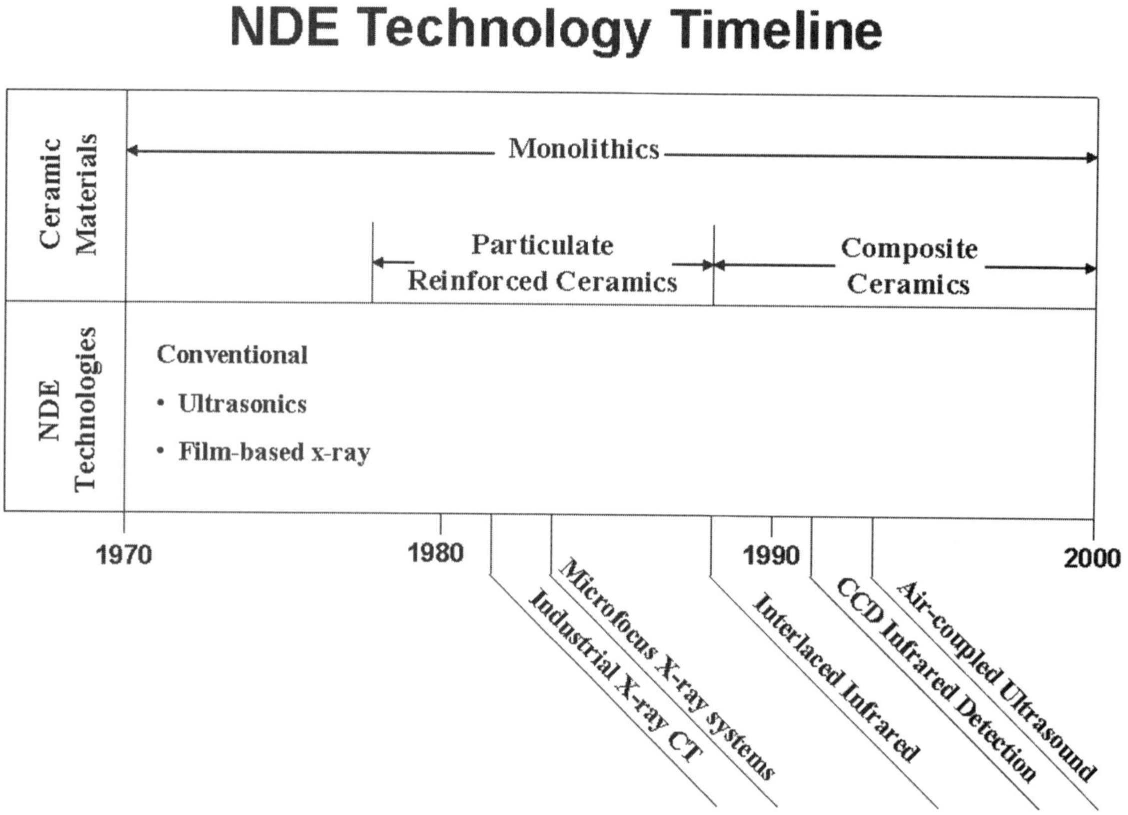

Figure 23.1 Diagram of Historical Development of NDE Technology for Ceramic Materials for Gas Turbines

some fluorescent penetrants. More recently, monolithic ceramics with higher fracture toughness and correspondingly larger critical flaw sizes of 50–80 μm. (many even as high as 100 μm) have been developed for turbines [16,17]. These material property advances, coupled with new advances in NDE/C technology, now allow detection of flaws in the size range of 50–80 μm.

A brief historical perspective of the development of NDE technology for ceramics used in gas turbines is shown in Figure 23.1. Please note that both monolithic structural ceramic materials and the more recently developed continuous-fiber ceramic matrix composite materials are shown on the diagram. The advent of desk-top personal computers is noted because it very significantly influenced the development of high-performance NDE technologies and automation. However, probability of detection (POD) still needs to be extensively studied.

During the development of very-high-fracture-toughness CMCs, in which defects or defect regions can be quite large (several mm), field engine tests have shown that these composites can contain relatively large defects and still perform under static-component conditions [18–20]. In rotating machinery applications with relatively short test times (<500 hrs) but rotation speeds that have exceeded 15,000 RPM, C/SiC composites have performed well, even with ≈1-mm defects [21]. One factor that has allowed NDE/C technology to advance in recent years, has been the development of high-performance computers, which allow sophisticated data acquisition and digital signal processing technologies, previously not possible. Almost all NDE/C technologies are now computer controlled and all data are presented through some form of digital signal and/or image processing.

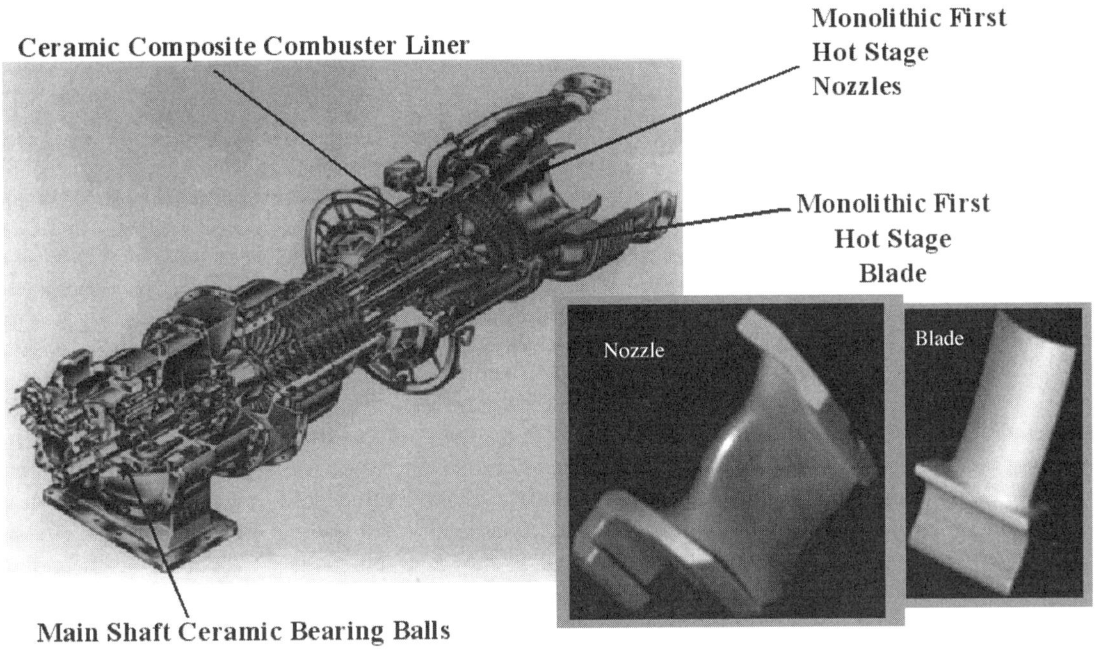

Figure 23.2 Cut Away Diagram Showing Location of Some Ceramic Components of Interest in a Typical Industrial Gas Turbine (Adapted with permission from Solar Turbines Incorporated)

The number of ceramic components under consideration for gas turbine engine applications varies, depending on the size of the engine. For one 300-kW-class engine that was under development in Japan [22], many smaller components were developed. For larger gas turbines, such as the Solar Turbines Centaur 50S, an engine with an output power of ≈4 MWe, the ceramic components included the combustor, the first-stage blade, and the first-stage nozzle [19–21]. Figure 23.2 is a cut-away diagram that shows the location of the components. Of these, only the first-stage blade is a rotating component. To date, the primary structural ceramic material of choice for blades and nozzles is Si_3N_4, because it is more resistant to oxidation and stronger than other monolithic ceramics, such as SiC. For the combustor, recent work has focused on the development of ceramic-matrix composites, because, in the combustion chamber, the concern is detection of material loss (recession rate) as well as burn-throughs and losses of thermal capacity that cause higher thermal stresses [23,24]. In the case of rotating monolithic blades and stationary vanes, where there is high surface stress, detection of surface flaws has been demonstrated by recent advances in fluorescent-penetrant technology as well as by the use of new methods based on elastic optical scatter that uses low-power lasers. Internal detection of critical flaw size in these complex shapes has also been demonstrated through the use of new very-high-spatial-resolution X-ray computed tomography. Detection of creep damage, and the onset of creep, in both blades and vanes is under study by ultrasonic and laser scatter methods that use a special reflectometer setup. These NDE/C developments and their historical backgrounds are explained in the sections that follow.

NDE/C TECHNOLOGY FOR MONOLITHIC TURBINE ENGINE COMPONENTS

Figure 23.3 shows an example of critical flaw size as a function of applied stress for two fracture modes: fast fracture and slow crack growth for a specific material. Such plots are very material dependent and provide direction for detection sensitivity of NDE/C. The following sections describe NDE/C technology for detecting internal cracks and porosity, creep, and surface-breaking cracks.

Detection of Internal Cracks and Porosity

Internal cracks and regions of porosity of monolithic materials are of interest because these can be locations for failure initiation. Various NDE methods for detecting such defects have been studied. The primary methods are X-ray radiography and ultrasonics.

Radiographic Imaging: Computed Tomography. Although conventional transmission X-ray imaging with film [25,26] is a fairly common practice, reliable use of film radiography to detect flaws in complex-shaped turbine components, such as blades and vanes, is very difficult. Thus, over the past 10 years, significant effort has been expended in the field of high-resolution X-ray tomographic imaging [27–29]. X-ray tomographic imaging is a desirable NDE method because it can provide detailed information about a component, including internal features. The first industrial X-ray computed tomography (CT) system was delivered in 1983. Data acquisition was slow and long times were necessary to reconstruct images. Recently, however, with new large-area solid-state X-ray detectors and fast computers, data can be acquired for an entire three-dimensional X-ray image in <6 min, and images of a cross-section, in <1 sec. Figure 23.4 shows

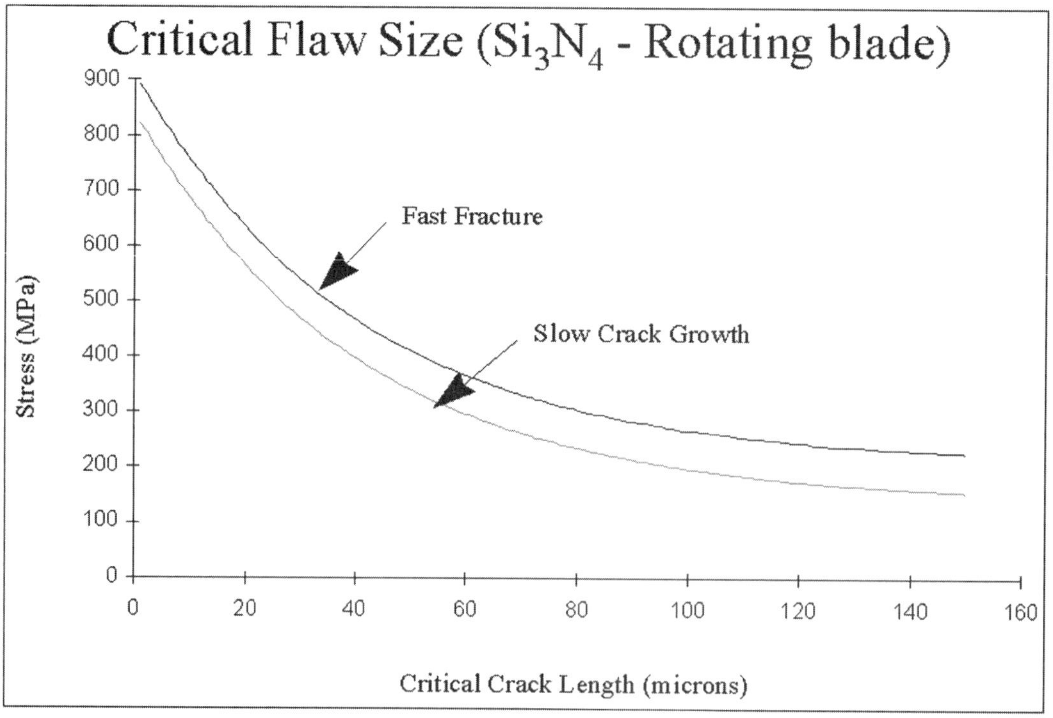

Figure 23.3 Relation between Critical Flaw Size and Stress for 2 Failure Modes for a Typical Silicon Nitride Material (Adapted with permission from Solar Turbines Incorporated).

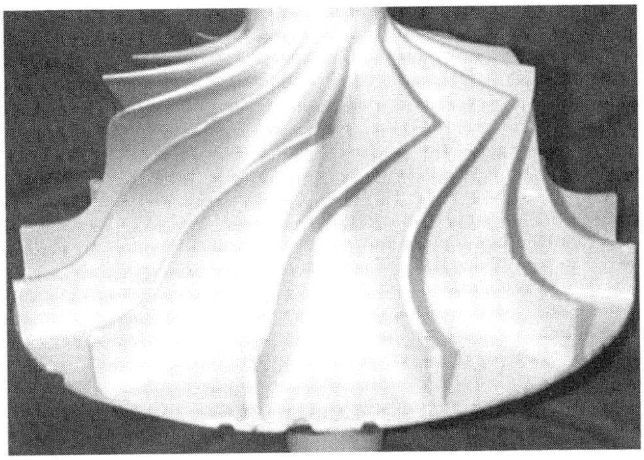

(a)

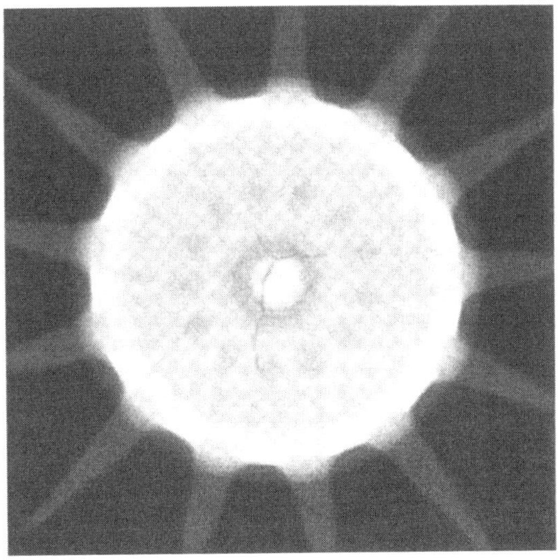

(b)

Figure 23.4 Gelcast 23 cm Diameter Rotor; (a) Photograph, (b) High-Speed X-Ray CT Image Showing Fine Internal Cracks

a 23-cm-diam radial-flow monolithic ceramic rotor and an X-ray tomographic image from fast data acquisition with one slice reconstructed in <1 sec.

Radiographic Imaging: Film and Digital Radiography. An important aspect of radiographic imaging of small defects in monolithic ceramics has been the development of small-focal-spot X-ray sources. These microfocus X-ray sources are important, because, to detect very small defects in monolithic ceramics, X-ray radiography with large magnification must be employed. Although film currently may allow higher spatial resolution, when compared with available solid-state detectors such as amorphous silicon or amorphous selenium [30–32], new solid-state detectors are under development and their spatial resolution may exceed that of film. Recent work with an image intensifier/TV camera system as a detector has been developed for improved X-ray imaging [33]. The method includes a unique digital image subtraction process that uses a translational shift of the object, with digital X-ray images acquired before and after shifting. The result is a pair of images from which a very small defect can be extracted. The resultant image can be used with very strong digital image contrast enhancement to extract small features. Figure 23.5 shows an example of the detectability of pores in a Si_3N_4 plate as a function of the ratio of pore size to plate thickness [34]. This method is now used by one fabricator of ceramic gas turbine parts in Japan.

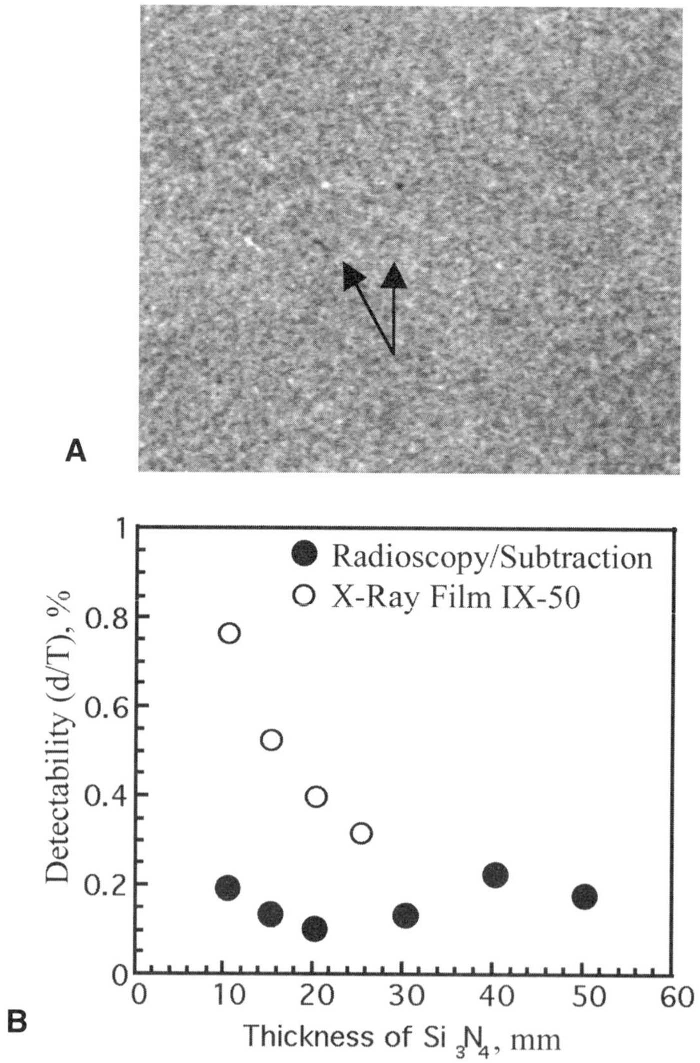

Figure 23.5 (a) Pair Image of a 20 μm pore (b) Detectability of Pore Defect in Si_3N_4 Ceramics (d/T = pore diameter/material thickness) (Adapted with permission from ref. [34]).

Focused Ultrasonics. Detection of internal porosity by ultrasonic methods has been studied by Ikeda et al. [35]. To demonstrate this method, reference blocks of silicon nitride were prepared, with several artificial pores that ranged in diameter from 50 to 500 μm. The depth positions of the pores varied from 0.5 to 17.2 mm. Figure 23.6 shows ultrasonic images of pore-embedded specimens obtained by using various transducers. The optimum probe to detect pores at various depths has been determined from echo-height data.

Another approach to the detection of porosity by ultrasonic methods was developed by Fujii and Kawashima [36] and Onda, et al. [37]. By using more than 1000 overlapped pulse echoes to eliminate noise, a highly precise (1 part in 10^5) measure of wave velocity was obtained.

Figure 23.7 shows the measured relationship between the porosity of silicon nitride and longitudinal wave velocity. For regions with >1% porosity, a linear relation was obtained.

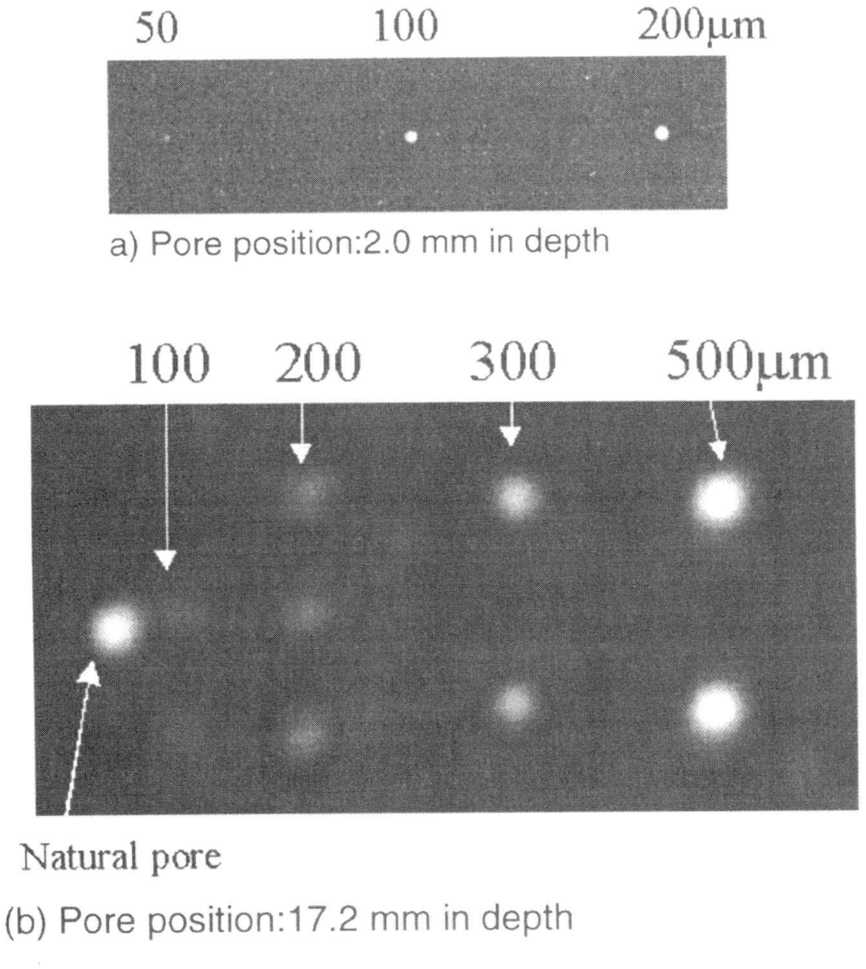

Figure 23.6 Detection of Porosity In Silicon Nitride by Focused Ultrasonic Methods: (a) Pores 2 mm Below the Surface, (b) Pores 17.2 mm Below the Surface. (Reprinted with permission from ref. [35]).

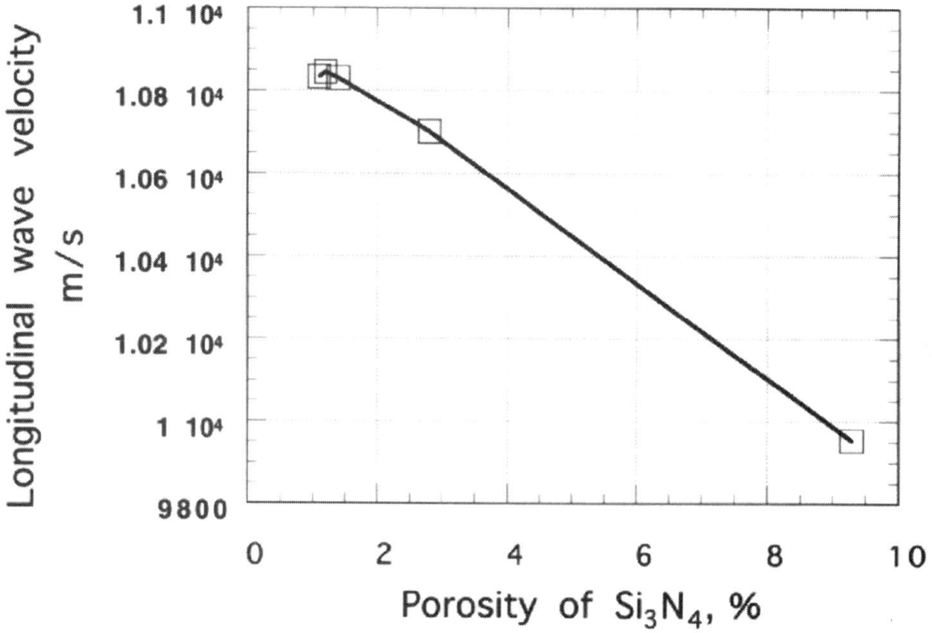

Figure 23.7 Sensitivity of Ultrasonic Velocity to Distributed Porosity Variation (Reprinted with permission from ref. [37])

Detection of Creep

Detection of creep is a concern in blades and vanes. Determination of the extent of creep in a component is an active area of study for NDE. Over the past few years two NDE/C methods of detecting creep have been investigated; one is based on scattered laser light and the other, on ultrasonics.

Elastic Optical Scattering. Creep detection based on back-scattered laser light is a new technological approach [38–40] that makes use of low-power laser-based elastic optical scatter from the surface and near-subsurface of structural ceramic materials. The elastic optical scatter method relies on the fact that many monolithic structural ceramic materials, including Si_3N_4 and SiC, are optically translucent at various optical wavelengths (see Figure 23.8). The technique is similar to classical ultrasonic technology, only rather than sound waves incident on the specimen, incident laser light probes the depth of the materials and reflected signals are detected. The data are presented in the form of a scanned area of the specimen under study.

Initial work to use polarized back-scattered laser light to detect creep was performed on a Si_3N_4 sample that was crept at 1400°C under 41.5 MPa stress. Elastic scatter data were collected; then, 50–60 μm of material was carefully polished away. Optical photomicrographs were taken of the surface, another laser scan was conducted, and the process was repeated until the laser data suggested no more creep damage. The optical-photomicrograph data and the prior laser scatter data consistently exhibited one-to-one correspondence. Figure 23.9 shows these correlations.

The detection sensitivity of the method has been further tested with a Si_3N_4 polished-surface flat specimen with subsurface Hertzian cone cracks created by using small-diameter balls with differing known loads. Figure 23.10 shows a schematic diagram of the specimen and also shows the resultant surface and subsurface laser scans. Note that, in the surface scan, only the known surface-breaking crack is detected whereas, in the subsurface scan, only the subsurface Hertzian cone cracks are detected.

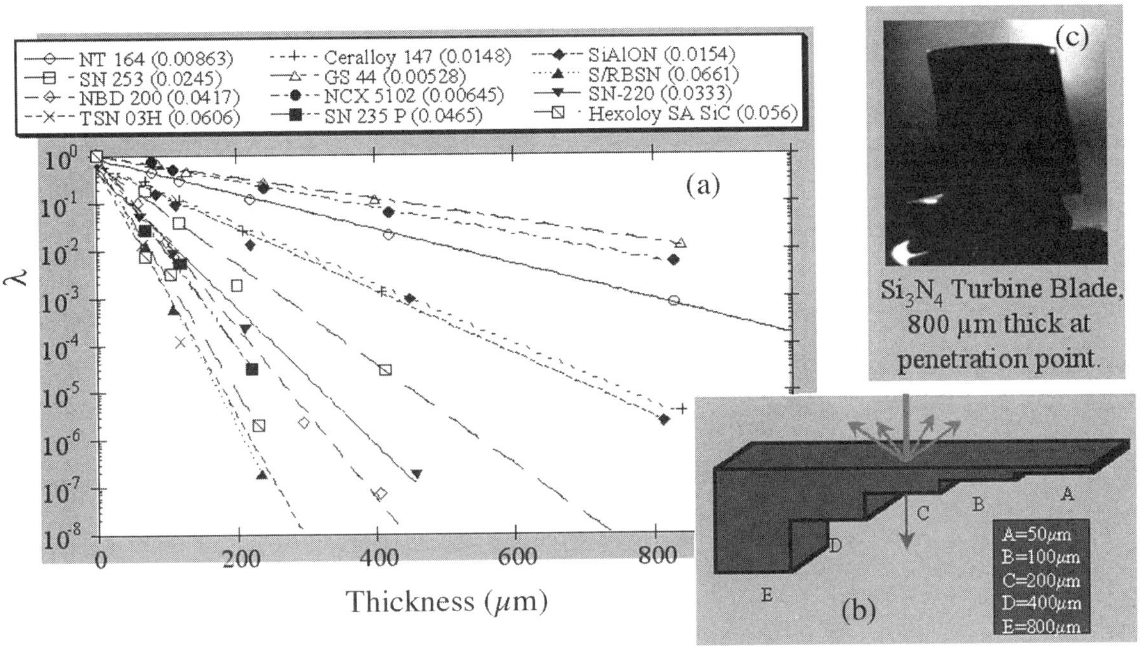

Figure 23.8 Optical Transmission. Properties of Several Structural Ceramics. (a) Transmission as a Function of Thickness. (b) Step Wedge Used to Obtain Data. (c) Photograph Showing Laser Light Through 1 mm Thick Trailing Edge of Blade.

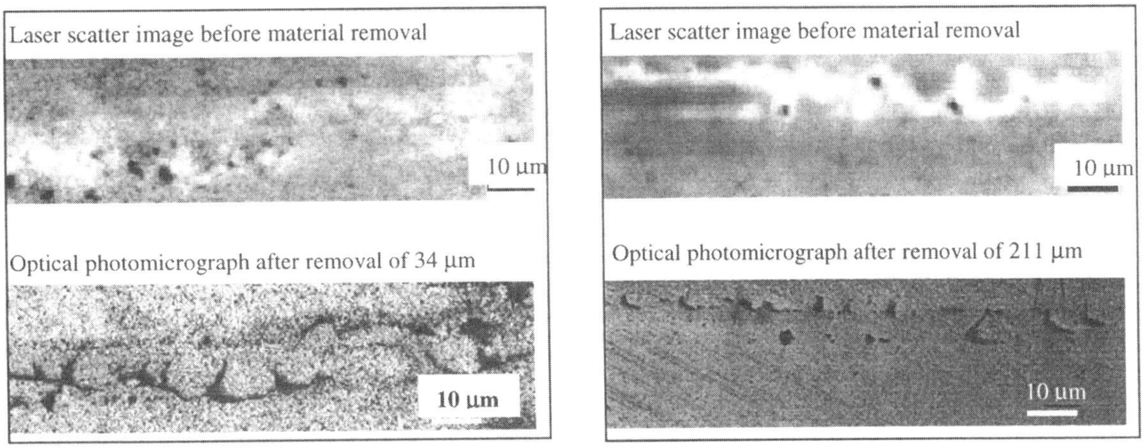

(a) (b)

Figure 23.9 Correlation between Elastic Optical Back-Scatter and Destructive Analysis for crept Si_3N_4 for Verification of Subsurface Creep Damage Detection. (a) Laser Scatter Image and Optical Photomicrograph After Removing 34 μm and (b) Laser Scatter Image and Optical Photomicrograph After Removing 211 μm.

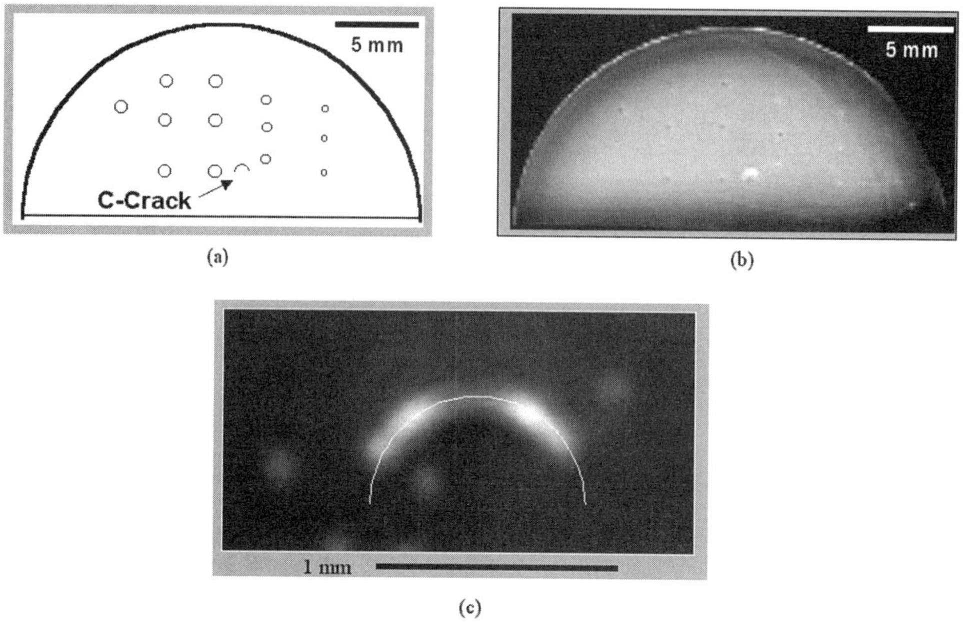

Figure 23.10 Example of Detection of Subsurface Cracks in 4 mm Thick Si_3N_4 Plate. (a) Schematic Diagram of Test Standard, (b) Sum Laser Scan Image of Standard, (c) Scaled Image of C-crack after Image Processing.

Ultrasonics. Detection of creep damage has also been studied by ultrasonic methods that measure ultrasonic velocity precisely [37]. Because creep changes the elastic property of a material, it also changes the acoustic velocity. Figure 23.11 shows typical profiles of ultrasonic longitudinal wave velocity in a Si_3N_4 crept at 1300°C for 200 hrs under various creep stress conditions [41]. The profiles show the velocity of two parts of the creep specimens: the gage part and the grip part. When compared with the virgin specimen, the velocities in the gage part are lower than those in the grip parts. The velocity decreases with increasing creep stress. Another specimen, crept under 200 MPa stress, shows a much larger reduction of the velocity in the gage part, which corresponds to larger creep strain and perhaps development of creep cavities. To try to obtain the net change of velocities caused by creep cavities, each difference at the grip part is subtracted from the original total wave velocity. It is assumed that this will then provide the net effect of creep cavities.

Detection of Surface-Breaking Cracks

Detection of surface-breaking cracks is an area of significant concern because crack crevice environmental conditions can significantly accelerate crack growth. Detection of surface-breaking cracks has primarily been addressed by dye-penetrant technology [27,42].

Fluorescent-Dye Penetrant Technology. Two penetrant methods are available: visible and fluorescent. For surface-breaking cracks, conventional visible penetrants have been [43–46] unsatisfactory because the surface tension of these penetrants will not allow penetration into the very tight cracks that are associated with modern structural ceramics. Several investigators [43–46] demonstrated detection of tight surface-breaking cracks in Si_3N_4 structural ceramics by fluorescent methods. The results suggest that observation/detection of the cracks must be performed with at least a 40–60X microscope that is fitted with

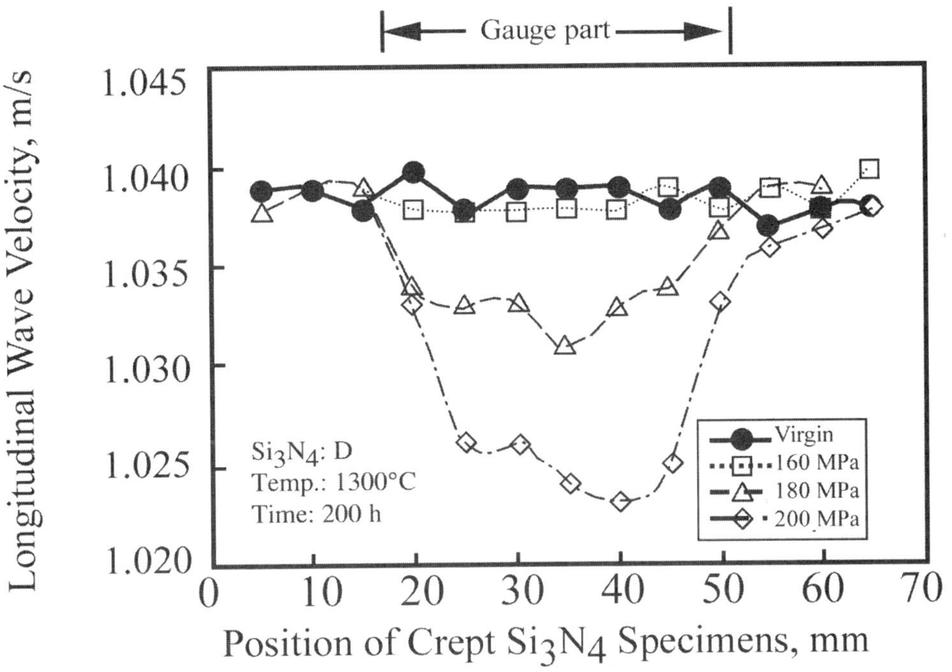

Figure 23.11 Corrected Profile of Longitudinal Wave Velocity for Creep Deformation of Si_3N_4 (Reprinted from ref. [41])

an ultraviolet light source. The ultraviolet light is incident on the sample directly under the viewing area of the microscope. It has been suggested that the ultraviolet-light power necessary is ≈1500 μW/cm^2 [44]. Two investigators [43,45] used Vickers indents with loads from 5 to 20 kg to establish known cracking patterns in several SiC and Si_3N_4 materials. Fluorescent penetrant inspections (FPSI) were then used to determine the length of the cracks across a Vickers indent. Figure 23.12 shows what is seen in ordinary light and what is seen with fluorescent penetrant across a typical 10–kg indent on a Si_3N_4 material. Examples of applying this penetrant method to a Si_3N_4 turbine blade are shown in Figure 23.13.

Focused Ultrasonics. Over the past 15 years, effort has also been devoted toward the use of water-coupled, focused ultrasound to detect surface-breaking cracks [26]. Results obtained by a normal beam technique with the transducers defocused below the ceramic surface have provided estimates of crack depth. Figure 23.14 shows the relationship between the depths of several artificially produced surface cracks and the reduction of the ultrasonic echo height from the defects [34,35]. Although this effort studied artificially created cracks with widths of 30 and 40 μm, this method has not been successful for actual components.

NDE/C TECHNOLOGY FOR CERAMIC MATRIX COMPOSITE TURBINE ENGINE COMPONENTS

Ceramic matrix composite materials present entirely different challenges to NDE. The complex microstructure of these materials makes definition of a "flaw" rather uncertain. Current work is underway to establish critical-flaw type relative to intended service. However, characteristics that might be classified as a flaw include delaminations, regions of open porosity, seams at plies or through-thickness cracks, and damage zones caused by factors such as foreign-object damage or accelerated oxidation [47]. Therefore, as

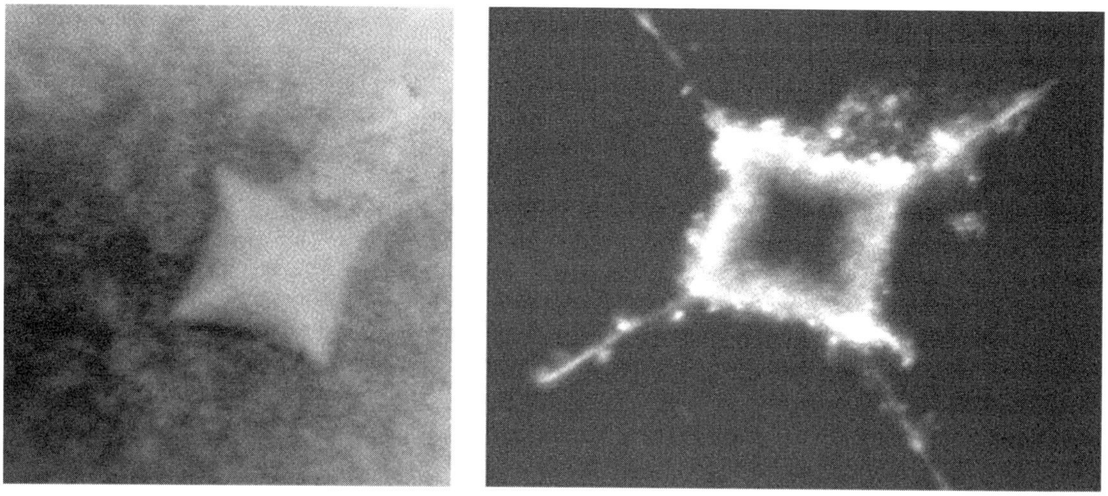

(a) (b)

Figure 23.12 Application of Fluorescent Penetrant. (a) Optical Photomicrograph of Si_3N_4 10 kg Vickers Indent without Penetrant (b) Typical Fluorescent Penetrant Image of Same Indent as (a).

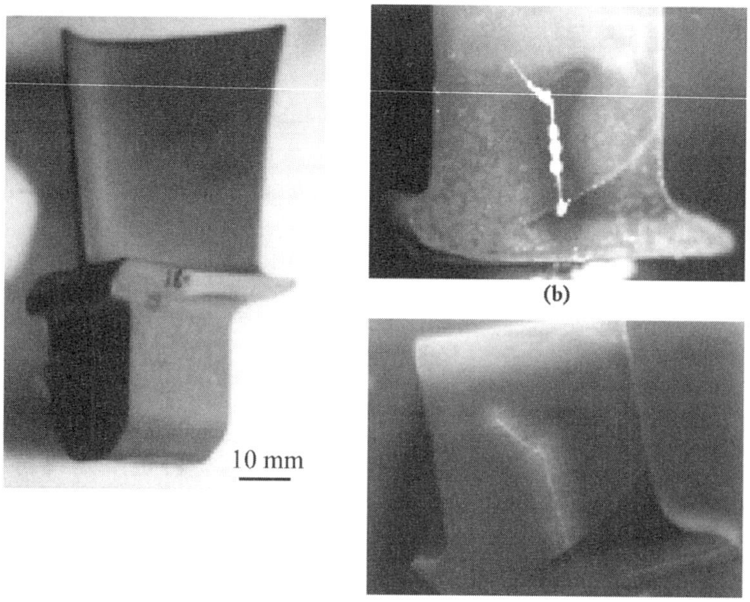

Figure 23.13 Example of Application of Fluorescent Penetrant to Surface Crack Detection on Silicon Nitride Turbine Blade

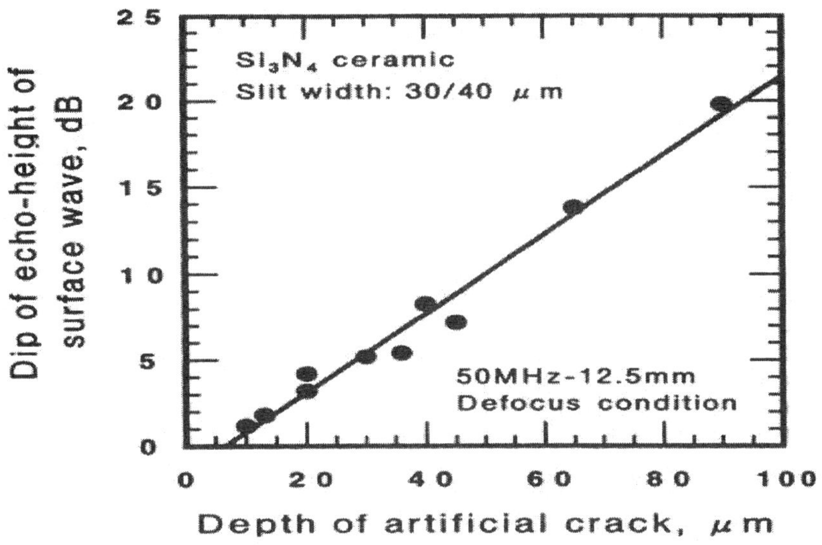

Figure 23.14 Ultrasonic Detection of Surface-Breaking Cracks in Si_3N_4 as a Function of Crack Depth (Reprinted with permission from refs. [34,35])

opposed to the case of monolithic ceramics, flaw zones in composites might better be described as distributed flaw regions. The development of nondestructive evaluation methods to detect these distributed flaw regions will be described in the following subsections.

Delamination Porosity Detection

Infrared Imaging. CMCs have been shown to degrade in turbine engine environments. Degradation occurs by oxidation-induced recession in SiC/SiC [47] and through creep and thermal shock in oxide-based materials [48]. Nondestructive characterization of the thermal properties over the entire combustor is important because large changes in thermal properties could lead to larger-than- anticipated thermal stresses. To assess CMC degradation, one NDE/C technology uses through-thickness thermal imaging technology coupled to computer-controlled time-dependent image capture to detect changes in thermal properties [23,24]. The early work focused on thermal-imaging detection of oxidation of pyrolytic carbon interfaces on SiC/SiC composites [49]. The basis of this thermal imaging method is determination of thermal diffusivity, wherein diffusivity is determined from a measurement of the time-dependent temperature rise on one side of a material test specimen after applying a thermal pulse on the opposite side of the specimen. The method has been fully described elsewhere [24,50] and consists of a high-intensity flash system, a thermal imaging system and an associated computer that is used to detect and measure time- dependent temperature change. The two-dimensional field of view of the camera covers sections of the component under study and is predetermined to cover a well-defined area, e.g., 50 x 50 or 100 x 100-mm. Thus, in the 50 x 50-mm case each pixel in the detector represents a spatial area of 195 x 195 μm and in the 100 x 100-mm case each pixel represents 390 x 390 μm. The established spatial resolution depends on the defect size that is considered important.

A delamination will cause a significantly different thermal conductivity in the delamination region than is present elsewhere in the specimen. The thermal imaging system shown in Figure 23.15 has been used to detect delaminations in combustor liners before use and as a way to measure the growth rate of a delamination region

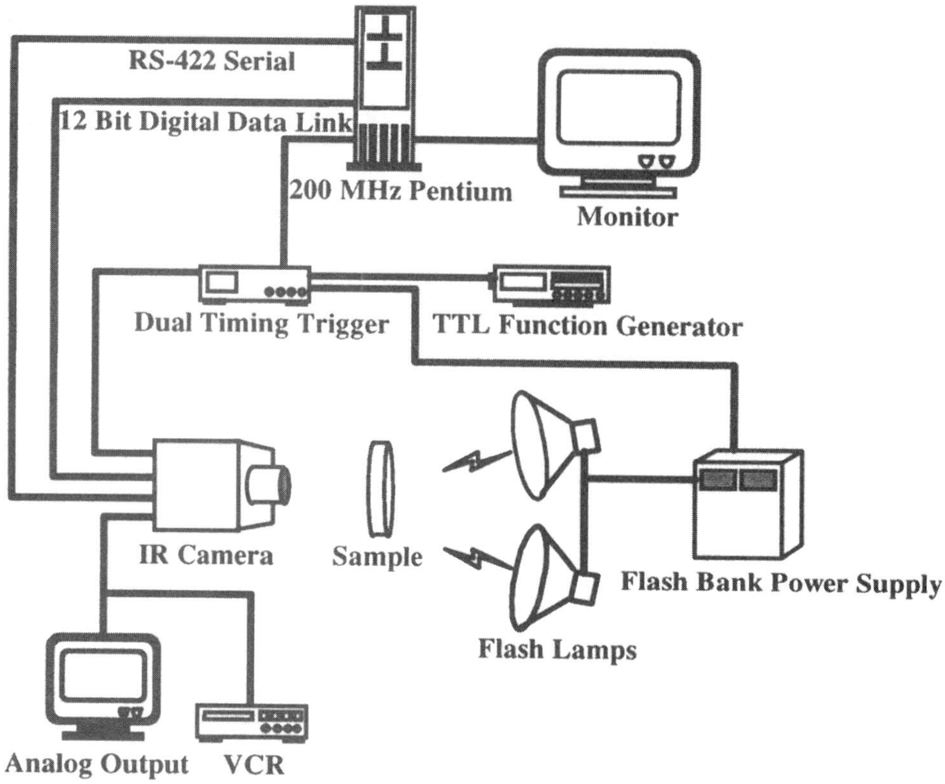

Figure 23.15 Schematic Diagram of Thermal Imaging System (Reprinted from ref. [50] with permission of the American Ceramic Society, www.ceramics.org, © 1998, all rights reserved)

during engine use. An example of tracking a delamination growth is shown in Figure 23.16, which shows a detected delamination in a SiC/SiC combustor liner and also shows how the delamination has grown after 25 h of rig testing. Both oxides as well as nonoxides can be studied by this method. Oxides present the special condition that they are optically translucent to many optical wavelengths; thus, attention should be given to the optical spectral characteristics of the thermal pulse that is used.

Air-Coupled Ultrasonics. SiC/SiC CMC materials that have been used to date have been produced by polymer impregnation processing (PIP), chemical vapor infiltration (CVI), and melt infiltration (MI). The MI method has been possible because of recent developments and the availability of new high-temperature SiC fibers that can withstand the high temperatures of the MI fabrication process. In any of the processes that are used to fabricate CMCs, several steps are involved. Between process steps, NDE/C data that provide information about completeness of infiltration as well as flaw presence provides an opportunity for rejection of parts in the early stages of the process. Immersion of ceramic composites in fluids during conventional ultrasonic methods is not desirable, because these fluids can cause uncertainties in subsequent processing steps. In addition, for periodic engine inspections, if an air-coupled ultrasonic systems could be operational, onsite ultrasonic inspection would be possible.

Over the past five years, air-coupled ultrasound systems [51,52] have been developed. The development has been principally advanced because of advances in piezoelectric devices. Coupling ultrasound across an air gap and into a test piece is inherently very inefficient, largely because the acoustic impedance between the transducer, the air, and the test piece is very large. The transducer designs used in almost all work done

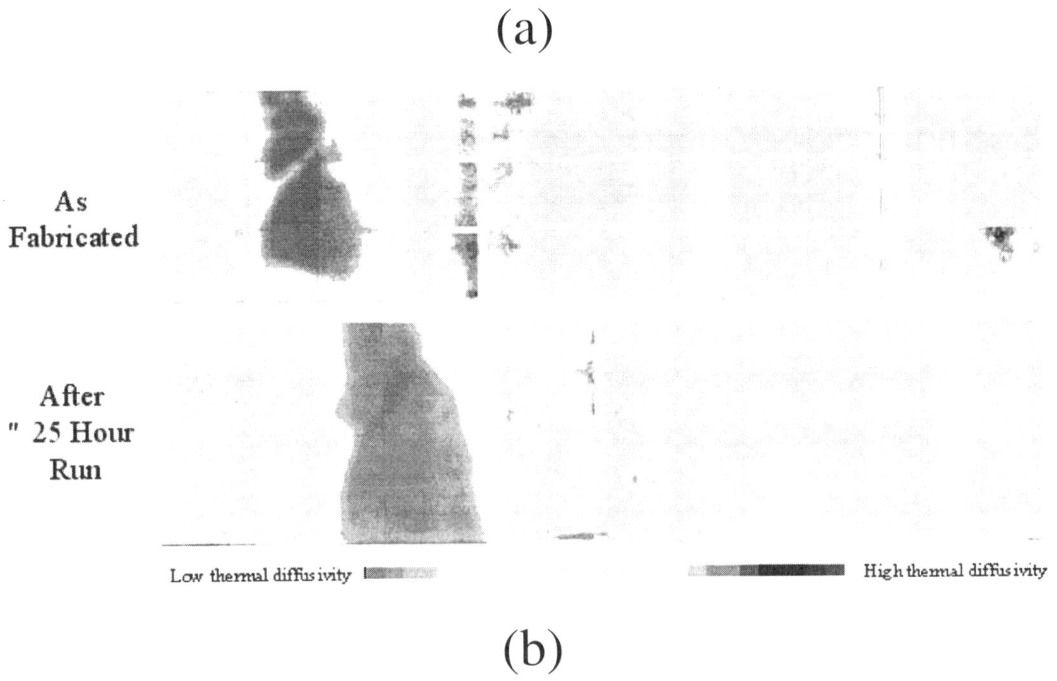

Figure 23.16 Thermal Imaging Detected Delamination in SiC/SiC Composite (a) As Produced, (b) After 25 Hours of Rig Testing

to date for CMC materials has used piezo elements. This has been primarily because the signal-to-noise ratio (S/N) is much better for narrow-band-width piezoelectric devices. Liquid-coupled ultrasonic systems often operate in the pulse-echo mode, using only one transducer; but high signal losses associated with air-coupled ultrasound do not currently permit this. Thus, most air-coupled ultrasonic systems use two transducers in through-transmission arrangement, as shown in Figure 23.17.

Using a through-transmission transducer arrangement, but with a special rotation stage added to handle large-diameter ceramic composite combustor liners, investigators have obtained air-coupled ultrasound data (see Figure 23.18). Attenuation of the ultrasound can be caused by delaminations, changes in porosity or other material-related differences, e.g., variations in fiber content. By synchronously scanning with the transducer pair and rotating the combustor liner, a complete map of ultrasound attentuation of the combustor liner is obtained.

An example of how air-coupled ultrasound data correlates with thermal image data and with destructive analysis is shown in Figure 23.19 for a sol-gel oxide/oxide CMC material. Variations in porosity have also been shown to be detectable with air-coupled ultrasonic systems. In one example, a MI SiC/SiC 35-cm-diameter test liner was known to have experienced a processing problem. Initially, the processing was thought to provide a uniform material. However, as noted in Figure 23.20, which shows water- and air-coupled ultrasound attenuation maps of the entire liner, and destructive optical photomicrographs, there indeed was an increase in the porosity between two stages in the process. It is this type of data that shows the strength and reliability of the air-coupled test method.

X-ray Computed Tomography. X-ray computed tomographic imaging of CMC materials has also been studied [53,54]. Data obtained by this technique have been used to establish the presence of delaminations and

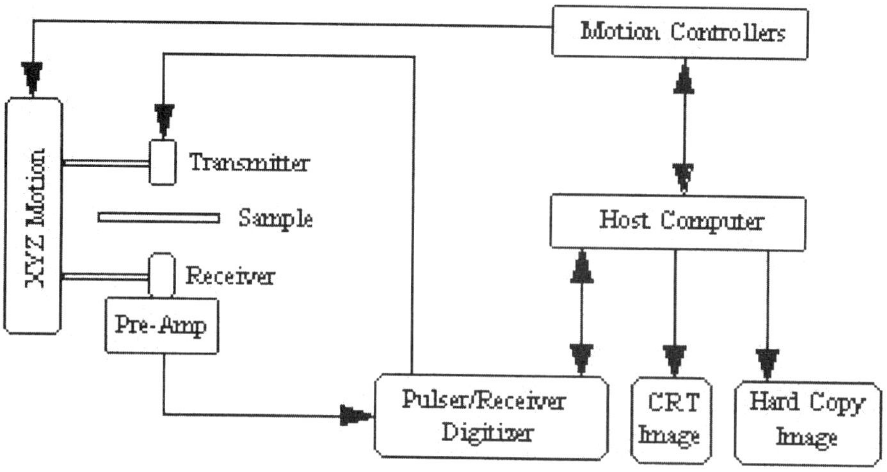

Figure 23.17 Schematic Diagram of Air-Coupled Ultrasonic System for Ceramic Composites, Shown in A Through-Transmission Configuration (Reprinted from ref. [50] with permission of the American Ceramic Society, www.ceramics.org, © 1998, all rights reserved)

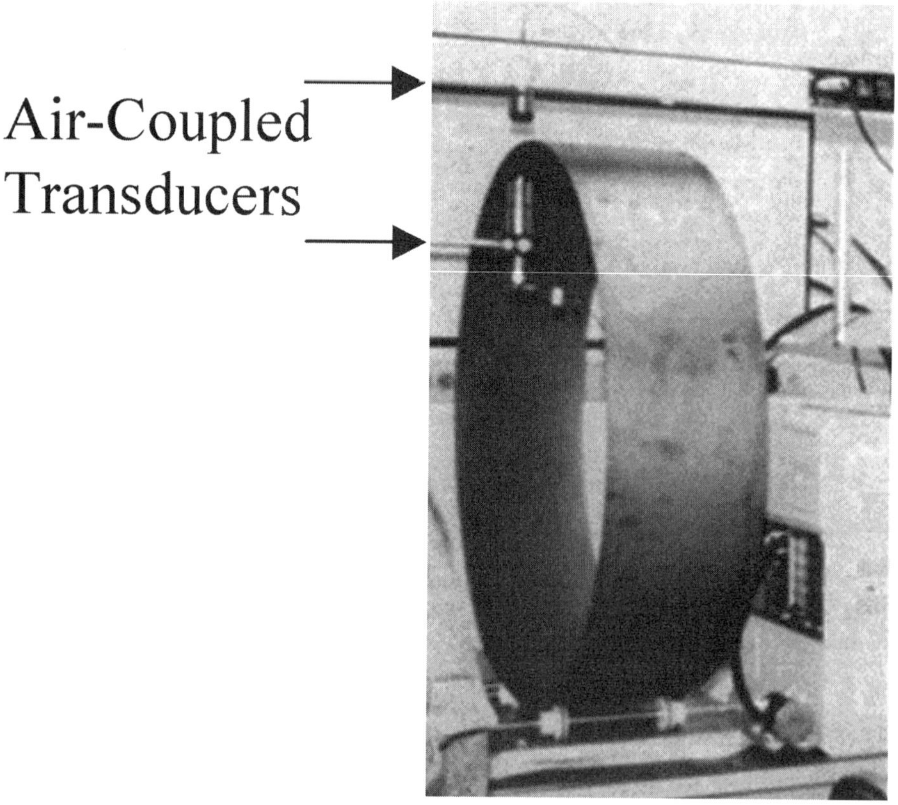

Figure 23.18 Photograph Showing Large 75-cm Diameter Combustor Liner on Rollers in Air-Coupled Ultrasound System

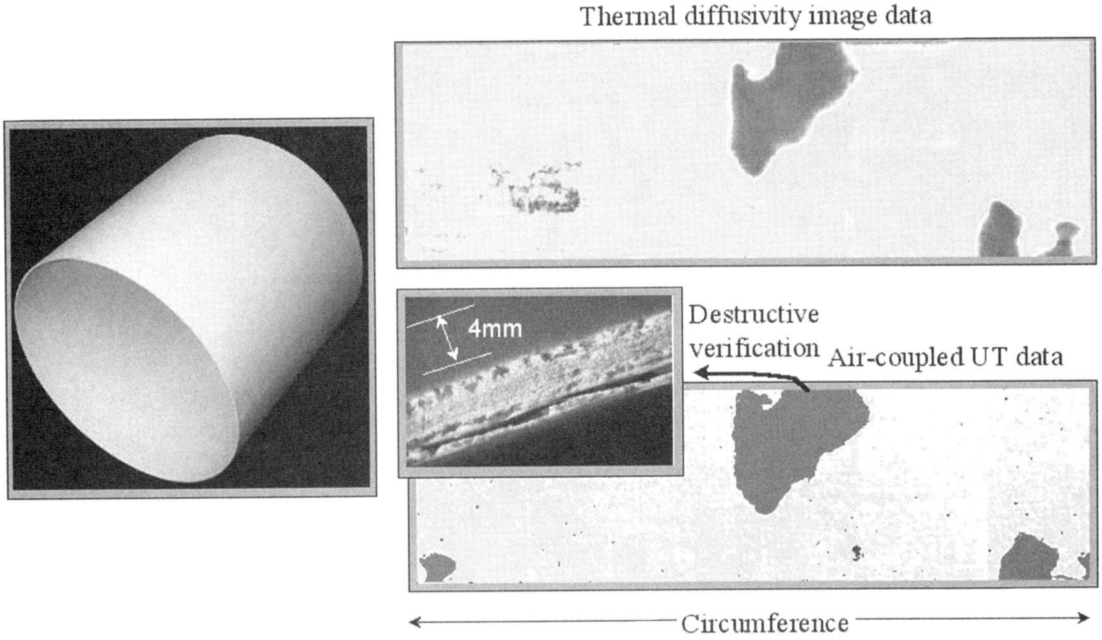

Figure 23.19 Correlation of NDE Data with Destructive Analysis for a 20-cm-Diameter, 20-cm-Long Oxide-Oxide Composite Combustor Liner

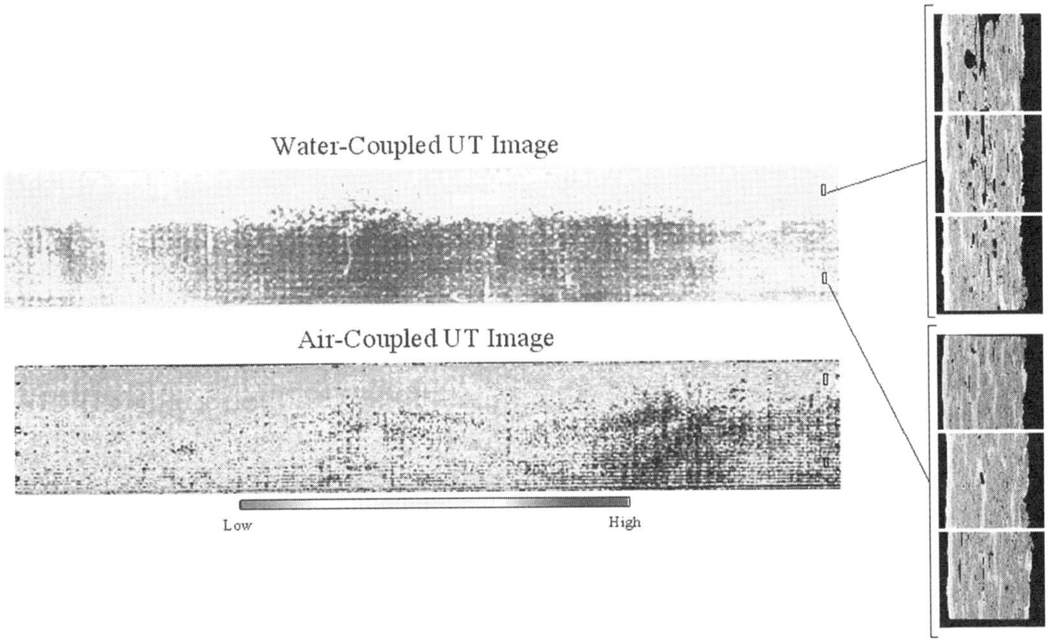

Figure 23.20 Correlation between Air-Coupled, Water Coupled, Thermal Image and Destructive Analysis for MI SiC/SiC Liner: a) Air Coupled Ultrasonic Attenuation Map, b) Water-Coupled Ultrasonic Attenuation Map and c) Optical Photomicrographs Showing Measured Differences in Porosity which Were Detected by UT

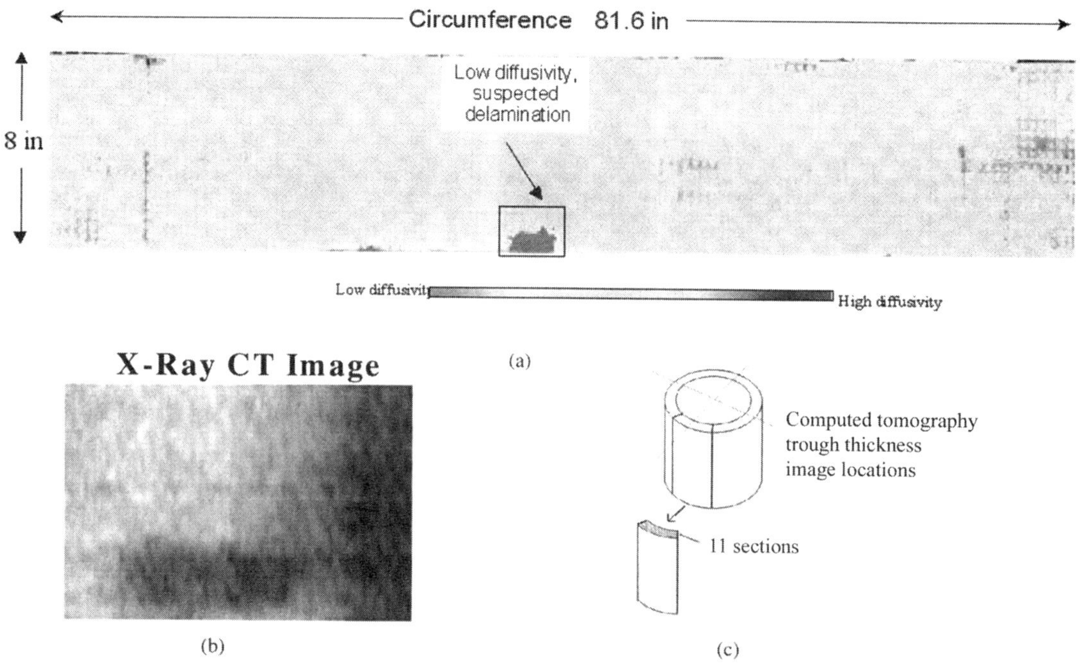

Figure 23.21 Depth of Delamination Detection Using X-Ray CT and One-Sided Thermal Imaging. a) Initial Thermal Image Showing Delamination. b) "Peeled" X-Ray CT Image at (c) 4th Section from Outer Surface.

to determine their depth in ceramic composites. Several sequential X-ray CT cross-sectional images are combined by digital imaging and, subsequently, "peeling" back through the stacked images (see Figure 23.21) which shows an example where one-sided thermal image data and X-ray CT data were used to detect and determine the depth of delamination.

Detection of Through-Thickness Crack

Stationary Components. Detection of through-thickness cracks by NDE/C technology is important to CMC materials because such cracks can provide a path for hot gases. One recent example occurred in a large, 75-cm-diameter combustor liner (see Figure 23.22) that was evaluated in an industrial gas turbine field test under the U.S. Department of Energy Solar Turbines Incorporated Ceramic Stationary Gas Turbine (CSGT) program [19,20]. The NDE methods used for data acquisition were thermal imaging and air-coupled ultrasonics. All NDE data suggested an axial separation of unknown depth. However, the liner was installed and run for 5016 h. After the engine test, the liner was removed for analysis that included destructive methods. What the destructive analysis suggested, see Figure 23.22, is that, at the time the liner was fabricated, the plies at the seam had been distorted so they no longer were concentric, but rather were orientated toward the outer surface. Thus, this destructive analysis confirmed the early NDE data in that there was a seam with a weakness.

Rotating Components. When ceramic matrix composites are used as rotating machinery components, it is important to be able to measure resonant frequencies as well as damping capacities. Uniformity of resonant frequencies ensures that the structure will not go into an unstable resonant frequency and thereby develop a potentially catastrophic condition.

Over the past 5 years, NDE/C technologies to measure resonant frequencies and damping capacity have been developed. Spohnholtz [55] has thoroughly described the various details involved in making these

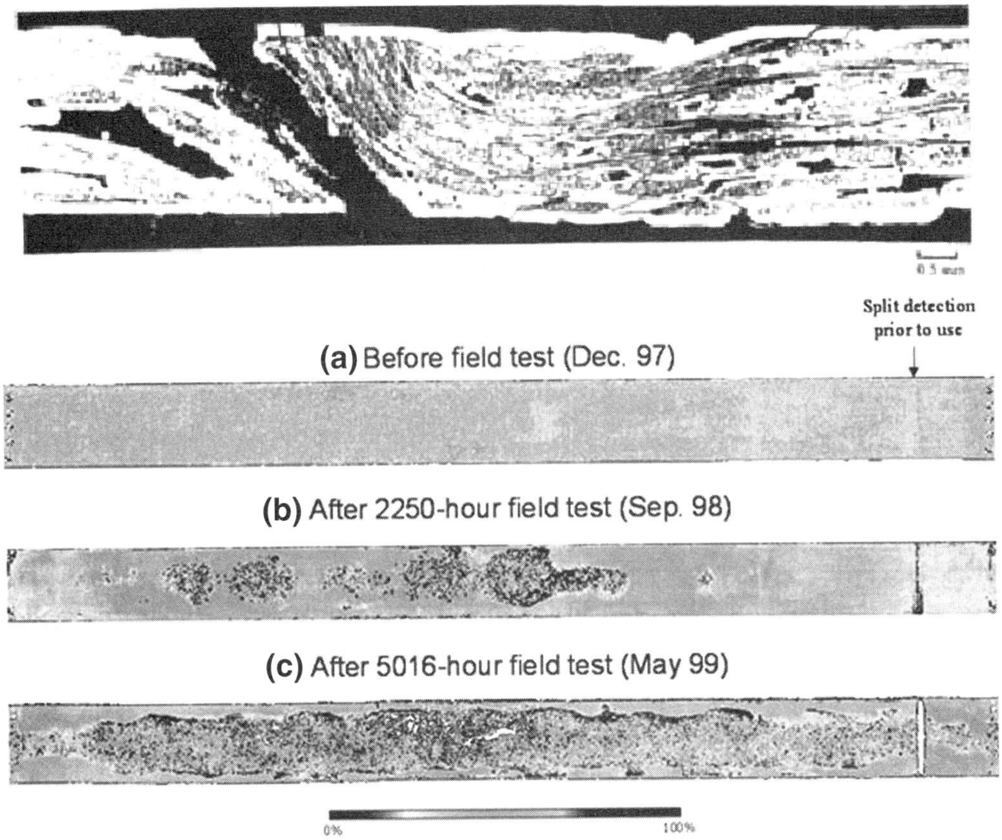

Figure 23.22 NDE Detection and Tracking of "split" in Large Diameter SiC/SiC Combustor Liner.(a) Before Field Test. (b) After 2250 Hours. (c) After 5016 Hours, and (d) Destructive Sectioning After 5016 Hours.

measurements. One component in his work was an integrally bladed disk, commonly referred to as a BLISK (see Figure 23.23). The work demonstrated that large voids could be detected in the blades by measuring resonant frequency. Figure 23.24 shows a plot of resonant frequency vs. blade number before spin testing. Note that Blade 21 shows a significantly lower resonant frequency. Subsequent X-ray tomographic imaging revealed that an ≈1-mm void was present. The value of using NDE data for predicting performance of rotating equipment has been discussed by Genge and Marsh [21].

Detection of Impact Damage. Although impact or foreign-object damage (FOD) is a reality in turbine engines, to date, very limited effort has been directed toward applying NDE methods to assess its extent. Detection of FOD has been approached by two primary methods: thermal imaging and air-coupled ultrasound. In a recent set of experiments [50], five 76 x 152-mm 8-ply (3-mm-thick) SYLRAMIC™ S200 CMC panels were studied. Impact damage was achieved with a 12.5-mm- diameter steel rod with a hemispherical head and a pendulum-type impact. Because visual inspection revealed some surface damage but could not assess the depth or extent of the damage, the latter was determined from NDE data. Typical retained strength of the panels that sustained impact damage is shown in Figure 23.25.

Although repair of CMC components is an area of interest and NDE/C methods are necessary to assess the quality of the repairs and determine the condition of repaired regions at inspection intervals, essentially

Figure 23.23 Photograph of Ceramic Matrix Composite BLISK

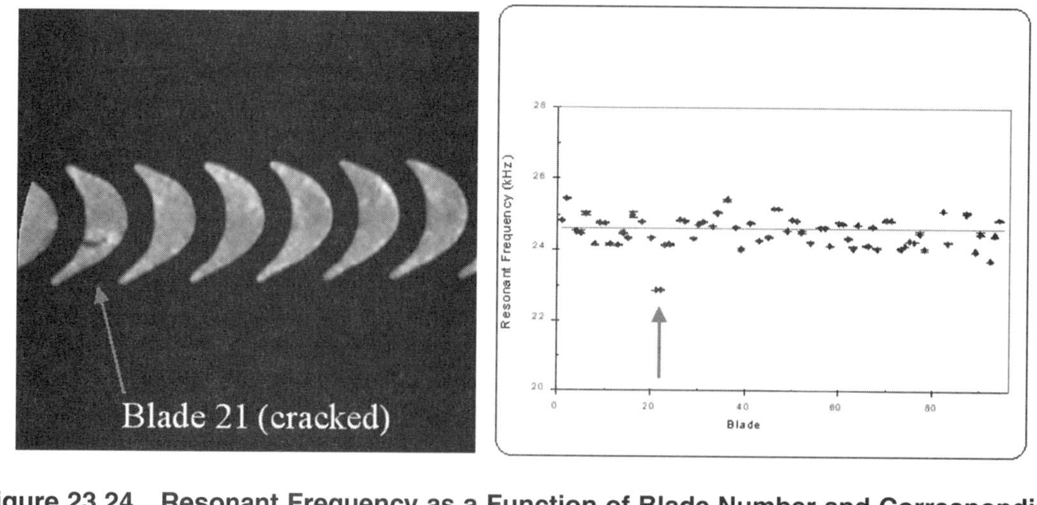

Figure 23.24 Resonant Frequency as a Function of Blade Number and Corresponding X-Ray Tomographic Image Showing Void in Blade with Low Resonant Frequency

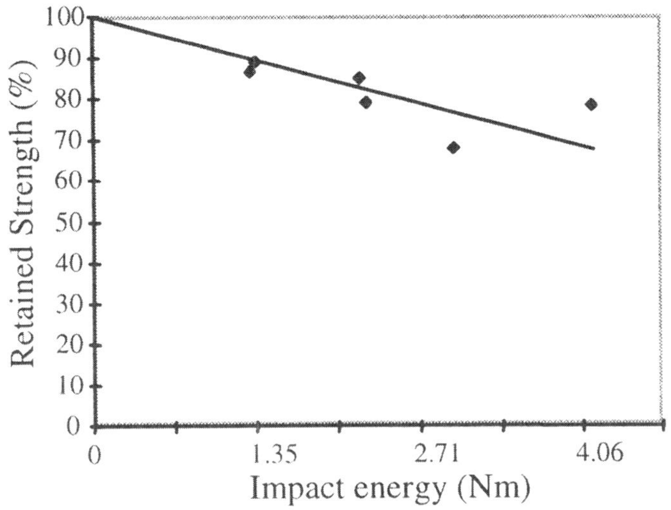

Figure 23.25 Percent Retained Strength of Impact-Damaged SYLRAMIC™ Composite Panels (Adapted from ref. [50] with permission of the American Ceramic Society, www.ceramics.org © 1998, all rights reserved)

no work has been done to evaluate NDE data for repair completeness. As part of the effort to study impact-damaged samples, and assess repair, we studied the impact-damaged panels by NDE methods before and after repair.

Through-transmission X-ray radiographs revealed few small, dark (low-density) lines that mark the impact. Delaminations (or cracks parallel to the surfaces) were not detected by the X-ray images. Figures 23.26 and 23.27 show through-thickness thermal diffusivity and air-coupled ultrasonic images, respectively, before and after repair. The correlation of the size and shape of the damaged region with the diffusivity and ultrasound images is good.

Detection of Flaws Associated with Environmental Barrier Coatings. Reduction of oxidation-induced recession in SiC/SiC CMC materials as well as reduced material temperature has recently been addressed by placing an environmental barrier coating (EBC) between the CMC and the path of the hot gas. The composition of the EBC depends on the CMC to which it is being applied and the temperature of operation. For SiC/SiC materials, current EBCs, typically 100–200 μm thick, are composed of barium-strontium-alumino-silicate (BSAS). For oxide/oxides, the composition of the EBC is still proprietary. The existing EBC for oxides is much thicker, i.e., >2 mm. These EBC materials present yet another challenge to defect detection by NDE/C methods. One issue to be addressed by NDE/C technology is assessment of the uniformity of adherence of the EBC; another is detection of flaw regions in the CMC below the EBC.

Recently, thermal imaging, air-coupled ultrasonics and X-ray tomography were shown to be able to detect a delamination in a BSAS (barium strontium alumino silicate) EBC applied to the liner of a SiC/SiC combuster. What was significant is that this particular delaminated part was subsequently installed in an operating engine. Borescope examination of the engine at various time intervals showed that the EBC had spalled off at the delaminated region that was detected by NDE. Furthermore, growth of the delaminated region as a function of engine run time corresponded to the NDE data that suggested a likely delamination, as shown in Figure 23.28 [56]. No similar data are yet available for oxide/oxide EBCs.

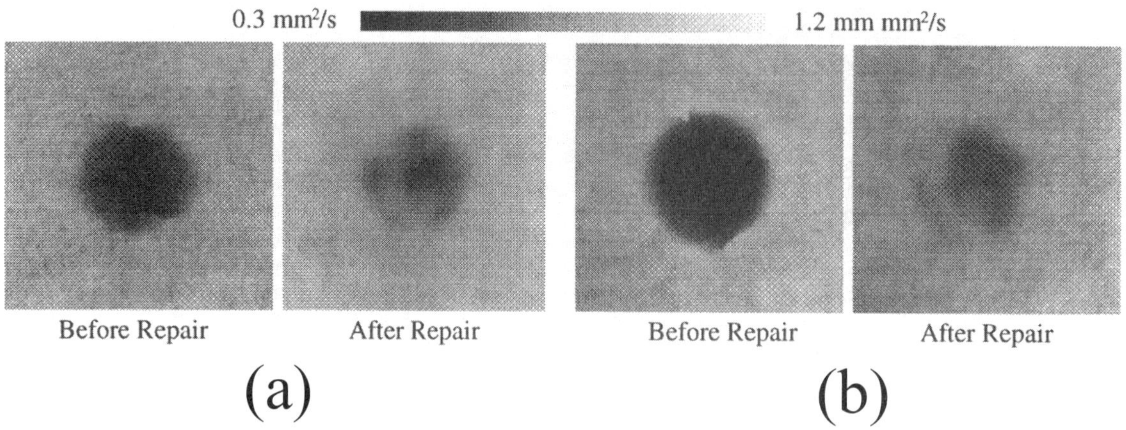

Figure 23.26 Thermal Diffusivity Images of Panels (a) E-3 and (b) A-2 (Reprinted from ref. [50] with permission of the American Ceramic Society, www.ceramics.org. © 1998, all rights reserved)

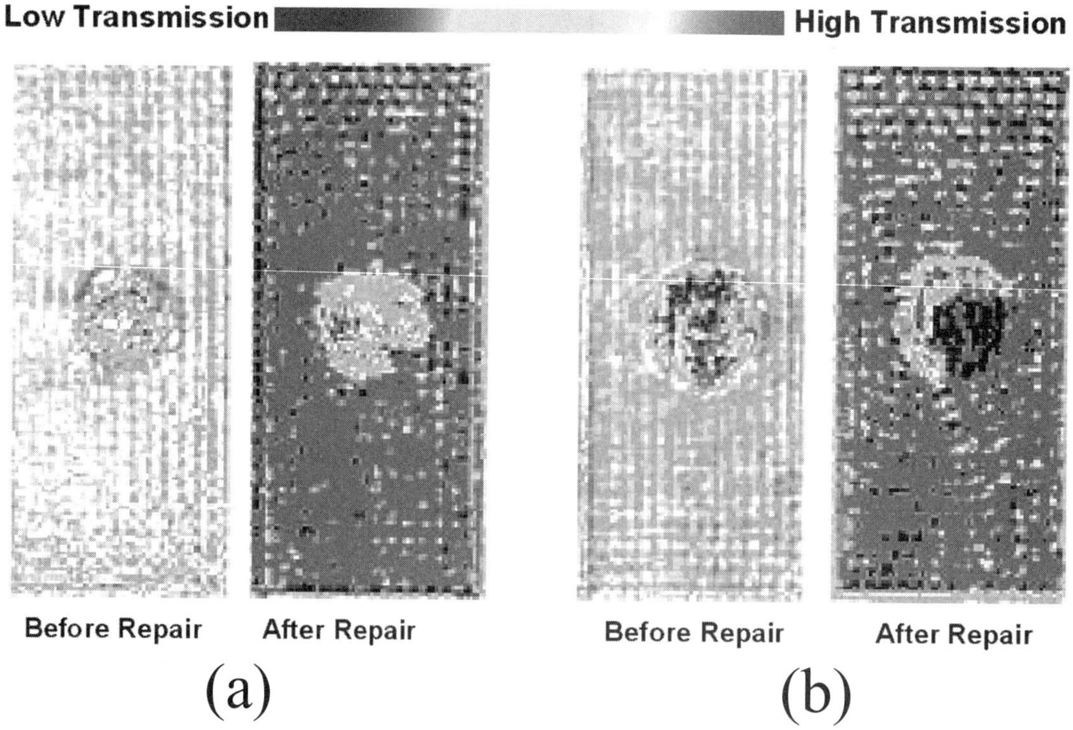

Figure 23.27 Air-Coupled UT Images Obtained Before and After Repair (a) E-3 and (b) A-2 (Reprinted from ref. [50] with permission of the American Ceramic Society, www.ceramics.org. © 1998, all rights reserved)

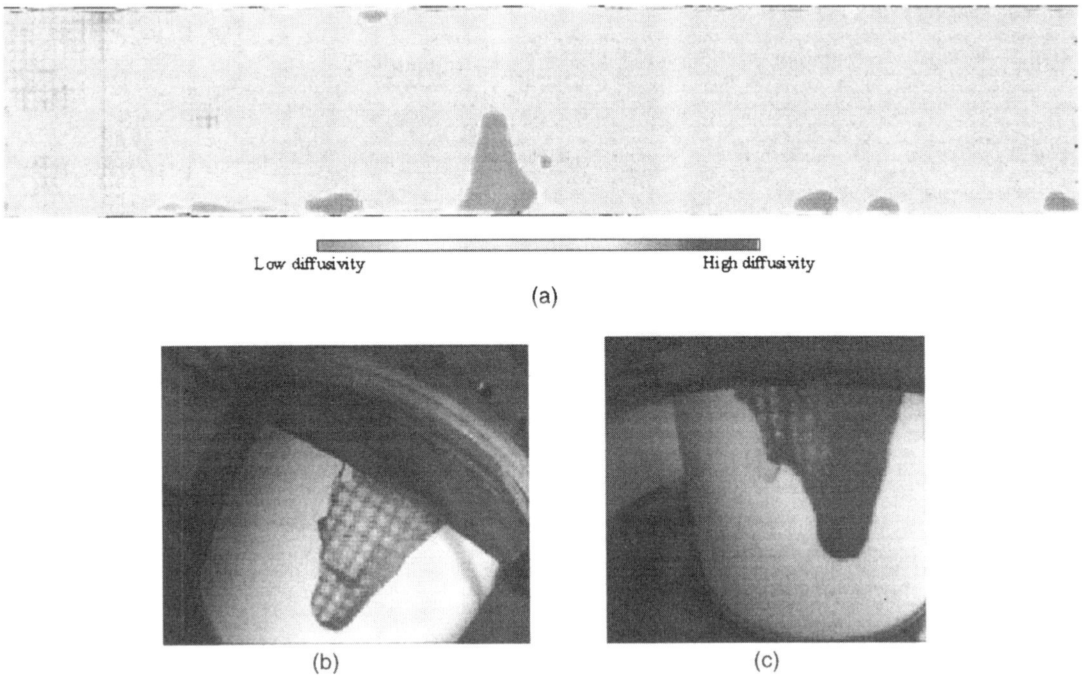

Figure 23.28 EBC on SiC/SiC (a) NDE Data Showing Likely Delaminated Regions. (b) Delamination after 2,897 hours (c) Delamination After 8,465 hours (Adapted from ref. [56])

Estimating Remaining Life. Current efforts couple NDE data to models that allow estimation of remaining useful life of components. The effects of oxidation, high- and low-cycle fatigue, thermal shock, thermal stresses, etc. all must be combined into a package that can be used for estimating remaining useful life. The challenge for NDE technology now is to couple with appropriate life prediction models or algorithms.

SUMMARY AND CONCLUSIONS

Over the past 20 years, NDE technology has been advancing in parallel with advances in ceramic materials for application to high-gas-firing-temperature gas turbine engines. For monolithic components, primarily Si_3N_4 blades and vanes, high-spatial-resolution X-ray CT can detect internal critical-sized cracks and subsurface voids. It has been shown that fluorescent-penetrant inspection, while relatively simple to apply with proper procedures, can be a useful and cost-effective method to detect surface and surface-breaking cracks down to the size of a few microns. New optical scatter methods that use low-power lasers and ultrasonic techniques can detect creep, as well as subsurface cracks. For ceramic matrix components, mainly combustor liners, NDE methods that use thermal imaging and air-coupled ultrasonic techniques have been shown to be very useful for detecting delaminations and porosity changes. Through-thickness cracks on composites can be detected by air-coupled ultrasonics as well as special application methods of thermal imaging. Verification of such defects and direct observation have been achieved through X-ray CT imaging technology. Damping capacity and resonant frequency measurements of CMCs have been correlated with changes in elastic modulus and damage.

ACKNOWLEDGEMENT

Unless otherwise indicated all figures were supplied by the authors' organizations.

REFERENCES

[1] Richerson, D. W., "Ceramics for Turbine Engines," Mech. Eng., 119[9], 80–83, Sept. 1997.

[2] McLean, A. F. and Fisher, E. A., *Brittle Materials Design, High Temperature Gas Turbine*, Final Report U.S. Army Materials and Mechanics Research Center, AMMRC- TR-81-14, March 1981.

[3] Goebbels, K. and Reiter, H., "Zerstörungsfreie Prüfverfahren für Komponenten der Keramik-Gasturbine—Entwicklungsstand, Möglichkeiten, Grenzen—(Nondestructive Testing of Ceramic Gas Turbine Components—Development Status, Potential, Limits -) in *Keramische Komponenten, für Fahrzeug-Gasturbinen II (Ceramic Components for Vehicular Gas Turbines II)*, W. Bunk and M. Böhmer, eds., Springer Verlag, Berlin 1981. pp. 163–193 (in German).

[4] Nakajima, K., "The Fine Ceramics Industry in Japan," J. Jpn Trade and Ind., 5, 10–14, 1983.

[5] Evans, A. G., "Structural and Microstructural Design in Brittle Materials," in *Proc. DARPA/AFMC Review of Prog. in NDE*, July, 1980. pp. 619–633.

[6] McClung, R. W. and Johnson, D. R., "Needs Assessment for NDT and Characterization of Ceramics: Assessment of Inspection Technology for Green State and Sintered Ceramics," in *Proceedings of the Joint Conference on Nondestructive Testing of High- Performance Ceramics*, August 25–27, 1987, Boston, MA, USA, Am. Ceram. Soc., 1987. pp. 33–51.

[7] Goebbels, K., Reiter, H., Arnold, W., and Hirsekorn, S., *A Concept for the Quality Assessment of Components of the Ceramic Gas Turbine by Nondestructive Testing*, Internal Report Fraunhofer Institute for Nondestructive Testing, Report T85-094, August 1985.

[8] Tanimoto, Y., Tanaka, S.-I., and Tomizawa,.M., "Application of X-ray CT Scanner to Nondestructive Testing," in *1987 Tokyo International Gas Turbine Congress and Exhibition*, October 26–30, 1987, Yu-Port, Japan. IGTC-119, pp. I-155–161, 1987.

[9] Minter, J., "The Role of NDE in Ceramic Turbine Engine Component Development," in *Proc. Third International Symposium on Ceramic Materials and Components for Engines*, V.J. Tennery, ed., November 27–30, 1988, Las Vegas, NV, USA, Am. Ceram. Soc., 1989, pp. 1289–1298.

[10] Schuldies, J. J. and Richerson, D. W., "NDE Approach Philosophy and Standards for the DARPA/NAVSEA Ceramic Turbine Program," in *Proc. 1977 DARPA/NANSEA Ceramic Gas Turbine Demonstration Engine Program Review*, Fairbanks, J. W., and Rice, R. W., eds., Metals and Ceramics Information Center, Batelle Laboratories, Columbus, OH, USA, 1978. pp. 381–402.

[11] Schuldies, J. J. and Spaulding, W. H., "Radiography and Image Enhancement of Ceramics," ibid., pp. 403–428.

[12] Schuldies, J. J. and Derkas, T., "Ultrasonic NDE of Ceramic Components," ibid., pp. 429– 448.

[13] Fate, W. A., "Pulsed Ultrasonic Measurements in Ceramic Materials at High Temperatures," in *Ceramics for High Performance Applications*, J.J. Burke, A.E. Gorum, and R.N. Katz, eds., Brook Hill Publishing Company, Chestnut Hill, MA, USA, 1974. pp. 687–605.

[14] Cassidy, D. J., "NDE Techniques Used for Ceramic Turbine Rotors," in *Ceramics for High Performance Applications—II*, J.J. Burke, E. Lenoe, and R.N. Katz, eds., Brook-Hill Publishing Company, Chestnut Hill, MA, USA, 1978. pp. 231–242.

[15] Baaklini, G. and Abel, P. B., "Flaw Imaging and Ultrasonic Techniques for Characterizing Sintered Silicon Carbide," in *Proceedings of the Joint Conference on Nondestructive Testing of High-Performance Ceramics*, August 25–27, 1987, Boston, MA, USA, Am. Ceram. Soc., 1987. pp. 304–321.

[16] Bower, J., "An Evaluation of Nondestructive Methods for Ceramic Heat Exchanger Applications," ibid., pp. 322–335.

[17] Ellingson, W. A., Sun, J. G., Pillai, T. A. K., and Forster, G. F. "Development and Application of NDE Methods for Monolithic and Continuous Fiber Ceramic Matrix Composites," in *Advances in Science and Technology*, Vol.16, Part D, P. Vincenzini, ed., Techna, Florence, Italy, 1999. pp. 331–343.

[18] Miriyala, N., Simpson, J. F., Parthasarthy, V. M., and Brentnall, W. D., "The Evaluation of CFCC Liners after Field-Engine Testing in a Gas Turbine," ASME paper 99-GT-395, presented at the International Gas Turbine and Aeroengine Congress and Exposition, Indianapolis, IN, USA, June 7–10, 1999.

[19] van Roode, M., "Ceramic Stationary Gas Turbine," in *Proceedings of the Advanced Turbine Systems Annual Program Review Meeting*, Abbie Layne, Patricia Hoffman, eds., DOE/OR-2048, November 7–8, 1996. pp. 633–657.

[20] Price, J.R., Jimenez, O., Parthasarthy, V., Miryala, N. and Leroux, D., "Ceramic Stationary Gas Turbine Development Program - Seventh Annual Summary," ASME paper 2000-GT-75, presented at the ASME TURBO EXPO 2000, Land, Sea &Air, Munich, Germany. May 8–11, 2000.

[21] Genge, G. G. and Marsh, M. W., "Carbon-Fiber-Reinforced/Silicon Carbide Turbine Blisk Testing in the SIMPLEX Turbopump," presented to the 49th JANNAF meeting, Tucson, AZ, USA, Dec 1999.

[22] Tatsuzawa, M., Sakida, T. and Yonekubo, Y., "Development of 300kW-Class Ceramic Gas Turbine (CST 301)," *Proceedings of the 1995 Yokohama International Gas Turbine Congress*, GTSJ, Yokohama, Japan, October 23–27, 1995. 95-YOKOHAMA-IGTC-124, Vol. I, I- 147–152, 1995.

[23] Ellingson, W. A., Rothermel, S. A. and Simpson, J. F., "Nondestructive Characterization of Ceramic Composites Used as Combustor Liners in Advanced Gas Turbines," ASME paper 95-GT-404, presented at the International Gas Turbine and Aeroengine Congress and Exposition, Houston, TX, USA, June 5–8, 1995. Transactions of the ASME, J. Eng. for Gas Turbines and Power, 118[3], 486–490, July 1996.

[24] Ellingson, W. A., Sun, J. G., More, K. L., and Hines, R. G., "Characterization of Melt- Infiltrated SiC/SiC Composite Liners Using Meso and Macro NDE Techniques," ASME paper, 2000-GT-0067, presented at the ASME TURBO EXPO 2000, Land, Sea &Air, Munich, Germany, May 8–11, 2000.

[25] Amin, K. E. and Leo, T. P., "Radiographic Detectability Limits for Seeded Defects in both Green and Densified Silicon Nitride," in *Proceedings of the Joint Conference on Nondestructive Testing of High-Performance Ceramics*, August 25–27, 1987, Boston, MA, USA, Am. Ceram. Soc., 1987. pp. 211–232.

[26] Cartz, L., *Nondestructive Testing*, ASM International, Materials Park, OH, USA, 1995.

[27] Goebbels, J., Heidt, H., Kettschau, A., and Reimers, P., "Computer Tomography," *NBS/BAM 1986 Symposium on Advanced Ceramics*, S.M. Hsu, and H. Czichos, eds., NIST Special Publication 766, 1989. pp. 97–100.

[28] Ikeda, Y., Mizuta, Y. and Tobita, H., "X-ray CT Test for Parts and Materials of Ceramic Gas Turbine," in *Proceedings of the 1995 Yokohama International Gas Turbine Congress*, GTSJ, Yokohama, Japan, October 23–27, 1995. 95-YOKOHAMA-IGTC-60, Vol. III, 24–36, 1995.

[29] Rabe, T., Goebbels, J., Kunzmann, A., and Riesemeier, H., "Quantitative Determination of Density Gradients by Three-Dimensional Computer Tomography (3D-CT)," *Proceedings of the 5th Conference and Exhibition of European Ceramic Society,* A. Bouquillon, ed., June 22–26, 1997, Versailles, France. Key Engineering Materials, 132– 136, 730–733, 1997.

[30] Antonuk, L. E., El-Mohri, Y., Huang, W., Lee, K. W., Maolinbay, M., and Siewerdsen, J. H., "Development of a High-Resolution, Active-Matrix, Flat-Panel Imager with Enhanced Fill Factor," in *Proc. 1997 Conf. on Medical Imaging*, R.L. Van Metter, and J. Beutel, eds., SPIE Vol. 3032, 2–13, 1997.

[31] Tsukamoto, A., Yamada, S., Tomisaka, T., Tanaka, M., and Ikeda, M., "Development and Evaluation of a Larger Radiography and Fluoroscopy," *in Proc. 1999 Conf. on Medical Imaging*, J.M. Boone, and J.T. Dobbins, eds., SPIE Vol. 3659, 14–23, 1999.

[32] Grantors, P. R. and Aufrichtig, R., "DQE (f) of an Amorphous-Silicon Flat-Panel X-ray Detector: Detector Parameter Influences and Measurement Methodology," in *Proc. 2000 Conf. on Medical Imaging*, J.T. Dobbins and J.M. Boone, eds., SPIE Vol. 3977, 2–13, 2000.

[33] Ikeda, Y., Mizuta, Y. and Onda, K., Rapid and Sensitive Radioscopy for Fine Ceramics Using an Image Subtraction Method," J.Am. Ceram. Soc., 82[4], 882–888, 1999.

[34] Ikeda, Y., Mizuta, Y., and Onda, K., "Nondestructive Testing for Ceramic Gas Turbine Project," in *Proceedings of the International Gas Turbine Congress 1999 Kobe*, T. Tamaru, ed., Kobe, Japan, November 14–19, 1999. Gas Turbine Society of Japan, 1999. IGTC'99 Kobe OS-105, pp. 87–92.

[35] Ikeda, Y., Onda, K. and Mizuta, Y., "Ultrasonic Testing of Silicon Nitride Ceramics Developed for Gas Turbine," *Proceedings of the 2nd Japan-US Symposium on Advances in NDE, JSNDI*, 1999. pp. 40–45.

[36] Fujii, I, and Kawashima, K., "Digital Measurement of Ultrasonic Velocity," Rev. Prog. in Quantitative NDE, Vol.14, 203–209, 1995.

[37] Onda, K., Mizuta, Y.and Ikeda, Y., "Nondestructive Evaluation of Porosity in Si_3N_4 Ceramics by using Ultrasonic Method," *Proceedings of the 4th Symposium on Ultrasonic Testing*, JSNDI, 1997. pp. 105–108,

[38] Ellingson, W. A., Todd, J. A., and Sun, J. G., "Optical Method and Apparatus for Detection of Defects and Microstructural Changes in Ceramics and Ceramic Coatings," U.S. Patent #6,285,449, issued September 2001.

[39] Ellingson, W. A. and Brada, M. P., "Optical Method and Apparatus for Detection of Surface and Near-Subsurface Defects in Dense Ceramics," U.S. Patent 5,426,506, 1995.

[40] Sun, J. G., Ellingson, W. A., Steckenrider, J. S., and Ahuja, S., "Application of Optical Scattering Methods to Damage Detection in Ceramics," in *Machining of Ceramics and Composites*, S. Jahanmir, M. Ramulu, and P. Koshy, eds., Marcel Dekker, Inc., New York, N.Y., USA, 1999. pp. 669–699.

[41] *Research and Development on Ceramic Gas Turbine, FY 1998 Annual Report*, NEDO, Tokyo, Japan, April 1999.

[42] Malins, R. J., McCall, D., and Rhodes, G. W., "Systems Level Considerations for Environmentally Responsive Fluorescent Penetrant Testing," Mater. Eval., 51[3], 338–351, 1993.

[43] Ellingson, W. A. and Forster, G. A., *An Investigation of Penetrant Techniques for Detection of Machining-Induced Surface-Breaking Cracks in Monolithic Ceramics*, Oak Ridge National Laboratory Report, ORNL/Sub93/-SN 893/1, 1996.

[44] Domon, H., Uemura, K., and Fujiwara, K., "Liquid Penetrant Test for Fine Ceramics," in *Proc. Eighth Asia-Pacific Conference on Nondestructive Testing*, December 11–14, 1995. Taipei, Taiwan, 1999. pp. 711–718.

[45] Kolker, H., Henze, P., Schwetz, K. A., and Lipp, A., "X-ray Microfocus and Dye Penetrant Techniques for Crack Detection in Ceramics," in Third International Symposium on Ceramic Materials and Components for Engines, J.V. Tennery, ed., Am. Ceram. Soc., Westerville, OH, USA, 1989. pp. 1122–1129.

[46] Horton, S. A., "Detection of Surface Defects in Ceramic Rolling Elements," in *Proceedings of the 4th International Symposium.on Ceramic Materials and Components for Engines*, R.L. Carlson, T. Johansson, and L. Kahlman, eds., Göteborg, Sweden, 10–12 June, 1991. Kluwer Academic Publishers, Dordrecht, The Netherlands. pp. 897–904.

[47] Eaton, H. E., Linsey, G.D., Sun, E.Y., More, K.L., Kimmel, J.B., Price, J.R., and Miriyala, N., "EBC Protection of SiC/SiC Composites in the Gas turbine Combustion Environment - Continuing Evaluation and Refurbishment Considerations," ASME paper 2001-GT-0513, presented at the ASME TURBO EXPO 2001, Land, Sea &Air, New Orleans, LA, USA, June 4–7, 2001.

[48] Razzell, A., Molliex, L., Holmquist, M., and Sudre, O., "Oxide/Oxide Ceramic Matrix Composites in Gas Turbine Combustors," ASME paper 98–GT-30, presented at the International Gas Turbine and Aeroengine Congress and Exhibition, Stockholm, Sweden, June 2–5, 1998.

[49] Winfree, W. P. and James, P. H., "Thermographic Detection of Disbonds," *Proc. 35th International Instrumentation Symposium,* 1989. pp. 183–188.

[50] Sun, J. G., Petrak, D. R., Pillai, T. A. K., Deemer, C., and Ellingson, W. A., "Nondestructive Evaluation and Characterization of Damage and Repair for Continuous Fiber Ceramic Composite Panels," Ceram. Eng. and Sci. Proc., 19[3], 615–622, 1998.

[51] Gradia, W. A. and Fortunko, C. M., "NDE Applications of Air-Coupled Ultrasonic Tansducers," in *Proc. IEEE Ultrasonic Symposium*, Seattle, WA, USA, Vol. 1, 697–709, 1995.

[52] Wykes, C., "Advances in Air-Coupled Ultrasonic Transducers," Nondestructive Testing and Evaluation, 12, 55–180, 1995.

[53] Ekenhorst, D., Riesemeier, H., and Goebbels, J., "Characterization of Ceramic Matrix Composites by Computerized Tomography, in *Proc. 6th European Conference on Composite Materials*, Sept. 20–24, 1993, Bordeaux, France, HT-CMC 1, Woodhead Publishing LTD, Cambridge, UK, 1993. pp. 743–750.

[54] Ekenhorst D., Goebbels, J., Riesemeier, H., and Illerhaus, B., "Nondestructive Inspection of Ceramic Matrix Composites Components by Computerized Tomography," in Proc. International Symposium on Advanced Materials for Lightweight Structures, March 22–25, 1994, Noordwijk, The Netherlands, ESA WPP-070, 1994. pp. 665–672.

[55] Spohnholtz, T. W., *Development of Impact-Resonant Spectroscopy for Characterizing Structural Ceramic Components*, MS Thesis, University of Illinois-Chicago, IL, USA, 1999.

[56] Brentnall, W.D., van Roode, M., Norton, P.F., Gates, S., Price, J.R., Jimenez, O., and Miriyala, N., "Ceramic Gas Turbine Development at Solar Turbines Incorporated," Chapter 7 in *Ceramic Gas Turbine Design and Test Experience, Progress in Ceramic Gas Turbine, Volume 1*, M. van Roode, M.K. Ferber, and D.W. Richerson, eds., ASME Press, New York, NY, USA, 2002. pp. 155–192.

CHAPTER

24

TEST STANDARDS AND DESIGN CODES FOR STRUCTURAL CERAMICS AND CERAMIC MATRIX COMPOSITES

Michael G. Jenkins
University of Washington, Seattle, Washington, USA

Yutaka Kagawa
Institute for Industrial Research, University of Tokyo, Tokyo, Japan

Marc Steen
Joint Research Centre, Petten, The Netherlands

INTRODUCTION

Demand for more efficient and environmentally-friendly gas turbines has fueled extensive worldwide investigations of the thermo-mechanical behavior (and its subsequent characterization) of structural ceramics (SCs) and ceramic matrix composites (CMCs) along with the important CMC subset, continuous fiber-reinforced ceramic-matrix composites (CFCCs). From an engineering perspective, determination of the properties and performance (e.g., mechanical, thermal, thermo-mechanical, physical, environmental, etc.) of SCs, CMCs and CFCCs is required for a number of reasons: 1) to provide basic characterization for purposes of materials development, quality control and comparative studies; 2) to provide a research tool for revealing the underlying mechanisms of mechanical performance; and 3) to provide performance-prediction data for engineering applications and component design [1]. As SC and CMC prototype and trial products begin to reach the marketplace, the paucity of standards (i.e., test methods, classification systems, unified terminology, and reference materials) for these materials and the lack of applicable design codes and their related data bases are limiting factors for their commercial diffusion and industrial acceptance [2].

Definition Standards

The term "standards" has multiple meanings for multiple users. In the research and technical communities, standards may be fundamental test methodologies and units of measure. For manufacturers or end-product users standards may be materials specifications and tests to meet requirements. Commercial standards are the rules and terms of information transfer among designers, manufacturers and product users [2]. Fundamental differences exist between levels of standards: company (internal use with only internal consensus); industry (trade/project use with limited organizational consensus); government (wide usage and varying levels of consensus); full-consensus (broadest usage and greatest consensus).

As of mid-2000, there are relatively few national or international full-consensus standards [3–6] for testing not only SCs but especially CMCs. Continued material development may be impaired by this limited ability to test on a common-denominator basis [2]. The lack of standards for SCs and CMCs may in part be due to the novelty of these advanced engineering materials. While initial speculative and actual applications of SCs and CMCs in gas turbines may be dated to Word War II in the mid 1940s along with follow-on work in 1950s [7], major efforts in Europe, Japan and the United States [8] to process, to characterize and to design with SCs and CMCs seem to have emerged from the 1970s push for more efficient heat engines. Indeed, while many different test methods and test results were being reported in the 1970s an oft-noted item on any list of hindrances to acceptance of SCs and CMCs in engineering applications was lack of standard test methods, design codes and data bases [9]. In 1981, the Japan Industrial Standards Committee (JISC) introduced a number of partial consensus standards for SCs including one for the flexural strength [10]. In due course other national standards bodies followed with their own standards for SCs and CMCs, eventually leading to the establishment of the Technical Committee TC206 on Fine Ceramics within the International Organization on Standards (ISO) in 1993 [4].

Definitions of Design Codes and Data Bases

The meaning of the term 'design code' is not generally well understood. As used in the following discussion 'design code' is not a design manual (i.e., a "cookbook" design procedure resulting in a desired component or system) or a computer program used in design (i.e., a programming 'code' for stress, strength or life analysis). Instead, 'design codes' are widely accepted, general rules for the construction of components or systems with emphasis on safety. A primary objective is the reasonably certain protection of life and property for a reasonably long safe-life of the design [11]. Although needs of the users, manufacturers and inspectors are recognized, the safety of the design can never be compromised.

By not imposing specific rules for design, the code allows flexibility for introducing new designs as required for performance, efficiency, usability, or manufacturability while still providing constraints for safety. The code must be wide ranging, incorporating figurative links between materials, general design (formulas, loads, allowable stress, permitted details), fabrication techniques, inspection, testing, certification by stamping and data reports, and finally quality control to insure that the code has been followed. Thus, implicit in the design codes may be many of the standards for materials testing, characterization, and quality control. In addition, unlike standards which provide no rules for compliance or accountability, codes require compliance through documentation, and certification through inspection and quality control.

A natural outcome of design codes is the incorporation of databases of material properties and performance 'qualified' for inclusion in the code. These data are 'qualified' because they have been attained through testing per the statistical requirement of the code as well as per the standards indicated in the code. 'Qualified' data bases often require a minimum numbers of tests for 1) a particular batch of material and 2) multiple batches of material. In addition, databases may include primary summary data (e.g., mean, standard deviation, and numbers of tests) along with secondary data from the individual tests. Some databases may contain only numerical information while others may include graphical information (e.g., stress-strain curves, temperature profiles, or test specimen geometry). Data bases are increasingly in electronic form to speed data retrieval and many are even web-based to provide instant access and frequent updatability.

Design codes and their data bases may even be backed as legal requirements for implementing an engineering design (e.g., certification and compliance with the American Society of Mechanical Engineers (ASME) Boiler and Pressure Vessel Code is a legal requirement in 48 of the 50 United States [11]). As of early-2003 there are no finalized national or international design codes relating to SCs, CMCs or CFCCs in any type of design, although efforts were initiated in 1995 to form a Task Group on Ceramic and Graphite Pressure Equipment within the ASME Boiler and Pressure Vessel Code and in 1996 to form a CMC effort within the Military Handbook 17 on Composites [6]. While progress in these two endeavors is noteworthy, introduction of these design codes is still several years off, although a limited consensus design code effort has been reported to be ready for release in the near future [12]. Currently, this lack of formalized design codes may be hampering utilization of SCs and CMCs because designers cannot use a material directly in new designs, but instead must first show evidence that the material meets the requirements of the code and then obtain special permission to use the material in the code design. In addition, further development of SCs and CMCs may be impaired because without a demand for a new material, there is no incentive for further refinement.

This chapter concentrates on standards, codes and databases for SCs, CMCs and CFCCs. First, current standards for SCs and CMCs are briefly reviewed, keying on important mechanical, thermal, and physical aspects of testing. Next, a similar brief review of current design codes and evolving databases for CFCCs is presented. Finally, the summary and conclusions are presented.

STANDARDS

Compared to metals and metal alloys, the number of standards for SCs and CMCs is limited. Hence, this number increases each year as these relatively new material systems mature. Table 24.1 lists key current full-consensus standards for SCs and CMCs. Of particular note for both SCs and CMCs are those full-consensus standards from the American Society for Testing and Materials (ASTM), the Committee for European Normalization (CEN), the Japan Industrial Standards Committee (JISC) and the International Organization for Standardization (ISO). In addition, Table 24.2 also lists some noteworthy industry/government standards for CMCs from the US High Speed Research/Enabling Propulsion Materials (HSR/EPM) program and the Japanese Petroleum Energy Center (PEC). The following subsections provide brief overviews of some of the key issues in standards for flexure, tension, compression, shear, hardness, fracture resistance, time-dependent behavior, physical/elastic constants, and other areas.

Flexure

Flexure testing is the "traditional" test method to evaluate the uniaxial strength of SCs. Either three- or four-point loading have been used and subsequently standardized as shown in Figure 24.1 for monolithic SCs [10]. While flexure testing is suitable for monolithic and particulate- and whisker-reinforced CMCs, the interpretation of the flexural test results for fiber-reinforced CMCs depends on the test configuration (e.g., uniaxial strength vs. interlaminar shear strength vs. compression failure vs. buckling) [13]. For both SCs and CMCs the test method may be applicable for material development, quality control purposes, and possibly design.

For monolithic SCs, flexural testing is often thought of as a method for evaluating surface-related phenomena such as sensitivity to surface preparation procedures (i.e., machining damage). However, studies have shown that with current standardized test methods (ASTM C1161, C1211, CEN.EN834-1, ENV820-1, ISO FDIS14704, JIS R1601, R1604) proper preparation, volume-related phenomena can be exposed as well [14]. Therefore, flexural test results for SCs (e.g., uniaxial strength, elastic constants, etc.), especially when obtained using standardized test methods, have been and are being used for design purposes. The flexure test for SCs has also become the basis of several other types of mechanical tests such as static (JIS R1612, R1632), dynamic (ASTM C1368, C1465) and cyclic fatigue (JIS R1621) as well as fracture toughness (ASTM C1421, JIS R1607, R1617).

TABLE 24.1 SOME CURRENT FULL-CONSENSUS STANDARD TEST METHODS FOR SCS, CMCS AND CFCCS

Organization/ Standard Number	Title	Completed
ASTM	**Committee C28 on "Advanced Ceramics"**	**Approved**
For SCs		
C1145-01	Terminology on Advanced Ceramics	1991
C1161-01	Test Method for Flexural Strength of Advanced Ceramics at Ambient Temperature	1990
C1175-99	Guide to Test Methods for Nondestructive Testing of Advanced Ceramics	1991
C1198-01	Test Method for Dynamic Young's Modulus, Shear Modulus, and Poisson's Ratio for Advanced Ceramics by Sonic Resonance	1991
C1211-98A	Test Method for Flexural Strength of Advanced Ceramics at Elevated Temperature	1992
C1212-98	Practice for Fabricating Ceramic Reference Specimens Containing Seeded Voids	1992
C1239-95	Practice for Reporting Uniaxial Strength Data and Estimating Weibull Distribution Parameters for Advanced Ceramics	1993
C1251-95	Guide for Determination of Specific Surface Area of Advanced Ceramics by Gas Adsorption	1993
C1259-98	Test Method for Dynamic Young's Modulus, Shear Modulus, and Poisson's Ratio for Advanced Ceramics by Impulse Excitation of Vibration	1994
C1273-95A	Test Method for Tensile Strength of Monolithic Advanced Ceramics at Ambient Temperatures	1994
C1274-95	Test Method for Advanced Ceramic Specific Surface Area by Physical Adsorption	1994
C1282-96A	Test Method for Determination of Particle Size Distribution of Advanced Ceramics by Centrifugal Photosedimentation	1995
C1286-94	Classification for Advanced Ceramics	1994
C1291-95E	Test method for Elevated Temperature Tensile Creep Strain, Creep Strain Rate, and Creep Time-to-Failure for Advanced Monolithic Ceramics	1995
C1322-96A	Practice for Fractography and Characterization of Fracture Origins in Advanced Ceramics	1996
C1323-96	Test Method for Ultimate Strength of Advanced Ceramics with Diametrally Compressed C-Ring Specimens at Ambient Temperature	1996
C1326-99 (96)	Test Method for Knoop Indentation Hardness of Advanced Ceramics	1996
C1327-97 (96)	Test Method for Vickers Indentation Hardness of Advanced Ceramics	1997
C1336-96	Practice for Fabricating Non-Oxide Ceramic Reference Specimens Containing Seeded Inclusions	1996
C1361-96	Practice for Constant-Amplitude, Axial, Tension-Tension Cyclic Fatigue of Advanced Ceramics at Ambient Temperatures	1996
C1366-97	Test Method for Tensile Strength of Monolithic Advanced Ceramics at Elevated Temperatures	1997
C1368-01	Test Method for Determination of Slow Crack Growth Parameters of Advanced Ceramics by Constant Stress-Rate Flexural Testing at Ambient Temperature	1997
C1421-01	Test Methods for the Determination of Fracture Toughness of Advanced Ceramics	1999
C1424-99	Test Method for Compressive Strength of Monolithic Advanced Ceramics at Ambient Temperatures	1999
C1465-00	Test Method for Determination of Slow Crack Growth Parameters of Advanced Ceramics by Constant Stress-Rate Flexural Testing at Elevated Temperature	2000
C1470-00	Guide for Testing the Thermal Properties of Advanced Ceramics	2000
C1499-01	Test Method for Monotonic Equibiaxial Flexural Strength Testing of Advanced Ceramics at Ambient Temperature	2001
For CMCs		
C1275-00	Standard Test Method for Monotonic Tensile Strength Testing of Continuous Fiber-Reinforced Advanced Ceramics with Solid Rectangular Cross Sections at Ambient Temperatures	1995
C1292-00	Standard Test Method for Shear Strength of Continuous Fiber-Reinforced Advanced Ceramics at Ambient Temperatures	1995
C1337-96	Standard Test Method for Creep and Creep Rupture of Continuous Fiber-Reinforced Advanced Ceramics under Tensile Loading	1996
C1341-00	Standard Test Method for Flexural Properties of Continuous Fiber-Reinforced Advanced Ceramics	1996
C1358-96	Standard Test Method for Monotonic Compressive Strength Testing of Continuous Fiber-Reinforced Advanced Ceramics with Solid Rectangular Cross Sections at Ambient Temperatures	1997
C1359-96	Standard Test Method for Monotonic Tensile Strength Testing of Continuous Fiber-Reinforced Advanced Ceramics with Solid Rectangular Cross Sections at Elevated Temperatures	1997

(continues)

TABLE 24.1 *(CONTINUED)*

Organization/ Standard Number	Title	Completed
For CMCs		
C1361-01	Standard Practice for Constant-Amplitude, Axial, Tension-Tension Cyclic Fatigue of Continuous Fiber-Reinforced Advanced Ceramics at Ambient Temperatures	1996
C1425-99	Test Method for Interlaminar Shear Strength of 1-D and 2-D CFCCs at Elevated Temperatures	1998
C1468-00	Standard Test Method for Transthickness Tensile Strength of Continuous Fiber-Reinforced Advanced Ceramics at Ambient Temperature	2000
C1469-00	Standard Test Method for Shear Strength of Joints in Advanced Ceramics at Ambient Temperature	2000
D3379-75	Standard Test Method for Tensile Strength and Young's Modulus for High-Modulus Single-Filament Materials	1975
CEN	**Committee TC184 "Advanced Technical Ceramics"**	**Completed**
For SCs		
ENV1071-1	Thickness of Ceramic Coatings by Probe Profilometer	1993
ENV1071-2	Thickness of Ceramic Coatings by Crater Grinding	1993
ENV1071-3	Adhesion by Scratch Test	1994
ENV1071-4	Chemical Composition of Ceramic Coatings by EPMA	1994
ENV1071-5	Porosity of Ceramic Coatings by Metallography	1995
ENV 12212	Classification	1997
EN 725-1	Impurities in Alumina	1997
ENV 725-2	Impurities in Barium Titanate	1994
EN 725-3	Oxygen in Non Oxides	1994
ENV 725-4	Oxygen in Aluminum Nitride	1994
EN 725-5	Particle Size Distribution	1996
EN 725-6	Specific Area	1996
EN 725-7	Absolute Density	1995
EN 725-8	Tap Density	1997
EN 725-9	Untapped Density	1997
EN 725-10	Compaction	1997
ENV 725-11	Sintering Curve	1993
ENV 1006	Sampling and Testing	1993
EN 623-1	Detection of Cracks by Dye Penetration	1995
EN 623-2	Density and Porosity	1993
ENV 623-3	Grain Size	1993
ENV 623-4	Surface Roughness	1993
ENV 820-1	Flexural Strength at Elevated Temperatures	1993
ENV 820-2	Deformation	1992
ENV 820-3	Thermal Shock	1993
EN 821-1	Thermal Expansion	1995
EN 821-2	Thermal Diffusivity	1997
ENV 821-3	Specific Heat	1993
EN 843-1	Flexural Strength	1995
ENV 843-2	Elastic Modulus	1995
ENV 843-3	Sub Critical Crack Growth	1996
ENV 843-4	Hardness	1994
ENV 843-5	Statistical Evaluation of Test Results	1996
ENV 12923-1	Laboratory Procedures for Chemical Corrosion Testing	1997
For CMCs		
EN 658-1	Tensile Strength of Continuous Fiber Reinforced Ceramic Composites	1998
EN 658-2	Compressive Strength of Continuous Fiber Reinforced Ceramic Composites	2000
ENV 658-3	Flexural Strength of Continuous Fiber Reinforced Ceramic Composites	1992
EN 658-4	Shear Strength (compression) of Continuous Fiber Reinforced Ceramic Composites	2000
ENV 658-5	Shear Strength (3-point) of Continuous Fiber Reinforced Ceramic Composites	1993
ENV 658-6	Shear Strength (double punch) of Continuous Fiber Reinforced Ceramic Composites	1993

(continues)

TABLE 24.1 (CONTINUED)

Organization/ Standard Number	Title	Completed
For CMCs		
ENV 1159-1	Thermal Expansion of Continuous Fiber Reinforced Ceramic Composites	1993
ENV 1159-2	Thermal Diffusivity of Continuous Fiber Reinforced Ceramic Composites	1993
ENV 1159-3	Specific Heat of Continuous Fiber Reinforced Ceramic Composites	1995
ENV 1159-4	Thermal Conductivity of Continuous Fiber Reinforced Ceramic Composites	1995
ENV 1389	Density of Continuous Fiber Reinforced Ceramic Composites	1994
ENV 1007-1	Size Level of Fibers for Continuous Fiber Reinforced Ceramic Composites	1993
ENV 1007-2	Linear Mass of Fibers for Continuous Fiber Reinforced Ceramic Composites	1993
ENV 1007-3	Filament Diameter of Fibers for Continuous Fiber Reinforced Ceramic Composites	1993
ENV 1007-4	Filament Strength of Fibers for Continuous Fiber Reinforced Ceramic Composites	1994
ENV 1007-5	Tensile Properties of Dry Fiber Tow at Room Temperature	1998
ENV 1892	Tensile Properties of Continuous Fiber Reinforced Ceramic Composites in Inert Atmosphere to 2000°C	1996
ENV 1893	Tensile Properties of Continuous Fiber Reinforced Ceramic Composites in Air to 1700°C	1996
ENV 1894	Shear Strength (interlaminar) of Continuous Fiber Reinforced Ceramic Composites in Inert Atmosphere to 2000°C	1996
EN 12289	Shear Strength (in-plane) of Continuous Fiber Reinforced Ceramic Composites	1996
ENV 12290	Compressive Properties of Continuous Fiber Reinforced Ceramic Composites in Inert Atmosphere to 2000°C	1996
ENV 12291	Compressive Properties of Continuous Fiber Reinforced Ceramic Composites in Air to 1700°C	1996
ENV 12788	Flexural Strength of Continuous Fiber Reinforced Ceramic Composites in Inert Atmosphere to 2000°C	1998
ENV 12789	Flexural Strength of Continuous Fiber Reinforced Ceramic Composites in Air to 1700°C	1998
ENV 13233	Notations and Symbols	1998
ENV 13234	Resistance to Crack Propagation of Continuous Fiber Reinforced Ceramic Composites	1999
ENV 13235	Resistance to Creep of Continuous Fiber Reinforced Ceramic Composites	1999
ISO	**Committee TC206 "Fine (Advanced Technical) Ceramics"**	**Approved**
For SCs		
ISO 14703	Sample Preparation for the Determination of Particle Size Distribution of Ceramic Powders	2000
ISO 14704	Test Method for Flexural Strength of Monolithic Ceramics at Room Temperature	2000
ISO 14705	Test Method for Hardness of Monolithic Ceramics at Room Temperature	2000
ISO 15490	Test Method for Tensile Strength of Monolithic Ceramics at Room Temperature	2000
FDIS 15732	Test Method for Fracture Toughness of Monolithic Ceramics at Room Temperature by Single Edge Precracked Beam (SEPB) Method	2002
ISO 15165	Classification Method	2001
ISO 17561	Test Method for Elastic Moduli of Monolithic Ceramics at Room Temperature by Sonic Resonance	2002
ISO 17562	Test Method for Linear Thermal Expansion of Monolithic Ceramics by Push Rod Technique	2001
For CMCs		
ISO 15733	Test Method for Tensile Stress-Strain Behavior of Continuous Fiber-Reinforced Composites at Room Temperature	2001
JISC	**Ceramics Division**	**Approved**
For SCs		
R1600-98	Terminology	1998
R1601-95	Flexural Strength	1995
R1602-95	Elastic Modulus	1995
R1603-94	Impurities in Silicon Nitride Part 1	1994
R1603-94	Impurities in Silicon Nitride Part 2	1994
R1604-95	Flexural Strength at Elevated Temperature	1995
R1605-95	Elastic Modulus at Elevated Temperature	1995
R1606-95	Tensile Strength at Elevated Temperature	1995
R1607-95	Fracture Toughness	1995
R1608-90	Compressive Strength	1990
R1609-90	Oxidation Testing	1990

(continues)

TABLE 24.1 *(CONTINUED)*

Organization/ Standard Number	Title	Completed
JISC	**Ceramics Division**	**Approved**
For SCs		
R1610-91	Hardness	1991
R1611-97	Thermal Diffusivity	1997
R1611-97	Specific Heat	1997
R1612-93	Flexural Creep	1993
R1613-93	Wear Resistance	1993
R1614-93	Chemical Corrosion	1993
R1615-93	Thermal Shock	1993
R1615-93	Erosion Resistance	1993
R1616-94	Chemical Analysis of Silicon Carbide	1994
R1617-94	Fracture Toughness at Elevated Temperature	1994
R1618-94	Thermal Expansion	1994
R1619-95	Particle Size Distribution	1995
R1620-95	Absolute Density	1995
R1620-95	Tap Density	1995
R1621-95	Bending Fatigue	1995
R1623-95	Hardness	1995
R1624-96	Bending Strength of Ceramic Joint	1996
R1625-96	Statistical Evaluation of Test Results	1996
R1626-96	Specific Area	1996
R1627-96	Dielectric Properties at Microwave Frequencies	1996
R1628-97	Untapped Density	1997
R1629-97	Particle Size Distribution	1997
R1630-97	Tensile Strength of Ceramic Joint	1997
R1631-98	Tensile Creep at Elevated Temperature	1998
R1632-98	Static Bending Fatigue	1998
R1633-98	Sample Preparation for SEM	1998
R1634-97	Density and Porosity	1997
R1635-98	Light Transmittance of Ceramic Films	1998
R1636-98	Coating Thickness by Probe Profilometer	1998
R1637-98	Resisivity of Conductive Thin Films, 4-Point Probe Array	1998
R1638-99	Iso-Electric Point	1999
R1639-99	Flowability of Powders	1999
R1639-1-99	Characterization of Granules: Part 1 Size Distribution	1999
R1639-2-99	Characterization of Granules: Part 2 Bulk Density	1999
R1639-3-99	Characterization of Granules: Part 3 Drying Loss	1999
R1639-4-99	Characterization of Granules: Part 4 Flowability	1999
R1640-02	Phase Composition Analysis of Silicon Nitride	2002
R1641-02	Dielectric Properties of Substrates:Microwave Frequency	2002
R1642-1-02	Internal Friction-Part 1: Torsion Pendulum Method	2002
R1642-2-02	Internal Friction-Part 2: Flexural Resonance Method	2002
R1645-02	Jet Water Erosion	2002
R1646-02	Cavitation Erosion	2002
R1647-02	Elution by Dipping in Water	2002
R1649-02	Chemical Analysis of Alumina Powders	2002
R1650-02	Thermoelectric Properties	2002
R1651-02	Pyroelectric Properties	2002
For CMCs		
R1643-02	Shear Strength (compression)	2002

* ASTM: American Society for Testing and Materials
CEN: Committee for European Normalization
ISO: International Organization for Standardization
JISC: Japan Industrial Standards Committee

TABLE 24.2 CURRENT INDUSTRY/GOVERNMENT STANDARD TEST METHODS FOR CMCS/CFCCS

Organization	Title	Completed
EPM (USA)		**Approved**
HSR/EPM-D-001-93	Monotonic Tensile Testing of Ceramic Matrix, Intermetallic Matrix and Metal Matrix Composite Materials	1993
HSR/EPM-D-002-93	Tension-Tension Load Controlled Fatigue Testing of Ceramic Matrix, Intermetallic Matrix and Metal Matrix Composite Materials	1993
HSR/EPM-D-003-93	Four Point Flexure Testing of Ceramic Matrix, Intermetallic Matrix and Metal Matrix Composite Materials	1993
HSR/EPM-D-004-93	Creep-Rupture and Stepped Creep Rupture of Ceramic Matrix, Intermetallic Matrix and Metal Matrix Composite Materials	1993
HSR/EPM-TSS-001-93	Measurement of Test System Alignment under Tensile Loading	1993
HSR/EPM-NDE-001-93	Measurement of the Bow and Warp of Continuous Fiber Reinforced Test Specimens	1993
Task A.5.4/A.5.5	CMC Pre-Cracking Standard	—
PEC (Japan)		**Approved**
PEC-TS CMC01	Test Method for Tensile Stress-Strain Behavior of Continuous Fiber Reinforced Ceramic Matrix Composites at Room and Elevated Temperatures	1997
PEC-TS CMC04	Test Method for Flexural Strength of Continuous Fiber Reinforced Ceramic Matrix Composites at Room and Elevated Temperatures	1997
PEC-TS CMC06	Test Method for Shear Strength of Continuous Fiber Reinforced Ceramic Matrix Composites at Room and Elevated Temperatures	1997
PEC-TS CMC08	Test Method for Fracture Toughness of Continuous Fiber Reinforced Ceramic Matrix Composites	1997
PEC-TS CMC09	Test Method for Fracture Energy of Continuous Fiber Reinforced Ceramic Matrix Composites	1997
PEC-TS CMC010	Test Method for Tensile-Tensile Cyclic Fatigue of Continuous Fiber Reinforced Ceramic Matrix Composites at Room and Elevated Temperatures	1997
PEC-TS CMC011	Test Method for Tensile Creep of Continuous Fiber Reinforced Ceramic Matrix Composites at Elevated Temperatures	1997
PEC-TS CMC013	Test Method for Elastic Modulus of Ceramic Matrix Composites at Room and Elevated Temperatures	1997
PEC-TS CMC014	Test Method for Oxidation Resistance of Non-Oxide Ceramic Matrix Composites at Elevated Temperatures	1997

HSR / EPM =High Speed Research / Enabling Propulsion Materials program (NASA, GE, Pratt and Whitney consortium) (USA)
PEC = Petroleum Energy Center (Japan)

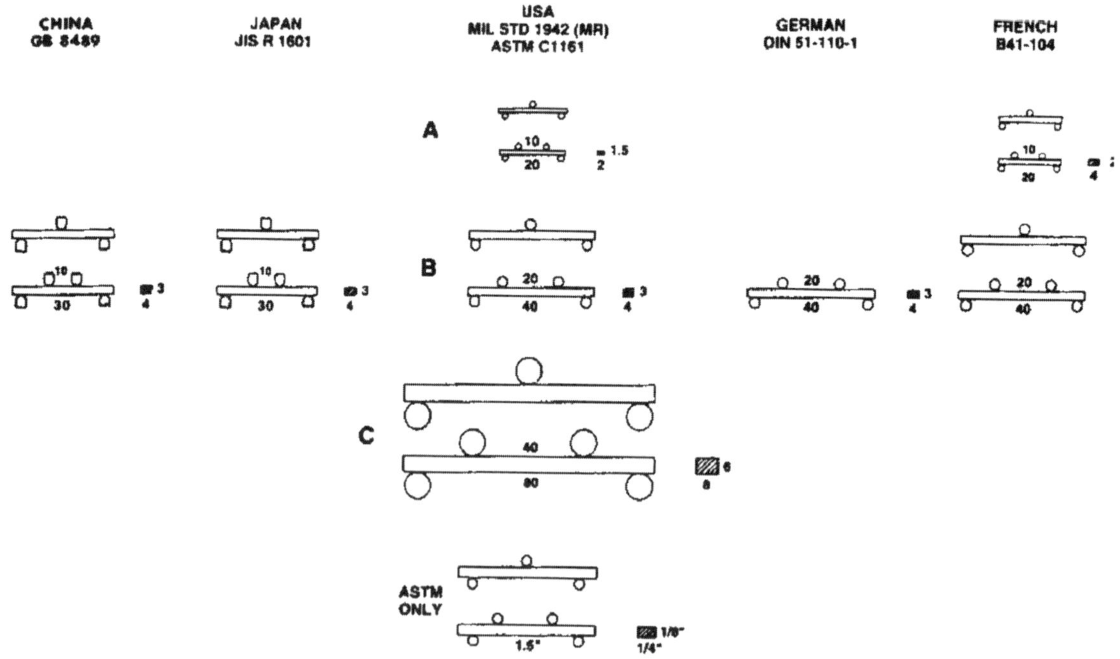

Figure 24.1 Some Test Specimen and Test Fixture Sizes of Various Flexural Strength Standards—Dimensions in mm Except Bottom Two Figures which are in Inches for Historical and Technical Consistency (Reprinted with permission from ref. [10], p. 361, by courtesy of Marcel Dekker, Inc.)

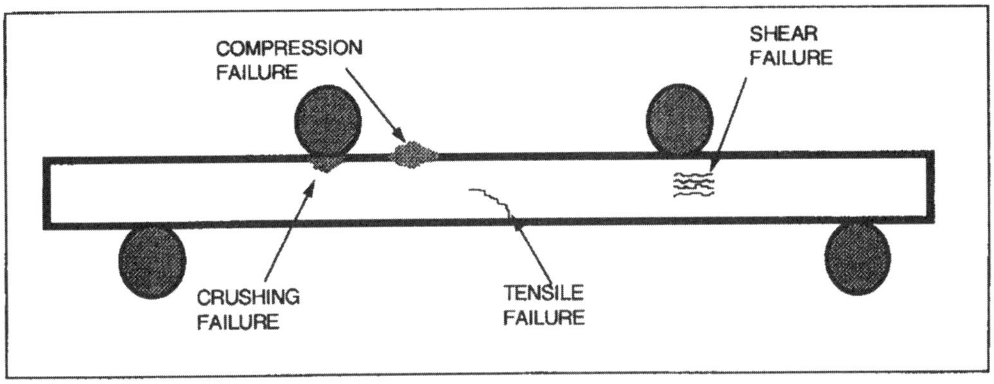

Figure 24.2 Various Failure Modes of CMCs in Flexure (Reprinted with permission from ASTM C1341)

For CMCs and especially CFCCs, depending on the reinforcement, the interphase, the matrix, and the ratio of test span to test specimen height (a.k.a span to depth ratio) among other parameters, a variety of failure modes may manifest themselves (see Figure 24.2). For example, the failure (i.e., loss of load carrying capability) may be due to shear stress if the interlaminar shear strength is weak compared to the in-plane tensile strength or if the span to depth ratio is small causing the transverse shear stress due to beam bending to dominate the in-plane tensile stress. In addition, because of differences in tensile and compressive stress-strain response in some CMCs, especially CFCCs (see Figure 24.3), an assumed non-linear tensile stress-strain response of a CFCC in flexure can be quite different (e.g., exhibiting "graceful" failure) compared to the tensile stress-strain response of a CFCC in tension (see Figure 24.4). This does not obviate the results from flexure tests of CMCs (ASTM C1341, CEN EN658-3, EN658-5, ENV12788,

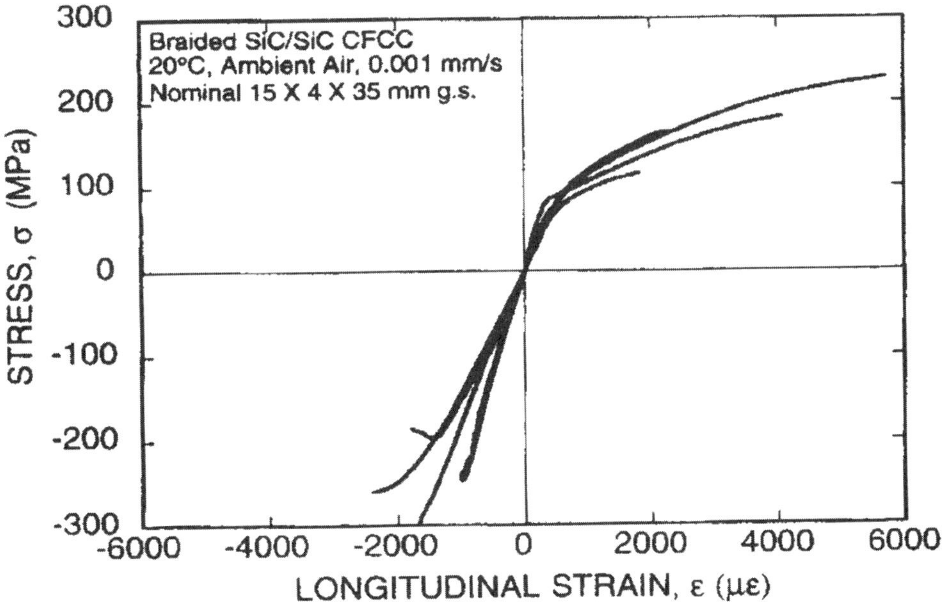

Figure 24.3 Difference in Tension and Compression for a CMC (Reprinted from ref. [15])

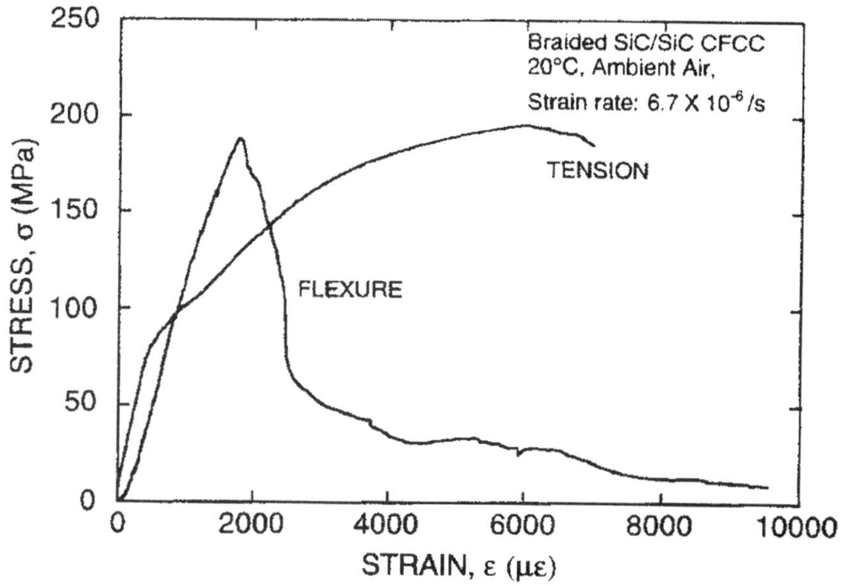

Figure 24.4 Comparison of Tensile and Flexural Behavior of a CMC (Reprinted from ref. [15])

ENV12789). Rather, use of the results requires meticulous attention to details, especially mode of failure to ascertain the utility of the results for a particular application. Often, flexural test results for CMCs are useful for screening purposes or for determining the response of the CMC to nonuniform stresses. However, for processing and design purposes, test methods resulting in uniform uniaxial stresses (e.g., uniaxial tension or compression) are often preferred [16].

Tension

Uniaxial tensile testing is the most fundamental means of measuring mechanical response of most engineering materials, and interest in it for SCs, CMCs and CFCCs is driven primarily by concern over the nonuniform stress state of flexure tests. This nonuniform stress state may overemphasize surface effects in SCs and may strongly affect failure mode in CMCs as just discussed.

While conceptually simple, the widespread implementation of the uniaxial tension test for ceramics has been limited by several factors [17]. A primary impediment has been the actual achievement of a uniform stress state (i.e., minimization of parasitic bending) which is a critical concern in brittle materials such as SCs because the test specimen is not able to "self align" by plastic deformation to accommodate any nonuniformity of the applied stress [17]. Despite the damage tolerance of CMCs and CFCCs, the brittle matrix has also been shown to be sensitive to nonuniform stress [18].

Because the tension test is the most fundamental test for mechanical characterization, the greatest number of in-plane monotonic tensile test standards (see Tables 24.1 and 24.2) exist for it. As with the flexure test, once developed, the monotonic tension test also becomes the basis for other types of tests, such as those for cyclic fatigue or creep/creep rupture. Recent efforts in both ASTM and CEN are moving toward developing elevated-temperature tensile tests for various environments other than ambient air.

Unlike the flexure test for SCs where standardization has led to a common $3 \times 4 \times 50$ mm prismatic beam test specimen geometry, no single preferred test specimen geometry has emerged for either SCs (ASTM C1273, C1366, ISO FCIS 15490, JIS R1606) or CMCs (ASTM C1275-C1359, CEN EN658-1, ENV

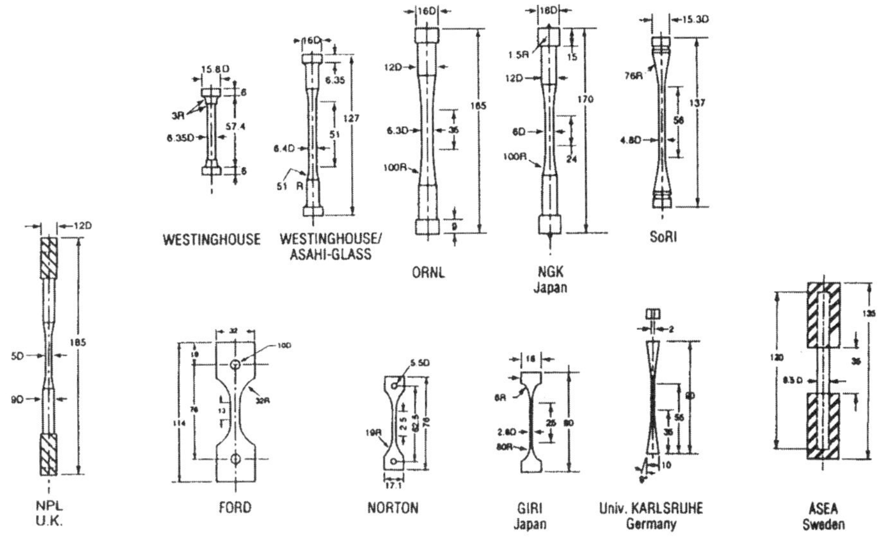

NOTE 1—All dimensions are in mm.
NOTE 2—Acronyms: NPL, U.K. = National Physical Laboratory, United Kingdom; ORNL = Oak Ridge National Laboratory; NGK = Nippon Gaishi Kaisha; SoRI = Southern Research Institute; ASEA = ASEA-Ceram; NIST = National Institute for Standards and Technology; GIRI = Government Industrial Research Institute.

Figure 24.5 Plethora of Tensile Test Specimens Geometries for SCs (ASTM C1273). Note that some of these test specimens are circular and others are rectangular in cross section (Reprinted with permission from ASTM C1273).

1892, ENV 1893, ISO DIS15733). Figures 24.5 and 24.6 show examples of test specimen geometries used for tensile testing SCs and CMCs, respectively. Of course, this wide range of test specimen geometries has also led to a range of gripping and alignment systems. Standards for tension testing have concentrated on normalizing test parameters (e.g., maximum allowable bending, rate of testing, etc.) rather than geometries or hardware. However, for elevated temperature tests (e.g., fast fracture strength, creep/creep rupture), concerns have arisen over the type of grip (hot, warm, or cold) and the design of the test specimen so as to minimize the imposition of undesirable thermal and imposed stresses (see Figure 24.7).

CMCs and two-dimensionally-reinforced CFCCs are unique in that the mechanical properties of the bulk material may be highly anisotropic. Thus, recently, demand has increased to determine the tensile response perpendicular to the in-plane reinforcements (see Figure 24.8). In two-dimensionally reinforced CFCCs, this direction can be the weakest since only matrix material with some interphase and no reinforcing fibers are present. Standardization activities produced a full consensus standard for "trans-thickness" tensile strength (ASTM C1468).

Compression

Although, it is generally accepted that SCs and CMCs achieve their greatest strength potential in compression, it is nearly impossible to create an advanced engineering design without subjecting the component to tensile stresses. Nonetheless, it is important to characterize the compressive strength of these materials for many design purposes. While it is accepted that elastic moduli of SCs and CMCs measured from compressive tests are independent of the test method employed, compressive strength results depend on the mode of failure and hence, the compression test method employed. Regardless of the method used, compressive performance is the most difficult to evaluate because test scatter is greater than that of any other mechanical test. The major source of scatter is the difference in failure modes [19,20] while the major experimental difficulties for compression tests are buckling instabilities and achieving good alignment to avoid load eccentricities that induce bending.

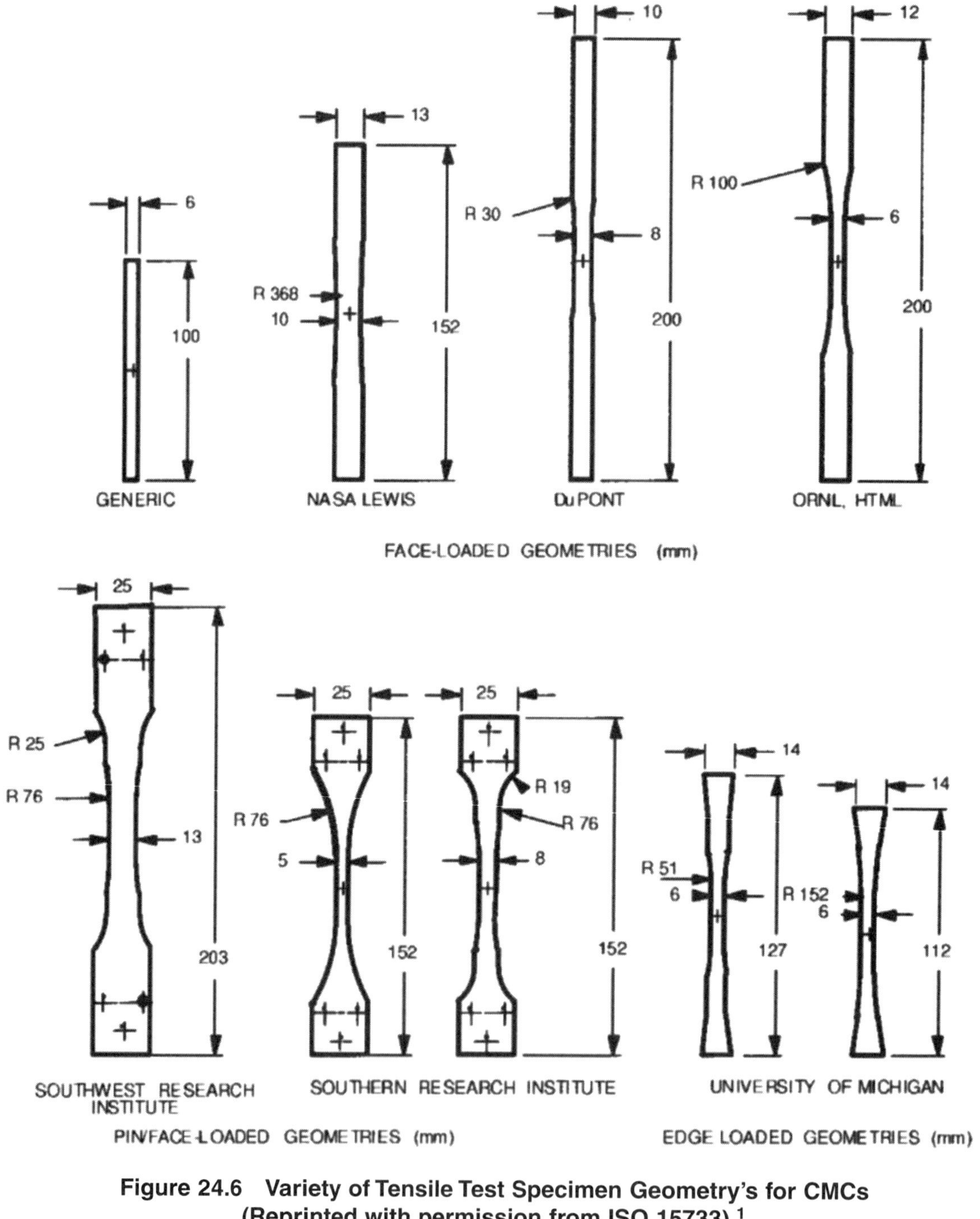

Figure 24.6 Variety of Tensile Test Specimen Geometry's for CMCs (Reprinted with permission from ISO 15733).[1]

[1]The terms and definitions taken from ISO 15733:2001 Fine Ceramics (advanced ceramics, advanced technical ceramics)—Test method for tensile stress-strain behavior of continuous fiber-reinforced composites at room temperature, Fig. B1, are reproduced with the permission of the International Organization for Standardization, ISO. This standard can be obtained from any ISO member and from the Web site of the ISO Central Secretariat at the following address: www.iso.org. Copyright remains with ISO.

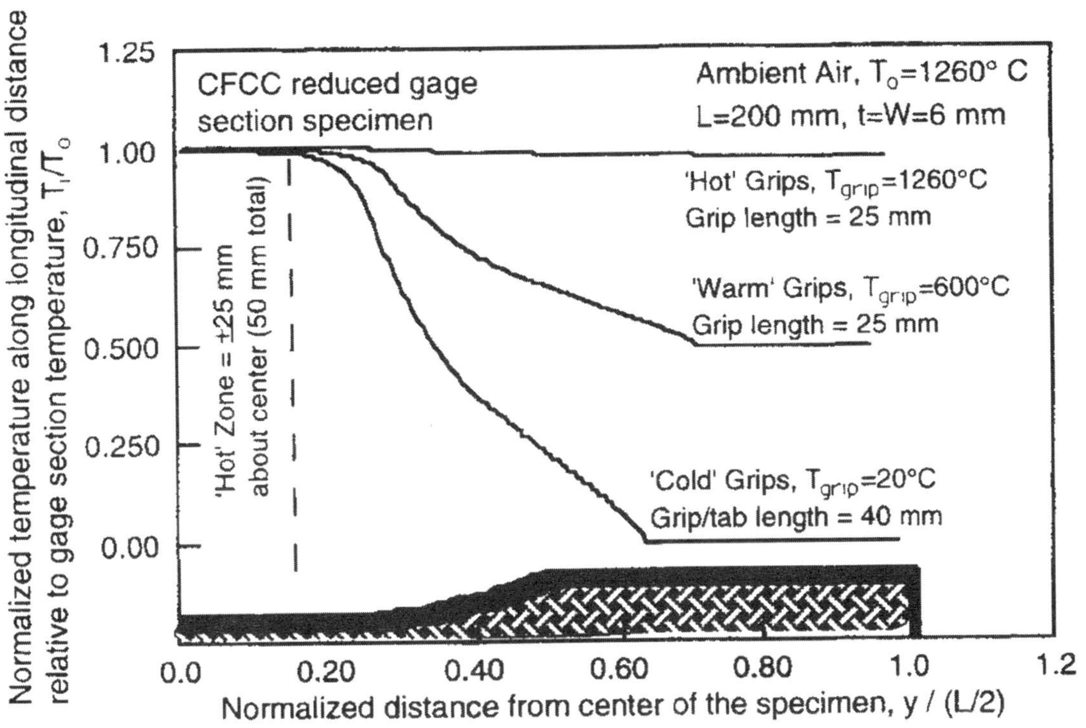

Figure 24.7 Distribution of Temperature and Applied Stresses in a High Temperature Test Specimen of a CMC (Reprinted with permission from ref. [6])

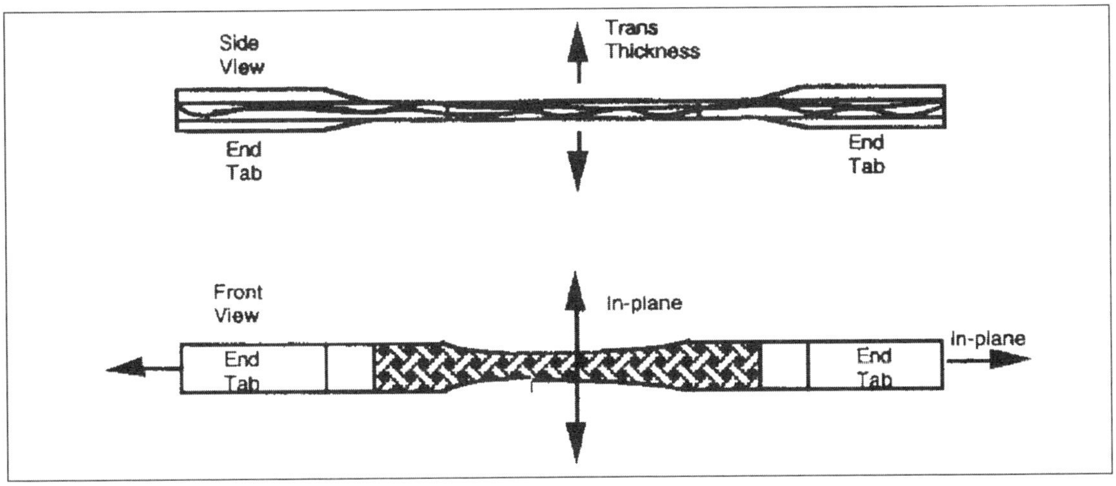

Figure 24.8 Transverse Tensile Testing Direction (Reprinted with permission from ref. [6])

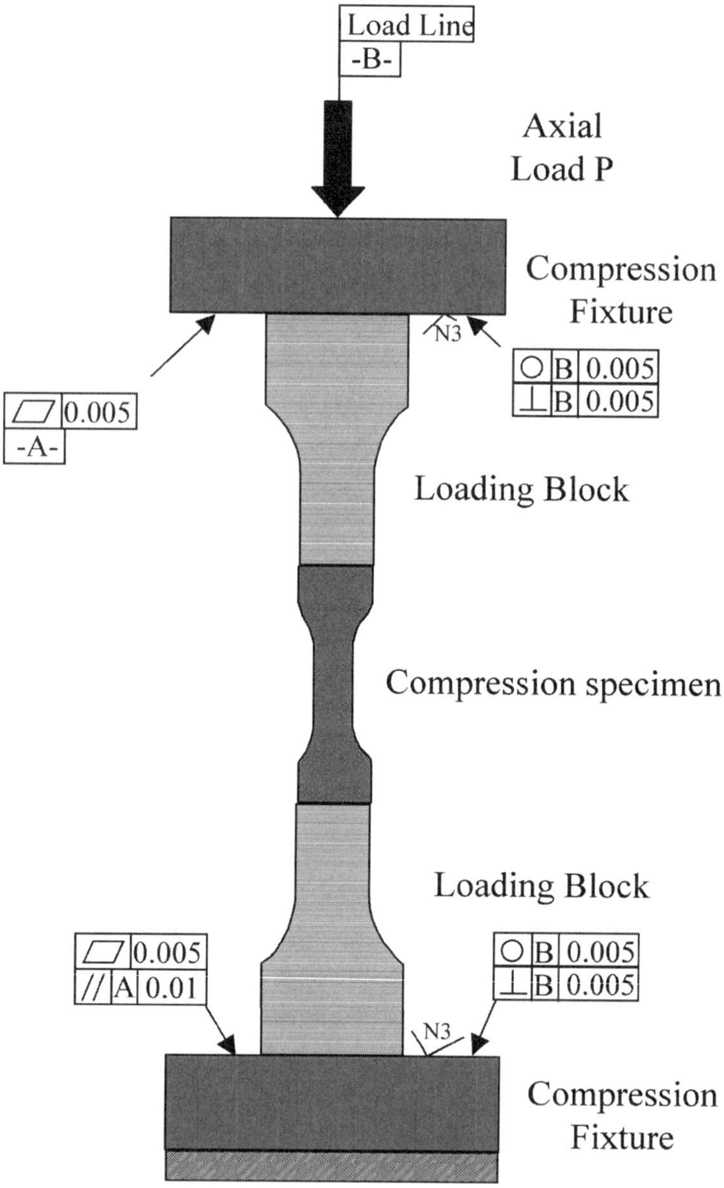

Figure 24.9 Compression Test Setup for SCs at Room Temperature (ASTM C1424) Using Cylindrical Hour-Glass Test Specimens (Reprinted with permission from ASTM C1424)

For SCs, successful compression tests are achieved by meticulous attention to detail especially in regard to test specimen design and fabrication as well as test fixture fabrication and maintenance (ASTM 1424, JIS 1608). An example of such a setup is shown in Figure 24.9. Although the test specimen geometry can be standardized to minimize the tendency to buckle by imposing dimensional tolerances so as to minimize eccentricities, the interface of the test specimen and test fixture may be the greatest contributor to nonuniform stresses and subsequent non-representative failures.

For CMCs, to avoid brooming and bending, robust tabs are usually bonded to the uniaxially-loaded test specimens. Buckling resistance can be increased if clamped end conditions are achieved and the bending

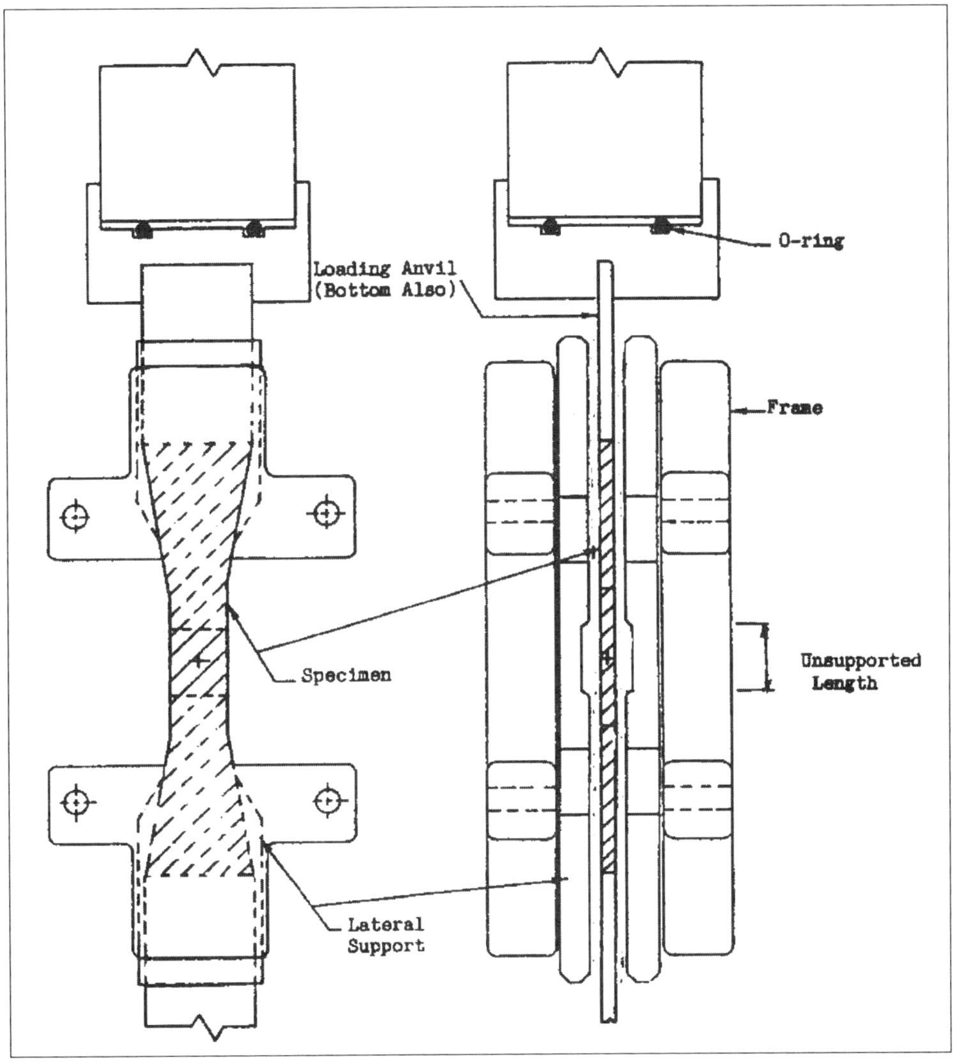

Figure 24.10 Compression Test Setup for CMCs at Room Temperature (ASTM C1358) Specimens (Reprinted with permission from ASTM C1358)

stiffness near the ends is increased. Finite element analyses (FEA) have been employed to optimize the loading configuration of compression test specimens which are based upon: preventing overall buckling, preventing debonding of end tabs and preventing failure at the gage region-tab juncture while insuring stress uniformity in the central region. Standardized compression tests for CMCs and CFCCs (ASTM C1358, CEN EN658-2, ENV12290, ENV 12291) have tended toward uniaxially-loaded test specimens (i.e., either with or without antibuckling guides) in anticipation of elevated temperature tests (see Figure 24.10). Results (see Figures 24.3 and 24.4) from these compression tests have been compared to tension and flexure test results to illustrate similarities (elastic modulus) and differences (strengths) between the different types of tests. Limited data is available in the open literature [15,21] on compressive strength of CFCCs although the standards currently in place (see Tables 24.1 and 24.2) should provide reliable and reproducible test results.

Another test available to determine the compressive performance of CFCCs is the bending of sandwich beams [17]. If the face in tension is deliberately made much stronger than the face in the compression side, a compressive failure will be induced. This test provides effective load transfer, ideal compressive strengths and data comparable to uniaxially loaded compression tests making the bending of composite sandwich beams highly recommended for design allowables. The major disadvantages for this test are that the test specimen is quite large and that the bonded core may artificially enhance the performance of the face in compression, although round robin tests with graphite/epoxy composites have dismissed the latter possibility [22].

Shear

The three primary testing modes are tension, compression and shear. For flat plates, there are three components of tensile and compressive stresses, axial, in-plane transverse and through-the-thickness transverse. Correspondingly there are also three components of shear stresses, (i.e., in-plane and two through-thickness components). Correspondingly, depending on the material, there are tests that only provide data on strength, or stiffness or both [22].

Because in general, SCs behave linear elastically up to fracture, constitutive relations (Hooke's Law) can be used to relate shear stress/strain and normal stress/strain behaviors. In addition, because dominant failure modes in SCs are related to normal stresses, tests involving shear for these materials are rare.

However, for CMCs failure modes can be multi-faceted and it is important to apply the appropriate normal or shear stress necessary to reveal the failure mode of interest. Most current shear tests methods for CMCs measure in-plane properties and only a few measure the interlaminar or through-thickness shear. The ideal quantitative shear test method should provide a region of pure and uniform shear stress. In addition there should be a unique relationship between the applied load and the magnitude of the shear stress in the test section and the test section should be one of maximum shear stress relative to all other regions of the test specimen. Although it can be argued that there are no good shear tests but only varying degrees of bad ones, the torsion of a thin tube comes closest to satisfying the criteria for a good in-plane shear test.

Nonetheless, shear test methods may be divided into two groups: those that use flat test specimens and those that use tubular or circular section specimens. For CMCs, standards (see Tables 24.1 and 24.2) that use flat specimens include (see Figure 24.11): the double notch compression test and the short flexure beam for interlaminar shear behavior and the double V-notch test (a.k.a. the Iosipescu test) for in-plane shear behavior (ASTM C1292, C1425, CEN EN658-4, EN658-5, EN658-6, EN 12289, ENV1894). Currently, there are no standards for CMCs and CFCCs, which use torsion for shear behavior although examples of tests using specimens with circular cross sections include the torsion of tubes and the fournsion, thermal conductivity and other non-mechanical results.

Hardness

Hardness by definition is a material's resistance to penetration. From a materials characterization standpoint, hardness is a resistance to penetration of particular indentor geometry and force in a prescribed manner (e.g., Brinell, Vickers, Rockwell, Knoop) that results in a well-defined and measurable indentation artifact on the materials surface (see Figure 24.12).

In SCs, Vickers and Knoop hardness tests have been standardized (ASTM C1326, C1327, CEN ENV 843, ISO FDIS14705, JIS R1610). Several issues are addressed in these standards, including surface preparation, maximum applied force and screening of indentations. One of the major problems (interferences) with hardness testing of brittle materials is cracking around the indentation. Knoop hardness tests are much less prone to cracking and are recommended for SCs [10]. Another major source of error in these measurements is the measurement of the indentation size, which can strongly influence the resulting hardness number. Interestingly, two of only few standard reference materials (SRM) for SCs exist for hardness [23]. These SRMs provide the means for testers to verify their test setup and their use of the standards for SCs.

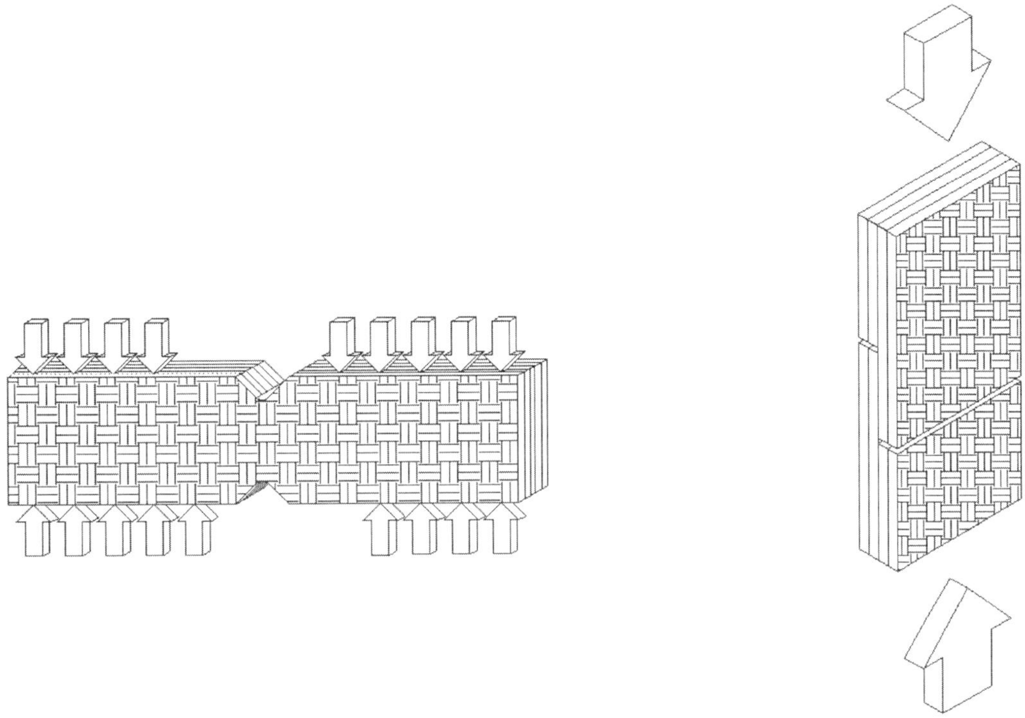

(a) (b)

Figure 24.11 Flat Shear Test Specimens for CMCs a) In-Plane Shear and b) Interlaminar Shear) (ASTM C1292) Specimens (Reprinted with permission from ASTM C1292)

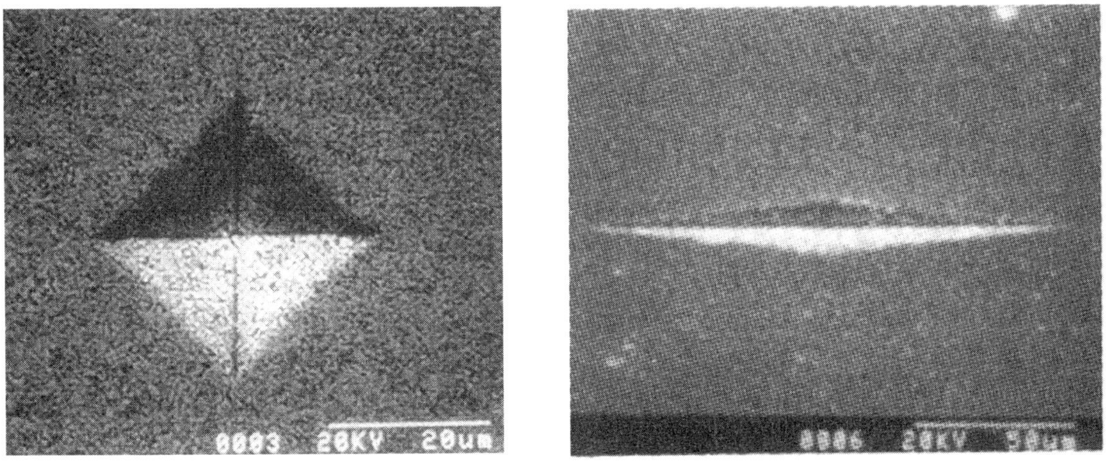

(a) (b)

Figure 24.12 Example of Vickers and Knoop Hardness Indentations in an SC a) Vickers in WC and b) Knoop in Si_3N_4 (Reprinted with permission from ref. [10], p. 370, by courtesy of Marcel Dekker, Inc.)

Hardness testing in CMCs is problematic. For dense matrix materials such as particulate and whisker-reinforced materials, the same standardized tests methods as for SCs can be used. However for non-dense matrix materials such as CFCCs hardness is not measurable in the usual sense and no standards have been forthcoming. Hardness test procedures may be useful for introducing prescribed damage into CFCCs in order to determine damage tolerance. However, this has not been a common practice.

Fracture Resistance

No doubt owing to the brittle nature of SCs and the potential for catastrophic failure in the presence of cracks, the determination of fracture toughness has met with considerable controversy over the years [24–27]. A variety of tests methods in use over the years sometimes gave results in agreement with one another and other times did not. Part of the controversy over the test results arose over the inconsistency of these various test methods.

Brittle fracture is accepted for SCs at room temperature because they rarely exhibit plastic deformation under those conditions. Two major problems with measuring fracture resistance at room temperature are environmentally assisted stable crack growth (so called slow crack growth) and increasing fracture resistance with increasing crack length (so called rising R-curve). In addition, it is generally agreed that three conditions need to be met for a valid fracture resistance test: atomistically sharp initial crack, valid stress intensity factor solution, and well-defined and measurable crack dimensions. These conditions and other issues have been addressed in recent standardization efforts (ASTM C1421, ISO DIS 15732, JIS R1607). Using standard test methods (see Figure 24.13) it is possible to actually discern the effects of processing variations in materials [28] such that an SRM for fracture toughness of SCs is now available for purchase.

Strictly-speaking because of the variety of constituents, the anisotropic nature of CMCs and CFCCs, and the lack of a well-defined macro crack tip, stress intensity factors used in linear elastic fracture mechanics (LEFM) are not defined. Therefore, the measurement of fracture toughness in the conventional sense in CMCs

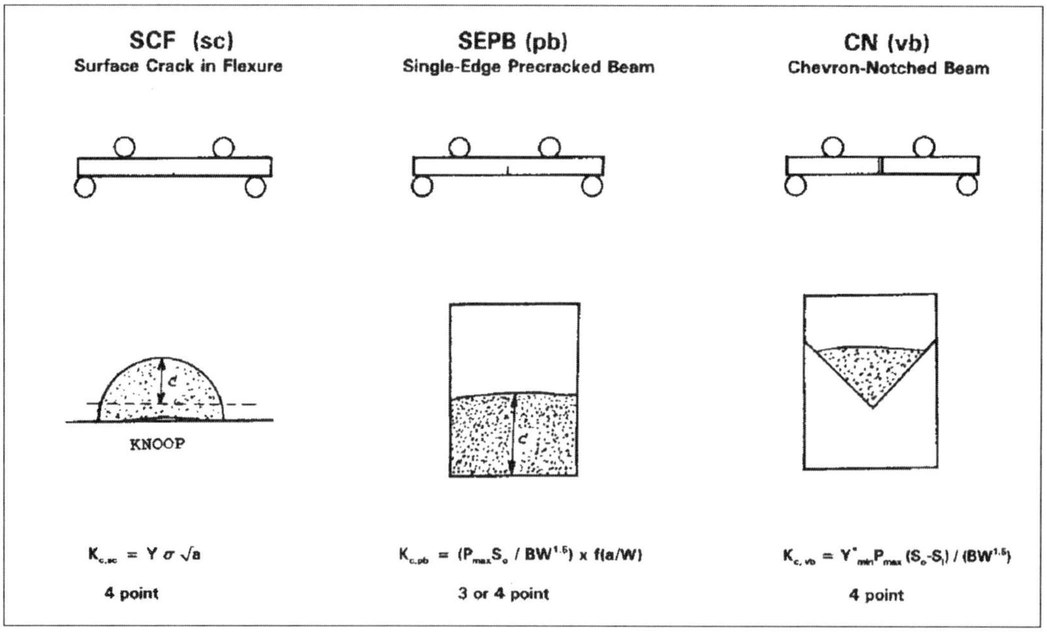

Figure 24.13 Examples of Some Standardized Fracture Toughness Testing Setups of SCs (Reprinted with permission from ref. [10], p. 376, by courtesy of Marcel Dekker, Inc.).

and CFCCs is inappropriate and meaningless. However, the inapplicability of LEFM for CMCs and CFCCs does not preclude the measurement of fracture resistance in these materials. Instead, more global energy parameters can be used such as variations on strain energy release rate or even modulus of toughness (area under the tensile stress-strain curve). Such energy-based parameters may be related to damage accumulation and damage tolerance, although only one standard has been forthcoming in this area (CEN ENV13234).

Time-Dependent Behavior

Time dependent degradation and failure of SCs and CMCs is of particular concern for long-term applications in aggressive environments (e.g., high moisture and/or high temperature). A particular concern in ceramics and glasses has been subcritical crack growth (a.k.a. slow crack growth or SCG). This phenomenon can manifest itself under constant or fluctuating loads and similar to fatigue crack growth in metals is known for its insidious nature and catastrophic results.

In SCs, numerous tests have been devised over the years to quantify SCG in SCs [29]. Some of the more common are static fatigue tests and dynamic fatigue tests. Strong evidence exists that the fundamental time-dependent failure mechanism in SCs is SCG and therefore even cyclic fatigue failure in SCs can be related to a SCG mechanism [29]. Several standards have been introduced to measure SCG parameters at various temperatures using flexure tests (ASTM C1368, ASTM C1465, CEN ENV 843-3, JIS R1632.). In addition, standards have also been introduced to measure cyclic fatigue behavior using flexure and tensile tests (ASTM C1361, JIS R1621).

SCG depends on the growth of preexisting crack-like flaws under large enough applied stress and hence for a large enough stress intensity factor. At elevated temperatures, a competing mechanism to SCG is creep cavity formation and coalescence. Because of a difference in creep response in tension and compression, it is impossible to deconvolute tensile or compressive creep behavior in SCs from flexure tests alone. Therefore meaningful creep tests of SCs have been standardized for uniaxial stress states (ASTM C1291, CEN ENV 820-2, JIS R1612, R1631).

While SCG may be a dominant time-dependent mechanism in many SCs, no single dominant mechanism reveals itself in CMCs. Instead, time-dependent phenomena are often tied to chemical kinetics with damage accumulating at the individual damage sites and eventually coalescing to cause failure of the material. Because of the nonlinear stress-strain response of CMCs, flexure tests are not used for measurements of time-dependent failure. Uniaxial tension or compression tests have been standardized to measure time-dependent degradation, creep and creep/stress rupture response (ASTM C1337, CEN ENV13235).

Physical/Thermal Properties and Performance

Physical and thermal properties and performance include density, elastic constants, thermal coefficient of expansion, thermal conductivity and other non-mechanical results. Some standards (see Tables 24.1 and 24.2) have been introduced for these properties. When available, a standard test method for SCs may be adapted to apply to CMCs. While application of the test method to the CMCs is experimentally straight forward, interpretation of the resulting information is far from resolved. Although some standards (see Tables 24.1 and 24.2) unique to SCs and CMCs have been introduced for these properties, in some cases a new standard guide may be introduced to interpret these general standards for application to SCs and CMCs

For thermal properties, standards for non-ceramics have been reinterpreted for applications to ceramics. Often, a major concern is extending the temperature range to include those temperatures of interest for ceramics. The concern may require modification of temperature sensors, fixturing and measurement equipment (CEN ENV820-3, EN821-1, EN821-2, ENV821-3, ISO DIS17562, JIS R1615, R1618, R1611, R1611).

For elastic constants, resonance methods (continuous and impulse) are often used to extract elastic modulus, shear modulus and Poisson's ratio (ASTM C1198, C1259, CEN ENV 843-3, JIS R1602). For specific test specimen sizes (bars, plates and discs) and isotropic materials, excellent repeatability and reproducibility (~1% coefficients of variation) are attainable using these methods. For anisotropic materials, there is some concern over the mode of excitation and subsequent interpretation of the test results although recent studies have shown applicability of the methods for CFCCs. As attention turns from the development of standards for characterizing basic mechanical properties and performance, standards for characterizing physical and thermal aspects of CFCCs are being further developed (CEN 1389, ENV1159-1, ENV1159-2, ENV1159-3).

Other Areas

Often characterization methods not related to thermal or mechanical behavior are required and standards exist for these methods as well. Microstructual characterization, non-destructive characterization, fractography and other similar areas reflect these methods.

For SCs, methods have been standardized for powder characterization prior to processing (ASTM C1251, C1274, C1282, CEN EN725-5, EN725-6, JIS R1619, R1629) and grain size characterization after processing (CEN ENV623-3, JIS R1639, R1616). Nondestructive methods involve ultrasonic, acoustic, infrared methods and dye penetrant (ASTM C1175, C1212, C1336, CEN EN623). Fractographic methods have been standardized to reflect the unique aspects of failure in brittle materials (ASTM C1322).

In CMCs and especially CFCCs there are three primary constituents: reinforcement/fiber, matrix and interphase. Understanding the properties and performance of the constituents is useful not only for predictive modeling but also for quality control. Standards (see Tables 24.1 and 24.2) exist for evaluating fibers (often single fibers, but also tows) (ASTM D3379, CEN ENV1007-1, ENV 1007-2, ENV1007-3, ENV1007-4, ENV1007-5). Standards (see Tables 24.1 and 24.2) also exist for evaluating matrices, albeit usually those listed for monolithic ceramics. However, no test standards currently exist for evaluating the interphase, although efforts have been initiated in ASTM and in CEN to develop a guide on how to conduct and analyze single fiber push-in/push-out tests. Another promising test method being developed is the unload/reload tensile test. The size and magnitude of stress-strain hysteresis loops (see Figure 24.14) can be

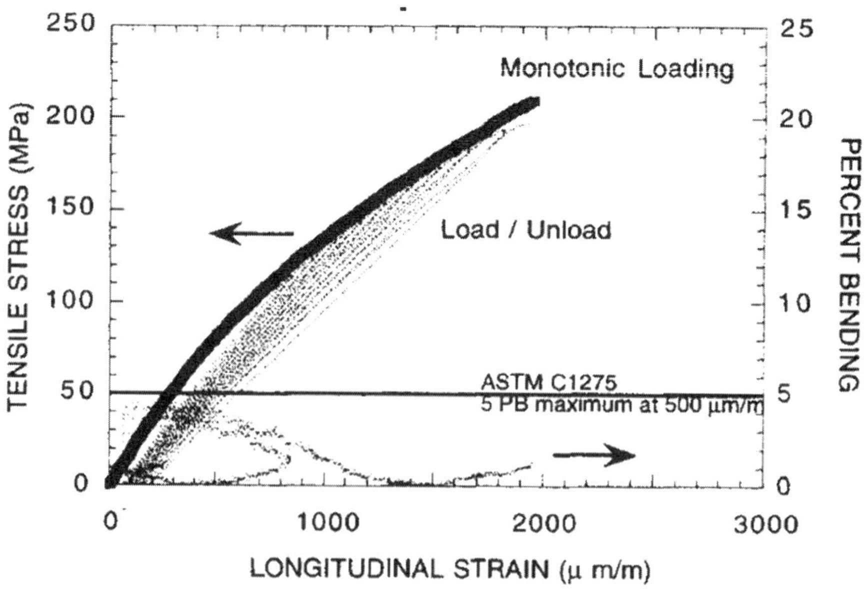

Figure 24.14 Example of an Unload-Reload Tensile Stress-Strain Curve for Extracting Additional Information from a Tensile Test of a CMC (Reprinted with permission from ref. [6])

used to extract information about residual stress state and interfacial shear strength [6, 30, 31]. Although this type of test has not been standardized, it does offer promise for a simple mechanical test for 'interrogating' the bulk CFCC to extract in-situ constituent properties and performance.

CFCCs are unique materials because the sum of their parts (fiber, matrix and interphase) can be greater or less than the composite itself depending on the environment or use conditions. For example, 'toughness' (i.e., area under the tensile stress-strain curve) is much greater for a CFCC than for any fiber, matrix or interphase (or their sums) because of the nonlinear energy dissipation that occurs during debonding and pullout of the fibers during the cumulative fracture process. Conversely embrittlement and degradation of fibers is accelerated during elevated temperature exposure in air due to the presence of excess oxygen in the interphase resulting in both strength and 'toughness' of the CFCC which are less than those of any fiber, matrix or interphase (or their sums). Therefore environmental tests (e.g., thermogravimetric analysis or differential gravimetric analysis) must be conducted under the actual use conditions. Another area needing standardization is component testing. For example, coupon test specimens cannot be extracted from a CFCC tube without disturbing the unique weaving for the fibers.

Prestandardization Collaborative Efforts

A number of collaborative efforts exist on an international level to develop experience, investigate methodologies, gather data for properties and performance, and draft prestandards via joint cooperative efforts such as multiple laboratory round robin programmes [10]. Some of these efforts are as follows:

- Versailles Advanced Materials and Standards (VAMAS) organization performs prestandardization research and consists of Canada, Commission of European Communities, France, Germany, Italy, Japan, United Kingdom and United States. VAMAS Technical Working Area #3 has objective of researching repeatability and reproducibility of test procedures for advanced ceramics.
- International Energy Agency (IEA) consists of representative companies and laboratories from Germany, Japan, Sweden and United States and has agreed to conduct research aimed at thermal and mechanical properties and performance of SCs selected by each country.
- The Technical Cooperation Program (TTCP) is defense research collaboration between Australia, Canada, New Zealand, United Kingdom and United States.

DESIGN CODES AND DATA BASES

Design codes contain in them not only the methodologies for successfully executing the steps in the design of a particular component or system, but also the information required to complete the design. This information is in the form of qualified data bases which contain material properties and performance obtained using standards referenced in the design code as well as statistical means for evaluating the "quality" of the data, also contained in the design code. Two efforts are currently underway to establish such design codes for SCs, CMCs and CFCCs: ASME Task Group on Ceramic and Graphite Pressure Equipment and Military Handbook 17 (Mil-Hdbk-17) CMC effort.

ASME Task Group on Ceramic and Graphite Pressure Equipment

The ASME Boiler and Pressure Vessel Code [32] is a good example of a design code where the material-specifics of SCs, CMCs and CFCCs must be incorporated before one of the potential applications (power generation) of these materials can be realized. As shown in Figure 24.15, the 'code' is divided into eleven sections, the two most important for CFCCs being Section I "Power Boilers" and Section II "Materials," each detailed in the following discussion.

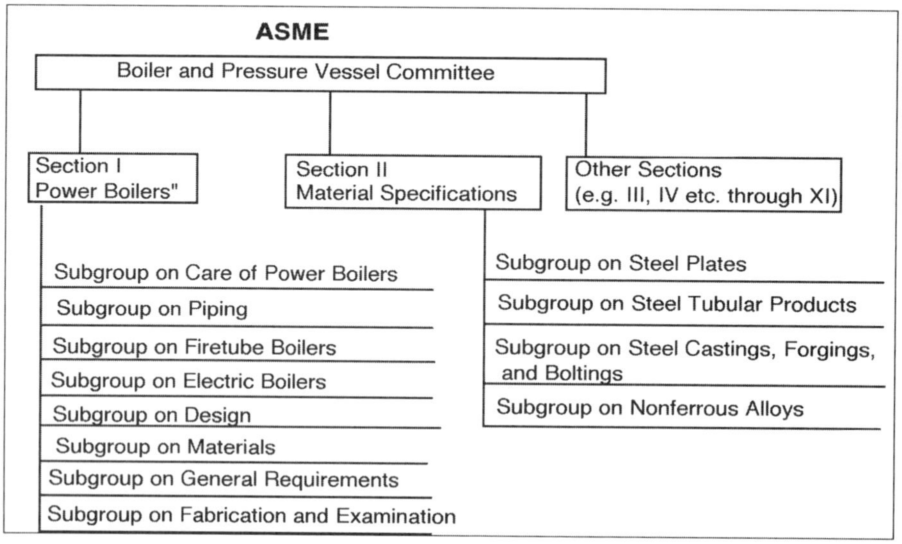

Figure 24.15 Partial Presentation of the Organization of the ASME Boiler and Pressure Vessel Effort

Section I dates to the adoption of the Code in 1914 and is divided into eight subgroups each dealing with specific aspects of the care, piping, design, fabrication and examination of power boilers [11]. In addition, subgroups also deal with particular types of boilers, including firetube and electric. Generally, Section I applies to boilers with >103 kPa pressure external to the boiler itself and >1103 kPa pressures and/or >121°C temperatures for high-temperature water boilers. The scope of jurisdiction of Section I is the boiler proper and the boiler external piping [11].

Of particular interest for SC and CMC producers is the subgroup on materials in Section I which limits materials to 1) those listed in Section II and 2) those listed in certain tables in Section I. In addition, certain materials are restricted in regard to usage. For example, an austenitic stainless steel might not be permitted for parts in water-wetted service, but may be permitted for steam-touched service. Also included in Section I, are provisions for Approval of New Materials for Code Construction [11,32]:

1) New materials shall have been previously adopted by ASTM (Adoption of a new material for the ASME Code means adoption of the ASTM specification for the material).
2) Items to be furnished for evaluation by the Code committee are:
   i) mechanical properties such as ultimate tensile strength, yield strength, creep and rupture strength, heat treatment, toughness, etc.
   ii) stress-strain data for vessels designed for external pressure
   iii) weldability, including data for establishing the requirements of Section IX "Welding"
   iv) physical changes and resistance to effects of both elevated temperature and cryogenic temperature where applicable
   v) availability of the material regarding patents and licensing

For design purposes, two approaches are prescribed in the code: design by rule and design by analysis. Section I relies primarily on design by rule which is basically an empirical approach (what has proven to work successfully in the past) setting limits on: factors of safety (typically 4 to 5); design pressures and temperatures, minimum thicknesses (e.g., 6.35 mm); maximum pressures (no greater than 1.06 times the maximum allowable working pressure (use of safety valves required)); loadings due to internal pressures (these set the minimum required thicknesses unless other loadings exceed 10% of the allowable working stress);

thickness of cylindrical components under internal pressure as determined from formulae; openings and reinforcements; fatigue, fast-fracture, creep and other failure mechanisms; and hydrostatic proof tests (1.5 times the maximum allowable working pressure).

Section III "Nuclear Power Plant Components" uses the more sophisticated, but less historically based, design by analysis. The use of design by analysis in Section III is a precedent important for acceptance of CMCs in the "Code" because CMCs require recently developed probabilistic/reliability techniques for determining their long-term suitability.

Two possible reliability methods have been considered. The first method employs information on the statistical distributions of the applied "load" and the "resistance" of the material to assess the factor of safety, FS, of the design such that:

$$FS = \frac{R}{L} = \frac{\bar{x}_R - 3(s_L^2)^{1/2}}{\bar{x}_L + 3(s_L^2)^{1/2}} \qquad (1)$$

where FS is the factor of safety, R is the "resistance" (for example, strength of the material as a function of time), L is the "load" (for example, stress acting on the component), $\bar{x}_R$ is the mean value of the distribution of the random variable R, $\bar{x}_L$ is the mean value of the distribution of the random variable L, s_R is the standard deviation value of the distribution of the random variable, R, and s_L is the standard deviation value of the distribution of the random variable, L. Note that the lower bound on the resistance, R, based on three standard deviations less than the mean is compared to the upper bound on the load, L, based on three standard deviations greater than the mean.

The second method employs computer algorithms, numerical stress analyses, and time-dependent material properties and performance distributions to iteratively determine the reliability (that is, probability of not failing). This method is illustrated schematically in Figure 24.16. Note that the numerical stress analysis can be carried out on commercial nonlinear finite element programs. However, the iterative routines must take into account the change in material properties and performance as functions of time (that is, empirical test information).

In a reliability-based approach the question arises as to the acceptable level of reliability. Table 24.3 illustrates probabilities of failure (fatality) for various types of accidents. Based on reliability design goals

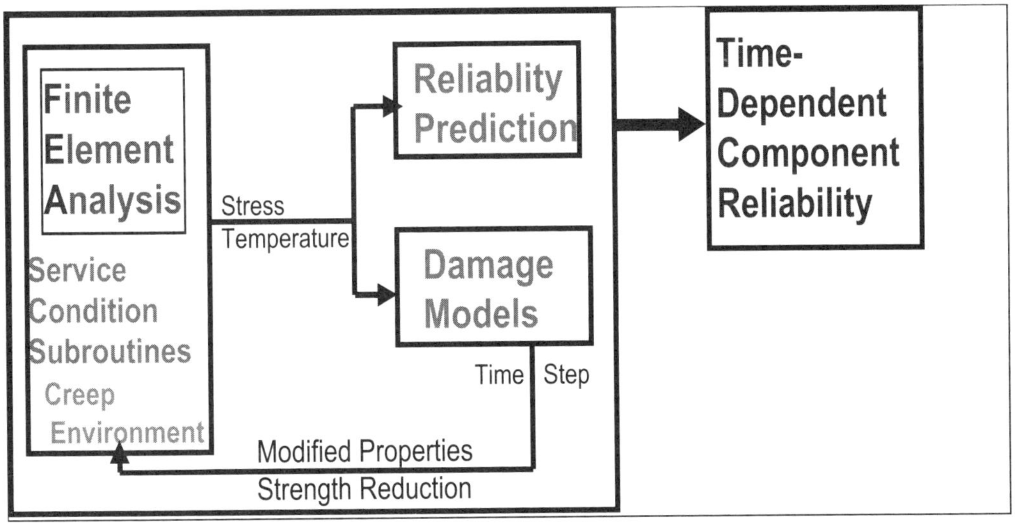

Figure 24.16 Schematic Illustration of Probabilistic Design by Analysis Approach for CMCs (Reprinted with permission from ref. [33] with permission from Interscience Enterprises Ltd.)

TABLE 24.3 SOME REPRESENTATIVE PROBABILITIES OF FAILURE (US DATA)

Accident Type	Probability of a Fatality per Year per Individual
Motor vehicle	3×10^{-4}
Falls	9×10^{-5}
Firearms	1×10^{-5}
Air Travel	1×10^{-6}
Hurricanes	8×10^{-7}
Lightning	5×10^{-7}

used in the nuclear industry, a probability of failure of $P_f = 10^{-6}$ has been proposed for CMCs resulting in a reliability of $R = 1 - P_f = 99.9999\%$.

However, rather than arbitrarily choosing values of probability of failure or reliability, it is important to note that probability of failure, P_f, can be combined with consequence of failure, C, to determine the risk, r, such that:

$$r = \Sigma C \times P_f \qquad (2)$$

Note from Eq. (2) that for the same risk, the probability of failure can increase if the consequence of failure decreases and vice versa. Thus, a risk analysis allows the imposition of different probabilities of failure on different components depending upon the consequence of failure. As a starting point, the overall system reliability (and the consequence of failure) can be used to determine the acceptable reliabilities of each component.

Note that Section I also requires affixing a special Code stamp and a proper nameplate to the components and resulting system, respectively after complying with all the Code requirements for design and construction. To document compliance, seven types of Manufacturers Data Report Forms must be completed.

Section II also has its roots in the original 1914 edition of the Code. The original materials specifications were developed in a joint effort of ASME and ASTM for ferrous and nonferrous materials. In addition, joint specifications with the American Welding Society (AWS) have since been developed for welding rods, electrodes and filler metals. Only those ASTM or AWS specifications required by ASME are addressed in Section II. The documentation of Section II is a four part compendium of materials data: Part A—Ferrous Materials Specifications; Part B—Nonferrous Material Specifications; Part C—Specification for Welding Rods, Electrodes, and Filler Metals; and Part D—Properties. Part D lists material properties for all materials accepted by Sections I, III, and VIII. Not only mechanical properties but also physical properties are contained in Part D. However, major portions of the data contained in Part D are tables of stresses as functions of temperature.

Because of increasing interest in SCs and CMCs from the power generation community along with similar continuing interest in impervious graphite from the chemical processing community, the Task Group on Graphite and Ceramic Pressure Equipment of the ASME Boiler and Pressure Vessel Code was formed in 1995. The task group was formed to address the possible explicit introduction of brittle materials (i.e., impervious graphite and ceramic composites) as acceptable, performance-based materials into the Code. Subsequent division of the task group resulted in two smaller groups, one of which was Subtask Group on Ceramic Pressure Equipment, specifically targeted to CFCCs but also including SCs where necessary. Two possible paths were considered for the sub task group: 1) development of a Code-case to demonstrate that brittle materials have been and are successfully being used in pressurized equipment and 2) development of a stand alone section of the 'Code.' The impervious graphite subtask group chose Path 1) because there are many current applications of this material in pressure equipment. SCs, CMCs and CFCCs are not currently used in pressurized equipment, therefore, Path 2) was chosen. A 100+ page draft of a stand-alone section (cloned from Section X on PMCs) was subsequently developed with the following organization. Note that this draft section also employs reliability-based design rather than design by rule or factor of safety design by analysis.

1. Introduction
2. General Requirements
3. Material Requirements
4. Design Requirements
5. Fabrication Requirements
6. Qualification Requirements
7. Pressure Relief Devices
8. Rules Governing Testing
9. Inspection Requirements
10. Marking, Stamping, Reporting
11. Appendices
12. Forms

Work on this draft section is ongoing although tangible results are still in the future by several years.

Mil-Hdbk-17 CMC Effort

Mil-Hdbk-17 is the outgrowth of a collaborative effort on the part of industry (i.e., defense contractors) and the U.S. Department of Defense to clarify and codify issues involving the use of polymer matrix composites (PMCs) in advanced designs. Mil-Hdbk-17 has been in existence in various forms since 1959 and has been relatively successful in creating common language, design philosophies, fabrication methods, maintenance approaches, and certification of advanced PMCs. Recent directives from U.S. Department of Defense have established the format for developing off shoots of Mil-Hdbk-17 for other advanced composites, namely, metal matrix composites (MMCs in 1993) and ceramic matrix composites (CMCs in 1996). The organization of the Mil-Hdbk-17 effort for CMCs is shown in Figure 24.17. The vision of the Mil-Hdbk-17 effort for CMCs is as follows:

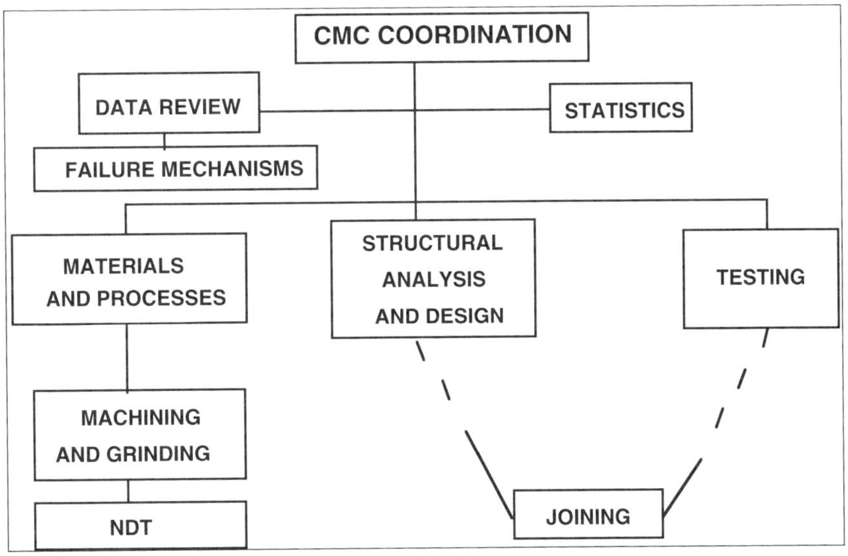

Figure 24.17 Organization of the Mil-Hdbk-17 Effort for CMCs

> Mil-Hdbk-17 is the primary and authoritative source for characterization, statistically based property and performance data of current and emerging ceramic matrix composites. It reflects the best available data and methodologies for characterization, testing, analysis and design, and usage guidelines in support of design methodologies for composites.

The objectives are:

- Development of a framework for the future, successful use of CMCs
- Provide guidance to industry for the collection of statistically meaningful critical data that designers need to utilize CMCs
- Based on the requirements from the design community, identify appropriate properties and broadly accepted testing procedures—including consideration of the designation of the precision level and prioritization of properties required
- Provide guidelines and recommendations for the characterization, testing, design, and utilization of CMC materials and structures
- Provide the primary and authoritative sources for characterization, property, and performance data of current and emerging CMC systems
- Provide recommendations for the statistical analysis of materials data and structures relativity

Note that imbedded in the organization shown in Figure 24.17 is an oversight activity for Data Review. Any data generated for inclusion in Mil-Hdbk-17 must satisfy requirements for confidence bounds and statistical sample size within a batch, and batch-to-batch. Because of the need to establish the CMC portion of the Mil-Hdbk-17 as quickly as possible, data base generation has been simultaneous with establishing design rules and guidelines. Note also in Figure 24.17 that while the Testing activity is separate, it can actually be thought as performing a service role to the other activities. Moreover, the stated mission of the Testing activity is not to develop standards (this is best left to the standards writing organizations already discussed), but instead to identify and recommend those existing standards, which are appropriate. Where appropriate standards do not exist, the Testing activity will assist standards writing organizations in the development of the standards.

It is important to note that work on the two design code efforts (Mil-Hdbk-17 CMC Effort and ASME Boiler and Pressure Vessel Code Subtask on Ceramic Pressure Equipment) are not being carried out separately. Rather, the same persons often fill similar roles in both efforts. Currently (mid-2000), progress in the Mil-Hdbk-17 effort is more advanced than in the ASME effort. A collection of Mil-Hdbk-17 "yellow pages" was circulated for review as a first "nearly complete" draft in early 2000 and as of early 2003 the first published edition of Volume 5 on CMCs was made publicly available by ASTM along with the other four volumes on PMCs and MMCs. This first edition will be refined into a final and complete partial-consensus design code within a limited time. The progress of the Mil-Hdbk-17 effort is also expected to stimulate continued progress in the ASME effort.

Data Bases

A logical outcome of design codes is the incorporation of databases of material properties and performance "qualified" for inclusion in the code. "Qualified" data have been obtained through testing per the standards and the statistical requirements indicated in the code. "Qualified" data bases often require minimum statistical samples (i.e., numbers of tests) for 1) a particular batch of material and 2) multiple batches of material. In addition, databases may include primary summary data (e.g., mean, standard deviation, and numbers of tests) along with secondary data from the individual tests. In addition, "qualified" databases also contain sufficient information about the materials pedigree to allow material to be easily obtained for replicating the test results.

Some databases contain only numerical information while others include graphical information (e.g., stress-strain curves, temperature profiles, or test specimen geometry). Traditional databases for SCs and CMCs are in the form of books [e.g., 34–37] with information contained in them compiled from various sources often with little or no regard as to whether these data is "qualified." While these traditional databases are useful for screening or comparison purposes, the lack of qualification often makes the information useless for anything except for preliminary design purposes. In addition, lack of specifications for SCs and CMCs causes the data in these books to rapidly fall out of date as formulation and processing changes obsolete the materials referenced in the database.

In response to some of the deficiencies of these "hard copy" databases are databases in electronic form for rapid data retrieval and relational data processing.. These electronic databases are often world wide web-based for ready access and frequent updating [e.g., 38,39] as illustrated by an example web homepage in Figure 24.18.

Some generally accepted requirements for data bases include the following:

1) Valid statistically significant data
   i) tested per recognized, preferably full consensus standard
   ii) various levels of significance such as screening, preliminary and fully qualified.
2) Sufficiently-described material (i.e., pedigree) and test parameters
3) User-friendly interface for ease of retrieval and input
4) Relational with graphics display capability
5) Web-friendly
6) Capable of storing information in non-proprietary format

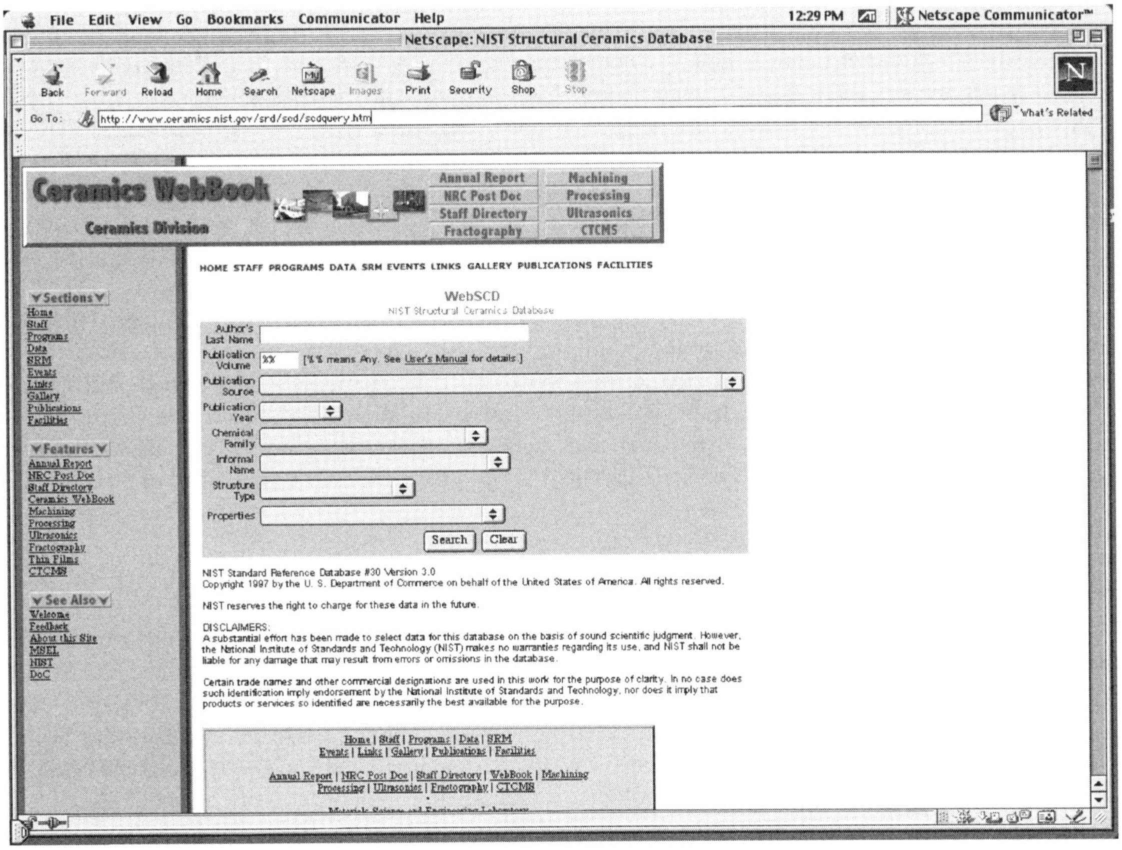

Figure 24.18 Example of the Home Page of a Web-Based Database for SCs and CMCs (Reprinted from ref. [38])

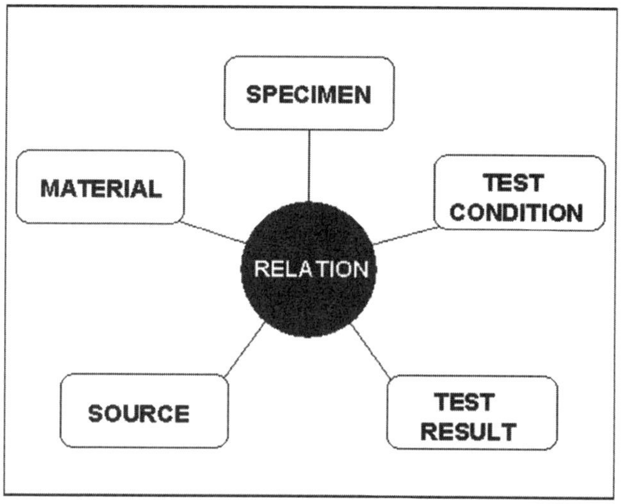

Figure 24.19 Example of a Web-Based Relational Database Targeted for CMCs

Design codes and their databases may be invoked as legal requirements for implementing an engineering design (for example, certification and compliance with the ASME Boiler and Pressure Vessel Code is a legal requirement in 49 of the 50 United States). Currently there are no national or international design codes and related "qualified" databases allowing SCs and CMCs in any type of code-based design. This situation may be hampering utilization of SCs and CMCs because designers are not allowed to use SCs and CMCs directly in new designs, but instead must 1) show evidence that the SCs and CMC meet the requirements of the code and 2) obtain special permission to use the SCs and CMC in the design. In addition, development of SCs and CMCs is impaired because without a demand for the SC and CMC, there is no incentive for refinement.

An example of an attempt to create an acceptable data base for a pre-commercial CMC has been completed as part of a multiple laboratory round robin program [40]. This round robin involved over 400 test specimens, 4 different types of tests (tension, flexure and shear), 13 different laboratories and 3 full-consensus ASTM standards for CMCs. An expanded effort to develop a relational web-based data base for CMCs has also addressed the issue of data retrieval and storage by using a commercially available system (Relational Data Base Management System (RDBMS)) but is still in its early stages (see Figure 24.19).

SUMMARY AND CONCLUSIONS

The main objective of this chapter was to provide information on the current state of standards and codes for structural ceramics and ceramic composites. While the number and diversity of test standards for SCs and CMCs is limited but growing, efforts are already underway to harmonize the various full-consensus industry/government standards through the international auspices of ISO. The success of ISO standards for SCs and CMCs will be instrumental in reducing redundancy and increasing productivity in the development of SC and CMC standards. Future directions include developing standards for components, sub components, environmental effects, electrical and electronic, thermo-mechanical, optical and biological characteristics.

Benefiting from the increasing number and diversity of test standards for SCs and CMCs are design codes and data bases for these materials. Two current efforts, ASME Task Group on Boiler and Pressure Equipment and Mil-Hdbk-17 CMC effort are helping coalesce the range of ideas of design with and design of SCs and CMCs. Various existing test methods are also being utilized to generate statistically significant databases for inclusion within the design codes. On-line, web-based databases are seen as necessary for keeping databases accessible and up to date.

In conclusion, the development and implementation of standards and codes for SCs and CMCs is ongoing and rapidly progressing. As SCs and CMCs move from materials in the development stage to prototype fabrication to final component production, not only standards but also the codes and attendant data will not only be essential but will also benefit from the maturation of this unique material system.

ACKNOWLEDGMENT

Research sponsored by the U.S. Department of Energy, Assistant Secretary for Energy Efficiency and Renewable Energy, Office of Industrial Technologies, Industrial Energy Efficiency Division and Continuous Fiber Ceramic Composites Program, under contract DE00OR22725 with UT-Battelle, LLC is acknowledged. Unless otherwise indicated the figures and tables were provided by the authors and their organizations.

REFERENCES

[1] Phillips, D., C., and Davidge, R., W., "Test Techniques for the Mechanical Properties of Ceramic Matrix Fibre Composites," Br. Ceram. Trans. J., 85, 123–130, 1986.

[2] Schneider, S., J., and Bradley, D. R., "The Standardization of Advanced Ceramics," Adv. Ceram. Matls., 3[5], 442–449, 1988.

[3] Jenkins, M.G. and Quinn, G.D., "ASTM Standards for Monolithic and Composite Advanced Ceramics: Industrial, Governmental and Academic Cooperation," ASME paper 96-GT-270, presented at the International Gas Turbine and Aeroengine Congress and Exposition, Birmingham, U.K., June 10–13, 1996.

[4] Kanno, T., "International Standardization Activities for Fine Ceramics—Status of ISO/TC206 on Fine Ceramics," ASME paper 96-GT-321, presented at the International Gas Turbine and Aeroengine Congress and Exposition, Birmingham, U.K., June 10–13, 1996.

[5] Steen, M., "European Standardization Efforts on Fiber-Reinforced Ceramic Matrix Composites," ASME paper 96-GT-269, presented at the International Gas Turbine and Aeroengine Congress and Exposition, Birmingham, U.K., June 10–13, 1996.

[6] Jenkins, M. G., "Standards and Codes for Ceramic Matrix Composites," *Advanced Composite Materials*, 8[1], 55–76, 1999.

[7] Glenny, E., "Ceramics and the Gas Turbine," in *Special Ceramics 1964: Proceedings of a Symposium held by the British Ceramic Research Association*, P. Popper, ed., Academic Press, London, UK, 1965. pp. 301–315.

[8] *Ceramics for High Performance Applications*, J.J. Burke, A.E. Gorum, and R. N. Katz, eds. Brook Hill Publishing Co., Chesnut Hill, MA, USA, 1974.

[9] Lenoe, E. M., "Recent Accomplishments and Research Needs in Structural Ceramics," in *Ceramics for High Performance Applications-III: Reliability*, E.M. Lenoe, R. N. Katz, and J. J. Burke, eds., Plenum Press, New York, USA, 1979. pp. 3–18.

[10] Brinkman, C.R., and Quinn, G. D., "Standardization of Mechanical Properties Tests for Advanced Ceramics," in *Mechanical Testing Methodology for Ceramic Design and Reliability*, D. C. Cranmer and D. W. Richerson, eds, Marcel Decker, Inc, New York, USA, 1998. pp. 353–386, © 1998 by Marcel Dekker, Inc., New York, NY, USA.

[11] Farr, J.R., "The ASME Boiler and Pressure Vessel Code: Overview," in *Pressure Vessel Codes and Standards*, Elsevier Applied Science, New York, NY, USA, 1983. pp. 1–34.

[12] Kawamoto, H. and Matsui, M., "Design Standard for Advanced Ceramic Materials and Components," *7th International Symposium on Ceramic Materials and Components for Engines*, J. G. Heinrich and F. Aldinger, eds., Goslar, Germany, June 19–21, 2000. pp. 201–212.

[13] Lewis, III, D., Bulik, C., and Shadwell, D., "Standardized Testing of Refractory Matrix/Ceramic Fiber Composites," Ceramic Engineering and Science Proceedings, 6[7–8], 507–523, 1985.

[14] Quinn, G.D. and Morrell, R., "Design Data for Engineering Ceramics: A Review of the Flexure Test," J. Am, Ceram. Soc., 74[9], 2037–2066, 1991.

[15] Jenkins, M.G. and Mello, M.D., "Processing and Characterization of 3-D Braided, Continuous SiC Fiber Reinforced/CVI Matrix Composites," *in Processing, Design, and Performance of Composite Materials*, MD-Vol. 52, American Society of Mechanical Engineers, New York, NY, USA,1994. pp.239–254.

[16] Cranmer, D., C., "Workshop on Assessment of Testing Methodology for Glass, Glass-Ceramic, and Ceramic Matrix Composites," Journal of Research NIST, 96[4], 493–501, 1991.

[17] Rudnick, A., Marschall, C.W., Duckworth, W.H., and Emrich, B.R., *The Evaluation and Interpretation of Mechanical Properties of Brittle Materials*, AFML-TR-67-316, Air Force Materials Laboratory, Wright Patterson Air Force Base, OH, USA, 1968.

[18] Jenkins, M.G., Piccola, Jr., J.P. and Lara-Curzio, E., "Onset of Cumulative Damage (First Matrix Cracking) and the Effects of Test Parameters on the Tensile Behavior of a Continuous Fiber-Reinforced Ceramic Composite (CFCC)," in *Fracture Mechanics of Ceramics*, Vol. 12, R.C. Bradt, D.P.H. Hasselman, D. Munz, M. Sakai, V.Y. Shevchenko, eds., Plenum Press, New York, NY, USA, 1996. pp. 267–282.

[19] Adams, M. and Sines, G., "A Statistical, Micromechanical Theory of the Compressive Strength of Brittle Materials," J. Am. Ceram. Soc., 61 [3–4], 126–131,1978.

[20] Rehfield, L. W., Armanios, E. A., and Changli, Q., "Analysis of Behavior of Fibrous Composite Compression Specimens," in *Recent Advances in Composites in the United States and Japan*, ASTM STP 864, J.R. Vinson and M. Taya, eds, ASTM, Philadelphia, Pennsylvania, USA, 1985. pp. 236–252.

[21] Barnett, T. R., and Starrett, H. S., *Room and Elevated Temperature Mechanical and Thermal Properties of Corning Nicalon CAS*, WRDC-TR-90-4131,1992

[22] Adams, D., F., "Mechanical Testing of Composite Materials for Quality Control," Proceedings Third Intl. Congress on Composites, Wiesbaden, Germany, October 1991.

[23] Gettings, R. J., Quinn, G. D., Ruff, A. W., and Ives, L. K., "Development of Ceramic Hardness Reference Materials," *Proceedings of the 8th CIMTEC-World Ceramics Congress and Forum on New Materials*, P. Vincenzini, ed., June 28–July 4, 1994, Florence, Italy. Techna Srl., Faenza, Italy,1995. pp. 617–624.

[24] *Fracture Mechanics Applied to Brittle Materials*, S.W. Freiman, ed., ASTM STP 678, ASTM, Philadelphia, PA, USA, 1979.

[25] Fujii, T. and Nose, T., "Evaluation of Fracture Toughness for Ceramic Materials, IJIS International, 29[9], 717–725, 1989.

[26] Sakai, M. and Bradt, R.C., "Fracture Toughness Testing of Brittle Materials," International Materials Review, 38[2], 53–58, 1993.

[27] Jenkins, M.G. and Salem, J.A, "Fracture Resistance Testing of Brittle Solids," in *Mechanical Testing and Evaluation*, ASM Handbook Vol. 8, ASM International, Metals Park, OH, USA, 2000. pp. 768–775.

[28] Quinn, G. D., Gettings, R, and Xu, K., "Standard Reference Materials 2100: Ceramic Fracture Toughness," Ceramic Engineering and Science Proceedings, 20[3], 513–520,1999.

[29] Salem, J. A. and Jenkins, M. G., "Fatigue Testing of Brittle Solids," in *Mechanical Testing and Evaluation*, ASM Handbook Vol. 8, ASM International, Metals Park, OH, USA, 2000. pp. 665–678.

[30] Steen, M. and Valles, J-L, "Unloading-Reloading Sequences and the Analysis of Mechanical Test Results for Continuous Fibre Ceramic Composites," in *Thermal and Mechanical Test Methods and Behavior of Continuous Fiber Ceramic Composites*, ASTM STP 1309, M. G. Jenkins S.T. Gonczy, E. Lara-Curzio, N. E. Ashbaugh, and L. P. Zawada, eds, ASTM, West Conshohocken, PA, USA, 1997, pp. 49–68.

[31] Campbell, C.X., *Thermomechanical Response of a Continuous Fiber Ceramic Composite (CFCC) Resulting from Oxidation, Temperature Degradation, and Residual Stress*, Master of Science Thesis, University of Washington, Seattle, WA, USA, 1998.

[32] American Society of Mechanical Engineers, *ASME Boiler and Pressure Vessel Code*, ASME, New York, NY, USA,1983.

[33] Jenkins, M.G., "Ceramic Matrix Composites and Enabling Supporting Technologies for Their Widespread Implementation," International Journal of Materials and Product Technology, 16 [1/2/3],1–11, 2001.

[34] Morrell, R., *Handbook of Properties of Technical and Engineering Ceramics*, HMSO Publications Centre, London, UK, 1989.

[35] Wachtman, J. B., *Mechanical Properties of Ceramics*, John Wiley & Sons, New York, NY, USA, 1996.

[36] Campbell, C. X., and El-Rahaiby, S. K., *Databook on Mechanical and Thermophysical Properties of Particulate-Reinforced Ceramic Matrix Composites*, Purdue Univ. Office Publications, W. Lafayette, IN, USA,1995.

[37] Campbell, C. X., and El-Rahaiby, S. K., *Databook on Mechanical and Thermophysical Properties of Fiber-Reinforced Ceramic Matrix Composites*, Purdue Univ Office Publications, W. Lafayette, IN, USA, 1997.

[38] *NIST Structural Ceramics Database*, www.ceramics.nist.gov/srd/scd/scdquery.htm, 2000.

[39] *Matec, Materials Properties Structural Ceramics Database, Version 3.0.* www.matec-softdat.de/materials_properties_structural_.htm, 2000.

[40] Jenkins, M.G., Gonczy, S. T., Lara-Curzio, E., and Zawada, L. P., "Multiple-Laboratory Round-Robin Study of the Flexural, Shear and Tensile Behavior of a Two-Dimensionally Woven Nicalon™/Sylramic™ Ceramic Matrix Composite," *Environmental, Mechanical, and Thermal Testing and Performance of Ceramic Composites*, ASTM STP1392, American Society for Testing and Materials, West Conshohocken, PA, USA, 2000, pp. 15–30.

CHAPTER

25

LIFE PREDICTION OF STRUCTURAL COMPONENTS

Stephen F. Duffy
Cleveland State University, Cleveland, Ohio, USA

Lesley A. Janosik
NASA Glenn Research Center, Cleveland, Ohio, USA

Andrew A. Wereszczak
Oak Ridge National Laboratory, Oak Ridge, Tennessee, USA

Bjoern Schenk
Honeywell Engines, Systems & Services, Phoenix, Arizona, USA

Akihiko Suzuki
Ishikawajima-Harima Heavy Industries Co., Ltd., Tokyo, Japan

Jacques Lamon
LCTS, University of Bordeaux, Pessac, France

David J. Thomas
Ohio Aerospace Institute, Cleveland, Ohio, USA

INTRODUCTION

This chapter will focus on the technical details as well as the historical development of life prediction methods and codes used in designing structural ceramic gas turbine engine components. Over the past several decades, significant efforts worldwide have focused on (1) improving powder composition and processing methods, (2) understanding the physics of deformation and failure of these materials, (3) designing experimental tests to characterize materials and verify analytical models, (4) developing analytical methods to accurately model the observed behavior and predict component lifetime reliability, and (5) manufacturing technology demonstrations and/or prototypes. These diverse efforts all have the common ultimate goal of utilizing advanced ceramic materials to potentially supplant metals and their alloys in commercial and military turbine engine applications because the use of such ceramic materials in components can potentially: operate at higher service temperatures and increase the thermodynamic efficiency of these engines, and; provide longer service lifetimes owing to the greater wear and erosion resistances afforded by those ceramics.

The envisioned widespread commercialization of ceramic engine components is, however, still limited by the fact that these materials exhibit low fracture toughness, which is usually defined by a critical stress intensity factor (typically quantified by K_{IC}). Lack of ductility (i.e., a low fracture toughness) leads to low strain tolerance. Combining low fracture toughness with high strength leads to large variations in observed fracture strength since critical crack sizes tend to be well below current detection levels. As a result of these flaws, components with a distribution of defects are produced which leads to an observed scatter in component strength. In addition, when subjected to elevated service temperatures, ceramics exhibit complex thermomechanical behavior that is both inherently time-dependent and hereditary in the sense that current behavior depends not only on current conditions, but also on thermomechanical history. Thus the design engineer must be cognizant that the ability of a ceramic component to sustain load degrades over time due to a variety of effects such as oxidation, creep, stress corrosion, cyclic fatigue, as well as erosion- and impact-induced damage under the application of a far field stress. Slow crack growth (SCG), which can be caused by any of the mechanisms just mentioned, enlarges cracks to a point where they propagate catastrophically. Time-dependent crack growth can be a function of chemical reaction, environment, debris wedging near the crack tip, and deterioration of bridging ligaments. Fracture mechanism maps, such as the one developed by Quinn [1] for ceramic materials, help illustrate the relative contribution of various failure modes as a function of temperature and stress. Thus there has been an on-going need for sound life prediction methods. Computational simulation of component performance based on these life prediction methods through the use of integrated computer algorithms allows for optimized designs before attempting expensive prototype testing.

Previous chapters focus on fabrication processes and testing methods for advanced ceramic materials. This chapter addresses the development of ceramic life prediction methods, as well as the inherent limitations and lessons learned from organizations at the forefront of this challenging endeavor. First, it presents an historical perspective of the parallel life prediction efforts undertaken in Asia, Europe, and North America. Next, it provides a review of accepted design methodologies and techniques for characterization of material properties of particular interest to the design engineer. Both fast fracture (time-independent) reliability and life prediction (time-dependent reliability) analyses are demonstrated. The chapter then highlights the evolution of design methods and tools (i.e., software algorithms) for ceramic component life prediction at the aforementioned organizations, and then illustrates the application of two life prediction examples involving monolithic ceramic components. Lastly, a brief description of (less advanced) life prediction methods used with ceramic matrix composite components is provided.

HISTORICAL OVERVIEW—CERAMIC COMPONENT DESIGN

Ceramics research in the USA has historically involved several organizations, representing government, industry, and academia. Federal agencies have supported numerous programs to foster the development of highly reliable structural ceramics for applications in advanced turbines. The goals of these programs included developing tougher, flaw-tolerant materials, advancing processing technology, mechanical testing methods, and developing and demonstrating ceramic life prediction methods. Highlights in the evolution of the life prediction efforts in these ceramic technology programs will be discussed in detail later in this chapter. Complementary topics mentioned above are addressed in other chapters of this publication. A timeline of the major programs and sponsors involved in the development of life prediction methods and tools for designing ceramic components in North America may be found in Ref. [2].

The Defense Advanced Research Projects Agency (DARPA) supported initial ceramic life prediction endeavors in the United States through the pioneering Brittle Materials Design program that was an effort initiated in 1971. That effort was a joint program that involved Ford Motor Company [3] in Dearborn, Michigan, and Westinghouse Corporation [4] in Pittsburgh, Pennsylvania, to encourage the use of brittle materials in engineering design. Several other federal-funded ceramic technology programs were established during the 1970s and 1980s [2]. Efforts supported by these federal research and development programs culminated in the development of ceramic life prediction capabilities at two engine manufacturers, Rolls-Royce USA (formerly Allison Engine Company, and Allison Gas Turbine Division of General Motors prior to that) [5] and Honeywell (formerly AlliedSignal, and Garrett prior to that) [6]. Concurrently, and somewhat independently, another government organization, the National Aeronautics and Space Administration Lewis Research Center (NASA LeRC, now known as NASA Glenn Research Center at Lewis Field, or simply NASA GRC) was undertaking its own endeavor to develop life prediction methods. Over the years these three major efforts served as "friendly competitors," i.e., crosschecking methods developed and verifying predictions obtained. Additionally, NASA GRC interacted with an international group, known as WELFEP (WEakest Link Failure Probability Prediction by Finite Element Postprocessors), which was organized in 1991 to provide a forum for discussion and evaluation of the life prediction programs being developed in Europe and the U.S. The WELFEP group conducted two international round robins to validate the various ceramic life prediction activities [7,8]. The ceramic life prediction efforts at NASA GRC and Honeywell have continued throughout the 1990s into the new millennium. Additionally, another organization, Sundstrand Power Systems (SPS), developed a proprietary ceramic life prediction algorithm (SPSLIFE) to support development of ceramic component designs [9]. SPSLIFE was used for SPS in house and funded programs and also for the DOE/Solar *Ceramic Stationary Gas Turbine* (CSGT) program [10].

In Asia, the Japanese have historically dominated ceramic design technology through government-sponsored programs. The Agency of Industrial Science and Technology (AIST) of the Ministry of International Trade and Industry (MITI) of the Japanese government was established in the late 1940s as a national research organization, superintending 15 research laboratories throughout the country. Scientific R&D initiatives at these research institutes served as the basis for technological innovations in Japan throughout the latter half of the 20th century. The majority of the ceramic life prediction activities throughout Japan have been (at least in part) funded through this organization. As a result of Japanese government restructuring, on January 6, 2001, the former MITI was transformed into METI, the Ministry of Economy, Trade, and Industry. On April 1, 2001, a newborn AIST, the National Institute of Advanced Industrial Science and Technology, emerged. The new AIST is an independent administrative organization that combines the 15 research institutes under the former AIST (in the former MITI) and the Weights and Measures Training Institute. The new AIST is one of Japan's largest public research organizations, and continues to sponsor ceramic life prediction activities in Japan. Significant efforts led by industry have resulted in improved ceramic material properties and processing, as well as improved experimental testing methods. Development of analytical methods and tools for ceramic component life prediction has evolved from these efforts. A timeline of the major programs and sponsors involved in the development of life prediction methods and tools for designing ceramic components in Asia may be found in Ref. [11].

The situation for life prediction programs in Europe is quite different to that in the United States and Asia. The interest in ceramics and ceramic structural components was very high in the 1970s and the early 1980s. It then started to decline from the mid-eighties as ceramic matrix composites were coming out particularly in a few countries like the UK and France. Additionally, there was no specific effort either at the European Commission level or at the national governments level to support or drive big programs for developing design methodologies for ceramic parts or ceramic structural components. Consequently, the number of large ceramic life prediction programs in Europe was small relative to those in the United States and Asia. Such life prediction efforts tended to be conducted individually either by companies in relation with universities or research laboratories, or by research institutions for a group of companies (e.g., Battelle Europe, Geneva) for their specific use. Therefore, most data are proprietary and cannot be found in the open literature. However, reports on activity devoted to reliability analysis problems, to reliability based design methodologies, and to developing computer codes for fracture prediction and for ceramic component design, are indeed available. A timeline of the major programs and sponsors involved in the development of life prediction methods and tools for designing ceramic components in Europe may be found in Ref. [12].

FAST FRACTURE (TIME-INDEPENDENT)

Of particular interest to the design engineer is the inherent scatter in strength exhibited by ceramic materials. Accounting for this phenomenon requires a change in philosophy on the design engineer's part that results in a reduced focus on the use of safety factors in favor of statistical reliability analyses. If a brittle material with an obvious scatter in tensile strength is selected for its high strength attributes or inert behavior, then components should be designed using an appropriate design method rooted in statistical analysis. However, the reliability approach demands that the design engineer must tolerate a finite risk of unacceptable performance. This risk of unacceptable performance is identified as a component's probability of failure (or alternatively, component reliability). The primary concern of the engineer is minimizing this risk in an economical manner.

Reliability analyses are typically segregated into two categories: time-independent (also referred to as fast fracture) and time-dependent. This classification is rooted both in the historic development of the reliability models presented here and also in a practical approach to the analysis of a component. In many instances, a component must perform in an adequate fashion over a predetermined service life. To accomplish this design goal, the component must survive the initial load cycle. Thus the calculated time-independent reliability value is used as a screening criterion, and is also used as an initial value for the time-dependent analyses discussed later. A fundamental premise of probabilistic analysis dictates that if the reliability of a component varies with time then it should never exceed the fast fracture reliability value (unless there exists some physical mechanism such as flaw healing that can account for this phenomenon). From a historical perspective time-independent models were developed first (hence they are presented here first).

System Reliability and the Use of the Weibull Distribution

The design of engineering components manufactured from structural ceramics requires a different approach relative to the design of metallic components. When metal components are subjected to applied loads the material has the capacity to generate dislocations that migrate through the microstructure. This property permits yielding and minor shape adjustments without catastrophic failure of the component. Structural ceramics do not have this property and therefore their failure behavior is very brittle in comparison to metals. As a result ceramic components are highly sensitive to stress concentrations, microstructural inhomogeneities, surface damage from processing, and/or dimensional tolerances that are not met.

Fracture mechanics is based on the assumption that failure takes place due to the presence of a defect in the microstructure. The approach describes the stress distribution in the near vicinity of the sharp leading edge of a defect, which results in expressions that indicate the existence of a stress singularity at the tip of the defect. Nearly all singularity expressions involve a $(1/\sqrt{r})$ term, where r is a radial distance from the crack tip. As r approaches zero, stresses become infinite. Since materials cannot sustain an applied stress that goes to infinity, the material fails near the crack tip and the defect advances through material. The advancing crack will move in a stable or unstable fashion depending on the material, applied load and geometry of the component.

The failure process just described can be easily quantified through the use of a stress intensity factor (K). The stress intensity factor defines the stress amplitude of the crack tip singularity, i.e., stresses in the singularity zone increase in proportion to K. Moreover, the stress intensity factor completely defines the crack tip condition. If K is known, then it is possible to determine all components of stress, strain and displacement. In order for the stress intensity factor to be useful, one must be able to determine *K* from applied loads (far-field stresses) that are remote to the crack and the geometry of the component under design. There are numerous stress intensity expressions that are functions of component geometry, far-field stress and a characteristic size of the defect. A component geometry factor, the applied far-field stress, and the current crack length can characterize most stress intensity relationships. When a critical combination of stress and crack length produces failure for a given component geometry, the critical stress intensity level (*Kc*) is attained. This value is a material property and is referred to as the fracture toughness of the material.

The stress intensity relationship can be used to design components in one of two ways:

- fracture toughness is specified *a priori*—materials are screened based on this value, and
- the material is pre-selected—critical combinations of applied stress and defect length are determined.

If ceramic materials are utilized to fabricate structural components, then component design falls under the second category. If fracture mechanics is used, then design engineers must determine critical combinations of applied far-field stress, defect length and defect orientation whereby the performance of the structural component would be deemed unacceptable. Here the design procedure based on fracture mechanics reduces to determining the defect length and orientation that will provide acceptable margins under mission loads. In practical terms this means characterizing a critical flaw, i.e., size and orientation. Once this characterization has taken place, all components must be inspected in a non-destructive fashion to determine the presence of this critical flaw. If the flaw is detected, then the component is removed from service. The design process just described seems simple at first; however, for ceramic materials complications arise. On relative terms, a material with high strength and low fracture toughness typically has critical flaw sizes that are quite small. These small defects can easily go undetected during inspection procedures.

Thus, the component fabricated from ceramic materials will contain a distribution of defects with variable lengths and orientations even if the material is inspected using the best available non-destructive evaluation (NDE) methods. Thus, performing strength tests makes fundamental sense and attempting to pursue a fracture mechanics approach is not a worthwhile endeavor. A distribution of flaw sizes and orientations is present in a ceramic material, the strength test will exploit the critical flaw. Since the critical flaw length and orientation will vary from sample to sample, a distribution in strength is expected. In addition, if the size and orientation of the critical flaw are random then the critical flaw causing failure may not be located in the region of a component subjected to the maximum applied stress. As a result the stress field throughout the entire component must be analyzed and this information must be combined with the statistical nature of the defects distributed throughout the component. Weibull analysis is well suited to this task.

A unique property of most brittle materials with high tensile strength and low fracture toughness is an apparent decrease in tensile strength as the size of the component increases. This is the so-called strength-size effect and is a direct result of the distribution of flaws present in the material. An illustration of strength-size-scaling is shown in Figure 25.1 for a variety of uniaxial and biaxial flexure tests on a 96%-pure alumina [13]. Design methods for ceramic components must admit size dependence. This is accomplished through the use of system reliability concepts. To account for size effects and deal with the probability of failure for a component in a general manner, the component should be treated as a system, and the probability of failure of the system must be computed.

The typical approach to designing structural components with varying stress fields involves discretizing the component in order to characterize the stress field using finite element methods. Since component failure may initiate in any of the discrete elements, it is convenient to consider a component as a system and utilize system reliability theories. A component is a series system if it fails when one discrete element fails. This type of failure can be modeled using weakest-link reliability theories. A component is a parallel system when failure of a single element does not cause the component to fail. In this latter case, the remaining elements sustain load through redistribution. This type of failure can be modeled with what has been referred to in the literature as "bundle theories." Weakest-link theories and bundle theories represent the extremes of failure behavior modeled by reliability analysis. They suggest more complex systems such as *"r out of n"* systems. Here a component (system) of n elements functions if at least r elements have not

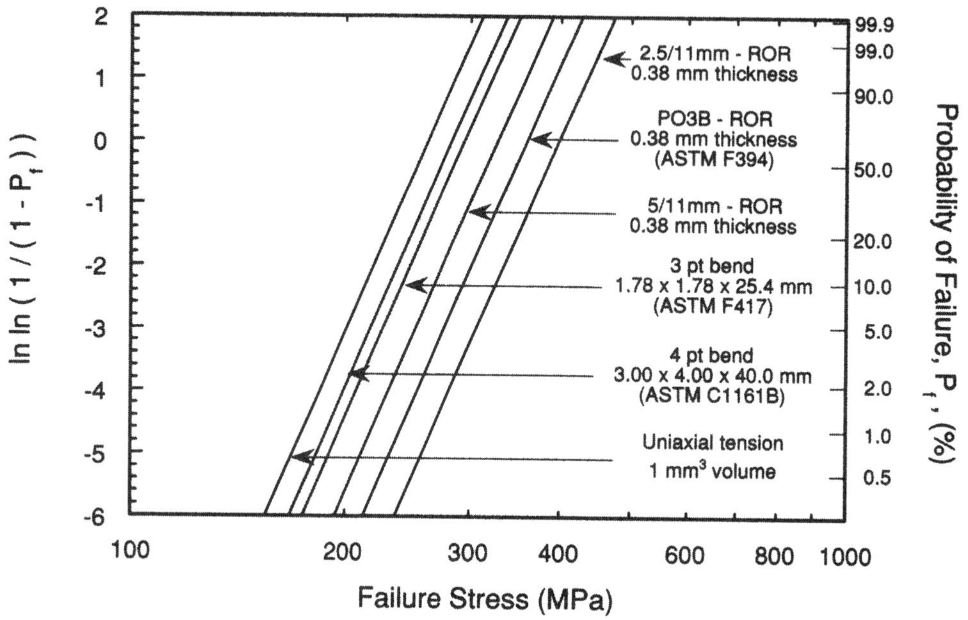

Figure 25.1 Example of Strength-Size-Scaling in an Alumina Ceramic. Smaller Effective Sizes Result in Higher Failure Stresses (Reprinted with permission from ref. [13])

failed. This type of system model has not found widespread application in structural reliability analysis. The historical perspective has been to assume that the failure behavior of the brittle materials is sudden and catastrophic. This type of behavior fits within the description of a series system, thus a weakest-link reliability system is adopted.

Now the probability of failure of a discrete element must be related to the overall probability of failure of the component. If the failure of an individual element is considered a statistical event, and if these events are independent, then the probability of failure of a discretized component that acts as a series system is given by the expression

$$P_f = 1 - \prod_{i=1}^{N} (1 - P_i) \tag{25-1}$$

where N is the number of finite elements for a given component analysis. Here P_i is the probability of failure of the ith discrete element. The next step involves adopting an expression for the probability of failure (or alternatively, the reliability) of the ith discrete element for a simplified state of stress, i.e., a uniaxial tensile stress. In the ceramic and glass industry the Weibull distribution [14,15] is universally accepted as the distribution of choice in representing the underlying probability distribution function for tensile strength. A two-parameter formulation and a three-parameter formulation are available for the Weibull distribution. However, the two-parameter formulation usually leads to a more conservative estimate for the component probability of failure. The two-parameter Weibull probability density function for a continuous random strength variable, denoted as Σ, is given by the expression

$$f_\Sigma(\sigma) = \left(\frac{\alpha}{\beta}\right)\left(\frac{\sigma}{\beta}\right)^{(\alpha-1)} \exp\left[-\left(\frac{\sigma}{\beta}\right)^{\alpha}\right] \tag{25-2}$$

for $\sigma > 0$, and

$$f_{\Sigma}(\sigma) = 0 \quad (25\text{-}3)$$

for $\sigma \leq 0$. The cumulative distribution is given by the expression

$$F_{\Sigma}(\sigma) = 1 - \exp\left[-\left(\frac{\sigma}{\beta}\right)^{\alpha}\right] \quad (25\text{-}4)$$

for $\sigma > 0$ and

$$F_{\Sigma}(\sigma) = 0 \quad (25\text{-}5)$$

for $\sigma \leq 0$. Here α (a scatter parameter, or Weibull modulus) and β (a central location parameter, or typically referred to as the Weibull scale parameter) are distribution parameters. In the ceramic literature when the two-parameter Weibull formulation is adopted then "m" is used for the Weibull modulus α, and "σ_0" or "σ_θ" is used for the Weibull central location parameter. A thorough description of established means for reporting Weibull distribution parameters can be found in American Society for Testing and Materials (ASTM) C1239 [16]. Thus if the random variable representing uniaxial tensile strength of an advanced ceramic is characterized by a two-parameter Weibull distribution, then the probability that a uniaxial test specimen fabricated from an advanced ceramic will fail can be expressed by the cumulative distribution function

$$P_f = 1 - \exp\left[-\left(\frac{\sigma_{\max}}{\sigma_\theta}\right)^{m}\right]. \quad (25\text{-}6)$$

Note that σ_{max} is the maximum normal stress in the component. When used in the context of characterizing the strength of ceramics and glasses, the central location parameter is referred to as the Weibull characteristic strength (σ_θ). The characteristic strength is dependent on the uniaxial test specimen (tensile, flexural, pressurized ring, etc.) used to generate the failure data. For a given material, this parameter will change in magnitude with specimen geometry (the so-called size effect alluded to earlier). The Weibull characteristic strength has units of stress and is not a material parameter as it is a function of geometry and the nature of the applied tensile stress. The scatter parameter m is dimensionless.

With the tensile strength characterized by the two-parameter Weibull distribution the discussion returns to the weakest link expression for component probability of failure. Let $\Re_i$ represent the reliability of the ith discrete element where

$$\Re_i = 1 - \mathrm{P}_i\,. \quad (25\text{-}7)$$

The reliability of this discrete element is then governed by the following expression

$$\Re_i = \exp\left(-\left(\frac{\sigma}{\sigma_0}\right)^{m}\Delta V\right) \quad (25\text{-}8)$$

where σ is the principal tensile stress applied to the discrete element, and ΔV identifies the volume of this arbitrary discrete element. In this expression, σ_0 is the Weibull material scale parameter and can be described as the Weibull characteristic strength of a specimen with unit volume loaded in uniform uniaxial tension. This is a material-specific parameter that is utilized in the component reliability analyses that follow. The dimensions of this parameter are based on the dimensional field in question and are: stress × (volume)$^{1/m}$ when volume flaws are the strength limiter; stress × (area)$^{1/m}$ when surface flaws are the strength limiter; and stress × (edge length)$^{1/m}$ when edge flaws are the strength limiter. Note—the use of established fractographical practices [17] allow the researcher to censor each ith strength value with respect to the dimensional field of the strength-limiting flaw.

Under the assumptions that the component consists of an infinite number of elements (i.e., the continuum assumption), considering an example when volume flaws are the strength limiter, and that the component is best represented by a series system, then

$$P_f = 1 - \lim_{k \to \infty} \left(\prod_{i=1}^{k} \Re_i \right). \qquad (25\text{-}9)$$

Substituting for $\Re_i$ yields

$$P_f = 1 - \exp\left(- \lim_{k \to \infty} \sum_{i=1}^{k} \left(\left(\frac{\sigma}{\sigma_0} \right)^m \Delta V \right)_i \right). \qquad (25\text{-}10)$$

Here ΔV once again represents the volume of an element. The limit inside the bracket is a Riemann sum, thus

$$P_f = 1 - \exp\left[- \int \left(\frac{\sigma}{\sigma_0} \right)^m dV \right]. \qquad (25\text{-}11)$$

Weibull [14] first proposed this integral representation for the probability of failure. The expression is integrated over all tensile regions of the specimen volume if the strength-controlling flaws are randomly distributed through the volume of the material, or over all tensile regions of the specimen area if flaws are restricted to the specimen surface. For failures due to surface defects the probability of failure is given by the expression

$$P_f = 1 - \exp\left[- \int \left(\frac{\sigma}{\sigma_0} \right)^m dA \right]. \qquad (25\text{-}12)$$

Equations 25-11 and 25-12 have historically served as the starting point for reliability analyses of components fabricated from ceramic materials. Note that an analogous expression to relate P_f to σ when edge flaws are the strength limiter can in turn be defined but it is not presented in this discussion. Component reliability analyses presented elsewhere representing different schools of thought from around the world find their genesis in these expressions.

Multiaxial Reliability Models

Over the years a number of reliability models have been presented that extend the uniaxial format of Eq. 25-11 and Eq. 25-12 to multiaxial states of stress. Only models associated with isotropic brittle materials are presented here. Anisotropic reliability models are beyond the scope of this chapter. Duffy and Arnold [18] as well as Duffy and Manderscheid [19] present reliability models for brittle composites. The monolithic models highlighted here include the principle of independent action (PIA) model [20,21], Batdorf's model [22–26], the Lamon-Evans model [27,28], and the Muest model [29]. A general discussion is presented for each though a detailed development is omitted for the sake of brevity. In order to simplify the presentation of each model recast Eq. 25-11 as

$$P_f = 1 - \exp\left[- \int \psi \, dV \right] \qquad (25\text{-}13)$$

where ψ is identified as a failure function per unit volume. What remains is the specification of the failure function ψ for each reliability model.

PIA—A Phenomenological Model. For the principle of independent action (PIA) model or the Barnett-Freudenthal approximation [20,21], the failure function per unit volume is

$$\psi = \left(\frac{\sigma_1}{\sigma_0}\right)^m + \left(\frac{\sigma_2}{\sigma_0}\right)^m + \left(\frac{\sigma_3}{\sigma_0}\right)^m, \tag{25-14}$$

where σ_1, σ_2 and σ_3 are the three principal stresses at a given point, and σ_0 is the Weibull material scale parameter. The PIA model is the probabilistic equivalent to the deterministic maximum stress failure theory. In this approach, the principal stresses are assumed to act independently in each principal direction. Failure probability is calculated from the product of the individual survival probabilities, in the direction of the tensile components. The PIA model has been widely applied in brittle material design; however, it does not specify the nature of the defect causing failure and has consequently been criticized by several authors because it ignores interaction of principal stresses and underestimates failure probabilities [24]. In essence, the model is phenomenological, which does not imply that the model is not useful; rather, the simplicity of a phenomenological model can often be a strength and not a weakness.

Batdorf's Theory—Mechanistic Model. The concepts proposed by Batdorf [22–26] are important in that the approach incorporates a mechanistic basis for the effect of multiaxial states of stress into the weakest-link theory. Here material defects distributed throughout the volume (and/or over the surface and/or along edges) are assumed to have a random orientation. In addition, the defects are assumed to be non-interacting discontinuities (cracks) with an assumed regular geometry. Consider a finite volume of material where the stress state is uniform throughout. Batdorf postulated that the probability of failure for this finite volume could be expressed by the following joint probability

$$P_f(\Sigma, \sigma_{cr}, \Delta V) = P_1 P_2. \tag{25-15}$$

Here Σ represents the uniform multiaxial state of stress throughout the finite volume. In addition, P_1 is the probability that a crack exists in ΔV (the finite volume) with an associated critical stress (i.e., fracture stress) in the range of σ_{cr} to $\sigma_{cr}+d\sigma_{cr}$. Note that σ_{cr} is defined as the critical far field normal stress for a given crack configuration under Mode I loading. The second marginal probability, P_2, denotes the probability that a crack with an associated critical stress of σ_{cr} will be oriented in a direction such that the effective far field applied stress, σ_e, equals or exceeds σ_{cr}. These two probabilities can be expressed as

$$P_1 = \Delta V \frac{d\eta(\sigma_{cr})}{\sigma_{cr}} d\sigma_{cr} \tag{25-16}$$

and

$$P_2 = \frac{\Omega}{4\pi}. \tag{25-17}$$

The function $\eta(\sigma_{cr})$ is referred to as the Batdorf crack density function. This function is independent of the multiaxial stress state Σ. Typically the function is expressed in a power law format, i.e.,

$$\eta(\sigma_{cr}) = k_B(\sigma_{cr})^m \tag{25-18}$$

where m is the Weibull modulus and k_B is a material parameter, both of which can be estimated or computed from failure data (see discussion below).

In Eq. 25-17, the solid angle Ω is mapped onto a unit sphere in the principal stress space and represents all crack orientations for which

$$\sigma_e \geq \sigma_{cr}. \tag{25-19}$$

Failure is assumed to occur when a far-field effective stress σ_e associated with the weakest flaw reaches a critical level, σ_{cr}. The effective stress σ_e is a predefined combination of the far field normal stress and the far field shear stress. It is also a function of the assumed crack configuration, the existing stress state, and the fracture criterion employed (hence the claim that the approach captures the physics of fracture). Accounting for the presence of a far-field shear stress reduces the far-field normal stress needed for fracture for special variations of the Batdorf model. The Batdorf model is defined by taking

$$P_f = 1 - \exp\left[-\int_V \left[\int_0^{(\sigma_e)_{\max}} \frac{\Omega}{4\pi} \frac{d\eta(\sigma_{cr})}{\sigma_{cr}} d\sigma_{cr}\right] dV\right] \tag{25-20}$$

Substituting for $\eta(\sigma_{cr})$ yields

$$P_f = 1 - \exp\left[-\int_V \left[mk_B \int_0^{(\sigma_e)_{\max}} \frac{\Omega}{4\pi} \sigma_{cr}{}^{m-1} d\sigma_{cr}\right] dV\right] \tag{25-21}$$

thus

$$\psi = mk_B \int_0^{(\sigma_e)_{\max}} \frac{\Omega(\Sigma, \sigma_{cr})}{4\pi} \sigma_{cr}{}^{m-1} d\sigma_{cr} \tag{25-22}$$

for Batdorf's model.

The solid angle Ω is defined by the following expression

$$\Omega = \int_{\beta=0}^{2\pi} \int_{\alpha=0}^{\pi} H(\sigma_e, \sigma_{cr}) \sin\alpha \, d\alpha \, d\beta \tag{25-23}$$

where α and β are azimuthal angles in the principal stress space, and the Heaviside step function equals one, i.e.,

$$H(\sigma_e, \sigma_{cr}) = 1 \tag{25-24}$$

if

$$\sigma_e \geq \sigma_{cr}. \tag{25-25}$$

The Heaviside step function in Eq. 25-23 is zero, i.e.,

$$H(\sigma_e, \sigma_{cr}) = 0 \tag{25-26}$$

if

$$\sigma_e < \sigma_{cr}. \tag{25-27}$$

Inserting Eq. 25-23 into Eq. 25-20 leads to the following expression for the component probability of failure

$$P_f = 1 - \exp\left[-\frac{k_{BV}}{2\pi} \int_V \int_{\beta=0}^{2\pi} \int_{\alpha=0}^{\frac{\pi}{2}} [\sigma_e(x,y,z,\alpha,\beta)]^{m_V} \sin\alpha \, d\alpha \, d\beta \, dV \right]. \tag{25-28}$$

An alternative expression for component failure probability can be derived from Eq. 25-28 by utilizing notation advocated by Johnson and Tucker [30]. Here a stress gradient factor (I_V) is identified as

$$I_V = \left(\frac{k_{BV}}{2\pi}\right)\left(\frac{\sigma_0}{\sigma_{\max}}\right)^{m_V}\left(\frac{1}{V}\right) \int_V \int_{\beta=0}^{2\pi} \int_{\alpha=0}^{\frac{\pi}{2}} \sigma_e^{m_V} \sin\alpha \, d\alpha \, d\beta \, dV. \tag{25-29}$$

Now

$$P_f = 1 - \exp\left[-I_V \, V \left(\frac{\sigma_{\max}}{\sigma_0}\right)^{m_V} \right]. \tag{25-30}$$

These expressions assume that flaws are distributed uniformly throughout the volume of the component. A similar development would lead to the following expression for failure due to flaws spatially distributed along the surface of the component (and analogously along edges), i.e.,

$$P_f = 1 - \exp\left[-\frac{k_{BA}}{\pi} \int_A \int_0^{\pi} [\sigma_e(x,y,\alpha)]^{m_A} \sin\alpha \, d\alpha \, dA \right]. \tag{25-31}$$

Once again a stress gradient factor associated with a surface area analysis (I_A) can be introduced, i.e.,

$$I_A = \left(\frac{k_{BA}}{2\pi}\right)\left(\frac{\sigma_0}{\sigma_{\max}}\right)^{m_A}\left(\frac{1}{A}\right) \int_A \int_0^{\pi} \sigma_e^{m_A} \sin\alpha \, d\alpha \, dA. \tag{25-32}$$

Then

$$P_f = 1 - \exp\left[-I_A \, A \left(\frac{\sigma_{\max}}{\sigma_0}\right)^{m_A} \right]. \tag{25-33}$$

The crack density coefficient k_B that appears in the expressions above is obtained from failure data. In the section that follows the methodology for obtaining estimates of the Weibull parameters m and σ_0 is outlined. Once these quantities are estimated k_B can be computed since

$$k_B = k_B(m, \sigma_\theta, \textit{Specimen Geometry}). \tag{25-34}$$

However, note that k_B is a material specific parameter in much the same manner that σ_0 (the Weibull material scale parameter) is a material specific parameter. The specimen geometry incorporated in the expression above is compensated by the fact that σ_θ is also dependent on specimen geometry. Thus the form of the expression above changes from specimen type, but the estimated value of σ_θ will also change, and the result is that the computed value of the crack density coefficient should stay consistent from test geometry to test geometry for a given material. In addition, the Weibull material scale parameter σ_0 is also a function of m, σ_θ, and the specimen geometry, i.e.,

$$\sigma_0 = \sigma_0(m, \sigma_\theta, Specimen\ Geometry). \tag{25-35}$$

Expressions for σ_0 associated with several simple bend geometries can be found in ASTM C1239 [16].

Numerous authors have discussed the stress distribution around cavities of various types, under different loading conditions, and proposed numerous criteria to describe impending fracture. A good compendium of fracture based reliability models can be found in Nemeth, *et al.* [31]. However, several issues must be pointed out. No prevailing consensus has emerged regarding a best probabilistic fracture theory. Most current models predict somewhat similar reliability results, despite the divergence of initial assumptions. Moreover, one must approach the mechanistic models with some caution. For example, the reliability models based on fracture mechanics incorporate the assumptions made in developing the fracture models on which they are based. One of the fundamental assumptions made in the derivation of fracture mechanics criteria is that the crack length is much larger than the characteristic length of the microstructure. This is sometimes referred to as the continuum principle in engineering mechanics. For ceramic materials that characteristic length is the grain size (or diameter). If one contemplates the fact that most ceramic materials are high strength with attending low fracture toughness, then the critical defect size can be quite small. If the critical defect size approaches the grain size of the material, then the Batdorf models discussed above should be based on the physics of cleavage fracture through or around single crystal microstructures.

Lamon-Evans Model. While the Batdorf approach in effect calculates the fraction of cracks for a given infinitesimal critical stress range that would lead to failure based on crack orientation and the relationship to σ_e, the Lamon-Evans model [27,28], also referred to as the elemental strength approach, in effect calculates the fraction of strengths for a given infinitesimal orientation range that would lead to failure as a result of that strength being less than σ_e. This approach considers flaws as physical entities as opposed to the Weibull theory that is fundamentally statistics and which does not address the flaw controlled physical processes producing fracture. Tucker and Johnson [32] proved that the Batdorf and Lamon-Evans approaches were equivalent when a common set of assumptions were considered. The Honeywell tandem set of life prediction codes, CERAMIC/ERICA, utilizes the Lamon-Evans multiaxial fracture criterion model, and it will be described in a later section.

Multiaxial Elemental Strength Model (Muest model). The Muest model [29] combines fracture mechanics concepts with the Lamon-Evans model, and incorporates concepts of non-coplanar crack extension. It is considered that fracture occurs from the pre-existing microstructural flaws in a direction depending upon the respective magnitudes of the normal and the shearing components operating on these flaws. This criterion is based upon the maximum in the strain energy release rate G_{max}, in the direction of crack propagation:

$$G_{\max} = \frac{(1+\nu)(1+x)}{4E}\left[K_I^4 + 6K_I^2K_{II}^2 + K_{II}^4\right]^{1/2} \tag{25-36}$$

where $x = 3 - 4\nu$ under plane strain conditions, $x = (3 - \nu)/(1 + \nu)$ under plane stress conditions, K_I and K_{II} are the Mode I and mode II stress intensity factors, E is the Young's modulus, and ν is the Poisson's ratio. Failure occurs when G_{max} reaches the critical value G_c.

The Lamon-Evans model is based upon the premise that the pre-existing flaws in the material can be characterized by their flaw extension stress or strength S. Failure probability is calculated from the following equation:

$$P_{fv} = 1 - \exp - \int_v dV \int_0^S g(S)d(s)ds \tag{25-37}$$

g(S)dS represents the number of flaws with a strength between S and S + dS. g(S) characterizes the distribution of flaws in the material, and it can be derived from strength data measured on specimens having well-defined geometry and stress state. Insertion of this fracture criterion and stress distribution in a mathematically convenient form of g(S) then permits the failure probability to be derived as follows in terms of the imposed loading, the specimen size and the flaw strength parameters (m_V and σ_{OMV}) [27, 33]:

$$P_{fv} = 1 - \exp\left[-\int_v \left(\frac{\sigma_1}{\sigma_{0v}} \right)^{m_v} I_v \left(m_v, \frac{\sigma_2}{\sigma_1}, \frac{\sigma_3}{\sigma_1} \right) dV \right] \qquad (25\text{-}38)$$

where I_V (m_V, σ_2/σ_1, σ_3/σ_1) accounts for shear sensitivity of volume flaws and their orientation relative to principal stresses. σ_1, σ_2, σ_3 are the principal stresses ($\sigma_1 > \sigma_2 > \sigma_3$) σ_2 and σ_3 may be compressive on the condition that σ_1, and the equivalent stress derived from Eq. 25-38 are tensile. The reader will later see that the life prediction code CERAM utilizes the Muest multiaxial fracture criterion model.

Fracture Strength Data and Parameter Estimates

As indicated earlier, the distribution of choice for characterizing the strength of advanced ceramic materials is the (two-parameter) Weibull distribution. In order to obtain point estimates of the unknown Weibull distribution parameters, well-defined functions that incorporate the failure data and specimen geometry are utilized. These functions are referred to as estimators. It is desirable that an estimator be consistent and efficient. In addition, the estimator should produce unique, unbiased estimates of the distribution parameters. Different types of estimators exist, including moment estimators, least-squares estimators, and maximum likelihood estimators (MLE). This discussion focuses on maximum likelihood estimators due to their efficiency and the ease of application when censored failure populations (see below) are encountered.

Strength Distributions. Uniaxial flexure, uniaxial tensile and equibiaxial flexure specimens are the most commonly used test configurations for determining statistical strength distributions in ceramic materials, and detailed test method descriptions for each may be respectively found in ASTM C1161 [34], ASTM C1273 [35], and ASTM C1499 [36]. Most ceramic material systems exhibit a decreasing trend in material strength as the test specimen geometry is increased (the so-called strength-size effect). The size effect phenomenon is dependent upon the Weibull modulus *m*. As *m* increases, the size effect diminishes. However, one should not surmise from this trend that the statistical approach advocated here does not apply in the absence of a pronounced size effect. The statistical design methods discussed here admit a diminishing size effect as a limiting case.

The total number of test specimens plays a significant role in estimating the distribution parameters. Initially, the uncertainty associated with parameter estimates decreases significantly as the number of test specimens increases. However a point of diminishing returns is reached when the cost of performing additional strength tests may not be justified. This suggests that a practical number of strength tests should be performed to obtain a desired level of confidence associated with a parameter estimate. This point cannot be overemphasized. However, quite often 30 specimens (a widely cited rule-of-thumb) are deemed a sufficient quantity of test specimens when estimating Weibull parameters. One should immediately ask why 29 specimens would not suffice. Or more importantly, why is 30 specimens sufficient? The answer to this is tied to the concept of computing confidence bounds for the maximum likelihood estimates. These bounds are a direct reflection of the precision of the parameter estimate.

Combining or pooling censored fracture data from multiple sizes and geometries can be very useful to the strength distribution analyst for several reasons [37]. The validity of strength-size-scaling can be assessed. Probabilities of survival can be determined with smaller confidence intervals. Fracture data may already exist from a different specimen size and geometry that could be combined with the analyst's current data. Initial component test data can be combined with laboratory specimen fracture data to determine self-consistency and to update probabilities of survival and interval estimates. Also, data pooling can be a confirmatory test of valid fractography/censoring for discrete strength-limiting flaws; namely, a valid censored pooled fracture data set should have the same Weibull modulus.

Advanced ceramics may contain two or more active flaw distributions, e.g., failures due to inclusions and failures caused by machining damage. Each flaw population will have its own "censored" strength distribution parameters. The censoring techniques presented in ASTM C1239 [16] for the two-parameter Weibull distribution require positive confirmation of multiple flaw distributions, which necessitates fractographic examination to characterize the fracture origin in each specimen. Multiple flaw distributions may also be indicated by a deviation from the linearity of the data from a single Weibull distribution. However, observations of approximately linear behavior should not be considered a sufficient reason to conclude that only a single flaw distribution is active. Lastly, the analyst needs to assess if dominant strength-limiting flaw populations are changing as a function of test condition (e.g., stressing rate and temperature [38]) and, if they are, to appropriately interpret those effects regarding the censoring of σ_o and m.

The data for each flaw distribution can also be screened for outliers. An outlying observation is one that deviates significantly from other observations in the sample. The reader should understand that an apparent outlying observation may be an extreme manifestation of the variability in strength. If this is the case, the data point should be retained and treated as any other observation in the failure sample. However, the outlying observation may be the result of a gross deviation from prescribed experimental procedure, or possibly an error in calculating or recording the numerical value of the data point in question. When the experimentalist is clearly aware that either of these situations has occurred, the outlying observation may be discarded, unless the observation (i.e., the strength value) can be corrected in a rational manner. The procedures for dealing with outlying observations are available elsewhere. For the sake of brevity this discussion omits any discussion on the performance of fractographic analyses, and omits any discussion concerning outlier tests.

Lastly, practical limits on the number of fracture strength specimens that can be tested leads to uncertainty in the estimate Weibull parameters, so it behooves the strength distribution analyst to determine and report confidence intervals about those parameters when possible. There are a variety of methods for determining confidence intervals about size-scaled Weibull parameters and they include the normal approximation method, likelihood ratio technique, nonparametric bootstrapping, and parametric bootstrapping methods. Chati, *et al.*, [39] recently assessed and compared these four methods using a simulated fracture experiment and concluded that the likelihood ratio technique and parametric bootstrapping method performed particularly well. Additionally, both those techniques were found to be readily extendable to confidence interval estimation of fast-fracture strength data from various test specimens (*i.e.*, pooled data).

Maximum Likelihood Estimators. With the above discussion serving as background, attention is now focused on how to compute values of the Weibull parameters m and σ_θ using maximum likelihood estimators. The MLE method is the most efficient in estimating parameters. Here efficiency is measured through the use of confidence bounds. It has been demonstrated that for an equivalent confidence level the confidence interval (defined by the associated upper and lower confidence bounds) for a MLE is always smaller than the confidence interval obtained using linear regression. For this reason the likelihood estimators should be used to compute parameters from failure populations characterized by a two parameter Weibull distribution.

The parameter estimates obtained using the maximum likelihood technique are unique (for a two-parameter Weibull distribution), and as the size of the sample increases, the estimates statistically approach the expected values of the true population parameter. Let $\sigma_1, \sigma_2, \ldots, \sigma_n$ represent realizations of the ultimate strength (a random variable) in a given sample, where it is assumed that the ultimate strength is characterized by the two-parameter Weibull distribution. The likelihood function associated with this sample is the joint probability density of the n random variables, and thus is a function of the unknown Weibull distribution parameters (m, σ_θ). The likelihood function for an uncensored sample under these assumptions is given by the expression

$$L = \prod_{i=1}^{n}\left(\frac{\tilde{m}}{\tilde{\sigma}_\theta}\right)\left(\frac{\sigma_i}{\tilde{\sigma}_\theta}\right)^{\tilde{m}-1}\exp\left[-\left(\frac{\sigma_i}{\tilde{\sigma}_\theta}\right)^{\tilde{m}}\right] \tag{25-39}$$

where the tilde overstrike denotes an estimation. The parameter estimates (the Weibull modulus $\tilde{m}$ and the characteristic strength $\tilde{\sigma}_\theta$) are determined by taking the partial derivatives of the logarithm of the likelihood function with respect to $\tilde{m}$ and $\tilde{\sigma}_\theta$, and equating the resulting expressions to zero. The system of equations obtained by differentiating the log likelihood function for an uncensored sample is given by

$$\frac{\sum_{i=1}^{n}(\sigma_i)^{\tilde{m}}\ln(\sigma_i)}{\sum_{i=1}^{n}(\sigma_i)^{\tilde{m}}} - \frac{1}{n}\sum_{i=1}^{n}\ln(\sigma_i) - \frac{1}{\tilde{m}} = 0 \tag{25-40}$$

and

$$\tilde{\sigma}_\theta = \left[\left(\sum_{i=1}^{n}(\sigma_i)^{\tilde{m}}\right)\frac{1}{n}\right]^{\frac{1}{\tilde{m}}}. \tag{25-41}$$

Equation 25-40 is solved numerically, since a closed form solution for $\tilde{m}$ cannot be obtained from this expression. Once $\tilde{m}$ is determined this value is inserted into Eq. 25-40 and $\tilde{\sigma}_\theta$ is calculated directly. Note that these expressions are used specifically for uncensored data. Data with two or more flaw populations are treated using similar MLE functions derived for censored data. These functions are presented in ASTM C1239 [16].

LIFE PREDICTION (TIME-DEPENDENT)

The ability of a brittle material component to sustain load degrades over time due to a variety of effects such as oxidation, creep, stress corrosion, and fatigue. Time-dependent probabilistic component design involves predicting the probability of failure for a thermomechanically-loaded component based on specimen rupture data. Typically experiments are performed using simple specimen geometries, i.e., flexural or tensile test specimens. A static (creep), monotonically increasing, or dynamic (cyclic) load is applied to a specimen until fracture occurs in a brittle fashion. Statistical strength parameters and SCG parameters are extracted from the data. Using these statistical parameters along with a multiaxial time-dependent reliability model, as well as the results obtained from a finite element analysis, an engineer can predict the life of a component. This is accomplished by generating the component's reliability as a function of time. When the component reliability falls below a predetermined reliability value, the associated point in time at which this occurs is assigned the life of the component. The methods outlined in this section combine the statistical nature of strength-controlling flaws presented earlier with the mechanics of crack growth to capture the effects of multiaxial stress states, concurrent (simultaneously occurring) flaw populations, and scaling effects. With a computer algorithm that integrates the items just mentioned an engineer can make appropriate design changes until an acceptable time to failure has been attained.

Subcritical Crack Growth

With load histories at stress levels that do not induce fast fracture, there is a regime defined by load and temperature where SCG occurs. Models for SCG that have been developed tend to be semi-empirical and approximate the behavior of crack growth phenomenologically. Experimental data indicates that crack

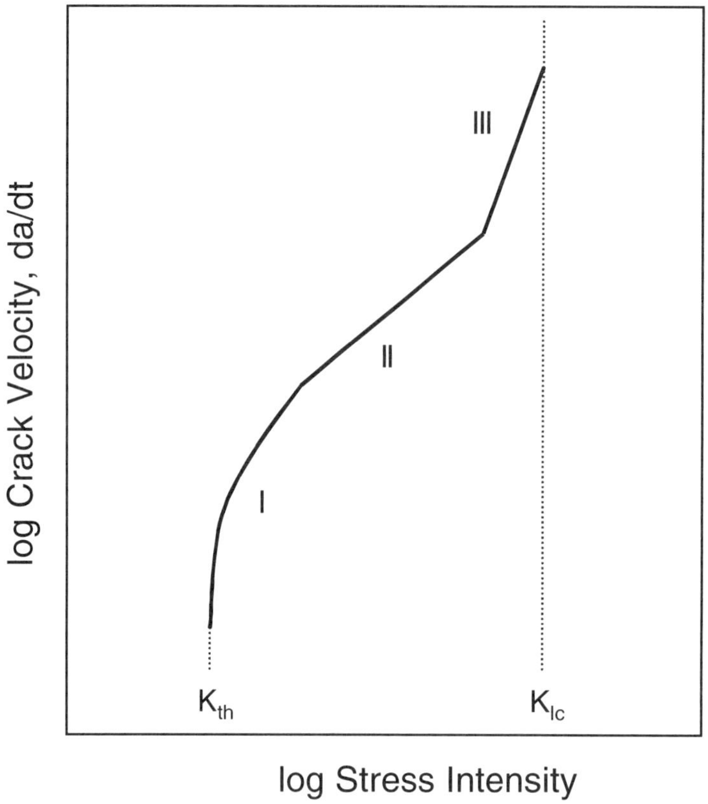

Figure 25.2 Crack Velocity as a Function of Stress Intensity (K) in Monolithic Ceramics. Crack Initation Does Not Occur When $K < K_{th}$. SCG Can Occur When $K_{th} < K < K_{1C}$ (Regions I, II, and III). Fracture Occurs When $K > K_{1C}$.

growth rate (*da/dt*) is a function of the applied stress intensity factor, K_I (or the range in the stress intensity factor), such as illustrated in Figure 25.2. Curves of experimental data show three distinct regimes of growth when the data is graphically depicted as the logarithm of the rate of crack growth versus the logarithm of the Mode I stress intensity factor. The first region indicates threshold behavior of the crack, where below a certain value of stress intensity (K_{th}) crack growth is zero. The second region shows an approximately linear relationship of stable crack growth. The third region indicates unstable crack growth as the material's critical stress intensity factor (K_{IC}) is approached. The second region typically dominates the life of the material. For the stress corrosion failure mechanism, these curves are material and environment sensitive.

The SCG models cited most often in the literature are based on power law formulations for the rate equations. Power law formulations are used to model the second region of the experimental data cited above. This power law formulation is expressed as

$$\frac{da(x,y,z,t)}{dt} = A\ K_{Ieq}^{N}(x,y,z,t) \tag{25-42}$$

where *A* and *N* (crack growth exponent) are material/environmental constants. Utilizing the following classical fracture mechanics expression

$$K_{Ieq}(x,y,z,t) = \sigma_{Ieq}(x,y,z,t)\ Y\ \sqrt{a(x,y,z,t)} \tag{25-43}$$

by inserting it into Eq. 25-42 leads to

$$\frac{da(x,y,z,t)}{dt} = A\, \sigma_{Ieq}{}^{N}(x,y,z,t)\, Y^{N}\, a^{N/2}(x,y,z,t) \tag{25-44}$$

where Y is a function of crack geometry, and $a(x,y,z,t)$ represents the crack length at time t. Separation of variables leads to the following integral

$$\int_{t=0}^{t_f} \sigma_{Ieq}{}^{N} dt = \left(\frac{1}{AY^{N}}\right) \int_{a=a_0}^{a_f} a^{-N/2} da. \tag{25-45}$$

Note that from Eq. 25-43 the initial critical flaw size can be expressed

$$a_0 = \left(\frac{K_{Ieq,C}}{\sigma_{Ieq,0}\, Y}\right)^2 \tag{25-46}$$

and the critical flaw size at failure can be expressed

$$a_f = \left(\frac{K_{Ieq,C}}{\sigma_{Ieq,f}\, Y}\right)^2. \tag{25-47}$$

Since component failure is the primary concern as the critical flaw grows from a_0 to a_f, then the stress intensity factors appearing in Eqs. 25-46 and 25-47 are critical stress intensities. These expressions for the flaw lengths a_0 and a_f are substituted for the limits appearing in Eq. 25-45. Integration of Eq. 25-45 yields the following expression

$$\sigma_{\text{Ieq},0}\,(x,y,z,t_f) = \left[\frac{\int_0^{t_f} \sigma_{\text{Ieq}}^{\text{N}}(x,y,z,t)\, dt}{\text{B}} + \sigma_{\text{Ieq},f}^{\text{N-2}}(x,y,z,t_f)\right]^{\frac{1}{(\text{N-2})}}. \tag{25-48}$$

Equation 25-48 provides the basis for computing time-dependent reliability. In this equation $\sigma_{Ieq,0}$ represents the equivalent applied far field stress that will fail a critical flaw with an initial length of a_0 under a fast fracture load rate. This far field stress is equal to the right hand side of Eq. 25-48 which is comprised of a load history that will produce failure when the critical flaw has grown to a length a_f. Since

$$a_0 < a_f \tag{25-49}$$

then

$$\sigma_{Ieq,0} > \sigma_{Ieq,f} \tag{25-50}$$

and the integral expression appearing in Eq. 25-48 is necessary for equality to hold. The integral expression represents the process of growing the critical flaw from a_0 to a_f in a subcritical fashion. If the right hand side of Eq. 25-48 can be evaluated, either in closed form or numerically, then the probability of failure at time t_f can be computed using fast fracture expressions along with the value of $\sigma_{Ieq,0}$. This is the computational approach adopted in the software algorithms (discussed later in this chapter) that are capable of evaluating time-dependent reliability. Using the PIA model or the Batdorf theory accommodates multiaxial stress fields. These multiaxial reliability expressions were outlined in previous sections on time-independent reliability analysis models. For the sake of brevity in this chapter we will focus on development of the SCG reliability model utilizing the Batdorf theory. Note that in Eq. 25-48

$$B = \frac{2}{A\ Y^2\ K_{Ieq,C}^{N-2}\ (N-2)} . \tag{25-51}$$

The parameter B has units of *stress² x time*. The crack geometry factor Y is assumed constant for subcritical crack growth. The parameters A, Y and K_{IC} are never computed directly, but are manifested through the parameter B via the expression above, and a value for the parameter B is estimated from experimental data.

Time-Dependent Batdorf Model. The determination of the time dependent component probability of failure using the Batdorf model parallels that presented earlier for fast fracture. Equation 25-28 is modified by taking

$$\begin{aligned}\sigma_e &= \sigma_{\mathrm{Ieq},0}(x,y,z,\mathrm{t_f}) \\ &= \left[\frac{\int_0^{\mathrm{t_f}} \sigma_{\mathrm{Ieq}}^{\mathrm{N}}(x,y,z,\mathrm{t})\ \mathrm{dt}}{\mathrm{B}} + \sigma_{\mathrm{Ieq},f}^{\mathrm{N\text{-}2}}(x,y,z,\mathrm{t_f})\right]^{\frac{1}{(\mathrm{N\text{-}2})}} .\end{aligned} \tag{25-52}$$

Substitution of this expression for σ_e into Eq. 25-28 yields

$$P_f = 1 - \exp\left[-\frac{k_{BV}}{2\pi}\int_V \int_0^{2\pi} \int_0^{\frac{\pi}{2}} \left[\sigma_{Ieq,0}(x,y,z,t_f)\right]^{m_V} \sin\alpha\, d\alpha\, d\beta\, dV\right] . \tag{25-53}$$

A similar substitution into Eq. 25-31 yields

$$P_f = 1 - \exp\left[-\frac{k_{BA}}{\pi}\int_A \int_0^{\pi} \left[\sigma_{Ieq,0}(x,y,t_f)\right]^{m_V} \sin\alpha\, d\alpha\, dA\right] . \tag{25-54}$$

The transformed equivalent stress $\sigma_{Ieq,0}$ is dependent on the load history, the selected fracture criterion, crack shape, and time to failure, t_f. The formulation of $\sigma_{Ieq,0}$ for creep and monotonically increasing load histories are presented below because closed form expressions can be derived. A similar development exists for the PIA model. For the sake of brevity it is omitted from this discussion. Expressions for cyclic fatigue are available—see, for example, Mencik [40].

Static Fatigue. Static fatigue is defined by the engineer as time-dependent strength degradation under a constant load. Integrating the right hand side of Eq. 25-48 with respect to time yields

$$\sigma_{Ieq,0}(x,y,z) = \sigma_{Ieq,f}(x,y,z,t=t_f)\left[\frac{t_f\ \sigma_{Ieq,f}^2(x,y,z,t=t_f)}{B} + 1\right]^{\frac{1}{N-2}} . \tag{25-55}$$

Dynamic Fatigue. Monotonically increasing uniaxial loading, referred to as dynamic fatigue, is defined as the application of a constant stress rate $\dot{\sigma}(x, y, z)$ over a period of time, t. Dynamic fatigue is an established [41], effective and valid accelerated static fatigue test (see Figure 25.3) to determine slow crack

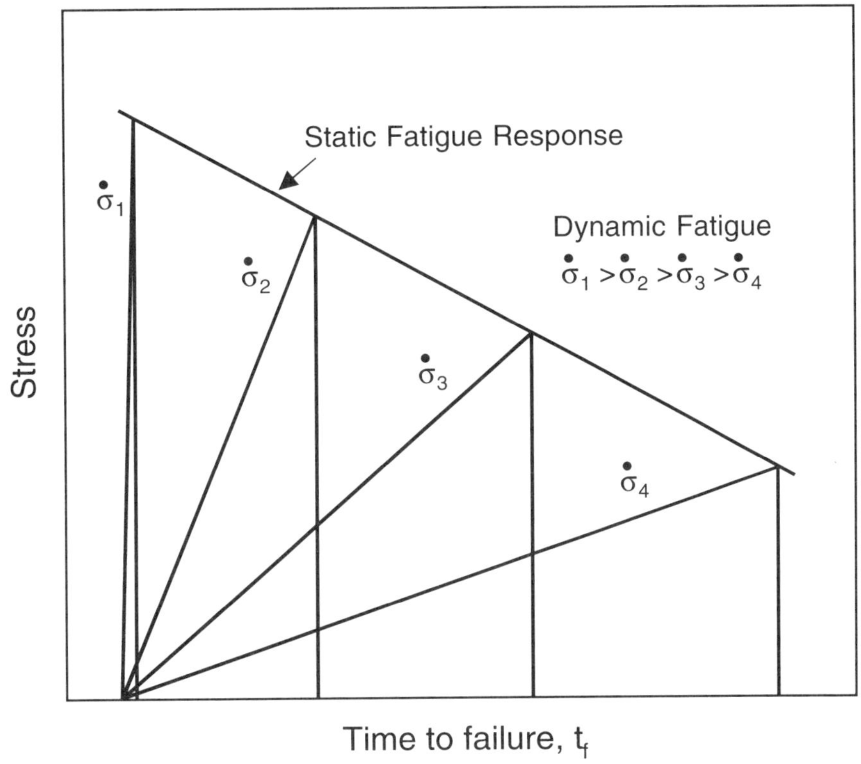

Figure 25.3 Dynamic Fatigue Testing is an Effective Accelerated Test for Static Fatigue Response if the Same Dominant Fatigue Mechanism is Operative

growth parameters if and only if the same dominant fatigue-limiting mechanism is operative under static and dynamic loadings. Assuming the applied stress is zero at time $t = 0$, then

$$\begin{aligned}\sigma_{\text{Ieq}}(x,y,z,\text{t}) &= \sigma(t)\\ &= t\,\dot{\sigma}(x,y,z).\end{aligned} \tag{25-56}$$

Substituting Eq. 25-56 into Eq. 25-48 results in the following expression for effective stress

$$\sigma_{Ieq,0}(x,y,z) = \left\{\left[\frac{1}{\dot{\sigma}(N+1)B}\right]\sigma_{Ieq,f}^{N+1}(x,y,z,t=t_f) + \sigma_{Ieq,f}^{N-2}(x,y,z,t=t_f)\right\}^{\frac{1}{N-2}}. \tag{25-57}$$

Note that only several load histories have been discussed above. These load histories correspond to analyses with closed form solutions. Several other load histories associated with cyclic fatigue have closed form solutions. These solutions are omitted for the sake of brevity.

Time-Dependent Strength Data and Parameter Estimates

The approach for computing the time dependent strength parameters B and N from failure data is presented in this section. A uniaxial creep test specimen is adopted in order to simplify the development presented below. Hence the stress state is uniform throughout the specimen, and is not spatially dependent. Begin by expressing the uniaxial formulation of Eq. 25-53 as

$$\begin{aligned} P_f &= 1-\exp\left[-\left(\frac{1}{\sigma_{0V}}\right)^{m_V} \int_V \left[\sigma_{1,0}\left(x,y,z,t_f\right)\right]^{m_V} dV\right] \\ &= 1-\exp\left[-\left(\frac{\sigma_{1,0}\left(t_f\right)}{\sigma_{0V}}\right)^{m_V} \int_V dV\right] \end{aligned} \tag{25-58}$$

for flaw distributions distributed through the volume of all test specimens (a similar expression exists for surface flaw distributions). Under the assumption that

$$\frac{\sigma_{Ieq,f}^{2}\, t_f}{B} >> 1 \tag{25-59}$$

then for creep tests (static fatigue) Eq. 25-55 yields

$$\sigma_{1,0}\left(t_f\right) = \left[\frac{\sigma_{Ieq,f}^{N}\, t_f}{B}\right]^{\frac{1}{N-2}}. \tag{25-60}$$

Substitution of Eq. 25-60 into Eq. 25-58 yields

$$P_f = 1-\exp\left[-\left(\frac{1}{\sigma_{0V}}\right)^{m_V} \left[\frac{\sigma_{Ieq,f}{}^{N}\, t_f}{B}\right]^{\frac{m_V}{N-2}} \int_V dV\right] \tag{25-61}$$

Introducing the parameter

$$t_{0V} = \frac{B(\sigma_{0V})^{N-2}}{\left(\sigma_{Ieq,f}\right)^{N}} \tag{25-62}$$

to Eq. 25-61 yields the following simplified expression for the probability of failure

$$P_f = 1-\exp\left[-V^{*}\left(\frac{t_f}{t_{0V}}\right)^{\frac{m_V}{N-2}}\right]. \tag{25-63}$$

Clearly the failure time t_f is a random variable described by a two parameter Weibull distribution. Here V^* is a volume parameter defined by the following expression

$$\begin{aligned} V^* &= Gage\ Volume \\ &= \int_V dV. \end{aligned} \tag{25-64}$$

For other test geometries this integral must be modified. Based on the development above the failure time t_f will be affected by the geometry and loading conditions imposed on the test piece being used to characterize the time dependent parameters B and N. Rearranging Eq. 25-63 and solving for t_f yields

$$t_f = \frac{t_{0V}}{\left[V*/\ln\left(1/1-P_f\right)\right]^{\frac{N-2}{m_V}}}. \tag{25-65}$$

Substituting Eq. 25-62 yields

$$t_f = \frac{B(\sigma_{oV})^{N-2}(\sigma_{Ieq})^{-N}}{\left[V*/\ln\left(1/1-P_f\right)\right]^{\frac{N-2}{m_V}}} \tag{25-66}$$

and taking

$$D = \frac{B(\sigma_{oV})^{N-2}}{\left[V*/\ln\left(1/1-P_f\right)\right]^{\frac{N-2}{m_V}}} \tag{25-67}$$

then

$$t_f = D(\sigma_{Ieq})^{-N}. \tag{25-68}$$

Taking the log of both sides of this last expression yields

$$\ln(t_f) = \ln(D) - N\ln(\sigma_{Ieq}). \tag{25-69}$$

Thus plotting the log of the time to failure (t_f) under creep load against the log of the applied creep failure stress σ_{Ieq} should yield a straight line with a slope of N. Typically linear regression techniques are used to determine the parameters N and D. The parameter D can be evaluated from the asymptotic nature of the creep data. Once these parameters are determined from the time dependent failure data this information would be combined with the Weibull distribution parameter estimates and the parameter B would be computed from Eq. 25-67.

Several issues should be noted carefully. First, in order to compute the parameter B from Eq. 25-67 a probability of failure must be utilized. The simplest approach would be to take

$$P_f = 0.50 \tag{25-70}$$

and use the median value of t_f from each data set (a data set would have a corresponding value of $\sigma_{Ieq,f}$).

Also note that the parameter N will be independent of the specimen geometry used to generate time to failure data, whereas the parameter D is strongly dependent on the specimen geometry. This is similar to the fast fracture Weibull parameters m and σ_θ, i.e., m is not dependent on the specimen geometry where σ_θ is. Although the formulations for a number of the equations above would change with specimen geometry (e.g., four point bend geometry) the derivation would remain intact.

Alternatively, another method of more accurately associating a probability of failure with a time to failure was recently developed by Schenk, *et al.,* [42]. The motivation for the development of this particular fatigue data analysis method was twofold and was a consequence of the existing limitations associated with the above-presented ordinary statistical treatment of fatigue data. Namely, those limitations could be overcome if the statistical interpretation of runouts and failures on loading (which are especially typical during fatigue testing of ceramic materials) were considered, and greater statistical efficiency could be introduced by pooling fast fracture and fatigue data sets generated from specimens of different size that are tested at different loading conditions, and temperatures. A runout describes a test that was interrupted before the specimen fails. Runouts do not contain the same information about the placement on the stress vs. time curve as failures do, and ignoring this introduces errors in estimated materials response. If runouts (and failures on loading) have to be included in the data analysis, censored data analysis using the maximum likelihood approach should be used to extract the most information out of the data available. The method developed and described in [42] is generic enough to be applied to cyclic fatigue data analysis and life prediction as well. The method was implemented into Honeywell's CERAMIC/ERICA probabilistic life prediction codes (described below) and currently allows the computation of maximum likelihood best estimates for all material parameters (Weibull modulus, characteristic strength, slow crack growth parameter B and n) including their likelihood ratio confidence bounds of all concurrently acting (competing) flaw populations in a single run using a pooled set of fast fracture and static fatigue data.

SOFTWARE ALGORITHMS

The development of life prediction methodologies and tools for designing ceramic components has progressed considerably over the past several decades. After an appropriate reliability model has been adopted, as described in the preceding sections, the primary task is the evaluation of the integral representation for the probability of failure. Closed form analytical solutions exist only for the simplest of component geometries and boundary conditions. Therefore, life prediction efforts worldwide have concentrated on developing integrated computer algorithms that enable a design engineer to predict the lifetime reliability of components subjected to thermomechanical loading. These algorithms typically couple finite element analysis programs with the probabilistic design models and parameter estimation techniques presented earlier in this chapter to determine the lifetime reliability of complex components. In the following three sections, the life prediction efforts in Asia, Europe, and North America will be discussed. Each section will discuss respective life prediction code developments on each continent.

Asia

A discussion of joint life prediction code development at Ishikawajima-Harima Heavy Industries (IHI) and Toyko Electric Power Company (TEPCO) is discussed. Several companies in Japan use CARES (described later in this chapter) or modified versions of CARES.

Ishikawajima-Harima Heavy Industries, Co. Ltd. The first research and development effort that focused specifically on the implementation of strùctural ceramic components began in Japanese fiscal year 1981 with a project entitled *Fine Ceramics*. This project was part of the *Research and Development of Basic Technologies for Future Industries* umbrella program administered by the MITI. The *Fine Ceramics* project was funded through Japanese fiscal year 1992 and focused on basic, fundamental technologies relating to the design and implementation of ceramic materials. Life prediction technologies were specifically developed under this work scope. Later follow-on programs sponsored by the Japanese government led to the development of a *Design Guide* that outlined engineering criteria developed under the *Fine Ceramics* program. A brief description of the *Design Guide* appears later in this section.

Utilizing the nomenclature advocated by researchers involved in the *Fine Ceramics* project, the fracture probability *(P)* of a ceramic component subjected to a uniaxial state of stress can be expressed as

$$P = 1 - \exp\left\{-\int_V \left(\frac{\sigma_{D,\max}}{S_0}\right)^m \frac{dv}{V_0}\right\}. \qquad (25\text{-}71)$$

Compare this expression with Eq. 25-11. Here σ_{Dmax} is the maximum value of σ_D during a given loading period. In the Japanese literature σ_D is referred to as the "effective loading stress." The value of σ_D is determined from the following expression

$$\sigma_D = \left\{\sigma^{n-2} + \frac{n-2}{2} BC^2 K_C^{n-2} \int_0^t \sigma^n dt\right\}^{\frac{1}{n-2}} \tag{25-72}$$

which is equivalent to Eq. 25-48 presented earlier. Again, Eq. 25-71 can be viewed as a time independent expression. However, Eq. 25-72 allows the use of Eq. 25-71 in the computation of time-dependent reliability.

For multiaxial stress states the Japanese defined an equivalent transformed normal stress Z. This parameter changes with the failure criterion utilized. If the defect population is comprised of Griffith cracks, then Z is defined by the expression

$$Z = \sqrt{\sigma_R^2 + \left(\frac{2}{2-\nu}\tau\right)^2} H(\sigma_R). \tag{25-73}$$

Here $H(\sigma_R)$ is the Heaviside step function. A double integration defined by solid angles associated with the orientation of the crack surface and accompanying constants is implicitly included in Eq. 25-73. As noted earlier, other multiaxial formulations exist. These formulations follow fracture mechanics failure criterion based on energy principles. A compilation of the most commonly used formulations can be found in the CARES Users Manual [31].

TEPCO and IHI conducted a research and development effort that ran in parallel with the *Fine Ceramics* project. This project specifically supported the development of ceramic components utilized in gas turbines. By 1985 a computer program entitled *GFICES* was developed under this program. This algorithm evaluates the reliability of ceramic components using most of the theoretical concepts mentioned throughout this chapter. An updated version of *GFICES* was released in 1988. From a functionality standpoint *GFICES* is quite similar to the *CARES/Life* algorithm developed by NASA (and discussed later in the chapter). The *GFICES* algorithm has been used successfully to design specific ceramic components that were the focus of the TEPCO project. Another modified version of *GFICES* was developed and utilized to design ceramic components in the Japanese Ceramic Gas Turbine (CGT) project, funded during fiscal years 1988 through 1998. This version of *GFICES* was developed by IHI and the modified algorithm was entitled *CCPRO.* GFICES and CCPRO have the same capability; however, CCPRO is a post processor for nonlinear (including creep deformation of ceramics) structural analysis and GFICES is the one for linear elastic structural analysis. This algorithm was utilized extensively in the evaluation of ceramic components inserted into test bed gas turbines. These test bed turbines were fabricated in the final stage of the *Fine Ceramics* project.

As part of the programs that produced the *GFICES* and *CCPRO* algorithms, design criteria were developed and subsequently incorporated into a published Design Guide. The Design Guide was developed specifically to aid engineers in the design of ceramic components in a manner consistent with their stochastic behavior. The intent of the Design Guide focused on the standardization of general design criteria, as well as design methods for stationary ceramic components, rotating ceramic components, contact/impact stress problems, and joining difficulties. An in-depth discussion of all facets of the Design Guide can be found in Ref. [43].

Based on the theoretical principles the design guide seeks to limit the applied stress in the following fashion. Define

$$\sigma_{\max} = \left(\frac{1}{K_0}\right)\frac{S_u \; or \; S_t}{K_1 \beta \gamma} \tag{25-74}$$

as the maximum tensile strength permitted within a ceramic component. Here S_u is the specified minimum strength of the material and is defined as the stress associated with a probability of survival equal to 99%. Alternatively, if the design engineer is conducting a time dependent analysis then S_t is utilized and this stress level is specified as the minimum time-dependent strength associated with a 99% probability of survival. In addition β and γ are design factors representing the volume effect present in ceramic materials. Finally, K_0

and K_I are safety factors that are dependent upon the component application. This design guide was used in the preliminary design of ceramic components for the test-bed gas turbine developed in the "Fine Ceramics" project and of the CGT-301 developed in the CGT project. At present the design guide is only useful in preliminary designs. The guide cannot account for local stresses that arise in actual service environments. These stresses are extremely detrimental since they facilitate fracture in ceramic components. These stresses cannot be ascertained a priori, and as a result, the final design of ceramic components is conducted using a trial and error procedure.

The Design Guide is now restricted to domestic use. To expand the utility of the design guide a new program has been recently completed. This program was started in Japanese fiscal year 1997. The effort was led by the Design Guide committee (Professor Kobayashi, Tokyo Institute of Technology—chair) as part of the *Techno-Infrastructure for Fine Ceramics* project. It was supported by the Japanese AIST, which is part of MITI. The Design Guide was modified to include lessons learned in the CGT projects as well as the design and development experience gained via the Japanese automotive industry. These automotive development projects focused on turbo chargers, swirl chambers, and cam rollers. The Design Guide also addresses fabrication issues associated with ceramic components fabricated for use in the Japanese automotive industry.

Europe

A variety of probabilistic life prediction codes were concurrently developed throughout Europe in the late 1980s and early 1990s, such as CERAM [44], STAU [45–47], CERITS-L [48], CERTUB [49], Daimler-Benz code [50], BRITPOST [51], FAILUR [52], FEM-LIFTAP [53], as well as various new codes developed for specific use at other academic institutions [54–56]. Most of these codes were compared in a round robin study as part of the WELFEP Group [8]. However, among these codes, CERAM and STAU have enjoyed continued maturation and evolution during the 1990s and have a presence in contemporary literature. Consequently, descriptions of only these two codes follows.

CERAM. The computer code CERAM [44] was developed in France and is based on both the well-known Weibull model and the Multiaxial Elemental Strength model (referred to here as the Muest model) [27–29]. The CERAM software package can consider any loading conditions and geometries that can be treated by finite element analysis codes. Ultimately, it identifies those regions prone to cracking and it indicates whether the component will fail or not.

CERAM provides detailed failure probabilities within the mesh elements and the overall failure probability for each population and for the concurrent effect of multiple flaw populations. Also displayed are various information including the time steps, the loading states, the dimensions, the location of flaw populations (volume, surface) and the failure probability. CERAM has broad capabilities by allowing the user to:

- perform 2- or 3-dimensional analyses
- evaluate failure from surface– or volume–located flaws or both
- specify up to 7 different surface– or volume–located flaw populations
- specify time-dependent flaw strength parameters to account for temperature or environmental effects, and analyze components made of one or several parts and of one or several materials.

Validation of the CERAM computer program was demonstrated by scaling strengths to different specimen sizes and configurations and ceramics. For this purpose, the flaw population characteristics (respectively, Weibull statistical parameters and the flaw strength parameters) were determined for a given specimen size and configuration selected as references. These flaw characteristics were subsequently incorporated in CERAM for computing the failure probability-strengths of different specimen sizes and/or configurations.

Comparison of the results with experimental data and theoretical calculations then allowed evaluation of the CERAM computer program [57–64]. In Ref. [57] in particular, 7 different loading geometries, two commercial ceramics and 715 specimens carefully prepared and tested were used. Silicon nitride, silicon carbide, and alumina were involved in the validation by considering the following:

- 3-point bending with variable spans [33,58–59,63],
- 4-point bending [28,63],
- traction [57],
- biaxial flexure of disks [28],
- bending of rings [57],
- compression of rings and C-rings [60],
- thermal shocks [61],
- thermal shock heat-up [62], and
- others [63].

Failure probabilities computed using CERAM agreed with experimental results in all cases. Additionally, the following factors that influence failure predictions were examined [57]:

- the estimator for experimental failure probability,
- the uncertainty and the variability in the statistical parameters,
- the probabilistic model,
- the finite element mesh,
- the results of stress analysis, and
- the stress-to-failure (stress at fracture origin, peak stress, etc.).

This analysis showed that the failure probabilities predicted with CERAM were in good agreement with experimental failure data when reproducible ceramics and the appropriate flaw strength parameters were considered. In certain cases involving shearing effects, the Weibull model led to very significant discrepancies between predictions and experimental results.

An alternative method based upon failure probability computations using the statistical post-processor CERAM can be used to derive the flaw strength parameters even from failures obtained on parts subjected to complex stress states. Failure predictions from statistical parameters extracted using conventional available methods (linear regression, maximum likelihood, mean strength) and the CERAM computations based method showed that the best fit to experimental data was obtained with those scale factors determined using the CERAM based method.

STAU. STAU is a finite element postprocessor developed by the Institute for Reliability and Failure Analysis at the University of Karlsruhe (Germany) that is used for the reliability analysis of ceramic components that are subjected to transient thermal and mechanical loadings. The reliability of multiaxially loaded monolithic ceramic components using STAU is evaluated based on a stress analysis, using the finite element method, and on a statistical analysis, using an extension of the weakest-link model. Failure is assumed to be caused by natural flaws. Both surface flaws and volume flaws are taken into consideration. The present version of the code, STAU 4, contains sub-critical crack growth, transient loads [65], temperature dependent material parameters as well as special routines for proof test evaluation. The effect of steep stress gradients can also be assessed [66,67]. Results on failure probability are available for the overall assessment of a component. Additionally, local risk of fracture results are available thus enabling identification of critical component regions. A graphical interface for Windows- and UNIX/LINUX-based systems is available. Additional features are devoted to the assessment of confidence intervals for the reliability (in terms of failure probability or lifetime distribution).

STAU was designed as a postprocessor package for the general-purpose finite element analysis code ABAQUS. A number of specific ABAQUS features such as element types are interfaced with STAU. Presently STAU is able to work with IDEAS, ANSYS, and ADINA, though these interfaces are not supported by default. The statistical analysis with STAU proceeds in three steps: (1) preparation of the input data, (2) statistical analysis based on the available FE stress analysis of the considered component, and 3) results processing for graphical presentation. Here, the default is a file format for the commercially available graphics postprocessor FEMVIEW, but other postprocessors can also be used if the corresponding interface routine is changed accordingly.

The STAU source code is written in standard FORTRAN 90 and requires about 1MB disk space. If a suitable compiler is available, STAU can be operated under Windows- and UNIX/LINUX-based systems. The graphical interface is based on the Tcl/Tk language and operates on both Windows- and UNIX/LINUX-based systems. No specific hardware requirements are necessary. The STAU code is currently maintained by a user group of several industrial and research partners under the overall control of the Forschungszentrum Karlsruhe GmbH.

North America

Descriptions of life prediction efforts or code-development at five US companies or US federal government agencies are presented. Ford Motor Company initiated efforts in the late 1960s and their progress set the foundation for much of the life prediction efforts that proceeded since then. NASA began developing life prediction codes for structural ceramics in the 1970s and have evolved over 30 years into being a worldwide distributor of their codes, while methods development and refinement has primarily been conducted by Honeywell and European academia. NASA has taken on more of a fast follower role in implementing methods developed elsewhere into their code over the past decade. Code developments at Honeywell, Rolls-Royce, and Sundstrand were somewhat contemporaneous, and have been used internally by each company for their respective gas turbine component designs.

Ford Motor Company. In 1967 Ford Motor Company began investing resources in ceramic development programs. The initial program advocated the design and manufacture of the Series 820 ceramic gas turbine engine for a passenger car. Additional details on the evolution of the major elements of this program are available in Ref. [3] and Chapter 2 of Volume II. By the end of 1970, investigations at Ford indicated that the Series 820 program should focus on developing a turbine with an all-ceramic hot flow path (including the turbine rotors), rather than the initially anticipated plan of using ceramics for stationary components of the hot flow path and air-cooled superalloy rotor blades. Coupled with this decision were the apparent needs for an approach to ceramic design, materials development, fabrication process development, and test and evaluation methodology. In 1971, DARPA sponsored a joint program with Ford to demonstrate and encourage the use of brittle materials in engineering designs. The basic design approach adopted for this government-supported program demanded both practical and theoretical aspects. From a practical perspective, it was necessary to develop general guidelines and design concepts for ceramic-ceramic and ceramic-metal attachments. From a theoretical perspective, it was essential to modify existing analytical codes and develop new analytical procedures to theoretically assess the structural integrity of ceramic components in a service environment (i.e., the critical flow path components in the 820 program). Ford's efforts in life prediction were a result of this latter effort. Analytical methods evolved for calculating both the fast fracture and the time-dependent reliability of structural ceramics. The objective was to develop new design tools and use them as a basis for the design iterations of ceramic components during the Series 820 program.

For fast fracture, a statistical approach was selected to determine the reliability of a ceramic component under engine loading conditions. A probabilistic analysis based on Weibull's model [14] of brittle strength was developed along the general guidelines suggested by Dukes [68] and incorporated into a computer algorithm. The algorithm captured a range of component responses. Material temperatures, stresses, and deformations were analyzed using finite element and/or finite difference computer codes. The reliability algorithm also required material strength data in the form of a characteristic modulus of rupture (i.e., the characteristic strength) and the Weibull modulus, both as a function of temperature. Using information from finite element analyses and a comprehensive material data base, the reliability algorithm numerically evaluates the integral formulation of the Weibull reliability equation given by Vardar and Finnie [69] for each volume element of the finite element mesh, assuming that the principal stresses are constant throughout a given element. Thus the algorithm computes the component survival probability under a specified load and thermal profile.

Additionally, if given a desired reliability level, the algorithm generates sets of Weibull strength parameters required of a material to meet a specified reliability goal.

For time-dependent reliability analysis, the reliability calculations are based on the experimental observation [70] that the velocity of a pre-existing crack can be described with a power law relationship as presented earlier in Eq. 25-42. Paluszny and Nicholls [71] combined the power law relationship for crack velocity with the Weibull relationship for fast fracture reliability to arrive at an equation for reliability versus time for a structure under constant load, and further extended the analysis to time-dependent load by approximating the load history with a simple step function. To determine the time-dependent reliability of multiaxially stressed ceramic turbine structures, Weibull's probabilistic statistical model of fracture was utilized with the premise that flaws which control strength are identical for both fast fractures and time-dependent fractures. Additional time-dependent effects, such as material instability and high-temperature oxidation are omitted from the analysis. Mode I stresses are utilized—Mode II and III loadings are omitted. If the load varies with time, the orientation of principal stresses throughout the structure is assumed to remain constant with time. As input, the time dependent algorithm utilized the output of the fast fracture reliability analysis. This information consisted of data on the elemental volume, fast fracture reliability, maximum principal stress, characteristic strength, Weibull modulus, and temperature of the element. The reader is directed to Paluszny and Wu [72] for comprehensive details. Additionally, the code required input to define the time-dependent characteristics of the material. Parameters required for time-dependent reliability calculations were obtained by analyzing data from available double torsion and stress rate testing databases. Optionally, a capability to directly input tables of either double torsion or stress rate material data as a function of temperature was incorporated. Using this information, the time-dependent reliability algorithm calculates a survival probability after the elapse of a specified service life or time at load. Also, the algorithm can calculate the service life available for a specified reliability. This becomes an iterative process, as strength degradation in a typical turbine structure is not uniform because stresses and temperatures are seldom uniform.

In addition to the life prediction efforts conducted under the DARPA-sponsored Series 820 program as described above, Ford also participated in other government programs to study the capability of the industry to manufacture ceramic components and to predict time dependent reliability. The reader is directed to Baker, *et al.*, [73], Caverly and Hartsock [74], and Hartsock [3] for additional information.

NASA Glenn Research Center. Ceramics research at NASA GRC began with the oil crisis in the 1970s and the establishment of the Environmental Protection Agency (EPA). After the end of the Apollo program, when President Nixon was downsizing NASA, the Department of Energy (DOE), seeking assistance in engine-efficiency research, approached NASA GRC with a proposal to do small gas turbine engine research. The push for lightweight, efficient, small-engine turbines could meet project goals and objectives sponsored by both research agencies, as well as the civilian sector. However, significant improvements in aerospace and terrestrial propulsion systems for the next century would require revolutionary advances in materials and structures. To achieve these goals, the development of a new class of high-temperature materials was identified as a critical national need. It was known that if monolithic ceramics could be used in engines, especially diesels, they could make the combustion process more efficient and reduce nitrous oxides, carbon monoxide, and other harmful pollutants. Ceramics could be simultaneously used in high temperature applications (e.g., gas turbines) or in machinery with high-speed parts such as engine valves. Ceramics were targeted for these applications because they potentially would last longer, offer less resistance, reduce friction, and weigh less. Additional aerospace technology applications include the design of turbine hot-section engine components, graphite and ceramic high temperature valves, ceramic bearings, spacecraft platforms (structural supports for imaging equipment), electronic chips and chip packages, and a variety of infrared windows and radomes.

As ceramics research progressed throughout the 1980s, NASA needed a general-purpose simulation tool to predict ceramic component durability. No such code existed at NASA, thus the goal was to develop an enabling technology to foster the introduction of ceramic materials in demanding engine environments,

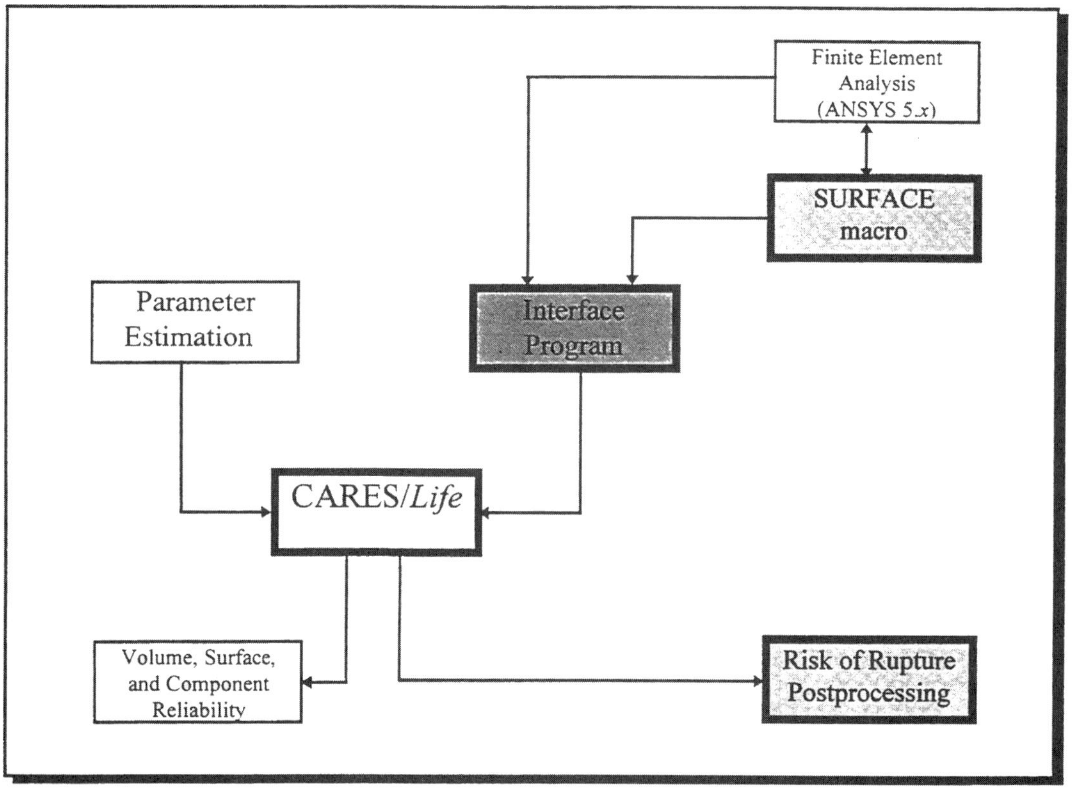

Figure 25.4 Flowchart of Life Prediction Algorithm for CARES/LIFE (Reprinted from ref. [75])

and to achieve breakthrough gains in energy efficiency and emissions reduction. A software algorithm was developed that combines the statistical nature of strength-controlling flaws with the mechanics of crack growth to allow for multiaxial stress states and predict the fast-fracture reliability of ceramic components due to flaws distributed throughout the material (as described throughout the "Fast Fracture" section of this chapter). The flowchart of the software algorithm is shown in Figure 25.4 [75]. The software was developed in the public domain under the *Ceramic Applications in Turbine Engines* program sponsored by the U.S. Department of Energy and NASA. The software was originally called SCARE (Structural Ceramics Analysis and Reliability Evaluation) [76,77] to warn engineers about the need to account for ceramic brittleness. Later the algorithm was renamed CARES (Ceramics Analysis and Reliability Evaluation of Structures). Details regarding additional features available in the CARES software but not presented here (for brevity) can be obtained in the CARES Users and Programmers Manual [31].

Over the past two decades, CARES has evolved into a series of general-purpose design tools that predict the probability of failure of a brittle material component as a function of its time in service. The next program in the CARES series was called CARES/*Life* [78,79]. CARES/*Life* retains all of the fast fracture capabilities of the original CARES program, and additionally assesses the probability that a structure will fracture prematurely due to SCG. CARES/*Life* accounts for the phenomenon of subcritical crack growth by utilizing the power law, Paris law, or Walker equation. The two-parameter Weibull cumulative distribution function is used to characterize the variation in component strength. The probabilistic nature of material strength and the effects of multiaxial stresses are modeled using either the PIA, the Weibull normal stress averaging method (NSA), or the Batdorf theory. Parameter estimation routines are available for obtaining inert strength and fatigue parameters from rupture strength data of naturally flawed specimens loaded in static, dynamic, or cyclic fatigue. Fatigue parameters can be calculated using the median value technique, a least squares regression technique, or a median deviation regression method that is somewhat similar to trivariant regression. In addition, CARES/*Life* can predict the effect of proof testing on component service probability of failure. Following STAU and CERAMIC/ERICA (described below), the most recent enhancement to CARES/*Life* is the ability to perform reliability analysis of ceramic components undergoing transient thermal and mechanical loading, (such as that encountered in engine startup, shutdown, and dynamic and vibrational loads) and to account for material properties that vary with time and temperature. Another program in the CARES series, C/CARES, was developed to predict the reliability of ceramic composite materials. The latest addition to the CARES family, CARES/*Creep*, has recently been released to account for effects in the creep stress-temperature operating regime.

Honeywell Engines, Systems & Services (formerly AiResearch Manufacturing Company of Arizona, Garrett Turbine Engine Company, and AlliedSignal Engines). DARPA/Navy initiated the Ceramic Turbine Engine Demonstration Program [80,81] in 1976 with AiResearch Manufacturing Company of Arizona (which later became Garrett Turbine Engine Company during the program) to retrofit ceramics into an existing turboprop engine. The primary goal of the program was to demonstrate a 40% increase in power output and a 10% decrease in fuel consumption during a 50 hour engine test. Under this program, Garrett modified the output processors of their existing finite element programs ISOPDQ and ISO3DQ to integrate stress fields over both component volume and surface area. The resulting post processor WESTAC used the two-parameter Weibull distribution to compare the calculated stress field with the material strength and strength distribution parameters. This allowed prediction of failure probability for each finite element and combination of the individual probabilities to obtain an overall probability for the component. Because planned engine testing was not to exceed 50 hours and since very little time-dependent data were available at the beginning of the program, the analyses were limited to fast fracture probability of failure calculations, including proof testing and transient operating conditions though. Garrett realized already in 1976 that in performing a probabilistic evaluation of a finite element model that a mesh size sufficient to represent the stresses is not sufficient to accurately represent the failure probability. To overcome this mathematically, the probabilistic evaluation programs WESTAC2 and WESTAC3 incorporated a refined integration calculation, which automatically increased the number of Gauss points per finite element during the probability calculation up to a maximum of 15, to accurately represent the probability of an element with even the most severe stress gradients. In the early 1980s, Garrett Automotive Products Companies developed their own in-house probabilistic life prediction code as part of their development efforts of ceramic turbine wheels for turbochargers [82]. The observed burst speeds and lifetimes agreed extremely well with the predictions.

In 1989, Honeywell Engines, Systems & Services (formerly AlliedSignal Engines), Phoenix, Arizona, started developing a new tandem set of life prediction codes, CERAMIC/ERICA [83,84], under the "Life Prediction Methodologies for Ceramic Components of Advanced Heat Engines" program funded by the U.S. Department of Energy/Oak Ridge National Laboratory (DOE/ORNL) under Contract No. DE-AC05-84OR21400, Subcontract No. 86X-SC674. A history of ceramic component development for gas turbines at Honeywell may be found in Ref. [6]. The emphasis of the Phases I and II of the program was to develop and demonstrate life prediction technologies that included the consideration of ceramic fast fracture, stress rupture, creep, and nondestructive evaluations. Significant advancements were made in the valid testing and characterization of monolithic ceramics, and the predictive capabilities of CERAMIC/ERICA were successfully demonstrated.

The objective of this life prediction program was to develop the methodology to address the failure modes that may be active in a given mechanical component. The general approach was to begin by addressing those modes which are likely to cause premature failure, followed by those that show promise of an easy solution, and culminating with the more difficult to analyze, complex modes. For ceramics, fast fracture (FF) and SCG are the most prevalent failure modes. Other failure modes, such as oxidation, corrosion, cyclic fatigue, impact damage, and contact damage, to name only a few, might also play an important role in the failure of ceramic components (see Figure 25.5), but most of these failure modes have yet to be

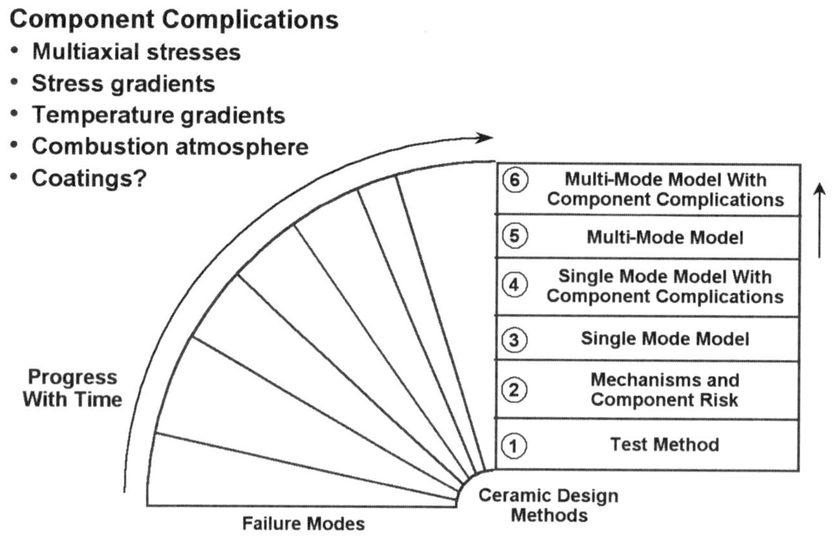

Figure 25.5 Comprehensive Design System Required For Reliability Assessment Of Ceramic Components In Gas Turbine Engine (Reprinted from ref. [83])

treated statistically. An outcome of the program is that Honeywell uses a building-block approach (increasing component simulation complexity as the program progresses) specifically designed for future industry and technology needs (see Figure 25.6).

Besides extensive database generation, Phase I of this program was focused on the development of advanced statistical concepts for the fast fracture failure mode prediction of brittle materials. Phase II emphasized primarily time-dependent failure modes (subcritical crack growth, cyclic fatigue), reliability enhancing methods (proof testing, NDE) and probabilistic life prediction codes development for distribution among the ceramics community within the US. As the logical next step, the Honeywell funded Phase III now focuses primarily on environmental effects on ceramic component reliability to address the currently most critical concern regarding ceramic applications in advanced high-temperature gas turbines. Additionally, Phase III

Increasing Component Simulation Complexity ↓

Phase	Primary Emphasis	Secondary Emphasis
I	• Fast fracture • NDE methods • Statistical methods	• Time-dependent failure modes (Stress rupture)
II	• Time-dependent failure modes (Subcritical crack growth) • Reliability-enhancing methods (Proof testing, NDE-POD) • Statistical methods	• Cycle-dependent failure modes (Cyclic fatigue)
III	• Probabilistic creep rupture • Environmental interaction – Oxidation & hot corrosion • Statistical methods	• Transient thermal & mechanical load • Thermomechanical fatigue • User-friendly codes enhancements

Figure 25.6 Honeywell Ceramic Life Prediction Program Utilizes A Building-Block Approach

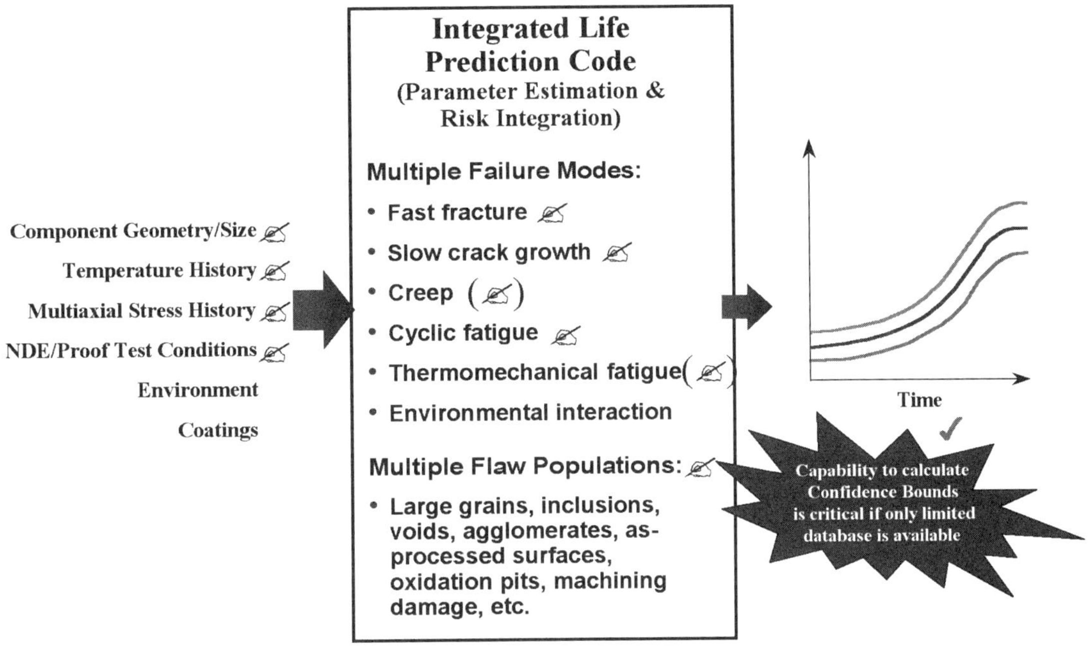

Figure 25.7 Honeywell's Phase III Program Will Provide An Integrated Ceramic Life Prediction Code

addresses the development of probabilistic creep life prediction methods, since none of the currently available creep rupture methodologies is capable of taking the scatter in observed specimen or component lives into account. The intent is that at the end of Phase III that an integrated ceramic life prediction code will be created (Figure 25.7).

In the process of assessing the reliability of ceramic components, two issues were deemed critical: generation of the Weibull and fatigue parameters of the material, and; risk integration of the component to determine the reliability. The basic theories that address these two issues are well developed, but in order to produce reliable designs with the typically small material properties databases and large extrapolations, more advanced statistical methods are required.

To this effect, Honeywell has been pursuing the development of probabilistic and statistical methods for extracting the most information out of a given set of data. The result has been the development of two computer codes, CERAMIC and ERICA, that are intended to be used in tandem, as shown in Fig. 25.8. These codes incorporate state-of-the-art methodologies for the design and life prediction of ceramic components. Statistical methods for the design of ceramic components for fast fracture and time-dependent failure modes have been developed which can significantly enhance component reliability and reduce baseline data generation costs. Use of the codes facilitates generation of material strength and fatigue parameters by pooling data from specimens that are of different sizes, stressed by different loading conditions, or tested at different temperatures, and even by including data from specimens or components that have been proof tested. The codes also include approaches for calculation of confidence bounds for the Weibull and fatigue parameters of censored data and for the predicted fast fracture or time-dependent reliability of ceramic components. The remainder of this section presents a summary of the methodology, and a synopsis of the codes features and capabilities.

The CERAMIC/ERICA codes developed at Honeywell benefited immensely from the implementation of advanced and previously unutilized statistical concepts outlined by Johnson and Tucker [30,32,37]. Because of that, CERAMIC/ERICA already had capabilities by the mid-1990s for maximum likelihood

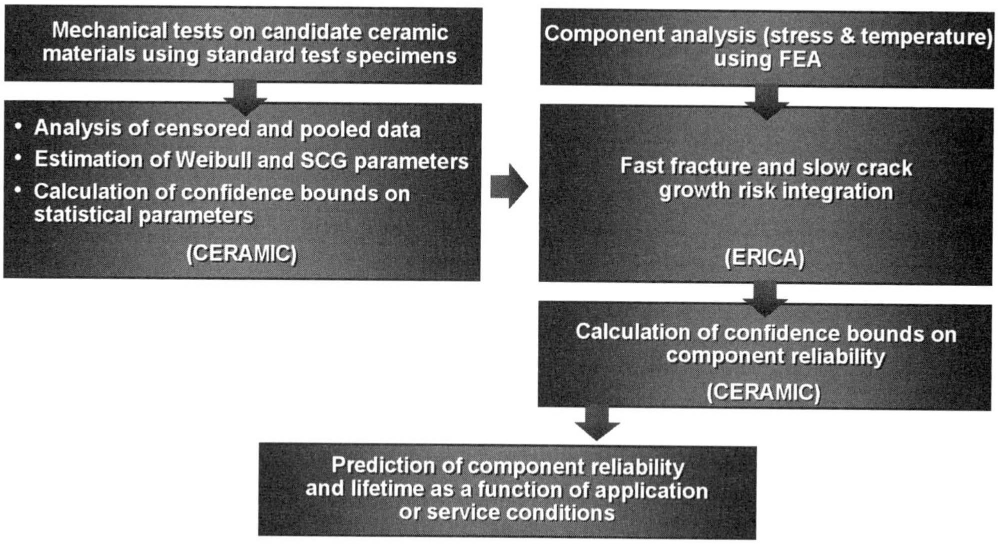

Figure 25.8 Honeywell's CERAMIC And ERICA Codes Are Intended For Use In Tandem To Develop Statistical Analysis Of Ceramic Failure Data And Reliability Evaluation Of Ceramic Components

estimation, censored data analyses, pooled data techniques, and likelihood ratio and bootstrap techniques for calculating confidence intervals. The result was a code that allowed consideration and interpretation of: (1) strength data with multiple specimen sizes, (2) multiple/competing strength distributions, and (3) multiple-temperature data to be pooled for estimating Weibull strength parameters and component reliability with confidence intervals. This was a breakthrough in ceramic component design because prior to this development, ceramic components could not be designed to levels of confidence, and the structural ceramic component designers/analysts were relegated to design to the best estimate of median strength and life properties without being able to account for the uncertainty in estimates due to limited sample sizes. The statistical methodology incorporated in CERAMIC can significantly reduce the number of specimens required to generate a comprehensive materials database. The data pooling capabilities and the ability to calculate confidence bounds for the Weibull parameters, as well as for the reliability based on censored data, are extremely useful in ensuring the reliable design of ceramic components. Used in tandem, CERAMIC and ERICA provide a complete package for the successful evaluation of the reliability of ceramic components.

The execution of the left-hand side of the flow chart in Fig. 25.8 is a consequence of the usage of CERAMIC and the consideration of censored mechanical test data. CERAMIC contains the advanced statistical concepts mentioned above. ERICA is a post finite element analysis post-processor that introduces risk integration using the Lamon-Evans multiaxial fracture criterion [27,28]. ERICA outputs effective volumes, areas, or edge-length strength gradient factors (e.g., see Eq. 25-29 for volumes and Eq. 25-32 for areas)) as a function of Weibull modulus. As a last step, the strength gradient factors for each censored flaw type (i.e., ERICA results) are combined with the censored material design data (of the ceramic material in question from the first run of CERAMIC) to calculate the POS in a second run of CERAMIC. CERAMIC also calculates the confidence bounds about that prediction.

Four key salient features of CERAMIC and ERICA are:

- The statistical methods consist of approaches that use censored data analysis techniques for the pooling of material data. By pooling, it is meant that specimens of different sizes that are loaded under different conditions, (see Figure 25.9), and at different temperatures can be analyzed together (in a single analysis), to

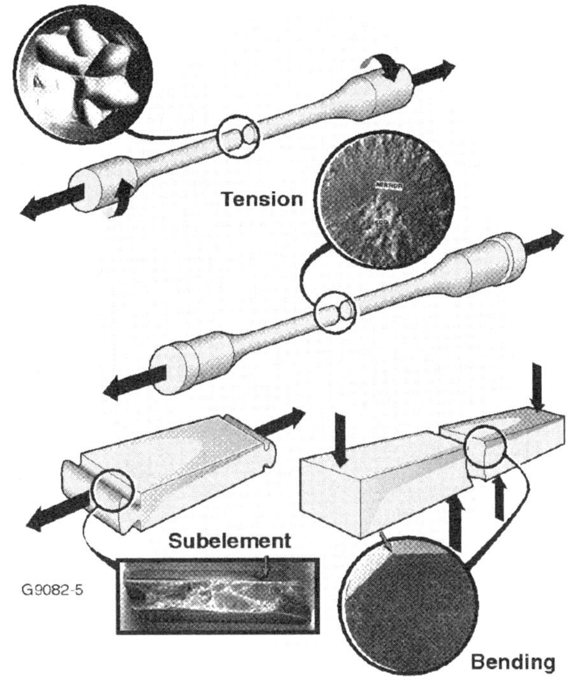

Figure 25.9 Material Data for Input into Probabilistic Design Is Generated Using a Variety of Specimen Types and Loading Conditions (Reprinted from ref. [83])

generate the required Weibull and, if applicable, fatigue parameters of the ceramic material. The material parameters generated in this way are more accurate than those obtained from the individual analyses of each data set. The pooling of specimens of different sizes and loading conditions can be used to provide an indication as to how well the chosen failure theory fits the given material. On the other hand, the pooling of specimens tested at different temperatures can be used to develop an approach to interpolate and/or extrapolate material behavior as a function of temperature (see Figure 25.10).

- The methodology includes approaches for the calculation of confidence bounds for the reliability prediction of a component. This is particularly important, because components are generally much larger and are loaded at significantly lower stress levels than test specimens, resulting in sizable extrapolations of the material properties. For actual components, the ratio of failures that originate from volume and surface flaws may also be much different than the ratio observed in the database specimens. Extrapolation away from the bulk of the stress levels where the data are generated results in significantly wide confidence bounds. One should, therefore, design ceramic components to a desired level of reliability using the upper-bound confidence limit (Figure 25.11).
- The risk integration approach has been modified to facilitate confidence bound calculation for components. The procedure in ERICA consists of the evaluation of the effective size of a component as a function of the Weibull slope in case of considering fast fracture, and as a function of Weibull slope, time, and subcritical crack growth parameter N and B in case of considering static fatigue [42]. This information is then used in CERAMIC to compute confidence bounds for the component reliability.
- An anisotropic risk integration approach has been developed for properly evaluating the risk integration of "transversely" machined surfaces.

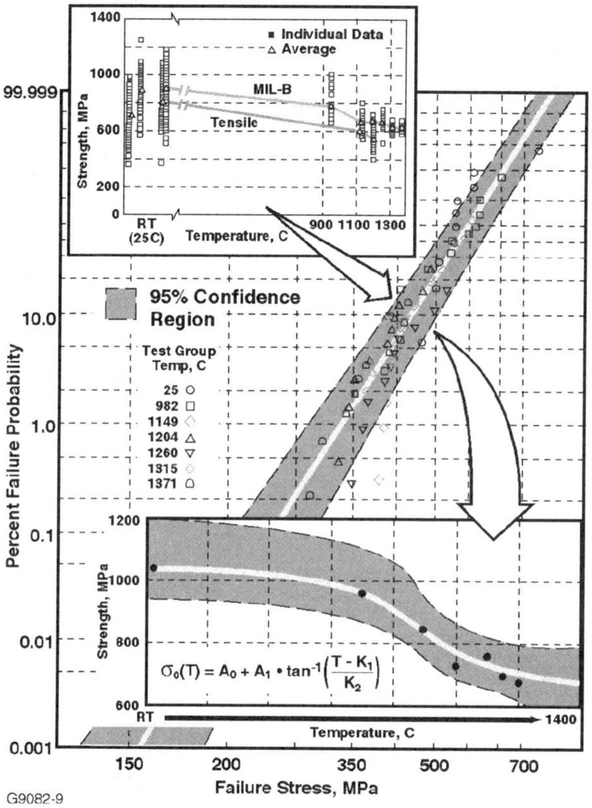

Figure 25.10 Specimen Data Points at Different Temperatures Are Used to Define Characteristic Strength as a Function of Temperature (Reprinted from ref. [83])

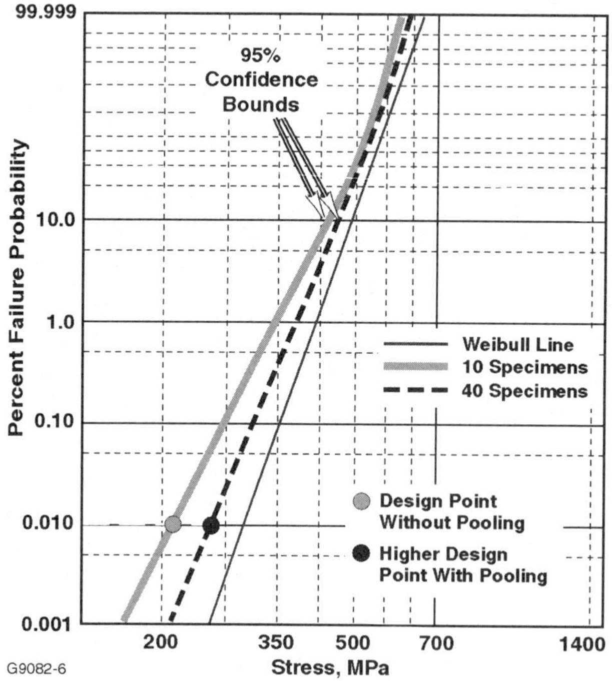

Figure 25.11 Increase In Specimen Quantity Due To Data Pooling Tightens Confidence Bounds and Significantly Increases Design Allowables (Reprinted from ref. [83])

In addition and most importantly, advanced statistical methods for the design of ceramic components for time-dependent failure modes (slow crack growth) have been developed which can significantly enhance component reliability, reduce baseline data generation costs, and lead to more accurate estimates of slow crack growth parameters. These methods are also incorporated into the CERAMIC and ERICA computer codes [42]. Use of the codes facilitates generation of material strength parameters and of slow crack growth parameters at the same time by pooling fast fracture data from specimens that are of different sizes, or stressed by different loading conditions, with data derived from static fatigue experiments. The codes also include approaches to calculation of confidence bounds for the Weibull as well as slow crack growth parameters of censored data and for the predicted reliability of ceramic components as a function of service time. Moreover, the codes' capabilities include performing reliability analysis of ceramic components undergoing transient thermal and mechanical loading, (such as that encountered in engine startup, shutdown, and dynamic and vibrational loads).

A large database was developed (NT154 Si_3N_4) to support and verify the life prediction development of CERAMIC/ERICA. The database included results from 1400 strength, stress rupture, and creep tests using 14 different specimen geometries (even after almost 10 years, this database is one of the largest for a single advanced structural ceramic). To ensure consistency, all NT154 specimens were fabricated from one powder lot. These improved probabilistic methods allowed 405 specimens with 4 different geometries to be combined into one data set (i.e., pooled) for estimating strength parameters and predicting confirmatory specimen failures. As described in an earlier section of this chapter, data pooling results in more confidence in predictions (narrow confidence bounds) that allows components to be designed closer to the true capability of the material. As a final test of these prediction methods, reliability estimates were made with confidence intervals and these were compared with measured confirmatory specimen strengths and lives [84]. An example is illustrated below that involves the use of CERAMIC/ERICA.

In closing, though CERAMIC/ERICA were developed to assist in the design optimization of structural ceramic components for gas turbine engines, the advanced statistical capabilities of CERAMIC have made it attractive for use for other ceramics components such as ceramic diesel exhaust valves and ceramic-containing electronic devices [85–87].

Rolls-Royce Corporation (North America). One of the major challenges involved in utilizing ceramic materials in advanced vehicular heat engines is ensuring adequate strength and durability. Consequently, the Rolls-Royce Corporation in North America (formerly Allison Engine Company and Allison Gas Turbine Division of General Motors prior to that) in Indianapolis, Indiana, has developed methodologies to predict the structural behavior of ceramic components [5]. Government-supported efforts have led to material characterization and analytical life prediction techniques for structural ceramics at Allison. Initial efforts resulted in a proprietary code called STRATA that performed fast fracture analyses. During the Ceramic Technology for Advanced Heat Engines (CTAHE) project sponsored by the DOE, Allison performed a study correlating results obtained using STRATA with those obtained using the public domain CARES code developed at NASA GRC (see previous section). This investigation precipitated a decision to incorporate future enhancements and modifications developed at Allison into a new algorithm that will be referred to here as Allison/CARES.

Analytical models for the fast fracture failure mode were developed based on the knowledge and information gained during the material characterization phase of the CTAHE effort. Techniques developed account for surface as well as internal (i.e., volume) component failures. The fast fracture model is based on a two-parameter Weibull distribution, which is the current industry standard. However, the proposed model utilizes a different distribution depending on the specimen/component failure origin. The design data input requires the Weibull modulus and the unit characteristic strength of the selected material. This data is obtained by testing a statistically significant number of specimens at each temperature in fast fracture, locating failure origin of each specimen under low-magnification microscopy, and conducting 2-parameter Weibull linear regression using suspended item analysis method. This in-house data analysis method has been very successful in predicting the rotor speeds of ultimate failure of blades and rotors. Both uniaxial and biaxial models have been developed and correlated with that material test data. Failure analysis of test specimen fracture surfaces revealed that in addition to volume and surface failure populations, a third type of failure originating at the specimen corners was present. Experimental results suggested that each type of failure plays a separate role in the assessment of the fast fracture strength and Weibull properties. The Allison/CARES program was modified to account for the corner flaw population [88]. As part of this process MOR failures are identified as surface, volume, or corner failures. The MLE method is used with the other two failure modes treated as suspended items for the third. This approach is used to determine the Weibull properties of the corner failures, which are applied during the fast fracture analysis of the component.

As the reader can ascertain from this discussion, the development of ceramic design protocols requires a significant testing program in order to obtain fast fracture Weibull properties at operational temperatures. One approach in reducing the requisite amount of test data involves consolidating the elevated temperature data using a single Weibull modulus. It was analytically demonstrated [89] that if the fast fracture failure mode is dominated by a Mode I failure, then the Weibull exponent is independent of temperature. As a result, the capability of determining a consolidated Weibull modulus and the characteristic strengths using this approach was incorporated in the Allison/CARES code.

The cornerstone of the later ceramic life prediction efforts at Allison has been the Allison/CARES algorithm. Under the DOE/ORNL sponsored program entitled *Ceramic Life Prediction Method for Vehicular Engines*, Allison expanded their capabilities in life prediction methodology. Efforts here included the incorporation of oxidation, slow crack growth, and creep/stress rupture failure modes into the Allison/CARES algorithm. Additionally, a creep deformation that couples the creep rate to the stress rupture information has been developed and incorporated into the Allison/CARES algorithm. Experimental characterization of subcritical crack growth is commonly performed by either the constant load test (i.e., static fatigue) or the constant loading rate test (i.e., dynamic fatigue). In the constant load test, the specimen rupture time is determined at various applied stress levels. In dynamic fatigue, the specimen failure strength is measured at various stressing rates. The equations for the two types of tests are derived from empirical crack velocity relationships in much the same manner that equations for creep tests were derived in an earlier section. Oxidation is modeled as a surface effect governed by a parabolic diffusion law and a thermal activation process of constant activation energy, with an Arrhenius type of temperature dependence. By assuming the Griffith flaw size to be proportional to the surface oxide thickness, the oxidation equations can be structured in a way to allow the time-dependent reliability of the component to be computed within the Allison/CARES program.

For components that are subjected to low stress at high temperature, and that also require a relatively long service life, creep rupture damage mechanisms tend to dominate failure. While there are a number of equations in the literature that adequately describe the creep rupture phenomenon, the creep model in the Allison/CARES program combines the classical Norton isothermal power law with a thermal component (which assumes a constant activation energy) exhibiting an Arrhenius type of temperature dependence to describe the relationship between creep rate and applied stress. This approach was advocated by Wiederhorn, *et al.* [90]. The classical Monkman-Grant equation is used to relate rupture time to creep rate. By combining the equations, a relationship between rupture time and applied load is obtained. Material creep data is analyzed using a least squares regression analysis. When both creep rate and rupture data are available, all of the available data should be used to obtain an optimum estimate of the material constants. The large amount of scatter observed in the time-to-failure in the tensile stress rupture data suggests that a prudent component design practice should incorporate a statistical "weakest link" concept. An approach by DeSalvo [91] based on Weibull theory was incorporated into the Allison/CARES algorithm.

Sundstrand Power Systems (SPS—currently Hamilton Sundstrand Power Systems). As new ceramic materials progress into advanced product design and development, the long-term performance of the ceramic components becomes increasingly important for the engineer who must design turbine systems. The analytical assessment of the expected life of ceramic components requires the statistical characterization of material properties in the various failure modes as well as the predicted stress-temperature distribution within the component as function of the structural thermal loads. The design and analysis of prototypical ceramic turbine components at SPS highlighted the need for an expeditious approach to the statistical life assessment analysis [9]. The design optimization of highly stressed ceramic components frequently requires several iterations, which can be time consuming. The design engineer can easily be overwhelmed with the detailed results obtained from finite element analyses, temperature profiles from heat transfer analyses and the corresponding life assessment analyses. In order to expedite the structural analysis of ceramic components SPS developed a proprietary computer code entitled SPSLIFE. This code was initially applied to the Sundstrand TJ-50 ceramic monorotor. Since then, the algorithm has been used in preliminary designs that evaluate the life of several ceramic components whose development was supported by the DOE/Solar Turbines Incorporated *Ceramic Stationary Gas Turbine Program.* This included two blade designs, one nozzle design and three combustor liner designs.

The life assessment computations for the various failure modes are performed using structural analysis and materials files as input data. The program links ANSYS results files and materials data files with the life assessment modules for different modes of failure to provide quantitative estimates of component life and survival probability. A number of menus are incorporated to request user input data and to guide the user through the problem definition process. The user has the capability to select existing material data files or create new ones. The computational modules are based on the analytical approaches currently used in the industry. Each computational module provides a guided analytical tool for the non-ceramic expert. Knowledgeable design engineers and stress analysts who are unfamiliar with the stochastic behavior of ceramic materials are quickly able to perform the life assessment analysis of ceramic components. Modularized construction of the algorithm facilitates expansion of analytical capabilities. Visual displays of the analyses are clear and easy to interpret. The output files are organized to facilitate detailed reviews of results in order to optimize designs. The SPSLIFE algorithm provides easy revision of data files, yet provides limited accessibility to different levels of users. In addition, the program produces presentation quality design assessment maps depicting stress and temperature distributions together with design envelopes for fast fracture, slow crack growth, creep and oxidation. Cyclic fatigue, proof testing, and wear modules are planned for future implementation.

SPSLIFE Design Assessment Maps (DAMs) are component-specific, and provide a graphical overview of the design evaluation results. The maps are based on the "fracture map" concept discussed in the ceramics literature [1]. Design envelopes for different failure modes are superimposed on a plot of maximum principle stress versus temperature for all the modes in the finite model. The fast fracture module uses Weibull theory that distinguishes between three different categories of flaws: volume flaws, as-fired surface flaws, and machined surface flaws. The Slow Crack Growth Module uses time dependent theory discussed earlier in this chapter. Here again the algorithm distinguishes between internal flaws, as-fired surface flaws, and machined surface flaws. The Creep Fracture Module uses Monkman–Grant analysis and provides a design envelope that is depicted on the DAM. Creep life corresponding to computed stress levels and temperatures are tabulated along the right hand side of the map. The Oxidation Module uses the fracture mechanics approach described by Soma, *et al.* [92]. A design envelope is depicted that corresponds to the selected design life, and the calculated oxidation life is tabulated on the right hand side of the map. An example of the demonstration and use of SPSLIFE is illustrated below. A summary description of the code can be found in the literature [93].

LIFE PREDICTION EXAMPLES

Two examples of ceramic component life prediction are provided. The first is the prediction of reliability of a Si_3N_4 spin disk tested at room temperature using Honeywell's CERAMIC/ERICA codes and that is excerpted from Refs. [84,94]. The second is the prediction of reliability of a Si_3N_4 nozzle by Solar Turbines Incorporared using SPSLIFE and NASA's CARES/LIFE [95,96].

Honeywell/AlliedSignal—Si_3N_4 Spin Disk

The strengths of many hundreds of ASTM C1161A, ASTM C1161B, and ASTM C1161C [34], and ASTM C1273 [35] NT154 Si_3N_4 test coupons were measured. All strengths were censored according to volume- and surface-strength-limiting flaws. The spin disk geometry had no edges, so edge-censored strength distributions from the ASTM C1161 bend tests were not considered. The pooled and censored strength-distribution for volume-based strength-limiting flaws had a $m_V = 8.3$ and a $\sigma_{0V} = 64.9$ ksi•in$^{3/m}$ = 1484 MPa•mm$^{3/m}$, and $m_A = 9.7$ and $\sigma_{0A} = 116.5$ ksi•in$^{2/m}$ = 1566 MPa•mm$^{2/m}$ for surface-based strength-limiting flaws. A Weibull plot of pooled surface strength from tensile and flexure specimens tested at multiple temperatures is shown in Figure 25.12, though only the room temperature data is used in the prediction. A Weibull plot of pooled volume strength from tensile and flexure specimens and multiple temperatures is

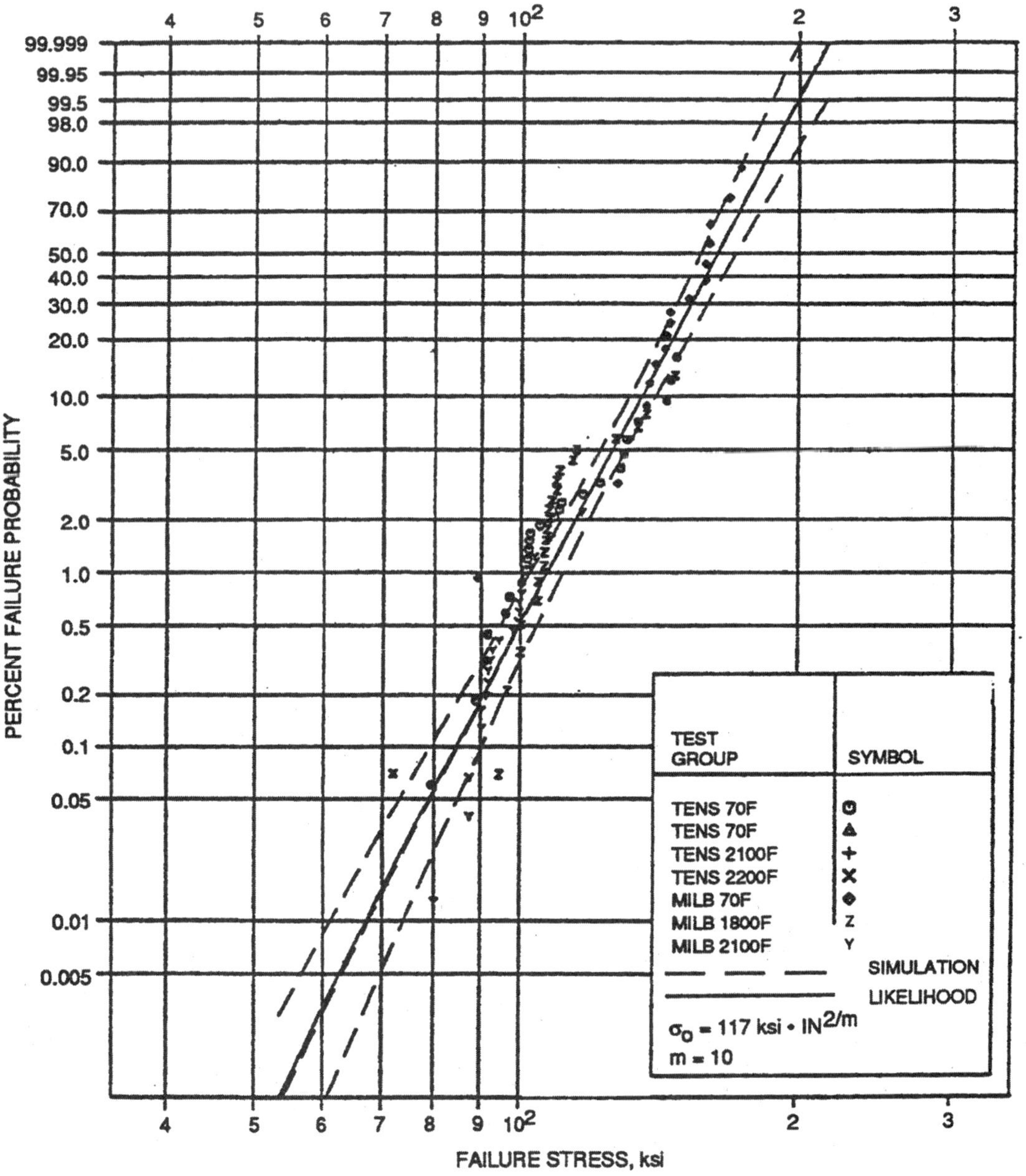

Figure 25.12 Weibull Plot of Pooled Surface Strength From Tensile and ASTM C1161-B Specimens Tested at Multiple Temperatures (Reprinted from ref. [84]).

shown in Figure 25.13, similarly, only the room temperature data is used in the prediction. These data constitute the left hand side of the life prediction algorithm shown in Figure 25.6.

The service stress state in question for the spin disk is shown in Fig. 25.14. Shown are the maximum principal stresses under centrifugal loading at 50000 rpm. The greatest stress is approximately 117 MPa (17 ksi). This step constitutes the upper right-hand box of the algorithm illustrated in Figure 25.6.

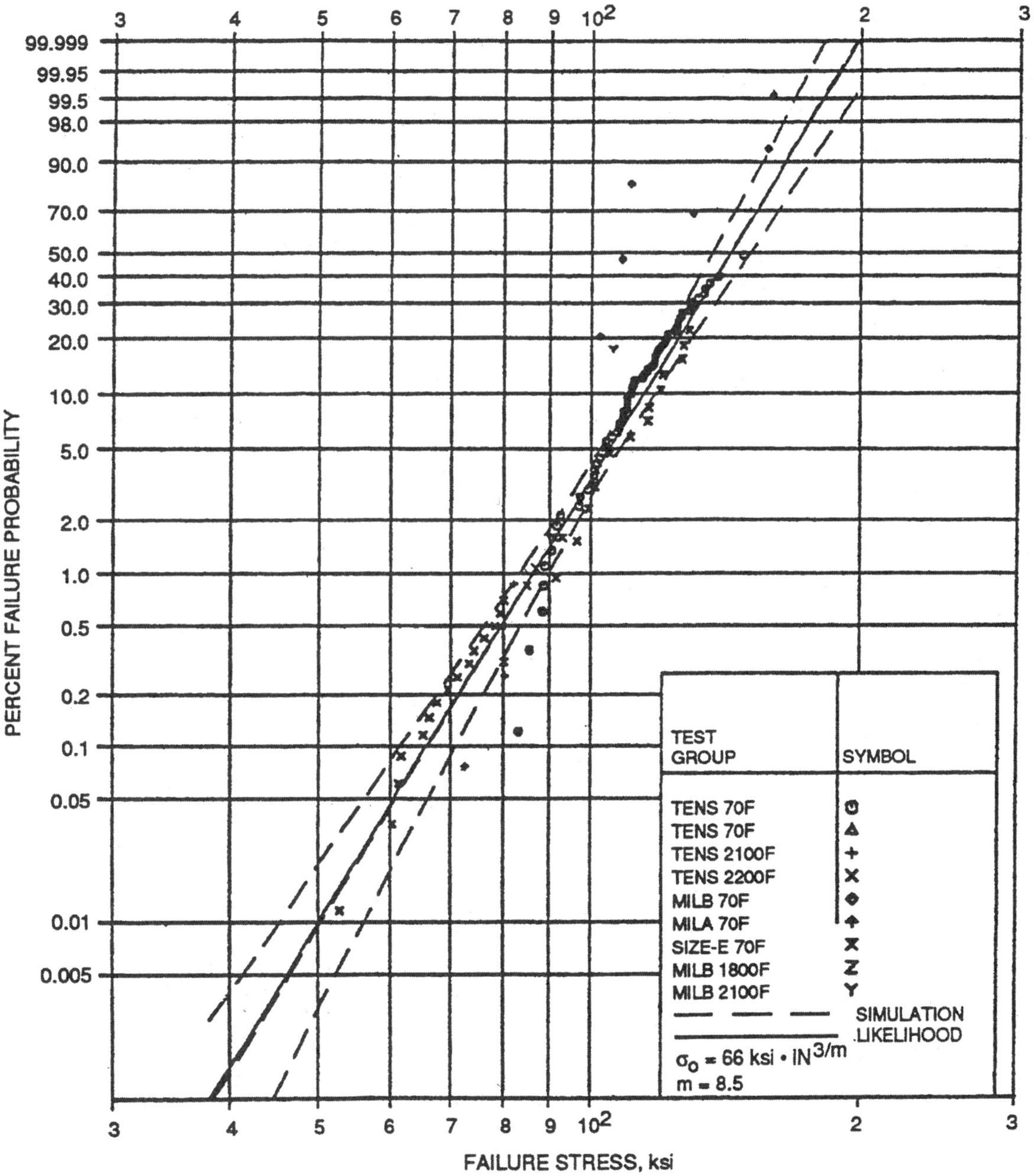

Figure 25.13 Weibull Plot of Pooled Volume Strength From All Specimens/Types and Multiple Temperatures (Reprinted from ref. [84])

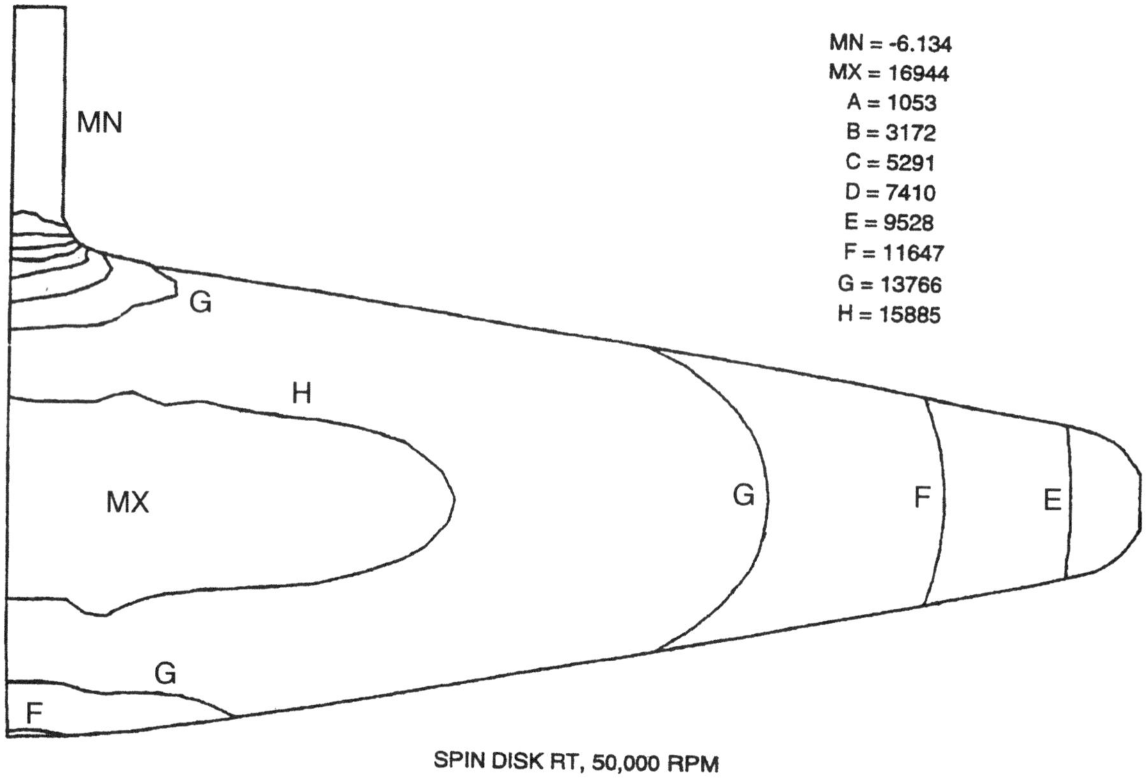

Figure 25.14 Spin Disk Maximum Principal Stress Distribution Under Centrifugal Loading Only, 50000 rpm (Units = psi, 1000 psi = 6.9 MPa) (Reprinted from ref. [84])

The predicted responses for surface-flaws and volume-flaws as the strength-limiters in the spin disk at the stress state shown in Figure 25.14 are shown in Figures 25.15 and 25.16, respectively. Also appearing in Figures 25.15 and 25.16 are censored data for spin disks tested to failure. The reader observes that poor correlation (non-conservative) between prediction and experiment resulted when surface-flaws limited spin disk performance whereas good correlation existed between the two when volume-flaws limited performance. The reason for lack of correlation when surface-flaws limited spin disk performance was that the machined surface produced on bend bars and tensile specimens did not mimic that surface on the spin disk; consequently, the surface-censored data from those test coupons provided incorrect input for the life prediction analysis. This observation represents a valuable finding in life prediction endeavors in general; namely, the quality of the life prediction is only as good as the representation of the censored input data.

Solar Turbines Incorporated—Si_3N_4 Nozzle

To illustrate how life prediction methodology can be used to guide ceramic component design is provided by the example of the development of SiC and Si_3N_4 nozzles under the US DOE sponsored Ceramic Stationary Gas Turbine (CSGT) program at Solar Turbines Incorporated, San Diego, California [10].

The initial concept nozzle design evolution in the CSGT program is schematically represented in Figure 25.17. The engine targeted for development in the CSTG program, the Centaur 50S metal engine (4 MW nominal output power) has 15 two-vane cooled first-stage FS-414 metal nozzles (Figure 25.17(a)). The nozzles have a Pt/Rh-aluminide coating for oxidation protection. The ceramic nozzle development started with

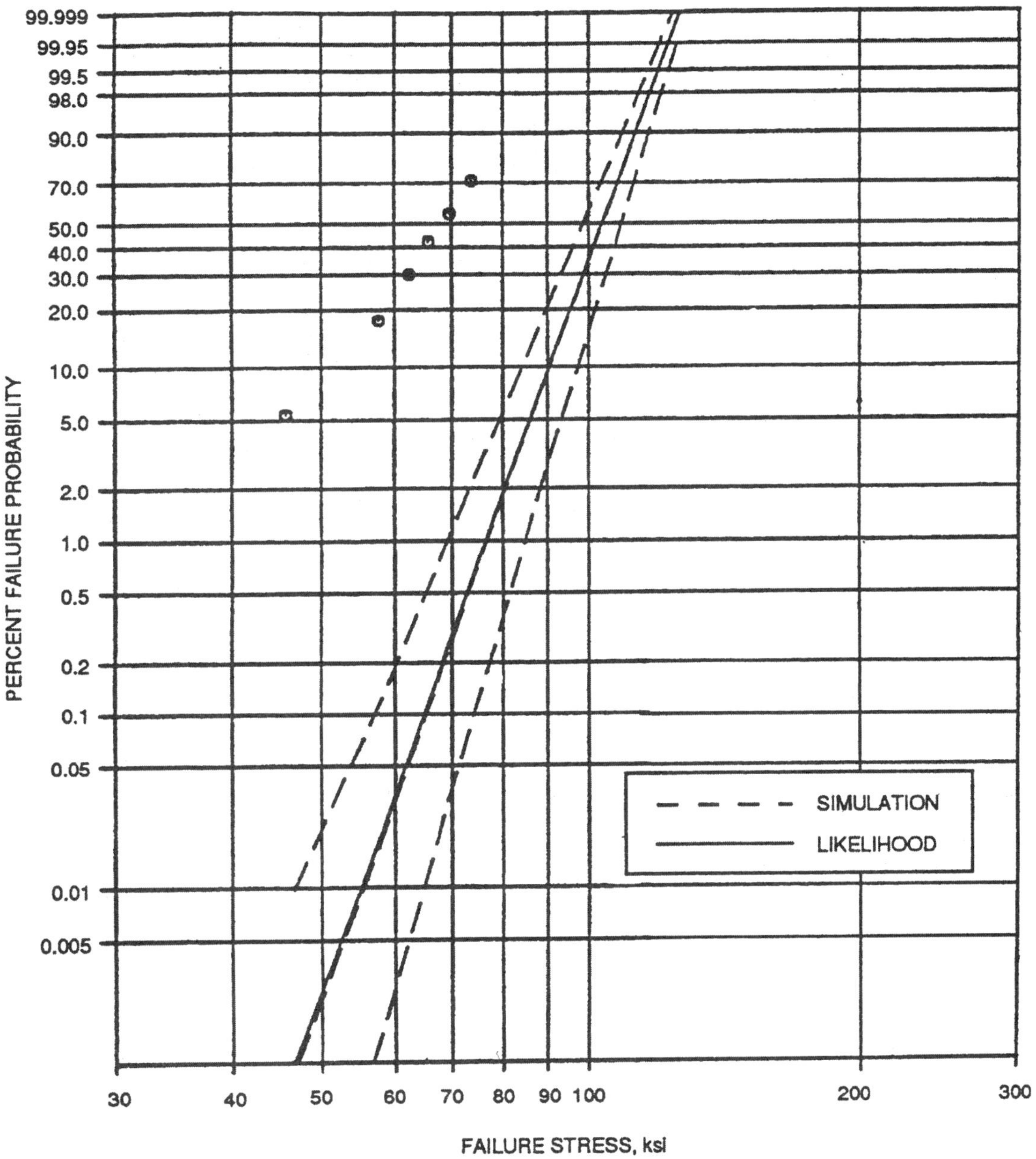

Figure 25.15 Spin Disk Room-Temperature Fast Fracture Surface Failure Predictions From Pooled Data Set (Units: 1 Ksi = 6.9 MPa) (Reprinted from ref. [84])

a single vane segment, representing one half of the current metal nozzle with tip shoe removed for ease of ceramic fabrication (Figure 25.17(b)). To maintain the basic aerodynamic performance of the turbine the initial design strategy was to retain the airfoil profiles of the metal nozzle.

Finite element studies revealed that replacing the metal airfoils with a ceramic vane of the same geometric configuration would result in an unacceptably high steady-state stress level (480 MPa) at the hottest temperature location (1288°C), which is incompatible with long service life (Figure 25.17). This high stress

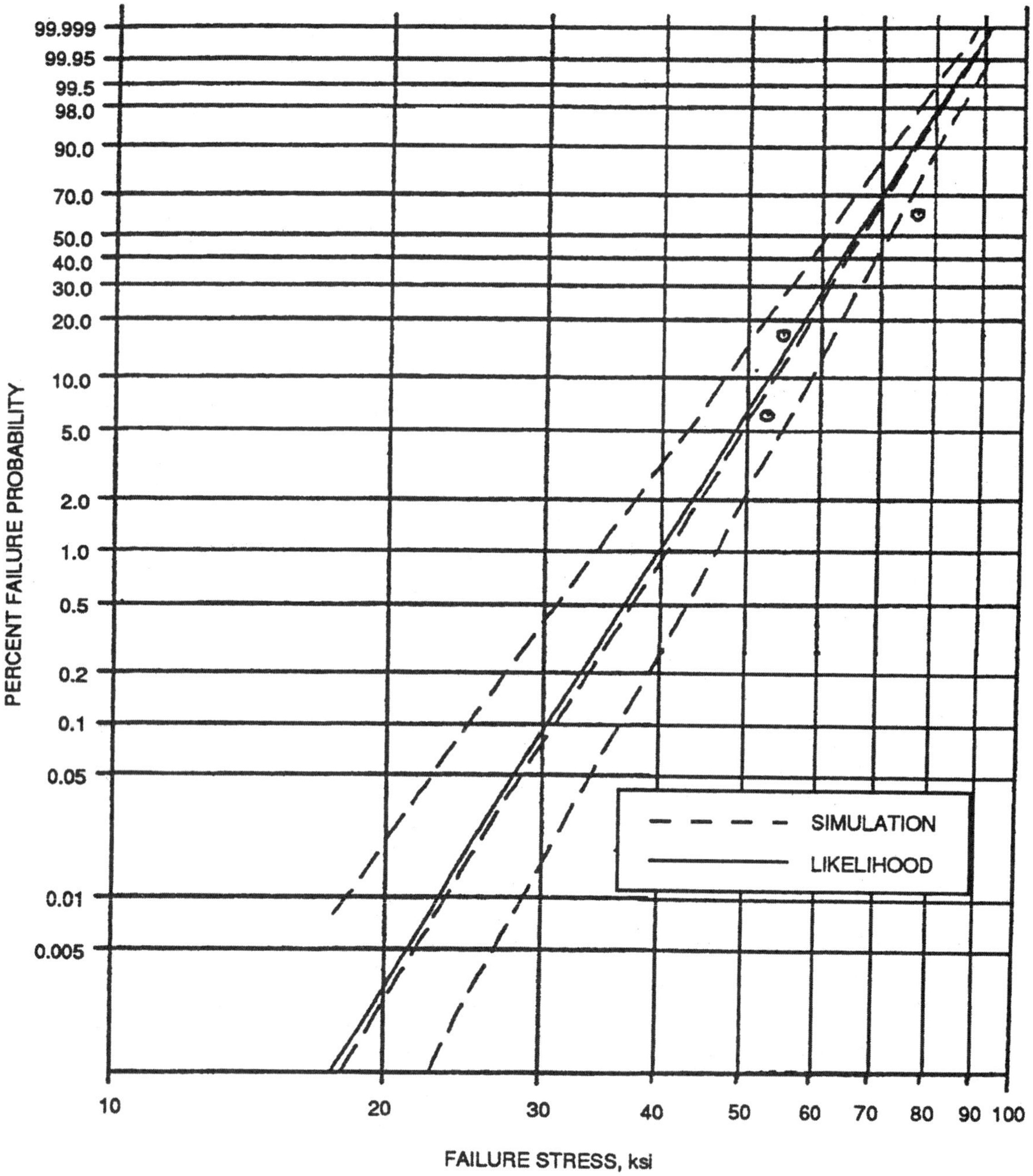

Figure 25.16 Spin Disk Room-Temperature Fast Fracture Volume Failure Predictions From Pooled Data Set (Units: 1 Ksi = 6.9 MPa) (Reprinted from ref. [84])

level is at the central portion of the airfoil trailing edge. The temperature of 1288°C was predicted at the nozzle hot spots (resulting from a combustor pattern factor of 0.20 and associated radial profile) for the engine operating at the design turbine rotor inlet temperature (TRIT) of 1121°C. To lower the stress level a concept design was generated in which the inner shroud was uncoupled from the airfoil. The outer shroud was retained and a pin feature was added for the attachment.

The initial life prediction study [95,96] was performed using the SPSLIFE life prediction code developed by Sundstrand Power Corporation [9,93] (currently Hamilton Sundstrand Power Systems), of San Diego, California. This code links structural (finite element) analysis files and materials property data files with life assessment modules for different modes of failure to provide quantitative estimates of component life and survival probability. The finite element map had 3522 brick elements and 5475 shell surface elements. Temperatures on the nozzle ranged from 834°C to 1291°C. Two commercially available materials, a silicon nitride and silicon carbide were evaluated in the study. The silicon nitride component was densified by glass encapsulation HIP-sintering and the silicon carbide was an alpha-sintered material. The maximum surface stress and internal stresses for the silicon nitride components in the study were predicted to be 234 MPa and 241 MPa, respectively. For the silicon carbide material the corresponding numbers were higher, 317 MPa and 324 MPa. The SPSLIFE life prediction analysis produces DAMs depicting stress and temperature distributions for the components together with design envelopes for fast fracture, SCG, creep and oxidation. The NASA CARES/LIFE routines are also embedded in the code and fast fracture and slow crack growth probability of survival (POS) are printed on the DAMs. SPSLIFE DAMs for the silicon nitride and silicon carbide single vane nozzles are shown in Figure 25.18(a) and Figure 25.18(b), respectively. The DAMs are plotted for a design life of 30,000 hrs [95,96].

Figure 25.18(a) shows that several data points are above the creep envelope (dashed-dotted line) indicating unacceptable creep life that was assessed at 4,100 hours. Predicted oxidation life (dotted line), in excess of 10^8 hours was excellent. It should be noted that the study was done in the 1993–94 time frame when the detrimental effects of accelerated oxidation in the gas turbine hot section (see chapter 26 in this volume) were not broadly recognized. The solid fast fracture lines, representing 90.0%, 95.0%, and 99.0% POS for each nozzle segment, are well above the data points which signals that the design is safe at these probability levels. NASA CARES predicts a fast fracture POS for each nozzle segment of 99.97%. However the SCG (dashed lines) show unacceptable life for this failure mode. CARES/LIFE calculates an SCG POS of only 1.76% [95,96].

The DAM for the SiC nozzle is shown in Figure 25.18(b). While the stresses were calculated to be higher than for the Si_3N_4 material creep life was significantly better. The SiC nozzle was assessed to have a creep life of 1,193,000 hours. An oxidation analysis was not performed for the SiC material due to lack of adequate data. The solid lines indicate that the SiC nozzle has somewhat less than 90% POS in the fast fracture mode (CARES computes 89.41%). NASA CARES/LIFE predicts unacceptable low POS for the SCG mode at 60.53% [95,96].

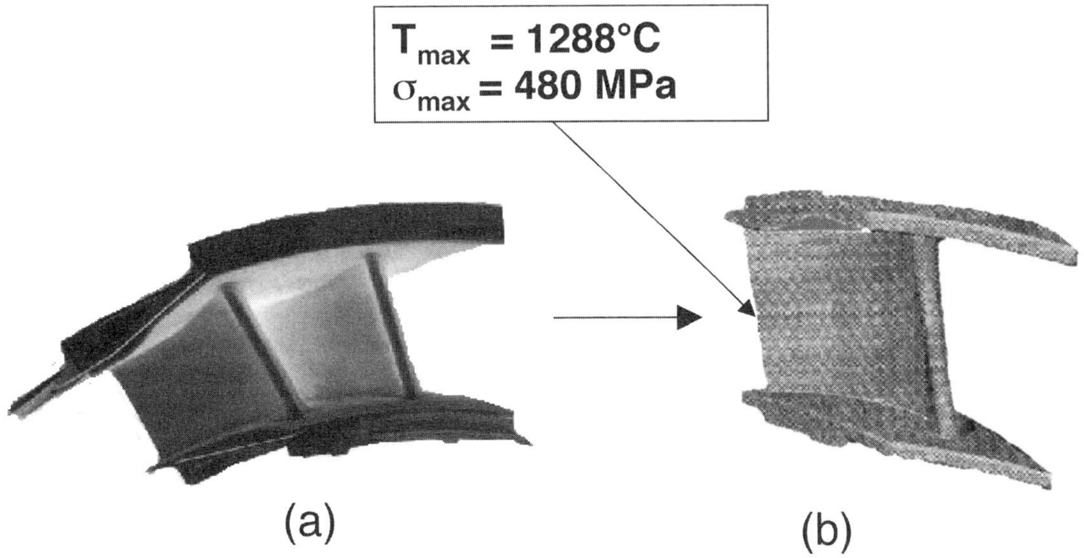

Figure 25.17 (a) FS-414 Metal Nozzle and (b) CSGT Straight Single Vane Nozzle. (Reprinted with permission from Solar Turbines Inc.)

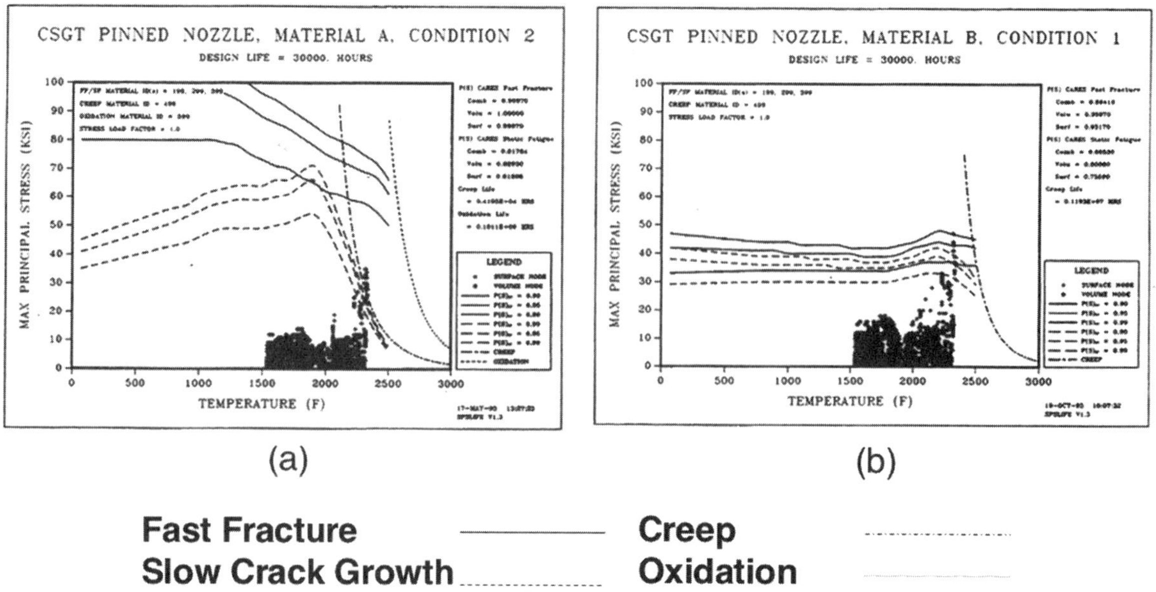

Figure 25.18 SPSLIFE Design Assessment Maps for (a) Si_3N_4 Nozzle and (b) SiC Nozzle. 1 ksi = 6.9 MPa, °C = {°F – 32}/1.8. (Reprinted with permission from refs. [95,96])

Lowering stress levels is one way to increasing the POS for a component. Therefore, an analysis was performed to determine the impact of lowering the maximum stress in the nozzle to 138 MPa. This was achieved by applying load factors of 0.57 and 0.43, to the stress spectra of the silicon nitride and silicon carbide nozzles, respectively. The impact on the component lives can be seen in the design assessment maps shown in Figures 25.19(a) and 25.19(b) for the silicon nitride and silicon carbide nozzles for a design life of 30,000 hours. For the silicon nitride component (Fig. 25.19(a) creep life was now assessed at 378,000 hours. Oxidation life increases to over one billion hours. Fast fracture POS approached 100%, but SCG life was still marginal since CARES/LIFE predicted 98.24% for each nozzle segment. Since there were 42 single vane nozzle segments, the POS for 30,000 hours of operation of the entire collection of stage one nozzles computes to only 47.44%, which poses an unacceptable risk to an engine in a commercial operation. For the silicon carbide component creep life increased further to 21,486,000 hours. Fast fracture POS approached 100% and SCG POS was predicted to be 99.98% by CARES/LIFE. The latter translates to an overall POS for the nozzle assembly of 99.16%, acceptable for the engine test duration planned for the CSGT program [96].

The life prediction study became a useful tool in the design process for the ceramic nozzle in the CSGT program. A number of candidate glass encapsulated HIP-sintered materials were eliminated as candidate materials for the fabrication of the ceramic nozzle, and a number of design strategies were examined for lowering component stresses. Alternative design concepts considered included airfoils detached from either the inner or the outer shroud, segmented/bolted airfoils (radial and axial), segmented shrouds, various pinned attachments and various clamped attachments, as well as a cooled nozzle. The design solution that was eventually adopted involved selecting materials with superior SCG and creep resistance, and a complete redesign of the nozzle airfoil. The airfoil chord was reduced by 50% and the airfoil was bowed, axially and tangentially, in order to reduce the steady-state stresses to the desired level. SN88 silicon nitride from NGK Insulators, Ltd. (NGK) was selected as the first generation nozzle material based on life prediction analysis, a POS of 99.34% for a set of 42 nozzles for 30,000 hours of operation, and creep and slow crack growth test data [10].

The data generated in the design process is shown in Figure 25.20. Shown are the outputs of a component finite element analysis, including steady state temperature (Figure 25.20(a)) and stress (Figure 25.20(b)) plots for the bowed nozzle design. Using these as inputs a CARES/LIFE life prediction plot was generated.

For a stress level of 179 MPa a POS of 99.34% for a set of 42 nozzles for 30,000 hours of operation was predicted for the SCG failure mode. Creep and oxidation data were also found to be acceptable. A summary of the CSGT nozzle design development can be found in Ref. [10].

LIFE PREDICTION OF CERAMIC MATRIX COMPOSITE COMPONENTS

The current state of life prediction of ceramic matrix composites (CMCs) is far less advanced than that of monolithic ceramics, and that is the reason why the present discussion involving CMC component life prediction is much less than that described in this chapter for the life prediction of monolithic components. The field of CMC life prediction is still very much a research field with, as yet, no consensus on the general lifing methodology that should be pursued. Research efforts are proceeding along a variety of paths including micro-mechanics, macro-level phenomenological, continuum damage, as well as empirical. At the same time, CMC components are being designed and fabricated for hot section applications, and are therefore in need of estimates of serviceable life. In the absence of a universally agreed upon lifing approach for CMCs, current practices tend to borrow from experience with metals and polymer matrix composites. This is the case with the three computer codes (CCLife, GENOA, NASALIFE) presently available in the United States.

CCLife is a computer code developed in the mid-1990s at the Virginia Polytechnic Institute (VPI) in an effort headed by Reifsnider. The endeavor was funded by NASA's Enabling Propulsion Materials program and is now in the public domain domestically [97] (a more recent version with technical support is available for an annual licensing fee [98]). CCLife grew out of VPI's experience with polymer matrix composites. The approach utilizes an evolutionary damage model that integrates cyclic and time-dependent phenomena and relates them back to residual strength. The public domain version of CCLife has been written as a post-processor to the research finite element package CSTEM [97] and runs on a Unix workstation. The more recent version is available on a PC platform with either an ANSYS or ABAQUS preference.

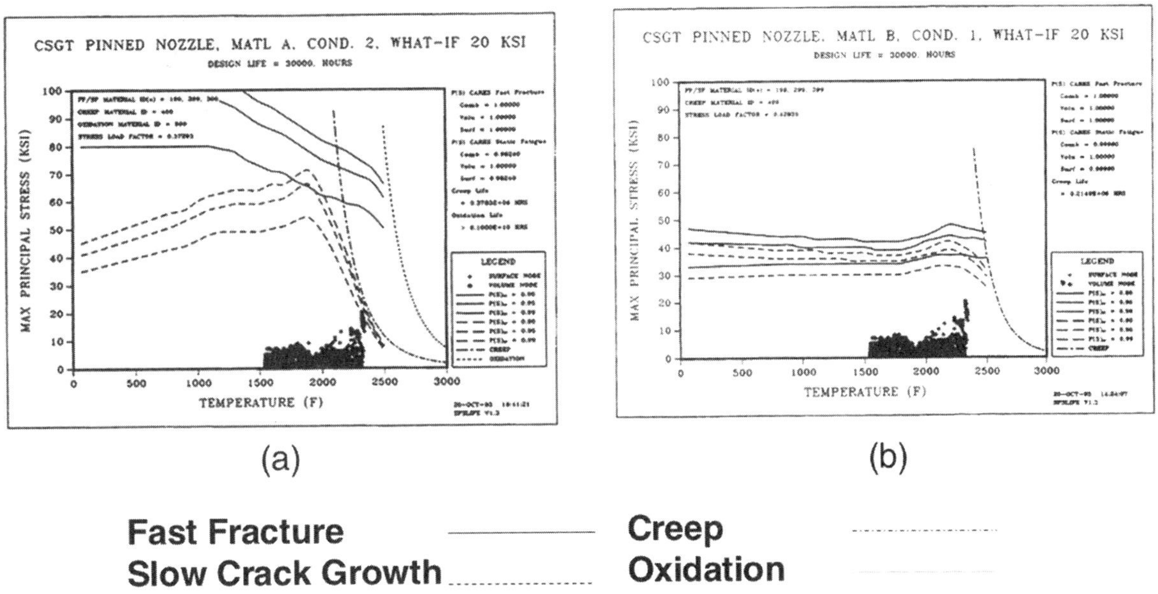

Figure 25.19 SPSLIFE Design Assessment Maps for (a) Si_3N_4 Nozzle (b) SiC Nozzle—Max. Stress Lowered to 138 MPa. 1 ksi = 6.9 MPa, °C = {°F − 32}/1.8. (Reprinted with permission from refs. [95,96].)

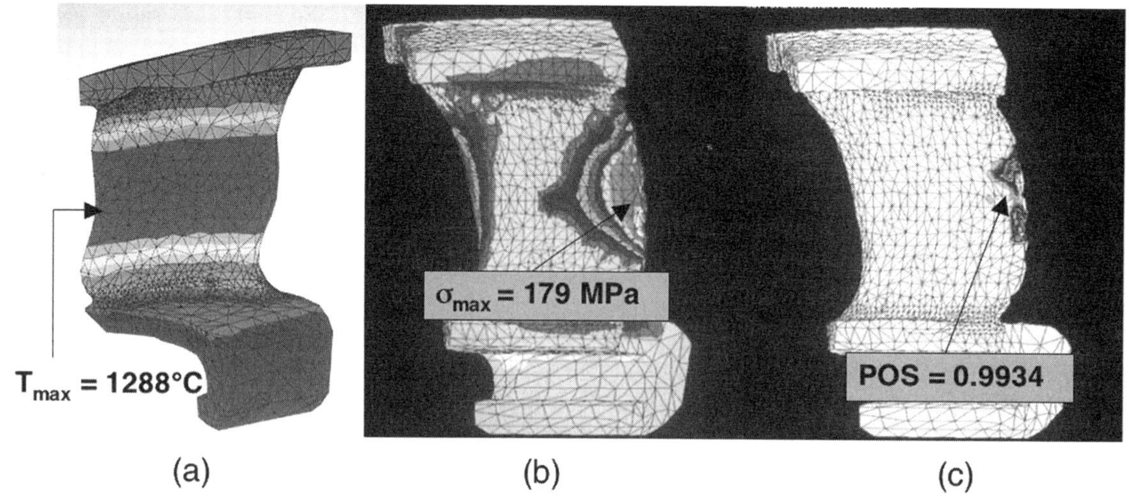

Figure 25.20 (a) Temperature Map, (b) Stress Map, (c) CARES/LIFE Map for Solar Turbines Inc. Bowed Si_3N_4 (SN88, NGK) Nozzle. POS for 42 Nozzle Assembly (Reprinted with permission from Solar Turbines Inc.)

Alpha Star Corp. commercially markets the computer code GENOA that is a complete finite element and progressive failure analysis package [99]. It has a Windows-based interface and runs on a PC platform. The software contains an adaptive mesher for automatic mesh refinement in the failure region, and allows for a graphical visualization of the failure process for a component. The failure criteria contained within the code are empirical; a curve fit to *S-N* data for fatigue, and a multi-factor exponential equation for stress rupture.

NASALIFE is another code that came into existence under NASA's Enabling Propulsion Materials program [97]. The software uses a damage rainflow approach. Required mission information (time, temperature, and six components of stress) is read by NASALIFE from an ASCII formatted input file. Creation of the input files generally requires manual manipulation of the finite element output files. Damage counting schemes identify the most damaging cycle based on the stress and temperature data. NASALIFE is a research level Fortran code, and as such can be used to post process results from any finite element package.

Additional proprietary lifing codes exist within industry at companies currently designing CMC components. Their technological state is similar to that just described. The lifing schemes are couched in terms of models the companies have engineering experience with, and that had been developed for, material systems other than CMCs. Many of these companies are also currently pursuing research in this field, either independently or in conjunction with various government agencies. While the next decade appears poised for great advances in this discipline, current practice relies heavily on past engineering experience with more traditional material systems.

CONCLUSION

Significant advancements have occurred over the last 30 years in life prediction of monolithic ceramic components in Europe, Asia, and North American, and the intent of this chapter was to describe that history and describe its current state-of-the-art. Understanding all the nuances of the mechanical performance of ceramic components is not trivial; however, many dedicated researchers and scientists have devoted their efforts toward that, and the tangible result is there are several sophisticated life prediction codes in existence today that can be used to dependably design structural ceramic components for service in harsh environments.

The future of probabilistic life prediction and ceramic component design has great potential. With successful application, the greater wear and erosion resistances and higher temperature capability of monolithic ceramic components can be realized and increased engine efficiency and life will result.

ACKNOWLEDGEMENTS

The authors would like to express their gratitude to the numerous contributors who have helped to prepare this chapter, including both those who have directly provided written contributions to this manuscript, as well as those whose efforts have generated the numerous references from which parts of this manuscript have been adapted. Additionally, the authors would like to recognize those whose efforts at the highlighted organizations have helped advance the ceramic life prediction efforts worldwide. Unless otherwise indicated the figures in this chapter were supplied by the authors.

NOMENCLATURE

A	surface area
a	crack length
ABAQUS	Commercial finite element analysis software
AIST	Agency of Industrial Science and Technology
ANSYS	Commercial finite element analysis software
ASTM	American Society for Testing and Materials
B	fatigue parameter
CARES	Ceramic Analysis and Reliability Evaluation of Structures; NASA life prediction code
CCLife	CMC life prediction code
CCPRO	life prediction code for nonlinear structural analysis
CGT	Japanese Ceramic Gas Turbine program
CERAM	Thermostructural life prediction code developed in France
CERAMIC	Honeywell code
CERITS	University of Karlsruhe life prediction code
CMCs	Ceramic Matrix Composites
CSGT	Ceramic Stationary Gas Turbine
CTAHE	Ceramic Technology for Advanced Heat Engines
D	fatigue constant
DAM	Damage assessment map
DARPA	Defense Advanced Research Projects Agency
DOE	US Department of Energy
EPA	Environmental Protection Agency (USA)
ERICA	Honeywell code
G	strain energy release rate
GENOA	CMC life prediction code
GFICES	TEPCO and IHI life prediction code
GRC	Glenn Research Center
H	Heaviside step function
I	Stress gradient factor
IHI	Ishikawajima-Harima Heavy Industries Co., Ltd
K	stress intensity factor
K_{IC}	Mode I critical stress intensity factor
k	crack density coefficient
L	likelihood function
LeRC	Lewis Research Center
m	Weibull modulus
MITI	Ministry of International Trade and Industry
MLE	maximum likelihood estimator
MPa	Mega Pascal
N	fatigue crack growth exponent
NASA	National Aeronautics and Space Administration
NASALIFE	CMC life prediction code
NDE	Non-destructive evaluation
NGK	NGK Insulators, Ltd.
NSA	Normal Stress Averaging
P	probability
PIA	Principal of Independent Action
POS	Probability of survival
$\Re$	reliability

SCARE	Structural Ceramics Analysis and Reliability Evaluation; original NASA life prediction code
SCG	Slow crack growth
SPS	Sundstrand Power Systems
SPSLIFE	Sundstrand Power Systems life prediction code
STAU	University of Karlsruhe life prediction code
STRATA	Rolls-Royce life prediction code
t	time
TEPCO	Tokyo Electric Power Company
TRIT	Turbine rotor inlet temperature
V	volume
VPI	Virginia Polytechnic University
WELFEP	WEakest Link Failure Probability Prediction by Finite Element Postprocessors
Y	crack geometry factor
α	Weibull modulus (scatter parameter)
β	Weibull scale parameter (central location parameter)
Δ	increment (Delta)
Σ	continuous random strength variable; summation operator
Π	product operator
Ψ	failure function per unit volume
σ	applied stress
σ_θ	Weibull characteristic strength
σ_o	Weibull scale parameter
Ω	solid angle in 3-dimensional principal stress space
∞	infinity

Subscripts:

A	area
e	effective
eq	equivalent
f	failure
i,k	observed values
max	maximum
C	critical
I	mode I
S	surface
V	volume
θ	Weibull characteristic strength

Overstrike Characters:

–	average
~	estimated

REFERENCES

[1] Quinn, G. D., "Fracture Mechanism Maps For Advanced Structural Ceramics: Part 1; Methodology and Hot-Pressed Silicon Nitride Results," Journal of Materials Science, 25, 4361–4376, 1990.

[2] Richerson, D. W. and Anson, D., "Evolution of Ceramic Gas Turbine Development Programs at Engine Manufacturers in the United States," Chapter 2 in *Ceramic Gas Turbine Design and Test Experience, Progress in Ceramic Gas Turbine Development, Volume 1*, M. van Roode, M. K. Ferber, and D. W. Richerson, eds., ASME Press, New York, NY, USA, 2002. pp. 11–16.

[3] Hartsock, D. L., "Ford's Development of the 820 High Temperature Ceramic Gas Turbine Engine," ibid., Chapter 3, pp. 17–75.

[4] Morrison, J., Lane, J., and Burke, M., "Progress in Ceramic Gas Turbine Development Westinghouse/Siemens Westinghouse," ibid., Chapter 8, pp. 193–223.

[5] Khandelwal, P. and Heitman, P, "Ceramic Gas Turbine Development at Rolls-Royce in Indianapolis," ibid., Chapter 5, pp. 111–132.

[6] Schenk, B., Easley, M. L., and Richerson, D. W., "Evolution of Ceramic Turbine Engine Technology at Honeywell Engines, Systems & Services," ibid., Chapter 4, pp. 77–109.

[7] Powers, L. M. and Janosik, L. A., "A Numerical Round Robin for the Reliability Prediction of Structural Ceramics," AIAA Paper 93-1498, in *34th AIAA/ASME/ASCE/AHS/ASC Structures, Structural Dynamics, and Materials Conference*, AIAA/ASME Adaptive Structures Forum, La Jolla, CA, USA, Apr. 19–22, 1993, Technical Papers, Pt. 3 (A93-33876 13-39), 1993. pp. 1647–1658.

[8] Dortmans, L., Thiemeier, T., and Brückner-Foit, A., "WELFEP : A Round Robin for Weakest-Link Finite Element Postprocessors," Journal of the European Ceramic Society, 11, 17–22, 1993.

[9] Bornemisza, T. and Jones, A., "Ceramic Gas Turbine Development at Hamilton Sundstrand Power Systems," Chapter 6 in *Ceramic Gas Turbine Design and Test Experience, Progress in Ceramic Gas Turbine Development, Volume 1*, M. van Roode, M. K. Ferber, and D. W. Richerson, eds., ASME Press, New York, NY, USA, 2002. pp. 133–154.

[10] Brentnall, W. D., van Roode, M., Norton, P. F., Gates, S., Price, J. R., Jimenez, O., and Miriyala, O., "Ceramic Gas Turbine Development at Solar Turbines Incorporated," ibid., Chapter 7, pp. 155–192.

[11] Itoh, T. and Sugawara, A., "Evolution of Ceramic Gas Turbine Development Programs at Engine Manufacturers in Japan," ibid., Chapter 12, pp. 277–281.

[12] van Roode, M., "Evolution of Ceramic Gas Turbine Development Programs at Engine Manufacturers in Western Europe," ibid., Chapter 20, pp. 445–452.

[13] Wereszczak, A. A., Kirkland, T. P., Breder, K., Lin, H.-T., and Andrews, M. J., "Biaxial Strength, Strength-Size-Scaling, and Fatigue Resistance of Alumina and Aluminum Nitride Substrates," International Journal of Microcircuits and Electronic Packaging, 22, 446–458, 1999. © 1999 by IMAPS—International Micoelectronics And Packaging Society.

[14] Weibull, W.A., "The Phenomenon of Rupture in Solids," Ingeniors Vetenskaps Akademien Handlinger, No. 153, 1939.

[15] Weibull, W. A., "A Statistical Distribution Function of Wide Applicability," Journal of Applied Mechanics, 18[3], 293–297, 1951.

[16] "Practice for Reporting Uniaxial Strength Data and Estimating Weibull Distribution Parameters for Advanced Ceramics," *ASTM C1239, Vol. 15.01, Annual Book of ASTM Standards*, American Society for Testing and Materials, West Conshohocken, PA, USA, 2002.

[17] "Standard Practice for Fractography and Characterization of Fracture Origins in Advanced Ceramics," *ASTM C1322, Vol. 15.01, Annual Book of ASTM Standards*, American Society for Testing and Materials, West Conshohocken, PA, USA, 2002.

[18] Duffy, S. F. and Arnold, S. M., "Noninteractive Macroscopic Statistical Failure Theory for Whisker Reinforced Ceramic Composites," Journal of Composite Materials, 24[3], 293–308, 1990.

[19] Duffy, S. F. and Manderscheid, J. M., "Noninteractive Macroscopic Reliability Model for Ceramic Matrix Composites with Orthotropic Material Symmetry," Transactions of the ASME, Journal of Engineering for Gas Turbines and Power, 112[4], 507–511, 1990.

[20] Barnett, R. L., Connors, C. L., Hermann, P. C., and Wingfield, J. R., *Fracture of Brittle Materials Under Transient Mechanical and Thermal Loading*, U.S. Air Force Flight Dynamics Laboratory, AFFDL-TR-66-220, March, 1967.

[21] Freudenthal, A. M., "Statistical Approach to Brittle Fracture. Fracture, An Advanced Treatise," in *Mathematical Fundamentals, Vol. 2,* H. Liebowitz, ed., Academic Press, New York, NY, USA, 1968. pp. 591–619.

[22] Batdorf, S. B. and Crose, J. G., "A Statistical Theory for the Fracture of Brittle Structures Subjected to Nonuniform Polyaxial Stresses," Journal of Applied Mechanics, 41[2], 459–464, 1974.

[23] Batdorf, S. B., "Weibull Statistics for Polyaxial Stress States," Journal of the American Ceramic Society, 57[1], 44–45, 1974.

[24] Batdorf, S. B., "Some Approximate Treatments of Fracture Statistics for Polyaxial Tension," International Journal of Fracture, 13, 5–11, 1977.

[25] Batdorf, S. B., "Fundamentals of the Statistical Theory of Fracture," in *Fracture Mechanics of Ceramics, Vol. 3,* Bradt, R. C., Hasselman, D. P. H. and Lange, F. F., eds., Plenum Press, New York, NY, USA, 1978, pp. 1–30.

[26] Batdorf, S. B. and Heinisch, H. L., "Weakest Link Theory Reformulated for Arbitrary Fracture Criterion," Journal of the American Ceramic Society, 61[7–8], 355–358, 1978.

[27] Lamon, J. and Evans, A. G., "Statistical Analysis of Bending Strengths for Brittle Solids : a Multiaxial Fracture Problem," Journal of the American Ceramic Society, 66[3], 177–182, 1983.

[28] Lamon, J., "Statistical Approaches to Failure for Ceramic Reliability Assessment," Journal of the American Ceramic Society, 71[2], 106–112, 1988.

[29] Lamon, J., "Probabilistic Failure Predictions in Brittle Materials Under Multiaxial Loading," in *Probabilities and Materials: Tests, Models and Applications,* D. Breysse, ed., NATO Asi Series E: Applied Sciences, Vol. 269, Kluwer Academic Publishers, The Netherlands, 1994. pp. 415–426.

[30] Johnson, C. A. and Tucker, W. T., "Derivation of Statistical Models," in *Life Prediction Methodology for Ceramic Components of Advanced Heat Engines—Phase I Final Report*, Volume I, Contract No. 86X-SC674C, 1994. pp. 5–60.

[31] Nemeth, N. N., Manderscheid, J. M., and Gyekenyesi, J. P., *Ceramics Analysis and Reliability Evaluation of Structures (CARES) Users and Programmers Manual*, NASA TP-2916, 1990.

[32] Tucker, W. T. and Johnson, C. A., "The Multiaxial Equivalent of Stressed Volume," in *Life Prediction Methodologies and Data for Ceramic Materials, ASTM STP 1201*, C. R. Brinkman and S. F. Duffy, eds., American Society for Testing and Materials, Philadelphia, PA, USA, 1994. pp. 265–279.

[33] Lamon, J., "Statistical Analysis of Fracture of Silicon Nitride (RBSN) Using the Short Span Bending Technique," ASME paper 85-GT-151, presented at the International Gas Turbine and Aeroengine Congress and Exposition, Houston, TX, USA, June 5–8, 1995.

[34] "Test Method for Flexural Strength of Advanced Ceramics at Ambient Temperature," *ASTM C1161, Vol. 15.01, Annual Book of ASTM Standards*, American Society for Testing and Materials, West Conshohocken, PA, USA, 2002.

[35] "Test Method for Tensile Strength of Monolithic Advanced Ceramics at Ambient Temperatures," *ASTM C1273, Vol. 15.01, Annual Book of ASTM Standards*, American Society for Testing and Materials, West Conshohocken, PA, USA, 2002.

[36] "Test Method for Monotonic Equibiaxial Flexural Strength of Advanced Ceramics at Ambient Temperature," *ASTM C1499, Vol. 15.01, Annual Book of ASTM Standards*, American Society for Testing and Materials, West Conshohocken, PA, USA, 2002.

[37] Johnson, C. A. and Tucker, W. T., "Weibull Estimators for Pooled Fracture Data," in *Life Prediction Methodologies and Data for Ceramic Materials, ASTM STP 1201*, C. R. Brinkman and S. F. Duffy, eds., American Society for Testing and Materials, Philadelphia, PA, USA, 1994. pp. 250–264.

[38] Wereszczak, A. A., Breder, K., Andrews, M. J., Kirkland, T. P., Ferber, M. K., "Strength Distribution Changes in a Silicon Nitride as a Function of Stressing Rate and Temperature," ASME paper 98-GT-527, presented at the International Gas Turbine and Aeroengine Congress and Exhibition, Stockholm, Sweden, June 2–5, 1998.

[39] Chati, M., Johnson, C., Kaya, A., and Schenk, B., "Confidence Intervals for Weibull Fast-Fracture Parameters," ASME paper 2002-GT-30501, presented at the ASME TUBO EXPO Land, Sea & Air, Amsterdam, The Netherlands, June 3–6, 2002.

[40] Mencik, J., "Rationalized Load and Lifetime of Brittle Materials," Communications of The American Ceramic Society, 67, 110–116, 1984.

[41] "Test Method for Determination of Slow Crack Growth Parameters of Advanced Ceramics by Constant Stress-Rate Flexural Testing at Ambient Temperature," *ASTM C1368, Vol. 15.01, Annual Book of ASTM Standards*, American Society for Testing and Materials, West Conshohocken, PA, USA, 2002.

[42] Schenk, B., Brehm, P. J., Menom, M. N., Tucker, W. T. and Peralta, A. D., "A New Probabilistic Approach for Accurate Fatigue Data Analysis of Ceramic Materials," Transactions of the ASME, Journal of Engineering for Gas Turbines and Power, 122[4], 637–645, 2000.

[43] Suzuki, A. and Hamanaka, J., "Design Guide for Fine Ceramics Components," IHI Engineering Review, 26[4], 133–137, 1993.

[44] Lamon, J., Pherson, D., and Dotta, P., *2D and 3D Ceramic Reliability Analysis Using CERAM, Statistical Post Precessor Software Technical Documentation*, Ceramic Design Final Report, Battelle-Geneva, May 1989.

[45] Heger, A., Brückner-Foit, A., and Munz, D., "STAU—A Computer Code to Calculate the Failure Probability of Multiaxially Loaded Ceramic Components," 2nd European Ceramic Society Conference, Euro-Ceramics, 1991.

[46] Heger, A., Brückner-Foit, A., and Munz, D., *STAU—ein Programm zur Berechnung der Ausfallwahrscheinlichkeit mehrachsig beanspruchter keramischer Komponenten als Postprozessor für Finite-Elemente-Programme (STAU—A Finite Element Analysis Postprocessor to Calculate the Failure Probability of Multiaxially Loaded Ceramic Components)*, Internal Report, Institute for Reliability and Failure Analysis, University of Karlsruhe, Germany, 1991 (in German).

[47] Brückner-Foit, A., Heger, A., and Munz, D., "Effect of Proof Testing on the Failure Probability of Multiaxially Loaded Ceramic Components," *Life Prediction Methodologies and Data for Ceramic Materials, ASTM STP 1201*, C.R. Brinkman and S.F. Duffy, eds., American Society for Testing and Materials, Philadelphia, PA, USA, 1994.

[48] Stürmer, G., Schulz, A., and Wittig, S., "Lifetime Prediction for Ceramic Gas Turbine Components," ASME paper 91-GT-96, presented at the 36th International Gas Turbine and Aeroengine Congress and Exposition, Orlando, FL, USA, 1991.

[49] Schenk, B., *Einsatz experimenteller und numerischer Methoden zur Untersuchung der Lebensdauer von keramischen Rotorbauteilen für Hochtemperatur-Kleingasturbinen (Experimental and Numerical Methods for Determining the Lifetime of Ceramic Rotor Components of Small High Temperatuer Gas Turbines)*, Wissenschaft & Technik Verlag, Berlin, Germany, 1996 (in German).

[50] Hempel, H., and Wiest, H., "Structural Analysis and Life Prediction for Ceramic Gas Turbine Components for the Mercedes-Benz Research Car 2000," ASME paper 86-GT-199, presented at the International Gas Turbine Conference and Exhibit, Düsseldorf, West Germany, June 8–12, 1986.

[51] Smart, J., "The Determination of Failure Probability Using Weibull Probability Statistics and the Finite Element Method," Res. Mechanica, 31, 205–219, 1990.

[52] Dortmans, L. and de With, G., "Weakest-Link Failure Prediction for Ceramics IV: Application of Mixed Mode Fracture Criteria for Multiaxial Loading," Journal of the European Ceramic Society, 10, 109–114, 1992.

[53] Kussmaul, K., Lauf, S., and Turan, K., "FEM-LIFTAP, A Finite Element Lifetime Analysis Postprocessor for Ceramic Components," 4th International Symposium on Ceramic Materials & Components for Engines, Göteborg, Sweden, 1991.

[54] Krüger, S., Jakel, R., Rubio, D., and Barth, H.-J., *Berechnung keramischer Bauteile mit dem neuen statistischen Software-Prozessor 'KerB' (Reliability Analysis of Ceramic Components Using the New Statistical Postprocessor 'KerB')*, Konstruktion, 51[3], Springer Verlag, Düsseldorf, Germany, 1999 (in German).

[55] Sander, D., *Bruchstatistischer FEM-Postprozessor zur Auslegung von keramischen Bauteilen (Fracture Statistics Based FEA-Postprocessor for Designing Ceramic Components)*, Ph.D. Dissertation, RWTH Aachen, Germany, 1996 (in German).

[56] Lüssgen, C., and Maier, R.H., "FEM-gestützte bruchstatistische Auslegung von Welle-Nabe-Pressverbindungen mit keramischer Hohlwelle (FEA and Fracture Statistics Based Design of Shaft-Hub Interference Fit Joints Using Hollow Ceramic Shafts), *Welle-Nabe-Verbindungen, VDI-Berichte 1384*, VDI Verlag GmbH, Düsseldorf, Germany, 1998 (in German).

[57] Lamon, J. and Melet, *J. CERAM Statistical Post Processor Software: Experimental Evaluation of Reliability Analysis by CERAM 2D and CERAM 3D, Ceramics Design Final Report*, Battelle-Geneva, May 1989.

[58] Lamon, J., "Ceramics Reliability: Statistical Analysis of Multiaxial Failure Using the Weibull Approach and the Multiaxial Elemental Strength Model," Journal of the American Ceramic Society, 73, 2204–2212, 1990.

[59] Lamon, J., "Ceramics Reliability: Statistical Analysis of Multiaxial Failure Using the Weibull Approach and the Multiaxial Elemental Strength Model," ASME paper 88-GT-147, presented at the Gas Turbine and Aeroengine Congress, Amsterdam, The Netherlands, June 6–9, 1988.

[60] Lamon, J., "Structural Reliability of Ceramics," in *Proceedings of the 11th Riso International Symposium on Metallurgy and Materials Science: Structural Ceramics-Processing, Microstructure and Properties*, J. J. Bentzen et al., eds., Riso National Laboratory, Roskilde, Denmark, September 3–7, 1990. pp. 39–56.

[61] Lamon, J., "Reliability Analysis of Ceramics Using the CERAM Computer Program," ASME paper 90-GT-98, presented at the Gas Turbine and Aeroengine Congress and Exposition, Brussels, Belgium, June 11–14, 1990.

[62] Lamon, J. and Pherson, D., "Thermal Stress Failure of Ceramics Under Repeated Rapid Heatings," Journal of the American Ceramic Society, 74, 1188–1196, 1991.

[63] Amar, E., Gauthier, F., and Lamon, J., "Reliability Analysis of a Si_3N_4 Ceramic Piston Pin for Automotive Engines," in *Proceedings of the 3rd International Symposium Ceramic Materials and Components for Engines*, Las Vegas, USA, V. J. Tennery, ed., The American Ceramic Society Inc., 1988. pp. 1334–1336.

[64] Lamon, J., "Probabilistic Failure Predictions in Ceramics and Ceramic Matrix Fiber Reinforced Composites—Methodologies and Data for Ceramic Materials," in *Life Prediction Methodologies and Data for Ceramic Materials, ASTM STP 1201*, C.R. Brinkman and S.F. Duffy, eds., American Society for Testing and Materials, Philadelphia, PA, USA, 1994. pp. 265–279.

[65] Brückner-Foit, A. and Ziegler, C., "Lifetime Prediction for Ceramic Components Subjected to Time-Dependent Loading," Ceram. Eng. and Sci. Proc., 19[4], 99–106, 1998.

[66] Brückner-Foit, A., Hülsmeier, P., Sckuhr, M., and Riesch-Oppermann, H., "Limitations of the Weibull Theory in Stress Fields with Pronounced Stress Gradients," ASME Paper 2000-GT-663, presented at the ASME TURBO EXPO, Land, Sea & Air, Munich, Germany, May 8–11, 2000.

[67] Brückner-Foit, A., Hülsmeier, P., Diegele, E., Rettig, U., and Hohmann, C., "Simulating the Failure Behaviour of Ceramic Components Under Gas Turbine Conditions," ASME paper 2002-GT-30502, presented at the ASME TUBO EXPO, Land, Sea & Air, Amsterdam, The Netherlands, 2002.

[68] Dukes, W. H., *Handbook of Brittle Material Design Technology*, AGARDograph 152, AGARD, Paris, France, 1971.

[69] Vardar, Ö. and Finnie, I., "An Analysis of the Brazilian Disk Fracture Test Using the Weibull Probabilistic Treatment of Brittle Strength," International Journal of Fracture, 11, 495–508, 1975.

[70] Evans, A. G. and Wiederhorn, S. M., "Crack Propagation and Failure Prediction in Silicon Nitride at Elevated Temperature," Journal of Materials Science, 9, 270–278, 1974.

[71] Paluszny, A. and Nicholls, P. F., "Predicting Time-Dependent Reliability of Ceramic Rotors," in *Ceramics for High Performance Applications II*, J. J. Burke, E. N. Lenoe, R.N. Katz, eds., Brook Hill Publishing Company, Chestnut Hill, MA, USA, 1978. pp. 95–112.

[72] Paluszny, A., and Wu, W., "Probabilistic Aspects of Designing with Ceramics," Transactions of the ASME, Journal of Engineering for Power, 99[4], 617–630, 1977.

[73] Baker, R. R., Swank, L. R., and Caverly, J. C., *Ceramic Life Prediction Methodology—Hot Spin Disc Life Program*, Army Materials and Mechanics Research Center TR 83-44, Interim Report. Contract No. DAAG 46-77-C-0028, Watertown, MA, USA, 1983.

[74] Caverly, J. C. and Hartsock, D. L., *Evaluation of Ceramics for Stator Applications—Gas Turbine Engines—Final Report—Reliability Analysis and Material Property Characterization*, National Aeronautics and Space Administration Report NASA-CR-165270, Cleveland, OH, USA, 1980.

[75] Powers, L. M. and Sokolowski, T. M., *ANSCARES 2.0 Users Guide*, NASA-GRC, March 21, 1997.

[76] Gyekenyesi, J. P., "SCARE: A Postprocessor Program to MSC/NASTRAN for Reliability Analysis of Structural Ceramic Components," Transactions of the ASME, Journal of Engineering for Gas Turbines and Power, 108[3], 540–546, 1986.

[77] Gyekenyesi, J. P. and Nemeth, N. N., "Surface Flaw Reliability Analysis of Ceramic Components With the SCARE Finite Element Postprocessor Program," Transactions of the ASME, Journal of Engineering for Gas Turbines and Power, 109[3], 274–281, 1987.

[78] Nemeth, N. N., Powers, L. M., Janosik, L. A., and Gyekenyesi, J. P., *Ceramics Analysis and Reliability Evaluation of Structures Life Prediction Program (CARES/LIFE), Users and Programmers Manual*, 1993.

[79] Powers, L. M., Janosik, L. A., Nemeth, N. N., and Gyekenyesi, J. P., "Lifetime Reliability Evaluation of Monolithic Ceramic Components Using the CARES/LIFE Integrated Design Program," paper presented at the 95th Annual American Ceramic Society Meeting and Exposition, Cincinnati, Ohio, April 19–22, 1993.

[80] Richerson, D.W. and Johansen, K.M., *Ceramic Gas Turbine Engine Demonstration Program," Final Report*, DARPA/Navy Contract N00024-76-C-5352, May 1982, Garrett Report 21-4410.

[81] *Proceedings of the 1977 DARPA/NAVSEA Ceramic Gas Turbine Demonstration Engine Program Review*, Fairbanks, J.W. and Rice, R.W., eds., Battelle Columbus Laboratories, Columbus, OH, USA, Report No. MCIC-78-36, March 1978.

[82] Egli, H., "Development of Ceramic Turbine Wheels for Turbochargers with Emphasis on Burst Speed Prediction Methodology," Aufladetechnische Konferenz, Aachen, Germany, 1984.

[83] Schenk, B., Brehm, P. G., Menon, M. N., Tucker, W. T., and Peralta, A. D., "Status of the CERAMIC/ERICA Probabilistic Life Prediction Codes Development for Structural Ceramic Applications," ASME paper 99-GT-318, presented at the International Gas Turbine and Aeroengine Congress and Exposition, Indianapolis, IN, USA, June 7–10, 1999.

[84] Cuccio, J. C., Brehm, P., Fang, H. T., Hartman, J., Meade, W., Menon, M. N., Peralta, A., Song, J. Z., Strangman, T., Wade, J., Wimmer, J., and Wu, D. C., *Life Prediction Methodology for Ceramic Components of Advanced Heat Engines, Phase 1*, AlliedSignal Engines, Report # ORNL/Sub/89-SC674/1/V1-2, March 1995.

[85] Wereszczak, A. A., Kirkland, T. P., Ferber, M. K., Corum, J. M., and Peralta, A., "Failure Probability Prediction of Fast-Fractured NT451 SiAlON Diesel Exhaust Valves," ASME paper 97-GT-318, pre-

sented at the International Gas Turbine and Aeroengine Congress and Exhibition, Orlando, FL, USA, June 2–5, 1997.

[86] Wereszczak, A. A., Kirkland, T. P., Ferber, M. K., and Andrews, M. J., "Prediction of SiAlON Diesel Valve Failure Probability Using AlliedSignal's CERAMIC and ERICA Life Prediction Codes," Ceram. Eng. and Sci. Proc., 18[3], 475–484, 1997.

[87] Wereszczak, A. A., Breder, K., Ferber, M. K., Bridge, R. J., Riester, L., and Kirkland, T. P., "Failure Probability Prediction of Dielectric Ceramics in Multilayer Capacitors," Multilayer Electronic Ceramic Devices, Ceramic Transactions, 97, 73–83, 1999.

[88] Khandelwal, P. K., Provenzano, N. J., and Schneider, W. E., "Life Prediction Methodology for Ceramic Components of Advanced Vehicular Heat Engines," in *Proceedings of the Annual Automotive Technology Development Contractors' Coordination Meeting*, 1993. pp. 199–211.

[89] Khandelwal, P. K., Provenzano, N. J., and Schneider, W. E., *Life Prediction Methodology for Ceramic Components of Advanced Vehicular Heat Engines-Final Report*, DOE Report# ORNL/SUB-88-SA920/1/V1-V3, Oak Ridge National Laboratory, Oak Ridge, TN, USA, 1996.

[90] Wiederhorn, S. M., Quinn, G. D., and Krause, R., "Fracture Mechanism Maps: Their Applicability to Silicon Nitride," in *Life Prediction Methodologies and Data for Ceramic Materials, ASTM STP 1201*, C.R. Brinkman and S.F. Duffy, eds., American Society for Testing and Materials, Philadelphia, PA, USA, 1994. pp. 36–61.

[91] DeSalvo, G. J., *Theory and Structural Design Application of Weibull Statistics*, WANL-TME-2688, Westinghouse Astronuclear Lab, 1970.

[92] Soma, T., Ishida, Y., Matsui, M., and Oda, I. "Ceramic Component Design for Assuring Long-Term Durability," Advanced Ceramic Materials, 2, 809–812, 1987.

[93] Bornemisza. T., and Saith, A., "SPSLIFE: A User Friendly Approach to the Structural Design and Life Assessment of Ceramic Components," Transactions of the ASME, Journal of Engineering for Gas Turbines and Power, 118[1], 179–183, 1996.

[94] Wu, D.C., Peralta, A.D., Menon, M.N., and Cuccio, J.C., "Subcritical Crack Growth Life Prediction For Ceramic Components of Advanced Heat Engines," ASME paper 95-GT-236, presented at the International Gas Turbine and Aeroengine Congress and Exposition, Houston, TX, USA, June 5–8,1995.

[95] Saith, A., *SPSLIFE: Application to Preliminary Design Evaluation and Life Assessment of CSGT Components*, Ceramic Stationary Gas Turbine Development, Topical Report Phase I, DOE Contract No. DE-AC02-92CE40960, Sep. 30, 1994.

[96] Saith, A., Norton, P. F., and Parthasarathy, V. M., "Application of SPSLIFE to Preliminary Design Evaluation and Life Assessment of CSGT Components," Transactions of the ASME, Journal of Engineering for Gas Turbines and Power, 117[3], 424–431, 1996.

[97] Contact information: Diane Chapman, Intellectual Property Officer at NASA Glenn Research Center, 21000 Brookpark Rd., Cleveland, OH 44135, USA.

[98] Reifsnider, K. L. and Case, S. W., *Damage Tolerance & Durability in Material Systems*, John Wiley & Sons, 2002.

[99] *GENOA User's Manual*, Alpha Star Corporation, Long Beach, CA, 90804, 2001.

CHAPTER

26

CORROSION ISSUES FOR CERAMICS IN GAS TURBINES

Nathan S. Jacobson, Dennis S. Fox, and James L. Smialek
NASA Glenn Research Center, Cleveland, Ohio, USA

Elizabeth J. Opila
Cleveland State University/NASA Glenn Research Center, Cleveland, Ohio, USA

Peter F. Tortorelli and Karren L. More
Oak Ridge National Laboratory, Oak Ridge, Tennessee, USA

Klaus G. Nickel
Eberhard-Karls-Universität Tübingen, Tübingen, Germany

Takehiko Hirata
Mitsubishi Heavy Industries, Ltd., Yokohama-shi, Kanagawa-ken, Japan

Makoto Yoshida
Kyocera Corporation, Kagoshima-shi, Kagoshima-ken, Japan

Isao Yuri
Central Research Institute of Electric Power Industry, Yokosuka-shi, Kanagawa-ken, Japan

INTRODUCTION

Historical and Programmatic Perspective

The requirements for hot-gas-path materials in gas turbine engines are demanding. These materials must maintain high strength and creep resistance in a particularly aggressive environment. A typical gas turbine environment involves high temperatures, rapid gas flow rates, high pressures, and a complex mixture of aggressive gases. Figure 26.1 illustrates the requirements for components of an aircraft engine and critical issues [1].

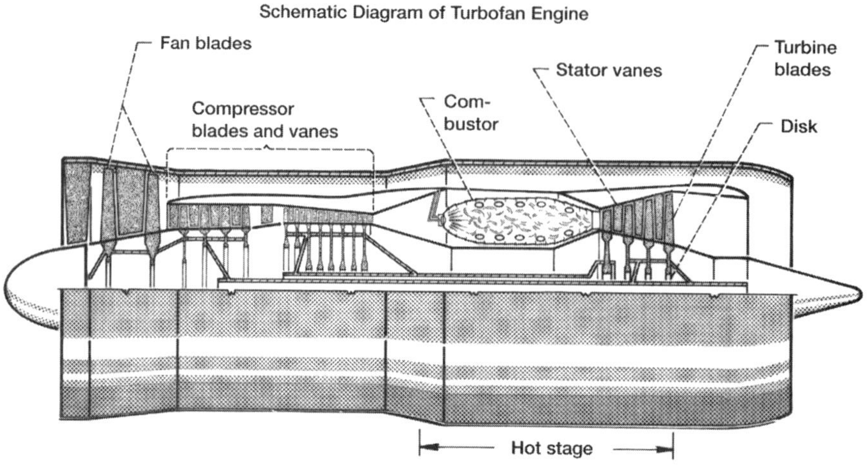

Component	Typical Operating Conditions: Temperature (°C)	Stress (MPa)	Life (hr)	Critical Problems
Blades	900 - 1050	140 - 210	5000	Creep strength, stability, oxidation, hot corrosion, thermal fatigue
Vanes	950 - 1100	35 - 70	5000	Thermal fatigue, oxidation, hot corrosion
Disks	400 - 650	420 - 1050	15 000	Low cycle fatigue
Combustors	850 - 1100	20 - 35	4000	Thermal fatigue, oxidation

Figure 26.1 Common Degradation Modes for High Temperature Turbine Components in An Aircraft Turbine Engine (Adapted from ref. [1] with permission of the American Ceramic Society, www.ceramics.org, © 1993, all rights reserved)

Currently, heat engines are constructed of metal alloys, which meet these requirements within strict temperature limits. In order to extend these temperature limits, ceramic materials have been considered as potential engine materials, due to their high melting points and stability at high temperatures. These materials include oxides, carbides, borides, and nitrides. Interest in using these materials in engines appears to have begun in the 1940s with BeO-based porcelains [2]. During the 1950s, the efforts shifted to cermets. These were carbide-based materials intended to exploit the best properties of metals and ceramics. During the 1960s and 1970s, the silicon-based ceramics silicon carbide (SiC) and silicon nitride (Si_3N_4) were extensively developed.

Although the desirable high-temperature properties of SiC and Si_3N_4 had long been known, consolidation of powders into component-sized bodies required the development of a series of specialized processing routes [3]. For SiC, the major consolidation routes are reaction-bonding, hot-pressing, and sintering. The use of boron and carbon as additives which enable sintering was a particularly noteworthy advance [4]. For Si_3N_4 the major consolidation routes are reaction-bonding and hot-pressing [5]. Reaction-bonding involves nitridation of silicon powder. Hot-pressing involves addition of various refractory oxides, such as magnesia (MgO), alumina (Al_2O_3), and yttria (Y_2O_3). Variations on these processes include a number of routes including Hot Isostatic Pressing (HIP), gas-pressure sintering, sinter-HIPing, and Encapsulation-HIPing. It is important to note that each process involves the addition of secondary elements, which later were shown to dramatically influence oxidation and corrosion behavior.

As dense bodies of silicon-based ceramics became more readily available, their desirable high temperature properties were confirmed. These materials retained strength to very high temperatures (i.e. 1300–1400°C).

Further, they were lightweight and made from abundant materials. SiC and Si_3N_4 therefore emerged as leading ceramic candidates for components in heat engines, designed to operate at higher temperatures for better performance and fuel efficiency. The first US programs for ceramics in heat engines have been reviewed [6]. Selected programs on ceramic engine parts are summarized here in regard to their contributions to understanding the corrosion behavior of ceramics in a heat engine environment.

The first major United States program was funded by DARPA (Defense Advanced Research Projects Agency) and involved the U.S. Army, Westinghouse Research Laboratories (Pittsburgh, PA), and Ford Motor Company (Dearborn, MI) in the early 1970s. The Westinghouse element involved power generation turbine vanes and the Ford element involved automotive engines. Several significant studies on the oxidation of additive-containing SiC and Si_3N_4 were done at Westinghouse [7,8]. These studies showed the additives controlled oxidation, as will be discussed in more detail later in this chapter.

During the 1970s, DARPA and NAVSEA (Naval Sea Systems Command) also funded a Ceramic Gas Turbine Engine demonstration program [9], which was conducted primarily at Garrett Turbine Engines (Phoenix, AZ). An AiResearch Turboprop Engine was selected for combustor and turbine redesign with ceramics. Numerous processing and property measurement studies were a part of this program. Oxidation and salt-induced corrosion tests were done on hot-pressed and reaction-bonded Si_3N_4. For hot-pressed Si_3N_4 in pure oxygen at temperatures below ~1100°C, strength actually increased due to surface flaw healing. However, higher temperature oxidation and salt-induced corrosion led to surface pitting and strength degradation [10].

Also in the 1970s, as part of the Department of Energy (DOE) Ceramic Technology Readiness (CTR) program, a corrosion/erosion study of various commercially available ceramic materials was conducted [11]. Various types of SiC and Si_3N_4, as well as AlN, Al_2O_3-ZrO_2, Al_2O_3-SiC, $MgAl_2O_4$ and SiAlON were examined after exposure to laboratory air environments and also coal-derived combustion environments. Damage was assessed from strength degradation and microscopic examination. Combustion environments led to more severe corrosion than laboratory air environments. This was attributed to water vapor and thermal cycling in the combustion environments. Unprotected AlN experienced particularly severe degradation and was determined to be unsatisfactory for combustion applications.

There was also an interest in the development of oxide ceramics for heat exchangers in automotive turbines. Lithium aluminum silicate (LAS) was selected due its low thermal expansion. A substantial testing program at Ford dating back to 1965 on LAS regenerators identified thermal stress and chemical attack as the two major modes of failure [12]. On the cold side of the regenerator, sulfuric acid forms and attacks the ceramic; on the hot side, sodium from fuel impurities attacks the ceramic. Other low thermal expansion oxides, such as magnesium aluminosilicate (MAS) and aluminosilicate (AS), were subsequently developed to minimize these problems.

Several programs were initiated that specifically focused on ceramics for automotive applications. The first was the Ceramic Automotive Turbine Engine (CATE) program in the late 1970s and early 1980s, which was funded by the National Aeronautics and Space Administration (NASA) and DOE and performed primarily at Detroit Diesel Allison. The emphasis here was on sintered SiC and Si_3N_4. Later, DOE and NASA funded the Automotive Gas Turbine (AGT) program [13]. This program involved development of both a MAS ceramic recuperator and SiC turbine blades. The AGT program later became the Advanced Turbine Technology Applications Program (ATTAP), which was aimed more at component development [14].

Related to these programs were 3500 hr durability tests at Garrett, funded by NASA. These involved cyclic oxidation tests on a variety of commercial SiC and Si_3N_4 materials [15,16]. Tests were conducted in a diesel fuel burner with exposures at 1204°C and 1371°C and cooling to below 204°C. Five cycles per hour were run with durations to 3500 hr. The materials tested showed good strength retention. As part of the AGT and ATTAP programs, salt corrosion tests were run at NASA Lewis Research Center (Cleveland, OH, USA). Particular care was taken to understand the deposition parameters. It was found that basic molten salt deposits led to severe surface pitting and strength degradation [17,18].

Concurrent with the engine development programs were programs at DOE/Oak Ridge National Laboratory (ORNL) in the 1980s and 1990s for ceramic heat exchangers in various types of elevated temperature industrial applications. These combustion environments are low-velocity as compared to heat engines. These studies again showed that pure oxidation was not a significant issue for ceramics, but deposit-induced corrosion and water vapor were quite important. Further a basic oxide deposit was particularly detrimental [19].

A significant international program from 1979–1993 was "Program of Research and Development on High Temperature Materials for Automotive Engines" of the International Energy Agency (IEA), which involved Germany, U.S., Sweden and Japan. The programs of the European Communities had ceramics for turbines as one focus within the larger program "Basic Research in Industrial Technologies for Europe/European Research on Advanced Materials" (Brite/EuRam) in the 1990s. The most significant national German program was funded from 1985–1994 [20]. It led to basic insight into the complex scale structures on additives containing Si_3N_4 and early attempts to model corrosion kinetics [21].

In Japan there have been a number of studies for development of ceramics in gas turbines, including corrosion studies. A 300 kW class Ceramic Gas Turbine (CGT) program (1988–1999), which was administered by the New Energy and Industrial Technology Development Organization (NEDO) under the sponsorship of the Ministry of International Trade and Industry (MITI), developed the ceramic gas turbines for cogeneration [22] and demonstrated excellent high temperature strength of a fine-grained Si_3N_4 with a stable rare earth disilicate grain boundary phase [23, 24]. In addition the Petroleum Energy Center (PEC) and the Japan Automobile Research Institute (JARI) developed a 100 kW ceramic gas turbine for automobiles [25], under a MITI-sponsored program (1990–1997). There was also a program for large-scale gas turbines conducted by the Central Research Institute of the Electric Power Institute (CRIEPI) and Hitachi, Ltd. [26,27]. Another program for large scale engines was sponsored by the Tokyo Electric Power Co. (TEPCO) [28]. Related to these programs were corrosion studies showing the deleterious effects of molten salts on the strength of SiC and Si_3N_4 [29].

In the late 1980s and 1990s several significant changes occurred which influenced the direction of ceramic corrosion studies.

1. The availability of pure, free-standing SiC and Si_3N_4 led to fundamental studies of oxidation of these materials. It was now possible to get accurate kinetic rates and product compositions without the influence of additives.
2. The importance of complex combustion environments was recognized. While simple isothermal oxidation tests provide useful fundamental information, actual engine environments are quite complex—consisting of water vapor, possible condensed phase deposits as well as high pressures and thermal cycling [1]. Resistance to pure oxygen at constant temperatures is not sufficient; other factors must be considered.
3. Ceramic matrix composites (CMCs) were developed. Continuous fiber reinforced ceramics offer a good deal of promise for improved fracture toughness. They are based on a carbon or SiC fiber in a SiC or Si_3N_4 matrix. For the SiC fiber composites, the fiber is coated with a carbon or boron nitride layer to allow proper load transfer to the fiber. The problem area is the carbon or boron nitride, which readily oxidize. Recent oxidation studies focus on these issues.

Major ceramic engine programs in the 1990s focused on CMCs were the NASA Enabling Propulsion Materials (EPM) program [30] and the DOE Continuous Fiber Ceramic Composite (CFCC) program [31]. The NASA EPM program centered on the use of a SiC-fiber/SiC-matrix for a high temperature, backside cooled combustor liner, for aircraft engine applications. In burner studies, which simulated the EPM environment, high pressure water vapor was shown to be a significant corrodent [32,33], resulting in increased recession of monolithic SiC. This was consistent with several Japanese studies which showed increased recession rates of SiC and Si_3N_4 in combustion environments [34,35]. This recession is related to water-enhanced volatilization and will be discussed in more detail in the next section.

The DOE CFCC program was more generic and directed at CMCs for a variety of industrial applications. In the later 1990s DOE together with Solar Turbines Inc. (San Diego, CA) began a demonstration program for a ceramic stationary gas turbine (CSGT) for cogeneration [36,37]. The field testing under the CSGT program involved testing of silicon nitride blades (AS800) and SiC/SiC CFCC combustor liners. The tests with blades ended in failure, very likely due to impact. Microstructural evaluation of uncoated SiC/SiC liners showed excessive material recession in hot, high velocity sections. However, SiC/SiC liners with an oxide environmental barrier coating were tested up to 13,937 hrs without failure. See Volume 1, Chapter 7 for more details on this test. In other DOE sponsored work, Rolls-Royce Allison (Indianapolis, IN) has tested an AS800 Si_3N_4 first stage vane assembly in a Model 501-K turbine (Exxon – Mobile, AL) [38]. Dimensional measurements found significant material recession at the mid-span of the trailing edge of the vanes with even greater losses at the vane leading edges. The recession in both field tests was attributed to volatility of the silica scale in the high water vapor partial pressure, high gas velocity environments as discussed later. See Volume I, Chapter 5 for more details of this Rolls-Royce test.

The development of CMCs in France for the Hermes program and in the U.S. for the EPM and CFCC programs led to several studies of the unique issues in CMC oxidation. At their current state of development CMCs are based on either a carbon fiber in a SiC matrix or a SiC fiber in a SiC or Si_3N_4 matrix. The SiC fiber must be coated with carbon or boron nitride. Thus all types of CMCs contain a second phase, which oxidizes easily. Initial studies were on composites of SiC matrices with carbon fibers [39, 40] and carbon-coated SiC fibers [41,42]. Later studies were on SiC matrices with BN-coated SiC fibers [43–45]. The final stage of CMC processing involves the application of a seal coating—typically CVD SiC—over the matrix. If this seal coat remains intact, the carbon or BN is not exposed to oxidation. However, in the event of seal coat damage in service, oxidation of the carbon or BN becomes a major problem.

In summary, oxidation/corrosion studies have been included in many ceramic engine development programs. While many of these are system specific, it is possible to draw some general conclusions:

1. SiC, Si_3N_4, and composites of these appear to be the most promising ceramics for hot stage applications. AlN undergoes substantial degradation in combustion environments and is not a suitable candidate.
2. SiC and Si_3N_4 show slow oxidation kinetics and good strength retention in purely oxidizing environments.
3. On some materials low temperature oxidation heals surface flaws and actually improves strength. However higher temperature oxidation creates surface pits and leads to strength degradation.
4. Basic molten salt or oxide deposits lead to surface pitting and strength degradation.
5. The water vapor present in a combustion environment enhances degradation.
6. The effects of temperature cycling also appear to be material dependent. In some cases cycling has little effect; in other cases it may lead to degradation.
7. The major issue with CMCs is oxidation of the fiber coating. This is due to an inadequate external coat—either not initially formed properly or damaged in service.

In the next sections, we focus on these key issues and the chemistry behind these issues.

Experimental Methods

Accurate simulation of engine environments in a laboratory environment is a major issue. Actual engine tests are expensive and it is difficult to accurately control all variables. It is important to simulate the high temperatures, pressures, flow rates, and gas compositions found in an engine. The most common method is the burner rig, shown schematically in Figure 26.2 [1]. Over the years burner rigs have evolved into more controlled test systems [46]. Although these rigs are well accepted for materials testing now, their very nature prevents them from offering the precise control of a laboratory furnace.

Laboratory tests in electric furnaces are substantially less expensive than burner rig tests. Yet they cannot fully simulate the engine environment. The investigator is often able to determine the critical issues from the burner and focus on this with a laboratory furnace [1].

Another important issue is the assessment of corrosion damage. Weight change is most commonly used, but this does not give an indication on the form of surface or internal damage. Examination of oxide and matrix morphologies with techniques such as optical microscopy as well as scanning and transmission electron microscopy is necessary. Measured reduction of load-bearing capability is another useful indication of corrosion damage.

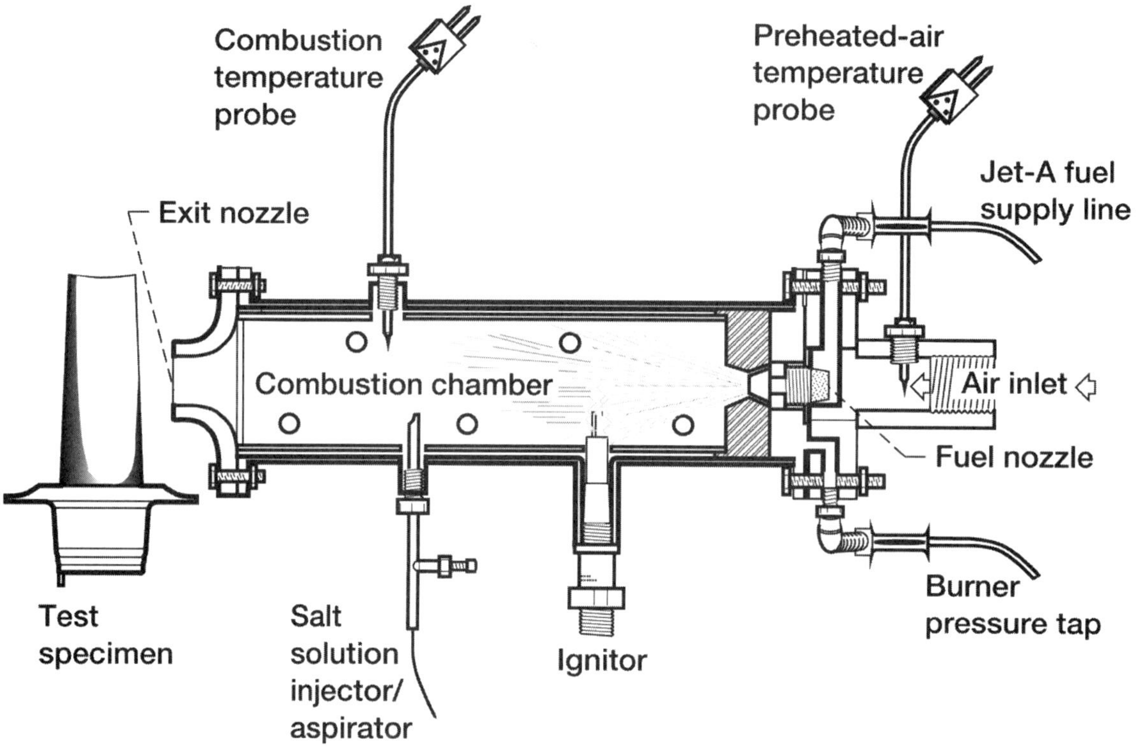

Figure 26.2 Schematic of NASA Glenn Research Center Burner Rig Used for Studying Exidation and Corrosion of High Temperature Materials in Actual Combustion Environments (Reprinted from ref. [1] with permission of the American Ceramic Society, www.ceramics.org, © 1993, all rights reserved)

CORROSION MECHANISMS FOR SILICON-BASED CERAMICS

Baseline Oxidation Data

Isothermal oxidation data for high purity SiC and Si_3N_4 in pure, dry oxygen provide a useful baseline of the slowest possible material consumption rates, as illustrated in Figure 26.3 [47]. The key oxidation reactions are:

$$\text{Oxidation of SiC: } SiC(s) + 3/2\ O_2(g) = SiO_2(s) + CO(g) \quad (1)$$

$$\text{Oxidation of } Si_3N_4\text{: } Si_3N_4(s) + 3/4\ O_2(g) = 3/2\ Si_2N_2O(s) + \_\ N_2(g) \quad (2)$$

$$Si_2N_2O(s) + 3/2\ O_2(g) = 2\ SiO_2(s) + N_2(g) \quad (3)$$

There is general agreement that for SiC and Si_3N_4, oxygen diffusion inward through the growing scale is rate controlling [1,48]. This leads to the familiar parabolic kinetics [49] where the rate slows as the oxide scale thickens:

$$x^2 = k_p t \quad (4)$$

Here x is scale thickness, k_p is the parabolic rate constant, and t is time.

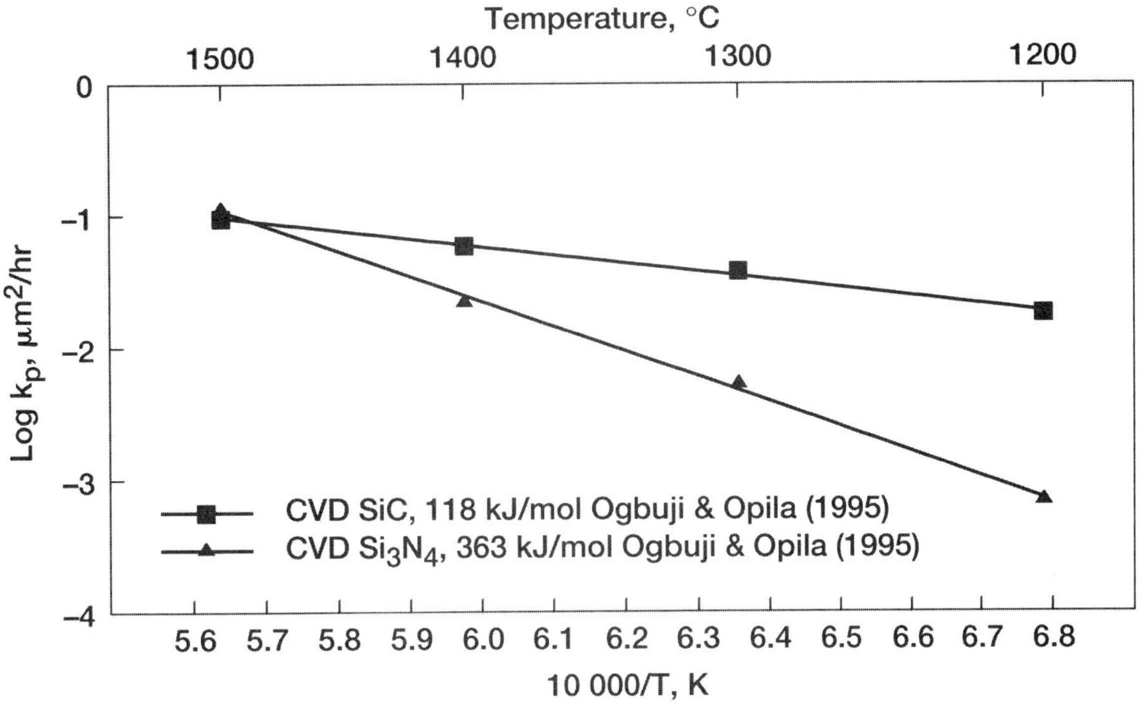

Figure 26.3 (Adapted from ref. [47], with permission from The Electrochemical Society, Inc., © 1995 and from ref. [48] with permission from the Wiley-VCH Verlag GmbH, © 2000.)

As shown in Figure 26.3 [47,48], the rates of oxidation for pure Si_3N_4 are slower than for pure SiC in pure oxygen. The formation of an inner layer of silicon oxynitride appears be a factor in the slower oxidation rates [50], although the exact mechanism for this is still controversial [1,48,51].

In practical situations, the differences between SiC and Si_3N_4 would not be observed [52]. Instead the additives to these ceramics and effects of other combustion gas components (e.g. water vapor, metallic impurities) tend to dominate oxidation behavior.

Material Effects on Oxidation

Additive-Containing SiC and Si_3N_4. Many of the early oxidation studies of SiC and Si_3N_4 were on additive-containing materials. For SiC materials, the most common additives are boron, carbon, and/or alumina. A common form of sintered α-SiC contains B_4C. This SiC material oxidizes at rates similar to CVD SiC [47, 53]. Hot pressed SiC generally requires alumina additives. The oxidation of alumina-containing SiC has been extensively studied. The alumina tends to migrate to the oxide scale. This leads to local crystallization and even mullite formation if the alumina concentration is sufficiently high. Oxidation rates tend to increase with increasing alumina content, as shown in Figure 26.4 [7].

Nearly all commercial forms of Si_3N_4 require refractory oxide additions to promote densification. Cubiciotti and Lau have demonstrated that these additives dominate oxidation behavior to the point where the metal cation diffusion outward is rate controlling [54,55]. These oxidation rates are substantially higher than those observed with pure Si_3N_4. Figure 26.5 illustrates the oxide film on a Si_3N_4 with Y_2O_3 additives—note the two phases. In multiphase materials, a variety of processes act simultaneously (e.g. internal and external oxidation, grain boundary widening, migration of additives to the scale, porosity formation, and

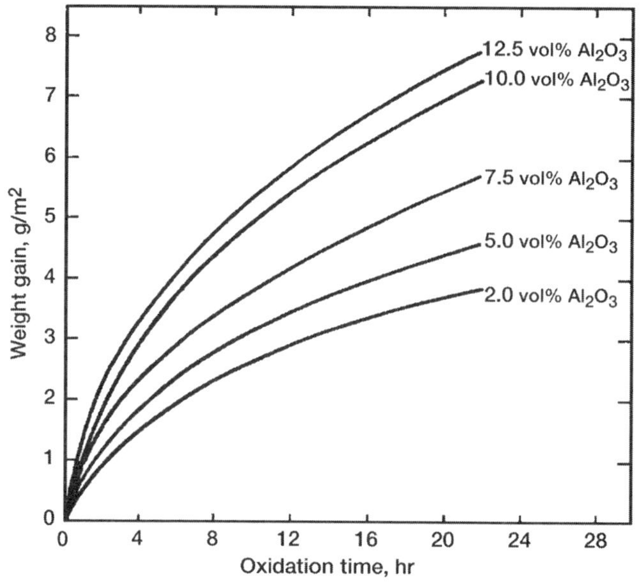

Figure 26.4 Oxidation Behavior at 1370°C for SiC Hot-Pressed with Varying Amounts of Al_2O_3 in 1 atm of Dry Oxygen (Reprinted from ref. [7] with permission of the American Ceramic Society, www.ceramics.org, © 1975, all rights reserved)

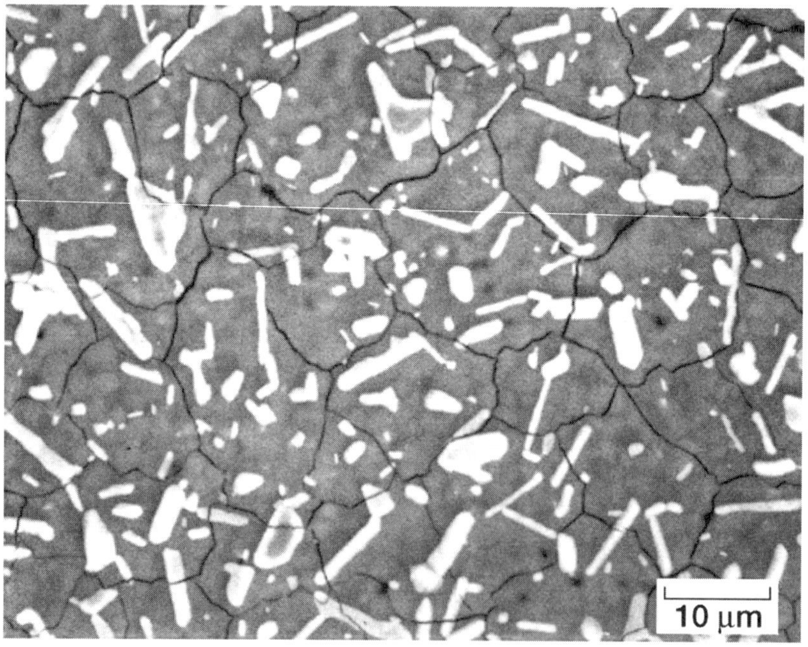

Figure 26.5 Y_2O_3 Containing Si_3N_4 Oxidized for 97 hrs at 1300°C in Dry Air. The Bright Particles are the Yttrium-Silicate Phase within the Silica. (Reprinted from ref. [1] with permission of the American Ceramic Society, www.ceramics.org, © 1993, all rights reserved)

scale crystallization). It may not be possible to model these with a simple rate law [21]. More complex expressions may be necessary [56]. Both the total amount and specific composition of the additives are critical. Some compositions of Si_3N_4 with Y_2O_3 sintering aids result in grain boundary phases which undergo a volume expansion on oxidation. This leads to disintegration of the sample [57].

Some processing routes lead to porous ceramics, such as reaction-bonded Si_3N_4. Although less favorable for candidate heat engine components, these materials may be less expensive than their dense counterparts. Oxidation studies indicate that internal oxidation is a critical issue and controlled by pore diffusion [58,59].

Continuous Fiber-Reinforced Ceramic Composites. As discussed in the introduction, continuous fiber-reinforced ceramic composites represent a large area of research in silicon-based ceramics. One area of major interest is in carbon fiber-reinforced SiC composites. This is due to the stability of carbon fibers to high temperatures in inert environments. The problem is that the carbon fibers oxidize quite readily in the event of a crack in the SiC matrix and subsequent exposure of these fibers to oxidizing gases. Lamouroux, et al. [39,40] conducted both an experimental and theoretical study of the oxidation of two dimensional carbon fiber-reinforced SiC composites. Basically they observe three oxidation regimes: I) Low temperature (T < 800°C): overall kinetics are limited by the reaction of carbon and oxygen, II) Intermediate temperatures (800°C < T < 1100°C): overall kinetics are limited by diffusion of CO and CO_2 from carbon oxidation out of pores and microcracks in the SiC matrix, and III) High temperatures (T > 1100°C): kinetics are dominated by sealing of the SiC matrix due to SiO_2 formation. More recently Halbig and Cawley [60] have modeled the oxidation of carbon fibers exposed to oxygen due to a matrix crack. They also see the important effect of reaction control at low temperatures and diffusion control at high temperatures. Although carbon fiber-reinforced SiC has attractive mechanical properties, the susceptibility of the carbon fiber to oxidation is a major limitation. Lamouroux, et al. [61] have proposed a multilayer matrix with improved oxidation resistance. More research is needed on improved oxidation resistance of carbon fiber-reinforced SiC to make this material viable.

A second area of interest is SiC-fiber/SiC or Si_3N_4 matrix reinforced composites. SiC fibers stable to high temperatures (T > 1200°C) have been developed. Although they are still quite expensive, they eventually are expected to decrease in price. A SiC fiber-reinforced SiC composite is shown schematically in Figure 26.6 [62]. A major problem with these types of composites is the rapid oxidation of the C or BN fiber coating. Several elegant models have been developed for this oxidation process [41,42,63], but there are currently no complete solutions to the problem of rapid oxidation of the fiber coating.

Consider first the SiC system with carbon-coated fibers. Figure 26.7 shows some typical oxidation curves for carbon coated SiC fiber in a reaction-bonded Si_3N_4 matrix [64]. At low temperatures the carbon coatings oxidize readily leading to the observed weight losses. At high temperatures a protective oxide film grows on the outside of the composite, leading to protection of the interior. However, protection would be negated if a large matrix crack forms when the ceramic is under load. Oxidation of an exposed fiber coating has been modeled by Filipuzzi and co-workers [41,42] and is shown schematically in Figure 26.8. The oxygen pressure dependence of SiC oxidation indicates that most of the oxidation occurs near the mouth of the pore. The process modeled includes these steps:

- Diffusion of oxygen into the pore.
- Two step oxidation of carbon to CO and CO_2.
- Diffusion of CO out of the pore.
- Consumption of oxygen by oxidation of the matrix.
- Consumption of oxygen by oxidation of the fiber.

Other investigators have extended this model to regimes where reaction control of carbon oxidation is important [63]. These models have been helpful in explaining the oxidation results and high-temperature environmental effects on mechanical properties.

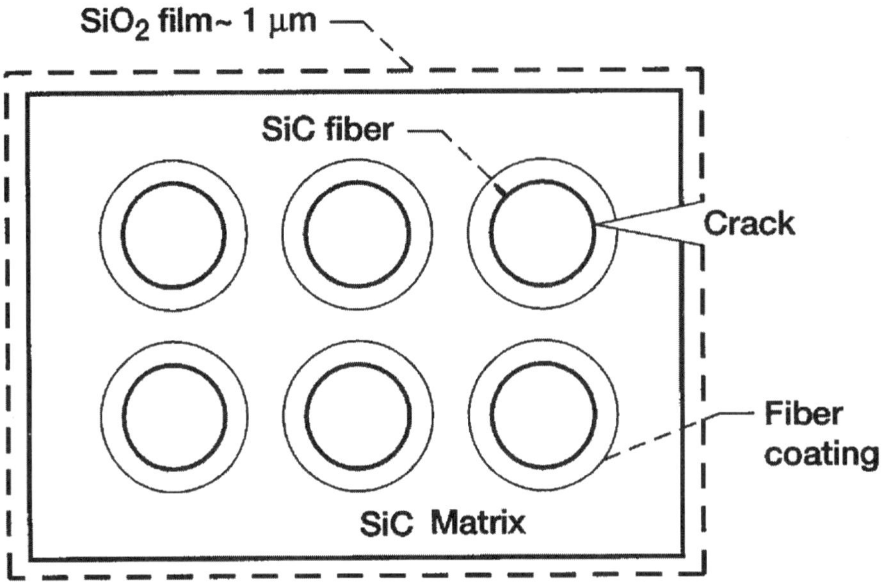

Figure 26.6 Schematic of Fiber Reinforced Ceramic Matrix Composite (Reprinted from ref. [62] with permission of the American Ceramic Society, www.ceramics.org, © 1995, all rights reserved)

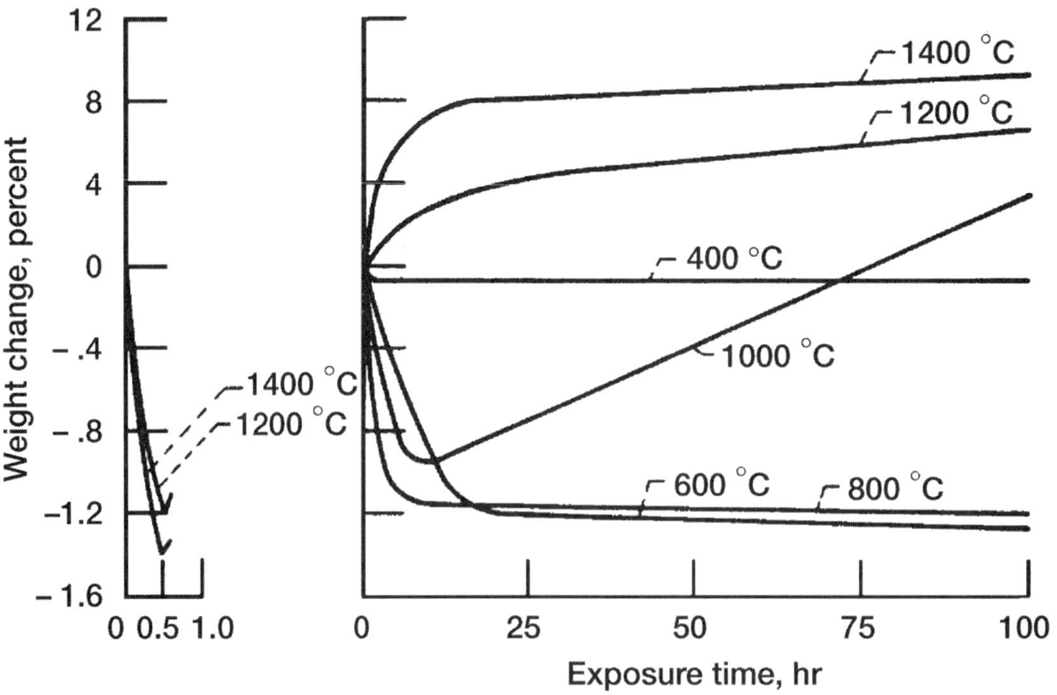

Figure 26.7 Weight Change Curves for Oxidation of SiC fiber/Si_3N_4 Matrix Composites (Adapted from ref. [64])

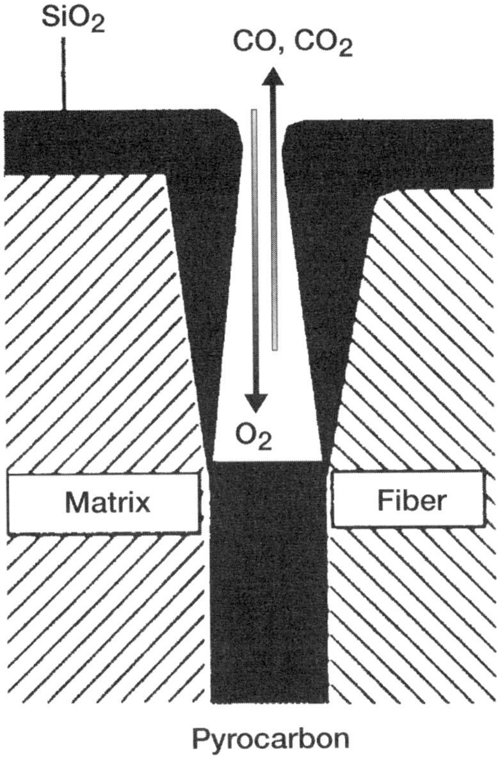

Figure 26.8 Schematic for Oxidation Model of a Carbon Coated SiC Fiber in a SiC Matrix (Reprinted from ref. [41] with permission of the American Ceramic Society, www.ceramics.org, © 1994, all rights reserved)

The other commonly used fiber coating is hexagonal boron nitride. Oxidation of this material is complex and several degradation routes are possible [45] as shown in Figure 26.9. As the BN and the SiC oxidize, both B_2O_3 and SiO_2 are formed. These react to form low melting borosilicates, as illustrated in Figure 26.9(a). Compare this to the second type of degradation behavior. Boria (B_2O_3), the oxidation product of BN, reacts readily with water vapor found in the combustion environment to form highly stable $H_xB_yO_z$(g) species. Thus the BN fiber coatings volatilize in a way analogous to carbon, but due to the reaction with water vapor. This explains the appearance of Figure 26.9(b). Moving down the annular region of Figure 26.9(b), into the composite, one would eventually reach the remaining BN coating. These fiber coating recession distances can be determined and related to simple diffusive fluxes through contracting annular pores [45]. This is illustrated in Figure 26.10.

Effect of Oxidation on Mechanical Properties

The effect of oxidation on strength is critical to determine the ceramic's suitability for an engine component. Consider first monolithic ceramics. As discussed in the Introduction, under some conditions mild oxidation can heal surface flaws and lead to a strength increase; whereas more severe oxidation can lead to pitting and strength decreases [10]. This has been verified and further explored with laboratory studies [65,66]. These issues have been discussed in detail in a review by Tressler [67].

In an actual engine environment, a component would be simultaneously exposed to both oxidation and mechanical stress. There are several studies of stress-enhanced oxidation of monolithic Si_3N_4 in the literature

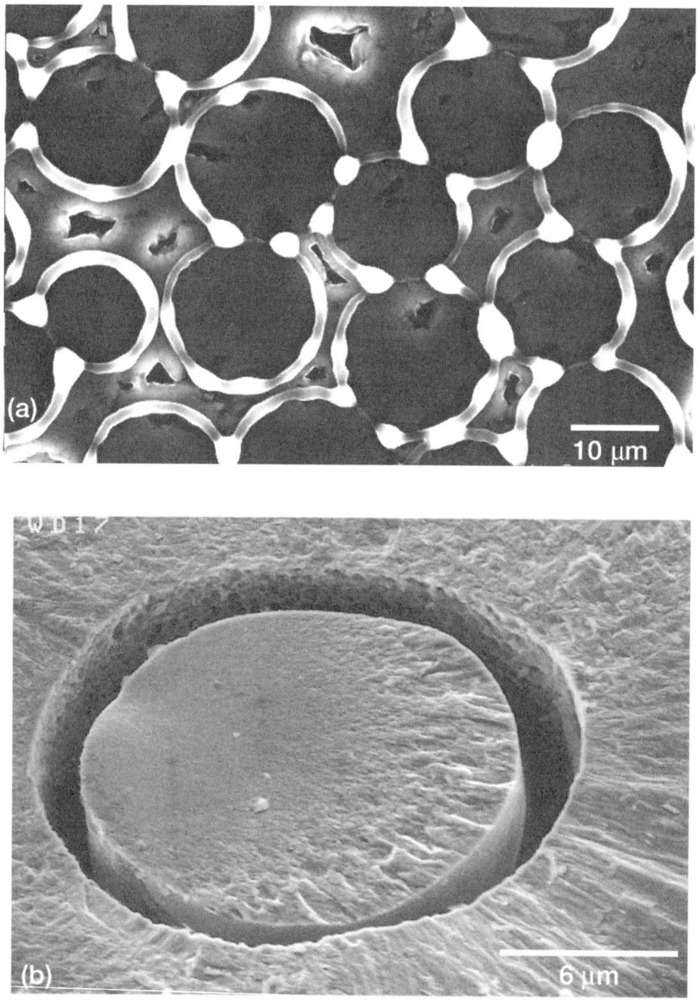

Figure 26.9 Two Oxidation Processes of a BN Coated SiC Fiber in a SiC Matrix (a) Borosilicate Glass Formation After Oxidation at 816°C in Oxygen for 100 h (b) Volatilization of BN Due to Water Vapor Interactions After Oxidation at 500°C in Humid Air (Reprinted from ref. [45] with permission of the American Ceramic Society, www.ceramics.org, © 1999, all rights reserved).

[68,69]. There is a complex interaction between stress and oxidation. Stresses form microcracks and cavities and also alter diffusivities. Each of these factors tends to enhance oxidation. Wereszczak, et al. [69] have observed the formation and growth of stress-corrosion cracks which ultimately led to failure. Thümmler and Gratwohl [70] have shown that oxidation of certain additive-containing Si_3N_4 compositions leads to formation of a softened grain boundary phase and consequently enhanced creep. SiC compositions without additives show better overall creep resistance and less susceptibility to oxidation effects on creep.

There is a good deal of literature on the mechanical properties of fiber-reinforced CMCs in oxidizing environments. Halbig, et al. [71] have examined oxidation of carbon fiber-reinforced CMCs under stress. Time-to-failure is dramatically reduced due to oxidation of the carbon fiber through an opened crack. The major problem occurs in the intermediate temperature (~500–1000°C) regime when the amount of SiO_2 formed is insufficient to seal cracks in the SiC matrix. Stress-rupture studies of SiC-fiber/C/SiC matrix [72]

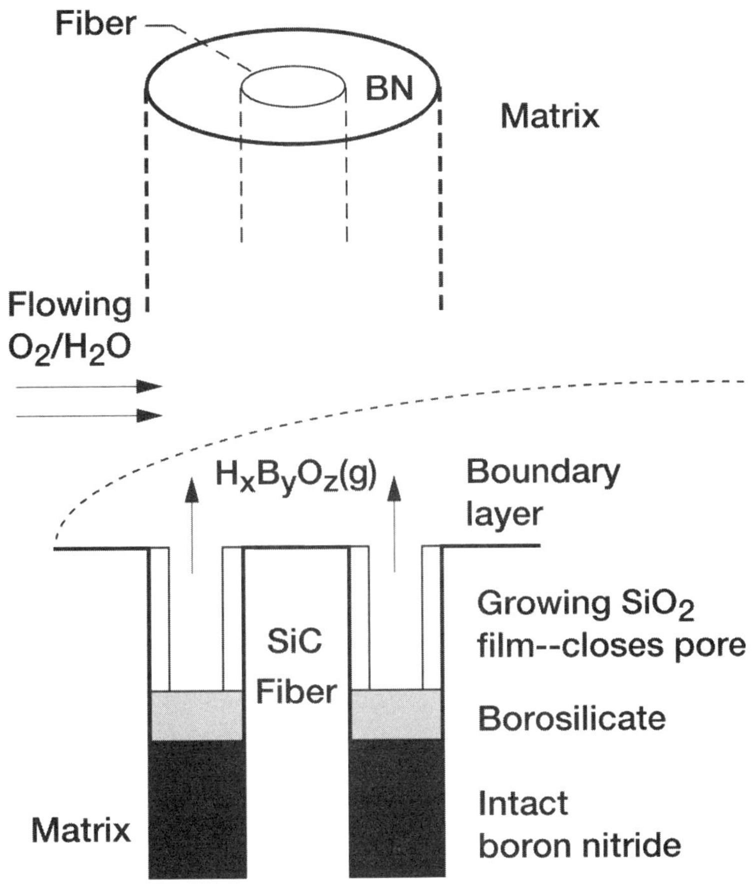

Figure 26.10 Schematic for Water Vapor Volatilization of a BN Fiber Coating in a SiC Fiber/SiC Matrix (Reprinted from ref. [45] with permission of the American Ceramic Society, www.ceramics.org, © 1999, all rights reserved)

and SiC-fiber/BN/SiC matrix [73] CMCs all indicate that oxidation of the fiber coatings (carbon or boron nitride) is the major cause of failure. Development of oxidation-resistant fiber coatings is a critical area of research and presents major challenges.

Complex Corrosion Environments

The discussion thus far has focused primarily on isothermal oxidation in pure oxygen. However, there are many other factors in the environment of a heat engine. These include thermal cycling, additional oxidants such as H_2O and CO_2, aggressive low-level gases such as SO_2 and HCl, and corrosive deposits such as sulfates and vanadates. To some degree, these are dependent on the type of heat engine. Aircraft gas turbines involve daily thermal cycling, whereas cycling would be more limited in utility and industrial turbines. A coastal-based turbine is more susceptible to salt attack than a turbine located inland. Nonetheless, some general comments can be made about combustion environments. Table 26.1 lists some properties of common hydrocarbon fuels [74–78]. Note that all contain impurities, which may be corrosive at high temperatures. Figure 26.11 is a plot of equilibrium gas composition and adiabatic flame temperature as function of equivalence ratio [1]. The fuel used for this calculation is a typical aviation fuel—Jet A, but the results are similar

TABLE 26.1 PROPERTIES OF SOME COMMON HYDROCARBON FUELS

Fuel	Boiling Range (K) [74]	H:C Molar Ratio [75]	S Content (w/o) [76,77]	Na + K content (ppm) [77]	V content (ppm) [78]
Unleaded Automotive Gasoline	300–375	2.02	0.15	3.6 <0.003	
Jet A (Commercial Aviation Fuel)	450–560	1.92	0.05	~10	0.06
Fuel Oils	450–615	1.61	0.1–1.0	10–20	<300

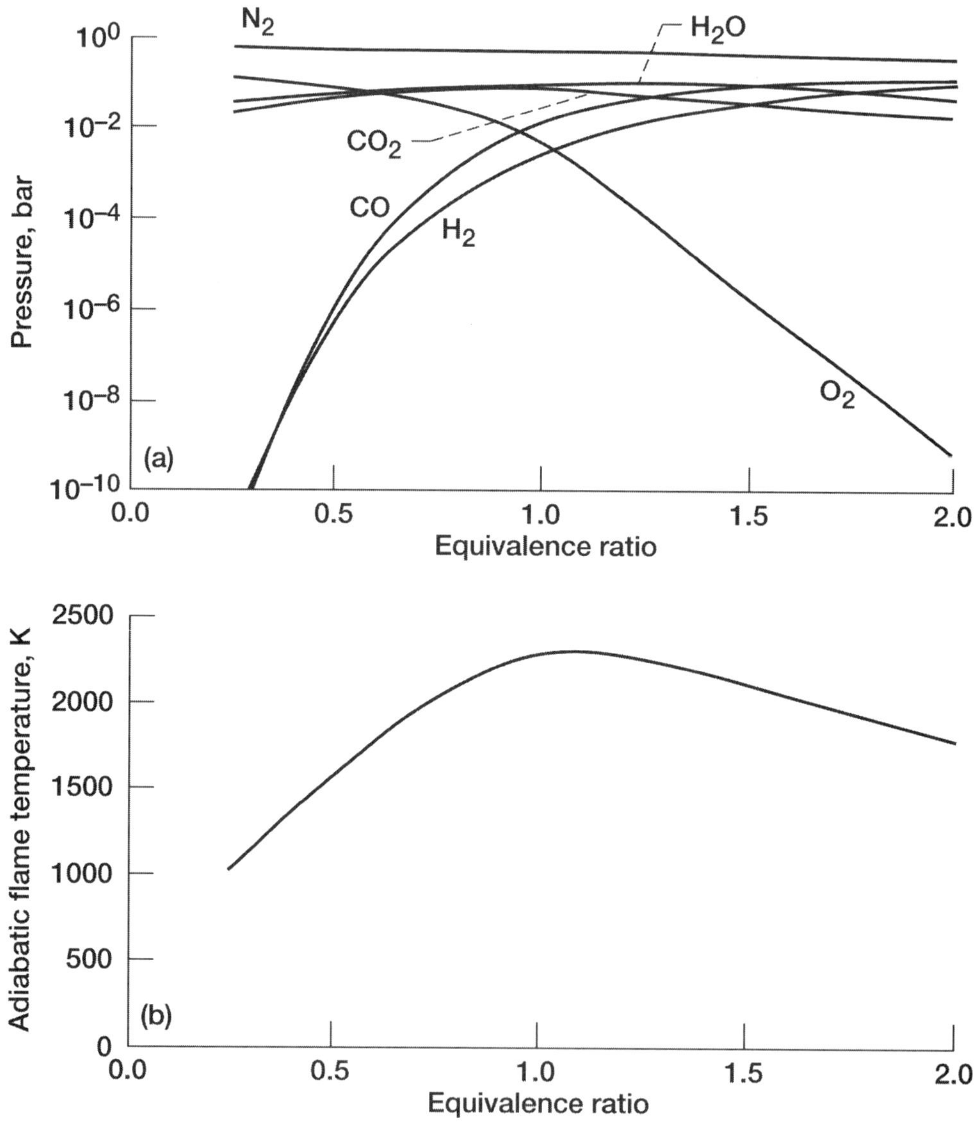

Figure 26.11 Combustion Products and Adiabatic Flame Temperature for the Combustion of Jet A Fuel in Air (Reprinted from ref. [1] with permission of the American Ceramic Society, www.ceramics.org, © 1993, all rights reserved)

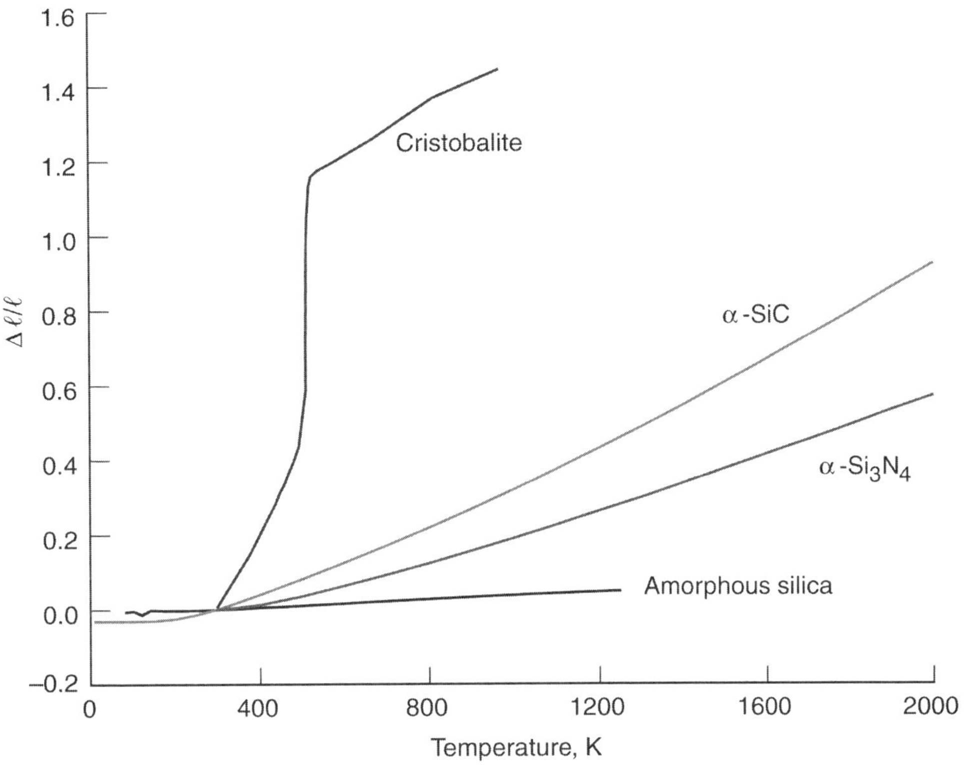

Figure 26.12 Thermal Expansion of SiC, Si_3N_4, Amorphous SiO_2, and Crystalline SiO_2 as a Function of Temperature.

for any type of hydrocarbon fuel. Note that H_2O and CO_2 are always present in amounts of about 10%. In a fuel lean environment (Equivalence Ratio < 1), which is most common, O_2 is also present. In a fuel-rich environment (Equivalence Ratio > 1), the O_2 content is very low and CO and H_2 become important.

In the following section, each of the additional factors, beyond isothermal oxidation, will be discussed for SiC and Si_3N_4.

Cyclic Oxidation. When samples oxidize during thermal cycling, stresses develop due to the coefficient of thermal expansion (CTE) mismatch between the silica scale and the SiC or Si_3N_4 substrate. These mismatches are illustrated in Figure 26.12 [48]. In actual engine environments, partially or fully crystalline scales are expected to form. Compressive stresses are expected on heating and then tensile cracks on cooling. On heating the compressive stresses can lead to crack healing. Hence SiC and Si_3N_4 show good behavior in cyclic oxidation, unlike their superalloy counterparts which often show scale buckling and spallation due to compressive scale stresses on cooling. Figure 26.13 illustrates the good oxidation behavior of some types of SiC and Si_3N_4 during thermal cycling [48].

Water Vapor. As discussed in the Introduction, the burner rig studies clearly suggest that water vapor in the combustion environment degrades silicon-based ceramics. Laboratory experiments have shown that water vapor has three major effects on the oxidation of silicon-based ceramics. First, water vapor enhances the transport of impurities to the ceramic surface. This may occur by formation of gaseous hydroxides, such as NaOH, from metallic impurities in the environment. Impurities become incorporated in the silica scale,

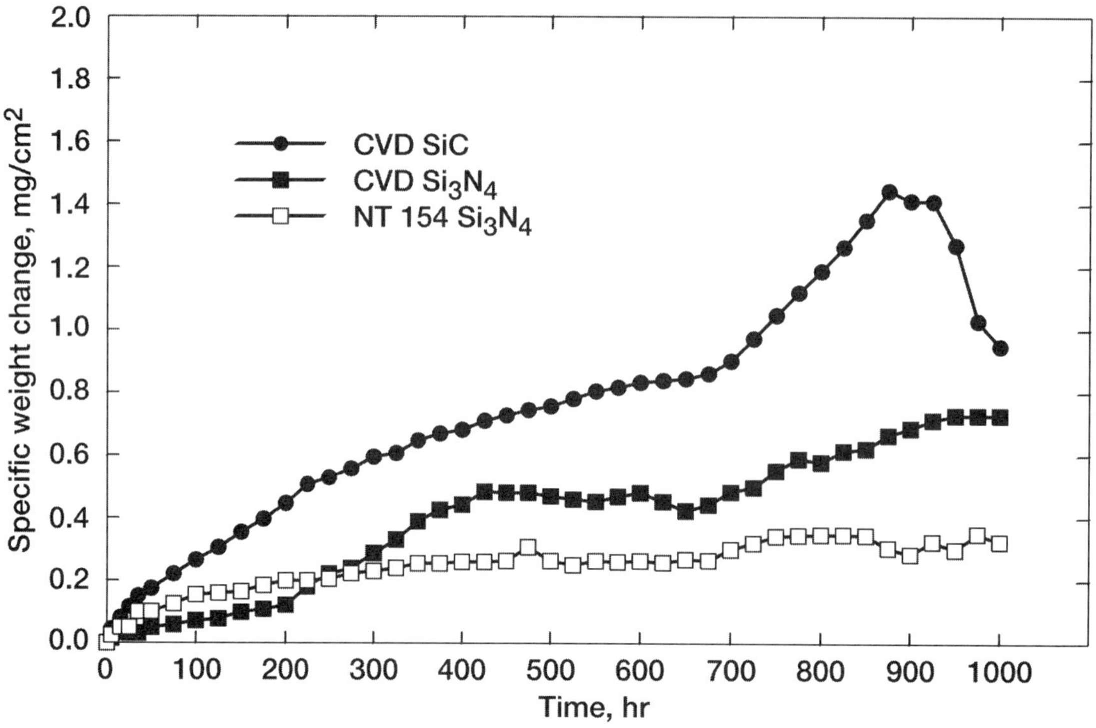

Figure 26.13 Cyclic Oxidation Weight Change Kinetics for Several Silica Forming Materials in Air at 1300°C in 5 hr Cycles (Reprinted from ref. [48] with permission from Wiley-VCH)

thereby increasing the oxidation rate substantially compared to the low rates observed under very clean conditions [79]. This is discussed further in the section on low-level metal cation impurities.

Second, water vapor increases the intrinsic oxidation rate of silica-formers compared to rates observed in dry oxygen. This mechanism has been described by Deal and Grove [80]. While the diffusivity of the water molecule in silica is about one order of magnitude lower than that of molecular oxygen, the solubility of water in silica is almost three orders of magnitude higher than that of oxygen. Since the parabolic rate constant is equal to the product of the diffusivity and the solubility, the parabolic rate constant for silica formation is higher in water vapor than in dry oxygen. Oxidation of silicon carbide in water vapor occurs by the reaction:

$$SiC(s) + 3H_2O(g) = SiO_2(s) + CO(g) + 3H_2(g) \tag{5}$$

One consequence of these higher oxidation rates is the generation of more product gases. The silica scale under these conditions often contains bubbles and pores formed by release of these gases [81]. Since the parabolic oxidation rate increases with water vapor partial pressure, high-pressure exposures of silicon carbide and silicon nitride in water vapor result in the formation of silica scales riddled with bubbles and pores. These bubbles and pores make the silica scale non-protective and very high oxidation rates are then observed [81,82].

Finally, water vapor causes volatilization of the silica scale by the reaction [83, 84]:

$$SiO_2(s) + 2H_2O(g) = Si(OH)_4(g) \tag{6}$$

Since this volatilization occurs simultaneously with the oxidation reaction (Equation 5), the resulting kinetics follow a paralinear rate law given by:

$$\frac{dx}{dt} = \frac{k_p}{2x} - k_l \tag{7}$$

where x is the scale thickness, t is time, k_p is the parabolic oxidation rate constant, and k_l is the linear volatilization rate constant. Paralinear kinetics can be monitored by either substrate recession measurements or weight change as shown in Figure 26.14 [48]. Typical paralinear weight change behavior is shown (Figure 26.15) for silicon-based materials exposed in a laboratory furnace containing water vapor [48]. At long times or high volatility rates, paralinear kinetics result in linear recession of the ceramic material, and a constant

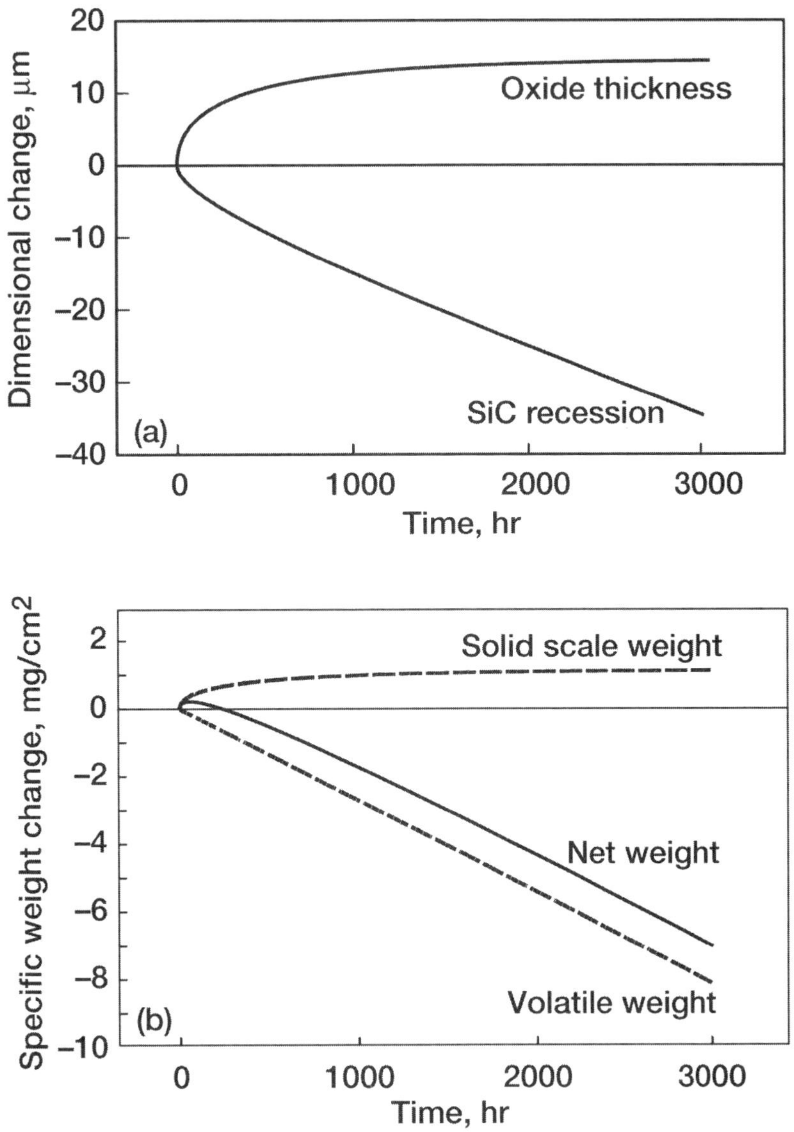

Figure 26.14 Model Paralinear Kinetics for SiC Oxidized in Water Vapor. Calculated Results are Typical of Exposures at 1200°C in 50% H_2O/50% O_2 at flow rates of 4.4 cm/s. (Reprinted from ref. [48] with permission from Wiley-VCH)

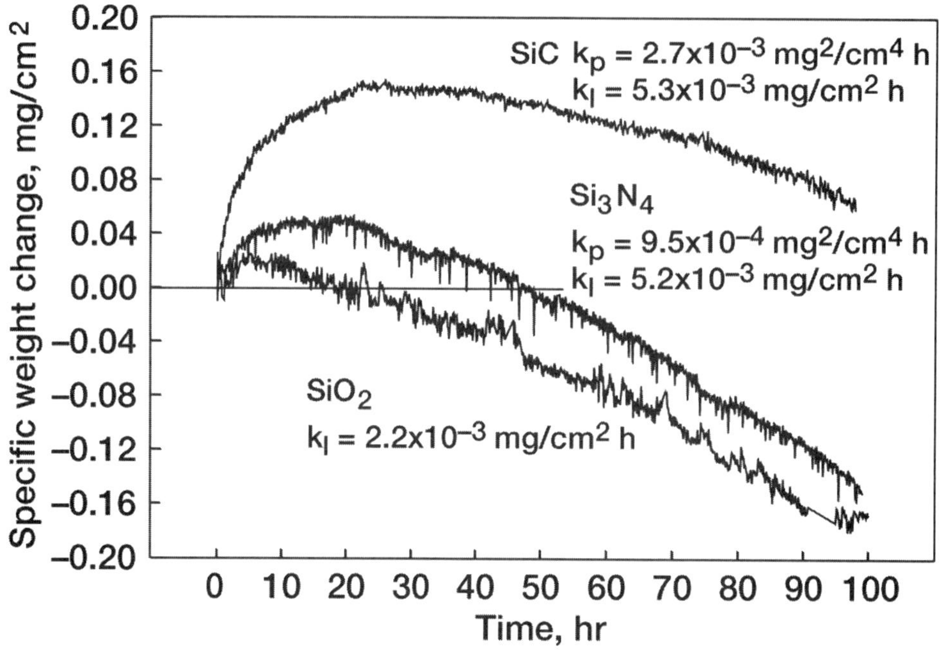

Figure 26.15 Oxidation/Volatilization Weight Change Kinetics for Silica, CVD SiC and CVD Si_3N_4 at 1200°C in 50% H_2O/50% O_2 at Flow Rates of 4.4 cm/s (Reprinted from ref. [48] with permission from Wiley-VCH)

oxide thickness, as well as linear weight loss [32,34,35]. Under these conditions, the recession and weight loss are limited by the transport of $Si(OH)_4(g)$ away from the ceramic through a gaseous boundary layer. This has been modeled using the following expression [33]:

$$k_l = 0.664\,Re^{1/2}\,Sc^{1/3}\,\frac{D\rho}{L} \tag{8}$$

where Re is the Reynold's number, Sc is the Schmidt number D, is the interdiffusion coefficient of $Si(OH)_4$ in the boundary layer gas, ρ is the concentration of $Si(OH)_4$ at the silica/gas interface and L is a characteristic length. This expression can be reduced to terms that describe the combustion environment such as pressure, P, gas velocity, v, and material temperature, T, so that

$$k_l \propto \exp\left(\frac{-DH_f}{RT}\right)P^{1.5}v^{0.5} \tag{9}$$

Here R is the gas constant. The temperature dependence arises from the enthalpy of formation (ΔH_f) of $Si(OH)_4$ according to the reaction given in Equation 6 [33]. The pressure dependence of SiC weight loss and recession has been demonstrated experimentally in a High Pressure Burner Rig at the NASA Glenn Research Center [32], as shown in Figure 26.16. The data, normalized by the $P^{1.5}v^{0.5}$ term fit the model based on equation (8) quite well for a variety of pressures and velocities, as shown in Figure 26.17 [32]. The high pressures, high temperatures, and high velocities present in gas turbine applications thus all promote water vapor attack of silicon-based ceramics.

Recession consistent with the volatilization mechanism has been observed in two recent engine field-tests as previously mentioned. Figure 26.18 shows recession and porous glass formation for a SiC/SiC combustor liner exposed for 2300 hrs in a land-based turbine (T_{max} = 1150°C, P = 10 atm, v = 61m/s). Figure 26.19 shows

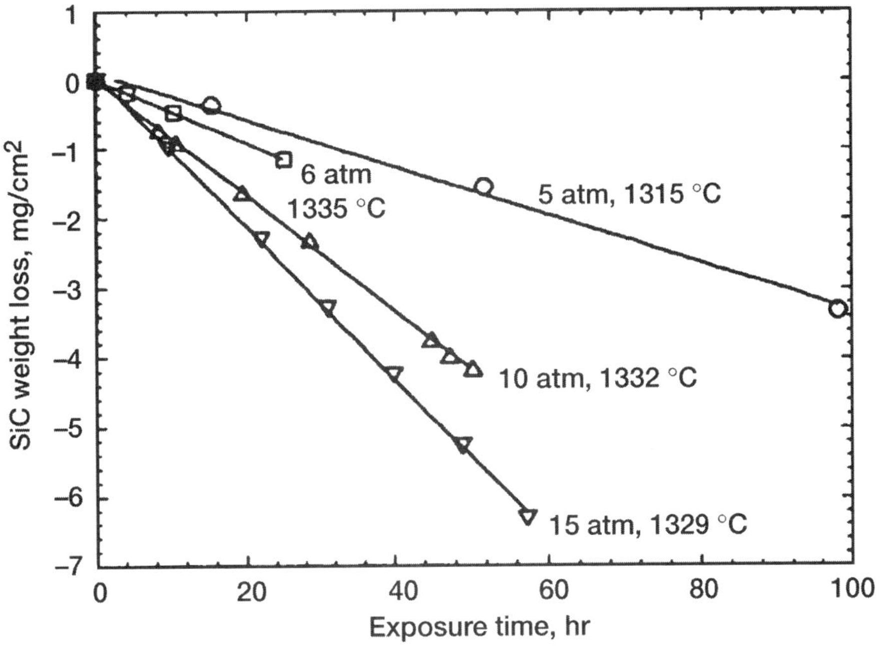

Figure 26.16 Weight Loss of CVD SiC in a Fuel-Lean High Pressure Burner Rig with Gas Velocities of 20 m/s (Reprinted from ref. [32] with permission of the American Ceramic Society, www.ceramics.org, © 1995, all rights reserved)

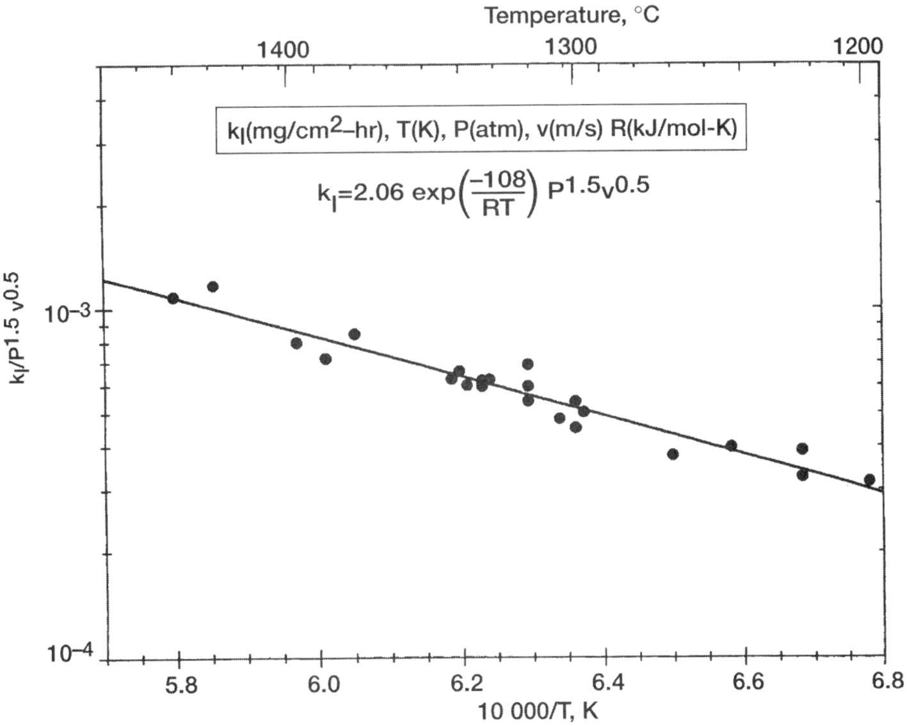

Figure 26.17 Normalized Weight Loss Results of CVD SiC in a Fuel-Lean High Pressure Burner Rig with Gas Velocities of 10–27 m/s and Total Pressures between 5 and 15 atm (Reprinted from ref. [32] with permission of the American Ceramic Society, www.ceramics.org, © 1999, all rights reserved).

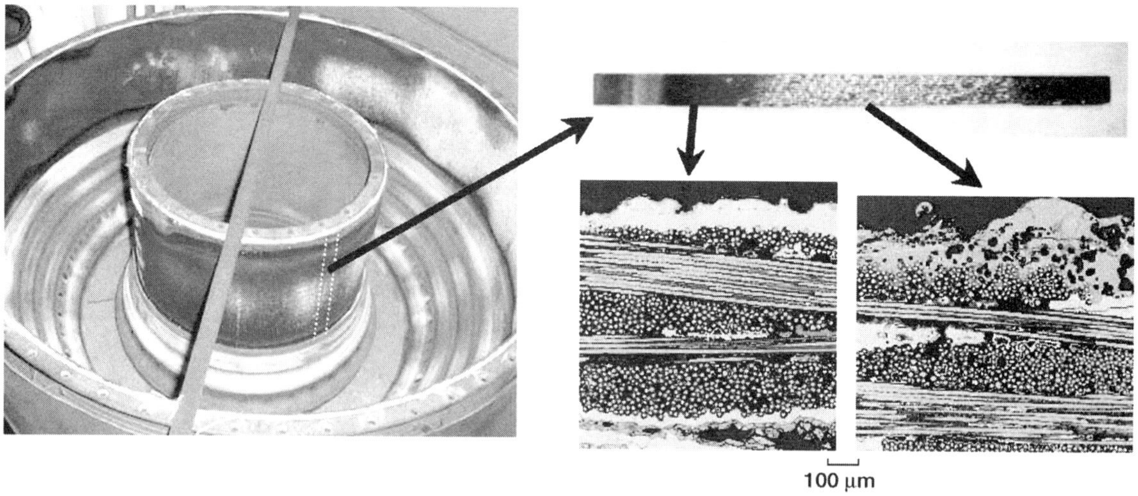

Figure 26.18 SiC/SiC Combustor Liner Utilized in an Industrial Gas Turbine Operated for About 2300h at a Total Pressure of 10 atm and a Peak Liner Temperature of 1150°C. The Micrographs Show an Intact SiC Seal Coat from the Cooler Portions of the Liner. The SiC Seal Coat is Breached Resulting in a Glassy Reaction Zone in the Hot Section of the Liner. The Micron Bar is Equal to 100μm (Reprinted with permission from M.K. Ferber, Oak Ridge National Laboratory)

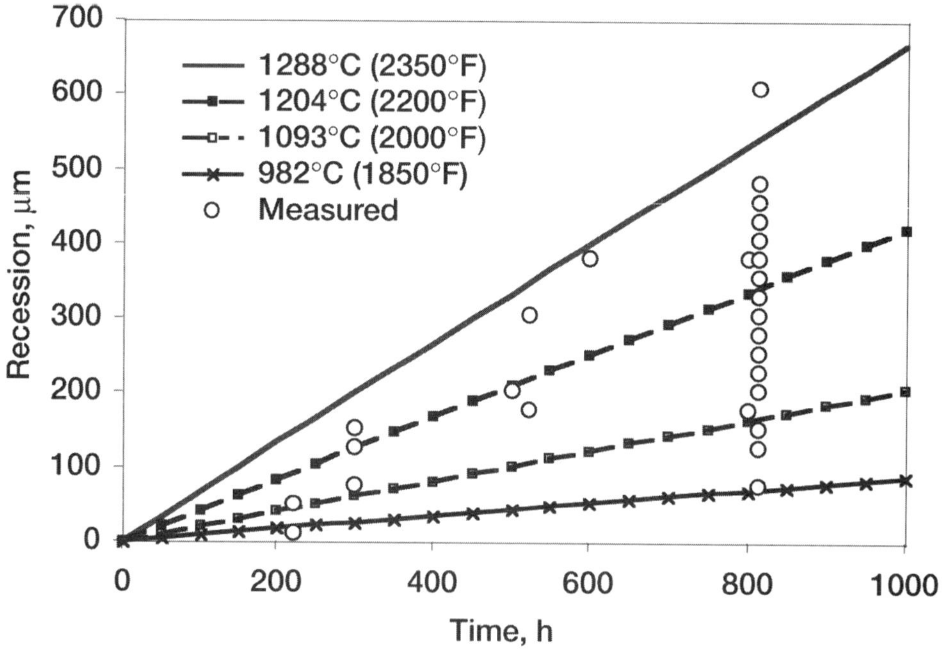

Figure 26.19 Measurement of Recession of AS800 Si_3N_4 Vanes at the Mid-Span of the Trailing Edge Tested in an Industrial Gas Turbine Operated for 815h at a Total Pressure of 8.9 atm and a Peak Temperature of 1260°C (open circles). The Lines Represent Predictions Based on the Model Developed in ref. [32] (Reprinted from ref. [38]).

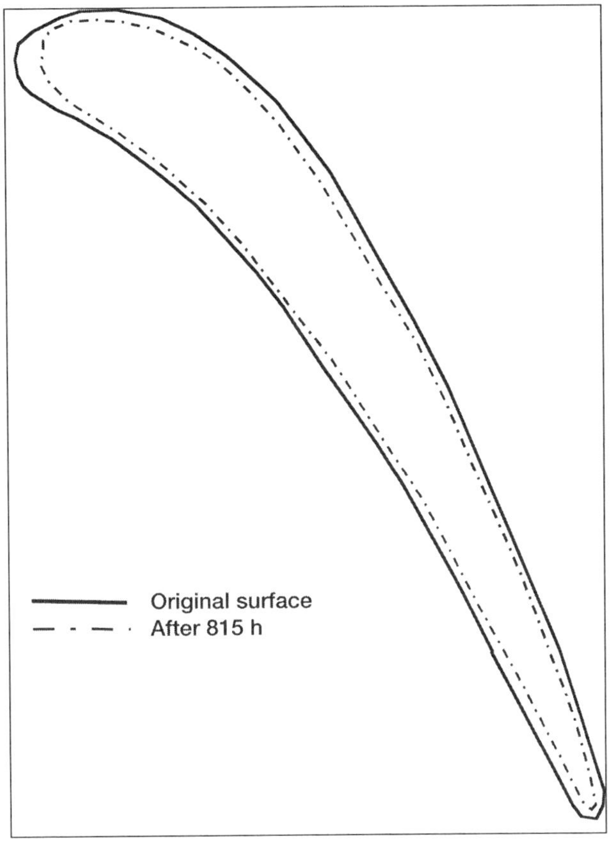

Figure 26.20 Profile of the Cross-Section at Mid-Span of an AS800 Si_3N_4 Vane Before and After Engine Testing in an Industrial Gas Turbine Operated for 815h at a Total Pressure of 8.9 atm and a Peak Temperature of 1260°C (Reprinted from ref. [38])

the measured recession at the mid-span trailing edge of AS800 Si_3N_4 vanes exposed for 815 h in a land based turbine (T_{max} = 1260°C, P = 8.9atm, v_{max} = 573m/s). The scatter is attributed to vane-to-vane temperature variation. The measured recession rates are consistent with rates calculated for a range of temperatures using a model of the type given by Equation (9). Figure 26.20 shows the dimensional change for a single vane at mid-span before and after engine testing for 815 h. Figure 26.21 shows that the recession at the leading edge exceeds the rate found elsewhere on the vane. Other mechanisms such as erosion or viscous flow may also be important here.

Because of the extensive degradation that occurs in turbine environments, as demonstrated in these field-tests, environmental barrier coatings (EBCs) are needed for SiC and Si_3N_4 turbine components. An EBC coated SiC/SiC combustor liner has undergone field-testing and had completed 13,937 hrs with limited recession problems (private communication, M.K. Ferber). EBCs therefore show promise for limiting recession due to silica volatility. Discussion of these coatings is included in chapter 27 in this volume.

Oxidation by CO_2. As previously shown in Fig. 26.11 [1], CO_2 is present in all combustion environments. The presence of this gas in the combustion environment has not been shown to affect the durability of silicon-based ceramics. Figure 26.22 shows that the effect of CO_2 as an oxidant is negligible compared to water vapor and even oxygen [85].

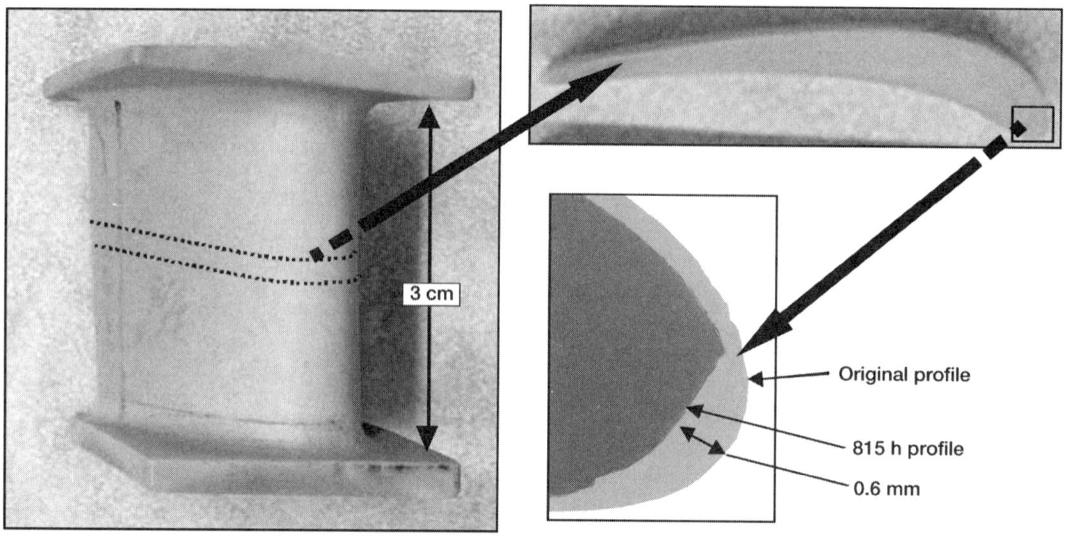

Figure 26.21 Profile of the Leading Edge of an AS800 Si_3N_4 Vane at Mid-Span Before and After Engine Testing in an Industrial Gas Turbine Operated for 815h at a Total Pressure of 8.9 atm and a Peak Temperature of 1260°C. Vane is ~2.5 cm Long (Reprinted with permission from M.K. Ferber, Oak Ridge National Laboratory)

Attack from Aggressive Gases such as HCl and SO_2. Although HCl and SO_2 are present in a combustion environment in small amounts, their effect appears minimal. Limited studies indicate the SiO_2 scale is sufficient to protect against $SiCl_x$(g) formation from HCl attack [86]. Attack of silicon-based ceramics is also negligible. Although SiS(g) formation is possible, this species is not very stable and the SiO_2 scale appears to protect against attack [87].

Low Level Metal Cation Impurities. In any type of combustion environment, low level metallic impurities are always present. These include sodium and potassium and also various transition metals such as iron. These are readily incorporated into the silica scale and generally open up the SiO_2 network leading to faster transport rates and enhanced oxidation rates [88, 89]. It appears that transport in SiO_2 is more readily affected by low level impurities than transport in the other common protective oxides, Al_2O_3 and Cr_2O_3.

Deposit-Induced Corrosion. As discussed in the Introduction, the burner rig studies have shown that salt-induced corrosion can lead to pitting and strength degradation. Salt deposits occur due to the interaction of sodium with sulfur fuel impurities [90]:

$$2\,NaCl(g) + SO_2(g) + 0.5\,O_2(g) + H_2O(g) = Na_2SO_4(l) + 2\,HCl(g) \tag{10}$$

The dewpoint for Na_2SO_4 is a function of sulfur content, sodium content, and pressure. This is shown in Figure 26.23 [91]. Above the dewpoint, deposit-induced corrosion is minimal. However, below the dewpoint, deposition and the resulting corrosion can be quite serious.

Deposit induced corrosion only occurs under specific conditions. Figure 26.24 shows a coupon of SiC oxidized in burner rig with and without the addition of sodium [92]. Note the substantial amount of corrosion due to the sodium. These processes are best interpreted with the acid/base theory of oxides [91, 93]. Na_2SO_4 decomposes to Na_2O, which is the key reactant:

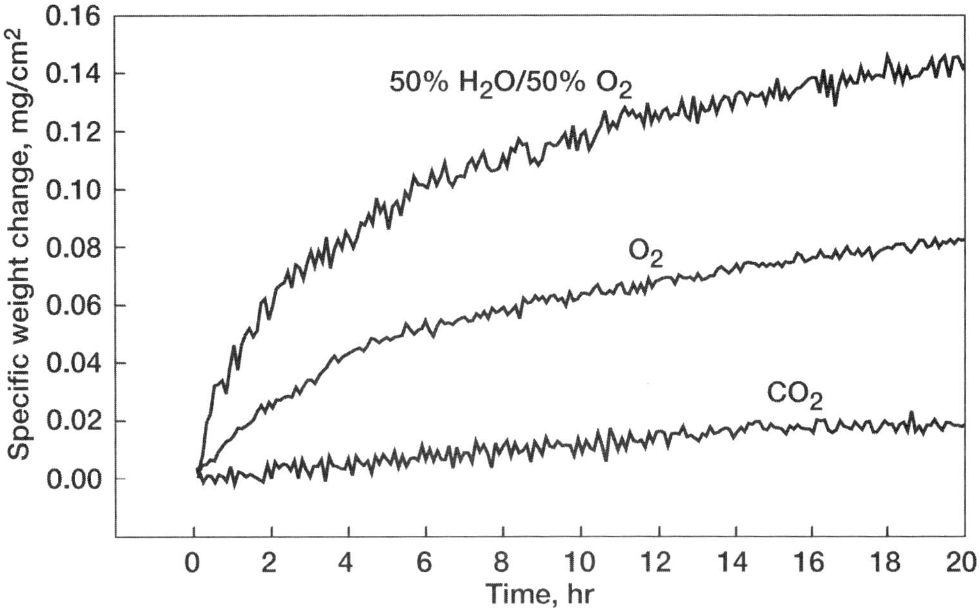

Figure 26.22 Comparison of CVD SiC Oxidation Kinetics in the Oxidants CO_2, O_2 and H_2O at 1200°C (Reprinted from ref. [85] with permission of the American Ceramic Society, www.ceramics.org, © 1998, all rights reserved)

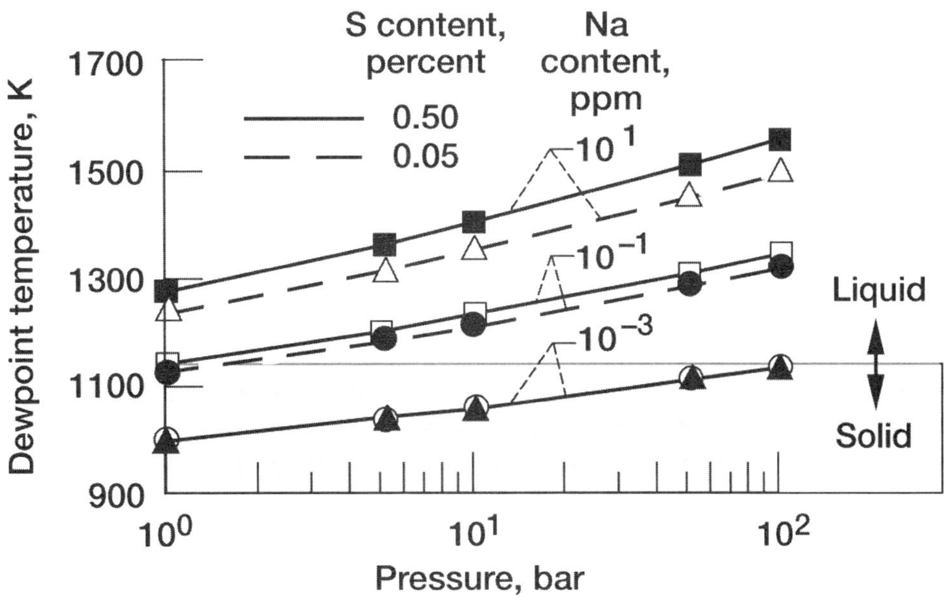

Figure 26.23 Dew Points for the Deposition of Na_2SO_4 (Reprinted from ref. [1] with permission of the American Ceramic Society, www.ceramics.org, © 1993, all rights reserved)

$$Na_2SO_4(l) = Na_2O(s) + SO_3(g) \quad (11)$$

The partial pressure of $SO_3(g)$ above the Na_2SO_4 deposit sets the chemical activity of Na_2O. A salt with a high activity of Na_2O is a basic molten salt; a salt with a low activity of Na_2O is an acidic molten salt. Since SiO_2 is an acidic oxide, it is readily attacked by a basic oxide to form sodium silicate:

$$2\ SiO_2(s) + Na_2O(s) = Na_2O \cdot 2(SiO_2)(l) \quad (12)$$

This is the key reaction in deposit-induced corrosion of SiC and Si_3N_4. A basic oxide readily attacks the acidic SiO_2 and the protective film is replaced by a liquid, non-protective oxide film. Transport rates through the liquid film are rapid and more SiO_2 forms. This immediately reacts with the Na_2O to form more liquid and the thick film observed in Figure 26.24 forms. Clearly, this process can continue until all of the Na_2O is consumed.

The kinetics of this process are complex and depend on a variety of external (e.g. basic/acidic deposits) and internal factors [94]. In some cases, Na_2SO_4 is continuously deposited and hence a substantial quantity of SiC or Si_3N_4 can be consumed. In other cases, the amount of deposit may be limited and a sub-layer of silica grows below the melt. This would slow corrosion, but the consumption of SiC or Si_3N_4 can be quite substantial before this occurs.

In practice, the deposit is not only Na_2SO_4. Sea salt leads to a mixed Na_2SO_4, K_2SO_4, Ca_2SO_4, and $MgSO_4$ deposit [95]. This, in turn, may lead to other silicates besides sodium silicate [96]. Sodium vanadates may also deposit from lower purity fuels. The chemistry of these mixed deposits can be quite complex, but

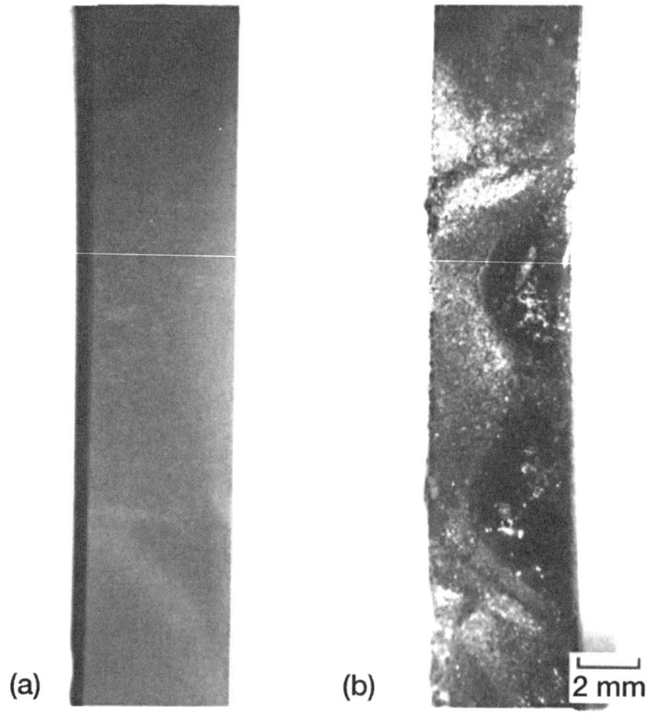

Figure 26.24 Optical Micrographs of Sintered SiC Coupons with Carbon and Boron Additives, Oxidized in a 4 atm Pressurized Burner Rig at 1000°C (a) 46 h with No Sodium (b) 13.5 h with a NaCl-Seeded Flame (Reprinted from ref. [1] with permission of the American Ceramic Society, www.ceramics.org, © 1993, all rights reserved)

in general the basic solutions are very corrosive to silica. In addition it has been shown that carbon can drive an acidic sulfate deposit more basic [91]. This is significant, since carbon is a common additive in monolithic SiC and composites.

An important issue is the form of attack of the SiC and Si_3N_4. Attack does not occur with even recession, but rather in the form of localized pitting and attack at structural discontinuities [97]. This is illustrated in Figure 26.25. In general SiC tends to show more pitting and attack than Si_3N_4. Attack of Si_3N_4 is primarily along grain boundaries [98].

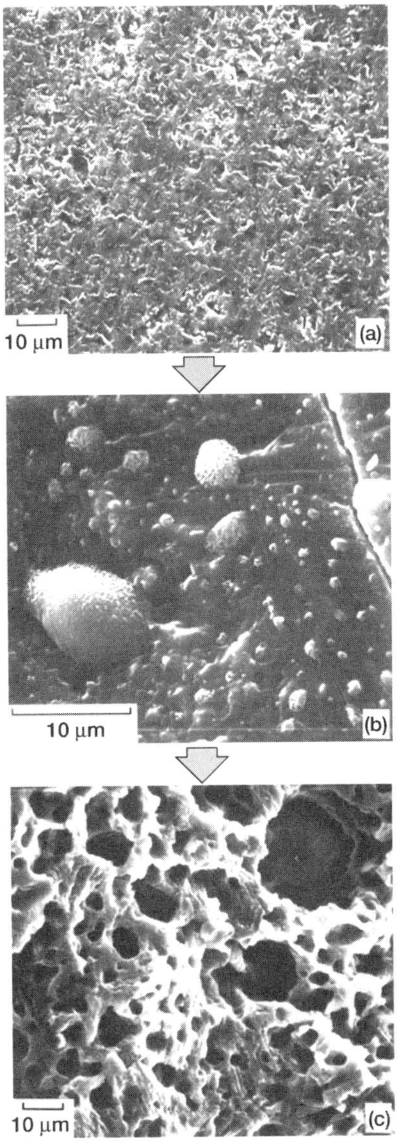

Figure 26.25 Sequence Showing Sintered SiC (a) Before Corrosion (b) After Corrosion with Na_2SO_4/(10^{-4} bar SO_3 + 1 bar O_2), 1273 K for 48 h (c) with Corrosion Products Removed, Revealing Pitting (Reprinted from ref. [1] with permission of the American Ceramic Society, www.ceramics.org, © 1993, all rights reserved)

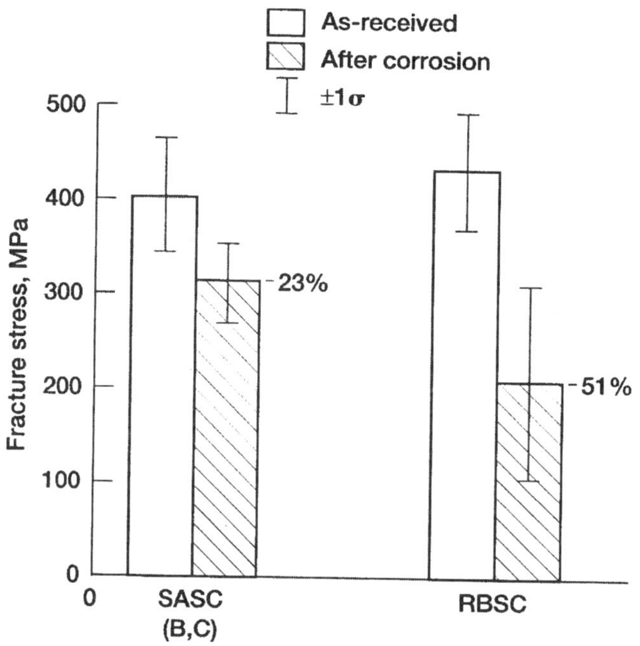

Figure 26.26 Strength Degradation of SiC After Burner Rig Corrosion. SASC (B,C) = Sintered α-SiC with Boron and Carbon Additives; RBSC = Reaction-Bonded SiC (Adapted from ref. [18] with permission of the American Ceramic Society, www.ceramics.org, © 1990, all rights reserved)

The type of surface attack described here is a problem with monolithic ceramics, since their strength is related to surface finish. In controlled laboratory studies significant strength degradation has been observed and fracture origins were often corrosion pits [17]. This is shown in Figures 26.26 and 26.27. This is consistent with the burner and engine testing programs, where similar observations have been made [10,18].

Upper Temperature Limits. The question of upper temperature limits for the use of silicon-based ceramics is often raised. Typically these are 1300–1500°C in pure oxygen and are quite dependent on the material and system. The major environmental factors which impose a temperature limit are:

1. Rapid oxidation rates, leading to extensive bubbling, and material consumption
2. Rapid volatility and material consumption
3. Scale/substrate reactions may lead to extensive bubbling in the scale. This is a particular issue for carbon-containing SiC [99].

Additive containing ceramics, which generally exhibit faster oxidation rates than pure materials in a high-velocity combustion environment would exhibit lower temperature limits than pure materials in pure oxygen. One approach to improve the oxidation behavior of additive-containing ceramics is a CVD coating of pure SiC or Si_3N_4.

As discussed, water vapor and salt deposits severely limit the upper use temperature for silicon-based ceramics. This attack is due to the nature of the SiO_2 film and hence a pure SiC or Si_3N_4 coating is not effective. For extended service at temperatures above about 1150°C in a turbine environment, which contains water vapor and the possibility of salt deposits, some type of environmental barrier coating (EBC) may be necessary. Several refractory oxides are being developed for this purpose. This is discussed in the next chapter of this book.

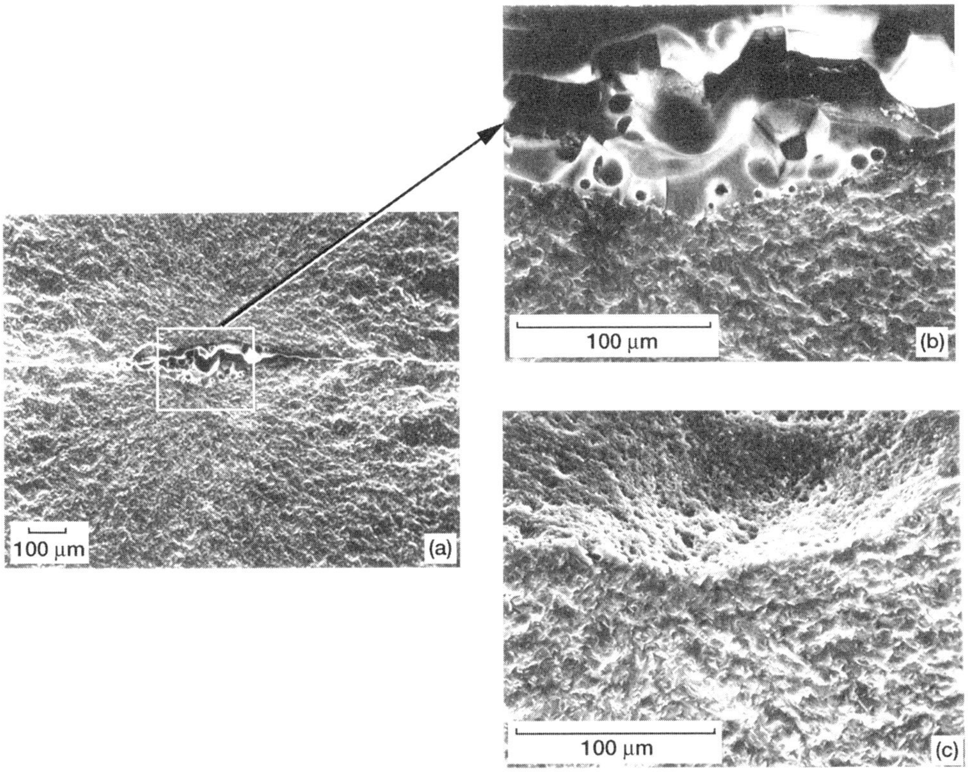

Figure 26.27 Typical Fracture Origin from Burner Rig Corrosion in SiC. The Enlar gements Show the Glass-Filed Pit and the Pit After an Hydrofluoric Acid Treatment to Remove the Glass (Reprinted from ref. [92] with permission of the American Ceramic Society, www.ceramics.org, © 1986, all rights reserved)

CORROSION MECHANISMS FOR OTHER CERAMICS

Aluminum Nitride

Aluminum nitride is often discussed as a potential high temperature material. It has a high thermal conductivity and forms an Al_2O_3 scale on oxidation. However, it failed catastrophically in some of the early engine tests [11]. It is known to rapidly oxidize in water vapor. Studies have shown that oxidation in water vapor leads to the formation of Al_2O_3 with micro-pores, which serve as rapid gas transport paths [100,101]. This problem appears to be intrinsic to AlN and thus renders it unsuitable for combustion applications.

Oxides

Oxides are very often the most chemically stable compounds containing metallic elements [102]. As mentioned, the first ceramic materials considered for turbine engines were porcelains. The problem is that oxides may not be microstructurally stable to high temperatures and are also susceptible to thermal shock and creep. However, application of certain refractory oxide ceramic coatings on both metals and silicon-based ceramics may be a particularly promising approach for using ceramics in gas-turbine applications. Ceramic coatings are the topic of another chapter in this book.

As discussed, monolithic oxide ceramics were prime candidates for heat exchangers in the automotive gas turbine (AGT) project [12]. Initially, lithium aluminum silicate (LAS) was proposed as a material for heat exchangers, however, LAS was quite susceptible to chemical attack. Lower temperature regions were attacked by small amounts of sulfuric acid in the combustion gas; higher temperature regions were attacked by the sodium or potassium impurities in the combustion gas. In both cases, ion exchange (H^+, Na^+, or K^+) occurred with the lithium, leading to an oxide with regions of different thermal expansion coefficient. This in turn led to cracking. As a result of these corrosion issues, magnesium aluminum silicate (MAS) was developed as a heat exchanger material. This material is substantially more chemically inert than LAS, but still appears to show some interactions with Na cations [103].

SUMMARY

From roughly 1970 to the present, there have been numerous programs to develop a ceramic heat engine. Although a commercially viable ceramic engine is not yet a reality, a good deal has been learned about the performance of ceramics in heat engine environments. Most of the emphasis has been on SiC and Si_3N_4 and, in recent years, on composites based on these materials. It has become apparent that the protective SiO_2 film which grows on these materials is not as inert as originally thought. While it is stable in an isothermal oxidizing environment, complications arise due to second phases in the ceramic substrate and exposure to complex combustion environments. These can be summarized as follows:

A. Material Issues
1. Refractory oxide additives migrate to the SiO_2 scale and weaken its otherwise protective properties.
2. Ceramic matrix composites contain an easily-oxidizable phase—either carbon fibers or carbon or boron nitride fiber coatings. Currently there is no adequate way to protect this phase in the event of a matrix crack. Some types of sealing schemes have been developed, but more research is needed.

B. Environment Issues
1. Thermal cycling is not a major problem for the pure materials, but can lead to substantial degradation
2. Water vapor can lead to substantial degradation both via enhanced oxidation and volatilization of the SiO_2 scale. In addition, boron-containing constituents are also quite susceptible to water vapor attack.
3. SiO_2 scales appear to be fairly impervious to HCl and SO_2.
4. Low level metallic impurities, such as Na and K, open the SiO_2 network and lead to faster oxidation rates. This is more of a problem for SiO_2 than other protective oxides.
5. Basic molten salt deposits readily dissolve the SiO_2 film, leading to a liquid, non-protective silicate. Attack of the SiC and Si_3N_4 is in the form of pitting and grain boundary attack, which leads to strength degradation.

Refractory oxide environmental barrier coatings may be a solution some of these issues and are discussed in more detail in the next chapter.

Other ceramics, such as AlN and MAS have also been considered for gas turbines. AlN is very reactive with water vapor and is therefore not suitable. MAS is a promising heat exchanger material, but reactivity with Na salts is a potential problem.

REFERENCES

[1] Jacobson, N. S., "Corrosion of Silicon-Based Ceramics in Combustion Environments," J. Am. Ceram. Soc., 76[1], 3–28, 1993.

[2] Conway, H. M., *The Possible Use of Ceramic Materials in Aircraft Propulsion Systems*, NACA Report CB No. 4D10, NACA, Washington, D.C., USA, 1944.

[3] Lay, L. A., *Corrosion Resistance of Technical Ceramics*, 2nd Edition, Her Majesty's Stationary Office, London, UK, 1991. pp. 18–23.

[4] Prochazka, S., "Sintering of Silicon Carbide,"in *Ceramics for High Performance Applications*, Proceedings of the Second Army Materials Technology Conference, Hyannis, MA, USA, November 13–16, 1973, J. J. Burke, A. E. Gorum, and R. N. Katz, eds., Brook Hill Publishing Company, Chestnut Hill, MA, USA, 1974. pp. 239–252.

[5] Riley, F. L., "Silicon Nitride and Related Materials," J. Am. Ceram. Soc., 83[2], 245–265, 2000.

[6] Rice, R. W., *An Assessment of the Use of Ceramics in Heat Engines*, Report 4499, Naval Research Laboratory, Washington, D.C., USA, 1981.

[7] Singhal, S. C. and Lange, F. F., "Effect of Alumina Content on the Oxidation of Hot-Pressed Silicon Carbide," J. Am. Ceram. Soc., 58[9–10], 433–435, 1975.

[8] Singhal, S. C., "Oxidation Kinetics of Hot-Pressed Silicon Carbide," J. Mater. Sci., 11, 1246–1253, 1976.

[9] Wallace, F. B., "Garrett-AiResearch Program Overview," in *Proceedings of the 1977 DARPA/NAVSEA Ceramic Gas Turbine Demonstration Engine Program Overview,* Fairbanks, J. W. and Rice, R. W., eds., Metals and Ceramic Information Center, Columbus, OH, USA, 1977. pp. 7–18.

[10] Richerson, D. W. and Yonushonis, T. M., "Environmental Effects on the Strength of Silicon-Nitride Materials," ibid., pp. 247–271.

[11] Adams, J. W. and Larsen, D. C., *Evaluation of Corrosion/Erosion Behavior of Various Ceramic Materials*, Report AFWAL-TR-84-4067, Air Force Wright Aeronautical Labs, Wright-Patterson Air Force Base, OH, USA, 1984.

[12] Rahnke, C. J. and Vallance, J. K., "Reliability and Durability of Ceramic Regenerators in a Gas Turbine," ASME paper 77-GT-59, presented at the ASME Gas Turbine Conference, Philadelphia, PA, USA, March 27–31, 1977. Transactions of the ASME, Journal of Engineering for Power, 100[1], 73–81, 1978.

[13] Boyd, G. L., Kidwell, J. R., and Kreiner, D. M., "A Technology Development Summary for the AGT 101 Advanced Gas Turbine Program," *Proceedings of the Twenty-Fourth Automotive Technology Development Contractors' Coordination Meeting*, Society of Automotive Engineers, Inc., Warrendate, PA, USA, P-197, 1987. pp. 115–135.

[14] Helms, H. E., Haley, P. J., Groseclose, L. E., Hilpisch, S. J., and Bell, A. H., "Advanced Turbine Technology Applications Project," *Proceedings of the Twenty-Sixth Automotive Technology Development Contractors' Coordination Meeting*, Society of Automotive Engineers, Inc., Warrendate, PA, USA, P-219, 1989. pp. 319–326.

[15] Carruthers, W. D., Richerson, D. W., and Benn, K. W., *3500-Hour Durability Testing of Commercial Ceramic Materials Interim Report*, DOE/NASA/0027-80/1, NASA Report CR-159785, NASA Lewis Research Center, Cleveland, OH, USA, 1980.

[16] Lindberg, L. J., "Durability Testing of Ceramic Materials for Turbine Engine Applications," Proceedings of the Twenty-Fourth Automotive Technology Development Contractors' Coordination Meeting, Society of Automotive Engineers, Inc., Warrendate, PA, USA, P-197, 1987. pp. 149–162.

[17] Smialek, J. L. and Jacobson, N. S., "Mechanism of Strength Degradation for Hot Corrosion of α-SiC," J. Am. Ceram. Soc., 69[10], 741–752, 1986.

[18] Fox, D. S. and Smialek, J. L., "Burner Rig Hot Corrosion of Silicon Carbide and Silicon Nitride," J. Am. Ceram. Soc., 73[2], 303–311, 1990.

[19] Federer, J.I., Ogle, J., Tennery, V.J., and Hensen, T., Characterization of Corrosion Mechanisms Occurring in a Sintered SiC Exposed to Basic Coal Slags," J. Am. Ceram. Soc., 68[4], 191–197, 1985.

[20] Seitz, E., "The German Materials Research Program 1985–1994," Advanced Performance Materials, 1, 87–117, 1994.

[21] Nickel, K. G., Danzer, R., Schneider, G., and Petzow, G., "Corrosion and Oxidation of Advanced Ceramics," Powder Metallurgy International, 21[3], 29–34, 1989.

[22] Tatsumi, T., Takehara, I., and Ichikawa, Y., "Development Summary of the 300kW Ceramic Gas Turbine CGT302," ASME paper 99-GT-105, presented at the International Gas Turbine and Aeroengine Congress and Exhibition, Indianapolis, IN, USA, June 7–10, 1999.

[23] Yoshida, M., Tanaka, K., Tsuruzono, S., and Tatsumi, T., "Development of Silicon Nitride Components for Ceramic Gas Turbine Engine (CGT302)," Ind. Ceram., 19[3], 188–192, 1999.

[24] Tanaka, K., Yoshida, M., Kubo, T., Terazono, H., and Tsuruzono, S., "Development and Evaluation of Ceramic Components for Small Gas Turbine Engines," ASME paper 2000-GT-531, presented at the ASME TURBO EXPO 2000, LAND, SEA, & AIR, Munich, Germany, May 8–11, 2000.

[25] Nakazawa, N., Sasaki, M., Nishiyama, T., Iwai, M., and Katagiri, H., "Status of the Automotive Ceramic Gas Turbine Development Program – Seven Years' Progress,"ASME paper 97-GT-383, presented at the International Gas Turbine and Aeroengine Congress and Exposition, Orlando, FL, USA, June 2–5, 1997.

[26] Yuri, I., Hisamatsu, T., Wanatabe, K., and Etori, Y., "Structural Design and High Pressure Test of a Ceramic Combustor for 1500°C Class Industrial Gas Turbine," ASME-Paper 96-GT-346, presented at the International Gas Turbine and Aeroengine Congress and Exposition, Birmingham, U.K., June 10–13, 1996. Transactions of the ASME, Journal of Engineering for Gas Turbines and Power, 119[4], 506–511, 1997.

[27] Machida, T., Nakayama, M., Wada, K., Hisamatsu, T., Yuri, I., and Watanabe, K., "Development of Ceramic Stator Vane for 1500°C Class Gas Turbine," ASME paper 96-GT-459, presented at the International Gas Turbine and Aeroengine Congress and Exposition, Birmingham, U.K., June 10–13, 1996.

[28] Hara, Y., Tsuchiya, T., Maeda, F., Tsuji, I., and Wada, K., "Development of Ceramic Components for a Power Generating Gas Turbine," ASME paper 91-GT-319, presented at the International Gas Turbine and Aeroengine Congress and Exposition, Orlando, FL, June 3–6, 1991.

[29] Kawamoto, H., "High Temperature Fracture Behavior of Structural Ceramics and the Evaluation Technologies," Japan Fine Ceramic Center, Nagoya, Japan, JFCC Review 8, 226–242, 1996 (in Japanese).

[30] Misra, A. K., Johnson, A. M., and Bartlett, B. J., "Progress Toward Meeting Material Challenges for High Speed Civil Transport Propulsion," in *Proceedings of the 14th International Symposium on Air Breathing Engines*, Waltrup, P. J., ed., International Society for Air Breathing Engines, Florence, Italy, 1999, ISABE Paper, Number 99-7018.

[31] Office of Industrial Technologies, *Continuous Fiber Ceramic Composite Program,* Report DOE/IET-91/8, Current Awareness Industrial Energy Technology, US Department of Energy, Washington, D.C., USA, 1991.

[32] Robinson, R. C. and Smialek, J. L., "SiC Recession Caused by SiO_2 Scale Volatility under Combustion Conditions: I, Experimental Results and Empirical Model," J. Am. Ceram. Soc., 82[7], 1817–1825, 1999.

[33] Opila, E. J., Smialek, J. L., Robinson, R. C., Fox, D. S., and Jacobson, N. S., "SiC Recession Caused by SiO_2 Scale Volatility under Combustion Conditions: II, Thermodynamics and Gaseous Diffusion Model," J. Am. Ceram. Soc., 82[7], 1826–1834, 1999.

[34] Etori, Y., Hisamatsu, T., Yuri, I., Yasutomi, Y., Machida, T., and Wada, K., "Oxidation Behavior of Ceramics for Gas Turbines in Combustion Gas Flow at 1500°C," ASME paper 97-GT-355, presented at the International Gas Turbine and Aeroengine Congress and Exposition, Orlando, FL, USA, June 2–5, 1997.

[35] Yuri, I., Hisamatsu, T., Etori, Y., and Yamamoto, T., "Degradation of Silicon Carbide in Combustion Gas Flow at High-Temperature and Speed," ASME paper 2000-GT-664, presented at the ASME TURBO EXPO 2000, LAND, SEA, & AIR, Munich, Germany, May 8–11, 2000.

[36] More, K. L., Tortorelli, P. F., Ferber, M. K., Walker, L. R., Keiser, J. R., Miriyala, N., Brentnall, W. D., and Price, J. R., "Exposure of Ceramics and Ceramic Matrix Composites in Simulated and Actual Combustor Environments," ASME paper 99-GT-292, presented at the International Gas Turbine and Aeroengine Congress and Exposition, Indianapolis, IN, USA, June 7–10, 1999.

[37] Miriyala, N., Simpson, J. F., Parathasarathy, V. M., and Brentnall, W. D., "The Evaluation of CFCC Liners after Field-Engine Testing in a Gas Turbine," ASME paper 99-GT-395, presented at the International Gas Turbine and Aeroengine Congress and Exposition, Indianapolis, IN, USA, June 7–10, 1999.

[38] Ferber, M. K., Lin, H. T., Parthasarathy, V., and Wenglarz, R. A., "Degradation of Silicon Nitrides in High Pressure, Moisture Rich Environments," ASME paper 2000-GT-0661, presented at the ASME TURBO EXPO 2000, LAND, SEA, & AIR, Munich, Germany, May 8–11, 2000.

[39] Lamouroux, F., Camus, G., and Thebault, J., "Kinetics and Mechanisms of Oxidation of 2D Woven C/SiC Composites: I, Experimental Approach," J. Am. Ceram. Soc., 77[8], 2049–2057, 1994.

[40] Lamouroux, F., Naslain, R., and Jouin, J-M., "Kinetics and Mechanisms of Oxidation of 2D Woven C/SiC Composites: II, Theoretical Approach," J. Am. Ceram. Soc., 77[8], 2058–2068, 1994.

[41] Filipuzzi, L., Camus, G., Naslain, R., and Thebault, J., "Oxidation Mechanisms and Kinetics of 1D-SiC/C/SiC Composite Materials: I, An Experimental Approach," J. Am. Ceram. Soc., 77[2], 459–466, 1994.

[42] Filipuzzi, L., and Naslain, R., "Oxidation Mechanisms and Kinetics of 1D-SiC/C/SiC Composite Materials: II, Modeling," J. Am. Ceram. Soc., 77[2], 467–489, 1994.

[43] Ogbuji, L. U. J. T., "A Pervasive Mode of Oxidative Degradation in a SiC-SiC Composite," J. Am. Ceram. Soc., 81[11], 2777–2784, 1998.

[44] More, K. L., Tortorelli, P. F., Lin, H.-T., Lara-Curzio, E., and Lowden, R. A., "Degradation Mechanisms of BN Interfaces in SiC/SiC Composites in Oxygen and Water-Vapor Containing Environments," in *Proceedings of the Symposium on High Temperature Corrosion and Materials Chemistry*, Hou, P. Y., McNallan, M. J., Oltra, R., Opila, E. J., and Shores, D. A., eds., San Diego, CA, USA, May 3–8, 1998, Electrochemical Society, Pennington, NJ, USA, 1998. pp. 382–394.

[45] Jacobson, N. S., Morscher, G. N., Bryant, D. R., and Tressler, R. E., "High-Temperature Oxidation of Boron Nitride: II, Boron Nitride Layers in Composites," J. Am. Ceram. Soc., 82[6], 1473–1482, 1999.

[46] Deadmore, D. L., Lowell, C. E., and Kohl, F. J., "The Effect of Fuel-to-Air Ratio on Burner Rig Hot Corrosion," Corrosion Science, 19, 371–378, 1979.

[47] Ogbuji, L. U. J. T. and Opila, E. J., "A Comparison of the Oxidation Kinetics of SiC and Si_3N_4," J. Electrochem. Soc., 142[3], 925–930, 1995.

[48] Opila, E. J. and Jacobson, N. S., "Corrosion of Ceramics, Corrosion and Environmental Degradation of Materials," in *Materials Science and Technology, Vol. 19*, Schutze, M., ed., Wiley-VCH, Weinheim, Germany, 2000. pp. 327–388.

[49] Kofstad, P., *High Temperature Corrosion*, Elsevier, London, UK, 1998. pp. 17–18.

[50] Du, H., Tressler, R. E., and Spear, K. E., "Thermodynamics of the Si-N-O System and Kinetic Modeling of Oxidation of Si_3N_4," J. Electrochem. Soc., 136[11], 3210–3215, 1989.

[51] Ogbuji, L. U. J. T. and Jayne, D. T., "Mechanism of Incipient Oxidation of Bulk Chemical Vapor Deposited Si_3N_4," J. Electrochem. Soc., 140[3], 759–766, 1993.

[52] Fox, D. S., "Oxidation Behavior of Chemically-Vapor-Deposited Silicon Carbide and Silicon Nitride from 1200° to 1600°C," J. Am. Ceram. Soc., 81[4], 945–950, 1998.

[53] Costello, J. A. and Tressler, R. E., "Oxidation Kinetics of Silicon Carbide Crystals and Ceramics: I, In Dry Oxygen," J. Am. Ceram. Soc., 69[9], 674–681, 1986.

[54] Cubiciotti, D. and Lau, K., "Kinetics of Oxidation of Hot-Pressed Silicon Nitride Containing Magnesia," J. Am. Ceram. Soc., 61[11–12], 512–517, 1978.

[55] Cubiciotti, D. and Lau, K. H., "Kinetics of Oxidation of Yttria Hot-Pressed Silicon Nitride," J. Electrochem. Soc., 126[10], 1723–1728, 1979.

[56] Nickel, K. G., "Multiple Law Modeling for the Oxidation of Advanced Ceramics and a Model-Independent Figure-of-Merit," in *Corrosion of Advanced Ceramics*, Nickel, K., ed., Kluwer Academic Publishers, Dordrecht, The Netherlands, 1994. pp. 59–72.

[57] Patel, J. K. and Thompson, D. P., "The Low-Temperature Oxidation Problem in Yttria-Densified Silicon Nitride Ceramics," Br. Ceram. Trans. J., 87, 70–73, 1988.

[58] Evans, A. G. and Davidge, R. W., "The Strength and Oxidation of Reaction-Sintered Silicon Nitride," J. Mater. Sci., 5, 314–325, 1970.

[59] Porz, F. and Thümmler, F., "Oxidation Mechanisms of Porous Silicon Nitride," J. Mater. Sci., 19, 1283–1295, 1984.

[60] Halbig, M. C. and Cawley, J. D., "Modeling the Environmental Effects on Carbon Fibers in a Ceramic Matrix at Oxidizing Conditions," Ceram. Eng. and Sci. Proc., 21[3], 219–226, 2000.

[61] Lamouroux, F., Bertrand, S., Pailler, R., and Naslain, R., "A Multilayer Ceramic Matrix for Oxidation Resistant Carbon-Fiber Reinforced CMCs," Key Engineering Materials, 164[5], 365–368, 1999.

[62] Jacobson, N. S., Smialek, J. L., Fox, D. S., and Opila, E. J., "Durability of Silicon-Protected Ceramics in Combustion Environments," in *High-Temperature Ceramic Matrix Composites I: Design, Durability, and Performance*, Evans, A. G. and Naslain, R., eds., American Ceramic Society, Westerville, OH, USA, Ceramic Transactions, 57, 157–170, 1995.

[63] Eckel, A. J., Cawley, J. D., and Parthasarathy, T. A., "Oxidation Kinetics of a Continuous Carbon Phase in a Nonreactive Matrix," J. Am. Ceram. Soc., 78[4], 972–980, 1995.

[64] Bhatt, R., *Oxidation Effects on the Mechanical Properties of SiC Fiber-Reinforced Reaction-Bonded Silicon Nitride Matrix Composites*, NASA TM-102360, NASA Lewis Research Center, Cleveland, OH, USA,1989.

[65] Becher, P. F., "Strength Retention in SiC Ceramics after Long-Term Oxidation," J. Am. Ceram. Soc., 67, C-120-1, 1983.

[66] Jakus, K., Ritter, J. E., Jr., and Rogers, W. P., "Strength of Hot-Pressed Silicon Nitride After High Temperature Exposure," J. Am. Ceram. Soc., 67[7], 471–475, 1983.

[67] Tressler, R. E., "Environmental Effects on the Long Term Reliability of SiC and Si_3N_4 Ceramics," in *Corrosion and Corrosive Degradation of Ceramics*, Tressler, R. E. and McNallan, M. J., eds., American Ceramic Society, Westerville, OH, USA, Ceramic Transactions, 10, 99–124, 1990.

[68] Gogotsi, Y. G. and Grathwohl, G., "Stress-Enhanced Oxidation of Silicon Nitride Ceramics," J. Am. Ceram. Soc., 76[12], 3093–3104, 1993.

[69] Wereszczak, A. A., Ferber, M. K., Kirkland, T. P., More, K. L., Foley, M. R., and Yeckley, R. L., "Evaluation of Stress-Failure Resulting from High-Temperature Stress—Corrosion Cracking in a Hot Isostatically Pressed Silicon Nitride," J. Am. Ceram. Soc., 78[8], 2129–2140, 1995.

[70] Thümmler, F. and Grathwohl, F., "High Temperature Oxidation and Creep of Si_3N_4 and SiC Based Ceramics and their Mutual Interaction," in *MRS International Meeting on Advanced Materials*, Materials Research Society, 1989. pp. 237–253.

[71] Halbig, M. C., Brewer, D. N., Eckel, A. J., and Cawley, J. D., *Stressed Oxidation of C/SiC Composites*, NASA TM 107457, National Aeronautics and Space Administration, Washington, DC, USA, 1997.

[72] Lara-Curzio, E., Tortorelli, P. F., and More, K. L., 1997, "Stress-Rupture of Nicalon^{TM}/SiC at Intermediate Temperatures," Ceram. Eng. and Sci. Proc., 18[4], 209–219, 1997.

[73] Morscher, G. N., Hurst, J., and Brewer, D.,"Intermediate-Temperature Stress Rupture of a Woven Hi-Nicalon, BN-Interphase, SiC-Matrix Composite in Air," J. Am. Ceram. Soc., 83[6], 1441–1449, 2000.

[74] Gary, J. H. and Handwerk, G. E., *Petroleum Refining: Technology and Economics*, Marcel-Dekker, New York, USA, 1975. p. 32.

[75] *North American Combustion Handbook*, 2nd Edition, North American Manufacturing Co., Cleveland, OH, USA, 1978.

[76] *Standard Specifications for Automotive Gasoline, ASTM Designation D," Book of ASTM Standards*, American Society for Testing and Materials, Philadelphia, PA, USA, 1991. pp. 439–489.

[77] Valkovic, V., *Trace Elements in Petroleum*, PPE Books, Tulsa, OK, USA, 1978.

[78] Jungers, R. H., Lee, R. E., Jr., and van Lehmden, D. J., "The EPA National Fuel Surveillance Network. I. Trace Constituents in Gasoline and Commercial Fuel Gasoline Additives," Environmental Health Perspectives, 10, 143–150, 1975.

[79] Opila, E. J., "Oxidation Kinetics of Chemically Vapor-Deposited Silicon Carbide in Wet Oxygen," J. Am. Ceram. Soc., 77[3], 730–736, 1994.

[80] Deal, B. E., and Grove, A. S., "General Relationship for the Thermal Oxidation of Silicon," J. Appl. Phys., 36[12], 3770–3778, 1965.

[81] Opila, E. J., "Variation of the Oxidation Rate of Silicon Carbide with Water-Vapor Pressure," J. Am. Ceram. Soc., 82[3], 625–636, 1999.

[82] More, K. L., Tortorelli, P. F., Ferber, M. K., and Keiser, J. R., "Observations of Accelerated Silicon Carbide Recession by Oxidation at High Water-Vapor Pressures," J. Am. Ceram. Soc., 83[1], 211–213, 2000.

[83] Hashimoto, A., "The Effect of H_2O Gas on Volatilities of Planet-Forming Major Elements: Experimental Determination of Thermodynamic Properties of Ca-, Al-, and Si-Hydroxide Gas Molecules and its Application to the Solar Nebula," Geochim. Cosmochim. Acta, 56, 511–532, 1992.

[84] Opila, E. J., Fox, D. S., and Jacobson, N. S., "Mass Spectrometric Identification of Si-O-H(g) Species from the Reaction of Silica with Water Vapor at Atmospheric Pressure," J. Am. Ceram. Soc., 80[4], 1009–1012, 1997.

[85] Opila, E. J., and Nguyen, Q. N., "Oxidation of Chemically-Vapor-Deposited Silicon Carbide in Carbon Dioxide," J. Am. Ceram. Soc., 81[7], 1949–1952, 1998.

[86] Marra, J. E., Kreidler, E. R., Jacobson, N. S., and Fox, D. S., "Reactions of Silicon-Based Ceramics in Mixed Oxidation Chlorination Environments," J. Am. Ceram. Soc., 71[12], 1067–1073, 1988.

[87] Nickel, H. and Förthmann, R., "Hot Gas Corrosion of Structural Ceramics," in *Corrosion of Advanced Ceramics*, Nickel, K., ed., Kluwer Academic Publishers, Dordrecht, The Netherlands, 1994. pp. 99–116.

[88] Pareek, V. and Shores, D. A., "Oxidation of Silicon Carbide in Environments Containing Potassium Salt Vapor," J. Am. Ceram. Soc., 74[3], 556–563, 1991.

[89] Zheng, Z., Tressler, R. E., and Spear, K. E., "The Effect of Sodium Contamination on the Oxidation of Single Crystal Silicon Carbide, Corrosion Science, 33, 545–556, 1992.

[90] Kohl, F. J., Stearns, C. A., and Fryburg, G. C., "Sodium Sulfate: Vaporization Thermodynamics and Role in Corrosive Flames," in *Proceedings, Metal-Slag-Gas Reactions and Processes*, Foroulis, Z. A. and Smeltzer, W. W., eds., The Electrochemical Society, Princeton, NJ, USA, 1975. pp. 649–664.

[91] Jacobson, N. S., 1989, "Sodium Sulfate: Deposition and Dissolution of Silica," Oxid. Met., 31, 91–103, 1989.

[92] Jacobson, N. S., Stearns, C. A., and Smialek, J. L., "Burner Rig Corrosion of SiC at 1000°C," Adv. Ceram. Mat., 1[2], 154–161, 1986.

[93] Rapp, R. A., "Chemistry and Electrochemistry of the Hot Corrosion of Metals," Corros. Sci., 42, 568–577, 1986.

[94] Berthold, C. and Nickel, K. G., "Hot-Corrosion Behaviour of Silica and Silica-Formers: External vs. Internal Control," Journal of the European Ceramic Society, 18[16], 2365–2372, 1998.

[95] Bornstein, N. M., "Reviewing Sulfidation Corrosion—Yesterday and Today," JOM, 11, 37–39, 1996.

[96] Fox, D. S., Cuy, M. D., and Nguyen, Q. N., "Sea-Salt Corrosion and Strength of a Sintered α-Silicon Carbide," J. Am. Ceram. Soc., 81[6], 1565–1570, 1998.

[97] Jacobson, N. S. and Smialek, J. L., "Corrosion Pitting of SiC by Molten Salts," J. Electrochem. Soc., 133[12], 2615–2621, 1986.

[98] Jacobson, N. S., Smialek, J. L., and Fox, D. S., "Molten Salt Corrosion of SiC and Si_3N_4," in *Handbook of Ceramics and Composites Volume I: Synthesis and Properties*, Cheremisinoff, N. P., ed., Marcel Dekker, New York, USA, 1990. pp. 99–137.

[99] Jacobson, N. S., Lee, K. N., and Fox, D. S., "Reactions of SiC and SiO_2 at Elevated Temperatures," J. Am. Ceram. Soc., 75[6], 1603–1611, 1992.

[100] Sato, T., Haryu, K., Endo, T., and Shimada, M., "High Temperature Oxidation of Hot-Pressed Aluminum Nitride by Water Vapor," J. Mater. Sci., 22, 2277–2280, 1987.

[101] Opila, E., Jacobson, N., Humphrey, D., Yoshio, T., and Oda, K., "The Oxidation of AlN in Dry and Wet Oxygen," in *Proceedings of the Symposium on High Temperature Corrosion and Materials Chemistry*, Hou, P. Y., McNallan, M. J., Oltra, R., Opila, E. J., and Shores, D. A., eds., San Diego, CA, USA, May 3–8, 1998, Electrochemical Society, Pennington, NJ, USA, 98[9], 430–437, 1998.

[102] Searcy, A. W., "Enthalpy and Predictions of Solid-State Reaction Equilibria," in *Chemical and Mechanical Behavior of Inorganic Materials*, A.W. Searcy, D.V. Ragone, and U. Colombo, eds., Wiley-Interscience, New York, USA, 1970. pp. 33–55.

[103] Bianco, R. and Jacobson, N., "Corrosion of cordierite ceramics by sodium sulfate at 1000°C," J. Mater. Sci., 24, 2903–2910, 1989.

CHAPTER

27

COATINGS FOR ENGINEERING CERAMICS

Kang N. Lee
NASA John H. Glenn Research Center, Cleveland, Ohio, USA

Holger Fritze
TU Clausthal, Institut fur Metallurgie, Clausthal-Zellerfeld, Germany

Yuzuru Ogura
Mitsubishi Heavy Industries, Ltd., Advanced Technology Research Center, Yokohama, Kanagawa, Japan

HISTORY

One key barrier to the application of advanced silicon-based ceramics and composites for hot section structural components in gas turbines is their lack of environmental durability. Silicon-based ceramics exhibit excellent oxidation resistance in clean, dry oxygen, by forming a slow-growing, dense silica scale [1]. However, the normally protective silica scale can be severely degraded by reacting with impurities, such as alkali salts [2–4] or water vapor [5–8], leading to accelerated degradation of Si-based ceramics. The application of oxide-based ceramic coatings is a promising approach to providing environmental protection for advanced heat engine components because oxide-based ceramics are in general more resistant to corrosive environments [9].

Coatings work for Si-based ceramics was first initiated at Garrett Turbine Engine Company in the late 1970s to seek solutions to the contact stress initiated fracture that was encountered under the DARPA/NAVY Ceramic Engine Demonstration Program [10]. Al_2O_3, ZrO_2, Cr_2O_3, mullite ($3Al_2O_3{\bullet}2SiO_2$), and mullite/ZrO_2 were applied by plasma spraying onto reaction-bonded silicon nitride (RBSN) and sintered silicon carbide (SSC). These coatings substantially reduced surface damage by contact stress, however, coating adherence was judged inadequate for long-term engine operation due to frequent spalling in thermal exposures. In 1984, under the support of DOE/Oak Ridge National Laboratories (ORNL), Garrett initiated a program, entitled "High Temperature Coating Study to Reduce Contact Stress Damage of Ceramics," to develop low friction, compliant ceramic coatings to reduce contact damage on Si_3N_4 and SiC ceramics [11,12]. Fully stabilized zirconia (ZrO_2–20 wt.% Y_2O_3) was deposited by electron beam physical vapor deposition (EB-PVD). The coating adherence was evaluated as processed and after oxidation exposures at 1200–1400°C for up to 100h. Initial results showed that the growth of silica layers at the ZrO_2/substrate interface degrades coating adherence and that mechanical bonding to rougher surfaces contributes to coating adherence. In subsequent study, various oxygen diffusion barriers, such as chemical vapor deposition (CVD) Al_2O_3, CVD AlN, EB-PVD mullite and sol-gel Al_2O_3, were investigated to minimize substrate oxidation as well as to increase chemical adherence. EB-PVD mullite was not successful because the silica in the mullite tends to evaporate more readily than the alumina, resulting in a silica-rich coating. CVD Al_2O_3 on RBSN resulted in excellent as-coated as well as post oxidation exposure adherence, while CVD AlN on RBSN showed excellent adherence only after oxidation exposure, indicating chemical interaction between ZrO_2 and AlN at temperature.

Coatings work in mid 1980s-early 1990s for Si-based ceramics was focused on protection from molten salt corrosion as well as contact stress damage. In 1986, Solar Turbines Inc. and GTE Laboratories independently initiated programs to develop coating systems for Si-based ceramics. Solar, under a GRI program, developed plasma-sprayed refractory oxide coatings for hot corrosion protection of Si-based ceramics [13–18]. Mullite or mullite + zircon ($ZrSiO_4$) was used either as a single layer coating or as a bond coat in multilayer coating systems in which the outer layers consisted of intermediate layers of compositionally-graded bond coat + Al_2O_3, YSZ (yttria-stablized zirconia) or Y_2O_3, and a top layer of Al_2O_3, YSZ or Y_2O_3. Single layer mullite coatings showed the best durability in hot corrosion tests at 760–1204°C. Mullite coatings, however, tended to develop cracks and corrosive species penetrated through the cracks and attacked the SiC substrates in long exposures. Solar's early work was augmented by a small study for Caterpillar in which mullite coatings were deposited onto silicon nitride glow plugs for corrosion protection. Solar also deposited mullite/alumina coatings on Nextel/SiC (3M syconex) which dramatically protected the substrate [18]. GTE developed a multilayer coating system using CVD to protect Si_3N_4 from contact stress damage [19–21]. The GTE coating consisted of AlN bond coat, graded intermediate layers of $Al_xO_yN_z$, and top layer of Al_2O_3 + YSZ. Despite the compositional grading, the coefficient of thermal expansion (CTE) mismatch stress led to the cracking and blistering of the outer layer on heating to temperatures above 1200°C.

In late 1980s, a research group at ORNL developed an aqueous slurry process to apply refractory oxide coatings, containing 50 to 90 wt.% alumina [22,23]. Coatings containing mullite as a major phase performed the best. All coatings, however, cracked and lost bonding with the SiC in thermal exposure. It was suggested that the coatings protected the SiC in a sacrificial manner in that the coating, combined with the silicate reaction products, made the reaction layer more refractory and protective. In early 1990s, the ORNL group developed a CVD Ta_2O_5 coating [24–27]. The Ta_2O_5-coated Si_3N_4 system was relatively stable at 1000°C in pure O_2. However, destructive interaction between the coating, salt, and the substrate occurred in an environment containing Na_2SO_4.

In 1991, a research group at NASA Glenn Research Center identified the crystallization of amorphous phase mullite in plasma-sprayed mullite, accompanied by a volumetric contraction, as the main source for the cracking of the coating [28–30]. Subsequently, a modified plasma spray process was developed at NASA Glenn to deposit fully crystalline mullite coatings. The second-generation plasma-sprayed mullite coatings exhibited dramatically improved thermal shock resistance and durability, remaining adherent out to 1200h at temperatures up to 1300°C under thermal cycling in air and out to 150h at 1000°C in a high pressure hot corrosion burner rig [31–34]. In the mid 1990s, a research group at Boston University developed a mullite coating by CVD [35–38]. Thin (5 ~ 10 μm), dense mullite coatings with the Al/Si ratio gradually increasing towards the outer surface were developed. The CVD mullite coating remained adherent after an exposure in O_2 at 1000°C for 100h and appeared to be effective in protecting Si_3N_4 substrates after 100h in Na_2SO_4 at 1000°C [39,40].

By mid 1990s, the volatilization of silica in water vapor and the resulting rapid recession of silicon-based ceramics emerged as a showstopper for the application of silicon-based ceramics in combustion environments [5,6]. This shifted the focus of coatings research to the protection from water vapor [41–44].

Mullite also lost silica through volatilization, leaving a porous alumina layer on the surface in simulated combustion environments. Mullite/YSZ duplex coatings, on the other hand effectively suppressed the silica volatilization. Mullite/YSZ duplex coatings, however, eventually suffered from water vapor attack as water vapor penetrated through cracks in the coating and attacked the substrate. A new crack-resistant coating system, consisting of silicon, modified mullite (mullite+BSAS) and BSAS (BaO-SrO-Al_2O_3-SiO_2), was developed under the HSR-EPM (High Speed Research-Enabling Propulsion Materials) program [45,46]. Today these coatings are referred to as EBCs (Environmental Barrier Coatings). A SiC/SiC composite combustor liner with the EPM EBC has undergone long-term engine testing in a Solar Turbines Inc. Centaur 50S turbine engine at Texaco (currently ChevronTexaco, Bakersfield, CA) under DOE's Ceramic Stationary Gas Turbine engine program [47]. At of end of the test in November 2000 the EBC coated SiC/SiC CFCC liners at Texaco had accumulated 13,937h of cumulative engine testing. A second Centaur 50S engine at the Malden Mills textile factory in Lawrence, Mass., was also retrofitted with EBC coated SiC/SiC CFCC combustor liners. This second engine, which has many more start/stop cycles, had accumulated 7,238h when the test ended in October 2000 (the coating was subsequently stripped and refurbished and the test is continuing at the ChevronTexaco site). A third engine also at the Malden Mills site had accumulated 15,144h in July of 2002 when the test was completed. Over 46,000h of engine test experience with EBC coated SiC/SiC combustor liners had been accumulated in the Solar field tests by January 2003.

In early 1990s in Japan, a group at Tokyo Electric Power Co. (TEPCO) applied oxide coatings, such as Al_2O_3, ZrO_2, and $ZrSiO_4$, on Si_3N_4 by plasma spraying. A significant reduction in weight loss at 1400°C in a combustion environment was observed. However, the coating began to peel off after 100h [48]. In the mid 1990s, the ORNL group developed plasma-sprayed $Ca_{1-x}Sr_xZr_4P_6O_{24}$ (CS50), a family of NZPs (named after their parent composition $NaZr_2P_3O_{12}$) [49,50]. The air plasma-sprayed CS50 coating on a Nicalon/SiC substrate remained adherent after exposure at 1000°C for 100h in air and protected the substrate from hot corrosion in Na_2SO_4 at 1000°C for 100h. In the late 1990s, the ORNL group modified the CVD mullite process [51,52], initially developed by the Boston University group. Unlike the mullite developed by the Boston University group, no significant variation in Al/Si ratio was observed across the thickness. Thin (~2μm), dense CVD mullite-coated SiC exhibited excellent resistance to Na_2SO_4.

Recently there have been some limited efforts to deposit EBCs on silicon nitride ceramics. NASA Glenn and Pratt &Whitney/URTC have deposited EPM type EBCs on silicon nitride and researchers at ORNL and Honeywell (New Jersey) have also deposited EBCs on silicon nitride (all unpublished researches). The research to develop EBCs for silicon nitride continues as the effort to develop monolithic silicon nitride ceramic turbine vanes is currently underway. In light of the critical role of EBCs for the realization of Si-based ceramic engine components, NASA Glenn initiated an EBC task under the UEET (Ultra Efficient Engine Technology) program in 2000. The goal of the EBC task is to optimize and refine the EPM EBCs and to develop new EBCs with low thermal conductivity and higher temperature capability, capable of 1482°C (2700°F) surface temperature and a 167°C (300°F) temperature gradient.

Carbon/carbon (C/C) composites are very attractive for structural applications because of their excellent mechanical properties at temperatures up to 2000°C. However, C/C composites gasify significantly above 500°C in an oxidizing environment, limiting their use to inert environments or short-term exposures in air [53]. Therefore the realization of the application of C/C composites for long-term high-temperature structural components in oxidizing environments requires the development of oxidation protection coatings.

The development of oxidation protection coatings for graphite was initiated more than 60 years ago. Early coatings had a boron glaze outer layer which was effective up to 1000°C. The effort to develop protective coatings for C/C composites was initiated in the late 1970s, driven by the need for components to be used on the shuttle orbiter vehicles. By the 1990s, the CVD SiC/boron oxide duplex coating system had been developed [54]. However, the use of boron oxide coatings for long term applications was limited by the volatility of the boron oxide above 1000°C. In the mid 1990s, further improvement was achieved by applying a thin (2.5 μm) CVD mullite overlay coating [55]. The mullite was dissolved in SiO_2, leading to improved crack resistance. In the early 1990s in Japan, CVD SiC/plasma-sprayed Y_2SiO_5 with a YSi_x intermediate layer coating system was developed which showed a life of over 1h at 1800°C in arc plasma [56–59]. Other coating systems investigated included HfC and HfN [60], Ir [61], Ir/$SrZrO_3$ [62], Ir/Al_2O_3 [63], and Ir/Y_2O_3 [64]. However, none of the coating systems developed to date appear to have the potential to protect C/C composites from long-term oxidation.

PROCESSING AND TESTING

Processing

The plasma spraying process involves melting of the powders of coating material, rapid transport of the molten particles to the substrate, and rapid solidification onto the substrate. The repeated melting and solidification process leads to the unique layered microstructure and a relatively high porosity. The microstructure and porosity can be controlled to some degree by controlling process parameters. The excellent strain tolerance of plasma-sprayed coatings is attributed to the layered microstructure and the high porosity. Other advantages include cost-effectiveness and scale-up capability. Because of these advantages, the plasma-spraying process has been widely used to apply environmental protection coatings for a wide range of heat engine components. One limitation is the difficulty of applying coatings on inner surfaces of complex structures, due to the line-of-sight nature of the plasma spray process. In some cases space restrictions (such as the small inner diameter of an inner cylinder) make spraying a coating difficult.

The chemical vapor deposition (CVD) process involves the deposition of a thin solid film onto a substrate from gaseous sources through chemical reactions. The substrate and/or the gas need to be heated to promote the deposition reaction. CVD is suitable for the fabrication of high density, high purity coatings. The non-line-of-sight nature of CVD allows the application of coatings on complex shapes. One limitation is the difficulty to process coatings of complex compositions.

Pulsed laser deposition (PLD) is mainly used in the laboratory. PLD allows the ablation of nonconductive and high melting target materials and the application of films with complex composition [65]. In addition, the substrate temperature remains low during the deposition, preventing the crack formation during the cooling after the deposition process. Sputtering is also mainly used for laboratory scale coating application because of limited deposition yield and high cost. Similar to CVD, this method requires extensive control of the deposition process. Other processing techniques include EB-PVD (electron beam physical vapor deposition), sol-gel and slurry coating deposition. The EB-PVD process has been widely used with a great success for the processing of YSZ TBCs (thermal barrier coatings) on metallic substrates, suggesting its potential for EBC processing. However, an attempt on EB-PVD mullite was not successful because the silica in the mullite tends to evaporate more readily than the alumina, resulting in a silica-rich coating [12].

Testing

Characterization of coatings includes the evaluation of environmental, physical, mechanical, and thermo-mechanical properties. Key properties evaluated during the course of coating development include coefficient of thermal expansion (CTE), phase stability, thermal conductivity, oxygen diffusivity, sintering, environmental durability, thermal shock resistance, adherence, creep, erosion, and fracture toughness.

Most commonly used test facilities are furnaces and burner rigs. Typical furnace tests include thermogravimetric analysis (TGA) and thermal cycling. Salt or water vapor can be incorporated in the furnace system to simulate combustion engine environments. The furnace test has advantages of low cost and easy maintainability, while its disadvantages include the difficulty to simulate high gas velocity. Therefore, the furnace test is suitable for the screening of multiple candidate coatings during the initial stage of coating development. The furnace test is also useful for long-term durability testing (over thousands of hours) due to its low operation cost. There are two types of burner rigs: the low-pressure burner rigs (LPBR) or atmospheric pressure burner rig, and the high-pressure burner rig (HPBR). Salt or oxide particles may be injected into the high velocity gas stream of the burner rig to simulate hot corrosion or erosion. A high-pressure burner rig can be set up to simulate the gas velocities as well as the pressures representative of gas turbines. In general the high-pressure burner rig test is suitable for evaluating the durability of mature coatings in simulated gas turbine environments since it requires a high operation and maintenance cost.

Typical furnaces and burner rigs do not have the capability to generate a thermal gradient across the thickness of the coating. Special rigs such as laser rigs or quartz lamp rigs can be used to simulate the thermal gradient. Mechanical testing devices such as tensile or bending test set ups can be incorporated in furnace or burner rig systems to evaluate thermo-mechanical properties. The other approach for evaluating materials in realistic environments is to characterize actual components. This approach is being used in the case of the Solar Turbines Inc. EBC/CFCC liners at ChevronTexaco and Malden Mills.

EVOLUTION OF COATING SYSTEMS AND TECHNOLOGY

There are several key issues that must be considered in designing coating systems [24,42,46]. Figure 27.1 schematically shows the key issues. They include the following. 1) The coating must possess the ability to resist reaction with aggressive environments and low oxygen permeability to limit the transport of oxygen. 2) The coating must possess a coefficient of thermal expansion (CTE) close to that of the substrate material to prevent delamination or cracking due to CTE mismatch stress. 3) The coating must maintain a stable phase under thermal exposure. Phase transformation is typically accompanied by a volumetric change, disrupting the integrity of the coating. 4) The coating must be chemically compatible with the substrate to avoid detrimental chemical interaction between the coating and the substrate. 5) Ideally, the coating should have a relatively low elastic modulus. The most ideal case is a single layer coating that meets all the requirements. However, such a material has not been identified to date. One way to address these key issues is a multilayer coating system, consisting of a bond coat with chemical compatibility and CTE match and a top coat with environmental durability and CTE match. If the top coat does not possess the matching CTE, compositional grading may be used to alleviate the CTE mismatch stress. Figure 27.2 compares the coefficients of thermal expansion of selected refractory oxides [66,67].

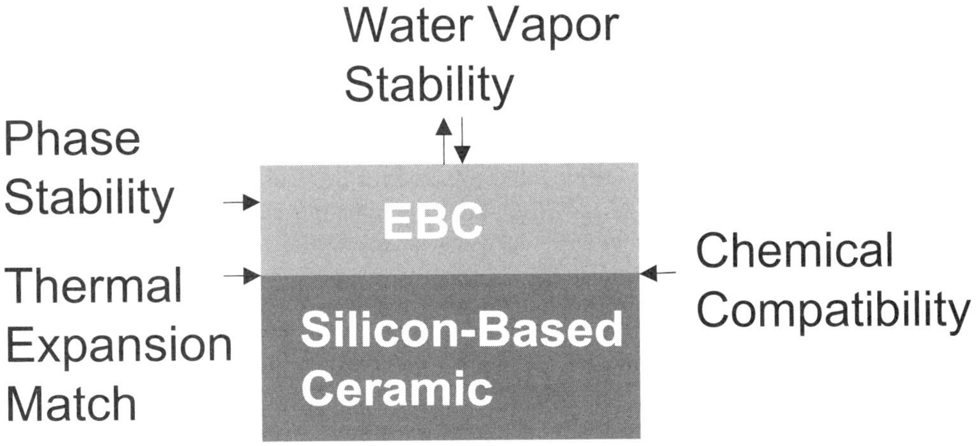

Figure 27.1 Key Issues in Selecting Coating Materials (Adapted from ref. [46] with permission from Elsevier Science)

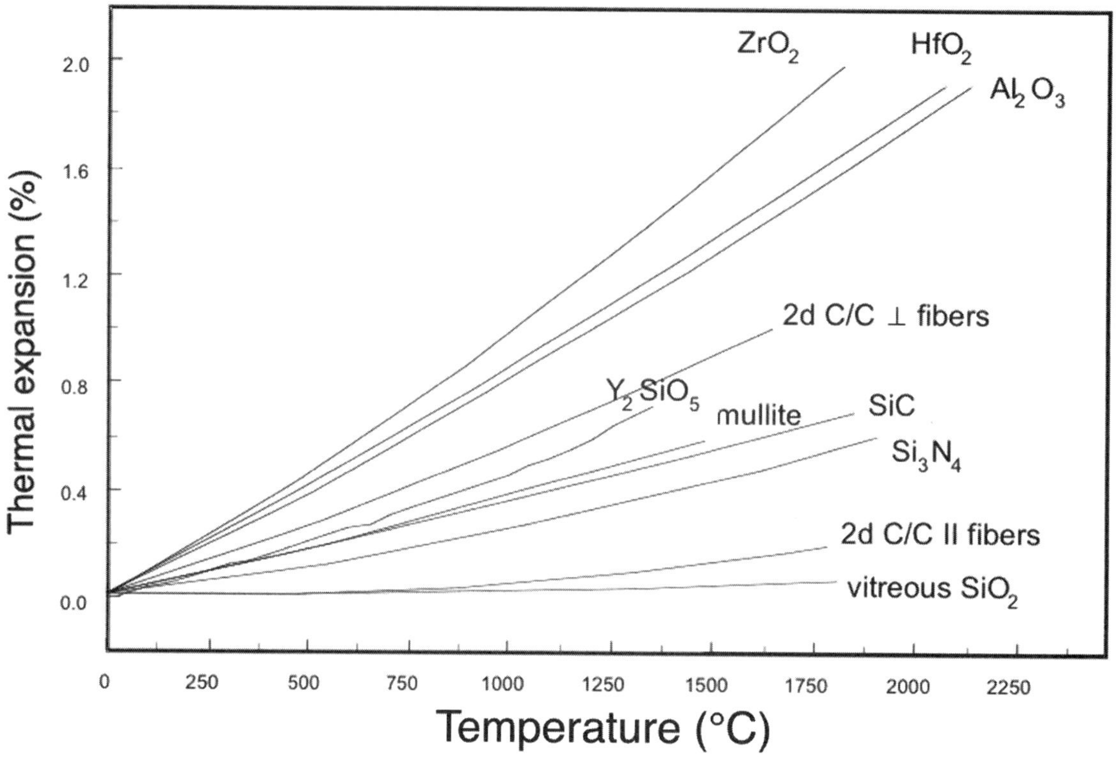

Figure 27.2 Thermal Expansion of Selected Refractory Oxides

Coatings for Si-Based Ceramics

Mullite and Mullite/Refractory Oxide Coating Systems. Mullite has attracted interest as a coating for Si-based ceramics because of its close CTE match with SiC. Major efforts have been invested in plasma spray process development that has led to the successful evolution of the current state-of-the-art EPM EBCs. CVD and aqueous slurry deposition are the other processes developed for mullite coating application. Figure 27.3 shows a chronological evolution of mullite and mullite/refractory oxide coating systems.

Plasma Spraying. Garrett Turbine Engine Company pioneered plasma spraying of refractory oxide coatings, such as Al_2O_3, ZrO_2, Cr_2O_3, mullite, and mullite/ZrO_2, on Si-based ceramic substrates, to provide protection from contact stress damage [10]. Mullite showed the best adherence and temperature capability in thermal shock test at temperatures up to 1427°C (2600°F). Although surface damage by contact stress was significantly reduced, coating adherence was judged inadequate for long-term engine operation due to frequent spalling in thermal exposures.

Solar Turbines Inc., sponsored by the Gas Research Institute (GRI), initiated a coatings program to develop protective coating systems for Si-based ceramics operating as heat recovery materials in highly corrosive environments [13–18]. Coatings were applied by conventional air plasma spraying onto monolithic SiC and SiC matrix composites. The SiC surface was treated by a proprietary process prior to the coating application to improve the adherence. Mullite or mullite + zircon ($ZrSiO_4$) was applied either as a single layer coating or as a bond coat in multilayer coatings. The top coat in multilayer coating systems was selected based on environmental durability, which included Al_2O_3, YSZ and Y_2O_3. Other coatings tested were cordierite, HfO_2, and Al_2TiO_5. Several modifications were examined to improve the durability, which

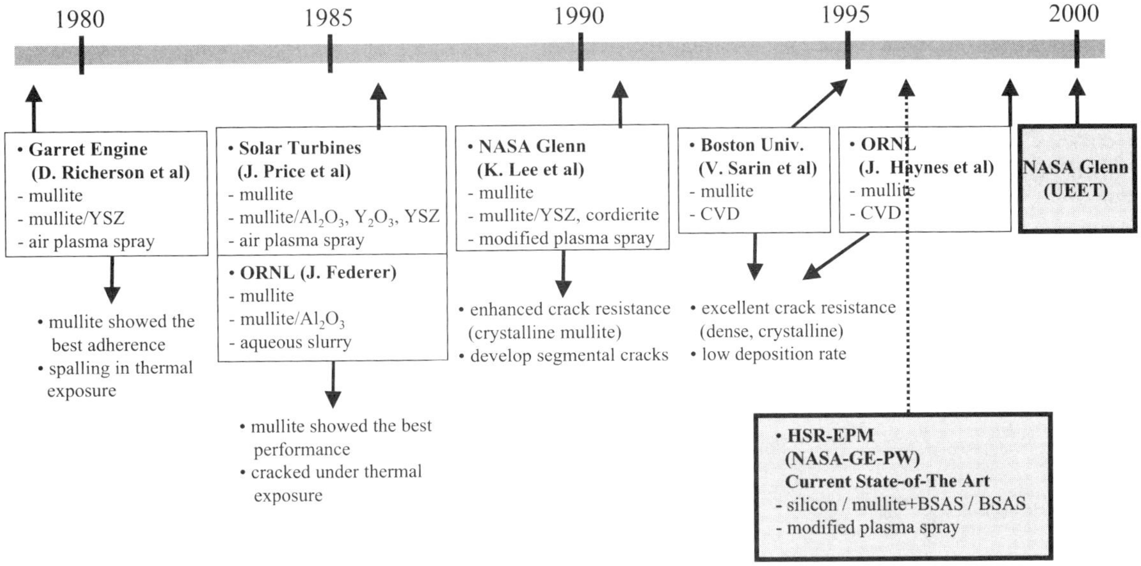

Figure 27.3 Evolution of Mullite and Mullite/Refractory Oxide-Based Coating Systems (Adapted from ref. [46] with permission from Elsevier Science)

included densified coatings, graded coatings (composition or density), and seal coatings. The purpose of the compositionally graded layers was to reduce the CTE mismatch stress between the low CTE bond coat and the high CTE top coat. The coatings were relatively thick, about 1.3 mm, since their function was to be sacrificial, i.e. it was assumed that the coatings would be consumed over time by the corrosive furnace atmosphere. Their consumption would postpone attack on the underlying Si-based substrate.

The coated specimens were exposed to a simulated aluminum reclamation environment at temperatures ranging from 760°C to 1204°C. The single layer mullite coatings provided the best corrosion resistance, followed by the coatings graded from mullite to alumina and mullite to yttria. After 500 hours of testing, however, the mullite-coated SiC began to exhibit cracks. The corrosive atmosphere penetrated through the cracks and attacked the SiC substrates. Various seal coats, such as Solaramic® glass, Er_2O_3, HfO_2 and Cr_2O_3, were applied to prevent the infiltration of corrosive species. Single layer mullite with a Solaramic® glass seal coat showed the best improvement. Adding zircon, which appeared to be incompatible with the SiC, was harmful to the mullite coating. Incorporation of a YSZ top coat resulted in eggshell cracking, with the damage increasing at higher temperatures. A cordierite coating completely spalled during the hot corrosion exposure, while a HfO_2 coating suffered cracking. The strength distributions of sintered α-SiC and siliconized SiC coated with various refractory oxides by plasma spraying were evaluated using a room temperature C-ring fracture test [17]. The strength of coated SiC was equivalent or superior to that of as-received material, despite the proprietary surface treatment which induced surface flaws on the SiC substrate. The good strength was attributed to the good mechanical bond between the coating and SiC, providing load transfer from failure-initiating defects to the coating, and to the relatively high strain tolerance of plasma-sprayed coatings due to their lamella type grain structure.

A research group at NASA Glenn Research Center discovered that plasma-sprayed mullites contained a large amount of amorphous phase mullite due to the rapid cooling during the mullite solidification [28–30]. The crystallization of amorphous phase mullite, accompanied by a volumetric contraction, was identified as the main source for the cracking in plasma-sprayed mullite coatings. Subsequently, a modified plasma spraying process was developed at NASA Glenn, which successfully eliminated most of the amorphous phase mullite from the coating. In the modified process, the substrate is heated above the crystallization temperature during the spraying using a specially designed furnace. The second-generation mullite coating exhibited dramatically improved thermal shock resistance and durability, remaining adherent for as long as 1200h at temperatures up to 1300°C under thermal cycling in air, and up to 150h in a 4 atm burner rig with jet A fuel containing 2 ppm sodium at 1000°C [31–33]. Figure 27.4 compares the first and second-generation mullite after two 24h-thermal cycles at 1000°C. Cracks disappeared from the second generation mullite except for some through-thickness-cracks. At higher temperatures (>1400°C) in air, a significant amount of pores developed at the mullite/SiC interface. EDS analysis showed that the scale was contaminated by Al_2O_3, Na_2O, K_2O, and CaO, leading to enhanced oxidation and reduced scale viscosity. The interfacial porosity was attributed to

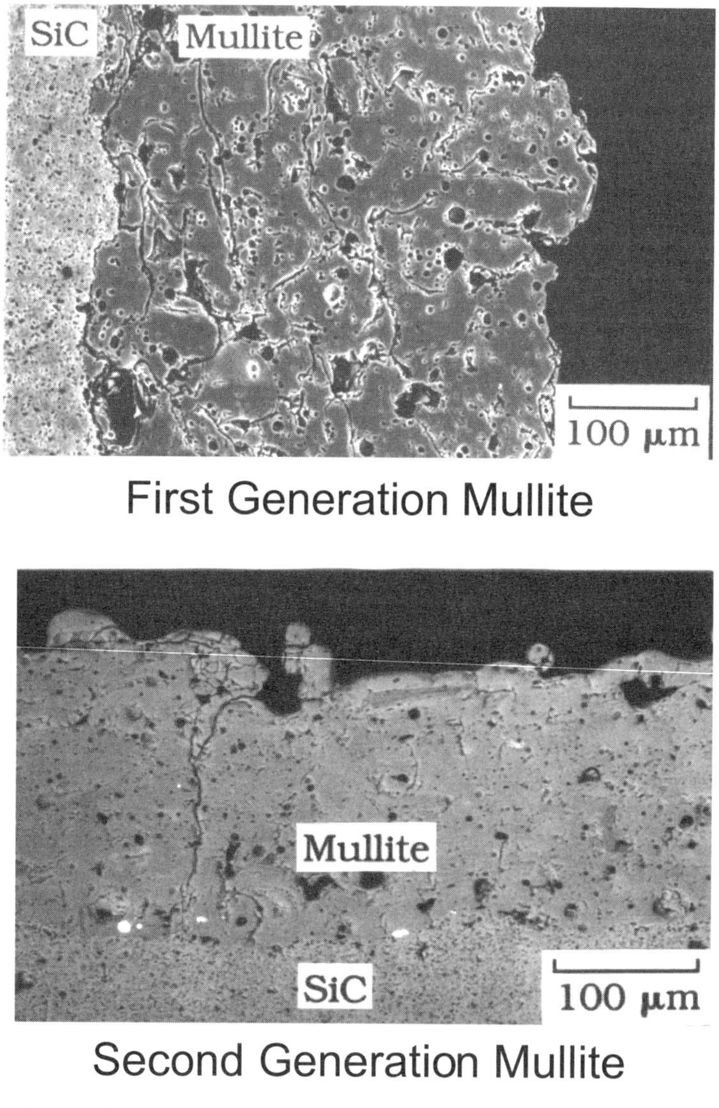

Figure 27.4 Comparison of First Generation vs. Second Generation Plasma-Sprayed Mullite (1000°C-2h Cycle, Air, 48h) (Reprinted from ref. [29] with permission of the American Ceramic Society, www.ceramics.org, © 1995, all rights reserved)

bubbling of gases, generated by the oxidation reaction, through the low-viscosity silicate scale [34]. Consequently, the interfacial porosity increased with increasing contamination and temperature. Post-hot corrosion examination revealed a fairly intact interface with only a trace amount of sodium silicate reaction products at the interface where cracks intersected the SiC interface. Formation of a solid Na_2O-Al_2O_3-SiO_2 compound was detected on the mullite surface, which was believed to act as a kinetic barrier to suppress further attack by salt. For longer exposures, continued transport of salt through cracks may become an issue.

Mullite coatings, however, do not have an adequate durability in water vapor, losing silica through selective volatilization due to high silica activity (~0.3) [41]. Figure 27.5 shows a porous alumina surface layer on the surface of a plasma-sprayed mullite coating after 50h in the NASA Glenn high pressure burner rig (HPBR) (1230°C, 6 atm, pH_2O = 0.6 atm, V_{gas} = 24 m/s). Therefore, an environmental overlay coating is necessary when the protection from water vapor is needed. At NASA Glenn, YSZ was selected as the baseline overlay coating because of its proven performance as a thermal barrier coating in combustion environments [41–43]. A typical thickness of the mullite and YSZ layers was 100–150 μm each. Despite the large CTE of YSZ, the mullite/YSZ-coated SiC showed excellent adherence under thermal cycling in air and effectively suppressed the silica volatilization in water vapor at 1300°C. Figure 27.5 shows the plot of weight change vs. time for SiC with and without a coating in the HPBR. Uncoated and mullite-coated SiC showed a weight loss due to the volatilization of silica, while the mullite/YSZ-coated SiC exhibited a fairly constant weight, indicative of suppressed silica volatilization. In long-term exposures, the mullite/YSZ system developed water vapor-enhanced oxidation as water vapor penetrated through the cracks and attacked the substrate. Figure 27.6 shows mullite and mullite/YSZ coated-SiC after 200h in a high steam thermal cycle (HSTC) (2h cycle, 1300°C, 1 atm, pH_2O = 0.9, V_{gas} = 11.5 cm/s). Water vapor, the predominant oxidant in a H_2O/O_2 environment, is known to enhance the oxidation of SiC. The silica scale formed in high water vapor was porous, allowing the oxidation to propagate readily along the interface. The porous scale was attributed to the generation of gaseous silicon hydroxide species. There was evidence of silica volatilization at the bottom of cracks (Figure 27.6).

Key durability issues in the mullite-based coating system include: (i) through-thickness cracking in the mullite; (ii) weak bonding of mullite onto silicon-based ceramics; and (iii) interface contamination [43]. Through-thickness-cracks provided the path for water vapor, leading to the eventual failure of the system. The YSZ overlay coating did not seal the cracks in mullite since YSZ also cracked due to the large CTE mismatch between the two layers and sintering. It should be noted that cracks in the coating do not affect the oxidation in air since silica scale formed in air is protective (Figure 27.6). It is believed that the development of through-thickness-cracks in mullite is due to stresses in the coating. The precipitation of second phases in the mullite coating, such as residual amorphous mullite and alumina, and the resulting volumetric shrinkage and CTE mismatch are suggested to be the major sources for the stresses in the coating. The mullite/SiC bond of as-sprayed coatings is mainly due to mechanical interlocking since mullite does not form a strong chemical bond with SiC. Interfacial contamination and the resulting chemical reaction can degrade the coating adherence by altering the physical or mechanical properties of the silica scale.

The crack resistance of the mullite coating can be improved by modifying the physical or mechanical properties of mullite or by sealing the cracks with a crack resistant overlay coating [43]. A new, low CTE, low silica activity overlay coat (BSAS: BaO-SrO-Al_2O_3-SiO_2) and a new, crack-resistant modified mullite bond coat (mullite+BSAS) were developed under the HSCT-EPM (High Speed Civil Transport-Enabling Propulsion Materials) program [45,46]. Today these coatings are referred to as EBCs (Environmental Barrier Coatings). Replacing the YSZ top coat in the mullite/YSZ system with BSAS delayed the onset of accelerated oxidation in water vapor by a factor of at least two to three. Further improvement was achieved by replacing the mullite bond coat with the modified mullite bond coat, while still further improvement was achieved in the modified mullite/BSAS duplex coating system. The improved durability of the EPM EBCs was attributed to their outstanding crack resistance. It has also been shown that EBC durability can be enhanced by improving the adherence through the modification of the mullite/SiC interface. Molybdenum silicide (2–10 μm) applied at the mullite/SiC interface delayed the onset of the accelerated oxidation in water vapor by a factor of at least two to three [43]. It was discovered in the HSR-EPM program that silicon can

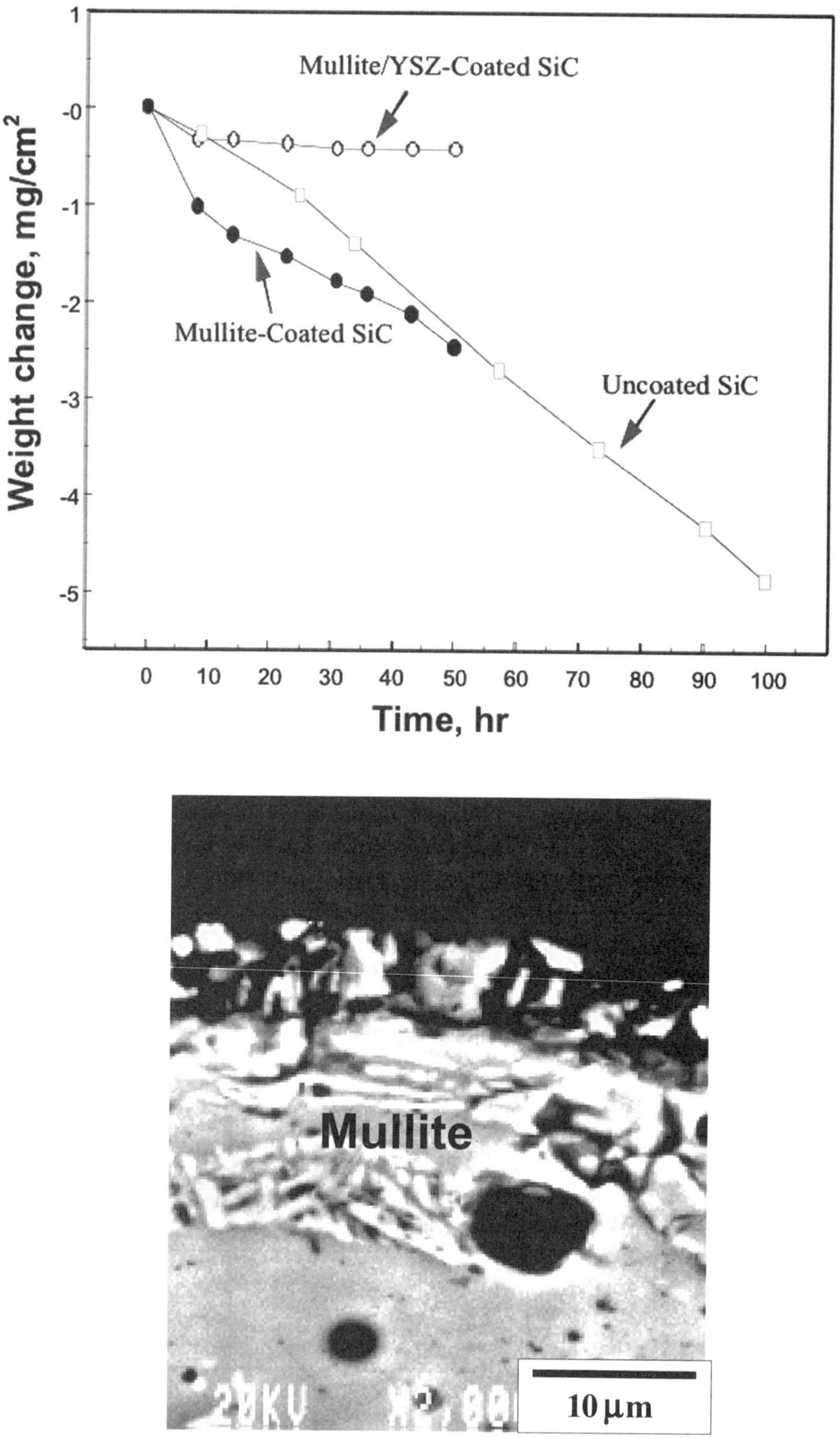

Figure 27.5 Weight Change of Mullite and Mullite/YSZ-Coated SiC in High Pressure Burner Rig (1230°C, 6 atm, pH_2O ~ 0.6 atm, gas velocity ~ 24 m/sec) (Reprinted with permission from NASA Glenn Research Center)

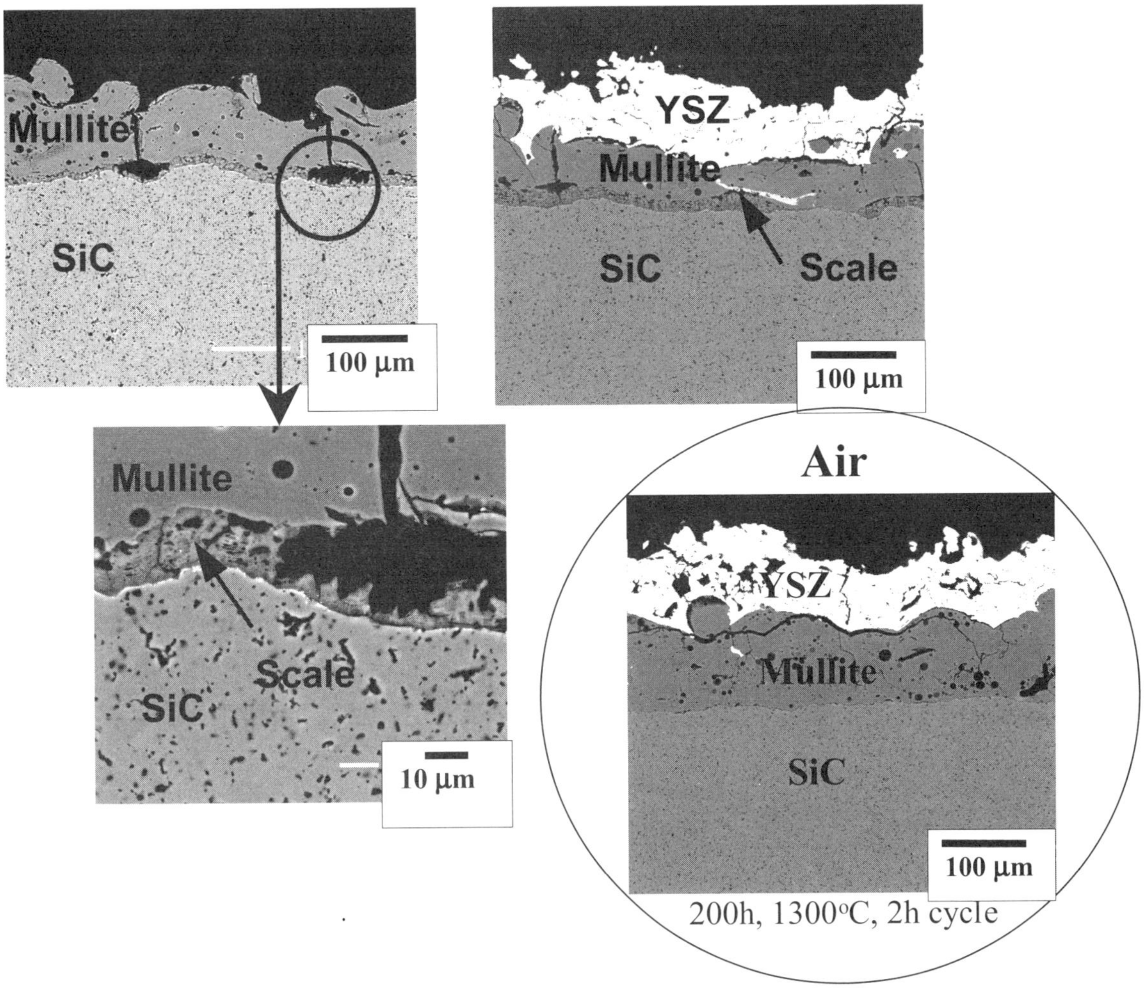

Figure 27.6 Mullite and Mullite/YSZ-Coated SiC in Cyclic Water Vapor Furnace (1300°C-2h Cycle, 1 atm, 90% H_2O/O_2, 200h) (Adapted with permission from ref. [45], and from ref. [46] with permission from Elsevier Science)

significantly improve the adherence of mullite onto SiC. EPM EBCs applied on the areas of a melt-infiltrated (MI) SiC/SiC CFCC with a silicon surface layer did not show any sign of water vapor-enhanced degradation out to 500h at 1300°C in the HSTC tests while significantly enhanced oxidation occurred on EBC/MI interfaces without the silicon surface layer [46]. EBCs with the silicon layer applied externally using plasma spraying (50–75 μm) showed a similar improvement in durability. Therefore, current state-of-the art EBCs consist of three layers, i.e. silicon, modified mullite, and BSAS. Figure 27.7 shows a plot of weight change vs. time for uncoated and EPM EBC-coated MI SiC/SiC CFCC exposed in the HPBR (1300°C, 6 atm, pH_2O ~0.6 atm, gas velocity ~24m/sec). Each datum in the plot represents one thermal cycle, as the test was interrupted to measure sample weight. The measured linear weight loss of uncoated CVD SiC and MI SiC/SiC CFCC was due to the volatilization of silica by water vapor. On the other hand, MIs coated with (mullite+BSAS)/BSAS or mullite/BSAS showed a much reduced weight loss, by an order of magnitude, demonstrating the effectiveness of EPM EBCs in limiting the silica volatilization.

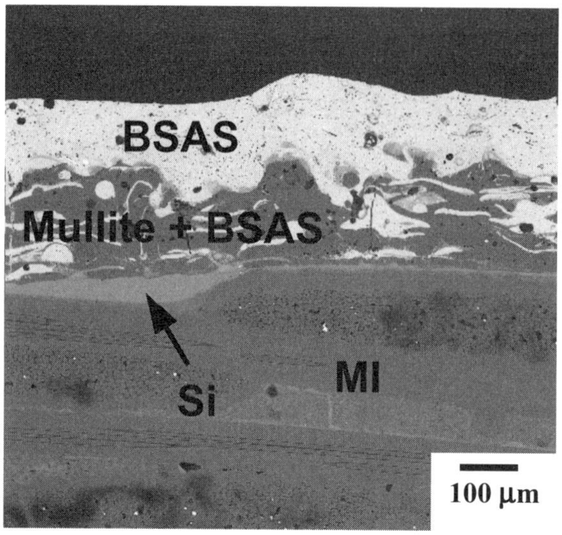

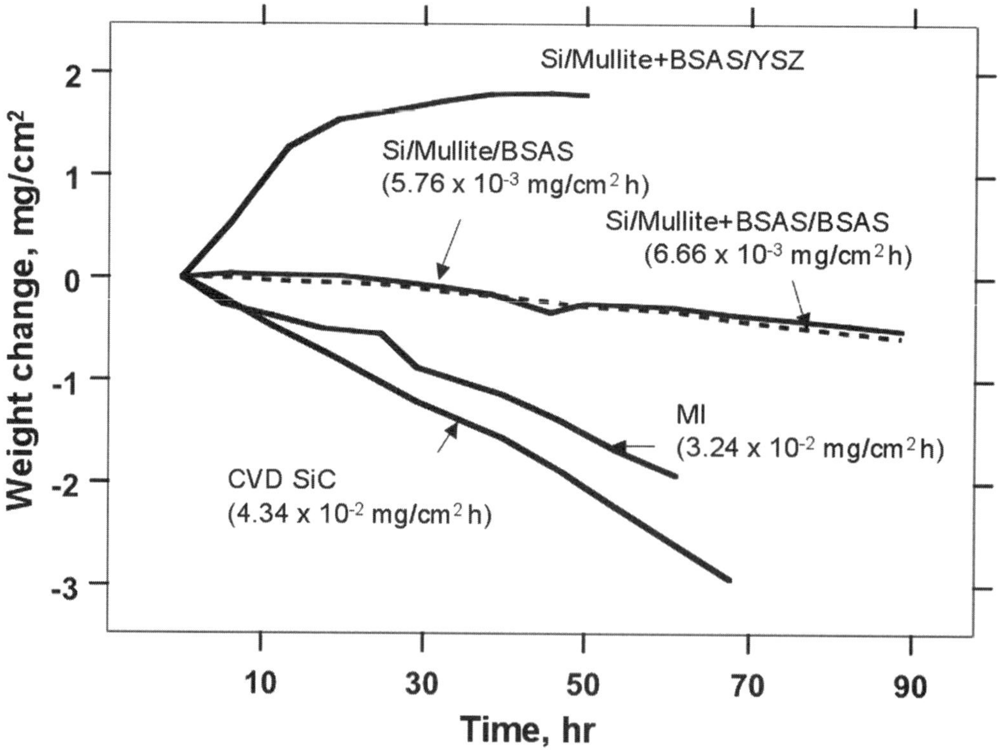

Figure 27.7 Weight Change of EPM EBC-Coated MI SiC CFCC in High Pressure Burner Rig (1300°C, 6 atm, pH_2O ~ 0.6 atm., Gas Velocity ~ 24 m/sec) (Adapted with permission from ref. [69])

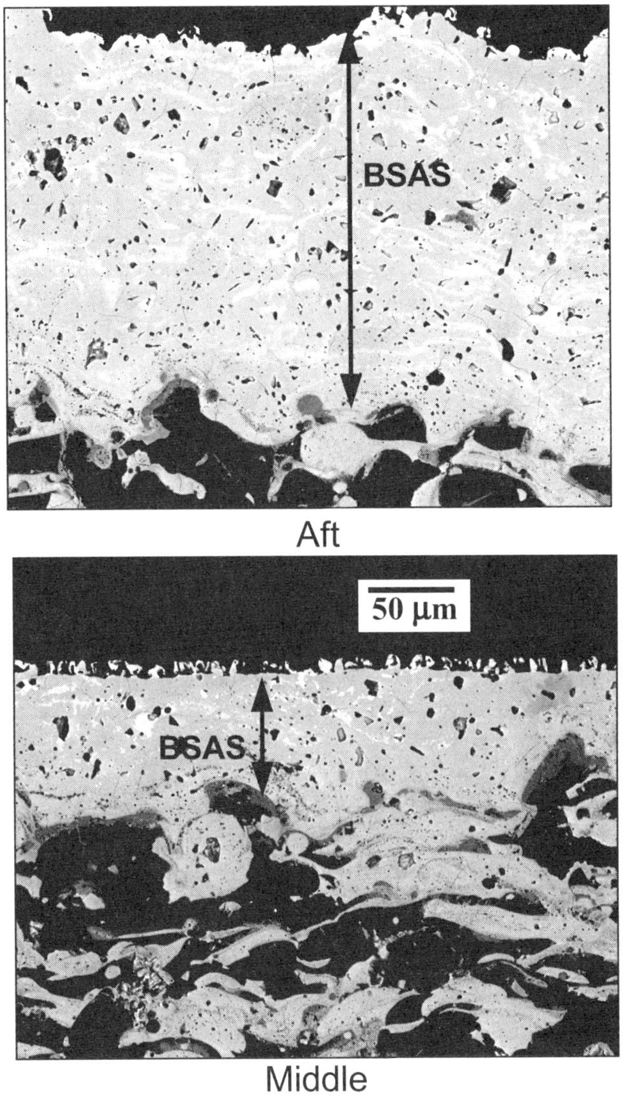

Figure 27.8 EBC-Coated SiC/SiC CFCC Liner After 13,937h of Cumulative Engine Testing at Texaco Test Site at Bakersfield, CA (Reprinted from ref. [68])

The EPM EBC was scaled up under the U.S. Department of Energy's Ceramic Stationary Gas Turbine (CSGT) development program [47]. SiC/SiC composite combustor liners coated with the EPM/DOE EBC applied by Pratt&Whitney/United Technologies Research Center have undergone long-term engine testing in Solar Turbines Inc. Centaur 50S turbine engines at the Texaco (currently ChevronTexaco), Bakersfield, CA, and Malden Mills industrial sites. The Texaco inner liner has the silicon/mullite bond coat and the outer liner has the silicon/modified mullite bond coat. As discussed in the previous sections, at the end of the test in November 2000 the EBC coated SiC/SiC CFCC liners at Texaco had accumulated 13,937h of cumulative engine testing. Evaluation of the tested liners at ORNL indicated three primary durability issues [68]. First, significant amount of BSAS recession was observed on some areas as shown in Figure 27.8, suggesting BSAS volatility in combustion environments. Second, pure alumina was detected at the BSAS/mullite interface, which was suggested as an indication of phase separation of mullite. Third, tool bumps on the surface of liners could be related to spalled EBC areas and deep holes into the CFCC, indicating deleterious effects of surface asperities on EBC durability. Subsequent testing at Malden Mills under DOE/Solar sponsored programs included a high time liner set (15,144h at conclusion of the test in July 2002) which showed essentially the same degradation pattern.

A task was undertaken in the UEET Program to determine the upper temperature limit of EPM EBCs by investigating environmental and chemical durability [69,70], silica volatility [69,70], BSAS phase stability [71], and thermal conductivity [72]. The BSAS-silica reaction formed a low melting silicate glass (mp ~1300°C) in short exposures (< 100h) at T > 1400°C. BSAS also exhibited a linear weight loss in water vapor at temperatures as low as 1200°C, as shown in Figure 27.9, which is a plot of weight change vs. time for hot pressed BSAS exposed to 50% H_2O-balance O_2 flowing at 4.4 cm/sec from 1200 to 1500°C at 1 atm total pressure. The initial weight loss is attributed to buoyancy. Table 27.1 is the recession of BSAS after 100h at 6 atm, pH_2O = 0.6 atm, gas velocity = 24m/sec, projected from the date in Figure 27.9 using the silica volatility model (equation 1) developed at NASA Glenn Research Center [6].

TABLE 27.1 PROJECTED RECESSIOON OF BSAS AFTER 1000H AT 6 ATM, PH_2O = 0.6 ATM, V_G = 24M/S, CALCULATED USING EQUATION (1) AND THE SILICA VOLITILITY DATA FROM TGA

Temperature (°C)	1300	1400	1500
Recession (mm)	28	67	268

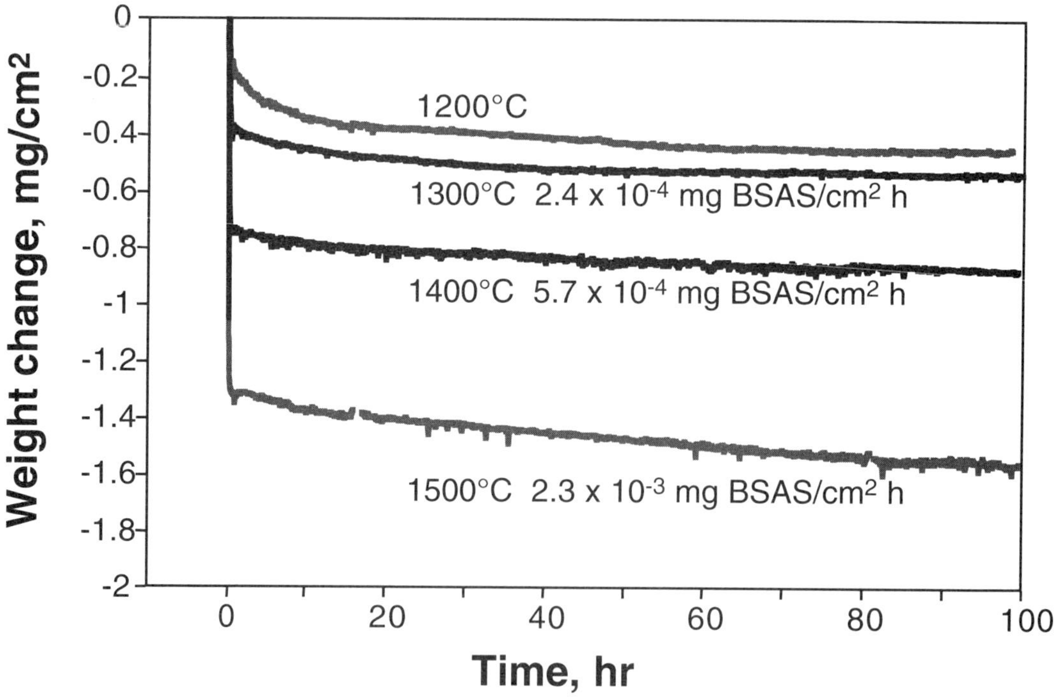

Figure 27.9 Weight Change of Hot Pressed BSAS in TGA (1200–1500°C, 1 atm, pH_2O = 0.5 atm, Gas Velocity = 4.4 cm/sec) (Reprinted with permission from ref. [69])

$$\text{Volatility} \propto \frac{v^{1/2} \times P(H_2O)^2}{(P_{TOTAL})^{1/2}} \tag{1}$$

Where v is gas velocity, $p(H_2O)$ is water vapor pressure, and P_{TOTAL} is total pressure. The BSAS recession becomes very significant at T > 1400°C. The glass formation and BSAS volatility suggest that the upper temperature limit of EPM EBCs for several thousand hours of life is ~1400°C [69,70]. EPM EBCs also showed a significant increase in thermal conductivity (~30%) after a 10h exposure at 1300°C in HSCT, which was attributed to sintering. These results indicate the need to develop EBCs with higher temperature capability and lower thermal conductivity than EPM EBCs to realize the full potential of Si-based ceramics in advanced gas turbine engines. BSAS was amorphous as sprayed but converted readily to hexagonal celsian phase after a furnace exposure at T > 1100°C. On the other hand, the conversion from high CTE hexagonal celsian ($\sim 8 \times 10^{-6}$/°C) to low CTE monoclinic celsian ($4\sim 5 \times 10^{-6}$/°C) was very sluggish, especially with the BSAS in the mullite+BSAS layer. The implications of the slow conversion from hexagonal to monoclinic celsian are under investigation in the UEET program.

Chemical Vapor Deposition. A research group at Boston University developed a CVD process to fabricate mullite coatings [35–38]. Two approaches were investigated: (i) diffusion of multilayers of Al_2O_3 and SiO_2 to form mullite; and (ii) co-deposition of Al_2O_3 and SiO_2. As-deposited multilayers had a definite interface with abrupt changes in composition, which resulted in high residual stresses. Extended oxidation resulted in crack formation, and eventual debonding of the coating. This was attributed to the stress due to the phase transformation from low CTE amorphous to high CTE crystalline SiO_2. In co-deposition, mullite coatings were grown in a CVD reactor at 950°C from the $AlCl_3$-$SiCl_4$-CO_2-H_2 system:

$$6AlCl_3 + 2SiCl_4 + 13CO_2 + 13H_2 = 3Al_2O_32SiO_2 + 13CO + 26HCl \tag{2}$$

Dense, crystalline mullite coatings with a preferential orientation in the (001) direction were deposited. The coatings were compositionally graded, with the Al/Si ratio less than 3 (stoichiometric mullite) at the coating/substrate interface and increasing to as high as 8 towards the outer surface. Under most circumstances, the coating started off as an intimate mixture of nanocrystalline γ-Al_2O_3 in a vitreous silica matrix. When the Al/Si ratio reached about 3.24, mullite grains nucleated and grew in a columnar fashion, typically with the c-axis corresponding to the growth direction. On annealing at 1200°C for 100h, the nanocrystalline region transformed to mullite without any cracking or adverse effects on the adhesion of the coating. The grading of the composition from Si-rich at the coating/substrate interface to Al-rich at the outer surface has potential advantages of higher environmental durability.

CVD mullite (5 ~ 10 μm) remained adherent on SiC after an exposure in O_2 at 1000°C for 100h with the growth rate of silica on SiC similar to that of uncoated SiC. CVD mullite-coated Si_3N_4 was subjected to a corrosive environment containing Na_2SO_4 and O_2 at 1000°C for 100h [39]. The coating appeared to be effective in protecting the underlying Si_3N_4 substrates. A small degree of Na penetration through mullite grain boundaries was observed to a depth of about 1μm. Continued penetration of Na in extended exposures may become an issue for long-term durability. Virtually no corrosion was observed after an exposure in a hot corrosion burner rig at 1050°C for 66h [40].

A research group at ORNL modified the mullite CVD process, initially developed at Boston University [51]. The modification included a higher reaction temperature (1050°C) and *in situ* formation of aluminum trichloride by flowing HCl through a vaporizer at 200°C to react with Al pellets:

$$Al + 3HCl = AlCl_3 + 1.5H_2 \tag{3}$$

Crystalline mullite coatings were deposited at $SiCl_4$ partial pressures of 19 Pa or less, while amorphous Al-Si-O coatings were deposited at higher $SiCl_4$ partial pressures. There was significant variation in the

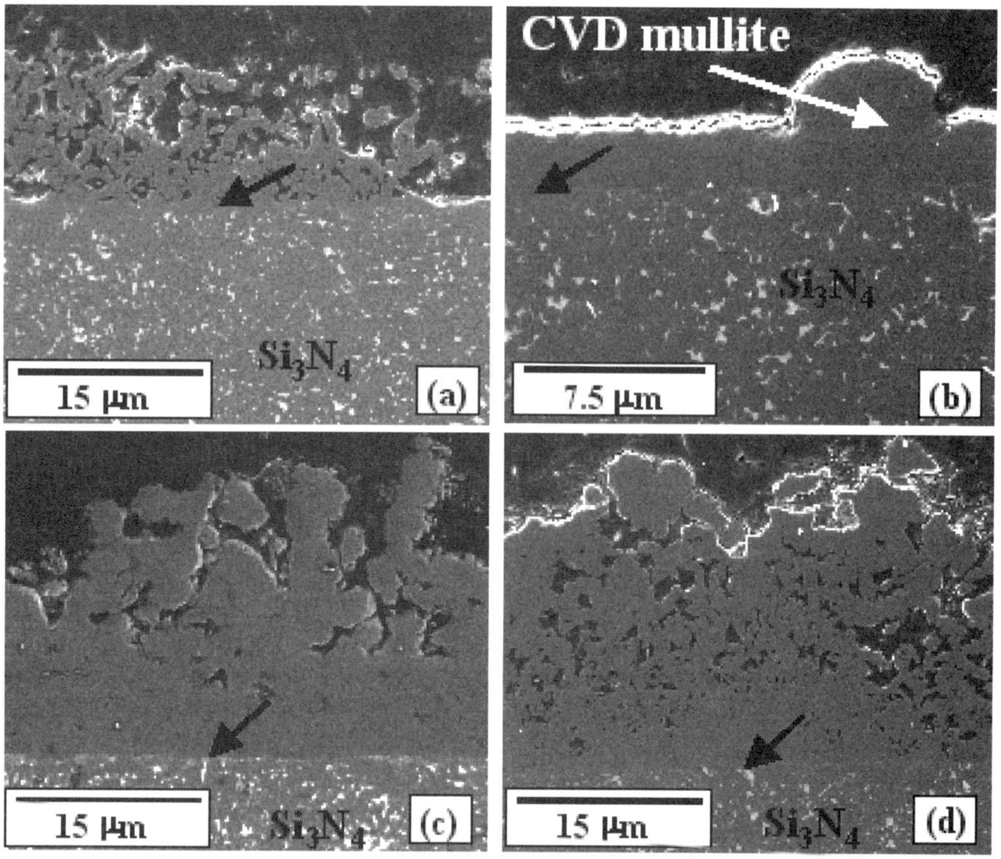

Figure 27.10 Microstructural Variations of CVD Mullite vs. Deposition Time and p($SiCl_4$): (a) 4 h, 19 Pa; (b) 2 h, 12 Pa; (c) 4 h, 12 Pa; (d) 4 h, 12 Pa (Reprinted from ref. [51] with permission of the American Ceramic Society, www.ceramics.org, © 1999, all rights reserved)

coating microstructure depending on the $SiCl_4$ partial pressure (Figures 27.10a–27.10d). Coatings deposited at a $SiCl_4$ partial pressure of 19 Pa (Figure 27.10a) were extremely porous and nonuniform, while coatings with improved density and microstructure were deposited at a lower $SiCl_4$ partial pressure (Figure 27.10b). Dense mullite coating (1.5 to 3.0 µm) grew during the first 2h after which a porous, nonuniform coating grew on top of the dense layer (Figures 27.10c and 27.10d). There was no significant variation in Al/Si ratio across the coating thickness. However, the Al/Si ratio could be controlled, ranging from 1 to as high as 6, by varying the partial pressure of $SiCl_4$.

In Na_2SO_4, CVD mullite with the Al/Si ratio 2.5 or higher displayed no evidence of damage, while coatings with the Al/Si ratio 1.5 exhibited a small amount of localized cracking and high Na concentration in the coating due to the reduced coating stability at higher SiO_2 activity. CVD mullite-coated SiC/SiC composites were exposed in the ORNL Keiser Rig (10 atm, 15 % H_2O, V_{gas} = ~ 3cm/minute) at 1200°C for 500h [52]. Dense, thin (~2µm) CVD mullite effectively suppressed the rapid oxidation of SiC/SiC with no microstructural evidence of mullite volatility (Figure 27.11). The absence of silica volatility from the mullite was in contrast to the plasma-sprayed mullite exposed in the NASA HPBR (6 atm, 10% H_2O, V_{gas} = 24 m/sec) [43]. The difference is attributed to the extremely slow gas velocity in the Keiser Rig and the resulting saturation of the reaction chamber with gaseous silicon hydroxides, which suppresses the volatilization reaction.

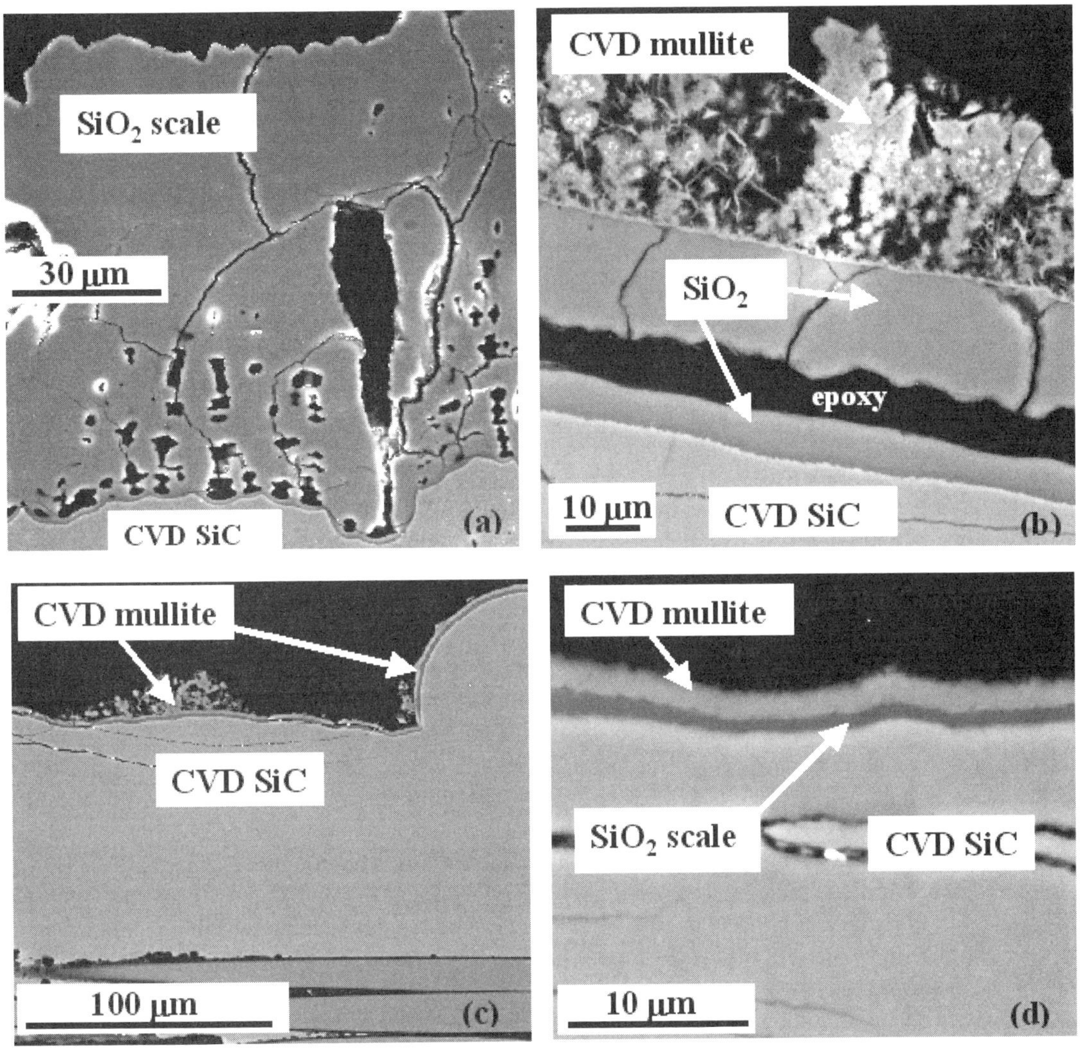

Figure 27.11 CVD Mullite-Coated SiC/SiC in High Pressure, Low Gas Velocity Rig (1204°C, 10 atm, V_{gas} = 0.05 cm/s, 15% H_2O/air, 500h): (a) Thick Silica Scale on Uncoated CVD SiC, (b) CVD SiC with a Bi-Layered Silica Scale Underlying Porous CVD Mullite, (c) Dense CVD Mullite on CVD SiC, (d) Thin, Dense Silica Scale Beneath Dense CVD Mullite (Reprinted from ref. [52] with permission of the American Ceramic Society, www.ceramics.org, © 2000, all rights reserved)

Aqueous Slurry. The ORNL group developed an aqueous slurry process to apply refractory oxide coatings [22,23]. Commercially available materials were used to form coatings containing 50 to 90 wt.% alumina. Initially thin (0.1 to 0.2 mm) coatings of alumina, silica (SiO_2), and zirconia-based ceramics were applied, most of which cracked and detached from the substrate during thermal cycling. Later work focused on thick (0.4 to 1.3 mm) alumina-based coatings, including mullite. Coated specimens were heated to 1200°C or higher to promote bonding within the coating, then thermally cycled to 1200°C as a test of integrity and adherence. Similar to Solar Turbines' results, coatings containing mullite as a major phase performed best. Subsequently, duplex coatings were developed which had a base layer (bond coat) containing mullite as the major phase and an outer layer containing alumina as the major phase (>85 wt.%). Hot corrosion resistance was evaluated at 1200°C for 55h in an oxidizing atmosphere containing Na_2CO_3 and H_2O vapor. Although cracked and debonding from the SiC was observed, coatings did not spall and the surface recession by hot corrosion was reduced by an order of magnitude. It was suggested that the coating protected the SiC in a sacrificial manner in that the coating combined with the silicate reaction products and became part of the corrosion product, making the reaction layer more refractory and protective.

Other Coating Systems. GTE sponsored a program to develop coatings to reduce the contact stress on Si-based ceramics [19–21]. A multilayer CVD coating system was developed, consisting of an AlN bond coat (0.5–2 μm), intermediate layers with compositional grading from AlN to $Al_xO_yN_z$ to Al_2O_3 + ZrO_2 (5–10 μm), and Al_2O_3 + ZrO_2 top coat (2–10 μm). The AlN bond coat was to promote the adherence by forming a SiAlON compound with Si_3N_4 substrates. The graded layer was to minimize the CTE mismatch between the low CTE bond coat and the high CTE top coat. Adherent and crack-free coatings were developed, which exhibited no apparent degradation under thermal cycling between room temperature and ~ 1100°C. However, the outer Al_2O_3 + ZrO_2 coating cracked on heating to temperatures above 1200°C. Penetration of oxygen through the cracks in the outer coating oxidized the AlN bond coat, leading to the blistering of the outer coating. It was concluded that the difference in CTE between SiC or Si_3N_4 and Al_2O_3+ ZrO_2 was too great to maintain an adherent, crack-free coating.

A group at Tokyo Electric Power Co (TEPCO) in Japan applied oxide coatings on Si_3N_4 by plasma spraying. Their coatings included Al_2O_3, ZrO_2, and $ZrSiO_4$. A graded coating consisting of $ZrSiO_4$ undercoat and ZrO_2 overcoat significantly reduced the weight loss of Si_3N_4 at 1400°C in a combustion environment. However, the coating began to peel off after 100h [48]. Solar Turbines reported the incompatibility of zircon with SiC and the resulting harmful effects on the coating's durability [18].

Ta_2O_5 has attracted interest as a coating, in part because of the ease of the CVD process [24]. CVD Ta_2O_5 was first applied as a coating at ORNL [24–25]. Crystalline Ta_2O_5 was deposited on Si_3N_4 and SiC substrates using $TaCl_5$ and O_2 at 1000–1300°C. Early coatings generally consisted of poorly coalesced, pebblelike grains. Improved Ta_2O_5 coatings with nonporous morphology were produced using $TaCl_5 + H_2 + CO_2$ at 1000–1300°C [26]. Microcracks were consistently observed on the as-deposited surface. The Ta_2O_5-Si_3N_4 system was relatively stable at 1000°C in pure O_2. However, the system became unstable in an environment containing Na_2SO_4 and O_2 at 1000°C [27]. Ta_2O_5 and Na_2SO_4 reacted rapidly to form $NaTaO_3$ which subsequently interacted destructively with the underlying Si_3N_4 substrate to form a molten phase.

The class of ceramics commonly referred to as NZPs have attracted interest as protective coating materials for Si-based ceramics because of their low CTE (~1 to 3×10^{-6} °C^{-1}) and low thermal conductivity [49]. NZPs in general exhibit anisotropic behavior. However, some NZP compositions such as $Ca_{1-x}Sr_xZr_4P_6O_{24}$ (CS50) can be tailored to obtain isotropic thermal expansion characteristics [49]. The interaction between several low-expansion bulk NZP materials and Na_2SO_4 at 1000°C in pure O_2 was investigated [50]. $Ca_{0.5}Sr0_{0.5}Zr_4P_6O_{24}$ remained intact visually and showed no significant weight change. The ion exchange between Na^+ and Ca^{++} was sufficiently rapid to allow the penetration of Na^+ into the specimen in 100h, causing the segregation of Ca on the specimen surface. NZP was first applied as a coating for SiC at ORNL using plasma spraying [49]. The air plasma-sprayed CS50 coating on Nicalon/SiC substrate remained adherent after 100h at 1000°C in air and protected the substrate from hot corrosion in Na_2SO_4 at 1000°C for 100h.

Other low CTE candidate coating materials include $ZrTiO_4$, $HfTiO_4$, $Ta_2O_5.6ZrO_2$, $Ta_2O_5.6HfO_2$, and Al_2TiO_5 [73]. $ZrTiO_4$ and $HfTiO_4$ crystallize in an orthorhombic structure that does not undergo polymorphic transformations. Al_2TiO_5 has been investigated as a protective coating for carbon/carbon (C/C) and as a corrosion resistant material for high-temperature particulate filters. However, the CTE is very anisotropic and the material exhibits significant microcracking. This material also tends to absorb water and may degrade significantly when exposed to an environment containing as little as 10% moisture at high temperatures. Not much is known about $Ta_2O_5.6ZrO_2$ and $TaO_2.6HfO_2$, except for the low CTE. The ORNL group performed a limited investigation on hot-pressed coupons of some of these materials with the main focus on the hot corrosion resistance [73]. Hot-pressed $ZrTiO_4$ and Al_2TiO_5, exposed to a combustion environment at 1000°C, showed reaction layers of ~ 4 and 2 μm thick, respectively. This was a significant improvement compared with 10 and 20 μm thick glassy reaction layers on mullite and sintered a-SiC, respectively [3]. More research needs to be done to determine the feasibility of these materials as EBCs.

Coatings for C/C

Early coatings with a boron glaze outer layer were effective only up to 1000°C. The CVD SiC/boron oxide system provided improved protection for extended exposure in the temperature range up to 1000°C because boron oxide sealed the microcracks in the SiC [54]. At temperatures above 1000°C, CVD SiC provided the protection by its self-healing characteristics, rapidly sealing pores and microcracks and thus limiting the oxidation to the vicinity of the surface. However, the use of the SiC/boron oxide system for long-term applications was limited by the volatility of boron oxide above 1000°C.

Further improvement was achieved by applying a mullite overlay coating [55]. A thin film (2.5 μm thick) of mullite was deposited by PLD on silicon-infiltrated C/C composite coated with CVD SiC. The SiC/mullite coated C/C composites exhibited a mass gain in the temperature range 1200°C ≤ T ≤ 1600°C due to the SiO_2 formation at the SiC/mullite interface. In extended oxidation, the mullite was dissolved in the newly formed SiO_2. In contrast to pure SiO_2 films, the mullite-containing SiO_2 layer did not crystallize for extended exposure at temperatures up to at least 1600°C, leading to improved crack resistance [55]. Under thermal cycling, the cracking in the inner SiC layer and spallation eventually led to the breakdown of oxidation resistance. The transport mechanisms in this system are described in detail by Lamouroux et al. [74,75].

Hafnium carbide and hafnium nitride coatings were deposited onto C/C and C/SiC composites by CVD [60]. These materials have very high melting points (HfC: 3887°C, HfN: 3310°C) and do not form volatile oxides. Therefore, these materials were suggested as candidates for intermediate layers despite their poor oxidation resistance. It was suggested that $HfO_{2-x}C_y$ may act as a bond layer in carbon containing systems.

In Japan the development of C/C was investigated in a research project of the New Energy and Industrial Technology Development Organization (NEDO) [62–64]. Iridium was suggested as an oxygen barrier for C/C because of the low oxygen permeability and chemical compatibility with carbon. However, iridium oxidizes rapidly in the oxidizing atmosphere above 1700°C. Therefore, $SrZrO_3$ (dipping + glazing), Al_2O_3 (CVD) and Y_2O_3 (plasma spray) were suggested as external layers to prevent the oxidation of iridium above 1700°C. The interface between these oxides and Ir was found to be chemically stable above 1800°C. The SiC/Y_2SiO_5 system was also proposed as an oxidation resistant coating system for C/C above 1700°C [56,57]. Y_2SiO_5 has a low CTE, similar to that of SiC, low vaporization rate and low oxygen permeability. A CVD SiC/plasma-sprayed Y_2SiO_5 duplex coating showed a life of over 1h at 1800°C in arc plasma, a simulated re-entry condition. The gas permeability can be further reduced by controlling the particle size of spray powder and the adherence of coatings to SiC can improved by applying an intermediate layer of YSi_x between SiC and Y_2SiO_5 [58,59].

SUMMARY AND CONCLUSIONS

Silicon-based ceramics need a robust, durable EBC to realize their full potential as hot section structural components in advanced gas turbines. The focus of early coatings research was on protection from contract stress damage and molten salt corrosion, which shifted to water vapor protection by the mid 1990s after the discovery of rapid silica volatilization in water vapor. Mullite has attracted most interest as a coating candidate because of the low CTE, chemical compatibility and environmental durability. In late 1970s, Garrett Turbine Engine Company pioneered the plasma spraying of mullite coatings, which was advanced by Solar Turbines in mid 1980s. The process was further advanced in the early 1990s at NASA Glenn Research Center, laying the foundation upon which the current-state-of-the-art EBCs were developed under the HSCT-EPM program (silicon/mullite+BSAS/BSAS). Crack-resistant CVD mullite coatings are promising and thus merit further research. Other low CTE materials, such as NZPs, deserve attention as potential EBC candidates.

Despite the remarkable progress during the last decade, EBCs are still far from being mature and excellent opportunities for further advancement are on the horizon. As recent researches indicate, there is a need to develop EBCs with higher temperature capability and lower thermal conductivity than EPM EBCs to realize the full potential of Si-based ceramics in next generation gas turbine engines. Future research should be directed to the exploration of low CTE, low thermal conductivity coating materials and coating design concepts to alleviate the stresses resulting from the complex nature of EBCs.

REFERENCES

[1] Jacobson, N. S., "Corrosion of Silicon-Based Ceramics in Combustion Environments," J. Am. Ceram. Soc., 76[1], 3–28, 1993.

[2] Jacobson, N. S., Smialek, J. L., and Fox, D. S., "Molten Salt Corrosion of SiC and Si_3N_4," *Handbook of Ceramics and Composites*, N. S. Cheremisinoff, ed., Marcel Dekker, New York, NY, USA, 1, 99–135, 1990.

[3] Federer, J. I., "Corrosion of SiC Ceramics by Na_2SO_4," Adv. Ceram. Mater., 76[1], 56–61, 1988.

[4] Pareek, V. and Shores, D. A., "Oxidation of Silicon Carbide in Environments Containing Potassium Salt Vapor," J. Am. Ceram. Soc., 74[3], 556–563, 1993.

[5] Opila, E. J. and Hann, R. E., "Paralinear Oxidation of CVD SiC in Water Vapor," J. Am. Ceram. Soc., 80[1], 197–205, 1997.

[6] Smialek, J. L., Robinson, R. C., Opila, E. J., Fox, D. S., and N. S. Jacobson, "SiC and Si_3N_4 Scale Volatility under Combustor Conditions," Adv. Composite Mater., 8[1], 33–45, 1999.

[7] More, M. K., Tortorelli, P. F., Ferber, M. K., Walker, L. R., Keiser, J. R., Miriyala, N., Brentnall, W. D., and Price, J. R., "Exposure of Ceramics and Ceramic Matrix Composites in Simulated and Actual Combustor Environments," ASME paper 99-GT-292, presented at the International Gas Turbine and Aeroengine Congress and Exposition, Indianapolis, IN, USA, June 7–10, 1999.

[8] Ferber, M. K., Lin, H. T., Parthasarathy, V., and Brentnall, W. D., "Degradation of Silicon Nitrides in High Pressure, Moisture Rich Environments," ASME paper 99-GT-265, presented at the International Gas Turbine and Aeroengine Congress and Exposition, Indianapolis, IN, USA, June 7–10, 1999.

[9] Lackey, W. J., Stinton, D. P., Cerny, G. A., Schaffhauser, A. C., and Fehrenbacher, L. L., "Ceramic Coatings for Advanced Heat Engines—A Review and Projection," Adv. Ceram. Mater., 2[1], 24–30, 1987.

[10] Garrett Turbine Engine Company, *Advanced Gas Turbine (AGT) Powertrain System Development for Automotive Applications*, Second Semiannual Progress Report (July-December, 1980), NASA CR-165329, July 1981, 67–82.

[11] Richerson, D. W. and Schienle J. L., "High Temperature Coating Study to Reduce Contact Stress Damage to Ceramics," *Proceedings of the Twenty-Second Automotive Technology Development Contractors' Coordination Meeting*, October 29–November 2, 1984, Dearborn, MI, USA. SAE P-155, 453–461, March 1985.

[12] Schienle, J. L., "High Temperature Coating Study to Reduce Contact Stress Damage to Ceramics," *Proceedings of the Twenty-Third Automotive Technology Development Contractors' Coordination Meeting*, October 21–24, 1985, Dearborn, MI, USA. SAE P-165, 249–257, March 1986.

[13] van Roode, M. and Price, J. R., *The Development of Protective Coatings for Ceramic Tubular Components*, Topical Report, GRI-88-0005, GRI Contract No. 5086-232-1233, 1987.

[14] Price, J. R. and van Roode, M., *Corrosion Resistant Coatings for Ceramic Heat Exchanger Tubes Operating in Highly Corrosive Environments*, Annual Report, GRI-90-0107, GRI Contract No. 5086-232, 1989.

[15] Federer, J. I., van Roode, M., and Price, J. R., "Evaluation of Ceramic Coatings on Silicon Carbide," Surface and Coatings Technology, 39–40, 71–78, 1989.

[16] Price, J. R., van Roode, M., and Stala, C., "Ceramic Oxide-Coated Silicon Carbide for High temperature Corrosive Environments," Key Engineering Materials, 72–74, 71–84, 1992.

[17] Butt, D. P., Mecholsky, Jr. J. J., van Roode, M., and Price, J. R., "Effects of Plasma-Sprayed Ceramic Coatings on the Strength Distribution of Silicon Carbide Materials," J. Am. Ceram. Soc., 73 [9], 2690–2696, 1990.

[18] Price, J. R. and van Roode, M., *Corrosion Resistant Coatings for Ceramic Heat Exchanger Tubes Operating in Highly Corrosive Environments*, Final Report, GRI-94-0353, GRI Contract No. 5086-232, 1994.

[19] Sarin, V. K., "Design Criteria for a Coating to Reduce Contact Stress Damage," Proceedings, *1987 Coatings for Advanced Heat Engines Workshop*, Castine, ME, USA, July 27–30, 1987, U.S. Department of Energy Report, Conf-870762, III-83, 1987.

[20] Rebenne, H. E. and Sarin, V. K., "Ceramic Coatings to Reduce Contact Stress Damage of Ceramics: Thermodynamic Modeling," *Proceedings, 25th Automotive Technology Development Contractors' Coordination Meeting*, Dearborn, MI, USA, Society of Automotive Engineers, Inc., Warrendale, PA, USA, P-209, 199–206, 1988.

[21] Rebenne, H. E. and Selverian, J. H., "Adherent Ceramic Coatings to Reduce Contact Stress Damage of Ceramics," *Proceedings, 28th Automotive Technology Development Contractors' Coordination Meeting*, Society of Automotive Engineers, Inc., Warrendale, PA, USA, 1991, P-243, 227–238, 1991.

[22] Federer, J. I., *An investigation of Ceramic Coatings for Protection of SiC from High-Temperature Corrosion*, ORNL/TM-10606, 1988.

[23] Federer, J. I., "Alumina Base Coatings for Protection of SiC Ceramics," J. Mater. Eng., 12, 141–149, 1990.

[24] Graham, D. W. and Stinton, D. P., "Chemical Vapor Deposition of Ta_2O_5 Corrosion Resistant Coatings," *Proceedings, 1992 Coatings for Advanced Heat Engines Workshop*, J. Fairbanks, ed., U.S. Department of Energy, Washington, DC, IV 65–71, 1992.

[25] Graham, D. W. and Stinton, D. P., "Development of Tantalum Pentoxide Coatings by Chemical Vapor Deposition," J. Am. Ceram. Soc., 77[9], 2298–2304, 1994.

[26] Bae, Y. W., Lee, W. Y., and Stinton, D. P., "Effects of Temperature and Reagent Concentration on the Morphology of Chemically Vapor Deposited β-Ta_2O_5," J. Am. Ceram. Soc., 78 [5], 1297–1300, 1995.

[27] Lee, W. Y., Bae, Y. W., and Stinton, D. P., "Na_2SO_4-Induced Corrosion of Si_3N_4 Coated with Chemically Vapor Deposited Ta_2O_5," J. Am. Ceram. Soc., 78[7], 1927–1930, 1995.

[28] Lee, K. N., Miller, R. A., and Jacobson, N. S., "Plasma-Sprayed Mullite Coatings on Silicon-Base Ceramics," *Proceedings, 5th Annual HITEMP Review, Volume III: Turbine Materials (Ceramic Matrix Composites)*, NASA-Lewis Research Center, Cleveland, OH, USA, 58-1-58-11, 1992.

[29] Lee, K. N., Miller, R. A., and Jacobson, N. S., "New Generation of Plasma-Sprayed Mullite Coatings on Silicon-Carbide," J. Am. Ceram. Soc., 78 [3], 705–710, 1995.

[30] Lee, K. N., Miller, R. A., and Jacobson, N. S., "Plasma-Sprayed Mullite Coatings on Silicon-Base Ceramics, " US Patent No. 5,391,404, 1995.

[31] Lee, K. N., Jacobson, N. S., and Miller, R. A., "Refractory Oxide Coatings on SiC Ceramics," MRS Bulletin, 14[10], 35–38, 1994.

[32] Lee, K. N. and Miller, R. A., "Oxidation Behavior of Mullite-Coated SiC and SiC/SiC Composites under Thermal Cycling between Room Temperature and 1200–1400°C," J. Am. Ceram. Soc., 79[3], 620–626, 1996.

[33] Jacobson, N. S., Lee, K. N., and Yoshio, T. Y., "Corrosion of Mullite By Molten Salts," J. Am. Ceram. Soc., 79[8], 2161–2167, 1996.

[34] Lee, K. N., "Contamination Effects on Interfacial Porosity during Cyclic Oxidation of Mullite-Coated Silicon Carbide," J. Am. Ceram. Soc., 81[12], 3329–3332, 1998.

[35] Sarin, V. K., "Systematic Development of Customized CVD Coatings," Surface and Coatings Technology, 73, 23–33, 1995.

[36] Mulpuri, R. P. and Sarin, V. K., "Synthesis of Mullite Coatings by Chemical Vapor Deposition," J. Mater. Res., 11[6], 1315–1324, 1996.

[37] Auger, M. L. and Sarin, V. K., "The Development of CVD Mullite Coatings for High Temperature Corrosive Applications," Surface and Coatings Technology, 94–95, 46–52, 1997.

[38] Basu, S. N., Hou, P., and Sarin, V. K., "Formation of Mullite Coatings on Silicon-Based Ceramics by Chemical Vapor Deposition," International Journal of Refractory Metals & Hard Materials, 16, 343–352, 1998.

[39] Lee, W. Y., More, K. L., and Bae, Y. W., "Characterization of Si_3N_4 Coated with Chemically-Vapor-Deposited Mullite after Na_2SO_4-Induced Corrosion," J. Am. Ceram. Soc., 79 [9], 2489–2492, 1996.

[40] Strangman, T. E., Honeywell Engines, unpublished research, 1996.

[41] Lee, K. N., Miller, R. A., Jacobson, N. S., and Opila, E. J., "Environmental Durability of Mullite/SiC and Mullite/YSZ Coating/SiC Systems," Ceram. Eng. and Sci. Proc., 16[5], 1037–1044, 1995. J.B. Watchman, ed., The American Ceramic Society, Westerville, OH, USA.

[42] Lee, K. N. and Miller, R. A., 1996, "Development and Environmental Durability of Mullite and Mullite/YSZ Dual Layer Coatings for SiC and Si_3N_4 Ceramics," Surface and Coatings Technology, 86–87, 142–148, 1996.

[43] Lee, K. N., 1999, "Key Durability Issues with Mullite-Based Environmental Barrier Coatings for Si-Based Ceramics," ASME paper 99-GT-443, presented at the International Gas Turbine and Aeroengine Congress and Exposition, Indianapolis, IN, USA, June 7–10, 1999.

[44] Lee, K. N. and Miller, R. A., "Durability of Mullite/YSZ-Coated SiC in 90% H_2O/O_2," Advances in Ceramic Matrix Composites IV, The American Ceramic Society, Westerville, OH, USA, 17–25, 1999.

[45] Lee, K. N., "Effect of Impurities and Coating Design on the Durability of EBC Coatings for Si-Based Ceramics," *Proceedings of the Per Kofstad Memorial Symposium, High Temperature Corrosion and Materials Chemistry*, M. McNallan, E. Opila, T. Maruyama, and T. Narita, eds., Electrochemical Society, Pennington, NJ, 417–428, 2000.

[46] Lee, K. N., "Current Status of Environmental Barrier Coatings for Si-Based Ceramics," Surface and Coatings Technology, 133–134, 1–7, 2000. © 2000, Elsevier Science.

[47] Eaton, H. E., Linsey, G. D., More, K. L., Kimmel, J. B., Price, J. R., and Miriyala, N., "EBC Protection of SiC/SiC Composites in Gas Turbine Combustion Environments," ASME paper 2000-GT-0631, presented at the International Gas Turbine and Aeroengine Congress and Exposition, Munich, Germany, May 8–11, 2000.

[48] Endoh, Y., Tsuchiya, T., and Furuse, Y., "Corrosion Behavior of Ceramics for Gas Turbine Application", *Proceedings of the 1993 PAC RIM Meeting*, November 7–10, Honolulu, HI, USA, 1993. pp. 319–326.

[49] Lee, W. Y., Cooly, K. M., Berndt, C. C., Joslin, D. L., and Stinton, D. P., "High-Temperature Chemical Stability of Plasma-Sprayed $Ca_{0.5}Sr_{0.5}Zr_4P_6O_{24}$ Coatings on Nicalon/SiC Ceramic Matrix Composite and Ni-Based Superalloy Substrates," J. Am. Ceram. Soc., 79 [10], 2759–2762, 1996.

[50] Lee, W. Y., Stinton, D. P., and Joslin, D. L., "Interaction of Low-Expansion NZP Ceramics with Na_2SO_4 at 1000°C," J. Am. Ceram. Soc., 79 [2], 484–486, 1996.

[51] Haynes, J. A., Cooly, K. M., Stinton, D. P., Lowden, R. A., and Lee, W. Y., "Corrosion-Resistant CVD Mullite Coatings for Si_3N_4," Ceram. Eng. and Sci. Proc., 20[4], 355–362, 1999. The American Ceramic Society, Westerville, OH, USA.

[52] Haynes, J. A., Lance, M. J., Cooley, K. M., Ferber, M. K., Lowden, R. A., and Stinton, D. P., "Protective Capacity of CVD Mullite Coatings in a High-Temperature, High-Pressure, Air-H_2O Environment," J. Am. Ceram. Soc., 83 [3], 657–659, 2000.

[53] Buchanan, F. J. and Little, J. A., "The Influence of Chemical Vapor Deposition Conditions on the Structural Stability and Cracking of Silicon-Carbide Coatings on Carbon-Carbon Composites," Surface and Coatings Technology, 53, 137–146, 1992.

[54] Gee, S. M. and Little, J. A., "Oxidation Behavior and Protection of Carbon/Carbon Composites," J. Mater. Sci., 26, 1093–1100, 1991.

[55] Fritze, H., Jojic, J., Witke, T., Rüscher, C., Weber, S., Scherrer, S., Weiß, R., Schultrich, B., and Borchardt, G., "Mullite Based Oxidation Protection for SiC-C/C Composites in Air at Temperatures up to 1900 K," J. Eur. Ceram. Soc., 18, 2351–2364, 1988.

[56] Morimoto, T., Ogura, Y., Kondo M., and Ueda, T., "Multilayer Coating for Carbon-Carbon Composites," Carbon, 33[4], 351–357, 1995.

[57] Ogura, Y., Kondo, M., and Morimoto, T., "Y_2SiO_5 as Oxidation Resistant Coating for C/C Composites," *Proceedings, 10th International Conference on Composite Materials*, A. Poursartip and K. Street, eds., Woodhead Publishing Limited, Cambridge, UK, Vol. IV, 1995. pp. 767–774.

[58] Sakakibara, N., Notomi, A., Ogura, Y., Kondo, M., Fujiwara, C., Sekigawa, T., Kouchiyama, J., and Miho, K., "Y_2SiO_5 High Temperature Oxidation Resistant Coating on C/C Composites by Plasma Spraying," J. Japan Inst. Metals, 63[1], 118–125, 1999 (in Japanese).

[59] Kondo, M., Ogura, Y., Morimoto, T., and Notomi, A., "Formation of Y_2SiO_5 Layer in Y_2SiO_5 Oxidation Protection Coating for Carbon/Carbon Composites," J. Japan Inst. Metals, 63[7], 851–858, 1999 (in Japanese).

[60] Ache, H. F., Goschnick, J., Sommer, M., Emig, G., Schoch, G. and Wormer, O., 1994, "Chemical Vapor Deposition of Hafnium Carbide and Characterization of the Deposited Layers by Secondary-Neutral Mass Spectrometry," Thin Solid Films, 241, 356–360, 1994.

[61] Treska, M., Hobbs, L. W. and Pemsler, J. P., "Carbon Fiber Reinforced Diffusion Barriers for Oxidation Protection of Carbon-Carbon Composites at 1600–2000°C," *Proceedings, Microscopy of Oxidation III*, S. Newcomb, ed., 1997. pp. 720–729.

[62] Kishi, K., Yamauchi, H., Hayashi, K., Miyagawa, K., and Kosaka K., *Development of Manufacturing Processes of C/C Composites and Oxidation Resistant Coatings*, NEDO Report No.010008475, R&D Institute of Metals and Composites for Future Industries, Tokyo, Japan, 15–26, 1995 (in Japanese).

[63] Ishizaki, M., *Development of Manufacturing Processes of C/C Composites and Oxidation Resistant Coatings*, NEDO Report No.010008475, R&D Institute of Metals and Composites for Future Industries, Tokyo, Japan, 85–99, 1995 (in Japanese).

[64] Enomoto, K. and Taguchi, M., *Development of Manufacturing Processes of C/C Composites and Oxidation Resistant Coatings*, NEDO Report No.010010129, R & D Institute of Metals and Composites for Future Industries, Tokyo, Japan, 1996. pp. 51–65 (in Japanese).

[65] Pompe, W., Scheibe, H. J., Siemroth, P., Wilberg, R. Schulze, D., Bucken, B., 1992, "Film Deposition by Laser-Arcs," Thin Solid Films, 208, 11–14, 1992.

[66] Strife, J. R. and Sheehan, J. E., "Ceramic Coatings for Carbon-Carbon Composites," Am. Ceram. Soc. Bull., 67 [2], 369–374, 1988.

[67] Ogura, Y., Kondo, M., Morimoto, T., Takeda, Y., and Notomi, A., "Thermal expansion of Plasma Sprayed Y2SiO5 Coating," J. Japan Inst. Metals., 63[10], 1295–1303, 1999 (in Japanese).

[68] More, K. L., Tortorelli, P. F., Walker, L. R., Kimmel, J.B., Miriyala, N., Price, J.R., Eaton, H. E., Sun, E. Y., and Linsey, G. D., "Evaluating Environmental Barrier Coatings on Ceramic Matrix Composite Combustor liners after Engine and Laboratory Exposures," ASME paper GT-2002-30630, presented at the ASME TURBO EXPO LAND, SEA & AIR, June 3–6, 2002, Amsterdam, The Netherlands.

[69] Lee, K. N., Fox, D. S., Robinson, R. C., and Bansal, N. P., "Environmental Barrier Coatings for Silicon-Based Ceramics," High Temperature Ceramic Matrix Composites, in *High Temperature Ceramic Matrix Composites*, W. Krenkel, R. Naslain, H. Schneider, eds., Wiley-VCH, Weinheim, Germany, 2001. pp. 224–229.

[70] Lee, K. N., Fox, D. S., Eldridge, J. I., Zhu, D., Robinson, R. C., Bansal, N. P., and Miller, R. A., *Upper Temperature Limit of Environmental Barrier Coatings Based on Mullite and BSAS*, NASA/TM-2002-211372, 2002.

[71] Eldridge, J. I., and Lee, K. N., "Phase Evolution of BSAS in Environmental Barrier Coatings," Ceram. Eng. and Sci. Proc., 22[4], 383–390, 2001. The American Ceramic Society, Westerville, OH, USA.

[72] Zhu, D., Lee, K. N., and Miller, R. A., "Thermal Conductivity and Thermal Gradient Cyclic Behavior of Refractory Silicate Coatings on SiC/SiC Ceramic Matrix Composites," Ceram. Eng. and Sci. Proc., 22[4], 443–452, 2001. The American Ceramic Society, Westerville, OH, USA.

[73] Stinton, D. P., McLaughlin, J. C., and Riester, L., "Fabrication and Testing of Corrosion Resistant Coatings," *Proceedings, 4th International Symposium on Ceramic Materials and Components for Engines*, R. Carlsson, T. Johansson, and L. Kahlman, eds, Göteborg, Sweden, 10–12 June, 1991.Elsevier Applied Science, New York, USA, 1992. pp. 1146–1153.

[74] Lamouroux, F., Camus, G., and Thébault, J., "Kinetics and Mechanisms of Oxidation of 2D Woven C/SiC Composites: I, Experimental Approach," J. Am. Ceram. Soc., 77[8], 2049–2057, 1994.

[75] Lamouroux, F., Naslain, R., and Jouin, J.-M., "Kinetics and Mechanisms of Oxidation of 2D Woven C/SiC Composites: II, Theoretical Approach," J. Am. Ceram. Soc., 77[8], 2058–2068, 1994.

CHAPTER

28

DESIGN OF IMPACT-RESISTANT CERAMIC STRUCTURAL COMPONENTS

Alonso D. Peralta

Mechanical Engineering Department, State University of New York, Stony Brook, New York, USA

Hiro Yoshida

Fluid Engineering Division, Department of Energy Engineering,
Mechanical Engineering Laboratory, AIST, METI, 1-2 Namiki, Tsukuba, Ibaraki, Japan

INTRODUCTION

The use of ceramic materials to fabricate structural components for applications in heat engines holds the promise of improving the thermodynamic efficiency of such engines. Efficiency improvements could be realized by increasing the operating temperatures of the thermodynamic cycle and by reducing the cooling flow requirements demanded by the metallic components that are currently used in such applications. Due to the brittle nature of the ceramic materials, design methodologies that are very different to those used for metallic materials are required. Issues that must be addressed are those of fast fracture, sub-critical crack growth, contact mechanics and impact mechanics. This chapter discusses the behavior of ceramic materials under impact conditions, also known as foreign object damage (FOD), and considerations that must be made to realize a successful impact resistant design. The material presented here is compiled from other sources, and it is packaged for the engineer that must design ceramic components to resist impact damage.

DISCUSSION

The introduction of ceramic materials into turbine engines has generally been for application in the first stage nozzle, turbine blades, combustor liners and scrolls. These components, in all likelihood, are designed to withstand failure under fast fracture (FF) and subcritical crack growth (SCG) modes. The methodologies to design against fast fracture and subcritical crack growth failure modes have been extensively developed as evidenced by the long publication list that exists in this area [1–8]. Chapter 25 in this volume also deals with probabilistic design methodologies. The fast fracture failure mode has received the most attention; it is currently possible to pool data from several different specimen types, under different loading conditions, and at different test temperatures to calculate the Weibull parameters [5,9]. Most importantly, it is possible to calculate the confidence limits on the reliability of the component for fast fracture under a given state of stress [5,10]. The subcritical crack growth failure mode has recently received more attention and new more powerful data analysis techniques have been developed [8,10]. These new techniques make it possible to pool failure data from fast fracture and subcritical crack growth tests into a single analysis that improves the calculation of the fast fracture and subcritical crack growth parameters. These advanced data analysis techniques allow the calculation of confidence bounds on the reliability of the time to failure predictions of components under SCG failure modes. The result has been the development of a very robust design methodology against fast fracture and SCG failure modes.

Unfortunately, the methodologies for designing to preclude failure under contact and particle impact conditions have not been developed to the extent that fast fracture methodologies have. Not to say that these failure modes have not been observed in real applications, contact and particle impact damage have been observed in programs funded (in the United States) by DARPA (Defense Advanced Research Projects Agency) in the 1970s [11], and DOE (Department of Energy) in the 1980s [12] and 1990s [13,14]. It is clear that these engine programs have had major setbacks as a result of failures due to contact and impact conditions. The prevalent approach has been the development of deterministic guidelines for contact and impact conditions, disregarding the probabilistic nature of the strength of ceramic materials. Out of contact and projectile impact failure modes, the guidelines for projectile impact conditions have been much more advanced than those for contact conditions. The remainder of this chapter will focus on guidelines that have been developed to preclude projectile impact damage.

Projectile impact conditions are experienced by static as well as rotating structural components. It is well known that projectile impact is more critical on rotating components since the intensity of the impact event depends on the relative velocity of the projectile. Since the velocity of the rotating component in the tangential direction is very large, i.e., in excess of 300 m/s, the relative velocity between a projectile traveling in the axial direction and the rotating component is also very large. The type of damage that may result from the impact event can be either structural or local damage. Structural damage is characterized by an immediate failure of the component as the impact event occurs. Local damage, in contrast, is characterized by the degradation of the strength of the ceramic material and is exacerbated as the numbers of impact events that produce local damage accumulate with time. The typical type of damage that may be encountered in ceramic structural components as a result of projectile impact can be characterized to be of the following modes [15], also shown in Figure 28.1:

(a) Chipping fracture mode
(b) Proximity cantilever beam fracture mode
(c) Flexure fracture mode, and
(d) Hertzian fracture mode at the impact location.

The first three are structural damage modes, whereas the Hertzian fracture mode causes local damage. In rotating components, the structural damage mode would likely initiate a cascade event where the chip that breaks off would impact against another ceramic component that would then cause major damage. It is important to understand that the above failure modes are illustrative and that depending on the type of component similar conditions of failure will arise. For example, Hara et al. [15] describe the type of structural damage that may be observed for a turbine blade as a result of projectile impact (see Figure 28.2). Similar results are cited in Chapter 7 of Volume I of this book series.

Given the above failure modes, considerations can be made for designing against impact damage. The following items should be included.

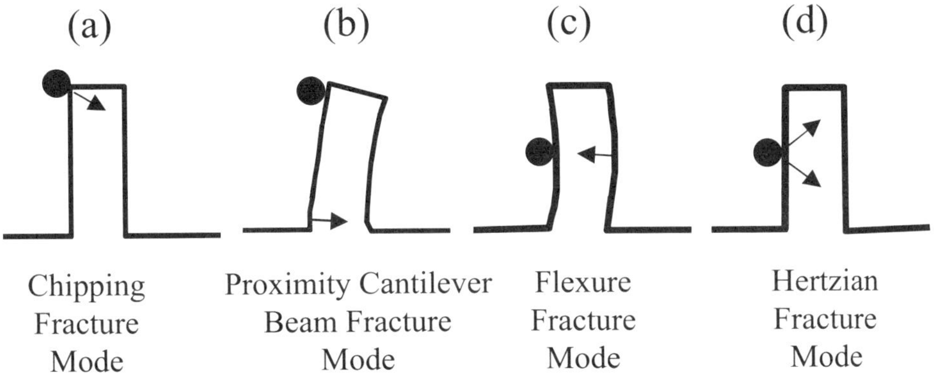

Figure 28.1 Typical Type of Damage in Ceramic Components Due to Projectile Impact (Adapted from ref. [15])

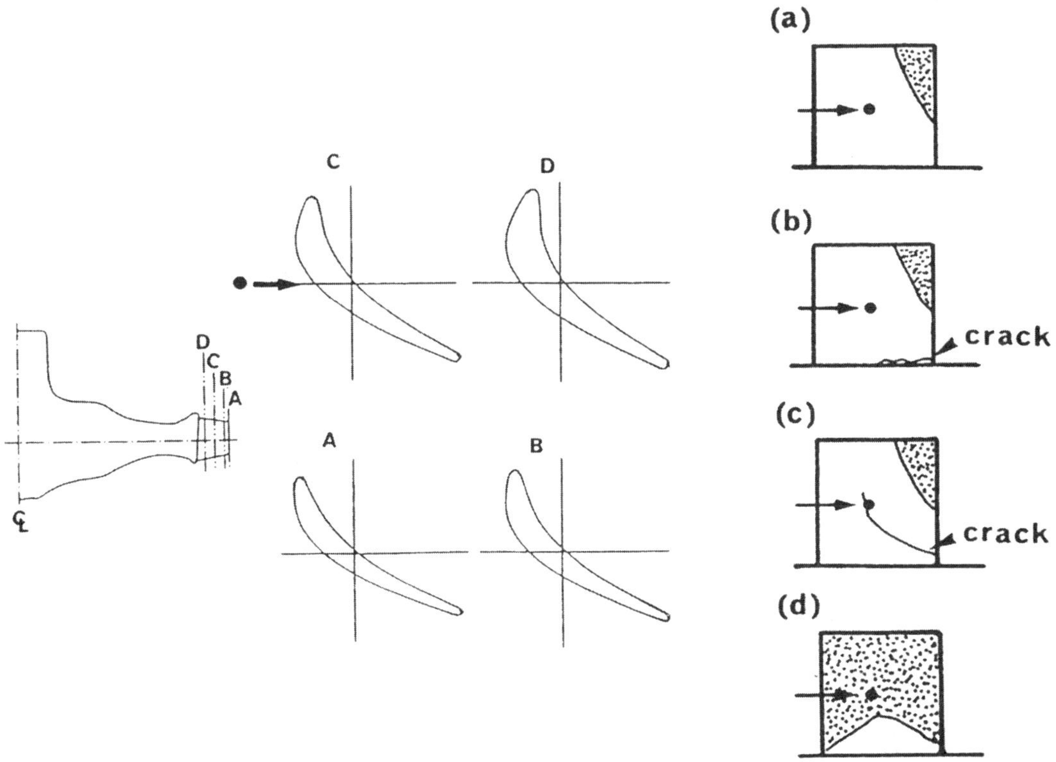

Figure 28.2 Fracture Modes of an Axial Turbine Blade as a Result of Projectile Impacts (Reprinted from ref. [15])

(a) The material of the component, which includes all of its material properties.
(b) The material of the projectile.
(c) The geometry of the projectile.
(d) The geometry of the component.
(e) The boundary conditions and temperature of the component.

(f) The angle of incidence of the impact event.
(g) The location of the impact on the component.
(h) The relative velocity of the projectile and the component.
(i) And, the magnitude of the background stress.

The influence of the above items may be best evaluated after a brief introduction to impact theory.

METHODS OF IMPACT

Before moving on to impact theory, a few typical methods of particle impact tests will be briefly described. These methods have been commonly used to study the impact resistance of ceramic materials and components. This is because foreign object damage seems to be deeply implicated with events that often appear in actual ceramic turbines [16, 17]. Understanding the impact behavior of the ceramic is inevitable to attain high reliability and to realize the reliable design of ceramic structural components in turbine engines. Suitable methods of launching particles differ depending on the mass of the particle and its velocity range. For example, in the FOD study for the Japanese small-scale ceramic gas turbines [18,19] particle sizes are assumed to be in the several-millimeter to sub-millimeter range. The impact velocity of the particle is supposed to be limited by the peripheral velocity of the turbine, i.e. 400 m/s–1,000 m/s at the highest. Therefore, the particle launchers for FOD experiments are expected to impart on the particle velocities in this range. A conventional gas gun with bursting diaphragm usually covers the impact velocities up to 400 m/s. And this is used most often in FOD experiments [20–23]. For velocities higher than 400 m/s and for relatively heavy particles, a powder gun is sometimes used [24]. In most cases impact particles are too small to launch directly, i.e. 1 millimeter to several millimeters in diameter, and are therefore mounted on a carrier, called sabot [25].

The operating conditions of ceramic component in turbine engines are typically at temperatures above 1,000°C, well above standard operating temperatures for most metal components. Moreover, the components can be subjected to tensile stresses, like in the rotating blades. Therefore, to simulate the impact behavior at the operating conditions of actual blade components, the control of the stress and temperature conditions is necessary. Recently, a launch system using an electro-thermal (ET) gun was developed (for a schematic of the design see Figure 1 of reference [25]). The working principle of the ET gun is as follows: when impulse current runs through the aluminum foil set in the gun, the foil is Ohmically heated and then evaporated explosively to generate a small amount of metal plasma. This plasma gas propels the sabot and the particle on it. The particle is separated from the sabot at the muzzle of the gun. Finally, only the particle hits the target. This system can launch a 1 mm diameter and 4 mg weight particle at velocities in the range of 200 m/s to higher than 1,000 m/s without using a large amount of driving gases like in the gas guns. Because of the low emission of gases, the test environment is scarcely contaminated. Using this system, therefore, one can conduct the impact tests under elevated temperatures and applied stress conditions.

In this section we referred to only the particle impact test methods though, there are other methods such as the Charpy and Izod impact tests and the Hopkinson bar method. They are so popular that they are described in many standard textbooks (for example, the book by Zukas et al., [26]).

IMPACT THEORY

In the impact event, the component (or target) and the projectile come into contact and a force develops between the two objects. The time history of the force, the location of the force, and the size of the contact area are ultimately responsible for the mode and the extent of damage generated by the impact event [27]. The structural damage mode usually occurs from a location that is removed from the impact location, whereas the local damage mode occurs at the impact location. The local damage mode has received the most attention, and for the most part under very idealized conditions. To evaluate the effect of particle impact two approaches have been used: a quasi-static method [28] and a dynamic method [29,30]. In the quasi-static method the laws of motion are used to relate the degree of penetration to the projectile kinetic energy in the local damage mode. For the structural damage mode, the kinetic energy of the projectile is transferred to the target, which stores it as strain energy as a result of the deformed configuration that it assumes after impact. In the dynamic method the relations between the particle velocity, the pressure between the projectile and target, and the wave velocity of the target and projectile characterize the transient contact between the impacting bodies.

The Force History, Contact Area and Duration of Contact

For the quasi-static method to be valid for calculating the contact force between the target and the projectile, the velocity of the projectile must be significantly below the velocity of the stress waves in the contacting materials. Furthermore, any plastic deformation, fracture or damage must absorb a negligible amount of the kinetic energy carried by the projectile [27]. The effect of this calculated dynamic force is then assumed to be the same as that of a static force. Hertz [31] first addressed the problem of the elastic contact between curved bodies compressed by a static force normal to the surface. The solution proceeds as follows:

For a sphere of radius R in contact with an elastic half space (an infinite flat surface) the maximum force that develops between the two bodies is calculated by equating the kinetic energy of the projectile before it impacts the flat surface to the work done by the applied load [32].

$$\frac{1}{2}\left(\frac{4}{3}\pi\rho R^3\right)V^2 = \int_0^{z_{max}} P(z)dz \tag{1}$$

Where ρ is the density of the sphere, V is the velocity of the sphere and z is the distance over which the load is applied and it is also known as the distance of mutual approach between the two bodies. The distance of mutual approach is given by [33]:

$$z = \frac{a^2}{R} \tag{2}$$

Where a is the radius of contact between the two bodies and can be calculated by the following equation [33]:

$$a^3 = \frac{3}{4}kPR \tag{3}$$

where $k = \left(\frac{1-\nu_1^2}{E_1} + \frac{1-\nu_2^2}{E_2}\right)$, E_1, E_2, ν_1, ν_2, are Young's modulus and Poisson's ratio respectively, the subscripts refer to the target and projectile, respectively. Substituting these variables into the above equation for the force one obtains:

$$P_{max} = \left(\frac{5}{3}\pi\rho\right)^{3/5}\left(\frac{3}{4}k\right)^{-2/5} V^{6/5}R^2 \tag{4}$$

In terms of the mass of the projectile, the above equation becomes:

$$P_{max} = 1.279\left(m^3V^6Rk^{-2}\right)^{1/5} \tag{5}$$

The duration of contact for this case is given by [33]:

$$t = 2.94\left(\frac{5}{4}\pi\rho k\right)^{2/5} \frac{R}{V^{1/5}} \tag{6}$$

For the case of a sphere impacting with a beam, the force can be calculated using a similar argument where the kinetic energy and the strain energy of the deformed beam are equated. For a cantilever beam and the sphere impacting at the free end, we have [34]:

$$P_{\max} = W\sqrt{V^2/g\Delta_{st}} \tag{7}$$

where Δ_{st} is the static deflection of the beam when loaded by a static force equal to W, the weight of the projectile. Hence, for the cantilever beam of length L and moment of inertia I:

$$\Delta_{st} = WL^3/3EI \tag{8a}$$

and

$$P_{\max} = \sqrt{3EIWV^2/gL^3} \tag{8b}$$

For the case of a simply supported beam,

$$\Delta_{st} = WL^3/48EI \tag{9a}$$

and

$$P_{\max} = \sqrt{48EIWV^2/gL^3} \tag{9b}$$

The stresses due to the contact force can be calculated so as to determine the effect the contact force has on the target. For the sphere on the elastic half space, the applied load is distributed as a hemisphere of compressive stresses within the diameter of contact area. This force depresses the surface of the flat plane and flattens the sphere. The result is tensile surface radial stresses σ_{rr} and compressive surface hoop stresses $\sigma_{\theta\theta}$. The variation of these stresses on the surface of the body are given by [32,35]:

$$\sigma_{rr} = \frac{(1-2\nu_1)}{2\pi}\frac{P_{\max}}{r^2} \tag{10a}$$

$$\sigma_{\theta\theta} = -\sigma_{rr} \quad \text{for} \quad r \geq a \tag{10b}$$

The stresses on the symmetry planes for this loading condition are plotted in Figure 28.3 and Figure 28.4 [36]. Hamilton and Goodman [35] obtained the solution to the problem of a sphere sliding along the surface. They found that at the edge of the contact area the stresses are intensified. Figure 28.5 shows the variation of the surface stress along the direction of the applied force. This figure shows that as the coefficient of friction increases the radial stresses are greatly intensified at the trailing edge, the position away from the direction of motion of the sphere relative to the surface.

In the dynamic method, the solution must account for the growing area of contact and the inertia effect of the material in the projectile and target. Tsai [30] solved the dynamic solution to the Hertz problem. Tsai remarked that the solution to the Hertz problem can be obtained by employing Hankel transforms operating

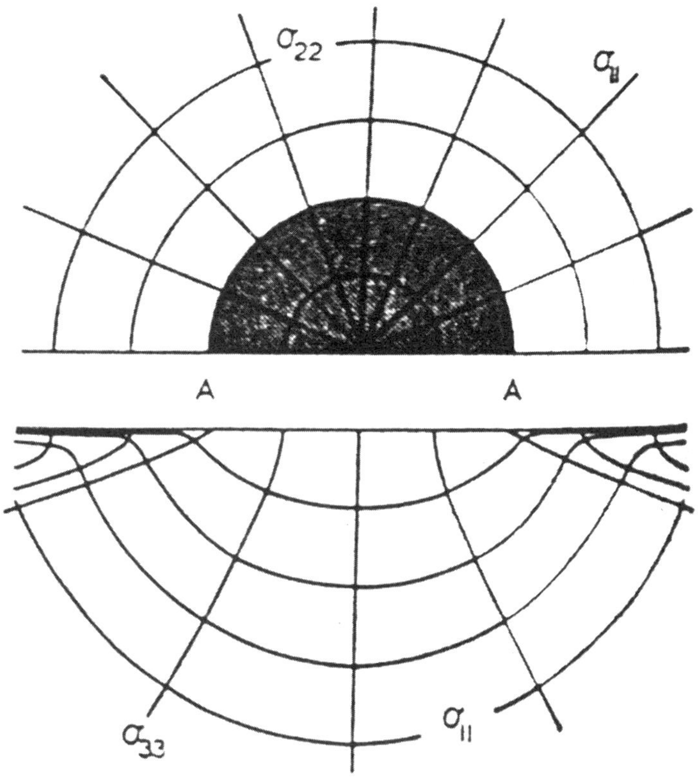

Figure 28.3 Stress Trajectories of the Principal Stresses in a Hertzian Loading Condition. Top View, Principal Stresses on the Surface and Bottom View Principal Stresses into the Material (Reprinted with permission from ref. [36]). Shaded Region A-A Denotes Area of contact.

on the variable r and Laplace transforms operating on the time t. Furthermore, the maximum load can be obtained from the solution to an integral equation. Tsai's main conclusions were that the Hertz theory is a good approximation for determining the maximum contact force produced during impact. Also, in the calculation of the radial surface stresses when the contact radius is a maximum, the Hertz theory applies only for moderate impact velocities where the contact time is more than 40 μs. The discrepancy between the radial stress in the Hertz problem and the full dynamic solution increases with decreasing contact time and increasing contact radius, and the discrepancy is larger immediately after impact than at the middle of the contact time when the load is largest.

Tsai [37] has calculated the effect of a finite body size on the magnitude of the dynamic stresses. For an elastic plate of finite thickness, Tsai has shown that the tensile stresses in a thick plate are larger than the associated tensile stresses for the Hertz problem, and these stresses increase as the thickness of the plate decreases. Furthermore, Bowden and Field [38] and Field [39] have shown that when a lead slug or high velocity water jet strikes a brittle solid, micro-cracks located well away from the impact site may be opened up by the surface Rayleigh waves. And Glenn [40] has proposed that the interaction of the impact stresses with the reflected compressive stress pulse play a significant role in the micro-fracture beneath the impact zone.

In the dynamic method for the case of structural damage, the approach is very much the same as that used by Tsai, except that now the displacements of the beam or plate must be taken into account.

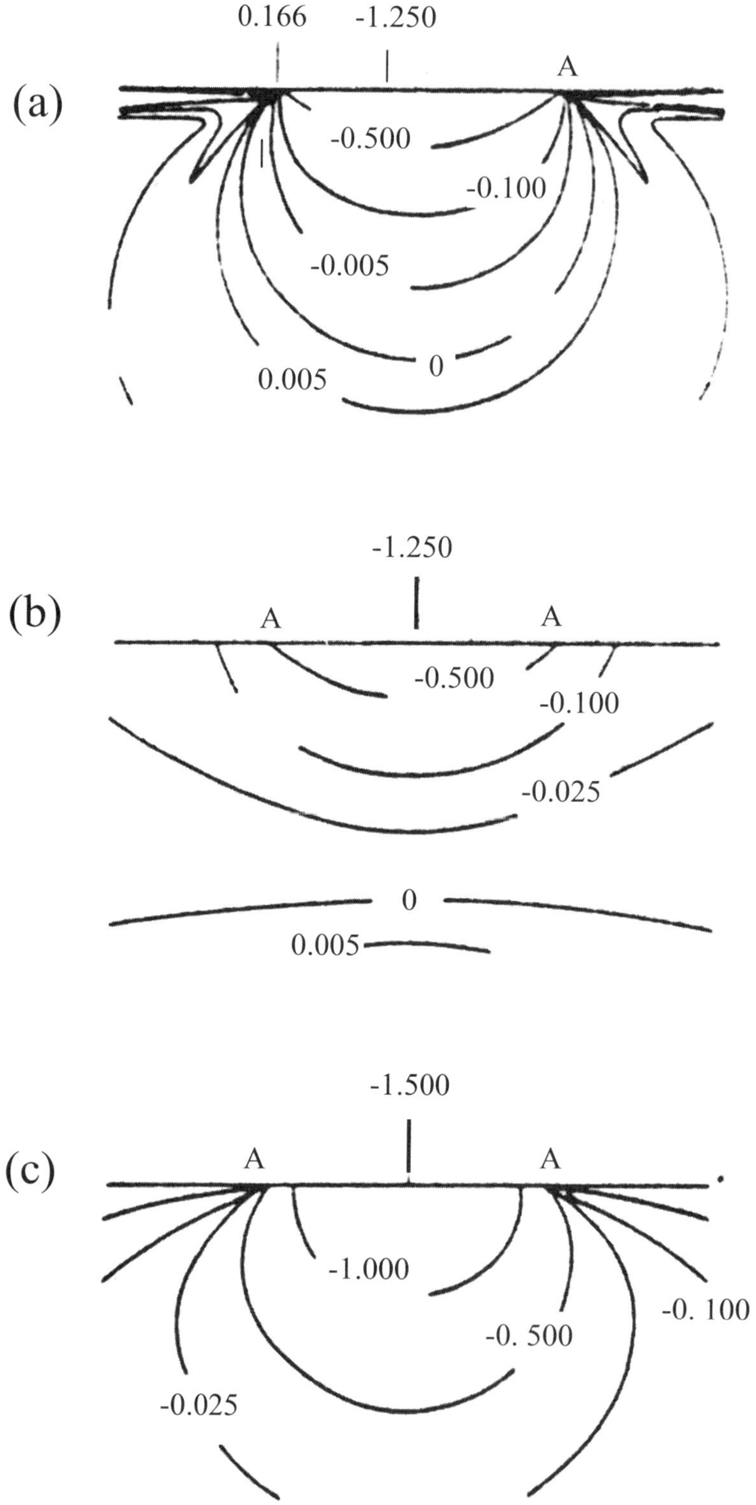

Figure 28.4 Contours of Principal Stress in a Hertzian Field, (a) (b) and (c) Stresses Normalized with Respect to Contact Pressure (Adapted with permission from ref. [36])

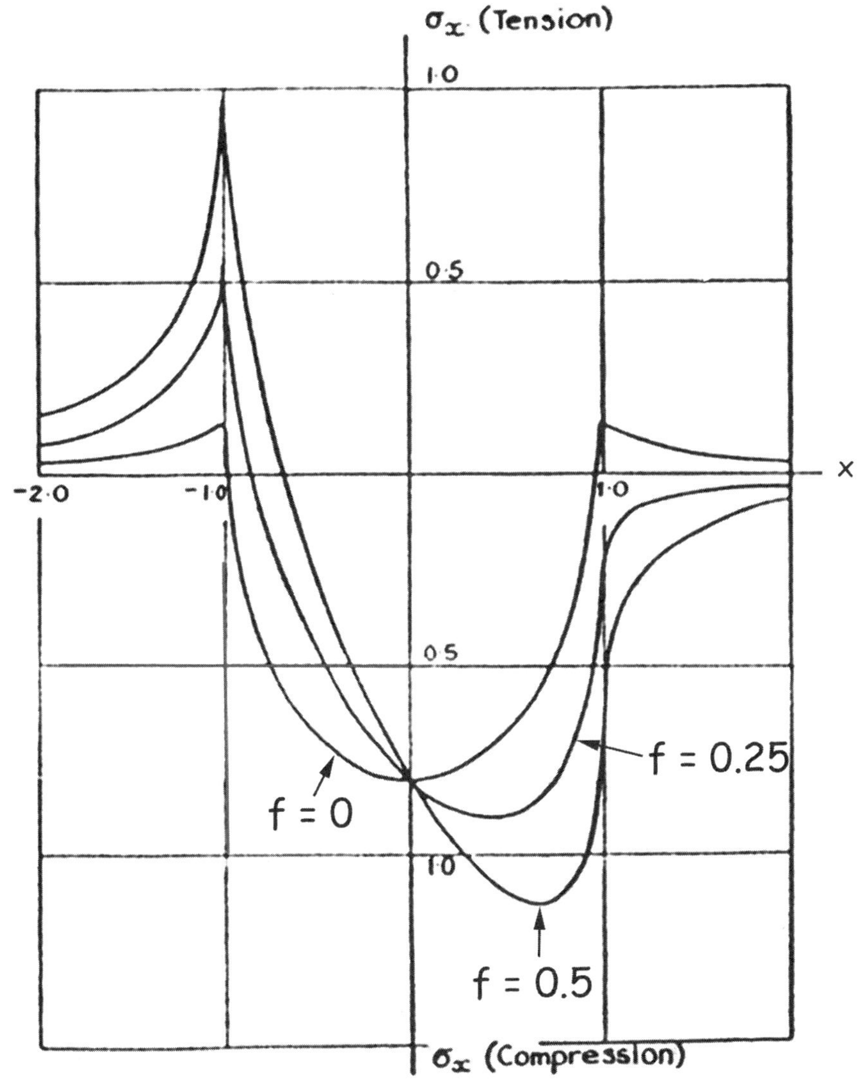

Figure 28.5 Stress Contours of the Surface Stress, σ_x, as a Function of Friction Factor. Note the Increased Stress as the Friction Factor, f, Increases. Stresses Normalized with Respect to Contact Pressure (Adapted from ref. [35])

Goldsmith (ref. [29], p.108) applies the dynamic approach to the impact of a simply supported beam by an elastic mass. Goldsmith finds that the contact force is the solution to an integral equation, as is the case for the dynamic solution to the Hertz problem. The beam displacements and the outer fiber stresses are then computed from the applied load and the dynamic response of the beam to such a load. The response is given in terms of the excited natural modes of vibration of the beam as a result of the time varying load. Goldsmith shows that the maximum contact force can be significantly reduced when the ratio of the mass of the beam to that of the projectile is much larger than one. This is due to the flexibility of the beam. When the ratio is close to one, the projectile and the beam will make contact for a second time. The first time the projectile and the beam make contact the applied force is of the order of the magnitude of the force computed for the dynamic Hertz problem. The second time the projectile and beam make contact, the magnitude of the force actually exceeds by a significant amount the magnitude of the force calculated for the dynamic Hertz problem.

These types of results are important for the impact analysis of turbine blades and stators. These structural members are very thin due to aerodynamic performance requirements and it is likely that conditions such as those described are encountered in the impact event, and must be accounted for.

The Local Damage Mode

Having established the magnitude of the forces generated by the impact of a spherical projectile on a flat surface, we proceed to describe the type of damage that is produced by the impact of a projectile on the target. In the more general case the projectile will not be a spherical object, instead the projectile may have an irregular shape with some sharp edges. In ceramic materials, contact and impact damage are characterized by complex patterns of cracks that are classified as ring, cone, median, radial and lateral cracks. Furthermore, pit-shaped permanent impressions are also found in the indentation or impact of ceramic materials similar to those found in metals. In metals, the formation of pits is due to the plastic deformation underneath the indenter and this may well happen in ceramics too. Furthermore, Liaw et al. [41] have shown numerically that the impressions in ceramic materials due to contact or impact conditions can be the result of microcracking.

The formation of the complex patterns of cracks that develop due to impact damage may evolve as follows. For a spherical projectile, the first cracks that may develop are ring cracks [36]. These cracks develop due to the radial stress field at the edge of the contact zone, and they are likely the result of the growth of preexisting flaws on the surface of the ceramic material. The crack extension around the circumference is due to the axisymmetric nature of the stress field. As the load between the projectile and the target increases, the radial cracks propagate into the target forming the cone cracks.

As the load increases, a zone of plastic deformation or microcracking forms under the projectile that gives rise to median cracks. Median cracks are penny-shaped and are oriented normal to the surface. They initiate below the plastic zone and propagate into the material. Upon unloading, radial cracks form within the surface about the area of contact at approximately half the maximum load. Just prior to complete unloading lateral cracks initiate. These cracks initiate from the subsurface tension about the deformed zone and propagate in a saucer–like shape towards the surface. See Figure 28.6 for a graphical representation of this sequence and the location and shape of the cracks and Figure 28.7 for an example of the generated flaws.

In the case of a sharp indenter, the crack formation process takes a slightly different path. The plastic region develops first followed by radial cracks, which emanate from the sharp edges of the indenter. Then, median cracks develop underneath the plastic zone, followed by lateral cracks in the unloading process.

The damage formation depends also on material properties besides the indenter shape, velocity of the projectile and the elastic properties of the material. Hardness and microstructure also play an important role. For example, Akimune et al. [20] have shown the difference in damage formation between two different silicon nitrides due to spherical impact damage. The first material is a gas pressure sintered silicon nitride (GPSSN), EC141, from NGK Spark Plug Co. Ltd., Japan; and the second material a pressureless sintered silicon nitride (SSN), SN60, from NGK Insulators Co. Ltd., Japan. For the GPSSN material Akimune et al. [20] observed that with increasing impact velocity, GPSSN first formed the pit impressions, then ring cracks and surface radial cracks, and last the lateral cracks (Figure 28.8). In contrast, SSN did not form lateral cracks as the impact velocity increased, but it did form the median cracks under the indenter. Furthermore, the GPSSN had a smaller permanent depression than SSN. Akimune et al. [20] propose that no plastic deformation formed under the impact location for the GPSSN, whereas plastic deformation or microcracks formed under the impression for the SSN as indicated by the presence of the median cracks.

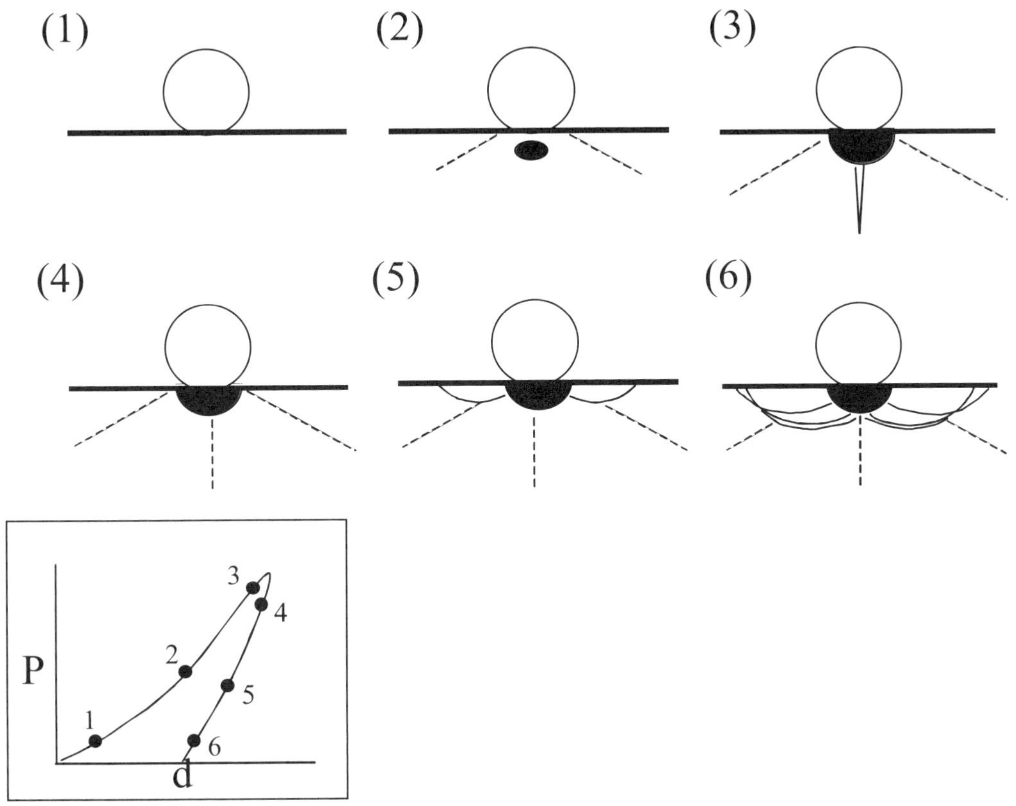

Figure 28.6 Representation of the Complex Patterns of Cracks that Develop During the Impact Process as Reflected by the Load, P, versus Displacement, d, Plot (Adapted with permission from ref. [32])

The microstructures of GPSSN and SSN, tested by Akimune et al. [20], are shown in Figure 28.9. This figure shows the denser microstructure for GPSSN, which gives GPSSN a hardness larger than the partially stabilized zirconia spheres that were used for the impact test. This figure also shows the more open microstructure and glassy phase of SSN, which gives SSN a hardness value similar to that of the impacting spheres. Akimune et al. [20] postulate that as a result of the higher hardness of GPSSN and the relatively low glassy phase, the GPSSN material behaves elastically up to the point of crack initiation. In contrast, SSN with hardness comparable to that of the indenter and the larger glassy phase develops microcracks that make it behave in an elastic/plastic fashion.

The Structural Damage Mode

The structural damage mode is the result of the excessive stresses that occur from the overall deformation of the structure due to the projectile impact event. This type of damage mode has received little attention in the literature [42], but it is probably the one responsible for the failures observed in the different programs. Song et al. [42] used finite element methods to calculate the structural response of a turbine blade to projectile impact conditions. This type of analysis includes the stresses due to the dynamic wave propagation superimposed on the stresses that arise due to the elastic deformation.

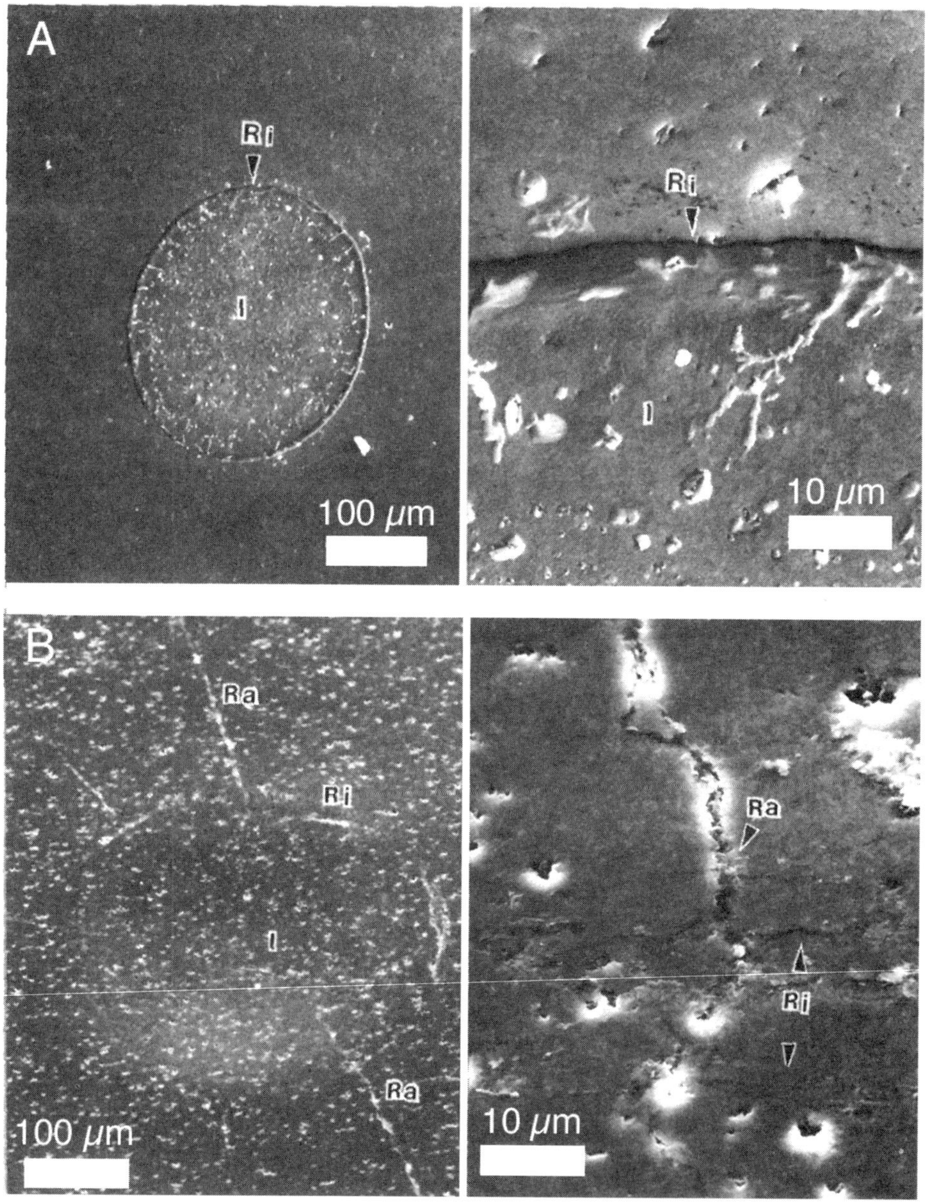

Figure 28.7 Optical Micrographs of Surface Damage for (A) GPSSN and (B) SSN Due to Projectile Impact (Adapted from ref. [20] with permission of the American Ceramic Society, www.ceramics.org, © 1989, all rights reserved)

STRENGTH DEGRADATION DUE TO PROJECTILE IMPACT

Normal Incidence of Attack

Strength degradation occurs as a result of the damage and the crack size that develops due to the impact event [16,21,43–45]. The first step in calculating the strength degradation is to develop a relationship between the maximum load due to the impact event and the resulting flaws that may be generated. We must

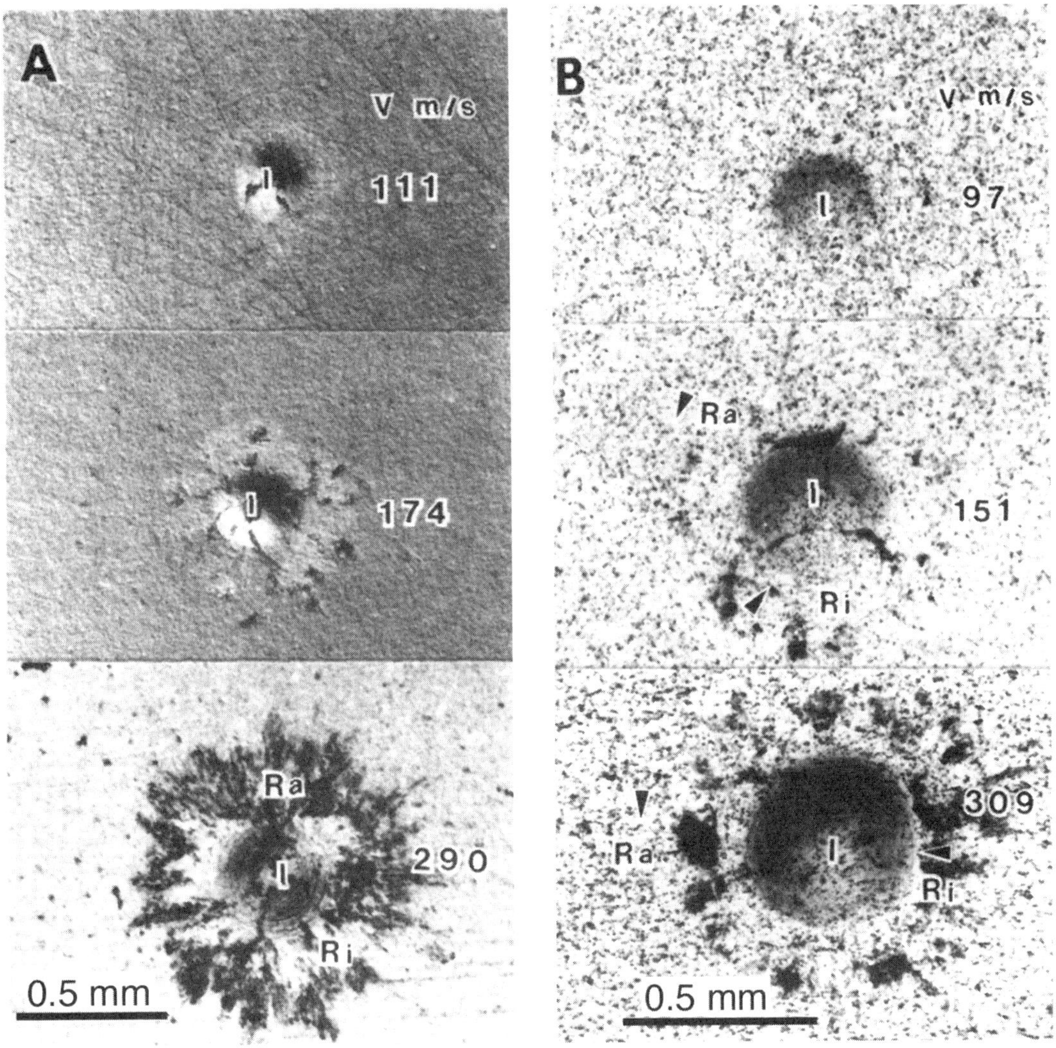

Figure 28.8 Optical Micrographs of Impact Damage on (A) GPSSN and (B) SSN. A Transition can be Observed where Cracks Begin to Appear (Adapted from ref. [20] with permission of the American Ceramic Society, www.ceramics.org, © 1989, all rights reserved)

keep in mind that the natural progression of crack generation for blunt and sharp indenters is somewhat different. In the case of a spherical (blunt) indenter, the ring cracks form first as a result of the extension of the preexisting inherent flaws. These ring cracks then flare downward and grow into the material to form Hertzian cone cracks, but not before a critical load is reached. Auerbach [46] observed that the critical load was related to the indenter radius in the following manner:

$$P_c = AR \tag{11}$$

where A is a constant of proportionality known as Auerbach's constant. Auerbach's law, as this relationship is known, has been shown to break down for small flaws and large spherical indenter radius [47]. The resulting critical velocity calculated from this equation is given by [32]:

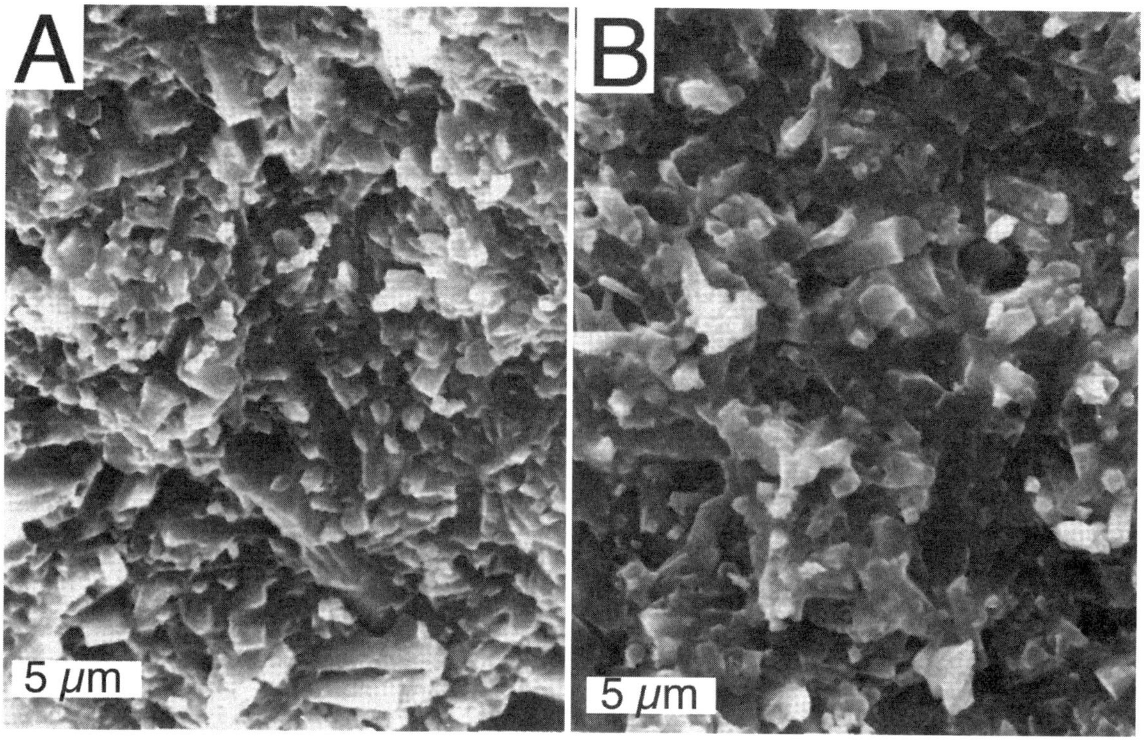

Figure 28.9 Transmission Electron Micrographs of (A) GPSSN and (B) SSN. The Denser Microstructure of GPSSN Yields an Elastic Behavior Up to where Cracking Occurs, Whereas the Less Dense Microstructure of SSN Yields an Elastic-Plastic Behavior (Adapted from ref. [20] with permission of the American Ceramic Society, www.ceramics.org, © 1989, all rights reserved)

$$V_c = \left\{ A \Big/ \left[\left(\frac{5}{3}\pi\rho \right)^{5/3} \left(\frac{3}{4}k \right)^{-2/5} R \right] \right\}^{5/6} \tag{12}$$

Wiederhorn and Lawn [44] derived a similar equation and it has the following form:

$$V_c \propto k^{7/6} / \rho^{1/2} R^{5/6} \tag{13}$$

The length of the cone crack produced by the impact must be determined in order to determine the extent of the strength degradation.

Evans [43] derived a relationship between the resulting impact load and the size of the Hertzian cone crack. The relation between the crack length, C, and the applied force, P, is shown in Figure 28.10. For an inherent crack size C_i an applied load P_o is required to propagate the crack. For a constant applied load P_o the inherent crack begins to grow. As the crack grows the required load to propagate the crack diminishes, as shown in the lower portion of the relationship of C vs. P in the figure, giving rise to unstable crack growth. As the crack grows, the crack jumps from the lower portion of the curve to the upper portion of the curve where the required load to propagate the crack increases as the crack increases. The crack then arrests as it reaches the stable crack size for the applied load P_o.

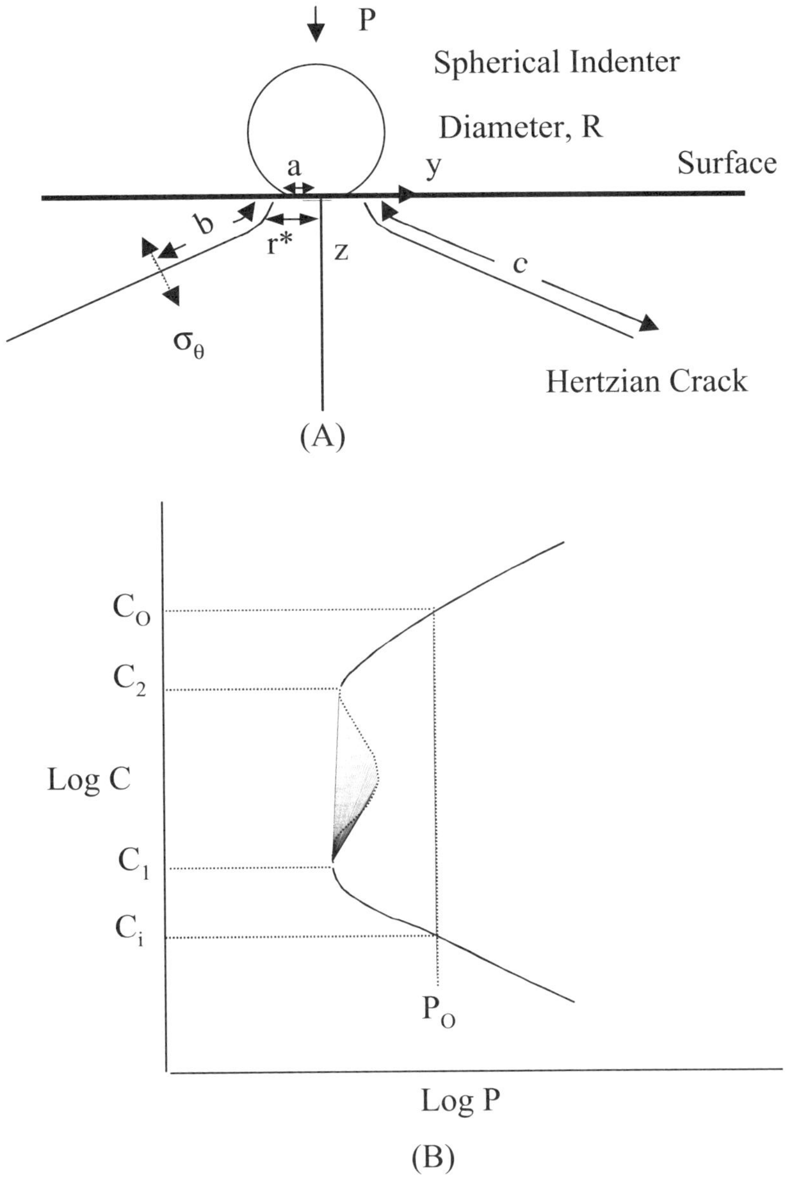

Figure 28.10 (A) Representation of a Section through a Hertzian Cone Crack Formed by a Blunt Indenter. (B) Relationship between the Applied Load and the Resulting Hertzian Crack Length (Adapted from ref. [43] with permission of the American Ceramic Society, www.ceramics.org, © 1973, all rights reserved)

The equation derived by Evans to relate the final size of the Hertzian cone crack to the applied load is given by:

$$P = kR\frac{\pi^2}{3\beta^2}\frac{a}{C}K_I(C)\mathrm{X}\left(r^*/a\right) \tag{14}$$

Where β is a geometry factor for a corner crack and is equal to 1.1, $K_I(C)$ is the stress intensity factor for an internal crack of length $2C$, and $X(r^*/a)$ is a dimensionless function. Again this relationship is not valid for small ratios of a/C (see Appendix I of Evans [43]).

Wiederhorn and Lawn [44] used an equation derived by Roesler [48] to derive a similar relationship. Roesler derived a relation between the load P and the radius φ of the base of the cone crack as follows:

$$\frac{P}{\varphi^{3/2}} = \beta_\varphi K_I \tag{15}$$

where β_ρ is an empirical constant. By substituting this equation into the equation for the maximum load observed during the impact Wiederhorn obtained:

$$\varphi = \left(\beta_\varphi K_I\right)^{-2/3} \left(\tfrac{5}{3}\pi\rho\right)^{2/5} \left(\tfrac{3}{4}k\right)^{-4/15} R^{4/3} V^{4/5} \quad \text{for} \quad V > V_c \tag{16}$$

Using a fracture mechanics approach to determine the residual strength for a given crack size one finds:

$$\sigma = K_{Ic}/Y\sqrt{c} \tag{17}$$

where c is given in terms of φ and a factor which is a function of the cone crack angle α for the case of Wiederhorn and Lawn [44]; and c is a function of the cone crack length C for the case of Evans. Note that this equation for the residual strength is valid only when the velocity of the projectile exceeds a critical velocity to initiate the Hertzian cone crack. The result is a failure envelope given by the form shown in Figure 28.11, where an initial region is shown where no strength degradation occurs, followed by a sharp change indicating the critical projectile velocity for crack initiation. A monotonically decreasing strength follows as the projectile velocity is increased. Figure 28.12 shows the results of residual strength testing of some SiC material and superimposed is the residual strength equation derived by Evans. Figure 28.13 shows the residual strength results for a soda lime glass that has been abraded with two different size grits. Superimposed is the strength degradation equation predicted by Wiederhorn and Lawn [44] which has the following form:

$$\sigma = \left[\left(48/125\pi^{10.5}\right)^{1/15} (9/16k)^{2/15} \beta_\varphi^{1/3} K_c^{4/3} / \Omega^{1/2} \rho^{1/5} R^{2/3}\right] V^{-2/5} \quad \text{for} \quad V > V_c \tag{18}$$

where Ω is a dimensionless function of the cone crack angle.

Similar trends in retained strength versus particle velocity were recently obtained by NASA researchers [49]. They evaluated foreign object damage in two commercial silicon nitride ceramics (SN282 and AS800) using steel ball projectiles having velocities in the range of 220 to 440 m/s. In both materials significant strength degradation occurred above a critical velocity due to the formation of cone and lateral cracks. The lower critical velocity obtained for the SN282 was attributed to its lower fracture toughness.

For the case of a sharp projectile impact the radial crack system is responsible for strength degradation [50]. By assuming that the quasi-static approach also applies for this type of impact event, one can relate the velocity or the kinetic energy of the projectile to the maximum load that develops in the impact event to obtain:

$$P_{max} = \varsigma H^{1/3} U^{2/3} \tag{19}$$

where ς is an indenter constant, H is the material hardness and U is the kinetic energy of the projectile. For values of ς for different sharp indenter geometry see Wiederhorn and Lawn [50].

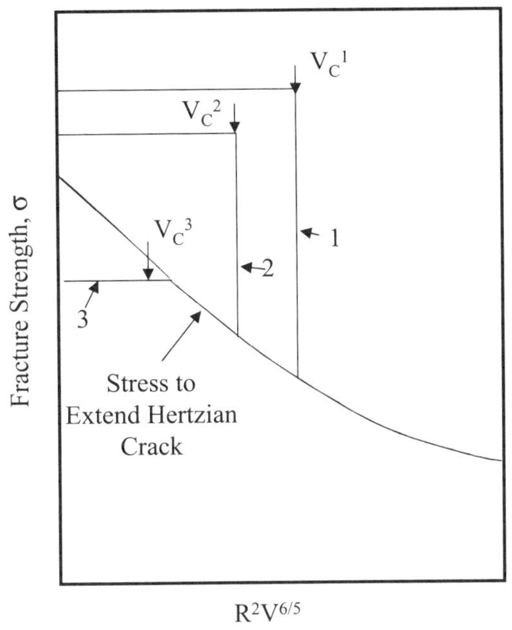

Figure 28.11 Relationship between the Projectile Momentum and Residual Strength. V_c is the Critical Projectile Velocity Required for Hertzian Cone Crack Formation (Adapted from ref. [43] with permission of the American Ceramic Society, www.ceramics.org, © 1973, all rights reserved). Curves 1, 2, and 3 Are for High, Intermediate, and Low Strength Materials, Respectively

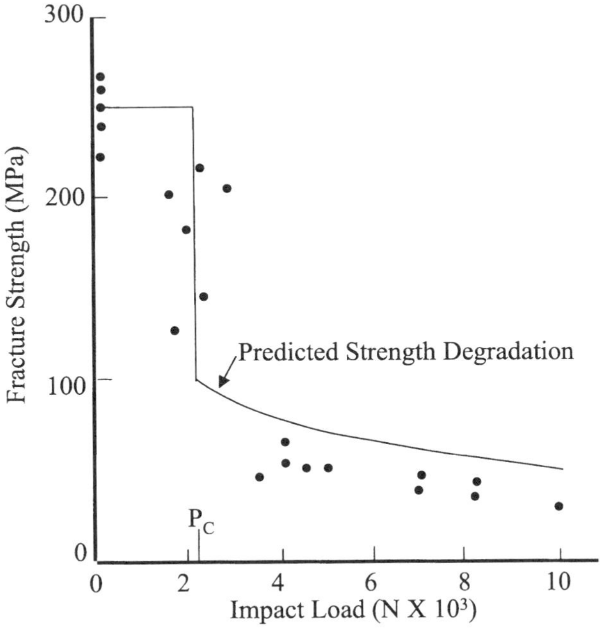

Figure 28.12 Comparison of SiC Residual Strength Data and the Analytical Equation Derived by Evans (1973) (Adapted from ref. [43] with permission of the American Ceramic Society, www.ceramics.org, © 1977, all rights reserved).

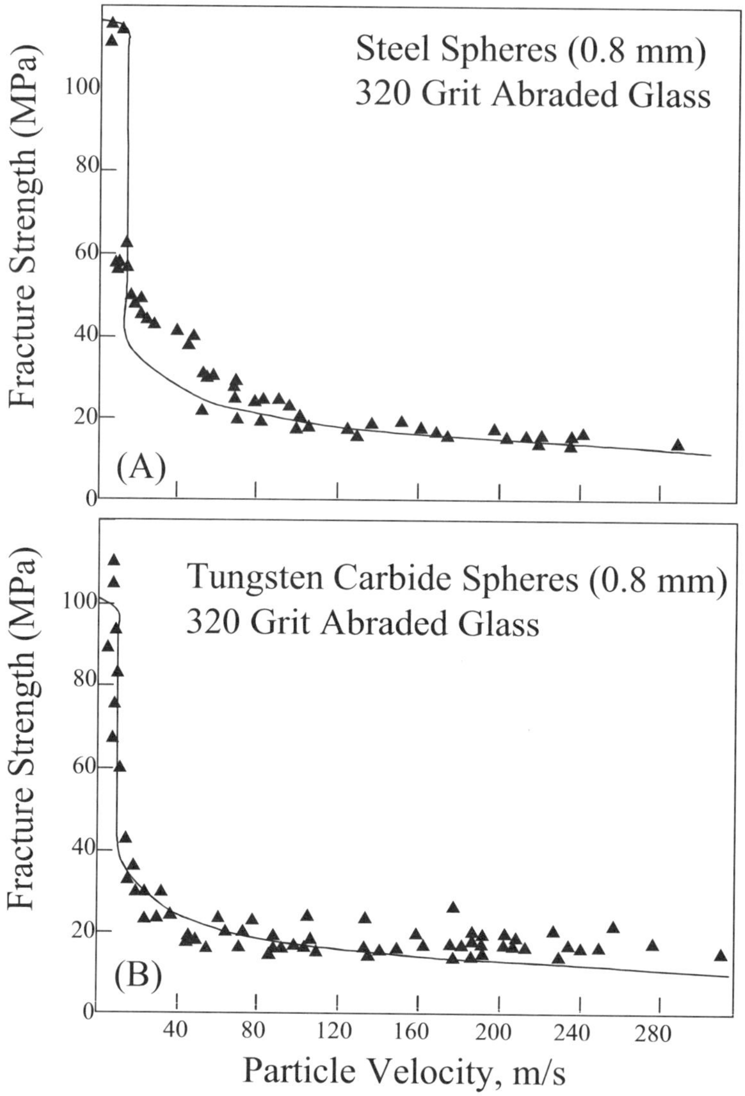

Figure 28.13 Comparison of Soda-Lime Glass Residual Strength Data and the Analytical Equation Derived by Wiederhorn and Lawn (Adapted from ref. [44] with permission of the American Ceramic Society, www.ceramics.org, © 1977, all rights reserved)

The relation between the impact load and the indentation crack size is given by:

$$\frac{P}{D^{3/2}} = \beta_P K_c \tag{20}$$

where D is the crack depth and β_p is a geometric factor, which is usually calculated empirically. K_c is the toughness of the material. Substituting into the usual fracture mechanics relation one obtains:

$$\sigma = \frac{\left(K^4 \beta_P \varsigma\right)^{1/3} \left(HU^2\right)^{-1/9}}{(\pi \Omega)^{1/2}} \tag{21}$$

For materials that exhibit crack resistance toughening [45], R-curve behavior, the relation between the impact load and the indentation crack size is given by:

$$c_o = \left[P\left(\frac{\chi}{\lambda} \right) \right]^{2/(3+2m)} \tag{22}$$

which yields the following residual strength relationship:

$$\sigma = \left(\frac{\lambda}{Y} \right)\left(\frac{3+2m}{4} \right)\left[\frac{4\chi\varsigma H^{1/3} U^{2/3}}{\lambda(1-2m)} \right]^{(2m-1)/(2m+3)} \tag{23}$$

Where χ is constant which is related to the residual stress, and λ and m are crack growth resistance constants.

Oblique Incidence of Attack

It is likely that on an impact event, the relative velocity of the projectile will not be normal to the surface of the target. In such a case, as the projectile contacts the target it wants to slide along the surface generating a stress field which is very different to the stress field for impact that is normal to the surface [35]. The important difference is the significant increase in the radial stress at the trailing edge of the contact area, as shown in Figure 28.5. Wiederhorn, et al. [51] proposed substituting the normal component of the velocity into the residual strength equation to give the following relationship:

$$\sigma_\alpha = \sigma_{90}(\sin\alpha)^\delta \tag{24}$$

where α is the angle of incidence, with $\alpha = 90$ being the normal incidence angle. The exponent δ depends on the type of indenter and it is usually negative. For a sharp indenter $\delta = -4/9$ [51] and for a spherical indenter $\delta = -2/5$ [21]. Akimune [21] performed studies of oblique impact of spherical particles onto a silicon nitride target and found that this relationship provides a very good fit to the residual strength of the impacted specimens for angles as small as 60 degrees. For angles less than 45 degrees a large discrepancy exists. He attributes this discrepancy to the change of the damage generated by the impact as the oblique angle is reduced. The damage transitions from the normal Hertzian cone crack at 90 degrees to a partial cone crack at 60 degrees followed by no cone crack at 45 degrees and lower as shown in Figure 28.14 [22].

Elevated Temperature Effects

The impact resistance of ceramic materials as a function of temperature was studied by Shockey et al. [52]. Shockey finds a deleterious effect of the increase in temperature on the impact resistance of ceramic materials. The negative effect is the result of the reduction in hardness of the ceramic material that produces a transition in behavior from an elastic response to an elastic/plastic response. The transition changes the damage generated from a Hertzian cone crack to the more damaging radial crack pattern. More recently, Yoshida [17] tried particle impact tests under elevated temperatures with tensile stresses. The impact tests under such complicated, but practically rather common conditions in the actual turbines became feasible by using a new electrothermal gun launching technique [25]. Data accumulation is in progress.

Effect of Background Stress

The effect of background stress on the impact resistance of structural ceramic materials is important for obvious reasons. Turbine blades as well as stators are highly stressed components and must survive impact

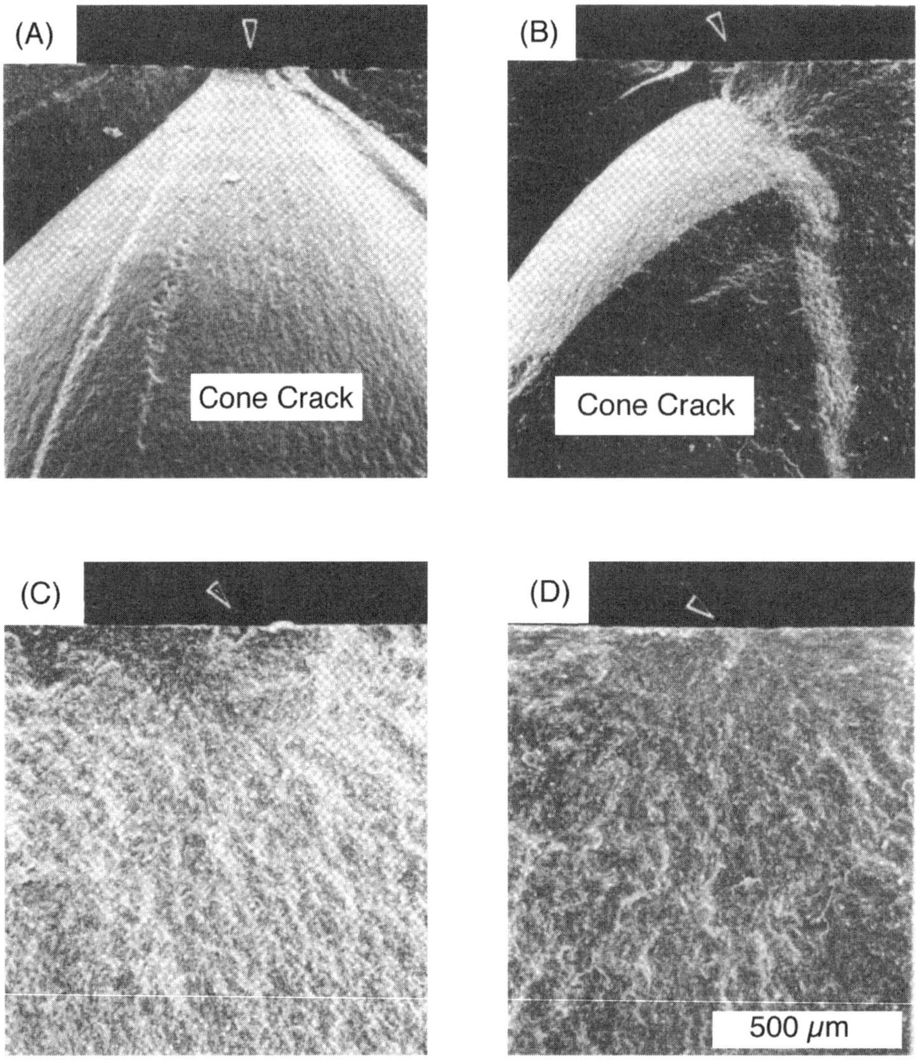

Figure 28.14 SEM Micrographs of Fractured Surfaces of Impacted Specimens Showing Crack Morphology Changes after Bend Testing. (A) 90°, (B) 60°, (C) 45°, (D) 30° (Adapted with permission from ref. [22]).

conditions given the applied loading. Yoshida, et al. [17,25,53] addressed this issue in a very limited form. They concluded that the effect of the background stress is detrimental to the residual strength, but did not quantify the amount due to the limited data they analyzed. One interesting observation they made involving the cone angle is that the presence of the background stress increased the cone angle. The cone angle is measured from the surface to the crack face.

Effect of Material Removal (Erosion)

Strength degradation of a ceramic component subjected to projectile impact loading can also occur as the result of erosion, or material removal. The degree of material removal occurs as the result of the

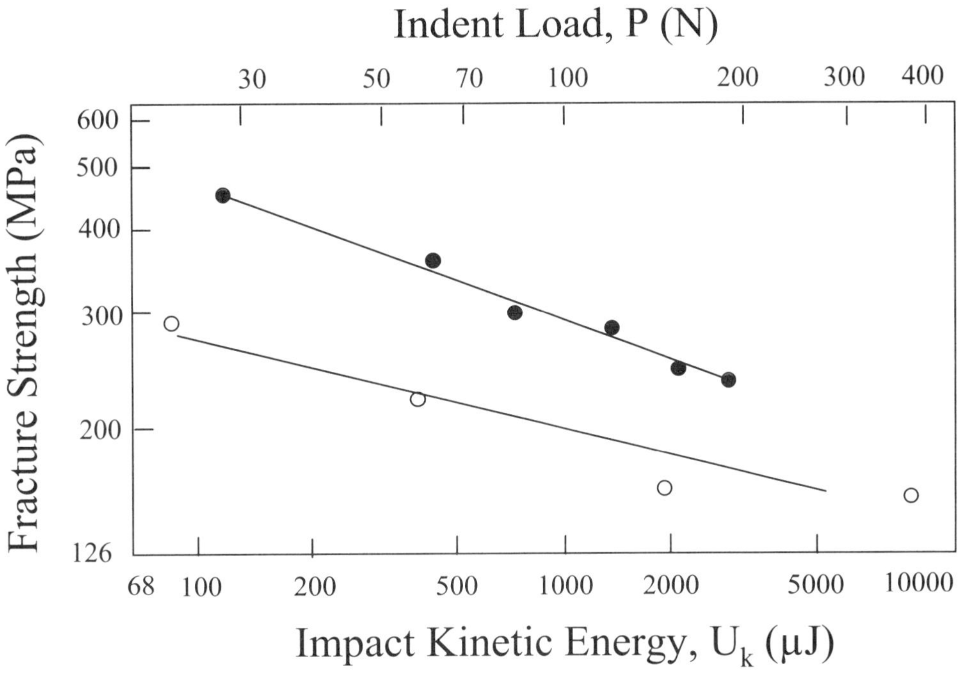

Figure 28.15 Comparison of Post Indentation (Closed Symbols) and Post Impact (Open Symbols) Residual Strength Data (Adapted with permission from ref. [56])

accumulation of multiple impacts with time. Evans, et al. [54] and Wiederhorn and Hockey [55] have proposed theories to predict the material removal rate. Both theories express the erosion process by a power law dependence on the particle velocity, radius and density, and on the target hardness and toughness, although with some difference in the value of the exponents. The equations have the following form:

$$\Theta \propto V^a R^b \rho^c H^d \qquad (25)$$

where Θ is the erosion or material removal rate with units of volume loss per impact event. The exponents in the right-hand side of the equation are $a = 3.2$, $b = 3.7$, $c = 1.3$, and $d = -1.25$ for the Evans, et al. formula and $a = 2.4$, $b = 3.7$, $c = 1.2$, and $d = -0.11$ for the Wiederhorn and Hockey formula.

Ritter, et al. [56] investigated the effect of multiple impacts on the residual strength of two silicon nitride materials, AS440 and AS44 from AlliedSignal Inc (now Honeywell Inc.). They found that the post-impact residual strength is lower (Figure 28.15) than the post-indentation residual strength of the material and attributed this effect to the interaction, or joining, of the cracks formed by multiple impacts. This is a further reduction in strength relative to the inherent strength of the material. Ritter, et al. [56] also investigated the effect of erosion temperature on the residual strength and find that the strength degradation is lower with an increase in temperature and they attribute this to a relief of the residual stresses during the elevated temperature testing.

Effect of the Mechanical Properties of the Material

Mechanical properties have a significant effect on the impact resistance of ceramic materials. We have presented equations that show that the magnitude of the contacting load during impact is proportional to the modulus of elasticity of the material, and inversely proportional to Poisson's ratio. The magnitude of this load plays an important role in determining the extent of damage due to impact. Wiederhorn and Lawn [50] showed that the magnitude of the force is proportional to the hardness of the material. One could assume that a lower hardness is preferable, but as discussed in the temperature effect section, an excessive reduction in hardness may in fact be deleterious since it can cause a transition in the behavior of the material under impact conditions. Last, it has been shown that the residual strength of the material after impact conditions is proportional to its fracture toughness [50], the larger the better since it serves to mitigate crack extension.

Impact Resistance of Ceramic Matrix Composites

The previous sections have addressed the behavior of monolithic ceramics during impact events. For completeness, a short discussion will be presented of the behavior of ceramic matrix composites during impact events. One of the main purposes of adding a second phase to the monolithic ceramic to form the composite is to increase the relatively low fracture toughness of the matrix material. Typical reinforcing structures are whiskers, particles and/or continuous fibers. Akimune's [57] work on the behavior during impact events of silicon nitride reinforced with silicon carbide particles showed that the hardness of the composite is lower than that of the matrix and that the toughness of the composite is relatively unchanged compared to that of the matrix. This resulted in an improved resistance to crack initiation and crack propagation during spherical impact. Furthermore, the composite exhibited an elastic response as its main behavior during the impact event and displayed a high resistance to Hertzian cone crack initiation and extension. In the case of a continuous fiber-reinforced ceramic composite, the study by van Roode, et al. [58] indicates similar trends as those discussed by Akimune [57] and describes that delamination is the prevalent failure mode. Ogawa, et al. [59] have shown that structural damage of continuously reinforced ceramic matrix composites can be significantly improved over those of the monolithic matrix. Observations of the fracture mode indicate that for a thick composite, a crack propagates into the specimen and then it bends to follow a path longitudinal to the fibers. In contrast, for a thin specimen, the crack just propagates through.

DESIGN CONSIDERATIONS

The structural integrity of a ceramic component subjected to projectile impact is influenced by many factors. The bulk of the literature regarding this issue addresses the local damage that occurs during impact, as we have presented in the prior sections. Only a few publications have addressed the structural damage mode [15,16,42]. In order for a structural ceramic component to survive the impact event, the component must be designed to reduce the effects of both the local and structural damage modes.

The local damage mode may be minimized by requiring that the critical velocity for damage formation and the residual strength be as large as possible. The critical velocity, presented earlier, for a blunt indenter has the following form:

$$V_c \propto \frac{k^a}{\rho^b R^c} \propto \frac{k^a R^{c'}}{m^b} \tag{26}$$

and the strength degradation equation has the form:

$$\sigma \propto \frac{1}{k^d \rho^e R^f V^g} \propto \frac{R^{f'}}{k^d m^e V^g} \tag{27}$$

where m is the mass of the projectile. From these two equations we find that the smaller the mass and the larger the radius of the projectile the larger the critical velocity and the residual strength. On the other hand, if the modulus of elasticity of the material is increased the critical velocity would be decreased, but the residual strength would be increased, and conversely for a decrease in the modulus of elasticity. For a sharp indenter the residual strength equation is, in addition, inversely proportional to the hardness of the material. Implying that a soft material would be more resistant to impact damage.

The effect of the strength degradation must also include the effect of multiple impact events. Ritter, et al. [56] showed that the coalescence of cracks due to multiple impacts could significantly reduce the strength of the material. But no functional relationship exists to evaluate such effect.

Considerations to minimize the structural damage mode must also involve the shape of the component, the boundary conditions and the impact location besides the considerations for the local damage mode. The shape and the boundary conditions of the component determine the natural modes of vibration and together with the impact location determine the modes that may be excited by the impact event. The strength of the impact determines the magnitude of the deformation of the component. Goldsmith [29] and Tsai [30] present analytic approaches that can be used to solve the problem.

Song, et al. [42], Hara, et al. [15] and Yoshida, et al. [16,53] have addressed the strength of components against the structural damage mode. Song, et al. used a numerical approach to determine the response of a turbine blade impacted at its tip with graphite spheres. They found that blade fracture occurred at the base of the blade as if by a bending cantilever beam mode. They concluded that the impact resistance of the blade was mostly affected by the velocity of the projectile and parameters that influence the stress concentration factor at the base of the blade such as the filet radius and blade thickness. They also found that the blade height had little effect on the fracture resistance, a result that is contrary to expectations given the strong dependence between the natural frequencies of the blade on the length of the blade.

Hara, et al. [15] and Yoshida, et al. [16,53] also addressed the flexural fracture mode. They observed that damage may occur when a critical velocity is exceeded. The critical velocity is calculated based on the assumption that the impact region of the component can be idealized as if it were a disk that is simply supported at its edges with a distributed load at the center of the disc. The critical velocity V_c is given as follows:

$$V_c = C\sigma_b^{5/6} t^{5/3} \tag{28}$$

where σ_b is the flexural bending strength of the material and t is the thickness of the component at the place of impact. C is a constant that depends on the geometry of the assumed disk but must be determined experimentally. Figure 28.16 shows some data from Yoshida et al. [53] with calculated values for C of 0.5 and 0.85.

Based upon the work in ref [60], the results in Figures 28.16 and 28.11 can be combined to generate an Impact Damage Map like that shown in Figure 28.17. Below the critical velocity for Hertzian cracking, failure occurs via a structural mode involving only flexural fracture. Low strength ceramics (path 1) will failure via a structural mode (Flexural Fracture Regime) once the velocity exceeds the critical value defined by Equation (28). In the case of high strength ceramics (path 2), strength degradation occurs due to particle impact damage when the velocity exceeds the critical velocity for Hertzian cracking. The extent of this strength degradation and thus the probability of failure will be dictated by Equation (18).

Examples of design modifications that have been performed as a result of impact considerations are highlighted by the work of Song, et al. [42] where the blade fillet radius was increased to reduce the dynamic stresses. A more drastic example can be rationalized when one compares the operating environment of an axial versus a radial turbine stage. In the case of an axial turbine stage, the rebound of a projectile that strikes against the turbine blades has a much greater probability of bouncing downstream than that in a radial stage. In a radial stage, the projectile could, for all practical purposes, rebound in a circumferential direction, and will eventually end hitting the stators as a result of the lack of the centrifugal loading that is required to maintain the projectile moving in a circumferential trajectory. In this case one may be inclined to choose a design with an axial stage rather than a radial stage. Finally, if a radial stage is chosen, one would have to consider the effect of projectile-rebounding on the design of the turbine blade and those of the stators. The probable

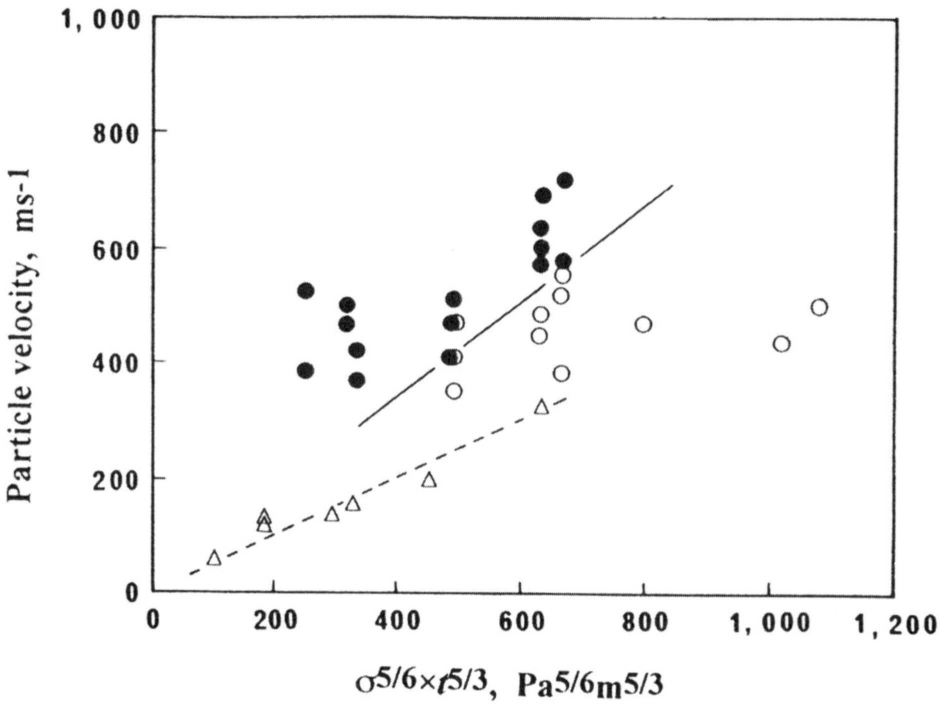

Figure 28.16 Critical Impact Velocity as a Function of the Bend Strength Parameter (Reprinted from ref. [53] with permission of Taylor & Francis, Ltd., www.tandf.co.uk/journals, © 1996, all rights reserved.). The Closed and Open Symbols Represent Damaged and Undamaged Specimens, Respectively.

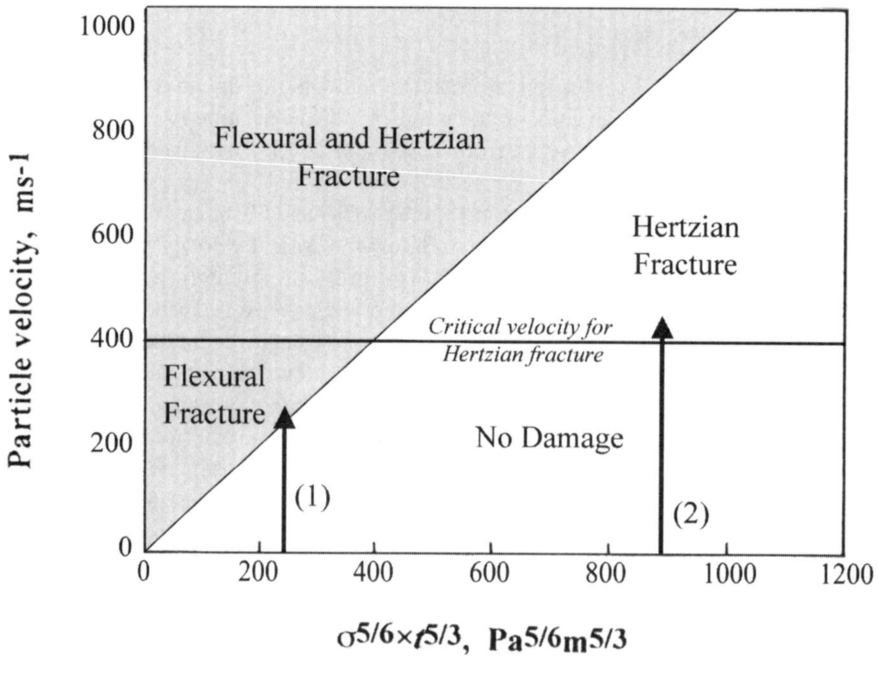

Figure 28.17 Critical Impact Velocity as a Function of the Bend Strength Parameter (Adapted from ref. [60]).

conclusion would be the thickening of the airfoils to increase the impact resistance, and possible erosion, of the components. This may result in a performance penalty over the ideal configuration.

CONCLUSION

A compilation has been presented of work that addresses the effect of projectile impact on the strength of ceramic materials. Most of the work that has been published focuses on the effect the impact event has at the local level and little has been published that addresses the structural level. A reason for this may be that the effect of impact at the structural level is very component specific. Furthermore, the structural damage level tends to be the result of the impact with a relatively large projectile, events that should be relatively rare in normal component operating conditions. On the other hand, impact with relatively small particles is expected to be more frequent and its damage would accumulate with time and in the long run could pose more reliability problems than the structural damage mode.

The design methodology that has evolved for designing ceramic components to resist impact damage is one where the component is designed to tolerate the worst possible crack generated by particle impact. As evidenced by the strength reduction equations that have been developed. What is lacking, and it was partially addressed by Ritter, et al [56], is a theory that can take into account the effect of multiple impacts on the effective crack length as a result of the linking of multiple cracks.

REFERENCES

[1] Weibull, W. W. "A Statistical Theory of the Strength of Materials," Royal Swedish Academy of Engineering Science Proceedings, 151, 1–45, 1939.

[2] Batdorf, S. B. and Heinisch H. L., "Weakest Link Theory Reformulated for Arbitrary Failure Criterion," Journal of the American Ceramic Society, 61 [7–8], 355–358, 1978.

[3] Evans, A. G. "A General Approach for the Statistical Analysis of Multiaxial Fracture," Journal of the American Ceramic Society, 61 [7–8], 302–308, 1978.

[4] Nemeth, N., Powers, L. P., Janosik, L. A., and Gyekenyesi, J., *Ceramic Analysis and Reliability Evaluation of Structures Life Prediction Program, Users and Programmers Manual*, NASA TM, 1993.

[5] Peralta, A. D., Wu, D. C., Brehm, P. J., Cuccio, J. C., and Menon, M. N., "Strength Prediction of Ceramic Components Under Complex Stress States," Journal of Engineering for Gas Turbines and Power, 118 [4], 856–862, 1996.

[6] Schenk, B., Brehm, P., Menon, M. N., Peralta, A. D., and Tucker W. T., "Status of the CERAMIC/ERICA Probabilistic Life Prediction Codes Development for Structural Ceramic Applications," ASME paper 99-GT-318, presented at the 44th ASME Gas Turbine and Aeroengine Technical Congress, Exposition and Users Symposium, June 7–10, 1999, Indianapolis, IN, USA.

[7] Evans, A. G., and Wiederhorn, S. M., "Crack Propagation and Failure Prediction in Silicon Nitride at Elevated Temperatures," Journal of Materials Science, 9, 270–278, 1974.

[8] Wu, D. C, Peralta, A. D., Menon, M. N., and Cuccio, J. C., "Subcritical Crack Growth Life Prediction for Ceramic Components of Advanced Heat Engines," Journal of Engineering for Gas Turbines and Power, 118 [2], 240–245, 1996.

[9] Schenk, B., "Ceramic Gas Turbine Technology Development," ASME paper 99-GT-315, presented at the 44th ASME Gas Turbine and Aeroengine Technical Congress, Exposition and Users Symposium, June 7–10, 1999, Indianapolis, IN, USA.

[10] Schenk, B., Brehm P., Menon, M. N., Tucker, W. T., and Peralta, A. D., "A New Probabilistic Approach for Accurate Fatigue Data Analysis of Ceramic Materials," ASME paper 99-GT-319, presented at the 44th ASME Gas Turbine and Aeroengine Technical Congress, Exposition and Users Symposium, June 7–10, 1999, Indianapolis, IN, USA.

[11] Richerson, D. W. and Johansen, K. M., *Ceramic Gas Turbine Engine Demonstration Program, Final Report*, Contract N00024-76-C-5352, 1982.

[12] Boyd, G. L, and Kreiner, D. M., "AGT101/ATTAP Ceramic Technology Development," in *Proceedings of the Twenty-Fifth Automotive Technology Development Contractors Coordination Meeting*, 1987. pp. 101.

[13] van Roode, M., Brentnall, W. D., Norton, P. F., and Boyd, G. L., "Ceramic Stationary Gas Turbine Development Program—First Annual Summary-," ASME paper 94-GT-313, presented at the International Gas Turbine and Aeroengine Congress and Exposition, The Hague, The Netherlands, June 13–16, 1994.

[14] Easley, M. L. and Smyth, J. R., "Ceramic Gas Turbine Technology Development," ASME paper 94-GT-485, presented at the International Gas Turbine and Aeroengine Congress and Exposition, The Hague, The Netherlands, June 13–16, 1994.

[15] Hara, Y., Makino, K., Mizuno, K., Shimamori, Y., and Yoshida, H., "Particle Impact Tests of Ceramic Gas Turbine Materials at Room Temperature," ASME paper 96-GT-504, presented at the International Gas Turbine and Aeroengine Congress and Exhibition, Birmingham, U.K., June 10–13, 1996.

[16] Yoshida, H., Nakashima, T., Yoshida, M., Hara, Y., and Shimamori, T. "Experiment on Foreign Object Damage of Gas Turbine-Grade Silicon Nitride Ceramic," ASME paper 98-GT-399, presented at the 43rd ASME Gas Turbine and Aeroengine Technical Congress, Exhibition and Users Symposium, Stockholm, Sweden, June 2–5, 1998.

[17] Yoshida, H. "FOD Tests in 300 kW class CGT R & D Project," Journal of the Gas Turbine Society of Japan, 28, 329–333, 1999 (in Japanese).

[18] *Report of Petroleum Energy Center*, Matsumoto, H., ed., PEC-91C04, March 1992 (in Japanese).

[19] Tsutsui, Y., Yoshida, H., Sasaki,S., Abe, H., Matsunuma, T., and Kurata, O. *Research on Ceramic Gas Turbine Engine Component Technologies—Final Report-*, Report of Mechanical Engineering Laboratory No.187, 2000.

[20] Akimune, Y., Katano, Y., and Matoba, K., "Spherical-Impact Damage and Strength Degradation in Silicon-Nitrides for Automobile Turbocharger Rotors," Journal of the American Ceramic Society, 72 [8], 1422–1428, 1989.

[21] Akimune Y., "Oblique Impact of Spherical Particles onto Silicon Nitride," Journal of the American Ceramic Society, 73 [12], 3607–3610, 1990.

[22] Akimune, Y., Akiba, T., and Ogasawara, T., "Damage Behavior of Silicon Nitride for Automotive Gas Turbine Use when Impacted by Several Types of Spherical Particles," Journal of Materials Science, 30, 1000–1004, 1995.

[23] Tsuruta, H., Masuda, M., Soma, T. and Matsui, M. "Foreign Object Damage Resistance of Silicon Nitride and Silicon Carbide," Journal of the American Ceramic Society, 73 [6], 1714–1718, 1990.

[24] Suzuki, A., Natsumura, T., Sugiyama, S., Teramae, T., and Hamada, S. "Particle Impact Damage in Silicon Carbide," Bulletin of JSME A, 56, 2431–2436, 1990 (in Japanese).

[25] Yoshida, H., Hoshi, Y., Uematsu, K., and Kitazawa, Y. "A Single, Small Particle Launch System by Electrothermal Gun and Microsabot," Review of Scientific Instruments, 68 [1], 178–183, 1997.

[26] Zukas, J.A., Nicholas, T., Swift, H.F., Greszczuk, L.B., and Curran, D.R., *Impact Dynamics*, John Wiley & Sons, New York, USA, 1982.

[27] Evans, A. G. and Wilshaw, T. R.. "Dynamic Solid Particle Damage in Brittle Materials: an Appraisal," Journal of Materials Science, 12, 97–116, 1977.

[28] Evans, A. G. and Wilshaw, T. R., "Quasi-Static Solid Particle Damage in Brittle Solids-I. Observations, Analysis and Implications," Acta Metallurgica, 24, 939–956, 1976.

[29] Goldsmith, W., *Impact, the Theory and Physical Behavior of Colliding Bodies*, Edward Arnold, London, UK, 1960. pp. 108–119.

[30] Tsai, Y. M. "Dynamic Contact Stresses Produced by the Impact of an Axisymmetrical Projectile on an Elastic Half-Space," International Journal of Solids and Structures, 7, 543–558, 1971.

[31] Hertz, H., *Hertz's Miscellaneous Papers*, chapters 5 and 6. Macmillan, London, UK, 1896.

[32] Knight, C. G., Swain, M. V., and Chaudhri, M. M.,"Impact of Small Steel Spheres on Glass Surfaces," Journal of Materials Science, 12, 1573–1586, 1977.

[33] Timoshenko, S. P. and Goodier J. N., *Theory of Elasticity*, 3rd ed., McGraw-Hill Book Company, New York, USA, 1951. pp. 412, 420.

[34] Popov, E. P. and Balan, T. A., *Engineering Mechanics of Solids*, Second Edition, Prentice Hall, Upper Saddle River, NJ, USA, 1998.

[35] Hamilton, G. M. and Goodman, L. E., "The Stress Fields Created by a Circular Sliding Contact," Journal of Applied Mechanics, 33, 371–376, 1966.

[36] Lawn, B. and Wilshaw, R., "Review Indentation Fracture: Principles and Applications," Journal of Materials Science, 10, 1049–1081, 1975.

[37] Tsai, Y. M. "Thickness Dependence of the Indentation in Elastic and Viscoelastic Plates Subjected to Symmetrical Rigid Indentation," Quarterly Applied Mathematics 27, 371–80, 1969.

[38] Bowden, F. P., F.R.S., and Field, J. E., "The Brittle Fracture of Solids by Liquid Impact, by Solid Impact, and by Shock," *Proceedings of the Royal Society of London,* A282, 331–352, 1964.

[39] Field, J. E., "Stress Waves, Deformation and Fracture Caused by Liquid Impact," Philosophical Transactions of the Royal Society of London A260, 86–93, 1966.

[40] Glenn, L. A., "The Fracture of a Glass Half-Space by Projectile Impact," Journal of the Mechanics and Physics of Solids, 24, 93–106, 1976.

[41] Liaw, B. M., Kobayashi, A. S., and Emery, A. F., "Theoretical Model of Impact Damage in Structural Ceramics," Journal of the American Ceramic Society, 67 [8], 544–548, 1984.

[42] Song, J., Cuccio, J. C., and Kington, H., "Impact Design Methods for Ceramic Components in Gas Turbine Engines," Journal of Engineering for Gas Turbines and Power, 115 [1], 83–90, 1993.

[43] Evans, A. G., "Strength Degradation by Projectile Impacts," Journal of the American Ceramic Society, 56 [8], 405–419, 1973.

[44] Wiederhorn, S. M. and Lawn, B. R., "Strength Degradation of Glass Resulting from Impact with Spheres," Journal of the American Ceramic Society, 60 [9–10], 451–458, 1977.

[45] Breder, K., Ritter J. E., and Jakus, K., "Strength Degradation in Polycrystalline Alumina Due to Sharp-Particle Impact Damage," Journal of the American Ceramic Society, 71 [12], 1154–1158, 1988.

[46] Auerbach, F., Annals of Physical Chemistry, 43, 61, 1891.

[47] Langitan, F. B., and Lawn, B. R., "Hertzian Fracture Experiments on Abraded Glass Surfaces As Definitive Evidence of an Energy Balance Explanation of Auerbach Law," Journal of Applied Physics, 4 [20], 4009–4017, 1969.

[48] Roesler, F. C. "Brittle Fracture Near Equilibrium," Proceedings Royal Society of London Series B, 69 [10], 981–992, 1956.

[49] Choi, S. R., Pereira, J. M., Janosik, L. A., and Bhatt, R. T., "Foreign Object Damage of Two Gas-Turbine Grade Silicon Nitrides at Ambient Temperatures, " Ceram. Eng. and Science Proc., 23 [3], 193–202, 2002.

[50] Wiederhorn, S. M. and Lawn, B. R., "Strength Degradation of Glass Impacted with Sharp Particles: I, Annealed Surfaces," Journal of the American Ceramic Society, 62 [1–2], 66–70, 1979.

[51] Wiederhorn, S. M., Lawn, B. R., and Hockey, B. J., "Effect of Particle Impact Angle on Strength Degradation of Glass," Journal of the American Ceramic Society, 62 [11–12], 639–640, 1979.

[52] Shockey, D. A., Erlich, D. C., and Dao, K. C.. "Particle Impact Damage in Silicon Nitride at 1400°C," Journal of Materials Science, 16, 477–82, 1981.

[53] Yoshida, H., Kano S., Hasegawa, Y., Shimamori, T., and Yoshida, M., "Particle Impact Phenomena of Silicon Nitride Ceramic," Philosophical Magazine A, 74 [5], 1287–1297, 1996.

[54] Evans, A. G., Gulden, M. E., and Rosenblatt, M., "Impact Damage in Brittle Materials in the Elastic-Plastic Response Regime," *Proceedings of the Royal Society of London Series A*, 361, 343–365, 1978.

[55] Wiederhorn, S. M. and Hockey, B. J., "Effect of Material Parameters on the Erosion Resistance of Brittle Materials," Journal of Materials Science, 18, 766–780, 1983.

[56] Ritter, J. E., Choi, S. R., Jakus, K., Whalen, P. J., and Rateick, R. G. Jr., "Effect of Microstructure on the Erosion and Impact Damage of Sintered Silicon Nitride," Journal of Materials Science, 26, 5543–5546, 1991.

[57] Akimune, Y., "Impact Damage and Strength Degradation in a Silicon Carbide Reinforced Silicon Nitride Composite," Journal of the American Ceramic Society, 73 [10], 3019–3025, 1990.

[58] van Roode, M., Jimenez, O., McClain, J., Price, J., Parthasarathy, V., Poormon, K.L., Ferber, M.K., and Lin, H.-T., "Ceramic Gas Turbine Materials Impact Evaluation," ASME paper 2002-GT-30505, presented at the 47th ASME IGTI ASME TURBO EXPO Land, Sea, and Air, June 3–6, 2002, Amsterdam, The Netherlands.

[59] Ogawa, K., Sugiyama, F., Pezzotti, G., and Nishida, T., "Impact Strength of Continuous-Carbon-Reinforced Silicon Nitride Measured by Using the Slip Hopkinson Pressure Bar," Journal of the American Ceramic Society, 81 [1], 166–172, 1998.

[60] Hara, Y., Matsubara, K., Mizuno, K., Shimamori, T., Aichi, K., and Yoshida, H., "Development and Evaluation of Silicon Nitride Components for Ceramic Gas Turbine," ASME paper 98-GT-498, presented at the 43rd ASME Gas Turbine and Aeroengine Technical Congress, Exhibition and Users Symposium, Stockholm, Sweden, June 2–5, 1998.

CHAPTER

29

THE CERAMIC GAS TURBINE— RETROSPECTIVE, CURRENT STATUS AND PROGNOSIS

David W. Richerson
University of Utah, Salt Lake City, Utah, USA

Mattison K. Ferber
Oak Ridge National Laboratory, Oak Ridge, Tennessee, USA

Mark van Roode
Solar Turbines Incorporated, San Diego, California, USA

INTRODUCTION

This volume and the companion volume of this treatise [1] present a detailed review of the worldwide development efforts to improve gas turbine engines through the use of ceramic materials. Difficult challenges have been encountered in developing design and life prediction methods, in the efforts to improve materials properties, in property characterization, in fabrication of complex ceramic parts, in quality assurance (such as non-destructive inspection, dimensional measurement, and proof testing), and in realistic rig and engine testing [2]. Enormous strides have been made in each of these areas. The objectives of this chapter are to quantify the magnitude of the advances, summarize the current status of ceramics implementation in turbines (including future technology needs), and identify important technology and application spin-offs.

TECHNOLOGY AND METHODOLOGY ADVANCES

Design/Development Process and Life Prediction

Design and life prediction techniques that were suitable for metals were quickly determined to not be suitable for ceramics [3]. Figure 29.1 summarizes the general timing of the development and evolution of

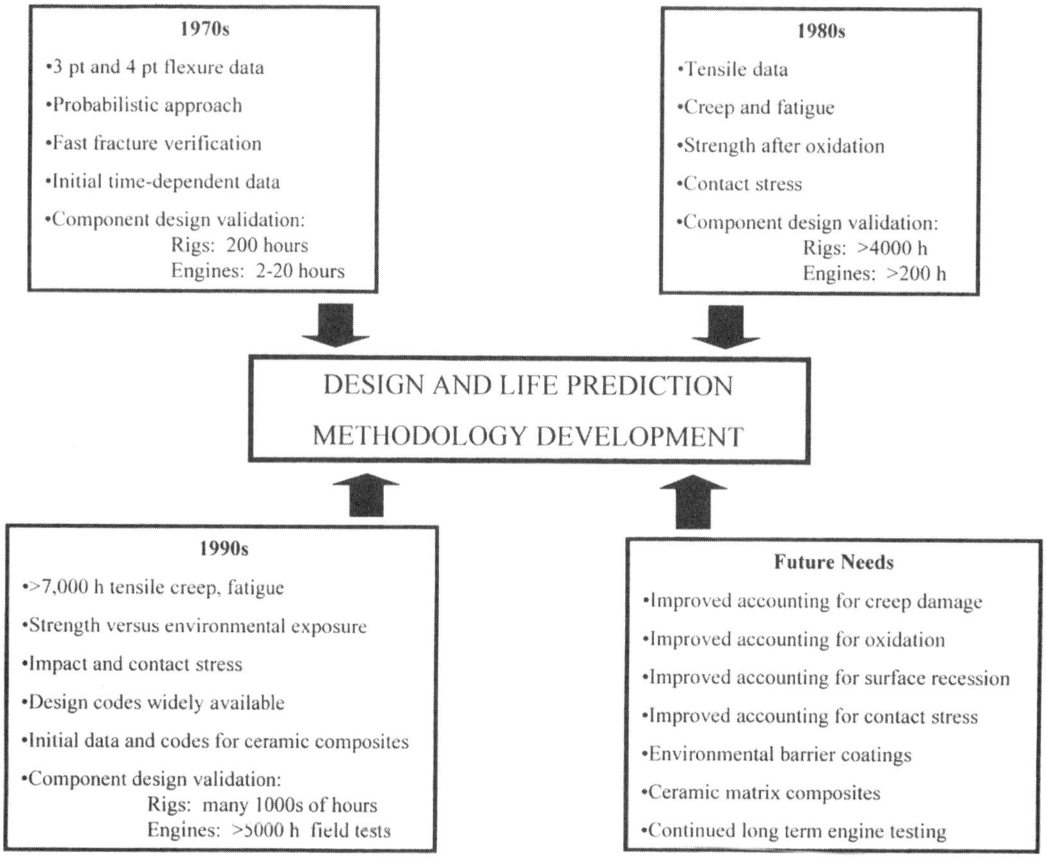

Figure 29.1 Timeline of the Development and Evolution of Techniques and Database for Monolithic Ceramics and Ceramic Matrix Composites

techniques and database, initially for monolithic ceramics, and later for ceramic matrix composites. Key factors have included transition from a flexure test database to a tensile database, progression from a fast fracture model (that was adequate for short life applications) to models that could predict effects of time-dependent materials behavior, and validation of design and life prediction codes through progressively longer term rig and engine tests.

Design methodology for brittle materials was at a very early stage prior to about 1970. Some work had been done in the 1950s at the US National Advisory Committee for Aeronautics NACA (later this became the National Aeronautics and Space Administration–NASA), but the ceramics and cermets at that time were not adequate and computers were cumbersome and very expensive to both purchase and use [4–6].

By 1967 when Ford Motor Company initiated their 820 ceramic engine design efforts, computers were still cumbersome and there were no established design tools and guidelines for the complex 3D analyses that would be required for brittle materials. Ford and other early pioneers adopted by necessity a "learn-by-doing" approach that required many iterations and experimental verification at each step of the way [7]. This involved a systems approach that became the model for future programs. It consists of iterations of the following sequence of activities (as illustrated in Figure 29.2):

- Define a platform for development, e.g. a specific engine or a generic engine concept.
- Select one or more conceptual component designs.
- Conduct detailed design analysis using finite element and finite difference analysis methodology.

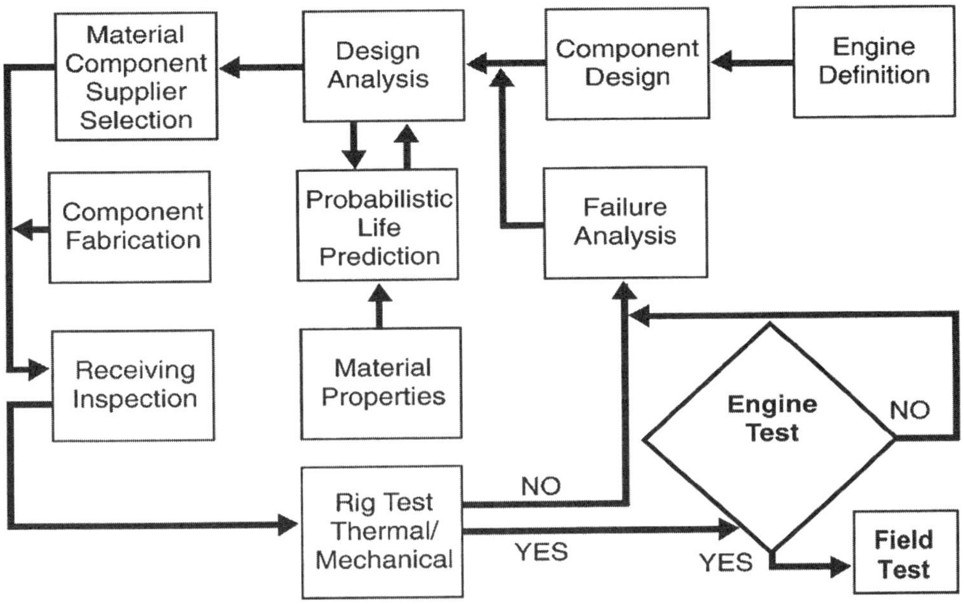

Figure 29.2 Iterative Ceramic Gas Turbine Design/Development Process (Reprinted with permission of Solar Turbines Incorporated)

- Conduct probabilistic life prediction for the components using materials properties and component boundary conditions as inputs.
- Select one or more configurations to fabricate.
- Select structural materials and suppliers for component fabrication.
- Fabricate prototype hardware.
- Inspect the hardware using dimensional inspection techniques and nondestructive evaluation (NDE) technology.
- Test the prototype hardware in a test rig to simulate the thermal and mechanical stresses expected in an engine.
- Conduct failure analysis to determine whether failure resulted from a design deficiency or material deficiency.
- Modify the component designs, materials, and/or fabrication process steps. Generate additional material property data as needed for guiding material improvement and for design analysis input.
- Repeat the complete design/development sequence or a specific loop within the sequence.
- The definitive test of design concepts is an engine test in which the new hardware is subjected to typical operating conditions. A long term engine endurance test, preferably at a commercial field site is the culmination of a successful development program.

The iterative design approach resulted in rapid development of a baseline design methodology and code for brittle materials [8]. The deterministic approach used for metals was found not to be suitable for ceramics. Instead, a probabilistic approach evolved [9,10]. It used finite element analysis and the best estimates of boundary conditions to define the temperature and stress distributions in the ceramic component. It then integrated this with a Weibull (weakest link) analysis that defined the statistical distribution of strength for the material based on both volumetric and surface strength-controlling flaw distributions. The result was a probability of failure estimate for the selected material for each finite element and a combined probability of failure for the whole component.

The initial design codes that evolved were for fast fracture reliability. These were verified by short term rig and engine tests, and the viability of the overall probabilistic design methodology was clearly demonstrated by the mid 1970s, i.e., it was clearly demonstrated that ceramic gas turbine components could be analytically designed and could survive the severe start-up, steady state, and shut-down conditions in a gas turbine engine [8,11]. However, most of the property data up to this time was fast fracture data, and all of the rig/engine tests were of short duration. It was recognized that long term uses in a turbine engine would involve changes in the material properties over time due to factors such as oxidation, slow crack growth and creep. As a result, efforts during the late 1970s and thereafter focused on gathering a time-dependent property database and establishing design codes that would calculate reliability both for fast fracture and time-dependent properties [12–17].

By the early 1980s, contact stress and foreign object damage were identified as key causes of failure during many engine tests [18]. Analysis tools and methods were refined to address these problems. Especially during the 1990s, engine components were redesigned to make them more resistant to impact stresses [19] and to minimize contact stresses at interfaces [20]. In spite of substantial improvements in design tools and designs, impact and contact stress are still key reliability challenges.

Many gas turbine engine companies worldwide initiated ceramic component design and test programs during the 1970s and early 1980s. Each adopted a probabilistic design approach and adapted it to its own computer programs. The design evolution at many of these companies is documented in the chapters in Volume 1 [1]. A major design life prediction effort was also conducted in the U.S. at NASA that has resulted in CARES/SCARE design codes that are available to researchers and industry [21,22]. The CERAMIC and ERICA codes developed at Honeywell with substantial support from DOE are available as well [23–25]. The Sunstrand Corporation developed the SPSLIFE reliability code [26,27]. In Japan workers at Tokyo Electric Power Company (TEPCO) and Ishikawajima-Harima Heavy Industries developed the G-FICES reliability analysis code [28,29]. And in Germany the CERITS and CERTUB codes were developed [30,31]. Information on the evolution of the design codes is reviewed in this volume in Chapter 25 [32].

Design and life prediction methods and tools have come a long way since 1970. Designers now have well-established codes for brittle material design and enormously improved computers. 3D analysis with high-resolution finite element grids is now routine. Furthermore, through years of experience we have gained much improved understanding of the boundary conditions in the engines and have assembled a large materials database. Thus the input available to the design analysis is much better now than in the 1970s, 1980s, and 1990s.

Even though much progress has been made in life prediction techniques, there are still some deficiencies. The techniques and databases still do not account adequately for creep damage, oxidation degradation, surface recession from time-dependent environmental effects, impact damage and contact stress. Furthermore, the successful application of the techniques is dependent upon having Weibull data that reflects the strength characteristics of the component. Generally, analyses are conducted with data from laboratory test samples that often are not adequately representative of the components.

Life prediction tools for ceramic matrix composites (CMCs) are lagging behind those for monolithics but some progress has been made in the US, mainly through efforts at NASA Glenn Research Center and work by Reifschneider and coworkers at Virginia Tech. Also, monolithic ceramic and composite materials continue to be refined to achieve longer life, so new design data must be generated for each new material. A brief summary on life prediction codes for CMCs can be found in ref. [32].

Material Characterization

Figure 29.3 illustrates the complex interactions of material characterization (including property measurement) with design and life prediction, material improvement, and in understanding the cause of a component failure. To meet these needs, advances were required in developing standardized tests, in designing and building reliable test fixtures and equipment, and in gathering and validating a wide range of property data. In addition, each new iteration and batch of ceramic materials and components required careful property measurement to certify that the ceramic met minimum standards. These material characterization efforts were (and continue to be) a very critical step in development of ceramic materials for gas turbine applications.

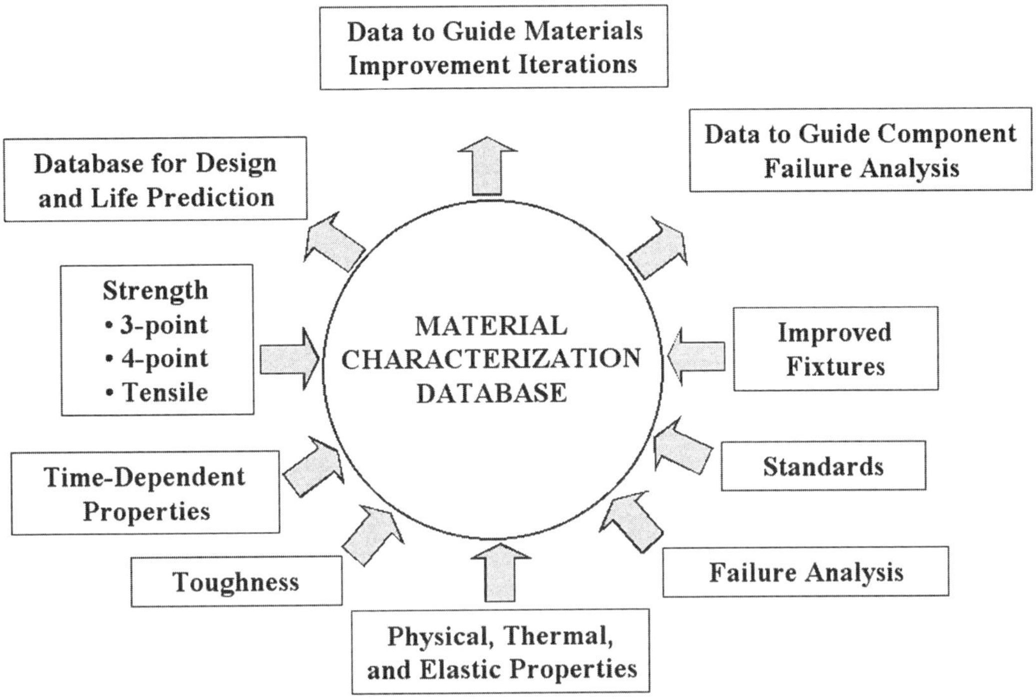

Figure 29.3 Interactions of Materials Characterization, and Design and Life Prediction

NACA and its successor NASA conducted or funded extensive screening of metals, ceramics, intermetallics, and cermets between 1950 and the mid-1960s for potential use in engine and aircraft applications [4]. By the late 1960s newly developed silicon nitride and silicon carbide materials were identified that appeared to have the best combination of properties to be potential ceramic candidates for turbine engine components. However, the screening data was not adequate for design analysis or for an adequate assessment of the materials' viability for turbine components. By 1970 there was only limited data on density, room temperature strength, thermal expansion, elastic modulus and even less data at elevated temperatures.

Considerable effort was initiated at ceramic companies, engine companies, and government laboratories during the 1970s to establish a statistical data base for the reaction bonded silicon nitride (RBSN), hot pressed silicon nitride (HPSN), and reaction sintered silicon carbide (RSSC) ceramics available at that time [33–35]. Not surprisingly, the properties of the early versions of these materials were determined to be inadequate, so the ceramic developers were continually making process modifications and offering new improved versions. Each new version then required characterization.

Up to about 1972 mechanical property characterization in the U.S. was conducted using 3-pont flexural loading of simple square or rectangular cross section samples. Each organization used a different size and shape of test sample and a different design of test fixture. Data was difficult to compare, especially since Weibull analysis was not yet in common use. Between about 1972 and 1974 substantial cooperative efforts were directed towards establishing standards for test bars and fixtures. For example, 4-point flexure strength testing with a test bar of specific dimensions and a self-aligning fixture was agreed to and adopted by several organizations in the U.S. To obtain further uniformity in data generation, the Air Force funded a program that lasted from June 1975 to July 1983 to the Illinois Institute of Technology Research Institute (IITRI) to conduct property characterization of a wide range of commercial and experimental silicon nitride and silicon carbide materials [36]. The program defined and followed specifications/standards for sample dimensions and surface preparation, for test methods, and for test equipment/fixtures. Similar characterization efforts were established and conducted during this time in Great Britain, Germany and Japan [37–40].

The ceramics and engine community recognized in the early 1970s that characterization of time-dependent properties was critical. Flexural stress rupture and creep, compressive creep, and static oxidation testing were conducted along with substantial flexural strength versus temperature testing. AMMRC [41], Norton [42,43], Westinghouse [33], and Battelle [44] were particularly active in the early 1970s in the U.S. Annawerk, DFVLR [45], and other organizations in Germany [37]and NGK in Japan conducted important efforts especially during the second half of the 1970s. NGK explored the effect of volume under tensile stress to assess the validity of the use of Weibull statistics and also initiated uniaxial tensile testing [46].

Flexural stress rupture and creep testing produced much data scatter. A new program was initiated in the U.S. in 1983 to characterize candidate silicon nitride and silicon carbide materials in uniaxial tension [47]. This program was funded by the US Dept. of Energy (DOE) and conducted primarily at Oak Ridge National Laboratory (ORNL), University of Dayton Research Institute (UDRI), and the National Bureau of Standards (NBS), the predecessor of the National Institute of Standards and Technology (NIST). Important goals of the program were to (1) gain an improved understanding of the mechanisms of crack growth and creep, (2) provide longer time tests to support design data and life prediction needs of engine programs, and (3) to provide feedback to ceramic material developers that might guide further material improvements.

Parallel to the DOE program, NASA funded a program to Garrett Turbine Engine Company to study the strength of advanced silicon nitride and silicon carbide ceramics as a function of exposure time and cycles in a burner rig [48,49]. NASA also conducted extensive inhouse studies on oxidation, especially under very high velocity burner conditions [50].

During the 1980s and 1990s considerable international cooperation has focused on establishing material and test standards. This has included round-robin testing, comparison of draft standards, and efforts to establish joint standards. Efforts have been closely linked to ASTM and ISO. As a result of these efforts, standards have been established for all of the mechanical and thermal properties of monolithic ceramics as described in Chapter 24 [51].

The ceramic materials characterization reviewed so far was for monolithic ceramics. Glass matrix composites from United Technologies Research Center (UTRC) became available for evaluation in the late 1970s [52,53] and SiC/SiC composites (fabricated by chemical vapor infiltration) became available from the Société Européenne de Propulsion (SEP) in France in the early 1980s [54]. New characterization methods and standards have been required to be developed for these ceramic matrix composites. Work is in progress [55].

In summary, much progress has been accomplished worldwide in establishing ceramic material tests and standards and in generating a large database suitable for design, material improvement, and quality assurance. The subsequent section highlights actual data that illustrates the dramatic improvements that have been achieved.

Materials Property Improvements

By 1970 four types of ceramic materials had been identified as important candidates for turbine components:

1. Reaction bonded silicon nitride (RBSN)
2. Reaction sintered silicon carbide (RSSC)
3. Hot pressed silicon nitride (HPSN)
4. Lithium aluminum silicate (LAS)

As gas turbine engine companies began to evaluate these 1970-vintage materials, key deficiencies were identified that limited the stress and temperature at which they could be used and the time that they would remain functional. Initial RBSN materials had about 30% porosity and were thus low strength and vulnerable to oxidation degradation. Early HPSN materials had excellent room temperature strength, but the

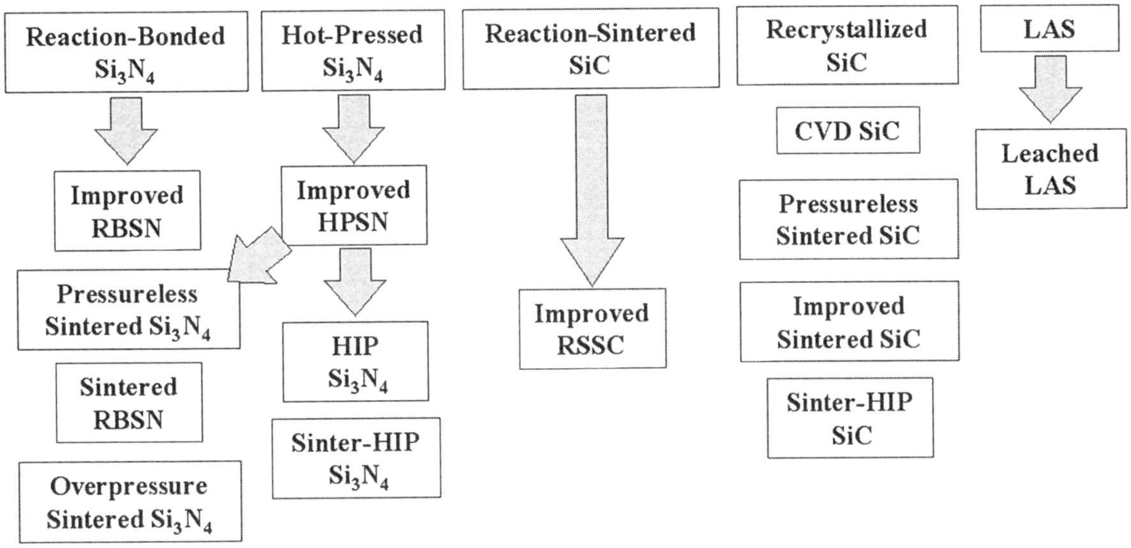

Figure 29.4 Progress in Monolithic Ceramic Materials Fabrication Development

strength, creep resistance, and stress rupture life were inadequate above about 1000°C. Furthermore, HPSN could not be fabricated directly to the complex shapes required for turbines, necessitating extensive and costly diamond grinding. 1970-vintage RSSC looked promising for low stress combustor components, but did not have adequate strength for other high stress components. LAS performed well as a honeycomb rotary regenerator in laboratory tests, but failed during on-road tests due to cracking caused by ion exchange of lithium ions with sodium ions (from road salt) and hydrogen ions (from sulfuric acid condensate from sulfur impurities in the fuel).

In spite of the deficiencies, the early 1970-vintage ceramic materials were good enough to demonstrate that ceramic turbine components could be designed to survive at least for a short time under gas turbine start, steady state, and shutdown conditions [8]. However, it was clear that improvements in properties and reduction in fabrication cost would be required before ceramics would be suitable for significant application in gas turbine engines. Since 1970 extensive worldwide efforts have been conducted that have resulted in large improvements in the baseline ceramic materials and in development of new categories of silicon nitride and silicon carbide ceramics that are fabricated by net-shape fabrication processes such as pressureless sintering, overpressure sintering, and hot isostatic pressing (HIP). This progression is illustrated in Figure 29.4 and reviewed in subsequent paragraphs.

RBSN. As described in earlier chapters, RBSN was fabricated by forming a compact of silicon particles into the desired shape and reacting with nitrogen in a high temperature furnace (about 1400°C) to convert to silicon nitride [56–65]. Because this requires a gas phase reactant, the resulting RBSN contains interconnected porosity. 1970-vintage RBSN typically had a density of about 2.2 g/cm^3, which means that the material contained about 30% porosity. This porosity limited the strength and also allowed ingress of oxygen to cause material degradation at high temperature. The typical strength of the 2.2 g/cm^3 RBSN in 3-point bending was under 100 MPa.

Through process development iterations during the 1970s [66–70], the density and strength of RBSN were progressively increased, as illustrated in Figure 29.5. By 1979 RBSN was commercially available with a density of 2.7–2.8 g/cm^3 and an average strength in 4-point bending greater than 350 MPa. This material

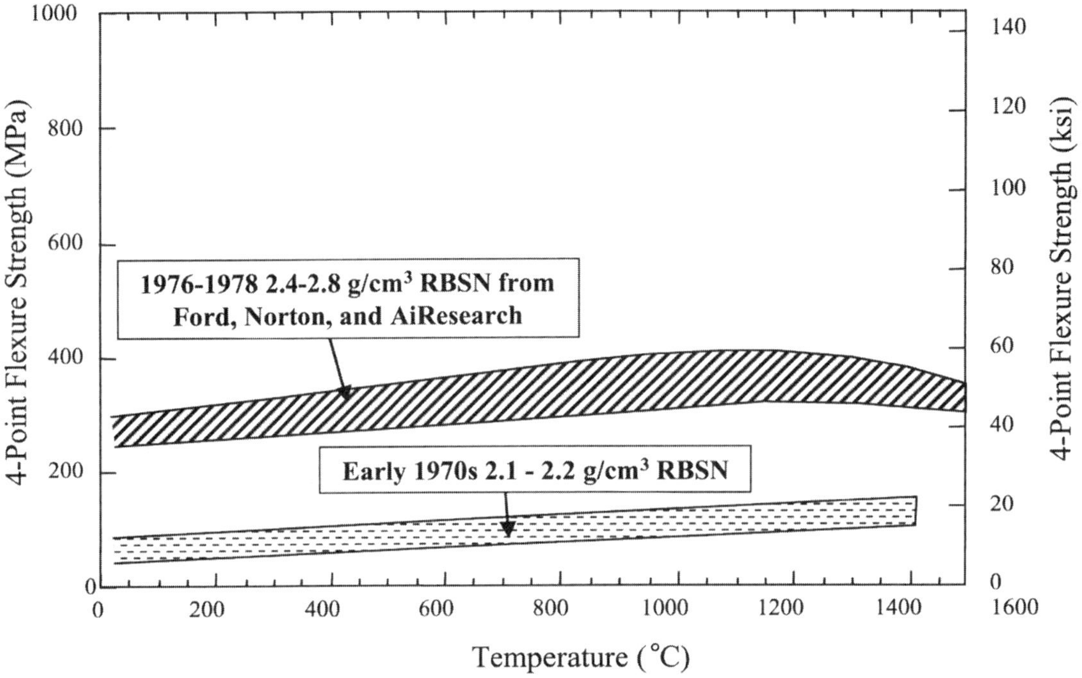

Figure 29.5 Improvements in Strength of RBSN

also had substantial improvement in resistance to oxidation. Another favorable factor was that the RBSN materials retained their strength to at least 1400°C.

RSSC. The second category of ceramics available in 1970 was RSSC, which was also referred to as siliconized silicon carbide (SiSiC) [71–76]. RSSC was fabricated by preparing a shaped preform from a mixture of silicon carbide powder and a source of carbon such as graphite powder. This preform was then infiltrated with molten silicon in a vacuum above 1405°C. The molten silicon reacted with the carbon to form additional SiC and also filled all of the remaining pores. The resulting RSSC had near-zero porosity, strength comparable to or better than RBSN, high thermal conductivity, and excellent oxidation resistance.

As shown in Figure 29.6, RSSC and recrystallized SiC (fabricated by firing a mixture of fine grain and coarse grain SiC particles at very high temperature) materials available in 1970 typically had flexure strength under 200 MPa. Improvements in density, uniformity, microstructure, and processing resulted in substantial increase in strength during the early 1970s [77–82]. One of the best early materials was Refel, developed by the United Kingdom Atomic Energy Authority and British Nuclear Fuels, Ltd. Later, Norton Company made some substantial improvements in RSSC ceramics.

HPSN. A second category of silicon nitride that received considerable development in Great Britain and was just becoming available by 1970 was HPSN [83–87]. HPSN was fabricated by adding an oxide such as MgO to silicon nitride powder and applying pressure through graphite punch and die tooling at about 1750°C. This resulted in a multiphase ceramic with near-zero porosity and very high strength (>700 MPa as measured in 3-point flexure) at room temperature, as illustrated by the Lucas HS-110 HPSN in Figure 29.7. In spite of the excellent room temperature strength, though, the strength dropped rapidly as the use temperature was increased above 1000–1100°C.

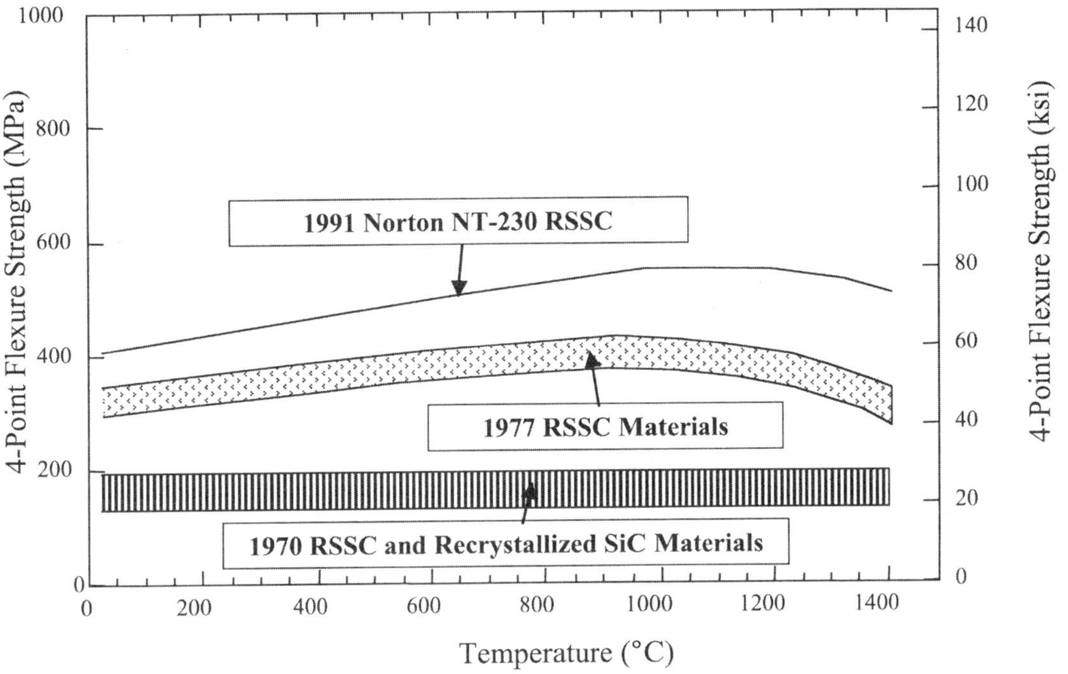

Figure 29.6 Improvements in Strength of RSSC

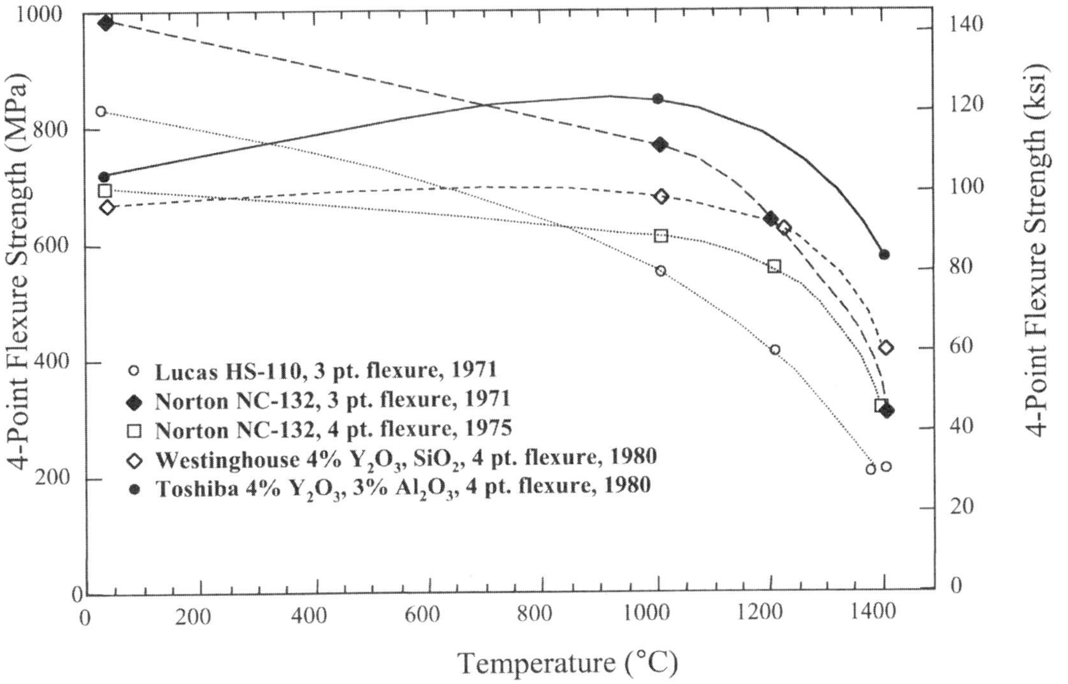

Figure 29.7 Improvements in Strength of HPSN

Studies during the early 1970s determined that the high temperature strength and rupture life were controlled by the chemistry of the phase (typically a complex silicate) at the grain boundaries and also by the degree of crystallinity of the grain boundary phase. A non-crystalline (glass) phase would soften above a critical temperature and lead to deformation and slow crack growth by grain boundary sliding. Efforts in the U.S. determined that the chemistry of the grain boundary phase could be controlled to decrease grain boundary sliding or to increase the threshold temperature and stress at which grain boundary sliding initiated. For example, minimizing the content of calcium for HPSN with MgO sintering aid resulted in the Norton NC-132 material with significant improvement in high temperature strength [88,89]. Note that both 3-point and 4-point flexure data are included for NC-132 in Figure 29.7 to show the large difference in strength value that is measured. Gazza [90] at the Army Materials and Mechanics Research Center (AMMRC) demonstrated that switching from MgO to yttria also improved the high temperature properties [42,43,45,91], as illustrated by the Westinghouse material in Figure 29.7 [36].

Efforts in Japan determined that a combination of chemistry selection and heat treatment could crystallize the glassy grain boundary phase and provide additional improvements [92]. An example is the Toshiba HPSN shown in Figure 29.7. Many other studies were conducted worldwide on hot pressed silicon nitride compositions [93–110].

Lithium Aluminum Silicate (LAS). The fourth material available in the early 1970s was lithium aluminum silicate. When fabricated as either glass or a polycrystalline ceramic, LAS had near-zero thermal expansion. It was studied extensively as a honeycomb rotary regenerator core [8], as a thick-walled flow separator housing [8], and later as a matrix for SiC fibers [52,53]. The honeycomb regenerator was tested successfully under a laboratory environment, but proved not to be successful during on-road testing. Ion exchange occurred between the lithium and sodium (from road salt) and hydrogen (from sulfuric acid condensate from sulfur impurities in the fuel) and resulted in cracking. A technique was developed to leach out the lithium from the LAS to form an aluminosilicate that was more stable. Considerable development also was conducted with cordierite-based compositions similar to those later developed for catalytic converter honeycomb catalyst substrates.

Hot Isostatic Pressing (HIP). Hot pressed silicon nitride was an important material during early turbine engine developments, especially for various rotor configurations. However, the high temperature properties, especially creep and stress rupture life, were still limiting. Also, complex shapes could not be achieved without extensive, expensive diamond grinding. HIP equipment and HIP of metals had been demonstrated by 1970. Asea in Sweden began in about 1971 to explore HIP of silicon nitride. By 1977 Larker and his coworkers reported direct fabrication of complex shapes of silicon nitride using a glass encapsulation process and achieving 3-point flexure strength of about 550 MPa at 1370°C [111,112].

Figure 29.8 shows the 1977 Asea data point compared to 1975 Norton NC-132 HPSN (measured by 4-point flexure) and HIP silicon nitride reported by Battelle in 1980 (also measured in 4-point flexure, but at a very low loading rate) [36,113]. The Asea and Battelle results demonstrated the potential for further improving the high temperature properties of silicon nitride at the same time as making possible complex shape fabrication. The property improvements were made possible by the much higher pressure that could be applied during HIP than by uniaxial hot pressing. The higher pressure allowed full densification to be achieved with decreased sintering aid and thus resulted in increased control of the composition of the grain boundary phase.

Other researchers began to study HIP in the late 1970s [114–123]. A number of companies licensed the Asea HIP technology and developed their own HIP silicon nitride materials during the 1980s [124,125]. As shown in Figure 29.8, a variety of HIP silicon nitride materials with excellent properties were available by the early 1990s. The NT154 and NT164 materials were developed at Norton Company [126,127], the GN10 at AlliedSignal Ceramics, and the CSN101 at AB Cerama (Asea). The data point at the upper left is for the experimental Norton NCX-5102 material [128,129]. It represents the average of 312 uniaxial tensile tests and illustrates the very high strength and reliability that is possible with HIP of silicon nitride.

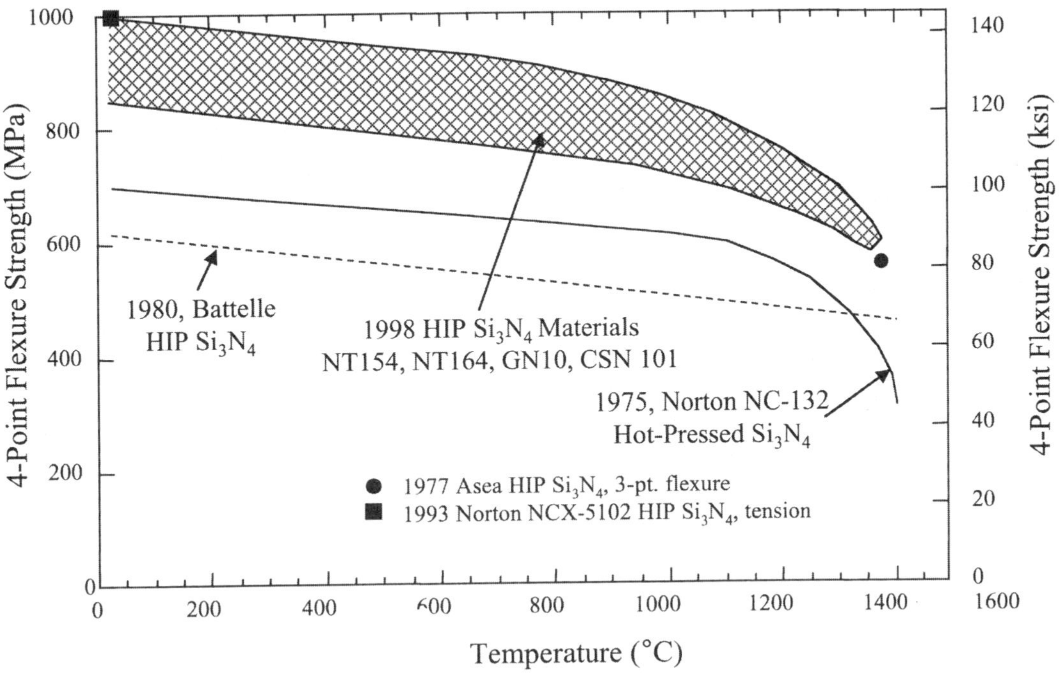

Figure 29.8 Improvements in Strength of HIP-Sintered Silicon Nitrides

Although HIP silicon nitride fabricated using glass encapsulation had excellent bulk properties, chemical interaction during the HIP cycle between the silicon nitride and the glass resulted in lesser properties at the surface. The overall probability of survival for candidate turbine components was decreased unless the surface material could be removed by machining. In most cases this was considered not to be cost effective, so the silicon nitride HIPed with glass encapsulation lost favor during the mid 1990s, especially for long life applications. However, "sinter-HIP" was still used whereby the ceramic was first densified by sintering to the point that the porosity was all internal (closed porosity) rather than connected to the surface. The component could then be HIPed without glass encapsulation to remove much of the closed porosity and achieve significant increase in strength. Even in this case, though, the HIP step represented an undesirable added cost.

Sintered Silicon Nitride. Extensive development efforts were conducted worldwide to achieve ceramics with improved properties and potentially reduced fabrication cost. Pressureless sintering was demonstrated for silicon nitride (and SiC) by the mid-1970s, but the properties were not much better than those of RBSN or RSSC. For example, the first reported sintered silicon nitride by Terwilliger [130] in 1974 was less than 77% of theoretical density. This material, which contained MgO as the sintering aid, exhibited high weight loss due to decomposition of silicon nitride and volatilization of MgO. Mitomo and coworkers [131] reported in 1976 achieving higher density with less weight loss by inhibiting the silicon nitride decomposition by a 1 MPa overpressure of nitrogen. Slowly, through use of different sintering aids, improved silicon nitride powders, and control of silicon nitride decomposition (by use of overpressure nitrogen, powder beds, and protective crucibles), the amount of porosity was decreased and the strength was improved at room temperature and high temperature [70,132–146]. The evolution of these improvements is illustrated in Figure 29.9.

As techniques were developed to minimize the decomposition of silicon nitride powder during sintering, especially overpressure sintering, higher sintering temperature became possible. This allowed reduced porosity, but also resulted in enhanced growth of elongated beta silicon nitride grains during sintering.

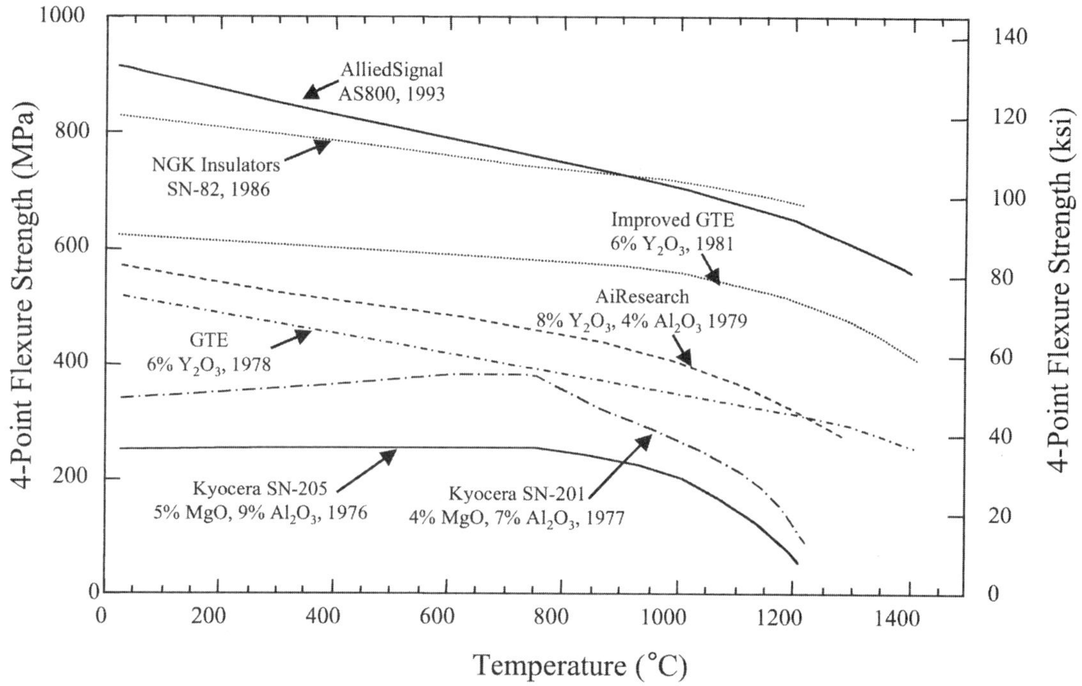

Figure 29.9 Improvements in Strength of SSN

These grains intertwined in such a way that fracture toughness was substantially increased [137–139]. 1975-vintage hot pressed silicon nitride had fracture toughness of about 4.5–5.0 MPa.m$^{1/2}$. Some of the overpressure sintered silicon nitride materials by 1990 had fracture toughness greater than 8.0 MPa.m$^{1/2}$. This unusually high toughness for a ceramic has been an important factor in the reliability and successful use of silicon nitride for turbine components and for other applications.

Sintered Silicon Carbide. SiC evolved similarly to silicon nitride. RSSC and recrystallized SiC were the primary SiC materials available up to the mid 1970s, followed sequentially by hot pressed SiC [147–149], pressureless sintered SiC [150–162], and HIP and sinter-HIP SiC. Hot pressed SiC became commercially available in the early 1970s, but never really caught on. In tests that provided a direct comparison, it did not have comparable thermal shock resistance to hot pressed silicon nitride. By the mid to late 1970s sintered SiC, especially from Carborundum Company, became an important candidate for turbine components and was evaluated extensively throughout the 1980s. As shown in Figure 29.10, 1977-vintage Hexoloy SA (sintered alpha-SiC) had 4-point flexure strength of 300 MPa at room temperature and about 350 MPa at 1450°C. By 1980 several sintered SiC materials were available with 4-point flexure strength between 400 and 500 MPa.

Ceramic Matrix Composites (CMCs). An alternate approach to monolithic SiC was pursued starting in the late 1970s when high-strength SiC-based fibers became available. SEP in France developed a composite consisting of SiC-based fibers (Nicalon fibers from Japan) in a chemical vapor infiltration (CVI) matrix [54]. This general category of composites became known as CVI SiC/SiC. Later General Electric Company developed a melt-infiltrated reaction sintered composite material (MI-SiC/SiC) and Dow Corning developed a polymer infiltrated and pyrolyzed (PIP-SiC/SiC) composite. Extensive development has been conducted on these ceramic matrix composite materials particularly since 1990 [163–172]. Parallel to

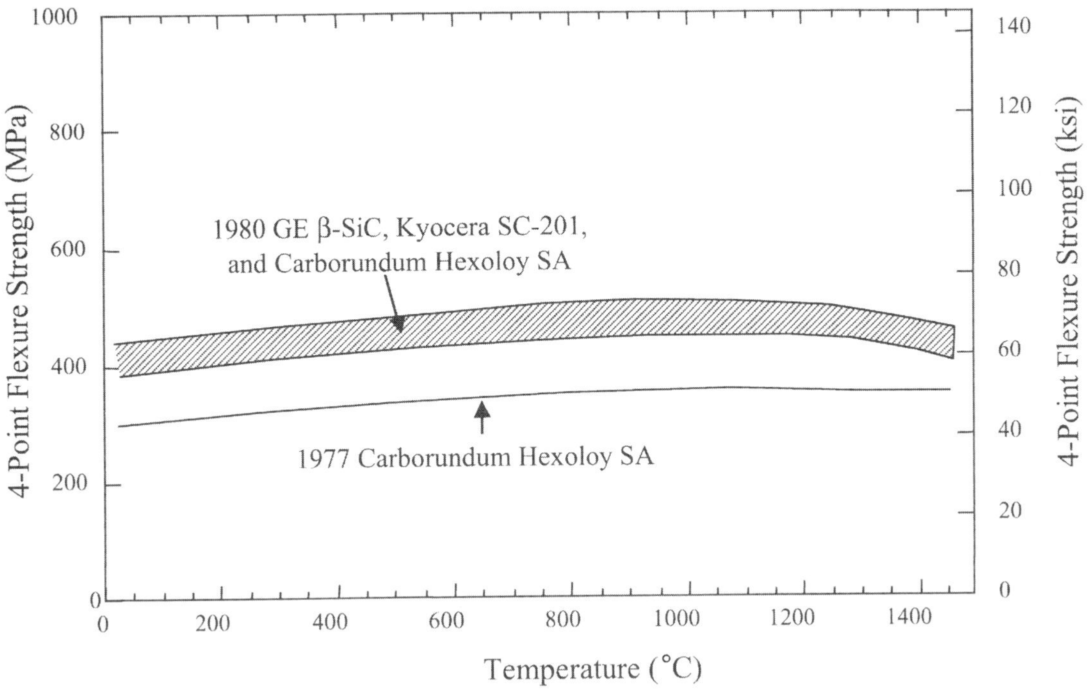

Figure 29.10 Improvements in Strength of SSC

development of these SiC-based composites, United Technologies Research Center developed a series of hot pressed composites with matrices of LAS, calcium aluminum silicate (CAS), and barium aluminum silicate (BAS) [52,53].

Prior to the introduction of the ceramic matrix composites, most development efforts were directed at making silicon nitride and SiC monolithic ceramics stronger to achieve increased reliability. However, these monolithic ceramics were still susceptible to brittle fracture initiated at relatively small material defects. The ceramic matrix composites offered a different fracture behavior, ie., ability to tolerate much larger material defects and to sustain substantial accumulation of structural damage prior to fracture. This resulted in a substantial increase in strain-to-failure compared to monolithic ceramics and generally a non-catastrophic failure mode.

Studies during the 1980s determined that the SiC/SiC ceramic matrix composites were susceptible to degradation due to interaction with the application environment (especially oxygen and water vapor) at high temperature [173]. Extensive developments on interface layers, fiber improvements, matrix modification, and surface coatings have been conducted during the 1990s and are continuing. Parallel efforts have focused on oxide/oxide ceramic matrix composites [174]. The oxide-based composites have the advantage of being less susceptible to environmental attack but mechanical strength limitations of the oxide-based fiber reinforcements limits their use temperature to below that of silicon-based CMCs.

Complex Component Fabrication

Learning to design with ceramics and achieving acceptable material properties were both difficult challenges. A third major challenge was to cost-effectively fabricate the required complex shaped, close tolerance turbine components. Fabrication of a ceramic component involves many steps, and each step requires a high level of reproducibility and careful controls to achieve adequate properties for successful turbine use. The flow chart for injection molding shown in Figure 29.11 gives an example of the complexity. If all of the

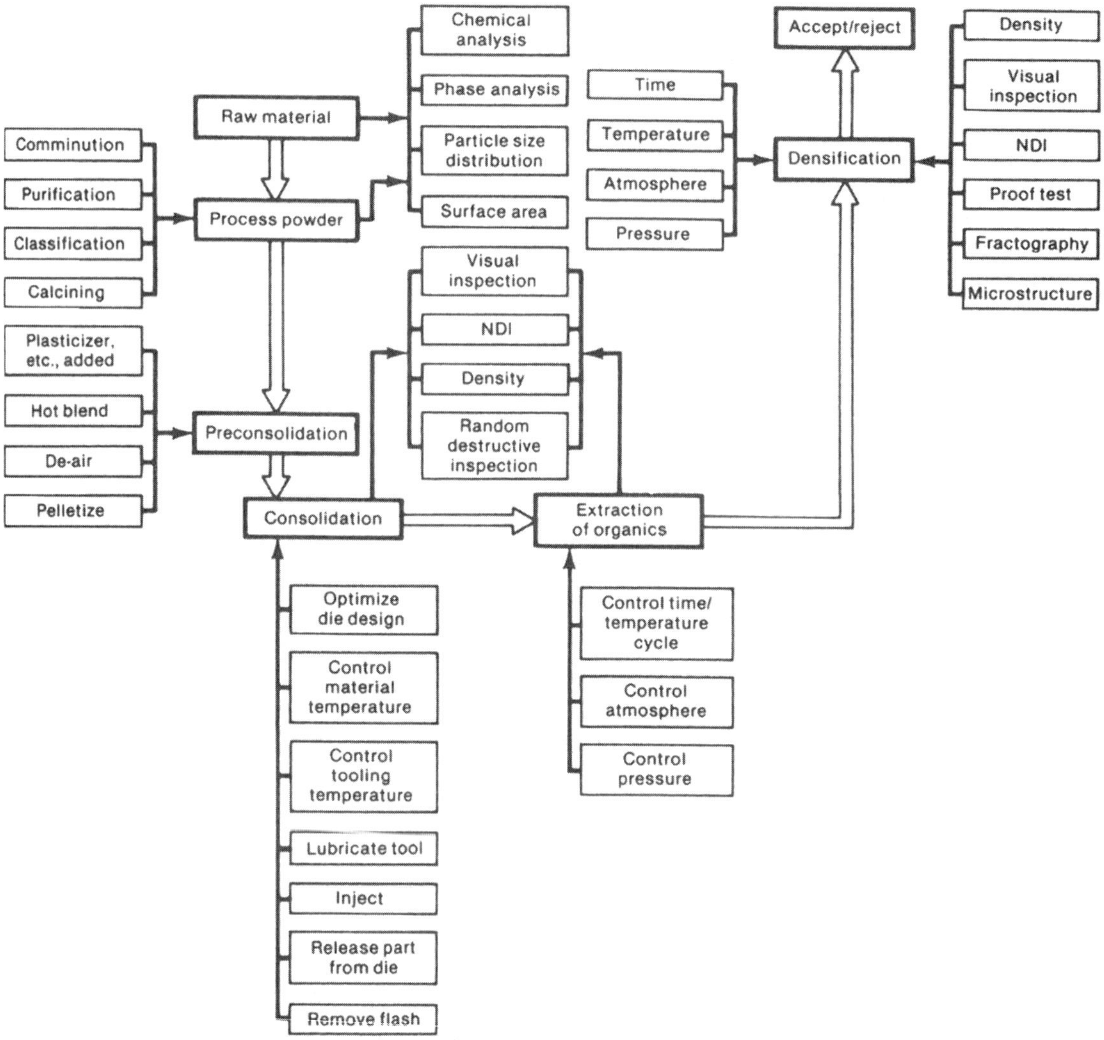

Figure 29.11 Flow Chart of Injection Molding Process

process parameters in each step are not done properly, the resulting component either will not have acceptable properties or will not meet dimensional tolerances.

The degree of process understanding and control necessary to produce acceptable ceramic gas turbine components did not exist in 1970. Each process for each material iteration required extensive long-term development and refinement.

As discussed in the prior section, early turbine components were fabricated from RBSN and RSSC. These methods had substantial merit from the fabrication standpoint. The raw materials were relatively inexpensive, and very little dimensional change occurred during the densification step. The major challenge was to form a uniform compact of the powder into the desired component configuration. Extensive development was conducted during the 1970s on three approaches: injection molding, slip casting, and cold isostatic pressing (CIP)/green machining.

Ford Motor Company in the US and AME in England conducted early development of injection molding [67,175]. Silicon powder was mixed with thermoplastic organic materials (such as paraffin and polyethylene) and forced under pressure and moderate temperature into a shaped mold. The mold tooling was cooler than the injected mixture, so the mixture would quickly cool and harden and allow removal of the part from the tool. The organic binder material could then be removed and the ceramic part densified by nitridation.

Ford fabricated during the early 1970s individual stator vanes, stator rings with continuous outer shroud and slotted inner shroud, and integral rotors. Although the yield of good components was low, injection molded RBSN stators performed well in 200 hour rig tests. This demonstrated that turbine-quality hardware could be produced by injection molding.

Efforts to develop slip casting also were conducted at Ford during the early to mid 1970s [176]. Of particular interest was the "fugitive wax" technique. The desired shape was prepared by injection molding water-soluble wax. This wax shape was then dipped in melted wax that was not water soluble. The water-soluble wax was then dissolved in water, leaving a wax mold with the interior in the shape of the desired part. This was placed with one edge in contact with plaster and filled with a slurry containing the ceramic particles. This technique was successfully used to produce engine-quality stator vanes.

The third method evaluated was CIP/green machining, especially at Norton Company with RBSN [177]. Silicon powder was compacted at high pressure into a simple block. This was heated in a furnace to a high enough temperature for the silicon particles to bond together only at their points of contact to provide enough strength to allow green machining. The desired shape was then achieved by cutting, grinding, and mounted-point profile grinding. This was followed by nitriding to convert the silicon to silicon nitride. The CIP/green machining process involved much more labor than injection molding or slip casting, but produced a much higher yield of acceptable parts.

Much was learned about injection molding, slip casting, and CIP/green machining of RBSN and also of RSSC. However, these materials did not have adequate strength for rotors. Efforts during the 1970s explored ways to make rotors using hot pressing. Ford demonstrated a "duo-density" integral axial rotor consisting of injection molded RBSN blades and a hot pressed silicon nitride hub [178,179]. Däimler-Benz prepared an integral axial rotor completely of hot pressed silicon nitride by a combination of ultrasonic grinding and profile grinding [180,181]. Garrett AiResearch prepared individual blades by diamond grinding and ultrasonic machining [182]. All of these methods were very expensive and had little chance of becoming commercially viable. However, they did provide high quality rotors of 1975-vintage hot pressed silicon nitride for rig and engine testing.

By the late 1970s emphasis was changing to HIP and sintering. Heavy focus returned to near-net-shape fabrication by injection molding and slip casting to prepare the preforms with high enough uniformity and green density to be successfully densified. What had been learned earlier with RBSN was a start, but much more needed to be learned. Specifically, sintering required very small particle size powders that behaved rheologically much differently than the silicon and silicon carbide powders that had been used for RBSN and RSSC. As a result, extensive study of powder processing and rheology was conducted during the 1980s and 1990s for both silicon nitride and silicon carbide material systems. A further challenge was control of shrinkage, typically 15–18 % linearly) and distortion during sintering to achieve the required dimensional tolerances in the final part. Many material and fabrication iterations were required to address these issues, each of which had to be accompanied by detailed property characterization.

Many companies demonstrated the technology to fabricate prototype engine-quality ceramic hardware [183–190]. Asea produced axial and radial rotors by HIP of injection molded performs of silicon nitride. Carborundum and GE used injection molding to prepare performs of SiC for sintering. Kyocera and NGK insulators developed undisclosed proprietary processes that produced high quality overpressure sintered silicon nitride turbine components. AlliedSignal explored injection molding, pressure casting, and CIP/green machining to prepare silicon nitride performs for both HIP and overpressure sintering densification. Later they collaborated with Oak Ridge National Laboratory to develop gel casting of silicon nitride.

In addition to the above examples of monolithic ceramic components, much progress has been made in development of ceramic matrix composite components and establishing controlled fabrication processes [168–172]. Success has particularly been notable for combustor liners and tip shrouds.

Composite fabrication development has focused mostly on preparation of fiber performs by layup of woven cloth. Some work has been done with filament winding. Besides preform fabrication, other key challenges have included uniform deposition of fiber-matrix interface coatings, achieving a high degree of uniformity of matrix infiltration, minimizing interlaminar defects, and applying an oxide environmental barrier coating.

The ceramics companies worldwide have made tremendous progress since 1970 in the fabrication of high quality ceramic engine components. They have demonstrated that they can fabricate the complex shapes, meet tight tolerances, and reproducibly achieve properties in the components that are comparable to the properties achieved in simple test bars. In spite of the progress, though, cost of fabrication and assurance of component reliability are still concerns that are impeding significant commercial introduction.

Quality Assurance

Quality assurance has been a concern and challenge from the earliest programs. A combination of certification samples fabricated (or cut from) each lot of components, non-destructive evaluation (NDE), proof testing, and rig or engine testing have been implemented to qualify ceramic turbine components.

NDE that was available in 1970 included dye penetrants, conventional X-ray radiography, and low frequency ultrasonic c-scan [191,192]. Microfocus X-ray radiography and image enhancement of digitized images were introduced during the 1970s [193]. A variety of different ultrasonic methods including high frequency and surface waves also were evaluated in the 1970s [194,195]. CT was explored in the 1980s and refined throughout the 1990s [196,197]. Thermal imaging and laser scattering were also developed during the 1990s [198].

Four primary challenges have been of concern for NDE: (1) ability to detect critical flaws (flaws that would lead to failure within the desired life of the component), (2) ability to examine the high-stress regions of the complex shaped turbine components, (3) ability to correlate NDE images with material and component behavior, and (4) the high cost of conducting the NDE. Although good progress has been made, turbine companies still do not trust NDE exclusively for qualification of ceramic components.

Virtually all of the turbine development programs have used a proof test for final accept/reject of highly stressed ceramic turbine components such as rotors and stators. Rotors or blades are proof tested by a room temperature spin test to a selected overspeed condition. Stators are typically proof tested through a thermal stress apparatus or rig. Considerable development has been conducted over the years to design, build, calibrate, and verify these proof test facilities, and they have been heavily relied upon for final qualification of ceramic components.

Rig and Engine Testing

As analytical design tools and an increasing number of ceramic materials and fabrication processes were developed, a wide variety of ceramic component designs have been explored. Many were discussed in Volume 1 [1]. Some of these are listed in Table 29.1.

Rig testing has been an important step in development of ceramics for turbines. A rig is a test facility designed to simulate some of the key conditions that a component will experience in an engine. In the earlier programs rig tests were used to validate the probabilistic design and life prediction approach and to essentially prove that the silicon nitride and silicon carbide ceramics could survive under both steady state and transient turbine operation conditions. The rigs were further used to compare design modifications, especially for static structure components, and to qualify ceramic components for additional rig and engine testing.

Table 29.2 summarizes some of the key issues that were addressed by rig and engine testing during the 1970s, 1980s, and 1990s. Efforts during the 1970s were focused on relatively small turbine engines for automotive or short life applications [7,19,20,199]. The automotive goal was 3500 hours life, but much of this time would be at part power decreased load conditions. As a result, a 200 hour rig test could be defined that

TABLE 29.1 EXAMPLES OF THE WIDE RANGE OF DESIGN CONCEPTS AND COMPONENTS THAT HAVE BEEN EXPLORED FOR CERAMICS

Component	Design Concepts
Rotors	• Hybrid axial (ceramic blades in a superalloy disk) with dovetail mechanical blade attachment • Hybrid axial with metal disk isothermally forged around ceramic blade dovetails • Inserted blades with dovetails ranging from 45 to 60°; variations of compliant layers and low friction coatings • Integral axial attached through curvic coupling • Integral axial with ceramic blade ring joined to ceramic disk core • Integral axial with attachment shaft • Integral radial • Ceramic shell on cooled metal core • Segmented ceramic shell on cooled metal core • Impact resistant design with thicker blades and reduced blade count • Hybrid axial with blades retained by a pin • Hybrid axial with rolling contact (rather than sliding contact) blade attachment • Blades with environmental barrier coatings (EBCs)
Nozzles (Stators)	• Individual vanes with integral inner and outer shrouds • Individual airfoils retained by individual spring-loaded slotted endwalls • Individual airfoils retained by slotted ring • Multi-vane segments • Integral ring consisting of vanes and shrouds • Integral with continuous outer shroud ring and slotted inner ring • Ceramic airfoil shell on cooled metal core • Segmented ceramic airfoil shell on cooled metal core • Vanes with EBCs
Combustor Liners	• Ceramic thermal barrier coating (TBC) on cooled metal • Tiles of low modulus, low strength thermal shock resistant alumina-mullite • Tiles of silicon carbide • Multiple ring segments • One-piece can • One-piece annular • Multi-piece annular • Above configurations constructed from a variety of monolithic ceramics and ceramic matrix composites, including SiC/SiC composites with EBCs and oxide/oxide composites
Heat Exchangers	• Plate-fin recuperators with and without integral manifolds • Primary surface recuperators • Tubular heat exchangers, especially for closed Brayton cycle engines • Rotary regenerator using low thermal expansion honeycomb ceramics; high temperature OD and ID regenerator seals
Other Components	• Scrolls • Transition liners • Seals (piston ring and others) • Bolts • Tip shrouds (monolithic, composite, with and without abradable coatings) • Static component support structures • Nose cones • Diffuser sections • Insulation • Bearing components

would give a suitable material and design assessment [8]. Short life applications could typically be assessed by two to ten hour rig tests.

Most of the testing during the 1970s was conducted in rigs. Some 2-hour engine tests were completed at Garrett Turbine Engine Company in the US, but these identified the need for better understanding of contact stress and ultimately for engine redesign. By the early 1980s engine test time was beginning to accumulate.

TABLE 29.2 EVOLUTION OF RIG AND ENGINE TESTING

1970s	1980s	1990 to Present
• Designed initial rigs and engines (form and function)	• Focused on automotive	• Focus shifted to industrial, APUs and microturbines
• Focused on short life and automotive	• Supported 2nd generation ceramic materials development	• Supported 3rd generation ceramic materials development
• Validated design and life prediction methods	• Identified design/material problems and support redesign	• Addressed long life issues such as oxidation, creep and fatigue
• Guided material development	• Addressed special problems like FOD and contact stress	• Conducted increased field testing • Completed >56,000 hours of engine field testing
• Guided component development	• Accumulated >5000 hours of rig testing	
• Qualified ceramic components	• Accumulated >1000 hours of engine testing, including first on-road tests	
• Assessed and refined engine designs		
• Completed 200-hour rig tests		
• Completed 2-hour engine tests		

Garrett successfully completed two 7.5-hour tests of hot pressed silicon nitride blades in a cooled metal engine [11]. Ford completed a 25-hour engine test with an all-ceramic hot section design and even briefly reached a turbine inlet temperature of 1370°C [7,8]. Allison Gas Turbine Division of General Motors accumulated several thousand hours of rig and engine test time, including on-road operation of a truck with a gas turbine engine containing key ceramic hot section components [19,200]. Later United Turbines in Sweden and Däimler-Benz in Germany conducted road testing of automobiles with ceramic engine components. In the early 1980s a modified Volvo 264 equipped with a 100 kW ceramic turbine was evaluated in road tests [201]. And in 1990, a W124 Mercedes-Benz (200 to 300 model E-range) car with a ceramic gas turbine successfully covered the distance from Stuttgart, Germany, to Brussels, Belgium, about 600 km. The total running time of the Daimler-Benz Research Gas Turbine equipped with two hot pressed silicon nitride rotors was more than 500 hrs on test rigs, in the car, and on the road [202].

New priorities were identified during the 1980s. The new sintered and HIPed ceramics appeared to have great potential for fabrication of near-net-shape components at reasonable cost. However, the materials and fabrication capabilities were at an early stage of development such that prototype components had insufficient or marginal properties and usually were not to the required dimensional tolerances. Rig testing and ceramic material/component development became closely integrated with the rig testing providing important support and guidance to the material/component development efforts.

Many ceramic component failures occurred during rig and engine testing during the 1980s [1]. The feedback from these failures led to improved materials, better understanding of design issues such as foreign object damage (FOD) and contact stress, and new designs. As confidence increased, testing began to transition during the 1990s to field testing. This was necessary to begin accumulating the thousands of hours required to evaluate durability and reliability of ceramic components and designs.

Another important change occurred during the 1990s. Emphasis shifted away from vehicular propulsion applications to industrial, auxiliary power (APUs), and short life (military) applications. Many thousands of hours of engine field testing have been accumulated by Honeywell Engines Systems & Services with silicon nitride ceramics (especially stator vanes) in military ground cart APUs and in commercial aircraft APUs. This has included more than 7000 hours on individual sets of ceramic hardware and has demonstrated that silicon nitride has superior life compared to the bill-of-materials metal alloys under standard engine operation conditions [20].

Thousands of hours have also been accumulated in industrial engines in field tests. Ceramic composite combustor liners for example have accumulated more than 56,000 hours in Solar Turbines Centaur engines between 1997 and the end of June 2003. High time so far on a single set of liners is over 15,000 hrs. These and parallel field tests with silicon nitride nozzles at Rolls-Royce [19,203], however, have identified an oxidation/volatilization problem for both silicon nitride and SiC in high pressure ratio engines with turbine inlet temperatures above about 1100°C when water vapor is present in the combustion atmosphere. This has led to the development of oxide-based environmental barrier coatings (EBCs). These are presently at an early stage of development and are the subject of substantial rig and engine testing.

Evolution of Component Designs

The evolution of component design took advantage of the improvements in materials properties and the body of "lessons learned" over almost 50 years of ceramic component and materials development. A number of examples have been listed here to show the progress in design capability and durability of the components tested.

Rotating Components. The design of a rotating ceramic component such as an integral turbine rotor or a turbine blade is particularly challenging because, in addition to the loads resulting from temperature gradients in the part, the component is subjected to the mechanical load imparted by the centrifugal force of the fast rotating turbine wheel in the engine hot section. Also, the contact area between the ceramic and the metal parts is always subject to contact stresses originating in differential thermal expansion between the components made of very different structural materials. The designer has to find ways to address these multiple challenges while ensuring the resulting stress and temperature profiles are within the capabilities of the selected ceramic materials.

Figure 29.12 shows some of the earliest concept designs for blade roots based on cermet structural materials [4,204,205]. The interface between the blade root and the disk cavity is critical for component durability because of contact stresses between the ceramic and metal parts. A conventional fir tree design resulted in blade root failures after start-up, because of unacceptably high stresses in the fir tree branches. The introduction of 4-pin and interlock designs improved the blade life to about 60 hrs. A dovetail design gave test lives of 100 hrs or more. The latter became a popular blade root design concept in subsequent ceramic gas turbine development programs.

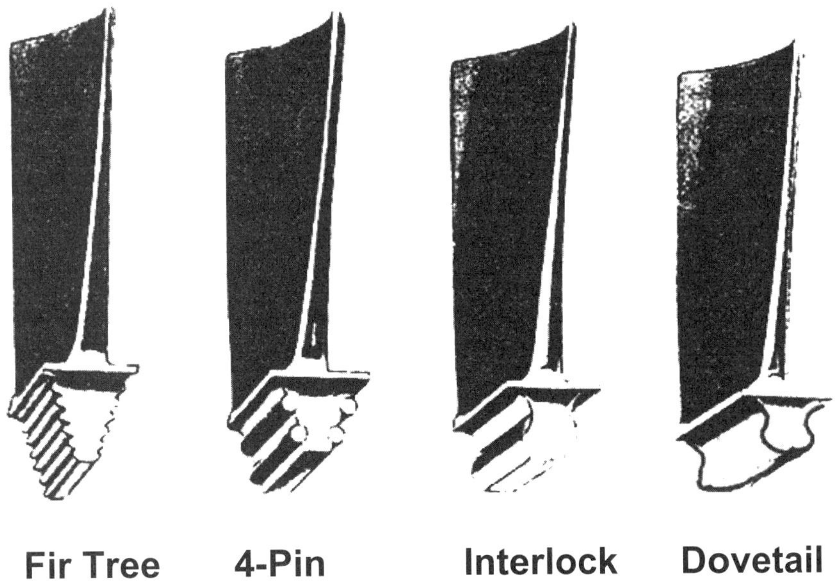

Figure 29.12 Various Early Blade Root Designs Based on Cermets (Adapted from ref. [204])

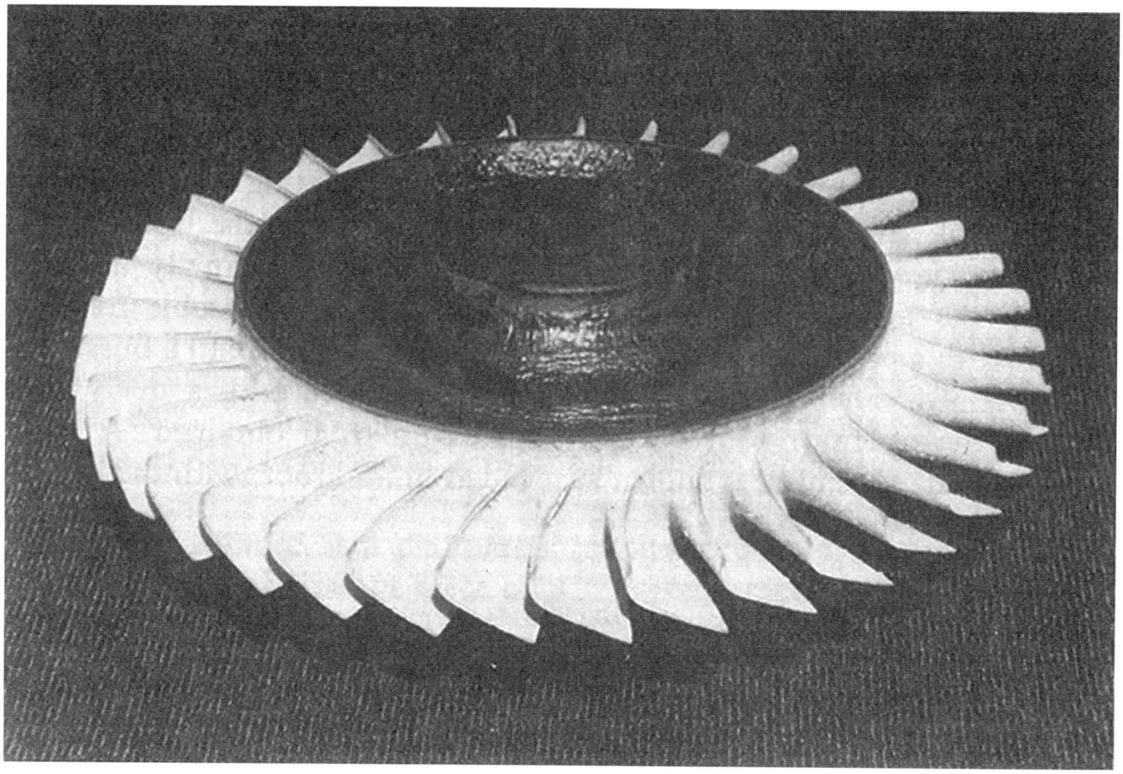

Figure 29.13 Duo-density Silicon Nitride Turbine Rotor Developed by Ford (Reprinted from ref. [206])

Design concepts for rotating components explored on both sides of the Atlantic Ocean were very similar. They reflected the limitations of the materials available. For instance, in the 1970s Ford in the US and Volkswagen AG and Daimler-Benz AG in West-Germany experimented with the duo-density rotor concept for an automotive rotor. Because of the high stresses in the disk hub an integrally bladed ceramic rotor had to be fabricated from a strong silicon nitride. The only material that met the strength requirements at that time was HPSN, but the cost of machining an all-ceramic rotor was prohibitive for an automotive application. The duo-density concept was a design compromise that involved joining a blade ring made from a relatively inexpensive but weak ceramic such as RBSN to a disk fabricated from a strong ceramic, HPSN. The complex shaped RBSN blade ring was readily formed using injection molding with minimal finish machining, while the HPSN disk didn't require elaborate machining either. Figure 29.13 shows a duo-density rotor developed by Ford. While some success was achieved in spin testing, the duo-density concept had limitations. Rotors failed either at the joint between the HPSN and RBSN components or in the RBSN ring at the blade roots, which is the area of highest stress in the ring. These failures led to the duo-density concept eventually being abandoned in favor of other more durable concepts.

Several alternative axial rotor concepts were developed concurrent with the duo-density rotor and they are amongst the most common ones today. One rotor concept that has been extensively evaluated is that of a hybrid design with ceramic blades inserted in a metal disk. The blade root is usually an axial dovetail design, similar to the one tested during the early development work at NACA/NASA on cermet inserted blades (see Figure 29.12). The design uses the ceramic where it is most needed, i.e. ceramic blades are placed in the hot gas path. Since the disk is shielded from immediate exposure to the hot gas stream it can be fabricated from

Figure 29.14 Turbine Rotor with RBSN Blades Inserted in a Metal Disk Developed by Volkswagen AG (Reprinted from ref. [202] with permission from Volkswagen AG)

a conventional disk alloy. Figure 29.14 shows a design of inserted RBSN blades in a metal disk developed by Volkswagen AG under the German National Automotive Ceramic Gas Turbine program in the 1970s.

In this design the critical design feature is the interface between the ceramic blade root and the disk cavity where contact stresses build up during engine operation, particularly during start-up and shut-down. Placing a compliant metal layer at the interface of the blade root and disk cavity is critical for this design concept to be functional. The purpose of the compliant layer is to mitigate the occurrence of unacceptably high contact stresses in the ceramic. The concept has been tested for hundreds of hours in spin test rigs and engines. Figure 29.15 shows a silicon nitride blade with metal compliant layer developed by Honeywell Engines, Systems & Services in the US which was tested for 200 hrs with 50 start-stop cycles in an engine test.

The use of a lenticular metal pad between the ceramic blade root and metal disk cavity is another example of a compliant layer system used to alleviate metal stresses. Such designs were used by Volkswagen AG in West-Germany [202] and Tokyo Electric Power Company (TEPCO) in Japan for its 20 MW component development program in collaboration with Hitachi [29].

The hybrid rotor design has been tested under numerous programs in the US, Japan, Western Europe and Russia. Figure 29.16 shows a hybrid rotor with AS800 silicon nitride blades (Honeywell Ceramic Components) inserted in a metal disk which was tested under the Ceramic Stationary Gas Turbines (CSGT) program sponsored by the US Dept. of Energy for 1052 hrs in a Solar Turbines Incorporated Centaur 50S engine, 948 hrs of which were at an industrial field test site. The rotor eventually failed because of impact damage, but post-test analysis showed the compliant layer system had remained intact during the testing.

Figure 29.15 Silicon Nitride Blade with Compliant Layer which Has Been Partially Pulled out for Illustration Purposes (Reprinted from ref. [20])

Figure 29.16 Hybrid Rotor with AS800 Silicon Nitride Blades in a First Stage Metal Disk Tested for 1052 hrs in a Solar Turbines Inc. Centaur 50S Engine (Reprinted from ref. [199])

Figure 29.17 Integral HPSN Rotors Evaluated by Daimler-Benz (Reprinted from ref. [202])

While it was realized during the early days of rotating ceramic component development that an integrally bladed HPSN rotor could never become a commercially viable component, this design continued to be evaluated because of the high strength of HPSN. It was believed that once the design was demonstrated, new process technologies would be developed that would lower the cost of fabrication of integral rotors. Figure 29.17 shows several HPSN rotors tested by Daimler-Benz AG in West-Germany. Rotors of this type were evaluated in rigs and eventually tested in automotive road tests, such as the successful 600 km trip from Stuttgart in Germany to Brussels in Belgium in 1990, mentioned previously.

Integral rotors have subsequently been fabricated using HIP-sintered and gas pressure sintered processes. Figure 29.18 shows an AGT-5 SN252 rotor of Kyocera that was evaluated in the early 1990s for more than 1000 hrs of durability testing at temperatures in excess of 1427°C under the General Motors-Allison Advanced Gas Turbine (AGT) program. The rotor remained a functional component notwithstanding the abuse it took of blade tip rubs into a shroud ring, impact of carbon particles and insulation, and of impact from melted superalloy combustor parts and vanes.While rotating parts have often failed in rig and engine testing because of foreign object damage (FOD), a combination of rugged component design and excellent materials properties has resulted in parts that are becoming increasingly more impact resistant. Figure 29.19 shows an AS800 silicon nitride rotor which is being field tested in a Honeywell Engines, Systems & Services Model ASE8 gas turbine an industrial plant in Salt Lake City, Utah, USA. As of May 2003 the rotor had accumulated over 10,000 field test hrs in addition to over 1,000 hrs during in-house testing at Honeywell.

The improvement in properties and processing and diagnostic techniques has enabled designers to design increasingly larger parts of monolithic silicon nitrides. An example is the CGT 301 silicon nitride power turbine rotor designed by Kawasaki Heavy Industry for the Japanese CGT302 program sponsored by the New Energy and Industrial Technology Development Organization (NEDO). Initially, the axial power

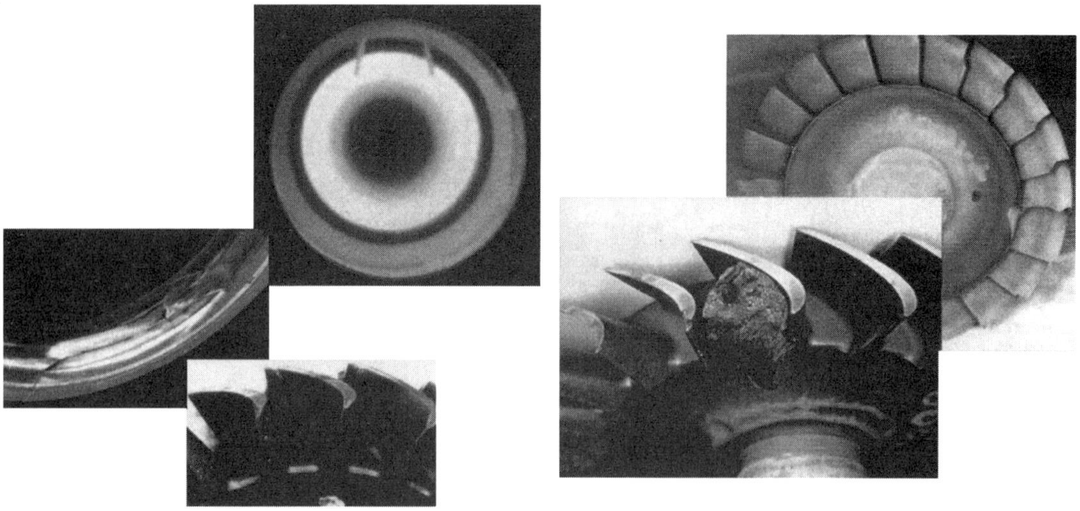

Figure 29.18 AGT-5 SN252 Gasifier Rotor after 1000 hrs of Durability Testing at Temperatures in Excess of 1427°C (Reprinted from ref. [19] with permission from Rolls-Royce)

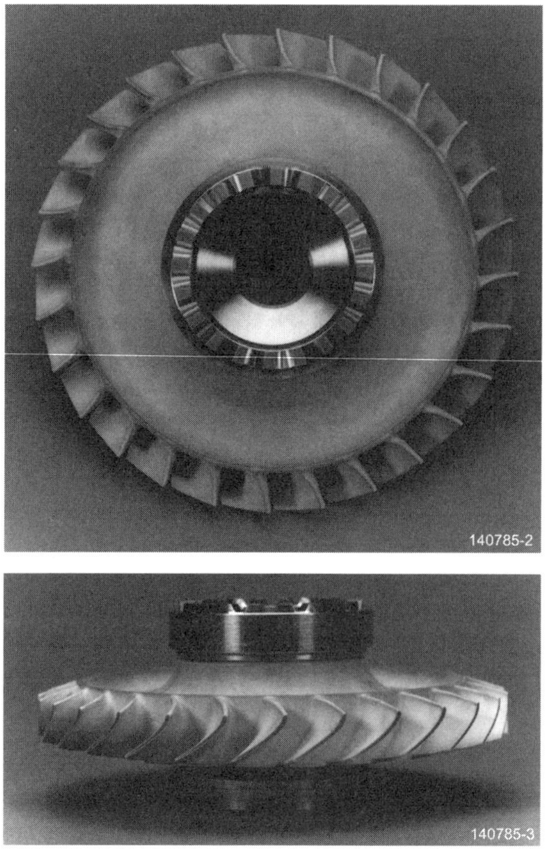

Figure 29.19 AS800 Rotor Tested for over 10,000 hrs in an ASE8 Industrial Gas Turbine (Reprinted with permission from Honeywell Engines, Systems & Services)

turbine rotor was a hybrid design with silicon nitride blades inserted in a metal disk. As the fabrication capabilities of the component supplier (Kyocera Corporation) improved, the final design was that of an integral rotor fabricated from SN282 silicon nitride. With a diameter of 192 mm this was one of the largest integral rotors fabricated. It operated in a short term engine test which achieved 42.1% thermal efficiency and 322 kW output power at a turbine inlet temperature (TIT) of 1396°C, easily meeting the performance targets for the program [207].

Stationary Parts. Stationary parts such as nozzles, vanes, shrouds, combustor liners, transition pieces and nose cones are not subject to the centrifugal loads of blades and rotors. The stresses in these parts are predominantly the result of thermal gradients or originate from the contact at the interfaces of the ceramic and adjacent metal components. For downstream hot section components such as nozzles and vanes, their fixed position makes them susceptible to differences in temperature from the combustor pattern factor and radial profile and they will experience higher temperatures than the rotating parts (blades see the average temperature profile of the hot gas path) in "hot spot" locations.

As for rotating components, a variety of designs have been explored for stationary parts (see Table 29.1). Fabrication development work in the US (Ford) and Germany (Volkswagen AG, MTU) during the 1970s involved stator vanes individual, stator rings with continuous outer shroud and slotted inner shroud, and integral rotors. The early components were fabricated from RBSN or RSSC (SiSiC) by slipcasting or injection molding. Figure 29.20 shows examples of stator designs developed by Ford and MTU. Although the yield of good components was low, injection molded RBSN stators performed well in 200 hour rig tests at Ford. This demonstrated that turbine-quality hardware could be produced by injection molding.

Cracks were frequently observed during the testing. Workers at MTU turned this observation into an advantage by inserting grooves in three locations in a nozzle ring, which forced the formation of localized cracks thereby relieving undesirable stresses. The benefits of this design approach were demonstrated at MTU during a 500 hr test in which a deliberately cracked SiSiC component retained its functionality.

It is only one step further from accommodating cracks by inserting grooves to fabricate the components of individual segments. The segmented component approach has been widely used as a design strategy for ceramic stationary components (and even for some rotating parts). The advantages are several. First, the segmented design consists of a number of constituting parts that have stresses that will be lower than for an integral component since temperature gradients are split up in smaller increments. Secondly, the segments have a smaller volume than an integral part and the probability of failure from a fabrication defect will be reduced. Figure 29.21 is an example of a segmented combustor design developed in the 1980s and early 1990s by Toshiba Corporation in Japan under a 20 MW ceramic gas turbine component development program sponsored by TEPCO. The liner consists of ceramic ring and tile segments fabricated from silicon carbide. The adjacent transition piece (not visible) is fabricated in a similar manner. A similar design was developed by the Japanese Central Research Institute of Electrical Power Generation (CRIEPI) in a parallel 20 MW engine component development program [208].

Segmented nozzle designs were also developed by TEPCO and CRIEPI [29,208]. The designs were hybrid nozzles consisting of a cooled metal support structure and heat resistant ceramic shield. The metal support structure provides the structural integrity of the component while the ceramic shield acts as a thermal barrier to the metal support. The cooling air requirement is about 10–20% of that of a cooled all-metal nozzle. Figure 29.22 is a schematic of a hybrid metal-ceramic nozzle developed by CRIEPI in the 1980s for its 20 MW ceramic gas turbine component development program. The ceramic shell is segmented in four pieces, two half-vane segments and an inner and outer shroud piece, to lower the stresses.

Segmentation is often the only way to design a part if the material has low strength. Figure 29.23 shows segmented designs for a turbine wheel, nozzle vanes, rotor blade and combustor liner developed for small (1.5–2.5MW) ceramic gas turbines at the St. Petersburg (Russia) based Bokyo Center (formerly the Research-Technological Institute for Power Engineering–NITI EM, and its affiliate NIZ KTD) [209]. The components were fabricated of "SUGRAV," an alumino-boron-nitride structural ceramic with a room temperature flexure

(a)

(b)

Figure 29.20 Examples of Early Stator Designs (a) Individual Vanes with Separate Integral Outer Shroud Ring Developed by Ford (Reprinted from ref. [8]) (b) Integral Nozzle Ring without Inner Shroud Develooped by MTU (Reprinted from ref. [202] with permission from MTU)

Figure 29.21 Segmented Combustor Liner Developed by Toshiba under the TEPCO 20 MW Ceramic Gas Turbine Component Development Program (Reprinted from ref. [29] with permission from TEPCO)

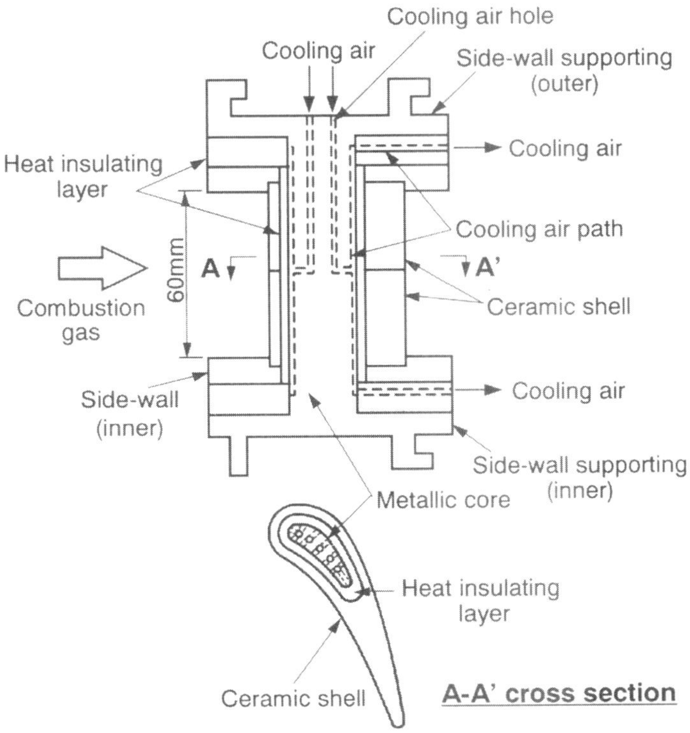

Figure 29.22 Schematic of a Hybrid Metal-Ceramic (Segmented) Nozzle Developed by CRIEPI for its 20 MW Ceramic Gas Turbine Component Development Program (Reprinted from ref. [208])

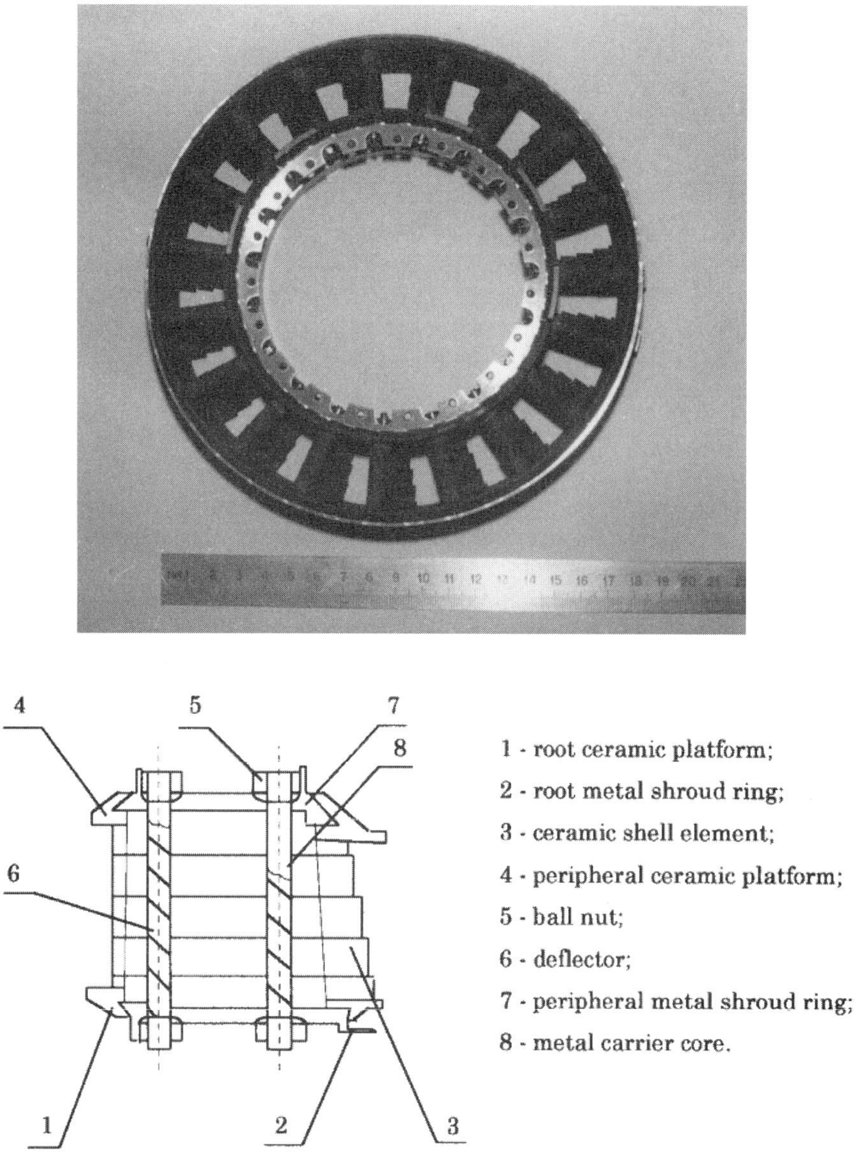

Figure 29.23 Metal-Ceramic Hybrid Components Developed by Boyko Center (Reprinted from ref. [209] with permission from JSC Boyko Center)

strength of 122 MPa. Designing with such a weak ceramic requires heavy segmentation. For instance, the blade and vane shells consist of four or five separate pieces that are held in compression. The mechanical load is carried by a metallic support structure that is cooled as in the TEPCO and CRIEPI nozzle designs.

Segmented components often have leakage problems and as the ceramic materials improved in mechanical properties, designers moved away from segmentation towards the design of integral vane shells for hybrid metal-ceramic nozzles. Figure 29.24 shows a hybrid metal-ceramic nozzle developed by Mitsubishi Heavy Industries in Japan under the TEPCO 20 MW program [29]. Silicon nitride and silicon carbide nozzles were tested in a vane cascade rig at a gas temperature up to 1500°C and a pressure of 6 atm. The nozzles shown

Figure 29.24 Air-Cooled Nozzles Fabricated from SN88 Silicon Nitride (Reprinted from ref. [210] with permission from NGK Insulators, Ltd.)

here were fabricated by NGK Insulators, Ltd. of Nagoya, Japan, using SN88, a gas pressure sintered silicon nitride [210]. The vane shell is a single piece and was air-cooled so that the maximum surface temperature could be maintained below 1300°C by a small amount of cooling air. The metal-ceramic hybrid nozzle concept was also evaluated by MTU in Germany [211].

An alternative way of lowering the stress on a ceramic stator part has been the bowing of the high temperature region of the airfoil. Figure 29.25 shows a bowed airfoil configuration for a stator designed for a 15 MW gas turbine by Toshiba Corporation [212,213]. The incorporation of heat shield plates ensures that the temperature gradient on the airfoil is minimized. The stator vane was tested in a cascade test rig in a wind tunnel to simulate both steady state and transient conditions in the engine. The average gas temperature was 1330°C and a maximum gas temperature of 1440°C was measured during the test. Vanes of sialon (SIN-E, Toshiba) and silicon nitride (SN252, Kycocera Corporation) were evaluated. A maximum thermal stress of 240 MPa was predicted 4 s after start up.

Another example of a bowed stator part is the bowed nozzle developed by Solar Turbines Incorporated under the US Dept. of Energy Ceramic Stationary Gas Turbine (CSGT) program [199]. The nozzle shown in Figure 29.26 had a calculated maximum steady state stress of 211 MPa, reducing by more than half the stress in a similar vane with straight airfoil. Silicon nitride nozzles of this design fabricated from SN88 (NGK Insulators, Ltd.) and SN282 (Kyocera Corporation) were evaluated in rig and engine testing. SN282 nozzles survived in a 104 hr engine test with a turbine rotor inlet temperature (TRIT) up to 1121°C. This part is predicted to have maximum temperatures of around 1288°C in hot spot locations at the airfoil trailing edge at this TRIT.

Integral nozzles for small auxiliary power units (APUs) have been successfully commercialized. The designs are simple, consisting of an airoil between two shrouds. Figure 29.27 shows a ceramic nozzle that was tested for 2,200 hrs in a comparative test with a metal nozzle. The silicon nitride ceramic nozzle which

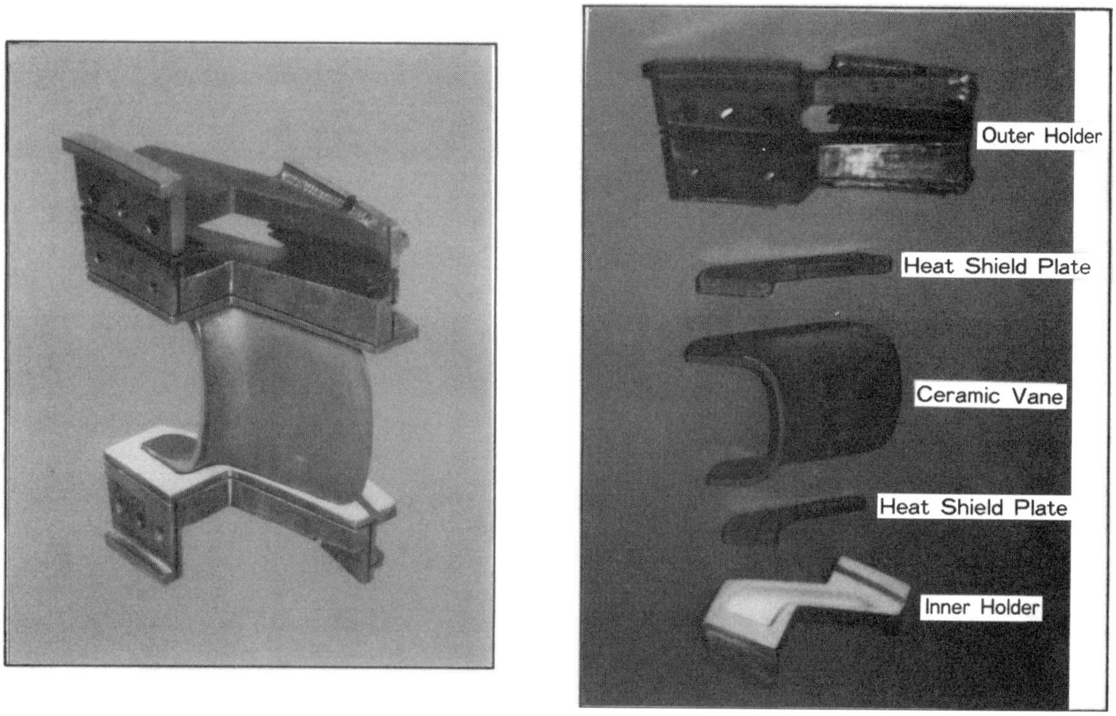

Figure 29.25 Assembled and Disassembled Components of Stator Vane (Reprinted from ref. [212] with permission of Toshiba Corporation)

Figure 29.26 Bowed Nozzle Developed under the CSGT Program (Reprinted with permission of Solar Turbines Incorporated)

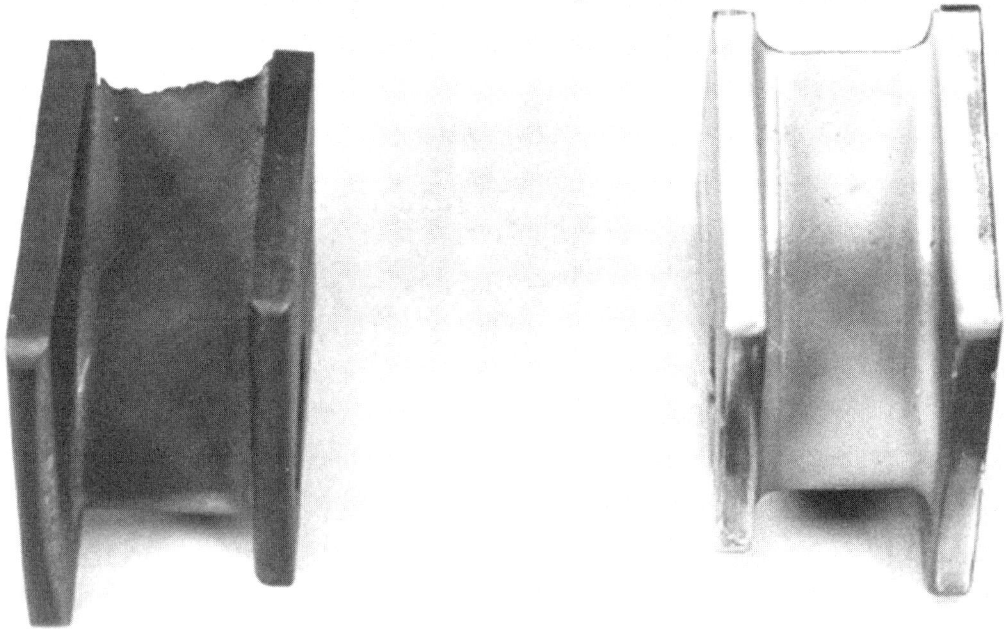

Figure 29.27 The Ceramic Nozzle (Right) is More Durable than Conventional Metal Nozzles (Left) in a 2,200-hour Comparative Test (Reprinted from ref. [20] with permission from Honeywell Engines, Systems & Services)

had accumulated over 5,000 hrs of previous engine testing showed virtually no wear or deterioration after 7,200 hrs operation while metal nozzles made of IN713 and MAR-M247 were visibly thinned and recessed on the trailing edge following testing in an erosive environment. The erosion was a severe problem caused by hard carbon particles generated in combustors and that passed through the hot gas path. Between 1994 and 2000 Honeywell and its predecessor AlliedSignal Engines installed ceramic nozzles in GTCP85-100 APUs and placed them in regular commercial service aboard MD-80 aircraft. Over 40,000 hrs of cumulative operating time had accrued during that period.

After 1990 ceramic matrix composites are being increasingly considered as structural materials for ceramic gas turbine components. The application is primarily for stationary components since mechanical properties limits the applications of CMCs to stresses not exceeding the proportional limit, i.e. about 140 MPa for the best commercially available silicon-based CMCs, Since rotating components have steady state stresses of 200 MPa or more, 2000 vintage CMCs cannot be considered for rotating components requiring long life, i.e. tens of thousand of hours. Currently silicon-based CMCs have been successfully tested for combustor liner applications. Noticeable is the work started under the CSGT program at Solar Turbines Incorporated. As of June 2003, SiC/SiC CMC combustor liners had been tested for over 56,000 hrs in the Solar Centaur 50S engine at industrial field sites with two tests having achieved 13,937 hrs at the ChevronTexaco industrial cogeneration site in Bakersfield, California, USA, and 15,144 hrs at the Malden Mills textile industrial plant in Lawrence, Massachusetts, USA, respectively. Liners tested were fabricated from Ceramic Grade, Hi-Nicalon or Tyranno ZM fiber-reinforced silicon carbide with the matrix being infiltrated using the CVI or MI process. To achieve improved durability, environmental barrier coatings developed by United Technologies Corporation and applied by conventional plasma spray techniques did increase liner life several-fold. Figure 29.28 shows a set of SiC/SiC CMC liners with protective EBC [199]. CMC applications for gas turbine hot section components including nozzle and shroud components, seal rings, and

Figure 29.28 Set of SiC/SiC CMC Combustor Liners with Protective EBCs Developed under thc US DOE/Solar Turbines Incorporated CSGT Program for Long Term Field Testing (Reprinted from ref. [199] with permission of Solar Turbines Incorporated)

combustor liners and transition pieces are currently under development at several companies in the US, Western Europe and Japan and it is expected that full commercialization of stationary CMC parts will be achieved in the first decade of the 21st century.

CURRENT STATUS

An enormous amount of funding has been expended since 1970 on the development of advanced ceramics for gas turbine engines, but a significant ceramic gas turbine business has not yet emerged. Exceptions are ceramic thermal barrier coatings and the Siemens refractory ceramic tile combustor liners. In addition, silicon nitride is in a moderate level of production for a seal runner application at Honeywell and in a low level of production for a few other components, particularly nozzles. The objective of this section is to briefly review key barriers that still need to be overcome and to identify sectors where production is likely in the future.

As of the writing of this text (i.e. 2003) perceived barriers are technological and economic. While they are not expected to impede the application of ceramic components in gas turbines in the long run, they are slowing down their implementation in the near term. The experience to-date indicates that the impediments are different for monolithic ceramics and CMCs.

Economies of scale play a major role. Unless the large automotive market starts to use ceramic gas turbines, annual part volumes will be aerospace quantities, i.e. the gas turbine market will need volumes of thousands of similar parts, and at best tens of thousands. This will be a key barrier to undercut the price of metal components significantly, and for some complex parts the price will be higher for the forseeable future.

From a cost perspective, monolithic ceramic components using state-of-the-art silicon nitrides and silicon carbides can in fact be fabricated at a cost that is comparable or even less than that of cooled single crystal superalloy parts. So, there exists no intrinsic cost barrier to the application of monolithics in the gas turbine hot section. Additionally, if there is incentive, part costs can be further reduced through advancements in part finishing (cost effective machining techniques).

This is currently not the case for CMCs. Stationary CMC components are expected to remain significantly more expensive than metallic parts for similar applications for the forseeable future because of the high cost of fibers, time consuming matrix infiltration processing, and inefficiencies in furnace loading. As more applications emerge, economies of scale will help to bring fiber and processing costs down, but this is not expected to occur before the second part of the first decade of the 21st century, at the earliest.

Technological hurdles are also substantial. While monolithics have greatly improved high temperature strength compared to superalloys, their fracture toughness is inferior which manifests itself in frequent failures because of impact of foreign objects in the gas turbine environment, particularly for rotating components. The best commercially available silicon nitrides have fracture toughnesses in the 8–9 $MPa.m^{1/2}$ range which does not compare favorably with superalloys. Failure of monolithic parts, especially rotating parts has been regularly observed in most of the ceramic gas turbine development programs.

CMCs have significantly improved fracture toughness compared to monolithics. K_{IC} values of 25 $MPa.m^{1/2}$ have been determined for SiC/SiC CMCs and studies have shown reduced catastrophic damage for this class of materials upon impact. However, compared to state-of-the-art silicon nitrides and carbides, CMCs have mechanical property limitations, primarily as a result of insufficient first matrix cracking strength. It is not clear at this time that this limitation can be overcome, although recent data show that the proportional limit for SiC/SiC CMCs can be as high as 200 MPa in laboratory testing. While this mechanical property limitation excludes CMCs for most rotating parts, mechanical strength is adequate for many stationary applications.

As firing temperatures of gas turbines continue to increase and emissions of NOx, CO, and unburned hydrocarbons are regulated to lower levels, and consequently the demand for cooling and dilution air is increased, ceramic materials strategies will become increasingly attractive. The high firing temperatures are already putting great demands on the component designs of aircraft, aeroderivative and utility gas turbines and with time these demands will also extend to industrial gas turbines. And emissions requirements will become more stringent for all classes of gas turbines, both for power generation and mechanical drive applications, initially in the developed world, but in the near future also in the developing world. The reduced need for cooling air for ceramic and composite components will enable gas turbine manufacturers to meet the rising fuel efficiency demands of the end users of their equipment and the regulatory emissions requirements. These more stringent demands can be expected to offset any existing cost disadvantages that monolithic ceramics and CMCs currently have.

Environmental durability has recently been noted as a critical barrier for the commercialization of ceramic components. During most of the development of these components testing times were rather short, on the order of tens and hundreds of hours, and surface recession of silicon-based ceramics was not considered to be a significant problem. However, with the extended field testing of the 1990s it has been observed that both silicon-based monolithic parts, particularly, nozzle components, and silicon-based monolithic and CMC combustor parts suffer accelerated degradation in the hot section of the gas turbine. The causes of the degradation are well understood, and are generally attributed to water vapor attack on the silicon-based structural materials, resulting in the formation of volatile silicon hydroxides.

The development of environmental barrier coatings is being extensively researched and successful EBC systems have been demonstrated on SiC/SiC CMC combustor liners. Still, continued development is needed to improve the durability of the EBCs, and to tailor them to monolithic silicon nitrides that require a lower coefficient of thermal expansion than the silicon carbide-based CMCs. The EBC systems are invariable oxide-based which suggests that oxide-based structural materials may someday become an alternative to the silicon-based monolithics and CMCs. The experience to-date has shown the limitations of oxides as primary structural materials: poor thermal conductivity and thermal shock resistance compared to silicon-based ceramics. Current oxide fibers also have inferior mechanical properties compared to silicon-based fibers, which limits the temperature range for the use of oxide-based CMCs. Nevertheless, there has been an upsurge in the interest in oxide-based systems recently because of the inherently superior environmental stability compared to silicon-based materials. There is also a perception that oxide CMC can be fabricated more cost-effectively than silicon-based CMCs, which should be an advantage in commercialization strategies.

In summary, substantial barriers do exist for commercialization of ceramic gas turbine components, but the efforts underway and the future requirements provide a strong incentive for the ceramic gas turbine to become a reality in the market place.

Microturbines

Over the past decade interest has grown to apply ceramic gas turbine technology to very small engine designs. Microturbines with output power between 25 kW and 500 kW for stationary power generation applications can be sited in locations constrained by space limitations for power production or where grid power is not easily available. They can also be used in direct mechanical drive markets such as compression and air conditioning.

The incorporation of ceramic components is believed to be critical for microturbines to achieve levels of fuel efficiency that makes them competitive with other sources of energy generation in their size range such as reciprocating engines. Ceramic gas turbine development programs for automotive and cogeneration applications in the US and Japan have shown that highly efficient designs of engines in the microturbine size range with ceramic components operating at high turbine inlet temperatures (TITs) are feasible. For instance, the ceramic gas turbine developed Kawasaki Heavy Industries under the Japanese 300 kW program achieved 42.1% thermal efficiency and 322 kW output power while operating at a TIT of 1396°C [207].

In the US the Department of Energy is sponsoring a six-year (FY 2000–2006) $ 60M advanced microturbine program for stationary power applications in industrial, commercial, and institutional sectors. The program encompasses conceptual design, development, and demonstration; component, sub-system development, and development of a technology base, in the areas of materials, combustion, and sensors and controls. Technology evaluations and demonstrations are also part of the program.

The program performance targets are very challenging [214]:

- High Efficiency—Fuel-to-electricity conversion efficiency of at least 40%
- Environment—NOx < 7 ppm (natural gas)
- Durability—11,000 hours of reliable operations between major overhauls and a service life of at least 45,000 hours
- Cost of Power—System costs < $500/kW, costs of electricity that are competitive with the alternatives (including grid) for market applications
- Fuel Flexibility—Options for using multiple fuels including diesel, ethanol, landfill gas, and bio-fuels

Contracts were awarded to Capstone Corp., General Electric Company, Ingersoll-Rand, and United Technologies Corporation for system development, and to Solar Turbines Inc. for recuperator development. The conceptual microturbine designs incorporate ceramics as structural materials for critical hot section components such as rotors, vanes and combustor liners.

Even smaller engine designs are envisioned. Allan Epstein and coworkers at the Massachusetts Institute of Technology are developing the technology for a micro-gas turbine which produces 10–20 W of electrical power or 0.05–0.1 Nt of thrust while consuming less than 10 g/hr of H_2. The concept is feasible because of advances in microelectrical and mechanical (MEMS) manufacturing techniques that have been pioneered in the semiconductor industry [215–217]. A prototype 4-mm diameter radial inflow turbine fabricated from silicon by deep reactive ion etching has been fabricated. Silicon is not adequate for the design temperatures of the application, but parts could conceivably be fabricated from SiC, or Si/SiC hybrid structures. An entire device is expected to weigh about 1 g. The concept is attractive for applications that are weight critical. Because the energy density of liquid hydrocarbon fuels is 20–30 times that of batteries, power sources can be a lot smaller. When produced in large quantities these low-cost heat engines could find applications as mobile power sources for propulsion of small unmanned aircraft, and for boundary-layer and circulation control and coolers for electronics and individuals.

TECHNOLOGY SPINOFFS

Even though the ceramic gas turbine market has not yet emerged, there have been numerous spinoffs of the technology advances stimulated by the gas turbine programs [218]. Some of these associated with the development of silicon nitride and silicon carbide materials are summarized in Table 29.3.

Cutting tool inserts was the first substantial application for hot pressed silicon nitride. The silicon nitride inserts could machine cast iron and superalloys five to six times faster than WC-Co inserts. This particularly had a big impact on productivity in the automotive industry. The second major success for silicon nitride has been bearings, especially for high-speed machine tools. Modern industry runs on machine tools, especially computer-controlled cutting and milling machines that shape the millions of metallic parts required for automobiles, jet engines, appliances, and for industrial machines. The silicon nitride bearings consume less energy, wear about 5 to 10 times longer than metal bearings, and allow the machines to run faster and with less vibration. Silicon nitride bearing are also being used in numerous aerospace applications (including gas turbine engines), in industrial equipment, in medical equipment, and even in sporting equipment.

Because of its unusually high fracture toughness for a ceramic and reasonable hardness, silicon nitride also has found numerous applications as a wear resistant material. Examples include sand blast nozzles, diesel cam follower rollers, coal and ore processing equipment, and papermaking machine liners. The combination of toughness and wear resistance with high strength, temperature resistance, and thermal shock resistance has made silicon nitride a prime candidate for automotive application. Starting in about 1988, about 30,000 silicon nitride turbocharger rotors were fabricated per month in Japan. Daimler-Chrysler has conducted extensive road testing of silicon nitride valves and clearly demonstrated their superiority over metal valves, but the cost has not yet come down enough for commercial introduction.

Silicon carbide also has been chosen for many applications. Being much harder than silicon nitride, it is inherently more wear resistant and is also very resistant to corrosion. Sintered silicon carbide and RSSC are used broadly for various types of seals and for pump parts. Because SiC can be fabricated to a high level of purity and can withstand high temperatures, it is an important material in our modern electronics industry in processing semiconductor devices. Finally, SiC is an important material for high temperature heat exchangers.

TABLE 29.3 APPLICATIONS THAT BENEFITED FROM TECHNOLOGY DEVELOPED UNDER THE GAS TURBINE PROGRAMS

Applications	Commercial Introduction
Silicon nitride and sialon cutting tool inserts	Late 1970s
Sintered SiC water pump seals for trucks	1980s
Sintered SiC for numerous other wear and corrosion resistance parts	1980s and 1990s
Silicon nitride diesel glow plugs	1981
Silicon nitride rocker arm wear pad for alternate fuel automobile engines	1984
Silicon nitride diesel swirl chamber	1986
Silicon nitride turbocharger rotors	1980s
Silicon nitride bearings	Late 1980s
Sintered SiC and RSSC for the electronics industry (integrated circuit fabrication)	1990s
Silicon nitride diesel cam follower rollers	1990s
Wide variety of silicon nitride parts for wear resistance applications	1990s
Silicon nitride for aluminum casting applications	1990s

Other materials technologies that were developed for the gas turbine programs have indirectly benefited other sectors. One of the most important was the rotary regenerator work at Corning. The technology developed for producing a thin-walled honeycomb structure was adapted by Corning in the mid-1970s to develop the catalytic converter. By 2000 the motor vehicle pollution control business had grown to $38 billion and had saved us from over 1.5 billion tons of air pollution [218].

Another indirect benefit of the ceramic gas turbine programs was the improvement in analytical design tools and codes. This has helped improve metal turbines. In addition, the potential competition posed by ceramics has stimulated significant improvements in metal turbines. Similarly, the potential competition of a ceramic turbine in the automotive industry has perhaps stimulated advances in reciprocating engines that have benefited all of us.

CONCLUSIONS

Achieving improvements in gas turbine engines through ceramic components has been a difficult challenge that has required advances across many fronts: development of new design and life prediction methods, improvement in the properties of ceramic materials, establishment of test methods and standards for characterization of ceramics, generation of an extensive database, development of NDE and proof test protocols, and implementation of extensive rig and engine testing. Much progress has been made in each of these categories and in integrating them. An impressive list of spinoff applications has resulted over the past 30 years, and turbine applications are starting to emerge, but there is still much to do before monolithic and composite ceramics will become routine bill-of-materials components in gas turbine engines. The editors hope that this two-volume treatise has provided the reader a comprehensive and useful review of the diverse work that has been done over the past 30 years and a good starting point in solving the remaining challenges.

ACKNOWLEDGEMENT

Unless otherwise indicated the figures and tables in this chapter were supplied by the authors.

REFERENCES

[1] *Ceramic Ceramic Gas Turbine Design and Test Experience, Progress in Ceramic Gas Turbine Development, Volume I*, van Roode, M., Ferber, M. K., and Richerson, D. W., eds, ASME Press, New York, 2002.

[2] Richerson, D. W., Iseki, T, Soudarev, A. V., and van Roode, M., "An Overview of Ceramic Materials Development and Other Supporting Technologies," Chapter 1 in this Volume.

[3] Dukes, W. H., *Handbook of Brittle Material Design*, AGARD AD719-712, Feb. 1971.

[4] Probst, H. B., "Structural Ceramics in Heat Engines—The NASA Viewpoint," in *Proceedings of the First International Symposium on Ceramic Components for Engines*, Hakone, Japan, Oct. 17–19, 1983, S. Sōmiya, E. Kanai and K. Ando, eds., KTK Sci. Pub. Tokyo, Japan, 1984, and Elsevier Applied Science Publishers Ltd., London, UK, 1986. pp. 45–58.

[5] Conway, H. M., *The Possible Use of Ceramic Materials in Aircraft Propulsion Systems*, NACA CB No. 4D10, 1944.

[6] Morgan, W. C. and Deutsch, G. C., *Experimental Investigation of Cermet Turbine Blades in an Axial-Flow Turbojet Engine*, NACA TN-4030, Oct. 1957.

[7] Hartsock, D. L., "Ford's Development of the 820 High Temperature Ceramic Gas Turbine Engine," Chapter 3 in ref. [1].

[8] McLean, A. F. and Fisher, E. A., *Brittle Materials Design, High Temperature Gas Turbine*, AMMRC TR 81–14, Final Report, March 1981.

[9] Paluszny, A. and Wu, W., "Probabilistic Aspects of Designing with Ceramics," Transactions of the ASME, Journal of Engineering for Power, 99[4], 617–630, 1977.

[10] McClean, A. F. and Hartsock, D. L., "Design with Structural Ceramics," in *Structural Ceramics*, J. B. Wachtman, Jr., ed., *Treatise of Material Science and Technology,* Academic Press, Inc., Boston, MA, USA, Vol. 29, 27–97, 1989.

[11] Richerson, D. W. and Johansen, K. M., *Ceramic Gas Turbine Engine Demonstration Program*, Final Report DARPA/Navy Contract N00024-76-C-5352, May 1982.

[12] Evans, A. G. and Wiederhorn, S. M., "Crack Propagation and Failure Prediction in Silicon Nitride at Elevated Temperatures," J. Mater. Sci., 9, 270–278, 1974.

[13] Quinn, G. D., *Characterization of Turbine Ceramics after Long-Term Environmental Exposure*, AMMRC-TR-80-15, Final Report, April 1980.

[14] Chuck, L., Goodrich, S. M., Hecht, N. L., and McCullum, D. E., "High-Temperature Tensile Strength and Tensile Stress Rupture Behavior of Norton/TRW NT-154 Silicon Nitride," Ceram. Eng. and Sci. Proc., 11[7–8],1007–1027, 1990.

[15] Ferber, M. K., Jenkins, M. G., and Tennery, V. J., "Comparison of Tension, Compression, and Flexure Creep for Alumina and Silicon Nitride Ceramics," Ceram. Eng. and Sci. Proc., 11[7–8],1028–1045, 1990.

[16] Liu, K. C. and Brinkman, C. R., "High Temperature Tensile Strength and Fatigue of Silicon Nitride," *Proc. 1989 Automotive Tech. Devt. Contractors Coord. Meeting*, Dearborn, Michigan, Oct. 23–26, 1989.

[17] Quinn, G. D. and Wiederhorn, S. M., "Structural Reliability of Ceramics at Elevated Temperatures," *ASTM Symposium on Life Prediction Methodologies for Ceramic Materials in Advanced Applications—A Basis for Standards*, Cocoa Beach, Florida, January 11–13, 1993; Cer. Eng. and Sci. Proc., 14[7–8], 305, 1993.

[18] Richerson, D. W., Finger, D. G., and Wimmer, J. M., "Analytical and Experimental Evaluation of Biaxial Contact Stress," in *Fracture Mechanics of Ceramics, Vol. 5, Surface Flaws, Statistics and Microcracking*, R. C. Bradt, A. G. Evans, D. P. H. Hasselman, and F. F. Lange, eds., Plenum Press, New York, NY, USA, 1983. pp. 163–184.

[19] Khandelwal, P. and Heitman, P., "Ceramic Gas Turbine Development at Rolls-Royce in Indianapolis," Chapter 5 in ref. [1].

[20] Schenk, B., Easley, M. L., and Richerson, D. W., "Evolution of Ceramic Turbine Engine Technology at Honeywell Engines, Systems & Services," Chapter 4 in ref. [1].

[21] Gyekenyesi, J. P., "SCARE: a Post Processor Program to MSC/NASTRAN for Reliability Analysis of Structural Ceramic Components," Transactions of the ASME, Journal of Engineering for Gas Turbines and Power, 108[3], 540–546, 1986.

[22] Nemeth, N. N., Powers, L. M., and Janosik, L. A., "Time-dependent Reliability Analysis of Monolithic Ceramic Components Using the CARES/LIFE Integrated Design Program, *ASTM Symposium on Life Prediction Methodologies for Ceramic Materials in Advanced Applications—A Basis for Standards*, Cocoa Beach, FL, Jan. 11–13, 1993; Cer. Eng. and Sci. Proc., 14[7–8], 318–319, 1993.

[23] Comfort, A. M. and Cuccio, J. S., "Life Prediction Methodology for Ceramic components of Advanced Heat Engines," in *Proc. 27th Automotive Tech. Devt. Contractors' Coord. Meeting*, Dearborn, MI, USA, Oct. 23–26, 1989, SAE P-230, Society of Automotive Engineers, Warrenville, PA, USA, 1990. pp. 275–284.

[24] Schenk, B., Brehm, P. G., Menon, M. N., Tucker, W. T., and Peralta, A. D., "Status of the CERAMIC/ERICA Probabilistic Life Prediction Codes Development for Structural Ceramic Applications," ASME paper 99-GT-318, presented at the International Gas Turbine and Aeroengine Congress and Exposition, Indianapolis, IN, USA, June 7–10, 1999.

[25] Schenk, B., Brehm, P. G., Menon, M. N., Tucker, W. T., and Peralta, A. D., "A New Probabilistic Approach for Accurate Fatigue Data Analysis of Ceramic Materials," Transactions of the ASME, Journal of Engineering for Gas Turbines and Power, 122[4], 637–645, 2000.

[26] Bornemisza, T and Saith, A., "SPSLIFE: A User Friendly Approach to the Structural Design and Life Assessment of Ceramic Components," ASME paper 94-GT-486, presented at the International Gas Turbine and Aeroengine Congress and Exposition, The Hague, The Netherlands, June 13–16, 1994. Transactions of the ASME, Journal of Engineering for Gas Turbines and Power, 118[1], 179–183, 1996.

[27] Bornemisza, T. and Jones, A., "Ceramic Gas Turbine Development at Hamilton Sundstrand Power Systems," Chapter 6 in ref. [1].

[28] Teramae, T., Hamada, S., and Hamanaka, J., "Probabilistic Structural analyses of Ceramic Gas Turbine Components," *IUTAM Symposium on Probabilistic Structural Mechanics—Advances in Structural Reliability Methods*, 1994. pp. 518–533.

[29] Furuse, Y., Teramae, T., Tsuchiya, T., Maeda, F., Tsukuda, Y., and Wada, K., "Application of Ceramics to a Power Generating Gas Turbine," Chapter 17 in ref. [1].

[30] Stürmer, G., Schultz, A., and Wittig, S., "Life Time Prediction for Ceramic Gas Turbine Components," ASME paper 91-GT-96, presented at the 36th ASME International Gas Turbine & Aeroengine Congress and Exposition, Orlando, FL, USA, June 3–6, 1991.

[31] Schulz, A., "New Design Principles for Ceramic Hot Gas Components of Gas Turbines," Chapter 25 in ref. [1].

[32] Duffy, S. F., Janosik, L. A., Schenk, B., Suzuki, A., Wereszczak, A. A., Lamon, J., and Thomas, D. J., "Life Prediction of Structural Components," Chapter 25 in this volume.

[33] Kossowsky, R., "Creep and Fatigue of Silicon Nitride as Related to Microstructure," in *Ceramics for High Performance Applications*, J. J. Burke, A. E. Gorum, and R. N. Katz, eds., Brook Hill Publishing Company, Chestnut Hill, MA, USA,1974. pp. 347–372

[34] Williams, R. M. and Uy, J. C., "Ceramic Material Characterization," in *Ceramics for High Performance Applications-II*, J. J. Burke, E. N. Lenoe, and R. N. Katz, eds., Brook Hill Publishing Company, Chestnut Hill, MA, USA,1978. pp. 151–176.

[35] Richerson, D. W., "Materials Characterization Program Overview," in *Proc. 1977 DARPA/NAVSEA Ceramic Gas Turbine Demonstration Engine Program Review*, J. W. Fairbanks and R. W. Rice, eds., MCIC-78-36, Battelle, Columbus, OH, USA,1978. pp. 187–191

[36] Larsen, D. C. and Adams, J. W., *Property Screening and Evaluation of Ceramic Turbine Materials*, Final Technical Report AFWAL-TR-83-4141, April 1984.

[37] Thümmler, F. and Grathwolh, G., "Creep of Ceramic Materials for Gas Turbine Application," in *Mechanical Properties of Ceramics for High Temperature Applications*, AGARD AD-A034 262, December 1976. pp. 1–26.

[38] Godfrey, D. J. and Lindley, M. W., "The Strength of Reaction-Bonded Silicon Nitride Ceramics," Proc. Brit. Ceram. Soc., 22, 229–252, 1973.

[39] *Progress in Nitrogen Ceramics*, Riley, F. L., ed., Martinus Nijhoff Pub., The Hague, The Netherlands, 1983.

[40] Davidge, R., Tappin, G., and McLaren, J., "Strength Parameters Relevant to Engineering Applications for Reaction bonded Silicon Nitride and REFEL Silicon Carbide," Powder Metall. Int., 8[3], 110–114, 1976.

[41] Quinn, G. D., "Review of Static Fatigue in Silicon Nitride and Silicon Carbide," Ceram. Eng. and Sci. Proc., 3[1–2], 77–98, 1982.

[42] Torti, M. L., Alliegro, R. A., Richerson, D. W., Washburn, M. E., and Weaver, G. Q., "High Temperature Properties of Silicon Carbide and Silicon Nitride," *in High Temperature Materials in Gas Turbines*, P. R. Sahn and M. O. Speidel, eds., Elsevier Technical Publishing Co., Amsterdam, The Netherlands, 1974. pp. 177–186.

[43] Weaver, G. Q. and Lucek, J.W., "Optimization of Hot-Pressed Si_3N_4–Y_2O_3 Materials," Am. Ceram. Soc. Bull., 57[12], 1131–1134, 1136,1978.

[44] Wills, R. R., Brockway, M. C., and Bansal, G. K., "Relationship Between Densification and High Temperature Mechanical Properties of HIP'ed Silicon Nitride," in *Ceramic Components for Engines*, S. Sōmiya, E. Kanai, and K. Ando, eds., Elsevier Applied Science, London, UK, 1984. pp. 321–332.

[45] Bunk, W., Gugel, E., and Walzer, P., "Overview of the German Ceramic Gas Turbine Program," in *Ceramics for High Performance Applications III Reliability*, E.M. Lenoe, R.N. Katz, and J.J. Burke, eds., Plenum Press, New York, NY, USA,1983. pp. 29–50.

[46] Oda, I., Matsue, M., Soma, T., Masuda, M., and Yamada, N., "Fracture Behavior of Sintered Silicon Nitride under Multiaxial Stress States," J. Ceram. Soc. Jpn. Inter. Ed., 96, 523–530, 1988.

[47] Johnson, D. R., Schaffhauser, A. C., Tennery, V. J., Long, E. L., and Schulz, R. B., "Ceramic Technology for Advanced Heat Engines Project," Am. Ceram. Soc. Bull., 64[2], 276–281,1985.

[48] Carruthers, W. D., Richerson, D. W., and Benn, K. W., "Combustion Rig Durability Testing of Turbine Ceramics," in *Ceramics for High Performance Applications III Reliability*, E. M. Lenoe, R. N. Katz, and J. J. Burke, eds., Plenum Press, New York, NY,USA, 1983. pp. 571–595.

[49] Schienle, J. L., *Durability Testing of Commercial Ceramic Materials*, Final Report DOE/NASA/0027-1, NASA CR-198497, Jan. 1996.

[50] Jacobson, N. S., Fox, D. S., Smialek, J. L., Opila, E. J., Tortorelli, P. F., More, K. L., Nickel, K. G., Hirata, T., Yoshida, M., and Yuri, I., "Corrosion Issues for Ceramics in Gas Turbines," Chapter 26 in this volume.

[51] Jenkins, M. G., Kagawa, Y, and Steen, M., "Test Standards and Design Codes for Structural Ceramics and Ceramic Matrix Composites," Chapter 24 in this volume.

[52] Holowczak, J., Jarmon, D. C., Mendelson, M. I., and Linsey, G. D., "United Technologies Pratt & Whitney/United Technologies Research Center Experiences with Ceramics and CMCs in Turbine Engines," Chapter 10 in ref. [1].

[53] Brennan, J. J. and Prewo, K. M., "Silicon Carbide Fiber Reinforced Glass-Ceramic Matrix Composites Exhibiting High Strength and Toughness," J. Mater. Sci., 17, 7371–7383, 1982.

[54] Lamicq, P. J. and Jamet, J. F., "Thermostructural CMCs: an Overview of the French Experience," in *High-Temperature Ceramic-Matrix Composites I: Design, Durability, and Performance*, A. G. Evans and R. Naslain, eds., American Ceramic Society, Westerville, OH, USA,1995. pp. 1–11.

[55] Lara-Curzio, E. "Properties of Continuous Fiber-Reinforced Ceramic Matrix Composites for Gas Turbine Applications," Chapter 22 in this volume.

[56] Collins, J. G. and Gerby, R. W., "New Refractory Uses for Silicon Nitride Reported," J. Metals, 7, 612–615, 1955.

[57] Parr, N. L., "Silicon Nitride, a New Ceramic for High Temperature Engineering and Other Applications," Res. (Lon.) 13, 261–269, 1960.

[58] Popper, P. and Ruddlesden, S. N., "The Preparation, Properties and Structure of Silicon Nitride," Trans. Brit. Ceram. Soc., 60, 603–626, 1961.

[59] Parr, N. L., Sands, R., Pratt, P.L., May, E. R. W., Shakespeare, C. R., and Thompson, D. S., "Structural Aspects of Silicon Nitride," Powder Met., 8, 152–163, 1961.

[60] Parr, N. L., Martin, G. F., and May, E. R. W., "Preparation, Microstructure, and Mechanical Properties of Silicon Nitride," in *Special Ceramics*, P. Popper, ed., Academic Press, New York, NY, USA, 1960. pp. 102–135.

[61] Pratt, P. L., "The Microstructure and Mechanical Properties of Silicon Nitride," in *Mechanical Properties of Engineering Ceramics*, W. W. Kriegel and H. Palmour, eds. III, Interscience Publishers, New York, NY, USA, 1961. pp 507–519.

[62] Noakes, P. B. and Pratt, P. L., "High-Temperature Mechanical Properties of Reaction-Sintered Silicon Nitride," in *Special Ceramics, Vol. 5, Proceedings of the 5th International Symposium on Special Ceramics*, BCRA, Stoke-on-Trent, England, UK, 14–16 July 1970, P. Popper, ed., British Ceramic Research Association, Manchester, England, UK, 1972. pp. 299–310.

[63] Cutler, I. B. and Croft, W. J., "Powder Met. Review 7: Silicon Nitride (Part 1)," Powder Met. Int. 6, 92–96, 1974; "(Part 2)," Powder Met. Int., 6, 144–146,1974.

[64] Davidge, R. W., Evans, A. G., Gilling, D., and Wilyman, P. R., "Oxidation of Reaction-Sintered Silicon Nitride and Effects on Strength," in *Special Ceramics, Vol. 5, Proc. 5th Int. Symp. on Special Ceramics*, BCRA, Stoke-on-Trent, England, UK, 14–16 July 1970, P. Popper, ed., British Ceramic Research Association, Manchester, England, UK, 1972. pp 329–344.

[65] "Nitrogen Ceramics," F. L. Riley, ed., Noordhoff Int. Publishing, Leyden, The Netherlands, 1977.

[66] Davidge, R. W., "Mechanical Properties of Reaction Bonded Silicon Nitride," in *Nitrogen Ceramics*, F. L. Riley, ed., Noordhoff Int. Publishing, Leyden, The Netherlands, 1977. pp 541–559.

[67] Mangels, J. A., "Development of Injection Molded Reaction Bonded Si_3N_4," in *Ceramics for High Performance Applications II*, Proceedings of the Fifth Army Materials Conference, Newport, RI, USA, March 21–25, 1977, J. J. Burke, E. N. Lenoe, and R. N. Katz, eds., Brook Hill Publishing Company, Chestnut Hill, MA, USA, 1978. pp. 113–130.

[68] Washburn, M. E. and Baumgartner, H. R., "High-Temperature Properties of Reaction-Bonded Silicon Nitride," in *Ceramics for High Performance Applications*, Proceedings of the Second Army Materials Conference, Hyannis, MA, November 13–16, 1973, J. J. Burke, A. E. Gorum, and R. N. Katz, eds., Brook Hill Publishing Company, Chestnut Hill, MA, USA, 1974. pp. 479–491.

[69] Gersch, H., Mann, D., Rorabaugh, M., and Schuldies, J. J., "Slip-Cast and Injection-Molding Process Development of Reaction-Bonded Silicon Nitride at AiResearch Casting Company," in *Proc. DARPA/NAVSEA Ceramic Gas Turbine Demonstration Engine Program Review*, J. W. Fairbanks and R. W. Rice, eds., Battelle Columbus Laboratories, Columbus, OH, USA, Report No. MCIC-78-36, March 1978. pp. 313–340.

[70] Richerson, D. W., Smyth, J. R., and Styhr, K., "Material Improvement Through Iterative Process Development," Cer. Eng. and Sci. Proc., 4 [9–10], 841–852, 1983.

[71] Andersen, J. C., "Method of Making Refractory Body," U.S. Patent 2,938,807, May 31, 1960.

[72] Taylor, K. M., "Cold Molded Dense Silicon Carbide Articles and Method of Making the Same," U.S. Patent 3,205,043, Sept. 7, 1965.

[73] Fredricksson, J. I., "Process of Making Recrystallized Silicon Carbide Articles," U.S. Patent 2,964,823, Dec. 20, 1960.

[74] Forrest, C. W., "Manufacture of Dense Bodies of Silicon Carbide (Refel)," U.S. Patent 3,495,939, Feb. 17, 1970.

[75] Popper, P., US Patent 3,275,722, "Production of Dense Bodies of Silicon Carbide," Sept. 27, 1966.

[76] Sawyer, G. R. and Page, T. F., "Microstructural Characterization of 'Refel' (Reaction–Bonded) Silicon Carbides," J. Mater. Sci., 13, 885–904, 1978.

[77] Alliegro, R. A. and Coes, S. H., "Reaction bonded Silicon Carbide and Silicon Nitride for Gas Turbine Applications," ASME paper 72-GT-20, presented at the Gas Turbine Conference, San Franciso, CA, USA, March 26–30, 1972.

[78] Lucek, J. W., Torti, M.L., Weaver, G. Q., and Olson, B. A., "Cast Densified Silicon Carbides," SAE Paper 790253, 1979.

[79] Whalen, T. J., Noakes, J. E., and Terner, L. L., "Progress on Injection-Molded, Reaction-Bonded SiC," in *Ceramics for High Performance Applications II*, Proceedings of the Fifth Army Materials Conference, Newport, RI, USA, March 21–25, 1977, J. J. Burke, E. N. Lenoe, and R. N. Katz, eds., Brook Hill Publishing Company, Chestnut Hill, MA, USA, 1978. pp. 179–189.

[80] Hucke, E.E., *Process Development for Silicon Carbide Based Structural Ceramics*, AMMRC TR 82–10, February 1982.

[81] Weaver, G. Q. and Olson, B. A., "Process for Fabricating Silicon Carbide Articles," U.S. Patent 4,019,913, April 26, 1977.

[82] Weaver, G. Q. and Logan, J. C., "Process for Forming Dense Silicon Carbide Bodies," U.S. Patent 4,127,629, Nov. 28, 1978.

[83] Deely, G. G., Herbert, J. M., and Moore, N. C., "Dense Silicon Nitride," Powder Met., 8, 145–151, 1961.

[84] Lumby, R. J. and Coe, R. F., "The Influence of Some Process Variables on the Mechanical Properties of Hot-Pressed Silicon Nitride," Proc. Brit. Ceram. Soc., 15, 91–101, 1970.

[85] Wild, S., Grieveson, P., Jack, K. H., and Latimer, M. J., "The role of Magnesia in Hot-Pressed Silicon Nitride," in *Special Ceramics, Vol. 5, Proc. 5th Int. Symp. on Special Ceramics*, BCRA, Stoke-on-Trent, England, 14–16 July 1970, P. Popper, ed., British Ceramic Research Association, Manchester, England, UK, 1972. pp. 377–384.

[86] Wild, S., Grieveson, P., and Jack, K. H., "The Crystal Chemistry of New Metal-Silicon-Nitrogen Ceramic Phases," ibid., pp. 289–298.

[87] Bowen, L. J., Weston, R. J., Carruthers, T. G., and Brook, R. J., "Mechanisms of Densification During the Pressure Sintering of Alpha-Silicon Nitride," Ceramurgia Int., 2[4], 173–176, 1976.

[88] Richerson, D. W., "Effects of Impurities on the High Temperature Properties of Hot-Pressed Silicon Nitride," Bull. Am. Ceram. Soc., 52 [7], 560–62, 1973.

[89] Richerson, D. W. and Washburn, M. E., "Hot Pressed Silicon Nitride," U.S. Patent 3,836,374, Sept. 17, 1974.

[90] Gazza, G. E., "Hot-Pressed Silicon Nitride," J. Am. Ceram. Soc., 56[12], 662, 1973.

[91] van Roode, M., "The German National Program for Ceramic Materials and Component Development," Chapter 9 in this volume.

[92] Tsuge, A. and Nishida, K., "High Strength Hot Pressed Silicon Nitride with Concurrent Y_2O_3 and Al_2O_3 Additions," Amer. Ceram. Soc. Bull. 57[4], 424–426, 431, 1978.

[93] Tsuge, A., Nishida, K., and Komatsu, M., "Effect of Crystallizing the Grain-Boundary Glass Phase on the High-Temperature Strength of Hot-Pressed Si_3N_4 Containing Y_2O_3," J. Am. Ceram. Soc., 58, 323–326, 1975.

[94] Wötting, G. and Ziegler, G., "Characterization of Microstructural Variables of Sintered, Hot-Pressed and Hot-Isostatically Pressed Si_3N_4 and their Correlation with Mechanical Properties," Science of Ceramics, 12, 361–370, 1984.

[95] Baumgartner, H. R. and Richerson, D. W., "Inclusion Effects on the Strength of Hot-Pressed Si_3N_4," in *Fracture Mechanics of Ceramics*, Vol. 1, R. C. Bradt, D. P. H. Hasselman, and F. F. Lange, eds.,Plenum Press, New York, NY, USA, 1974. pp. 367–386.

[96] Mitomo, M., "Pressure Sintering of Silicon Nitride," J. Mater. Sci., 11,1103–1107, 1976.

[97] Osipova, I. I. and Pogorelova, D. A., "Recrystallization of Silicon Nitride During Hot Pressing," Sov. Powder Met. Metal Ceram., 14, 1015–1017,1975.

[98] Iskoe, J. L., Lange, F. F., and Diaz, E. S., "Effect of Selected Impurities on the High Temperature Mechanical Properties of Hot-Pressed Silicon Nitride," J. Mater. Sci., 11, 908–912, 1976.

[99] Colquhoun, I., Thompson, D. P., Wilson, W. I., Grieveson, P., and Jack, K. H., "The Determination of Surface Silica and Its Effect on the Hot-Pressing Behavior of Alpha-Silicon Nitride Powder," Proc. Brit. Ceram. Soc., 22, 181–195,1973.

[100] Knoch, H. and Ziegler, G., "Influence of MgO Content and Temperature on Transformation Kinetics, Grain Structure and Mechanical Properties of Hot-Pressed Silicon Nitride," Science of Ceramics 9, 494–501,1977.

[101] Weston, R. J. and Carruthers, T. G., "Kinetics of Hot-Pressing of Alpha-Silicon Nitride Powder with Additives, " Proc. Br. Ceram. Soc., 22, 197–206,1973.

[102] Fickel, A. F. and Kessel, H., "Technology and Applications of Dense Silicon Nitride," Glas. Email. Keram. Tech., 25, 193–196, 1974.

[103] Gugel, E., Fickel, A. F., and Kessel, H., "Developments in the Production of Hot-Pressed Silicon Nitride," Powder Met. Int. 6, 136–140, 1974.

[104] Oyama, Y. and Kamigaito, O., "Sintered Silicon Nitride—Magnesium Oxide System," Yogyo Kyokai Shi, 81, 290–293,1973.

[105] Bowen, L. J., Carruthers, T. G., and Brook, R. J., "Hot-Pressing of Si_3N_4 with Y_2O_3 and Li_2O Additives," J. Amer. Ceram. Soc., 61 [7–8], 335–339, 1978.

[106] Tsuge, A., Kudo, H., and Komeya, K., "Reaction of Silicon Nitride and Yttrium Oxide in Hot-Pressing," J. Amer. Ceram. Soc., 57 [6], 269–270, 1974.

[107] Rice, R. W. and McDonough, W. J., "Hot Pressed Silicon Nitride with Zr Based Additives," J. Amer. Ceram. Soc., 58 [5–6], 264,1975.

[108] Huseby, I. C. and Petzow, G., "Influence of Various Densifying Additives on Hot-Pressed Silicon Nitride," Powder Met. Int., 6, 17–19, 1974.

[109] Weston, J. E., Pratt, P. L., and Steele, B. C. H., " Crystallization of Grain boundary Phases in Hot-Pressed Silicon Nitride Materials, Parts 1 and 2," J. Mat. Sci., 13, 2137–2156,1978.

[110] Lange, F. F., "Relation Between Strength, Fracture Energy and Microstructure of Hot-Pressed Silicon Nitride," J. Am. Ceram. Soc., 56[10], 518–522, 1973.

[111] Larker, H. T., "Hot Isostatic Pressing of Ceramic Turbine Components to Shape—Experience at Asea High Pressure Laboratory, Asea Cerama AB, ABB Cerama AB and AC Cerama AB," Chapter 10 in this volume.

[112] Larker, H., Adlerborn, J., and Bohman, H., "Fabricating of Dense Silicon Nitride Parts by Hot Isostatic Pressing," SAE Paper 770335, 1977.

[113] Wills, R., Brockway, M. C., McCoy, L., and Niesz, D., "Preliminary Observations on the Hot Isostatic Pressing of Silicon Nitride," Ceram. Eng. and Sci. Proc., 1[7–8B], 534–539, 1980.

[114] Richerson, D. W. and Wimmer, J.M., "Properties of Isostatically Hot-Pressed Silicon Nitride," J. Am. Ceram. Soc., 66[9], C-173–C-176, 1983.

[115] Heinrich, J. and Böhmer, M., Science of Ceramics, 11, 439–446, 1981.

[116] Moritoki, M., "HIP Activities in Japan," Metal Powder Report, 38[7], 393–396, 1983.

[117] Ishii, T., Tsuzuki, H., and Inoue, Y., "Modular-type, Hot-loading-type and Compact Hot Isostatic Pressing Systems," *Proc. 2nd Int. Conf. On Isostatic Pressing*, Vol. 2, 1982. pp. 20.1–20.24

[118] Fujikawa, T., Moritoki, M., and Homma, K., "The High Temperature HIP Furnace and Its Application in Densification of Ceramics," ibid., Vol. 1, 1982. pp.12.1–12.25.

[119] Homma, K., Tatuno, T, Okada, H., Kawai, N., and Nishihara, M., "HIP Treatment of Sintered Silicon Nitride," Zairyo-Journal of the Soc. of Mater. Sci. Japan, 30, 1005–1011, 1981 (in Japanese with English summary).

[120] Larker, H. T., "On Hot Isostatic Pressing of Shaped Ceramic Parts," in *Ceramic Components for Engines*, S. Sōmiya, E. Kanai and K. Ando, eds., Elsevier Science Publishing, New York, NY, USA, 1986. pp. 304–310.

[121] Fujikawa, T., Moritoki, M., Kando, T., Homma, K., and Okada, H., "Hot Isostatic Pressing: Its Application in High Performance Ceramics," ibid., pp. 425–433.

[122] Hirota, K., Ichikizaki, T., Hasegawa, Y. T., and Suzuki, H., "Densification of Silicon Nitride by Hot Isostatic Press in N_2 Atmosphere," ibid, pp. 434–441.

[123] Ziegler, G. and Wötting, G., "Post-treatment of Pre-sintered Silicon Nitride by Hot Isostatic Pressing," Int. J. High Tech. Ceram., 1, 31–58, 1985.

[124] Richerson, D. W., Bright, E., and Licht, R., "Ceramic Component Development at Saint-Gobain/Norton Advanced Ceramics," Chapter 3 in this volume.

[125] Pollinger, J. P., "The Development of Monolithic Silicon Nitride Structural Ceramics at Honeywell," Chapter 5 in this volume.

[126] McEntire, B. J., Taglialavore, A. P., Heichel, D. N., Johnson, J. W., Bright, E., and Yeckley, R. L., "ATTAP Ceramic Component Development Using Taguchi Methods," *Proc. 26th Auto. Tech. Dev't Contractors Coord. Meeting*, Dearborn, MI, USA, Oct. 24–27, 1988, SAE Publication P-219, April 1989. pp. 289–300.

[127] Hengst, R. R., Heichel, D. N., Holowczak, J. E., Taglialavore, A. P., and McEntire, B. J., "A Comparison of Forming Technologies for Ceramic Gas-Turbine Engine Components," ASME paper 90-GT-184, presented at the Gas Turbine and Aeroengine Congress and Exposition, Brussels, Belgium, June 11–14, 1990.

[128] Pujari, V. K., Amin, K. E., and Tewari, P. H., "Development of Improved Processing and Evaluation of Silicon Nitride," ASME paper 91-GT-317, presented at the Int. Gas Turbine and Aeroengines Congress and Exposition, Orlando, FL, USA, June 3–6, 1991.

[129] Pujari, V. K. and Tracey, D. M., "Processing Methods for High Reliability Silicon Nitride Heat Engine Components," ASME paper 93-GT-319, presented at the Int. Gas Turbine and Aeroengine Congress and Exposition, Cincinnati, OH, USA, May 24–27, 1993.

[130] Terwilliger, G. R., "Properties of Sintered Silicon Nitride," J. Am. Ceram. Soc., 57[1], 48–49, 1974.

[131] Mitomo, M., Tsutsumi, M., Bannai, E., and Tanaka, T., "Sintering of Silicon Nitride," Am. Ceram. Soc. Bull., 55[3], 313, 1976.

[132] Masaki, H. and Kamigaito, O., "Pressureless Sintering of Silicon Nitride with Additions of MgO, Alumina, and/or Spinel," Yogyo Kyokai Zasshi (J. Ceram. Soc. Japan), 84[10], 508–512, 1976 (in Japanese).

[133] Oda, I., Kaneno, M., and Yamamoto, N., "Pressureless Sintered Silicon Nitride, in *Progress in Nitrogen Ceramics, Proc. NATO-ASI Nitrogen Ceramics Conf.*, Canterbury, England, UK, Aug. 16–27, 1976. pp. 359–365.

[134] Greskovich, C. and Rosolowski, J. H., "Sintering of Covalent Solids," J. Am. Ceram. Soc., 59[7–8], 336–343, 1976.

[135] Priest, H. F., Priest, G. L., and Gazza, G. E., "Sintering of Si_3N_4 under High Nitrogen Pressure," J. Am. Ceram. Soc., 60[1–2], 81, 1977.

[136] Rowcliffe, D. J. and Jorgensen, P. J., "Sintering of Silicon Nitride, in *Proc. Workshop on Ceramics for Heat Engines*, Energy Res. & Dev. Adm., CONRT, Orlando, FL, USA, Jan. 24–26, 1977. pp. 191–196.

[137] Tani, E., Umebayashi, S., Kishi, K., Kobayashi, K., and Nishijima, M., "Gas-Pressure Sintering of Silicon Nitride with Concurrent Addition of Alumina and 5 wt.% of Rare Earth Oxide: High Fracture Toughness Silicon Nitride with Fiber-like Structure," Am. Ceram. Soc. Bull., 65[9], 1311–1315, 1986.

[138] Matsuhiro, K. and Takahashi, T., "The Effect of Grain Size on the Toughness of Sintered Silicon Nitride," Ceram. Eng. and Sci. Proc., 10[7–8], 807–816, 1989.

[139] Li, C. W. and Yamanis, J., "Super-Tough Silicon Nitride with R-Curve Behavior," Ceram. Eng. and Sci. Proc., 10[7–8], 632–645, 1989.

[140] Smith, J. T. and Quackenbush, C. L., "A Study of Sintered Silicon Nitride Compositions with Yttria and Alumina Densification Additives," in *Proc. Int. Symp. on Factors in Densification and Sintering of Oxide and Nonoxide Ceramics*, Hakone, Japan, Oct. 3–6, 1978, S. Sōmiya and S. Sayto, eds., Association of Science Documents Information, Tokyo Institute of Technology, Japan, 1979. pp. 426–443.

[141] Mangels J. A., "Sintered Reaction-Bonded Silicon Nitride," Ceram. Eng. and Sci. Proc., 2 [7–8], 589–593, 1981.

[142] Giachello, A. and Popper, P., "Post-Sintering of Reaction-Bonded Silicon Nitride," Ceramurgia Int., 5 [3],110, 1979.

[143] Takahashi, T., "Sintering Character of Yttria-Magnesia-Zirconia Doped and Yttria-Ytterbia Doped Silicon Nitrides," *Proc. of the Symposium on the Sintering of Advanced Ceramics*, May 3, 1988, American Ceramic Society, Westerville, OH, USA, 1989.

[144] Hampshire, S., *The Sintering of Nitrogen Ceramics*, Parthenon Press, Lancashire, England, UK, 1986.

[145] Lumby, R. J., North, B., and Taylor, A. J., "Properties of Sintered Sialons," in *Ceramics for High Performance Applications II*, J. J. Burke, E. N. Lenoe, and R. N. Katz, eds., Brook Hill Publishing Company, Chestnut Hill, MA, USA, 1978. pp. 893–906.

[146] Lumby, R. J., "The Preparation, Structure and Properties of Commercial Sialon Ceramic Materials," Ceram. Eng. and Sci. Proc., 3[1–2], 50–66, 1982.

[147] Alliegro, R. A., Coffin, L. B., and Tinklepaugh, J. R., "Pressure-Sintered Silicon Carbide," J. Am. Ceram. Soc., 39[11], 386–389, 1956.

[148] Marshall, R. C., Faust, J. W., and Ryan, C. E., "Silicon Carbide—1973," Univ. of South Carolina Press, Columbia, SC, 1974.

[149] Lange, F. F., "Hot-Pressing Behavior of Silicon Carbide Powders with Additions of Aluminum Oxide," J. Mater. Sci., 10, 314–320, 1975.

[150] Prochazka, S., "Dense Polycrystalline Silicon Carbide," U.S. Patent 3,954,483, May 4, 1976.

[151] Prochazka, S., "Silicon Carbide Sintered Body; Boron Additive," U.S. Patent 4,041,117, Aug. 8, 1977.

[152] Coppola, J. A. and Smoak, R. H., "Method of Producing High Density Silicon Carbide Product; Sintering Silicon Carbide with Boron or Boron Compounds Present as Densification Aids," U.S. Patent 4,080,415, March 21, 1978.

[153] Stroke, F., "Submicron Beta Silicon Carbide Powder and Sintered Articles of High Density Prepared Therefrom," U.S. Patent 4,133,689, Jan. 9, 1979.

[154] Lipp, A. and Schwetz, K. A., "Dense Sintered Shaped Articles of Polycrystalline Alpha-Silicon Carbide and Process for Their Manufacture," U.S. Patent 4,230,497, Oct. 28, 1980.

[155] Coppola, J. A., Hailey, L. N., and McMurtry, C. N., "Process for Producing Sintered Silicon Carbide Ceramic Body," U.S. Patent 4,124,667, Nov. 7, 1978.

[156] Boecker, W., Landfermann, H., and Hausner, H., "Sintering of Alpha Silicon Carbide with Additions of Aluminum," Powder Met Int., 11[2], 83–85, 1979.

[157] Tanaka, H. and Inomata, Y., "Normal Sintering of Al-Doped Beta SiC," J. Mater. Sci. Letters, 4, 315–317, 1985.

[158] Omori, M. and Takei, H., "Pressureless Sintering of SiC," J. Am. Ceram. Soc., 65 [6], C-92, 1982.

[159] Omori, M. and Takei, H., "Method for Preparing Sintered Shapes of Silicon Carbide," U.S. Patent 4,564,490, Jan. 14, 1986.

[160] Negita, K., "Effective Sintering Aids for Silicon Carbide Ceramics: Reactivities of Silicon Carbide with Various Additives," J. Am. Ceram. Soc., 69[12] C-308–C-310, 1986.

[161] Prochazka, S., "The Role of Boron and Carbon in the Sintering of Silicon Carbide," Special Ceramics 6, 171–181, 1974.

[162] Schwetz, K. A. and Lipp, A., "The Effect of Boron and Aluminum Sintering Additives on the Properties of Dense Sintered Alpha Silicon Carbide," Science of Ceramics, 10, 149–158, 1980.

[163] Johnson, D. D. and Sowman, H. G., "Ceramic Fibers," in *Composites, Engineered Materials Handbook,* Vol 1, ASM Int., Metals Park, OH, USA, 1987. pp. 60–65.

[164] Vasilos, T., "Structural Ceramic Composites," ibid, pp. 925–932.

[165] Richerson, D. W., "Ceramic Matrix Composites," Chapter 19 in *Composites Engineering Handbook*, P. K. Mallick, ed., Marcel Dekker, New York, NY, USA, 1997. pp. 983–1038.

[166] Belitskus, D., *Fiber and Whisker Reinforced Ceramics for Structural Applications*, Marcel Dekker, New York, NY, USA, 1993.

[167] *High Temperature / High Performance Composites*, F. D. Lemkey, S. G. Fishman, A. G. Evans, and J. R. Strife, eds., MRS Proc., Vol. 120, Mater. Res. Soc., Pittsburgh, PA, USA, 1988.

[168] *High-Temperature Ceramic-Matrix Composites, I and II*, A.G. Evans, and R. Naslain, eds., American Ceramic Society, Westerville, OH, USA, 1995.

[169] Landini, D. J., Fareed, A. S., Wang, H., Craig, P. A. and Hemstad, S., "Ceramic Matrix Composites Development at GE Power Systems Composites, LLC," Chapter 14 in this volume.

[170] Corman, G. S., Luthra, K. L. and Brun, M. K., "Silicon Melt Infiltrated Ceramic Composites—Processes and Properties," Chapter 16 in this volume.

[171] Hiromatsu, M, Nishide, S., Nishio, K., Igashira, K., Matsubara, G., and Suemitsu, T., "Development of Hot Section Components Composed of Ceramic Matrix Composites (CMC) at the Research Institute of Advanced Materials Gas Generator (AMG)," Chapter 17 in this volume.

[172] Kameda, T., Suyama, S., Itoh, Y., Nishida, K., Ikeda, I., Hijikata, T., and Okamura, T., "Development of Continuous Fiber-Reinforced Reaction Sintered Silicon Carbide Matrix Composites for Gas Turbine Hot Parts Applications," Chapter 18 in this volume.

[173] More, K. L., Tortorelli, P. F., Ferber, M. K., Walker, L. R., Keiser, J. R., Brentnall, W. D., Miriyala, N., and Price, J. R., "Exposure of Ceramics and Ceramic Matrix Composites in Simulated and Actual Combustor Environments," ASME paper 99-GT-292, presented at the International Gas Turbine and Aeroengine Congress and Exposition, Indianapolis, Indiana, USA, June 7–10, 1999.

[174] Szweda, A., Easler, T. E., Jurf, R. A., and Butner, S. C., "Ceramic Matrix Composites for Gas Turbine Applications," Chapter 15 in this volume.

[175] Johnson, C. R. and Mohr, T. G., "Injection Molding 2.7 g/cc Silicon Nitride Turbine Rotor Blade Rings Utilizing Automatic Control," *in Ceramics for High Performance Applications-II*, J. J. Burke, E. N. Lenoe, and R. N. Katz, eds., Brook Hill Publishing Company, Chestnut Hill, MA, USA, 1978. pp. 193–206.

[176] Ezis, A., "The Fabrication and Properties of Slip-Cast Silicon Nitride," in *Ceramics for High Performance Applications*, J. J. Burke, A. E. Gorum, and R. N. Katz, eds., Brook Hill Publishing Company, Chestnut Hill, MA, 1974. pp. 207–222.

[177] Torti, M. L., Baker, S. H., Tuffs, S, and Olson, B. A., "Norton Process Optimization Studies for Hot Pressed Silicon Nitride Blades and Reaction Sintered Silicon Nitride Stator and Static Components," in *Proc. 1977 DARPA/NAVSEA Ceramic Gas Turbine Demonstration Engine Program Review*, J. W. Fairbanks and R. W. Rice, eds., MCIC-78-36, Metals and Ceramics Information Center, Battelle Columbus Laboratories, Columbus, OH, USA, 1978. pp. 341–363.

[178] Uy, J. C., Williams, R. M. and Goodyear, M. U., "Processing of Hot Pressed Silicon Nitride," in *Ceramics for High Performance Applications—II*, J. J. Burke, E. N. Lenoe, and R. N. Katz, eds., Brook Hill Publishing Company, Chestnut Hill, MA, USA, 1978. pp. 131–149.

[179] Baker, R. R., Ezis, A., Hartsock, D. L., and Goodyear, M. U., "Developments in Press Bonding of Duo-Density Rotors," ibid., pp. 207–230.

[180] Bunk, W., "Overview on the German Program on Ceramic Componoents for Vehicular Gas Turbines," in *Ceramic Components for Engines*, Proc. 1st Int. Symp. 1983, Japan, S. Sōmiya, E. Kanai, and K. Ando, eds., Elsevier Applied Science, London, 1986. pp. 9–20.

[181] Mörgenthaler, K. D. and Neubrand, F., "Thermal Shock Testing of a Hot-Pressed Silicon Nitride Turbine Rotor," in *Ceramics for High Performance Applications III Reliability*, E. M. Lenoe, R. N. Katz, and J. J. Burke, eds., Plenum Press, New York, NY, USA, 1983. pp. 805–810.

[182] Robare, M. W. and Richerson, D. W., "Rotor Blade Machining Development," in *Proc. 1977 DARPA/NAVSEA Ceramic Gas Turbine Demonstration Engine Program Review*, J.W. Fairbanks and R. W. Rice, eds., MCIC-78-36, Metals and Ceramics Information Center, Battelle Columbus Laboratories, Columbus, Ohio, 1978. pp. 291–311.

[183] McEntire, B. J., Taglialavore, A. P., Heichel, D. N., Bright, E., Yeckley, R. L., Holowczak, J. E., and Quackenbush, C. L., "Silicon Nitride Component Development for Advanced Gas Turbine Engines," in *Proc. 27th Automotive Tech. Devt. Contractors' Coord. Meeting*, Dearborn, MI, USA, Oct. 23–26, 1989, SAE P-230, Society of Automotive Engineers, Warrendale, PA, USA, 1990. pp. 341–356.

[184] Pollinger, J. P. and Busovne, B. J., "Development of Pressure Slip Cast Silicon Nitride Rotors for ATTAP," ibid., pp. 357–364.

[185] Funk, J. E., Lawler, H. A., Ohnsorg, R. W., and Storm, R. S., "ATTAP SiC Ceramic Component Development Year 2," ibid., pp. 319–326.

[186] Neil, J., Bandyopadhyay, G., Sordelet, D., and Mahoney, M., "Fabrication of Silicon Nitride ATTAP Components at GTE Laboratories," ibid., pp. 303–310.

[187] Pasto, A. E. and Natansohn, S., "Advanced Processing of High-Performance Silicon Nitride Ceramics," ibid., pp. 197–206.

[188] McEntire, B. J., Hengst, R. R., Collins, W. T., Taglialavore, A. P., and Yeckley, R. L., "Ceramic Component Processing Development for Advanced Gas Turbine Engines," ASME paper 91-GT-120 presented at the Int. Gas Turbine and Aeroengine Congress and Exposition, Orlando, FL, USA, June 3–6, 1991. Transactions of the ASME, J. Engineering for Gas Turbines and Power, 115[1], 1–8, 1993.

[189] Pollinger, J. P., "Improved Silicon Nitride Materials and Component Fabrication Processes for Aerospace and Industrial Gas Turbine Applications," ASME paper 95-GT-159, presented at the International Gas Turbine and Aeroengine Congress and Exposition, Houston, TX, USA, June 5–8, 1995.

[190] Bright, E., Burleson, R., Dynan, S. A., and Collins, W. T., "NT 164 Silicon Nitride Gas-Turbine Engine Turbine Blade Manufacturing Development," ASME Paper 95-GT-74, presented at the Int. Gas Turbine and Aeroengine Congress and Exposition, Houston, TX, USA, June 5–8, 1995.

[191] Cassidy, D. J., "NDE Techniques Used for Ceramic Turbine Rotors," in *Ceramics for High Performance Applications II*, Proceedings of the Fifth Army Materials Conference, Newport, RI, USA, March 21–25, 1977, J.J. Burke, E. N. Lenoe, and R. N. Katz, eds., Brook Hill Publishing Company, Chestnut Hill, MA, USA, 1978. pp. 231–242,

[192] Schuldies, J. J. and Richerson, D. W., "NDE Approach, Philosophy, and Standards for the DARPA/NAVSEA Ceramic Turbine Program," in *Proc. 1977 DARPA/NAVSEA Ceramic Gas Turbine Demonstration Engine Program Review*, J. W. Fairbanks and R. W. Rice, eds., MCIC Rep. MCIC-78-36, Battelle Laboratories, Columbus, OH, USA, 1978. pp. 381–402.

[193] Schuldies, J. J. and Spaulding, W. H., "Radiography and Image Enhancement of Ceramics," ibid., pp. 403–428.

[194] Schuldies, J. J. and Derkacs, T., "Ultrasonic NDE of Ceramic Components," ibid., pp. 429–448.

[195] Kino, G. S., Khuri-Yakub, B. T., Murakami, Y., and Yu, K. H., "Defect Characterization in Ceramics Using High Frequency Ultrasonics," in *Proceedings of the ARPA/AFML Review of Progress in Quantitative Nondestructive Evaluation,* July 17–21, 1978, Scripps Institute of Oceanography, La Jolla, CA, USA, 1979. pp. 242–245.

[196] Ellingson, W. A. and Vannier, M. W., *X-Ray Computed Tomography for Nondestructive Evaluation of Advanced Structural Ceramics*, Argonne National Laboratory Report ANL-87-52, Argonne National Laboratory, Argonne, IL, USA, 1987.

[197] Ellingson, W. A., Ikeda, Y., and Goebbels, J., "Nondestructive Evaluation/Characterization," Chapter 23 in this volume.

[198] Sun, J. G., Shirber, M., and Ellingson, W. A., "Laser-Based Optical Scattering Detection of Surface and Subsurface Defects in Machined Silicon Nitride Components," Ceram. Eng. and Sci. Proc.,18[4], 273–280, 1997.

[199] Brentnall, W. D., van Roode, M., Norton, P. F., Gates, S., Price, J. R., Jimenez, O., and Miriyala, N, "Ceramic Gas Turbine Development at Solar Turbines Incorporated," Chapter 7 in ref. [1].

[200] Helms, E. H., Lindgren, L. C., Heitman, P. W., and Thrasher, S. R., *Ceramic Applications in Turbine Engines*, Noyes Publications, Park Ridge, NJ, USA, 1986.

[201] Lundberg, R. and Gabrielsson, R., "Ceramic Components for Automotive Gas Turbines—Experience at Volvo," Chapter 22 in ref. [1].

[202] van Roode, M., "German Automotive Ceramic Gas Turbine Development," Chapter 21 in ref. [1].

[203] Ferber, M. K., Lin, H. T., Parthasarathy, V., and Wenglarz, R. A., "Degradation of Silicon Nitrides in High Pressure, Moisture-Rich Environments," ASME paper 2000-GT-0661, presented at the ASME Turbo Expo 2000, Land, Sea & Air, Munich, Germany, May 8–11, 2000.

[204] Deutsch, G. C., Meyer, Jr., A. J., and Morgan, W. C., *Preliminary Investigation in J33 Turbojet Engine of Several Root Designs for Ceramal Turbine Blades*, NACA RM E52K13, January 1953.

[205] Meyer, Jr., A. J., Deutsch, G. C., and Morgan, W. C., *Preliminary Investigation of Several Root Designs for Cermet Turbine Blades in Turbojet Engine II—Root Design Alterations*, NACA RM E53G02, October 1953.

[206] McLean, A. F., Fisher, E. A., and Bratton, R. J., *Brittle Materials Design, High Temperature Gas Turbine*, Interim Report, January 1, 1973–June 30, 1973, Ford Mortor Company, Dearborn, MI, USA, AMMRC CTR 73-32, Sept. 1973. pp. 13–65.

[207] Tatsumi, T., "Development of the 300 kW Ceramic Gas Turbine CGT302, Chapter 15 in ref. [1].

[208] Hisamatsu, T., Yuri, I., Machida, T., and Wada, K., "Development of Ceramic Gas Turbine for Power Plants," Chapter 18 in ref. [1].

[209] Soudarev, A. V. and Tikhoplav, Yu.V., "Research-Engineering Center 'Ceramic Heat Engines (NIZ KTD) at the Research Technological Institute for Power Engineering (NITI EM),'" Chapter 32 in ref. [1].

[210] Yamamoto, T. and Masuda, M., "Research & Development of Ceramic Materials And Components For Gas Turbines," Chapter 7 in this volume.

[211] Krüger, W., and van Roode, M., "Ceramic Gas Turbine Component Development at Motoren Turbinen Union," Chapter 24 in ref. [1].

[212] Okamura, T., Hijikata, T., and Koga, A., "A Design and Experimental Evaluation of Ceramic Stator and Rotor Blades for 1300°C Class Medium Gas Turbine," Chapter 19 in ref. [1].

[213] Hijikata, T., Koga, A., and Asakuma, K., "Reliability of Ceramic Stator Blade Tested in Hot Wind Tunnel," (IGTC-79), *Proceedings of the 1995 International Gas Turbine Congress*, Yokohama, Japan, October 23–27, 1995, Vol. III, 161–167, 1995.

[214] U.S. Department of Energy, distributed energy Resources website "http://www.eere.energy.gov/der/microturbines/microturbines.html, 2000.

[215] Epstein, A. H., Senturia, S.D., Al-Midani, O., Anathasuresh, G., Ayon, A., Breuer, K., Chen, K.-S., Ehrich, F. E., Esteve, E., Frechette, L., Gauba, G., Ghodssi, R., Jacobson, S., Kerrebrock, J.L., Lang, J.H., Lin, C.-C., London, A., Lopata, J., Mehra, A., Mur Miranda, J. O., Nagle, S., Orr, D. J., Piekos, E., Schmidt, M. A., Shirley, G., Spearing, S. M., Tan, C. S., Tzeng, Y.-S., and Waitz, I. A., "Microheat Engines, Gas Turbines, and Rocket Engines—The MIT Project," American Institute of Aeronautics and Astronautics, 1997, reprint, 12 pages.

[216] Ashley, S., "Turbines on a Dime," Mechanical Engineering, 119 [10], 78–81, October 1997.

[217] Epstein, A. H., "Millimeter-Scale, MEMS Gas Turbine Engines," ASME paper GT2003-38866, presented at the ASME TURBO EXPO 2003 and International Joint Power Generation Conference, Atlanta, GA, USA, June 16–19, 2003.

[218] Richerson, D. W., *The Magic of Ceramics*, American Ceramic Society, Westerville, OH, USA, 2000.

CHAPTER

30

Future Needs and Developments in Structural Ceramic Materials for Gas Turbines

W. Dave Carruthers

Dave Carruthers & Associates, Battle Ground, Washington, USA

John Pollinger

Tetralux Engineering, Los Angeles, California, USA

Paul F. Becher and Mattison K. Ferber

Oak Ridge National Laboratory, Oak Ridge, Tennessee, USA

Mark van Roode

Solar Turbines Incorporated, San Diego, California, USA

INTRODUCTION

Over the last 30 years, a number of programs in the United States, Japan, and countries of Western Europe and the former Soviet Union have sought to introduce monolithic and composite ceramic components into gas turbines with the goals of increasing thermal efficiency and output power, and lowering emissions. A historical summary of these efforts is presented in previous chapters of this volume as well as in volume 1 of this series, *Ceramic Gas Turbine Design and Test Experience*. The primary purpose of the present chapter is to discuss how materials technology advancements might proceed in the future to solve some of the challenges that remain.

The superior high temperature strength and durability and acceptable physical properties make silicon-based monolithic ceramics (silicon carbide, silicon nitride) and ceramic matrix composites (CMCs) attractive candidate materials for gas turbine hot section components. Programs in various areas of the world to develop and demonstrate the application of these silicon-based ceramics have not yet led to broad-based commercial application. Some successes have been achieved in specific applications, such as silicon nitride turbochargers in cars, silicon nitride vanes in auxiliary power units (APUs), silicon nitride seal rings in small gas turbines, and SiC/SiC CMCs for combustor liners. However, long-term (20,000–30,000 hr) applications are not yet assured because of fundamental materials property deficiencies. Three critical material property issues that limit the application of the silicon-based ceramics for gas turbine hot section applications include:

1. Fracture toughness (K_{IC}) and strength properties of silicon carbide, silicon nitride and CMCs, also referred to as continuous fiber reinforced ceramic composites (CFCCs) limit successful application. Early impact studies on monolithic ceramics showed that fracture toughness is a major factor affecting impact resistance, indicating that post-impact retained strength is proportional to K_{IC} [1]. Subsequent impact studies support these conclusions [2,3]. For example, in early field tests conducted under the Ceramics for Stationary Gas Turbine (CSGT) Program of Solar Turbines Incorporated silicon nitride first-stage blades failed after 948 hours, most likely due to impact damage from a metallic locating-pin associated with the inner combustor liner [4]. Although the tensile strengths of the monolithics are relatively high, ~700 MPa, the fracture toughness values obtained with long cracks are typically in the 4–7 $MPa{\cdot}m^{0.5}$ range. A K_{IC} of 7 $MPa{\cdot}m^{0.5}$ is very good for monolithic ceramics, but is low relative to that of metallic turbine alloys. The low fracture toughness results in poor resistance to impact and slow crack growth. On the other hand, the CMCs have the characteristic of graceful rather than catastrophic failure, but the mechanical strengths of CMCs are quite limited (e.g., the stress required for the generation of matrix cracks is 150 MPa or less). This constrains the use of these materials to stationary applications with relatively low stress such as combustor liners, transition pieces, seal rings, and shrouds, and excludes their use for stationary applications with significant temperature gradients (nozzle vanes) and rotating applications where stresses are typically in the 200–300 MPa range.
2. Very high temperature oxidation results in surface and subsurface microstructural damage, including the formation of pits or voids, subcritical crack growth, or creep. In ambient air, these damage mechanisms are influenced by reactions of silicon nitride and its sintering additives with an oxidative atmosphere. Besides the formation of a surface oxide, alterations in composition and distribution of the grain boundary phases and the formation of gas can occur after long-term exposures [5]. After long exposures to non-corrosive high temperature conditions, newly formed oxidation damage controls the strength and creep resistance of the material [6–9]. Microstructures and compositions that are highly stable in an oxidizing environment are needed for long term use.
3. Environmental degradation in water vapor bearing combustion atmospheres generally leads to unacceptable degradation of both silicon-based monolithics and CMCs. This problem is discussed more completely in Chapters 26 and 27. A key problem with water vapor induced oxidation is that the otherwise protective silica scales are non protective. Therefore the silicon nitride or silicon carbide continues to oxidize (lose material) at relatively high rates. The problem is further aggravated by high velocity gas streams which remove the silica scale. Loss of function can occur simply by the reduction in the component dimensions. Recession has been observed after 500 to 1000 hours of service in gas turbines with pressure ratios over 8:1 for parts operating at 1000°C and above [10]. The latter problem is currently being addressed with environmental barrier coatings (EBCs), and some measure of success has been achieved for EBC coated SiC/SiC CMCs in the CSGT program sponsored by the U.S. Dept. of Energy [11,12]. However, EBCs capable of higher temperature use, and with improved stability in gas turbine environments are needed.

Notwithstanding the steady improvement in properties over the past few decades engine durability testing worldwide has shown that fundamental material limitations need to be addressed to assure that ceramics achieve their potential as enabling materials for component fabrication for advanced gas turbine systems with improved performance compared to state-of-the-art all-metal engines. In addition to the research investigating near-term refinements to further advance self-reinforced silicon nitride properties, a number of efforts around the world are investigating the tailoring of ceramic compositions and microstructures ranging from the nanoscale to the macroscale, and incorporating multiple levels of composition and structural design (hierarchical structures) to generate radically improved properties. There is now an opportunity to incorporate the variety of new material composition/microstructure and fabrication concepts that have been developed over the past ten years into a new generation of more sophisticated structural ceramics, effectively combining the best properties of structural monolithic ceramics and CMCs.

In this chapter we review a number of technology developments that are establishing the foundation for these potential improvements. This chapter is an elaboration of ideas previously presented at an ASME conference [13].

NEXT GENERATION SILICON NITRIDE CERAMICS: POTENTIAL FOR HIGHER TOUGHNESS AND STRENGTH

Improvements made in silicon nitride during the past three decades make them serious candidates for use in gas turbines. For example, the creep resistance of Si_3N_4 has steadily improved with the advent of more refractory and viscous grain boundary phases [14,15], as shown in Figure 30.1. A real progression has taken place in silicon nitride and carbide ceramics with respect to strength and toughness, as seen in Figure 30.2.

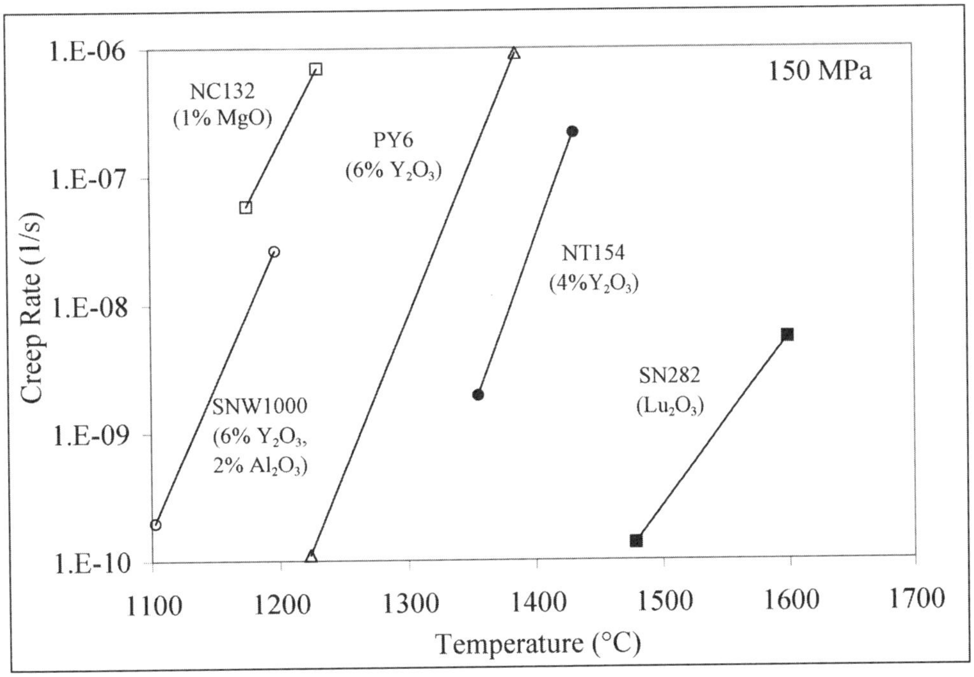

Figure 30.1 Creep Rate Improvements Have Come with More Refractory Grain Boundaries (Adapted from ref. [14] with permission from Wiley-VCH Verlag GmbH)

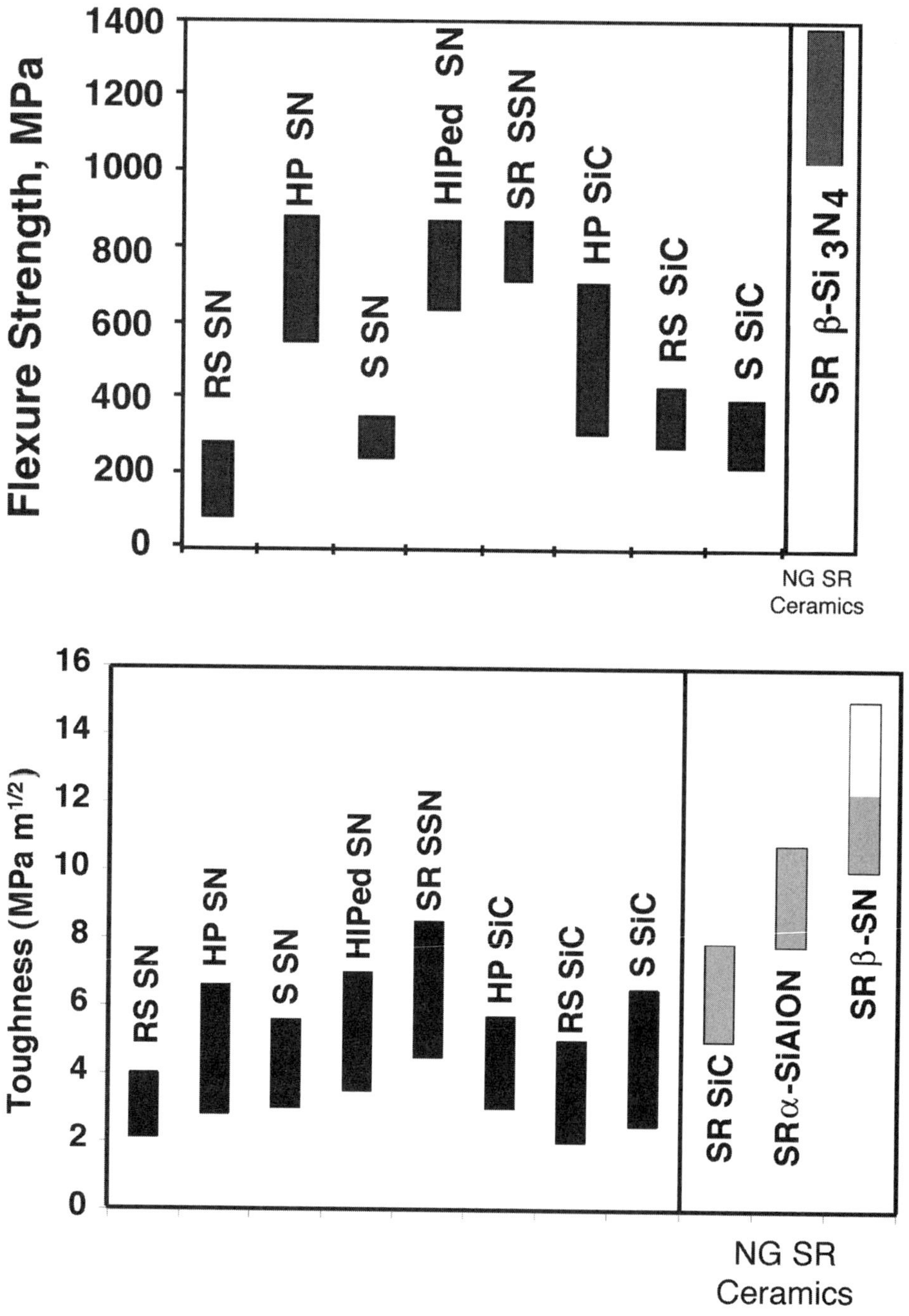

Figure 30.2 Progression in Strength (Top) and Toughness (Bottom) Values versus Time (X-Axis) Achieved in Silicon-Based Ceramics Including: Reaction Sintered Si_3N_4 (RS SN) and SiC (RS SiC), Hot Pressed Si_3N_4 (HPSN) and SiC (HPSiC), Sintered Si_3N_4 (Early Vintages-S SN), HIPed Si_3N_4 (HIPed SN), Sintered Self-Reinforced Si_3N_4 (SR SSN) and Next Generation (NG) (Lined Bars Lighter Shading) Ceramics Including Self-Reinforced Beta- Si_3N_4 (SR β-Si_3N_4) [16,17,19,23,24], SiC (SR SiC) [25] and α-SiAlON (SR α-SiAlON) [26] (Reprinted from ref. [13])

Both charts in the figure reflect progress with time in silicon nitride and silicon carbide mechanical properties, from left to right, with the self-reinforced and next generation ceramics representing the most recent results. Since the properties of Si_3N_4 are determined by the microstructure, engineering the microstructure is key to achieving property improvements. For Si_3N_4, the critical structural properties are controlled by the grain size, orientation, and shape, as well as by the properties of the grain boundary, i.e., additive composition, grain boundary viscosity, phase stability, oxygen permeability, and chemistry, as illustrated in Figure 30.3. Each of these characteristics is interrelated; changes to improve one aspect could adversely or beneficially affect other properties.

In the case of the toughness for the next generation self-reinforced ceramics (NGSR β-Si_3N_4), the unfilled portion of this bar is representative of material with highly aligned elongated reinforcing grains [16,17] (e.g., similar to the effects of uniaxially aligned fiber-reinforced ceramics). The properties of NGSR ceramics (e.g., beta-silicon nitrides, alpha-SiAlONs, SiC), exceed those of current commercial materials. The fundamental understanding gained during the evolution of self-reinforced silicon nitride ceramics is now being utilized to develop the next generation silicon nitride (NGSN) materials with higher toughness and strength to address the property requirements for a number of application areas, e.g. in advanced diesel engines and heat exchangers (note that bulk Si_3N_4 ceramics can achieve thermal conductivities that are now nearing the 150 W/m.K level) [18]. Such thermal conductivity will contribute to lower thermal stresses. Studies have now established a basis for controlling both the development of microstructures containing larger elongated grains and the tailoring of the sintering additive composition to improve the fracture toughness [19,20]. The elongated grains serve as the reinforcing elements much like whiskers in whisker-reinforced ceramics [20,21], or fibers in fiber-reinforced ceramics [22].

Control of the microstructure is key because the toughening achieved by crack bridging processes is dependent upon the size and volume fraction of these larger elongated grains. In commercial self-reinforced ceramics, the elongated beta-phase grains are generated during the sintering of the ceramic, which greatly simplifies the processing and avoids the use of expensive ceramic fibers or whiskers. Such reinforcing grains form in silicon nitride ceramics and larger grains form due to the initial presence of some beta-phase in the alpha-phase silicon nitride powders typically used. Unfortunately, when sintering these powders it is difficult, if not impossible, to regulate the size of the larger reinforcing grains and to obtain a uniform spatial distribution of

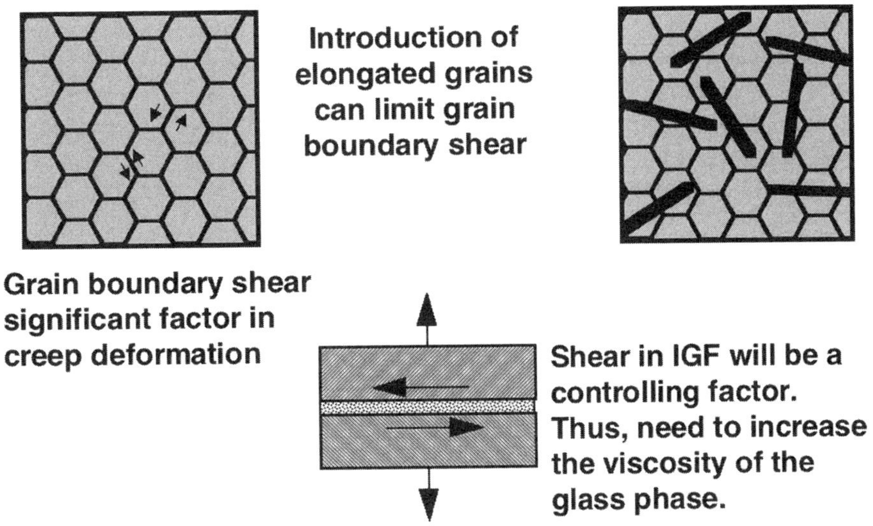

Figure 30.3 Key Contributors to Ceramic Mechanical and Oxidation Properties are Interrelated

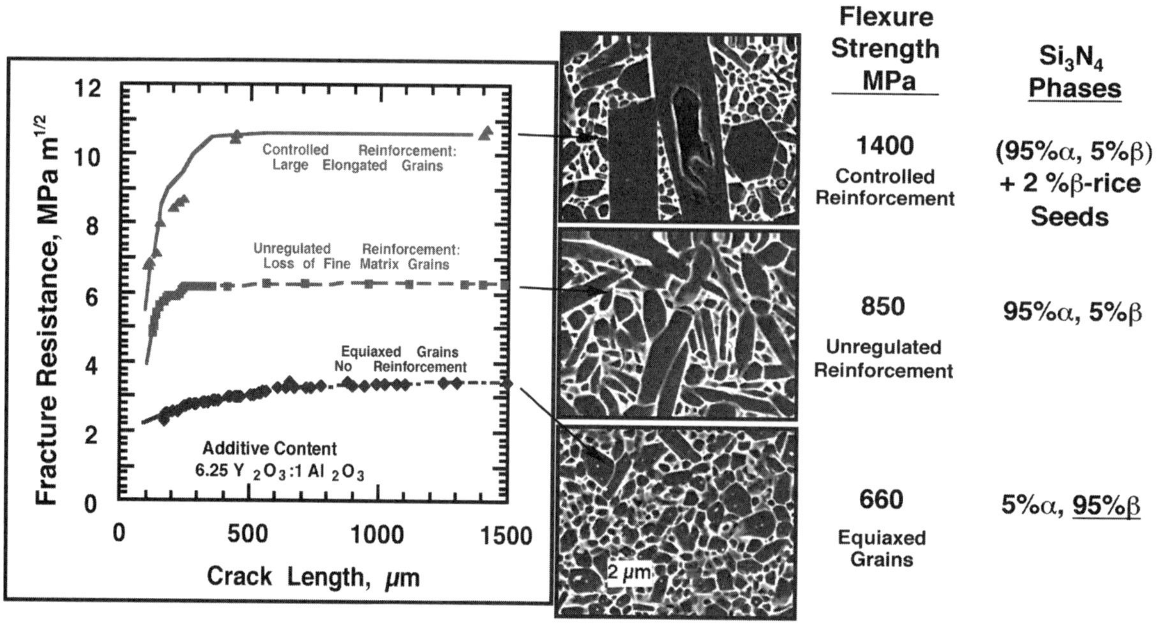

Figure 30.4 Very High Strength and Toughness Values are Achieved by Controlled Generation of Large Elongated β-Si_3N_4 Reinforcing Grains into a Fine-Grained β-Si_3N_4 Matrix. (Reprinted from ref. [19] with permission of the American Ceramic Society, www.ceramics.org, © 1998, all rights reserved).

them. Both factors must be satisfied in order to achieve both high toughness and high strength. Control of the size and distribution of reinforcing grains is one of two major factors that have limited the achievement of high toughness, high strength self-reinforced silicon nitride ceramics. Studies have shown that control can be achieved by the incorporation of a small fraction of rice-shaped beta seeds into the silicon nitride powder. The "beta-rice" seeds used to date are typically 3 to 4 μm long by ~1 μm in diameter. However, the optimum size and shape of such seeds has not been determined as yet. These "beta-rice seeds" serve as the nuclei for generating well dispersed elongated reinforcing grains with controlled size(s) [19,23,24]. Recent studies suggest that toughening by self-reinforcing grains can also be developed in alpha-SiAlON ceramics using a similar seeding approach [26]. Results using these principles show that next generation self-reinforced silicon nitride (NGSRSN) ceramics can have high strengths (1 to 1.4 GPa) combined with high fracture resistance, which yield plateau values (i.e. long crack toughness values) of up to 11 MPa·$m^{0.5}$ (Figure 30.4).

Control of the sintering additive compositions, which dictate the intergranular films typically formed in the sintering step, is the second important factor in the development of self-reinforced silicon nitride ceramics. The additive composition is important because both the growth of the reinforcing grains and their subsequent ability to act as reinforcements, are controlled by the intergranular material typically found in silicon nitride ceramics. (They are also a major factor in the intermediate and elevated temperature oxidation, strength, and creep properties). These intergranular films (typically < 2 nm thick) are a result of the reaction between the native oxygen in silicon nitride powders and the specific chemical additives used to promote sintering. This reaction results in the formation of a viscous glass that allows liquid phase sintering to occur and acts as a medium for the growth of the elongated reinforced beta grains. The resultant intergranular films (IGFs) also control the debonding of the reinforcing grains from the matrix material in the crack tip region [27,28]. This debonding process prevents the crack from cutting through the reinforcing grains and allows them to bridge across the crack. This is similar to the steps that occur in fiber-reinforced ceramics. Recent studies show that the composition of the sintering additives and, hence, the IGF can be tailored to optimize the toughening effects of the elongated reinforcing grains in silicon nitride ceramics. As an example, the fracture toughness of SRSN ceramics with ~7 wt. % additives of yttria plus alumina was raised by ~50 % simply by increasing the yttria to alumina ratio in the additives [29,30]. Tailoring of the additive composition plus the microstructure is now recognized as the critical factor in obtaining both high fracture toughness and high fracture strength. Similar controlled and elongated grain microstructure approaches apply to oxide ceramics as well. A $LaAlO_3$-Al_2O_3 composition with strength over 600 MPa and high fracture toughness of 6 MPa.m$^{0.5}$ was achieved by a combination of elongated Al_2O_3 grains and $LaAl_{11}O_{18}$ platelets. The increase in toughness is achieved by the crack bridging effect of anisotropically grown alumina grains, while the lanthanum aluminate platelets play a role in hindering the alumina growth [17]. Compared with silicon-based materials these results are not impressive but they are for oxides where insufficient toughness is problematic.

Alternate or combinatorial approaches have also demonstrated to improve strength and toughness. Low elastic modulus materials developed by Ishikawajima-Harima Heavy Industries (IHI) have shown high strength (>1000 MPa) combined with high K_{IC} (15 MPa.m$^{0.5}$). IHI has made high strength and toughness components using tape casting and a technique based on alternating dense and porous layers [31]. When research emphasis was on reducing the elastic modulus of silicon nitride to improve its strain to failure, a porous (14%) silicon nitride with aligned fibrous grains was evaluated. Using a chevron-notched beam technique a K_{IC} ranging from ~11 to as high as 20 MPa·m$^{0.5}$ was achieved for crack sizes from ~0 to 2.6 mm, respectively, The material has a flexural strength of 1081 MPa, and an elastic modulus of 246 GPa vs. ~320 GPa for dense Si_3N_4. The controlled oriented porosity material [32,33] has a surprisingly high work of fracture of ~490 J/m^2 versus ~70 J/m^2 for dense Si_3N_4, or 96 J/m^2 for SiC_w reinforced Si_3N_4 [34], or 246 J/m^2 for highly anisotropic forged Si_3N_4 [35].

An early study showed that a 7-fold improvement in impact resistance could be achieved with only a 1 mm thick silicon rich (20–25%) reaction bonded silicon nitride (RBSN) coating on hot pressed NC132. The coating also survived thermal shock and gas erosion testing. [36]. Controlled but random porosity has also been achieved by adding carbon black to Si_3N_4 powder. The carbon reacted with the SiO_2 or Si_3N_4 to form SiC, which limited the sintering shrinkage to 2–3% resulting in porosity [37].

Based on these results, one could consider the benefit in an impact event of a surface layer on airfoil engineered to be strong in the direction of highest stress, with a low elastic modulus and high work of fracture. The surface layer could consist of aligned and controlled porosity, or, instead of porosity, a low elastic modulus oxide, e.g., Si_2N_2O (E = 83 MPa) or SAS. Surface layers could be applied by tape application [38,39,40], spraying or dip coating.

Porous materials will likely suffer from a higher oxidation rate. Also achievement of these properties in thin coatings is unproven, and incorporation of a porous layer on complex shapes will require significant effort. However, the potential benefits may justify further study of these materials as coatings or structural materials.

Hitachi, Ltd. has developed nanocomposite silicon nitrides with high strength (1400 MPa) and K_{IC} (10.5 MPa·m$^{0.5}$) [41]. Silicon nitride composites containing nano-size second phase particulates have demonstrated improved creep resistance, but there continues to be disagreement on the occurrence and mechanism of improved fracture toughness [42,43]. It has become clear that the degree of improvement in properties is dependent on the type of composite system involved, i.e., whether the nano-particulate thermal expansion is higher or lower that the matrix grains. [44,45]

A summary of some of the results obtained indicates that engineered microstructures have the ability to provide both improved strength and K_{IC}, as summarized in Table 30.1

What can we expect in the future from such toughened ceramic materials? The object must be to generate more robust materials that exhibit improved mechanical properties and resistance to damage during handling or in-service. As an example, the fracture strength should not be altered by surface damage. The resistance to damage of toughened silicon nitride ceramics has been evaluated by examining how the fracture strength is influenced by more severe (i.e., coarser diamond grit size) surface grinding [55]. A very significant benefit of the toughened self-reinforced silicon nitrides can be seen in Figure 30.5 where the high strengths are unaffected by the severity of the surface damage. In this figure, the x-axis represents the size,

TABLE 30.1 A SUMMARY OF PROPERTIES ACHIEVED BY TAILORED MICROSTRUCTURES IN Si_3N_4

Approach	K_{IC} ($MPa.m^{0.5}$)	Room Temperature Strength (MPa)
Aligned β-seeded grains via tape casting w/a-Si_3N_4 [19,29,46]	~11	>1100 (stress // to grain alignment)
Aligned grains via Superplastic forging or sinter-forging [47,48]	~13	1650
	~8	~2100
Aligned Porosity[34]	~11–19 w/ a ~ 7x increase in fracture energy	1080 w/ low elastic modulus, E = 246 GPa
HIPed SRBSN w/SiC_P, (nanocomposite)[49]	~6.5	~1200 (3pt)
$MoSi_2$ in Si_3N_4 [50, 51]	>8	~800 (40 v/o)
TiN in Si_3N_4, [52]	~10	~850
Si_2N_2O in-situ reinforcement, bi-modal distribution [53]	~5+	1000
HP, β-seeded, fine α-Si_3N_4 powder, bimodal [54]	NA	NA

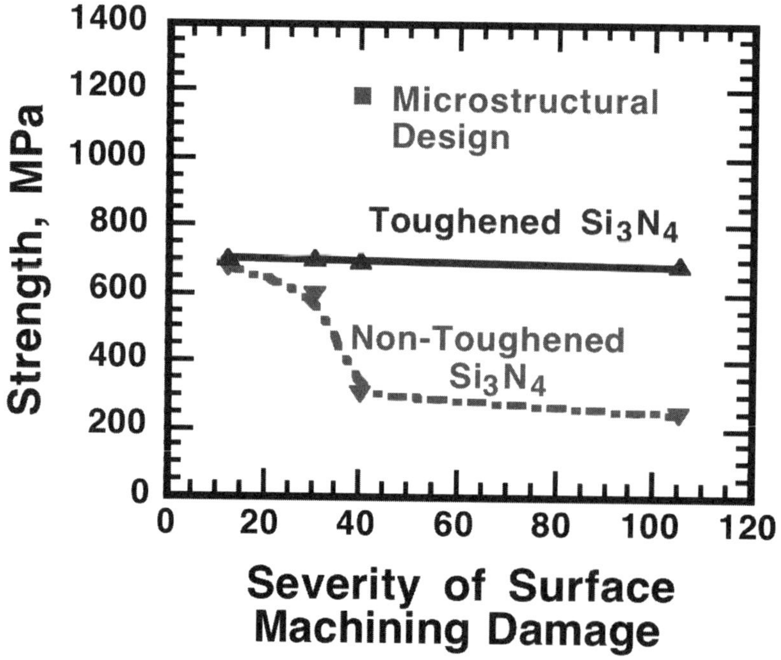

Figure 30.5 The Self-Reinforced Silicon Nitride Ceramics Exhibit Fracture Strengths That Are Not Degraded by Increasingly Severe Surface Damage While the Unreinforced Silicon Nitrides Undergo Substantial Strength Loss. Data for Toughened and Non-Toughened Silicon Nitride are Taken from ref. [55] (Adapted from ref. [13])

in microns, of the diamond particles used in the grinding wheel used to finish the tensile surface of four point flexure bars. In this case, the depth of the residual grinding grooves and associated damage zone as well as the size of the surface cracks generated all increase as the size of the diamond particles in the grinding wheel increases. These results suggest that such self-reinforced silicon nitride ceramics will have both greater resistance to local impact damage and less flaw size dependent strengths (e.g., narrow strength distributions). Indeed, the self-reinforced silicon nitride ceramics typically exhibit a substantial narrowing of the fracture strength distributions with Weibull moduli of up to 40 as compared to ~10 in the absence of toughening effects [56].

POTENTIAL FOR IMPROVED OXIDATION AND CREEP RESISTANCE

The mechanical reliability of silicon nitride ceramics at intermediate and elevated temperatures is also dominated by the intergranular phases present. Amorphous and or crystalline intergranular films control the temperature and stress for the onset of crack growth at intermediate temperatures and creep deformation at elevated temperatures. As the demand for sinterable Si_3N_4 with improved creep and oxidation resistance grew, rare earth (RE) additives were selected for their high melting points and oxidation resistance.

Recent work reveals that substitution of smaller rare earth (RE) ions into Si-RE-Al oxynitride glasses can be used to increase the softening point of such glasses and, hence, raise the glass viscosity at elevated temperatures [57–59]. A caveat to the selection of grain boundary constituents is that if metastable phases are present that undergo volumetric expansion with oxidation, severe cracking can occur. Catastrophic oxidation occurs when the volume expansion exceeds ~1%. [60].

The diffusion of oxygen along intergranular films typically increases as the glass viscosity at a given temperature is lowered. This increases the susceptibility of ceramics to degradation by oxidation processes. The results of these glass studies provide directions for tailoring the sintering additives for silicon nitride to improve the mechanical reliability and reducing the transport of oxygen through the grain boundaries of these ceramics at intermediate and high temperatures. In addition, studies at Oak Ridge National Laboratory (ORNL) reveal that the reinforcing microstructure with the larger elongated grains can be used to gain additional increases in creep resistance [61]. In fact, the creep resistance of commercial silicon nitrides has been improved by taking advantage of these approaches while eliminating deleterious additives like alumina, which tend to lower glass viscosity. The highest reported tensile creep resistance has been obtained in a self-reinforced silicon nitride that utilizes Lu_2O_3 as a sintering additive [62]. An additional aspect of oxidation resistance has included the evaluation of SiC or $MoSi_2$ particles in a Si_3N_4 matrix [5]. With the addition of 10% SiC or $MoSi_2$ in a Y_2O_3 doped Si_3N_4, after 1000 hours of oxidation testing from 1400 to 1500°C the composites had a residual strength of 650 to 700 MPa versus a level of ~450 MPa for the unreinforced material. The improvement is attributed to the increased formation of a layer of crystalline Si_2N_2O caused by a modified oxidation mechanism. The layer, formed below the upper surface of the material, prevents both the diffusion of oxygen into the bulk material, and the formation of typical oxidation defects such as cracks, pores, or glassy-rich regions. The formation of the Si_2N_2O rather than SiO_2 also reduces the driving force for grain boundary cations to diffuse into the silica rich glass at the surface of the material. With the formation of the Si_2N_2O during oxidation, the degradation of the composite materials was reduced. Further, hot pressed $Si_3N_4/MoSi_2$ composite materials were exposed for 10,000 hours at 1500°C with only minimal alterations in the microstructure while maintaining ~65% of their baseline room temperature strength [63].

While it is critical to improve the room and elevated temperature properties of silicon nitride ceramics, it is also important to understand and control other properties that will make them more reliable. Studies have shown that the presence of oriented large elongated grains, the substitution of rare earth ions, or elimination of alumina in the sintering aids can greatly increase, or alter the thermal conductivity of silicon nitride ceramics (e.g., up to 120 W/m.K, which approaches that for AlN) [18,64–67]. As an example, Nissan Motor Co., Ltd. is developing silicon nitride with exceptionally high thermal conductivities. In a 2-D structure a value of 128 W/m.K has been achieved, while in a 1-D structure values in the range of 140–160 W/m.K have been measured [31]. Increases in thermal conductivity in brittle materials can improve thermal shock resistance. As an example, Honeywell Ceramic Components' AS800 silicon nitride achieves a doubling of thermal conductivity (to 80 W/m.K) over fine-grained silicon nitrides with its combination of large elongated grains and grain boundary phase composition [68]. The result is impressive thermal shock resistance. Tests at NASA Glenn Research Center using rocket engine exhaust impingement [69] have shown that AS800 can withstand repeated 1700°C/s upshock rates, and conventional thermal shock quench tests [70] demonstrate

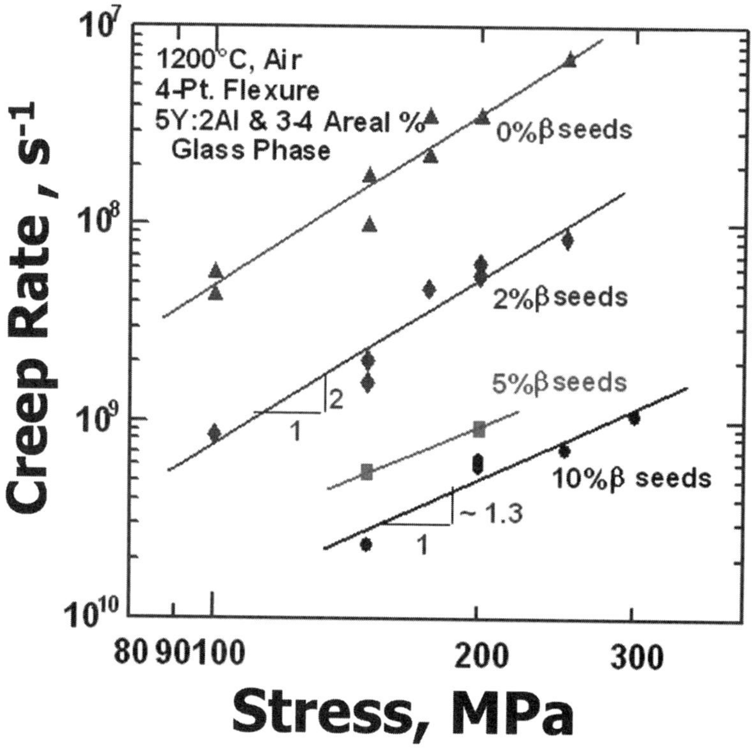

Figure 30.6 Reduced Creep was Achieved Through the Growth of β-Si_3N_4 in a 2% Alumina Containing Si_3N_4 (Reprinted with permission from ref. [61])

that it has the ability to survive greater than 1200°C downshocks without strength loss. Even greater thermal shock resistance capability should be achievable as these ultrahigh thermal conductivity structural ceramics are optimized for fabrication into actual components.

A noteworthy observation has been made regarding a small amount of alumina in the grain boundary. If a small amount of alumina is present in the powder blend during sintering and grain growth, the grain boundary content of alumina can decrease as it is consumed in the growth of SiAlON on the β-Si_3N_4 seeds. The growth process reduces the Al_2O_3 content of the IGF, and increases its viscosity. Thus, improved creep resistance in alumina containing Si_3N_4 can be improved by increasing the number of large, elongated grains, or increasing their size. Results are shown in Figure 30.6.

Several researchers have reported that gas pressure sintered or hot pressed Si_3N_4-SiC_P composites (nano- or other) show creep resistance equal to or better than the constituent monolithics[71–76]. The creep rate is affected by both the sintering additive content, as well as the SiC content. The additive quantity will affect creep, but not simply because a large quantity will produce more glassy phase. With Y_2O_3 as the additive, a peak in creep rate occurs at approximately 5 weight percent and reduces as more additive is used, to a minimum creep at approximately 8% [71]. This is illustrated in Figure 30.7. The explanation for this behavior is the formation of eutectics at intermediate additive contents.

The creep rates are also affected by the SiC content. In a study of the creep behavior of hot pressed Si_3N_4 containing 0 to 30 % SiC nano particles and 8 or 10 weight % Y_2O_3, it was found that a minimum creep rate occurred near the 10% SiC content. Creep behavior was evaluated in tensile and bending conditions. The results are shown in Figure 30.8.

The oxynitride glass transition temperature (Tg), and thus, the viscosity, increase with smaller RE ions, as illustrated in Figure 30.9, and increases with the nitrogen content. However, the nitrogen solubility can

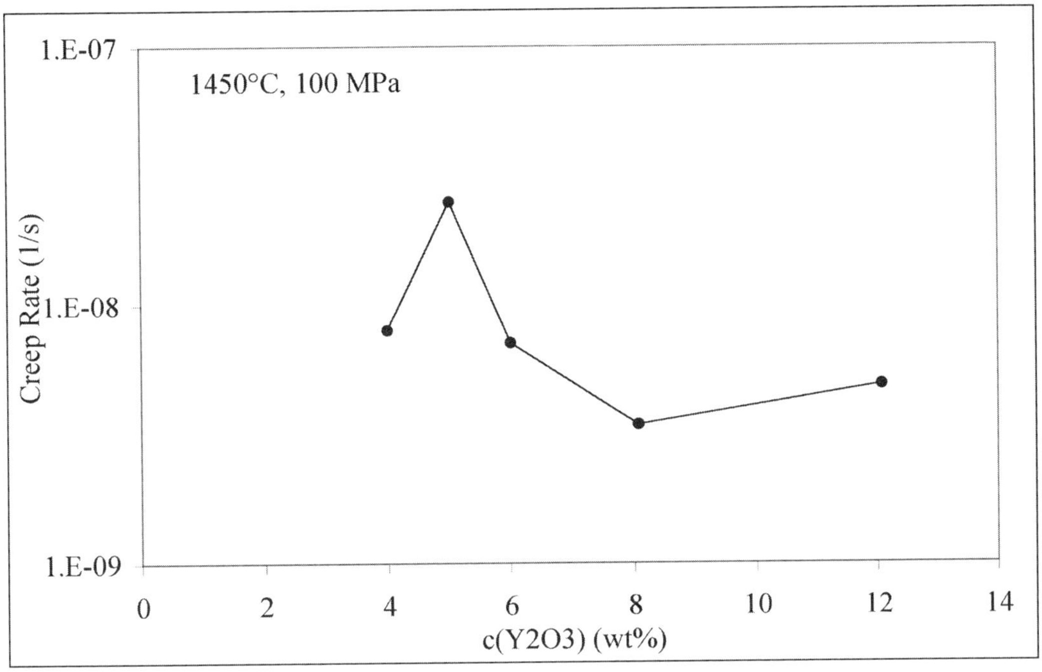

Figure 30.7 Varying Additive Contents Can Result in the Formation of Eutectics that Increase the Creep Rate (Adapted from ref. [71] with permission of the American Ceramic Society, www.ceramics.org, © 2000, all rights reserved)

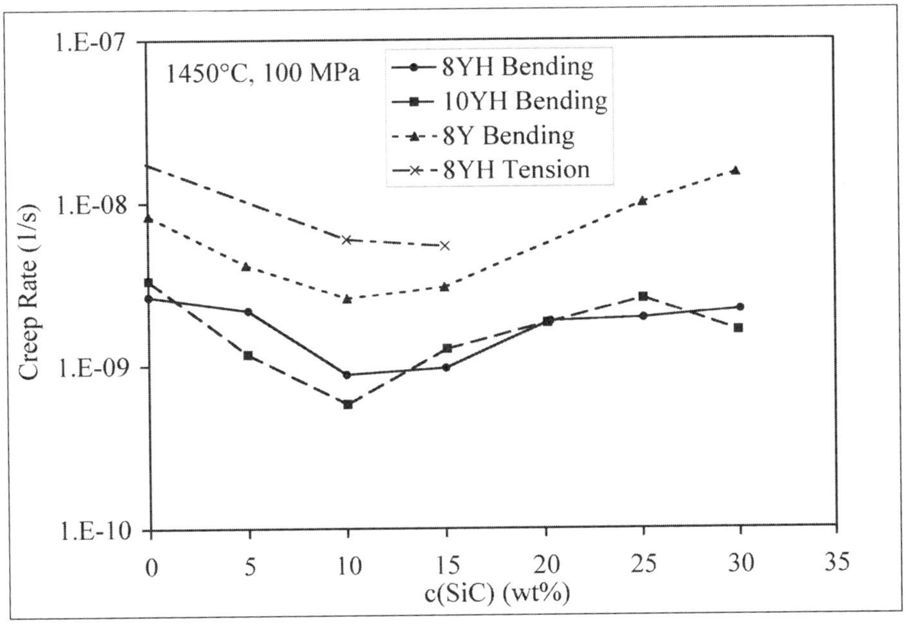

Figure 30.8 Effect on the Steady-State Creep Rate of SiC Nano Particle Content Tested in Bending and Tension (Adapted from ref. [71] with permission of the American Ceramic Society, www.ceramics.org, © 2000, all rights reserved)

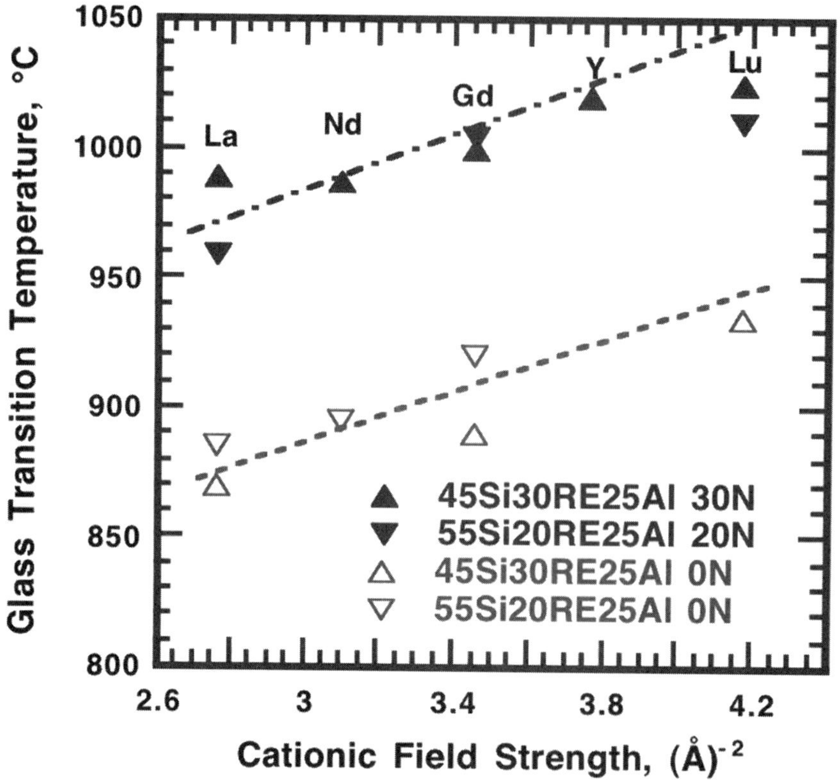

Figure 30.9 The Amorphous Grain Boundary Phase May Be Made More Refractory by the Use of Smaller Rare Earth Ions

also be affected by the ion size (e.g., in the case of Lu). One can tailor the additives to increase the IGF viscosity and, thus, the creep resistance of the silicon nitride.

The additions of SiC or $MoSi_2$ to Si_3N_4 have shown to be a benefit for property retention in oxidizing conditions. The addition of $MoSi_2$ results in an increased weight gain during initial oxidation, and both additives promote the formation of Si_2N_2O below the silica oxide surface layer [5]. A partial-fiber microstructure of Si_2N_2O occurs. Si_2N_2O is responsible for the improved oxidation of Si_3N_4 over SiC [77] because its oxygen diffusion is low [78,79]. It has been suggested that the presence of the subsurface Si_2N_2O reduces the cation driving force to the surface, which results in a cleaner SiO_2 layer [5]. Monolithic Si_2N_2O has good oxidation resistance and high temperature strength (~600–800 MPa) [53,80]. Figure 30.10 shows that the fracture toughness of Si_3N_4 sintered with cordierite increases as bi-modal content of Si_2N_2O increases [81,82]. The presence of Si_2N_2O may also be helpful for improving erosion resistance of Si_3N_4 [53].

Si_3N_4-$MoSi_2$ showed improved strength retention after 10,000 hr of exposure in ambient air at 1500°C [83]. The damage processes seen in monolithic Si_3N_4 was not observed in the composite.

IMPROVED ENVIRONMENTAL RESISTANCE

With the advent of more extensive silicon nitride and SiC/SiC CFCC component testing in industrial gas turbines in the late 1990s, it became obvious that the combination of significant amounts of water vapor with high temperature, high pressure and high velocity air resulted in excessive surface recession [10] due

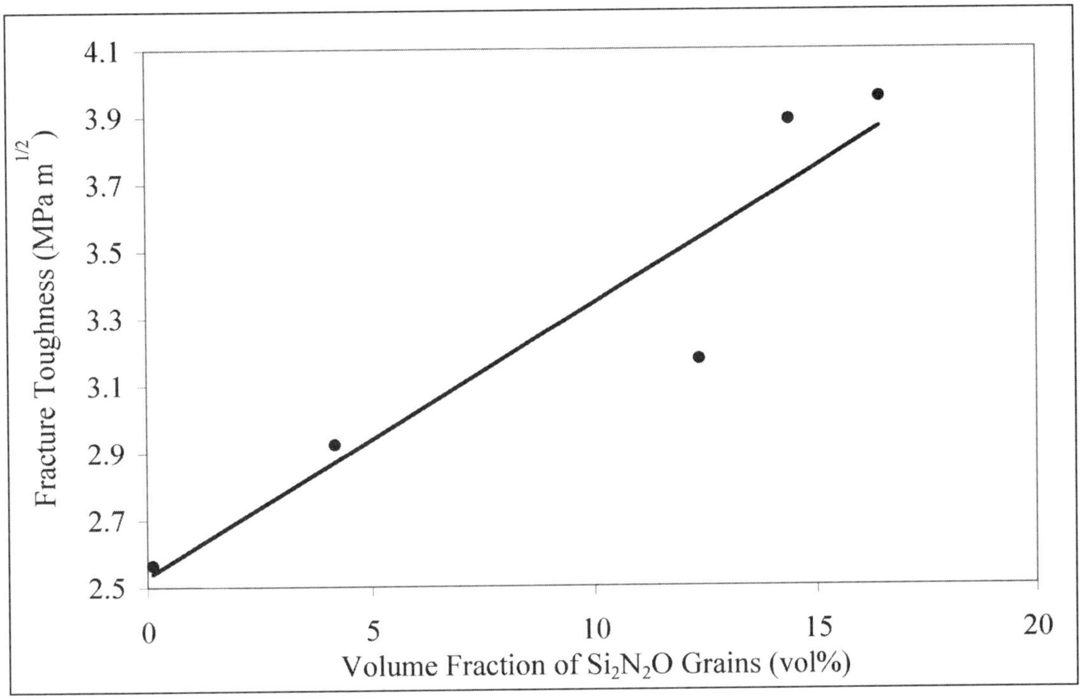

Figure 30.10 Fracture Toughness of Si_3N_4 Materials Sintered with Cordierite as a Function of Volume Fraction of Si_2N_2O Grains (Reprinted from ref. [81] with permission from Elsevier, © 1998)

to rapid volatilization of the surface protective SiO_2 layer. Recently, coatings incorporating mullite and celsian have been applied to silicon nitride monolithic materials and SiC/SiC CMCs to inhibit water vapor and oxygen penetration to the silicon nitride or carbide beneath. The coatings have been selected to closely match the coefficient of thermal expansion (CTE) of the ceramic substrate to prevent spalling, and for the ability to be applied by conventional methods on dense parts (e.g. plasma flame spraying, chemical vapor deposition—CVD, electron beam-physical vapor deposition—EB-PVD). The coatings show some improvements in protection but they still have significant oxygen permeability and contain silicon, which makes them susceptible to the water vapor induced volatilization mechanism. There are obvious concerns about coating integrity, adherence, and performance over the thousands of hours of life needed for these components to be accepted for commercial implementation. In addition, engine manufacturers have identified applications, which require higher temperatures (1450–1500°C) than the current coatings being evaluated can handle. It is also likely that effective surface EBCs will need to be multi-component, because many materials being considered for their resistance to water vapor attack have significant oxygen permeability. Ideally, a self-healing protection system would be designed into the material itself. This system would function much the same way the Si does in silicon nitride when it forms a protective SiO_2 coating in an oxidizing atmosphere. The protective species could be uniformly distributed throughout the ceramic or localized near the surface. The challenge to this self-healing approach would be to maintain the integrity of both the protection system, and the grain boundary phase so that the material properties are not degraded. The benefit of such an approach is that the protection system would be an integral part of the component, would not add dimensionally to critical gas path surfaces, and would not require additional expensive processing steps. Studies in the $Si_3N_4/MoSiO_2$ composite systems using an Yb_2O_3 sintering additive have been shown to behave in such a manner [84]. After hot pressing the composite, an Yb_2O_3 coating was applied, and the material was exposed in a natural gas combustion atmosphere at 50 m/s, 1400°C, 5 bar pressure, with steam introduction. Although EBC adhesion was poor, the functionality of the EBC was demonstrated, and below the EBC a stable layer of amorphous SiO_2, cristobalite and $Yb_2Si_2O_7$ was observed, which prevented substrate oxidation. Many potential EBC materials being considered as oxygen/water vapor protectors have significantly higher CTE than silicon nitride or silicon carbide, which currently prevents their use. There are two potential methods being investigated to allow application of new candidate surface oxygen/water vapor protective coatings.

(a) Incorporate post-densification displacement or diffusion reactions, where the densified component is processed in a gas, liquid metal, or powder bed to generate a near-surface composition gradient of the desired surface protection composition. The graded composition would have the potential to minimize CTE mismatch stresses and maximize adherence to prevent spalling.

(b) Generate graded compositions on part surfaces during fabrication using solid freeform processes such as Direct Photo Shaping (DPS) or Fused Deposition. The surface compositions would obviously need to be compatible with the ceramic densification conditions. This type of technology has already been demonstrated with DPS on silicon nitride turbine nozzles where a CeO_2-doped surface layer was applied as part of the net-shape forming step, and then co-sintered and oxidized to form a compressive surface stress [85], taking advantage of the volume change expansion of the CeO_2-doped layer.

Ideally, surface EBC protection mechanisms should be combined with damage resistant surface coatings/mechanisms to create a holistic approach to surface protection—addressing both environmental and impact resistance. A key problem is that, currently, EBCs on monolithics tend to reduce mechanical properties.

THE SYNERGY CERAMICS PROJECT: A MODEL FOR MATERIAL DEVELOPMENT

Recently, in Japan materials development programs have been conducted which focus on routes to solve the inadequacies of current silicon based ceramic materials. Specifically, the Synergy Ceramics project has been ongoing at the National Industrial Research Institute of Nagoya (NIRIN) (now known as National Institute of Advanced Industrial Science and Technology, Synergy Materials Research Center) and other organizations in Japan. By focusing on the micro- and macro-scale aspects of processing new materials many potential improvements have been demonstrated. The Synergy Ceramics project was inaugurated in 1994 as a collaborative effort among national research laboratories under the Industrial Science and Technology Frontier Program of the Japanese Agency of Industrial Science and Technology (AIST) of the Ministry of International Trade and Industry (MITI) currently renamed as the Ministry of Economy, Trade and Industry (METI). A part of the program is being conducted under the sponsorship of New Energy and Industrial Technology Development Organization (NEDO) through entrustment to the Fine Ceramics Research Association (FCRA). The project is being conducted in two phases. Phase I, conducted between 1994 and 1998, emphasized basic research in materials structure control and supportive analysis and evaluation method development. The mode of operation during Phase I was free-flow, open ended R&D without focus on specific applications. Phase II, which started in 1998 and will be conducted through 2003, is much more focused on taking the successful areas of Phase I research to the applications stage. At this stage collaborations with industry are an important aspect of the program. Participants in the Synergy Ceramics project are currently national research laboratories, universities, industries, and overseas research institutes. Of the Japanese gas turbine Original Equipment Manufacturers (OEMs) Hitachi, Ltd., Toshiba Corp. and Ishikawajima-Harima Heavy Industries (IHI) are involved. NGK Insulators, Ltd., Kyocera Corporation, and Asahi Glass Co. are supplier participants. Toyota Motor Corp. and Nissan Motor Co. represent the automotive industry. Dow Chemical Co. is also a participant. Researchers from the Synergy Ceramics Project and ORNL are currently collaborating under a program titled, "Assessment of Key Technologies Required for the Long-Term Reliability of Advanced Ceramics for Turbine Engine Application," which was initiated by the US-Japan High-level Consultations on Climate Change Science and Technology Working Group.

The basic premise of the Synergy Ceramics project is to develop materials that have multi-functionality and multi-capability, combining properties that are currently only present in different classes of materials, e.g.

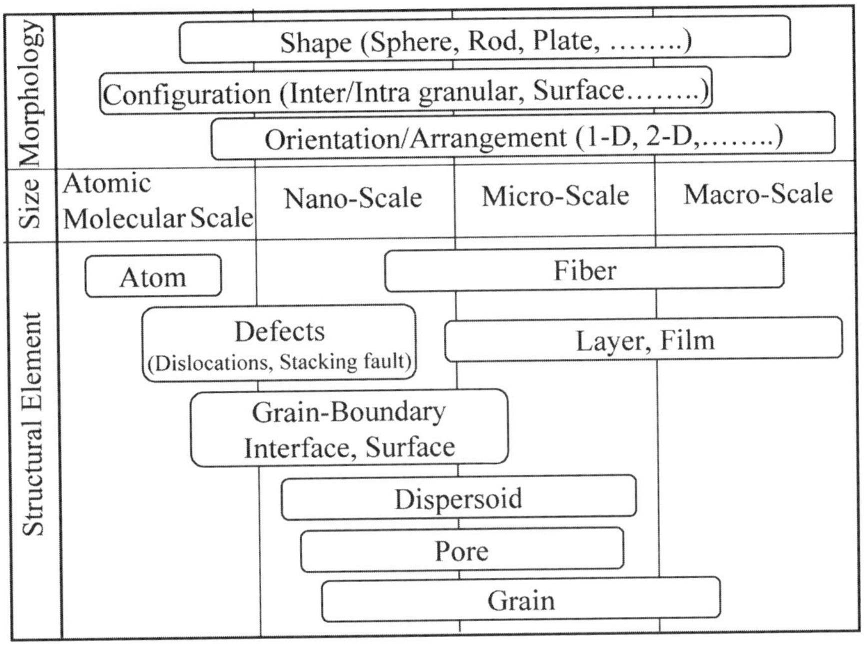

Figure 30.11 Classification of Material Structures (Reprinted with permission from ref. [17])

monolithic vs. CMC, often with antagonistic properties. An example is the development of materials that combine high mechanical strength with high fracture toughness, or materials that have good strength and good thermal conductivity. The objective of synergistic properties is being achieved by a strategy broadly termed "Hyper-Organized Structure Control". In the Phase I core research area the four predominant themes were: precursor design, structure development simulation design, nano-structure processing, and in-situ structure controlling process. One objective of the Project is to produce ceramics with excellent mechanical strength (1000 MPa flexure strength) and superior fracture toughness (15 $MPa.m^{0.5}$) [86]. Key concepts of the project to-date are illustrated in Figures 30.11 and 30.12. Figure 30.11 classifies material structures based on the scale (atomic-molecular-, nano-, micro-, macro-scale). To attain compatibility of antagonistic properties and a synergistic effect of functions, it is necessary to select the elements conforming at each level, and exert simultaneous control of different structural elements, such as shape and size, at plural scale levels. Figure 30.12 illustrates how two antagonistic properties, strength and toughness can be combined in the same material. Strength benefits from a homogeneous microstructure while toughness is favored by a heterogeneous microstructure; both of these are characteristics at the micro-scale level. It is rather difficult to achieve simultaneous improvement of both properties with traditional techniques of structural control operating at a single scale level. The solution proposed by the Synergy Ceramics project is to perform structural control at two different scale levels. For example, combining an intra-granular dispersion of nano particles (supplying strength) with orientation control of anisotropic grains (providing toughness) will have the synergistic result of a hybrid material with both properties.

In reviewing the work addressing reliability issues, several precautions in the results should be noted:

- Most work has been conducted by Hot Pressing or HIPing, which are not desirable processes for low cost parts.
- Most reported work included RE + Alumina as a densification aid. Other additive combinations need to be explored.
- Assessments of intermediate temperature stability and cyclic properties have typically not been included in the studies. Some of the work used compositions that may result in grain boundary phases outside of the "compatibility triangle."

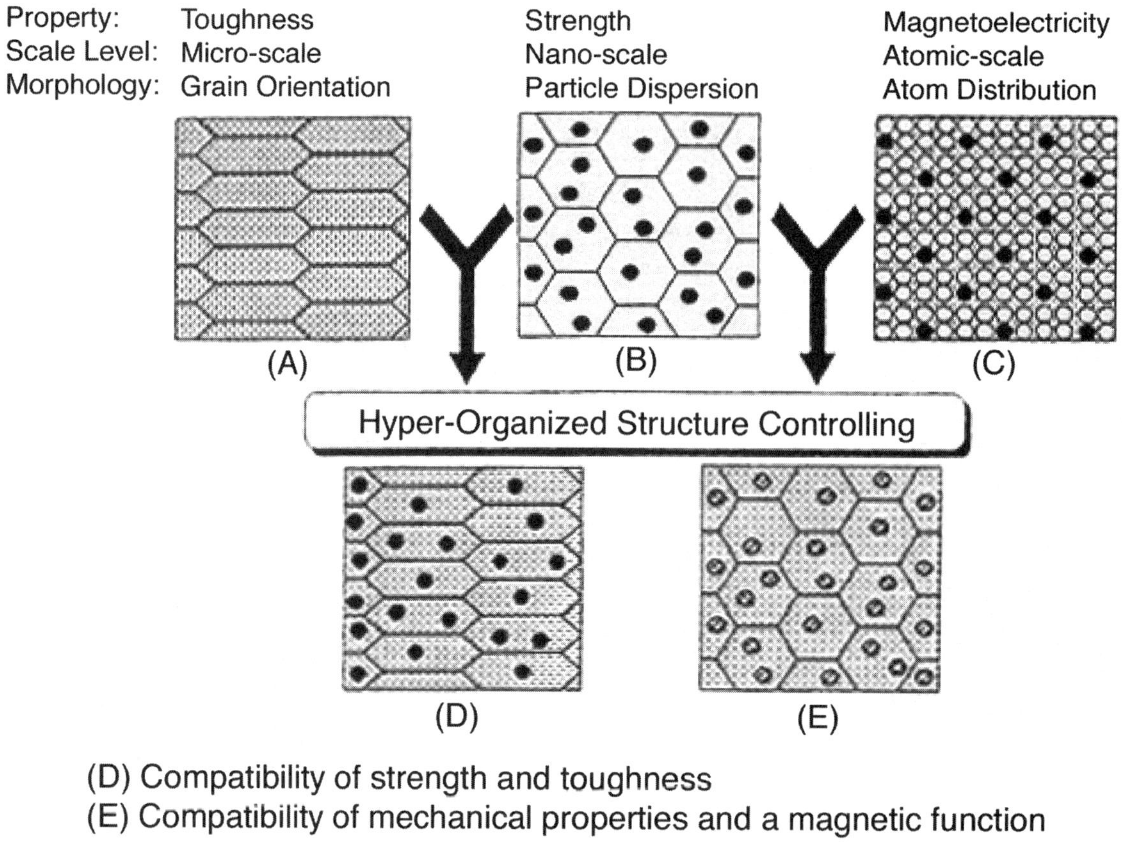

Figure 30.12 Conceptual Diagram of Hyper-Organized Structure Control (Reprinted with permission from ref. [17])

- 3pt bend testing is being used for some strength results.
- Several different methods are used for K_{IC} measurements.
- Porous structures are vulnerable to increased oxidation.
- References to high K_{IC} may be for large cracks, not small cracks. R curves should be included (Standards needed).
- Obtaining meaningful fracture resistance values for crack lengths less than 50 microns in length remains a concern.
- Component properties can be much lower than those of baseline materials, e.g., creep rates can vary 4 to 100X [87]. Typically, this issue has not been addressed.

INNOVATIVE MATERIALS DEVELOPMENTS THAT MAY HAVE FUTURE IMPLICATIONS FOR STRUCTURAL Si_3N_4

The U.S. does not have an analogous coordinated synergy ceramics program but a small number of individual research efforts are investigating similar aspects of nanoscale structural ceramic composition and structure. Additionally, two other specific areas of ceramic composition/ structure research hold the potential to significantly improve structural ceramic properties—biomimetics and hierarchical structured materials.

One potential path to improving structural ceramic material properties is to simulate structures in nature—referred to as biomimetics [88]. A number of examples exist in nature of strong structures being generated from weak starting materials due to ingenious nanoscale and hierarchical structures (e.g. seashells being made from calcium carbonate). One of the key tools used in nature is to "self-assemble" inorganic nanoscale structures using biological (organic) processes [89]. Researchers now are able to generate a myriad of nanoscale structures with materials such as SiO_2. The self-assembled SiO_2 structures have been shown to demonstrate unique elastic modulus properties in bulk material, compared to SiO_2 fabricated from traditional processes such as melting and sintering [90]. This is a very nascent field, but one that holds promise for providing tools to allow generation of desired nanoscale structures in structural ceramics.

At the other end of the size scale, since the mid-1980s a number of U.S. researchers have been developing structural ceramic materials technologies that generate "macrostructures" in monolithic materials to provide improved toughness, an example being fibrous monoliths [91]. The scale of structure is one to two orders of magnitude greater (e.g. 200–500 μm) than the ceramic grain size. With careful tailoring of interface properties between these macroscopic structures, effective debonding can be generated and toughnesses of 15–20 $MPa{\cdot}m^{0.5}$ can be demonstrated. Although current fibrous monolith materials exhibit some significant limitations, such as temperature limited interfaces (e.g. BN) and relatively low strengths (150–300 MPa), there is obvious potential for improvement in temperature and strength capability. The extension of biomimetic or fibrous monolith technologies may enhance the durability of advanced structural silicon nitrides.

A key need for optimization of these synergy/hierarchical materials concepts is the ability to model and predict material properties from composition and structure designs. The National Institute of Science and Technologies' (NIST) Object Oriented Finite Element Analysis of Real Material Microstructures (OOF) program [92] is a useful first generation tool. While currently limited to 2-D, one can input virtual or actual microstructures to establish the grid. Although its strength is in solving elasticity and thermal diffusion problems, one can gain some useful insights about microstructural effects [93]. Ideally, a tool is needed that allows plug-in of material properties (e.g. for silicon nitride—silicon nitride grain properties, second phase properties, and grain boundary interface properties) and predicts resulting properties for a given nano/micro/macrostructure based on applied external conditions—mechanical load, environment, temperature, etc. Even more desirable, given a set of material properties and microstructural constraints, is a model that would predict optimum nano/micro/macrostructure and compositions for a desired set of properties and component geometries.

ADVANCES IN FABRICATION TECHNOLOGY

Concomitant with the needed refinements in microstructure and material composition is the ability to fabricate functional complex shaped components. Structural ceramic component fabrication continues to be time consuming and expensive. The fabrication process for gas turbine parts typically requires iterations in the forming process to accommodate shrinkage and distortion during densification, because they are not robustly predictable. Forming large parts such as turbine blisks presents a challenge to obtaining uniform green and sintered densities without flaws such as voids and cracks. Advanced forming technologies such as gelcasting and hybrid molding have shown the ability to generate engine quality complex shaped silicon nitride and silicon carbide components but are still immature in regard to production volume and quality capability.

The development of structural ceramics with spatially complex microstructures and compositions at multiple scales adds even further complexity to fabrication processes. Although they are in early development, a variety of rapid prototyping and solid freeform fabrication approaches such as direct photo shaping [85], fused deposition [94], and shape deposition manufacturing [95] have the potential to fabricate these spatially complex materials. They require substantial further development to improve part quality, spatial and size capability, and throughput rate to meet gas turbine component requirements. Densification of these materials presents equally significant challenges including differential densification of the various materials and structures and the resulting possibility of distortion. Effective shrinkage stress/densification models and resulting predictive software (now available for metallic superalloy components) are a critical need to support the fabrication of complex shape and large structural components implementing these advanced microstructures and compositions.

ROADMAP FOR IMPROVED STRUCTURAL RELIABILITY

The goals for the development of improving the reliability of Si_3N_4 should be to develop economical and robust methods to optimize and control microstructures to provide:

- High strength (>1000 MPa), fracture toughness (> 10 MPa.m$^{1/2}$), and Weibull modulus (>20).
- Oxidation and creep resistance that meets or exceeds that of SN281.
- Strength retention and microstructural stability at both intermediate and elevated temperatures in static and cyclic conditions.
- Resistance to degradation by environmental corrosion/oxidation.

The overall approach to accomplishing these goals is to further the understanding and optimization of:

- Bimodal microstructures
- Grain boundary phase that promote the desired grain growth, IGF bond strength, viscosity, nitrogen content
- The use of Si_2N_2O formers such as $MoSi_2$, SiC or Mo_2C particulates or other reducing additives
- The use and benefits of polymer precursors
- The use and benefits of nanocomposites
- Energy absorbing surfaces that have high strength and toughness
- Application of the optimized material to complex components

CONCLUSIONS

Although great progress has been made in the development of structural ceramic materials for gas turbine hot section applications, a number of limitations remain that limit their effective implementation, including strength, toughness, and oxidation and environmental resistance properties. We have presented possible near-term (e.g. silicon nitride microstructure and grain boundary phase composition refinement, EBC coatings) and longer-term solution paths. New approaches emphasizing "synergy" ceramics and hierarchical structures, although at a very early concept and development stage, are showing that antagonistic properties such as toughness and strength can be simultaneously optimized. Effective component fabrication processes must be developed simultaneously with these new materials to assure achievable shape fabrication, part quality, and cost competitiveness. Unless all the critical property limitations are addressed and overcome simultaneously, new materials will always fall short of the demands of gas turbine engines. The Japanese Synergy Ceramics program offers a model of that approach.

ACKNOWLEDGEMENT

Unless otherwise indicated the figures and tables in this chapter were supplied by the authors.

REFERENCES

[1] Wiederhorn, S. M., and Lawn, B. R., "Strength Degradation of Glass Resulting from Impact with Spheres," J. Am. Ceram. Soc., 60[9–10], 451–458,1977.

[2] Choi, S. R., Pereira, J. M., Janosik L. A., and Bhatt, R. T., "Foreign Object Damage of Two Gas-Turbine Grade Silicon Nitrides at Ambient Temperature," Cer. Eng. and Sci. Proc., 23[3], 193–202, 2002.

[3] van Roode, M., Jimenez, O., McClain, J., Price, J., Parthasarathy, V., Poormon, K. L., Ferber, M. K., and Lin, H.-T., "Ceramic Gas Turbine Materials Impact Evaluation," ASME paper GT-2002-30505, presented at the ASME TURBO EXPO Land, Sea & Air, Amsterdam, The Netherlands, June 3–6, 2002.

[4] Price, J. R., Jimenez, O., Faulder, L., Edwards, B., and Parthasarathy, V., "Ceramic Stationary Gas Turbine Development—Fifth Annual Summary," ASME paper 98-GT-181, presented at the International Gas Turbine and Aeroengine Congress and Exhibition, Stockholm Sweden, June 2–5, 1998.

[5] Klemm, H., Herrmann, M., and Schubert, C., "Silicon Nitride Composites with an Improved High Temperature Oxidation Resistance," Ceram. Eng. Sci. Proc. 18[3], 615–623, 1997.

[6] Wereszczak, A. A., Ferber, M. K., Kirkland, T. P., More, K. L., Foley, M. R., and Yeckly, R. L., "Evolution of Stress Failure Resulting from High-Temperature Stress Corrosion Cracking in a Hot Isostatically Pressed Silicon Nitride," J. Am. Ceram. Soc., 78 [8], 2129–2140, 1995.

[7] Tressler, R. E., "Environmental Effects on Long Term Reliability of SiC and Si_3N_4 Ceramics," Ceram. Trans., 10, 99–124, 1990.

[8] Van der Biest, O., Weber, C., and Garguet, L. A., "Role of Oxidation on Creep and High Temperature Failure of Silicon Nitride," *Proc. 3rd Int. Symp. Ceram Mater. & Components for Engines*, V. Tennery, ed., Las Vegas, Nevada, USA, Nov. 27–30, 1988, American Ceramic Soc., Columbus, OH., USA, 1989. pp. 729–738.

[9] Nishimura, N., Masuo, E., and Takita, K., "Effect of Microstructural Degradation on the Strength of Sintered Silicon Nitride after High Temperature Exposure," *Proceedings of the 4th International Symposium on Ceramic Materials and Components for Engines*, Göteborg, Sweden, June 10–12, 1991, R. Carlson, T. Johansson, and T. Kahlman, eds., Elsevier Science Publishers, Ltd., Barking, Essex, England, UK. 1992. pp. 1139–1145.

[10] Smialek, J.L., Robinson, R.C., Opila, E.J., Fox, D.S., and Jacobson, N.S., "SiC and Si_3N_4 Recession Due to SiO_2 Volatility under Combustor Conditions," Adv. Composite Mater., 8 [1], 33–45, 1999.

[11] Eaton, H. E., Linsey, G. D., Sun, E. Y., More, K. L., Kimmel, J. B., Price, J. R., and Miriyala, N., "EBC Protection of SiC/SiC Composites in the Gas Turbine Combustion Environment—Continuing Evaluation and Refurbishment Considerations", ASME Paper 2001-GT-0513, presented at the 46th ASME TURBO EXPO 2001, LAND, SEA, & AIR, New Orleans, LA, USA, June 4–7, 2001.

[12] Price, J., Jimenez, O., Miriyala, N., Kimmel, J.B., Leroux, D.R., and Fahme, T., "Ceramic Stationary Gas Turbine Development Program—Eighth Annual Summary," ASME paper 2001 GT-0517, presented at the 46th ASME TURBO EXPO 2001, LAND, SEA, & AIR, New Orleans, LA, USA, June 4–7, 2001.

[13] Carruthers, W. D., van Roode, M., Becher, P. F., Ferber, M.K., and Pollinger, J., "Advances in the Development of Silicon Nitride and Other Ceramics," ASME paper GT-2002-30504, presented at the 47th ASME TURBO EXPO 2002, LAND, SEA, & AIR, Amsterdam, The Netherlands, June 3–6, 2002.

[14] Lofaj, F., Wiederhorn, S. M., Long, G. G., Jemian, P. R., and Ferber, M. K., "Cavitation Creep in the Next Generation Silicon Nitride", in *Ceramic Materials and Components for Engines, 7th International Symposium*, Goslar, Germany, June 19–21, 2000, J. G. Heinrich and F. Aldinger, eds., Wiley VCH Verlag GmbH, Weinheim, Germany, 2001. pp. 487–493.

[15] Wiederhorn S. M. and Ferber, M. K., "Silicon Nitride for Gas Turbines," Current Opinion in Solid State & Mater. Sci., 5, 311–316, 2001.

[16] Hirao, K., Ohashi, M., Brito, M., and Kanzaki, S., "Processing Strategy for Producing Highly Anisotropic Silicon Nitride," J, Am. Ceram. Soc., 78 [6], 1687–1690, 1995.

[17] Kanzaki, S. and Shimada, M., Komeya, K., and Tsuge, A., "Recent Progress in the Synergy Ceramics Project" in Key Engineering Materials Vols. 161–163, 437–442. Copyright © Trans Tech Publications Inc., Uetikon-Zürich, Switzerland, 1999.

[18] Watari, K. and Shinde, S. L., "High Thermal Conductivity Materials," MRS Bulletin, 26[6], 440–441, 2001.

[19] Becher, P. F., Sun, E. Y., Plucknett, K. P., Alexander, K. B., Hsueh, C-H, Lin H-T, Waters, S. B., Westmoreland, C. G., Kang, E-S, Hirao, K., and Brito, M., "Microstructural Design of Silicon Nitride with Improved Fracture Toughness, Part I: Effects of Grain Shape and Size," J. Am. Ceram. Soc., 81[11], 2821–2830, 1998.

[20] Becher, P. F., Hsueh, C. H., Alexander, K. B., and Sun, E. Y., "The Influence of Reinforcement Content and Diameter on the R-Curve Response in SiC Whisker-Reinforced Alumina," J. Am. Ceram. Soc., 79[2], 298–304, 1996.

[21] Becher, P. F., "Microstructural Design of Toughened Ceramics," J. Am. Ceram. Soc., 74[2], 255–269, 1991.

[22] Evans, A. G., "Perspective on the Development of High-Toughness Ceramics," J. Am. Ceram. Soc., 73[1], 187–206, 1990.

[23] Hirao, K., Nagaoka, T., Brito, M. E., and Kanzaki, S., "Microstructure Control of Silicon Nitride by Seeding with Rodlike β-Silicon Nitride Particles," J. Am. Ceram. Soc., 77[7], 1857–1862, 1994.

[24] Hirao, K., Nagaoka, T., Brito, M. E., and Kanzaki, S., "Mechanical Properties of Silicon Nitrides with Tailored Microstructures by Seeding," J. Ceram. Soc. Jpn, 104, 55–59, 1996.

[25] Cao, J. J., Moberly Chan, J., De Jonghe, L. C., and Gilbert, C., "Carbide with Al-B-C Additions," J. Am. Ceram. Soc., 79[2], 461–469, 1996.

[26] Kim, J., Rosenflanz, A., and Chen, I. W., "Microstructure Control of In-Situ Toughened α'-SiAlON Ceramics," J. Am. Ceram. Soc., 83[7], 1819–1821, 2000.

[27] Becher, P. F., Sun, E. Y., Hsueh, C. H., Alexander, K. B., Hwang, S. L., Waters, S. B., and Westmoreland, C. G., "Debonding of Interfaces Between Beta-Silicon Nitride Whiskers and Si–Al–Y Oxynitride Glasses," Acta Metall., 44[10], 3881–3893, 1996.

[28] Sun, E. Y., Becher, P. F., Hsueh, C. H., Painter, G. S., Waters, S. B., Hwang, S. L., and Hoffmann, M. J., "Debonding Behavior between β-Si_3N_4 Whiskers and Oxynitride Glasses with and without an Epitaxial β-SiAlON Interfacial Layer," Acta Mater., 47[9], 2777–2785, 1999.

[29] Sun, E.Y, Becher, P. F., Hsueh, C-H, Waters, S. B., Plucknett, K. P., Hirao, K., and Brito, M., "Microstructural Design of Silicon Nitride with Improved Fracture Toughness, Part II: Effects of Additives," J. Am. Ceram. Soc., 81[11], 2831–2840, 1998.

[30] Becher, P. F., Painter, G. S., Sun, E. Y., Hsueh, C. H., and Lance, M. J, "The Importance of Amorphous Intergranular Films in Self-Reinforced Si_3N_4 Ceramics," Acta Mater., 48[12], 4493–4499, 2000.

[31] Sasa T., "Research and Development of Engineering Ceramics, High Strength and High Toughness Advanced Ceramics with Novel Multilayered Structure," presented at the French-Japanese Scientific Symposium, Materials Science and Technology Session, December 16–17, 1997.

[32] Inagaki, Y., Shigegaki, Y., Kondo, N., Suzuki, Y., and Ohji, T., "Enhanced Fracture Resistance of Highly Anisotropic Porous Silicon Nitride," Ceram. Trans., 118, 355–362, 2000.

[33] Inagaki,. Y., Ohji, T., Kanzaki, S., Shigagaki, Y., “Fracture Energy of an Aligned Porous Silicon Nitride,” J. Am. Cer. Soc., 83 [7], 1807–1809, 2000.

[34] Ohji, T., Goto, Y., and Tsuge, A., “High-temperature Toughness and Tensile Strength of Whisker-Reinforced Silicon Nitride,” J. Am. Ceram. Soc., 74[4], 739–745, 1991.

[35] Kondo, N., Inagaki, Y., Suzuki, Y., and Ohji, T., “High Temperature Fracture Energy of Superplastically Forged Silicon Nitride,” J. Am. Ceram. Soc., 84[8], 1791–1796, 2001.

[36] Probst, H. B., “Structural Ceramics in Heat Engines—the NASA Viewpoint”, *Proc. of International Symposium on Ceramic Components for Engines*, Hakone, Japan, Oct. 17–19, 1983, S. Sōmiya, E. Kanai, and K. Ando, eds., Elsevier Applied Science, London, UK, 1984. pp. 45–58.

[37] Yang, J-F., Zhang, G-J., Ohji, T., “Fabrication of Low-Shrinkage, Porous Silicon Nitride Ceramics by Addition of a Small Amount of Carbon,” J. Am. Ceram. Soc., 84[7], 1639–1641, 2001.

[38] Zhang, G., Yang, J-F., Ohji, T., ”Fabrication of Porous Ceramics with Unidirectionally Aligned Continuous Pores.” J. Am. Ceram. Soc., 84[6], 1395–1397, 2001.

[39] Shigegaki, Y., Brito. M. E., Kirao, K., Toriyama, M., and Kanzaki, S., ”Strain Tolerant Porous Silicon Nitride,” J. Am. Ceram. Soc., 80[2], 495–498, 1997.

[40] Shigegaki, Y., Brito. M. E., Kirao, K., Toriyama, M., and Kanzaki, S., ”Processing of a Novel Multilayered Silicon Nitride,” J. Am. Ceram. Soc., 79[8], 2197–2200, 1998.

[41] van Roode, M., “Ceramic Stationary Gas Turbine Development Program, Japan Trip Report,” DOE Contract DE-AC02-92CE40960, December 10, 1999.

[42] Pezzotti, G., Nishida, T., Sakai, M., “Physical Limitations of the Inherent Toughness and Strength in Ceramic-Ceramic and Ceramic-Metal Nanocomposites,” J. Ceram. Soc. Jpn., 103[9]. 901–909,1995. Also, Pezzotti, G., Nishida, T., and Sakai, M., “Reply to “Comments on ‘Physical limitations of the Inherent Toughness and Strength in Ceramic-Ceramic and Ceramic-Metal Nanocomposites’”,” J. Ceram. Soc. Jpn., 104[1210], 583–585, 1995.

[43] Sternitzke, M., “Review: Structural Ceramic Nanocomposites,” J. Eur. Ceram. Soc., 17, 1061–1082, 1997.

[44] Niihara, K., “New Design Concept of Structural Ceramics—Ceramic Nanocomposite,” J. Ceram. Soc. Jpn., 99[10], 974–982, 1991.

[45] Ohji, T., Jeong, Y-K., Choa, Y. H., and Niihara, K., “Strengthening and Toughening Mechanisms of Ceramic Nanocomposites,” J. Am. Ceram. Soc., 81 [6], 1453–1460, 1998.

[46] Hirao, K., Ohashi, M., Brito, M.E., and Kanzaki, S., “Processing Strategy for Producing Highly Anisotropic Silicon Nitride,” J. Am. Ceram. Soc., 78[6],1687–1690, 1995.

[47] Kondo, N., Ohji., T., and Wakai, F., “Strengthening and Toughness of Silicon Nitride by Superplastic Deformation,” J. Am. Ceram. Soc., 81[3], 713–716, 1998.

[48] Kondo, N., Suzuki, Y., and Ohji, T., “Superplastic Sinter-Forging of Silicon Nitride with Anisotropic Microstructure Formation,” J. Am. Ceram. Soc., 82[4], 1067–1069, 1999.

[49] Yang, J-F., Sekino, T., Choa, T-H., Niihara,, K., and Ohji, T., “Microstructural and Mechanical Properties of Sinter-Post-HIPed Si_3N_4-SiC Composites,” J. Am. Ceram. Soc., 84[2], 406–412, 2001.

[50] Petrovic, J. J., Pena, M. I., Reimanis, I. E., Sandlin, M. S., Conzone, S. D., Kung, H. H., and Butt, D. P., “Mechanical Behavior of $MoSi_2$-Reinforced Si_3N_4 Matrix Composites,” J. Am. Ceram. Soc., 80[12], 3070–3076, 1997.

[51] Filsinger, D., Schultz, A., Wittig, S., Taut, C., Klemm, H., and Wötting, G., "Model Combustor to Assess the Oxidation Behavior of Ceramic Materials Under Real Engine Conditions," ASME paper 99-GT-349, presented at the 44th ASME Gas Turbine and Aeroengine Technical Congress, Exposition and Users Symposium, June 7–10, 1999, Indianapolis, IN, USA.

[52] Yoshida, A., Isozaki, K., and Nakajima, Y, *Particulate-Reinforced Silicon Nitride Materials*, Synergy Ceramics Project Report, Japan. Publication source unknown.

[53] Park, D-S., Han, B-D., Kim, H-D., Choi, H-J., and Lim, D-S., "Erosion of Silicon Nitride Based In-Situ Composites", paper presented at the 104th Annual Am Ceram. Soc. Meeting and Exposition, St Louis, MO, April 28–May 1, 2002

[54] Emoto, H. and Mitomo, M., "Control and Characterization of Abnormally Grown Grains in Silicon Nitride Ceramics," J. Eur. Ceram. Soc., 17, 797–804, 1997.

[55] Tajima, Y. and Urashima, K., "Improvement of Strength and Toughness of Silicon Nitride Ceramics," in Tailoring of Mechanical Properties of Si_3N_4 Ceramics, M. J. Hoffmann and G. Petzow, eds., Kluwer Academic Publishers, Dordrecht, The Netherlands, 1994. pp. 101–109.

[56] Hoffmann, M. J., 1994, "Analysis of Microstructural Development and Mechanical Properties of Si_3N_4 Ceramics," in Tailoring of Mechanical Properties of Si_3N_4 Ceramics, M. J. Hoffmann and G. Petzow, eds., Kluwer Academic Publishers, Dordrecht, The Netherlands, 1994. pp. 59–72.

[57] Sun, E.Y., Becher, P. F., Hwang, S. L., Waters, S. B., Pharr, G. M., and Tsui, T. Y., "Properties of Silicon-Aluminum—Yttrium Oxynitride Glasses," J. Non-Cryst. Solids, 208 [11], 162–169, 1996.

[58] Becher, P. F., Waters, S. B., Westmoreland, G., and Riester, L., "Compositional Effects on the Properties of Si-Al-RE-Based Oxynitride Glasses (RE = La, Nd, Gd, Y, or Lu)," J. Am. Ceram. Soc., 85[4], 897–902, 2002.

[59] Yoshida, M, Tanaka, K., Kubo, T., Terazona, H, and Tsurazono, S, "Development of Ceramic Components for Ceramic Gas Turbine Engine (CGT302)", ASME paper 98-GT398, presented at the International Gas Turbine & Aeroengine Congress & Exhibition, Stockholm, Sweden, June 2–5, 1998.

[60] Patel, J. K., Thompson, D. P., "Further studies of the Low-Temperature Oxidation Problem in Yttria-Densified Silicon Nitride Ceramics," *Third International Symposium on Ceramic Material and Components for Engines*, Las Vegas NV, November 27–30, 1988, V. J. Tennery, ed., American Ceramic Society, Columbus, OH, USA, 1989. pp. 987–996.

[61] Lin, H.-T., "Creep Response of Silicon Nitride with a Highly Anisotropic Microstructure," To Be Published.

[62] Wiederhorn, S. M. and Ferber, M. K., "Silicon Nitride for Gas Turbines," Current Opinions in Solid State and Mater. Sci., 5, 311–316, 2001.

[63] Klemm, H., Schubert, C., "Silicon Nitride/Molybdenum Disilicide Composite with Superior Long Term Oxidation Resistance at 1500°C," J. Am. Ceram. Soc., 84[10], 2430–2432, 2001.

[64] Hirosake, H., Okamoto, Y., Ando, M., Munakata, F., and Akimune, Y., "Thermal Conductivity in Gas—Pressure-Sintered Silicon Nitride," J. Am. Ceram. Soc., 79[11], 2878–2882, 1996.

[65] Hirao, K., Watari, K., Brito, M. E., Toriyama, M., and Kanzaki, S., "High Thermal Conductivity in Silicon Nitride with Anisotropic Microstructure," J. Am. Ceram. Soc., 79[9], 2485–2488, 1996.

[66] Samsonov, G. V., *The Oxide Handbook*, IFI/Plenum Press, New York, NY, USA, 1973.

[67] Hampshire, S., Drew, R. A. L., and Jack, K. H., "Oxynitride Glasses," Phys. Chem. Glasses, 26[5], 182–186, 1985.

[68] Pollinger, J. P., "The Development of Monolithic Silicon Nitride Structural Ceramics at Honeywell," Chapter 5 in *Ceramic Gas Turbine Component Development and Characterization, Volume II of Progress in Ceramic Gas Turbine Development*, M. van Roode, M. K. Ferber, D. W. Richerson, eds, ASME PRESS, New York, NY, USA, 2003.

[69] Pollinger, J. P. "Improved Silicon Nitride Materials and Component Fabrication Processes for Aerospace and Industrial Gas Turbine Applications," ASME Paper 95-GT-19, presented at the International Gas Turbine and Aeroengine Congress and Exposition, Houston, TX, USA, June 5–8, 1995.

[70] Boeing Rocketdyne, Unpublished Test Results, Canoga Park, CA, 2000.

[71] Rendtel, A., Hubner, H., "Creep Resistant Silicon Nitride Ceramics—Approaches of Microstructural Design," Ceram. Eng. and Sci. Proc., 21[4], 515–526, 2000.

[72] Rendtel, P., Rendtel, A., and Hubner, H., "Mechanical Properties of Gas Pressure Sintered Si_3N_4/SiC Nanocomposites," J. Eur. Ceram Soc., 22, 2061–2070, 2002.

[73] Krause, R. F., Wiederhorn, S. M., and Li, C-W., "Tensile Creep Behavior of a Gas-Pressure-Sintered Silicon Nitride Containing Silicon Carbide", J. Am. Ceram. Soc., 84[10], 2394–2400, 2001.

[74] Yamada, K. and Kamiya, N., "High Temperature Mechanical Properties of Si_3N_4-$MoSi_2$ and Si_3N_4-SiC Composites with Network Structures of Second Phases," Materials Science and Engineering, A261, [70], 270–277, 1999.

[75] Hirano, T., Niihara, K., Ohji, T., and Wakai, F., "Improved Creep Resistance of Si_3N_4/SiC Nanocomposites Fabricated from Amorphous Si-C-N Precursor Powder", J. Mater. Sci. Letters, 15, 505–507, 1996.

[76] Woetting, G., Frassek, L., and Gugel, E., "New Developments on Silicon Nitride-Based Ceramics for High-Temperature Applications," *Proceedings of the 3rd European Workshop on High Temperature Materials*, Stuttgart, Germany, October 23–27, 1995. pp. 40–45.

[77] Du, H., Tressler, R. E., Spear, K. E., and Pantano, C. G. "Oxidation Studies of Crystalline CVD Silicon Nitride," J. Electrochem. Soc., 136, 1527–1536, 1989.

[78] Du, H., Tressler, R. E., and Spear, K. E., "Thermodynamics of the Si-N-O System and Kinetic Modeling of Oxidation of Si_3N_4," J. Electrochem. Soc., 136[11], 3210–3215, 1989.

[79] Tressler, R. E., Spear, K. E., Zheng, Z., and Di, H., "Fundamental Studies of the Oxidation of Silicon Carbide Crystals and CVD Silicon Nitride," in High Temperature Corrosion of Technical Ceramics, R. J. Fordham, ed., Elsevier Applied Science, 1990. pp 69–89.

[80] Ohashi, M., Kanzaki, S., and Tabata, H., "Processing, Mechanical Properties, and Oxidation of Silicon Oxynitride Ceramics," J. Am. Ceram Soc., 74[1], 109–114, 1991.

[81] Emoto, H., Mitomo, M., Wang, C.-M., Hirosturu, H., and Inaba, T., "Fabrication of Silicon-Nitride Oxynitride In-Situ Composites," J. Eur. Cer. Soc. 18, 527–533, 1998.

[82] Park, D.-S., Choi, H.-J., Han, B.-D., Kim, H.-D., and Lim, D.-S., "Effect of Si_2N_2O Content on the Microstructure, Properties, and Erosion of Silicon Nitride—Si_2N_2O in Situ Composites," J. Mater. Res., 17[9], 2275–2280, 2002.

[83] Klemm, H., Taut, C., and Wötting, G., "Long Term Stability of Nonoxide Ceramic in an Oxidative Environment at 1500°C," J. Eur. Ceram Soc., 23, [4], 619–627, 2003.

[84] Klemm, H., "Corrosion of Silicon Nitride Materials in Gas Turbine Environment," J. Eur. Ceram. Soc., 22[14–15], pp. 2735–2740, 2002.

[85] Ventura, S., Narang, S., Guerit, P., Liu, S., Twait, D., Khandelwal, D., Cohen, E., and Fish, R."Freeform Fabrication of Functional Silicon Nitride Components by Direct Photo Shaping," Materials Research Society Proceedings, Warrendale, PA, USA, Materials Research Society, 625, 81–90, 2000.

[86] "Synergy Ceramics," Joint Research Consortium of Synergy Ceramics, Fine Ceramics Research Association, 1997.

[87] Lin, H.-T., Becher, P. F., Ferber, M. K., and Parthasarathy, V., "Verification of Creep Performance of a Ceramic Gas Turbine Blade," *The Science of Engineering Ceramics II, Proceedings of the 2nd International Symposium on the Science of Engineering Ceramics (EnCera'98)*, K. Niihara, T. Sekino, E. Yasuda, and T. Sasa, eds., Osaka, Japan, September 6–9, 1998, CSJ Series, Vol. 2, 1998. Trans Tech Publications Inc., Zürich, Switzerland, 1998. 671–674. Key Engineering Materials, Vol. 161–163, 1999.

[88] Aksay, I. A., "Nanostructured Ceramics through Self-Assembly," in *R&D Status and Trends in Nanoparticles, Nanostructured Materials, and Nanodevices in the United States*, R. W. Siegel, E. H. Hu, and M. C. Roco, eds., International Technology Research Institute, WTEC, Baltimore, MD, USA, 1998. pp. 79–82.

[89] Kresge, C. T., Leonowicz, W. J., Roth, W. J., Vartuli, J. C., and Beck, J.S., "Ordered Mesoporous Molecular Sieves Synthesized by a Liquid Crystal Template Mechanism," Nature, 359, 710–712, 1992.

[90] Wu, J., Liu, X., and Tolbert, S.H., "High Pressure Stability in Ordered Mesoporous Silicas—Rigidity and Elasticity through Nanometer Scale Arches," J. Phys. Chem. B104, 11837–11841, 2000.

[91] Lienard, S. Y., Kovar, D., Moon, R. J., Bowman, K. J., and Halloran, J. W., "Texture Development in Silicon Nitride/Boron Nitride Fibrous Monolith Ceramics," J. Mater. Sci. 35[13], 3365–3371, 2000.

[92] Langer, S.A., Fuller, E. R., and Carter, W. C., "OOF: An Image-Based Finite-Element Analysis of Material Microstructures," Computing in Science and Engineering, 3[3], 15–23, 2001.

[93] Hsueh, C. H., Becher, P. F., and E.Y. Sun, "Analyses of Thermal Expansion Behavior of Intergranular Two-Phase Composites," J. Mater. Sci., 36[1], 255–261, 2001.

[94] Agarwala, M. K., Bandyopadhyay, A., van Weeren, R., Safari, A., Danforth, S. C., Langrana, N. A., Jamalabad, V. R., and Whalen, P. J., "Rapid Fabrication of Structural Components by Fused Deposition" Am. Ceram. Soc. Bull., 75[11], 60–65,1996.

[95] Kang, S., Stampfl, J., Cooper, A. G., and Prinz, F. B., "Application of the Mold SDM Process to the Fabrication of Ceramic Parts for a Micro Gas Turbine Engine." in *Ceramic Materials and Components for Engines, 7th International Symposium*, Goslar, Germany, June 19–21, 2000. J. G. Heinrich and F. Aldinger, eds., Wiley VCH Verlag GmbH, Weinheim, Germany, 2001. pp. 359–364.

INDEX

A

AC Cerama 20
Acoustic damping 285
Acoustic emission 46
Admiralty Materials Laboratory 16
Advanced Ceramic Manufacturing Technology Program (ACMT) 66
Advanced Gas Turbine Project 8
Advanced Gas Turbines for Automobiles (AGATA) 20
Advanced Materials Engineering (AME) 16
Advanced Technology Program (ATP) 111
Advanced Turbine Technology Applications Project (ATTAP) 60, 88
Aerospatiale 20
AGATA 212, 219
AGT 84
AGT-5 67,88,133
AGT101 67, 115
Air-coupled ultrasonics 507
AiResearch Casting Company 3, 97
AiResearch Manufacturing Company of Arizona 7
Air Force 5
AlliedSignal Ceramic Components 4, 98
AlliedSignal Composites Inc. 263
AlliedSignal Corporation 97
Allison 67
Allison/CARES 588
Allison scroll 63
Allison Turbine Engines 7
Alpha alumina 267
Alstom 21
Alumino-boron-nitride 246
Aluminum 267
Aluminum-boron-nitride compositions 15
AMG 314
ANB 248
Annawerk-Ceranox 17, 179
Applications 269
Aqueous injection molding 114
Aqueous slurry process 657
Argonne National Laboratory (ANL) 7
Army Materials and Mechanics Research Center (AMMRC) 3, 5, 58
Asahi Glass Co., Ltd. 165
Asea 4, 19
Asea Cerama AB 200
AS800 101, 116, 680, 751
AS800 Si_3N_4 371, 627
ASME Boiler and Pressure Vessel Code 541
Atomic Energy Laboratory at Harwell 16
AVCO Specialty Materials Division 62
Axial turbine wheels 203
AY-6 7

B

Batdorf's Theory 561
Battelle 4
Bayer 195
Bayer AG 21
Bearings 208, 727
Beta seeds 748
Biaxial flexure 377
Binder 41
Biomimetics 759
Bisque firing 109
Bisque machining 109, 134
Blade fillet radius 687
Blades 418
BN 262, 336
BN/SiC 332
Boostec Industries 20

BOOSTEC SA 211
Borosilicon-nitride fiber 21
Boston University 642
Boyko Research-Engineering Ceramic Heat Engines Center 245
Braze 93
Brazing 419
Brite/Euram project 21
British Ceramics Research Association 16
British Nuclear Fuels, Ltd. 16
BSAS (BaO-SrO-Al_2O_3-SiO_2) 643
B-SiC 7
Burner rig 612, 648, 698
Burst speed 86, 89
Bustor afterburner cone 20

C

Carborundum Company 6, 77
Calcined block machining 159
Calendaring 220
Carbon-carbon grips 427
CARES/Life 580, 596
CARES Users Manual 575
Catalytic igniter 237
CATE 78
Cavitation 366
C/C composites 643
CCLife 597
Censoring techniques 566
Central Institute of Aviation Motors (CIAM) 425
Ceradyne Inc. 36, 49
CERAM™ 576
Ceramic Applications for Turbine Engines Program 78
Ceramic Components for Turbine Engines Program 98
Ceramic Engine Demonstration Program 641
CERAMIC/ERICA 583
CERAMIC/ERICA probabilistic life prediction codes 574
Ceramic Gas Turbine Engine Demonstration 106
Ceramic Life Prediction Method for Vehicular Engines 588
Ceramic Matrix Composite (CMC) 12, 21, 259, 314, 441, 686, 704, 725
Ceramic powder 178
Ceramic Stationary Gas Turbine (CSGT) 444
Ceramic Technology for Advanced Heat Engines 62
Ceramic Turbine Engine Demonstration Program 581
Céramiques et Composites (C&C) 20, 214
CeramTec AG 195
Ceraset™ 262
Cerbec 64
CFCC inner scroll support 478
CFI 195
CGT301 161
CGT302 engine 138
Chemical Vapor Infiltration (CVI) 263
ChevronTexaco 273
ChevronTexaco, Bakersfield, CA 643
Coatings 641
Coatings for C/C 659
Code 2 silicon nitride 99
Coefficient of friction 670
Coefficient of thermal expansion 65, 148, 166, 178, 205, 225, 226, 230, 428, 468, 621, 645
COI ceramics 277
Cold isostatic pressing (CIP) 158, 167, 706
Colloidal processing 64
Combustion chambers 418
Combustor 267
Combustor liners 57, 270, 307, 314, 344, 443, 724
Compliant layer 713
Component development 67
Component fabrication 40, 106, 157, 179, 270, 280, 705
Composites 414
Compressive strength 531
Compressive surface stress 756
Computed tomography 7, 90, 496
Contact damage 642
Contact stress 6, 696
Contact zone 674
Continuous Fiber Ceramic Composite (CFCC) program 299, 615
Continuous Fiber-reinforced Ceramic Matrix Composites (CFCCs) 441
Controlled matrix porosity 279
Cordierite 20, 221
Corrosion 607, 619
Corrosion kinetics 610
Corrosion mechanisms 612
Cost 66, 141, 272, 298, 309, 725
Crack bridging 101, 460, 749

Creep 172, 365, 389
Creep rate 19, 129
Creep resistance 103, 745, 749, 751
Creep rupture 204, 283, 389
Critical flaw size 496
Critical velocity for damage 686
CRM process 298
CSGT 69
C/SiC 21
CSN 101 204
CSN 101 HIPed Si_3N_4 20
CT 7
Cutting tool 39, 46
Cutting tool inserts 727
CVD 63, 332, 642, 644, 655
CVD mullite 655
CVI 314
CVI C/SiC 260
CVI process 264
CVI SiC/SiC 260, 265
Cyclic fatigue 388, 469
Cyclic oxidation 621
Cyclic oxidation tests 609

D

Daimler-Benz AG, Volkswagen AG 17
Data bases 546
Data pooling 586
Defense Advanced Research Projects Agency (DARPA) 4, 111
Deflocculants 41
Degradation 476, 632, 744
Degussa 17, 184
Delamination 505
Density 34, 81
Department of Energy 299, 444
Design by analysis 542
Design by rule 542
Design codes 522, 541
Design concepts 709
Design guide 575
Detroit Diesel Allison 78
DFVLR 17, 175
Dilatometer 428
DIMOX 260
Direct metal oxidation 260, 266
Direct photo shaping 759
DOD Military Handbook 17, 447
DOE 64, 111, 654
DOE/ORNL Ceramic Technology Project 64
Dog-bone tensile specimen 376
Duo-density rotor 712
Duo-density turbine rotor 44
DuPont Lanxide Composites 260
Dye-penetrant 7, 502
Dynamic fatigue 365
Dynamic wave propagation 675

E

EBC 272, 287
EB-PVD 642, 644
Effects of molten salts 610
8-Harness Satin Weave 449
Elastic modulus 65, 148, 178
Electro-Thermal (ET) gun 668
Enabling Propulsion Materials (EPM) program 268
Engine components 418
Enhanced CVI SiC/SiC 266
Environmental Barrier Coating (EBC) 105, 390, 643, 649, 725
Environmental effects 371
ESK 17, 191
Evolution of component designs 711
Exhaust flaps 283
Exhaust panels 285
Extrusion 220

F

Failure modes 529
Feldmühle 17
Feldmühle AG 182
Fiber coating 442
Fiber debonding 442
Fiber pull-out 460
Fiber-reinforced silicon nitride 62
Fibrous monoliths 759
Field testing 273
Fine Ceramics Project 8
First matrix cracking 725
Flame swirler 240
Flaps and seals 445
Flexural strength 57, 65, 127, 148, 205, 316

Flexure strength 19, 63, 84, 99, 191, 248, 381
Flow lines 87
Focused ultrasonics 499
Ford Motor Company 3, 31
Foreign Object Damage (FOD) 138, 665, 696
Forged Si_3N_4 749
4-Point Bending Strength 12
Fracture behavior of CFCCs 464
Fracture mechanism maps 370
Fracture modes 667
Fracture surfaces 340
Fracture toughness 60, 65, 79, 102, 129, 148, 166, 205, 230, 359, 387, 406, 420, 430, 433, 686, 744
Fracture toughness testing 438
French Ministry of the Defense 217
French National Aerospace Institute (ONERA) 20
Fugitive wax slip casting 43
Fused deposition 759

G

Garrett Auxiliary Power Division (GAPD) 60
Garrett Corporation 97
Garrett Turbine Engine Company 3, 58, 97, 581, 641, 646
GE CRD 7
GE Global Research Center (GEGRC) 291
Gel casting 110, 707
GENOA 597
GE Power Systems Composites, LLC 259
German Aerospace Research Establishment 17
German Federal Ministry for Research and Technology 17, 175
German National Program 18
GFICES 575
Glass encapsulation 202
Glass frit encapsulation 6
Glass matrix composites 698
GN-10 6, 100
Graded coating 658
Graded compositions 756
Grain boundary engineering 57
Grain boundary phase 154, 411
Grain growth 472
Green density 34
GS-44 101
GTE 7, 99, 658
GTE Laboratories 642

H

Hardness tests 536
H.C. Starck Ceramics 195
Heat exchanger 161, 188, 211
Herman Starck Company 4
Hertzian cone cracks 677
Hertzian loading 671
Hexoloy SA 6
Hexoloy® SP 95
Hierarchical structures 759
High Pressure Burner Rig 624
High-temperature mechanical tests 425
Hi-Nicalon™ 12, 265, 303, 306, 332, 471
Hi-Nicalon™/MI-SiC 444
HIP 19, 59, 89, 100, 168, 191, 200, 702
HIPed turbine wheels 207
Honeywell Advanced Composites Inc. 263
Honeywell Ceramic Components (HCC) 98
Honeywell Engines, Systems & Services 6, 581
Hot corrosion 283, 658
Hot Isostatic Pressing (HIP) 4, 6
Hot Pressed Silicon Nitride (HPSN) 2, 57
Hot pressing 401
HPSN 17, 39, 181, 230, 700
HSCT-EPM (High Speed Civil Transport-Enabling Propulsion Materials) 649
HS-110 57
HS-130 57
Hybrid metal-ceramic nozzle 720
Hybrid molding 131
Hybrid rotor 713
Hyper-organized structure control 758

I

Illinois Institute of Technology Research Institute (IITRI) 5, 697
Impact 666
Impact damage 674
Impact Damage Map 687
Impact resistance 744, 749
Impact theory 668
Inex Inc. 93

Injection molding 40, 84, 91, 106, 108, 184, 203, 706
In-plane shear strength 463
Institute for Problems of Materials Science 398
In-situ reinforced 101
Intergranular phases 751
Ishikawajima-Harima Heavy Industries (IHI) 749
Isopressing 108
Iterative design approach 695
Image-enhancement 7
Improvements in strength 701, 703, 704
Infrared imaging 505
INSTRON 428
Integral HPSN Rotors 715
Interlaminar shear strength 458, 461
Iosipescu test method 458
IzfP Saarbrücken 17

J

Japanese Ministry of International Trade and Industry (MITI) 8
Joining 160, 192, 251
Joseph Lucas, Ltd. 16

K

Kawasaki Heavy Industries Limited (KHI) 315
Keiser rig 656
Keramische Komponenten für Fahrzeug-Gasturbinen 17, 175
KX01 silicon infiltrated SiC 82
Kyocera Corporation 125

L

Lamon-Evans model 564
Lanxide 262
Larson-Miller 367
Larson-Miller Parameter 316
Laser scattering 7
Lateral cracks 674
Lessons learned 71, 143
Life prediction 555
Life prediction codes 8, 576
Life prediction methodology 374
for ceramic components of advanced heat engines 581
Lifetime prediction 151
Linear elastic limit 21
Linear regression 573
Liquid Polymer Impregnation (LPI) and Pyrolysis 21
Lithium aluminum silicate (LAS) 609, 702
Low elastic modulus 749
Low-pressure injection molding 42
Lucas 3, 57
Lu_2O_3 751

M

Machining damage 750
Malden Mills 273, 643
MAN Technology AG 21
Material characterization 696
Matrix cracking 442, 452, 469
Matrix cracking stress 300
Maximum likelihood estimators 566
Max Planck Institut 17
Mechanical characterization 354
Median cracks 674
Melt Infiltrated Ceramic Matrix Composite (MI-CMC) 294
MEMS 726
MI-CMC fabrication process 296
Microcracked matrix 268
Microfocus X-ray 497
Microfocus X-ray radiography 7
Microturbines 726
Mil-Hdbk-17 545
Ministry of Aircraft Industry (MAP) 14
Ministry of Economy, Trade and Industry (METI) 8
MI SiC/SiC 268, 269
MI SiC/SiC CFCC 651
MITI 8
Model 501-K turbine 371
Modulus of elasticity 225, 226
Molybdenum silicide 649
Monkman-Grant relationship 366, 389
Monofilament SiC fibers 294
Moonlight Project 8
MTU 17
Mullite 642, 646, 655

Mullite+BSAS 643
Multiaxial reliability models 560
Multilayer coating system 645
Multiple flaw distributions 566

N

Nanocomposite silicon nitrides 749
NASA 8, 100
NASA Enabling Propulsion Materials (EPM) program 105
NASA Glenn Research Center 371, 579, 642, 647
NASALIFE 598
National Bureau of Standards 5
National Industrial Research Institute of Nagoya (NIRIN) 756
National Institute of Standards and Technology (NIST) 111
Na_2SO_4 628
NC-machining 203
NC-430 57
NC-433 57
NC-132 58
NC-350 RBSN 56
NC-203 57
NCX-5102 64
NCX-34 58
NDE 7, 270, 493
New Energy and Industrial Technology Development Organization (NEDO) 8
Nextel 720 283
Nextel™720/AS 465
NGK Insulators, Ltd. 147
Nicalon fiber 12
Nicalon™ 265
NIID-Research-Engineering Institute for Engines 15
Nitriding 36
Nondectructive Evaluation (NDE) 45, 493
Noralide NC-400 57
Norton Company 3, 55
Norton-TRW Ceramice 59
Notch sensitivity 464
Novel Oxide Ceramic Composites (NOCC) 444
Nozzle guide vanes 63
Nozzle vanes 235
Nozzles 105, 107, 187, 418, 592, 717
NT154 6, 60, 63, 67
NT154 Si_3N_4 587
NT164 6, 67
NT230 Siliconized SiC 63
NT235 63
NZP 287
NZPs 658

O

Oak Ridge National Laboratory (ORNL) 5, 105, 642, 655
Oblique incidence 683
100 kW Automotive CGT Development Project 12
ONERA 217
Optical scattering 500
OR&PE "Technologiya" 229
Oxidation 5, 59, 104, 129, 149, 173, 204, 224, 283, 295, 300, 305, 371, 414, 475, 613, 744, 754
Oxidation model 617
Oxide CMC 279
Oxide fibers 472
Oxide/oxide CFCCs 442
Oxide-oxide CMCs 279

P

Partial size distribution 34, 60
Particle impact tests 668
Petroleum Energy Center (PEC) 10, 447
Plasma spray 642, 644, 646
Polycarbosilane 12
Polymer-derived SiC-based fibers 13
Powder processing 707
Powder synthesis 4
Pratt & Whitney 58
Pratt & Whitney/URTC 643
Pre-ceramic polymer infiltration and pyrolysis (PIP) process 277, 278, 314
Prepreg 298
Pressure slip casting 63, 67, 108
Pressureless sintering 3, 78, 167, 399
Probabilistic design 543
Probabilistic design methodology 695
Probability of failure 558

Projectile impact 666
Projectile-rebounding 687
Proof testing 140
Proof tests 16, 188
Properties 35, 148, 166, 178, 190, 193, 205, 225, 226, 230, 248, 265, 267, 269, 281, 299, 303, 317, 338, 345, 379, 403, 419, 448, 750
Properties of CFCCs 451
Proportional limit 725
Proportional stress limit 477
PSC 259
PY6 Silicon Nitride 369

Q

Quality assurance 708

R

Radial cracks 674
Radial rotor 63, 67
Radial turbine rotor 86
Rapid prototyping 759
RBN-104 98
RBN-101 98
R-curve behavior 358
Reaction Bonded SiC (RBSiC) 78, 81, 291
Reaction Bonded Silicon Nitride (RBSN) 2,3,17, 34,56, 98, 106, 180, 184, 188, 230, 699
Reaction-Sintered Silicon Carbide (RSSC) 2, 3, 700
Recession 373, 610, 624
Reliability 66
Reliability methods 543
Reproducibility 309
Research-Engineering Center 16
Research Institute of Advanced Materials Gas-Generator Co., Ltd. (AMG) 313
Residual properties 478
Residual strength 680
Retained strength 377
Rheology 707
Rig and engine testing 708
Rolls-Royce 21
Rolls-Royce Allison 105
Rosenthal Technik 17
Rosenthal Technik AG 188
Rotor blades 187, 234
Rotors 40
Rybinsk Motors Works 15

S

Saint Gobain 20
Saint–Gobain Ceramics & Plastics, Inc. 55
Salt corrosion tests 609
Scroll 70, 92, 137, 193
SCS-9 fiber 299
SCS-6 291
SCS-6 fiber 299
SCS-6 SiC monofilaments 62
Seal runners 118
Seals 727
Self-reinforced ceramics 747
SEP 21, 261, 443
Shape deposition manufacturing 759
Shear tests 536
Sherby-Dorn 367
Shroud Segment 345
Sialons 17, 412
SiC-Al_2O_3 165
SiC-based fibers 472
SiC-B_4C 416
SiC_f/Al_2O_3 DIMOX™ 268
SiC fiber-reinforced reaction sintered SiC 332
SiC fibers 322
SiC^f/SiC 315
SiC/SiC 21
SiC-TiB_2 407, 416
SiC whiskers 63, 101
Siemens Westinghouse Power Corporation 283
SIGRI 17, 192
Silcomp 292
Silicon 34, 56, 649
Silicon carbide 727
Silicon-doped BN 295, 306
Silicon nitride bearings 64
Silicon nitride powder 99, 398
Silicon nitride ring rotors 119
SiNC 279
Sintered Alpha Silicon Carbide (SASC) 78, 84
Sintered Reaction-bonded Silicon Nitride (SRBSN) 4, 36, 180, 230
Sintered silicon carbide 224, 230, 704

Sintered silicon nitride 115, 703
SiSiC 182, 188, 700
Si_3N_4 11
$Si_3N_4/MoSi_2$ 751
Si_2N_2O 751
Slip casting 42, 60, 91, 136, 707
Slow crack growth 84, 172, 363, 412
Slurry casting 298
Snecma Moteurs 21 148
SN-88 148, 160
SN88 Silicon Nitride 370
SN281 127
SN282 127, 680
SN253 127
SN252 127
Société Européenne de Propulsion 261
Solaramic® glass 647
Solar Turbines Inc. 5, 132, 272, 594, 642, 646
Sol-gel 279
Solid freeform fabrication 115, 759
Spin testing 322
Spring-loaded elements 235
SPSLIFE 589
SRBSN 49
$SrZrO_3$ 659
Standards 9, 522
Standard test methods 447, 524
Standard test methods for CMCS/CFCCS 527
Static fatigue 65, 570
Stationary parts 717
Stator 40, 63, 80
Stator vane 722
STAU 577
S300 CMC 286
Stoichiometric SiC fibers 306
Strain to failure 304
Strength 417
Strength degradation 676
Strength distributions 565
Strength-size effect 557
Stress-relaxation 475
Stress rupture 13, 128, 266, 363, 472
Stress-strain behavior 452
Structural Ceramics Program 18
Structural damage mode 675
S200 CMC 281
Subcritical crack growth 567
SUGRAV 247
Sundstrand Power Systems 64, 67
Surface recession 105, 119
Swedish Board for Technical Development (STU) 19
Swedish Institute for Silicate Research 19
Swedish National Program 19
SYLRAMICTM™ 306
SYLRAMICTM™ SiC fiber 279
Synergy ceramics project 756

T

Taguchi experimental design 60
Tailored microstructures 750
Ta_2O_5 658
$Ta_2O_5.6HfO_2$ 658
$Ta_2O_5.6ZrO_2$ 658
Technical University-Berlin 17
Technical University-Clausthal-Zellerfeld 17
Technical University-Karlsruhe 17
Technologiya 15
Tensile creep 471
Tensile strength 21, 65, 453
Tensile stress-strain 448
Tensile testing 204, 362
Tensile tests 428
Textron 62, 299
Thermal barrier coating 649
Thermal conductivity 102, 148, 166, 178, 192, 205, 225, 226, 230, 281, 300, 345, 467, 747, 751
Thermal imaging 7
Thermal shock 267
Thermal shock resistance 103, 751
Thermoplastic slip casting 399
3-D Woven preform 3212
300kW Industrial Ceramic Gas Turbine Research and Development Project 138
300 kW NEDO CGT Project 10
331-200 APU 114
3M Corporation 279
Time-dependent Batdorf model 570
Time-dependent behavior 388, 539
Time-dependent reliability 569
Toshiba Corporation 331
Tokyo Electric Power Co. (TEPCO) 643, 658
Transfer-molded 80
Transthickness tensile strength 456

Tubular recuperator 215
Turbine blade 85
Turbine blisk 321
Turbine rotor 711
Turbocharger rotors 11, 42, 46, 727
Tyranno 12
Tyranno SA™ 306
Tyranno ZMI 319

U

Ube Industries Ltd. 12, 319
Ultrasonic c-scan 7
Ultrasonic inspection 45
Uniaxial tensile testing 530
Uniaxial tension 698
United tensile strength 337
United Turbine 19
University of Dayton Research Institute (UDRI) 5, 361
University of Leeds 16
University of Newcastle upon Tyne 16
University of Warwick 16

V

Vacuum-compression treatment 246
Valves 727
Virginia Polytechnic Institute 597
Volatilization 104, 266, 295, 618, 649
Volvo 19
Volvo Aero Corporation 21

W

Warm compression molded 80
Water vapor 621, 649
Wear resistant 727
Weibull characteristic strength 559
Weibull distribution 558
Weibull modulus 19, 65, 102, 186, 205, 386, 559
WESTAC 581
Westinghouse 58
Whisker-reinforced silicon nitride 431
Williams International 267

X

X-ray radiography 7, 45

Y

Yogyo Taikan 8
Young's modulus 345
Y_2SiO_5 659

Z

$ZrSiO_4$ 658